Lecture Notes in Networks and Systems

Volume 204

The series "Lecture Notes in Networks and Systems" publishes the latest developments in Networks and Systems—quickly, informally and with high quality. Original research reported in proceedings and post-proceedings represents the core of LNNS.

Volumes published in LNNS embrace all aspects and subfields of, as well as new challenges in, Networks and Systems.

The series contains proceedings and edited volumes in systems and networks, spanning the areas of Cyber-Physical Systems, Autonomous Systems, Sensor Networks, Control Systems, Energy Systems, Automotive Systems, Biological Systems, Vehicular Networking and Connected Vehicles, Aerospace Systems, Automation, Manufacturing, Smart Grids, Nonlinear Systems, Power Systems, Robotics, Social Systems, Economic Systems and other. Of particular value to both the contributors and the readership are the short publication timeframe and the world-wide distribution and exposure which enable both a wide and rapid dissemination of research output.

The series covers the theory, applications, and perspectives on the state of the art and future developments relevant to systems and networks, decision making, control, complex processes and related areas, as embedded in the fields of interdisciplinary and applied sciences, engineering, computer science, physics, economics, social, and life sciences, as well as the paradigms and methodologies behind them.

Indexed by SCOPUS, INSPEC, WTI Frankfurt eG, zbMATH, SCImago.

All books published in the series are submitted for consideration in Web of Science.

More information about this series at http://www.springer.com/series/15179

V. Suma · Joy Iong-Zong Chen · Zubair Baig ·
Haoxiang Wang
Editors

Inventive Systems and Control

Proceedings of ICISC 2021

Springer

Editors
V. Suma
Research and Industry Incubation Center
Dayananda Sagar College of Engineering
Bengaluru, Karnataka, India

Joy Iong-Zong Chen
Department of Electrical Engineering
Dayeh University
Changhua, Taiwan

Zubair Baig
School of Information Technology
Deakin University
Geelong, VIC, Australia

Haoxiang Wang
GoPerception Laboratory
Cornell University
Ithaca, NY, USA

ISSN 2367-3370 ISSN 2367-3389 (electronic)
Lecture Notes in Networks and Systems
ISBN 978-981-16-1394-4 ISBN 978-981-16-1395-1 (eBook)
https://doi.org/10.1007/978-981-16-1395-1

This Springer imprint is published by the registered company Springer Nature Singapore Pte Ltd.
The registered company address is: 152 Beach Road, #21-01/04 Gateway East, Singapore 189721, Singapore

We are honored to dedicate the proceedings to all the participants, organizers, technical program chairs, technical program committee members, and editors of ICISC 2021.

Foreword

On behalf of the conference committee of ICISC 2021, it is my great pleasure to write a foreword to the 5th International Conference on Inventive Systems and Control (ICISC 2021), which has been organized by JCT College of Engineering and Technology on January 7–8 , 2021.

ICISC continues the research tradition that has been established over the past four years, to bring together the scientific and research community that are interested in sharing the research ideas and developments along with the technological breakthroughs in the areas of computing, communication, and control. The main aim of ICISC 2021 is to promote discussions about recent research achievements and trends to encourage the exchange of technological and scientific information by covering a wide range of topics ranging from data communication, computing networks to control technologies that encompasses all the relevant topics in design, modeling, simulation, and analysis.

The proceedings of ICISC 2021 includes a total of 71 papers. The submission of papers originated from validating the international nature of the conference. The quality of papers accepted for presentation at the conference and publication in the proceedings is guaranteed by the Paper Review Board and Technical Program Committee members. ICISC is constantly encouraging research scholars' participation and their interaction with leading professional experts and scientists from research institutes, universities, and industries. The technical sessions of ICISC 2021 are complemented by a special keynote session from the renowned research experts.

I would like to thank all of the authors, technical programme committee members, review board members and session chairs for their research contributions and efforts in support of the 5th ICISC. I hope that the conference program provides you with valuable research insights and will further stimulate research in the fields of computing, communication, and control.

Dr. K. Geetha
Dean Academics and Research
JCT College of Engineering and Technology
Coimbatore, India

Preface

We are pleased to introduce you to the proceedings of the 5th International Conference on Inventive Systems and Control (ICISC 2021), which was successfully held on January 7–8, 2021. ICISC 2021 has brought together research experts, scholars, and industrialists from around the world to disseminate and explore the most recent advanced research works in the field of intelligent systems, as well as to establish innovative academic exchange among computing and communication researchers.

The geographically distributed committee of ICISC 2021 consists of experts, reviewers, and authors in the areas of computing, communication, and control from different parts of the world. With its professional and significant research influence, ICISC is honored to invite three renowned research experts as keynote speakers. We are pleased to report that we received 382 submissions, from which 71 high-quality papers were chosen and compiled into proceedings after each manuscript was stringently reviewed. Moreover, the committee has always ensured that all the papers have gone through the peer-review process in order to meet the international research publication standard.

We would like to extend our gratitude to the organizing committee, distinguished keynote speakers, internal/external reviewers, and also the authors for their continual support for the conference. We would also be very thankful to the Springer publications for publishing this proceedings. The readers will be highly benefitted by gaining state-of-the-art research knowledge from the ICISC 2021 proceedings. We would also expect the same overwhelming response and more number of scholars and experts across the globe to join the international conference, which will be organized in the upcoming years.

Bengaluru, India — V. Suma
Changhua, Taiwan — Joy Iong-Zong Chen
Geelong, Australia — Zubair Baig
Ithaca, USA — Haoxiang Wang

Acknowledgments

We would like to express our gratitude to everyone who helped make this 5th edition of the conference a success in the midst of the global pandemic. Moreover, we are very pleased to thank our institution JCT College of Engineering and Technology, Tamil Nadu, India, for their support and timely assistance during the conference event.

The conference organizers are very grateful to all the internal/external reviewers, faculties, and committee members for delivering an extended support to the authors and participants, who dedicated their research efforts to submitting high-quality manuscripts to the conference. Furthermore, we particularly acknowledge the efforts made by Dr. K. Geetha, who leveraged immense support to all the stages of the conference event right from submission, review to publication. We are also grateful to the conference keynote speakers Dr. Mohanmmad S Obaidat from the University of Sharjah, UAE; Dr. Wang Haoxiang from Cornell University, USA; and Dr. Abul Bashar from Prince Mohammad Bin Fahd University, Saudi Arabia, who delivered valuable research insights and expertise to assist the future research directions of the conference attendees. We also extend our gratitude to the advisory and review committee members for their valuable and timely reviews that greatly improved the submitted manuscripts to ICISC 2021.

Further, we also extend our appreciation to all the potential authors, who contributed their research works to enhance the publication quality of the 5th ICISC 2021. Also, we would like to thank all the session chairs and organizing committees for contributing their tireless continuous efforts to this successful conference event. Finally, we are pleased to thank Springer publications for their guidance throughout this publication process.

Conference Team
ICISC 2021

Contents

Editors and Contributors

About the Editors

Dr. V. Suma has obtained her B.E. in Information Science and Technology, M.S. in Software Systems, and her Ph.D. in Computer Science and Engineering. She has a vast experience of more than 17 years of teaching. She has published more than 183 international publications which include her research articles published in world class international journals such as ACM, ASQ, Crosstalk, IET Software, and international journals from Inderscience publishers, from journals released in MIT, Darmout, USA, etc. Her research results are published in NASA, UNI trier, Microsoft, CERN, IEEE, ACM portals, Springer, and so on.

Dr. Joy Iong-Zong Chen is currently a full professor of the Department of Electrical Engineering, Dayeh University at Changhua, Taiwan. Prior to joining the Dayeh University, he worked at the Control Data Company (Taiwan) as a technical manager since September 1985 to September 1996. His research interests include wireless communications, spread spectrum technical, OFDM systems, and wireless sensor networks. He has published a large number of SCI journal papers in the issues addressed physical layer for wireless communication systems. Moreover, he also majors in developing some applications of the IoT (Internet of Thing) techniques, and Dr. Joy Iong-Zong Chen owned some patents authorised by the Taiwan Intellectual Property Office (TIPO).

Dr. Zubair Baig works in the School of Science and a member of the ECU Security Research Institute. He completed Doctor of Philosophy, Monash University, 2008. He received many Research Grants which includes Academic Centre of Cyber Security Excellence, Department of Education and Training, Academic Centres of Cyber Security Excellence, 2017–2021, Investigation of social media for radicalisation and terrorism in the Maldives, Edith Cowan University, Edith Cowan University, School of Science Collaborative Research Grant Scheme, and Authentication and Authorisation for Entrusted Unions (AU2EU), European Commission, Grant—Seventh Framework Programme (FP7).

Dr. Haoxiang Wang is currently the director and a lead executive faculty member of GoPerception Laboratory, NY, USA. His research interests include multimedia information processing, pattern recognition and machine learning, remote sensing image processing, and data-driven business intelligence. He has co-authored over 60 journal and conference papers in these fields on journals such as Springer MTAP, Cluster Computing, SIVP; IEEE TII, Communications Magazine; Elsevier Computers and Electrical Engineering, Computers, Environment and Urban Systems, Optik, Sustainable Computing: Informatics and Systems, Journal of Computational Science, Pattern Recognition Letters, Information Sciences, Computers in Industry, Future Generation Computer Systems; Taylor & Francis International Journal of Computers and Applications and conference such as IEEE SMC, ICPR, ICTAI, ICICI, CCIS, and ICACI.

Contributors

Taghreed Abdullah Department of Studies in Computer Science, Mysore, India

K. Abirami Department of Computer Science and Engineering, Amrita School of Engineering, Coimbatore, Amrita Vishwa Vidyapeetham, Coimbatore, India

Md. Abu Khaer Bangladesh Atomic Energy Commission, Dhaka, Bangladesh

Subrata Kumar Aditya University of Dhaka, Dhaka, Bangladesh

Neha Adnekar Department of Computer Engineering and Applications, GLA University, Mathura, India

G. Aghila Department of Computer Science and Engineering, National Institute of Technology Puducherry, Karaikal, India

Reema Agrawal Department of Computer Engineering and Applications, GLA University, Mathura, India

Syed Abdul Basit Andrabi Department of CS and IT, Maulana Azad National Urdu University, Hyderabad, India

G. Anjana Department of Electronics and Communication, Amrita Vishwa Vidyapeetham, Amritapuri, India

R. Annie Uthra Department of CSE, College of Engineering and Technology, SRM Institute of Science and Technology, Kattankulathur, Chennai, India

Akhil Antony Applied Electronics and Instrumentation, Rajagiri School of Engineering and Technology, Ernakulam, India

Joseph Antony Applied Electronics and Instrumentation, Rajagiri School of Engineering and Technology, Ernakulam, India

Barbadekar Aparna Department of Electronics and Telecommunication, AISSMS IOIT, Pune, India;
Department of Electronics and Telecommunication, VIIT, Pune, India

M. Suresh Babu Gouthama Budha Society, Anantapur, A.P, India

Vipin Balyan Cape Peninsula University of Technology, Cape Town, South Africa

Malti Bansal Department of Electronics and Communication Engineering, Delhi Technological University (DTU), Delhi, India

Prachi Bansal Department of Computer Engineering and Applications, GLA University, Mathura, India

Kamaladevi Baskaran School of Management and Commerce, Amity University, Dubai, UAE

Pradeep Bedi Lingayas Vidyapeeth, Faridabad, Haryana, India

Oshin Benny Department of Computer Science, CHRIST (Deemed to Be University), Lavasa, Pune, India

Teresa Benny Applied Electronics and Instrumentation, Rajagiri School of Engineering and Technology, Ernakulam, India

J. Bhagya LBS Institute of Technology for Women, Poojapura, Kerala, India

Neeraj Bhargava Department of Computer Science, School of Engineering and Systems Sciences, MDS University, Ajmer, Rajasthan, India

Sunandan Bhunia ECE Department, Central Institute of Technology Kokrajhar, Kokrajhar, Assam, India

Joel S. Biyoghe Cape Peninsula University of Technology, Cape Town, South Africa

Prudhvi Raj Budumuru Vishnu Institute of Technology, Bhimavaram, India

Rekha Chakravarthi Faculty, School of Electrical and Electronics, Sathyabama Institute of Science and Technology, Chennai, India

Pragati Chandankhede Sir Padampat Singhania University, Udaipur, Rajasthan, India

Balina Surya Chandra Department of Computer Science and Engineering, Koneru Lakshmaiah Education Foundation, Vaddeswaram, A.P, India

K. Ramesh Chandra Vishnu Institute of Technology, Bhimavaram, India

Abhishek Chauhan Institute of Technology, Gopeshwar, Uttarakhand, India

Aswathy K. Cherian SRM Institute of Science and Technology, Kattankulathur, Chennai, India

Sarat Kr. Chettri Department of Computer Applications, Assam Don Bosco University, Guwahati, India

N. R. Chilambarasan PG & Research Department of Computer Science, Government Arts College (Autonomous), Salem, India

Apoorva Choudhary Department of Electronics and Communication Engineering, Delhi Technological University (DTU), Delhi, India

Vishwas Dalal Center for Creative Cognition, S R Engineering College, Warangal, India

P. S. Deepthi LBS Institute of Technology for Women, Poojapura, Kerala, India

Arpan Desai Charotar University of Science and Technology, Anand, India

Dharmavaram Asha Devi Sreenidhi Institute of Science and Technology, Hyderabad, India

Mallisetti Dileep School of Electrical and Electronics, Sathyabama Institute of Science and Technology, Chennai, India

Madhusudan Donga Vignan's Institute of Information Technology, Visakhapatnam, India

Jigyasu Dubey Shri Vaishnav Vidyapeeth Vishwavidyalaya, Indore, India

S. Emalda Roslin Faculty, School of Electrical and Electronics, Sathyabama Institute of Science and Technology, Chennai, India

R. Eswari Department of Computer Applications, National Institute of Technology, Tiruchirappalli, Tamilnadu, India

V. V. Gana Department of Electrical and Electronics Engineering, Amrita School of Engineering, Amrita Vishwa Vidyapeetham, Bengaluru, India

Apuroop Sai Ganesh Engineering Product Development Pillar, Singapore University of Technology and Design, Singapore, Singapore

Gourav Garg Jaypee Institute of Information Technology, Noida, India

Yash Vardhan Gautam Department of Computer Engineering and Application, GLA University, Mathura, U.P., India

Jossy George Department of Computer Science, CHRIST (Deemed to Be University), Lavasa, Pune, India

Sharvari Govilkar Pillai College of Engineering, Mumbai, India

Apoorva Goyal Department of Electronics and Communication Engineering, Delhi Technological University (DTU), Delhi, India

S. B. Goyal City University, Petaling Jaya, Malaysia

Rishu Gulati Department of CSE, Manav Rachna International University Faridabad, Faridabad, India

Tushar Gupta Department of Computer Engineering and Application, GLA University, Mathura, U.P., India

H. S. Guruprasad Department of ISE, B.M.S. College of Engineering, Bangalore, Karnataka, India;
Visvesvaraya Technological University, Belagavi, Karnataka, India

Mazin Kadhum Hameed Department of Software, University of Babylon, Babylon, Iraq

Prasad Naik Hamsavath Department of Computer Applications, NMIT, Bengaluru, Karnataka, India

Durlav Hazarika EE & IE Department, Assam Engineering College Jalukbari, Guwahati, Assam, India

R. Hema PVC Office - ECE, BIHER, Chennai, India

V. M. Hemanth Department of Computer Science and Engineering, M. S. Ramaiah Institute of Technology (Affiliated to VTU), Bangalore, India

Ali Kadhum Idrees Department of Computer Science, University of Babylon, Babylon, Iraq

A. Jackulin Mahariba Department of CSE, SRMIST, Chennai, India

Ravichander Janapati ECE Department, S R Engineering College, Warangal, India

Jacob Jeevamma Department of Electrical Engineering, National Institute of technology Calicut, Calicut, India

S. Kamalakkannan Department of Computer Science, School of Computing Sciences, Vels Institute of Science, Technology and Advanced Studies (VISTAS), Chennai, India

Ormila Kanagasabapathy Department of Electrical and Electronic Engineering, A.M.K. Technological Polytechnic College, Chennai, India

Anita Kanavalli Department of Computer Science and Engineering, M. S. Ramaiah Institute of Technology (Affiliated to VTU), Bangalore, India

A. Kangaiammal Department of Computer Applications, Government Arts College (Autonomous), Salem, India

K. Karibasappa Deptartment of ECE, Dayananda Sagar Academy of Technology and Management, Bangalore, India

Anatoly Karpenko Bauman Moscow State Technical University, Moscow, Russia

M. Karthika Gurubarani Department of Computer Science and Engineering, Amrita School of Engineering, Coimbatore, Amrita Vishwa Vidyapeetham, Coimbatore, India

Javed Ahmad Khan Department of IT, Government Girls Polytechnic Ballia, Ballia, Uttar Pradesh, India

Vaibhav Khanna Department of Computer Science, Dezyne E'Cole College, Ajmer, Rajasthan, India

B. Kirubagari Department of Computer Science and Engineering, Annamalai University, Chidambaram, India

G. Shyam Kishore ECE Department, University College of Engineering, Osmania University, Hyderabad, Telangana, India

T. Kosalai University College of Engineering Kancheepuram, Tamilnadu, India

Avinash Hegde Kota Department of Electronics and Communication Engineering, Amrita Vishwa Vidyapeetham, Amritapuri, India

V. S. S. Krishna Mohan Department of Electronics and Instrumentation Engineering, Dayananda Sagar College of Engineering, Bengalore, India

Arun Kumar Sir Padampat Singhania University, Udaipur, Rajasthan, India

G. Sathish Kumar SCSVMV University, Kanchipuram, Tamilnadu, India

Jugnesh Kumar St. Andrews Institute of Technology and Management, Gurgaon, India

K. Senthil Kumar Rajalakshmi Institute of Technology, Chennai, Tamil Nadu, India

Vinay Kumar Department of Computer Engineering and Applications, GLA University, Mathura, India

Inna Kuzmina Bauman Moscow State Technical University, Moscow, Russia

B. N. Lakshmi Narayan Department of Computer Applications, NMIT, Bengaluru, Karnataka, India

H. N. Latha Department of ECE, BMS College of Engineering, Bangalore, India; Department of CSE, IIT, Kharagpur, India

Abdeldjalil Ledmi Department of Computer Science, ICOSI Lab, University of Khenchela, Khenchela, Algeria

Toufik Messaoud Maarouk Department of Computer Science, ICOSI Lab, University of Khenchela, Khenchela, Algeria

V. Madhivanan School of Electrical and Electronics, Sathyabama Institute of Science and Technology, Chennai, India

Kishan Chaitanya Majji Amity University, Dubai, UAE

R. Manasa Department of ECE, Dayananda Sagar College of Engineering, Bangalore, India

M. V. Manoj Kumar Department of Information Science and Engineering, Nitte Meenakshi Institute of Technology, Bangalore, India

Ephron Martin Applied Electronics and Instrumentation, Rajagiri School of Engineering and Technology, Ernakulam, India

Juby Mathew Amal Jyothi College of Engineering, Kanjirapally, Kerala, India

Yash Mathur Department of Computer Engineering and Applications, GLA University, Mathura, India

Rajesh Kannan Megalingam Department of Electronics and Communication Engineering, Amrita Vishwa Vidyapeetham, Amritapuri, India

Shashi Mehrotra Department of Computer Science and Engineering, Koneru Lakshmaiah Education Foundation, Vaddeswaram, A.P, India

Divya Midhunchakkarvarthy Lincoln University College, Kaula Lampur, Malaysia

S. K. Mona Sweata Department of Computer Science and Engineering, Amrita School of Engineering, Coimbatore, Amrita Vishwa Vidyapeetham, Coimbatore, India

A. Monika Department of Computer Applications, National Institute of Technology, Tiruchirappalli, Tamilnadu, India;
Department of Computer Science, Shaheed Rajguru College of Applied Sciences for Women, University of Delhi, Delhi, India

Pranav Moorthy Sri Sivasubramaniya Nadar College of Engineering, Kalavakkam, India

Akhil M. Nair Department of Computer Science, CHRIST (Deemed to Be University), Lavasa, Pune, India

N. M. Nandhitha Faculty, School of Electrical and Electronics, Sathyabama Institute of Science and Technology, Chennai, India

B. Nandini Department of Computer Applications, NMIT, Bengaluru, Karnataka, India

Md. Julkar Nayeen Mahi Jahangirnagar University, Dhaka, Bangladesh

Pornima Niranjane Babasaheb Naik College of Engineering, Pusad, Maharashtra, India

Vaishali B. Niranjane Yashwantrao Chavan College of Engineering, Nagpur, Maharashtra, India

T. K. Niriksha Department of Computer Science and Engineering, National Institute of Technology Puducherry, Karaikal, India

Potnuru Sai Nishant Department of Computer Science and Engineering, Koneru Lakshmaiah Education Foundation, Vaddeswaram, A.P, India

P. V. Nithara Department of Electrical and Electronics Engineering, CMR Institute of Technology, Bengaluru, India

Raja Shekar P. V. Center for Creative Cognition, S R Engineering College, Warangal, India

Ketan Pandey Jaypee Institute of Information Technology, Noida, India

Chiranjit R. Patel Electronics and Communication, RNS Institute of Technology, Bengaluru, Karnataka, India

Ishita Patel Sardar Patel University, Anand, India

Ketan D. Patel Department of Computer Science, Ganpat University, Kherva, Gujarat, India

Bonani Paul Department of Computer Science, St. Mary's College, Shillong, India

V. Poornima Department of Computer Science and Engineering, M. S. Ramaiah Institute of Technology (Affiliated to VTU), Bangalore, India

E. Poovammal SRM Institute of Science and Technology, Kattankulathur, Chennai, India

Patıl Pradeep Department of Electronics and Telecommunication, JSPM/TSSM'S COE, Pune, India

N. H. Prasad Department of MCA, Nitte Meenakshi Institute of Technology, Bengaluru, India

Ravi Pratyusha Sri Sivasubramaniya Nadar College of Engineering, Kalavakkam, India

S. Priya Department of CSE, College of Engineering and Technology, SRM Institute of Science and Technology, Kattankulathur, Chennai, India;
Applied Electronics and Instrumentation, Rajagiri School of Engineering and Technology, Ernakulam, India

Vijaya Krishna Tejaswi Puchakayala Department of Electronics and Communication Engineering, Amrita Vishwa Vidyapeetham, Amritapuri, India

Krushil Punwatkar Babasaheb Naik College of Engineering, Pusad, Maharashtra, India

Rachit Department of Computer Engineering and Applications, GLA University, Mathura, India

Rahul Department of Computer Science and Engineering, Delhi Technological University, Delhi, India

Rojin Alex Rajan SAINTGITS College of Engineering, Pathamuttom, Kerala, India

Hemalatha Rallapalli ECE Department, University College of Engineering, Osmania University, Hyderabad, Telangana, India

Sandesh Ramesh Department of Information Science and Engineering, Nitte Meenakshi Institute of Technology, Bangalore, India

Neha Rane Pillai College of Engineering, Mumbai, India

Lalitha Rangarajan Department of Studies in Computer Science, Mysore, India

Yash Rathi SRM Institute of Science and Technology, Chennai, India

Kritesh Rauniyar Department of Computer Science and Engineering, Delhi Technological University, Delhi, India

R. Rengaraj Sri Sivasubramaniya Nadar College of Engineering, Kalavakkam, India

Bokkisam Rohit Department of Computer Science and Engineering, Koneru Lakshmaiah Education Foundation, Vaddeswaram, A.P, India

Ishita Sagar Department of Electronics and Communication Engineering, Delhi Technological University (DTU), Delhi, India

Rajiv R. Sahay Department of EEE, IIT, Kharagpur, India

Himanshu Saini Jaypee Institute of Information Technology, Noida, India

M. S. Sangeeetha Faculty, School of Electrical and Electronics, Sathyabama Institute of Science and Technology, Chennai, India

H. A. Sanjay Department of Information Science and Engineering, Nitte Meenakshi Institute of Technology, Bangalore, India

Tirumala Satya Savithri JNTUH, Hyderabad, India

M. Selvaganesh Sri Ramakrishna Engineering College, Coimbatore, India

Rakesh Sengupta Center for Creative Cognition, S R Engineering College, Warangal, India

S. Senthilarasi Department of Computer Science, Vels Institute of Science, Technology and Advanced Studies (VISTAS), Chennai, India

Md. Shaiful Islam Babu Changchun University of Science and Technology, Jilin, China

Ashutosh Shankhdhar Department of Computer Engineering and Application, GLA University, Mathura, U.P., India

Aditi Sharma Jaypee Institute of Information Technology, Noida, India

Jitendra Sharma Shri Vaishnav Vidyapeeth Vishwavidyalaya, Indore, India

K. S. Shivakumar Department of Computer Science and Engineering, M. S. Ramaiah Institute of Technology (Affiliated to VTU), Bangalore, India

K. Sindhu Department of ISE, B.M.S. College of Engineering, Bangalore, Karnataka, India;
Visvesvaraya Technological University, Belagavi, Karnataka, India

Manoj Kumar Singh Manuro Tech Research Pvt. Ltd, Bangalore, India

Vejay Karthy Srithar Department of Computer Science and Engineering, Amrita School of Engineering, Coimbatore, Amrita Vishwa Vidyapeetham, Coimbatore, India

Mohammed El Habib Souidi Department of Computer Science, ICOSI Lab, University of Khenchela, Khenchela, Algeria

K. Sowmya Department of Computer Applications, NMIT, Bengaluru, Karnataka, India

Gavini Sreelatha Lincoln University College, Kuala Lampur, Malaysia

K. Sreeram Department of Computer Science and Engineering, M. S. Ramaiah Institute of Technology (Affiliated to VTU), Bangalore, India

Mayank Srivastava Department of CEA, GLA University, Mathura, UP, India

S. Suganthi Department of Electrical Engineering, National Institute of technology Calicut, Calicut, India

P. Suguna Department of Computer Science, Annamalai University, Chidambaram, India

Pramod Sunagar Department of Computer Science and Engineering, M. S. Ramaiah Institute of Technology (Affiliated to VTU), Bangalore, India

M. Sundararajan PVC Academics, BIHER, Chennai, India

Amit B. Suthar AMPICS, Ganpat University, Kherva, Gujarat, India

Dipankar Sutradhar IE Department, Central Institute of Technology Kokrajhar, Kokrajhar, Assam, India

S. Tamilsenthil PG & Research Department of Computer Science, Government Arts College (Autonomous), Salem, India;
Department of Computer Science, Padmavani Arts and Science College for Women, Salem, India

Polly Thomas SAINTGITS College of Engineering, Pathamuttom, Kerala, India

Divya Tripathi Institute of Engineering and Technology, Dr. APJ. Abdul Kalam Technical University, Lucknow, India

S. S. Tyagi Department of CSE, Manav Rachna International University Faridabad, Faridabad, India

R. Umamaheswari Gnanmani College of Technology, Namakkal, India

Chandni Upadhyaya Sardar Patel University, Anand, India

Trushit Upadhyaya Charotar University of Science and Technology, Anand, India

Jyoti Vaishnav Department of Computer Application, Presidency College Hebbal, Kempapura, Bengaluru, India

V. Vaishnavi Sri Ramakrishna Engineering College, Coimbatore, India

S. Varalakshmi Aadhi College of Engineering and Technology, Tamilnadu, India

K. Sukanya Varshini SRM IST, Chennai, India

K. Veena Sri Sivasubramaniya Nadar College of Engineering, Kalavakkam, India

G. R. Venkatakrishnan Sri Sivasubramaniya Nadar College of Engineering, Kalavakkam, India

E. Esakki Vigneswaran Sri Ramakrishna Engineering College, Coimbatore, India

V. Vimal Kumar Applied Electronics and Instrumentation, Rajagiri School of Engineering and Technology, Ernakulam, India

A. Vinaya Babu Stanley College of Engineering and Technology for Women, Hyderabad, India

K. Vishal Vinod Department of Computer Science and Engineering, Amrita School of Engineering, Coimbatore, Amrita Vishwa Vidyapeetham, Coimbatore, India

H. L. Viswanath Department of Electronics and Communication Engineering, Christ University, Bengalore, India

Vivek B. A Electronics and Communication, RNS Institute of Technology, Bengaluru, Karnataka, India

Abdul Wahid Department of CS and IT, Maulana Azad National Urdu University, Hyderabad, India

Subodh Wairya Institute of Engineering and Technology, Dr. APJ. Abdul Kalam Technical University, Lucknow, India

Md. Ziaul Hasan Majumder Bangladesh Atomic Energy Commission, Dhaka, Bangladesh

Knowledge Graphs in Recommender Systems

T. K. Niriksha and G. Aghila

Abstract Recommender systems are used to overcome the information overload problem and provide a personalized recommendation to the user. Recommender systems suffer from several challenges, like data sparsity and cold start problems. Knowledge graphs are proven to benefit recommender systems in multiple ways, like elevating cold start problems, tackling data sparsity problems, increasing the accuracy of recommendations, and can also be used to provide explanations for recommendations. Knowledge graphs are heterogeneous information graph that is made up of entities as nodes and relationships as edges; they store rich semantic information. Many researchers have used a knowledge graph for recommendations of movies, news, music, fashion, etc. Knowledge graph embedding or knowledge graph representational learning is the trending approach for the usage of knowledge graphs in recommender systems. Translational models like TransE, TransH are widely used and neural network models like graph convolutional network and graph attention network are seen to perform efficiently. The article, gives a background of recommender systems and knowledge graphs and then talks about some different approaches available for their integration and also presents a unique classification of available methods.

Keywords Recommender systems · Knowledge graph · Translational model · Knowledge graph embedding

T. K. Niriksha (✉) · G. Aghila
Department of Computer Science and Engineering, National Institute of Technology Puducherry, Karaikal 609609, India

G. Aghila
e-mail: aghila@nitpy.ac.in

V. Suma et al. (eds.), *Inventive Systems and Control*,
Lecture Notes in Networks and Systems 204,
https://doi.org/10.1007/978-981-16-1395-1_1

1 Introduction

1.1 Recommender Systems

Recommender systems are a subset of information filtering algorithms, which reduce millions of options into hundreds, which are relevant to the user. Recommender systems provide tailor-made options to the user. Using recommender systems is advantageous for both the user and the provider. This survey literature will discuss the broad classification of recommender systems and common problems or research areas of recommender systems.

The broad classification of recommender systems is as follows:

A. *Content-based filtering*:

This kind of recommender system depends on the content of the item (i.e., description of the items) and also considers the user's choice based on the past. In this kind of filtering, there is no cold start problem, but the disadvantage is the performance does not increase as the user transaction increases. If someone watches cute cat videos, then the recommender system recommends videos of cute animals.

B. *Collaborative filtering*:

Collaborative filtering uses the user's similarity and the item's similarity. The main advantage is it works even if there is no domain knowledge. But it suffers from a cold start problem. If user A is similar to user B, and user B watches cricket videos, then the cricket videos are recommended to user A, even though user A did not watch any videos related to cricket before.

C. *Hybrid filtering*:

Hybrid models are built to use the advantage of both models and as well discard the disadvantage of the two models. Hybrid filtering is a combined model of both collaborative filtering and content-based filtering.

Below are some of the common problems faced in recommender systems, and they are also the research areas of recommender systems:

(a) *Cold Start*:

There exist two types of cold start problem

1. *Cold start for new items*:

The problem of cold start for fresh things or items is, it is difficult to decide if the item is relevant to a user until it has a proper description of the item or user ratings.

2. *Cold start for new users*:

Cold start problem for new users is it is difficult to recommend items to the new user because the system does not know what the user likes or does not like. Later, the user can give a rating or feedback about any items.

(b) *Scalability*:

Since recommender systems deal with a huge amount of data, scalability is a major problem. Scalability is the capacity to change in size.

(c) *Data sparsity*:

Data sparsity is handling a user-item dataset that is mostly empty.

(d) *Privacy*:

In demographic models, user demographic data like sex, age, location, hobbies, and so on are possible effects of the user's privacy.

There have been many methods to overcome the problems faced by recommender systems using social network information and tagging methods. The overfitting problem can be tackled by introducing diversity and surprise in the list of recommended items. Knowledge graphs are proven to solve the cold start and data sparsity issue and also helps to improve the quality of recommendation.

1.2 Knowledge Graph

A knowledge graph in simple words is knowledge represented in graph form. A knowledge graph is a directed heterogeneous multi-graph, made up of nodes as entities/concepts and edges represent the relations between the concepts. A knowledge graph is a type of semantic network representation of knowledge. Knowledge graphs are easy to understand and interpret.

$$\text{Knowledge graph } G = (V, E)$$

V = Vertices representing entities or concepts
E = Edges represented by a triple (h, r, t) [(head, relation associating head and tail, tail)].

Google announced its knowledge graph on May 16, 2012. Google's knowledge graph is the knowledge base used by Google to answer user queries efficiently. Knowledge graphs can be used in search engines, recommender systems, question answering systems, and relation prediction systems.

Figure 1 shows an example of a knowledge graph, here the entities are Bill Gates, Microsoft, Apple, Steve Wozniak, USA, and Harvard.

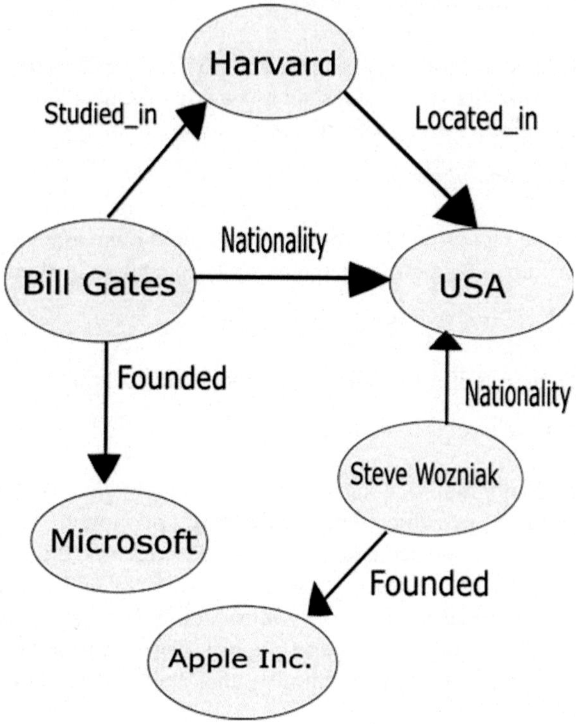

Fig. 1 Knowledge graph example

The relationship is represented as triples (h, r, t), and the following are the triples in Fig. 1. (Bill Gates, Founded, Microsoft), (Bill Gates, Studied_In, Harvard), (Harvard, Located_In, USA), (Bill Gates, Nationality, USA), (Steve Wozniak, Nationality, USA), (Steve Wozniak, Founded, Apple Inc).

Note that the entities are not of a single type and relationships are also of many types, hence these are heterogeneous. There exist general knowledge graphs like DBpedia, Freebase, Wikidata, which are public, and Satori knowledge graph, Google knowledge graph which is private. Domain-specific knowledge graphs are also available like Microsoft's academic graph, LinkedIn economic graph, etc.

Some of the advantages of knowledge graphs are:

(a) Knowledge graph representations are highly similar to how humans organize knowledge in their brain
(b) The clear structure of knowledge graphs aids in the easy explanation
(c) Highly beneficial for personalization and recommendations
(d) Knowledge graphs can be used to support business decisions.

Knowledge graphs have many benefits since it stores semantic information of the entities and relationships. A concise introduction to the structure of knowledge graphs is seen. A brief literature survey on the existing knowledge graph-based recommender systems is discussed in the next section.

2 Knowledge Graphs in Recommender System

Knowledge graphs in recommender systems are being used as an auxiliary or add-on information for a few years, and it is known to benefit the recommender systems in multiple ways like elevating cold start problem, solving the data sparsity problem, and also increase the breadth of recommendations [1]. Using a knowledge graph for the generation of recommendations can generate both accurate predictions and semantically rich explanations to justify the predictions [2].

Knowledge graphs have been used in a wide variety of applications of recommender systems like movie recommender system [3–5], news recommender systems [6], fashion recommendation system [1], Web API recommender system [7], tweet and followee recommender system [8], etc. Zhang et al. used a knowledge graph along with the topic analysis model to deal with the "long tail" words in the collaborative news recommender system [6].

Yan et al. proposed a fashion recommendation system using data augmentation and knowledge graph in the factorization machine (FM) model to generate high-quality recommendations on the Amazon fashion dataset. Data augmentation was used to filter irrelevant data and label items with fine-grained tags. Using a knowledge graph as auxiliary information can enrich the description of items and users, which effectively help fight the data sparsity problem. A differentiated fashion recommendation strategy was employed for active users and inactive users. The classification of active and inactive users was done manually on the dataset. Inactive users are those who have rated less than ten items (cold start problem); for these users, recommendations were generated by depth search on the knowledge graph, whereas for active users, the author used both the knowledge graph and the FM model to provide high-quality recommendations [1]. Table 1 shows the literature survey of a recommender system based on knowledge graphs in different application domains.

The usage of knowledge graphs in recommender systems can be classified in different ways. Sun et al. have classified the recommender systems that utilize knowledge graphs into three types, graph-based, meta-path-based, and knowledge graph embedding-based methods [9]. Guo et al. classified knowledge graph-based recommender systems into the following categories, embedding-based, path-based methods, and unified methods. Liu et al. classified them as recommendations based on ontology, recommendations based on knowledge graph embedding, and recommendations based on linked open data [10]. This survey literature classified the KG based recommender systems into two main categories knowledge graph embedding-based methods and other methods which include the path-based, graph-based and other non-embedding-based methods. The knowledge graph embedding methods can be further classified into translational models like TransE, TransH, etc., and the neural network models like convolution neural networks, attention networks, etc. The upcoming sections discuss these methods in detail. Figure 2 shows the classification of methods to use knowledge graphs in recommender systems.

Table 1 Summarized literature survey of recommender systems based on knowledge graphs in different application domains

Authors	Title of the paper	Source and year of publication	Recommendation domain	Method used	Limitations
Cairong Yan, Yizhou Chen, Lingjie Zhou	Differentiated fashion recommendation using knowledge graph and data augmentation	IEEE Access (2019)	Fashion	Factorization model, knowledge graph (depth search)	Do not explore deep connections between entities since it uses a general graph algorithm
Benjamin A. Kwapong and Kenneth K. Fletcher	A knowledge graph-based framework for web API recommendation	IEEE world congress on services (2019)	Web API	Collaborative filtering, graph walking approach (GraphSAGE)	The attention mechanism is not used while aggregating the feature information from the node's neighborhood
Danae Pla Karidi, Yannis Stavrakas, Yannis Vassiliou	Tweet and followee personalized recommendations based on knowledge graphs	Springer, J Ambient Intell Human Comput (2018)	Social Media (Twitter)	Content-based filtering, Knowledge graph (DFS)	Lower efficiency compared to machine learning-based methods
Weizhuang Han and Quanming Wang	Movie recommendation algorithm based on the knowledge graph	IICSPI (2019)	Movie	Collaborative filtering, translational method(TransH)	Potential relationships between users are not added to the knowledge graph

(continued)

Table 1 (continued)

Authors	Title of the paper	Source and year of publication	Recommendation domain	Method used	Limitations
Minghong Cai and Jinghua Zhu	Knowledge-aware graph collaborative filtering for recommender systems	IEEE International Conference on MSN (2019)	Movie	Neural network collaborative filtering	Computation cost is high
Kuai Zhang, Xin Xin, Pei Luo, Ping Guo	Fine-grained news recommendation by fusing matrix factorization, topic analysis, and knowledge graph representation	IEEE International Conference on SMC (2017)	News	Collaborative filtering, translational method (TransR)	Cannot work with the composite type of relationships
Zhang Yang and Zhang Guiyun	Collaborative filtering recommendation algorithm Fuses Semantic Nearest Neighbors based on the knowledge graph	Asia-Pacific Conference on IPEC (2020)	Movie	Collaborative filtering, translational method(TransE)	Performs well only if the relationship type is one-to-one
Xiao Yu, Xiang Ren, Yizhou Sun, Bradley Strut, and Urvashi Khandelwal	A recommendation in heterogeneous information networks with implicit user feedback	7th ACM conference on recommender systems (2013)	Movie	Collaborative filtering, meta-path	System performance depends on the types and length of meta-paths defined

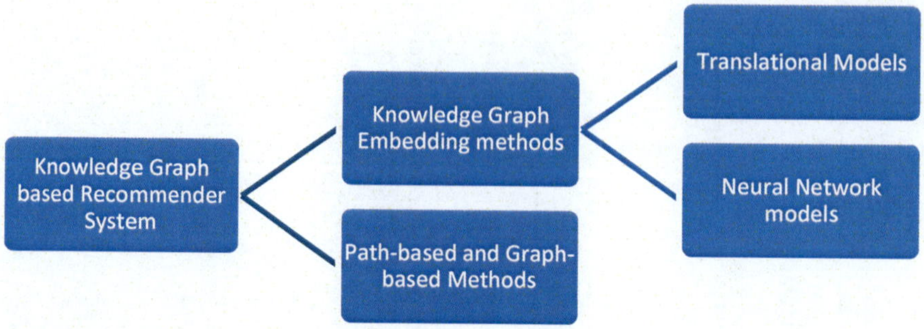

Fig. 2 Classification of methods available to use knowledge graphs in recommender systems

3 Knowledge Graphs Embedding

Knowledge graph embedding means translating the graph which is of higher dimension into a lower-dimensional space like a vector. Knowledge graph embedding is a lower-dimensional representation of entities and relations in a knowledge graph that attempts to preserve the structure and semantics of the knowledge graph [11]. Graph representational learning is used to automatically learn the latent feature matrix rather than handcrafting the feature matrix which is a very expensive and time-consuming task. Knowledge graph embedding helps learn the latent feature matrix automatically.

Liu et al. have classified the relationship between knowledge graph embedding and recommendation algorithm in two ways, jointly learning and independent learning [12]. Knowledge graph embedding methods define a scoring function to calculate the plausibility of a triple (h, r, t) [10]. Another way of classifying the knowledge graph embedding methods is the type score function the method uses. In this literature, classification is also based on a similar method, and knowledge graph embedding can be divided into translational models and neural network models.

In this section, a brief introduction to translational models and neural network models is given; later, some existing recommender systems which employ translational and neural network models are discussed.

3.1 Translational Models

Translational models are accurate and efficient for the knowledge graph embedding tasks in the field of machine learning [11]. Translational models are said to be the special case of distance models. The recommendation task is seen as a special type of knowledge graph completion problem, and the scoring function used for ranking the recommendations can be derived from translational models. There exist many translational models like TransE, TransH, TransR, TransD, TransA, etc. This proposed literature discusses translational models TransE, TransH, TransR, and talks about a few existing recommender systems which have employed translational models.

3.1.1 Translating Embeddings (TransE) Model

Bordes et al. in 2013 proposed translating embeddings (TransE) for modeling multi-relational data. TransE model represents the relation as a translational vector r connecting embedding entities *h* and *t* [13]. Figure 3 shows the illustration of the TransE model. Here *h* and *t*, entities are represented as vectors and *r* is the relation vector connecting them. The TransE model believes that the golden triples (h, r, t) have lower energy when compared to incorrect triples. The error $|h + r - t|$ is low or equal to zero if the triple is golden triple [14].

Score function of TransE is $f(h, r, t) = D(h + r, t)$. The TransE model works well if the relation r is one-to-one and if it is non-reflexive but fails to perform well if the relationship is many-to-one, one-to-many, many-to-many, or reflexive.

Han and Wang in 2019 proposed a movie recommendation algorithm based on a knowledge graph, and they effectively implemented TransH with a collaborative filtering algorithm on dataset MovieLens-1M and built a knowledge graph by utilizing the data obtained from the Baidu encyclopedia. After constructing the knowledge graph, they matched the entities between the knowledge graph and the MovieLens dataset to produce 3223 entity data; here, the entities are movies, actors, directors, genres, etc. The embedded knowledge graph into low-dimensional space using TransH generated the semantic similarity matrix using cosine similarity. Collaborative filtering was used to generate the item similarity matrix. The weighted fusion method was applied to the semantic similarity and item similarity matrix to generate the fusion similarity matrix. User score prediction is done on the fusion similarity matrix to generate the recommendation list of movies [3].

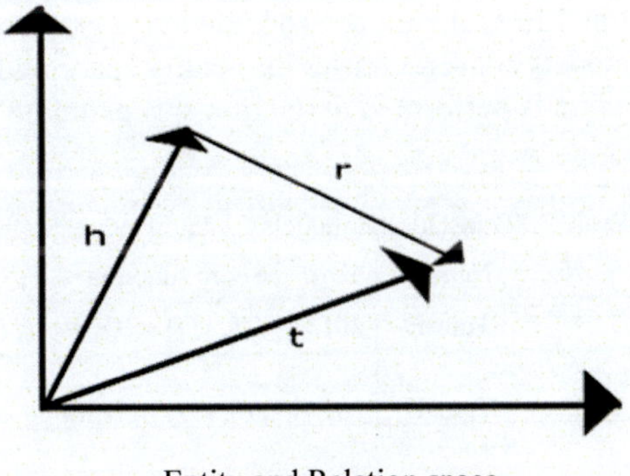

Fig. 3 Simple illustration of TransE

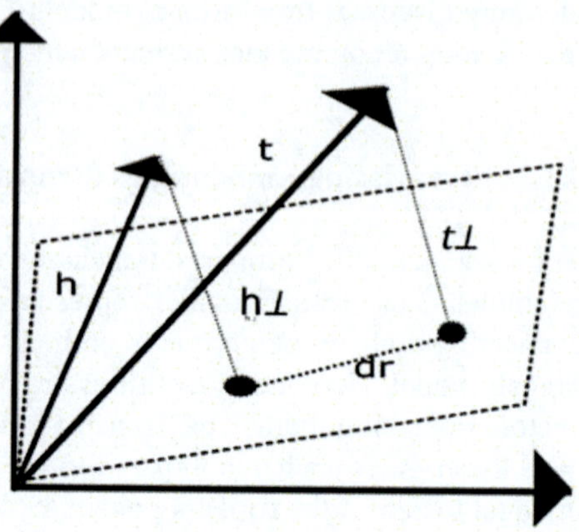

Fig. 4 Simple illustration of TransH

3.1.2 Translating Hyperplanes (TransH)

Wang, Zhang et al. in 2014 proposed knowledge graph embedding by translating hyperplanes (TransH) [14] and tried to overcome the drawbacks of the TransE model by inheriting the efficiency. TransH model handles reflexive, one-to-many, many-to-one, and many-to-many relations by multiple representations of different relations projecting on hyperplanes [14]. Figure 4 depicts the illustration of the TransH model.

Here, the vectors h and t are projected to in the hyperplane as $h\perp$and $t\perp$ and dr connect both of them in the hyperplane. The score function is $f(h, r, t) = D(h\perp + dr, t\perp)$ [14], the TransH model is seen to overcome the drawbacks of the TransE model.

3.1.3 TransR Model

Lin, Liu et al. proposed a knowledge graph embedding method called TransR. These models convert entities into entity space and relations into relation space, and translation is performed in relation space, unlike TransE and TransH [15]. Table 2 shows

Table 2 Translational models for knowledge graph embedding

S. No.	Name	Year	Score function	Limitations
1	TransE	2013	$f(h, r, t) = D(h + r, t)$	Faces issues if relationship is M-to-1, 1-M, M-to-M or reflexive
2	TransH	2014	$f(h, r, t) = D(h\perp + dr, t\perp)$	Projects both entities and relationships into the same space
3	TransR	2015	$f(h, r, t) = D\ (hr + r, tr)$	Unique relation vectors may be under-representative to fit all entity pairs under the relation

different translational models available for knowledge graph embedding and their score functions along with the year of publication and its limitations.

Zhang, Xin et al. used the TransR model for the news recommender system. Their model combines topic analysis, knowledge graph representation, and matrix factorization, to generate fine-grained news, and they have shown how long tail of words caused by topic analysis can be taken care of by using a knowledge graph [6]. Here, the authors did not mention anything about the usage of knowledge graph to model similar users, and this model can generate more accurate recommendations using semantic information from the knowledge graph to produce much better recommendations.

There are few other translational models available like CtransR, TransD, TransA, etc.

3.2 Neural Network Models

Neural network models operate by mimicking the working of the human brain, and they can adapt to the changes in the input. In neural network models, representation of knowledge graph is how to express entities or concepts and relations using a neural network. The neural network-based models are affected by the problem of high complexity and computational cost, but their triples have the strong expressive ability.

Wang, Zhao et al. proposed a knowledge graph convolutional network (KGCN) [16] in 2019, the proposed method is claimed to discover both higher-order structure and semantic information of the graph. It is recognized that both the node and its neighbors are considered to build the computation graph. They have applied KGCN on three datasets namely movies, music, and books. The KGCN is designed for a single layer. Aggregation is the key step and final step of the algorithm; they have used three types of aggregators, sum aggregator, concat aggregator, and neighbor aggregator. They showed that KGCN-sum performs the best on all three datasets [16].

Wang, Li et al. in 2019 proposed a knowledge graph embedding via graph attenuated attention network (GAAT) [17]. Traditional neural network models assign the same weights to all the neighbors of a node, and attention networks assign different attention to different nodes in the network. Attention networks perform better than traditional convolutional networks since they have the inductive capacity and are storage and computationally efficient. GAAT assigns different weights in different relation paths and acquires information from their neighbors. Graph attention networks are the improved or advanced version of graph convolutional networks. GAAT can be used for applications like relation prediction, triple classification, and link prediction [17].

Sun et al. presented recurrent knowledge graph embedding (RKGE) for an effective recommendation [9]. RKGE automatically semantic paths are mined between entity pairs, then different paths are encoded using a recurrent network, salient paths

Table 3 Comparison of different methods

S. No	Method name	Pros	Cons
1.	Translational models	Does not depend on handcrafted features	Does not work for certain types of relations
2.	Neural network models	Triples have high expressive ability	High complexity and computational cost
3.	Path-based methods	Semantic similarities can be captured using meta-paths	Depends on handcrafted features and domain knowledge
4.	Graph-based methods	Inductive approaches generalize to new nodes	Do not consider the semantic structure of the graph Depend on handcrafted features

are determined further by pooling operation. The author found that shorter paths in knowledge graphs indicate clearer semantics and stronger connectivity. Pooling operation is used to highlight important features. Two kinds of pooling strategies max-pooling and average-pooling have been used, and it is found that average-pooling is better than max-pooling [9].

Neural network (NN) models are perceived to perform better than translational models, but the computation cost of neural network models is higher when compared to translational models. Table 3 shows the brief advantages and disadvantages of the methods.

4 Path-Based and Graph-Based Models

In this section, path-based methods to use knowledge graphs in recommender systems are discussed initially on the basic terminologies and then graph-based methods are discussed.

4.1 Path-Based

These methods essentially make use of the meta-path/meta-graphs. Path-based methods have been developed since 2013 [10]. Path-based methods build a user-item knowledge graph and leverage the connectivity patterns of users and items in the graph for the generation of recommendations [10]. This method is used to find the semantic similarity between entities in the path, meta-paths. Meta-path-based methods rely on handcrafted features, unlike embedding-based methods.

4.1.1 Meta-Path

Meta-path is a path defined on a network schema, it is a relation sequence between two entities, and it defines an existing or a new relationship between those objects [10].

4.1.2 Meta-Graph

Meta-graph is a meta-structure that connects two entities in a heterogeneous network. Meta-graph is a combination of meta-paths [10].

Yu et al. in 2013 proposed HeteRec [18], they used meta-path in heterogeneous information networks (HIN) to get the semantic similarity between items in the network. The recommendations were generated by using implicit feedback along with the semantic similarity. The implicit feedback tells the preference of the user toward the item and the semantic similarity is used to find similar items for that particular item. Bayesian ranking optimization is used to estimate the model. HeteRec was compared with other recommendation strategies like popularity, non-negative matrix factorization, hybrid SVM, etc., and found that HeteRec performs better than these methods.

Yu et al. in 2014 [19] generalized a framework for entity recommendation uses a heterogeneous information network approach. Matrix factorization was used to generate an implicit feedback matrix and meta-paths define how two entities kinds could be linked by diverse kinds of paths. Meta-paths are used to mine latent features. The recommendation module is defined for both global recommendations and personalized recommendations and found that the proposed method is of high computational cost; in the future, these methods can be found to decrease the complexity.

4.2 Graph-Based

The graph walking-based methods consider the topological structure of the graph ignoring the semantics. It encodes the relationship between the concepts and relations of the graph. This kind of method is based on the handcrafted features and domain knowledge of the knowledge graph. Graph walking approaches are of two types namely, transductive and inductive. Transductive approaches attempt to predict a known set of unlabeled samples, i.e., fail to work with unseen data whereas the inductive approach works efficiently even with previously unseen data. This literature will discuss random tracks that are transductive by nature and graphSAGE that is inductive by nature.

Fig. 5 Structure of graph G

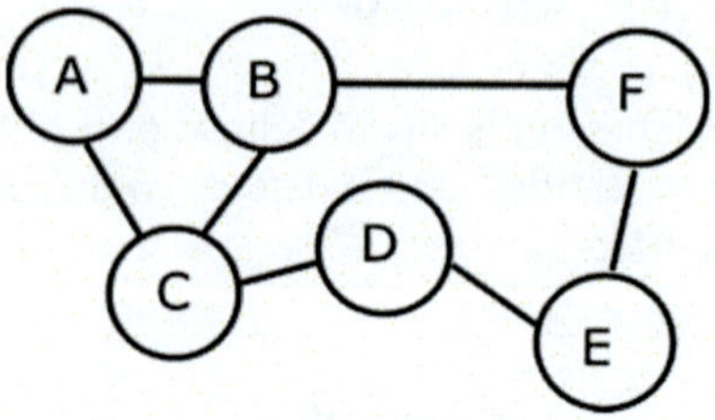

Fig. 6 Random walks on G

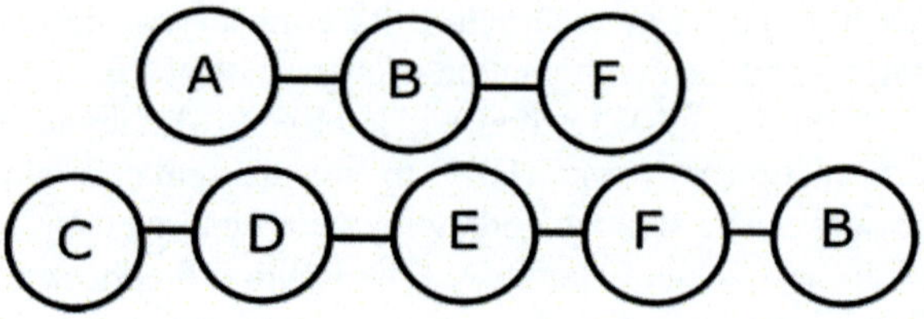

4.2.1 Random Walk

Given a graph, the random walk is starting from a node, at random selects a neighbor and moves to its neighbor, and repeats the process. In this way, a random sequence of nodes is represented as a random walk on the graph.

Figure 5 depicts the structure of graph G and Fig. 6 shows some random walks done on graph G.

4.2.2 GraphSAGE

GraphSAGE aka Graph SAmple and aggreGatE is a graph walking approach. The main idea in this method, is it determines how to aggregate feature information from a node's local neighborhood.

Kwapong and Fletcher in 2019 proposed a knowledge graph framework for the recommendation of web API [7]. They used a knowledge graph to elevate cold start problems, data sparsity problem, and also to generate a high-quality recommendation. The knowledge graph was built from a dataset in three steps, entity extraction, relation extraction, and the final step is triple extraction and integration component. They used graph walking algorithm graphSAGE which is better than the random walk approach, and graphSAGE is an inductive approach and can work with newly seen nodes, unlike random walk which is transductive and cannot generalize to unseen nodes [7].

In graph walking-based methods, inductive approaches are proven to perform better than the transductive approach.

5 Conclusion

This survey presented some techniques of usage of knowledge graphs in recommender systems. Knowledge graphs store rich semantic information that can be used to generate semantically correct recommendations. The item recommendation problem can be seen as a problem of completing the knowledge graph. This literature has classified the usage of knowledge graphs in recommender systems into knowledge graph embedding approaches and path and graph-based approaches. The knowledge graph embedding approach is further classified into the translational model and neural network models. Each method has its advantages and disadvantages. By applying the knowledge graph embedding approach for the integration of knowledge graphs in recommender systems, it is perceived to be better than meta-path-based or graph walking-based approaches since the model learns the features automatically, and it need not be handcrafted. Knowledge graph embedding methods and neural network models are similar to the attention network and convolutional network models automatically mine the features and can work with variable-sized inputs, unlike the translational models. They also generate feature embeddings with strong expressive ability. Further improvements can be done to reduce the computational cost of neural networks by maintaining efficiency.

References

1. C. Yan, Y. Chen, L. Zhou, Differentiated fashion recommendation using knowledge graph and data augmentation. IEEE Access **7**, 102239–102248 (2019)
2. M. Alshammari, O. Nasraoui, S. Sanders, Mining semantic knowledge graphs to add explainability to black box recommender systems. IEEE Access **7**, 110563–110579 (2019)
3. W. Han, Q. Wang, Movie recommendation algorithm based on knowledge graph, in *2019 2nd International Conference on Safety Produce Informatization (IICSPI)* (IEEE, 2019), pp. 409–412
4. Z. Yang, Z. Guiyun, Collaborative filtering recommendation algorithm fuses semantic nearest neighbors based on knowledge graph, in *2020 Asia-Pacific Conference on Image Processing, Electronics and Computers (IPEC)* (IEEE, 2020), pp. 470–474
5. M. Cai, J. Zhu, Knowledge-aware graph collaborative filtering for recommender systems, in *2019 15th International Conference on Mobile Ad-Hoc and Sensor Networks (MSN)* (IEEE, 2019), pp. 7–12
6. K. Zhang, X. Xin, P. Luo, P. Guot, Fine-grained news recommendation by fusing matrix factorization, topic analysis and knowledge graph representation, in *2017 IEEE International Conference on Systems, Man, and Cybernetics (SMC)* (IEEE, 2017), pp. 918–923
7. B. Kwapong, K. Fletcher, A knowledge graph based framework for web API recommendation, in *2019 IEEE World Congress on Services (SERVICES)*, vol. 2642 (IEEE, 2019), pp. 115–120
8. D.P. Karidi, Y. Stavrakas, Y. Vassiliou, Tweet and followee personalized recommendations based on knowledge graphs. J. Ambient Intell. Humaniz. Comput. **9**(6), 2035–2049 (2018)
9. Z. Sun, J. Yang, J. Zhang, A. Bozzon, L.K. Huang, C. Xu, Recurrent knowledge graph embedding for effective recommendation, in *Proceedings of the 12th ACM Conference on Recommender Systems* (2018), pp. 297–305
10. Q. Guo, F. Zhuang, C. Qin, H. Zhu, X. Xie, H. Xiong, Q. He, A survey on knowledge graph-based recommender systems. arXiv preprint arXiv:2003.00911 (2020)

11. E. Palumbo, G. Rizzo, R. Troncy, E. Baralis, M. Osella, E. Ferro, Translational models for item recommendation, in *European Semantic Web Conference* (Springer, Cham, 2018), pp. 478–490
12. C. Liu, L. Li, X. Yao, L. Tang, A survey of recommendation algorithms based on knowledge graph embedding, in *2019 IEEE International Conference on Computer Science and Educational Informatization (CSEI)* (IEEE, 2019), pp. 168–171
13. A. Bordes, N. Usunier, A. Garcia-Duran, J. Weston, O. Yakhnenko, Translating embeddings for modeling multi-relational data, in *Advances in Neural Information Processing Systems* (2013), pp. 2787–2795
14. Z. Wang, J. Zhang, J. Feng, Z. Chen, Knowledge graph embedding by translating on hyperplanes. AAAI **14**(2014), 1112–1119 (2014)
15. H. Lin, Y. Liu, W. Wang, Y. Yue, Z. Lin, Learning entity and relation embeddings for knowledge resolution, Procedia Comput. Sci. **108**, 345–354 (2017)
16. H. Wang, M. Zhao, X. Xie, W. Li, M. Guo, Knowledge graph convolutional networks for recommender systems, in *The World Wide Web conference* (2019), pp. 3307–3313
17. R. Wang, B. Li, S. Hu, W. Du, M. Zhang, Knowledge graph embedding via graph attenuated attention networks. IEEE Access **8**, 5212–5224 (2019)
18. X. Yu, X. Ren, Y. Sun, B. Sturt, U. Khandelwal, Q. Gu, B. Norick, J. Han, Recommendation in heterogeneous information networks with implicit user feedback, in *Proceedings of the 7th ACM Conference on Recommender Systems* (2013), pp. 347–350
19. X. Yu, X. Ren, Y. Sun, Q. Gu, B. Sturt, U. Khandelwal, B. Norick, J. Han, Personalized entity recommendation: a heterogeneous information network approach, in *Proceedings of the 7th ACM International Conference on Web Search and Data Mining* (2014), pp. 283–292

Cyberbullying Detection on Social Media Using SVM

J. Bhagya and P. S. Deepthi

Abstract Social media has become a part of human day-to-day life. With the increasing use of social media, cyberbullying has become most common on social networking sites. This article focuses on a more extensive method for detecting cyberbullying comments which appear in online social media. Cyberbullying detection using support vector machine (SVM) along with TF-IDF for word vectorization is implemented in this work. The SVM performs classification on the target corpus into bully or non-bully comments, after training with text data extracted from the Wikipedia dataset. A set of English text has been collected from available social media platforms and labeled as either bullied or not bullied for training the machine learning based classification model. The support vector machine-based algorithm achieved an accuracy of 92%.

Keywords Social media · Machine learning · Support vector machine · Term frequency-inverse document frequency

1 Introduction

Cyberbullying through social networks has become a major threat nowadays. It is very easy to bully somebody through any of the social media as there is no monitoring of accounts and activities. Cyberbullying may be done using e-mails, messages, chat rooms, blogs, images, video clips, text messages, etc. The person affected by bullying is the victim, and the person doing cyberbullying is a bully. Bullying through the Internet makes a negative impact on human minds. It is easy to bully a person before an entire online network, which can end up in the emotional and psychological breakdown of the victim resulting in depression, stress, lack of self-confidence, anger, sadness, loneliness, health degradation, and even suicides. This has become a reason for the decline of the psychological and physical condition of an individual because

J. Bhagya (✉) · P. S. Deepthi
LBS Institute of Technology for Women, Poojapura, Kerala, India

V. Suma et al. (eds.), *Inventive Systems and Control*,
Lecture Notes in Networks and Systems 204,
https://doi.org/10.1007/978-981-16-1395-1_2

of which an expansive number of suicides and discouragement are being reported. Women and children abuse also has increased a lot through such illegal activities.

A study [1] of 1034 tweens in the USA on bullying, cyberbullying behaviors, social media, and app usage in June and July of 2020 says around 80% had faced bullying. 57% had been targeted in one environment or another. In this survey, about half of the tweens were bullied at school and 15% were cyberbullied. The study says more than two-thirds of cyberbullied tweens were negatively impacted concerning their feelings about themselves. One-third of the tweens had affected their friendships due to cyberbullying. Cyberbullying affected the physical health of 13.1% and schoolwork of 6.5%. It was also found that one in five tweens has been cyberbullied, cyberbullied others, or seen cyberbullying. Tweens use a variety of strategies to stop cyberbullying. Nine out of ten tweens used social media and gaming apps. Two-thirds of tweens are helpers to support, to step into defend, or otherwise assist those who are bullied at school and online when they see it. 90% of the tweens have their own devices and they used more than one gaming apps and most popular social media in the last year.

The survey results indicate that cyberbullying is growing at an alarming rate, with teenagers being the most affected. The most common way of cyberbullying through social media platforms is in the form of textual comments and messages. The bullying comments make a toxic effect on social media. So it is very important to identify those bullying comments. The cybercrime rates can be reduced to a limit by controlling cyberbullying. The aim is to classify the textual data into bully or non-bully and evaluate the performance measures with the help of a machine learning model.

Machine learning has a major role in identifying cyberbullying in online social media. The machine learning algorithms help in text classification. Bullying messages consist of insulting, toxic, and even hate speech. These features have major importance in text processing. Machine learning-based automatic cyberbullying detection contains two parts such as finding these inappropriate bullying words and classifying those messages. It is necessary to preprocess the textual comments, in which the message is first converted into a vector of fixed-length followed by classification. This is followed by training the machine learning model and then testing that model to detect cyberbullying. At last, a trained model can detect the bullied messages in each new data contributed.

This work intends to investigate cyberbullying on social media English textual comments using the SVM [2]. The experiments were carried on the Wikipedia dataset. The comments in the form of text data are preprocessed and then converted into vectors using term frequency-inverse document frequency (TF-IDF) [3]. Finally, the data is trained and the performance is evaluated using an SVM classifier.

This article is organized as follows. In Sect. 2, some related works for cyberbullying detection were discussed. Then, the proposed model and architecture for cyberbullying detection were presented in Sect. 3. In Sect. 4, experimental results on the Wikipedia dataset are illustrated. Finally, concluding remarks are provided in Sect. 5.

2 Related Works

Cybersecurity has attracted the focus of researchers with increasing cybercrime rates. A risk assessment strategy for critical infrastructure which involves identification, analysis, and mitigation of risks to enhance cybersecurity is proposed in [4]. In the case of cyberbullying detection, machine learning techniques and deep learning methods are widely used for text classification. The existing text classification techniques use support vector machine (SVM), Naive Bayes (NB), decision tree, neural networks, nearest neighbor, logistic regression (LR), etc. The dataset used for cyberbullying detection is from Twitter, Formspring, Wikipedia, YouTube, Amazon mechanical Turk.

Rosa et al. [5] proposed a cyberbullying detection model with the machine learning classifiers like SVM, logistic regression, and random forests. The proposed work used 10 textual, 21 sentiment features, and word embedding for feature extraction in the Form Spring dataset. Kumar et al. [6] give the details of soft computing techniques such as machine learning, neural networks, and fuzzy logic for cyberbullying detection. It gives an idea about the different types of datasets used for cyberbullying detection. Zaho et al. [7] proposed cyberbullying detection with different feature extraction techniques such as word embedding with bag-of-words (BoW) and latent semantic features in the Twitter dataset. A linear SVM classifier is used for cyberbullying detection. The embedding-enhanced BoW model is compared with the BoW model, semantic-enhanced BoW model, latent semantic analysis, and latent Dirichlet allocation model.

A literature review on different approaches using supervised learning, lexicon-based, mixed-initiative, and other approaches is reported by Salawu et al. [8]. Cyberbullying detection based on features like content, sentiment, user, and network on various researches is discussed in depth. The group of features that is listed under content-based features is the cyberbullying keywords, document length and spelling, n-grams, BoW, and TF-IDF. Van Hee et al. [9] proposed an automatic cyberbullying detection method on social media site ASK.fm dataset with English and Dutch corpus was evaluated with binary classification using linear SVM. For feature extraction tokenization, n-gram extraction, sentiment lexicon matching, and stemming were used. Finally, performance measures are evaluated.

A review on various machine learning models for cyberbullying prediction in social media by Al-Garadi et al. [10] discusses the use of support vector machine, Naïve Bayesian, random forest, decision tree, logistic regression, and K-nearest neighbor. Agrawal et al. [11] used real-world datasets such as Twitter, Wikipedia, and Formspring. The deep learning-based models and transfer learning were applied to detect cyberbullying. Four deep neural network models such as CNN [12], LSTM [13], BLSTM [14] and BLSTM with Attention were compared with the machine learning models such as logistic regression, SVM, random forest, and Naive Bayes. The DNN models with transfer learning achieved the best result.

Van et al. [15] introduced a multiplatform dataset that contains a set of text from posts that are collected from seven social media platforms. In this work, a multistage and multitechnique annotation system is used. Initially, it used crowd-sourcing for representing posts and hashtags. Later it was used in machine learning methods for cyberbullying detection. The VISR dataset consists of 603,379 posts from different social media platforms such as Instagram, Facebook, Pinterest, Twitter, Gmail, YouTube, Tumblr, and other media. The SVM with TF-IDF for feature selection and RBF kernel is used in this work. The stemmed word unigrams, bigrams, and trigrams were used in SVM for cyberbullying detection. The model is then evaluated with convolutional neural networks and XGBoost model and an accuracy of 84.6 was obtained.

Chavan et al. [16] proposed a model to detect cyberaggressive comments in social media. The dataset collected from Kaggle and for features extraction the techniques such as TF-IDF score, Ngrams, bad word count, and stemming is used to model supervised machine learning algorithms like support vector machines and logistic regression. Isa et al. [17] used the Naive Bayes method and SVM. For each method, they applied n-gram 1–5 for the number of classes 2, 4, and 11. Naive Bayes got an average accuracy of 92.81%, SVM with a poly kernel obtained an average accuracy of 97.11% for the cyberbullying detection conversations in Formspring.me collected from Kaggle.

The cyberbullying detection in Turkish messages was proposed by Özel et al. [18]. SVM, decision tree (C4.5), Naïve Bayes multinomial, and K-nearest neighbors (KNN) classifiers were used for cyberbullying detection. Wu et al. [19] proposed an improved TF-IDF-based fastText (ITFT) for cyberbullying detection. The TF-IDF is added with the position weight and the keywords are extracted. It is then used as input to get filter noise data to improve accuracy. Balakrishnan et al. [20] used Twitter to analyze the user's psychological features like personalities, sentiment, and emotion for cyberbullying identification. The machine learning models random forest, J48, and Naïve Bayes were used to detect bullying.

In [21], Al-garadi et al. proposed a set of unique features derived from Twitter such as network, activity, user, and tweet content, based on these features, and developed a supervised machine learning technique to detect cyberbullying in Twitter. Dani et al. [22] proposed a sentiment-informed cyberbullying detection model to detect cyberbullying behaviors in social media. They used Twitter and MySpace datasets for the proposed framework.

Though there are several works in the literature using SVM, the accuracy of prediction is very less. The highest accuracy reported with SVM is 84.6 in [15]. Motivated by these findings our aim was to make a cyberbullying detection model with higher accuracy than the related works.

3 Proposed System

The proposed work aims at automatic cyberbullying detection using a machine learning technique. An SVM classifier along with the TF-IDF feature selection method is used for cyberbullying text identification from the Wikipedia dataset [23].

The major steps involved in this work include preprocessing [24], feature extraction [25], and classification. The data preprocessing helps in achieving a better result in cyberbullying analysis.

The data is collected from Wikipedia, and it consists of binary labels 0 and 1. The bullying comments are labelled by 1 and non-bullying are labeled with 0. Some words which are good are labeled and the words which have the bullying property are labeled separately. A set of 723 texts is good which are under non-bully. A bad list consists of a group of 426 bullying words. The next step of cyberbullying detection is preprocessing. In this work, for preprocessing tokenization, stop words removal, and stemming are used. After preprocessing, the next phase is the feature selection using TF-IDF. Then the dataset is split into train and test datasets. The train set is used for training the model, and the test set is used for testing.

The SVM classifier is used for classification. The SVM is trained with a set of training data and the classifier is then tested with a set of training data. Here, an SVM with radial basis function kernel is used. The SVM classifier predicts the comments into bully or non-bully. The overall methodology is depicted in Fig. 1.

Data preprocessing stages used in the proposed model are removing the brackets, tokenization, stop words removal, and stemming [26]. The special symbols are removed; they have no role in detecting the bullying texts.

Tokenization is the splitting up of sentences into tokens. In this work, sentence tokenization and word tokenization are used. In sentence tokenization, the text is split into individual sentences. The sentence tokenization works in the sentences "Good morning all. Have a nice day." as "Good morning all.", "Have a nice day.". The Word tokenization splits a large sample of text into words. The word tokenization of "Have a nice day" is "Have", "a", "nice", "day".

The words like "as", "is", "to", "an", "a", "the" do not make any impact on cyberbullying detection, these are the stop words. The process of removing such stop words is called stop words removal. Stemming is done in a text to get the extract root words. This reduces the length of words and thus it can be easily used for the next stage. In this work, TF-IDF creates its own dictionary and each word in the dictionary has given a TF-IDF value. The TF-IDF values are assigned for each bully and non-bully words. The features selected are bad word value, good word value, good word count, bad word count, second-person pronoun, third-person pronoun, second-person words count, third-person words count. The second-person pronoun consists of words such as "yourself", "u", "you", "ur", "your". Also, the third-person pronoun consists of words such as "she", "he", "him", "her", "them", "his".

The term frequency is calculated by,

$$\text{TF} = \frac{\text{Frequency of a word in the document}}{\text{Total words in the document}} \tag{1}$$

The inverse document frequency, IDF, is calculated by,

$$\text{IDF} = \log\left(\frac{\text{Total number of documents}}{\text{Number of documents containing the word}}\right) \tag{2}$$

Fig. 1 Proposed methodology

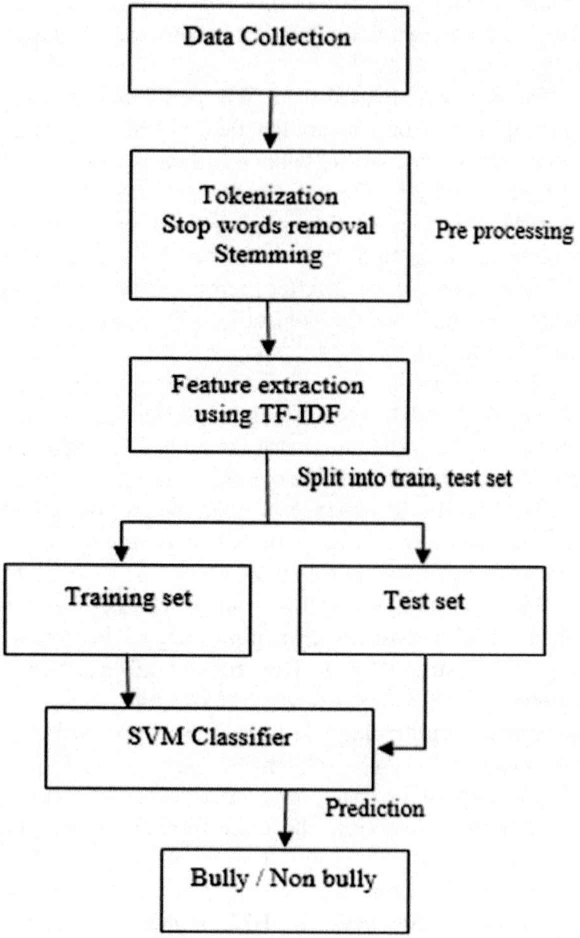

In this work, SVM is a supervised machine learning algorithm that is used as one of the most commonly used algorithms for classification. A kernel trick technique in SVM is used to transform the data and an optimal boundary is found between the outputs. The support vector machine plays well when there exists a margin of separation between the outputs. The efficiency of SVM is more in high-dimensional spaces. It is difficult to determine the separating hyperplane in many real-time applications. The SVM is capable to specify a better hyperplane. SVM has advantages like high speed, efficiency, and scalability.

SVM effectively separates the two classes (bully or non-bully) in a 2D plane contain linearly separable objects. In real-time applications, the SVM determines a maximum margin hyperplane for separating different classes. SVM is a capable classification algorithm because of its characteristics. SVM for cyberbullying prediction models and found to be effective and efficient.

The Wikipedia dataset which consists of 100 k which is split into train and test set in the ratio 80:20. The proposed model selects the test dataset size as 20 in which the parameter test_size is set to 0.20. Then the datasets are ready to be fed into the SVM classifier. SVM is an optimal classifier that learns a classification hyperplane in the feature space having the maximal distance to all the training examples. A plane that can divide the set of objects into different classes is known as a hyperplane. Here, a set of objects is split into a bully or non-bully classes.

SVM takes a labeled training data, the bullying comments are labeled by 1, and non-bullying are labeled with 0 and outputs an optimal hyperplane which can then be used to group the set of another example. The set of training data is provided to the SVM; it generates a boundary to differentiate between the classes learned from the training. The SVM parameters are C and kernel. SVM with a radial basis function (rbf) kernel is used in the proposed work. The C parameter is set to 1.0 in this work. C is a regularization parameter. The strength of regularization is inversely proportional to C.

4 Results and Discussion

For evaluating the performance of the model, the dataset was split into training and test set in the ratio 80:20. The tenfold cross-validation is performed with each of the binary averages, microaverage, macroaverage, and weighted average to evaluate the performance. The performance measures accuracy, precision, recall, and F-score are then evaluated for the model. The model achieved 92% accuracy.

Accuracy is the percentage of correctly classified instances.

$$\text{Accuracy} = \frac{(\text{TP} + \text{TN})}{(\text{TP} + \text{TN} + \text{FP} + \text{FN})} \tag{3}$$

Precision is the ratio of data elements that are correctly classified to a total number of classified instances.

$$\text{Precision} = \frac{\text{TP}}{(\text{TP} + \text{FP})} \tag{4}$$

The recall is the ratio of the minority class instances that are correctly classified to the total number of actual minority class instances.

$$\text{Recall} = \frac{\text{TP}}{(\text{TP} + \text{FN})} \tag{5}$$

where TP is the number of true positives, FN is false negatives, FP is the false positives, and TN represents the true negatives.

F-score is calculated by taking the harmonic mean of precision and recall.

$$\text{F Score} = 2 * \frac{\text{precision} * \text{Recall}}{\text{precision} + \text{Recall}} \tag{6}$$

Table 1 shows the outcome of the proposed model with the binary average with tenfold cross-validation. It gives an accuracy of 0.9198, the precision is 0.8327, recall of 0.4017, and an F-score of 0.542. The mean and standard deviation in binary average for accuracy, precision, recall, and F-score were calculated.

Table 2 shows the outcome of the proposed model with the microaverage with tenfold cross-validation. It gives accuracy, precision, recall, and F-score for 10 runs.

Table 1 Performance measure with binary average

Run #	Accuracy	Precision	Recall	F-score
1	0.921	0.826	0.408	0.546
2	0.920	0.820	0.401	0.538
3	0.921	0.839	0.404	0.546
4	0.919	0.840	0.387	0.530
5	0.918	0.829	0.395	0.535
6	0.917	0.825	0.400	0.539
7	0.921	0.831	0.412	0.551
8	0.920	0.838	0.413	0.554
9	0.922	0.842	0.403	0.545
10	0.919	0.837	0.394	0.536
Mean	0.9198	0.8327	0.4017	0.542
Standard deviation	0.001549	0.007514	0.008193	0.007601

Table 2 Performance measure with microaverage

Run #	Accuracy	Precision	Recall	F-score
1	0.921	0.921	0.921	0.921
2	0.920	0.920	0.920	0.920
3	0.921	0.921	0.921	0.921
4	0.919	0.919	0.919	0.919
5	0.918	0.918	0.918	0.918
6	0.917	0.917	0.917	0.917
7	0.921	0.921	0.921	0.921
8	0.920	0.920	0.920	0.920
9	0.922	0.922	0.922	0.922
10	0.919	0.919	0.919	0.919
Mean	0.9198	0.9198	0.9198	0.9198
Standard deviation	0.001549	0.001549	0.001549	0.001549

Table 3 shows the outcome of the proposed model with the macroaverage. A tenfold cross-validation is done with macroaverage and it gives an accuracy of 0.919, precision of 0.879, recall 0.6952, and F-score 0.7489.

Table 4 shows the outcome of the proposed model with the weighted average with tenfold cross-validation. It gives an accuracy of 0.9198, the precision is 0.9143, recall of 0.9198, and F-score of 0.907.

Table 5 shows the comparison with some cyberbullying detection models using support vector machine. The highest accuracy with SVM is reported in [15] with 84.6 for cyberaggression and 86.7 for cyberbullying.

Table 3 Performance measure with macroaverage

Run #	Accuracy	Precision	Recall	F-score
1	0.921	0.876	0.698	0.751
2	0.920	0.873	0.694	0.747
3	0.921	0.883	0.697	0.751
4	0.919	0.882	0.688	0.743
5	0.918	0.877	0.692	0.745
6	0.917	0.874	0.694	0.746
7	0.921	0.879	0.700	0.754
8	0.920	0.881	0.701	0.755
9	0.922	0.884	0.696	0.751
10	0.919	0.881	0.692	0.746
Mean	0.9198	0.879	0.6952	0.7489
Standard deviation	0.001549	0.00383	0.003994	0.00404

Table 4 Performance measure with weighted average

Run #	Accuracy	Precision	Recall	F-score
1	0.921	0.915	0.921	0.909
2	0.920	0.913	0.920	0.907
3	0.921	0.916	0.921	0.908
4	0.919	0.914	0.919	0.906
5	0.918	0.913	0.918	0.905
6	0.917	0.911	0.917	0.904
7	0.921	0.916	0.921	0.909
8	0.920	0.914	0.920	0.907
9	0.922	0.917	0.922	0.909
10	0.919	0.914	0.919	0.906
Mean	0.9198	0.9143	0.9198	0.907
Standard deviation	0.001549	0.001767	0.001549	0.001764

Table 5 Comparison with related works

Paper	Year	Dataset	Model used	Metric used	Metric value
Dinakar [27]	2012	YouTube	SVM	Accuracy	66%
Waseem [28]	2015	Ask.fm	SVM	F1-score	53.39%
Di Capua [29]	2016	Formspring	SVM	Recall	67%
Zhao [7]	2016	Twitter	SVM	Recall	79.4%
Van Hee [9]	2018	ASKfm	SVM	F1-score	64.32% (English) 58.72% (Dutch)
Van Bruwaene [15]	2020	VISR	SVM	Accuracy	84.6% (Cyberaggression) 86.7% (Bullying)
Proposed work		Wikipedia	SVM with TF-IDF	Accuracy, precision, recall, F1-score	92%

5 Conclusion

In this article, the SVM is used for analyzing cyberbullying and the dataset used in Wikipedia. Data collected from Wikipedia dataset is preprocessed by removing brackets, tokenization, stop words removal, and stemming. Here, TF-IDF is used for word vectorization. The classification of cyberbullying text is executed with SVM and a machine learning algorithm. The results are measured in terms of accuracy, precision, recall, and F-score. The experimental results show that our method obtains an accuracy of 92%. Thus, SVM with TF-IDF is the best-suited model for cyberbullying detection.

References

1. J.W. Patchin, S. Hinduja, TWEEN CYBERBULLYING in 2020. www.cyberbullying.org/2020-tween-data
2. X.-L. Liu, S. Ding, H. Zhu, L. Zhang, Appropriateness in applying SVMs to text classification. Comput. Eng. Sci. **32**, 106–108 (2010)
3. A. Aizawa, An information-theoretic perspective of tf–idf measures. Inform. Process. Manag. **39**(1), 45–65 (2003)
4. Z. Baig, S. Zeadally, Cyber-security risk assessment framework for critical infrastructures. Intell. Autom. Soft Comput. **25**(1), 121–129 (2019)
5. Hugo Rosa, Automatic cyberbullying detection: a systematic review. Comput. Hum. Behav. **93**(2019), 333–345 (2019)
6. A. Kumar, N. Sachdeva, Cyberbullying detection on social multimedia using soft computing techniques: a meta-analysis. Multimedia Tools Appl. **78**(17), 23973–24010 (2019)
7. R. Zhao, A. Zhou, K. Mao, Automatic detection of cyberbullying on social networks based on bullying features, in *Proceedings of the 17th International Conference on Distributed Computing and Networking* (2016)
8. S. Salawu, Y. He, J. Lumsden, Approaches to automated detection of cyberbullying: a survey. IEEE Trans. Affec. Comput. (2017)

9. C. Van Hee, Automatic detection of cyberbullying in social media text. PloS one **13**(10), e0203794 (2018)
10. M.A. Al-Garadi, Predicting cyberbullying on social media in the big data era using machine learning algorithms: review of literature and open challenges. IEEE Access **7**(2019), 70701–70718 (2019)
11. S. Agrawal, A. Awekar, Deep learning for detecting cyberbullying across multiple social media platforms.in *European Conference on Information Retrieval*. (Springer, Cham, 2018)
12. Y. Kim, Convolutional neural networks for sentence classification. *arXiv preprint* arXiv:1408.5882 (2014)
13. R. Johnson, T. Zhang, Supervised and semi-supervised text categorization using LSTM for region embeddings. *arXiv preprint* arXiv:1602.02373 (2016)
14. P. Zhou, Text classification improved by integrating bidirectional LSTM with two-dimensional max pooling. *arXiv preprint* arXiv:1611.06639 (2016)
15. D. Van Bruwaene, Q. Huang, D. Inkpen, A multi-platform dataset for detecting cyberbullying in social media. Lang. Resour. Eval. 1–24 (2020)
16. V.S. Chavan, S.S. Shylaja, Machine learning approach for detection of cyber-aggressive comments by peers on social media network, in *2015 International Conference on Advances in Computing, Communications and Informatics (ICACCI)*. (IEEE, 2015)
17. S.M. Isa, L. Ashianti, Cyberbullying classification using text mining, in *2017 1st International Conference on Informatics and Computational Sciences (ICICoS)*. (IEEE, 2017)
18. S.A. Özel, Detection of cyberbullying on social media messages in Turkish, in *2017 International Conference on Computer Science and Engineering (UBMK)*. (IEEE, 2017)
19. J. Wu, Toward efficient and effective bullying detection in online social network. Peer-to-Peer Netw. Appl. 1–10 (2020)
20. V. Balakrishnan, S. Khan, H.R. Arabnia, Improving cyberbullying detection using Twitter users' psychological features and machine learning. Comput. Sec. **90**, 101710 (2020)
21. M.A. Al-garadi, K.D. Varathan, S.D. Ravana, Cybercrime detection in online communications: the experimental case of cyberbullying detection in the Twitter network. Comput. Human Behav. **63**, 433–443 (2016)
22. B. Haidar, M. Chamoun, A. Serhrouchni, Multilingual cyberbullying detection system: detecting cyberbullying in Arabic content, in *2017 1st Cyber Security in Networking Conference (CSNet)*. (IEEE, 2017)
23. https://figshare.com/articles/Wikipedia_Talk_Labels_Personal_Attacks/4054689
24. S. García, Big data preprocessing: methods and prospects. Big Data Anal. **1**(1) (2016)
25. D. Kim, Multi-co-training for document classification using various document representations: TF–IDF, LDA, and Doc2Vec. Inform. Sci. **477**, 15–29 (2019)
26. J. Atwan, M. Mohd, G. Kanaan, Enhanced arabic information retrieval: light stemming and stop words, in *International Multi-Conference on Artificial Intelligence Technology*. (Springer, Berlin, Heidelberg, 2013)
27. K. Dinakar, Common sense reasoning for detection, prevention, and mitigation of cyberbullying. ACM Trans. Interact. Intell. Syst. (TiiS) **2**(3), 1–30 (2012)
28. Z. Waseem, D. Hovy, Hateful symbols or hateful people? Predictive features for hate speech detection on Twitter, in *Proceedings of the NAACL Student Research Workshop* (2016)
29. M. Di Capua, E. Di Nardo, A. Petrosino, Unsupervised cyber bullying detection in social networks, in *2016 23rd International Conference on Pattern Recognition (ICPR)*. (IEEE, 2016)

Ultraviolet Radiation in Healthcare Applications: A Decades-Old Game Changer Technology in COVID-19

Abhishek Chauhan

Abstract Ultraviolet (UV) radiation and its prominence are always explicitly observed in various applications, since its incarnation. In this paper, UV-based products and its prominence during COVID-19 are highlighted and critically reviewed. Moreover, the constructional steps and function of the proposed 40 and 120 W smart UV chamber, which holds the innovations added to make it more functional as a product are also underlined.

Keywords Ultraviolet (UV) rays · Innovation · Disinfectants · Coronavirus

1 Introduction

The UV rays are one of the most prominently used forms of electromagnetic wave with a wavelength range of 10–400 nm, which is shorter than the wavelength of visible lights but having a long wavelength when compared with X-rays [1–12]. In contrast to its presence, the UV is well available in the sunlight, i.e., about 10% of the consolidated electromagnetic radiation are observed from the sun. Moreover, UV rays can also be produced artificially through specially designed UV lamps, i.e., mercury vapor, blacklight, and tanning lamps. Electrical arcing and lighting are also considered as one of the sources of UV, which is not visible through the naked human eye [1, 2].

The UV light was discovered in the early eighteenth century by a German physicist Johann Wilhelm Ritter. The discovery is completely revolving around his experiment and finally concludes the presence of invisible light beyond the visible light spectra, which is capable of tanning the paper soaked in silver chloride (AgCl) more swiftly than the violet light. In early times, UV rays are also known as oxidation rays and chemical rays, where it is important to know that the breakthrough in the application of UV that comes into its eminence, whereas in 1878, the effect of short-wavelength UV on bacteria is analyzed and studied [9, 12].

A. Chauhan (✉)
Institute of Technology, Gopeshwar, Uttarakhand 246424, India
e-mail: abhishek.chauhan@itgopeshwar.ac.in

V. Suma et al. (eds.), *Inventive Systems and Control*,
Lecture Notes in Networks and Systems 204,
https://doi.org/10.1007/978-981-16-1395-1_3

During the exposure of UV-C rays, the energy is absorbed by the double bonds in the DNA of the bacterial cells [12] which the opening of the bond,where these bonds are reformed with the bases in adjacent, results in the mutation in the DNA. These mutations are responsible for the repair mechanism in the cell, if the bacteria cell gets sufficient UV-C exposure the mutation rate is increased which cannot be repaired by this mechanism and inhibits the reproduction of the bacteria or viruses [12] which in turn manifests its ability of disinfection and still in this COVID-19 pandemic UV-based disinfection plays a vital role [1].

In this paper, application of UV-C in health care especially during the COVID-19 scenario is critically underlined. Moreover, 40 and 120 W smart UV chamber is also proposed and designed for disinfecting the surfaces like documents, currency notes, vegetables, and eatables with smart connected features for safe operation along with voice control strategy.

1.1 Classification of UV as UV-A, UV-B, and UV-C

UV lights are classified based on their wavelength, i.e., UV-A, UV-B, and UV-C, also illustrated in Fig. 1.

Different UV rays are responsible for different applications which are listed below as,

(a) UV-A rays (315–400 nm): It is the wave which is also known as the blacklight wave and is considered the safest among all UV waves. UV-A enables several objects or surfaces to emit fluorescence. It is most commonly used for the identification of several precious metals, in banks for fake currency identification, and also for artistic designs [2].

(b) UV-B rays (280–315 nm): In general, about 90–95% of UV-B from the sun re absorbed by the ozone layer of earth. The UV-B upsurges the probability of skin cancer and to minimize the risk of sunburn, and sunscreens are introducing in the market with different sun protection factor (SPF) values [2].

(c) UV-C rays (100–280 nm): It is the most harmful, short-wavelength UV rays which cause serious damage to the eyes and skin if exposed directly, whereas

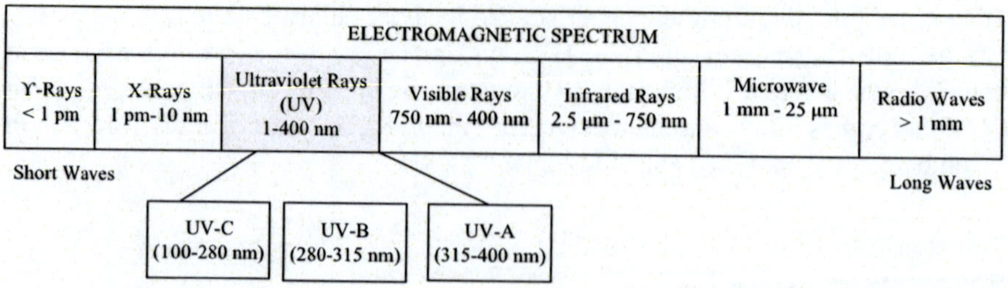

Fig. 1 Electromagnetic wave spectrum diagram

it can disinfect air, water, food, and even the surfaces. UV-C breaks the nucleic acid links of the DNA of bacteria and viruses which leads to disinfection [3].

It is important to underline that UV holds some health hazards, specifically over a wavelength of 180 nm. Short-duration exposure to UV leads to sunburn on the skin which leads to the risk of skin cancer. Moreover, prolonged exposure to UV results in permanent damage to the retina. Hence, it is important to follow the safety instruction given on the UV product under usage.

2 Related Work: COVID-19, UV-C and the World

In the COVID-19 scenario, it is observed that the most common problem is to disinfect the surfaces, where the disinfection on the surfaces through liquid disinfectant is not possible, and for this UV-C-based disinfections leads the world. Currently, across the world, public transport, hospitals, stores, laboratories, food, and packaging industry are working with various UV-C products. Hence, innovations along with new products come into existence during the COVID-19 pandemic era.

2.1 UV-C in Public Transport

The design of a disinfection system for public transport is the problem whose holistic solution, and for the new normal after lockdown, situation in many countries is exercised, and few challenges as mentioned in Fig. 2 are to be overcome while selecting UV as a tool for the same.

(a) *Bus disinfection*: In the recent COVID-19 situation, national health commission issued a guideline, which refers to the usage of UV-C for disinfecting the interior and exterior of the public busses by Shanghai public transport company, 210 UV tubes are employed to disinfect about 250 busses daily, which takes

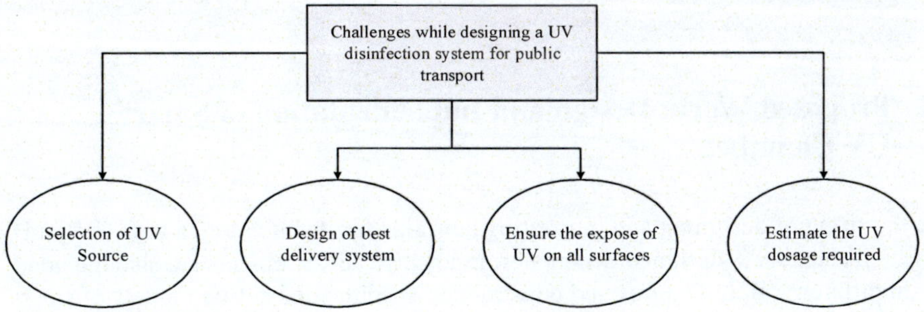

Fig. 2 Challenges for designing a UV-based disinfection system

5–7 min for each bus to kill about 99% of the viruses [4]. Disinfecting public busses are prominently used in the COVID-19 scenario, whereas it is studied as well as implemented in the pre-COVID era; also, in [4], the implementation of ultraviolet germicidal irradiation (UVGI) in airconditioning systems of the transit buses to evaluate the air quality improvement and reduction of harmful pathogens in the bus are asserted.

(b) *Aerospace industry*: Disinfection of a complete airplane and keep it ready for the next operation in a petite period is a challenge after the resumption of lockdown throughout the world. So, to keep the industry operated, there is a need for a safe and fast disinfection protocol. Autonomous UV-based robots are the innovation developed and implemented at various stages in aerospace industries that include the disinfection of airplane interiors and to the baggage area.

The GR Ford international airport is the first airport in the US which tests and employs autonomous UV robots for disinfection at the airport, which takes 10–15 min for zone disinfection independently without any human intervention, whereas footwear sanitization station and luggage trolley sanitization station took only 8 s for disinfection. Delhi international airport limited also installed UV tunnels for the disinfection of the luggage at New Delhi airport.

2.2 *UV-C in the Health and Food Industry*

In the health industry, UV is prominently used since the disinfection property of UV comes into existence, which includes the sanitization of the operation wards and disinfection of surgical instruments. In the COVID-19 era, the disinfection of protection kits is also done through UV, as in the initial stages, the availability of protection kits is scarce; hence, reuse of protection kits is important and only possible after proper disinfection, which is done through UV-based disinfection. Moreover, the food and packaging industry also employed UV tunnel through which the food items are crossed and eliminate the possibility of any contamination of bacteria and viruses in the food items and on the packaging surfaces.

3 Proposed Work: Design and Implementation of Smart UV Chamber

In the string of designing and innovating new aids for disinfection, a new UV sanitization chamber is designed, where new innovative advancements are also included with different safety features and construction specifications. Two models of smart UV chambers are designed, i.e., 40 and 120 W as shown in Figs. 3 and 4, respectively.

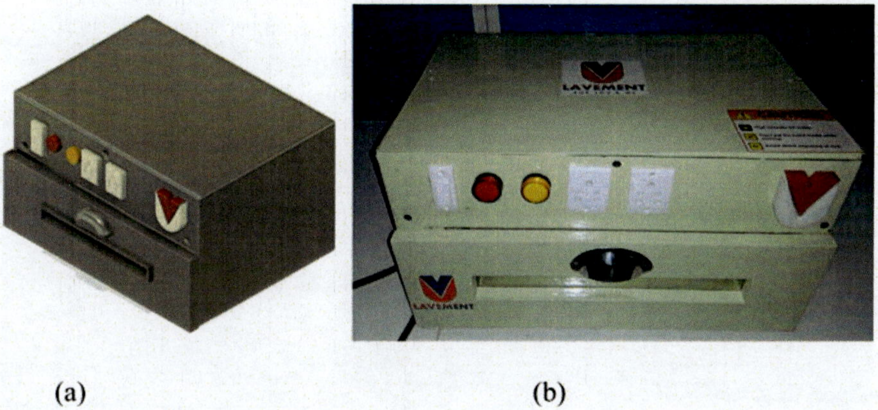

Fig. 3 **a** CAD design and **b** actual 40 W smart UV chamber

(a) *40 W UV chamber*: The chamber is designed with a 2.5 mm non-reflective metallic sheet (inner side of the chamber is reflective), where two Hg, 20 W each Philips UV-C lamps are employed, where each UV lamp gives an output of 12 W of UV-C. It is important to understand that the 20 W tube or lamp does not mean the UV-C output is 20 W, as the lamp also generated other secondary rays including the visible light. Hence, the total UV-C output watt depends upon the efficiency of that particular lamp. Moreover, the time of UV exposure is calculated based on UV dosage and the watt power of the lamp along with the area underexposure. This chamber can be remotely controlled over the Internet (by using mobile applications like Google home, Google assistant, or controlled through voice commands with Alexa), manually and through the four-channel relay module, which enables the minimum human intervention. It takes approximately 40 s for the disinfection process.

(b) *120 W UV chamber*: A 120 W UV chamber, with an additional safety feature and an automatic timer, is also designed. The time in the timer switch can be changed as per the calculated exposure period of the UV, which can be calculated by the dosage required for different viruses and bacterias. As the UV is not safe for the skin and the eye, so to minimize the error while operating the chamber a safety switch is also employed in the door, which ensures the disconnection of the circuit when anyone tries or erroneously opens the door of the chamber while operating. To ensure uniform exposure of the UV inside the chamber, the UV tubes are so arranged that the objects placed inside the chamber observe a uniform exposure as shown in Fig. 5c.

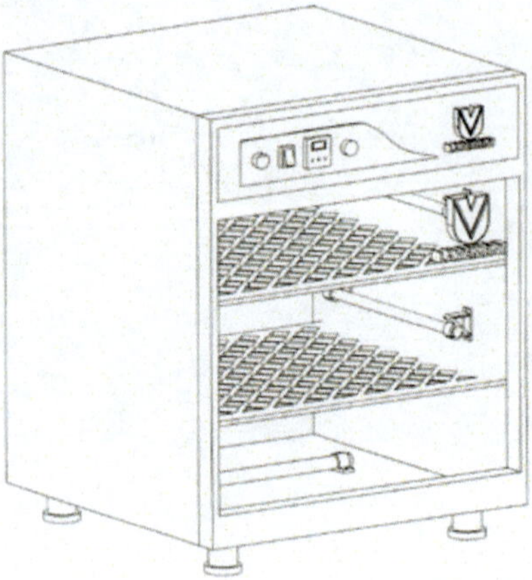

(a) Door Less CAD Design

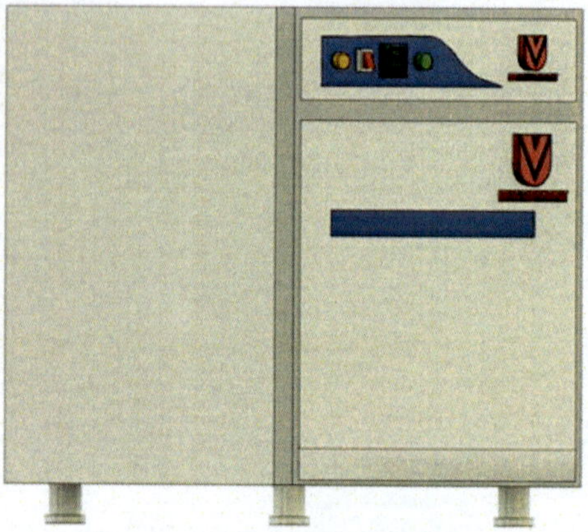

(b) With Door CAD Design

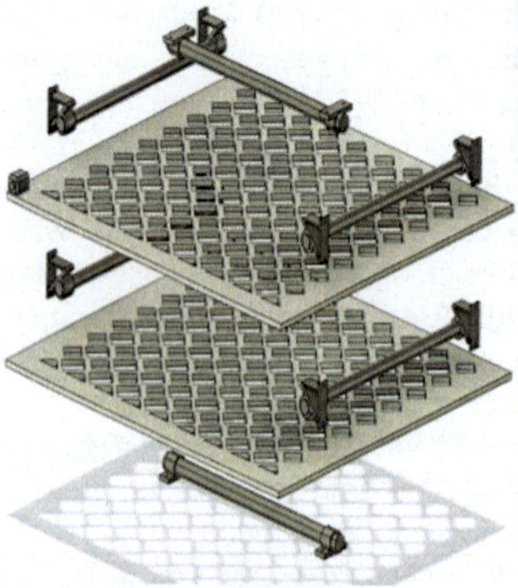

(c) UV-C Tube Arrangement

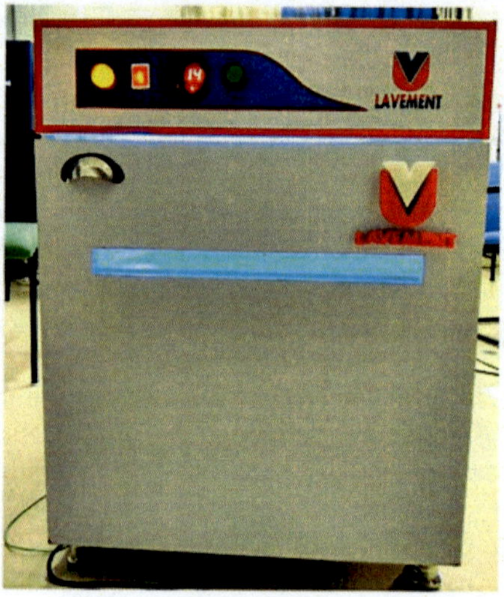

(d) Original Product

Fig. 4 Different views of UV chamber

4 Conclusion

In this paper, applications of UV rays are critically studied and reviewed; moreover, the importance of UV-based products in COVID-19 situation is also underlined, where the prominence is given to the disinfection products used in hospitals, offices, public transports, and aviation. A 40 and 120 W UV chamber is also proposed and

Fig. 5 Interior of original product

designed for the hospitals and offices, where prominence is given to the smart and safety features.

Acknowledgements I would like to convey my gratitude to the Director, faculty, and staff of the Institute of Technology, Gopeshwar for the support during this project. This project is funded under TEQIP-III, NPIU, Ministry of HRD.

Design Registration

Applied for design registration, application number is 331145-001, pending in Controller General of Patents, Design and Trademarks.

References

1. S.I. Ahmad, L. Christensen, E. Baron, History of UV lamps, types, and their applications. Adv. Exp. Med. Biol. **996**, 3–11 (2017). https://doi.org/10.1007/978-3-319-56017-5_1
2. X. Li, M. Cai, L. Wang, F. Niu, D. Yang, G. Zhang, Evaluation survey of microbial disinfection methods in UV-LED water treatment systems. Sci. Total Environ. **659**, 1415–1427 (2019). https://doi.org/10.1016/j.scitotenv.2018.12.344
3. M. Hanoon, UV light disinfection robots help to overpower pathogens. OR Manag. **31**(6), 24–25 (2015)
4. Bus disinfection through UV lights. www.sustainable-bus.com. Accessed on 26/10/2020
5. L.R. Dougall, J.B. Gillespie, M. Maclean, I.V. Timoshkin, M.P. Wilson, S.J. MacGregor, Pulsed ultraviolet light decontamination of artificially-generated microbiological aerosols, in *2017 IEEE 21st International Conference on Pulsed Power* (PPC), Brighton (2017), pp. 1–5. https://doi.org/10.1109/ppc.2017.8291260

6. M. Gostein, B. Stueve, Accurate measurement of UV irradiance in module-scale UV exposure chambers, including spectral and angular response of sensor, in *2016 IEEE 43rd Photovoltaic Specialists Conference* (PVSC), Portland, OR (2016), pp. 0863-0866. https://doi.org/10.1109/PVSC.2016.7749731
7. M.D. Kempe, Accelerated UV test methods and selection criteria for encapsulants of photovoltaic modules, in *2008 33rd IEEE Photovoltaic Specialists Conference*, San Diego, CA, USA (2008), pp. 1–6. https://doi.org/10.1109/PVSC.2008.4922771
8. H. Yu, S. Pang, R. Liu, Effects of UV-B and UV-C radiation on the accumulation of scytonemin in a terrestrial cyanobacterium, Nostoc flagelliforme, in *2011 4th International Conference on Biomedical Engineering and Informatics* (BMEI), Shanghai (2011), pp. 1424–1427. https://doi.org/10.1109/bmei.2011.6098596
9. V. Soldatkin, L. Yuldashova, A. Shardina, A. Shkarupo, T. Mikhalchenko, Device for water disinfection by ultraviolet radiation, in *2020 7th International Congress on Energy Fluxes and Radiation Effects* (EFRE), Tomsk, Russia (2020), pp. 870–873. https://doi.org/10.1109/EFRE47760.2020.9242002
10. C. Purananunak, S. Yanavanich, C. Tongpoon, T. Phienthrakul, A system for ultraviolet monitoring, alert, and prediction, in *2018 10th International Conference on Knowledge and Smart Technology* (KST), Chiang Mai (2018), pp. 237–241. https://doi.org/10.1109/kst.2018.8426146
11. Y. Cao, W. Chen, M. Li, B. Xu, J. Fan, G. Zhang, Simulation based design of deep ultraviolet LED array module used in virus disinfection, in *2020 21st International Conference on Electronic Packaging Technology* (ICEPT), Guangzhou, China (2020), pp. 1–4. https://doi.org/10.1109/icept50128.2020.9202924
12. S. Ravichandran et al., Advances in the development of ultraviolet sterilization system for specific biological applications, in *2009 International Conference on Biomedical and Pharmaceutical Engineering*, Singapore (2009), pp. 1–5. https://doi.org/10.1109/icbpe.2009.5384081

Multi-agent Ludo Game Collaborative Path Planning based on Markov Decision Process

Mohammed El Habib Souidi, Toufik Messaoud Maarouk, and Abdeldjalil Ledmi

Abstract The multi-agent path planning problem is usually addressed via the use of reinforcement learning methods. In this article, a new Lludo game path planning algorithm based on Q-learning was proposed, and a set of cooperation proprieties that the agents have to follow in order to reach the goal location. To accomplish that a Lludo game environment modeling is introduced according to the Markov decision process (MDP) principles. The main objective of this work is to increase agents' cooperation rate with the aim of decreasing the execution time of the assigned tasks. To demonstrate the improvement produced by this proposal, the path planning algorithm is applied in comparison with a greedy strategy also based on Q-learning. During this comparison, the execution time as well as the agents' reward acquisition were examined in different cases. The simulation results reflect the advantages provided by this new algorithm in comparison with related work.

Keywords Multi-agent games · Ludo game · Collaborative path planning · Markov decision process · Reinforcement learning

1 Introduction

Board games [1] are multi-player games that can be considered as a type of tabletop game. For each participating player, the main objective is to choose the best playing strategy allowing the achievement of the goal game before the other players. The playing strategies usually depend on a random number sequence obtained via the use of dice or on cards' draw containing ambiguous instructions. Among the board games, Ludo [2] is one of the most popular games of this kind. This non-deterministic game can be played through the participation of two or four players. In this derivative of Pachisi [3], each player manages a group of four pawns or agents with the aim of providing them a collaborative path planning strategy that allows them to reach the goal location before the concurrent groups.

M. E. H. Souidi (✉) · T. M. Maarouk · A. Ledmi
Department of Computer Science, ICOSI Lab, University of Khenchela, Khenchela, Algeria

V. Suma et al. (eds.), *Inventive Systems and Control*,
Lecture Notes in Networks and Systems 204,
https://doi.org/10.1007/978-981-16-1395-1_4

Multi-agent path planning [4] can be defined as the movement coordination of the mobile agents to reach the goal location. This coordination allows the cooperation of the agent in a centralized or a decentralized way. Regarding the centralized path cooperation [5], the trajectory of each agent depends on the trajectory of the other agents such as in the Ludo game. The decentralized path cooperation [6] concerns the cases where the agents are independently moving without taking into consideration the trajectories of each other. Path planning mechanisms are usually related to the multi-player games with the aim of providing adequate displacement strategies for the game's components (agents). Pursuit-Evasion game (PEG) [7] was one of the most targeted cases regarding the application of the path planning methods during the last years. In [8], the authors proposed an environment modeling according to MDP principles in order to guide the agents' motion. Precisely, the distributed payoffs in the environment were proportionally inversed in relation to the distance between the evader and the concerned environment's cells. Furthermore, they are based on Q-learning to allow the pursuers to choose their directions during the pursuit iterations. In [9], they introduced a new extension of Bug algorithms [10] to avoid the encountered obstacles. Even in this work, they have based on the reward function to detect the obstacle's leaving points in less time in relation to the precedent Bug algorithms proposed.

In relation to the important works regarding the multi-agent Ludo game [11], they attempted to improve the domain-specific knowledge for Ludo and its variant race games. In fact, they estimated the state-space complexity of the Ludo game, proposed and analyzed strategies based on four basic moves. Furthermore, they experimentally compared the pure and mixed versions of the studied strategies. In relation to reinforcement learning [12], they proposed an expert player based on improving their proposed basic strategies on Ludo. Specifically, they implemented a TD (λ) based Ludo player and used the expert player to train it. Moreover, they implemented a Q-learning based Ludo player using the achieved knowledge from building the expert player. The obtained results showed a slight advantage to the benefit of the Q-learning and TD (λ) players in relation to the expert player. In [13], they presented a new strategy to enhance the performance of swarm algorithms [14] based on the rules of Ludo. Based on these rules, two-Ludo game-based swarm intelligence was introduced (LGSI2, LGSI4) in order to simulate the cases of two and four Ludo players. In LGSI2, the two players are the moth flame optimization (MFO) [15] and grasshopper optimization algorithm (GOA) [16]. Meanwhile, the sine cosine algorithm (SCA) [17] and gray wolf optimization (GWO) [18] are used in addition to the two previous algorithms in the case of LGSI4. The main objective of this work was to make the two/four algorithms compete to find the optimal solution. Recently, neural networks [19, 20] are hugely used in combination with reinforcement learning in order to drive optimal policies for the agents. In [21], they introduced an important aspect of deep reinforcement learning in relation to the situations requiring a set of agents to communicate and collaborate with each other to resolve complex problems.

In this article, the application of MDP principles is focused to model the Ludo game environment. This modeling permits the application of the new multi-agent path planning proposed. This collaborative path planning is based on Q-learning, and a set of playing proprieties that the agents have to follow in order to block the motion strategies undertaken by the adversary players and achieve the goal cell.

The article is organized as follows: Sect. 2 describes the Ludo problem and also how the environment is modeled to allow the motion of the agents. Section 3 introduces the new collaborative path planning proposed. Section 4 reflects the main simulation results obtained in comparison with another multi-agent path planning strategy. Finally, Sect. 5 concludes the research work.

2 Problem Description

The Ludo problem can be considered as a multi-agent game, in which different groups of agents are trying to achieve their goal cells according to the values furnished by a dice. Recognizing that the groups are using only one dice in turn, the objective of the agents belonging to the same group is to reach their concerned goal cell before the agents belonging to other groups.

In this kind of game, two different agent's behaviors were analyzed such as collaborative and concurrent behaviors. On the one hand, the collaborative behavior concerns the agents belonging to the same group, which are always cooperating according to a specific path strategy until the end of the game. The concurrent behaviors of the agents belong to different groups. In other words, an agent can bring back another agent of another group to its respective initial position if the first agent achieves the cell occupied by the second one (killing case). Moreover, an agents' group can reuse the dice in two different cases, if the dice returns the number six or in the case of killing.

Figure 1 represents our simulation environment. This figure shows the initial state of the Ludo game. Note that the agents belonging to the same group have the same color. The environment is composed of different cells. The features of each cell's type can be summarized as follows:

(a) The initial cells: The cells containing the agents until they obtain the dice number (number six) bringing them to the beginning cell. Each initial cell is accessible to only one agent.
(b) The beginning cells: These are the cells containing arrows. Each beginning cell receives the agents of the respective group from their initial cells. Furthermore, these cells are considered neutrals because they can contain more than one agent of the same or different groups at the same time.
(c) The star cells: These cells are also considered neutrals.
(d) The black cells: On the one hand, this kind of cell can contain more than one agent if they belong to the same group. On the other hand, if the cell is already occupied by an agent, and another agent from another group will access the black cell, and the first agent will return to its initial cell.

Fig. 1 Ludo game environment

(e) The colored cells: These cells can be accessed only by the agents of the same color as the concerned cells. Each cell can contain more than one agent. It can be assumed that these cells are also neutrals.

(f) The colored triangle: Each colored triangle represents the goal cell of the agents having the same color as the concerned triangle. This kind of cell has the same priorities in comparison with the colored cells.

In order to implement our path planning method, the usefulness in modeling the environment game as follows.

2.1 Multi-agent Task Planning

In MAS, a task planning method could be defined as an automatic mechanism that answers the question "which agent or set of agents should be assigned to resolve the concerned task or the set of tasks." In the Ludo problem, the agents are coordinated according to the agent-group-role organizational model. In fact, each agent belongs to a specific group and plays a specific role. In this problem, there exist four groups formed by four agents each. The role proposed by each group is the motion planning strategy that the agents have to undertake to achieve the goal cell.

2.2 *Multi-agent Path Planning*

MDP is a discrete-time stochastic control process often used in MAS in order to allow the displacement of the agents in the environment. This stochastic process is based on the quadruplet (S, A, R, T) where:

S: The set of environment states.
A: The set of possible actions that the agents can execute.
R: The reward function that determines the different payoffs returned to the agents according to the undertaken actions.
T: The transition function that allows the agents to switch from a state to another one through the execution of a specific action. This switch is totally based on the payoffs returned via the application of the reward function in the environment.

To allow the motion of the agents, the environment is modeled according to the MDP principles. In fact, the values of a reward function are distributed in each environment cell regarding each agent's group. This reward function is proportionally inverse to the distance separating the position of each agent from its goal cell according to the circuit to cross. The distance is recursively calculated as follows

$$\text{dist}_i^k = \begin{cases} \text{dist}_{i-1}^k - 1 \\ \text{dist}_0^k = \alpha \end{cases} \tag{1}$$

$$r_i^k = \alpha - \text{dist}_i^k \tag{2}$$

α: The number of cells between the beginning and the goal cell regarding the group k.

Moreover, each cell contains an index detailing if the cell is free or occupied by another agent. The cell's values can be represented by the vector $[x, y, z, w, n]$ where:

x: The reward value could be obtained by the agents belonging to the first group if they achieve the concerned cell.
y: The reward value could be obtained by the agents belonging to the second group if they achieve the concerned cell.
z: The reward value could be obtained by the agents belonging to the third group if they achieve the concerned cell.
w: The reward value could be obtained by the agents belonging to the fourth group if they achieve the concerned cell.
n: The cell index. This variable can reflect five different value:

$$n = \begin{cases} 0 \text{ if empty(cell)} = \text{true or neutral(cell)} = \text{true} \\ 1 \text{ if agent } \in \text{Group}_1 \text{ and agent} \subset \text{cell and neutral(cell)} = \text{false} \\ 2 \text{ if agent } \in \text{Group}_2 \text{ and agent} \subset \text{cell and neutral(cell)} = \text{false} \\ 3 \text{ if agent } \in \text{Group}_3 \text{ and agent} \subset \text{cell and neutral(cell)} = \text{false} \\ 4 \text{ if agent } \in \text{Group}_4 \text{ and agent} \subset \text{cell and neutral(cell)} = \text{false} \end{cases} \quad (3)$$

3 Path Planning Algorithm

In this section, the new MAS path planning proposed is introduced through its application to the Ludo problem. Knowing that in the Ludo game, only one agent can move at each iteration, the path planning of the agent must be centralized. Furthermore, the main objectives of the algorithm are described and also explain the necessity of each motion planning's step:

Algorithm 1: the proposed Ludo game algorithm

```
While end(game) = false do
        For each group do
          Use-dice(group_i);
          If Use-dice(group_i) = 6 then
                if killing(group_i) = true then
                   Killing-best-choice(group_i);
                   Ask-killed-agent(move-to(waiting-line));
                   Goto(line 3);
                Else-if empty(waiting-line(group_i)) = false then
                   Ask-one-of-group_i(move-to(initial-cell));
                   Goto(line 3);
                Else-if neutral(next-cell(group_i)) = true then
                   Move-best-neutral(group_i);
                   Goto(line 3);
                Else
                   Move-best-agent(group_i);
                   Goto(line 3);
                End-if;
          Else
                if killing(group_i) = true then
                   Killing-best-choice(group_i);
                   Ask-killed-agent(move-to(waiting-line));
                   Goto(line 3);
                Else-if neutral(next-cell(group_i)) = true then
                   Move-best-neutral(group_i);
                Else
                   Move-best-agent(group_i);
                End-if;
        End-if;
      Update(cells-index);
     counter ← 0;
    For each agent do
      If agent_i ⊂ Goal-cell then
        counter ← counter + 1;
      End-for;
      If counter = 16 then
        End(game) ← true;
   End-for;
  End-while;
```

At the beginning of the game, the reward values of each cell are statically calculated according to equation (2). However, the index of each cell is dynamically updated at each game the iteration are according to the group of the agent situated in the concerned cell. The objective of each agents' group in the Ludo problem is to reach the concerned goal cell before the other groups. Algorithm 1 details how the agents belonging to the same group should cooperate to achieve this goal. This algorithm is based on the reward function generated through the application of MDP principles. Knowing that the groups are using the dice in turn, each group must make a decision in order to determine which agent has to change their position. In order to deal with this problem, our algorithm is based on a set of priorities.

The first priority concerns the adversary killing. If the agents have the possibility to kill opponent agents, the algorithm classifies the importance of the targets according to the rewards detected in the environment. This importance is predicted in relation to the reward's degree of each target's cell calculated according to equation (2). In other words, the agents must kill the target belonging to the cell with the highest reward (the closest target to the concerned goal cell). Realizing that our algorithm is based on the collaboration of the agents, the second priority is based on the increase of the playing agents' number in order to increase the playing possibility. This priority is reflected in the algorithm by asking the agents belonging to the waiting line to move from the initial to the beginning cell until the emptying of the concerned line. The agents can only participate in the game if they are not placed in their initial cells. This propriety is executed if the is no killing possibility, and the waiting line is still containing agents. The third priority is to avoid the decrease of the playing agents (avoid to be killed). To do that, the agents have to move to the best possible neutral cell. The best neutral cell is the neutral cell returning the highest reward's degree. If the conditions of preceding priorities are not met, the agent situated in the cell returning the highest reward degree will be asked to move:

$$Q_{i+1}(\mathrm{s},\mathrm{a}) = (1-\alpha)Q_i(\mathrm{s},\mathrm{a}) + \alpha\left[r + \lambda \max_{a'}[Q_i(S_{ag1},a'),Q_i(S_{ag2},a''),Q_i(S_{ag3},a'''),Q_i(S_{ag4},a'''')]\right] \quad (4)$$

Knowing that the Ludo environment is deterministic, then $\alpha = 1$:

$$Q_{i+1}(\mathrm{s},\mathrm{a}) = r + \lambda \max_{a'}[Q_i(S_{ag1},a'),Q_i(S_{ag2},a''),Q_i(S_{ag3},a'''),Q_i(S_{ag4},a'''')] \quad (5)$$

Before passing the dice from a group to another one, the algorithm checks if all the existing agents have already achieved their cells destination for further processing. If this condition is met, the game is considered as finished. Each group automatically transmits the dice to the next group if all its agents have already achieved the goal cell. This priority is checked according to the following equation:

$$\mathrm{Max}_r = \frac{1}{4} \times \sum_{i=1}^{4} r_s^{Ai} \quad (6)$$

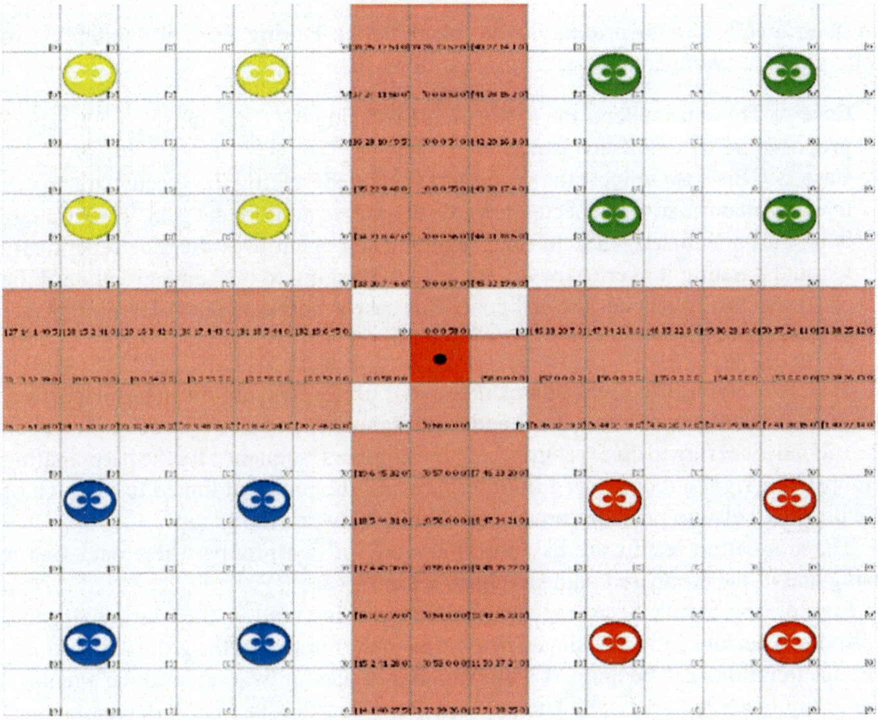

Fig. 2 Ludo problem modeling via MDP principles

As shown in Fig. 2, the maximum reward of each agent is reflected by the number of cells across in order to achieve the goal cell (58 cells).

The novelty part of this proposed algorithm is the introduction of MDP principles in the Ludo problem. Moreover, this algorithm is based on a set of proprieties allowing the increase of the collaboration degree between the agents belonging to the same group. This collaboration is reflected by the algorithm's defense strategy allowing the selection of the best target to kill. In addition, this collaboration is also reflected by the increase of the playing possibility through the increase of the participating players. The path planning proposed can also be used in order to solve engineering problems such as robot path control.

4 Simulation Results

In order to implement the Ludo problem, the application to use the NetLogo platform was analyzed [22]. This agent-oriented platform proposes predefined methods that facilitate our implementation. From Fig. 1, each agent must browse 58 cells with the aim of reaching the concerned goal cell. This agents' circuit is composed of

14 neutral cells and 44 ordinary cells (black cells). During these simulations, two different cases were considered:

Case A: This case reflects the group of agents using the cooperative path planning proposed in Sect. 3 of this paper.
Case B: This case reflects the group of agents undertaking the greedy strategy or trying to maximize the expected rewards in order to achieve the goal. This strategy is based on selfish agents. In other words, each agent uses the dice repetitively without handing it over to the other agents belonging to the same group until the concerned agent reaches the cell goal. This case is based on the Q-learning player proposed in [12].

Moreover, two different kinds of dice were used. The first dice (ordinary dice) returns a random number between 1 and 6 throughout each game iterations. However, the second dice (rigged dice) returns the same numbers' sequence for the participating groups. The rigged dice is used in order to avoid the problem linked to the luck of the draw and also to provide certain equitability between the groups.

The simulation results are based on the study of two groups where each one is using one of the compared path strategies (case A, case B).

Figure 3 represents the average game duration using random dice during 50 game episodes. The duration is calculated in relation to the number of the groups' iterations. Groups iterations can be defined as the number of dice utilization until the agents of the group reach the goal cell. The average game duration in case A decreases until 29.23% in relation to case B. This fact is due to the intelligent behavior of the agents provided by the collaborative path planning proposed. In other words, the priority

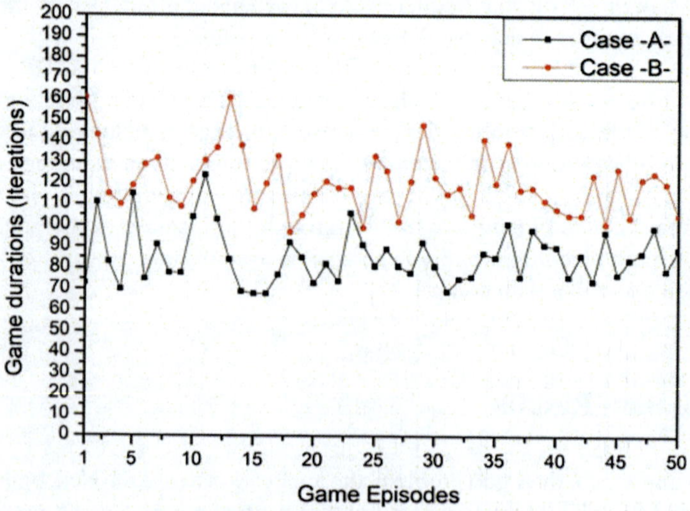

Fig. 3 Average game duration using a random dice

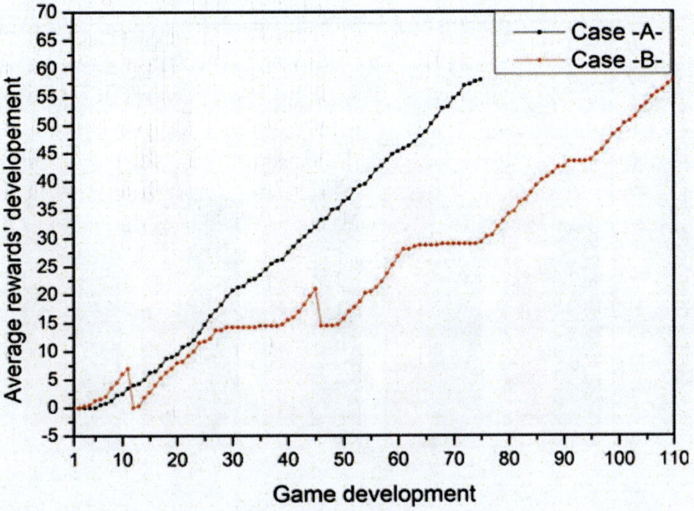

Fig. 4 Average rewards development during a game episode

aiming to empty the waiting line allows the increase of the collaboration rate between the agents (the increase of the decision-making possibility).

Figure 4 showcases the average reward's development of the agents during a complete game episode using random dice. The agents of case A reach the maximum average reward after 75 game iterations. However, in the case of B, the agents achieve the maximum reward after 110 iterations. Also, there are two important decreases regarding case B (iteration 11 and iteration 46). These decreases mean that the agents using the collaborative path planning realized two killings during this game episode. This fact showcases the advantages of the killing priority of the proposed algorithm. This priority provokes a certain imbalance in the opponent group leading to a substantial delay regarding the task execution.

Figure 5 reflects the average game duration in five-game episodes using rigged dices. In each episode, the rigged dice generates a repetitive sequence of numbers as follows.

In relation to the average game duration obtained in case B, the average decreases to 24.34% in case A. This result confirms the results shown in Fig. 3. Remembering that in this case, the players belonging to the different groups draw equitably the dice numbers.

Figure 6 shows the average rewards obtained by the agents per iteration during a complete game episode. In these cases, the rigged dice number 2 reflected in Table 1. The average reward per iteration is calculated as follows:

$$\Omega_s = \frac{\left(r_s^{A1} + r_s^{A2} + r_s^{A3} + r_s^{A4}\right) - \left(r_{s-1}^{A1} + r_{s-1}^{A2} + r_{s-1}^{A3} + r_{s-1}^{A4}\right)}{4}$$

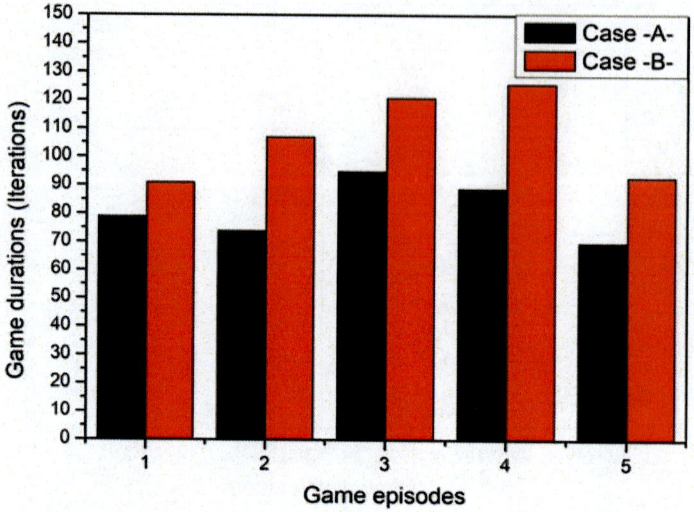

Fig. 5 Average game duration using a rigged dice

$$\Omega_s = \frac{1}{4} \times \sum_{i=1}^{4} (r_s^{Ai} - r_{s-1}^{Ai}) \tag{7}$$

In case B of this figure, there are three negative decreases in the iterations 12, 21 and 26 meaning that the agents of case A realized three killings during this game episode. Furthermore, the average reward obtained in case A is 0.84. However, this average decreases to 0.29 in case B. Furthermore, the results shown in this figure aim to validate the results shown in Fig. 2, in which the rewards are based on random dice.

The limitation of the compared algorithm and addressed in this new proposition is the greedy behavior of the agents belonging to the same group. This selfish strategy provokes the inexistence of cooperation between the agents. Furthermore, the compared strategy does not undertake a defense strategy regarding the kill of adversary players. Table 2 summarizes all the obtained results during the simulation part. From these results, the superiority of the collaborative path planning proposed in comparison with the greedy strategy. This superiority concerns the agents' development during the game (reward acquisition) and also the game duration (task execution time).

5 Conclusion

In this article, a new collaborative path planning is based on the payoffs generated through the application of MDP principles. This path planning aims to increase the

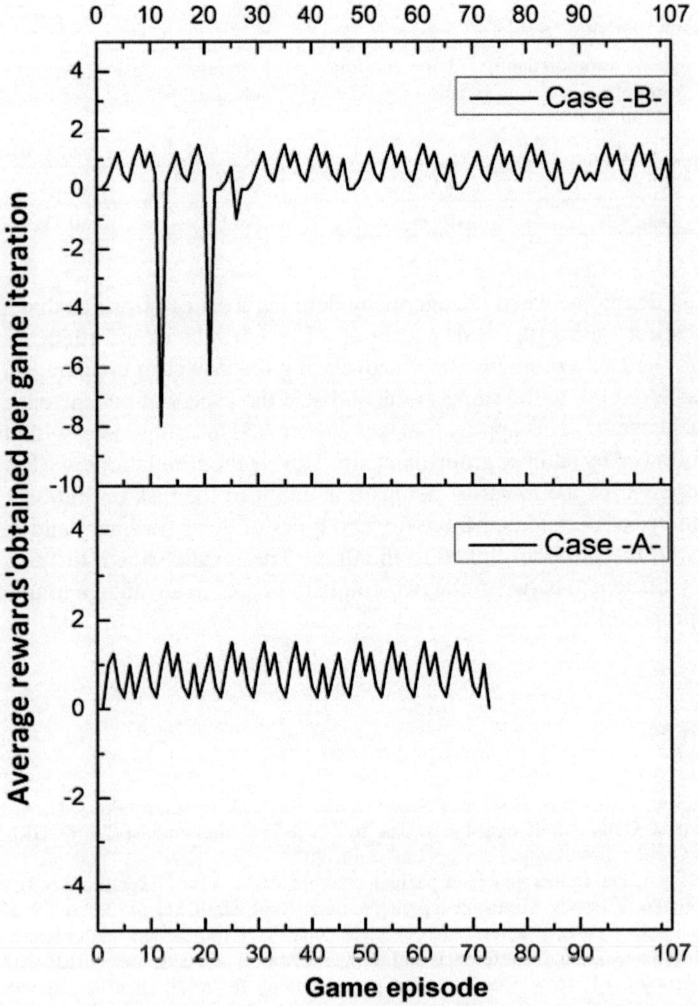

Fig. 6 Average reward's obtained per iteration during a complete game episode

Table 1 Static cases studied

Game episodes	Numbers' sequence
1	[1, 2, 3, 4, 5, 6]
2	[6, 3, 5, 2, 1, 4]
3	[5, 6, 2, 1, 4, 3]
4	[1, 3, 6, 4, 5, 2]
5	[3, 4, 6, 5, 1, 2]

Table 2 Main simulation results

	Game duration using ordinary dice (iterations)	Game duration using rigged dices (iterations)	Average reward obtained per iteration using ordinary dice	Average reward obtained per iteration using rigged dices
Case A	85.44	81.4	0.75	0.84
Case B	120.74	107.6	0.093	0.29

cooperation degree between the agents by defining a set of priorities that the agents have to respect during the task execution. The advantages afforded by this new proposal to the Ludo problem allow showcasing the aspect of cooperation between the agents belonging to the same group and also the aspect of concurrence between the opponent agents. This application was effectuated in comparison with the greedy strategy followed by another group of agents. During the simulations, we have studied the development of the rewards' acquisition and also the task execution's duration via the number of iterations. Moreover, two types of dices (ordinary and rigged) are used to avoid the problem linked to the draw. The results shown in Sect. 4 reflect the glaring difference between the two compared cases in advantage to the new path planning proposed.

References

1. D.S. Parlett, *The Oxford History of Board Games* (Oxford University Press, USA, 1999)
2. M. Kroll, R. Geiß, A ludo board game for the agtive 2007 tool contest (2007). URL: http://gtcases.cs.utwente.nl/wiki/uploads/ludokarlsruhe.pdf
3. W.N. Brown, The Indian games of pachisi, chaupar, and chausar. Expedition **6**(3), 32 (1964)
4. A. Gorbenko, V. Popov, Multi-agent path planning. Appl. Math. Sci. **6**(135), 6733–6737 (2012)
5. R. Luna, K.E. Bekris, Efficient and complete centralized multi-robot path planning, in 2011 IEEE/RSJ International Conference on Intelligent Robots and Systems. (IEEE, 2011)
6. V.R. Desaraju, J.P. How, Decentralized path planning for multi-agent teams with complex constraints. Autonomous Rob. **32**(4), 385–403 (2012)
7. M.El.H. Souidi, A. Siam, Z. Pei, Multi-agent pursuit coalition formation based on a limited overlapping of the dynamic groups. J. Intell. Fuzzy Syst. Preprint 1-13 (2019)
8. M.El.H. Souidi, et al., Multi-agent pursuit-evasion game based on organizational architecture. J. Comput. Inform. Technol. **27**(1), 1-11 (2019)
9. M.El.H. Souidi, S. Piao, G. Li, Mobile agents path planning based on an extension of Bug-Algorithms and applied to the pursuit-evasion game. Web Intell. **15**(4) (2017)
10. S. Rajko, S.M. LaValle, A pursuit-evasion bug algorithm, in *Proceedings 2001 ICRA. IEEE International Conference on Robotics and Automation (Cat. No. 01CH37164)*, vol 2. (IEEE, 2001)
11. F. Alvi, M. Ahmed, Complexity analysis and playing strategies for Ludo and its variant race games, in *2011 IEEE Conference on Computational Intelligence and Games (CIG'11)*. (IEEE, 2011)
12. M. Alhajry, F. Alvi, M. Ahmed, TD (λ) and Q-learning based Ludo players, in *2012 IEEE Conference on Computational Intelligence and Games (CIG)*. (IEEE, 2012)
13. P.R. Singh, M.A. Elaziz, S. Xiong, Ludo game-based metaheuristics for global and engineering optimization. Appl. Soft. Comput. **84**, 105723 (2019)

14. J. Kennedy, C.E. Russell, A discrete binary version of the particle swarm algorithm, in *1997 IEEE International conference on systems, man, and cybernetics. Computational cybernetics and simulation*, vol. 5. (IEEE, 1997)
15. S. Mirjalili, Moth-flame optimization algorithm: a novel nature-inspired heuristic paradigm. Knowl. Based Syst. **89**, 228–249 (2015)
16. S. Saremi, S. Mirjalili, A. Lewis, Grasshopper optimisation algorithm: theory and application. Adv. Eng. Softw. **105**, 30–47 (2017)
17. S. Mirjalili, SCA: a sine cosine algorithm for solving optimization problems. Knowl. Based Syst. **96**, 120–133 (2016)
18. S. Mirjalili, S.M. Mirjalili, A. Lewis, Grey wolf optimizer. Adv. Eng. Softw. **69**, 46–61 (2014)
19. J.S. Raj, J. Vijitha Ananthi, Recurrent neural networks and nonlinear prediction in support vector machines. J. Soft Comput. Paradigm (JSCP) **1**(1), 33–40 (2019)
20. P.P. Joby, Expedient information retrieval system for web pages using the natural language modeling. J. Artif. Intell. **2**(02), 100–110 (2020)
21. T.T. Nguyen, N.D. Nguyen, S. Nahavandi, Deep reinforcement learning for multiagent systems: a review of challenges, solutions, and applications. IEEE Trans. Cybern. (2020)
22. U. Wilensky, *Center for Connected Learning and Computer-Based Modeling* (Northwestern University, NetLogo, 1999)

An Analysis of Epileptic Seizure Detection and Classification Using Machine Learning-Based Artificial Neural Network

P. Suguna, B. Kirubagari, and R. Umamaheswari

Abstract Epileptic seizure appears owing to the disorder in brain functioning that damages the health of the patients. Identification of seizures in the earlier stage is helpful to prevent seizure from treatment. Machine learning (ML) and computational models are applied to predict the seizures from electroencephalograms (EEG) signals. EEG signal-based epileptic seizure identification is a hot research topic that proficiently identifies the non-stationary development of brain activities. Basically, epilepsy is identified by physicians based on the visual reflection of EEG signals which is a tedious and time-consuming task. This article presents an efficient epileptic seizure detection and classification using the ML-based artificial neural network (ANN) model. The ANN is a biologically evolved computational model that activates a system for learning tracking details. It is utilized commonly for developing the prediction results. The performance of the proposed ANN model undergoes validation using EEG signal dataset, and the experimentation outcome verified the superior performance of the ANN model.

Keywords Classification · EEG · Machine learning · Seizure detection · Artificial neural network

1 Introduction

Several patients are suffering due to seizures which are caused due to the abnormal activity of the brain which is named epilepsy [1]. A maximum number of people were diagnosed with epilepsy, especially in the US where millions of peoples were infected by epilepsy. Epilepsy is one of the common brain infections [2], which is

P. Suguna (✉)
Department of Computer Science, Annamalai University, Chidambaram, India

B. Kirubagari
Department of Computer Science and Engineering, Annamalai University, Chidambaram, India

R. Umamaheswari
Gnanmani College of Technology, Namakkal, India

V. Suma et al. (eds.), *Inventive Systems and Control*,
Lecture Notes in Networks and Systems 204,
https://doi.org/10.1007/978-981-16-1395-1_5

caused by several reasons like a molecular mutation that leads to immigration of neurons. Even though the major cause of epilepsy is unknown, primary analysis is highly helpful to consume the treatment accordingly. Epilepsy patients are treated with high dosage medicines or surgical events [3]. Therefore, these medicines are not so effective. But, seizures that are treated completely would enhance the mortal rate of a patient. At this point, the patients become highly inactive and need the help of other people. The previous detection of epileptic seizures takes sufficient time before its occurrence where the early treatment would save the life of a patient.

Epileptic seizures are composed of four phases like a preictal state: It is the beginning stage of seizure; secondly, ictal state: It is invoked with the onset of seizure; thirdly, postictal state: It emerges after the first phase; finally, interictal state: It is initialized after third stage of seizure and results in a subsequent seizure. Moreover, seizures are detected at the initial stage of the preictal state. Inspection of preictal state [4] detects the seizure. Hence, the main objective of this examination is to analyze the presence of a preictal state for epileptic seizures. Here, ML methods have been employed to diagnose seizures. Also, EEG signal acquisition, signal pre-processing, feature extraction, and classification [5] are performed for various seizure conditions. The main theme of the detection method with ML is to examine preictal states for enough duration before starting the seizure onset. Therefore, sufficient time for the detective preictal state of a gland and higher sensitivity is significant, and it is considered as a performance problem of epileptic seizures.

Pre-processing and feature extraction from EEG signal highly impact the detection time as well as a true positive rate (TPR). Initially, Pre-processing is carried out to remove the artifacts from signals and to enhance the signal-to-noise ratio (SNR). Massive developers [6] defined pre-processing for transforming various channels of EEG signals as a surrogate channel, and filters are utilized for enhancing the SNR. EEG signals are obtained under the application of several electrodes which are converted as a surrogate channel by applying an averaging filter, common spatial filtering (CSP), large Laplacian filter, as well as optimized spatial pattern (OSP) filtering. Moreover, the extractions of linear and nonlinear features are essential for this detection of seizures [7].

An attempt is made to predict the initialization of the preictal state using EEG signals [8]. Therefore, some considerable prediction of the preictal state of epilepsy has been performed. Pre-processing of the EEG signal is to enhance SNR, and features extraction is highly essential in predicting the seizure. The concatenation of various exclusive features into a feature vector is utilized for examining the preictal state of epileptic seizures. Rasekhi et al. [9] projected a method for detecting seizures using univariate linear features. Developers have applied 6 EEG channels and obtained 22 univariate linear characteristics.

Also, it has utilized the support vector machine (SVM) classification method for classifying preictal as well as ictal states of signals. Researchers employed univariate linear features under the application of window size. Secondly, pre-processing is performed, and finally, a review was carried out on the EEG signal, with specific normalization. There are 3 EEG channels are obtained by fixing electrodes in the scalp of a patient, whereas three electrodes are placed on the external surface. Once the conversion process is completed, the Butterworth filter [10] is applied for reducing the artifacts and irregularities. The developer is obtained four statistical moments as features. To resolve the predefined problems, the developers have normalized these features, so that the complete elimination of noise is attained. Moreover, smoothing is processed on EEG signals for noise elimination.

This article presents an efficient epileptic seizure detection and classification using ML-based ANN model. The ANN is a biologically evolved computational model that activates a system for learning tracking details. It is utilized commonly for developing the prediction results. The performance of the proposed ANN model undergoes validation using EEG signal dataset, and the experimentation outcome verified the superior performance of the ANN model.

2 ANN-Based Classification Model

In general, neural networks (NNs) include a set of techniques modeled by reflecting the actions of the human brain. It is mainly used for performing nonlinear statistical modeling. ANN is a computational model used to activate a system for learning tracking details. It is utilized commonly for developing the prediction results. A fundamental element in the network structural design is known as neurons that perform a specific operation according to the activation function. A fundamental structural design is composed of three types of neuron layers, namely input, hidden, and resultant layers as depicted in Fig. 1.

where aj implies the outcome of a jth neuron in the hidden layer, o_k refers to the network result, $\omega_{i,j}$ indicates the network weight to a path (i, j), and b_j denotes jth neuron bias. In feed-forward networks, signal flow is driven from input to outcome units. The center premises of training NN are to reduce the cumulative error as provided below:

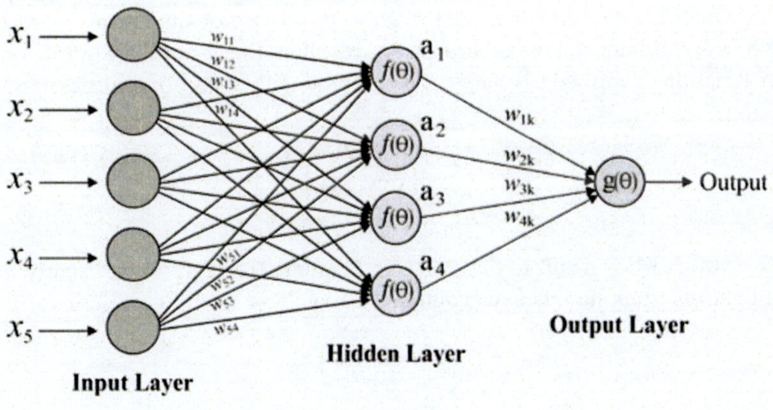

Fig. 1 Structure of artificial neural network

$$E = \frac{1}{2}\sum_{\mathrm{i=1}}^{\mathrm{N}}(y_j - o_k)^2 \tag{1}$$

where

N represents the overall count of neurons, y_i signifies the desired result, and o_k denotes the exact neuron outcome. Based on the system error, weights w is changed into:

$$w_{ij} = w_{ij} \pm ax_i \tag{2}$$

where α implies a learning parameter, and the sign is +ve or −ve which depends upon the outcome and errors. Several researchers have stated that NNs are applied for the classification process. ANN is stimulated by inducing the input vector to the input layer and propagate the activations in a feed-forward manner, through weighted connections, with the whole network. The applied input x_k is a state of i-th neurons (s_i) is calculated by:

$$s_i = f\left(w_{i,0} + \sum_{j \in P_l} w_{i,j} \times s_j\right) \tag{2}$$

where P_i signifies the group of nodes obtaining node; i; f refers the activation function; $w_{i,j}$ implies the correlation among nodes j and i; and $s_1 = x_{k,1}, \ldots, s_I = x_{k,I}$. The $w_{i,0}$ associations are known as bias and frequently enhance the MLP learning flexibility. But, many hidden layers are applied for complex operations (2-spirals), the common application for one hidden layer of H hidden nodes with the logistic $\left(f(x) = \frac{1}{1+\exp(-x)}\right)$ activation function. In order to compute binary classification, one resultant node with logistic function in a pattern which permits the result interpretation which is equal to LR model if $H = 0$. In case of multi-class operations (with $N_G > 2$ resultant classes), there are N_G resultant linear nodes $(f(x) = x)$, and softmax function is applied to change the outcome with class possibilities:

$$p(G_c|x_k) = \frac{\exp(\hat{y}_{k,c})}{\Sigma_{j=1}^{N_G}\exp(\hat{y}_{k,j})} \tag{3}$$

where $\hat{y}_{k,c}$ refers MLP result to class G_c. To compute the regression, usually linear form of neurons is applied, because outcomes stay away from linear series ([0, 1]).

3 Performance Validation

The experimental results of the ANN model are examined using the EEG signal dataset. It holds a set of 2300 instances under class 0 and 9200 instances under class 1 as shown in Table 1. The results are examined in terms of precision, recall, F-score, and accuracy.

Table 2 performs a detailed comparative study of the ANN with existing models in terms of different measures. Figures 2 and 3 investigate the classifier results analysis of the proposed ANN model in terms of distinct performance measures. On analyzing the results in terms of precision, the ANN model has reached a maximum precision of 80.15, whereas the KNN and MLP models have exhibited lower precision values of 70% and 77%, respectively. At the same time, on determining the classifier outcome in terms of recall, the ANN model has achieved significant classification performance with the recall of 80.26%, whereas the KNN and MLP models have exhibited lower precision values of 76% and 79%, respectively.

Table 1 Dataset description

Class name	Class label	No. of instances
EEG signals having seizure activity	0	2300
EEG signals not having seizure activity	1	9200

Table 2 Result analysis of existing with proposed method on applied dataset

Methods	Precision	Recall	F-score	Accuracy
ANN	80.15	80.26	80.44	80.21
KNN	70.00	76.00	72.00	76.00
MLP	77.00	79.00	78.00	78.00
Linear SVM	–	–	–	77.10

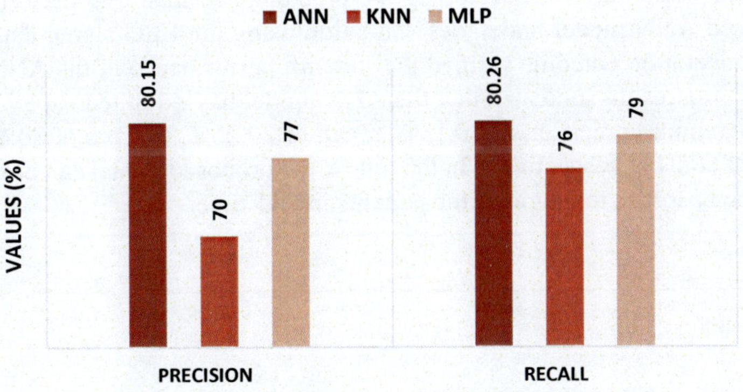

Fig. 2 Comparative analysis of linear SVM model

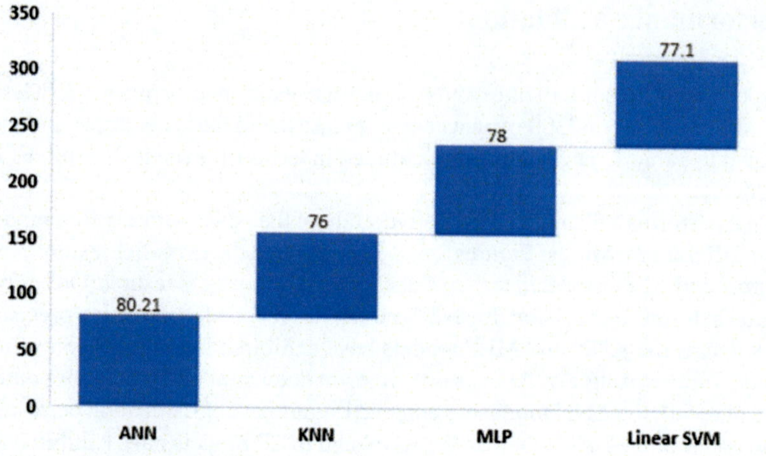

Fig. 3 Accuracy analysis of linear SVM model

Similarly, on determining the classifier outcome in terms of F-score, the ANN model has obtained betterment with the F-score of 80.44%, whereas the KNN and MLP models have exhibited lower precision values of 72% and 78%, respectively. Finally, on evaluating the classification performance in terms of accuracy, the ANN model has found to be superior classification over the other methods with the accuracy of 80.21%, whereas the KNN, MLP, and linear SVM models have exhibited lower precision values of 76%, 78%, and 77.10%, respectively.

4 Conclusion

This article has introduced a novel epileptic seizure detection and classification using ML-based ANN model. The ANN comprises a set of input, hidden, and output layers. It is utilized commonly for developing the prediction results. The performance of the proposed ANN model undergoes validation using the EEG signal dataset, and the experimentation outcome verified the superior performance of the ANN model. The simulation results indicated that the ANN model has exhibited superior results with the maximum precision of 80.15%, recall of 80.26%, F-score of 80.44%, and accuracy of 80.21%, respectively. In the future, the proposed model can be deployed in real-time hospitals to diagnose the patients in real-time.

References

1. U.R. Acharya, S.V. Sree, G. Swapna, R.J. Martis, J.S. Suri, Automated EEG analysis of epilepsy: a review. Knowl. Based Syst. **45**, 147–165 (2013)

2. L.E. Hebert, P.A. Scherr, J.L. Bienias, D.A. Bennett, D.A. Evans, Alzheimer disease in the US population: prevalence estimates using the 2000 census. JAMA Neurol. **60**(8), 1119–1122 (2003)
3. M. Guenot, Surgical treatment of epilepsy: outcome of various surgical procedures in adults and children. Rev. Neurol. **160**(5), S241–S250 (2004)
4. Y. Wang, W. Zhou, Q. Yuan et al., Comparison of ictal and interictal eeg signals using fractal features. Int. J. Neural Syst. **23**(6) (2013). Article ID 1350028
5. J. Engel, ILAE classification of epilepsy syndromes. Epilepsy Res. **70**(1), 5–10 (2006)
6. H. Ramoser, J. Muller-Gerking, G. Pfurtscheller, Optimal spatial filtering of single trial EEG during imagined hand movement. IEEE Trans. Neural Syst. Rehabil. Eng. **8**(4), 441–446 (2000)
7. K.C. Chua, V. Chandran, U.R. Acharya, C.M. Lim, Application of higher order statistics/spectra in biomedical signals—a review. Med. Eng. Phys. **32**(7), 679–689 (2010)
8. A.S. Zandi, R. Tafreshi, M. Javidan, G.A. Dumont, Predicting epileptic seizures in scalp EEG based on a variational bayesian gaussian mixture model of zero-crossing intervals. IEEE Trans. Biomed. Eng. **60**(5), 1401–1413 (2013)
9. J. Rasekhi, M.R.K. Mollaei, M. Bandarabadi, C.A. Teixeira, A. Dourado, Preprocessing effects of 22 linear univariate features on the performance of seizure prediction methods. J. Neurosci. Methods **217**(1–2), 9–16 (2013)
10. R. Palaniappan, D.P. Mandic, EEG based biometric framework for automatic identity verification. J. VLSI Sig. Process. Syst. Sig. Image Video Technol. **49**(2), 243–250 (2007)

Improving Image Resolution on Surveillance Images Using SRGAN

Aswathy K. Cherian, E. Poovammal, and Yash Rathi

Abstract The process of generating an image from a single low-resolution input to a high-resolution image is known as single image super-resolution. Image resolution finds its application in images where there involves a blurriness or high brightness in the pictures. The same can be corrected to produce a higher resolution image using generative adversarial networks (GAN), specifically super-resolution GAN (SRGAN). This method uses perceptual losses instead of the traditional peak signal-to-noise ratio (PSNR). The application of SRGAN in the field of CCTV footage is immense as the video from a CCTV camera produces low-resolution images. The low-resolution images are passed through a window where it is sliced into smaller images, overlapping, and then passed through the SRGAN. All the SRGAN produced images are then stitched together to make the complete image. Finally, the contrast limited adaptive histogram equalization (CLAHE) filter is applied to the created image with contrast equalization. The final produced images have high resolution and increase visibility through the pictures' hazy, foggy parts.

Keywords Generative adversarial network · SRGAN · Peak signal-to-noise ratio · CLAHE · Descriptor · Generator

1 Introduction

A high-resolution (HR) image is achieved by enhancing the low-resolution (LR) image, which is better insight and properties are now an area of focus by science fiction (Sci-fi) films and scientific literature. This area is known as single image super-resolution (SISR) [1], a subject that has attracted significant attention and advancement in recent years. With super-resolution generative adversarial network

A. K. Cherian (✉) · E. Poovammal · Y. Rathi
SRM Institute of Science and Technology, Kattankulathur, Chennai, India
e-mail: aswathy2205@gmail.com

E. Poovammal
e-mail: poovamme@srmist.edu.in

V. Suma et al. (eds.), *Inventive Systems and Control*,
Lecture Notes in Networks and Systems 204,
https://doi.org/10.1007/978-981-16-1395-1_6

(SRGAN) [2, 3], state of the high-resolution art images can be produced, making them easier to understand and comprehend. Similar resolution enhancement models like ESRGAN and EDSR are also present. The process of enhancing and recovering an HR image from an LR image is called super resolution. The state-of-art techniques used for comparison purposes are the following: regularity-preserving image interpolation [4], new edge-directed interpolation (NEDI) [5], hidden Markov model (HMM) [6], HMM-based image super-resolution (HMM SR) [7], WZP and cycle-spinning (WZP-CS) [8], WZP, CS, and edge rectification (WZP-CS-ER) [9], DWT-based super-resolution (DWT SR) [10], complex wavelet transform-based super-resolution (CWT SR) [11], discrete and stationary wavelet decomposition [12]. The experimental results claim that the proposed method performs in a better way when compared to all other state-of-art techniques for image resolution enhancement.

The super-resolution models' extensive training is undertaken to use low-resolution images as input images and high-resolution images as target images. The central part of mapping the images is the transformation and is taken care of by the inverse of a downgrade function. To produce large datasets, downgrading functions like bicubic down sampling [13] are applied to the images. This processes pipelines to generate low-resolution images from the high-resolution images. Such an approach enables self-supervised learning by the production of extensive dataset from readily available HR images. Figure 1 describes the sample CCTV footage images of a fighting scene happening in the pathway.

Usually, the CCTV footage produced or captured at crime scenes is either very blurry or hazy and at a shallow resolution. Even the CCTV camera is usually placed at angles that generate very noisy images. At every crime scene CCTV, footage plays the primary role and is always referred to, but often due to low quality or noisy images, it becomes difficult to analyze the pictures. Merging the SRGAN produces high-resolution images and low-resolution images of the CCTV. The results are astonishing and have increased the readability and desirability of the images manifold. The SRGAN produces better and more effective outputs when the image input is smaller than the standard size. The entire image can be reduced in size and then made to pass through the SRGAN, or the partitioned image using the sliding window library of Python can be used. This library essentially forms smaller images

Fig. 1 Sample input and output CCTV images

or windows from the main image with suitable overlap, and then these individual images can be made to pass through the SRGAN. Finally, the images are stitched to get the original images. The difference between the three images such as the original image, SRGAN produced images, and the stitched SRGAN images are massive and in ascending order. The stitched SRGAN images can have a variable number of windows and depend on multiple factors. The stitching is done using the OpenCV [14] library of python. Contrast limited adaptive histogram equalization (CLAHE) filter is used to equalize the varying contrasts of the images. Hence, the application of SRGAN in CCTV footages has many uses and prospects.

2 Dataset

The dataset used the DIV2K dataset [15], consisting of both low-resolution images and high-resolution images. It is further divided into a set of training and testing sets of images. The training set images are further divided into the train and validation datasets.

This dataset has 800 images, which consist of high definition high-resolution images. Corresponding images with downscaling factors of 2, 3, and 4 are also generated from the high-resolution images. For validation purposes, 100 images have been taken, which are useful in tuning the proposed models' hyperparameters. The testing data has more than 100 images that consist of the LR images and its counterpart, which is the HR images generated by the model.

3 Literature Survey

Before super-resolution images, high-resolution images were the benchmark for the images. Many conventional techniques like upsampling techniques were used to make the resolution better of the images, but the approach of the SRGAN adds a new super-resolution benchmark to change the entire dynamics of image resolution. The application of SRGAN on CCTV images has not been explored much and "CCTV Surveillance Camera's Image Resolution Enhancement using SRGAN" [16] has tried the implementation of SRGAN on CCTV images. The authors implemented SRGAN on CCTV images without coupling them with any other sort of image enhancing techniques or filters. The proposed model produces super-resolution images but suffers from the varying contrasts to the CCTV images. Since visually the images are of a low resolution along with uneven contrasts and blurriness, it can lead to distortions in the final image. Another article, "A Development of Image Enhancement for CCTV Images," [17] has used different types of homomorphic filters and then compared them on various metrics. The produced images are not super-resolution images and hence lack the clarity which SRGAN produced images have. The possibility of coupling SRGAN with filters has not been explored before as this very useful in generating super-resolution images with less distortion and blurriness. The main problem in the SRGAN generated image is that of blurriness and contrast; hence, filters have to be applied to enhance the SRGAN output. Many SOTA filters are available in the market today for contrast correction like histogram equalization (HE), adaptive histogram equalization (AHE), and contrast limited adaptive histogram equalization (CLAHE). These methods are based on normalizing the contrast histograms to a median value. AHE and CLAHE also take care of the blurriness along with contrast. As discussed in [18, 19], since images of the eye require a good resolution and attention to details, implementation of SRGAN in this field looks promising.

4 Proposed Method

The flow diagram of the proposed SRGAN method is described in Fig. 2. The input image is initially passed through a sliding window process where the images are partitioned into smaller images. These are then passed through the SRGAN to improve the resolution of the images and stitched to get the images. Finally, it is made to transpire through the CLAHE filter to get the final output image. A detailed explanation of each process is summarized below.

4.1 *Using the Sliding Window*

The sliding window is the process in which larger images are broken into smaller images. The input image of a CCTV camera contains a lot of haziness and blurriness. Dividing an image into smaller images and then passing through the SRGAN helps ineffective handling of the unwanted factors, i.e., noise, blurriness, uneven contrast, and distortion present in the image. This leads to a decrease in the number of pixels underscan, and an in-depth analysis of the small area can be done. The images are divided with a certain degree of overlapping in both vertical and horizontal directions, as this facilitates the stitching of the high-resolution output images. The distance matrix for a window, which represents each pixel's position in a window (relative to its center), needs to be taken into account. A tuple represented by (*x*, *y*, *w*, *h*) helps the window objects to be converted from and to rectangles. These tuples of (*x*, *y*, *w*, *h*) give the window dimensions for further manipulations. The pixel channel length is altered according to the input image and is generally half the input

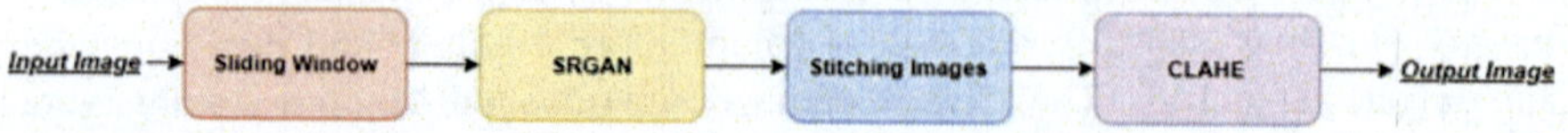

Fig. 2 Flowchart of the proposed SRGAN method

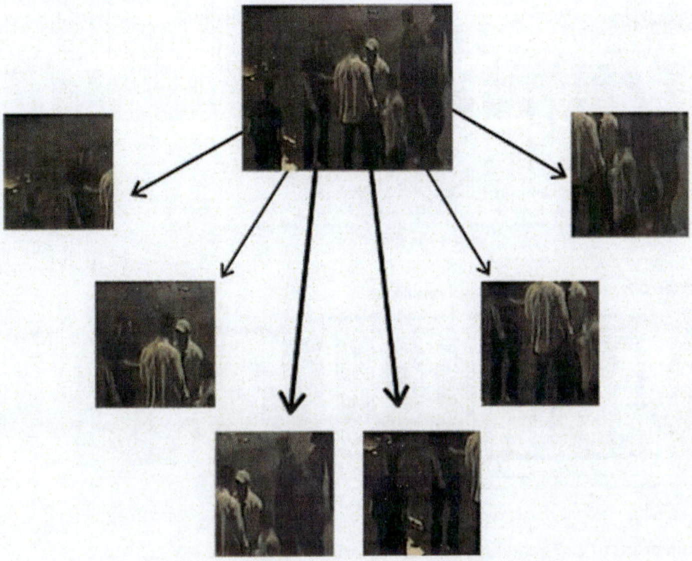

Fig. 3 Images obtained after the process of sliding

image's length to get even several windows and easier stitching. Figure 3 depicts a sample image to show how exactly the sliding works. Here, the image is split into six smaller fragments with overlapping which helps the descriptor functions to map the features while stitching back the images when passing it through the SRGAN. The overlapping can be controlled by the parameters that depend on the input image. The value of the overlapping varies in proportion to the number of frames the image is sliced.

4.2 GAN Model

Generative adversarial network (GAN) consists of two neural networks that work together to make the predictions more accurate and reliable. GAN is a machine learning model. The GAN includes a convolutional neural network generator, a discriminator, and a deconvolutional neural network. A discriminator network is defined to optimize the adversarial min-max problems in an alternating manner.

$$\min_{\theta_G} \max_{\theta_D} E_{I^{\mathrm{HR}} \sim p_{\mathrm{train}}(I^{\mathrm{HR}})}\left[\log D_{\theta_D}\left(I^{\mathrm{HR}}\right)\right] + E_{I^{\mathrm{LR}} \sim p_G(I^{\mathrm{LR}})}\left[\log\left(1 - D_{\theta_D}\left(G_{\theta_G}\left(I^{\mathrm{LR}}\right)\right)\right)\right] \quad (1)$$

The formula helps us the discriminator to become better at discriminating between the real and the generated images. Both the discriminator and the generator are

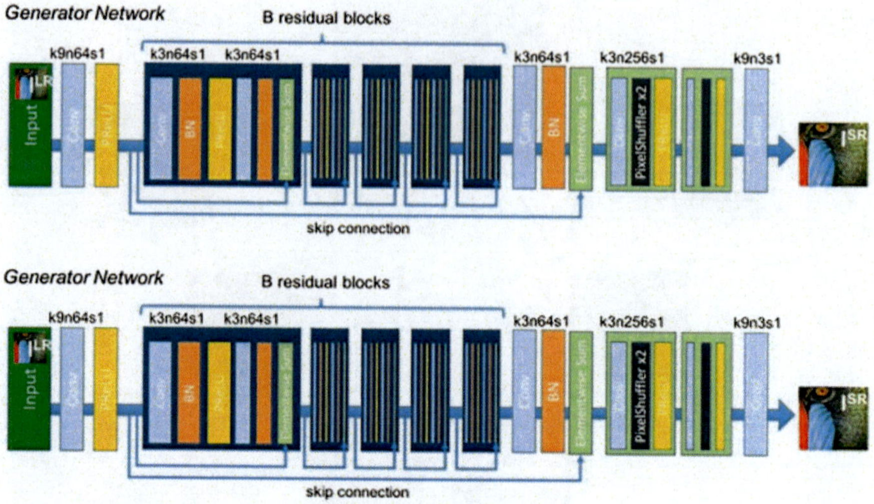

Fig. 4 Architecture of the generator and the discriminator

provided with the best possible images in terms of resolution and quality to output the best possible product from the model. The output image is thus perceptually superior and resides in the subspace of natural images. The end product is an image produced by a generator that cannot be acquired or classified as fake or unnatural by the discriminator. The architecture of the GAN is shown in Fig. 4.

A similar thought process has been implemented within the G and the D networks. The networks have been designed to facilitate deeply layered models and, with the help of residual block in the G network, several erratic behaviors. The residual blocks' structure inside the ResNet consists of two convolutional layers with small 3 × 3 kernels for capturing the minute details of the images along with 64 feature maps. The batch-normalization layers are added for faster computation and normalization. This is followed by PReLU as an activation function to perform the major transformation and learning of the next layers' alpha parameter. Upsampling is done, increasing the resolution, and the two convolution layers take that care. The next important part of the model is the discriminator D network, whose main job is to differentiate the generated SR images from the original HR images.

The guidelines govern the design of the discriminator. It consists of eight convolutional layers with a monotonic increase in the number of 3 × 3 filter kernels, starting from 64 till 512, increasing by 2. The kernel movement is stridden to reduce the image's resolution as the features within the layer get doubled. The activation function used within the discriminator network is leaky ReLU. No max pool layer is used with this network. Figure 5 shows the architecture of the generator and discriminator network.

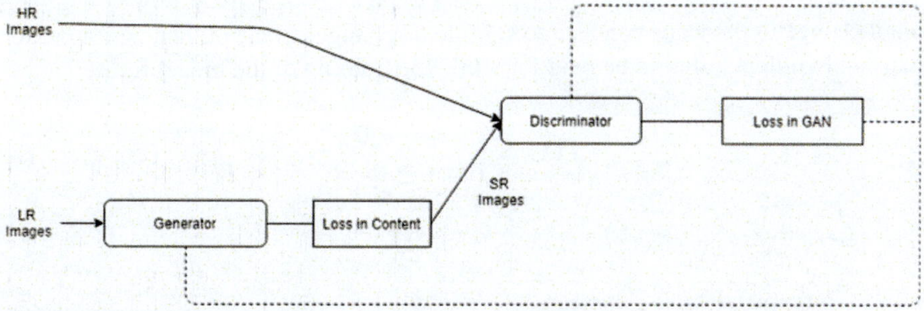

Fig. 5 Training process of GAN

4.2.1 Residual Blocks

The residual blocks [20] or identity blocks are referred to as the building blocks of a ResNet [21]. Theoretically, a model's training error should show a monotonic decrease in its values as the number of layers increases within a model. But in practice, there is a deviation from the expected behavior as the neural networks reach a point where the training error starts to increase. However, ResNets are immune to such erratic behavior as there is a monotonic decrease in the error as the number of layers increase. This inherent property of the ResNets has enabled the training of hundreds and even thousands of layers without any problems. This is possible because of the unique structure of the ResNets, which gives them improved speed and makes them substantially deeper. Since these layers possess such unique properties, they play an essential role in the proposed models.

4.2.2 PixelShuffler ×2

This layer is essentially responsible for up sampling [22] and the feature map upscaling. An inbuilt Keras function has been used for the same. The main job of pixel shuffler is to rearrange the elements of $H \times W \times C \cdot r^2$ tensor to form $rH \times rW \times C$ tensor. There are two sub-pixel CNNs are applied in the generator.

4.2.3 PRelu (Parameterized Relu)

A small change is done in the activation function by changing it from ReLU [23] to PReLU [24]. PReLU is much more efficient as compared to the simple ReLU as the adaptable parameter (α) takes into consideration the negative values of x as well, hence, the corresponding information on the negative axis is not lost, unlike ReLU in which the negative value of x as is taken as 0. This is how the problem of dying ReLU is resolved by using PReLU and retaining the information in deep networks is automatically calculated by the model while training the model. Even a small change in the value can lead to significant changes in the output, hence proper steps and vigilance should be paid for while the training of the model for α.

$$f(x) = \begin{cases} \alpha x, x < 0 \\ x, x \geq 0 \end{cases} \tag{2}$$

4.2.4 Perceptual Loss Function

The perceptual loss function l^{SR} of the proposed method is different and enhanced as it is designed to assess the solution concerning perceptually relevant characteristics, whereas l^{SR} is based on mean square error (MSE) [25]. Instead of the normal approach, a weighted sum approach for content loss (l_X^{SR}) has been used. Hence, the adversarial loss components can be represented as:

$$l^{\text{SR}} = l_X^{\text{SR}} + 10^{-3} l_{\text{Gen}}^{\text{SR}} \tag{3}$$

In the following sections, all different options are explored and examined for content loss $\left(l_X^{\text{SR}}\right)$ and the adversarial loss ($l_{\text{Gen}}^{\text{SR}}$).

4.2.5 Content Loss

Calculation of the pixel-wise MSE loss can be done as:

$$l_{\text{MSE}}^{\text{SR}} = \frac{1}{r^2 WH} \sum_{x=1}^{rW} \sum_{y=1}^{rH} \left(I_{x,y}^{\text{HR}} - G_{\theta_G}\left(I^{\text{LR}}\right)_{x,y} \right)^2 \tag{4}$$

The state-of-the-art approaches rely on this target optimization and is, hence, most widely accepted for image super-resolution. But a severe problem with MSE optimized solutions while achieving high PSNR [26] is the generation of perceptually unsatisfied solutions due to lack of high-frequency content.

The approach of THR loss function is toward perceptual similarity rather than the conventional pixel-wise losses. The VGG loss is coupled and reliant on the ReLU activation layers of the pre-trained 19 layer VGG network [27]. With $\emptyset_{i,j}$ indicates the feature map obtained by the jth convolution (after activation) before the ith max-pooling layer within the VGG19 network. Now, for the definition of the VGG loss between the features, define the VGG loss as the Euclidean distance between the feature representations of a reconstructed image $G_{\theta_G}(I^{\text{SR}})$ and the reference image I^{HR}:

$$l_{VGG/i,j}^{SR} = \frac{1}{W_{i,j}H_{i,j}} \sum_{x=1}^{rW} \sum_{y=1}^{rH} \left(\varphi_{i,j}(I^{HR})_{x,y} - \varphi_{i,j}(G_{\theta_G}(I^{LR}))_{x,y} \right)^2 \quad (5)$$

Here, $W_{i,j}$ and $H_{i,j}$ describe the dimensions of the respective feature maps within the VGG network.

4.2.6 Adversarial Loss

The main reason for adding the adversarial loss to the perceptual loss is the ability of the loss to bring a more natural look and feel to the images as it favors the real images and hence attempts to mislead the discriminator. The generative loss l_{Gen}^{SR} is defined based on the probabilities of the discriminator $D_{\theta_D}(G_{\theta_G}(I^{LR}))$ overall training samples as:

$$l_{Gen}^{SR} = \sum_{n=1}^{N} -\log D_{\theta_D}(G_{\theta_G}(I^{LR})) \quad (6)$$

Here, $D_{\theta_D}(G_{\theta_G}(I^{LR}))$ is the probability that the reconstructed image $G_{\theta_G}(I^{LR})$ is a natural HR image. For better gradient behavior, minimize $-\log D_{\theta_D}(G_{\theta_G}(I^{LR}))$ instead of $\log\left[1 - D_{\theta_D}(G_{\theta_G}(I^{LR}))\right]$.

4.3 Stitching

The main function of stitching of images back is the descriptor detection function [28]. The better the descriptor function, the better is the stitched images. Stitching takes place in pairs, as it yields optimum results. Descriptors and key points for a pair of images are found out, which are in turn dependent on the degree of overlapping performed in the sliding window function. Laplacian of Gaussian function (LoG) [29] is the main parameter that dictates the entire working in stitching the images. LoG works on the blob detection system where different sizes of blobs are detected in an image as σ changes. σ is called the scaling parameter. Since the computation of LoG is a little costlier, the difference of Gaussian [30], which is an approximation of LoG can be used. This is calculated with two different σ, say σ and $k\sigma$, which is the difference of Gaussian blurring [31] of an image. This process is repeated for all octaves of the image in the Gaussian Pyramid [32]. Accurate results can be generated by refining the potential keypoint locations. The Taylor series expansion [33] of scale space is implied to get a more accurate location of extrema, and if the intensity at these extrema is less than a threshold value (0.03 as per the execution), it is rejected. Distance between every descriptor for the pair of images is computed, and the best result is chosen. Finally, holography matrix is estimated which results in proper alignment of the image and is then finally stitched. Figure 6 shows the sample output images after passing the input image through the proposed network. It contains the low-resolution image, high-resolution image, and the CLAHE optimized image. In the following table of images, images 1(a), 2(a), 3(a), 4(a) are the low-resolution images. Images 1(b), 2(b), 3(b), 4(b) are the high-resolution images and images 1(c), 2(c), 3(c), 4(c) are the CLAHE optimized images. Images 1(a) and 4(a) have relatively higher resolution than 2(a) and 3(a). Since the 2(a) and 3(a) have relatively low resolution, their high-resolution images that are 2(b) and 3(b) show

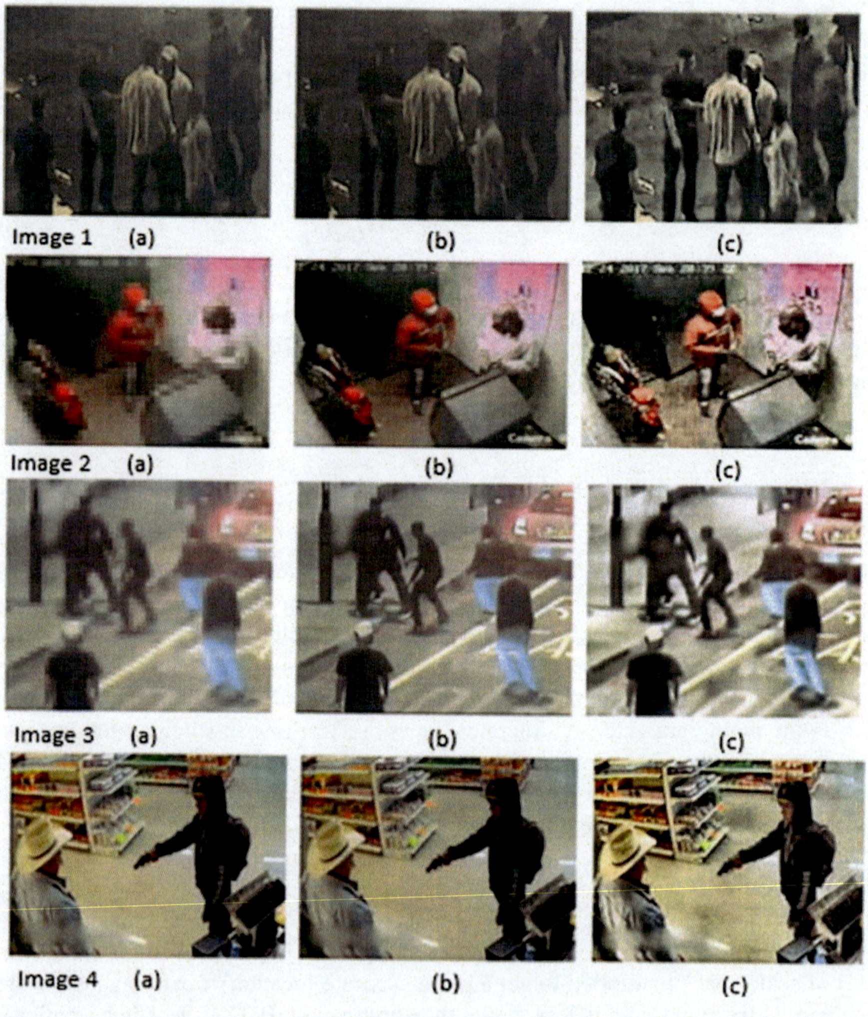

Fig. 6 Images reconstructed through the proposed method. **a** Low resolution, **b** high resolution, **c** CLAHE optimized

massive enhancement. The images 1(a) and 4(a) have better resolution hence 1(b) and 4(b) show a minute difference in the form of smoother edges.

4.4 Contrast Limited Adaptive Histogram Equalization (CLAHE)

Stitched images are now equalized to get a better contrast in the images, which further increases the image's readability using CLAHE [34]. It works on a bottom-up approach by considering smaller areas from the image, identified as tiles, and responsible for the over-amplification of the contrast. The artificial boundaries have to be handled with care because they might lead to loss of information if not appropriately handled. Hence, bilinear interpolation can be used to take the liability of the same. The transformation function slope depicts the contrast amplification of a given pixel. This generated slope is directly proportional to the cumulative distribution function (CDF) [35] in the neighborhood and the histogram value at that pixel. The clipping of the histogram at a defined value contributes to the amplification concerning the CDF. This limits the CDF slope and, thus, the transformation function. The value to which the histogram is clipped is known as the limit of the clip. It depends on the histogram normalization and, therefore, the size of the neighborhood area.

Comparing CLAHE to its other popular counterparts such as histogram equalization (HE) and adaptive histogram equalization (AHE), the produced images have better contrast, less blurriness, and less distortion. CLAHE is an upgrade on the AHE which in turn is an upgrade of the HE technique. The concept of clip limit and block size helps CLAHE to overcome the drawback of the noise problem of AHE. The clip limit takes care of the value at which the histogram has to be clipped, a pre-defined number after it is normalized. Hence, CLAHE is a more dynamic and suitable technique since the input images are low-resolution CCTV images in which the contrasts can be very varying.

5 Experimental Results

The proposed model produces an output image that is less hazy, less blurry, with better contrasts and higher readability of the image compared to the input low-resolution images. CLAHE make the edges more defined by doing an adaptive contrast equalization. The SRGAN removes the unwanted and noisy elements present in the images and also produces a higher resolution image.

The signal noise ratio (SNR) and structural similarity (SSIM) are used to validate the images. The similarity of the two images can be understood by its SSIM index value. This value ranges from 0 to 1, where closer to 1 represents a perfect match and closer to 0 represents a complete mismatch between the input and the reconstructed image. Some of the output images are shown in Fig. 7.

The SNR of the output image with comparison to the input image is calculated as shown as (7)

$$\text{SNR} = 10 * R \tag{7}$$

where

$$\text{R} = \text{Variance}(v_{\text{ref}})/\text{MSE}(v_{\text{ref}}, v_{\text{cmp}}) \tag{8}$$

Here, v_{ref} refers to the ground image or the reference image or the ground truth image and v_{cmp} refers to the reconstructed/noise image.

Fig. 7 Images reconstructed through the proposed method. **a** Low resolution, **b** high resolution, **c** CLAHE optimized

Table 1 Difference in resolution of input and output images (pixels)

Images	Resolution of input image (pixels)	Resolution of output image (pixels)	Percentage of increase in resolution (%)
Image 1	2,674,763	152,820,483	98.24
Image 2	1,923,313	97,814,243	98.03
Image 3	2,122,683	121,311,533	98.25
Image 4	1,813,343	103,514,373	98.24

Table 2 The SSIM and PSNR of images

Image	SSIM	PSNR
Image 1	0.811	43.4
Image 2	0.91	19.56
Image 3	0.97	35.15
Image 4	0.89	20.68

Table 1 shows the resolution of the input image and the output image after passing the images through the SRGAN. The quantitative results show that there exists a major increase in the resolution of the images. The average increase in resolution is about 98%.

Table 2 depicts the value of the similarity index, and the PSNR value of the output image regarding the input images represented in Fig. 6. The value of the PSNR provides obvious evidence to understand the efficiency of the proposed method. The PSNR value between 20 and 40 dB is said to claim to be a good image resolution. It also shows the increase in resolution of the input and output images. The difference in the values proves the improvement over the resolution manifold and the image by removing the noise.

The stated values prove that the proposed method is highly efficient in producing high-resolution images which help in the application of CCTV footages.

Hence, the entire process yields an image that has almost a 98% increase in its resolution, and a significant improvement in the PSNR value, along with better contrast, less noise, less distortion, better-defined edges, and less blurriness. All these are the basic components of an image and hence rearranging them to proper values helps to retrieve more information from the image generated visually. Since the image is broken down into smaller segments using the sliding window for each segment, greater attention can be paid to the minute details such as the edges of objects to generate a much better image (Table 3).

Table 3 PSNR (DB) results for resolution enhancement of the proposed technique compared with the conventional and state-of-art image resolution enhancement techniques

Techniques/Images	PSNR (DB)			
	Image 1	Image 2	Image 3	Image 4
Bilinear	35.45	12.8	26.34	12.88
Bicubic	35.89	12.58	26.86	12.93
Regularity-preserving image interpolation [4]	37.36	14.01	28.81	14.28
NEDI [5]	37.38	14.07	28.81	14.3
HMM [6]	37.48	14.19	28.86	14.57
HMM SR [7]	37.56	14.23	28.88	14.55
WZP-CS [8]	38.82	15.69	29.27	14.58
WZP-CS-ER [9]	38.91	15.88	29.36	14.87
DWT SR [10]	42.62	19.60	34.79	19.95
CWT SR[11]	41.23	18.97	33.74	19.5
DSWD [12]	42.8	19.49	34.82	20.12
Proposed method	43.4	19.56	35.15	20.68

6 Conclusions and Future Work

In this study, the specifically super-resolution GAN framework was developed for improving the surveillance images successfully. This model mainly focuses on enhancing the images and conversion of low-quality images to high-resolution images. This is an area of very high potential, as this can be used to convert low-quality videos to high-quality videos or high definition videos. This model shows little promise with the video conversion, but proper and suitable addons can show more significant results. The aim was to work on the perceptual quality of super-resolved images rather than type computational efficiency. The efficiency had to be given a backseat to generate better and more realistic looking images, but finding an optimization between the two is an area to be explored. Also, SRGAN with CLAHE and other filters is a very viable and realistic option that can be used in a variety of fields such as medical images. Generating super-resolution images of MRI images can be a breakthrough, as it can help the doctors to understand the problems better.

References

1. W. Yang, X. Zhang, Y. Tian, W. Wang, J.-H. Xue, Q. Liao, Deep learning for single image super-resolution: a brief review. IEEE Trans. Multimedia **21**(12), 3106–3121 (2019)
2. C. Ledig, L. Theis, F. Huszar, J. Caballero, A. Cunningham, A. Acosta, A. Aitken, A. Tejani, J. Totz, Z. Wang, W. Shi, Photo-realistic single image super-resolution using a generative adversarial network

3. I. Goodfellow, J. Pouget-Abadie, M. Mirza, B. Xu, D. Warde-Farley, S. Ozair, A. Courville, Y. Bengio, Generative adversarial networks, in *Advances in Neural Information Processing Systems*, vol. 3 (2014)
4. W.K. Carey, D.B. Chuang, S.S. Hemami, Regularity-preserving image interpolation. IEEE Trans. Image Process. **8**(9), 1295–1297 (1999)
5. X. Li, M.T. Orchard, New edge-directed interpolation. IEEE Trans. Image Process. **10**(10), 1521–1527 (2001)
6. K. Kinebuchi, D.D. Muresan, R.G. Baraniuk, Wavelet-based statistical signal processing using hidden Markov models, in *Proceedings of international Conference on Acoustics, Speech, and Signal Processing,* vol. 3 (2001), pp. 7–11
7. S. Zhao, H. Han, S. Peng, Wavelet domain HMT-based image super resolution. Proc. IEEE Int. Conf. Image Process. **2**, 933–936 (2003)
8. A. Temizel, T. Vlachos, Wavelet domain image resolution enhancement using cycle-spinning. Electron. Lett. **41**(3), 119–121 (2005)
9. A. Temizel, T. Vlachos, Image resolution upscaling in the wavelet domain using directional cycle spinning. J. Electron. Image **14**(4) (2005)
10. G. Anbarjafari, H. Demirel, Image super resolution based on interpolation of wavelet domain high frequency subbands and the spatial domain input image. ETRI J. **32**(3), 390–394 (2010)
11. H. Demirel, G. Anbarjafari, Satellite image resolution enhancement using complex wavelet transform. IEEE Geosci. Rem. Sens. Lett. **7**(1), 123–126 (2010)
12. H. Demirel, G. Anbarjafari, Image resolution enhancement by using discrete and stationary wavelet decomposition. IEEE Trans. Image Process. **20**(5) (2011)
13. S. López-Tapia, et al., A single video super-resolution GAN for multiple downsampling operators based on pseudo-inverse image formation models. Dig. Sig. Process. **104** (2020)
14. N. Mahamkali, A. Vadivel, OpenCV for computer vision applications, in *National Conference on Big Data and Cloud Computing,* March 2015
15. E. Agustsson, R. Timofte, NTIRE 2017 challenge on single image super-resolution: dataset and study, in *IEEE Conference on Computer Vision and Pattern Recognition Workshops (CVPRW)*, July 2017, pp. 1122–1131
16. H. Ha, B.Y. Hwang, Enhancement method of CCTV video quality based on SRGAN. J. Korea Multimedia Soc. **21**(9), 1027–1034 (2018)
17. M. Sodanil, C. Intarat, A development of image enhancement for CCTV images, in *2015 5th International Conference on IT Convergence and Security (ICITCS)* (IEEE, 2015), pp. 1–4
18. I.J. Jacob, Capsule network based biometric recognition system. J. Artif. Intell. **1**(2), 83–94 (2019)
19. J.D. Koresh, Quantization with perception for performance improvement in HEVC for HDR content. J. Innov. Image Process. (JIIP) **2**(1), 55–64 (2020)
20. X. Li, Z. Hu, X. Huang, Combine Relu with Tanh, in *IEEE 4th Information Technology, Networking, Electronic and Automation CCOntrol Conference (ITNEC)*, June 2020, pp. 51–55
21. S. Hayou, E. Clerico, B. He, G. Deligiannidis, A. Doucet, J. Rousseau, Stable ResNet (2020)
22. R. Fattal, Image upsampling via imposed edge statistics. ACM SIGGRAPH 2007 papers. (2007), p. 95-es
23. J. Si, S. Harris, E. Yfantis, A dynamic ReLU on network network, in *IEEE 13th Dallas Circuits and Systems Conference (DCAS)*, November 2018, pp. 1–6
24. F. Zuo, X. Liu, DPGAN: PReLU used in deep convolutional generative adversarial networks, in *International Conference on Robotics Systems and Vehicle Technology*, October 2019, pp. 56–61
25. M. Tuchler, A.C. Singer, R. Koetter, Minimum mean squared error equalization using a priori information. IEEE Trans. Sig. Process. **50**(3), 673–683 (2002)
26. J. Lian, Image sharpening with optimized PSNR (2019), pp. 106–110
27. H. Hassannejad, G. Matrella, P. Ciampolini, I. De Munari, M. Mordonini, S. Cagnoni, Food image recognition using very deep convolutional networks, in *2nd International Workshop on Multimedia assisted Dietry Management*, October 2016, pp. 41–49

28. J. Chen, S. Shan, C. He, G. Zhao, M. Pietikainen, X. Chen, W. Gao, WLD: a robust local image descriptor. IEEE Trans. Pattern Anal. Mach. Intell. **32**(9), 1705–1720 (2009)
29. J.S. Chen, A. Huertas, G. Medioni, Fast convolution with Laplacian-of-Gaussian masks. IEEE Trans. Pattern Anal. Mach. Intell. **PAMI-9**(4), 584–590 (1987). https://doi.org/10.1109/tpami.1987.4767946
30. S. Wang, et al., An improved difference of gaussian filter in face recognition. J. Multimedia **7**(6), 429–433 (2012)
31. J. Flusser, S. Farokhi, C. Höschl, T. Suk, B. Zitová, M. Pedone, Recognition of images degraded by Gaussian Blur. IEEE Trans. Image Process. **25**(2), 790–806 (2016). https://doi.org/10.1109/TIP.2015.2512108
32. Z. Lan, et al., Beyond gaussian pyramid: multi-skip feature stacking for action recognition, in *Proceedings of the IEEE Conference on Computer Vision and Pattern Recognition* (2015)
33. R. Tehrani, L.C. Ludeman, Use of generalized Taylor series expansion. **2**, 979–982 (2021)
34. A.M. Reza, Realization of the contrast limited adaptive histogram equalization (CLAHE) for real-time image enhancement. J. VLSI Sig. Process. Syst. Sig. Image Video Technol. **38**(1), 35–44 (2004)
35. M.H. Chun, S.J. Han, N.I. Tak, An uncertainty importance measure using a distance metric for the change in a cumulative distribution function. Reliab. Eng. Syst. Saf. **70**(3), 313–321 (2000)

Smart City: Recent Advances and Research Issues

Bonani Paul and Sarat Kr. Chettri

Abstract Smart cities use technological innovations to improve urban services and people's livelihoods to develop sustainably. Different technology solutions and technologies like IoT sensors, big data analytics, communication networks, and applications are being used to collect and analyze data to boost various services in smart cities, including public services, transport, and various other utilities. The article intends to discuss the state-of-the-art technologies of smart cities and their roles and applications. It also seeks to analyze the current research trend in the smart city domain and its key enabling technologies. It also pursues to identify some of the open issues and challenges facing efficient use of energy, smart decision-making systems, privacy and security of data, and effective and secure communication technologies in smart cities.

Keywords Smart city · Internet of Things · Information and communication technologies · Big Data · Issues and challenges

1 Introduction

Urbanization is an unending phenomenon. Rapid urbanization requires services and physical infrastructure in urban cities that meet the growing needs of urban residents and promote sustainable development. A significant role in the development of smart cities has been played by the rise of cloud computing, big data analytics, and the development of the Internet of things (IoT) technologies. Big data analytics enables smart cities to gain significant insight from data collected from various sources in

B. Paul (✉)
Department of Computer Science, St. Mary's College, Shillong, India
e-mail: b.paul@smcs.ac.in

S. Kr. Chettri
Department of Computer Applications, Assam Don Bosco University, Guwahati, India
e-mail: sarat.chettri@dbuniversity.ac.in

V. Suma et al. (eds.), *Inventive Systems and Control*,
Lecture Notes in Networks and Systems 204,
https://doi.org/10.1007/978-981-16-1395-1_7

Fig. 1 Overview of smart cities and their key enabling technologies

large volumes. IoT allows the use of Internet services in the physical environment to integrate sensors, actuators, radio-frequency identification (RFID), and Bluetooth technologies.

IoT and big data integration is an emerging research field that has provided new, innovative, and fascinating insights into the future of smart cities. However, there are various challenges focusing primarily on business and technological issues that enable cities to update the vision, principles, and needs of smart city applications by taking advantage of the intelligent environment's core characteristics. A smart city has several components, including smart infrastructure, smart citizens, smart security, smart grid, and smart technology (e.g., big data, IoT, sensors, 5G connectivity, robotics, geospatial technology, etc.). Smart infrastructures in a smart city comprise several operators from different sectors, such as smart energy, intelligent transport options, public safety, and so on. Smart infrastructures control and monitor the cyber-physical systems, which are essentially data-controlled systems interacting with the physical world. Due to the various smart city initiatives, the burden on the environment and the natural resources of urban areas is increasing and thus the smart citizen empowers communities to better understand their environment and address the various environmental problems in cities. Figure 1 presents an overview of smart cities and technological innovations.

The prime objective of the mission of smart cities is to foster economic growth and improve the quality of life of people by facilitating local area development by applying different tools and technologies that produce smart outcomes.

The remaining part of the article is organized accordingly: Section 2 provides an overview of the current trend in smart city research. Section 3 describes how the use of big data in the context of smart cities, together with IoT, ICT and smart-based applications, geospatial technologies, machine learning, and artificial intelligence (AI), can fundamentally improve urban lifestyles on a day-to-day basis. Some of the open issues and challenges in this area with possible future research directions are highlighted in Sect. 4, and Sect. 5 concludes the research work.

2 Current Research Trend

By 2025, the smart city sector will be a $1712 billion market with 600 cities worldwide. According to McKinsey research, by 2025, these smart cities will contribute 60% to the global GDP. The upsurge in the adoption of green technology, AI, and IoT continues to drive the market. Industry and academia are carrying out numerous research work and even collaborative work for future research directions

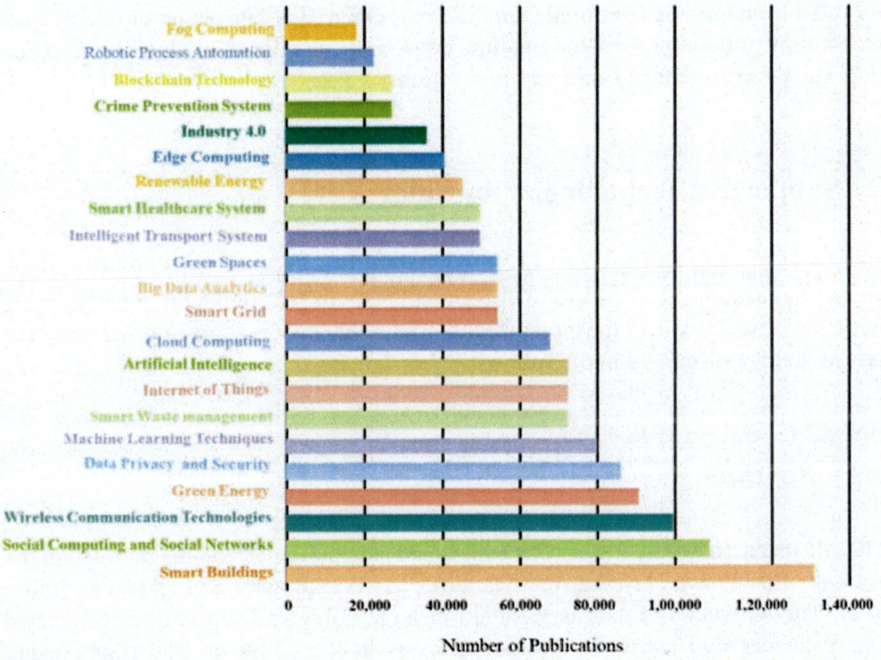

Fig. 2 Number of research publications made in 2016–2020 for various applications and technologies driving smart cities

to understand and provide practical applications of various technologies and their performance measurement in the smart city domain.

To get an idea of the research trend in smart cities, we attempted to find the number of publications made by the scientific community in terms of scientific articles, books, or chapters of a book. Figure 2 shows the trend for various applications and technologies that drive smart cities over the last five years (2016–2020) as per the number of paper publications indexed in Google Scholar. Based on the findings, the most researched topics in the scientific community over the last five years have been smart buildings using AI, machine learning techniques and IoT, social computing and social networks and wireless communication technologies, green energy and data privacy and security in smart cities. The proliferation of AI and IoT in buildings makes it smarter through optimization in energy use, detection of faults, improvement of occupant comfort, automated and more adaptive buildings, and so on. Social computing and social networks have become an integral part of the population of smart cities, not only for business interaction and product promotion but also for opinion mining, urban management, and planning to accommodate the growing population. In fact, social computing tools offer a wide range of applications for smart cities, from traffic management to pollution control, solid waste management and public safety, etc. Such applications constitute the backbone of smart cities; however, wireless communication technology plays a crucial role in meeting the connectivity requirements between different devices and networks and the ever-growing demand for reliable and secure communication in smart cities. Big data analytics, blockchain technology, robotic process automation, edge, and fog computing are becoming an important component of smart city innovations.

3 Supportive Technologies for Smart City

To improve urban services and residents' living experience, smart cities leverage technology solutions. The state-of-the-art and evolving technologies are supporting these initiatives. Some of the typical technologies that lay the foundation of the smart city model are discussed in the following subsections.

3.1 Big Data

The advancement of the Internet of things (IoT) technologies and the increase in the volume of data have played a significant role in the exploitation of smart city initiatives. The unstructured data is gathered in large volumes from various connected smart devices and stored for further analysis in the cloud or data center using distributed fault-tolerant databases such as NoSQL to implement smart city services. Health care, transport services, and other smart city services are included in some of these services [1, 2]. The programming model for processing large datasets with parallel algorithms, such as Parallel Optimization, Jacobi Method, LASSO [3], can therefore be used for data analytics to obtain value from stored data.

Big data applications can serve a wide range of sectors in a smart city [4]. It helps to improve customer experience and services (e.g., higher returns or increased market shares) that contribute to the growth and performance of companies. Big data applications make it possible to improve health and well-being through better emergency services, proper diagnosis, medication, recovery, and cure to ensure optimal health outcomes. Big data can be of significant benefit for transportation, not only through smart parking but also through congestion control [5] and routing by using the navigation system to find the best route on a real-time basis. Improved water and waste management can also be harnessed through technological innovation to predict the likelihood of waste levels in trash bins [6].

The deployment of big data applications requires the support of a good information and communication technology (ICT) infrastructure. Otherwise, it would not have been possible for smart education and smart governance to mention some useful and unique solutions.

3.2 Smart Sensors and IoT

The IoT fitted with smart sensors and actuators is the fundamental building block of a smart city. The Internet of things has the potential for the exploitation of sustainable information and communication technologies [7]. Smart city applications based on IoT can be categorized based on the type of network, flexibility, coverage, scale, repeatability, heterogeneity, and end-user involvement and impact [8].

With the Internet revolution, the IoT framework provides a platform in which many smart and self-configuring objects, such as watches, cars, tablets, wearable technology, and home appliances, are connected. IoT is a broadband network that uses standard communication protocols [9] to exchange, process, and analyze data generated from different smart devices to provide smart cities with more efficient, cost-effective, secure, and real-time services.

Smart cities can deliver scalable and secure solutions by exploiting the full potential of IoT standard protocols and networking technologies. Radio-frequency identification (RFID) is the major IoT enabling technologies and protocols. Some others include near-field communication (NFC), LTE-A, low-energy Bluetooth, wireless sensor networks (WSN), and so on.

RFID and NFC are short-range wireless connectivity that is used wirelessly to track and identify objects. Low-energy Bluetooth technology is a low-power wireless communication technology that allows smart machines to communicate within a few kilometers. Low-energy wireless communication not only reduces battery consumption but also increases the life of the device through reduced use.

Wireless radio-frequency communication protocols such as ZigBee, Z-Wave, and Thread enable remote monitoring and control of connected devices in a private area network. These technologies consume less energy but deliver high throughput. Similarly, LTE-A or LTE Advanced, an upgrade to existing LTE, is a standard mobile communication technology that provides a smoother, higher bandwidth experience with low power consumption. LTE-A is robust and has lesser dropped connections and delivers larger and faster wireless data payloads. Wi-Fi Direct can create a Wi-Fi network by enabling peer-to-peer (P2P) connections between smart devices without the need to connect to an access point. WSNs are self-configured, finite sets of sensor devices distributed geographically over wireless networks to monitor and collect environmental data such as temperature, sound, pressure, and forward data to sink nodes for analysis.

3.3 Information and Communication Technology

The use of information and communication technology (ICT) in the smart city domain makes it smart and allows resources to be used effectively [10]. The ICT tools enhance the concept of smart living, reflecting mainly improved public safety, smart buildings [11], smart health care, assured electricity supply, smart transportation, and tourism, smart education, etc. Information is the key to the smart city concept used to provide a better quality of life for citizens through transparency, efficiency, and active involvement [12]. An ICT-based smart city empowers people to interact effectively with one another and share knowledge and experiences about mutual interests. ICT is the basic foundation for a smart city, where enormous data collection is carried out for effective decision making by local government and various stakeholders in the city administration. A robust and scalable ICT infrastructure is therefore required to manage such an enormous volume of data and provide citizens with real-time responses for a better quality of living. Different technologies are used for data acquisition purposes which include wireless sensor networks (WSNs), vehicular ad hoc networks (VANETs), unmanned aerial vehicles (UAVs), mobile ad hoc networks (MANETs), and 5G networks. These technologies are mainly used for activities relating to data transfer, communication, monitoring, and tracking. Monitoring is used to analyze and control systems that include agriculture, industry, smart homes, and so on and also tracking is used to record any incidental change.

With the use of various devices and technologies, there is immense energy consumption. The multiple energy requirements and energy consumption of the ICT infrastructure need to be optimized and some work has been done in the literature in this direction. Kumar et al. [13] have proposed a new nanocloud-based grid that stores and converts energy generated from multiple sources in smart buildings, such as PV panels and wind power generation to DC power to be consumed by smart devices. Besides, a cloud controller is used by a variety of geographically separate smart buildings to make optimum use decisions. In the same context, Bukhechba et al. [14] have introduced a prototype called NomaBlue, where Bluetooth technologies are used to collect data and collaborate with users.

3.4 *Geospatial Technology*

Cities are producing many forms of real-time geospatial data. Geospatial technology is central when it comes to providing a technology platform that forms the backbone of a city. Geospatial technology is an IT field devoted to the collection, mapping, and analysis of spatial data. Technology simplifies the distribution of spatial data with ease and efficiency. It assists cities with a variety of applications, from finding the fastest route to identifying the exact location of an emergency call, to simply wrapping up information to improve health care, law enforcement, and government services. Indeed, the smart city concept is synchronized with advances in geospatial technology that move toward more real-time data inputs, 3D visualization [15], and the ability to track any changes over time. For instance, in agriculture, the LiDAR remote sensing technology [16] helps in identifying the types of soil, forecasting yields, and monitoring crops. It also helps to track air pollution and to ensure safe navigation for self-driving cars. Smart cities are aided by geographic information systems to plan, develop, and operate. There are several advantages of GIS in smart cities, such as helping to create public maps to improve public transport systems, helping to respond to emergencies, such as mapping disease spread to a search and rescue zone. GPS technology in smart cities actually contributes to the management of smart waste [17], fleet tracking, etc. To enable all of these services in smart cities using geospatial technology, geospatial information is communicated to various stakeholders of smart cities. Geospatial information also flows from all the stakeholders to centrally controlled information systems. However, there are major issues as to which data is shared, by whom, where it is stored and how it is used. Various checks and balances should be carried out to preserve the privacy, security, and integrity of sensitive data.

3.5 *Artificial Intelligence and Machine Learning*

With the increasing urban population, the challenges of providing all residents with resources and energy and, at the same time, avoiding environmental deterioration are increasing in many ways. Besides, administration and management to avoid sanitation problems, providing intelligent transportation system (ITS), crime prevention, cyber-security, energy-efficient use of smart grids, smart healthcare system, and so on are other critical challenges in a smart city. Artificial intelligence (AI) and machine learning (ML) techniques play a key role in facilitating ITS by monitoring and estimating the flow of city traffic data in real time [18]. The operating structure of smart grids and the efficient use of energy are being revolutionized by machine learning and big data [19]. Cyber-attack is one of the key issues in smart grids and to mitigate it a deep reinforcement learning-based intrusion detection system [20] is suggested in the literature where the proposed model uses short signatures and hash functions to create blocks. In smart cities, AI and ML can be used to assess spatio-temporal data to analyze trends and heterogeneities in a city that contributes to the deployment of infrastructure, proper planning, mobility, neighborhood relationships, etc. With the advancement of sensors and the amalgamation of IoT devices, edge computing, advanced data analytics with AI, ML, and deep reinforcement learning (DRL) technologies is making a paradigm shift in disease diagnosis and treatment in the healthcare sector [21]. In smart cities, there are numerous areas for AI-powered applications that can improve the lives of citizens and residents' businesses. The applications range from smart farming to smart parking, smart police to smart governance, smart manufacturing, etc. There are, however, many issues and challenges where the academic and industry-based professional can focus solely on the efficient use of the smart city's AI and ML approaches [22]. Some of the significant issues include developing machine learning models for more precise and accurate decision making, efficient collection, storage, and analysis of huge amounts of real-time data, detection of anomalies and prevention of security breaches, etc.

4 Issues and Challenges

Innovations in smart city novelties are increasing; there remain several issues and challenges facing smart city solutions today. The subsequent sections highlight some of the issues and challenges that will need further investigation and work for future generations of smart cities.

4.1 Data Privacy and Security

Data security includes unauthorized access and attacks causing physical disruption of services available. Collecting and storing digital data for citizens increasingly collect data about their location and activities that foster citizens concerning their privacy protected privacy systems [21] which collect and cause emergency response if necessary are the technological challenges to meet the data security challenges. It is often a challenge to achieve optimal security and privacy, but it is necessary for a smart city. Table 1 provides a summary of some of the approaches in smart city applications to address the issue of data privacy and security.

Table 1 Approaches to data privacy and security issues in smart city applications

References (Year)	Issues	Approach	Summary
[23] (2015)	Data over-collection	Mobile-cloud framework	To eliminate the data over-collection in the mobile-cloud framework has been presented
[24] (2019)	Data privacy protection	Edge computing and DIKW (Data, Information, Knowledge, Wisdom) framework	Developed a system for privacy protection by classifying privacy resources (data, information, and knowledge) as typed resources in the hierarchy of DIKW
[25] (2018)	Location privacy protection	Blockchain and cloud computing	To provide location privacy protection in crowd sensing networks, an incentive mechanism based combined approach with blockchain technology was used
[26] (2020)	Spatio-temporal data protection	Locality-sensitive hashing (LSH) technique	A new privacy-aware data fusion and prediction approach was proposed using the LSH technique for the industrial environment of smart cities
[27] (2017)	Protection of data confidentiality and integrity in big data analysis	Selective encryption (SEEN) method	A selective encryption (SEEN) method was proposed to secure the big data streams in real-time data analytics to achieve data integrity and confidentiality

4.2 Accurate and Precise Decision-Making Systems

Intelligent decision making based on data combined with analytics is used to forecast the likely effects of major decisions, whether they are planning choices in new areas, focusing on improving public services, or maximizing the impact of investments in smart cities, etc. Decision support system (DSS) components provide decisionmakers with the opportunity to make decisions at three main levels: strategic, tactical, and operational. One of the necessary tools to plan and complete the set of projects needed for the development of a smart city is the DSS. Several techniques have been widely proposed and developed to support the process of decision making, such as evolutionary algorithms, neural networks, machine learning, artificial intelligence, and fuzzy systems. Decision support system can be divided into five main categories [28]: model-driven DSS, communication-driven DSS, data-driven DSS, document-driven DSS, and knowledge-driven DSS. The application domain of DSS includes medical, agriculture, academics, inventory management, transportation, business planning, and manufacturing.

There are various issues related to smart city decision-making systems that hinder the process of precise and accurate decision-making systems, such as lack of information, lack of adequate resources, lack of pro-professional decision-making systems, and so on. Table 2 summarizes some of the approaches to addressing the issue of accurate and precise decision-making systems in smart cities.

4.3 Energy Consumption

Smart cities require continuous energy supply for their industrial and commercial activities, transportation, infrastructure, manufacturing, and different production activities. So, practically all activities in a smart city require a lot of energy. Thus, energy efficiency is becoming a challenge for urban life.

Increasing demand and consumption of a huge amount of energy not only raises the issue of efficient energy utilization but also looks into the issues that could be detrimental to the environment. For instance, the increasing demand for electricity leads to more energy and heat production causes greenhouse gas, followed by agriculture, industry, transportation, and other sectors. Table 3 summarizes some approaches to addressing energy consumption and production issues to meet the demands of smart cities.

4.4 Efficient and Secure Communication Technologies

Telecom infrastructures in smart cities, using a broad range of digital appliances, provide quality services and deliver information efficiently by using a range of technologies such as IoT, cloud computing, blockchain, WSN, communication gateways, and network virtualization among others. To develop machine-to-machine (M2M), machine-to-human (M2H), and human-to-machine (H2M) interaction to make effective and secure communication technologies in smart cities, researchers are currently fusing different technologies with IoT.

Table 2 Approaches to decision support systems in smart city applications

References (Year)	Issues	Approach	Summary
[29] (2020)	Decision support system (DSS) for disaster management in smart cities	Big Data and artificial intelligence	A new IDSS disaster management conceptual framework was proposed. Wildfires and cold/heat waves have received special attention
[30] (2020)	Smart healthcare decision support system	Genetic algorithm with efficient evolutionary techniques	A decision support system (DSS) has been developed to help hospital staff allocate beds to patients over a given time horizon, taking into account the availability of beds, medical facilities, etc
[31] (2016)	Smart healthcare decision support system	Markov logic network (MLN)	A smart home intelligent decision support system has been proposed to help people having mental disabilities with several types of emergency aid
[32] (2017)	Decision support system in smart meters	Internet of Things	To improve the cost forecast for meter field operations the DSS uses advanced smart smart meter network network communication and quality data analysis to provide actionable decision-making recommendations for sending technologists where a customer needs to solve an Electric Smart Meter (ESM) problem
[33] (2018)	DSS for real-time road trafiic network management system	Genetic algorithm	Presented a framework for modelling the decisions in real-time for robust traffic network administration

Table 3 Approaches to balance the demand and supply of energy in smart cities

References (Year)	Issues	Approach	Summary
[34] (2019)	Balancing electricity generation and consumption	Deep neural network and reinforcement learning	To help suppliers balance the energy variation and increase the reliability of smart girds with their customers buying energy resources, a demand response algorithm has been proposed which is incentive-based and works in real-time
[35] (2019)	Managing storage unit operation within a household to maximize the cost savings in purchasing power on the demand side	Deep reinforcement learning	Proposed a data-driven technique based on deep reinforcement learning for efficient use of the energy storage system with different tariff structures to maximize demand-side cost savings in power purchase
[36] (2019)	Designing energy-efficient demand-driven smart buildings	Internet of Things	Developed an IoT-based system for an active infrared-based occupancy counting system in smart buildings, including data processing and visualization
[37] (2019)	Designing an energy-efficient network of wireless sensors in smart cities	Wireless Sensor Networks and 3D geo-clustering algorithm	To reduce cluster overlap and prevent redundant data transmission, a 3D geo-clustering algorithm was proposed for the Wireless Sensor Network (WSN) and the sensor head node for energy saving was determined
[38] (2019)	Reducing the consumption of energy and improving the use of materials in smart plants throughout the manufacturing process	Cyber-physical system	A green manufacturing model for future smart factories enabled by the cyber-physical system has been developed

The communication standards for information exchange and communication could be long-range communication such as WiMAX, 2G/3G/4G, NB-IoT, NWave, Satellite Communication, etc., as well as short-range communication [] such as RFID, Bluetooth, Z-Wave, NFC, and 6LowPAN. However, several issues need to be resolved

Table 4 Approaches for efficient and secure communication technologies in smart cities

References (Year)	Issues	Approach	Summary
[39] (2018)	İntrusion detection system in smart city buildings	Big data analytics	Developed a secure smart city system with a strong system-level intrusion detection and access control system and a secure and effective data transmission security protocol
[40] (2018)	Public smart lighting and building-monitoring	IoT, IEEE 802.15.4 short-range communication technology, and LoRa technology	A pilot project for smart public lighting and to monitor temperature, humidity, lighting, etc. in smart buildings with secure data transmission for analysis with minimal packet loss in multipoint-to-point communication
[41] (2016)	Secure communication platform	Blockchain technology	Integrated blockchain technology with smart city devices to build a common platform that allows all devices to communicate securely in a decentralized environment
[42] (2017)	Remote health care	Internet of Medical Things (IoMT) and Wi-Fi	Presented a cloud-based IoMT platform for remote health monitoring and information transfer to a cloud for data analysis and storage
[43] (2019)	Intelligent transportation system	Wi-Fi and Android-based smartphone	An ad hoc vehicle network was used to implement an intelligent route-selection transport system based on information received from nearby vehicles in real-time

to address the expected requirements for effective and secure ICT in smart cities. Table 4 summarizes some of these issues and their potential solutions.

5 Conclusion

Smart cities across the world are using state-of-the-art technologies to provide cleaner air and water, better mobility, and efficient public services to their citizens, promoting greener and safer urban areas. The world is being transformed with the emergence of big data and IoT. The convergence of two independent technologies is taking the world to new heights and creating a digitized ecosystem. In addition to collecting, transmitting, and analyzing IoT-generated data from connected smart devices in real time, smart decision making is also possible for a sustainable society in real time. Collectively, IoT, big data, and ICT are reshaping the next-generation healthcare system, optimizing traffic by controlling congestion and routing passengers to their destination on time, apart from smart parking, providing virtual assistance, managing e-waste, improving data security and privacy platforms, among others. However, the power consumption in smart buildings has increased several folds as the use of Internet-enabled smart gadgets has increased. Smart decisions need to be taken on the efficient use of various smart appliances in smart buildings to meet demand and supply requirements. Cloud-based infrastructure, together with machine learning and artificial intelligence, can make intelligent decisions on the use of energy by various appliances by monitoring and estimating the flow of traffic data in real-time. Big data analytics, AI, edge computing, and blockchain amalgamation are now an important component of smart city innovations. This study explored the current research trends in smart cities and focuses primarily on supporting technologies driving the smart cities, highlighting the open issues and potential research opportunities for smart innovations of the next-generation smart innovative cities.

References

1. Z. Khan, A. Anjum, S.L. Kiani, Cloud based big data analytics for smart future cities, in *Proceedings of the 2013 IEEE/ACM 6th International Conference* (IEEE, 2013)
2. I. Yaqoob, V. Chang, A. Gani, S. Mokhtar, I. Abaker, et al., WITHDRAWN: information fusion in social big data: foundations, state-of-the-art, applications, challenges, and future research directions (2016)
3. F. Facchinei, S. Simone, S. Gesualdo, Flexible parallel algorithms for big data optimization, in *2014 IEEE International Conference on Acoustics, Speech and Signal Processing (ICASSP)* (IEEE, 2014)
4. M. Boukhechba, A. Bouzouane, S. Gaboury, C. Gouin-Vallerand, S. Giroux, B. Bouchard, A novel Bluetooth low energy based system for spatial exploration in smart cities. Exp. Syst. Appl. **77**, 71–82 (2017)
5. A. Kramers, M. Höjer, N. Lövehagen, J. Wangel, Smart sustainable cities—exploring ICT solutions for reduced energy use in cities. Environ. Model Softw. **56**, 52–62 (2014)
6. P. Neirotti, A. De Marco, A.C. Cagliano, G. Mangano, F. Scorrano, Current trends in smart city initiatives: some stylised facts. Cities **38**, 25–36 (2014)
7. A.H. Alavi, P. Jiao, W.G. Buttlar, N. Lajnef, Internet of Things—enabled smart cities: state-of-the-art and future trends. Measurement **129**, 589–606 (2018)
8. J. Gubbi, R. Buyya, S. Marusic, M. Palaniswami, Internet of Things (IoT): a vision, architectural elements, and future directions. Future Gener. Comput. Syst. **29**, 1645–1660 (2013)

9. I. Ruiz-Mallén, Co-production and resilient cities to climate change, in *Participatory Research and Planning in Practice*, ed. by J. Nared, D. Bole. (SpringerOpen, Cham, Switzerland, 2020)
10. L. Cilliers, S. Flowerday, Factors that influence the usability of a participatory IVR crowd-sourcing system in a smart city. South African Comput. J. **29**, 16–30 (2017)
11. K.L. Terence, R. Hui, S. Sherratt, D.D. Sánchez, Major requirements for building smart homes in smart cities based on Internet of Things technologies. Future Gener. Comput. Syst. **76**, 358–369 (2017)
12. T. Kim, C. Ramos, S. Mohammed, Smart city and IoT. Future Gener. Comput. Syst. **76**, 159–162 (2017)
13. N. Kumar, A.V. Vasilakos, J.P.C. Rodrigues, A multi-tenant cloud-based DC nano grid for self-sustained smart buildings in smart cities. IEEE Commun. Mag. **55**, 14–21 (2017)
14. M. Boukhechba, A. Bouzouane, S. Gaboury, C. Gouin-Vallerand, S. Giroux, B. Bouchard, A novel Bluetooth low energy based system for spatial exploration in smart cities. Exp. Syst. Appl. **77**, 71–82 (2017)
15. S. Ortega, J.M. Santana, J. Wendel, A. Trujillo, S.M. Murshed, Generating 3D city models from open LiDAR point clouds: advancing towards smart city applications, in *Open Source Geospatial Science for Urban Studies*. (Springer, 2020), pp. 97-116
16. L. Pantoli, G. Barile, A. Leoni, M. Muttillo, V. Stornelli, Electronic interface for lidar system and smart cities applications. J. Commun. Softw. Syst. **15**, 118–125 (2019)
17. S. Joshi, U.K. Singh, S. Yadav, Smart dustbin using GPS tracking. Int. Res. J. Eng. Technol. **6**, 165–170 (2019)
18. J. Zhang, Y. Zheng, D. Qi, R. Li, X. Yi, Deep spatio-temporal residual networks for citywide crowd flows prediction, in *Thirty-First AAAI Conference on Artificial Intelligence*, vol. 259 (2017), pp. 147-166
19. B.P. Bhattarai, S. Paudyal, Y. Luo, M. Mohanpurkar, K. Cheung, R. Tonkoski, R. Hovsapian, K.S. Myers, Big data analytics in smart grids: state-of-the-art, challenges, opportunities, and future directions. IET Smart Grid. **2**, 141–154 (2019)
20. M.A. Ferrag, L. Maglaras, Deepcoin: a novel deep learning and blockchain-based energy exchange framework for smart grids. IEEE Trans. Eng. Manag. **67**, 1285–1297 (2019)
21. E.J. Topol, High-performance medicine: the convergence of human and artificial intelligence. Nat. Med. **25**, 44–56 (2019)
22. A.S. Elmaghraby, M. Losavio, Cyber security challenges in smart cities: safety, security and privacy. J. Adv. Res. **5**, 491–497 (2014)
23. L. Yibin, W. Dai, Z. Ming, M. Qiu, Privacy protection for preventing data over-collection in smart city. IEEE Trans. Comput. **65**,1339-1350 (2015)
24. Y. Duan, Z. Lu, Z. Zhou, X. Sun, J. Wu, Data privacy protection for edge computing of smart city in a DIKW architecture. Eng. Appl. Artif. Intell. **81**, 323–335 (2019)
25. B. Jia, et al., A blockchain-based location privacy protection incentive mechanism in crowd sensing networks. Sensors **18**(11) (2018)
26. L. Qi, C. Hu, X. Zhang, M.R. Khosravi, S. Sharma, S. Pang, T. Wang, Privacy-aware data fusion and prediction with spatial-temporal context for smart city industrial environment. IEEE Trans. Ind. Inform. (2020)
27. D. Puthal, X. Wu, S. Nepal, R. Ranjan, J. Chen, SEEN: a selective encryption method to ensure confidentiality for big sensing data streams. IEEE Trans. Big Data **5**, 379–392 (2017)
28. D.J. Power, R. Sharda, Model-driven decision support systems: concepts and research directions. Decis. Supp. Syst. **43**, 1044–1061 (2007)
29. D. Jung, et al., Conceptual framework of an intelligent decision support system for smart city disaster management. Appl. Sci. **10** (2020)
30. K. Dorgham, et al., A decision support system for smart health care, in *IoT and ICT for Healthcare Applications*. (Springer, Cham, 2020), pp. 85-98
31. K.S. Gayathri, K.S. Easwara Kumar, Intelligent decision support system for dementia care through smart home. Procedia Comput. Sci. **93**, 947–955 (2016)
32. J. Siryani, B. Tanju, T.J. Eveleigh, A machine learning decision-support system improves the Internet of Things' smart meter operations. IEEE Internet Things J. **4**, 1056-1066 (2017)

33. K. Abdelghany, H. Hashemi, M.E. Khodayar, A decision support system for proactive-robust traffic network management. IEEE Trans. Intell. Transp. Syst. **20**, 297-312 (2018)
34. R. Lu, S.H. Hong, Incentive-based demand response for smart grid with reinforcement learning and deep neural network. Appl. Energ. **236**, 937–949 (2019)
35. H. Kumar, P.M. Mammen, K. Ramamritham, Explainable AI: deep reinforcement learning agents for residential demand side cost savings in smart grids (2019)
36. Q. Huang et al., Rapid Internet of Things (IoT) prototype for accurate people counting towards energy efficient buildings. J. Inform. Technol. Constr. **24**, 1–13 (2019)
37. S. Azri, U. Ujang, A. Abdul Rahman, 3D geo-clustering for wireless sensor network in smart city, in *International Archives of the Photogrammetry, Remote Sensing and Spatial Information Sciences*, vol. 42.4/W12 (2019)
38. S. Ma et al., Energy-cyber-physical system enabled management for energy-intensive manufacturing industries. J. Clean. Prod. **226**, 892–903 (2019)
39. M.M. Rathore et al., Real-time secure communication for smart city in high-speed big data environment. Future Gener. Comput. Syst. **83**, 638–652 (2018)
40. G. Pasolini, et al., Smart city pilot projects using LoRa and IEEE802. 15.4 technologies. Sensors **18**(4) (2018)
41. K. Biswas, V. Muthukkumarasamy, Securing smart cities using blockchain technology, in IEEE *18th International Conference on High Performance Computing and Communications* (IEEE, 2016)
42. A. Rashed, et al., Integrated IoT medical platform for remote healthcare and assisted living, in *2017 Japan-Africa Conference on Electronics, Communications and Computers (JAC-ECC)* (IEEE, 2017)
43. T. Zaheer, et al., A vehicular network-based intelligent transport system for smart cities. Int. J. Distrib. Sens. Netw. **15**(11) (2019)
44. S. Zeadally, F. Siddiqui, Z. Baig, A. Ibrahim, Smart healthcare: challenges and potential solutions using Internet of Things (IoT) and big data analytics. PSU Rev. J. 1-17 (2019)

HOUSEN: Hybrid Over–Undersampling and Ensemble Approach for Imbalance Classification

Potnuru Sai Nishant, Bokkisam Rohit, Balina Surya Chandra, and Shashi Mehrotra

Abstract Classification is an essential technique and is omnipresent in our day-to-day life. In real time, the datasets are often imbalanced, i.e., the majority class is dominating the minority class. Classification of data becomes tough due to the imbalanced nature of data. Identifying the impact of imbalance techniques for classification techniques, this article presents a model named the HOUSEN model. It uses a novel hybrid class imbalance sampling method and an ensemble classification approach. Based on our experimental analysis, the oversampling and undersampling techniques are used to design a hybrid sampling method. The other techniques like random forest, support vector machine (SVM), gradient boost, and AdaBoost algorithms are also used for our analysis. The performance evaluation of the HOUSEN model is conducted over five imbalanced datasets. The HOUSEN model achieved a promising result when compared to existing techniques.

Keywords Imbalance classification · DBSMOTE · Ensembling · Hybrid sampling · Random forest · AdaBoost · Gradient boost

1 Introduction

In the classification problems, the distribution of classes plays a vital role in prediction [1]. If the class distribution is not in proper order, then it leads to a class imbalance problem. The word imbalanced refers to the uncertainty in the distribution of the class variable. This issue is confronted more often in binary classification than multilevel classification issues, considering a dataset consists of two classes positive and negative. The positive class is 95% of the dataset, and the other 5% is a negative class. Therefore, the positive class is called the majority class and the negative class is called the minority class. Due to the domination majority class over the negative class, this algorithm overfits the majority class. This leads to the poor prediction of the minority class. There are more changes that occur to this problem in real-world

P. S. Nishant (✉) · B. Rohit · B. S. Chandra · S. Mehrotra
Department of Computer Science and Engineering, Koneru Lakshmaiah Education Foundation, Vaddeswaram, A.P, India

V. Suma et al. (eds.), *Inventive Systems and Control*,
Lecture Notes in Networks and Systems 204,
https://doi.org/10.1007/978-981-16-1395-1_8

datasets such as fraud detection, cancer detection, online adds conversion, oil-spill detection in radar images, intrusion detection, and fault monitoring of gearbox in helicopters [2].

There exist some methods to deal with imbalanced datasets such as anomaly detection technique, cost-sensitive-based learning, sampling-based approaches, and ensemble-based approaches. These methods may be used along with machine learning algorithms to obtain better results.

This article addresses class imbalance issue for classification. The study first conducted experiments to observe the impact of the sampling-based techniques on imbalanced datasets with machine learning algorithms and then proposed a hybrid sampling model to improvise the results. Further, ensembling is proposed for classification. The study used five datasets banana, Haberman, glass0, ecoli1, yeast4, which are collected from KEEL repository for our experiments.

1.1 Contribution

The study proposed a novel hybrid model named: "HOUSEN" intending to improve classification results for imbalance dataset. The main contribution of our study is as follows:

i. Conducted experimental analysis of the imbalance data techniques.
ii. Designed a novel hybrid approach for imbalance data sampling.
iii. Introduced a novel classifier ensembling method.
iv. Exhaustive experiments are conducted to evaluate the performance of our proposed model.

The layout of this article is organized as follows. Section 2 discusses some related research. Section 3 describes the proposed model such as HOUSEN. Section 4 presents the experimental results and analysis. Finally, Sect. 5 concludes the research work along with the future scope, respectively.

2 Related Work

In recent years, a lot of research is conducted in this area. Imbalance classification problem mainly appears in real-time datasets such as fraud detection and medical diagnosis. Hu et al. applied SMOTE on the wine dataset to handle the unequal distribution of wine quality [4]. This approach gave the right prediction on this dataset using random forest. Galar et al. discussed the review on the ensemble approaches for the imbalance classification problem [1]. Chawla et al. discussed the cost-sensitive methods and kernel-based procedures, sample-based ways for class imbalance problem [7]. Gosain et al. applied sampling techniques over the following datasets: Pima India Diabetes, Breast Cancer Wisconsin, Statlog (Heart),

Ionosphere, etc. [6]. Deng et al. proposed an ensemble model which is based on automatic and clustering and under sampling named ACUS [8]. Milaré et al. applied evolutionary algorithms to select rules which maximize the AUC [11]. Khoshgoftaar et al. experimented with hybrid sampling by taking different combinations of undersampling and oversampling techniques. This article took different proportions of majority and minority classes for oversampling and undersampling compared the results. Oversampling techniques like SMOTE and borderline smote and undersampling techniques like Wilson's editing and random undersampling were used for their experiment [13]. Hanskunatai proposed a new algorithm based on DB-scan algorithm and SMOTE to balance the datasets which improved the accuracies of decision trees and Naïve Bayes machine learning models [14]. Popel et al. proposed a new model called HUSBoost which used random undersampling technique for the data and then classification is done based on ensembling of boosting algorithms which gave them better results [15]. Seiffert et al. came up with a new hybrid boosting algorithm called RUSboost which is based on SMOTEBoost and AdaBoost algorithm, and the results are compared with the same algorithms individually and they random undersampling at different distributions to balance the data [16].

2.1 Background

This section elucidates some of the most popular approaches for imbalance classification problem. The class imbalance problem is handled in two ways. The first method is the cost function-based approach where the cost attached to the imbalance data is adjusted, whereas the second method is resampling the data to balance them [10]. Figures 1 and 2 show a glimpse of undersampling and oversampling, respectively.

This section also entails about the oversampling techniques that mainly focus on SMOTE, ADASYN, DBSMOTE by applying various algorithms on the result of these oversampling techniques.

Fig. 1 Before and after undersampling

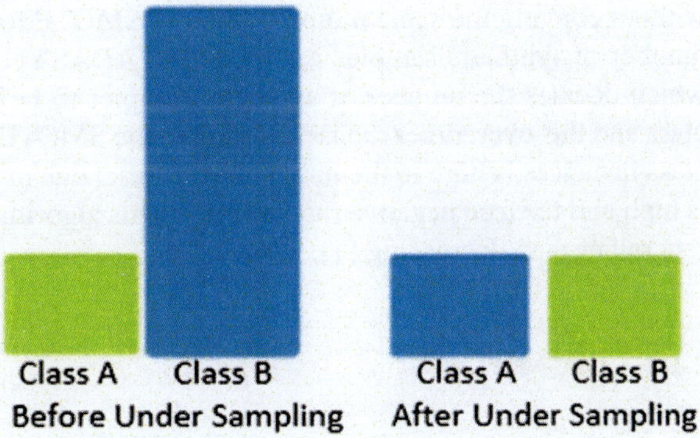

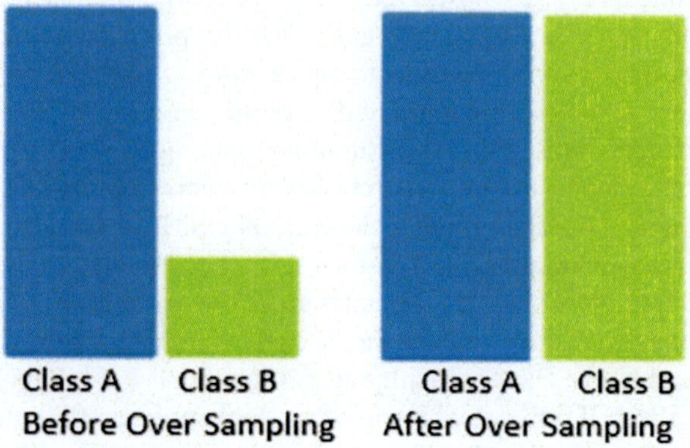

Fig. 2 Before and after oversampling

2.1.1 SMOTE

Synthetic minority oversampling technique is one of the oversampling techniques that is used to handle the imbalanced data. The basic idea behind this SMOTE algorithm is to generate the synthetic samples of the minority class [4]. The working process of SMOTE is as follows. It selects the p minority class from the set of minority classes M which are available in the original dataset D. It iterates for each minority class p it identifies some K-nearest neighbors based on some distance measures. Out of these nearest neighbors, it selects n points and generates a synthetic point s by interpolating between p and n. Then the new point s attaches to the minority class [12].

2.1.2 ADASYN

In the oversampling technique, the instances of minor variables are replicated x times, where x is the number. Due to this, the model overfits the minority class because of the replication of the minority class. This kind of problem can be avoided with the help of ADASYN. It is an adaptive synthetic algorithm that generates synthetic data without copying the same minority data. In SMOTE for each minority class, a same number of synthetic samples are made [5]. ADASYN uses the density distribution which decides the number of synthetic samples to be generated for every minority class and this overcomes the issue found in the SMOTE. The distribution of majority class instances is high in the imbalanced dataset due to which the actual positive rate is high and the true negative rate is lesser. This algorithm improves the true negative rate rather than decreasing [1].

2.1.3 DBSMOTE

It is a combination of DB-scan and SMOTE algorithm. It finds the density of the instances plotted over a graph. DBSMOTE generates more artificial instances around a dataset core than the boundary of a dataset and does not work within a noisy environment. Therefore, in a protected region, these instances are dense and sparse in an area that overlaps [17].

3 The Proposed Model: HOUSEN

The proposed model named HOUSEN intends to improve classification results for the imbalance dataset. The model uses a hybrid sampling method for imbalanced datasets and an ensemble classification model. The HOUSEN model works in two phases. In Phase 1, hybrid sampling method is executed, and in Phase 2, an ensemble classification approach is executed.

3.1 Phase I

Initially, the experiments conducted are to observe the impact of the sampling-based techniques on imbalanced datasets with machine learning algorithms and then analyze the various imbalance techniques by experimenting with three oversampling approaches: SMOTE, ADASYN, and DBSMOTE. The DBSMOTE achieved the best result among all the three experimented techniques. HOUSEN is a model that uses a hybrid sampling approach. It used DBSMOTE for oversampling and random undersampling (RUS) technique for undersampling. The mentioned approaches is selected and it is observed that the DBSMOTE oversampling technique outperformed all other oversampling techniques and then combined it with RUS. By integrating both will overcome the drawbacks of each other sampling techniques.

Initially, the HOUSEN model will take imbalanced datasets for hybrid sampling. At this step, the minority class of the target attribute will undergo oversampling by the DBSMOTE algorithm and then the majority class will undergo undersampling by the random undersampling technique. Subsequently, new hybridized and balanced dataset in obtained from which it could be used for classification.

3.2 Phase II

There are three classifiers namely gradient boost, AdaBoost, and random forest that were selected out of four because these three were way better in almost all the cases of our datasets and then the obtained results for each classifier on the new hybridized dataset are ensembled by the voting majority to get better and efficient results. Figure 3 shows the workflow of the proposed method. The concept of ensembling can be achieved in the majority voting in many ways.

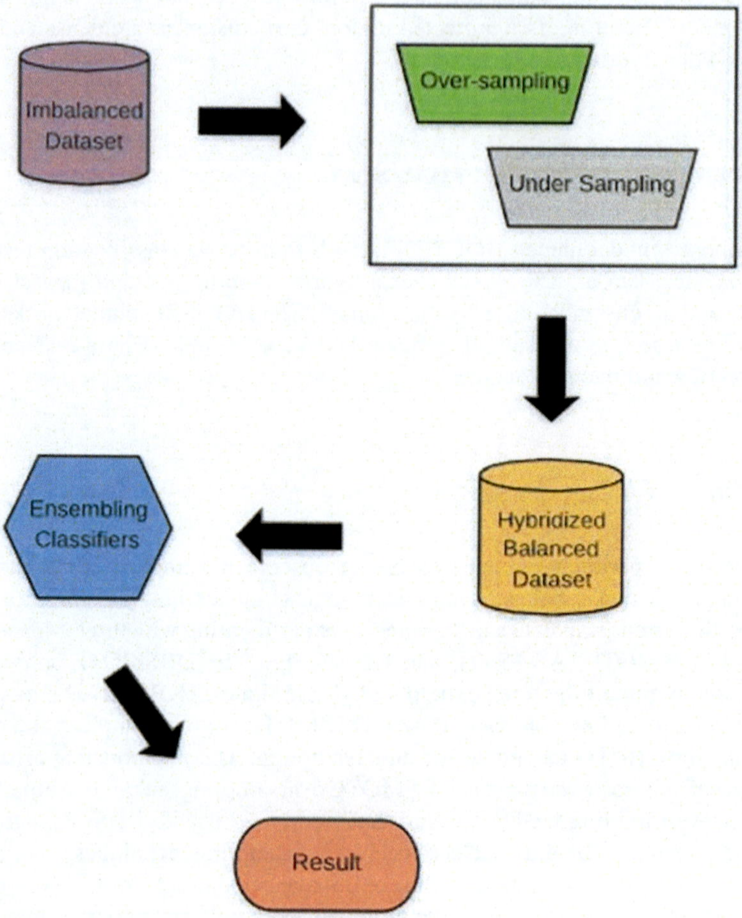

Fig. 3 Framework of HOUSEN model

```
Algorithm: Pseudo code of HOUSEN
HybridSampling ()
{
    Input: Imbalanced dataset
    Oversampling: DBSMOTE
    Under-sampling: Random Undersampling
    Output: Hybridized and Balanced Dataset
}
Classification ()
{
    Input: Hybridized and Balanced dataset
    Classifier_1: Random Forest
```

```
    Classifier_2: AdaBoost (CART)
    Classifier_3: Gradient Boost
    Output: The Results of three classifiers
}
EnsembleVoting()
  {
    Input: The results of three classifiers
    Prediction
    #Binary Classification - 1 or 0
    If (Classifier_1 & Classifier_2 == 1)
     {
      Prediction = 1
     }
Else if (Classifier_1 & Classifier_3 == 1)
    {
     Prediction = 1
    }
   Else if (Classifier_2 & Classifier_3 == 1)
    {
      Prediction = 1
    }
   Else Prediction = 0
   Output: Final Prediction
}
```

The algorithm takes an imbalance dataset as input and after applying DBSMOTE and RUS output is the balanced dataset. Classification function uses three machine learning algorithms: random forest, AdaBoost, gradient boost, and used predictions of three classifiers for ensembling. If any two of the three classifiers predicted the same class, then the final prediction would be the same class. For an instance, if random forest predicted "positive", AdaBoost predicted "negative" and gradient boost predicted "negative" then our final prediction would be "negative". In this way, the majority of voting works for ensembling the results and below that represents the pseudocode of the HOUSEN Model. In our initial analysis, many classification algorithms like logistic regression, C4.5, decision tree—Gini and ID3, gradient boost, AdaBoost, random forest, and support vector machine (SVM) were taken and out of these, the best four were selected where other algorithms are SVM, random forest, AdaBoost, and gradient boost. Another appealing reason is that all these algorithms work the best when the data is numeric, and our analysis is confined solely to numeric kind of data. Moreover, gradient boost overcomes overfitting and handles larger data, whereas AdaBoost has the ability to combine poor classifiers in the forest of decision trees to improve the accuracy in a way pruning the trees. Furthermore, random forest and support vector machines are flexible with regression as well as classification. Random forests give the benefit of handling missing values and maintains in terms of accuracy.

Table 1 Banana data

Banana	Original data	Hybridized data
Negative class	2376	2268
Positive class	264	2192

Table 2 Haberman data

Haberman	Original data	Hybridized data
Negative class	2250	1526
Positive class	810	1545

4 Experiment and Result Analysis

4.1 Data Description

The study used five datasets banana, Haberman, glass0, ecoli1, yeast4 for the experiment. The datasets are collected from the KEEL repository for our experiment.

4.1.1 Banana

An artificial dataset that includes instances belongs to several banana-shaped clusters. There are two attributes, respectively, corresponding to the x, which consists of 2640 observations. One of the two banana shapes in the dataset is depicted by the class label (positive and negative) (Table 1).

4.1.2 Haberman

This is a survival dataset of patients who experienced an operation for breast cancer. It consists of 306 instances and four features age, year of operation, the number of positive auxiliary nodes detected and predictor class, the survival status whether the patient has survived five years or longer or died within five years of operation (Table 2).

4.1.3 Glass0

A glass dataset that is used for imbalance binary classification to identify the variety of glass. It consists of 214 instances and nine features. The features are refractive index (RI), weight percent of sodium in the component (NA), weight percent of magnesium in the component (Mg), weight percent of aluminum in the component (Al), weight percent of silicon in the component (Si), weight percent of potassium in the component (K), weight percent of the calcium in the component (Ca), weight

Table 3 Glass0 data

Glass0	Original data	Hybridized data
Negative class	1440	1240
Positive class	700	1330

Table 4 Ecoli1 data

Ecoli1	Original data	Hybridized data
Negative class	2590	2100
Positive class	770	2070

percent of barium in the component (Ba), weight percent of iron in the component (Fe) and class attribute with two possible glass type positive if it is building windows and float processed, negative for all rest (Table 3).

4.1.4 Ecoli1

The dataset which is used for imbalanced binary classification to identify the protein traits which contains 336 observations and eight attributes. The features are mcg, gvh, Lip, chg, Aac, Alm1, alm2, and class attribute for predicting either positive or negative (Table 4).

4.1.5 Yeast4

It is an imbalanced dataset consisting of 1484 observations and nine attributes. The features include mcg, gvh, alm, mit, erl, pox, vac, nuc, and class attribute (either positive or negative). It is used to identify the proteins in the yeast (Table 5).

4.2 Preprocessing

The preprocessing step is to improve the performance of the model. **Missing Values**. In this step, check the data for missing values. Missing values are eliminated using KNN missing value imputation.

Table 5 Yeast4 data

Yeast4	Original data	Hybridized data
Negative class	1433	1369
Positive class	510	1374

Table 6 Confusion matrix

Actual	Predicted	
	Positive	Negative
Positive	True positive (TP)	False negative (FN)
Negative	False positive (FP)	True negative (TN)

Outlier Detection. In this step, check the data for outliers and apply the normalization techniques to bring them into the range by that handled few outliers.

Multicoloniality. In this step, identify the data for multicollinearity and the variables which are having the highest collinearity.

4.3 Performance Metrics

In the following metrics, accuracy and F1-score are used to measure the performance evaluation of the proposed model (Table 6).

4.3.1 Accuracy

Accuracy is a performance measure (statistical measure) that requires true positives and true negatives to estimate the proposed model.

True positive (TP): No. of cases are classified correctly.
True negative (TN): No. of cases are identified as negative values correctly.
False positive (FP): No. of cases are wrongly predicted as positive
False negative (FN): No. of cases are incorrectly predicted as negative

$$\text{Accuracy} = (\text{TP} + \text{TN})/(\text{TP} + \text{FN} + \text{FP} + \text{TN}) \quad (1)$$

4.3.2 F1-Score

F1-score is the harmonic mean of precision and recall which is better than accuracy because accuracy determines the perfectly classified cases whereas this score represents the more important and incorrectly classified cases.

$$\text{Precision} = \text{TP}/(\text{TP} + \text{FP}) \quad (2)$$

$$\text{Recall} = \text{TP}/(\text{TP} + \text{FN}) \quad (3)$$

Precision is the exact number of positive cases identified from the predicted positive cases, and it is quite useful when FP is more. The recall is the number of positive cases observed from the actual positive cases in the data, and it is significant when FN values are very high. The reason for choosing F1-score in certain cases where the imbalanced datasets could have good accuracy but low F1-score due to precision and recall values that are incorrectly classified classes and better metric when the data has imbalanced nature. From Eqs. 2 and 3, it is possible to calculate the F1-score.

$$\text{F1 - score} = ((\text{Precision} * \text{Recall})/(\text{Precision} + \text{Recall})) * 2 \quad (4)$$

It can be observed from Table 7 that after the hybridization of DBSMOTE and random undersampling better accuracy is achieved by the overall datasets except for the dataset yeast 4. Further, it can be noticed from Table 8, after hybridization better F1-score is obtained for the overall datasets except for the dataset glass0. These two tables produce the results after the hybridization of data. Remember, the last column in both the tables are not the results of the HOUSEN model, they are only the results of each algorithm in all combinations that were chosen before and after the hybridization of data for those four machine learning models.

Table 7 Accuracy measure of classification models

Datasets	Algorithm		SMOTE	ADASYN	DBSMOTE	
Banana	RANDOMFOREST	95.83	95.83	90.32	95.64	**96.25**
	GRADIENTBOOST	95.54	96.02	86.91	95.64	**96.47**
	ADABOOST	96.20	93.93	90.13	96.20	**96.47**
	SVM	95.64	95.64	85.77	96.20	95.34
glass0	RANDOMFOREST	90.48	90.48	90.48	90.48	**93.29**
	GRADIENTBOOST	85.71	85.71	88.10	**90.48**	88.89
	ADABOOST	80.95	88.10	85.71	80.95	**88.89**
	SVM	**83.33**	80.95	76.19	**83.33**	**83.33**
Yeast4	RANDOMFOREST	96.96	94.26	94.26	95.61	**97.62**
	GRADIENTBOOST	**97.30**	93.92	93.58	92.23	94.33
	ADABOOST	**97.30**	95.27	93.24	94.93	96.89
	SVM	**96.62**	89.86	88.18	91.55	93.78
Haberman	RANDOMFOREST	68.85	68.85	65.57	70.49	**75.44**
	GRADIENTBOOST	68.85	67.21	65.57	70.49	**77.19**
	ADABOOST	67.21	60.66	65.57	67.21	**71.93**
	SVM	70.49	65.57	63.93	**72.13**	68.42
Ecoli1	RANDOMFOREST	90.91	89.39	92.42	93.94	**94.23**
	GRADIENTBOOST	80.30	89.39	93.94	92.42	**94.23**
	ADABOOST	89.39	92.42	90.91	92.42	**93.40**
	SVM	89.39	89.39	83.33	86.36	**94.23**

Table 8 F1-Score of classification models

Datasets	Algorithm	F1-Score before applying sampling techniques	F1-score after applying oversampling techniques			F1-score after hybridization of DBSMOTE and RUS
			SMOTE	ADASYN	DBSMOTE	
Banana	RANDOMFOREST	78.80	89.39	86.72	88.25	**95.49**
	GRADIENTBOOST	73.06	83.92	85.87	90.32	**95.22**
	ADABOOST	83.99	87.46	87.57	90.59	**95.59**
	SVM	71.60	83.77	88.49	90.59	**94.47**
glass0	RANDOMFOREST	89.14	**91.04**	**91.04**	86.59	88.36
	GRADIENTBOOST	77.15	80.75	87.46	**89.14**	88.36
	ADABOOST	77.92	87.46	83.59	81.98	**92.36**
	SVM	79.37	77.92	78.42	**85.12**	82.69
Yeast4	RANDOMFOREST	56.97	73.68	45.77	56.74	**97.62**
	GRADIENTBOOST	57.03	73.58	65.54	73.05	**93.98**
	ADABOOST	66.43	81.02	33.09	45.85	**96.47**
	SVM	69.44	78.96	78.29	65.04	**93.47**
Haberman	RANDOMFOREST	38.57	45.29	44.59	67.48	**72.94**
	GRADIENTBOOST	30.83	44.94	50.12	69.91	**73.42**
	ADABOOST	44.94	52.84	58.71	69.32	**70.64**
	SVM	30.97	38.10	54.60	61.54	**66.21**
Ecoli1	RANDOMFOREST	79.37	85.65	92.74	91.13	**94.12**
	GRADIENTBOOST	61.51	78.72	68.04	87.31	**93.95**
	ADABOOST	82.43	92.74	91.74	90.24	**93.82**
	SVM	74.44	78.72	82.10	83.92	**93.82**

Figures 4 and 5 represent a comparative analysis of our HOUSEN model with SVM, random forest, gradient boost, and AdaBoost and the overall five datasets. The figures in this section depict the results of the three models being ensembled after the hybridized data has been generated which is called the HOUSEN model.

After analyzing Figs. 4 and 5, it is observed that our proposed HOUSEN model has achieved the highest accuracy and F1-score among all the experimented models such as SVM, random forest, AdaBoost, gradient boost, and the overall datasets, respectively.

5 Conclusion

The proposed HOUSEN model is used to address the imbalance data class issue. Based on the experimental analysis, some sampling techniques were designed using a

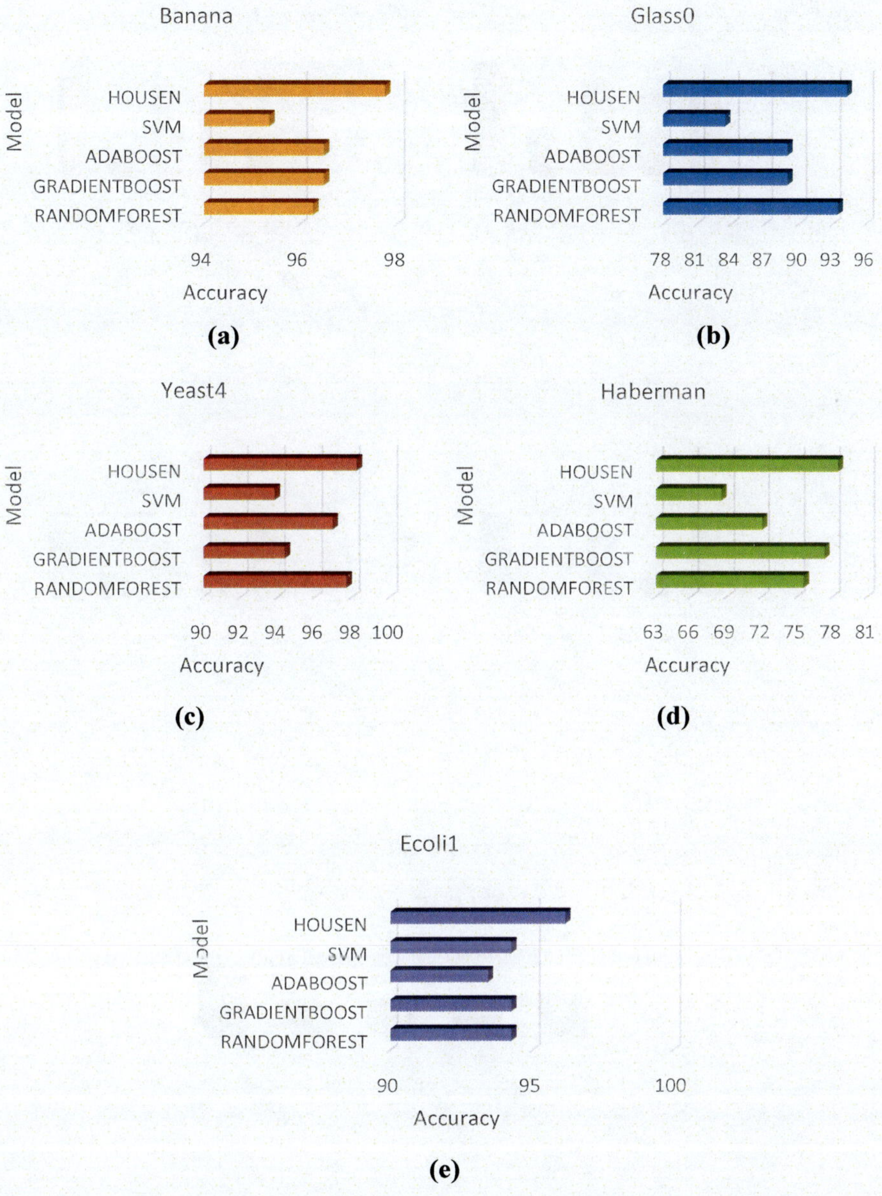

Fig. 4 Accuracy of models: **a** banana, **b** glass0, **c** yeast4, **d** Haberman, and **e** ecoli1

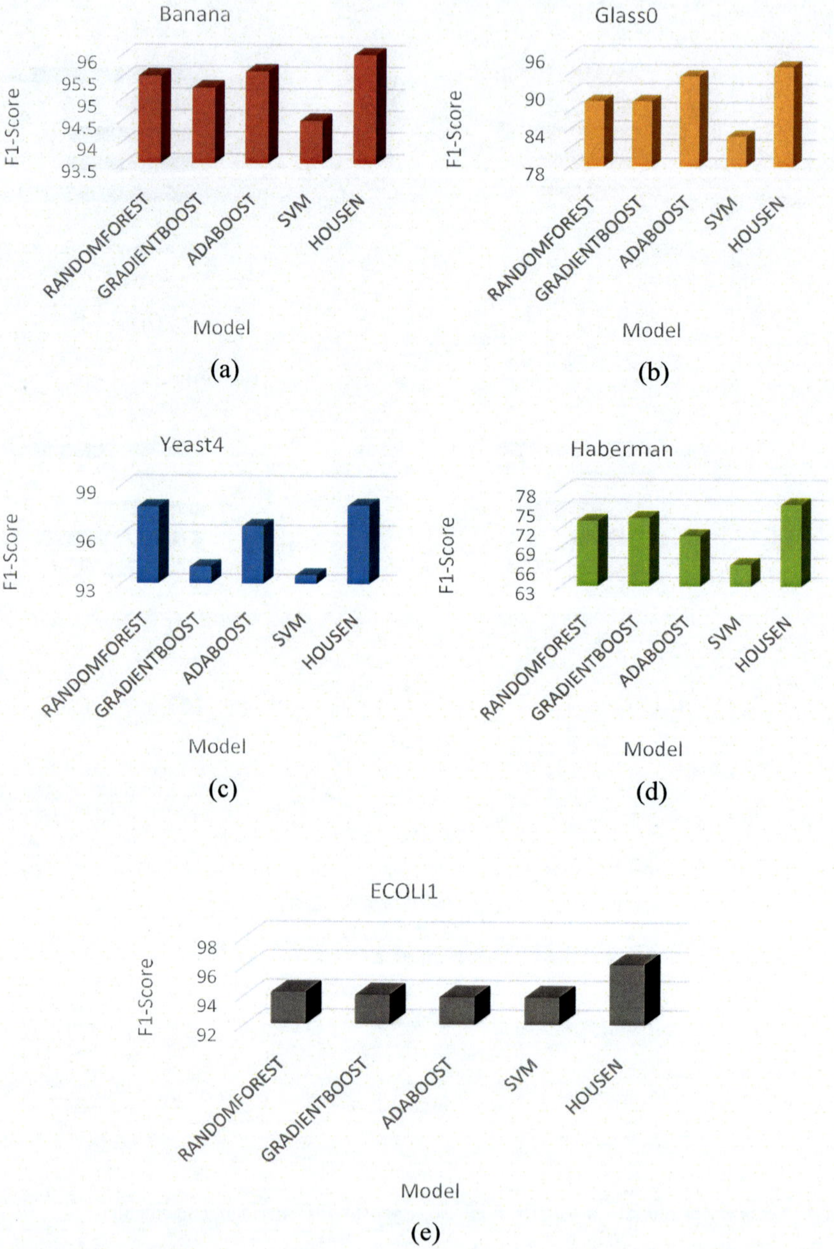

Fig. 5 F1-Score of Models over different datasets: **a** banana, **b** glass0, **c** yeast4, **d** Haberman, and **e** ecoli1

sampling method by hybridizing DBSMOTE and random undersampling technique. Our novel designed sampling method outperformed all other sampling methods used for the experiment. The HOUSEN model used our designed hybrid sampling method and ensembling of the classifiers—random forest, gradient boost, and AdaBoost. It works in two phases: in Phase 1, a hybrid sampling method is applied over the data. Further, an ensemble classification model is executed to obtain the final result. A total of five datasets such as banana, glass0, yeast4, ecoli1, and Haberman is used for the experiments. The comparative evaluation of our proposed HOUSEN model is performed with SVM, random forest, gradient boost, and AdaBoost. The experimental results indicate that the HOUSEN model has achieved a promising result.

6 Future Work

In the future, nature-inspired evolutionary algorithms are used for better results.

References

1. M. Galar, A. Fernandez, E. Barrenechea, H. Bustince, F. Herrera, A review on ensembles for the class imbalance problem: bagging, boosting, and hybrid-based approaches. IEEE Trans. Syst. Man Cybern. Part C (Applications and Reviews) **42**(4), 463–484 (2012)
2. X. Guo, Y. Yin, C. Dong, G. Yang, G. Zhou, On the class imbalance problem, in *2008 Fourth International Conference on Natural Computation*, vol. 4 (IEEE, 2008), pp. 192–201
3. https://www.analyticsvidhya.com/blog/2016/03/practical-guide-deal-imbalanced-classification-problems/
4. G. Hu, T. Xi, F. Mohammed, H. Miao, Classification of wine quality with imbalanced data, in *2016 IEEE International Conference on Industrial Technology (ICIT)* (IEEE, 2016), pp. 1712–1217
5. H. He, E.A. Garcia, Learning from imbalanced data. IEEE Trans. Knowl. Data Eng. **9**, 1263–1284 (2008)
6. A. Gosain, S. Sardana, Handling class imbalance problem using oversampling techniques: a review, in *2017 International Conference on Advances in Computing, Communications, and Informatics (ICACCI)* (IEEE, 2017), pp. 79–85
7. N.V. Chawla, K.W. Bowyer, L.O. Hall, W.P. Kegelmeyer, SMOTE: synthetic minority over-sampling technique. J. Artif. Intell. Res. **16**, 321–357 (2002)
8. X. Deng, W. Zhong, J. Ren, D. Zeng, H. Zhang, An imbalanced data classification method based on automatic clustering under-sampling, in *2016 IEEE 35th International Performance Computing and Communications Conference (IPCCC)* (IEEE, 2016), pp. 1–8
9. V. Ganganwar, An overview of classification algorithms for imbalanced datasets. Int. J. Emerg. Technol. Adv. Eng. **2**(4), 42–47 (2012)
10. A. Sonak, R. Patankar, N. Pise, A new approach for handling imbalanced dataset using ANN and genetic algorithm, in *2016 International Conference on Communication and Signal Processing (ICCSP)* (IEEE, 2016), pp. 1987–1990
11. C.R. Milaré, G.E. Batista, A.C. Carvalho, A hybrid approach to learn with imbalanced classes using evolutionary algorithms. Logic J. IGPL **19**(2), 293–303 (2010)

12. https://www.datasciencecentral.com/profiles/blogs/handling-imbalanced-data-sets-in-supervised-learning-using-family
13. T. Khoshgoftaar, C. Seiffert, J. Van Hulse, Hybrid sampling for imbalanced data, in *Proceedings of IRI*, vol. 8 (2008), pp. 202–207
14. A. Hanskunatai, A new hybrid sampling approach for classification of imbalanced datasets, in *2018 3rd International Conference on Computer and Communication Systems (ICCCS)* (IEEE, 2018), pp. 67–71
15. M.H. Popel, K.M. Hasib, S. Ahsan Habib, F. Muhammad Shah, A hybrid under-sampling method (HUSBoost) to classify imbalanced data, *2018 21st International Conference of Computer and Information Technology (ICCIT)* (2018)
16. C. Seiffert, T.M. Khoshgoftaar, J. Van Hulse, A. Napolitano, RUSboost: a hybrid approach to alleviating class imbalance. IEEE Trans. Syst. Man Cybern. Part A Syst. Humans **40**(1), 185–197 (2009)
17. C. Bunkhumpornpat, K. Sinapiromsaran, C. Lursinsap, DBSMOTE: density-based synthetic minority over-sampling technique. Appl. Intell. **36**(3), 664–684 (2012)

E-Pro: Euler Angle and Probabilistic Model for Face Detection and Recognition

Sandesh Ramesh, M. V. Manoj Kumar, and H. A. Sanjay

Abstract The high priority to give facial appearance is common nature among human as it reflects the person nature on first appearance. Every individual has specific and unique facial appearance and this unique feature is considered as vital information. This research work proposed a face detection and recognition approach termed as E-Pro identifies the person based on the facial features obtained from the images. Proposed model can be utilized in various domains such as surveillance, authentication, attendance and crowd monitoring etc. In this research work, the proposed model is developed as a mobile application which helps the lectures to collect the presence of students in a classroom by recognizing the students face obtained through application. Google Firebase face recognition API is used to develop the E-Pro application considering the Euler's Angles and probabilistic models. Experimental results are obtained by testing the proposed approach using stock images and observed the better detection and recognition performance.

Keywords Face detection · Face recognition · Neural networks · Firebase · Euler angle · Prob. model

1 Introduction

Face recognition is one of the key thrust area in image processing domain [1], and due to technology development it has wide proliferation in various applications. With its introduction on smartphone devices, millions of people around the world have this technology at the palm of their hands, storing and protecting valuable data. But to recognize a face, the device must first detect it. Face detection refers to the ability of the device or computer to identify the presence of a human entity within a digitized image or video.

S. Ramesh (✉) · M. V. Manoj Kumar · H. A. Sanjay
Department of Information Science and Engineering, Nitte Meenakshi Institute of Technology, Bangalore 560064, India

V. Suma et al. (eds.), *Inventive Systems and Control*,
Lecture Notes in Networks and Systems 204,
https://doi.org/10.1007/978-981-16-1395-1_9

Face detection has plenty of applications, most prominent applications are

- Facial recognition—attempts to establish identity. In order to establish person identity, digital image captures through computer applications are used to obtain the individual face. A comparison process is performed with the captured image and existing image stored in database to match the features [2].
- Prevention of retail crime—where shoplifters with a history of fraud enter a retail establishment are identified instantly or smarter advertising—where companies make use of advertisements to educated class to guess peoples age.
- Forensic investigations or Gender identification—by identifying dead individuals through video surveillance footage data [3].

Another application that the paper focuses on is marking students' attendance. Objectives of the proposed model discussed in this paper are as follows:

1. Detection of face/s in the image by considering vital factors such as lighting, facial expression, facial posture, etc. for accurate recognition.
2. Calculation of 3-Dimensional (3D) geometrical orientation of the face using Euler Angle.
3. Calculation of percentage recognition by probabilistic decision.
4. Obtain identification results.
5. Calculate total count of faces in image.

The upcoming sections are ordered in the following pattern.

Section 2 discusses a condensed overview of the work carried out by previous researchers in the domain of facial recognition and requirements needed for the same. Section 3 illustrates and describes the framework of E-Pro and Sect. 4 tabulates the readings obtained through experimentation of the model. Section 5 concludes with the contribution of the paper with concise summary.

2 Literature Survey

According to the research paper titled "Face recognition/ detection by probabilistic decision-based neural network", the authors have utilized probabilistic decision-based neural networks (PDBNN) comprising of a tri-methodology for facial recognition system—the first step involves a face detector that detects the human face location in an image using an eye localizer. To evaluate and present the essential features the position of eyes is analyzed. Other than eyes, eyebrows and nose are also used to represent the facial features. Mouth could not be used to represent the facial features as it differs for every time. The final module was the face recognizer. Recognition system adopts network structures in a hierarchical network along with nonlinear basis functions. The whole recognition process consumed approximately one second, without the use of hardware accelerator or co-processor [4].

Raphaël Féraud in their paper "A Fast and Accurate Face Detector Based on Neural Networks", approached the task of detecting faces in complex backgrounds using generative approach. The learning model evaluates the generated system probability from the input data and utilized constrained rules to improve the performance of estimation model. False alarm occurrence and side view detection, conditional mixture of networks are employed in the system [5]. In early developmental stages of facial detection algorithms, focus was mainly directed towards the frontal part of the human face. According to the research paper titled "A Neural Basis of Facial Action Recognition in Humans", the authors state that by combining various facial muscular actions, called action units, humans have the ability to produce an large number of facial expressions [6]. With the use of functional magnetic resonance imaging and certain machine learning technique, the authors were able to identify a consistent and differential coding of action units by the brain. Most of the attempts by Hjelmås and Low, have aimed to improve the ability to cope with alterations, but are still restricted to certain body parts such as frontal face, shoulder and head. Detection of faces among multiple faces is essential. Therefore, the proposed model goes on to ignore the basic understanding of the face and tends to use stock images facial patterns. This process is considered as a training stage in the detection process [7]. They go a step further to put the system to experiment, based on a series of combination of various classifiers to help in obtaining more reliable results. The obtained results are used to compare with single classifier. The authors have presented a model which is efficient to handle different face patterns as a multiple face classifier using three classifier approaches such as Gradient feature, Texture Feature, and Pixel Intensity Feature. Gradient feature classifier considers the features which has information. Pixel distribution and invariability for facial features are considered in this classifier. In Texture Feature approach texture features by correlation obtained through occurrence of joint probability, local distribution of features, image disorder features are considered while extracting the features. Pixel Intensity classifier extracts the eye, mouth and nose pixel intensity to extract face patterns.

3 Framework

Functional requirements are defined as those requirements the system must deliver. In the case of E-Pro, it was vital that we collect certain parameters to attain the desired objective of the proposed model. Both functional and non-functional requirements were collected. However, the collected functional requirements were important, as in, without these, the entirety of this system would be a failure. These requirements were also based on the users' feedback about the overall functioning of the system. The following algorithm defines a detailed overview of the objectives we aim to achieve through E-Pro.

Step 1: Start
Step 2: Capture face images via camera.
Step 3: Detection of faces within an image must be confirmed.
Step 4: Bounding boxes must be used around each face.

Step 5: Complete attendance must be marked based on number of faces obtained.
Step 6: All detected faces must be cropped.
Step 7: Cropped images can be resized to meet mapping requirements.
Step 8: All cropped images must be stored onto a folder.
Step 9: The database must be loaded with images of faces.
Step 10: These images can be used for training the model.
Step 11: Capture image and recognize faces over and again.
Step 12: Compare image stored in the database along with the input image.
Step 13: Display the name or ID of the student over the image captured
Step 14: Stop.

Other features of E-Pro include detailed measures to judge E-Pro system operations. They cover ease, protection, availability of support, speed of operation, and considerations for implementation. To be more specific, it is more convenient and user friendly to capture photographs and inform students regarding their facial positions. The system is very secure and can be easily installed. With a response time of less than 10 s, E-Pro is fast and reliable (Fig. 2).

One needs to prioritize the importance of certain requirements that are vital for the functioning of the system. To make this process easy to understand we classify the requirements under Must Have, Should Have, Could Have, and Will Not Have. "Must Have"—These are requirements without which the system ceases to exist. The "Should Have"—These are requirements with a slightly higher priority and must be implemented if possible. "Could Have"—These are features that are desirable in the system but not a necessary requirement. "Not Have"—These are features that could be implemented in the future to enhance the overall working of the system.

3.1 Must-Have

Considering the approach, "Must Have" defines the identified user requirements to obtain desired output. Absence of these requirements, the ultimate outcome will not be achieved.

- Bounding boxes must surround the faces of people in the image.
- Images of all faces detected must be cropped.
- Captured image sizes must be cropped to meet the image size of the database.
- The total attendance of the class must be calculated depends on the detected faces.
- Pictures must be trained rigorously for recognition.
- The input and output images must be displayed side-by-side.
- Name of image outputted must be displayed above it.

3.2 Should Have

These features will be implemented if possible. These features form a priority to the system. However, even without these features, the system will continue to perform its functions.

- Ensure the names of both output and input search images are displayed.
- Calculate the recognition percentage of an image captured by the system to that of its database counterpart.
- Calculate the rate at which the system effectively recognizes faces.

3.3 Could Have

These are features, if added, could make the application much more interactive and fun to use. However, without these, the app doesn't stop functioning and continues to perform its duty.

- An enhanced and easy to use Graphical User Interface (GUI).
- High definition Camera to capture quality images.

3.4 Will not Have

Under this header we have features that have not been included in the current system, as we don't see much use of them. Since the system is college/university specific, we find it easy to maintain the data records of students and teachers. However, should this system be used in a much larger commercial scale, the database and servers must be revamped to meet the demands of every college/university simultaneously.

4 Methodology

Table 1 contains experimental readings of people, individually and in a group of greater than two. The following abbreviations are used:

Zzwy—getBoundingBox ()—Returns the axis-aligned bounding rectangle of the detected face.
Zzxq—getTrackingId ()—Returns the tracking ID if the tracking is enabled.
Zzxu—getHeadEulerAngleY ()—Returns the rotation of the face about the vertical axis of the image.
Zzxv—getHeadEulerAngleZ ()—Returns the rotation of the face about the axis pointing out of the image.

Table 1 Quantitative attributes of facial recognition

Observation	Number of people under consideration	Individual classification	zzwy	zzxq	zzxu	zzxv	zzxt	zzxs	zzxr
1	1	a. P1	Rect (270, 488–469, 687)	0	−5.2812	−2.9454	−1.0	−1.0	−1.0
		b. Misc							
2	2	a. P1	Rect (134, 4 94–234, 593)	0	−3.5252	−3.9695	−1.0	−1.0	−1.0
		b. P2	Rect (117, 872–202, 9580)	0	−3.1259	−0.4368	−1.0	−1.0	−1.0
3	3	a. P1	Rect (43, 438–296, 697)	0	36.8327	−21.325	−1.0	−1.0	−1.0
		b. P2	Rect (283, 443–495, 655)	0	1.4248	−2.4146	−1.0	−1.0	−1.0
		c. P3	Rect (459, 491–715, 757)	0	−20.905	9.9750	−1.0	−1.0	−1.0
4	6	a. P1	Rect (86, 528–214, 655)	0	−1.7599	−17.752	−1.0	−1.0	−1.0
		b. P2	Rect (480, 505–610, 636)	0	−4.8044	−7.6158	−1.0	−1.0	−1.0
		c. P3	Rect (528, 662–675, 813)	0	−5.0763	−16.526	−1.0	−1.0	−1.0
		d. P4	Rect (305, 691–459, 844)	0	−11.528	−7.7964	−1.0	−1.0	−1.0
		e. P5	Rect (303, 691–420, 608)	0	−1.4715	−12.268	−1.0	−1.0	−1.0
		f. P6	Rect (117, 679–278, 836)	0	5.7566	−33.651	−1.0	−1.0	−1.0
5	8	a. P1	Rect (581, 573–665, 658)	0	−12.467	0.38612	−1.0	−1.0	−1.0
		b. P2	Rect (498, 561–575, 640)	0	−9.00484	7.73211	−1.0	−1.0	−1.0
		c. P3	Rect (73, 595–151, 672)	0	5.829581	−18.794	−1.0	−1.0	−1.0
		d. P4	Rect (279, 594–343, 659)	0	−1.99653	−0.5076	−1.0	−1.0	−1.0
		e. P5	Rect (365, 567–440, 642)	0	−3.63065	−3.3822	−1.0	−1.0	−1.0
		f. P6	Rect (212, 665–288, 741)	0	8.367999	−10.364	−1.0	−1.0	−1.0
		g. P7	Rect (163, 570–234, 642)	0	1.052805	−5.9284	−1.0	−1.0	−1.0
		h. P8	Rect (394, 665–475, 747)	0	−8.54884	−8.5854	−1.0	−1.0	−1.0

(continued)

Table 1 (continued)

Observation	Number of people under consideration	Individual classification	zzwy	zzxq	zzxu	zzxv	zzxt	zzxs	zzxr
6	10	a. P1	Rect (233, 683–300, 750)	0	−3.98541	2.441709	−1.0	−1.0	−1.0
		b. P2	Rect (416, 627–465, 676)	0	0.487392	0.209005	−1.0	−1.0	−1.0
		c. P3	Rect (295, 604–348, 657)	0	−0.80602	−5.32229	−1.0	−1.0	−1.0
		d. P4	Rect (182, 610–237, 665)	0	−7.00699	−1.65343	−1.0	−1.0	−1.0
		e. P5	Rect (65, 693–135, 764)	0	1.865780	−3.20102	−1.0	−1.0	−1.0
		f. P6	Rect (510, 612–565, 668)	0	−1.32773	−11.9088	−1.0	−1.0	−1.0
		g. P7	Rect (61, 611–119, 670)	0	6.605692	−0.36110	−1.0	−1.0	−1.0
		h. P8	Rect (496, 678–570, 753)	0	−0.28733	−1.65976	−1.0	−1.0	−1.0
		i. P9	Rect (605, 618–658, 672)	0	6.210866	−10.6227	−1.0	−1.0	−1.0
		j. P10	Rect (380, 691–438, 749)	0	3.595529	−5.46549	−1.0	−1.0	−1.0

Zzxt—getIsSmilingProbability ()—Returns a value between 0.0 and 1.0 giving a probability that the face is smiling.
Zzxs—getIsLeftEyeOpenProbability ()—Returns a value between 0.0 and 1.0 giving a probability that the face's left eye is open.
Zzxr—getIsrightEyeOpenProbability ()—Returns a value between 0.0 and 1.0 giving a probability that the face's right eye is open.

Vital factors such as lighting, facial expression, facial posture, etc. need to be considered for accurate recognition [8] (Fig. 3).

5 Results and Discussions

The attributes obtained by detecting faces in images are tabulated below. For the purpose of experimentation, we do not take into consideration the Smiling Probability (zzxt), LeftEyeOpenProbability (zzxs), RightEyeOpenProbability (zzxr) and TrakingId (zzxq) of the images and are assigned the values −1.0, −1.0, −1.0 and 0 respectively. The rectangular bounding box around faces is represented by Rect ((left, top), (width, height)). This method indicates the size of boxes that need to be created during the process of facial identification. Groups of people are sub-divided into individual entities and their values are recorded. Observation 1 (Fig. 1) takes into account, a person and a miscellaneous object, in this case a dog. The E-Pro system successfully generates a box around the person but not the object. Euler head angles (zzxu and zzxv) are those that describe the orientation of a rigid head with respect to a fixed coordinate system. Negative zzxu and zzxv values indicate that the heads are pointed in the negative direction of the axes. The system also successfully identifies

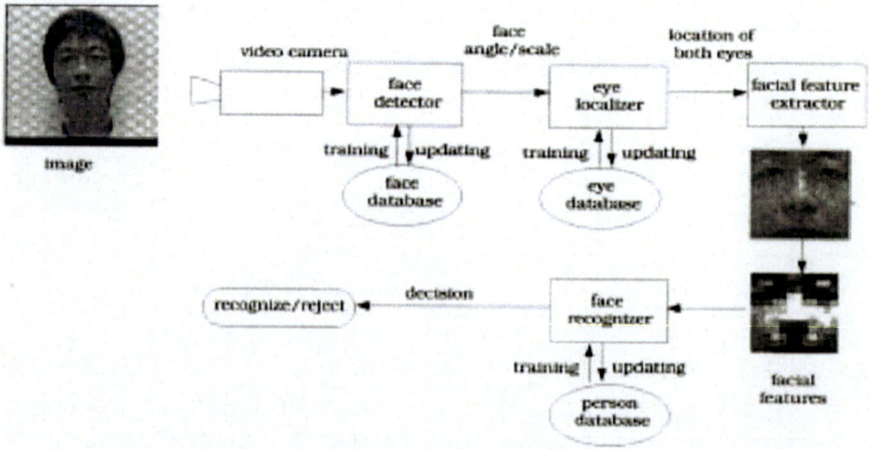

Fig. 1 Face recognition system framework as suggested by Shang-Hung Lin (2000, p. 2)

a large group of people as well. Observations 5 (Fig. 8) and 6 (Fig. 9) demonstrate this (Figs. 4, 5, 6, and 7).

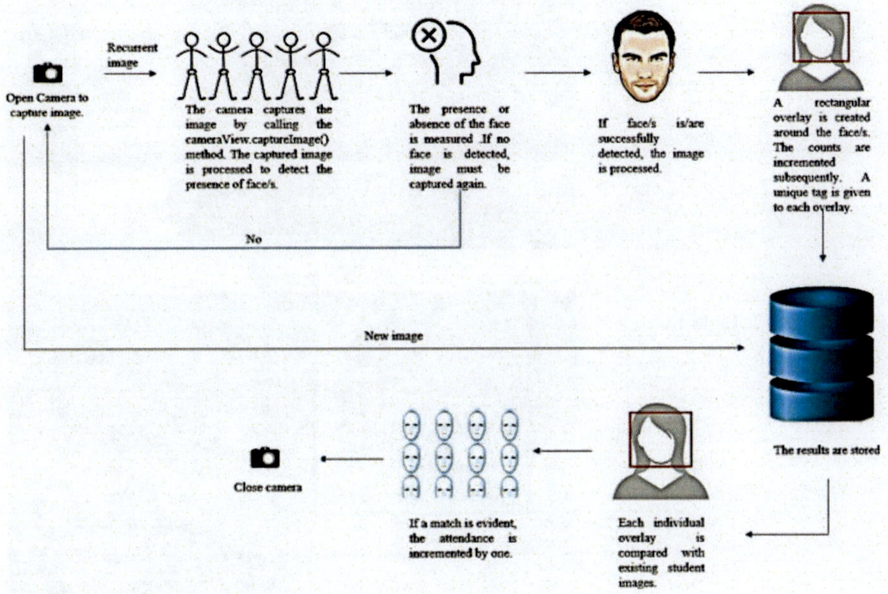

Fig. 2 Proposed E-Pro system framework

Fig. 3 Head pose angles and axes as illustrated by Jerry (2018, p 22)

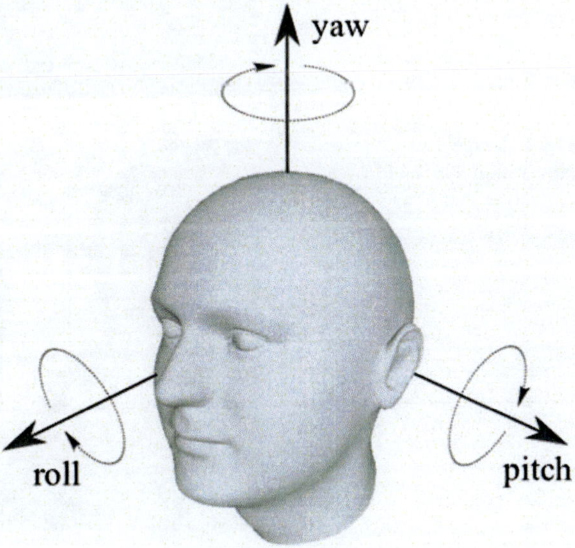

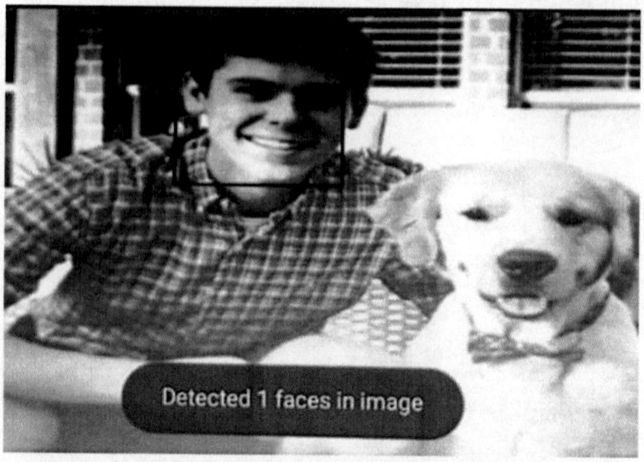

Fig. 4 Image w.r.t observation 1 in Table 1

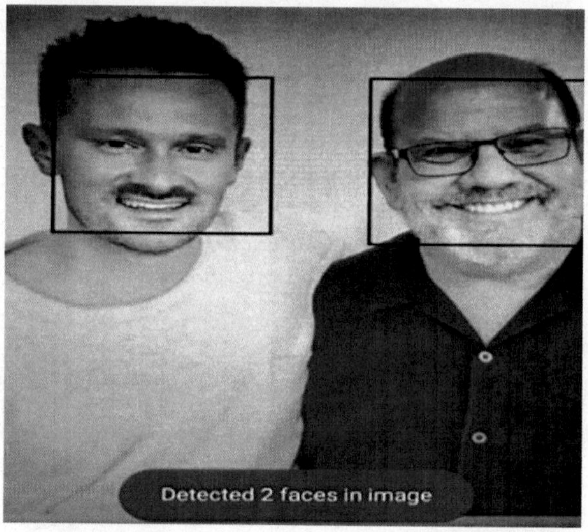

Fig. 5 Image w.r.t observation 2 in Table 1

Fig. 6 Image w.r.t observation 3 in Table 1

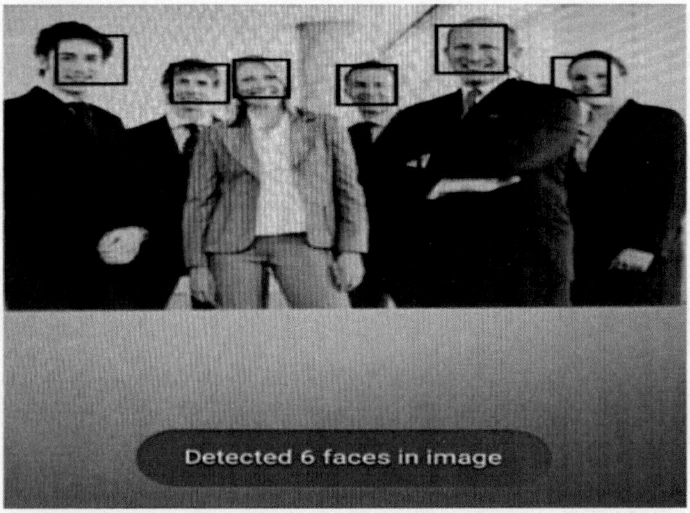

Fig. 7 Image w.r.t observation 4 in Table 1

Fig. 8 Image w.r.t observation 5 in Table 1

Fig. 9 Image w.r.t observation 6 in Table 1

6 Conclusion

This paper has successfully demonstrated the use of a facial detection and recognition system named E-Pro on a dataset composing of varied number of faces. Face detection is rapidly growing in the technology sphere. While the system finds its usefulness in the attendance domain, it can also be used in the surveillance of video footages, access and security, criminal identification and payments. Any personal information can become sensitive information. With the technology's constant development, one can expect better tools for law enforcement, customized advertisements and pay-by-face authentication in the future.

References

1. H. Sharma, S. Saurav, S. Singh, A.K. Saini, R. Saini, Analyzing impact of image scaling algorithms on Viola-Jones face detection framework, in *2015 International Conference on Advances in Computing, Communications and Informatics (ICACCI)* (IEEE, 2015), pp. 1715–1718
2. S.H. Lin, S.Y. Kung, L.J. Lin, Face recognition/detection by probabilistic decision-based neural network. IEEE Trans. Neural Netw. **8**(1), 114–132 (1997)
3. R. Srinivasan, J.D. Golomb, A.M. Martinez, A neural basis of facial action recognition in humans. J. Neurosci. **36**(16), 4434–4442 (2016)
4. R. Feraund, O.J. Bernier, J.E. Viallet, M. Collobert, A fast and accurate face detector based on neural networks. IEEE Trans. Pattern Anal. Mach. Intell. **23**(1), 42–53 (2001)
5. E. Hjelmås, B.K. Low, Face detection: a survey. Comput. Vis. Image Underst. **83**(3), 236–274 (2001)

6. H. Ryu, S.S. Chun, S. Sull, Multiple classifiers approach for computational efficiency in multi-scale search-based face detection, in *International Conference on Natural Computation* (Springer, Berlin, Heidelberg, 2006), pp. 483–492
7. J.D. Wright, K. Wright, G.P. Israel, D.M. Thornock, L.L. Hofheins, *Face Protector* (Save Phace Inc, 2006). U.S. Patent Application 29/214,330
8. J. Gonion, D.R. Kerr, *Personal Computing Device Control Using Face Detection and Recognition* (Apple Inc, 2013). U.S. Patent 8,600,120

Classification of Covid-19 Tweets Using Deep Learning Techniques

Pramod Sunagar, Anita Kanavalli, V. Poornima, V. M. Hemanth, K. Sreeram, and K. S. Shivakumar

Abstract In this digital era, there is an exponential growth of text-based content in the electronic world. Data as texts exist in the form of documents, social media posts on Facebook, Twitter, etc., logs, sensor data, and emails. Twitter is a social platform where users express their views on various aspects in a day to day life. Twitter produces over 500 million tweets daily that is 6000 tweets per second. Twitter data is, by definition, very noisy and unstructured in nature. Text classifications based on the machine learning techniques have problems like poor generalization ability and sparsity dimension explosion. Classifiers based on deep learning techniques are implemented to improve accuracy to overcome shortcomings of machine learning techniques and to avoid feature extraction processes and have high prediction accuracy and strong learning ability. In this work, the classification of tweets is performed on Covid-19 dataset by implementing deep learning techniques namely Convolution Neural Network (CNN), Recurrent Neural Network (RNN), Recurrent Convolution Neural Network (RCNN), Recurrent Neural Network with Long Short Term Memory (RNN+LSTM), and Bidirectional Long Short Term Memory with Attention (BI-LSTM + Attention). The algorithms are implemented using two-word embedding techniques namely Global Vectors for Word Representation (GloVe) and Word2Vec. RNN with Bidirectional LSTM model has performed better than all the classifiers considered. It has classified the text with an accuracy of 93% and above when used with GloVe and Word2Vec.

Keywords Text classification · Deep learning · Text Pre-processing · Word embedding techniques · LSTM · Attention · Word2vec · GloVe

---

P. Sunagar (✉) · A. Kanavalli · V. Poornima · V. M. Hemanth · K. Sreeram · K. S. Shivakumar
Department of Computer Science and Engineering, M. S. Ramaiah Institute of Technology (Affiliated to VTU), Bangalore 560054, India
e-mail: pramods@msrit.edu

A. Kanavalli
e-mail: anithak@msrit.edu

V. Suma et al. (eds.), *Inventive Systems and Control*,
Lecture Notes in Networks and Systems 204,
https://doi.org/10.1007/978-981-16-1395-1_10

1 Introduction

Text classification is the process of classifying or categorizing the text into one or more predefined sets of classes according to its contents. The key applications of text classification are sentiment analysis, topic labeling, intent recognition, language detection, and spam detection. Large collections of documents or social media posts require improved data handling strategies for searching, recovering, and sorting. Text classification is supervised learning which is fundamental to these data preparing techniques. In this digital age, around 80% of the data generated is in an unstructured format. The unstructured data is available as social media posts, emails, web pages, chats, survey responses, and so on. The unstructured data contains valuable information and mining insights from it is hard and time-consuming. Machine learning and deep learning algorithms are measured for classifying the texts to improve the decision-making processes in many applications.

In the existing system, the classification of the text is only based on the Sentiment Analysis, Movie Review, etc. Due to the Covid-19 pandemic, a lot of tweets are being generated on various topics like, safety measures, social distancing, advisories, etc by Government agencies, WHO, Scientists, NGOs, and individuals. Classifying these tweets into different categories like Social Distancing, Vaccination, Advisories, etc was one of the motivations for this work. The primary aim of the project is to categorize the tweets and classify according to the category which it belongs. The scope of the project is to build a classifier and provide all the necessary tweets which are needed to perform the classification. The model accepts at least one classifier, and brokers between them, ensuring all parameters are forwarded to the next classification activities in the model.

Classification of Twitter data on Covid-19 is the primary objective of this work. The scope of the article is to build a classifier to perform text classification on the Covid-19 Twitter data. Covid-19 Dataset used in this article was collected from the Kaggle website (https://www.kaggle.com/markb98/tweets-related-to-coronavirus-trends/). This dataset has 260,000 tweets collected from February 1, 2020, to April 15, 2020. The dataset has 2 columns namely Tweets and Labels. The dataset has pre-defined labels and the numbers of labels were reduced to 15. The list of 15 Labels are Entertainment, Essential Workers, Facts, General, Government Action, Medical Test/Analysis and Supply, Tribute, Pandemic, Panic Shopping, Political, Self-Care, Social Distancing, Stay At Home, and Telecommunicating Life. The layout of this article is structured as follows. Section 2 presents a broad description of the related work on text classification techniques. Section 3 depicts the design and implementation part where data cleaning and data preprocessing techniques are used for text classification. Section 4 illustrates the different deep learning algorithms implemented in this work. Section 5 explains the comparative analysis of these deep learning algorithms. Finally, Sect. 6 concludes the work with findings.

2 Literature Survey

Kowsari et al. [1] explains the different feature extraction methods like Term Frequency (TF), Term Frequency-Inverse document frequency (TF-IDF), different Word-Embedding techniques, dimensionality reduction techniques. The article also presents a comprehensive study of existing algorithms etc. Wang and Song [2] explains the classification of hot news by embedding the deep learning algorithm BiGRU (Bi-Gated Recurrent Unit). The author suggested that Bi-GRU with the attention of deep learning models performed better than traditional modeling. The author had suggested the BERT word vector model for a more complex deep learning network. Luan and Lin (2019) [3] describes that performing text classification with deep learning models like CNN and LSTM can yield good accuracy. The author proposes these models NA_CNN_LSTM (Non-Activation function with Convolution Neural Network and Long Short Term Memory) and NA_CNN_COIF_LSTM (Non-Activation function with Convolution Neural Network Coupled of Input and forgotten gates with Long Short Term Memory).

Hallac et al. [4] describes the tweet classification by using CNN, LSTM, Bi-LSTM, and RNN algorithms and by handling noisy data due to special symbols, tweets with missing letters, etc. The author proposed an approach for Fine-Tuning the model with different sources and extracting functionality from these models. Cai et al. [5] describes the use of hybrid deep learning models like AT-BIGRU-CNN (Bidirectional Gated Recurrent Unit with Attention) and Convolution neural network model. The authors intend to enhance the contextual semantic information through BIGRU and the Attention mechanism. Then the authors explain how to obtain preliminary features and to obtain deeper features through the CNN model. Lakhotia and Bresson [6] have demonstrated several machine learning models and deep learning models for text classification on 3 benchmark datasets for comparisons. The authors have used the pre-trained word vector such as word2vec, Glove, encode the semantic relationship between words and allow the model to explore the associations between the semantics of a text and its assigned label. Song and Geng [7] explain the TextCNN model based on CNN for classification. The classified text is converted to the word vector. Then this word vector is used in TF-IDF for training. The observation based on the CNN curve mentioned the weight vector and word vector.

Anand [8] explained the classification of offensive words from social blogging sites. The author had discussed having the Kaggle toxic dataset, which was trained with deep learning concepts. An efficient model was proposed by having a comparative study between CNN, LSTM with and without vector conversion (GloVe embedding). With these comparative studies, the author concluded that Recursive Neural Network performs better than others, as the model compositionality in NLP using parse trees-based structural representation. Aslam Parwez [9] had performed classification on social media blogging site data. The classification of the social media blogging site data is performed by passing it to the CNN-based architectures. Later word embedding techniques have been used and the model which they had proposed is examined on the Twitter classification dataset. A comparative study on the other traditional and advanced machine learning models has been presented. Marwa et al. [10] summarized the online harassment in social media where Twitter is one of the most widely used microblogging sites. Online harassment is common on the Twitter site and it leads to the user causing low self-esteem and depression. The author demonstrated the word embedding technique and deep learning models like CNN, LSTM, and BLSTM which were compared with other classification model and the results which were obtained from them was very encouraging.

This section summarizes that all the above-mentioned authors have implemented their knowledge in classifying the dataset with different deep learning techniques. Few authors have used traditional models like CNN and RNN, and other authors have used the hybrid approach with a traditional model like CNN + LSTM, NA + CNN + LSTM, etc. and few have used separate models other than these with BI-LSTM, TextCNN, etc. In this work, the hybrid approaches like Recurrent Convolution Neural Network (RCNN), Recurrent Neural Network with Long Short Term Memory (RNN + LSTM), and Bidirectional Long Short Term Memory with Attention (BI-LSTM + Attention) is implemented to study the text classification process.

3 Design and Implementation

The Covid-19 dataset consisted of 260,000 tweets collected in the period of 1st February to 15th April 2020. The entire dataset was divided into training and testing sets. The split ratio is 90–10%. Hence, the training set was comprised of 234,000 tweets, and the testing set comprised of 26,000 tweets. The same training and testing datasets are used for all the deep learning techniques in this work. The sample tweets in the dataset are as shown in Fig. 1.

3.1 Data Cleaning

Twitter data is considered to be noisy as users tend to add emojis, special symbols, misspelling, acronyms, and slang words while expressing their views. Due to this, the cleaning and pre-processing of the Twitter data are required to get good accuracy

	label	tweet
0	Facts	Bernie needs to get to Michigan and Florida kick out Biden because he might just choose Kamalaâ...
1	Stay At Home	Yep. Iâm grateful for some of what heâs doing now, esp wrt acquiring vents/masks, but he too...
2	Social Distance	SO...why are the faith healers social distancing?\r\n\r\nCan't they heal each other of Coronavirus?
3	Pandemic	Again he has lied to the Nation and People of #US \r\n\r\nCan we really keep him as #POTUS in 20...
4	Social Distance	not sure why this hashtag exists, when there are plenty of people across the UK who aren't socia...

Fig. 1 Dataset of Covid-19 tweets

when algorithms are implemented on the processed data. In this work, the removal of NaN values and duplicates, converting text to ASCII format with the help of Unidecode word, removing URLs, Hashtags, emoji's, smiley, converting the text to lowercase, removing punctuation and whitespace from text and converting numbers to words.

3.2 Data Pre-processing

The data needs to be processed after it has been cleaned. This step includes Tokenization where the tweets are converted to separate sentences and later these sentences are divided into words. The next step is Stemming, it is one of the normalization techniques where there is only one root word for example eat, eating, eaten these entire were considered to eat. Then Lemmatization is performed as it brings the content of the words to its similar meaning in the tweets. The final step in the data pre-processing technique is removing stop word sand single-word sentence.

3.3 Feature Extraction

The Covid-19 dataset is unstructured in nature. For applying text classification algorithms on the dataset, the unstructured text sequences must be converted into a structured feature space. The common techniques of feature extractions are Term Frequency-Inverse Document Frequency (TF-IDF), Term Frequency (TF) [11], Word2Vec [12], and Global Vectors for Word Representation (GloVe) [13]. The gloVe is used to obtain the vector representations for words from the text documents. Training is performed on aggregated global word-word co-occurrence statistics from a corpus. The resulting representations showcase interesting linear substructures of the word vector space. Word2vec is a 2-layer neural network. It processes the text by transforming the words present in the text documents into a vector form. It takes text corpus as input and outputs a set of vectors. These vectors are nothing but feature vectors that symbolizes the words from the corpus. The Word2Vec and GloVe are utilized to translate the words into meaningful vectors. In feature extraction, N-gram and Unigram TF-IFD techniques are performed on the processed data. In N-gram TF-IFD features hashing and filter-based feature selections are performed on the Covid-19 dataset. In Unigram TF_IFD features based on the maximum, minimum word length, and frequency of words is performed. In this work, Word2Vec and GloVe are used for performing a comparative analysis of the deep learning algorithms for the classification of Covid-19 tweets. Once the dataset is pre-processed and the feature extraction step is performed, the deep learning algorithms are implemented to classify the tweets. The system architecture is illustrated in Fig. 2.

During the pre-processing stages, the entire dataset was cleaned by removing the URL, username, whitespace, and punctuation. Then calculated the word count

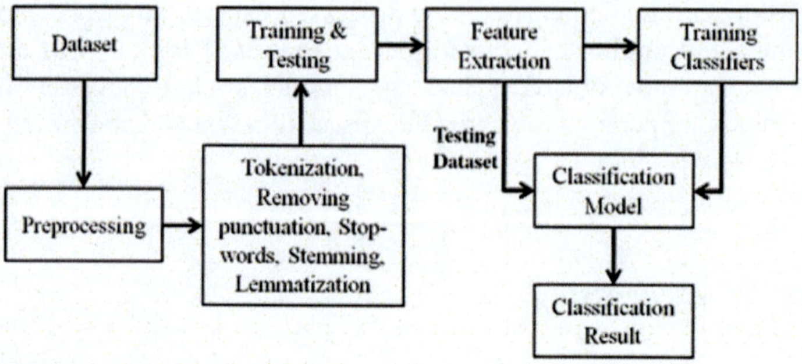

Fig. 2 System architecture

which helped in classifying the labels manually and also applied the tokenization on the Covid-19 dataset where the tweets were converted to sentences and then from sentences to words. The tokenized words were input to the process of normalization, where the Stemming and Lemmatization were applied. Before that, there was a need to remove the stop words in the dataset, and then Normalization was applied. The need to use Stemming and Lemmatization is to modify all the grammatical categories like gender, person, voice, tense, number, case, and aspect, by the normalization process. Despite the expression of inflection like infix, prefix, suffix, and the other internal modification, vowel change, etc., will also be handled perfectly. Based on the analysis, it was observed that these 15 labels are converted from categorical to the numerical format. The reason for this translation is to make it easier for the model to understand the labels in the form of numerical rather than categorical labels.

4 Deep Learning Techniques

4.1 Convolution Neural Network (CNN)

CNN is one of the traditional deep learning models used for predicting and classifying the text data [14, 15]. CNN is a powerful deep learning model in spatial data. CNN takes the input and has assistance members like learnable weights and bias. The purpose of using CNN is to lower the input form so that the text data can be easily processed without losing the features of predicting and classifying the dataset. The limitations of CNN are exploding gradient and long-term dependency.

4.2 Recurrent Neural Network (RNN)

RNN is an internal memory feedback neural network [16, 17]. The nature is recurrent where the function is performed on every input whereas the output is dependent on the current execution of the input. After the output is obtained it is sent back to the recurrent network or the decision making which it learned from the previous input. The limitations of the RNN model are gradient dimension and vanishing.

4.3 Recurrent Neural Network with Long Short Term Memory (RNN + LSTM)

RNN has some disadvantages like the vanishing gradient and it is difficult to train RNN. To overcome this, the LSTM is used with RNN [18, 19]. LSTM works on the back-propagation and works well in prediction, classifying, and processing in the unknown duration of the time delay. The limitations of this model are to overfit and are affected by random weight initialization.

4.4 Recurrent Convolution Neural Network (RCNN)

RCNN is a mixture of two largely used deep learning algorithms namely CNN and RNN [20, 21]. RCNN is used to overcome the problems like eradicating the dimension problem and optimizing the dataset existing in CNN and RNN. The limitations of the RCNN are that the training time is slow and is expensive.

4.5 BIdirectional LSTM with Attention

Bi-LSTM is used for relational classification and captures the most important semantic information in a sentence [22]. Bi-LSTM is used to access the output of preceding and succeeding layers in the process. Bi-LSTM with attention uses high-level feature word embedding. The attention model gives better accuracy when compared to other models.

4.6 Remarks

Deep Learning Techniques exhibit some disadvantages when applied to classification jobs. One of the primary limitations of deep learning is that it is difficult to understand the intricacies of the method that generated the output. Unlike machine learning, there is a need for abundant data to train deep learning techniques. The computational complexity increases during training due to the massive data used.

5 Results and Analysis

In this work, the GloVe and Word2Vec word are used for embedding techniques for tweet classification. The traditional deep learning algorithms like CNN, RNN, and RCNN have demonstrated the accuracy of 79%, 81%, and 82% respectively when implemented with the GloVe technique. RNN + LSTM and RNN + Attention models have displayed the accuracy of 92 and 93.5% in classifying the texts when used with the same feature extraction technique. The CNN, RNN, RCNN, RNN + LSTM, and RNN + Attention algorithms are tested for prediction accuracy using the Word2Vec method and the results are 78.1%, 81%, 83%, 91%, and 93.1%, respectively. During the implementation of the models, considered 10 epochs as common for all the models. The comparisons of prediction accuracy of different classifiers are shown in Table 1.

The accuracy of the different classifiers using GloVe is shown in Fig. 3 and the accuracy of the different classifiers using Word2Vec is shown in Fig. 4. In both cases, the RNN + Attention model has demonstrated better accuracy. Training time for RNN + LSTM is high compared to other models. Prediction time for RCNN is high compared to other models. The RNN + LSTM is the slowest model for training and prediction. The LSTM works using the backpropagation method. The iterations in LSTM make it slow for training and prediction tasks. Figures 5 and 6 display the training time of the different classifiers using GloVe and Word2Vec respectively.

In Fig. 3, the accuracy of the traditional models and hybrid models are compared when the GloVe technique is used. The BI-LSTM with ATTENTION has shown the highest accuracy when compared to others.

In Fig. 4, the accuracy of the traditional models and hybrid models are compared when the Word2Vec technique is used BI-LSTM with Attention holds the highest accuracy when measured with other deep learning models.

Table 1 Prediction accuracy for deep learning classifiers

Deep learning models implemented	Word embedding techniques	
	GloVe (%)	Word2Vec (%)
CNN	79	78.1
RNN	81	81
RCNN	82	83
RNN+LSTM	92	91
RNN with BI-LSTM and attention	93.5%	93.1

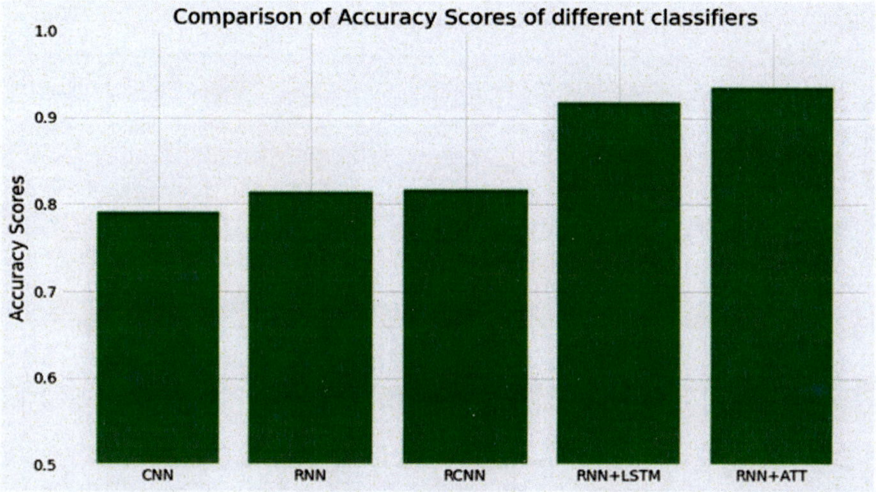

Fig. 3 Accuracy of the classifiers using GloVe method

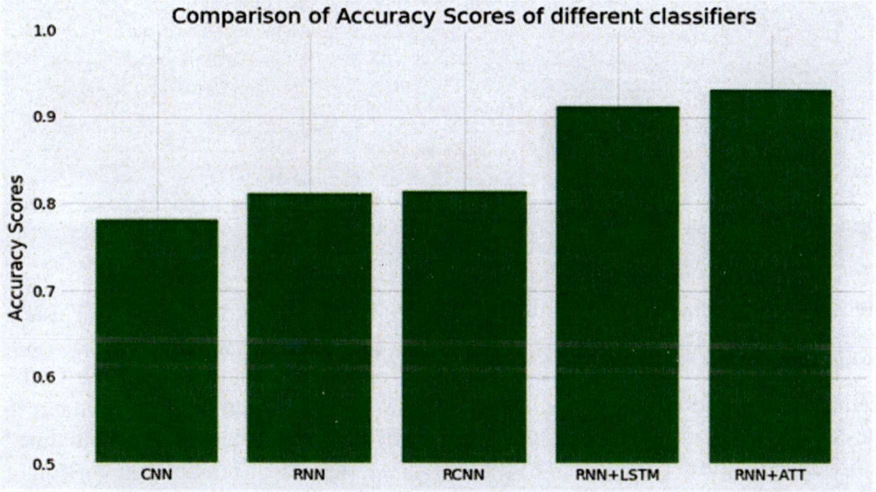

Fig. 4 Accuracy of the classifiers using Word2Vec method

When the deep learning models were implemented and compared according to the prediction time, the RNN+LSTM performed worst due to its backpropagation nature. In Figs. 5 and 6, the prediction time of the different classifiers with Glove and Word2Vec methods are presented respectively.

The confusion matrix is one of the ways to show the percentage of labels correctly classified for a particular algorithm. A confusion matrix is a summary of results from

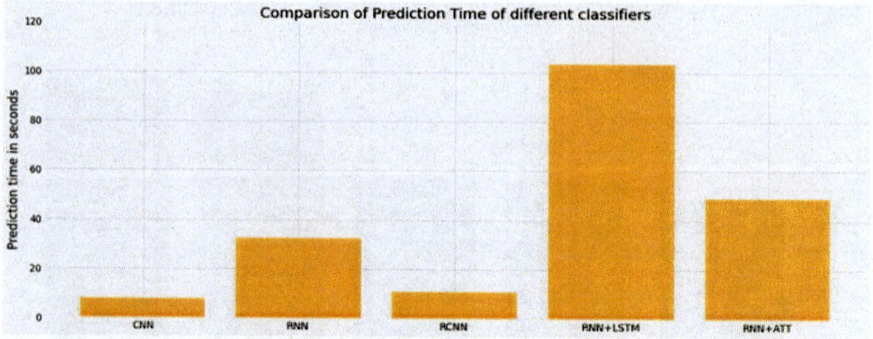

Fig. 5 Prediction time of the classifiers using GloVe method

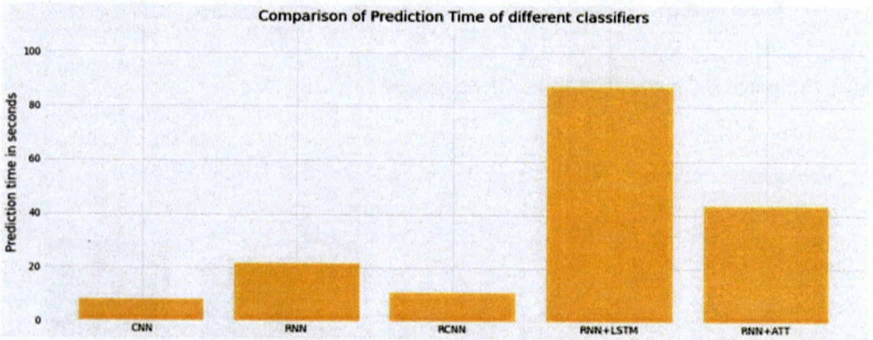

Fig. 6 Prediction time of the classifiers using Word2Vec method

predictions on a classification problem. The confusion matrix for RNN+LSTM and RNN with Bi-LSTM + Attention is shown in Figs. 7 and 8 respectively.

The work is carried out by considering 8 epochs. The accuracy of the CNN, RNN, RCNN, and RNN+LSTM algorithms has demonstrated the improvement in the accuracy after every epoch. The RNN with Bi-LSTM + Attention has maintained consistency in the accuracy above 90% in every epoch. The accuracy comparison of all the models is illustrated in Fig. 9. For the optimization of deep learning algorithms, a loss function is used. On training and testing, the loss is calculated and its analysis is based on how well the model is doing in these two sets. In Fig. 10, the model loss of traditional and hybrid models are compared and it is found that the RNN + Attention model has the lowest model loss compared to other models. Loss value implies how a process performs good or bad after every optimization step.

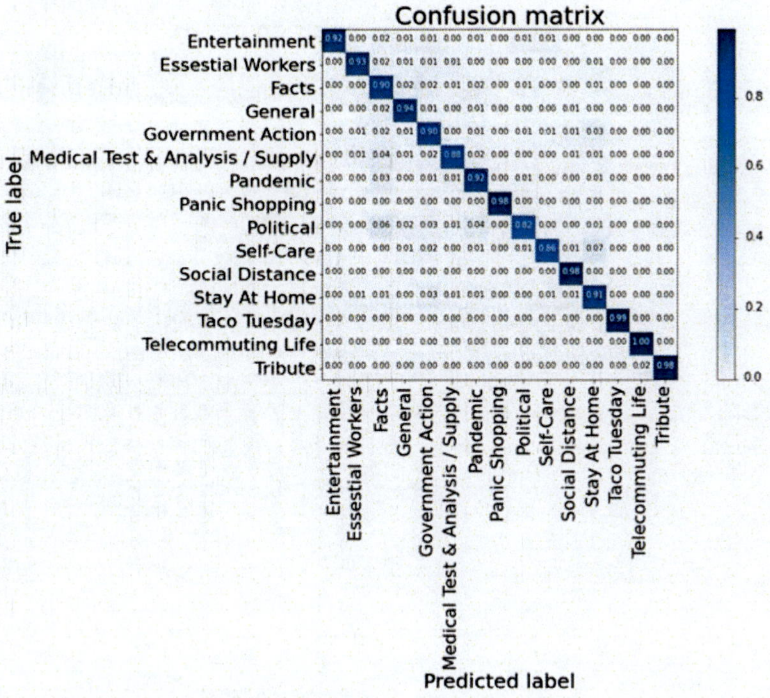

Fig. 7 Confusion matrix for the RNN + LSTM

6 Conclusion

Text classification is one of the vital tasks in the machine learning field. In this work, the classification of Covid-19 tweets using deep learning techniques were focused. Through this work, it can be accomplished three objectives, namely, (1) to implement GloVe and Word2Vec for word embedding techniques, (2) to implement deep learning techniques for text classification, and (3) to perform analysis of results obtained. In the first objective, the effects of word embedding techniques were examined based on the accuracy of the classifiers. In the second objective, scrutiny of traditional as well as hybrid deep learning models were performed on the Covid-19 Twitter dataset and noted that the GloVe technique gives an improved accuracy compared to Word2Vec. The deep learning model RNN with Bidirectional LSTM demonstrated better accuracy while classifying the texts compared to other classifiers considered in this work. The future scope will include the comparison of different word embedding techniques to assess the improvement in deep learning. Character level embedding is one more approach for improving the accuracy of text classification tasks. Compared to word-level embedding, the character level embedding can give improved results as it can build any word as long as the character is included.

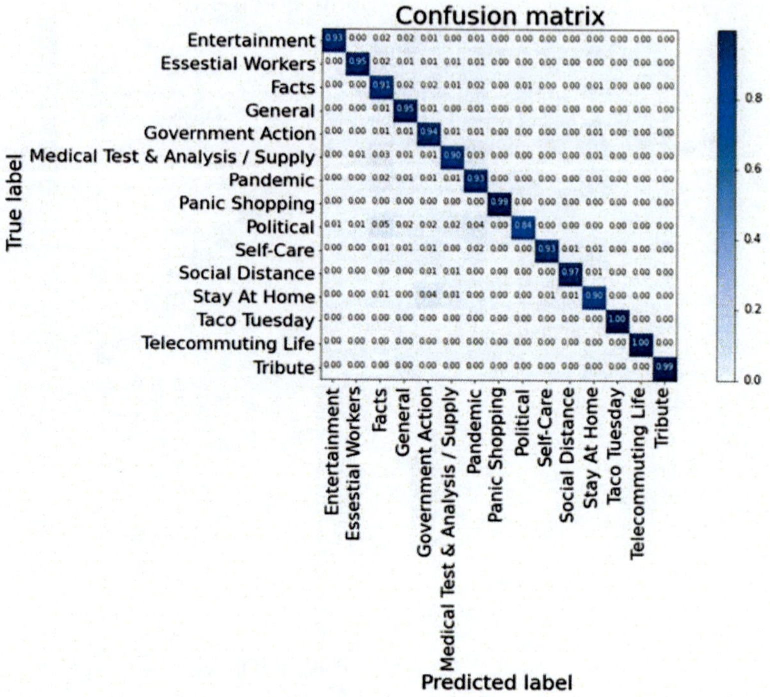

Fig. 8 Confusion matrix for the RNN with Bi-LSTM + attention

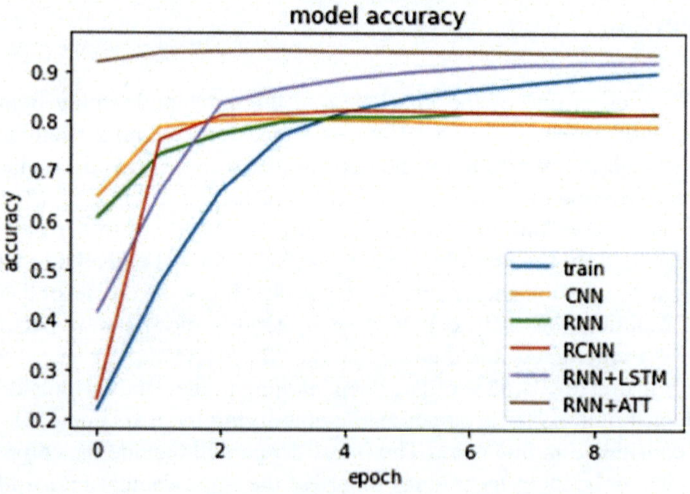

Fig. 9 Accuracy comparison of different models

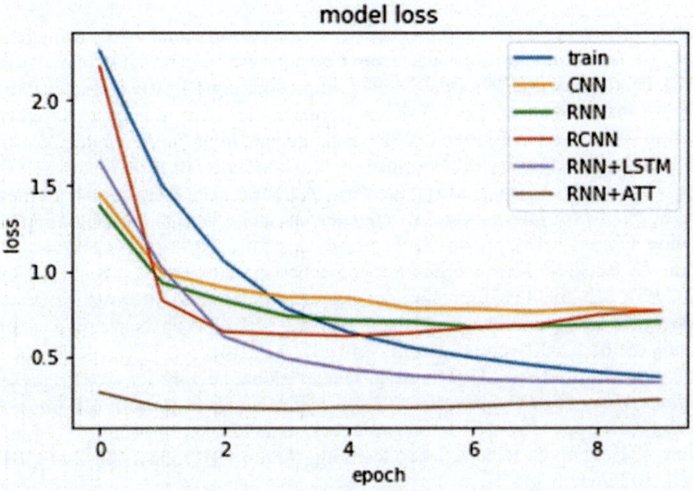

Fig. 10 Model loss comparison of different models

Acknowledgements This work was supported by Ramaiah Institute of Technology, Bangalore-560054, and Visvesvaraya Technological University, Jnana Sangama, Belagavi-590018.

References

1. K. Kowsari, K. JafariMeimandi, M. Heidarysafa, S. Mendu, L. Barnes, D. Brown, Text classification algorithms: a survey. Information **10**(4), 150 (2019). https://doi.org/10.3390/info10040150
2. Z. Wang, B. Song, Research on hot news classification algorithm based on deep learning, in 2019 IEEE 3rd Information Technology, Networking, Electronic and Automation Control Conference (ITNEC) (IEEE, 2019), pp. 2376–2380. https://doi.org/10.1109/itnec.2019.8729020
3. Y. Luan, S. Lin, Research on text classification based on CNN and LSTM, in *2019 IEEE International Conference on Artificial Intelligence and Computer Applications (ICAICA)* (IEEE, 2019), pp. 352–355. https://doi.org/10.1109/icaica.2019.8873454
4. I.R. Hallac, B. Ay, G. Aydin, Experiments on fine tuning deep learning models with news data for tweet classification, in *2018 International Conference on Artificial Intelligence and Data Processing (IDAP)* (IEEE, 2018), pp. 1–5. https://doi.org/10.1109/idap.2018.8620869
5. J. Cai, J. Li, W. Li, J. Wang, Deep learning model used in text classification, in *2018 15th International Computer Conference on Wavelet Active Media Technology and Information Processing (ICCWAMTIP)* (IEEE, 2018), pp. 123–126. https://doi.org/10.1109/iccwamtip.2018.8632592
6. S. Lakhotia, X. Bresson, An experimental comparison of text classification techniques, in *2018 International Conference on Cyberworlds (CW)* (IEEE, 2018), pp. 58–65. https://doi.org/10.1109/cw.2018.00022
7. P. Song, C. Geng, Z. Li, Research on text classification based on convolutional neural network, in *2019 International Conference on Computer Network, Electronic and Automation (ICCNEA)* (IEEE, 2019), pp. 229–232. https://doi.org/10.1109/iccnea.2019.00052

8. M. Anand, R. Eswari, Classification of abusive comments in social media using deep learning, in *2019 3rd International Conference on Computing Methodologies and Communication (ICCMC)*, Erode, India (2019), pp. 974–977. https://doi.org/10.1109/iccmc.2019.8819734
9. T. Saha, S. Saha, P. Bhattacharyya, Tweet act classification: a deep learning based classifier for recognizing speech acts in Twitter, in *2019 International Joint Conference on Neural Networks (IJCNN)*, Budapest, Hungary (2019), pp. 1–8. https://doi.org/10.1109/ijcnn.2019.8851805
10. X. Chen, C. Ouyang, Y. Liu, L. Luo, X. Yang, A hybrid deep learning model for text classification, in *2018 14th International Conference on Semantics, Knowledge and Grids (SKG)*, Guangzhou, China (2018), pp. 46–52. https://doi.org/10.1109/skg.2018.00014
11. G. Salton, C. Buckley, Term-weighting approaches in automatic text retrieval. Inf. Process. Manag. **24**(5), 513–523 (1988). https://doi.org/10.1016/0306-4573(88)90021-0
12. Y. Goldberg, O. Levy, word2vec Explained: deriving Mikolov et al.'s negative-sampling word-embedding method. arXiv preprint arXiv:1402.3722 (2014)
13. J. Pennington, R. Socher, C.D. Manning, Glove: global vectors for word representation, in *Proceedings of the 2014 Conference on Empirical Methods in Natural Language Processing (EMNLP)* (2014), pp. 1532–1543. https://doi.org/10.3115/v1/d14-1162
14. Y. LeCun, Y. Bengio, G. Hinton, Deep learning. Nature **521**(7553), 436–444 (2015). https://doi.org/10.1038/nature14539
15. Y. LeCun, L. Bottou, Y. Bengio, P. Haffner, Gradient-based learning applied to document recognition. Proc. IEEE **86**(11), 2278–2324 (1998). https://doi.org/10.1109/5.726791
16. I. Sutskever, J. Martens, G.E. Hinton, Generating text with recurrent neural networks, in *ICML* (2011)
17. D. Mandic, J. Chambers, *Recurrent Neural Networks for Prediction: Learning Algorithms, Architectures and Stability*. (Wiley, 2001). https://doi.org/10.1002/047084535X
18. J. Schmidhuber, S. Hochreiter, Long short-term memory. Neural Comput. **9**(8), 1735–1780 (1997)
19. A. Graves, J. Schmidhuber, Framewise phoneme classification with bidirectional LSTM and other neural network architectures. Neural Netw. **18**(5–6), 602–610 (2005). https://doi.org/10.1016/j.neunet.2005.06.042
20. S. Lai, L. Xu, K. Liu, J. Zhao, Recurrent convolutional neural networks for text classification, in *Twenty-Ninth AAAI Conference on Artificial Intelligence* (2015)
21. B. Wang, J. Xu, J. Li, C. Hu, J.S. Pan, Scene text recognition algorithm based on faster RCNN, in *2017 First International Conference on Electronics Instrumentation and Information Systems (EIIS)* (IEEE, 2017), pp. 1–4. https://doi.org/10.1109/eiis.2017.8298720
22. J. Zheng, L. Zheng, A hybrid bidirectional recurrent convolutional neural network attention-based model for text classification. IEEE Access **7**, 106673–106685 (2019). https://doi.org/10.1109/ACCESS.2019.2932619

Applied Classification Algorithms Used in Data Mining During the Vocational Guidance Process in Machine Learning

Pradeep Bedi, S. B. Goyal, and Jugnesh Kumar

Abstract Recent developments in information management and corporate computerization of company processes have made data processing faster, easier, and more accurate. Data mining and machine learning techniques have been used increasingly in different areas, from medical to technological, training, and energy applications to analyze data. Machine learning techniques allow significant additional knowledge to be deducted from data processed by data mining. This critical and practical knowledge allows companies to develop their plans on a sound basis and reap substantial time and expense benefits. This article implements the classification methods used for data mining and computer training for the collected data during technical advice processes and aims to find the most powerful algorithm.

Keywords Data mining · Machine learning · Classification algorithm

1 Introduction

Data mining is a logical method used to look for valuable knowledge across the vast volumes of data. The purpose of this methodology is to discover designs that were already unknown. At the point when these patterns are found, they can be additionally used to settle on those market planning decisions [1]. There are three steps involved such as Exploration, Pattern Identification, and Deployment.

1. Exploration: Data is cleaned and translated into another type in the first phase of data exploration, and essential variables are calculated and then the essence of the data depending on the problem is determined.

P. Bedi
Lingayas Vidyapeeth, Faridabad, Haryana, India
e-mail: pradeep.kcu21@gmail.com

S. B. Goyal (✉)
City University, Petaling Jaya, Malaysia

J. Kumar
St. Andrews Institute of Technology & Management, Gurgaon, India

V. Suma et al. (eds.), *Inventive Systems and Control*,
Lecture Notes in Networks and Systems 204,
https://doi.org/10.1007/978-981-16-1395-1_11

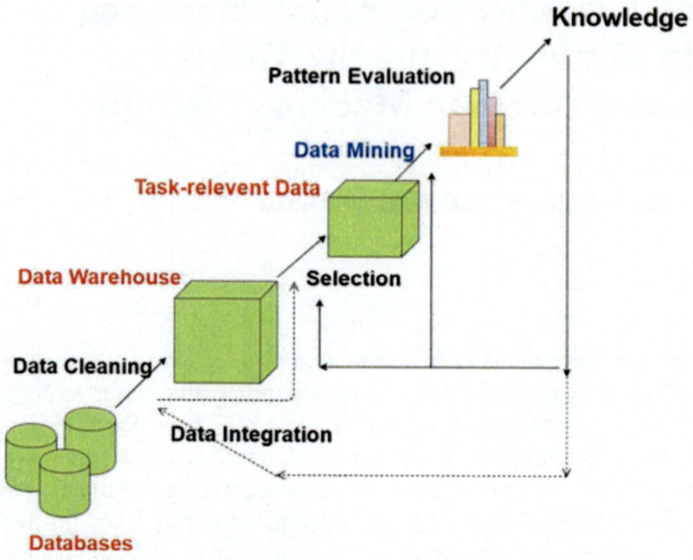

Fig. 1 Illustration of data mining concept [1]

2. Pattern Identification: The second step is to establish pattern recognition until knowledge is explored, optimized, and described for the particular variables. Identify the trends which make the best prediction for the selected patterns.

1.1 Data Mining Algorithm and Techniques

Different calculations and strategies, for example, Sorting, Clustering, Regression, Artificial Intelligence, Neural Networks, Correlation Rules, Decision Trees, Genetic Algorithms, Nearest Neighbor Processes, and so forth, are utilized to find data from information bases [2] (Fig. 1).

1.2 Classification

The most usually utilized information mining strategy is an arrangement, which utilizes a set of pre-grouped examples to develop a standard that can classifications the record strength on the large area. For this type of analysis, fraud identification and credit risk applications are especially important. This approach also uses a decision tree or classifying algorithms based on neural networks. Learning and sorting are part of the data classification process. Training data is validated by the Learning Classification Algorithm. The consistency of the classification rules is measured in the outcome of the classification exam. Guidelines for new data tuples would be introduced where the consistency is acceptable. In the case of an application for fraud identification, This includes complete accounts for both bogus and real record-by-register activities. These pre-classified examples are used to evaluate the parameters required for proper discrimination by the classifier-training algorithm. The algorithm codes these parameters to a classifier model [2].

1.3 Types of Classification Models:

1.3.1 Clustering

A collection of related entity types may be believed to be clustering through the clustering approaches, which can be distinguished as dense and sparse regions in object space, and discover general distribution patterns and associations between data attributes. The grouping method can also be used to easily isolate classes or categories of objects, but it is expensive to use clustering as a mechanism for obtaining and classifying subsets of attributes. The development of customers based on shopping favorites into genes of similar characteristics. [3]. Types of clustering methods are described in subsections.

Predication

The process of regression can be modified for mitigation. The relation between one or more independent and dependent variables can be estimated by a regression study. For data mining functions, independent variables are already known, and reaction variables are used for modeling. Sales, inventory costs, and commodity are the failure rates, for example, are all highly difficult to forecast since they can be related to several predictor variables in dynamic correlations [4]. Additional complex approaches may be used in modeling potential values (e.g., logistic regression, policy-making trees, or neural networks). The same models can also be used for regression and classification [4], for example, to construct the classification tree (classify categories) and the regression tree (to predict continuous response variables). Neural networks may also build models for classification and regression [5].

Association Rule

The discovery of frequent observations between large datasets is generally amalgamation and association. This finding form helps corporations to make choices such as the development of catalogs, cross-marketing, and user model research. Algorithms of Association Law should be able to generate laws of less than one trust value. However, the number of possible link rules for a data set is typically very large and a substantial proportion of the rules is usually very small (if any).

A neural network is a group of linked entrances and outputs for each connection with a weight present. The network will learn by weight-adjustment during the study process to assume the right labels in the class of the input tuples. Neural networks have an extraordinary ability to focus on complicated or unsuitable data to identify trends and innovations that individuals or other mathematical approaches find too difficult to use. They are suitable for inputs and outputs which are continuously valued. Many industries have already effectively introduced hand-written reorganization of character, Tech training in English, and many market challenges in the modern world. Neural networks may be used to recognize trends or data flows and to estimate or foresee needs [6].

Data Mining Application

Data mining has not been fully developed and is a comparatively new technology. Nevertheless, a host of businesses now use it every day. Some have drug stores, pharmacies, banks, and insurance. Many of these organizations introduce data processing into the study, model recognition, and other primary techniques. This technology is popular with many businesses because it makes them learn and make effective marketing decisions for their customers. This is a description of the problems and opportunities of businesses using the techniques of data mining [7].

1.4 Machine Learning

Machine learning is a series of techniques that allow "teaching" computers to perform tasks with examples of how they are accomplished. For example, a program is needed to distinguish between genuine e-mails and unwanted spam [7]. An attempt is made to write a variety of universal rules such as flagging messages with certain functionality (like viagra or, clearly, bogus headers). However, it can be very difficult to write rules to distinguish a text contributing either to a lot of missed spam messages or, worse, a lot of misplaced emails. Worse, spammers will purposely adjust their method of delivering spam to trick these strategies. It is likely to become an insurmountable challenge to write good rules and keep them up to date. Fortunately, a solution took place on artificial learning. Different models are used to 'practice' the current spam filters to provide a research algorithm with examples of e-mails that manually call 'ham' or 'spam,' and algorithms automatically discriminate between them. Machine learning is a fascinating and complex field, and it is defined in several ways [7]:

1. Artificial Intelligence Vision: The invention of intelligent machinery is therefore important to learn and is central to people's intellect and understanding. A lot of efforts taken in AI have to show that artificial intelligence is developed by programming and cannot render any rule. Automatic learning is necessary. For example, humans don't have the ability to interpret the language that is to be

learned because it is meant not to have to program anything but an attempt to find computers to learn the language.

2. In Info Engineering View: Machine learning can be used as an example to program the devices that can be easier than typing code.
3. Statistical View: The synthesis of computer science and statistics in machine learning computational techniques are used to solve the problems.
4. Mathematics: Machine learning has in many contexts, including conventional statistic questions that have been used on a wide range of topics. Machine learning also relies on variables different from statistics.

Often, machine learning methods are broken into two phases:

1. Training: A model is derived from a training dataset [8].
2. Application: The model is used to settle on new test outcomes. When filtering spam, for example, the training data will be email messages that are categorized as spam or ham, and test data will be used for any new email message received (and are to be classed). However, there are other types of machine learning [8].

A comparison of different data mining techniques was discussed in Table 1.

Table 1 Comparison of different data mining [9]

Algorithm	Findings	Drawbacks
Decision tree	It can work with both consistent and discreet details. This results easily in the classification of unknown records [9]. It's going to be perfect with a small measure tree. The findings should not affect deviations. It doesn't require a planning procedure such as normalization. It fits best for numeric details [8]	It is not possible to predict the value of a constant class attribute. It sends an error message where a huge number of classes is used that are unrelated attributes which leads to a poor decision tree. Also, minor modifications made to the data that can modify the full decision tree
Naive Bayesian	Compared to other classifiers, the error rate is smaller. Fast to adapt. It can manage continuous data in a positive way. When dealing with a large database it provides high precision and speed. It can handle discrete values [7]	Provides less precision as it relies more on independent features
Neural Networks	Used to define the pattern for untrained data and fits well for continuous values [10]	Less interpretability and Long training time is expected
K-Means	Easy and efficient algorithm. It's reasonably fast. It gives a better outcome when different data is used [6]	Does not work with noisy data and non-linear data sets

1.5 Types of Machine Learning

Some of the main types of machine learning are:

1. Supervised learning in which the training outcomes are tagged with suitable answers, such as spam or ham. Classification (where distinct outputs are marking, for example, spam filtering) and regression (where outputs are tested in practice) are the two most common types of supervised learning.
2. Uncontrolled learning, with a series of unmarked findings given to us, which is required to dig at and discover patterns inside. The reduction and clustering of dimensions are the two major examples.
3. To improve the awareness that an individual (for example, a robot or a controller) try to learn the most relevant steps based on past actions.

There are many other types of machine learning as well, for example:

1. Semi-conduct learning with a classification of just a part of training data
2. Forecast time series on financial markets, for example.
3. Anomaly identification, as seen in the plant tracking.
4. Active learning, which costs data acquisition and the method also has to decide on the testing phase and some other data [8].

2 Literature Survey

Buczak [11] the authors published a literature study on machine learning and data mining approaches for cyber analytics in favor of intrusion detection. It addresses numerous ML/DM approaches which are described on well-known cyber data sets and the sophistication of ML/DM. Many issues related to the use of ML/DM are also discussed.

Sharma et al. [12] a variety of classification methods were studied and a comparative analysis of various classification algorithms was carried out. Different grouping approaches include Decision Trees (DT), Support Vector Machine (SVM), Nearest Neighbour (NN), etc.

Huang [13], the authors presented two algorithms with category domains such as mixed numerical and categorical domains in this article on the extension of the K-Medium Algorithm. The authors have used popular soy disease and credit acceptance data sets to demonstrate the efficiency of the clustering of two algorithms.

Fayyad [14] this article offers an overview of the database mining and information discovery field, explaining the interconnection between data mining and knowledge discovery in databases and related fields such as computer study, statistics, and databases. The study focuses on different actual implementation steps, particularly data mining techniques, barriers to the certified use of data exploration in the current and future fields of science.

Alpayd [15] Machine learning field has been growing for several years and has produced a variety of classifying algorithms, e.g., neural networks, decision trees, rule inducers, nearest neighbor, support vector machine (SVM), etc. The consumer must select the right algorithm that is ideally suited to his mission. This algorithm selection problem is difficult, as no algorithm is performing better than all the others, regardless of unique problem characteristics, as has been found in various empiric comparisons.

Amasya [16] later, the empiric findings of the non-free lunch theorems were verified in this article. They state, among others, that their success averaged overall learning disabilities identified over a particular training set would be the same. For any field of problems where one learner outperforms the other, there will be another category where the reverse is the case.

3 Proposed Methodology

Machine learning helps the computer program to learn with similar situations based on previous experience. This approach tends to generalize the interpretation of the situation. It can be used as a learning aid. The ties between machine learning and data mining are obvious. The use of machine learning algorithms for large databases is data mining. The implementation phase of the data mining method is machine learning. Data mining deals with the information obtained and its review. Machine learning applies, however, to the manner in which this knowledge is acquired and to the self-expanding application of these methods for computers. Computer tech along with the accumulated knowledge in the area of machine learning can predict new circumstances that can arise in the future. These estimates provide substantial advantages and opportunities in many vital practices, including time, expenses, and human life. The assessment of progress or achievement of the student in the chosen profession would be a sample in the ground of education and would lead the student to the right technical sector rather than encouraging being qualified in the wrong segment. In medicine, early diagnosis of cancer, viewed as a terminal disease, can be vital to the health of a person and may be another reason for it. Predicted in high-cost applications including natural disasters, security networks, or threat evaluations, machine learning may also be used which can certainly have major implications in terms of time, costs, and labor (Fig. 2).

4 Result Analysis

The goal is to collect data based on the common characteristics in the required framework. Classification allows assessment of the unknown type or grouping of associates. Methods of classification are categorized as models that generate different results. Both of these methods can analyze and distinguish according to examples from the compilation of training results. In several areas, the classification methodology is used. For example, pharmacy cancer risk identification, banking credit risk

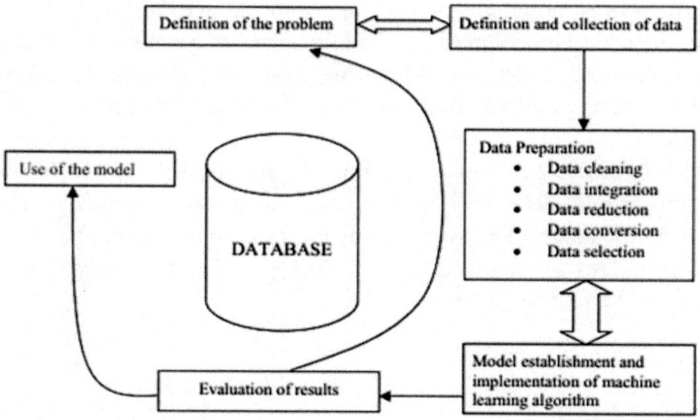

Fig. 2 Classification of application process

assessment, quality management analysis of industrial processes in a range of energy applications, building material stability determinations, school student performance standards, weather forecasts, medication marking, and so on are considered for adopting classification. An analysis algorithm is a basis for the classification to construct main model. The algorithm is used for both this data package and the other undefined grouped data package to evaluate the group of these research data. The machine learning classification uses decision-making trees, regression trees, vector assist machines, and methods of mathematical classification and each shape has a separate algorithm. Models of estimators are grouping methods. Besides, these models are rule-based learning model, the decision tree model, and so on. The root node will then be identified at the beginning of this model and the subnodes will then be generated based on a decision on the quality status of each node. Since any node is minimized to a single quality state called leaf node, and at the end of the node, the class is determined. This transaction is resource-based until at the end of each node the class is defined. To build machine learning decision-tab modeling algorithms like ID3, C4.5, CHAID, CRT, and QUEST were created.

The conditions for the creation of sub-nodes after the classification of root nodes are varying in those algorithms. Regression trees must be classified by obtaining sub-nodes and leaves, as in the Decision Tree, from the root node. In the regression trees, however, each node is divided into only two sub-nodes such as left and right. The algorithms developed for this kind of technique can be CART, Towing, and Gini. SVM approach classifies with the aid of linear or non-linear functions. The Support Vector Machine (SVM) approach is based on estimating the most suitable data separation function. The goal of the SVM approach is to find a special linear line dividing group. During classification, this linear line can be drawn more than once. For all classes, the SVM decides the farthest line and thus estimates the highest error tolerance.

Where training data and boundary lines are established, the estimated data can be classified in terms of their border location. A technique to quantify the probability that new data is put in one of the existing categories based on known, open, and currently confidential information is the Bayes classification approach. Bayes theorem is further developed and extended to a wide range of fields from medicine to economics, statistics, archaeology, law, climate science, research and measurement and genetics, and physics. The single basic algorithm used to classify data is the Naive Bayes algorithm based on the Bayesian data mining classification. The Bayes theorem depends on the likelihood. If event B frequency depends on event A, a conditional probability may be explained.

$$P(B|A) = \frac{P(B)P(A|B)}{P(A)} \tag{1}$$

where *P*(*B*) is a later likelihood, *P*(*B*) is a former likelihood, *P*(*A*) represents conditional chance and *P*(*A*) is an isolated likelihood of occurrence *A*.

$$P(c_1|x_1) = \frac{P(x_1|C_1)P(C_1)}{P(x_1|C_1)P(C_1) + P(x_1|C_2)P(C_2)} \tag{2}$$

According to the theorem of the Bayes, the hypothesis of which class the X value belongs to should be created:

$$P(C|X) = \frac{P(X|C_1)}{P(X|C_2)} \tag{3}$$

The MSSQL 2005 database was used to gather data using questionnaires or the internet and to store data in four (4) main categories classified according to 31 parameters, belonging to 100 professionals in various fields of energy usage even participating in vocational guidance. Data Mining methods were used to view objects generated in t-sql query languages to make them able to be used in the classification techniques. Open-source data mining software called Weka has been used for algorithms which are explained in Table 2.

Table 2 Results of classification algorithms

Algorithm/results	Naïve bayes	Oner	JRIP	KSTAR
Correctly classified instance	83	72	76	79
Incorrectly classified instance	17	28	24	21
Kappa statistic	0.297	0.0151	0	0.1906
Mean absolute error	0.1254	0.1205	0.1476	0.1068
Root mean squared error	0.2033	0.2919	0.2451	0.2384

5 Conclusion

Presently, the growing number of computerized business operations and the data are evaluated parallelly are likely to demonstrate and use data mining techniques. It can be used through machine learning techniques by accessing the data for data-mining analysis in order to predict or estimate future results accurately. This study used algorithms for a group of people who are in the process of vocational training and found that a Bayes mathematical estimate model is the best algorithm in this field. The findings associated with substantial time and cost savings were reliable from the use of computer teaching technologies in classification studies. It is strongly advised that algorithms used in databases mining and technique of machine learning for applications can be used to re-build. This research is considered to support organizations and people working on all facets of their jobs, through their business processes computerization.

References

1. M. Gera, S. Goel, Data Mining—techniques, methods and algorithms: a review on tools and their validity international. J. Comput Appl. **113**(18), 0975–8887 (March 2015)
2. Ö. Ünsal, *Determination of Vocational Fields With Machine Learning Algorithm*, M.Sc. Thesis (Gazi University, Ankara, Turkey, 2011)
3. I.H. Witten, E. Frank, *Practical Machine Learning Tools and Techniques*, 2nd edn. (Morgan Kaufmann Publications, USA, 2005).
4. S. Özekes, Data mining models and their applications. J. Istanbul Commerce Univ. **3**, 65–82 (2003) (Stanbul, Turkey)
5. M.T. Mitchell, *Machine Learning* (McGraw-Hill, USA, 2016).
6. Y. Özkan, *Data Mining Methods* (Papatya Publications, Stanbul, Turkey, 2008)
7. A.S. Albayrak, K. Ylmaz, Data mining: decision tree algorithms and an application on MKB data. J. Suleyman Demirel University, Facul of Econom Administrative Sci. **14**(1), 31–52 (2009) (Isparta, Turkey)
8. M.F. Amasyal, *Introduction to Machine Learning* (2010). https://www.ce.yildiz.edu.tr/mygetfile.php?id=868
9. G. Silahtarolu, *Basic Concepts and Algorithms of Data Mining* (Papatya Publishing, Stanbul, Turkey, 2008)
10. N. Murat, *The Use Of Bayesian Approaches To Model Selection*, M.Sc. Thesis (Ondokuz Mays University, Samsun, Turkey, 2007)
11. A.L. Buczak, E. Guven, A survey of data mining and machine learning methods for cyber security intrusion detection (2019)
12. S. Sharma, J. Agrawal, S. Agarwal, S. Sharma, Machine Learning techniques for data mining: a survey (2018)
13. Z. Huang, *Extensions to the k-Means Algorithm for Clustering Large Data Sets with Categorical Values*. (Acsys CRC, CSIRO, 2018)
14. U. Fayyad, G.P. Shapiro, P. Smyth, The KDD process for extracting useful knowledge from volumes of data. Commun. ACM **39**, 27–34 (2017)
15. N.E. Alpayd, *Introduction to Machine Learning* (The MIT Press, London, England, 2016)
16. M.F. Amasyal *New Machine Learning Methods and Drug Design Applications*, Ph.D. Thesis (Yldz Technical University, Stanbul, Turkey, 2016)

Meta-Heuristic Algorithm for the Global Optimization: Intelligent Ice Fishing Algorithm

Anatoly Karpenko and Inna Kuzmina

Abstract The Intelligent Ice Fishing Algorithm, (IIFA), is encouraged by the logic of ice fishing. The algorithm is related to the class of "tracking" algorithms when during the evolution of the agent's population (fishermen), coordinates and results of several previous tests are recorded. The *IIFA* algorithm implies that each fisherman has devices and equipment (e.g., a GPS navigator, binoculars, and pad) that allow determining the exact coordinates of holes, and fishing neighbors holes, to determine the amount of fish caught in the hole to form the approximate model of fish distribution in the considered area of ice to find coordinates of local and global maxima of this model. The IIFA algorithm is meta-heuristic, which allows forming on its basis a large number of specific heuristic population-based global optimization algorithms. The article describes the algorithm and presents the results of computational experiments.

Keywords Global conditional maximization · Tracking algorithms · Meta-heuristic algorithms

1 Introduction

Global conditional optimization is a branch of applied mathematics and numerical analysis that attempts to find the global minima or maxima of a function or set of functions on a given set [1–5]. These include many socio-economic, technical, organizational and managerial, combinatorial problems, game theory problems, and so on. For most of these problems, deterministic methods are unacceptable or do not provide the necessary accuracy [6, 7]. Therefore, an alternative approach is needed to use of evolutionary methods of global optimization and the deliberate introduction of an element of randomness in the search algorithm [8]. The main advantages of such methods are increased performance; high reliability; relatively simple internal

A. Karpenko · I. Kuzmina (✉)
Bauman Moscow State Technical University, 105005 2-Nd Baumanskayast, 5, Moscow, Russia

V. Suma et al. (eds.), *Inventive Systems and Control*,
Lecture Notes in Networks and Systems 204,
https://doi.org/10.1007/978-981-16-1395-1_12

implementation; low sensitivity to the growth of the dimension of the optimization set, etc.

The development of new methods for solving the global optimization problem is an urgent task [9–11]. New methods are aimed at improving the accuracy of the obtained solution and the speed of obtaining the result. This article presents a new algorithm developed by the authors for solving global optimization problems, the effectiveness of which is proved by experiments.

2 Problem Statement

2.1 The Basic Definitions and Notation

Consideration of problem in global conditional maximization [12–15]

$$\max_{X \in D} f(X) = f\left(X^*\right) = f^*, \tag{1}$$

where X—is $|X|$—dimensional vector of variable parameters; $D \subset R^{|X|}$—is search area, depending on limited, closed and simply connected; $f(X) \in R^1$—is an objective function; X^*, f^*—is required optimal vector X and the objective function value. The boundary line of area D (ice edge) is referred Γ_D.

The computational complexity of objective function $f(X)$ indicates identical at all points in the area D, and greatly exceeding total computational complexity of bounding line functions [16]. Introduced in the formula (1) notation, gives the following meanings: vector X—is coordinates of points (holes) on the ice surface; $D \subset R^{|X|}$—is ice floe; $f(X)$—is number of fish in hole with coordinate $X \in D$.

1. **Ice floe** is the ice cover of the reservoir in which fishing is carried out. It also maintains a constant position relative to boundary lines of the reservoir that is stationary. The ice floe retains on its surface traces of holes left by fishermen during the fishing. The ice block does not have holes before the fishing process.
2. **Fish**. The total number and position of fish in the reservoir is a priori unknown. The fish do not migrate during the fishing (stationary global optimization problem is considered), i.e. their position relative to the ice floe is unchanged. As a result of the fishing, the number of fish does not change at a given point or nearby locations in the ice floe.
3. **Fishermen**. The outfit of each fisherman includes the devices and equipment that make it possible to localize without errors the exact coordinates of all their holes and the holes of fishermen, located in near and far visibility zones (*close* and *distant* fishermen); accurately to determine the amount of fish, caught by him and all his close fishermen in each of holes separately.

At the beginning of the fishing process, the fishermen (their *starting holes*) are distributed over the ice floe so that distance between them is limited from below by suitable "*gentlemanly*" distance.

The potential arsenal of each fisherman has the following possible variants of behavior.

1. *Fishing*. The fisherman can make several holes in his *developed* ice block and make a *fishing session* in them. The fishing is subject to some local maximization algorithm that is in the process of fishing the fisherman makes holes in such way as to try to approach the maximum concentrations of fish. In the process of fishing, the fisherman watches his close neighbors and knows the coordinates of their holes and the number of fish caught in each of them.
2. *Near Relocation* is carried out by fisherman if one of the conditions is managed during the fishing process:
   (a) Stagnation of caught fish number;
   (b) Progress of one of the closest fishing neighbors "significantly" exceed successes of given fisherman.

In the process of near relocation, the fisherman forms a surrogate model of fish distribution (based on information, collected from the current fishing session) and gets positions of maxima, moves to the new site in the vicinity of one of the found maxima for the surrogate model, selects starting hole in a new area and starts new one fishing session.

The *fishing series* of a given fisherman is a set of fishing sessions, divided by one or more close relocation of a given fisherman. The fishing series of fishermen are separated by his distant relocation.

3. *Far Relocation* is carried out by fisherman if one of the conditions is managed during the fishing process:
   (a) In given series of fishing, it is stagnation of the number of fish caught;
   (b) Progress of the closest fishing neighbors is "slightly" higher than given fisherman's successes.

In the process of long-distance relocation, the fisherman based on information about distant fishing neighbors determines the direction and distance of the movement, relocating to the selected point and chooses a new developed area to start a new fishing session.

4. *The end of fishing*. The fishing process continues until the end of fishing conditions.

Introduce the following notation.

$S = \{s_i, i \in [1 : |S|]\}$—is many fishermen by number $|S|$.

$t \in \left[0 : \hat{t}\right]$— is the current number of the hole for the fisherman in a *given fishing* session;

$\hat{t}$— is the maximum allowable number of holes *in one fishing session.*

$t_{\max}$—is the total maximum number of holes for each fisherman *during the whole fishing.*

$\Delta t > 0$—is a number of holes after which the fisherman in the fishing session can get their tracing and corresponding values of the objective function from his nearest neighbors.

$X_i(t) \in D$—is a vector of current coordinates for fisherman (i.e., coordinates of his hole with a number in given fishing session).

$D_i^n(t), D_i^f(t) \subset D$—is close and distant visibility zones of fisherman $s_i \in S$ in his current position $X_i(t)$, having a center at point $X_i(t)$ and radius, consequently;$D_i^n(t) \subseteq D_i^f(t) \neq \emptyset; r_f \geq r_n > 0$.

$d_i(t) \in D_i^n(t)$—is developed fishery spot $s_i \in S$, having a radius r.

$d_i^\varepsilon(t) \subset D$—is ε-*forbidden* subzone of region D, having radius ε and center at a point $X_i(t)$.

$d_i^\rho(t) \subset D$—is ρ-*forbidden* subzone, centered at the point of the current position $X_i(t)$ of fisherman $s_i \in S$, radius of which is equal $\rho > 0$ and makes sense of "gentlemanly" distance.

$S_i^n = \{s_{i_j} | X_{i_j}(t) \in D_i^n(t); i_j \in [1 : |S|], j \in [1 : |S_i^n(t)|], i_j \in i\}$—is the current set of fisherman's *close neighbors* $s_i \in S$.

$S_i^f = \left\{s_{i_j} | X_{i_j}(t) \in D_i^f(t); i_j \in [1 : |S|], j \in \left[1 : \left|S_i^f(t)\right|\right], i_j \in i\right\}$—is the current set of fisherman's *distant neighbors* $s_i \in S$.

$\mathbf{X}_i(t) = (X_i(0), \ldots, X_i(t))$—is current *tracking* of fisherman in his given fishing session, that is, coordinates of all previous and current holes.

$\mathbf{f}_i(t) = (f_i(0), \ldots, f_i(t))$—is values, corresponding to tracking $X_i(t)$ of fish caught number (values of the objective function).

$e(d_i(t)) \in e_i(t)$—is an *elaboration* of spot $d_i(t)$, equal, for example, to density of holes in given area, owned by fisherman $s_i \in S$.

$p(d_i(t)) \in p_i(t)$—is prospects of spot $d_i(t)$. If the site $d_i(t)$ is located in the vicinity of local or global maximum for surrogate model of function $f(X)$, then value $p_i(t)$ can be determined by relative value of given maximum.

$a_i(t)$—is an attractiveness of spot $d_i(t)$. It can be calculated, for example, as additive convolution of development and prospects of given site:

$$a_i(t) = \lambda_e e_i(t) + \lambda_p p_i(t). \tag{2}$$

Here λ_e, λ_p—is weight multipliers; $\lambda_e, \lambda_p \in [0; 1]$.

$\delta_t > 0, \delta_f > 0$—are values that determine the stagnation of fishing process: stagnation situation occurs if during δ_t holes, value of the objective function could not be increased by more than δ_f.

$\Delta_f > 0$—is value that determines progress of neighbor fisherman: fisherman is considered to be significantly more successful than given if his current catch exceeds the catch of last one by an amount Δ_f.

$\varepsilon > 0$—is required accuracy for localization of function maximum $f(X)$ in search space.

The values $|S|$, Δt, $\hat{t}$, λ_e, λ_p, r, r_n, r_f, ρ, δ_t, δ_f, Δ_f are *free parameters* of the algorithm, that is, its values are set by the decision maker (DM) [17].

3 The General Scheme and Main Algorithm Procedures

The basic ones are the following algorithm procedures:

1. *Initialization* (initial placement of fishermen on an ice floe);
2. *Series of fishing in a given subzone of the ice:*
   (a) *Fishing session* at a developed site;
   (b) *Short-range relocation* (choice and movement of fisherman to new development spot);
3. *distant relocation* (selection and movement of fisherman to new subzone of the ice);
4. *the end of fishing.*

For the given fisherman, the fishing series consists of one or more fishing sessions, separated by near basing procedures. The fishing series is interspersed with long-range relocation procedures (Fig. 1). Generally, for each of fishermen, the following quantities values are different such as the number of fishing episodes; the number of fishing sessions within each fishing series; the number of holes in each of fishing sessions. Thus, the *IIFA* algorithm is asynchronous.

A. **Initialization**: The specifics of the initialization procedure have resulted from the requirement of fishermen placing s_i, $i \in [1 : |S|]$ in the region D so, that distance between them is not less than a "gentlemanly" one.

The procedure scheme is as follows (Fig. 2).

1. Cover each of the edges of the definition domain of uniform net with a step so that $h\sqrt{X} \geq \rho$.
2. Find the coordinates for centers of all cubes, whose centers are in the region (belong to given region). Denote given cubes Π_k, $k \in [1 : n]$, where $n \geq |S|$.

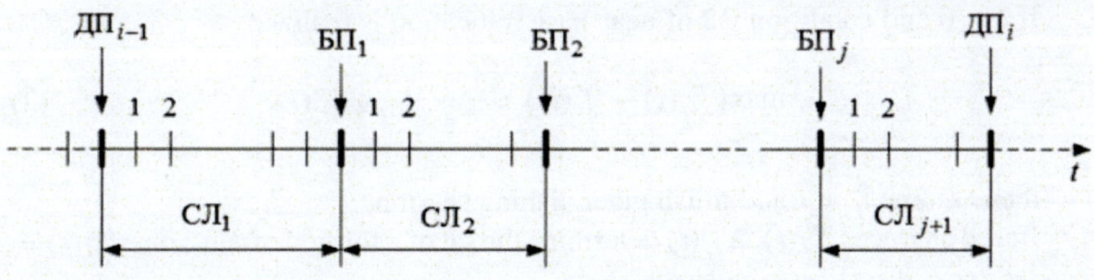

Fig. 1 Scheme of fishing series with number $i > 0$: LR—long-distance relocation; NR—close relocation; FS—fishing session

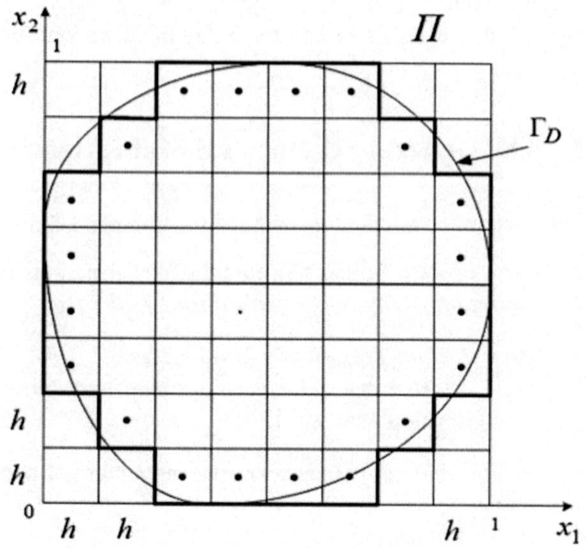

Fig. 2 To scheme of the initialization procedure: $|X| = 2$; by points are showed the centers of boundary cubes of the set $\{\Pi_k, k \in [1 : n]\}$

3. If $n = |S|$, in the center of each cubes place fisherman and go to step 5).
4. If $n > |S|$, then generate $|S|$ of non-repeating random numbers $k_1, \ldots, k_{|S|}$, uniformly distributed in the interval $[1 : n]$, in the center of each of cubes $\Pi_{k_1}, \ldots, \Pi_{k_{|S|}}$ place fishermen.
5. At all points $X_i(0), i \in [1 : n]$ compute values of the objective function and complete the initialization process.

B. **Session of fishing:** Let $X_i(0) \in d_i(0) \subset D_i^n(0), \subset D$—be coordinates of starting hole of fisherman, $s_i \in S$, obtained after the initialization, close or distant relocation.

The general scheme of fishing session procedure for fisherman has the following form (Fig. 3).

1. Assume $t = 0$.
2. Determine the set of close neighbors $S_i^n(t)$ of fisherman and get from all these fishermen their current traces $\boldsymbol{X}_{i_j}(t)$ and corresponding values of the objective function $\boldsymbol{f}_{i_j}(t)$.
3. If $t > 0$ and condition B2 of near-base relocation is realized

$$\max_{i_j}\left(f_{i_j}(t) - f_i(t)\right) > \Delta_f, s_{i_j} \in S_i^n(t), \tag{3}$$

   then assume $\hat{t}_i = t$ and finish given fishing session.
4. Based on traces $\boldsymbol{X}_i(t), \boldsymbol{X}_{i_j}(t)$ determine the set of ε forbidden sub-zones $T_i(t) = \left\{d_{i_k}^{\varepsilon}(t) \in d_i(t)\right\}$ (*many restrictions*).
5. Using one or another local conditional optimization algorithm, perform one step of solving the maximization problem

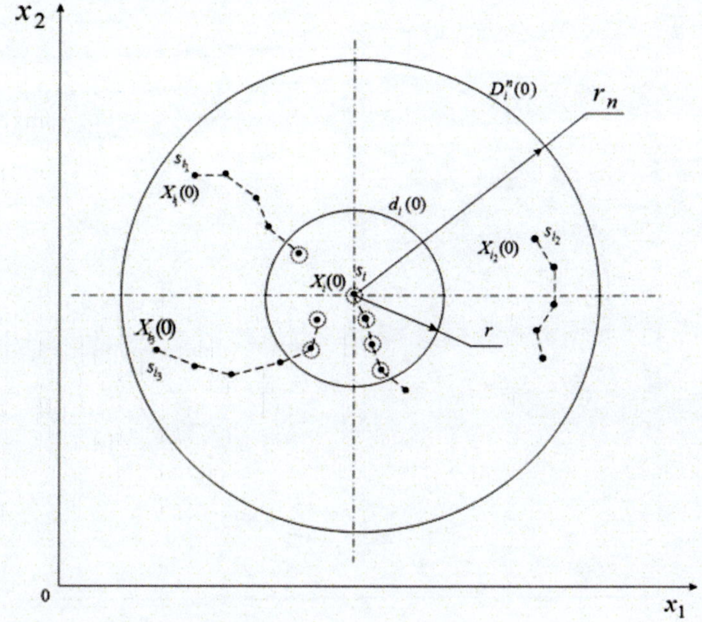

Fig. 3 Scheme of procedure for fishing session for fisherman s_i:$t = 0$; $|X| = 2$; $|S_i^n| = 3$;– traces of fishermen; ◌—ε-forbidden zones

$$\max_X f(X) = f(X_i(t+1)), X \in d_i(t) \backslash T_i(t). \tag{4}$$

6. Assume $t = t + 1$.
7. If t is not multiple to value and condition B1 of near base is not realized, having form

$$\tilde{f}_i^*(t) - \tilde{f}_i^*(t - \delta_t) \le \delta_f. \tag{5}$$

then return to step (4). Here $\tilde{f}_i^*(t)$—is the best value of the objective function found by this fisherman $s_i \in S$ in his all fishing sessions.
8. If t is multiple to Δt, then go to step 2.
9. If condition (5) is realized, then assume $\hat{t}_i = t$ and finish given fishing session.

C. **Near relocation:** The near relocation is based on the information of fisherman about the closest neighbors, obtained in the current fishing session. With the displacement of fishermen on the ice, many close neighbors can change during the fishing.

The scheme of procedure for near relocation of a fisherman is as follows (Fig. 4).

1. Form subzone $\hat{D}_i \in D$, covering the tracks $X_i(\hat{t}_i)$, $X_{i_j}(\hat{t}_i)$ center of which is at point $X_i(\hat{t}_i)$, and the radius is

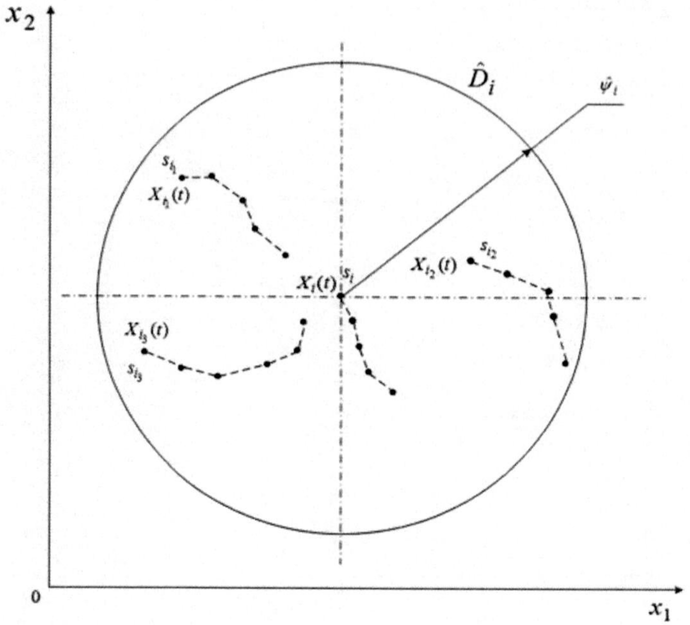

Fig. 4 Scheme of the procedure of near relocation for fisherman s_i:$|X| = 2$; $|S_i^n| = 3$; •—traces of fishermen

$$\hat{\psi}_i = \max\left(\max_{\tau \in [0:\hat{t}_i - 1]} \mu\left(X_i(\tau), X_i\left(\hat{t}_i\right)\right), \max_{\tau \in [0:\hat{t}_i - 1]} \mu\left(X_{i_j}(\tau), X_i\left(\hat{t}_i\right)\right) \right),$$

where $s_{i_j} \in S_i^n\left(\hat{t}_i\right)$.

2. On the basis of traces $\boldsymbol{X}_i\left(\hat{t}_i\right), \boldsymbol{X}_{i_j}\left(\hat{t}_i\right)$ and corresponding values of the objective function $\boldsymbol{f}_i\left(\hat{t}_i\right), \boldsymbol{f}_{i_j}\left(\hat{t}_i\right)$, form some surrogate model $\tilde{f}_i(X)$, of given function in the subzone $\hat{D}_i$.
3. Using one or another optimization algorithm, numerically solve the problem

$$\tilde{f}_i(X) \to \max_{X \in \hat{D}_i} \tag{6}$$

and find approximate values of maximum (local and global) $\left\{\tilde{f}_{i_k}\right\} = \left\{\tilde{f}_i\left(X_{i_k}\right)\right\}$ of function $\tilde{f}_i(X)$ in the subzone $\hat{D}_i$.
4. Based on results in solving problem (6), *determine new development area* $d_i(0)$ and the coordinates of *starting hole* $X_i(0) \in d_i(0)$.

C.1 Far relocation: The scheme of the procedure for distant relocation of fisherman $s_i \in S$ has the following form.

1. The current set of distant neighbors $S_i^f(t)$ are defined to obtain their current positions from all these fisherman $X_{i_j}(t)$; $s_{i_j} \in S_i^f(t)$.
2. Based on the given information, the target subzone $\tilde{D}_i \subset D$ and *new development area* $d_i(0) \subset \hat{D}_i$ (starting hole $X_i(0) \in d_i(0)$ and radius *are determined*).

D. **The end of fishing**: In the IIFA basic algorithm, termination condition is that each of fishermen achieves number of processed holes, equal $t_{\max}$.

4 Research of the Algorithm Efficiency

The IIFA algorithm is implemented in the object-oriented language C++. The research of the algorithm efficiency and developed software is performed for the Shekel function [18, 19]

$$f(X) = \sum_{i=1}^{k} \frac{1}{c_i + \sum_{j=1}^{|X|}\left(x_j - a_{i,j}\right)^2}.$$

Here k—is a number of maximum, estimated to be 10 in the study;$a_i = \left(a_{i,1}, a_{i,2}, \ldots, a_{i,|X|}\right)$—is a vector that defines the coordinates of i maximum; c_i—is a constant that sets the value of the given maximum.

The peculiarity of the Shekel function is that it allows setting the number and the intensity of maximum [20–22]. The range of valid values of function is set equal to $D = \{X| - 10 \le x_i \le 10\}$. For two-dimensional case ($|X| = 2$) landscape of the function Shekel is illustrated in Fig. 5, where it is accepted:

$$\begin{aligned} c =& (1, 620, 782, 21, 10, 841, 90, 550, 410.8);\\ a_i =& \{(5, 89; 4, 53)(-5, 71; -9, 98)(-8, 23; 4, 57)(3, 9; -7, 05)\\ & (-3, 76; -2, 34)(0, 87; 0, 35)(-9, 52; -3, 87)(4, 23; -1, 87)\\ & (-5, 35; 6, 85)(3, 49; 4, 75)\}. \end{aligned}$$

The available values of the algorithm parameters are assumed equal $|S| = 30$, $\Delta t = 3$, $\hat{t} = 5$, $\lambda_e = 0.05$, $\lambda_p = 0.05$, $r = 1$, $r_n = 2$, $r_f = 4$, $\rho = 0, 5$, $\delta_t = 10$, $\delta_f = 0, 01$, $\Delta_f = 1$.

Figure 6 shows an example of a single run of an intelligent ice fishing algorithm, where the initial and final positions of fishermen are superimposed on the terrain of the target function, $|X| = 2$. Figure 6b shows that all fishermen reached the maxima of the target function as a result of the solution.

The algorithm efficiency is evaluated with the indicators $\xi_1, \xi_2, \ldots, \xi_{10}$, which make sense of the estimates for localization probabilities of one (global) maximum, any two local maximums,…, all ten local maximums, consequently [23]. As a criterion for localization of l maximum use the condition

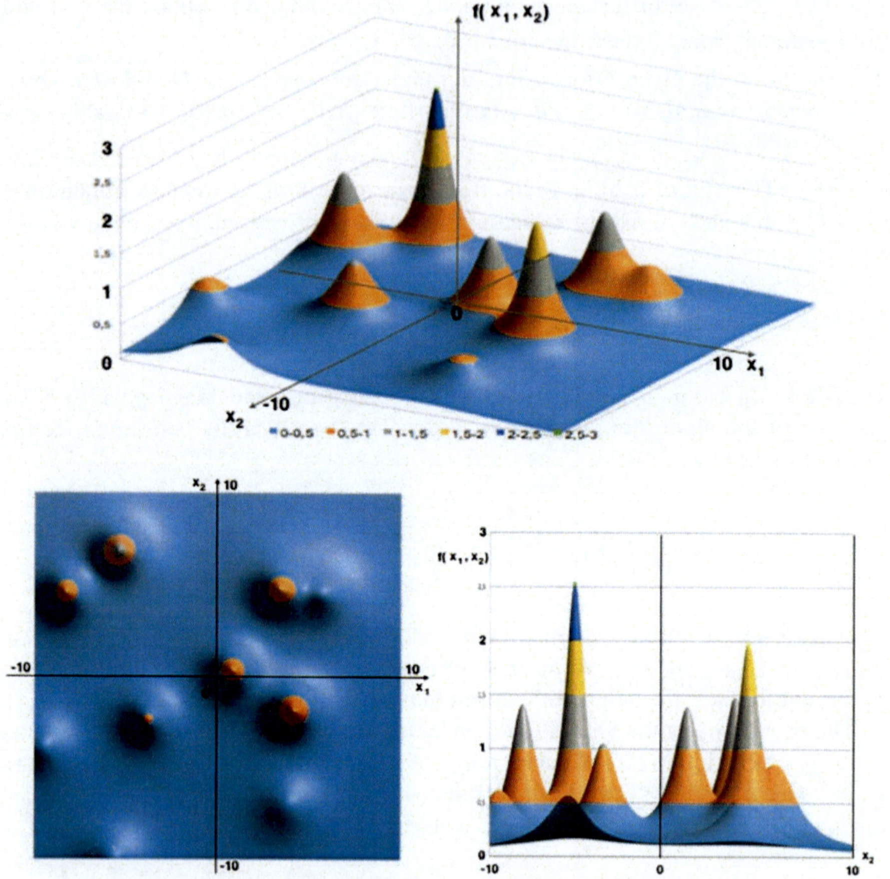

Fig. 5 Landscape of the Shekel function for given parameters:$|X| = 2$

$$\tilde{X}_l - X_l < \delta_f, l \in [1 : 10],$$

where $\tilde{X}_l$, X_l—is found by the algorithm and the exact global or local solution, consequently; δ_f—is the required localization accuracy.

Since the algorithm efficiency can significantly depend on the initial location fishermen, the computational experiments are performed 100 times (multistart method) [24–26].

Figure 7 shows the frequency of localization of the global maximum of the target function for $|X| = 2, 4, 8, 16, 32, 64$.

Figure 8 shows the results of computational experiments. Here, each chart column corresponds to a single program launch, and the column height reflects the number of maximums found.

Some experimental results are presented in Table 1.

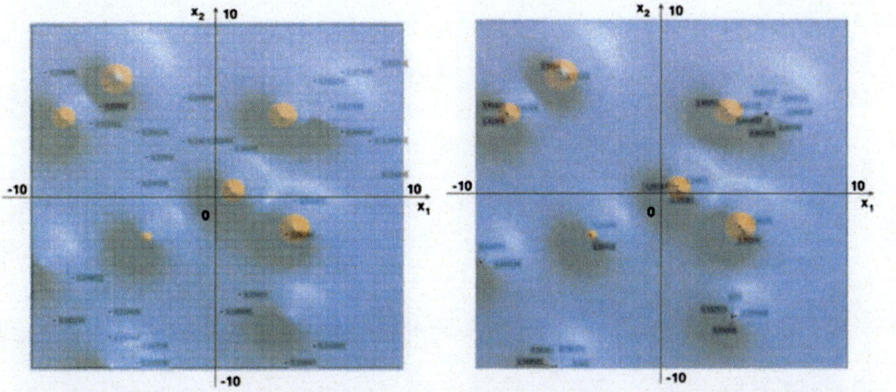

a) initial positions of fishermen b) final positions of fishermen

Fig. 6 Result of one run of the ice fishing algorithm

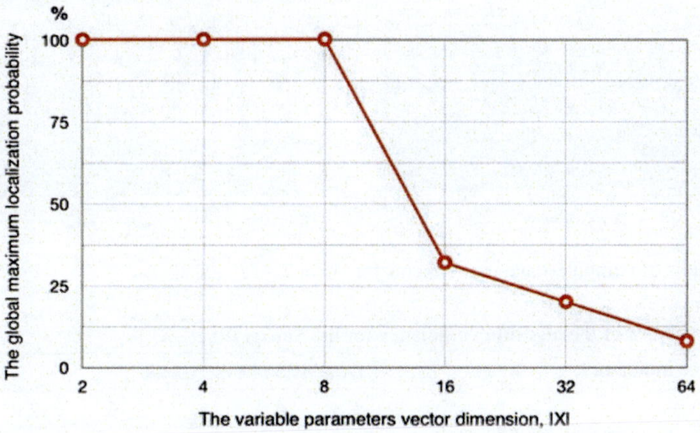

Fig. 7 Frequency localization of the global maximum

5 Conclusion

The article presents an algorithm for intelligent ice fishing, and a wide computational experiment is performed to study the effectiveness of the algorithm. The Shekel test function was selected for analysis, the dimension of the X vector varied from 2 to 64, and the probability of localization from 1 to 10 maxima was estimated. The analysis of the obtained data showed that the frequency of localization of the global maximum significantly decreases and increasing the problem dimension. This is due to a decrease in the efficiency of the local optimization method. Besides, with an increase in the dimension of the vector X from 2 to 64, a significant (more than 300 times) increase in the calculation time was observed. Analysis of the frequency

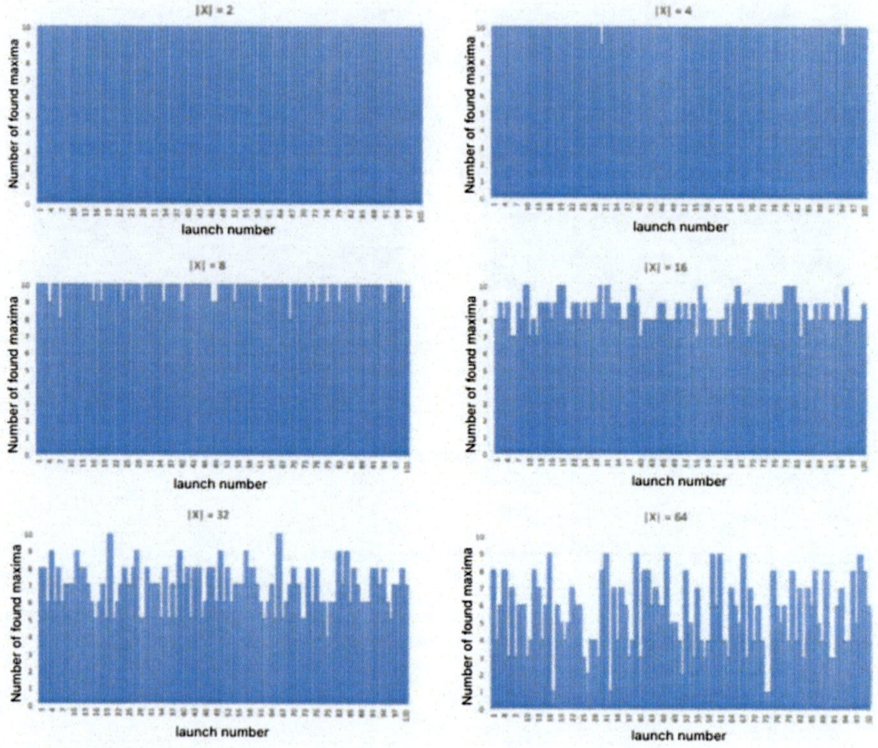

Fig. 8 Results of computational experiments for $|X| = 2, 4, 8, 16, 32, 64$

Table 1 Estimates of the algorithm efficiency for the Shekel function

Performance indicator	Dimension of the vector X					
	2	4	8	16	32	64
ξ1	100	100	100	100	100	100
ξ2	100	100	100	100	100	96
ξ3	100	100	100	100	100	94
ξ4	100	100	100	100	100	80
ξ5	100	100	100	100	98	62
ξ6	100	100	100	100	82	52
ξ7	100	100	100	100	60	33
ξ8	100	100	100	90	34	21
ξ9	100	100	98	52	10	8
ξ10	100	98	80	12	2	0
Frequency of finding the global maximum	100	100	100	32	20	8

of calling algorithm functions showed that this is due to a significant increase in the calculation time of the target function. This disadvantage can be eliminated by replacing the real target function with its surrogate model at some stages of calculations. In the future, it can be planned to compare the algorithm with other optimization algorithms, modify the method, and parallelize calculations.

References

1. S. Shan, G.G. Wang, Survey of modeling and optimization strategies to solve high-dimensional design problems with computationally-expensive black-box functions. Struct Multi. Optim. **41**(2), 219–241 (2010)
2. P.J.M. Van Laarhoven, E.H.L. Aarts, *SiMulated Annealing. Simulated Annealing: Theory and Applications* (Springer, Dordrecht, 1987), pp. 7–15
3. Z. Michalewicz, M. Schoenauer, Evolutionary algorithms for constrained parameter optimization problems. Evol. Comput. **4**(1), 1–32 (1996)
4. A.H. Wright, Genetic algorithms for real parameter optimization. Found Genetic Algor. **1**, 205–218 (1991) (Elsevier)
5. J. Kennedy, Particle swarm optimization. Encycl. Mach. Learn. 760–766 (2010)
6. D. Karaboga, B. Basturk, *Artificial Bee Colony (ABC) Optimization Algorithm For Solving Constrained Optimization Problems. International Fuzzy Systems association World congress* (Springer, Berlin, Heidelberg, 2007), pp. 789–798
7. M. Dorigo, C. Blum, Ant colony optimization theory: a survey. Theoret. Comput. Sci. **344**(2–3), 243–278 (2005)
8. A.P. Karpenko, Z.O. Svianadze, Meta-optimization based on self-organizing map and genetic algorithm. Opt. Mem. Neu. Netw. **20**(4), 279–283 (2011)
9. A.I.J. Forrester, A.J. Keane, Recent advances in surrogate-based optimization. Prog. in Aerosp. Sci. **45**(1–3), 50–79 (2009)
10. P. Kerschke, H. Trautmann, Automated algorithm selection on continuous black-box problems by combining ex-ploratory landscape analysis and machine learning. Evol. Comput. **27**(1), 99–127 (2019)
11. H. José Antonio Martín, J. de Lope, D. Maravall. Adaptation, anticipation and rationality in natural and artificial systems: computational paradigms mimicking nature. Nat. Comput. **8**(4). 757 775 (2009)
12. J. Branke, J.A. Elomari, Meta-optimization for parameter tuning with a flexible computing budget, in *Proceedings of the 14th Annual Conference on Genetic and Evolutionary Computation.* (ACM, 2012), pp. 1245–1252
13. M.S. Nobile et al., Fuzzy self-tuning PSO: a settings-free algorithm for global optimization. Swarm Evol. Comput. **39**, 70–85 (2018)
14. O. Mersmann et al. Exploratory landscape analysis, in *Proceedings of the 13th Annual Conference on Genetic and Evolutionary Computation.* (ACM, 2011), pp. 829–836
15. V. Beiranvand, W. Hare, Y. Lucet, Best practices for comparing optimization algorithms. Optim. Eng. **18**(4), 815–848 (2017)
16. E.D. Dolan, J.J. Moré, Benchmarking optimization software with performance profiles. Math. Prog. **91**(2), 201–213 (2002)
17. N. Hansen, A. Ostermeier, Completely derandomized self-adaptation in evolution strategies. Evol Comput. **9**(2), 159–195 (2001)
18. Á.E. Eiben, R. Hinterding, Z. Michalewicz, Parameter control in evolutionary algorithms. IEEE Trans. Evol. Comput. **3**(2), 124–141 (1999)
19. Y.-J. Gong, J.-J. Li, Y. Zhou, Y. Li, H.S.-H. Chung, Y.-H. Shi, J. Zhang, Genetic Learning Particle Swarm Optimization. IEEE Trans. Cybern. **46**(10), 2277–2290 (2016)

20. J. Kavetha, Coevolution evolutionary algorithm: a survey. Int. J. Adv. Res. Comput. Sci. **4**(4), 324–328 (2013)
21. A.K. Qin, P.N. Suganthan, Self-adaptive differential evolution algorithm for numerical optimization, in *2005 IEEE Congress on Evolutionary Computation,* vol. 2 (IEEE, 2005), pp. 1785–1791
22. V. Popov, Genetic algorithms with exons and introns for the satisfiability problem. Adv. Stud. Theo. Phys. **7**(5–8), 355–358 (2013)
23. B. Xing, W.-J. Gao, *Innovative Computational Intelligence: A Rough Guide to 134 Clever Algorithms* (Springer International Publishing, Switzerland, 2014), p. 450
24. X. Koua, S. Liua, J. Zheng, W. Zheng, Co-evolutionary particle swarm optimization to solve con-strained optimization problems. Comput. Math. Appl. **57**, 1776–1784 (2009)
25. Q. Chen, , B. Jiao, S. Yan, *A Cooperative Co-evolutionary Particle Swarm Optimization Algorithm Based on Niche Sharing Scheme for Function Optimization/Advances in Computer Science, Intelligent System and Environment* (Springer Berlin Heidelberg, Verlag, 2011), pp 339–345
26. Е.Ю. Воробьева, А.П. Карпенко, Ко-эволюционный алгоритм глобальной оптимизации на основе алгоритма роя частиц // Наука и образование: электронное научно- техническое издание, 2013. №11, http://technomag.bmstu.ru/doc/619595.html

Progression of EEG-BCI Classification Techniques: A Study

Ravichander Janapati, Vishwas Dalal, Rakesh Sengupta, and Raja Shekar P. V.

Abstract The introduction of various neuroimaging methods, opened the doors to the field of brain–computer interfaces (BCI). Suddenly, controlling the elements of the world with your brain seemed to be possible, and that is what happened, given a few decades. The field of BCI does not simply aim to create a futuristic world, but also help people with neurodegenerative disorders. EEG-based BCIs (EEG-BCI) are the most common of all the BCI out there today. EEG-BCI consists of the hardware to record data, the experimental paradigm, and the signal processing pipeline. In this review, we will discuss a specific part of the processing pipeline, i.e., the classification techniques. More specifically the advancements, complexity, accuracy, and sensitivity in classification technique. We will also discuss the issues with the current trends of EEG-BCI.

Keywords Electroencephalogram · Brain–computer interface · Classification techniques · Usability

1 Introduction

If we as humans have learned something during the course of evolution is to use our brain to solve the complexities of the world. We without the processing capacities of our brains would not be able to survive as a species. But our brain, no matter how logical, powerful, creative it is, is limited to our body. Humanity recognized these limitations long ago. Attempts to control things that are not part of our body, using our mind, can be found in traces of human history. But none were successful as far as the science is concerned. It was the introduction of various neuroimaging

R. Janapati (✉)
ECE Department, S R Engineering College, Warangal, India
e-mail: ravi_chander_j@srecwarangal.ac.in

V. Dalal · R. Sengupta · Raja Shekar P. V.
Center for Creative Cognition, S R Engineering College, Warangal, India

V. Suma et al. (eds.), *Inventive Systems and Control*,
Lecture Notes in Networks and Systems 204,
https://doi.org/10.1007/978-981-16-1395-1_13

methods that brought the possibility of controlling things using our minds. A brain–computer interface is the combination of both hardware and software that allows the person using it, the ability to interface with the computer to pass on commands using the brain. This newly found capacity through the BCI had endless possible implementations. BCI where it stands today is the result of its evolution from the first BCI, where Philip Kennedy using neural implants was able to give a paralysis patient control of their external using on/off signals from brain implants [1]. This evolution was brought about due to advances in other fields of research like machine learning, computing, signal processing, and neuroimaging. Now, we have hardware that can process data very quickly. The high computing power reduces the lag of the modern BCI systems (latency). The latency of BCI does not simply depend on the computing power of hardware but also on the data processing pipeline that is being used. A typical signal processing pipeline of BCI includes a filter, feature extractor, and a classifier. All the parts of the pipeline play a role in making a BCI faster or slower. In this review, we discuss the most important part of the signal processing pipeline, the classifiers. We will discuss the advances that happened in the classification technique over the past few decades and what advantages or disadvantages these advances bring along with them.

2 Advances in Classification Techniques

As we know now that BCI converts our thoughts into an action/command in the computer domain, but how does this exactly happen? Sure there is a person who is imagining opening or closing of their hand (motor-imagery paradigm), or looking at a flickering screen (SSVEP paradigm), or doing some other task, but the signal processing pipeline does the actual job of interfacing the brain with the computer. The pipeline, from the EEG data, that we have collected finds out what these signals mean. The pipeline filters the raw data then extract different features from the preprocessed data followed by the classification of the data [2, 3]. Simple as it may sound these processing steps are the most complex and require high computational power. Features are signal properties that are different for a different types of brain processes. The extracted feature are then used to first train a classifier. A classifier without training will not be able to distinguish between two signals originating from different mental tasks [4–6]. After the classifier is trained, it can now be used to classify different signals based on their features. The first step is called offline training, and the second is online testing. Only a trained classifier is used in online testing [7–9]. The above-mentioned method was used in early BCI systems. The modern BCI does not require offline standardization [10], and this was made possible due to the advances in the field of machine learning and deep learning. The early classifier includes classifiers like LDA and SVM. These classifiers required less computational capacity. With good pre-processing steps, these classifiers could get a good classification accuracy of 75–95%. The era of machine learning and deep learning came newer more complex classifiers like deep learning classifiers, LSTM classifiers, CNN, and adaptive classifiers. These modern age classifiers, without any doubt, improved the classification accuracy of the BCI. Deep learning classifiers could get 80–97% accuracy. It can be noted that the margin of improvement in accuracy was not as high as was the increase in computational power required to execute the classifier [11–13]. With a slow processor, a deep learning classifier may take hours to execute. The apparent problem is that the focus of BCI research in the last decade was just to improve classifiers. No regard as to, what the end-user of BCI wants, was given [14–16]. This kind of tunnel vision did lead to some very innovative classifiers such as LSTM+ attention [1] classifier, but in doing so steered away from the goal, which was to make BCI work and accessible for all. Table 1 contains a list of modern and old classifiers with some of their properties.

As talked about before, there is no rejecting that an astounding measure of work went into the classification strategies over the most recent twenty years or so. The classifier research was done so often that there are now more classifiers than there are BCI systems. With such a high number of classifiers available, we need a way to access them. Most of the research in the last three decades does not talk about any other parameters of evaluation other than classification accuracy. Such a high number of classifiers, which are getting more and more complex every year, simply being evaluated on just the merits of their classification accuracy seems unreasonable. Some of the papers do mention information transfer rate (ITR), latency, sensitivity as another measure, but these parameters are simply mentioned in some of the papers, others do not mention them at all, they are not still used as an evaluation parameter.

3 Conclusion

BCI makes our thoughts valid in a real sense. It pumps hope in the individuals who lost any or the entirety of their body's control. BCI ought to be an innovation for all; people should get it and use it without any hitch. And the work that has happened in improving the BCI's core (the classification techniques) has been unreal.

There is no denying that without it BCI would not be where it is today. Contribution from multiple fields have further enhanced BCI as we knew it, a decade back. With all this positive work, there are certain issues that were overlooked, and the issue of an affordable and usable BCI is one of the biggest issues today. BCI needs to be evaluated from a user-centric point of view and evaluation parameter such as latency, usability, etc., must be taken in account.

Table 1 Classifiers used in EEG-BCI

Paper	Classifier name and type	Validation protocol (No. of subjects)	Complex processing power	Accuracy	Sensitivity
Zhang et al. [4]	LSTM + Attention	Intr subject	The network consists of three layers. Each layer had 7 cells in it	98.3 ± 0.9	95.9 ± 1.7
	Deep learning	Cross subject	This was followed by the attention layer the attention layer was followed by a fully connected layer with sigmoid activation function. This makes the proposed network computationally powerful and costly at the sometime	83.2 ± 1.2	82.2 ± 2.1
Ma et al. [5]	Channel-correlation	Intra-subject	For feature extraction, they first found Pearson correlation coefficient which was then used to estimate the PSD	87.03	Sensitive to sliding window length and not sensitive to number of filters in classifier
	CNN classification	Cross-validation	This was followed by a CNN model with two 1D convolution layers where the filter size was same as number of electrodes. CNN given here is bit less complex and faster	83.93 ± 9.21	Nothing is reported to deter-mine the sensitivity

(continued)

Table 1 (continued)

Paper	Classifier name and type	Validation protocol (No. of subjects)	Complex processing power	Accuracy	Sensitivity
Ko et al. [6]	CFP + KNNC	Cross-validation	Although there is not training needed the execution time of the KNNC depends on the size of data. As the data size increase the speed of algorithm goes down very fast	97.3 ± 1.0	Cannot determine (Lack of data)
	CFP + PARZENDC		Time complexity and space/memory complexity	97.3 ± 1.0	
	CFP + LDC		Depend on m, n, i.e., number of samples and number of features, respectively, and $t = \min(m, n)$	97.2 ± 1.1	
	CFP + SVC		Training time depend on number of support vectors (training set size * error rate) hence they are not ideal for larger training set	97.4 ± 1.1	
Yu et al. [7]	SWLDA	Intra-subject	Step-wise LDA is a very unstable classifier it out performs LDA in most cases but step-wise discrimination takes more time	93.6 ± 1.2	Cannot deter-mine (Lack of data)
Yuan et al. [17]	MSI	Intra-subject	No comment	91.46	Cannot determine (Lack of data)

(continued)

Table 1 (continued)

Paper	Classifier name and type	Validation protocol (No. of subjects)	Complex processing power	Accuracy	Sensitivity
	FBCCA			88.38	
	CCA			83.3	
	FWHT-NBC			87.34	
Park et al. [18]	EMSI	Intra-subject	No comment	82–90.8	Sensitive to duration of stimulus. Accuracy saturates at 2.5 s
Li et al. [19]	CSP	Intra-subject	No comment	84.04	Cannot determine (Lack of data)
Sadiq et al. [20]	LR	Cross-subject	No comment	91.33 ± 7.4	89.67 ± 88
	MLP			92.33 ± 6.7	90.67 ± 83
	LS-SVM			90.67 ± 8.0	90.00 ± 10.2
	LMT			88 ± 3.7	87.00 ± 7.1
	NB			92.33 ± 6.7	90.67 ± 83
Kwon et al. [21]	CSP + CNN	Intra-subject	No comment	74.15	Number of epochs and number offrequency indices
Stawicki et al. [22]	MEC	Intra-subject	No comment	89.77	Cannot determine (Lack of data)

(continued)

Table 1 (continued)

Paper	Classifier name and type	Validation protocol (No. of subjects)	Complex processing power	Accuracy	Sensitivity
Nakanishi et al. [8]	CCA	Intra-subject	$O(M^2 (M + Dt_x t_y K))$. Where X and Y be the two datasets, and number of features in X and Y are denoted by p and q. $M = \max(p, q)$ $K = \min(p, q)$ D represents number of extracted features	89.83 ± 6.07	Cannot determine (Lack of data)
MartinezCagigal et al. [9]	SWLDA	Intra-subject	Step-wise LDA is a very unstable classifier it out performs LDA in most cases but step-wise discrimination takes more time	77.46 (MS)	Cannot determine (Lack of data)
Wang and Bezerianos [23]	CSP + SVM	Intra-subject	No comment	94.23 (contro l)	Cannot determine (Lack of data)
Riechmann et al. [24]	OCSVM	Intra-subject	No comment	81	Cannot determine (Lack of data)
Leeb et al. [25]	CVA + Gaussian classifier	Intra-subject	No comment	87	Cannot determine (Lack of data)
Chae et al. [26]	IAC + MDC	Intra-subject	No comment	80.9	Cannot determine (Lack of data)

(continued)

Table 1 (continued)

Paper	Classifier name and type	Validation protocol (No. of subjects)	Complex processing power	Accuracy	Sensitivity
Galan et al. [10]	CVA + LDA		LDA is one of the most straight forward classifiers, it requires less training time and give fast and reliable results. It is highly unstable	74–92	Cannot determine (Lack of data)
Scherer et al. [11]	CFR1(DSLVQ + LDA)	Intra-subject	O (mnt + t^3) time complexity where m and n are number of samples and number of features, respectively, and t = min (m, n)	83.67	Cannot determine (Lack of data)
	CFR2(DSLVQ + LDA)		O (mn + mt + nt) Space/memory complexity	79.67	
Wang et al. [12]	ICA	Intra-subject	O (K^2N^2) O (3 k N + K^2 N) where K and N are number of sources and number of samples, respectively	95	Cannot determine (Lack of data)

Acknowledgements The authors acknowledge Science & Engineering Research Board (SERB) a statutory body of Department of Science and Technology (DST), Government of India for financial support vide Reference No: EEQ/2019/000624under the scheme of Empowerment and Equity Opportunities for Excellence in Science to carry out this work. The authors also acknowledge the Management and Principal of S R Engineering College, Warangal Urban for their continuous support by providing all the necessary facilities.

References

1. G. Zhang, V. Davoodnia, A. Sepas-Moghaddam, Y. Zhang, A. Etemad, Classification of hand movements from EEG using a deep attention-based LSTM network. IEEE Sens. J. **20**(6), 3113–3122 (2020). https://doi.org/10.1109/jsen.2019.2956998
2. R. Janapati, V. Dalal, R. Gupta, P. Anuradha, P. Shekar, Various signals used for device navigation in BCI production. IOP Conf. Ser. Mater. Sci. Eng. **981** 032003 (2020). https://doi.org/10.1088/1757-899X/981/3/032003
3. .R. Ravi Kumar, M. Babu Reddy P. Praveen, A review of feature subset selection on unsupervised learning, in *2017 Third International Conference on Advances in Electrical, Electronics, Information, Communication and Bio-Informatics (AEEICB)* (Chennai, 2017), pp. 163–167. https://doi.org/10.1109/AEEICB.2017.7972404
4. P.U. Anitha, C.V.G. Rao, S. Babu, Email spam classification using neighbor probability based Naïve Bayes algorithm, in *Proceedings—7th International Conference on Communication Systems and Network Technologies, CSNT 2017* (2018), pp. 350
5. R. Ravi Kumar, M. Babu Reddy, P. Praveen, An evaluation of feature selection algorithms in machine learning. Int. J. Sci. Technol. Res. **8**(12), 2071–2074 (2019)
6. M.A. Iqbal, K. Devarajan, S.M. Ahmed, A brief survey of asthma classification using classifiers. Int. J. Adv. Sci. Technol. **28**(15), 717–740 (2019)
7. X. Ma, S. Qiu, W. Wei, S. Wang, H. He, Deep channel-correlation network for motor imagery decoding from the same limb. IEEE Trans. Neural Syst. Rehabil. Eng. **28**(1), 297–306 (2020). https://doi.org/10.1109/tnsre.2019.2953121
8. P.R. Kennedy, R.A.E. Bakay, Restoration of neural output from a paralyzed patient by direct brain connection. NeuroReport **9**, 1707–1711 (1998)
9. L. Ko, O. Komarov, S. Lin, Enhancing the hybrid BCI performance with the common frequency pattern in dual-channel EEG. IEEE Trans. Neural Syst. Rehabil. Eng. **27**(7), 1360–1369 (2019). https://doi.org/10.1109/tnsre.2019.2920748
10. O. Kwon, M. Lee, C. Guan, S. Lee, Subject-independent brain-computer interfaces based on deep convolutional neural networks, in *IEEE Transactions on Neural Networks and Learning Systems,* 1–14 (2020). https://doi.org/10.1109/tnnls.2019.2946869
11. Y. Yu, Y. Liu, E. Yin, J. Jiang, Z. Zhou, D. Hu, An Asynchronous hybrid spelling approach based on EEG–EOG signals for chinese character input. IEEE Trans. Neural Syst. Rehabil. Eng. **27**(6), 1292–1302 (2019). https://doi.org/10.1109/tnsre.2019.2914916
12. M. Nakanishi, Y. Wang, X. Chen, Y. Wang, X. Gao, T. Jung, Enhancing detection of SSVEPs for a high-speed brain speller using task-related component analysis. IEEE Trans. Biomed. Eng. **65**(1), 104–112 (2018). https://doi.org/10.1109/tbme.2017.2694818
13. V. Martinez-Cagigal, J. Gomez-Pilar, D. Alvarez, R. Hornero, An Asynchronous P300-based brain-computer interface web browser for severely disabled people. IEEE Trans. Neural Syst. Rehabil. Eng. **25**(8), 1332–1342 (2017). https://doi.org/10.1109/tnsre.2016.262338
14. F. Galán, M. Nuttin, E. Lew, P. Ferrez, G. Vanacker, J. Philips, J.D. Millán, A brain-actuated wheelchair: asynchronous and non-invasive Brain–computer interfaces for continuous control of robots. Clin. Neurophysiol. **119**(9), 2159–2169 (2008). https://doi.org/10.1016/j.clinph.2008.06.001

15. R. Scherer, F. Lee, A. Schlogl, R. Leeb, H. Bischof, G. Pfurtscheller, To- ward Self-paced brain-computer communication: navigation through virtual worlds. IEEE Trans. Biomed. Eng. **55**(2), 675–682 (2008). https://doi.org/10.1109/tbme.2007.903709
16. Y. Wang, R. Wang, X. Gao, B. Hong, S. Gao, A practical VEP-based brain–computer interface. IEEE Tran. Neural Syst. Rehabil. Eng. **14**(2), 234–240 (2006). https://doi.org/10.1109/tnsre.2006.875576
17. Y. Yuan, W. Su, Z. Li, G. Shi, Brain–Computer Interface-Based Stochastic Navigation and Control of a Semiautonomous Mobile Robot in Indoor Environments. IEEE Trans. Cognitive Develop. Syst. **11**(1), 129–141 (2019). https://doi.org/10.1109/tcds.2018.2885774
18. S. Park, H.-S. Cha, C.-H Im, Development of an online home appliance control system using augmented reality and an SSVEP-based Brain–Comput. Inter. IEEE Access, **7**, 163604–163614 (2019). https://doi.org/10.1109/access.2019.2952613
19. Z.Li, Y. Yuan, L. Luo, W. Su, K. Zhao, C. Xu, ... M. Pi, Hybrid brain/muscle signals powered wearable walking exoskeleton enhancing motor ability in climbing stairs activity. IEEE Trans. Med. Robotic. Bionics **1**(4), 218–227 (2019). https://doi.org/10.1109/tmrb.2019.2949865
20. M. T. Sadiq, X. Yu, Z. Yuan, F. Zeming, A.U. Rehman, I. Ullah, ... G. Xiao, Motor imagery EEG signals decoding by multivariate empirical wavelet transform-based framework for robust brain–computer interfaces. IEEE Access **7**, 171431–171451 (2019). https://doi.org/10.1109/access.2019.2956018
21. O.-Y. Kwon, M.-H. Lee, C. Guan, S.-W. Lee, (2020). Subject-independent brain–computer interfaces based on deep convolutional neural networks. IEEE Trans. Neural Net. Learn. Syst. **31**(10), 3839–3852 (2020). https://doi.org/10.1109/tnnls.2019.2946869
22. P. Stawicki, F. Gembler, A. Rezeika, I. Volosyak, A novel hybrid mental spelling application based on eye tracking and SSVEP-based BCI. Brain Sci. **7**(12), 35 (2017). https://doi.org/10.3390/brainsci7040035
23. H. Wang, A. Bezerianos, Brain-controlled wheelchair controlled by sustained and brief motor imagery BCIs. Electron. Letter. **53**(17), 1178–1180 (2017). https://doi.org/10.1049/el.2017.1637
24. H. Riechmann, A. Finke, H. Ritter, Using a cVEP-based brain-computer interface to control a virtual agent. IEEE Trans. Neural Syst. Rehab. Eng. **24**(6), 692–699 (2016). https://doi.org/10.1109/tnsre.2015.2490621
25. R. Leeb, L. Tonin, M. Rohm, L. Desideri, T. Carlson, J. del. Millan, Towards independence: A BCI telepresence robot for people with severe motor disabilities. Proc. IEEE, **103**(6), 969–982 (2015). https://doi.org/10.1109/jproc.2015.2419736
26. Y. Chae, J. Jaeseung, S. Jo, Toward brain-actuated humanoid robots: asynchronous direct control using an EEG-based BCI. IEEE Trans. Robot. **28**(5), 1131–1144 (2012). https://doi.org/10.1109/tro.2012.2201310

Cuckoo Scheduling Algorithm for Lifetime Optimization in Sensor Networks of IoT

Mazin Kadhum Hameed and Ali Kadhum Idrees

Abstract The network topology control represents an essential factor during designing the Wireless Sensor Networks (WSNs) due to its primary role in the lifetime optimization of the WSNs. This article proposes a Cuckoo Scheduling Algorithm (CSA) for lifetime optimization in Sensor Networks of Internet of Things (IoT). Here, the sensor devices have clustered using the DBSCAN method, and then the CAS technique is applied at each cluster head. The CSA provides the best schedule of the sensor nodes inside each cluster to monitor the cluster region with a minimum number of nodes while keeping a suitable level of coverage. The cluster head pooling and the CSA have been executed periodically. The experimental results confirm that the CSA technique enhances the network lifespan and whilst preserving an acceptable ratio of coverage in WSNs.

Keywords Sensor networks · DBSCAN clustering · Cuckoo algorithm · Performance evaluation · Energy saving

1 Introduction

Sensors could be defined as small, economizing, and low-power operating devices that can sense, process, and transmit any acquired data over wireless networks. Wireless Sensor Network(WSN) is an essential factor in a wide variety of uses, including security, battlefield surveilling, air traffic controlling, bio-detection, environmental monitors, industrial automation, and smart grids [1, 2]. Monitor such uses demand the deployment of several sensors in the environment. The area that could be reached by the sensor during monitoring or sensing is known as its coverage, as a sensor is capable of monitoring multiple targets in its coverage [3].

M. K. Hameed
Department of Software, University of Babylon, Babylon, Iraq
e-mail: it.mazen.kadhum@uobabylon.edu.iq

A. K. Idrees (✉)
Department of Computer Science, University of Babylon, Babylon, Iraq
e-mail: ali.idrees@uobabylon.edu.iq

V. Suma et al. (eds.), *Inventive Systems and Control*,
Lecture Notes in Networks and Systems 204,
https://doi.org/10.1007/978-981-16-1395-1_14

The relatively smaller size of sensors, as well as their uses in hardly accessible regions, make it rather impractical to replace or change their batteries [4]. Therefore, expanding the network lifespan is considered to be an important issue in Wireless Sensor Networks. Among the main aspects to be taken into consideration when increasing the lifespan are deploying sensors optimally [5–7], sleep scheduling (i.e.) changing the sensor mode from active to sleep [8–11], and maintaining the balance of load [12].

Hence, efficient use of the energy could be realized by relying on the scheduling of sensor nodes for monitoring the target, as this would help in avoiding network congestions and packet retransmitting, thereby reducing any additional energy consumption for communication within networks [13]. As for the target coverage, sensor scheduling remarkably extends the network lifespan. Multiple sensors could be clustered in sensor sets, which could either have a duplicate or not. Whenever such a set is in the active mode, the element sensor node may be active or inactive, meanwhile, the remainder (the elements of other sets) are idle. Maintaining the balance of load represents another essential variable in achieving a maximal network lifespan. The extent to which the energy is available for every sensor differs remarkably according to its distance from the base station as well as its sense range and the quantity of the node's transmitting power consumption. According to the usable residual energy, a balance could be drawn among sensor loads [14, 15].

This article introduces the following contributions.

1. This article suggests the Cuckoo Scheduling Algorithm (CSA) for Enhancing Lifetime of Cluster-based WSNs. This technique achieves two efficient algorithms: clustering and then scheduling the sensor devices in WSN. WSN is clustered into clusters using the distributed DBSCAN approach. The scheduling phase splits the lifespan into periods, and it achieves its goal in three steps. In the first step, the cluster head polling is implemented in a distributed way inside each cluster. After that, the cluster head will execute Cuckoo Algorithm in the second step to optimize the coverage and the lifespan of the network to produce the optimal schedule of sensor devices that are responsible for monitoring the cluster region in the next step. In the third step, each sensor device will receive a packet from the cluster head to inform it to stay active or sleep until the beginning of the next period.
2. The Cuckoo Algorithm (CA) is employed to optimize the network lifespan while maintaining acceptable coverage for the monitoring area. CA is applied to solve the proposed model of optimization to reduce the centers' number of devices that are not covered and decrease the number of active sensor device in each period. The scheduling algorithm-based CA is to provide the best schedule of sensor devices in each period instead of using optimization solvers. This can decrease the execution time and maximize network lifespan.
3. Extensive experiments are performed using the C++ custom simulator. The conducted simulation results prove that the proposed CSA technique can improve the lifespan of the network and maintaining an acceptable level of

coverage in the area of interest in comparison with some existing methods such as DESK [16], GAF [17], and PeCO [49].

The remainder of the work could be sketched as follows: Sect. 2 discusses the related works, Sect. 3 presents a description of the proposed algorithm. Section 4 presents and discusses the results of the experiments. Section 5 concludes the research remarks.

2 Related Works

Yaxiong and Jie (2010) introduced a generic duty-cycling scheduling technique that depends on the stochastic theory [18]; whereas Chih-fan and Mingyan (2004) proposed a precise scheduling algorithm as part of every approach, along with the analysis of the coverage and duty cycle characteristics[19]. Both Nath and Gibbons (2007) and Zhuxiu et al., created two common Sleep Scheduling algorithms, further identified as the Connected K-Neighbourhood (CKN) and the enhanced Energy Consumed uniformly—Connected K-Neighbourhood (EC-CKN) algorithms, respectively [20]. Any of these two scheduling algorithms could close the lower power nodes meanwhile each of the network's connectivity and its sufficient routing latency is maintained. As for the CKN algorithm, every node is K connected, implying that whenever a node has over K active neighbors (the k-connectivity is satisfied), it will decide on terminating itself, otherwise (if the active adjacent neighbors are under K), the node will remain active. Being a distributed SS algorithm, it could efficiently extend the lifespan of every single node and thereby eventually the whole network's lifespan. The rank value could be used in identifying the active nodes within CKN algorithms. The initial phase of performing CKN algorithms is provided for every period in a random manner, where the set of active nodes differs from one period to another. The main issue of this type of algorithm lies in the fact that their energy could not be guaranteed to have a uniform consumption [21]. Alternatively, the EC-CKN algorithm considers the remainder of the energy for the nodes based on the CKN, thereby balancing the amount of energy consumed by the network as a whole and at the same time keeping it K-connected.

In [22] a scheduling technique is suggested according to generic duty-cycling by use of the stochastic theory, meanwhile [23] introduced the coverage and duty cycle feature analyses that are determined by scheduling algorithms. Based on where the neighbouring nodes are located, Sun et al. [24] employed redundancy algorithms in judging the sensor node status, as any redundant node will be turned into sleep mode. The Sleep-awake Energy-Efficient Distributed (SEED) algorithm was introduced by Ahmed et al. [25]. It divides any network field into three different regions; the high-energy region cluster heads take the responsibility of communicating with the base station and conserving the energy of low energy region cluster heads. The SEED outperformed various previously proposed energy-efficient mechanisms in the field.

The work in [16] proposed a scheduling algorithm that can schedule the sensor devices in subdivided grids of the area of interest based on their geographical positions of sensor devices. A distributed scheduling algorithm named DESK is proposed by [17]. The authors executed the DESK algorithm at each sensor device and the decision is based on local information from the neighboring devices using perimeter coverage model. The authors in [26, 27] proposed two coverage algorithms to extend the lifespan of the WSNs. The first algorithm is named DiLCO which uses an optimization solver to a coverage optimization model based on primary points to produce the optimal schedule in each round. The second algorithm proposed an optimization model to maintain the coverage and improve the lifespan of the WSNs based on the perimeter coverage model.

Several algorithms have been proposed to solve the scheduling problem in WSNs. Some of them distributed scheduling approaches, which are fast because they have based on local information, but they cannot give an optimal schedule of sensor devices. The other types of scheduling algorithms are centralized, and they can provide an optimal solution but with a high execution time in the case of large WSN. Some proposed works proposed scheduling algorithms that are globally distributed but locally centralized and using optimization solvers. These methods can enhance network lifespan, but they consume a high execution time.

In this article, a Cuckoo Scheduling Algorithm (CSA) is used for improving the lifespan of Cluster-based WSNs. This technique partitions the WSN lifespan into periods. Every period performs a distributed network lifespan optimization based on distributed DBSCAN clustering, distributed cluster head polling, and sensor scheduling based on CA. This approach extends the network lifespan while maintaining a suitable level of coverage for the monitored area of interest.

3 Proposed CSA Technique

In this article, a Cuckoo Scheduling Algorithm for Enhancing Lifespan of Cluster-based WSNs, called (CSA) technique is suggested. It is composed of two phases: Clustering and scheduling. Figure 1 illustrates the proposed CSA technique.

3.1 Clustering Phase

The sensor devices in the network are clustered based on our proposed distributed DBSCAN clustering algorithm introduced in [28]. This algorithm will group the nodes into clusters, after that the nodes in each cluster cooperate and share their information for choosing a cluster head periodically.

This algorithm is selected for the following different reasons:

Fig. 1 CSA technique

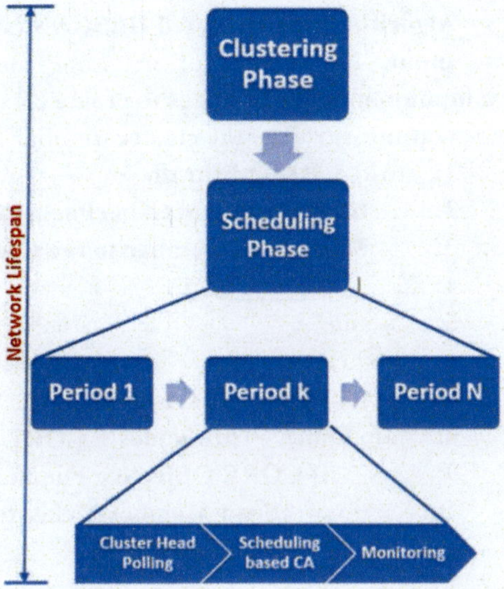

1. DBSCAN does not require one to specify the number of clusters in the data a priori, as opposed to k-means.
2. DBSCAN can find arbitrarily shaped clusters. It can even find a cluster surrounded by (but not connected to) a different cluster.
3. DBSCAN has a notion of noise and is robust to outliers.
4. The parameters minPts and ε can be set by a domain expert if the data is well understood.

According to the Dbscan algorithm, the algorithm for each sensor to execute DbScan is as follows:

1. According to the principle of the Dbscan algorithm, each sensor node will perform the same test, regardless of whether it is a core point because it scans the surrounding area to find out the number of sensors that are within the sensing range and must be larger or equal to specific parameters.
2. For sensor nodes within the sensing range of the core point, they will become its members.
3. If this core point does not belong to any cluster, it will form a new cluster, otherwise it will remain in the same cluster.
4. The core point sends a message to all its members to be included in the same cluster.
5. Repeat steps 1–5 until all sensor nodes pass.

Algorithm 1. Distributed DBSCAN (sj)

Input: N: number of neighbor nodes, Sr: sensing range, minNodes: minimum number of nodes to create cluster.

Output: sj.rejon: the cluster number for node sj.

```
while REj ≥ Ethr do
    If sj Receive MemberPacket from si then
    Mark sj as member to the Core si ;
        Update REj;
    end
    sj.rejon ← 0;
    for each node si in N do // i ∈ N and i ≠j
nbrNodes ← nbrNodes + CORE Objective Function (sj, si, Sr);
        if CORE Objective Function return 1 then
            Send MemberPacket to the sensor node i;
             Update REj;
        end
        if nbrNodes ≥ minNodes then
           save the information
if (((sj.rejon = 0) Or (sj.rejon ≠ 0)) and (r==0)) then
               sj.rejon ← sj.rejon +1;
                Call Cluster(sj);
           end
               else if ((sj.rejon = 0) Or (r ≠ 0)) then
                       sj.rejon ← sj.rejon +1;
                       Call Cluster1(sj);
           end
                else if ((sj.rejon ≠ 0) Or (r ≠ 0)) then
                        Call Cluster2(sj);
            end
               end
     end for
    end while
returen sj.rejon;
```

CORE Objective Function **return** 1 and r = **0** if the sensor node i is **within the** sensing **range** Sr and it **isn't always** a member in **other** clusters. Otherwise, CORE Objective Function returns **0** and r = 1. The **function** Cluster **placed** any neighbor node **inside the** sensing **variety** of sj in the **same** cluster and sj **send** MemberPacket to the sensor node i **to inform** it that it **turns into** a member in the **identical** cluster of sj. The **function** Cluster1 **placed** any neighbor node **in the** sensing **variety** of sj and it has **no longer** assigned to any cluster in the **identical** cluster of sensor node j. The **function** Cluster2 **positioned** any neighbor node **within the** sensing **range** of sj and it has **no longer** assigned to any cluster (or **it is** assigned to the cluster of sensor node j) in the **same** cluster of sensor node j. After **reaching** the **capabilities** Cluster, Cluster1, and Cluster2, the remaining **electricity** of the sensor node j **might be** updated **because of** sending a MemberPacket to the sensor node i **to inform** it that it **becomes** a member in the **equal** cluster of sj.

3.2 Scheduling Phase

The scheduling phase will be started periodically after the clustering phase. It includes three steps in each period such as cluster head selection, sensor node activity scheduling based Cuckoo Algorithm (CS) optimization, and monitoring.

A. Cluster Head Selection

After producing the clusters, the exchange of information among the core points (sensor nodes) is achieved inside each cluster, where each core point sends a message to all the core points inside the cluster. It includes the whole essential information like rested power, status, position, members' number, the entire number of devices in the cluster, etc.

In this step, every sensor node inside each cluster will involve the information of other nodes in the same cluster. Therefore, each node in the same cluster will achieve Eq. 1 using the information of each member inside the node. The node that gives the better value of the Eq. 1 will be chosen as a cluster head in the current cluster for this period. The whole device inside the cluster will perform the same calculation and will produce the same results for the winner device. This will be executed in a distributed way and every node will know if it is a cluster head or not.

$$\mathrm{FitVal}_j = \frac{E_{\mathrm{remaining}}}{E_{\mathrm{initial}}} + \left(1 - \sum_{j \in N} \left|S_j(x, y) - S_i(x, y)\right|\right) + \frac{S_j(\mathrm{Members})}{\mathrm{Cluster}(\mathrm{Members})} \tag{1}$$

where $E_{\mathrm{remaining}}$ is the residual energy of the node j, E_{initial} is the initial energy value of node j, N is the number of nodes in the current cluster, S_j (x, y) and S_i (x, y) refer to the locations of nodes S_j and S_i respectively. S_j (Members) indicates the number of nodes members of node j, Cluster (Members) indicates the whole number of nodes in the cluster.

In all the clusters of the WSN, the cluster heads are chosen in an independent, asynchronous, and distributed way.

B. Activity Scheduling BAsed Cuckoo Algorithm (CS)

In this part, the optimization model of the scheduling problem will be formulated, and then a Cuckoo Algorithm will be employed because it is easy to implement and has fewer tuning parameters. The Cuckoo Algorithm can solve this model to find the optimal/near-optimal solution by producing the best schedule of sensor devices to take the mission of monitoring in the next step in the current period.

A scheduling mathematical model is used to optimize the network lifetime and the coverage of the WSN.

In this article, two objectives are considered into account during formulating the scheduling optimization model: minimizing the uncovered region inside the cluster region and minimizing the number of active sensor devices after the decision by the CA. This model is inspired by the work in [29] with some modifications by considering decreasing the number of active devices per period as another objective to reduce the consumed energy and enhance the lifespan of WSN. Let parameter A be an indicator for covering the center points of sensor nodes inside each cluster. Parameter A can be defined as follow:

$$Aji = \begin{cases} 1 \text{ if point center } ji \text{ scover ed by sensori} \\ 0 \text{ Otherwise} \end{cases} \tag{2}$$

for $1 \leq j \leq N$ and $1 \leq i \leq N$, where N is the number of sensor devices (or center points) inside the cluster. Let S refers to the solution parameter that can be either 0 or 1 according to the status of the sensor device. It can be defined as follows

$$S_i = \begin{cases} 1 \text{ if Sensor } i \text{ is Active} \\ 0 \text{ Otherwise.} \end{cases} \tag{3}$$

The coverage probability P_j of the center point j is defined as follows

$$P_j = 1 - \prod_{i=1}^{N} \left(1 - \left(A_{ji} * S_i\right)\right) \tag{4}$$

The first principal objective of the suggested scheduling optimization model is to increase the coverage ratio over the area of interest by decreasing the uncovered ratio ($1-P_j$) of this area as follows.

$$P_j = 1 - P_j \tag{5}$$

The second objective is to minimize the number of those that cover the same center point in the sensing field. This can be defined as follows

$$L_j = \begin{cases} 0 \text{ if point } j \text{ is not covered} \\ \sum_{i=1}^{N} \left(A_{ji} * S_i\right) \end{cases} \tag{6}$$

Hence, the problem of scheduling optimization is modeled as follows:

$$\text{minimize} \quad \delta \cdot \sum_{j=1}^{N} P_j + \vartheta \cdot \sum_{j=1}^{N} L_j \tag{7}$$

$$\text{Subject to} \sum_{i=1}^{N} \left(A_{ji} * S_i\right) = 1 + L_j - P_j \,\forall j \in N \tag{8}$$

$$\mathrm{S_i} \in \{0, 1\} \quad \forall \mathrm{i} \in \mathrm{N} \tag{9}$$

$$P_j \ \geq 0 \,\forall j \in N \tag{10}$$

After that, in each cluster head in the clusters, and energy-efficient activity scheduling mechanism-based CS optimization is achieved. Depending on the sensor nodes focuses only the CS optimization yields the ideal cover set of active sensor nodes, which are responsible for sensing throughout the monitoring step in the current period.

The local random walk is as follows

$$xi^{(k+1)} = \begin{cases} xi^{(k)} + ra\text{if}\,\acute{r}\,a > P\alpha \\ xi^{(k)}\text{otherwise.} \end{cases} \tag{11}$$

where ra and$\acute{r}a$ are two random numbers in range (0,1).

As for the suggested algorithm, the assumptions below concerning the sensor nodes will remain constant:

Sensor coverages take the form of circles, where every sensor has a similar coverage with radius Rs.

A sensor is not capable of sensing through or moving across boundaries or obstacles that are regarded to be walls.

The sensing quality remains fixed within Rs and equals zero when outside it (following a binary model).

Algorithm 2 illustrates the Scheduling based CS for providing the optimal or near-optimal schedule of sensor devices to stay active in the monitoring step in the current period.

```
Algorithm 2 Scheduling based CS
Input: POP_S: is the population size
Output: Gbest:is the best solution(nest)
    Initialize population of POP_S solutions(nests);
    Population transformed into 0 or 1 by (7);
    Evaluate each solution(nest) using (7);
    Update best solution Gbest;
    While Stopping criteria is not satisfied do
        For i = 1 to POP_S
            Generate new solution(nest) yi(new) via (2);
            Evaluate yi(new) via (7);
            If fit(y(new) < fit (yi(k));
                yi(k+1) = yi(new);
            Else
                yi(k+1) = yi(k);
            End If;
        End For;
        For i = 1 to POP_S
            Generate a new solution yi(new) via (11);
            Evaluate Individual yi(new) via (7);
            If fit(yi(new) ) < fit (yi(k) );
                yi(k+1) = yi(new)
            Else
            yi(k+1) = yi(k);
            End If;
        End For;
        Update the best solution Gbest;
    End While;
return Gbest;
```

The Scheduling based CS can be identified as an evolutionary algorithm of global optimization whose basis is the distinct breeding behavior [15] of the cuckoo bird, in addition to the Levy flight, a pattern used by a bird for searching food. Initially, a population is initialized which consists of several candidate solutions that are generated randomly. These solutions undergo improvement generation after another until the maximal number of generations is reached or a certain condition is satisfied. The improving process which the first solution is subjected to takes place through applying the levy flight on a population, after which the least useful solution is disposed of and the better ones are kept.

The initial population includes a constant number of nests; as it takes the form of a matrix with k rows and n columns, representing the number of nests and sensors, respectively. The scheduling-based CS algorithm can be explained in more detail as follows.

1. Generating the initial cuckoo Population (nests): The initial population has been created with nests that represent possible sensor schedule solutions for covering the whole cluster region. According to the CS algorithm, the type of the initial population's values will be real.
2. Representation of Solution: Given the fact that the aim of the suggested CS is finding the (near) optimum sensor node scheduling that can be responsible for monitoring the cluster region throughout the following step. A nest is identified as a schedule of the sensor nodes, and every nest includes several eggs. Therefore, the egg will include a value 1 that corresponds to the active device or it has a value 0 that refers to the sleep mode of the sensor device. The sigmoid function is used to convert the continuous population to a discrete population with 0 or 1.
3. Fitness function: Next, all individuals undergo evaluation, as fitness values are assigned based on their fitness functionality presented in Eq. (7). The proposed CS indicates that solutions with minimal fitness values tend to be considered as optimal candidates. A negative correlation seems to exist between the fitness value and an individual's opportunity to survive. The function tends to reward a decline in the number of sensor nodes covering the same sensor device center, meanwhile, a decline to zero is penalized.
4. Generate a new nest (solution) $xi(k+1)$: The standard CS makes use of both a global and local random position combined randomly. The former is indicated by Eq. 2, whereas the local random walk is given by Eq. 11.
5. Evaluate New Nests (solutions): each new individual is evaluated using Eq. 7.
6. Immigration: As the new individuals (solutions) have been evaluated, the algorithm is iterated through two steps: first, based on the CS algorithm, is replacing all nests (with exception of the ideal one) with a new solution produced through a random walk with Lévy flight around the (so far) optimal nest in terms of quality, and secondly, calculating the Pα fraction of the worst nest to replace them by newer ones.
7. Update Gbest: The global best is update according to the new best solution reached by this algorithm in this iteration.

C. Monitoring

After producing the best schedule of sensor devices by the Scheduling-based CS algorithm, the cluster head will send messages to all sensor devices in the cluster to inform them of their status in the next step (monitoring). Each sensor device inside the cluster will receive this message. If the message contains 0, it means that it must stay in sleep mode until starting the next period. If the message includes 1, this means that the sensor device must stay active to perform all the tasks during this step.

4 Performance Evaluation and Analysis

In this section, the evaluation is made regarding the efficiency of the CSA technique by executing multiple experiments with the use of a C++ custom simulator. Table 1 presents the parameters applied in the simulating process. Fifty executions have been performed using different WSN topologies. The presented results indicate the average rate of these executions. Five network sizes from100 to 300 nodes have been used in the simulation process, deploying nodes in a controlled manner over a sensing area of (50 × 25) m^2 for ensuring full coverage for the present area of interest. The suggested protocol uses the energy model discussed in [49].

The power of all sensor nodes was initialized randomly in the range [500–700]. Similar performance metrics that had been applied by [49] have been employed in evaluating the performance of the CSA technique, which include Coverage Ratio, Active Sensors Ratio, Network Lifetime, and Energy Consumption. Additionally, three other methods were used in drawing a comparison, namely DESK [47], GAF [46], and PeCO [49].

4.1 Coverage Ratio

The average coverage ratios of each of these 5 methods for 200 nodes are presented in Fig. 2. During the first periods, DESK, GAF, and PeCO result in slightly better coverage rates (99.99%, 99.96%, and 98.76% respectively), as compared to the (97.1%) provided by CSA. The reason behind this is that CSA turns off relatively more redundant nodes than DESK, GAF, and PeCO. After the 65th period, the CSA tends to result in a more favorable coverage performance than the alternative methods, maintaining a coverage rate of over 80% for many rounds. Such an increase in efficiency is the result of the large quantity of energy that has been saved by CSA during the initial rounds.

Table 1 The variables applied in the simulating process

Parameter	Value
Field of the sensing	(50 * 25) m^2
WSN size	100, 150, 200, 250 and 300 nodes
Range of the initial energy	500–700 J
Rs	5 m
Rc	10 m
POP_S	30
δ	0.05
ϑ	0.95

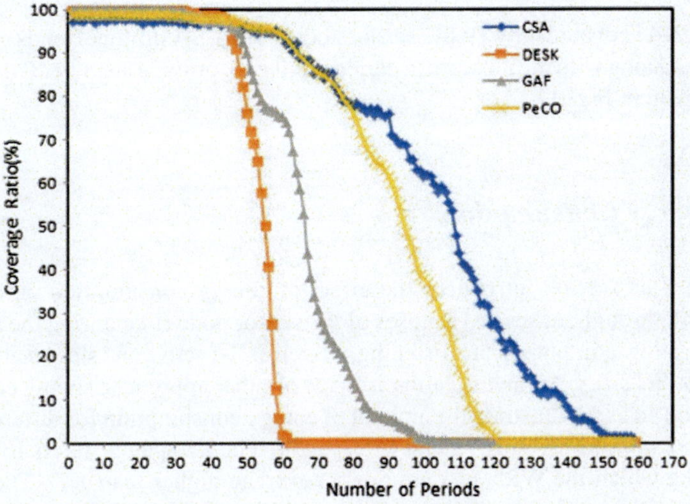

Fig. 2 Coverage ratio for WSN size of 200 deployment nodes

4.2 Active Sensors Ratio

Decreasing the number of active nodes during each round is important for conserving more power, hereby maximizing the WSN lifespan. Figure 3 illustrates the average ratio of active nodes for every 200 deployed ones. During the first fifteen periods, DESK, GAF, and PeCO activated 30.68%, 34.5%, and 20.18% nodes respectively,

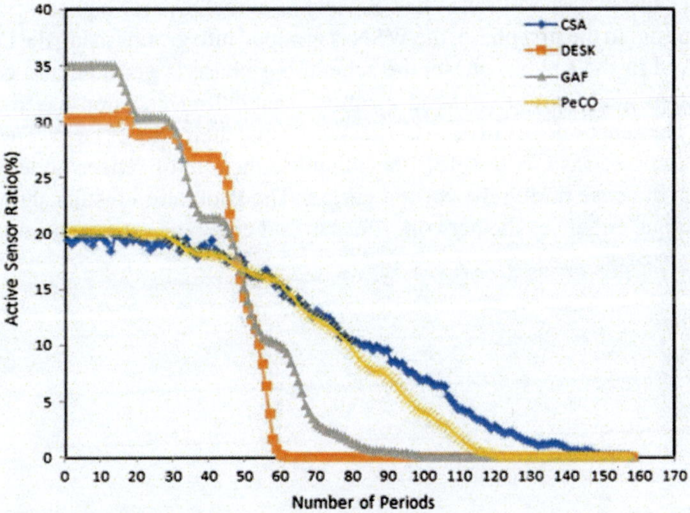

Fig. 3 Active sensors ratio for WSN's size of 200 deployed nodes

whereas CSA activated only 19.8% sensor nodes. The CSA protocol tends to activate more nodes along with the increase in period number to provide an increased coverage rate as shown in Fig. 3.

4.3 Energy Consumption

The current subsection introduces the effect of energy consumption on behalf of the network throughout several statuses of the sensor node (like during the modes of communicating, computing, and listening, as well as the active and sleep statuses), for different WSN sizes. An investigation is made of other approaches being compared. Each of Fig. 4a and b illustrate the amount of energy consumption for different WSN sizes, in addition to Lifespan95 and Lifespan50. The lifespan X refers to the total time during which the WSN can provide a coverage higher than X%. The relative superiority of CSA in terms of economizing could be concluded from the illustration. Both Figures indicate the reduction in the amount of energy consumed by CSA in comparison with other methods. The rate of energy consumption hits relatively lower for Lifespan95 and Lifespan50.

5 Conclusion

Network lifespan optimization is one of the important factors in designing WSNs. This article proposed a Cuckoo Scheduling Algorithm (CSA) for ımproving the lifespan of Cluster-based WSNs. The CSA technique achieves two phases: clustering and scheduling. In the first phase, the WSN is grouped into groups using the DBSCAN method. And in the second phase, the scheduling phase is periodic and composed of three steps such as cluster head election, scheduling decision-based CA, and covering. The sensor nodes in each cluster determine their cluster head. The selected cluster head performs CA to select the suitable schedule of sensor nodes that take the mission of sense during the current period. The simulation results show that the CSA technique enhances the network lifespan and coverage ratio, and improves the lifespan of WSNs.

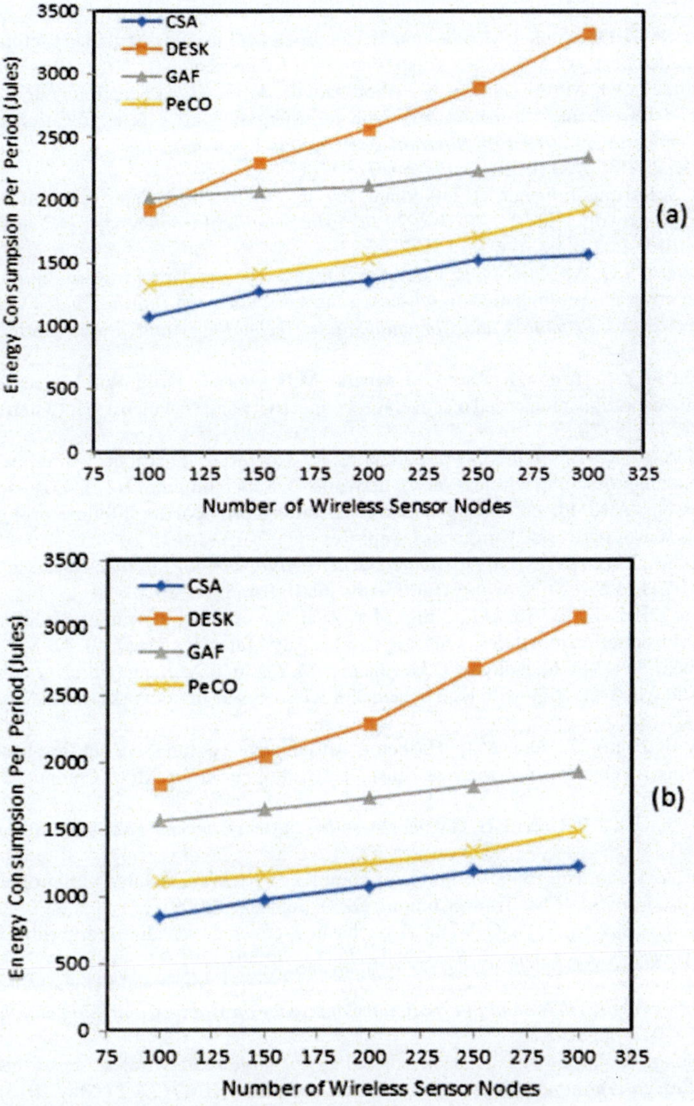

Fig. 4 Energy consumption per round for **a** Lifetime95 and **b** Lifetime50

References

1. A.K. Idrees, A.K.M. Al-Qurabat, Energy-efficient data transmission and aggregation protocol in periodic sensor networks based fog computing. J. Netw. Syst. Manage. **29**(1), 1–24 (2020)
2. A.K. Idrees, R. Alhussaini, M.A. Salman, Energy-efficient two-layer data transmission reduction protocol in periodic sensor networks of IoTs. Personal and Ubiquitous Comput. (2020)

3. A.K. Idrees, K. Deschinkel, M. Salomon, R. Couturier, Multiround distributed lifetime coverage optimization protocol in wireless sensor networks. J. Supercomput. **74**(5), 1949–1972 (2018)
4. A.K. Idrees, A.K.M. Al-Qurabat, C. Abou Jaoude, W.L. Al-Yaseen, Integrated divide and conquer with enhanced k-means technique for energy-saving data aggregation in wireless sensor networks, in *2019 15th International Wireless Communications & Mobile Computing Conference (IWCMC)* (IEEE, 2019), pp. 973–978
5. J. Guo, S. Karimi-Bidhendi, H. Jafarkhani, Energy-efficient node deployment in wireless ad-hoc sensor networks, in *ICC 2020–2020 IEEE International Conference on Communications (ICC)* (IEEE, 2020), pp. 1–6
6. A.K. Idrees, S.O. Al-Mamory, R. Couturier, Energy-efficient particle swarm optimization for lifetime coverage prolongation in wireless sensor networks, in *International Conference on New Trends in Information and Communications Technology Applications* (Springer, Cham, 2020), pp. 200–218
7. S.E. Bouzid, Y. Serrestou, K. Raoof, M. Mbarki, M.N. Omri, C. Dridi, Wireless sensor network deployment optimisation based on coverage, connectivity and cost metrics. Int J Sens Networks **33**(4), 224–238 (2020)
8. A.K. Idrees, K. Deschinkel, M. Salomon, R. Couturier, Distributed lifetime coverage optimization protocol in wireless sensor networks. J. Supercomput. **71**(12), 4578–4593 (2015)
9. W. Hussein, A.K. Idrees, Sensor activity scheduling protocol for lifetime prolongation in wireless sensor networks. Kurdistan J Appl Res **2**(3), 7–13 (2017)
10. A.K. Idrees, W.L. Al-Yaseen, *Distributed Genetic Algorithm for Lifetime Coverage Optimization in Wireless Sensor Networks* (Int. J. Adv. Intell, Paradigm, 2020).
11. S.G. Jia, L.P. Lu, L.D. Su, G.L. Xing, M.Y. Zhai, An efficient sleeping scheduling for save energy consumption in wireless sensor networks. Adv Mater Res **756**(759), 2288–2293 (2013)
12. C.P. Chen, S.C. Mukhopadhyay, C.L. Chuang, M.Y. Liu, J.A. Jiang, Efficient coverage and connectivity preservation with load balance for wireless sensor networks. IEEE Sens. J. **15**(1), 48–62 (2015)
13. J. Hao, B. Zhang, Z. Jiao, M.R. Hashemi, An adaptive compressive sensing-based sample scheduling mechanism for wireless sensor networks. Pervasive Mob. Comput. **2**, 113–125 (2015)
14. A. Silberschatz, P.B. Galvin, G. Gagne, *Operating System Concepts* (Addison-Wesley, Boston, 1998).
15. X. Yuan, Z. Duan, Fair round-robin: a low complexity packet scheduler with proportional and worst-case fairness. IEEE Trans. Comput. **58**(3), 365–379 (2009)
16. J. Yu, S. Ren, S. Wan, D. Yu, G. Wang, A stochastic k-coverage scheduling algorithm in wireless sensor networks. Int. J. Distrib. Sens. Netw. **8**(11), 746501 (2012)
17. Y. Xu, J. Heidemann, D. Estrin, Geography-informed energy conservation for ad hoc routing, in *Proceedings of the 7th Annual International Conference on Mobile Computing and Networking.* (ACM, 2001), pp. 70–84
18. Z. Yaxiong, W. Jie, Stochastic sleep scheduling for large scale wireless sensor networks, in *2010 IEEE International Conference on Communications (ICC)* (23–27 May 2010). pp. 1–5.
19. H. Chih-fan, L. Mingyan, Network coverage using low duty-cycled sensors: random & coordinated sleep algorithms, in *IPSN 2004. Third International Symposium on Information Processing in Sensor Networks* (26–27 April 2004). pp. 433–442
20. S. Nath, P.B. Gibbons, Communicating via fireflies: geographic routing on duty-cycled sensors, in *IPSN 2007. Sixth International Symposium on Information Processing in Sensor Networks* (25–27 April 2007). pp. 440–449
21. Y. Zhuxiu , W. Lei, S. Lei, T. Hara, Q. Zhenquan, A balanced energy consumption sleep scheduling algorithm in wireless sensor networks, in *Seventh International Conference on Wireless Communications and Mobile Computing Conference (IWCMC)* (4–8 July 2011) pp. 831–835
22. H. Chih-Fan, L. Mingyan, Network coverage using low duty- cycled sensors: Random and coordinated sleep algorithms, in *IPSN 3rd International Symposium on Information Processing in Sensor Networks* (26–27 Apr 2004) pp. 433–442.

23. Z. Yaxiong, W. Jie, Stochastic sleep scheduling for large scale wireless sensor networks, in *2010 IEEE International Conference on Communications (ICC)* (23–27 May 2010), pp. 1–5
24. L.J. Sun, J. Wei, J. Guo et al., Node scheduling algorithm for heterogeneous wireless sensor networks. Acta Electron Sinica **42**(10), 1907–1912 (2014)
25. Ahmed et al., Sleep-awake energy efficient distributed clustering algorithm for wireless sensor networks. Comput. Electr. Eng. 1–14 (2015)
26. A.K. Idrees, K. Deschinkel, M.R. Salomon, Couturier, Distributed lifetime coverage optimization protocol in wireless sensor networks. J Supercomput **71**, 4578–4593 (2015)
27. A.K. Idrees, K. Deschinkel, M. Salomon, R. Couturier, Perimeter-based coverage optimization to improve lifetime in wireless sensor networks. Eng. Optim. **48**(11), 1951–1972 (2016)
28. M.K. Hameed, A.K. Idrees, Distributed clustering-based DBSCAN protocol for energy saving in IoT networks, in *2nd International Conference on Communication, Computing and Electronics Systems (ICCCES 2020) on 21–22, October 2020, Lecture Notes in Electrical Engineering Series* (Springer, 2020), ISSN: 1876–1100
29. J.E. Sander, K. Martin, X. Hans-Peter, Xiaowei, Density-based clustering in spatial databases: the algorithm GDBSCAN and its applications. Data Mining Knowl. Disc **2**(2), 169–194. (Springer-Verlag, Berlin)

Design and Development of an Automated Snack Maker with CNN-Based Quality Monitoring

Akhil Antony, Joseph Antony, Ephron Martin, Teresa Benny, V. Vimal Kumar, and S. Priya

Abstract This article discusses the automation of the cooking process of a deep-fried food item called Unniyappam. It involves the design of a machine capable of mass production along with the provision for monitoring the quality of the cooked product. The working of the machine involves pouring a fixed volume of the prepared batter into a mold immersed in boiling oil. The mold remains immersed in the boiling oil for a preset time, and the fried food products are automatically removed from the mold. The quality monitoring discussed here refers to a deep neural-network-based computer vision system to check the fried products for partially, optimally, or over fried conditions. The computer vision system uses a GoogLeNet, with its last fully connected layer and subsequent soft-max layer and classification layer modified to classify input images into three classes. The preset time of frying is changed to optimal value the output of the computer vision system.

Keywords Automated cooking · Temperature control · Quality monitoring · Deep CNN · Machine learning · GoogLeNet · MATLAB

1 Introduction

Kerala cuisine offers a wide variety of dishes from three-course meals to tea-time snacks. Unniyappam is the traditional snack of South Indians. Unniyappam is a small round snack made from rice, jaggery, banana, roasted coconut pieces, roasted sesame seeds, and cardamom powder fried in oil. In many religious places like Guruvayur temple, Unniyappam is produced in bulk quantities for the congregation. It also

---

A. Antony (✉) · J. Antony · E. Martin · T. Benny · V. Vimal Kumar · S. Priya
Applied Electronics and Instrumentation, Rajagiri School of Engineering and Technology, Ernakulam, India

V. Vimal Kumar
e-mail: vimalk@rajagiritech.edu.in

S. Priya
e-mail: priyas@rajagiritech.edu.in

V. Suma et al. (eds.), *Inventive Systems and Control*,
Lecture Notes in Networks and Systems 204,
https://doi.org/10.1007/978-981-16-1395-1_15

involves a lot of manpower and is very time consuming. As technology develops day after day, food processing industry is getting automated. As a result of the past research and development currently, the automated appam [1] and dosa maker [2] became a huge success in the mass catering and hotel industry. The main objective of the project is to automate the time consuming and tedious processes in the food industry. Following this trend attempting to focus on an unexplored area of automating, the Unniyappam making process has a huge potential in south Indian food culture. To design and fabricate, an automatic Unniyappam maker is planned for mass catering and households by considering cost, usability, safety, easy handling, and hygiene. It allows the user to obtain ready to eat Unniyappam at the press of a button without any skilled labor.

Quality checking of the cooked Unniyappams has a major role in maintaining grades and standards. Adding to that, the cooking time of the Unniyappams can be varied after monitoring as there will be changes in the fermentation amount of batter, hence the cooking time. The manual checking of the Unniyappams requires time and skilled labor. By proposing a model of real-time, quality monitoring using image processing techniques can dramatically increase the quality and grade of the snack.

Deep learning is a subcategory of machine learning in artificial intelligence that has networks capable of learning from the given data. The network is pre-trained to learn from the raw data prescribed to them [3, 4]. The data is collected from the proposed machine and categorized as overcooked, perfectly cooked, and partially cooked Unniyappams. The data collected in this manner is used to train the last layers of a pre-trained Googlenet to create a deep neural network (DNN) suitable to implement classification for any domain of interest. By this process, the single batch of Unniyappams (4 Unniyappam in each batch in our model) is monitored just before it is deposited in the collecting chamber.

2 Literature Review

Food industries often require mass production, so the constituting parts should be selected to provide the best possible performance. Food manufacturing applications may use permanent magnet DC motors [1], as they provide low torque ripple and have low heat generation. In the paper of appam maker [1], proximity sensors are used for limiting the movements wherever necessary. Since the food processing machines, high temperatures are given out; this can alter the readings of the sensor. In this project, limit switches are used instead of proximity sensors which are not affected by the altering temperature.

As the article discussing dosa maker [2] comes up with different designs to improve the overall performance of a food production machine. The removal mechanism is carried out using a DC motor which is much simpler, though it adds up the power consumption. Hence in the Unniyappam machine, the removal mechanism is made much better without added power consumption. In the article on automatic food making machine [5], a special motor driver known as L298N is used for interfacing the motor with the controller. When compared to other motor drivers, the L298N motor driver provides highly precise and easy for controlling [6]. A relay module is used to control the various temperatures of an induction cooktop [5]. In this project, the temperature is kept constant by a temperature checking loop, with the aid of a thermocouple and relay [7].

3 Project Design

3.1 Hardware Module

3.1.1 Batter Reservoir

The batter reservoir is where the batter for Unniyappam is stored, and it is made airtight. The flow of the batter is regulated with the help of a metal ball valve of size 1 inch. The valves are made up of metal due to the large amount of heat produced from the boiling oil. Ball valve further splits into four sub-outlets which make sure that an equal amount of batter is supplied to the die. The smooth flow of batter through the valves is maintained by the sufficient pressure head provided by the compressed air supplied to the reservoir. There is an actuator that helps in the opening and closing of the ball valves. The period of the ball valve kept open is transferred into the controller (Fig. 1).

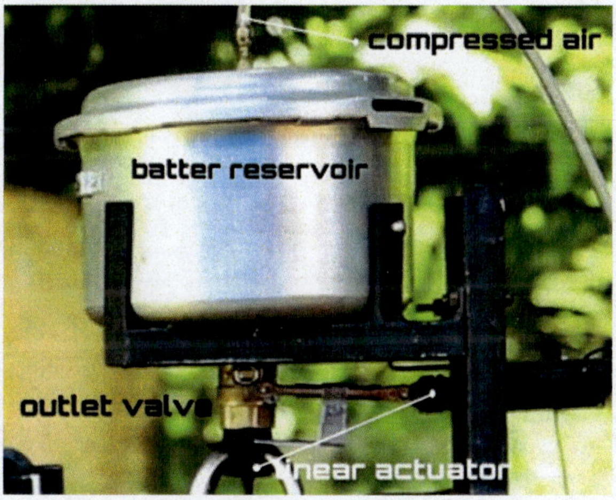

Fig. 1 Proposed die setup

3.1.2 Die Setup

The die setup is where the Unniyappam is cooked in its specific shape. A square-shaped die is used here rather than a normal round one for the easiness of manufacturing and running of the machine. Brass is used to making the die as it an ideal metal used by the utensil makers. The die is fixed into a rectangular frame such that it is free to rotate on its axis. This feature is incorporated in the machine for the sole purpose of removing the cooked Unniyappam from the die. A sprocket is coupled to the die which when intercepts the chain it is rotated, and in turn, the die is also rotated. The rectangular frame which holds the die is further suspended from a trolley that runs on a track. The horizontal motion of the trolley is achieved with the help of a PMDC motor. This PMDC motor is mounted on the trolley and coupled to the wheels of the trolley with the help of a chain (Fig. 2 and 3).

Fig. 2 Complete die setup

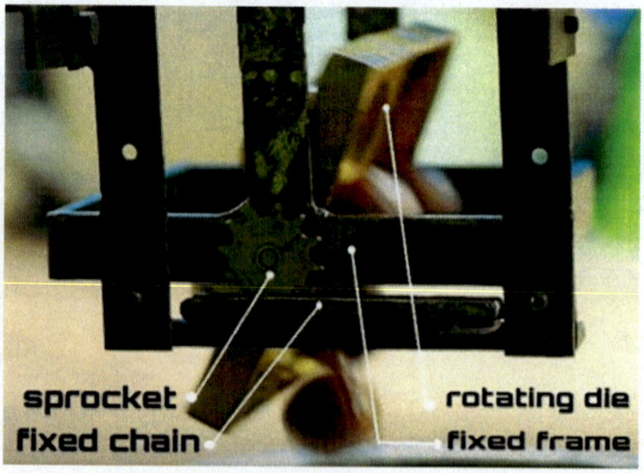

Fig. 3 Die rotating in the frame with the action of sprocket and chain

Fig. 4 Oil reservoir, jack, and motor setup

3.1.3 Oil Container

The oil reservoir is where the required amount of oil for cooking the Unniyappam is being heated, and it has a capacity of 6 L in total with 8*9*6 inches of dimensions. It is made up of stainless steel, as it is light in weight and can withstand the high temperature produced by the boiling oil due to its higher melting point property. An electrical heating coil is used to heat the oil, and the temperature of the oil is maintained at 300 °C. The microcontroller cuts off the supply using a relay-based input received from the temperature sensor. A simple jack used to lift cars is used here to move the reservoir in the upward and downward direction. The movement of the reservoir is controlled by a PMDC motor which is coupled to the car jackscrew. The rotation of the shaft present in the PMDC motor makes the jack wind and unwinds which leads to the upward and downward direction movement as needed. To start cooking the Unniyappam, initially, the batter stored in the die is immersed in the boiling oil for almost 5 min. This is done by moving the oil reservoir in an upward direction until the die is fully submerged in the boiling oil. Next, when the Unniyappam is cooked completely, the oil is removed from the die by moving the oil reservoir in the downward direction (Fig. 4).

3.1.4 Controlling Module

Arduino UNO open-source microcontroller board based on microchip AT mega328p is the heart of the automated food processor [5]. The machine is controlled by the program fed into the board. Figure 6 shows the flowchart of the process.

The microcontroller controls the following:

1. Movement of the permanent magnet DC motor.
2. Driving the valve opening actuator and the solenoid valve.

3. Cooking time and temperature.

All the movements are performed using a permanent magnet DC motor arrangement. It can control the heavy die set up along with the oil container which needs special care, as the DC motor runs at 60 rpm with higher torque. Using a circuit based on 2 relays, rotation of the motor in clockwise and anticlockwise motion could be performed, respectively. Due to this, the horizontal motion of the die setup and vertical motion of the oil container could be controlled accordingly.

The uses of five-limit switches with higher response time operated by the motion of a machine part or the presence of an object. They are mechanical relay type switch which contains heavy-duty contacts capable of switching high current than normal proximity sensor. Two pairs of switches (L1 & L5) are placed at both extreme ends of the horizontal and vertical racks, such that it limits the motion of the motor exceeding the boundaries. The fourth limit switch (L4) is placed in the pathway of the die movement, which is used to stop the die for 2 s. This switch is placed in such a way to align the die under the IP camera for quality monitoring. The output from the limit switch helps the program to control the DC motor when and where required. For the constant flow of batter, a fixed pressure is provided at the inlet of the solenoid valve using a compressor. For a particular period, the valve actuator and the solenoid valve outlet is opened simultaneously using a motor driver. Due to the applied external pressure, only a particular amount of batter falls onto the die, thus maintains a similar shape for all the appams.

The valve actuator along with the solenoid valve does not require high input power. Instead of a relay circuit, a normal motor driver could be used instead. L298N is a dual-channel h bridge motor driver that works on 12v dc, capable of driving a pair of DC motors [8]. By directly connecting the motor driver to the Arduino, the batter flow can be regulated based on the program fed to it. The change in pressure and time delay could result in the uneven quantity of the appam. Based on the experimental results, when a small external pressure of 1.2 psi is applied, the valve actuator along with the solenoid valve needs to be open only for 2 s to achieve perfectly shaped appams considering the die hole area.

Controlling oil temperature plays an important role in perfectly cooking the snack. As the oil is boiled at an approx. 300 °C, a thermocouple that has the capability of withstanding this condition is used to measure temperature. The max 6675 performs cold junction compensation and digitalize the signal from a K-type thermocouple [6] which could be submerged. Max6675 is a thermocouple–digital module with an inbuilt analog to digital convertor. The K-type thermocouple's hot junction can be read from 0 °C to + 1023.75 °C, so the hot boiling oil could be sensed. Using a temperature sensing diode, it converts the ambient temperature reading into a voltage. This voltage value is fed to the Arduino, and thus comparing the limits set by the program, the input to the heating coil is controlled. Thus, by turning the heater coil ON and OFF using a relay-based switch with required delay, the temperature is compensated. Through in the experimental method [9], the response time of the thermocouple used was found to be 2.2 s. The average time taken by the module to detect a 1 °C rise in temperature is 4.4 s. The resolution of the module is 0.25 °C. So, according to the program, the temperature can vary between 299.75 °C and 300 °C, which keeps up the optimum temperature for best cooking.

Still, after controlling most of the parameters, the perfect cooking period for the snack may vary due to many unpredictable external factors. So, by monitoring the results of the image recognition techniques, the cooking period could be predicted based on the feed forwarded data. Concerning the data collected, the period could be manually adjusted using a potentiometer which is calibrated in such a way to provide a delay for cooking (Figs. 5 and 6; Table 1).

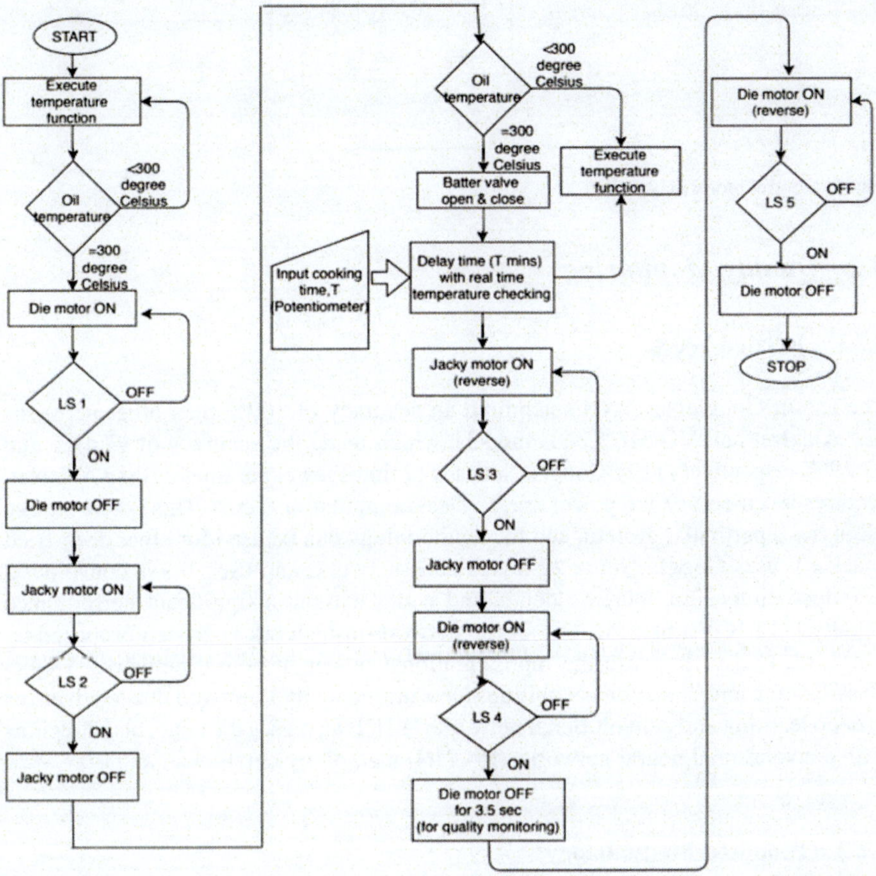

Fig. 5 Flowchart

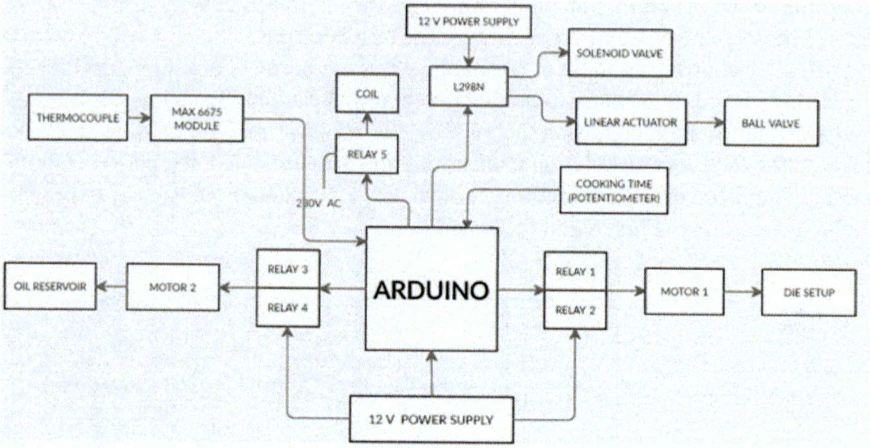

Fig. 6 Control block diagram

3.2 *Quality Monitoring*

3.2.1 Related Work

The use of GoogleNet CNN's acquired an accuracy of 100% than other networks like AlexNet and VGG16 CNNs models, which made the accuracy of 97.49% and 97.29%, respectively [10]. GoogLeNet uses 12 times fewer parameters than AlexNet, requires less memory and power use, and less computational cost. This article though discusses a particular system, and the methodology can be used for other deep-fried snacks. It uses GoogLeNet, a 22 layers deep network, and uses 1×1 convolution to reduce dimension, increase depth, and width without a significant performance penalty [11]. It is a modification of the network-in-network approach proposed by Lin et al. [12] to increase the representational power of neural networks. Object classification and detection capabilities have dramatically improved due to advances in deep learning and convolutional networks [13]. It also uses the concepts of regions with convolutional neural networks (R-CNN) method by Girshick et al. [14].

3.2.2 Proposed Methodology

The GoogleNet is used for the Unniyappam's quality monitoring process. The MATLAB program uses a DNN/CNN to classify objects. Transfer learning was carried out on Googlenet to obtain the required classification for three categories of cooked Unniyappams, namely perfectly cooked, overcooked, and not cooked. The pc used for the processing is an Asus laptop with Intel corei5 8th gen, with NVIDIA GTX 1050 4 GB graphics. The project work is carried out in different steps.

Table 1 Specifications

S. No	Item	Specifications	No's
1	PMW DC motor	Rated torque: 53 in-lb Stall torque: 177 in-lb High speed: 50 rpm, 1.5 A(12 V DC) Low speed: 35 rpm, 1 A(12 V DC) Maximum wattage: 50 W	
2	Linear actuator	Motor input: 12 V DC Current consumption: 0.15–2.22 A Maximum pull range: 22 mm	
3	Solenoid valve	Voltage: 12 V DC Orifice: 2.5 mm Working pressure: 0–120 PSI	
4	Heating coil	Voltage: 220 V AC Maximum wattage: 3000 W	1
5	Microcontroller	Arduino UNO	1
6	Ball valve	1.25 inch	1
7	Car jack	3.5–13.8 inch, 1.5 tons	1
8	Relay module	1 channel 5 V 10 A relay control module	2
9	L298N	Input voltage: 3.2–40 V DC Power supply: 5–35 V DC Peak current: 2 A Maximum power consumption: 20 W	1
10	IRF520	Output load voltage: 0–24 V DC Input voltage: 3.3–5 V Maximum load current: less than 5 A	1
11	MAX 6675 temperature module	Operating voltage: 3–5.5 V DC Temperature resolution: 0.25 C	1
12	Thermocouple	Type K, Temperature range 0–1024 °C	1
13	Relay	Voltage: 12 V Current rating: 40 A	1

Steps in Monitoring the Cooked Unniyappams

I. *Collecting Images Using GUI*

The sample images are collected from the proposed Unniyappam maker by manually checking the quality of the snack and categorizing them into the three grades. The network was trained on 300 images collected by using the special GUI shown in Fig. 7. The GUI is made using MATLAB APP DESIGNER. The collected images were separated into three categories, namely not cooked, perfectly cooked, and overcooked with each category with 100 images.

Fig. 7 GoogLeNet architecture with the part used for retraining labeled

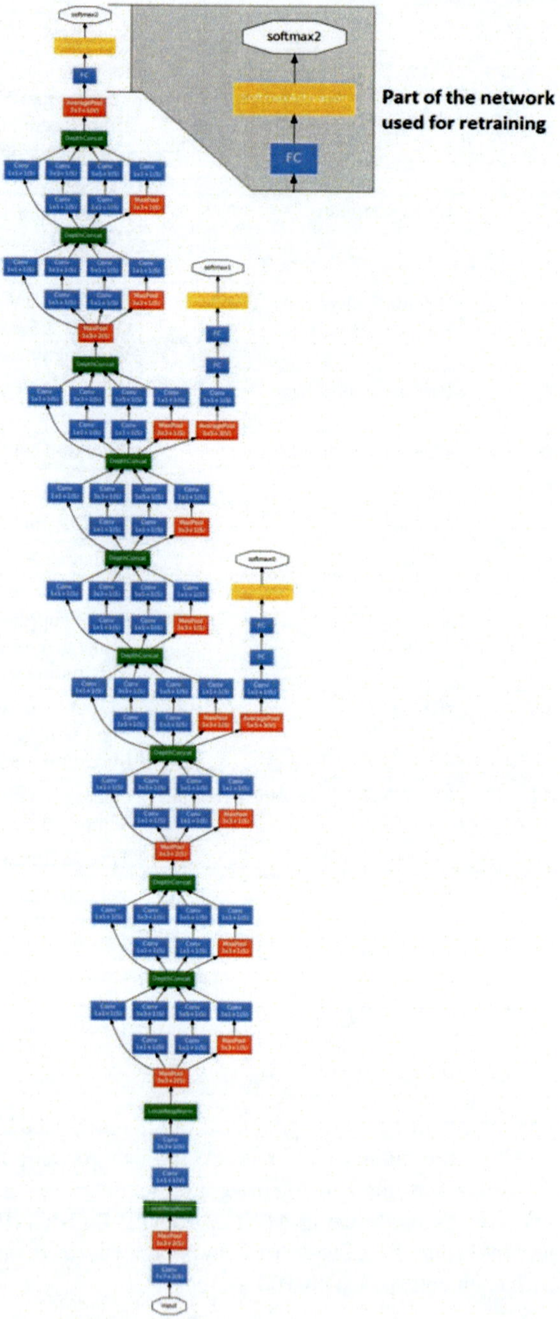

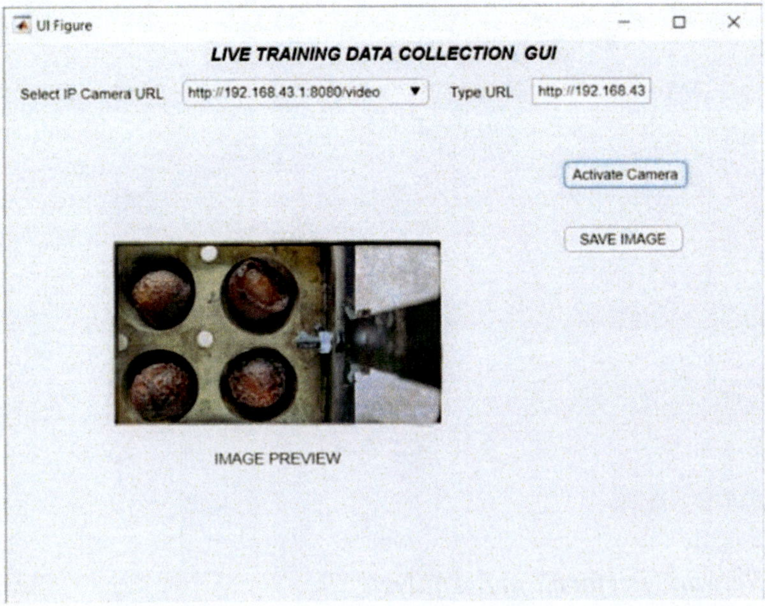

Fig. 8 GUI for collecting data

II. *Deep Neural Network for Transfer learning (GoogLeNet)*

The architecture of *GoogLeNet* [11] was modified at the final fully connected layer to change the number of classifications from 1000 to 3, and this fully connected layer of network was trained on the new 300 images collected to train the network for classification of the cooked food product. Figure 8 shows the architecture of *GoogLeNet* with its part used for retraining highlighted.

4 Network Training

The modified Googlenet was trained on 300 images, 80% of the images were used for training, and the rest used for validation. The images provided as input to the network should be 224 × 224 × 3 in size. However, the images collected may vary in size. Image resizing was carried out by data augmentation. Additional augmentation like flipping and scaling is carried out, on the images collected for the training network.

A stochastic gradient descent with momentum (SGDM) optimizer was used to update the weights of the layers from the last fully connected layer during training. The graph in Fig. 9 shows the plot of the accuracy of the network on both training and validation data. It was observed that both training and validation accuracies were around 96% which indicated good network performance without overfitting.

.

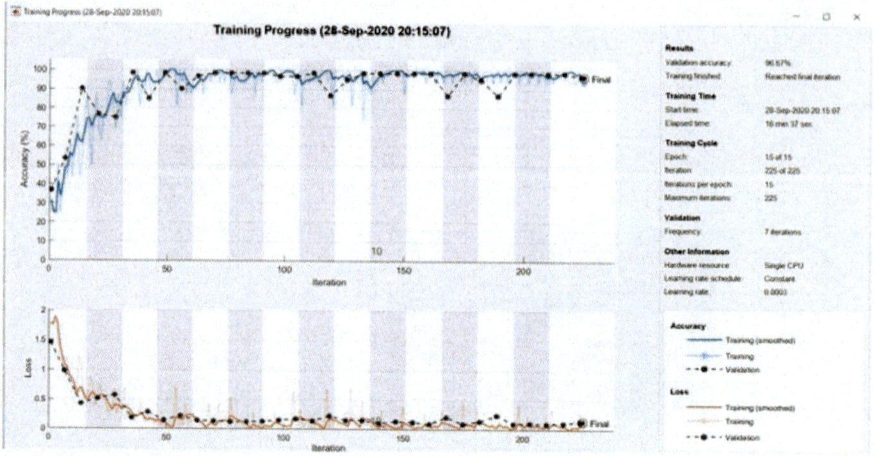

Fig. 9 Training progress

4.1 Network Performance Analysis

The training was performed with 80% of data for training and 20% for validation. The performance of the DNN was analyzed using the confusion matrices. The confusion matrix of output class and target class category of training data and validation data is classified using GoogleNet CNN, shown in Figs. 10 and 11, respectively.

80 images from each category were classified by using the confusion matrix for training data. In class 1(not cooked), the diagonal element shows 80 which specifies that each image in this category was correctly classified. For class 2(overcooked), 78 out of 80 were correctly classified-two overcooked Unniyappam were misclassified as perfectly cooked. In class 3(perfectly cooked), one was misclassified as not cooked item. In the same manner, 20 images from each category were classified by the validation data confusion matrix. For class 1 and class 3, one from each set was misclassified. For class 2, all of the sets were perfectly classified.

The analysis of confusion matrixes on training and validation data proved similar performance on the training and validation data confirming that the network was not overfitting.

The network was reasonably accurate with the worst case of misclassification of 5%. (One among the 20 items in a class was misclassified.)

4.2 GUI for Quality Checking

An image testing program is also developed using MATLAB for capturing and monitoring real-time images. It provides a display platform for the operator to monitor the cooking grade. An IP Camera positioned at a suitable location streams the video

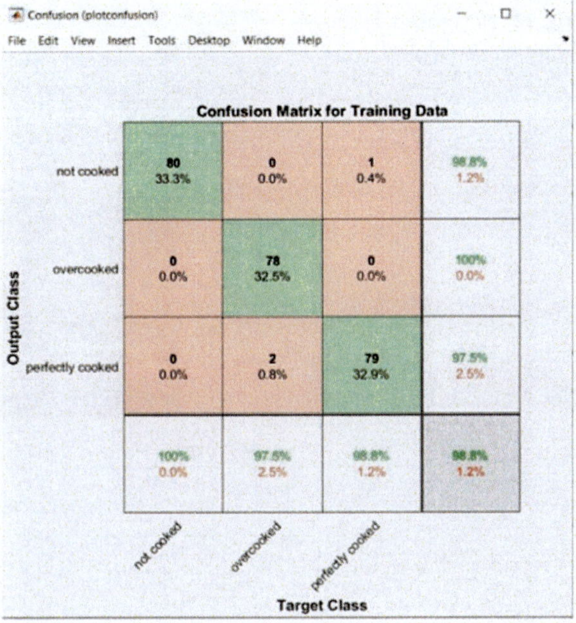

Fig. 10 Training data confusion matrix

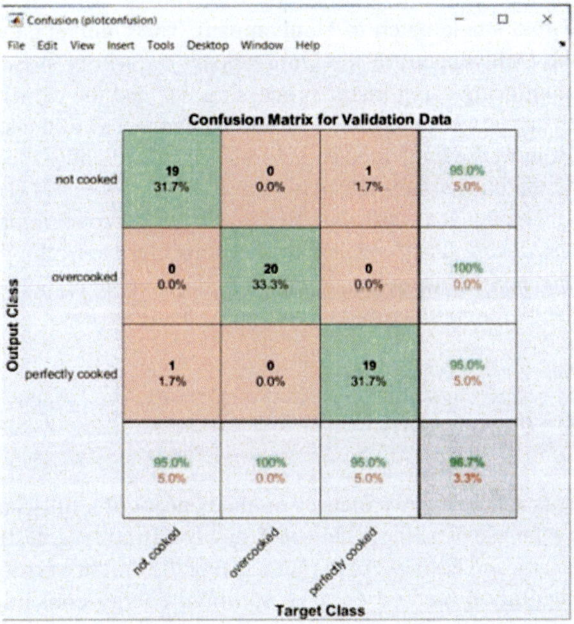

Fig. 11 Validation data confusion matrix

Fig. 12 Not cooked Unniyappam detected

with a single batch of Unniyappam. The Unniyappam dye setup before unloading the Unniyappam to the Unniyappam repository stops for a second for the quality monitoring using image processing. PC and the camera (an Android phone with an IP camera installed in it) should be connected to the same network. Figures 12 and 13 show the live video streaming of the different grades of the Unniyappam detected by the image classifier platform.

The images were classified within five seconds. Implementing the response from classification as feedback to the controller controlling the cooking time shall entail this delay. During the experiment, the average response time of the CNN to classify the captured image was found to be 3 s (Fig. 14).

5 Conclusion

This article mainly focuses on the concept of a fully automated Unniyappam maker by the use of a single button. A newly customized design is used to increase the efficiency and easiness of manufacturing the unit in a cost-effective method. The unique integration method ensures optimum energy consumption, safety, easy handling, and hygiene throughout the whole process. The concept has successfully proven that with a completely new idea and innovative design the automated Unniyappam maker can make a bulk number of Unniyappams in a short period. In the future, the Unniyappam maker can be modified with features like an oil level sensor, a different

Fig. 13 Overcooked Unniyappam detected

Fig. 14 Proposed Unniyappam maker

die structure for a variety of snacks, and an expanding module to increase the count of Unniyappam produced.

A GoogLeNet CNN was modified to change the number of classifications, and only a small part of it was retrained on a small set of fresh images to apply the capability of pre-trained Googlenet for the application of food quality inspection. The pre-trained network proved reasonably accurate in detecting if a food product is uncooked, partially cooked, or overcooked. The response of the CNN was suggested to be feedback to the controller to optimize the cooking time. Future works may include improvement on CNN performance and modifying CNN structure to reduce response time.

References

1. Design and development of automated appam maker. IJIRST—Int. J. Innovat. Res. Sci. Technol. **3**(11), (2017) ISSN (online): 2349-6010. https://202.88.229.59:8080/xmlui/handle/123456789/815
2. Design and fabrication of automatic dosa maker. Int. J. Eng. Adv. Technol. (IJEAT) **9**(5), (2020). ISSN: 2249-8958. Retrieval Number: E1151069520/2020©BEIESP https://doi.org/10.35940/ijeat.E1151.069520
3. H. Al Hiary, S. Bani Ahmad, M. Reyalat, M. Braik, Z. ALRahamneh, Fast and accurate detection and classification of plant diseases. Int. J. Comput. Appl. **17**, 31–38 (2011). https://doi.org/10.5120/ijca
4. U. Mokhtar, A.E. Hassenian, E. Emary, M.A. Mahmoud, SVM-Based detection of tomato leaves diseases, in *Advances in Intelligent System and Computing* (Springer, 2015), pp. 641–652
5. M.J. Rakshitha, G. Madan, K.R. Prakash, C.S. Shivaraj, Design and development automated food maker. Int. Res. J. Eng. Technol. (IRJET) **06**(06) (2019). E-ISSN: 2395-0056. P-ISSN: 2395-0072
6. L. Yin, F. Wang, S. Han, Y. Li, H. Sun, Q. Lu, C. Yang, Q. Wang, Application of drive circuit based on L298N in direct current motor speed control system, in *Proceeding SPIE 10153, Advanced Laser Manufacturing Technology*, 101530N (19 October 2016). https://doi.org/10.1117/12.2246555
7. Dalian Jiaotong University, Dalian Jiaotong University, Dalian 116028, Liaoning, Dalian 116028, Liaoning; Design of temperature controller based on MAX6675"
8. Y.A. Badamasi, The working principle of an Arduino, *in 2014 11th International Conference on Electronics, Computer and Computation (ICECCO)* (2014). https://doi.org/10.1109/icecco.2014.6997578
9. Determination of the response time of thermocouples to be used for the measurement of air or gas phase temperature in reaction to fire testing. European Group of *Organisations* for Fire Testing, SM1:1995, 2008. (Revision of EGOLF SM1:1995).
10. A.K. Rangarajan, R. Purushothaman, A. Ramesh, Tomato crop disease classification using pre-trained deep learning algorithm. Procedia Comput. Sci. **133**, 1040–1047 (2018)
11. C. Szegedy, W. Liu, Y. Jia, P. Sermanet, S. Reed, D. Anguelov, D. Erhan, V. Vanhoucke A. Rabinovich "Going deeper with convolutions." IEEE Conference on Computer Vision and Pattern Recognition (CVPR) (2015)
12. M. Lin, Q. Chen, S. Yan. Network in network. CoRR, abs/1312.4400 (2013)
13. Y. LeCun, B. Boser, J.S. Denker, D. Henderson, R.E. Howard, W. Hubbard, L.D. Jackel, Backpropagation applied to handwritten zip code recognition. Neural Comput. **1**(4), 541–551 (1989)
14. R.B. Girshick, J. Donahue, T. Darrell, J. Malik, Rich feature hierarchies for accurate object detection and semantic segmentation, in IEEE *Conference on Computer Vision and Pattern Recognition, CVPR 2014* (2014)

An Effective Deep Learning-Based Variational Autoencoder for Zero-Day Attack Detection Model

S. Priya and R. Annie Uthra

Abstract In recent times, machine learning (ML) and deep learning (DL) models are commonly employed to design effective intrusion detection systems (IDS). A significant increase in the number of recent unknown cyberattacks necessitates respective improvement in the performance of the IDS solution for the identification of zero-day attacks. Consequently, it is needed to develop effective IDS for the detection and classification of zero-day attacks. In this view, this article presents a novel DL-based variational autoencoder (VAE) model for zero-day attack detection. The goal of this study is to design a new IDS model with a maximum detection rate with minimal false-negative rate. The DL-VAE model involves pre-processing to convert the raw data into a compatible format. Then, the preprocessed data is fed into the VAE model to detect the existence of zero-day attacks in the networking data. To validate the effective performance of the DL-VAE model, a series of experiments were conducted, and the results are determined under several aspects. The obtained simulation values ensured the betterment of the DL-VAE model with the sensitivity of 0.985, the specificity of 0.977, accuracy of 0.989, F-score of 0.982, and kappa of 0.973.

Keywords Deep learning · Zero-day attack · Intrusion detection · Machine learning · Variational autoencoder

1 Introduction

Recently, a zero-day attack prediction system is a significant study in intrusion detection systems (IDS) and cybersecurity, managing the progressive increase in cyberattacks [1]. To develop an effective IDS system, machine learning (ML) method has

S. Priya · R. Annie Uthra (✉)
Department of CSE, College of Engineering and Technology, SRM Institute of Science and Technology, Kattankulathur 603203, India

S. Priya
e-mail: priyas3@srmist.edu.in

V. Suma et al. (eds.), *Inventive Systems and Control*,
Lecture Notes in Networks and Systems 204,
https://doi.org/10.1007/978-981-16-1395-1_16

been applied widely due to its massive benefits and applications. Even though the recently developed IDS are applicable in gaining maximum prediction accuracy for well-known attacks, it still fails in predicting novel and zero-day attacks. This is because of the limitations involved in present IDS which depend upon the previous patterns as well as signatures. Besides, it experiences maximum false-positive rates (FPR) by reducing the working function and real-time application of IDS in actual scenarios. Finally, the zero-day attacks remain undetected by enhancing the serious issues. Based on the study of [2], a zero-day attack is meant to be "a traffic pattern of interest where no matching patterns are presented in malware prediction in a system." The representations of zero-day attacks are defined by Smys et al. [3]. The central premises of this study are to learn the effect of zero-day attacks with better significance. Several studies have pointed out that zero-day attacks have been a prominent one which implies that out of already predicted attacks, some of them were unknown zero-day attacks. Moreover, the findings have represented that a zero-day attack can be identified within a limited time interval (around 10 months). Also, the number of zero-day attacks has been increased periodically in the last decades [4].

From the above-mentioned statement, it is apparent that an effective zero-day attack prediction method has to be developed. The zero-day attacks can be predicted by forecasting the noises like instances or occurrences which differs from the benign traffic. But, the major limitations of the outlier-examination depend upon the minimum accuracy values where the maximum FPR (that spends a valid time of cybersecurity) as well as false-negative rates (FNR) (the reduced system performance). An approach is proposed to identify zero-day attacks on the Internet of things (IoT) network [5]. It relies on a distributed diagnosing system for zero-day prediction. A Bayesian probabilistic method is developed for detecting zero-day attack paths [6]. Also, they have visualized attacks in a graph-like structure and developed a prototype for identifying zero-day attacks.

Diverse supervised ML approaches by applying CIC-AWS-2018 dataset [7]. Researchers have applied decision tree (DT), random forest (RF), K-nearest neighbor (KNN), multi-layer perceptron (MLP), and quadratic discriminant analysis (QDA), as well as Gaussian naive Bayes (NB) classification models. Moreover, developers have not provided a brief definition of supervised ML training on benign traffic that has been applied for predicting unknown attacks. Also, transfer learning (TL) has been applied for predicting zero-day attacks. A deep transudative TL for examining zero-day attacks in [8]. Also, ML has been utilized for addressing zero-day malware examination, for instance, the application of deep-convolutional generative adversarial network (DCGAN) [9].

Here, an effective DL-based variational autoencoder (VAE) model for zero-day attack detection was introduced. This research work aims to design a new IDS model with a maximum detection rate with minimal false-negative rate. The DL-VAE model involves pre-processing to convert the raw data into a compatible format. Followed by, the preprocessed data is fed into the VAE model to distinguish the presence of zero-day attacks in the networking data. For ensuring the better results of the DL-VAE model, a series of experiments were conducted, and the results are determined under several aspects.

2 The Proposed DL-VAE Model

The overall working process involved in the DL-VAE model is shown in Fig. 1. As depicted in the figure, the DL-VAE model incorporates two major processes such as pre-processing and VAE-based classification. The detailed working of these processes is provided in the succeeding sections.

2.1 Pre-processing

Primarily, the raw data is preprocessed, and it is divided into different types of attacks along with the timestamp provided. Moreover, the bidirectional flow features are generated, and the feature with maximum correlation gets discarded for improvising the stability of the model. Finally, the scaling of the features takes place by the use of a standard scalar.

2.2 VAE-Based Zero-Attack Detection

Once the feature is preprocessed, the VAE model gets executed to determine the existence of zero-day attacks. In general, VAE is referred to as a directed probabilistic graphical method that is attained by the approximation of artificial neural network (ANN) to the posterior. In VAE, the secondary variable z where the generative operation is initialized is considered as a greater layer of the graphical method. The difficult procedure of data generation results in data x, as depicted by $g(z)$ which is labeled in developing ANN. As the marginal likelihood is intractable, the variational lower boundary of the marginal likelihood of input data is the major objective function of VAE. Therefore, the marginal likelihood is accomplished by consolidating the

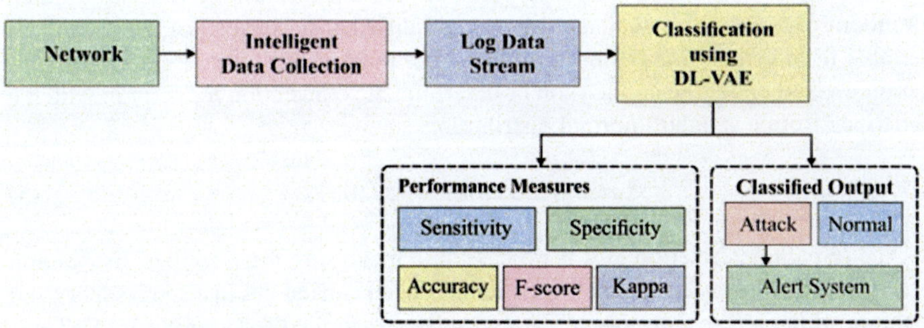

Fig. 1 Block diagram of proposed model

marginal likelihood of diverse data points from Eq. (1). Equation (2) is gained when the marginal likelihood of various data points is formulated again.

$$\log p_{\theta}\left(x^{(1)}, \ldots, x^{(N)}\right) = \sum_{i=1}^{N} \log p_{\theta}\left(x^{(i)}\right) \tag{1}$$

$$\log p_{\theta}\left(x^{(i)}\right) \geq \mathcal{L}\left(\theta, \phi; x^{(i)}\right) \tag{2}$$

$$= E_{q_{\phi}(z|x^{(i)})}\left[-\log q_{\phi}(z|x) + \log p_{\theta}(x|z)\right] \tag{3}$$

$$= -D_{\mathrm{KL}}\left(q_{\phi}\left(z|x^{(i)}\right)||p_{\theta}(z)\right) + E_{q_{\phi}(z|x^{(i)})}\left[\log p_{\theta}(x|z)\right] \tag{4}$$

In Eq. (4), D_{KL} means the Kullback–Leibler divergence from approximate posterior as well as a previous latent parameter z. Therefore, the likelihood of input data x from the latent parameter z is depicted as $p_{\theta}(x|z)$.

The variables of approximate posterior $q_{\phi}(z|x)$ are attained using a neural network (NN) by VAE. Also, the directed probabilistic graphical method $p_{\theta}(x|z)$ is referred to as a decoder, and the approximate posterior $q_{\phi}(z|x)$ is called an encoder. Here, it has to be emphasized that VAE is used for modeling the distribution attributes instead of the original value. Followed by, $f(x, \phi)$ in the encoder generates a parameter of approximate posterior $q_{\phi}(z|x)$ and to gain the original value of a latent variable z, that samples from $q(z, f(x, \phi))$ is mostly required. Also, the general choice is isotropic that is normal for the distribution of a secondary variable z, where $p_{\theta}(z)$ and $q_{\phi}(z|x)$, as the relationship has variables from latent variable space should be simple when compared with actual data space. This likelihood $p_{\theta}(x|z)$ differs based on data behavior. Especially, multivariate Gaussian distribution has been used if the input data is in a continuous format, and when it is binary, Bernoulli distribution has been utilized.

Here, the VAE training is implemented under the application of the backpropagation (BP) model. Then, Eq. (4) is processed under the application of Monte Carlo gradient approaches along with a re-parameterization technology that has applied a random parameter from remarkable normal distribution instead of using random variable from actual distribution. Moreover, the random variable $z \sim q_{\phi}(z|x)$ is re-parameterized under the application of deterministic conversion $h_{\phi}(\epsilon, x)$ in which ϵ is derived from a standard normal distribution.

$$\tilde{z} = h_{\phi}(\epsilon, x) \text{with} \epsilon \sim \mathcal{N}(0, 1) \tag{5}$$

The re-parameterization task is processed to make sure the $\tilde{z}$ applies distribution of $q_{\phi}(z|x)$. The zero-day attack prediction has been carried out in a semi-supervised fashion, which refers that normal data samples are applied for training VAE. Thus, the probabilistic decoder g_{θ}, as well as encoder f_{ϕ}, parameterizes isotropic normal distribution from actual input variable space and latent variable space, correspondingly. The testing operation has been performed by deciding various samples from the probabilistic encoder of the trained VAE approach.

3 Performance Validation

The performance of the presented DL-VAE model has been assessed using NSL-KDD 2015 dataset [10]. It includes a set of 125,973 instances with the existence of 41 attributes under two classes. A total of 67,343 instances come under normal class and the remaining 58,630 instances under anomaly class. For experimentation, tenfold cross-validation process is applied. The details related to the dataset are shown in Table 1. The measures used to examine the results are sensitivity, specificity, accuracy, F-score, and kappa [11–13].

Table 2 and Fig. 2 show the classification outcome of the DL-VAE model with existing methods [14, 15] such as radial basis function (RBF) network, logistic regression (LR), random forest (RF), random tree (RT), and decision tree (DT). The resultant values stated that the RBFNetwork model has obtained poor classification outcomes with the minimum sensitivity of 0.934, the specificity of 0.924, accuracy of 0.929, F-score of 0.934, and kappa of 0.858. Also, the RF model has resulted in slightly better results with a sensitivity of 0.924, the specificity of 0.938, accuracy of 0.930, F-score of 0.936, and kappa of 0.860. Similarly, the DT model has accomplished moderate performance with a sensitivity of 0.957, the specificity of 0.954, accuracy of 0.955, F-score of 0.958, and kappa of 0.911. Along with that the reasonable results are achieved by the RT model with a sensitivity of 0.957, the specificity of 0.954, accuracy of 0.956, F-score of 0.958, and kappa of 0.911. Though the LR model has obtained competitive classification results with a sensitivity of 0.973,

Table 1 Dataset description

Dataset	No. of instances	No. of attributes	No. of classes	Normal/anomaly
NSL-KDD 2015	125,973	41	2	67,343/58,630

Table 2 Performance evaluation of existing with proposed DL-VAE method

Methods	Sensitivity	Specificity	Accuracy	F-score	Kappa
DL-VAE	0.985	0.977	0.989	0.982	0.973
RBF network	0.934	0.924	0.929	0.934	0.858
LR	0.973	0.969	0.971	0.973	0.942
RF	0.924	0.938	0.930	0.936	0.860
RT	0.957	0.954	0.956	0.958	0.911
DT	0.957	0.954	0.955	0.958	0.910

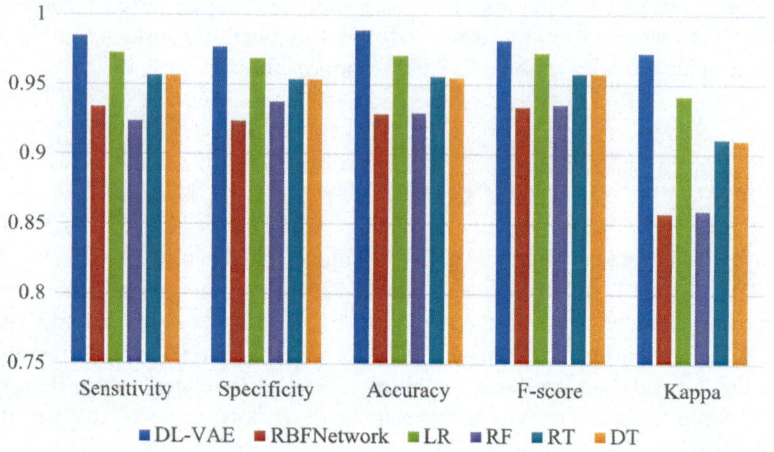

Fig. 2 Result analysis of DL-VAE model with different measures

the specificity of 0.969, accuracy of 0.971, F-score of 0.973, and kappa of 0.942, the presented DL-VAE model has led to a maximum classification performance with the sensitivity of 0.985, specificity of 0.977, accuracy of 0.989, F-score of 0.982, and kappa of 0.973.

Table 3 and Fig. 3 examine the comparative analysis of the DL-VAE method using the present model's operating with accuracy. The figure demonstrated that the CS-PSO, gradient boosting, and Gaussian process methods have implied limited accuracy values of 0.755%, 0.843%, and 0.911% correspondingly. Additionally, the DNN + SVM and Fuzzy C-means methods have shown moderate accuracy measures of 0.920% and 0.953% respectively. Followed by, the LSTM, GA + Fuzzy, and

Table 3 Performance evaluation of proposed DL-VAE with recent methods

Methods	Accuracy
DL-VAE	0.989
DNN	0.987
LSTM	0.962
Cuckoo optimization (2018)	0.969
CS-PSO (2019)	0.755
PSO-SVM (2019)	0.991
Behavior-based IDS (2019)	0.989
Gaussian process (2015)	0.911
DNN + SVM (2018)	0.920
GA + Fuzzy (2018)	0.965
Fuzzy C-means (2018)	0.953
Gradient boosting (2018)	0.843

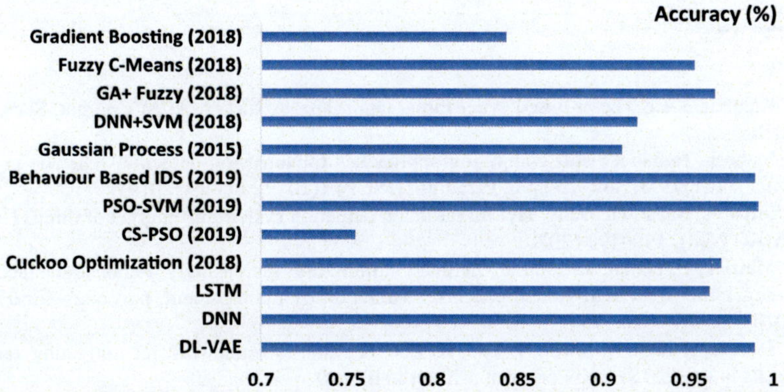

Fig. 3 Accuracy analysis of DL-VAE model with existing methods

Cuckoo optimization schemes have attained considerable accuracy scores of 0.962%, 0.965%, and 0.969% respectively. On the other hand, the DNN, behavior-based IDS, and PSO-SVM approaches have shown competing results with an accuracy of 0.987%, 0.989%, and 0.991%. However, the newly developed DL-VAE technology has exhibited supreme outcomes with a maximum accuracy of 0.989%.

4 Conclusion

This article has presented a novel DL-based VAE model for zero-day attack detection. The study aims to develop an effective IDS model with a maximum detection rate with minimal false-negative rate. The DL-VAE model incorporates two major processes such as pre-processing and VAE-based classification. The DL-VAE model involves pre-processing to convert the raw data into a compatible format. Followed by, the VAE model gets executed to determine the existence of zero-day attacks in the networking data. To validate the effective performance of the DL-VAE model, a series of experiments were conducted, and the results are determined under several aspects. The obtained simulation values ensured the betterment of the DL-VAE model with the sensitivity of 0.985, the specificity of 0.977, accuracy of 0.989, F-score of 0.982, and kappa of 0.973. The proposed method finds useful for real-time applications such as banking transactions, e-commerce, etc. In the future, the performance can be further improvised using metaheuristic-based parameter tuning algorithms.

References

1. N. Kaloudi, J. Li, The ai-based cyber threat landscape: A survey. ACM Comput. Surv. **53**(1) (2020)
2. A. Sathesh, Enhanced soft computing approaches for intrusion detection schemes in social media networks. J. Soft Comput. Paradigm (JSCP) **1**(2019), 69–79 (2019)
3. S. Smys, A. Basar, H. Wang, Hybrid intrusion detection system for internet of things (IoT). J. ISMAC **2**(04), 190–199 (2020)
4. K. Metrick, P. Najafi, J. Semrau, Zeroday exploitation increasingly demonstrates access to money, rather than skill—intelligence for vulnerability management, part one—FireEye inc (2020)
5. V. Sharma, J. Kim, S. Kwon, I. You, K. Lee, K. Yim, A framework for mitigating zero-day attacks in iot (2018) arXiv preprint arXiv:180405549
6. X. Sun, J. Dai, P. Liu, A. Singhal, J. Yen, Using bayesian networks for probabilistic identification of zero-day attack paths. IEEE Trans. Inf. Forensics Secur. **13**(10), 2506–2521 (2018)
7. Q. Zhou, D. Pezaros, Evaluation of machine learning classifiers for zero-day intrusion detection– an analysis on cic-aws-2018 dataset (2019) arXiv preprint arXiv:190503685
8. N. Sameera, M. Shashi, Deep transductive transfer learning framework for zeroday attack detection (2020)
9. J.Y. Kim, S.J. Bu, S.B. Cho, Zero-day malware detection using transferred generative adversarial networks based on deep autoencoders (2018)
10. NSL-KDD (2019) Dataset of NSL-KDD University of new Brunswick. https://www.unb.ca/research/iscx/dataset/iscx-NSL-KDD-dataset.html
11. J. Uthayakumar, T. Vengattaraman, P. Dhavachelvan, Swarm intelligence based classification rule induction (CRI) framework for qualitative and quantitative approach: an application of bankruptcy prediction and credit risk analysis. J. King Saud Univ.-Comput. Inf. Sci. (2017)
12. J. Uthayakumar, N. Metawa, K. Shankar, S.K. Lakshmanaprabu, Financial crisis prediction model using ant colony optimization. Int. J. Inf. Manage. **50**, 538–556 (2020)
13. M. Bhattacharya, Datta, A. Uthra, Comparative analysis of bayesian methods for estimation of locally-invariant extremes, in *IEEE International Conference on Electrical, Control and Instrumentation Engineering, ICECIE 2019*, (Malaysia, 2019)
14. U. Sabeel, S.S. Heydari, H. Mohanka, Y. Bendhaou, K. Elgazzar, K. El-Khatib, Evaluation of deep learning in detecting unknown network attacks, in *2019 International Conference on Smart Applications, Communications and Networking (SmartNets)* (IEEE, 2019) pp. 1–6
15. Z. Chiba, N. Abghour, K. Moussaid, M. Rida, Intelligent approach to build a deep neural Network based IDS for cloud environment using combination of machine learning algorithms. Comput. Secur. **86**, 291–317 (2019)

Image-Text Matching: Methods and Challenges

Taghreed Abdullah and Lalitha Rangarajan

Abstract Image-text matching has gained increasing popularity, as it bridges the heterogeneous image-text gap and plays an essential role in understanding image and language. In the recent few years, there have been extensive studies of matching visual content and textual data with deep architectures. This article reports an overview of image-text matching, especially the most relevant and recent image-text methods and presents a taxonomy of these methods based on alignment level as well as describes the main approaches. Further, we identify the benefits and limitations of each method in this review and provide readers with prominent challenges that can be helpful for researchers in the cross-modal community.

Keywords Image-text matching · Deep learning · Embedding · Attention

1 Introduction

Single-modal matching, such as image–image matching and text–text matching, has been performed conventionally. However, these methods only perform matching on the same modality. Image and text are two essential elements that help to understand the real world. It is easy to connect an image with text and vice versa, but due to the gap between image and text, such a connection is still difficult in computer vision. In fact, different modalities have different representations and distributions, and these heterogeneous characteristics make it difficult to directly measure the similarities of vision and language. Currently, with the development of deep learning technologies, exploring the association between visual and textual contents has garnered great interest from researchers because of its significance in several applications including image-text matching [1], cross-modal retrieval [2], image captioning [3], and visual question answering (VQA) [4]. In this article, the main focus is on the bidirectional image-text retrieval task, i.e., image-text matching, which is considered one of the most common topics in the cross-modal field.

T. Abdullah (✉) · L. Rangarajan
Department of Studies in Computer Science, Mysore, India

V. Suma et al. (eds.), *Inventive Systems and Control*,
Lecture Notes in Networks and Systems 204,
https://doi.org/10.1007/978-981-16-1395-1_17

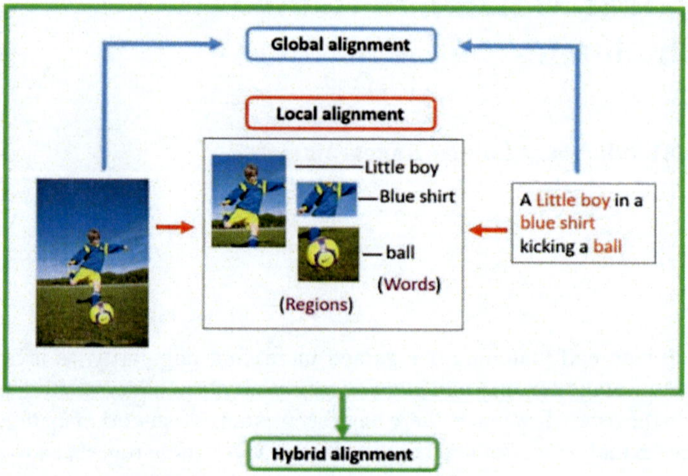

Fig. 1 Illustration of the different image-text matching methods

The core issue in image-text matching is how to perfectly find and associate common semantics in image and text, where semantically related images and text pairs posse a higher matching score than unmatched ones. Over the last ten years, several earlier studies have made significant strides in matching image-text. According to the alignment level, existing image-text matching approaches based on deep learning can be categorized into global, local, and hybrid matching methods (Fig. 1).

Global matching methods learn joint embeddings for whole images and text, while local matching methods focus on the local-level correlation, image regions, and text words. Hybrid matching methods combine the global and local alignment for a more accurate matching score. Most of the prior works fall under global matching methods [5–7], which aim to learn neural networks to map whole images and entire sentences, global-level, into a common semantic space, where the similarity between image-text pairs can be directly measured. Although these methods achieved considerable improvements in the image-text matching task, the major drawback of these methods is that they are unable to benefit from the fine-grained interaction between image and text to concentrate on finding the common semantics due to their focus on the whole image and text for global alignment, which hinders digging into the specifics of the regions of the image or words of the sentence. Besides, sampling the useful triplets and choosing suitable margins is a challenging problem in real applications.

Consequently, many studies are starting to concentrate on a fine-grained analysis of the local-level correlation [8–10]. The attention mechanism, which can concentrate on salient features as needed, is used for image-text embedding models [11] to correlate image regions with words, and the local similarity between region-word pair is computed. Several studies have confirmed that adopting attention is useful for modelling a more consistent relationship between visual and textual data [9, 11, 12].

While these approaches have accomplished satisfactory results, they still suffer from component-level mismatching and massive computation burden. Hybrid matching methods [13, 14] combine both global and local level (multi-level) for obtaining more accurate performance. Currently, in some cross-modal matching works, it is possible to find some reviews about image-text matching. However, most of them have classified image-text matching approaches based on the type of objective function used for the learning procedure. Unlike most of the works, we focus on the type of alignment level at which the learning occurs, where learning to match image and text basically relied on object co-occurrence. Therefore, we divide the existing methods into three categories: global, local, and hybrid matching methods.

The layout of this article is structured as follows: In Sect. 2, the results of recent image-text matching works were discussed. In Sect. 3, a taxonomy of the main deep learning-based image-text matching methods is presented and each method is described. Finally, Sect. 4 concludes the research work and introduces the main challenges of image-text matching.

2 Related Work

In this section, the results of the latest image-text matching articles are reviewed in chronological order, image-text matching methods based on deep-learning techinques can be classified into global, local, and hybrid, which are shown in Table 1.

3 Deep Learning Image-Text Matching Methods

Existing works learn semantic similarity between image and text by using different deep-learning techniques. Based on alignment level these works can be categorized as shown in Fig. 2.

3.1 Global Matching Methods

The goal of global methods is to learn joint semantic embedding space where images and text embeddings are comparable directly. Specifically, these methods are learning two mapping functions that map whole image and full text into a joint space $f : V \to E$ and $g : T \to E$, where V and T visual and textual feature spaces, respectively, and E joint embedding space (see Fig. 3). This type of method usually learns these embeddings by designing different loss functions. The most widely used functions

Table 1 Some recent works in image-text matching

Methods	References	Year	Proposed method	Image-to-text			Text-to-image		
				R@1	R@5	R@10	R@1	R@5	R@10
Global	[1]	2018	Two-branch neural networks	54.0	84.0	91.2	43.3	76.8	87.6
	[15]	2018	VSE++	52.0	84.3	92.0	64.6	90.0	95.7
	[16]	2017	RRF	56.4	85.3	91.5	43.9	78.1	88.6
Local	[17]	2018	GXN	68.5	–	97.9	56.6	–	94.5
	[12]	2018	SCAN	72.7	94.8	98.4	58.8	88.4	94.8
	[11]	2017	DAN	55.0	81.8	89.0	39.4	69.2	79.1
Hybrid	[18]	2020	Instance loss	65.6	89.8	95.5	47.1	79.9	90.0
	[13]	2020	GSLS	68.9	94.1	98.0	58.6	88.2	94.9
	[19]	2019	MDM	54.7	84.1	91.9	44.6	79.6	90.5
	[20]	2018	CRAN	23.0	52.0	66.0	21.1	48.9	64.5
	[21]	2018	JGCAR	52.7	82.6	90.5	40.2	74.8	85.7

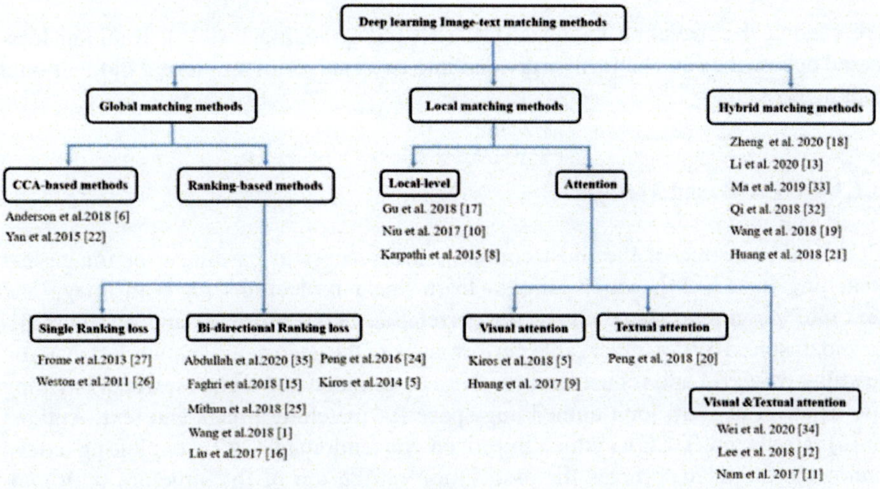

Fig. 2 Taxonomy of image-text matching methods

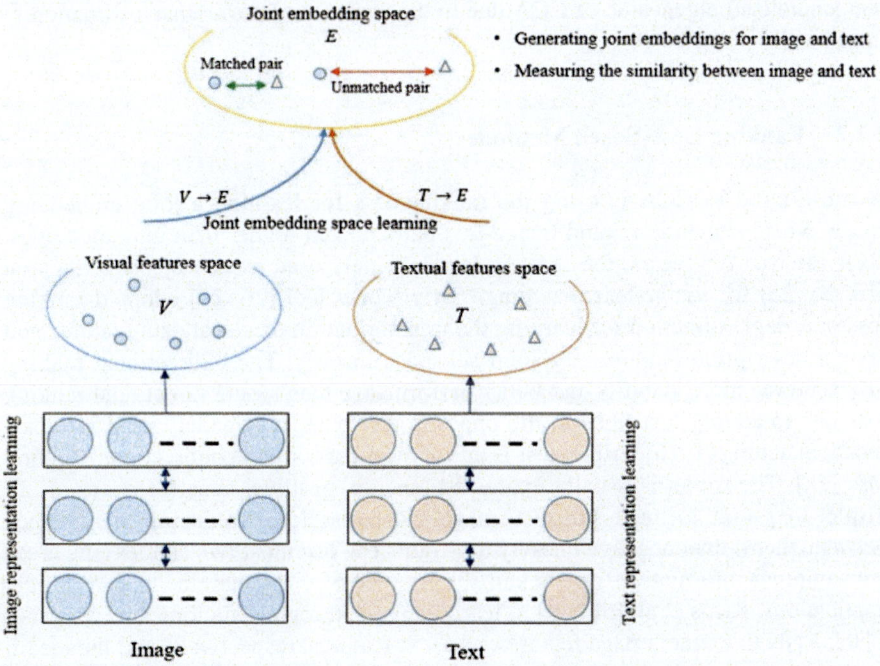

Fig. 3 Overview of image-text matching

are canonical correlation analysis (CCA) [6] and ranking loss [5]. Ranking loss-based approaches can be further divided into two categories single and bidirectional ranking loss.

3.1.1 CCA-Based Methods

CCA has been one of the most common and successful baselines for image-text matching [6, 22, 23], which aims to learn linear projections for both image and text into a common space where the correlation between image and text is maximized. Inspired by the remarkable performance of the deep neural network (DNN) on learning powerful and robust features from both image and text, researchers making use of DNN to learn joint embedding space for matching image and text. Authors in [6] introduced (DCCA) which improved conventional CCA by exploiting a deep architecture that maximizes the association on the top of the structure with two-branch neural networks. DCCA can be implemented on large and trained by using stochastic gradient descent (SGD). However, SGD unable to tackle the problem of the generalized eigenvalue of CCA due to the variance of covariance estimation in mini-batch.

3.1.2 Ranking Loss-Based Methods

A substitution to CCA is using the ranking loss for learning a joint embedding space, where semantic-related image-text pairs ranked higher than unrelated ones. Over the past few years, the ranking loss is widely used as an objective function [16, 24, 25] for image-text matching. Early works like [26, 27] adopted ranking loss in a single direction for learning the transformation of visual data features and textual description features to a joint subspace linearly. The bidirectional ranking loss achieves more stability and better performance than single-directional ranking loss due to adding lost links in the opposite direction. It is widely used in cross-modal matching [1, 16]. Triplet loss is one of the most common bidirectional ranking loss [28]. The triplet is usually expressed (anchor, positive, negative) or (a, p, n). Triplet loss aims to make positive image-text pairs closer (reducing the distance between them) than negative ones by a margin. The foremost two challenging areas are sampling informative triplets and selecting convenient margins in several real applications. Kiros et al. [5] tried to learn image-text representations through using CNN, RNN to extract image features, and text features, respectively, and then learn a joint semantic embedding space with a triplet loss. Authors in [22] utilized hard negative in the function of triplet loss to enhance embedding learning. In [17], the image-text feature embedding method is introduced for cross-retrieval to combine generative models into classical cross-modal feature embedding. Wang et al. [29] constructed a simple matching network with two layers to learn common space that preserves the structure relations between image and text. Liu et al. [16] adopted residual blocks to correlate image and text densely.

3.2 *Local Matching Methods*

In addition to the traditional image-text matching feature embedding at the global semantic level, image-text matching at the local level is introduced in many works [8, 10]. Karpathy et al. [30] adopted R-CNN as the encoder of image regions (region-level) and then performed local similarity learning between the regions of image and sentence words by assembling the similarity scores of all region-word pairs. Niu et al. [10] used a tree-structured LSTM for learning the hierarchical relations between image objects and text words, further to learn the relations between sentences and images at the phrase level. Recently, the attention mechanism, which is one of the recent neural network techniques, has made significant enhancements in different multimodal tasks including image-text matching, by using deep learning structures like RNNs [31] and CCNs [18]. The attention mechanism makes the model focus on the important fine-grained parts of the inputs (image or text), i.e., it can make alignment between image and text, which can focus on the prominent features as required for more accurate semantic correlations. The most recent image-text matching models tend to be fine-grained image region-text word matching methods, the similarities for each image region or text word are being measured and then aggregating to obtain the global image-text similarity. Nam et al. [11] proposed a dual attention (visual and textual attention) network for image-text matching by applying a self-attention technique to concentrate on specific regions in images and words in a text to capture the fine-grained information from image and sentence for more accurate matching. In [20], an RNN is constructed to independently find the modality-specific properties in text and image space.

3.3 *Hybrid Matching Methods*

Although the goal of local methods is finding local alignment among all region-words pairs, they ignore significant-related information in fine-grained regions, which can help to give rich additional indications for image-text matching learning. Recently, the idea of integrating the global and local alignment has been adopted by some works [13, 21, 30, 32–35]. Qi et al. [32] developed a cross-media relation attention network with three branches to find global, local, and relation alignment (multi-level alignment) across image and text data for learning more precise cross-modal correlation. In [21], a spatial-semantic attention mechanism is adopted to build a bidirectional network, which strengthens the relationship of the word to regions and object in visual content to words within a deep structure for further efficient matching. Wang et al. [19] proposed to learn representation globally to improve the semantic consistency of image/text content representations and developed a co-attention learning mechanism to completely leverage varying levels of image-text relations. Li et al. [13] developed a network with two levels to obtain more accurate image-text matching by fusing local and global similarity. Combining local and

Table 2 Comparison between different image-text matching methods

Methods	Benefits	Limitations
Global	• The most direct way to judge whether image and text are similar	• The main limitation of these methods is that they deal with the whole image and the entire text for global alignment, which make digging into the image and text details is very coarse • The global level may bring irrelevant or noisy information • blends some redundant information (useless regions • Requiring high memory for computing the covariance matrix of the whole images and texts as well as the inability to discover the nonlinear relation between image and text
Local	• There is interest in distinct objects/regions in an image • The visual attention mechanism adaptively focuses on specific discriminative local regions rather than spreads evenly over the whole image	• Lack of emphasis on relations between objects and non-object elements like the background, the surroundings, or the environment
Hybrid	• Semantically investigate the best matching between visual and textual contents	• Most of these methods calculate the final similarity by incorporating global and local similarities and even relationship similarity, which is very complex for real-time applications, as it not only increases computation complexity but also memory usage

global similarity helps to obtain more accurate matching than adopting only one level of alignment. Table 2 summarizes the main benefits and limitations of image-text methods.

4 Conclusion

In this article, we have presented a review of image-text matching methods from the level of alignment, provided a taxonomy of these methods. From the perspective of the authors, the challenges of image-text matching include the following three aspects: (1) how to measure accurately image-text semantic similarity, which requires a thorough understanding of both modalities, (2) how to learn appropriate joint embeddings of image and text content, (3) how to sample useful triplets and choosing suitable margins, and (4) how to learn robust features of both modalities.

References

1. L. Wang, Y. Li, S. Lazebnik, Learning two-branch neural networks for image-text matching tasks. IEEE Trans. Pattern Anal. Mach. Intell. **41**, 394–407 (2018). https://doi.org/10.1109/TPAMI.2018.2797921
2. X. Xu, H. Lu, J. Song, Y. Yang, H.T. Shen, X. Li, Ternary adversarial networks with self-supervision for zero-shot cross-modal retrieval. IEEE Trans. Cybern. **50**, 2400–2413 (2020). https://doi.org/10.1109/TCYB.2019.2928180
3. K. Xu, J. Ba, R. Kiros, K. Cho, A. Courville, R. Salakhutdinov, R. Zemel, Y. Bengio, Show, attend and tell: neural image caption generation with visual attention. arXiv:1502.03044 [cs] (2015)
4. V. Kazemi, A. Elqursh, Show, ask, attend, and answer: a strong baseline for visual question answering. arXiv:1704.03162 [cs] (2017)
5. R. Kiros, R. Salakhutdinov, R.S. Zemel, Unifying visual-semantic embeddings with multimodal neural language models. arXiv:1411.2539 [cs] (2014)
6. P. Anderson, X. He, C. Buehler, D. Teney, M. Johnson, S. Gould, L. Zhang, Bottom-up and top-down attention for image captioning and visual question answering, in *2018 IEEE/CVF Conference on Computer Vision and Pattern Recognition* (IEEE, Salt Lake City, UT, 2018), pp. 6077–6086
7. Y. Guo, H. Yuan, K. Zhang, Associating images with sentences using recurrent canonical correlation analysis. Appl. Sci. **10**, 5516 (2020). https://doi.org/10.3390/app10165516
8. A. Karpathy, L. Fei-Fei, Deep visual-semantic alignments for generating image descriptions, in *Presented at the Proceedings of the IEEE Conference on Computer Vision and Pattern Recognition* (2015)
9. Y. Huang, W. Wang, L. Wang, Instance-aware image and sentence matching with selective multimodal LSTM, in *Presented at the Proceedings of the IEEE Conference on Computer Vision and Pattern Recognition* (2017)
10. Z. Niu, M. Zhou, L. Wang, X. Gao, G. Hua, Hierarchical multimodal LSTM for dense visual-semantic embedding, in *Presented at the Proceedings of the IEEE International Conference on Computer Vision* (2017)
11. H. Nam, J.-W. Ha, J. Kim, Dual attention networks for multimodal reasoning and matching, in *2017 IEEE Conference on Computer Vision and Pattern Recognition (CVPR)* (2017), pp. 2156–2164
12. K.-H. Lee, X. Chen, G. Hua, H. Hu, X. He, Stacked cross attention for image-text matching, in *Presented at the Proceedings of the European Conference on Computer Vision (ECCV)* (2018)
13. Z. Li, F. Ling, C. Zhang, H. Ma, Combining global and local similarity for cross-media retrieval. IEEE Access **8**, 21847–21856 (2020). https://doi.org/10.1109/ACCESS.2020.2969808
14. X. Xu, T. Wang, Y. Yang, L. Zuo, F. Shen, H.T. Shen, Cross-modal attention with semantic consistence for image-text matching. IEEE Trans. Neural Netw. Learn. Syst. 1–14 (2020). https://doi.org/10.1109/TNNLS.2020.2967597
15. F. Faghri, D.J. Fleet, J.R. Kiros, S. Fidler, VSE++: Improving visual-semantic embeddings with hard negatives. arXiv:1707.05612 [cs] (2018)
16. Y. Liu, Y. Guo, E.M. Bakker, M.S. Lew, Learning a recurrent residual fusion network for multimodal matching, in *Presented at the Proceedings of the IEEE International Conference on Computer Vision* (2017)
17. J. Gu, J. Cai, S.R. Joty, L. Niu, G. Wang, Look, imagine and match: improving textual-visual cross-modal retrieval with generative models, in *Presented at the Proceedings of the IEEE Conference on Computer Vision and Pattern Recognition* (2018)
18. Z. Zheng, L. Zheng, M. Garrett, Y. Yang, Y.-D. Shen, Dual-path convolutional image-text embedding with instance loss. arXiv:1711.05535 [cs] (2018)
19. S. Wang, Y. Chen, J. Zhuo, Q. Huang, Q. Tian, Joint global and co-attentive representation learning for image-sentence retrieval, in *Proceedings of the 26th ACM international conference on Multimedia* (Association for Computing Machinery, New York, NY, USA, 2018), pp. 1398–1406

20. Y. Peng, J. Qi, Y. Yuan, Modality-specific cross-modal similarity measurement with recurrent attention network. IEEE Trans. Image Process. **27**, 5585–5599 (2018). https://doi.org/10.1109/TIP.2018.2852503
21. F. Huang, X. Zhang, Z. Li, Z. Zhao, Bi-directional spatial-semantic attention networks for image-text matching. IEEE Trans. Image Process. (2018). https://doi.org/10.1109/TIP.2018.2882225
22. F. Yan, K. Mikolajczyk, Deep correlation for matching images and text, in *2015 IEEE Conference on Computer Vision and Pattern Recognition (CVPR)* (2015), pp. 3441–3450
23. W. Wang, X. Yan, H. Lee, K. Livescu, Deep variational canonical correlation analysis. arXiv:1610.03454 [cs] (2017)
24. Y. Peng, X. Huang, J. Qi, Cross-media shared representation by hierarchical learning with multiple deep networks, in *IJCAI* (2016)
25. N.C. Mithun, R. Panda, E.E. Papalexakis, A.K. Roy-Chowdhury, Webly supervised joint embedding for cross-modal image-text retrieval, in *Proceedings of the 26th ACM international conference on Multimedia.* (Association for Computing Machinery, New York, NY, USA, 2018), pp. 1856–1864
26. J. Weston, S. Bengio, N. Usunier, Wsabie: scaling up to large vocabulary image annotation, in *Proceedings of the International Joint Conference on Artificial Intelligence, IJCAI* (2011)
27. A. Frome, G.S. Corrado, J. Shlens, S. Bengio, J. Dean, M.A. Ranzato, T. Mikolov, DeViSE: A deep visual-semantic embedding model, in *Advances in Neural Information Processing Systems*, vol. 26, ed. by C.J.C. Burges, L. Bottou, M. Welling, Z. Ghahramani, K.Q. Weinberger (Curran Associates, Inc., 2013), pp. 2121–2129
28. A. Hermans, L. Beyer, B. Leibe, In defense of the triplet loss for person re-identification. arXiv:1703.07737 [cs] (2017)
29. B. Wang, Y. Yang, X. Xu, A. Hanjalic, H.T Shen, Adversarial cross-modal retrieval, in *Proceedings of the 25th ACM international conference on Multimedia* (Association for Computing Machinery, New York, NY, USA, 2017), pp. 154–162
30. A. Karpathy, A. Joulin, L.F. Fei-Fei, Deep fragment embeddings for bidirectional image sentence mapping, in *Advances in Neural Information Processing Systems*, vol. 27, ed. by Z. Ghahramani, M. Welling, C. Cortes, N.D. Lawrence, K.Q. Weinberger (Curran Associates, Inc., 2014), pp. 1889–1897
31. W. Zaremba, I. Sutskever, O. Vinyals, Recurrent neural network regularization. arXiv:1409.2329 [cs] (2015)
32. J. Qi, Y. Peng, Y. Yuan, Cross-media multi-level alignment with relation attention network. arXiv:1804.09539 [cs] (2018)
33. L. Ma, W. Jiang, Z. Jie, X. Wang, Bidirectional image-sentence retrieval by local and global deep matching. Neurocomputing **345**, 36–44 (2019). https://doi.org/10.1016/j.neucom.2018.11.089
34. K. Wei, Z. Zhou, Adversarial attentive multi-modal embedding learning for image-text matching. IEEE Access **8**, 96237–96248 (2020). https://doi.org/10.1109/ACCESS.2020.2996407
35. T. Abdullah, Y. Bazi, M.M. Al Rahhal, M.L. Mekhalfi, L. Rangarajan, M. Zuair, TextRS: deep bidirectional triplet network for matching text to remote sensing images. Remote Sens. **12**, 405 (2020). https://doi.org/10.3390/rs12030405

Fault Location in Transmission Line Through Deep Learning—A Systematic Review

Ormila Kanagasabapathy

Abstract In a power system, transient stability is very essential. Huge disturbances such as faults in the transmission line need to be separated as rapidly as probable to replace transient stability. In a transmission network, the faulty voltage regulator along with current signals is used for fault location, classification, and detection. Detecting the location of the fault on transmission lines precisely can save the effort of labor and enhance the restoration and repairing process essentially. On transmission lines, accurate pinpointing reduces labor costs and outage time. The location of the fault is independent of resistance of fault and the approach does not need any knowledge of source impedance. Relay predicts an unusual sign and then the breaker of the circuit separates the transmission line which is unhealthy from the remaining health system. This study uses a systematic and explicit technique of deep learning algorithms to determine the location of faults in the transmission line. Deep learning which offers a hierarchy of feature which can study experiences and identify raw data automatically as the human brain performs. It shows greater importance to solve the issues of location in power transmission systems. The deep learning algorithms train neural networks effectively and hinder fundamental issues in overfitting. Deep learning training depends mainly on measurements of electricity. Since deep learning is independent of factors of system namely parameters of line and topology which will have the essential prospect of the application when it is used in the location of the fault. Extensive researches have explained the efficiency and accuracy of deep learning approaches based on fault location in transmission lines.

Keywords Fault location · Deep learning · Artificial neural network · Deep neural network · Decision tree · Transmission lines · Stacked autoencoder · Convolutional neural network (CNN) · Support vector machine deep belief network

O. Kanagasabapathy (✉)
Department of Electrical and Electronic Engineering, A.M.K. Technological Polytechnic College, Chennai to Bangalore Highway, Sembarambakkam, Chennai 600123, India

V. Suma et al. (eds.), *Inventive Systems and Control*,
Lecture Notes in Networks and Systems 204,
https://doi.org/10.1007/978-981-16-1395-1_18

1 Introduction

Nowadays, transmission lines are much complex as huge networks owing to developing electrical power demand. In an interlinked system of transmission, it is important to have an accurate scheme of protection to assure extreme system reliability [1–4]. Transmission lines are considered as an important connection between customers and power stations which undertake a huge volume of power to the required premises [5]. Transmission lines form a connection in the interlinked operation of the system for bi-directional power flow. Transmission lines carry out nearly 100s of km to provide electrical supply to customers [6]. They are exposed to surroundings; hence, the opportunity of fault occurrence in a transmission line is greater which has to be taken immediately to reduce the destruction caused. It has been a concern for engineers to detect the power system faults as early as possible [7]. However, fault identification is not often a simple job. If the fault exists it must be separated rapidly to secure the system stability. For transmission lines, protective relays usually employ input signals of voltage and current to locate, categorize, and detect faults in a secured line. In fault case, the relay requires a trip indication to the circuit breaker to detach every line which is at fault [8–10]. In an interlinked system, the remaining network continues working usually or under usual conditions [11]. In a transmission line, the fault distracts the electric power from its intended way [11]. A fault produces an abnormal circumstance in an electric system [12]. The fault is classified into two kinds namely short circuit faults and series faults. The series fault is categorized into one, two, or three open fault conductors [12]. Similarly, the short circuit fault is categorized into two different types, i.e., symmetrical and unsymmetrical fault. Fault clearance is essential for the reliable operation of an electric system [13]. In power system control and operation, the recognition of faults is an important part of transmission lines. It satisfies an essential function in dealing with the condition of the power system and enhances the power system functions safely [14]. The exact recognition of faults models the power system protection along lines of transmission and enhances speedy faults of power system forecast and assures failure analysis associated with components of the power system. Power transmission lines develop in size and difficulty and the necessity to recognize faults of transmission line have become much essential [15]. Transients generate electrical system currents that destruct the power system relying on its occurrence severity. To avoid repetitions of the fault and greater cost related to predicting utilities of line faults endeavor for evolving exact methods of fault location [16]. The systems of transmission protection are designed to recognize the fault's location and separate only the faulted network part. The main barrier in the protection of transmission lines depends on isolating and detecting faults co-operates the system security with essential accuracy [17]. For power system operators, it is a challenging task to provide uninterrupted control to relevant users. However, when the intrusion of fault is beyond the control of humans, it is essential to locate, perceive, and categorize the fault accurately [18]. The location and detection of transmission lines are important to provide dependable and proficient power flow. Several authors proposed different schemes for the fault location evaluation in transmission lines. Along with the various fault location evaluation methods on transmission lines, this research used a deep learning algorithm for the location of a fault on the power transmission line Kapoor [19]. The neural network research has accomplished many developments that have led to what is referred to as deep learning [20]. On account of these developments, it is possible that the application of a neural network is not restricted to an individual hidden layer [21–23]. Deep learning permits models of computation that are comprised of numerous layers of processing to study data representation with numerous abstraction levels [24]. These deeper NN have been recognized to be good systemically at valuation issues in many applications due to their good generalization properties [25]. When developing DNN and other deep learning algorithms, it is essential to regard optimization algorithms, regularization techniques, and activation functions to acquire accurate networks that can be trained effectively by Guo et al. [26]. To isolate the faulty components of the power system and the safety of the power system is designed whenever a disturbance exists. Coordination, quickness, and selectivity are the most essential features of the protection system [27–30]. Faulted components must be isolated rapidly in case of disturbances to reduce the damage risk in power device system. The energy supply interruption must be avoided or minimized [31]. In a coordinated scheme, the protection device must perform to assure that only faulted processes are detached. Backup protection is possible when the protection device liable for separating the defected component fails to perform properly, whereas other devices of protection should perform to remove the fault [32]. Proper schemes of protection must be implemented and designed for distribution systems in order to identify the defective part from the healthy part to lessen the power blackout time based on deep learning algorithm [33]. Thus, it can be inferred that automatic fault location can highly develop the reliability of the system because quicker the system is restored a huge amount of money and the valuable time is saved [34].

2 Related Work

This section reviews various deep learning techniques used for fault location in transmission lines. Each technique is explained below briefly:

2.1 Deep Learning Techniques

Yadav and Dash [35] concentrate on the concept of classification of fault, phase selection of fault, fault location, and difference of fault direction by using the ANN technique. For applications of the power system, ANN is valuable as it can be qualified with disconnected data. Hessine and Sabir [36] present two algorithms of fault classification where the first algorithm utilizes the single artificial neural network method and the next one uses the modular artificial neural network method. The two classifiers relative research is performed in order to select which artificial neural network structure of fault classifier leads to good performance. The three locators of fault are proposed, and comparative research is carried out of three locators of fault to decide which structure of locator of fault leads to the exact location of the fault. Sanjay Kumar [37] discussed the BPNN structure as an alternative technique for detection, classification, and separation of fault in the transmission line. In addition, the main aim of the research is the whole scheme implementation of the transmission line system for the security of distance. To carry out this, the task of distance protection is categorized into different NNs for fault detection, recognition of fault as well as the location of the fault in various regions. These relevant faults were explained, namely 1 phase to G, 2 phase faults, and 2 phases to G faults. The study of Jamil et al. [38] concentrates on the classification and detection of fault transmission in electrical power using ANN. The results of the research infer that present methods based on NN are effective in classifying and detecting the faults of transmission lines with satisfactory performances [7]. The research of Koley et al. [39] presents a hybrid modular ANN and wavelet transform-based locator of fault, classification, and detection for six lines of phase by employing separate end data. The approximate coefficients standard deviation of current and voltage signals acquired using DWT are considered as input to modular ANN for fault detection and classification. Patil and Prajapati [40] develop a combined ANN-based scheme of protection that can solve the issues related to the conventional distance protection approach. In this research, ANN based on a backpropagation algorithm will be developed. Hatata et al. [41] proposed a hybrid protection scheme of transmission lines based on ANN. The hybrid transmission lines comprise two parts namely underground cable and overhead transmission line. Similarly, Thwe et al. [42] used ANN applications to analyze the fault classification and identification in transmission lines of voltage for greater protection of speed that could be employed in power system protection digitally. In the proposed research, the end side of three-phase currents is considered as inputs. Along with the BP algorithm, the ANN algorithm such as the feed-forward neural network has been employed for classification and detection of fault for the examination of three phases included in the process. In the transmission line system, the research of Sarathkumar et al. [43] presents ANN architecture used as a substitute approach for isolation, detection, and classification of the fault. In terms of the transmission line, the key role of architecture is the whole system implementation for the security of distance. The protection of distance is classified into varied NN for identification, location as well as detection of a fault in varied areas. Contrary to that Mbamaluikem et al. [44] employ a feed-forward network with a backpropagation neural network in developing the classifier of fault detector. Also, the instantaneous values of voltage are used and extracted to direct the fault detector classifier. The simulation results of transmission lines provide the developed intelligent systems effectiveness for fault classification and detection. Contrary to that Swetapadma and Yadav [45] propose an ANN resolution to locate multi-location faults in dual circuit SCCTLs, unlike earlier researches that find the fault at only one location. Though different fault schemes location has been suggested for faults of normal shunt existing at one place in series capacitor compensated transmission lines, but identifying the faults of multi-location in dual circuit SSCTLs has not yet been solved. The proposed ANN-based technique determines the location of the multi-location fault by employing voltage signals of one end of the line, therefore, avoiding the connection of communication requirement. Elnozahy et al. [46] discuss fault classification, detection, and determining the location of the fault as possible through the ANN algorithm. The testing, evaluation, and training of intelligent locator techniques are carried out on the basis of a multilayer perceptron feed-forward neural network with the algorithm of backpropagation neural network [47]. Table 1 shows the use of ANN for the fault location in transmission line.

2.2 *Stacked Autoencoders*

The study of Chen et al. [44] presents a study for classification and identification of faults in transmission power lines on the basis of CSAE. The proposed approach provides precise and fast results in classifying and identifying faults and is considered effective for the protection of online transmission lines for its strength and generalizability. To examine the incipient faults in power cables accurately, the research of Liu et al. [48] integrates a deep belief network and a sparse autoencoder which consists of a DNN that depends on the strong ability of NN learning to identify and classify different signals of cable fault without needing operations of pre-processing for the fault signals. The research of Luo et al. [49] proposed a technique for fault location on transmission lines of greater voltage direct current. Varied from traditional approaches which depend on the interaction between varied units of measurement or retrieval of post fault transients feature, the proposed algorithm considers locally predicted traveling surges of raw data or current as fault location input and output directly. The stacked autoencoder is used to structure the association between fault locations and currents. Relatively, Luo et al. [50] introduced an intelligent location technique to denote the location of the fault based on the data of phasor measurement units which are allocated properly. The section of the fault is decided by comparing the current waveforms of 0 sequences on both fault parts. Then, a stack autoencoder is structured to offer an end to end way to denote the fault point with the phasors of voltage and current. The performance of the proposed technique is authenticated by a simulated network of distribution on the PSCAD platform. Table 2 shows the reviews of the use of stacked autoencoder for the location of the fault in a transmission line.

2.3 *Support Vector Machine*

Saber et al. [51] present a classification scheme of defects on a double circuit parallel line of transmission using the integration of SVM and discrete wavelet transform. The proposed technique considers the mutual combination between parallel lines of transmission and defects randomness on transmission line regarding the occurrence

Table 1 Review of use of ANN for fault location in transmission line

Technique used	Advantages of technique	Author	Year
Artificial neural network	A robust, efficient, and accurate method for fault classification, detection, and discrimination of direction and selection of faulty phase and localization transmission line	Yadav and Dash	2014
Artificial neural network	Recognizes all types of faults and estimates the accurate location of the fault in transmission lines with greater accuracy	Hessine and Sabir	2014
Artificial neural network	A reliable and attractive method for the growth of secure relaying system for systems of power transmission	Sanjay Kumar	2014
Artificial neural network	Effective in classifying and detecting transmission lines fault with fulfilled performances	Jamil et al.	2015
Artificial neural network and wavelet transform	Brings the possibility of reliability and effectiveness for real-time execution	Koley et al. Kandekar and Khule Lala et al.	2015 2018 2018
Artificial neural network-based backpropagation algorithm	Resolves the issues related with conventional distance protection approaches	Patel and Prajapati	2015
Artificial neural network	Indicate the existence of fault and locate it accurately	Hatata et al. Kale and Pandey	2016 2016
Artificial neural network	Used for precise classification of fault on transmission line effectively	Thwe et al.	2016
Artificial neural network	A reliable and alternative method for the growth of secure relaying system for power transmission system	Sarathkumar et al.	2017
Feed-forward artificial neural network	Classifies and detects faults accurately and the configuration used is effective	Mbamaluikem et al. Maheshwari et al.	2018
Artificial neural network	Classifies the fault categories and determines the fault location precisely	Elnozahy et al. Okwudili et al. Rosle et al.	2019 2019 2020

time, location of the fault, type of fault, the resistance of fault, and conditions of loading [52, 53]. Similarly, Singh and Chopra [54] emphase on defect location, classification, and identifying power transmission line faults. The fault classification and detection can be achieved using SVM. The classifier based on SVM is trained on the database of faults to categorize one phase of the transient to faults in the ground.

Table 2 Review of use of stacked autoencoder for fault location in transmission line

Technique used	Advantages of technique	Author	Year
Convolutional sparse autoencoder	Accurate and fast and practical for protection of online transmission line for its generalizability and robustness	Chen et al.	2016
Sparse autoencoder and deep belief network	Greater reliability and recognition accuracy than the traditional pattern recognition approach	Liu et al.	2019
Stacked autoencoder	Efficient in locating the points of fault and robust against attenuation, traveling surges overlapping and different resistances of ground	Luo et al.	2020
Stacked autoencoder	Efficient in fault location and withstand the impacts of fault type, the resistance of transition, noise, and better accuracy of the location	Luo et al.	2020

Ray et al. examine SVM-based distance estimation and fault type method in a long transmission line and the study examined ten various kinds of short circuit faults [48]. The result of simulation denotes that the expected method for classification of fault provides 99.21% accuracy and least error of fault distance estimation for entire discussed cases. Babu and Mohan [55] propose a power system fault classification using SVM and empirical mode decomposition (EMD) [49]. Cho and Hoang [56] proposed a PSO-based SVM classifier in order to categorize faults in electricity [50]. The proposed method is capable of choosing the proper features of the input and optimizes the parameters of SVM to enhance the classification accuracy. Gururajapathy et al. [57] developed a technique using SVM to recognize the fault types, a section of the fault, and distance regarding the phase of fault. The proposed approach uses the magnitude voltage of the distribution system as the main element for SVM to recognize faults [58]. Wani and Singh [59] proposed an SVM approach to predict a method with which it is accessible to classify and detect the type of faults [51]. The voltage signals are usually generated on the basis of extensive fault simulation beside the central line of transmission that has a test system. Table 3 shows the reviews of SVM used for fault detection in the transmission line.

2.4 Decision Trees

The study of Ray et al. offers an improved classification of power quality PQ disturbances which are related to environmental factors and changes in load [52]. Several characteristics are acquired through S-transform hyperbolic by means of which optimal characteristics are taken using a genetic algorithm. On the other side, these optimal characteristics are utilized for PQ disturbances classification using decision trees and SVM classifiers. Niyas and Sunitha [60] recommend classification and

Table 3 Review of use of stacked autoencoder for fault location in transmission line

Technique used	Advantages of technique	Author	Year
Support vector machine and discrete wavelet transform	Classify entire faults on parallel lines of transmission properly	Saber et al. Zakri et al. Hosseini	2015 2018 2015
Support vector machine	Locate faults in power transmission lines effectively and accomplish fulfilled performances	Singh and Chopra Kasinathan and Kumarapan Gopakumar et al.	2014 2015 2015
Support vector machine	Removes redundant features by enhancing the prediction accuracy	Ray and Mishra Johnson and Yadav	2016 2016
Support vector machine and empirical mode decomposition	Effective classifier with acceptable accuracy levels	Babu and Mohan	2017
PSO-based support vector machine	Improves the performance by choosing proper kernel parameters and feature subset	Cho and Hoang Thom et al.	2017 2018
Support vector machine	Used to identify, phase, section, and distance of the fault and yields better accuracy	Gururajapathy et al Cho and Thom	2018 2017
Wavelet transform and support vector machine	Classify and find out the fault type effectively	Wani and Singh Livani and Everensoglu	2018 2012

identification of fault at power swing time using the DT method for the network transmission system. During switching, the swing power which exists in and out of extreme loads after the condition of defect clearance causes alteration in active and as well reactive power [53]. The article of Jana and De [61] proposes a smart and a planned classifier of fault with the integration of a reliable method of pre-processing for the selection of major attribute from power system recorded waveforms and uses a dependable DT-based classification algorithm to obtain precision of fault detection even when the network power is huge [54]. Zhang et al. [62] study use an algorithm such as a decision tree (DT) to realize the detection of fault type of transmission line. The DT algorithm has a better effect on classification and it is easy and simple to implement [63]. The research of Wasnik et al. [64] presents a semi-supervised ML method-based decision tree and K-nearest neighbor classifiers designed for the identification and categorization of defects on the transmission power lines [65]. The aim is to compare the two ML algorithms, decision tree and K-nearest neighbor, to acquire the most appropriate technique for analysis of fault. Table 4 shows the reviews of DT used for fault detection in transmission line.

Table 4 Reviews of DT employed for fault location in transmission line

Technique used	Advantages of technique	Author	Year
Decision tree and support vector machine	Performed with harmonics and noise in disturbance signals offering complete outcomes	Ray et al.	2014
Decision tree	Retrieves the data about the condition of fault and power swing and then recognizes the fault	Niyas and Sunitha Upendar et al.	2017 2012
Decision tree	Proficient highly in categorizing power transmission network faults with rapidity and precision	Jana and De	2017
Decision tree	Reduces the operation workload and maintenance staff	Zhang et al Mohanty et al.	2019 2020
Decision tree and *K*-nearest neighbor	Decision tree outperforms KNN with reduced time of testing providing similar accuracy of classification	Wasnik et al.	2020

2.5 Convolutional Neural Network

Fan et al. article propose a single-ended approach of fault detection for power transmission lines employing various current techniques of deep learning. A combination of CNN with LSTM framework is trained to find the distance of fault given the present measurements and single-ended voltage [66]. Bai et al. [4] proposed an approach based on NLP and an integrated structure that combines CNN and long and short-term memory for the recognition of alarm events grid monitoring [55]. A model of monitoring alarm event detection on the basis of integration of CNN and long short-term memory has been set up for the features of alarm data. Contrary to that Rudin et al. [8] discuss the feasibility of using the DL framework using CNN for fault classification of the real-time power system [56]. The fault classification system aims to categorize the power system signal samples in real-time and decide either they are in a faulted or non-faulted condition. Wang et al. [1] propose a CNN into the power line fault detection field [67]. In this study, a novel detection approach integrates the sliding window method and output map data. Li et al. [22] have proposed an approach of faulted line localization on the basis of CNN classifiers using voltages of the bus. With the physical representation features, the developed method proposes the strength of location performance. In order to further develop the location performance, a dual PMU placement method is validated and suggested against other techniques [57]. The study of Ekici and Unal [27] presents a new approach based on CNN for the automatic categorization of faults of the transmission line. The proposed method is inspired by the concept of how CNN is used for fault signal classification. For this purpose, colorized images of fault voltage signals from the

Table 5 Reviews of DT employed for fault location in transmission line

Technique used	Advantages of technique	Author	Year
Convolutional neural network	Resolves the power line fault detection issue in complex background effectively	Wang et al. Chen et al.	2017 2019
Natural language processing, CNN, and long short-term memory	Better identification impact for every fault type	Bai et al.	2019
CNN with long short-term memory	Predicts the fault distance with effectiveness and accuracy	Fan et al.	2019
CNN classifier	Determine fault features of the power system and classify them properly	Rudin et al.	2017
CNN classifier	Improves the location performance robustness	Li et al. Paul and Mohanty	2019 2019
CNN algorithm	Particularly used for detection and classification of fault signal	Ekici and Unal	2020

sending end of the power system were acquired by using continuous wavelet transform (CWT) [59]. Table 5 shows the reviews of CNN used for identification of fault in the power transmission line.

2.6 *Deep Belief Network*

Srinivasa Rao et al. [10] article proposes DBN in classifying and detecting signals, namely sag, swell, and transient in the transmission line [68]. In this article, signals of wavelet decomposed are retrieved, and the defect is identified on the basis of the decomposed signal by deep belief network. Hong et al. [21] proposed a novel fault classification approach based on deep belief networks. The fault voltage and current samples are pre-processed by the standardization of min-max and splicing of the waveform, and then it trains the deep belief network along with the fault type label [69]. The voltage and current amount will be extracted by a well-skilled deep belief network model automatically [60]. Table 6 shows the reviews of deep belief network used for fault location in transmission line.

Table 6 Reviews of use of deep belief network for fault location in transmission line

Technique used	Advantages of technique	Author	Year
Deep belief network	Comparatively, it classifies and detects the fault signals efficiently in power distribution system than that of other traditional models	Srinivasa Rao et al.	2018
Deep belief network	Provides greater accuracy of fault classification and better adaptability	Hong et al.	2020

2.7 *Other Deep Learning Algorithms Used for Fault Detection in Power Transmission*

Guomin et al. [28] proposed fault location-based depth learning for distribution networks of direct current [70]. The first distribution network of direct current with radiant topology is structured and the faults are added with varied parameters to pretend different situations in practical developments. Then, a DNN is trained and generated with normalized fault currents [70]. Mirzaei et al. [29] presented a DNN algorithm for fault detection in a three-terminal line of transmission with the existence of parallel flexible AC transmission systems (FACTS) device [71]. This study concentrates on both parallel compensated and multi-terminal lines. The created features are employed to train a DNN which decides faulted distance and line part simultaneously [62]. The performance of the algorithm is approved for both unsymmetrical and symmetrical types of faults, little inception angles of the fault, and greater resistance of fault. Muzammel evaluated the direct current fault in multi-terminal increased voltage transmission line on the basis of RBM. RBM is a resourceful stochastic ANN, wherein probability distribution learning is conducted over input sets [72]. Three high voltage direct current transmission system stations are simulated in both faulty and normal situations to examine the differences in parameters of electricity. These differences serve as a parameter of learning of RBM [64]. Table 7 shows the reviews of other techniques used for fault detection in transmission line.

Table 7 Reviews of use of ANN for fault location in transmission line

Technique used	Advantages of technique	Author	Year
Deep neural network	Minimize the reconstruction error	Guomin et al.	2018
Deep neural network	Provides a robust, accurate, and fast tool for the location of the fault in parallel compensated three-terminal lines of transmission	Mirzaei et al.	2019
Restricted Boltzmann machine	Follows the gradient of divergence differences	Muzzammel et al.	2020

3 Conclusion

Deep learning has been developed in the present decades. An architecture of deep learning that can be changed to new issues with ease is using techniques, namely RNN, CNN, LSTM, etc. Deep learning is not economical and as well consumes lots of time to train through several equipped machines with costly GUIs. Deep learning poses certain issue that essentially surpasses other solutions in several domains. In this research study, several authors have discussed different deep learning techniques for fault location in the transmission line. Swetapadma and Yadav [45] propose an ANN solution to detect faults of multi-location in the double circuit. The proposed ANN-based approach decides the faults of multi-location sing voltage and current signals of one line end, thus, avoiding the communication link requirement. Contrary to that the research of Luo et al. [49] proposed a technique for fault location on HVDC transmission lines. The stacked autoencoder is used to structure the association between locations of fault and fault currents. Fan et al. propose a single-ended approach of fault detection for transmission lines with the help of modern deep learning techniques. A combination of CNN with LSTM framework is trained to find the distance of fault given the present measurements and a voltage that is single-ended. Contrary to that the study of Mirzaei et al. [29] developed a DNN algorithm for fault detection in a three-terminal line of transmission with the existence of parallel flexible alternate current transmission systems device. This study concentrates on both parallel compensated and multi-terminal lines.

The safety of transmission line against fault is a difficult task in the security of the electrical power systems. Secure relaying can be used to recognize the abnormal signals indicating power transmission system faults. So the location of the fault is essential for high speed and reliable secure relaying. A systematic analysis of deep learning algorithms for fault identification in transmission line has been proposed in the current research. The proposed deep learning technique of fault location is predicted resistant to variation impact in the location and type of fault. Deep learning is an efficient technique that reduces the need for feature engineering and it is considered as an essential time-consuming factor of machine learning. It can be concluded that a deep learning algorithm is necessary for fault location in the power transmission system. Since the transmission of the network is progressing with certain difficulties and inadequate sizes are expected to be similar, vast area approaches would be used mainly for fault location and identification in predicting the near future.

References

1. M. Wang, W. Tong, S. Liu, Fault detection for power line based on convolution neural network, in *Proceedings of the 2017 International Conference on Deep Learning Technologies* (2017), pp. 95–101
2. J. Chen, X. Xu, H. Dang, Fault detection of insulators using second-order fully convolutional network model. Math. Probl. Eng. **2019** (2019)

3. R. Fan, T. Yin, R. Huang, J. Lian, S. Wang, Transmission line fault location using deep learning techniques, in *2019 North American Power Symposium (NAPS)* (2019), pp. 1–5
4. Z. Bai, G. Sun, H. Zang, M. Zhang, P. Shen, Y. Liu, Z. Wei, Identification technology of grid monitoring alarm event based on natural language processing and deep learning in China. Energies **12**(17), 3258 (2019)
5. A.S. Neethu, T.S. Angel, Smart fault location and fault classification in transmission line, in *Proceedings of IEEE International Conference on Smart Technologies and Management for Computing, Communication, Controls, Energy and Materials* (2017), pp. 339–343
6. H. Mahajan, A. Sharma, Various techniques used for protection of transmission line-a review. Int. J. Innov. Eng. Technol. (IJIET) **3**(4), 32–39 (2014)
7. G.P. Ahire, N.U. Gawali, Fault classification and location of series compensated transmission line using artificial neural network. Int. J. Adv. Electron. Comput. Sci. **2**(8), 77–81 (2015)
8. F. Rudin, G.J. Li, K. Wang, An algorithm for power system fault analysis based on convolutional deep learning neural networks. Int. J. All Res. Edu. Sci. Methods **5**(9), 11–17 (2017)
9. B. Singh, O.P. Mahela, T. Manglani, Detection and classification of transmission line faults using empirical mode decomposition and rule based decision tree based algorithm, in *2018 IEEE 8th Power India International Conference (PIICON)* (2018), pp. 1–6
10. T.C. Srinivasa Rao, S.S. Tulasi Ram, J.B.V. Subrahmanyam, Fault signal recognition in power distribution system using deep belief network. J. Intell. Syst. **29**(1), 459–474 (2018)
11. L. Teklić, B. Filipović-Grčić, I. Pavičić, Artificial neural network approach for locating faults in power transmission system, in *Eurocon 2013* (2013), pp. 1425–1430
12. S. Kirubadevi, S. Suthan, Wavelet based transmission line fault identification and classification, in *Proceedings of IEEE International Conference on Computation of Power, Energy Information and Communication* (2014), pp. 737–741
13. P.B. Singh, R. Sharma, N.K. Swarnkar, G. Kapoor, A review on fault detection, classification and its location evaluation methodologies in transmission lines. Gyan Vihar Univ. **5**(1) (2019)
14. S. Ghimire, Analysis of fault location methods on transmission lines, University of New Orleans (2014), pp. 1–79
15. P. Nonyane, *The application of artificial neural networks to transmission line fault detection and diagnosis*, Doctoral dissertation, 2016
16. R. Fan, Y. Liu, R. Huang, R. Diao, S. Wang, Precise fault location on transmission lines using ensemble Kalman filter. IEEE Trans. Power Deliv. **33**(6), 3252–3255 (2018)
17. V. Venkatesh, Fault classification and location identification on electrical transmission network based on machine learning methods, Virginia Common Wealth University, 2018
18. A. Raza, A. Benrabah, T. Alquthami, M. Akmal, A review of fault diagnosing methods in power transmission systems. Appl. Sci. **10**(4), 1312 (2020)
19. G. Kapoor, Evaluation of fault location in three phase transmission lines based on discrete wavelet transform. ICTACT J. Microelectr. **6**(1), 897–890 (2020)
20. N. Papernot, P. McDaniel, S. Jha, M. Fredrikson, Z.B. Celik, A. Swami, The limitations of deep learning in adversarial settings, in *2016 IEEE European symposium on security and privacy (EuroS&P)* (2016), pp. 372–387
21. C. Hong, Y.Z. Zeng, Y.Z. Fu, M.F. Guo, *Deep-Belief-Networks Based Fault Classification in Power Distribution Networks* (Wiley, 2020).
22. W. Li, D. Deka, M. Chertkov, M. Wang, Real-time faulted line localization and PMU placement in power systems through convolutional neural networks. IEEE Trans. Power Syst. **34**(6), 4640–4651 (2019)
23. D. Paul, S.K. Mohanty, Fault classification in transmission lines using wavelet and CNN, in *2019 IEEE 5th International Conference for Convergence in Technology (I2CT)* (2019), pp. 1–6
24. Y. LeCun, Y. Bengio, G. Hinton, Deep learning. Nature **521**(7553), 436–444 (2015)
25. N. Sapountzoglou, J. Lago, B.D. Schutter, B. Raison, A generalizable and sensor-independent deep learning method for fault detection and location in low-voltage distribution grids. Appl. Energy **276** (2020)

26. J. Guo, Y. Jiang, Y. Zhao, Q. Chen, J. Sun, Dlfuzz: differential fuzzing testing of deep learning systems, in *Proceedings of the 2018 26th ACM Joint Meeting on European Software Engineering Conference and Symposium on the Foundations of Software Engineering* (2018), pp. 739–743
27. S. Ekici, F. Unal, Classification of Energy Transmission Line Faults using Convolutional Neural Networks, IZDAS (2020)
28. L. Guomin, T. Yingjie, Y. Changyuan, L. Yinglin, H. Jinghan, Deep learning-based fault location of DC distribution networks. J. Eng. **2019**(16), 3301–3305 (2019)
29. M. Mirzaei, B. Vahidi, S.H. Hosseinian, Accurate fault location and faulted section determination based on deep learning for a parallel-compensated three-terminal transmission line. IET Gener. Transm. Distrib. **13**(13), 2770–2778 (2019)
30. R. Muzzammel, Restricted Boltzmann machines based fault estimation in multi terminal HVDC transmission system, in *Intelligent Technologies and Applications* (2020), pp. 772–790
31. N. Srivastava, G. Hinton, A. Krizhevsky, I. Sutskever, R. Salakhutdinov, Dropout: a simple way to prevent neural networks from over fitting. J. Mach. Learn. Res. **15**(1), 1929–1958 (2014)
32. A. Azriyenni, M.W. Mustafa, D.Y. Sukma, M.E. Dame, Backpropagation neural network modeling for fault location in transmission line 150 kV. Indonesian J. Electr. Eng. Inf. (IJEEI) **2**(1), 1–12 (2014)
33. S.V. Khond, G.A. Dhomane, Fault classification accuracy measurement for a distribution system with artificial neural network without using signal processing technique. Int. J. Innov. Technol. Exploring Eng. **9**(3), 1523–1526 (2020)
34. R.K. Goli, A.G. Shaik, S.T. Ram, A transient current based double line transmission system protection using fuzzy-wavelet approach in the presence of UPFC. Int. J. Electr. Power Energy Syst. **70**, 91–98 (2015)
35. A. Yadav, Y. Dash, An overview of transmission line protection by artificial neural network: fault detection, fault classification, fault location, and fault direction discrimination. Adv. Artif. Neural Syst. **2014** (2014)
36. M. Ben Hessine, S. Ben Saber, Accurate fault classifier and locator for EHV transmission lines based on artificial neural networks. Math. Probl. Eng. **2014** (2014)
37. K. Sanjay Kumar, R. Shivakumara Swamy, V. Venkatesh, Artificial neural network based method for location and classification of faults on a transmission lines. Int. J. Sci. Res. Publ. **4**(6), 1–5 (2014)
38. M. Jamil, S.K. Sharma, R. Singh, Fault detection and classification in electrical power transmission system using artificial neural network. SpringerPlus **4**(1), 1–13 (2015)
39. E. Koley, K. Verma, S. Ghosh, An improved fault detection classification and location scheme based on wavelet transform and artificial neural network for six phase transmission line using single end data only. Springerplus **4**(1), 551 (2015)
40. F. Patil, H.N. Prajapati, A review on artificial neural network for power system fault detection. Indian J. Res. **4**(1), 52–54 (2015)
41. A.Y. Hatata, Z.M. Hassan, S.S. Eskander, Transmission line protection scheme for fault detection, classification and location using ANN. Int. J. Mod. Eng. Res. **6**(8), 1–10 (2016)
42. E.P. Thwe, M.M. Oo, Fault detection and classification for transmission line protection system using artificial neural network. J. Electr. Electron. Eng. **4**(5), 89–96 (2016)
43. M. Sarathkumar, S. Pavithra, V. Gokul, N. Prabhu, Automatic fault detection and fault location in power transmission lines using ANN algorithm with labview. S. Asian J. Eng. Technol. **3**(3), 112–117 (2017)
44. P.O. Mbamaluikem, A.A. Awelewa, I.A. Samuel, An artificial neural network-based intelligent fault classification system for the 33-kV Nigeria transmission line. Int. J. Appl. Eng. Res. **13**(2), 1274–1285 (2018)
45. A. Swetapadma, A. Yadav, An artificial neural network-based solution to locate the multilocation faults in double circuit series capacitor compensated transmission lines. Int. Trans. Electr. Energy Syst. **28**(4), e2517 (2018)
46. A. Elnozahy, K. Sayed, M. Bahyeldin, Artificial neural network based fault classification and location for transmission lines, in *IEEE Conference on Power Electronics and Renewable Energy* (2019), pp 140–144

47. M. Jamil, A. Kalam, A.Q. Ansari, M. Rizwan, Generalized neural network and wavelet transform based approach for fault location estimation of a transmission line. Appl. Soft Comput. **19**, 322–332 (2014)
48. N. Liu, B. Fan, X. Xiao, X. Yang, Cable incipient fault identification with a sparse autoencoder and a deep belief network. Energies **12**(18), 3424 (2019)
49. G. Luo, C. Yao, Y. Liu, Y. Tan, J. He, K. Wang, Stacked auto-encoder based fault location in VSC-HVDC. IEEE Access **6**, 33216–33222 (2018)
50. G. Luo, Y. Tan, M. Li, M. Cheng, Y. Liu, J. He, Stacked auto-encoder-based fault location in distribution network. IEEE Access **8**, 28043–28053 (2020)
51. A. Saber, A. Emam, R. Amer, Discrete wavelet transform and support vector machine-based parallel transmission line faults classification. IEEJ Trans. Electr. Electron. Eng. **11**(1), 43–48 (2016)
52. K. Hosseini, Short circuit fault classification and location in transmission lines using a combination of wavelet transform and support vector machines. Int. J. Electr. Eng. Inf. **7**(2), 353 (2015)
53. A.A. Zakri, S. Darmawan, J. Usman, I.H. Rosma, B. Ihsan, Extract fault signal via DWT and penetration of SVM for fault classification at power system transmission, in *2018 2nd International Conference on Electrical Engineering and Informatics (ICon EEI)* (2018), pp. 191–196
54. R. Singh, T. Chopra, Fault classification in electric power transmission lines using support vector machine. Int. J. Innov. Res. Sci. Technol. **1**(12), 388–399 (2015)
55. N.R. Babu, B.J. Mohan, Fault classification in power systems using EMD and SVM. Ain Shams Eng. J. **8**(2), 103–111 (2017)
56. M.Y. Cho, T.T. Hoang, Feature selection and parameters optimization of svm using particle swarm optimization for fault classification in power distribution systems. Comput. Intell. Neurosci. **2017** (2017)
57. S.S. Gururajapathy, H. Mokhlis, H.A.B. Illias, Classification and regression analysis using support vector machine for classifying and locating faults in a distribution system. Turk. J. Electr. Eng. Comput. Sci. **26**(6), 3044–3056 (2018)
58. C.D. Prasad, N. Srinivasu, Fault detection in transmission lines using instantaneous power with ED based fault index. Procedia Technol. **21**, 132–138 (2015)
59. N.S. Wani, R.P. Singh, A novel approach for the detection, classification and localization of transmission line faults using wavelet transform and support vector Machinenclassifier, Int. J. Eng. Technol. **7**(2) (2018)
60. M. Niyas, K. Sunitha, Identification and classification of fault during power swing using decision tree approach, in *International Conference on Signal Processing, Information and Communication and Energy Systems* (IEEE Publisher, India, 2017)
61. S. Jana, A. De, Transmission line fault pattern recognition using decision tree based smart fault classifier in a large power network, in *2017 IEEE Calcutta Conference (CALCON)* (2017), pp. 387–391
62. W. Zhang, Y. Wang, X. Wang, J. Wang, *Decision Tree Approach for Fault Type Identification of Transmission Line*, vol. 477 (IOP Publishing, 2019)
63. G. Kasinathan, N. Kumarappan, Double circuit EHV transmission lines fault location with RBF based support vector machine and reconstructed input scaled conjugate gradient based neural network. Int. J. Comput. Intell. Syst. **8**(1), 95 (2015)
64. P.P. Wasnik, N.J. Phadkule, K.D. Thakur, Fault detection and classification in transmission line by using KNN and DT technique. Int. Res. J. Eng. Technol. **7**(4), 335–340 (2020)
65. P. Ray, D.P. Mishra, Support vector machine-based fault classification and location of a long transmission line. Eng. Sci. Technol. Int. J. **19**(3), 1368–1380 (2016)
66. J.M. Johnson, A. Yadav, Complete protection scheme for fault detection, classification and location estimation in HVDC transmission lines using support vector machines. IET Sci. Meas. Technol. **11**(3), 279–287 (2016)
67. H.T. Thom, C.H.O. Ming-Yuan, V.Q. Tuan, A novel perturbed particle swarm optimization-based support vector machine for fault diagnosis in power distribution systems. Turk. J. Electr. Eng. Comput. Sci. **26**(1), 518–529 (2018)

68. H. Livani, C.Y. Evrenosoğlu, A fault classification method in power systems using DWT and SVM classifier, in *PES T&D 2012* (IEEE, 2012), pp. 1–5
69. P.K. Ray, S.R. Mohanty, N. Kishor, J.P. Catalão, Optimal feature and decision tree-based classification of power quality disturbances in distributed generation systems. IEEE Trans. Sustain. Energy **5**(1), 200–208 (2014)
70. J. Upendar, C.P. Gupta, G.K. Singh, Statistical decision-tree based fault classification scheme for protection of power transmission lines. Electr. Power Energy Syst. **36**, 1–12 (2012)
71. M.M. Taheri, H. Seyedi, B. Mohammadi-ivatloo, DT-based relaying scheme for fault classification in transmission lines using MODP. IET Gener. Transm. Distrib. **11**(11), 2796–2804 (2017)
72. S.K. Mohanty, A. Karn, S. Banerjee, Decision tree supported distance relay for fault detection and classification in a series compensated line, in *2020 IEEE International Conference on Power Electronics, Smart Grid and Renewable Energy (PESGRE2020)* (2020), pp. 1–6
73. K. Chen, J. Hu, J. He, Detection and classification of transmission line faults based on unsupervised feature learning and convolutional sparse autoencoder, in *2017 IEEE Power & Energy Society General Meeting* (2017), p. 1

A Cost-Efficient Magnitude Comparator and Error Detection Circuits for Nano-Communication

Divya Tripathi and Subodh Wairya

Abstract Quantum-dot Cellular Automata (QCA) is a leading computing paradigm for nano-communication. QCA is appreciated due to its less power consumption, fast speed, and a minor dimension, and so it is an encouraging substitute to CMOS technology. One of the latest and emerging nanotechnologies used nowadays is QCA which is based on Coulomb repulsion. Presently, circuit designers are leading toward an advanced applied science dependent on the electron polarization is named Quantum-dot Cellular Automata, which contribute productive results about area decrement and quantum cost to the excessive low horizon. In this article, an efficient, less complex XOR gate (N11) and with the help of this XOR gate, designed a magnitude comparator, parity generators, and parity checker circuitry for nano-communication applications. The proposed circuitry for nano-communication illustrates the competency of these designs. The competency of the planned architecture is checked by the QCA designer simulation environment. The simulation outcomes illustrate that the planned circuitry outperforms the parameters of quantum cell count, area, latency, and quantum cost as compared to its best existing counterpart. The cost function of the proposed QCA architecture is rising to the finest as related to its previous best counterpart.

Keywords XOR · QCA · Nano-communication

1 Introduction

Quantum-dot Cellular Automata (QCA) suggests high device density, less power consumption, and rapid switching speed structures [1, 2]. Traditional CMOS has conquered our fabrication in recent years and it holds prove to be an improved alternative than previous automation methods. QCA is an innovative technology beyond the CMOS paradigm that may be used for executing nanoscale circuitry. The impediments of CMOS methodology are such high density in a very small arena due to the

D. Tripathi (✉) · S. Wairya
Institute of Engineering and Technology, Dr. APJ. Abdul Kalam Technical University, Lucknow, India

V. Suma et al. (eds.), *Inventive Systems and Control*,
Lecture Notes in Networks and Systems 204,
https://doi.org/10.1007/978-981-16-1395-1_19

expanding numbers of transistors [3, 4]. In QCA, the logical data will be transmitted through Coulombic repulsion between the adjacent QCA cells. Here, the optimization of QCA cells is used for reducing the area of any digital logic circuitry, and the quantum cost of the function is also reduced [5]. Quantum computing, QCA, optical computing, spin electronics, DNA computing, molecular computing, nano-communication, and power-efficient nano-computing are few evolutionary nanotechnologies implemented by reversible logic [6, 7]. Nano-communication design of the proposed circuitry illustrates the effectiveness of the QCA designs. Distinctive parameters like quantum cell counts, area, latency, and quantum cost are utilized to estimate the proposed designs that affirm the speedier processing at the nanoscale. In this paper, a cost-efficient design of a 2-input QCA XOR gate has been introduced. To calibrate the dominance of our projected QCA XOR gate, error detection circuitry has been proposed later. Here, the cell optimization and realization of digital logic circuitry were used in nano-communications. The proposed QCA designs are modest and conquer a fraction of the area, the number of cells used and quantum cost as associated with previously existing designs. The design approach is very apt in designing various arithmetic and other such circuits with very less power utilization and lesser surface which is useful in nano-communication.

In digital communication, the identification of fault in the receiving information, there is a major concern of key. The parity bit is utilized in the detection of such sort of errors, amid transfer of binary data via network an additional bit is cushioned with the data to identify the mistake inside the message. This additional bit is called the parity bit. The parity bit is cushioned to convey a total number of 1's (counting parity bits) in a binary message either even or odd. Other than, at the nanoscale, the complexity of fault location and rectification is the foremost daring viewpoint at the hardware level in terms of circuitry area and energy dissipation [8, 9]. So, in this article, the cell optimization and realization of magnitude comparators, parity generators, and parity checkers using QCA are explained. This article has the basic contribution are as follows;

(a) Designing a cost-efficient XOR gate using QCA.
(b) Designing of an efficient magnitude comparator using a proposed cost-efficient XOR gate in QCA.
(c) Designing of numerous parity generators and parity checkers using a proposed cost-efficient XOR(N11) gate.
(d) The proposed architectures are compared to existing designs based on quantum cell count, area, latency, and the quantum cost which confirms that proposed designs have a lesser area and faster speed as related to their previous best counterpart.

The layout of the article is organized in the following manner. Section 2 deals with a brief overview of QCA. Section 3 shows the proposed architectures of QCA XOR gate, parity generator, and parity checker. Section 4 executed simulation results and discussion of the proposed architecture and comparison with its previous best counterparts. Finally, Section 5 draws the conclusion of the research work.

2 A Brief Oversight of Quantum-Dot Cellular Automata

QCA formations are put together as an arrangement of quantum cells within which each cell has communicated electrostatically with its neighboring cells. QCA technology has built-in capacities for digital circuitry scheme without ensuring any Boolean function [6, 7]. QCA cell plays a vital role in QCA methodology, which allows us to implement a couple of the computation and transfer the data through the overlap. A simple QCA cell contains a speculative squarish space that holds the four energy sites there in which electrons can absorb. Each cell is invaded by two electrons. Also, the Coulombic interaction between electrons can create two distinct cell states with different charge arrangements [8]. This arena for an electron is meant by a dot in the cubicle cell. Quantum mechanical tunneling barriers are used to couple the phases so that electrons can tunnel via them determined on the structural state.

Columbia's repulsion compels the electrons to involve the furthermost dots in a QCA cell which resembles the nethermost energy state-owned of the circuitry. Cell polarization refers to the relative locations of the electrons in a cell and it selects whether it is signifying binary '1' or '0'. Two categories of QCA cells are there, the first is 90° and the next is 45°. Figure 1 presents a 90° cell with the state of polarization of $P = +1$ which signifies binary 1 [9]. The clock signal delivers the desirable force to do the calculation. A simple QCA cell contains a speculative squarish-shaped arena in which there are four energy sites in which electrons can absorb. This arena for an electron is signified through a dot in the cubicle cell.

QCA cells are organized continually to compose a QCA wire. The attractiveness and repulsiveness taking place because of the Columbia force between the end-to-end cellular sources the cell polarization so that it can arrange in a line according to its neighboring cells so the communication of data besides the cell array takes place. QCA wires transmitting binary '1' using 90° QCA cells. The input cells are enforced by an outside source and are powerfully polarized in one way. The input cells drive another QCA cell in the NULL state which tends to bring into line them to the polarization of the input cell to attain the system's ground states [10].

2.1 *Majority Gate*

The arrangements of the majority gate are offered in Fig. 2. The output Y is well-defined as $Y = \text{PN} + \text{PO} + \text{NO}$. The output of the QCA cell of the gate diverges affording to the calculations of a QCA cell in the medium of the gate. The output 'Y' as it may be driven out along the help of QCA wire whichever can execute as an input to other gates. The majority gate plays a significant role to construct the OR/AND gates [10, 11]. If one input is stable to 1/0, the subsequent function Y is the OR/AND of the rest of the two inputs.

$$M(P, N, O) = F = (P * N) + (N * O) + (P * O) \quad (1)$$

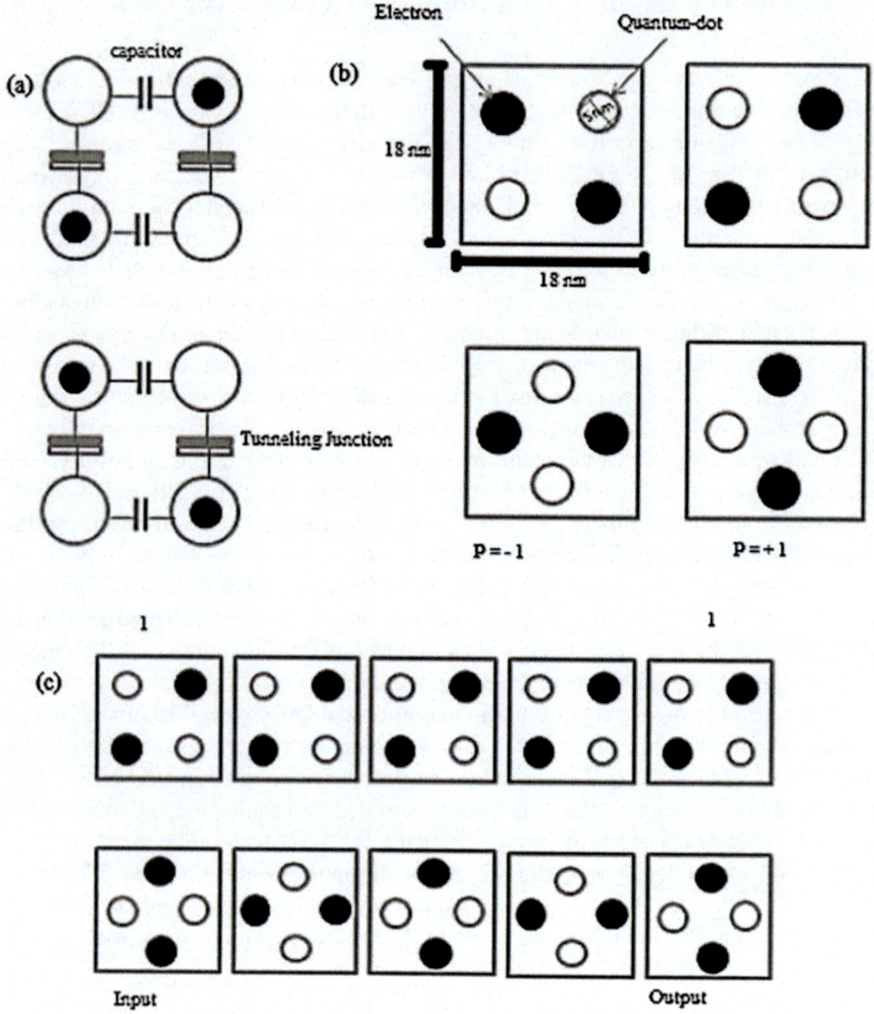

Fig. 1 QCA cells **a** functional outline, **b** polarizations in cells, **c** QCA wire

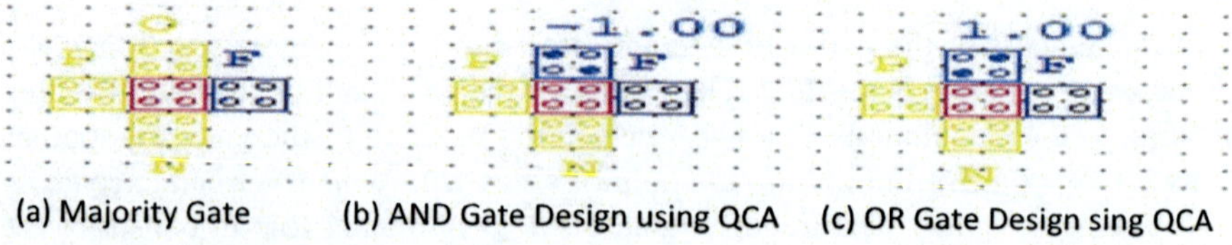

Fig. 2 Basic QCA logic gates architecture

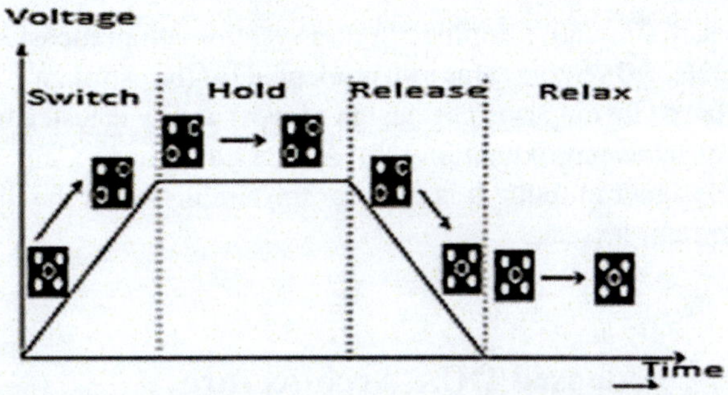

Fig. 3 Clocking concept in QCA: four phase

where P, N, and O are 3 inputs of the majority gate and F is the one output of the majority gate. The basic architecture of the majority gate contains 5 quantum cells in which 3 cells used as input 1 cell are used as output and 1 cell is fixed normal cell, and it can change the polarization of any input cell if it is assigned polarization of one input cell is -1 so whole architecture works as AND gate and if the polarization of the cell is $+1$ then whole architecture become OR gate as presented in Fig. 2.

2.2 *QCA Clocking*

The data streaming in the QCA architecture is managed and coordinated via the clocking mechanism. Clocking offers the ability to make and keeps off the metastable phase [11, 12]. The clocking in QCA technology is not similar as they are customary CMOS circuits, the QCA clocking arrangements contain quadruple phases: Switch (unpolarized cells determined by several inputs and become polarized dependent on their neighbors' polarization), control (cells contained in the same binary state so that it can apply as per the input to other cells), release (barriers are taken down and cells remain unpolarized) and relax (cell remains unpolarized) [13, 14]. This is the phase divergence of the quarter cycle in all these clocking phases as it may be supported through generating 4 clocks each one with $\pi/2$ phase difference from the previous one as shown in Fig. 3.

2.3 *QCA Designer*

QCA designer tool used for simulation of the complex QCA architectures. It was produced in the ATIPS Laboratory of the University of Calgary. The newest version of QCA designer has three diverse simulation engines included. A piece of the three engines has a dissimilar and significant band of profits and disadvantages to boot, and each simulation engine can do a systems comprehensive confirmation or a set of user-selected vectors. This tool is adapted for the estimation of energy dissipation in QCA-based layout also. The design of workability is tested under the simulation engine of bistable approximation [15]. AITPS Laboratory, Calgary University developed QCA designer initially. It is used for the simulation of the digital circuit by creating QCA circuit layouts.

3 Proposed QCA Architecture

3.1 Proposed QCA XOR Gate

In this section, an efficient and optimum design of a 2-input QCA XOR gate has been introduced. It is a main digital circuitry that is practiced in many diverse kinds of combinatorial circuitry such as comparators and error detector circuits. The conventional and less complex QCA XOR layout had been suggested by so many researchers. But our proposed design is optimum as compared to its best previous counterparts. To check the dominance of our proposed QCA XOR gate, numerous complex QCA designs have been proposed later. The QCA layout of the 2-input QCA XOR layout is presented in Fig. 4 here '*A*', '*B*' is the input of the QCA XOR gate and 'XOR' is the output simulation waveform is presented in Fig. 5.

In this literature, an efficient and optimum design of a 2-input XOR layout using the QCA designer tool is suggested. It is defined by very little quantum cell count and higher density as compared to previous designs. The new 2-input XOR gate design contains 11 numbers of quantum normal cells, an area of 0.018 μm^2.

Some prior QCA layout designs of 2-inputs QCA "XOR" gate are used for literature survey and summarized in Table 1. Niemier [8] aimed QCA XOR gate contains 60 number of cells, 0.011 μm^2 area, and 1.5 clocking zone latency. This architecture delivers a significant number of cells and delivers a great field. To solve these difficulties, Hashemi et al. [9] planned a novel 2-input QCA XOR gate publicized with only 51 cells, 0.092 μm^2 area, and 2 clocking latency. To diminution the cell count, one more configuration is proposed by Chabi et al. [10] contains 29 cells,

Fig. 4 QCA layout of the proposed XOR gate (N11)

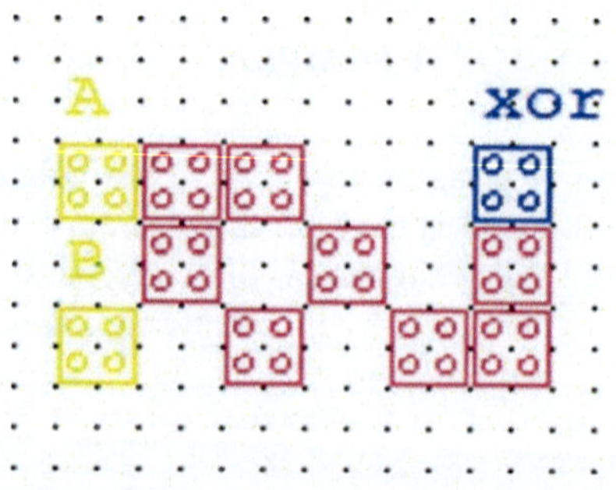

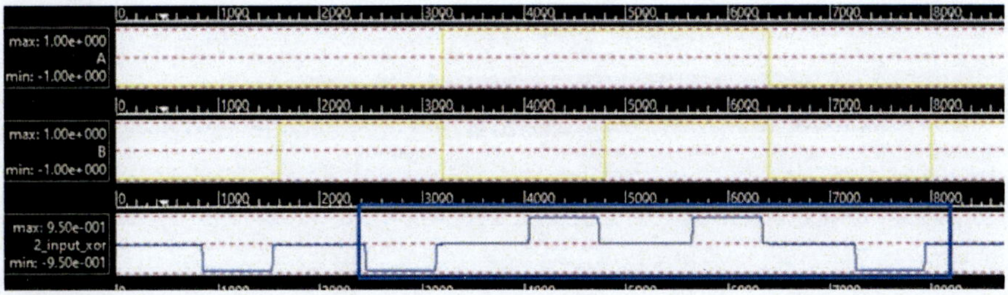

Fig. 5 Simulation waveform of the proposed XOR gate (N11)

Table 1 QCA XOR designs: a comparison

QCA XOR gate	Year	No. of QCA cells	Area (μm^2)	Latency (clocking cycles)	Quantum cost (area * latency)
[8]	2004	60	0.113	1.50	0.016
[9]	2013	51	0.092	2.00	0.184
[10]	2014	29	0.041	0.25	0.010
[11]	2016	28	0.035	0.75	0.026
[12]	2017	12	0.012	0.50	0.005
[13]	2017	14	0.015	0.50	0.008
[14]	2019	27	0.034	0.75	0.025
Proposed XOR gate PD (N11)		11	0.010	0.25	0.004

0.041 μm^2 area, and 0.25 clock latency. One more design offered to cut the expense of the QCA XOR gate by Singh, et al. [11], using two QCA inverters and five input QCA majority gate with 28 cells, an area of 0.035 μm^2, and 0.75 clocking zones latency. The embodiment according to Bahar et al. [12] has compacted the cell count to 12 cells for 2-input and 3-input XOR gate, area 0.021 μm^2, and latency of 0.05. According to Roohi et al. [13] introduced XOR gate has 14 cells, an area of 0.034 μm^2, and 0.5 clock latency. A new XOR gate was introduced by Das et al. [14] having 27 cells and 0.034 μm^2.

3.2 *Proposed QCA Binary Magnitude Comparator*

In this section, an efficient 1-bit magnitude comparator has been suggested with less number of cells and area. The comparators can be designed as a combinatorial circuitry that relates to the assessment of the two binary numbers and acknowledges if one of the inputs is identical, big, or lesser than the other input. The QCA layout of the suggested magnitude comparator is presented in Fig. 6 and the simulation waveform is presented in Fig. 7.

Fig. 6 QCA layout of the proposed magnitude comparator (N24)

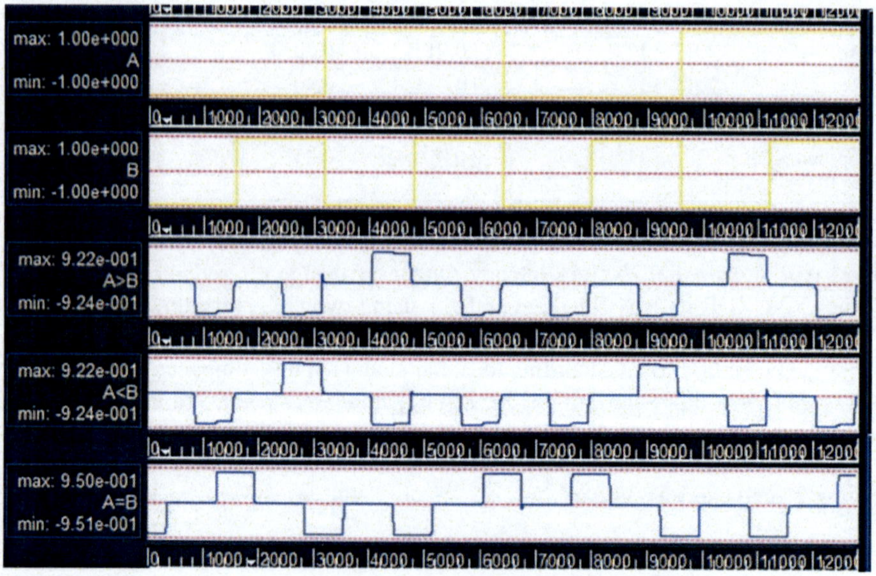

Fig. 7 Simulation waveform of the proposed magnitude comparator (N24)

Some prior QCA layout designs of QCA magnitude comparator are used for the literature survey and summarized in Table 2. Shiri et al. [15] aimed QCA magnitude comparator contains 38 cells, 0.060 μm^2 area, and 1.5 clocking zone latency. This architecture delivers a significant number of cells and delivers a great field. To solve these difficulties, Mokhtari et al. [16] planned a novel QCA magnitude comparator publicized with only 29 cells, 0.052 μm^2 area, and 2 clocking latency. Ghosh et al.

Table 2 QCA magnitude comparator: a comparison

QCA magnitude comparator	Year	No. of QCA cells	Area (μm^2)	Latency (clocking cycles)	Quantum cost (area * latency)
[15]	2019	38	0.060	1.50	0.090
[16]	2018	29	0.052	2.00	0.104
[17]	2012	73	0.063	0.25	0.016
[18]	2017	37	0.043	0.75	0.032
[19]	2017	42	0.067	0.50	0.033
[20]	2020	58	0.083	0.50	0.041
[21]	2018	100	0.150	0.25	0.037
[22]	2020	31	0.056	0.25	0.014
[23]	2020	27	0.051	0.25	0.013
PD (N24)		24	0.035	0.25	0.008

[17] introduced the novel magnitude comparator layout having only 73 numbers of cells and 0.063 μm^2 and 0.25 clocking latency. Roy et al. [18] proposed an optimum design of magnitude comparator having only 37 quantum cells and 0.43 μm^2 area and 0.75 clocking latency which was the optimum design. Deng et al. [19] 42 quantum cells, 0.067 μm^2, and 0.50 clocking latency. Mira et al. [20] introduced an efficient design having 58 numbers of quantum cells, 0.083 μm^2, and 0.50 clocking latency. Jun-wen et al. [21] proposed a design having 100 quantum cells having 0.015 μm^2 and 0.25 clocking cycle. To diminution the cell count, two more recent configurations are proposed by Gao et al. [22], and Tripathi et al. [23] contain 27 and 24 quantum cells, 0.051 μm^2 and 0.035 μm^2 area, and 0.25 clocking latency.

3.3 *Parity Generator*

The parity generator is a basic component of the data processing system in which the precise coordination of all received and transmitted information must be verified. Because of the popularity of the operation, numerous endeavors have been made to execute this basic component in QCA technology [24–26]. In this segment, an efficient parity generator using QCA designer was proposed to have less number of QCA cell count as compared to its best previous counterpart. Similarly, for designing more bit, parity generators have to cascade 4-bit parity generators as per the bit so can conclude that this circuitry can be effortlessly expanded up to the n-bit QCA parity generator. Figures 8 and 9 show the QCA layout of the planned 4-bit and 8-bit parity generator circuits, and the simulation waveform is presented in Fig. 10.

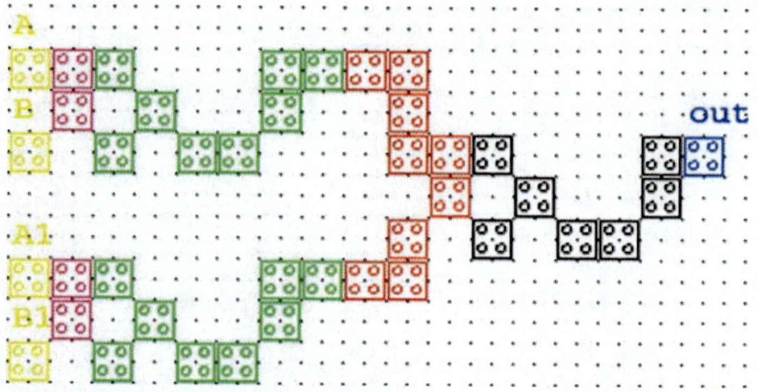

Fig. 8 QCA layout of proposed 4-bit parity generator

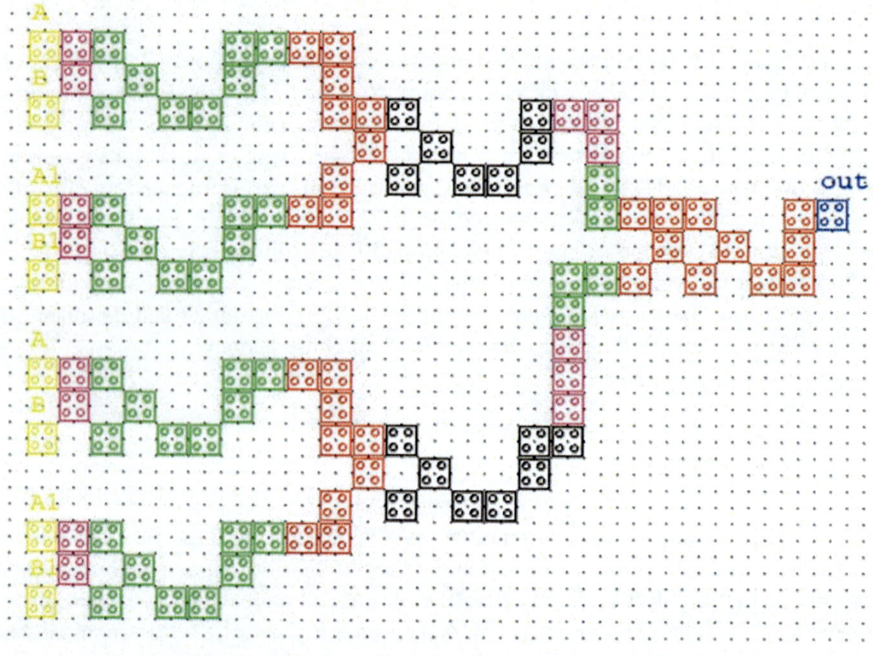

Fig. 9 QCA layout of proposed 8-bit parity generator

The QCA layouts of 4-bit and 8-bit parity generators were demonstrated. The number of quantum cells, area, and latency of the proposed 4 and 8-bit parity generator circuits is significantly enhanced as related to 4-bit and 8-bit parity generator circuitry as shown in Table 3.

Some prior QCA layout designs of parity generators are used for the literature survey and summarized in Table 3. Singh et al. [24] aimed QCA parity generators 4-bit having 87, 8-bit having 213 quantum cell, 16-bit having 480, and 32-bit having

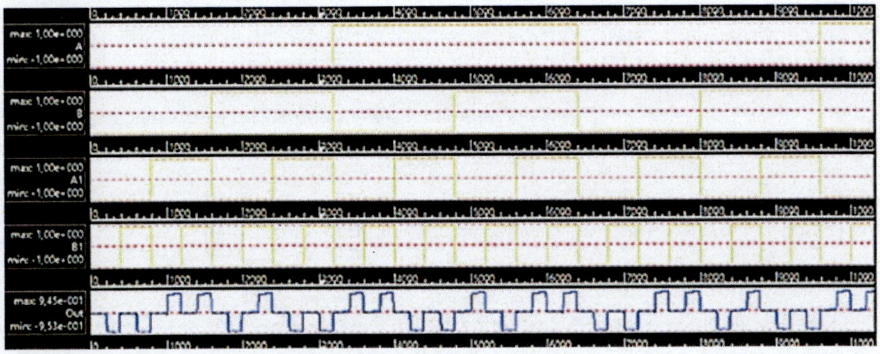

Fig. 10 Simulation waveform of 4-bit parity generator

Table 3 Even parity generator circuits: a comparison

Parity generator	Year	No. of bits	Cell count	Area (μm^2)	Latency (clocking cycles)	Quantum cost (area * latency)
[24]	2016	4	87	0.10	1.75	0.17
		8	213	0.30	2.75	0.82
		16	480	0.81	3.75	3.03
		32	1,044	2.08	4.75	9.88
[25]	2018	4	111	0.14	2	0.28
		8	269	0.43	3	1.29
		16	603	1.13	4	4.52
		32	1,312	2.81	5	14.0
[26]	2017	4	37	0.05	1.5	0.07
		8	97	0.18	2.5	0.45
		16	227	0.50	3.5	1.75
		32	511	1.31	4.5	5.85
Proposed QCA parity generator		4	30	0.06	1.5	0.09
		8	104	0.23	2.5	0.57
		16	218	0.46	4.0	1.84
		32	463	1.02	4.5	4.59

1044 quantum cells. Poorhosseini et al. [25] planned an efficient design containing 111 quantum cells in 4-bit, 269 number of cells in 8-bit, 603 cells in 16-bit, and 1312 quantum cells in 32-bit parity generators design. To diminution the cell count, one more configuration is proposed by Kumar et al. [26] contains 37 quantum cells in 4-bit, 97 number of cells in 8-bit, 227 quantum cells in 16 bit, and 32-bit parity generator contains only 511 cells. Our proposed design is optimum in several cells as well as in area also as depicted in comparison with Table 3.

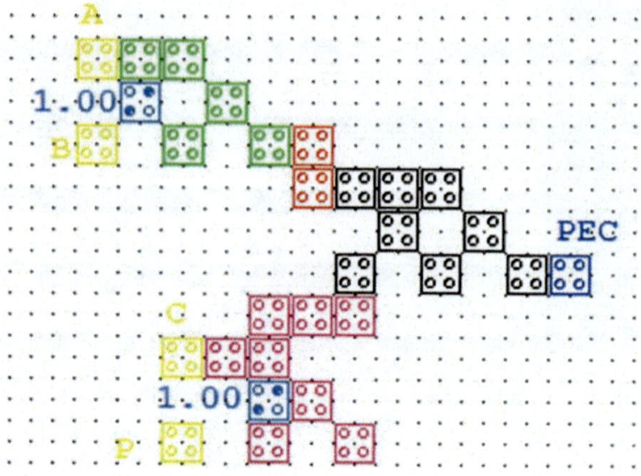

Fig. 11 QCA layout of proposed 4-bit parity checker

3.4 Parity Checker

Parity bit, *P* produced over will be transmitted along with the message bits. These four inputs *A*, *B*, *C*, and *P* will be the input to the parity checker circuitry which checks the plausibility of error with inside the information. Since the information with even parity is transmitted, the received message must have an even number of 1 s. The output of the parity detector, PEC will be 0 in case of error-free data transfer and 1 if there occurs error, i.e., the four bits received have four numbers of bits. The basic logic expression for error check bit PEC is written as XOR operation of the inputs as mentioned in Eq. (2). The proposed QCA layout of the parity checker is shown in Fig. 11 and the simulation waveform is shown in Fig. 13 (Fig. 12).

$$\text{PEC} = A \oplus B \oplus C \oplus P \tag{2}$$

Some prior QCA layout designs of parity checkers are used for the literature survey and summarized in Table 4. Agarwal et al. [29] aimed QCA parity checker 4-bit having 220 quantum cells and 0.38 μm^2. Das et al. [14] planned an efficient design containing 96 quantum cells and a 0.08 μm^2 area. Our proposed design is optimum in the number of cells as well as in area also as depicted in comparison with Table 4.

4 Simulation Result and Discussion

QCA the designer is taken for simulation of proposed QCA architectures (PD). The simulation parameter like the number of cells, area, latency, and quantum cost has been calculated and matched after optimizing the number of cells and further reducing area so the latency will also be decreased. The proposed methodology is used in

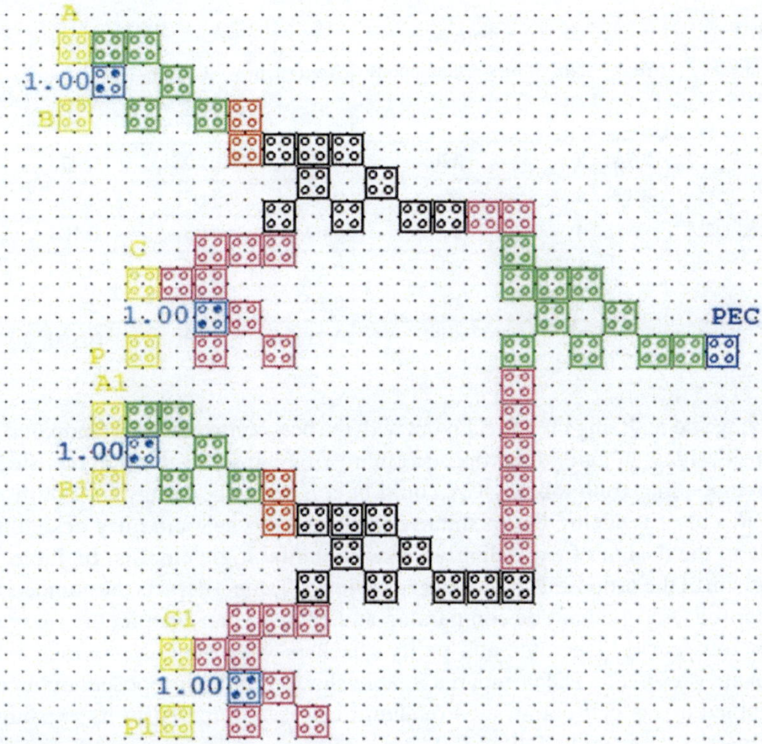

Fig. 12 QCA layout of proposed 8-bit parity checker

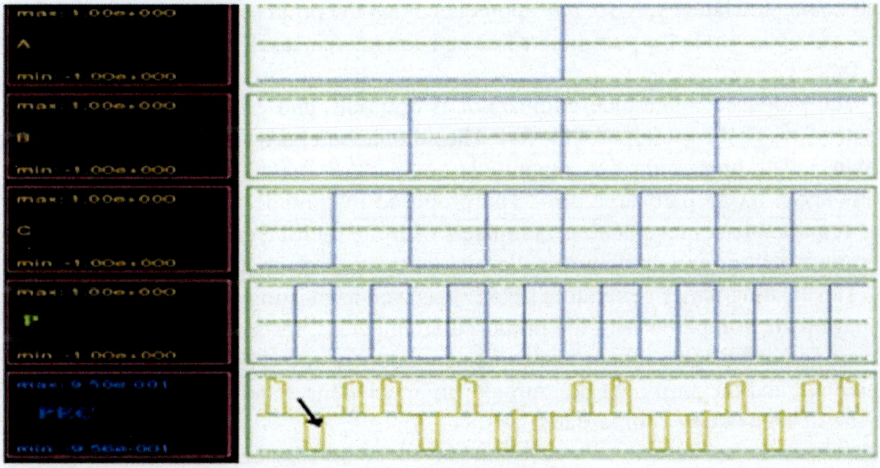

Fig. 13 Simulation waveform of proposed QCA parity checker

Table 4 Even parity checker: a comparison

Parity checker designs	Year	No. of bits	Cell count	Area (μm^2)	Latency (clocking cycles)	Quantum cost (area * latency)
[29]	2016	4	220	0.38	1.25	0.475
[14]	2019	4	96	0.08	1.75	0.146
Proposed QCA Parity Checker		4	34	0.07	1.5	0.105
		8	80	0.19	3.5	0.665
		16	184	0.50	4.0	2.000
		32	405	0.71	5.0	3.550

optimizing and realizing the QCA layout with the best competency as compared to the previous designs. In gain, the proposed circuit simulation results were verified with the corresponding truth table. QCA could be a new methodology for optimization of the area after reducing the quantum cell for a nanoscale circuitry design that is suitable to the design of exceedingly versatile digital logic circuits. In Table 1, the comparison of the proposed QCA XOR is depicted, Table 2 depicts the comparison of the proposed QCA magnitude comparator, and Tables 3 and 4 depict the comparison of numerous parity generator and parity checkers with various previous existing designs are presented. The comparison table shows that the proposed architectures have gained a less cell count, area quantum cost as compared to its existing best counterpart as depicted in Tables 1, 2, 3 and 4.

The projected architecture of the QCA XOR gate represents the exponential presentation than the multilayer XOR gate design in different parameters like area, cell count, and latency. Also, it is appreciated that the proposed XOR gate is competent and optimum in the parameters like cell count, area, and quantum cost as offered in Table 1

The proposed magnitude comparator is optimum and efficient as compared to its previous best counterpart [15–23]. The comparison between designs is shown in Table 2. The proposed QCA magnitude comparator contains only 24 numbers of cells and a 0.035 μm^2 area also. The proposed magnitude comparator represents the remarkable achievement as compared with the multilayer designs in numerous parameters like area, number of cells, and latency which are presented in Table 2.

The existing parity generators [24–27], as well as existing parity checker circuits [28, 29], are compared with the proposed architecture, respectively. The results are investigated through Table 4. The comparison demonstrates that the proposed parity generator and the parity checker circuits are much denser and speedier as compared to their best existing counterparts.

5 Conclusion

Quantum-dot Cellular Automata can attain higher density, fastest switching speed, and room temperature performances. Recently, this new technology is attracting so many scientists worldwide. It could be a new methodology for optimization of the area after reducing the quantum cell for a nanoscale circuitry design that is suitable to the design of exceedingly versatile digital logic circuits. The proposed design of the optimum and efficient XOR gate, magnitude comparator, parity generator, and parity checker is executed and compared with the existing layout. The proposed QCA layouts are executed and simulated on the QCA designer simulation environment with all default parameters. The functionality and performance of the proposed designs are better as compared with the best existing previous designs. The cell optimization and the realization of the magnitude comparator, parity generator, and parity checker were analyzed using the proposed XOR (N11) gate. The suggested architectures can be utilized as a primary component in designing of QCA-based transceiver for nano-transmission. The proposed XOR (N11) used only normal cells, and a comparison of numerous XOR gate, magnitude comparator, 4-bit, 8-bit, 16-bit, and 32-bit parity generator and parity checker circuits concerning cell count, area, latency, and the quantum cost is explained in this article. The proposed architecture represents a remarkable achievement as compared to its best previous counterparts. Nevertheless, a bunch of research and consideration is required in this course to record for the period of the nano-communication paradigm.

References

1. S.R. Kassa, R.K. Nagaria, A novel design of quantum dot cellular automata 5-input majority gate with some physical proofs. J. Comput. Electr. **15**, 324–334 (2015)
2. M.A. Shafi, A.N. Bahar, M.R.B. Mohammad, S.M. Shamim, K. Ahmed, Average output polarization dataset for signifying the temperature influence for QCA designed reversible logic circuits, in *Data in Brief*, vol.19 (Elsevier Inc., 2018), pp. 42–48
3. M. Khakpour, M. Gholami, S. Naghizadeh, Parity generator and digital code converter in QCA nanotechnology. Int. Nano Let. **10**, 49–59 (2020)
4. M. Balali, A. Rezai, H. Balali, S. Emadid, Towards coplanar QCA adders based on efficient three-input XOR gate. Result Phys. **7**, 1389–1395 (2017)
5. K. Sridharan, V. Pudi, *Design of Arithmetic Circuits in Quantum Dot Cellular Automata Nanotechnology*, vol. 599 (Springer, 2015), pp. 1–71
6. S. Seyedi, A. Ghanbari, N.J. Navimipour, New design of a 4-bit ripple carry adder on a nano-scale QCA. Moscow Univ. Phys. **74**, 494–501 (2019)
7. S. Kidwai, D. Tripathi, S. Wairya, Design of full adder with self-checking capability using QCA. Adv. VLSI, Comm. Sig. Process. 719–731 (2020)
8. M.T. Niemier, Designing digital systems in QCA. M.S. thesis. Uni. of Notre Dame. (2004)
9. S. Hashemi, Farazkish, R.K. Navi, New QCA cell arrangements. J. Comput. Theor. Nanosci. **10**, 798–809 (2013)
10. A. Chabi, S. Sayedsalehi, S. Angizi, K. Navi, Efficient QCA XOR and multiplexer circuits based on a nanoelectronic- compatible designing approach. **9** (2014)
11. G. Singh, R.K. Sarin, B. Raj, A novel robust exclusive-or function implementation in qca nanotechnology with energy dissipation. J. Comp. Elect. **15**, 455–465 (2016)

12. A.N. Bahar, S. Waheed, N. Hossain, Md. Asaduzzaman, A Novel 3-input XOR function implementation in quantum dot- cellular automata with energy dissipation analysis. Alexandria Eng. J. **57**, 729–738 (2017)
13. A. Chabi, A. Roohi, H. Khademolhosseini, S. Sheikhfaal, S. Angizi, K. Navi, R.F. DeMara, Towards ultra- efficient QCA reversible circuits. Microprocess. Microsyst. **49**, 127–138 (2017)
14. J.C. Das, D. De, S.P. Mondal, A. Ahmadian, F. Ghaemi, N. Senu, QCA based error detection circuit for nano communication network. **7**, 67355–67366 (2019)
15. A. Shiri, A. Rezai, H. Mahmoodian, Design of efficient coplanar comparator circuit in QCA technology. Facta Uni Ser. **32**, 119–128 (2019)
16. R. Mokhtari, A. Rezai, Investigation and design of novel comparator in quantum-dot cellular automata technology. J. Nano Electron. Phys. **10** (2018)
17. B. Ghosh, S. Gupta, S. Kumari, Quantum dot cellular automata magnitude comparators, in *Paper presented at: 2012 IEEE International Conference on Electron Devices and Solid State Circuit (EDSSC)* (2012), pp. 1–2
18. S.S. Roy, C. Mukherjee, S. Panda, A.K. Mukhopadhyay, B. Maji, Layered T comparator design using QCA. Devices Integr. Circuit (DevIC) **2017**, 90–94 (2017)
19. F. Deng, G. Xie, Y. Zhang, F. Peng, H. Lv, A novel design and analysis of comparator with XNOR gate for QCA. Microprocess. Microsyst. **55**, 131–135 (2017)
20. S. Umira, R. Qadri, Z.A. Bangi, B.M. Tariq, A novel comparator-A cryptographic design in QCA. Int. J. Dig. Signals Smart Syst. **4**, 1–10 (2020)
21. L. Jun-wen, X. Yin-shui, A novel design of quantum-dots cellular automata comparator using five-input majority gate, in *Paper presented at. 14th IEEE International Conference on Solid-State and Integrated Circuit Technology (ICSICT)*, (2018), pp. 1–3
22. M. Gao, J. Wang, S. Fang, J. Nan, L. Daming, A new nano design for implementation of a digital comparator based on QCA. Int. J. Theor. Phys. (2020)
23. D. Tripathi, S. Wairya, Energy Efficient Binary Magnitude Comparator Nanotechnol. Appl. **8**, 430–436 (2020)
24. G. Singh, R.K. Sarin, B. Raj, A novel robust exclusive-OR function implementation in QCA nanotechnology with energy dissipation analysis. J. Comput. Electron. **15**, 455–465 (2016)
25. M. Poorhosseini, A.R. Hejazi, A fault-tolerant and efficient XOR structure for modular design of complex QCA circuits. J. Circ. Syst. Comput. **27** (2018)
26. D. Kumar, C. Kumar, S. Gautam, D. Mitra, Design of practical parity generator and parity checker circuits in QCA, in *2017 IEEE International Symposium on Nanoelectronic and Information Systems (iNIS)* (Bhopal, 2017), pp. 28–33
27. T. Sasamal, A. Singh, U. Ghanekar, Design and analysis of ultra-low power QCA parity generator circuit, in *Advances in Power Systems and Energy Management* (2018), pp. 347–354
28. I. Gassoumi, L. Touil, B. Ouni, A. Mtibaa1, An ultra-low power parity generator circuit based on QCA technology. J. Electr. Comput. Eng. **2019**, 1–8 (2019)
29. P. Agrawal, S.R.P. Sinha, N.K. Misra, S. Wairya, Design of quantum dot cellular automata based parity generator and checker with minimum clocks and latency. Int. J. Mod. Edu. Comput. Sci. **8**, 11–20 (2016)

A Survey of Existing Studies on NOMA Application to Multi-beam Satellite Systems for 5G

Joel S. Biyoghe and Vipin Balyan

Abstract The non-orthogonal multiple access (NOMA) technology and the satellite communication systems have been identified as key enabling technologies for the achievement of the fifth-generation (5G) networks. There are ongoing researches about the implementation of 5G networks, and more precisely, employing NOMA technology. While most existing survey papers on NOMA-based 5G networks' studies focus mainly on terrestrial networks, this paper provides a comprehensive survey of NOMA-based satellite network studies for 5G. More precisely, this article gives an up-to-date survey of studies that have investigated the performances (ergodic capacity, outage probability, secrecy rate, etc.) of NOMA-based satellite networks. It also gives a survey of studies that have designed network components such as power allocation algorithms or pre-coding algorithms for NOMA-based satellite networks. The surveys presented reveal that the field of NOMA application to satellite networks for 5G is still relatively new, and therefore, holds many interesting topics opened for future research.

Keywords Non-orthogonal multiple access (NOMA) · Orthogonal multiple access (OMA) · Satellite · User clustering · Power allocation · Pre-coding · 5G · 4G

1 Introduction

Mobile networks are radio access communication networks (RANs) in which a mobile user is granted access for a designated period of time to utilize the radio resources of the network (frequency spectrum) to either transmit or receive information [1]. The network access point can be a terrestrial base station (in which case, the network is called terrestrial network or cellular network) or an earth orbit satellite (in which case, the network is called a satellite network) [2]. General performance

J. S. Biyoghe (✉) · V. Balyan
Cape Peninsula University of Technology, Cape Town 7530, South Africa

V. Balyan
e-mail: balyanv@cput.ac.za

V. Suma et al. (eds.), *Inventive Systems and Control*,
Lecture Notes in Networks and Systems 204,
https://doi.org/10.1007/978-981-16-1395-1_20

Table 1 Overview of the evolution of the mobile network's generations [5–8]

Generation	Year	Radio access technology	Analog/digital technologies	Applications
1G	1980s	FDMA	Analog technology	Voice calls only
2G	1990s	TDMA CDMA	GSM, GPRS, EDGE	Voice calls, SMS, packet-switched data services
3G	2000s	CDMA	CDMA-2000, EV-DO, WCDMA	Services from 2G + video calls
4G	2010s	OFDMA	LTE, LTE-A	Services from 3G + enhanced quality video calls and live streaming

parameters of mobile networks includes their total capacity, reliability (quality of service (QoS), and user fairness), latency as well as geographical coverage or span [3, 4]. Based on their achievable performances as well as the technologies used, mobile networks have been classified in various generations over the years. Table 1 below summarizes the evolution of mobile networks, which applies to both the terrestrial and the satellite networks.

The fourth-generation (4G) networks have been the most advanced form of mobile networks throughout the 2010 decade, and they deliver performances such as high connectivity, high capacity, high end user speed, intermediate latency, low reliability, and a geographical coverage still relatively small (about 20% of the earth surface is covered by the mobile network). With such performances, 4G networks offer services such as video calls and live streaming. From the 2020 decade, the mobile communication industry seeks to achieve the Internet-of-Things (IoTs) which include services such as industrial Internet, remote surgery, and the whole- world connection [9]. These services require network performances such as massive connectivity, ultra-low latency, ultra-high reliability ultra-high speed, and ubiquity [5, 6] and cannot, therefore, be supported by 4G networks [10].

To address this need, the fifth generation (5G) of networks has been envisaged by the international telecommunication union (ITU) as a technological advancement of 4G networks. The expected performance metrics of 5G networks have been summarized by [6]. The enhanced mobile broadband communication (eMBB), the ultra-reliable and low-latency communication (URLLC), and the massive machine-type communication (mMTC) are three main communication concepts that should be supported by any 5G network [8, 11]. To enable the development of 5G networks, the ITU has proposed some emerging technological concepts including multiple-input multiple/single-output (MIMO/MISO), non-orthogonal multiple access (NOMA), multicasting, beamforming (BF), and the multi-beam satellite system. The NOMA and the MBSS technologies are regarded as the key enabler of the 5G achievement [12, 13].

Various ongoing researches have investigated the development of 5G networks using the above-mentioned technological concepts, independently or inclusively.

However, since NOMA is regarded as a key technology enabler of 5G and has captured much attention from researchers, this paper therefore, focuses on 5G development studies that are based on NOMA technology. In this regard, some existing papers have given a survey of NOMA-based researches for 5G [1–3, 8]; but these papers focused purely on works related to terrestrial networks, and none on works related to satellite networks.

This article accordingly seeks to fill this gap and provides a comprehensive and upgraded survey of existing researches that investigated the usage of NOMA technology on MBSS for 5G applications. The surveys presented can serve as a guide for researchers to know the state of the art in the field as well as the gap areas opened for future possible researches. The contribution of this article includes:

1. A survey of all studies that have investigated performances (ergodic capacity, outage probability, asymptotic outage probability, secrecy rate) of NOMA-based satellite networks.
2. A survey of all studies that designed a power allocation algorithm for the MA system of NOMA-based satellite networks.
3. A survey of all studies that designed a pre-coding allocation algorithm for the MA system of NOMA-based satellite networks.

The rest of the article is organized as follows: Sect. 2 gives a brief overview of the NOMA technology concept. Then, in Sect. 3, the comprehensive surveys of NOMA-based satellite networks studies are given. Important observations made from these surveys, and the possible opened research areas identified are discussed in Sect. 4. Finally, conclusion is given in Sect. 5.

2 Overview of NOMA Concept

The non-orthogonal multiple access (NOMA) is a novel multiple access (MA) technology for mobile networks in which more than one users are allowed to utilize a designated sub-frequency resource simultaneously over a given time period [8, 14]. This is unlike the orthogonal multiple access (OMA) technology traditionally used in previous generations of mobile networks (1G to 4G) in a form of time-division multiple access (TDMA), frequency-division multiple access (FDMA), or offset-FDMA (OFDMA), in which only one user is allowed to utilize a designated sub-frequency resource [15, 16], and Fig. 1 illustrates two scenarios of an MBSS, one with OMA technology where only one user is served for a given frequency slot; and the other with NOMA technology (case of two users per beam), where two users are served simultaneously using the same frequency slot. In the NOMA technology, the information of respective users utilizing the same frequency resources is separated in either power level, hence, power-domain NOMA (PD-NOMA) or in code-level, hence, code-domain NOMA (CD-NOMA) [3, 6]. Figure 2 gives an illustration of the power diagram of OMA and PD-NOMA technologies. These diagrams show that in the case of OMA, the one user served at the designated frequency slot

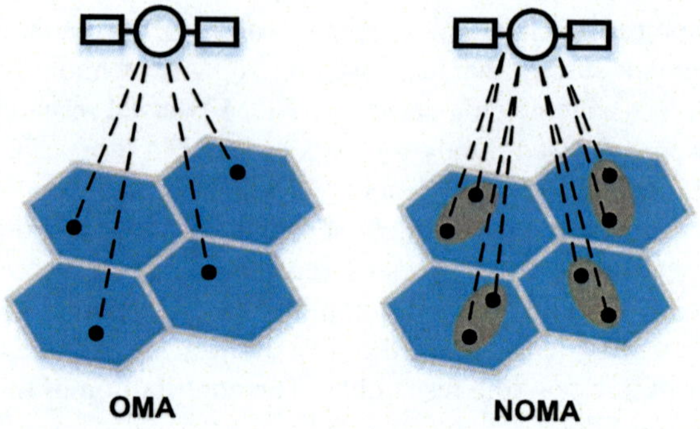

Fig. 1 MBSS illustration with OMA and NOMA

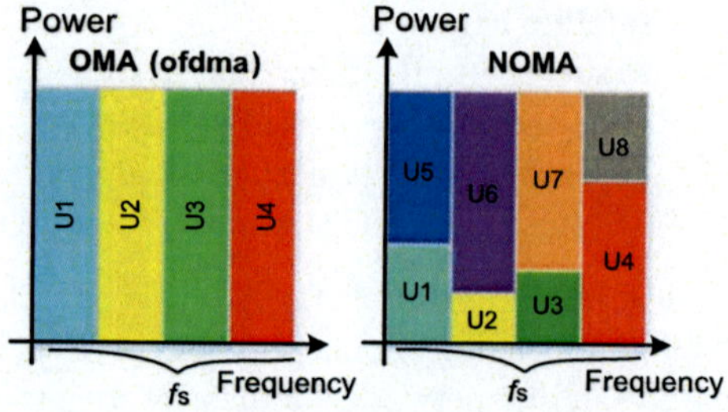

Fig. 2 Illustration of power-domain diagram: OMA versus NOMA

utilizes all the power available for that frequency slot, whereas in the PD-NOMA concept, the transmitted signal is a superposition of multiple signals from respective users that share the same frequency, with different power levels. These users share the total power available for the designated frequency slot, and power levels are assigned to users based on their respective channel conditions, where the order of power coefficients is inverse to the order of channel gains. The users with the best channel gains are assigned less power, and users with the poorer channel gains are assigned more power. At the receiver side, the concept of successive interference cancelation (SIC) is used to decompose the superposed signals. The user with the highest power level decomposes its signal directly as it sees other signals as noise. Subsequently, users with lower power levels first have to successively decompose signals of users with higher powers [17, 18]. A few important observations can be made: (a) NOMA delivers better system performances than its predecessor OMA in most network aspects such as achievable connectivity, capacity, latency, and fairness [12, 19]; (b) NOMA is compatible with OMA and can therefore be used in a system where OMA has been employed without having to upgrade the system considerably [20, 21]; (c) due to the implementation of the SIC concept, the complexity of NOMA receiver is high and keeps on growing with the increase in the number of users which may constitute a considerable limitation of the NOMA technology [8, 11, 22].

3 Survey of Existing NOMA-Based Satellite Networks' Studies

3.1 Type of Network Studies

The study of a RAN network generally involves two types of works, namely system analysis and system design. The system analysis investigates statistical performances of the system such as the outage probability (OP), the asymptotic outage probability (ASOP), the ergodic capacity (EC), and the secrecy rate (SR). The system design on the other hand consists of designing blocks of the multiple access (MA) encoder including the user-clustering (UC) process to group users to be served, the power allocation (PA) process to assign power levels to respective users and possibly the pre-coding (PC) process to do execute beamforming and mitigate inter-cell-interferences (ICIs). It also consists of designing blocks of the MA decoder, which in the case of NOMA would include the SIC process to decompose the superposed signals. The SIC forms part of the multi-user-detection (MUD) techniques, which are used to mitigate the ICIs at the receiver side. The design goal for the user-clustering algorithm is to find the best matching pairs of users to be served together for either intra-beam NOMA power allocation or beamforming. For the power allocation, the design goals usually include maximizing the system capacity or user fairness. The pre-coding algorithm design goal is to mitigate inter-cell-interference and therefore to maximize the system capacity. For the three systems, the design considerations for all these design processes generally include the number of users to be paired for NOMA, the users' channel conditions, the type of channel state information (CSI) estimation, whether perfect or imperfect and the channel impairment, wither considered or assumed negligible. In the survey below, all identified works regarding NOMA application to satellite networks for 5G be it system analysis or system design are presented.

3.2 Survey of Network Analysis Studies

The following system analysis studies presented in Table 2 have been identified for NOMA-based satellite systems. The following abbreviation should be noted: ergodic capacity (EC), outage probability (OP), secrecy rate (SR), inter-channel-interference (ICI), channel impairments (CI), channel state information (CSI).

3.3 Survey of network's Power Allocation Design Studies

The following power allocation (PA) design studies listed in Table 3 have been identified for NOMA-based satellite systems with the following abbreviations such as

Table 2 Summary of system's analysis studies for NOMA-based MBSS

Paper	Analysis study done	Technical details	Outcome
[23]	Compared the performances of a MBSS's downlink with NOMA and OMA	Evaluated EC and OP for both cases	Demonstrated that NOMA > OMA in performances
[24]	Investigated the impact of ICI on the performances of a NOMA-based MBSS	Evaluated EC and OP with ICI	The ICI deteriorates the system performances
[25]	Performed a SR analysis of satellite communications with frequency domain NOMA	SR evaluation	The SR is affected by the level of spectral overlapping
[26]	Investigated the impact of both imperfect CSI and CI on the performance of a NOMA-based MBSS downlink	Evaluated OP with imperfect CSI and CI	Demonstrated that imperfect CSI and CI deteriorate system's performances
[27]	Investigated the physical layer security challenges in mm wave, with NOMA and MIMO downlink	Evaluated secrecy capacity for both NOMA and MIMO	Secrecy capacity depends on richness of the radio frequency
[28]	Investigated the ergodic capacity of NOMA-based uplink satellite networks with randomly deployed users	Evaluated EC	Location information, link performance, transmitted power, and imperfection of CSI all impact on system performance

power allocation (PA), admission control (AC), particle swarm optimization (PSO), deep learning search (DLS).

3.4 *Survey of network's Precoding Design Studies*

The following precoding (PC) design studies listed in Table 4 have been identified for NOMA-based satellite systems and the following abbreviations such as multiple-user-detection (MUD) and beamforming (BF).

Table 3 Identified PA algorithm's design studies for NOMA-Based satellite systems

Paper	Design study done	Technical details	Outcome
PA algorithm's design for single MBSS			
[29]	Designed a PA algorithm for a NOMA-based MBSS	PA algorithm designed to improve total system transmission rate	Achievable system rate is above that of OMA
[30]	Designed a PA algorithm for a MISO-NOMA satellite system	PA algorithm designed to minimize the system's total power consumption	Performances of proposed algorithm are superior to those of OMA and SISO systems
[31]	Designed a PA algorithm for a NOMA-based MBSS	PA algorithm (NOMA + PC) designed to maximize the system's fairness	Performances of NOMA + PC are superior to those of some algorithms such as OMA, NOMA1, and NOMA2
[32]	Designed a PA algorithm for a NOMA-based MBSS	PA algorithm designed to maximize the weighted sum rate system transmission rate	Performances of designed algorithm are superior to those of OMA
Combined PA and AC algorithm's design for single MBSS			
[33]	Developed PA and admission control (AC) algorithm for a NOMA-based MBSS	A joint PA and AC algorithm (NOMA-JOPT) designed to maximize long-term network utility of the system	Performances of NOMA-JOPT are superior to those of other algorithms such as OMA, NOMA-PRO-G
[34]	Developed PA and admission control (AC) algorithm for a NOMA-based MBSS	A combined PA and AC algorithm (NOMA-PSO) designed to maximize long-term network utility of the system, based on DLS concept	Performances of NOMA-DLS are superior to those of another algorithm called NOMA-FuS
[35]	Developed PA and admission control (AC) algorithm for a NOMA-based MBSS	A combined PA and AC algorithm (NOMA-PSO) designed to maximize long-term network utility of the system, based on PSO concept	Performances of NOMA-PSO are superior to those of other algorithms such as OMA, NOMA-PRO-G

(continued)

Table 3 (continued)

Paper	Design study done	Technical details	Outcome
[36]	Developed PA and admission control (AC) algorithm for a NOMA-based MBSS	A combined PA and AC algorithm (NOMA-algo1) designed to maximize the system's number of users	Performances of NOMA-Algo1 are superior to those of some existing algorithms such as OMA-OFDMA, dynamic exchange, static-exchange
PA algorithm's design for an integrated 2-MBBSs			
[37]	Designed a PA algorithm for a NOMA-based integrated 2 satellites systems	The single PA algorithm for both MBSSs (NOMA-PSO) designed to maximize the total sum rate of the system	Performances of NOMA-PSO are superior to those few other algorithms such as OMA and NOMA-uniform
[38]	Proposed a NOMA-based irregular repetition slotted ALOHA (IRSA) scheme for satellites networks	The designed NOMA-based IRSA differ from usual OMA-based IRSAs	Performances of NOMA-IRSA proved superior to those of some OMA-IRSAs

4 Observations and Opened Research Areas Identification

4.1 Observations

Table 5 presents some observations made (relevant findings) from a careful interpretation of the above-presented surveys.

4.2 Opened Research Areas

Based on the above observations from the surveys presented, the following points can constitute opened areas for future research in the field of NOMA application to satellite networks for 5G:

1. System performance analyses to be done with the scenario of more than two users per NOMA-beam.
2. System performance analyses to be done by considering channel impairments.
3. More investigations of the possibility to cooperate with multiple LEO/MEO satellites operating with NOMA.
4. Design of user-clustering algorithms to be done for the scenario of more than 2 users per NOMA-beam.

Table 4 Identified PC-algorithm's design studies for NOMA-Based satellite systems

Paper	Design study done	Technical details	Outcome
PC-algorithm's design for single MBSS			
[39]	Designed a precoding algorithm for a MBSS that employs NOMA and beamforming technologies	Precoding algorithm (overlay-coding-BF) designed to maximize the system's spectral efficiency	Performances of overlay-coding-BF are superior to those of few other algorithms such as FR4-BF, FR2-BF
[40]	Designed a precoding algorithm for a MBSS that employs NOMA and beamforming technologies	Precoding algorithm (Geo-NOMA-BF) designed to maximize the system's sum rate	Performances of Geo-NOMA-BF are superior to those of few other algorithms such as OMA-BF, NOMA-BF-ZF, NOMA-BF user scheduling
Combined PC and UC/MUD algorithm's design for single MBSS			
[41]	Investigated the effect of combining precoding at transmitter and MUD at receiver for a MBSS	Evaluated the performance of the MBSS with only precoding, only MUD and both precoding and MUD	The performance of joint precoding and MUD outperform those of the system when only one of the two is used
[42]	Combined a geographical user scheduling and precoding for a NOMA-based MBSS	The system (combined GUS and precoding) was designed to maximize the system throughput	The proposed system's performance are superior to those of using precoding alone
PC-algorithm's design for an integrated 2-MBBSs			
[43]	Designed a precoding algorithm for an integrated 2-MBSS that employs NOMA and beamforming technologies	single precoding algorithm (IPF-Based-BF) for both MBSSs designed to maximize the total sum rate of the system	Performances of IPF-Based-BF are superior to those of few other algorithms such as OMA-BF, SDR-NOMA, and Maxeigen-NOMA

5. Design of power allocation algorithms to be done for the scenario of more than 2 users per NOMA-beam.
6. Design of power allocation algorithms to be done by considering imperfect CSI.
7. Design of power allocation algorithms to be done by considering channel impairments.
8. More research on the design of a power allocation algorithm for maximization of system's user fairness.

Table 5 List of important observations made from the surveys: relevant findings

General observation
• The research is still quite young as most research range from about the end of the year 2016 to date (2020); and as a logical consequence, there is only a handful of studies that have been reported so far
Observations on system analysis's studies
• While most presented studies have investigated system performances of NOMA-based satellite systems under conditions of perfect and imperfect CSIs, no identified papers have reported analyzing the system by considering channel impairments • In most of the researches presented, the number of users per beam for NOMA has been limited to 2, and no paper has reported system analysis for the case of more than 2 users per beam for NOMA • Thus far, only one study has investigated the cooperation of multiple LEO or MEO satellites employing NOMA
Observations on system design's studies
• Most user-clustering, power allocation, or precoding algorithms design considered only the scenario of two user-per-beam for NOMA. None of the identified studies considered the scenario of more than two user per beam during their system's design • All reported design studies assumed the scenario of perfect CSI and no channel impairments; no reported design work considered a scenario of imperfect CSI and/or channel impairment during their system's design • All reported design studies have considered operating in the microwave-wave frequency spectrum; no identified design works considered millimeter-wave (mmWave) frequency spectrum for their system's design • The reported user-clustering algorithm designs, mostly based their designs either on the channel gain, for the intra-beam (NOMA) power allocation process, or on channel-correlation, for the beamforming process • For the case of power allocation algorithm designs, an outstanding observation is that most reported works design their PA algorithm to maximize the total system capacity; the only one reported work designed its PA algorithm to maximize system fairness

9. Design of precoding algorithms to be done for the scenario of more than 2 users per NOMA-beam.
10. Design of precoding algorithms to be done by considering imperfect CSI.
11. Design of precoding algorithms to be done by considering channel impairments.
12. Do the above system's design by considering operating in the mmWave frequency spectrum.

It is to be noted that the above-listed topics are to be added to a non-exhaustive list of challenges, and possible research areas for NOMA-based satellite networks already are mentioned in some papers [3, 8, 19, 23, 44, 45], and which include:

1. The challenge of the non-stationary ground beam for the case of non-geostationary earth orbit satellites.
2. The satellite to gateway link limitation is considered as a bottleneck challenge.
3. Inter-cell interference management in case of multiple gateway networks.

4. Very limited work on the physical layer security which requires more attention as satellite communications are very vulnerable to eavesdroppers.

5 Conclusion

A comprehensive survey of existing studies on NOMA application to satellite networks for 5G was discussed. Initially, a survey of studies that have investigated performances (ergodic capacity, outage probability, secrecy rate, etc.) of NOMA-based satellite networks was presented. Then, a survey of system design studies including power allocation algorithm studies and precoding algorithm studies for NOMA-based satellite networks was presented. A few important observations have been drawn from these surveys. In the global view, only a few researches exist in this field of NOMA application to satellite networks for 5G; and all of them being very recent therefore means that the field is still very opened for future research. Subsequently, from a technical observation point of view, most studies have limited themselves to the case of two users per NOMA-beam; and most studies considered the case of perfect CSI and no channel impairments during their designs. Also, most design studies focused on maximization of system capacity with no much consideration for the user fairness aspect. These observations constitute opened areas that give way to possible future researches.

References

1. S.M.R. Islam, N. Avazov, O.A. Dobre, K. Kwak, Power-domain non-orthogonal multiple access (NOMA) in 5G systems: potentials and challenges. IEEE Commun. Surv. Tutorials **19**(2), 721–742 (2017)
2. W. Ejaz, S.K. Sharma, S. Saadat, M. Naeem, A. Anpalagan, N.A. Chughtai, A comprehensive survey on resource allocation for CRAN in 5G and beyond Networks. J. Netw. Comput. Appl. **160**, 1–24 (2020)
3. M. Aldababsa, M. Toka, S. Gökçeli, G.K. Kurt, O. Kucur, A tutorial on non-orthogonal multiple access for 5G and beyond. Wirel. Commun. Mob. Comput. 1–24 (2018)
4. L. Bai, L. Zhu, X. Zhang, W. Zhang, Q. Yu, Multi-satellite relay transmission in 5G: concepts, techniques, and challenges. IEEE Netw. 38–44 (2018)
5. G. Liu, D. Jiang, 5G: vision and requirements for mobile communication system towards year 2020. Chin. J. Eng. 1–9 (2016)
6. Y. Wang, B. Ren, S. Sun, S. Kang, X. Yue, Analysis of non-orthogonal multiple access for 5G. Chin. Commun. **2**, 52–66 (2016)
7. L. Dai, B. Wang, Z. Ding, Z. Wang, S. Chen, L. Hanzo, A survey of non-orthogonal multiple access for 5G. IEEE Commun. Surv. Tutorials **20**(3), 2294–2322 (2018)
8. A. Anwar, B. Seet, M.A. Hasan, X.J. Li, A survey on application of non-orthogonal multiple access to different wireless networks. Electronics **8**, 1–46 (2019)
9. R. Khan, D.N.K. Jayakody, An ultra-reliable and low latency communications assisted modulation based non-orthogonal multiple access scheme. Phys. Commun. 15758–15761 (2020)
10. X. Zhu, C. Jiang, L. Kuang, N. Ge, S. Guo, J. Lu, Cooperative transmission in integrated terrestrial-satellite networks. IEEE Netw. 204–210 (2019)

11. V.K. Trivedi, K. Ramadan, P. Kumar, M.I. Dessouky, F.E. Abid El-Samie, Enhanced OFDM-NOMA for next generation wireless communication: a study of PAPR reduction and sensitivity to CFO and estimation errors. Int. J. Electron. Commun. (AEÜ) **102**, 9–24 (2019)
12. L. Dai, Z. Wang, Y. Yuan, Non-orthogonal multiple access for 5G: solutions, challenges, opportunities, and future research trends. IEEE Commun. Mag. 74–81 (2015)
13. X. Zhang, D. Guo, K. An, Z. Chen, B. Zhao, Y. Ni, B. Zhang, Performance analysis of NOMA-based cooperative spectrum sharing in hybrid satellite-terrestrial networks. IEEE Access **7**, 172321–172329 (2019)
14. V. Balyan, R. Daniels, Resource allocation for NOMA based networks using relays: cell centre and cell edge users. Int. J. Smart Sens. Intell. Syst. **13**(1), 1–18 (2020)
15. V. Balyan, D.S. Saini, Call elapsed time and reduction in code blocking for WCDMA networks, in *IEEE International Conference on Software Telecommunication and Computer Networks* (SoftCom, 2009), pp. 141–145
16. V. Balyan, D.S. Saini, Integrating new calls and performance improvement in OVSF based CDMA Networks. Int. J. Comput. Commun. **2**(5), 35–42 (2011)
17. Y. Saito, A. Benjebbour, Y. Kishiyama, T. Nakamura, System-level performance evaluation of downlink non-orthogonal multiple access (NOMA), in *2013 IEEE 24th International Symposium on Personal, Indoor and Mobile Radio Communications: Fundamentals and PHY Track* (IEEE, 2013), pp. 611–615
18. M. Caus, M.A. Vazquez, A.I.P. Neira, NOMA and interference limited satellite scenarios. IEEE Asilomar, 497–501 (2016)
19. S. Kumar, K. Kumar, Multiple access schemes for cognitive radio networks: a survey. Phys. Commun. **38**, 1–31 (2020)
20. D. Wan, M. Wen, F. Ji, H. Yu, F. Chen, Non-orthogonal multiple access for cooperative communications: challenges, opportunities, and trends. IEEE Wirel. Commun. 109–117 (2018)
21. I. Baig, N.U. Hasan, M. Zghaibeh, I.U. Khan, A.S. Saand, A DST precoding based uplink NOMA scheme for PAPR reduction in 5G wireless network. IEEE Access, 1–4 (2017)
22. V. Balyan, Outage probability of cognitive radio network utilizing non orthogonal multiple access, in *2020 7th International Conference on Signal Processing and Integrated Networks (SPIN)* (Noida, India, 2020), pp. 751–755
23. X. Yan, K. An, T. Liang, G. Zheng, Z. Ding, S. Chatzinotas, Y. Liu, The application of power-domain non-orthogonal multiple access in satellite communication networks. IEEE Access **7**, 63531–63539 (2019)
24. S. Xie, B. Zhang, D. Guo, B. Zhao, Performance analysis and power allocation for NOMA-based hybrid satellite-terrestrial relay networks with imperfect channel state information. IEEE Access **7**, 136279–136289 (2019)
25. Z. Yin, M. Jia, W. Wang, N. Cheng, F. Lyu, Q. Guo, X. Shen, Secrecy rate analysis of satellite communications with frequency domain NOMA. IEEE Trans. Veh. Technol. **68**(12), 11847–11858 (2019)
26. F. Zhou, R. Wang, J. Bian, Performance analysis of non-orthogonal multiple access based-satellite communication networks with hardware impairments and channel estimations. Electron. Lett. **56**(1), 52–55 (2020)
27. K. Xiao, S. Zhang, K. Michel, C. Li, Study of physical layer security in mmwave satellite networks. IEEE Access, 1–6 (2018)
28. X. Yan, H. Xiao, K. An, G. Zheng, S. Chatzinotas, Ergodic capacity of NOMA-based uplink satellite networks with randomly deployed users. IEEE Syst. J. 1–8 (2019)
29. X. Liu, X.B. Zhai, W. Lu, C. Wu, QoS-guarantee resource allocation for multibeam satellite industrial Internet of Things with NOMA. IEEE Trans. Indus. Inf. J. Latex Class Files **14**(8), 1–10 (2015)
30. M. Alhusseini, P. Azmi, N. Mokari, Optimal joint subcarrier and power allocation for MISO-NOMA satellite networks. Phys. Commun. **32**, 50–61 (2019)
31. A. Wang, L. Lei, Lagunas, E., A.I.P. Neira, S. Chatzinotas, B. Ottersten, On fairness optimization for NOMA-enabled multi-beam satellite systems, in *2019 IEEE 30th Annual International Symposium on Personal, Indoor and Mobile Radio Communications (PIMRC): Track 3: Mobile and Wireless Networks* (IEEE, 2019), pp. 1–6

32. T. Ramırez, C. Mosquera, Resource management in the multibeam noma-based satellite downlink, in *IEEE Conference of ASSP* (IEEE, 2020), pp. 8812–8816
33. Y. Sun, J. Jiao, S. Wu, Y. Wang, Q. Zhang, Joint power allocation and rate control for NOMA-based space information networks. IEEE Access, 1–6 (2019)
34. Y. Sun, Y. Wang, J. Jiao, S. Wu, Q. Zhang, Deep Learning-based long-term power allocation scheme for NOMA downlink system in S-IoT. IEEE Access **7**, 86288–86296 (2019)
35. J. Jiao, Y. Sun, S. Wu, Y. Wang, Q. Zhang, Network utility maximization resource allocation for NOMA in satellite-based Internet of Things. IEEE Internet Things J. 1–13 (2019)
36. R. Wang, W. Kang, G. Liu, R. Ma, B. Li, Admission control and power allocation for NOMA-based satellite multi-beam network. IEEE Access **8**, 33631–33643 (2020)
37. R. Wan, L. Zhu, T. Li, L. Bai, A NOMA-PSO based cooperative transmission method in satellite communication systems. IEEE Access, 1–6 (2017)
38. X. Shao, Z. Sun, M. Yang, S. Gu, Q. Guo, NOMA-based irregular repetition slotted ALOHA for satellite networks. IEEE Commun. Lett. **23**(4), 624–627 (2019)
39. N.A.K. Beigi, M.R. Soleymani, Interference management using cooperative NOMA in multi-beam satellite systems. IEEE Access, 1–6 (2018)
40. Y. Zhu, T. Delamotte, A. Knopp, Geographical NOMA-beamforming in multi-beam satellite-based Internet of Things. IEEE Access, 1–6 (2019)
41. M.A. Vazquez, M. Caus, A. Preze-Neira, Performance analysis of joint precoding and MUD techniques in multibeam satellite systems. IEEE Access, 1–5 (2016)
42. A. Guidotti, A.V. Coralli, Geographical scheduling for multicast precoding in multi-beam satellite systems, in *2018 9th Advanced Satellite Multimedia Systems Conference and the 15th Signal Processing for Space Communications Workshop (ASMS/SPSC)* (IEEE, 2018), pp. 1–8
43. R.P. Sirigina, A.S. Madhukumar, M. Bowyer, NOMA precoding for cognitive overlay dual satellite systems. IEEE Access, 1–5 (2019)
44. K. An, X. Yan, T. Liang, W. Lu, NOMA based satellite communication networks: architectures, techniques and challenges, in *2019 IEEE 19th International Conference on Communication Technology* (IEEE, 2019), pp. 1105–1110
45. A.I.P. Neira, M. Caus, M.A. Vázquez, Non-orthogonal transmission techniques for multibeam satellite systems. IEEE Commun. Mag. Mob. Commun. Netw. 58–63 (2019)

Luhn Algorithm in 45 nm CMOS Technology for Generation and Validation of Card Numbers

Vivek B. A and Chiranjit R. Patel

Abstract Nowadays, a safe and easy way to check the validity of credit card numbers is important with the growing use of online banking and online shopping. We have proposed a circuit that integrates the use of Luhn algorithm developed by Hans Peter Luhn in the 1950s. The circuit requires the generation of a check digit appended to the end of a card number and further a verification circuit is used to verify the correctness of the number. The circuit design was done in 90 and 45 nm technologies. A comparison of the design's power consumption and propagation delay was carried out between two technologies. Simulations are performed in all process corners by using the Spectre simulator, to account for the variance on the chip during fabrication. All performance measurement results are tabulated and represented in graphs.

Keywords Luhn algorithm · Card number validation · CMOS · GPDK45 · GPDK90

1 Introduction

Hans Peter Luhn, who created the Luhn algorithm, indexing KWIC (Key Terms In Context) and Selective Dissemination of Information, was a computer science researcher and IBM Library and Information Science researcher (SDI). His technologies have been used in numerous areas such as computer science, garment industry, linguistics, and information science. There were over 80 patents given to him.

Luhn algorithm is used all the time from the viewpoint of a customer, without even realizing it. This algorithm has been used by researchers in the past to develop advanced techniques for validating the International Mobile Equipment Identity (IMEI) and Credit Card numbers [1]. Computer systems will easily say whether we have made an error by entering our details when placing orders online or using the point of sale terminal of a merchant. This is because the Luhn algorithm has been implemented into the programming of such systems. Without it, before understanding

Vivek B. A (✉) · C. R. Patel
Electronics and Communication, RNS Institute of Technology, Bengaluru, Karnataka, India

V. Suma et al. (eds.), *Inventive Systems and Control*,
Lecture Notes in Networks and Systems 204,
https://doi.org/10.1007/978-981-16-1395-1_21

if the transaction was accepted, we would need to wait for the whole purchase order to be submitted. The algorithm helps to immediately fix user mistakes, thereby making transactions quicker.

For credit cards, one digit at the end of the credit card number is written. Instead the check digit is dynamically calculated by the Luhn algorithm, depending on the previous numbers in the list, rather than being selected by the credit card company directly. The payment processing program will use the Luhn algorithm to detect whether the specified number is exact and dependent in part on the check digit when entering your credit card numbers for full transactions.

The Luhn algorithm is now incorporated into common programming languages and code libraries, by enabling it to be integrated into modern software apps in a reasonably simple manner.

The algorithm works by applying a series of calculations to the number provided by the credit card, adding the results of those calculations, and verifying if the resulting number corresponds to the expected result. If it does then the credit number will be assumed to be valid. If not the credit card number will be rejected by the algorithm and implying that the user has made a mistake when entering the number.

2 Literature Survey

To distinguish legitimate devices, GSM networks use the IMEI number and can stop a stolen phone from accessing the network. For example, the owner may make their network provider use the IMEI number to block the phone if a cell phone is stolen. The IMEI numbers were studied by Krishan Kumar and Prabhpreet Kaur [2] and noted that while being used worldwide for specific mobile equipment recognition, attackers have developed multiple techniques to alter the IMEI of mobile phones so that robbed or missing mobile equipment is not discovered by its manufacturers or intelligence agencies. They ended by emphasizing the need for improved, non-hackable security measures.

CARDWATCH, a database mining method is used for credit card fraud detection, which is proposed by Aleskerov et al. [3]. The framework offers an interface to a range of industrial, neural learning module-based databases.

Kim described skewed data distribution and the mixture of fraudulent and legal transactions as the two key explanations for the difficulty of detecting bank card fraud [4]. The fraud density of actual transaction data is used as a trust value based on this study and the weighted fraud score is generated to minimize the number of incorrect detection.

Brause et al. illustrate how advanced data mining techniques and neural network algorithms can be effectively combined to achieve high coverage of frauds coupled with a low false alarm rate [5].

The main aim of this paper is to reflect technologies that can be used for credit card fraud detection [6]. Such strategies can aid in the detection and acceptance of credit card fraud. The authors have been working on an unsupervised learning process in which their network is trained to identify fraudulent transactions.

The Verhoeff algorithm is a tested formula developed by the Dutch mathematician Jacobus Verhoeff for the detection of errors and was first published in 1969 [7]. It was the first digit decimal search algorithm to identify all single-digit errors with two joining digits. Verhoeff designed his algorithm using the properties of the order 10 dihedral group, combined with a permutation, he argued that it was the first practical use of the dihedral group, confirmed the theory that all beautiful mathematics would eventually find use, although the algorithm would be implemented in practice using simple search tables without knowing how to produce those [8].

The survey done in [9] gives the reader a broad perspective on the types of credit card frauds and ways on how these frauds can be reduced.

The transposition error of the digits 09 and 90 due to the reduction of items into single terms is highlighted in a study by L.W.Wachira [10]. A modulo 13 algorithm is developed, and it is demonstrated how the modulo 13 algorithm outperforms the Luhn algorithm in a few areas.

The Luhn algorithm has a few shortcomings [11], including the fact that it fails to detect errors in judging the length of the card number and twin errors, as shown by the authors in [12, 13].

3 The Luhn Algorithm

On several e-commerce pages, the Luhn algorithm, shown in Fig. 1, is the first line of defense. It is used to validate a set of identity numbers, such as numbers for Master and Visa cards. It was designed to protect businesses and customers from unintended mistakes and typing mistakes.

An example of the Luhn algorithm applied for decimal base numbers is shown in Fig. 2.

$$\left(\sum_{j=0}^{m} 2k_{2j} + \sum_{j=0}^{m} 2k_{2j+1}\right) \mod 10 \tag{1}$$

$$k \in Z_{10} \ , \ m \in N \ , \ \exists\, k \geq 10 \tag{2}$$

Referring to Eq. (1) the alternative terms are multiplied by 2. The check is then computed. The numbers resulting with two digits are added with their respective digits to obtain a single digit number lesser than 10 in decimal.

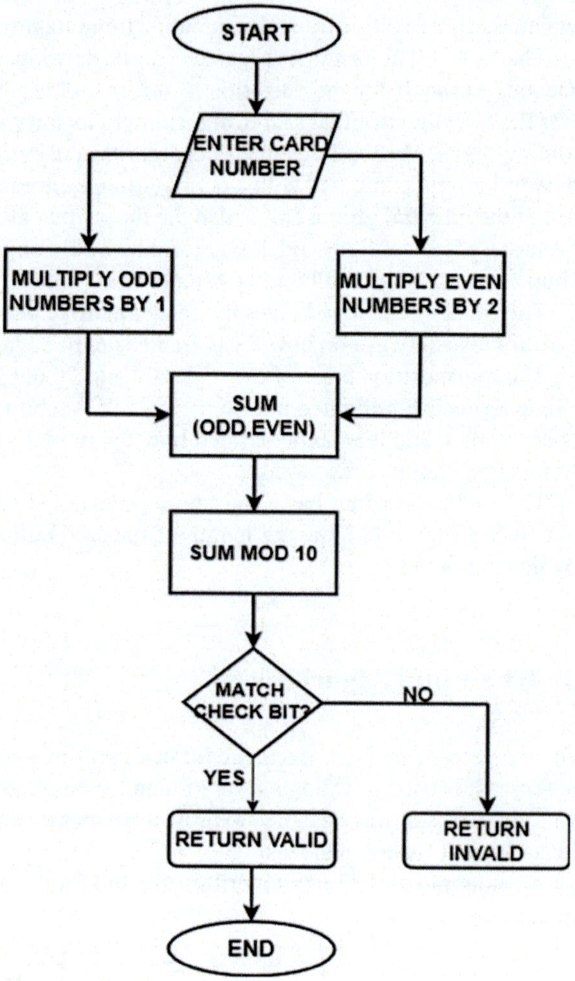

Fig. 1 Flowchart of the Luhn algorithm

4 Generation System for Hex Card Numbers

The intrinsic advantage of using hexadecimal numbers is that more card numbers can be generated. Hexadecimal numbers reduce the complexity in performing decimal-based conversions and computations on the transistor level. The same example from the decimal card number in Fig. 2 is considered in Fig. 3 for hexadecimal based computations.

- Double every even digit. This is achieved by performing the function of the left shift once in binary notation.
- After doubling, if two digits result from the product obtained, add the digits to obtain a single digit.

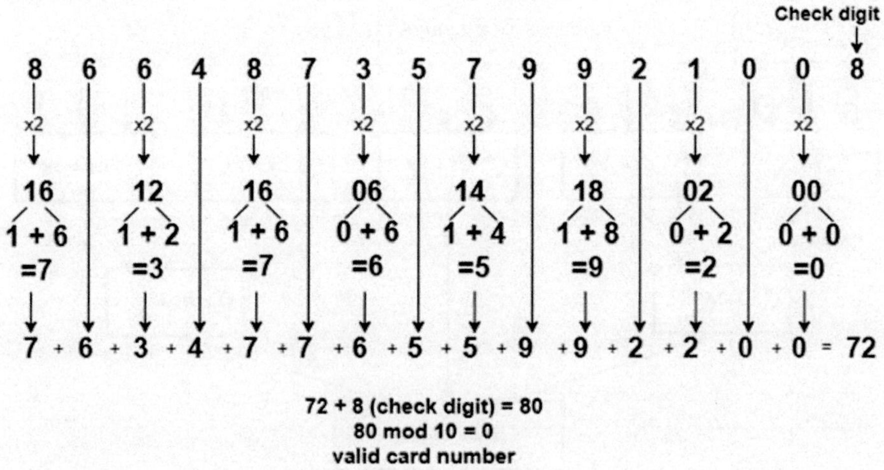

Fig. 2 Example of Luhn algorithm for regular card numbers

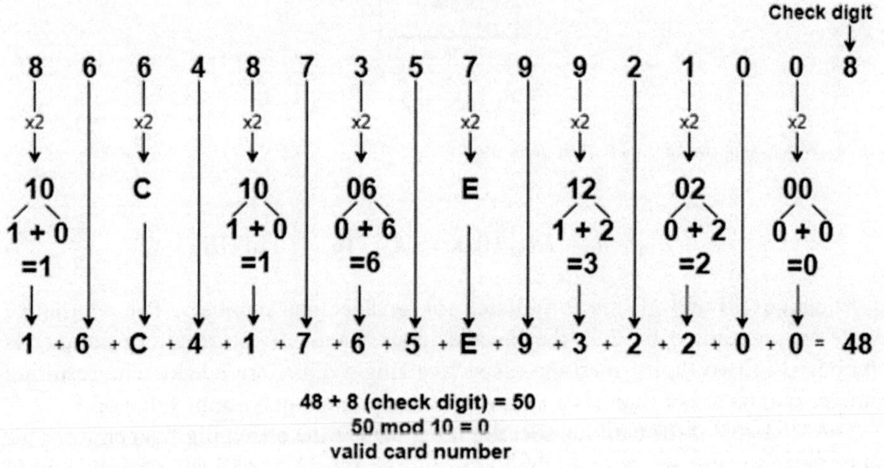

Fig. 3 Example of Luhn algorithm for hexadecimal card number

- With the exception of the check digit attached at the end of the card number, add all the digits.
- Divide the total by $(10)_{16}$, and then subtract $(10)_{16}$ from the remainder. The number resulting from this is the check digit. This computational step is improvised by just reading the last nibble of the checksum.

$$\left(\sum_{j=0}^{m} 2k_{2j} + \sum_{j=0}^{m} 2k_{2j+1}\right) \quad \text{mod } 16, \{\text{mod } 10 \text{ in HEX}\} \tag{3}$$

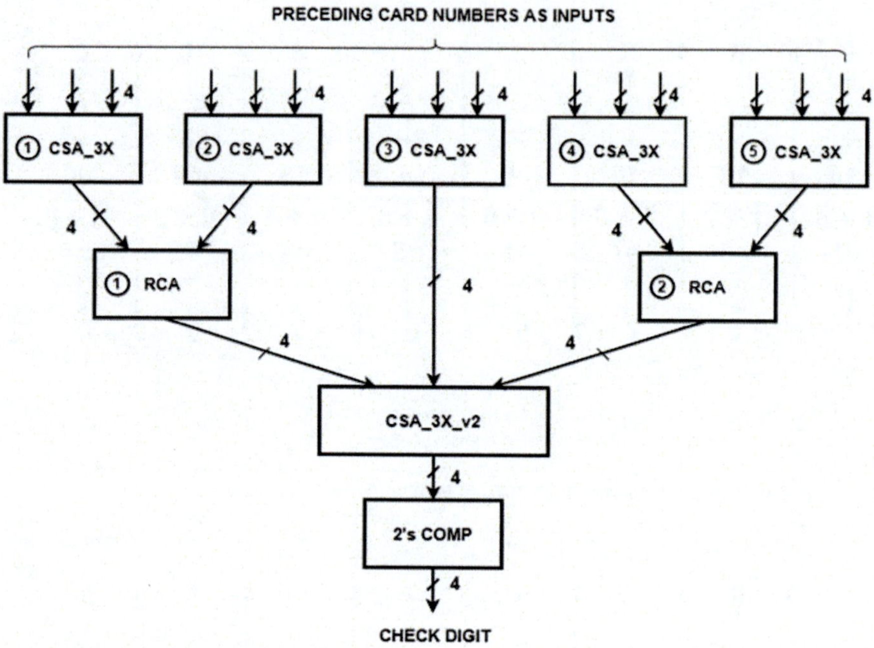

Fig. 4 Block diagram of check digit generator

$$k \in Z_{16}\ ,\ m \in N \in \mathrm{HEX},\ \exists\, k \geq 16,\ \{10 \text{ in HEX}\} \tag{4}$$

Equations (3) and (4) are formulated for hexadecimal numbers. The alternative digits are multiplied by 2 in the hexadecimal system. If the resulting number is composed of two digits, then the respective single digits are added. The resulting number will be lesser than $(10)_{16}$, that is, it can attain a maximum value of F.

The 15 digits of the card number are the inputs to the check digit generator. This block performs the operation of the Luhn algorithm for Hex and outputs one digit (4 bits in binary). At the end of the 15 digits, to form the 16 digits of the card number, this output is appended. As shown in Fig. 4, a 60-bit card number is processed to compute the check bits if 4-bit length. The Hex card number is then formed with 64-bits.

In Fig. 4, the 60 bits are read as input. These bits represent hexadecimal numbers. They are then to be multiplied by '2' which can be performed with a simple left shift. To accommodate for the number overflowing in case of numbers represented by all the four bits (numbers above 7), the multiplication by '2' is performed with left rotating the 4-bits once instead of left shift.

As it can be inferred, with left shift a zero is placed in the LSB of the 4 bits. If the left shift or multiply by '2' overflows 4 bits, then the number is represented by two digits in hexadecimal. The successive step is to add the two digits. In hexadecimal, it will bring us to add the LSB of the 4 bits which is zero, and the overflowing bit which will be '1'. The left rotation can simplify this step thereby enhancing the performance.

The 60-bit sequence is then added with CSA adders in three stages. The resultant checksum is then taken modulo from $(10)_{16}$ in base-16 (hexadecimal). The remainder of the modulo is concatenated with the 60 bits to generate the 64-bit card number with check bits.

To overcome such computationally intensive operation, the nibble containing the LSB is taken as the 4-bit parity that is to be appended. As the 60-bit number is divided by $(10)_{16}$, the remainder will be the last nibble. This adds to the improvements in performance. The check bits or parity added will aid in the verification of the card number. These check bits are tailored for the card number, which would satisfy the condition of remainder zero from modulo by $(10)_{16}$.

4.1 Schematic and Layouts of Generator Circuit

Figure 5 depicts the schematic of the checksum computing block. The input of the 60-bit card number is drawn to the carry-save adders (CSA) blocks. The final stages are designed to compute the modulus of checksum in base-16 or hexadecimal. The generator provides a 4-bit check number to be appended to the 60-bit card number. The resulting 64-bit card number is used on the ATM cards. The verification block derives these numbers to perform checking on the validation of the card number.

Figure 6 shows a version of carry-save adder. A carry-save adder is a type of digital adder that is used to compute the sum of three or more binary numbers efficiently. For our application, the carry-save adder tends to be the most effective adder. Typically, a broad adder implemented using this approach would be much quicker than the standard addition of those numbers.

As shown in Fig. 4 there is a requirement for three numbers to be taken as input at once. The benefit of CSA is that three numbers can be added and the total can be created instantly [14]. Uma and others studied various adder topologies [15], in order to select the adder topology with the trade-off between delay, power consumption, and area, which presents the appropriate option [16].

The floor planning of the check digit generator, validator, and the sub-blocks was carried out in such a way that the area occupied by the layout is minimal. Each alternate metal is laid out at a 90° angle to prevent interlayer parasites between different metal layers. The layouts were designed in 45 nm process node. Figures 7 and 8 show the floor plan and routing interconnects between blocks. For the routing of the interconnections, four metal layers were used.

Higher metals were used to route connections between blocks. A strong power mesh is maintained in metal2 horizontally and metal3 vertically. The power mesh lines are wider and designed to minimize resistance, and the power mesh is present in all metal layers.CSA_3x refers to the three input carry-save adders, followed by RCA which is a two input ripple carry adder. The cells are linked together so that the VDD and VSS terminals of neighboring cells short together to ensure that each cell gets power and ground connection. The gaps between cells are connected with the help of filler cells.

5 Verification System for Hex Card Numbers

The card number obtained from the generator is verified as depicted in Fig. 9. Equation (3) is incorporated in the verification process. The block of verification circuit is shown in Fig. 10. The circuit requires an additional CSA as the input bits to the verification circuit is 64-bit unlike the 60-bit input for the generator circuit.

5.1 Schematic and Layouts of Verification Circuit

Figure 11 shows the schematic of the card number verification circuit. The input to this block will be all 16 digits of the card number. The output is a single bit which indicates the card number's validity. The result may be one of two values, high or

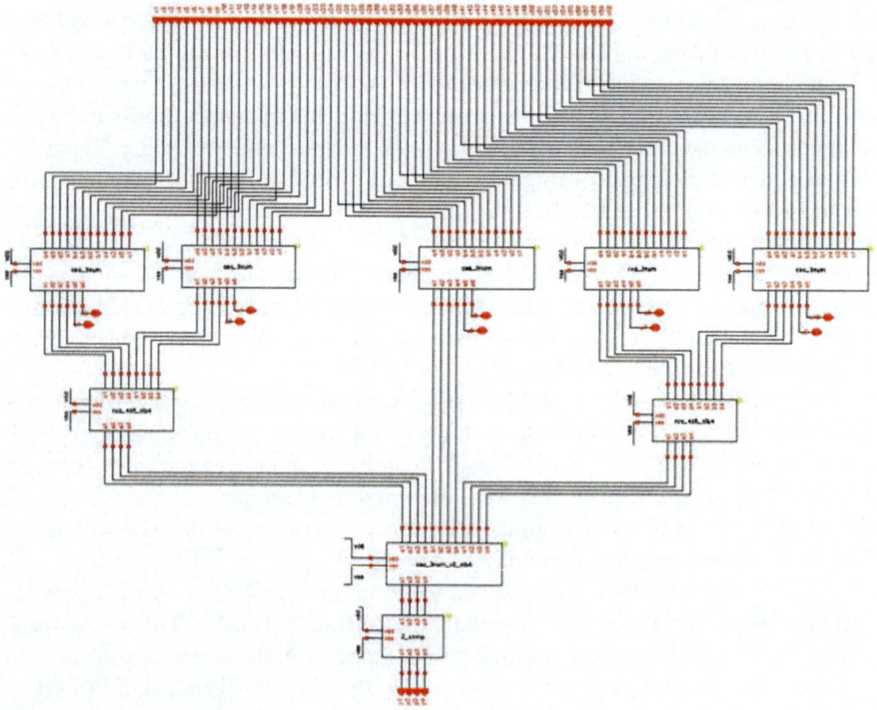

Fig. 5 Schematic of check digit generator

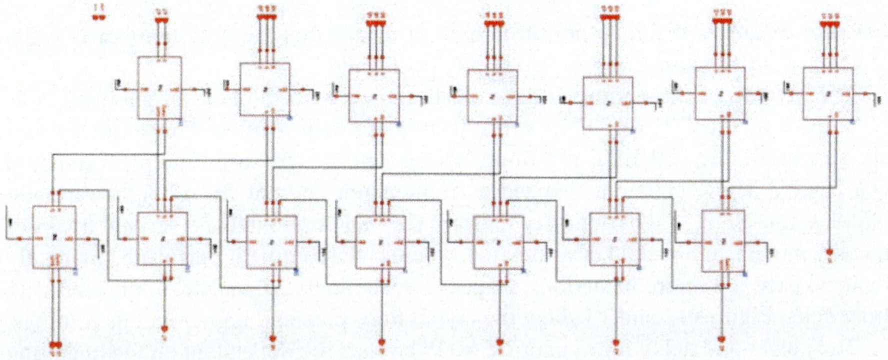

Fig. 6 Schematic of carry-save adder v2

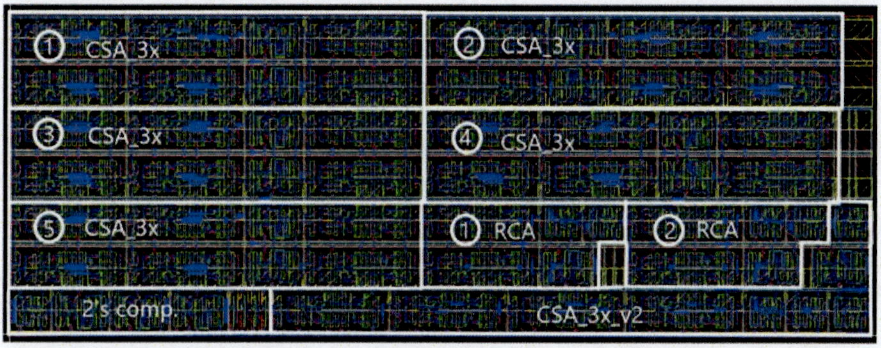

Fig. 7 Layout design and floor plan of the generation circuit in 45 nm

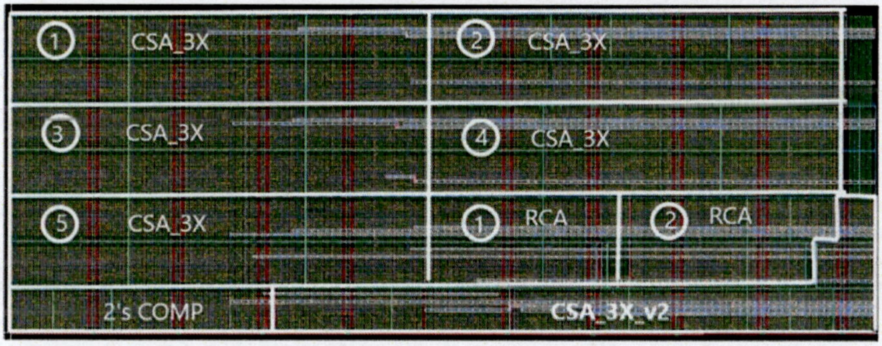

Fig. 8 Higher metal connections of generation circuit (Metal2, Metal3, and Metal4)

low. For example, if the output bit is high, it means that the card number is valid. When low, this means that the card number is invalid.

The layouts of the various blocks are designed with the help of standard cells obtained from the GPDK45 library. A selection of low-level electronic logic features such as AND, OR, INVERT, flip-flops, latches, and buffers forms a typical standard cell library. These cells are realized as full-custom cells of fixed-height variable-width. Using SKILL, the routing of uniform parallel horizontal and vertical tracks of metal2, metal3, and metal4 is automated, whereas the routing is performed manually by hand between interconnections. Gaps between blocks, if any, are connected with filler cells. Figures 12 and 13 show the layout floor plan and higher metals routings.

The power and delay for dynamic CMOS circuits are dependent on the operating frequency and quadratically dependent on the supply voltage. Power reduction is done by reducing the power supply voltage.

Graphs are shown in Figs. 14 and 15 depict the behavior of the circuit (power consumption and propagation delay) with different supply voltages.

5.2 *Results and Performance Analysis*

To verify the robustness of the circuit, it has been simulated through three even corners. Process angles reflect the limits of these differences in parameters in which the circuit grafted into the wafer would operate properly. A circuit that runs on devices created in such process corners can run slowly or faster than stated and at lower or higher temperatures and voltages, but the design is considered to have an insufficient design margin if the circuit does not work at any one of those extremes. The circuit proposed successfully works in all process corners.

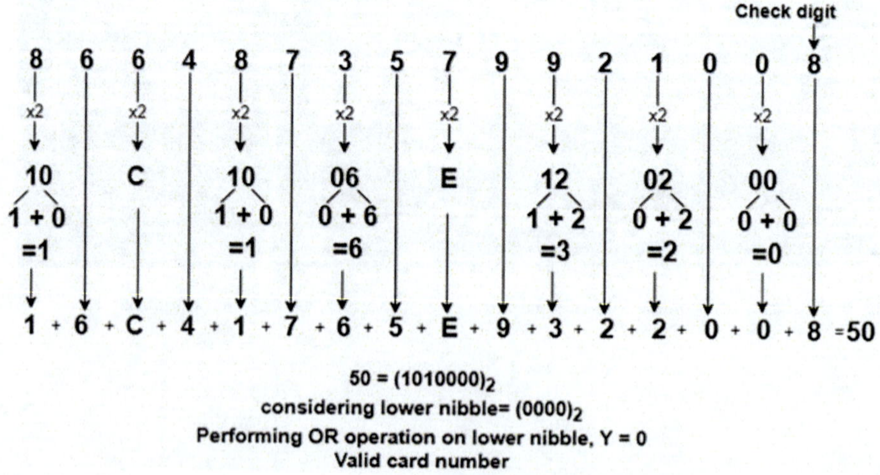

Fig. 9 Example of Luhn algorithm for hexadecimal card number verification

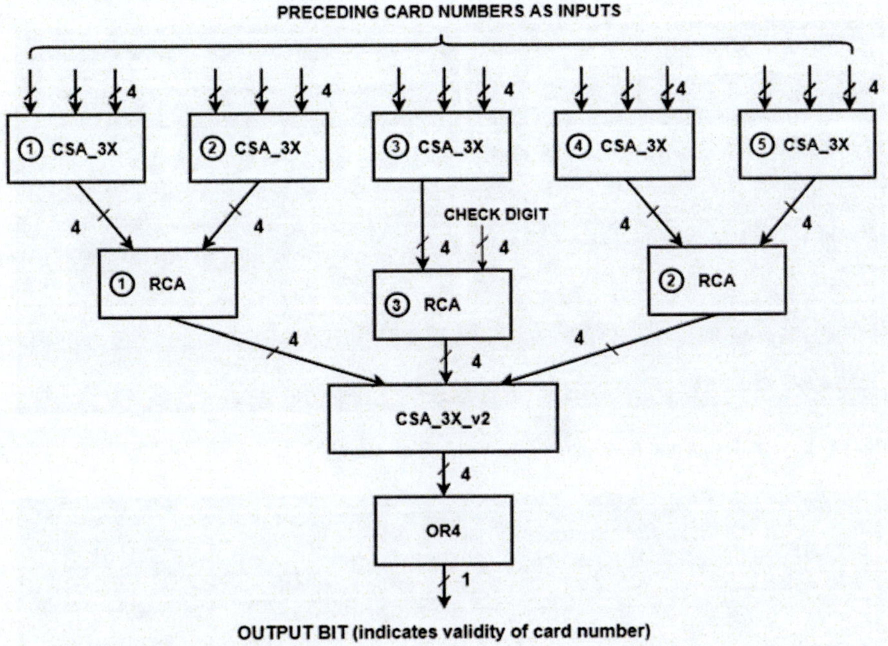

Fig. 10 Block diagram of the verification circuit

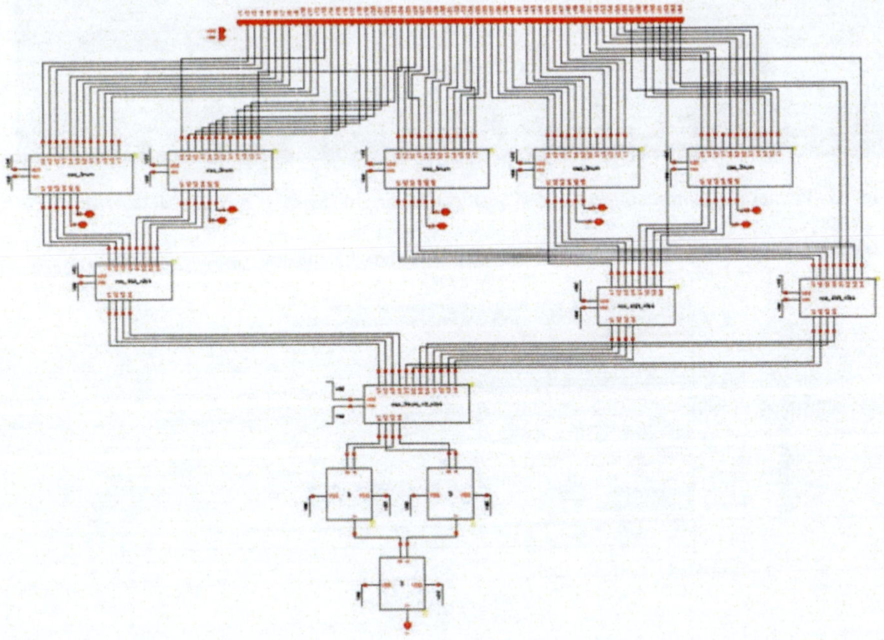

Fig. 11 Layout design and floor plan of the verification circuit

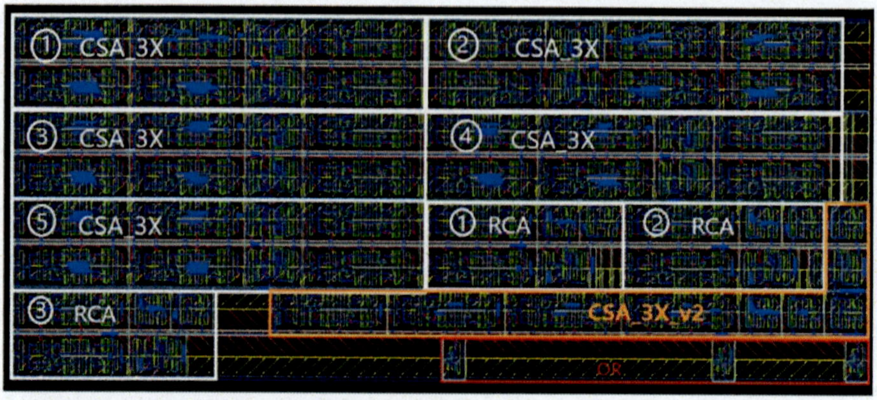

Fig. 12 Layout design and floor plan of the verification circuit

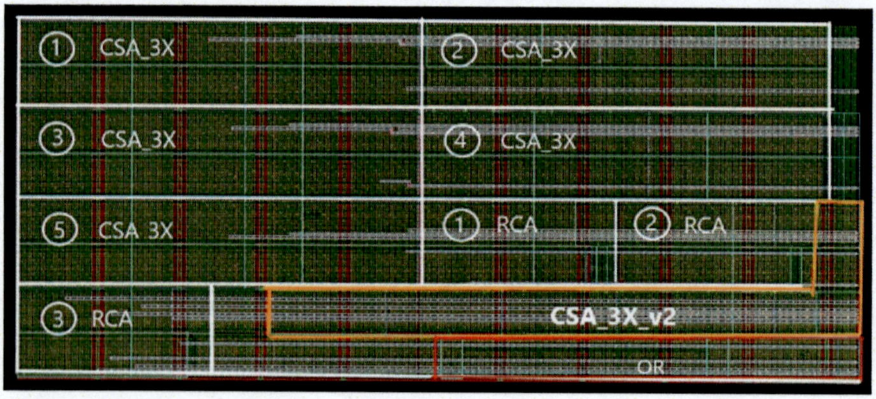

Fig. 13 Higher metal interconnections of verification circuit (Metal2, Metal3, Metal4)

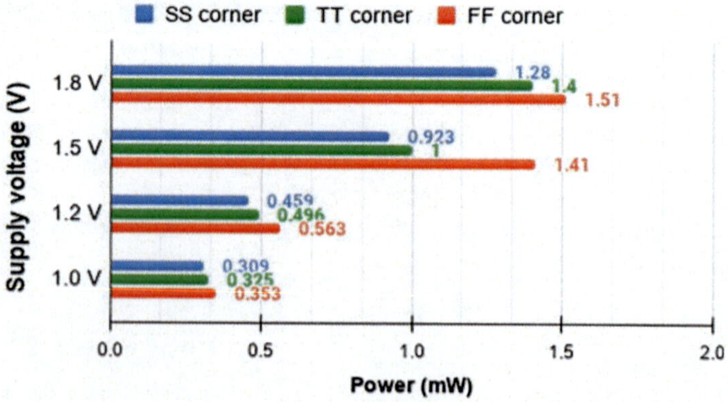

Fig. 14 Average power consumed at various supply voltages and process corners

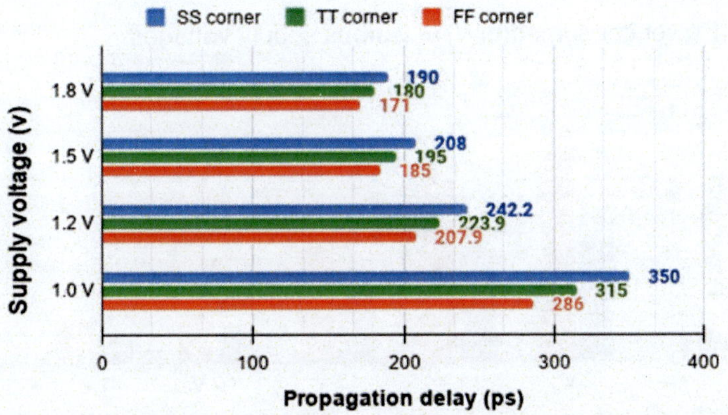

Fig. 15 Propagation delay(ps) at various supply voltages and process corners

Table 1 Power and delay comparison between 90 and 45 nm technologies

Supply voltage (V)	Power consumed (mW)		Propagation delay (ps)	
	90 (nm)	45 (nm)	90 (nm)	45 (nm)
2	2.9	1.9	323	170
1.8	2.33	1.4	336	180
1.5	1.41	1.0	370	195
1.2	0.804	0.496	439	223.9
1.0	0.512	0.325	528	315

Table 1 summarizes the power consumed (in mW) and propagation delay (in ps) of the circuits at different supply voltages. The results of the performance analysis is obtained with the help of Cadence Spectre simulator. The analysis is done for both 90 and 45 nm technology nodes.

Both the GPDK90 and GPDK45 technologies were used to conduct a thorough comparison. In the graphs shown in Figs. 16 and 17, the data obtained is represented. Scaling down the technology from 90 to 45 nm makes the circuit 60% more power efficient and also 55% efficient in terms of speed. The red color in the graphs indicated the power/delay consumed by the 90 nm circuit and the blue color indicated the power/delay consumed by the 45 nm circuit.

The 45 nm architecture consumes a delay of 315 ps at the supply voltage of 1 V, which means that it can be worked at a 3 GHz frequency.

The area occupied (in μm^2) by the architecture of all blocks built at 45 nm is summarized in Table 2. The complete card number check digit generator occupies 385 μm^2 and the card number validator occupies a total of 440μm^2, designed in 45 nm technology.

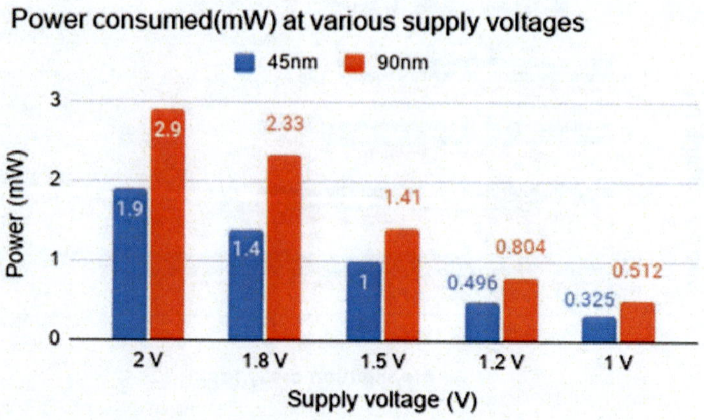

Fig. 16 Power consumed comparison between 90 and 45 nm

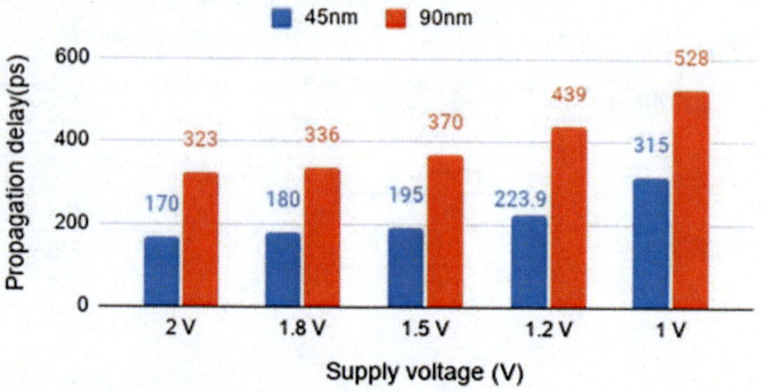

Fig. 17 Delay comparison between 90 and 45 nm

Table 2 Area occupied by different blocks designed in 45 nm

Blocks	Area (μm^2)
CSA_3X	52.668
RCA	25.992
CSA_3X_v2	44.802
2_COMP	16.7
Generator	385.434
Validator	440.496

6 Conclusion

There must be a secure and easy way to verify the correctness of the card number with the ever-growing growth in population and credit/debit cards. An algorithm developed by Hans Peter Luhn in the 1950s is still being used to check card numbers by many banks. A digital hardware circuit is proposed that can be manufactured into a physical chip that implements this algorithm to validate the card numbers. Not only does scaling down the production technology make the circuit quicker, but it consumes less power at the same time. The comparison of the efficiency of 90 and 45 nm technologies has proven this. Scaling down the technology from 90 to 45 nm makes the circuit 60% more power efficient and also 55% efficient in terms of speed. The 45 nm technology architecture can be operated with a 1 V supply at a frequency of 3 GHz. This means that it is possible to validate 3 billion valid card numbers in hexadecimal form per second. With the population of the world being 7.8 billion and believing everybody has a card, verifying them all takes only 2.6 s [11].

References

1. Dr. V. Ilango, N. Nadeem Akram, Advanced IMEI and credit card validation techniques using layered based LUHN's algorithm. Int. J. Recent Innov. Trends Comput. Commun. (2018)
2. K. Kumar, P. Kaur, Vulnerability detection of international mobile equipment identity number of smartphone and automated reporting of changed IMEI number. Int. J. Comput. Sci. Mob. Comput. **4**, 527–533 (2015)
3. E. Aleskerov et al., CARDWATCH-a neural network based database mining system for credit card fraud detection
4. M.-J. Kim, T.-S. Kim, *A Neural Classifier with Fraud Density Map for Effective Credit Card Fraud Detection* (Springer, Berlin, Heidelberg, 2002), pp. 378–383
5. R. Brause et al., Neural data mining for credit card fraud detection, in *Proceedings of the 7th International Conference on Machine Learning and Cybernetics* (2008)
6. R. Kumar et al., Payment card fraud identification. Int. J. Eng. Adv. Technol. (IJEAT) **2**(4) (2013)
7. J. Verhoeff, Error detecting decimal codes. *Mathematics Centrum Amsterdam* (1969)
8. N.R. Wagner, The Laws of Cryptography with Java Code (2002)
9. M.B. Suman, Survey paper on credit card fraud detection. Int. J. Adv. Res. Comput. Eng. Technol. (IJARCET) **3** (2014)
10. L.W. Wachira, Error detection and correction on the credit card number using Luhn algorithm, Jomo Kenyatta University of Agriculture and Technology (2016)
11. A. Maheshwari, S.K. Saritha, Flaws and frauds hindering credit cards security. Int. J. Eng. Trends Technol. (IJETT) **24** (2015)
12. K.W. Hussein et al., Enhance Luhn algorithm for validation of credit cards numbers. Int. J. Comput. Sci. Mobile Comput. **2**(7), 262–272 (2013)
13. W. Kamaku, W. Wachira, Twin error detection in Luhn's algorithm. Int. J. Sci. Res. (IJSR) (2015)
14. C.R. Patel et al., Inverted gate vedic multiplier in 90nm CMOS technology. Am. J. Electr. Comput. Eng. Sci. (Publishing Group) 2020)

15. R. Uma et al., Area, delay and power comparison of adder topologies. Int. J. VLSI Des. Commun. Syst. (VLSICS) **3**, (2012)
16. R. Mahalakshmi, Dr. T. Sasilatha, A power efficient carry save adder and modified carry save adder using CMOS technology, in *IEEE International Conference on Computational Intelligence and Computing Research* (2013)

Stability Analysis of AFTI-16 Aircraft by Using LQR and LQI Algorithms

V. S. S. Krishna Mohan and H. L. Viswanath

Abstract The stability analysis of the dynamical system of linearized plant model of Advanced Fighter Technology Integration (AFTI)-16 aircraft was proposed along with the optimal control methods by applying linear quadratic regulator (LQR) and linear quadratic algorithm (LQI) algorithms. The LQR and LQI algorithms results were compared with state-space model analysis results. The state-space methods like pole placement method, without using the LQR algorithm the negative feedback system were found to be unstable. By the application of LQR and LQI algorithms to the linearized plant AFTI-16 aircraft open-loop system having negative feedback found to be stable. The stability parameters were verified by using MATLAB programming software. The eigenvalues play a key role in finding closed-loop system stability analysis. MIMO dynamical system with state feedback gain matrices is calculated by using MATLAB programming software.

Keywords AFTI-16 aircraft · Optimal control systems · LQR · CARE · LQI · LQRD · DARE · MIMO dynamical systems

1 Introduction

In this article, the stability of the linear plant model of AFTI-16 aircraft [1] was employed by using the LQR algorithm [2]. Conventional state-space stability techniques like finding poles of closed-loop system by using eigenvalues of system matrix may not give all the times having negative feedback found to be stable [2]. By applying conventional state-space methods [3] and LQR technique, the AFTI-16 linear plant model having negative feedback was found to be stable. State feedback gain matrix K of given MIMO dynamical system found by using MATLAB program using optimal

V. S. S. Krishna Mohan (✉)
Department of Electronics and Instrumentation Engineering, Dayananda Sagar College of Engineering, Bengalore, India
e-mail: krishnamohan-eie@dayanandasagar.edu

H. L. Viswanath
Department of Electronics and Communication Engineering, Christ University, Bengalore, India

V. Suma et al. (eds.), *Inventive Systems and Control*,
Lecture Notes in Networks and Systems 204,
https://doi.org/10.1007/978-981-16-1395-1_22

control system toolbox [4]. The AFTI-16 aircraft linearized plant model contains both state equation and output equation [1].

AFTI-16 aircraft linearized plant model
Inputs: Elevators and flaperon angle
Outputs: Attack and pitch angle
Sampling Time: 0.05 s.
Constraints: Maximum 25° on both angles.
Open-loop response: Unstable.
State-space representation mode

$$\dot{X} = \begin{bmatrix} -0.0151 & -60.5651 & 0 & -32.174 \\ -0.0001 & -1.3411 & 0.9929 & 0 \\ 0.00018 & 43.2541 & -0.86939 & 0 \\ 0 & 0 & 1 & 0 \end{bmatrix} X + \begin{bmatrix} -2.516 & -13.136 \\ -0.1689 & -0.2514 \\ -17.251 & -1.5766 \\ 0 & 0 \end{bmatrix} U \tag{1}$$

$$Y = \begin{bmatrix} 0 & 1 & 0 & 0 \\ 0 & 0 & 0 & 1 \end{bmatrix} X \tag{2}$$

The above state-space model indicates two equations. The first equation is the state equation and the second equation is the output equation. The system matrix A which is having an order 4×4 square matrix having four state variables $\times 1$, $\times 2$, $\times 3$, and $\times 4$, and matrix B is an input matrix having order 4×2. This gives information above system can be modelled as multi-input and multi-output (MIMO) dynamical control system [4].

2 Literature Survey

In a classical control system [5], there are many stability techniques are available like RH criteria, root locus, Nyquist stability [6, 7] criteria, and bode plot. The drawbacks of classical control techniques mentioned above are computations are very tedious and handling for control systems having higher-order characteristic equation are time-consuming, and results are not very accurate. Classical control systems are mainly based on hit and trail methods [5]. The major drawback of the classical control system is all initial conditions are made to be zero. The classical control systems mainly applicable to the single input and single output (SISO). For multi-input and multi-output (MIMO) control system [3] problems classical control system approach is very difficult to find the amenable solutions. To analyze the complex control systems like aircraft, missiles, unmanned air vehicles (UAV), and space aircraft, the best approach is modern control systems [2, 4]. The modern control systems

[20] approach is completely based on state-space analysis. If the control system has physical constraints, the best possible approach is the optimal control system.[8] Optimal control system is mainly based on state-space approach technique and can easily solve the MIMO dynamical control system modeled and can be solved real-time problems based on matrix analysis and linear algebra mathematical techniques. In the state-space approach [3], results can be verified by using simulation programs that will be developed in the MATLAB platform.

3 Stability Analysis with Conventional State-space Method

From the system dynamics of AFTI-16 mentioned above, the plant model can be compared with standard state equations as $\dot{X} = AX+BU$ and $Y = CX+DU$. Here, matrix A as system matrix, B as Input matrix, C as output matrix and D as transmission matrix. The determinant of $(\lambda I - A)$ gives the roots of negative feedback of the AFTI-16 aircraft open-loop control system. By using standard MATLAB command $eig(A)$. It was observed that one of the closed-loop poles of the characteristic equation was found to be the right-hand side of s plane. In the state-space technique, one of the eigenvalues was found to be positive. It was concluded that the closed-loop system was unstable [3].

The following eigenvalues of the system matrix are obtained from the MATLAB program output.

$$e = \begin{matrix} -7.6636 + 0.0000i \\ 5.4530 + 0.0000i \\ -0.0075 + 0.0556i \\ -0.0075 - 0.0556i \end{matrix}$$

$$fx >>$$

From Table 1, it was observed that one of the eigenvalues λ2 was found to be positive. It was concluded that closed-loop system was unstable.

Table 1 Eigenvalues of the system using conventional state-space approach

Parameter	Eigen value
λ1	−7.6636
λ2	5.4530
λ3	−0.0075 + 0.00556*i*
λ4	−0.0075 − 0.00556*i*

4 Stability Analysis with Optimal Control Systems (LQR Technique)

LQR algorithm [9, 10] solves the continuous-time, linear quadratic regulator problem, and the associated Riccati equation. The following MATLAB command calculates the optimal feedback gain matrix K such that the feedback control law

$U = -KX$ Reduces the performance index J.[8]

Where

$$J = \int_0^\infty [X^T QX + U^T RU]\mathrm{d}t \tag{3}$$

Subject to the physical restrictions

$$\dot{X} = AX + BU \tag{4}$$

With MATLAB command $[K, P, E] = lqr(A, B, Q, R)$

Returns the gain matrix K, eigenvalue vector E, and matrix P, the unique positive definite solution to the associated matrix Riccati equation. If matrix $A - BK$ is a stable matrix, such a positive definite solution P always exists. The eigenvalue vector E gives the roots of the $A - BK$ matrix.

The stability analysis with LQR [11] was tested in the MATLAB program by choosing Q as a positive definite matrix such that 4 × 4 unit matrix and R as $\begin{bmatrix} 1 & 0 \\ 0 & 1 \end{bmatrix}$. After execution of the MATLAB program, the following eigenvalues of matrix $A - BK$ are displayed (Table 2).

E closed loop Eigenvalues of the system.

It was observed from the above table that all the eigenvalues of matrix $A - BK$ have a negative real part. It was concluded that the open-loop system having a negative feedback was stable [4]. The feedback gain matrix K is a 2 × 4 matrix. The elements of matrix K are given below.

K Feedback gain Matrix

Table 2 Eigenvalues of the system after LQR algorithm

Parameter	Eigen value
$E1$	−0.5769
$E2$	−9.7283 + 1.1415i
$E3$	−9.7283 + 1.1415i
$E4$	−19.5068

$$K = \begin{bmatrix} 0.3271 & -8.0640 & -1.4689 & -3.1618 \\ -0.9437 & 3.1732 & 0.1039 & 1.7342 \end{bmatrix}$$

$$K = \begin{matrix} 0.3271 & -8.0640 & -1.4689 & -3.1618 \\ -0.9437 & 3.1732 & 0.1039 & 1.7342 \end{matrix}$$

$$P = \begin{matrix} 0.0831 & -0.4187 & -0.0270 & -0.1866 \\ -0.4187 & 6.3317 & 0.4665 & 1.6298 \\ -0.0270 & 0.4665 & 0.0845 & 0.1945 \\ -0.1866 & 1.6298 & 0.1945 & 2.5075 \end{matrix}$$

$$E = \begin{matrix} -0.5769 + 0.0000i \\ -9.7283 + 1.1415i \\ -9.7283 - 1.1415i \\ -19.5068 + 0.0000i \end{matrix}$$

$fx >>$

4.1 Solution of Care

The output of the CARE algorithm [12] for the given linear system can be computed by using proper MATLAB command and by choosing proper values of Q, R, and S matrices. $[X, L, G] = \text{care}(A, B, Q, R, S, E)$ calculates the output value X of the continuous-time algebraic Riccati equation [11]. The MATLAB program output gives the following values:

X solution set of CARE

$$X = \begin{bmatrix} 0.0945 & -0.9373 & -0.0045 & -0.8644 \\ -0.9373 & 29.5216 & 0.14300.1430 & 31.7044 \\ -0.0045 & 0.1430 & 0.0583 & 0.2145 \\ -0.8644 & 31.7044 & 0.2145 & 38.3246 \end{bmatrix}$$

Command Window

$$L = \begin{matrix} -0.5769 + 0.0000i \\ -9.7283 + 1.1415i \\ -9.7283 - 1.1415i \\ 19.5068 + 0.0000i \end{matrix}$$

$$G = \begin{matrix} 0.3271 & -8.0640 & -1.4689 & -3.1618 \\ -0.9437 & 3.1732 & 0.1039 & 1.7342 \end{matrix}$$

Table 3 Eigenvalues of the system using CARE algorithm

Parameter	Eigen value
$L1$	$-0.5769 + 0.0000i$
$L2$	$-9.7283 + 1.1415i$
$L3$	$-9.7283 - 1.1415i$
$L4$	$-19.5068 + 0.000i$

$fx >>$

It was observed that all Eigenvalues have a negative real part. So, the open-loop system having negative feedback was found to be stable (Table 3).

G Gain Matrix

$$G = \begin{bmatrix} 0.3271 & -8.0640 & -1.4689 & -3.1618 \\ -0.9437 & 3.1732 & 0.1039 & 1.7342 \end{bmatrix}$$

5 Stability Analysis with Optimal Control Systems (LQI) Technique

Linear quadratic integral [10, 7] calculates an optimal state-feedback control law for the tracking loop as shown in Fig. 1.

For an open-loop system plant system with the state-space equations, the state-feedback control is of the form $U = -K[x; x_i]$ Where x_i is the integrator output. This control law ensures that the output y tracks the reference command r. For MIMO systems, the number of integrators equals the dimension of the output y. [13]

MATLAB command $[K, S, E] = lqi(SYS, Q, R, N)$ calculates the optimal gain matrix K, given a state-space model SYS for the plant and weighting matrices Q, R, N [14].

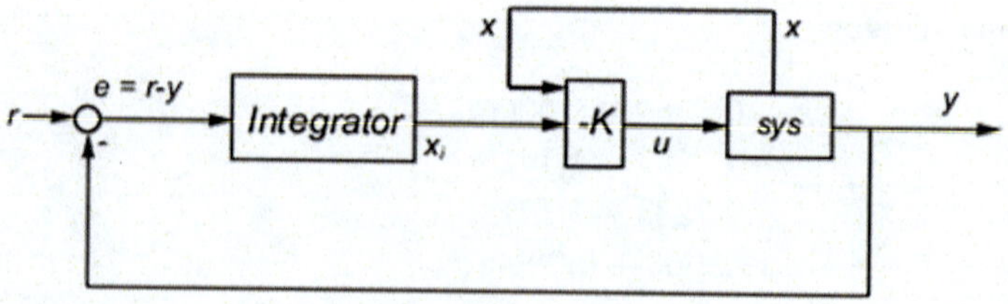

Fig. 1 Block diagram of linear quadratic integrator

Table 4 Eigen values of system after LQI algorithm

Parameter	Eigen value
$E1$	−0.0002
$E2$	−0.1647
$E3$	−0.5780
$E4$	−9.7273 + 1.1399i
$E5$	−9.7273 − 1.1399i
$E6$	−19.5069

The MATLAB program output for the AFTI-16 aircraft dynamical system produced the following results:

$$K = \begin{bmatrix} 0.0145 & -5.8956 & -1.0247 & -7.8746 & 0.7244 & 0.6893 \\ -0.9760 & 3.4205 & -0.0597 & 1.8199 & 0.6893 & 0.7244 \end{bmatrix}$$

The closed-loop eigenvalues of the above systems are

$$e = \begin{matrix} -0.0002 + 0.0000i \\ -0.1647 + 0.0000i \\ -0.5780 + 0.0000i \\ -9.7273 + 1.1399i \\ -9.7273 - 1.1399i \\ 19.5069 + 0.0000i \end{matrix}$$

$$E = [-0.0002, -0.1647, -0.5780, -9.7273 + 1.1399i, -9.7273 - 1.1399i, -19.5069]$$

The entire closed-loop eigenvalues were found to be negative real numbers. It was concluded that the closed-loop system was stable (Table 4).

6 Stability Analysis with Optimal Control Systems (Discrete LQR) Technique

In this section, the stability analysis were performed by using discrete linear-quadratic (LQ) regulator [15] for continuous plant AFTI-16. *lqrd* MATLAB command designs a discrete full-state-feedback regulator that has response characteristics similar to a continuous state-feedback regulator designed using LQR [7]. This command is useful to design a gain matrix for digital implementation after a satisfactory continuous state-feedback gain has been designed.

$[Kd, S, E] = lqrd(A, B, Q, R, T_S)$ calculates the discrete state-feedback law. $U[n] = -KdX[n]$ minimizes a discrete cost function equivalent to the continuous cost function.

$$J = \int_0^{\infty} [X^T QX + U^T RU] \mathrm{d}t$$

The matrices A and B specify the continuous plant dynamics [16, 17].

$$\dot{X} = AX + BU$$

and $Y = CX + DU$.

Matrix Q is a positive definite matrix having order same as system matrix A and matrix R is a positive definite matrix as several columns as matrix B. The MATLAB results were obtained based on the command $[Kd, S, E] = lqrd(A, B, Q, R, T_S)$. T_s specify the sample time of a discrete regulator. From the reference [1] $T_s = 0.05$ s. T_s specifies sample time of discrete regulator. From the result, the closed-loop discrete eigenvalues are $E = [0.4172, 0.6467, 0.7369, 0.9407]$. All eigenvalues lie inside the unit circle in the Z plane. By using a discrete linear quadratic (LQ) regulator for continuous plant AFTI-16, the closed-loop system was stable [7].

6.1 Solution of Dare Algorithm

The output of the DARE algorithm for the given linear system can be computed by using proper MATLAB command and by choosing proper values of Q, R, and S matrices. $[X, L, G] = \text{dare}(A, B, Q, R, S, E)$ Calculates the output X of the discrete-time algebraic Riccati equation [7]. The MATLAB program output gives the following values.

X solution set of DARE algorithm

$$X = \begin{bmatrix} 1.000 & 0.0026 & -0.0052 & 0.0070 \\ 0.0026 & 48.4592 & 16.0142 & 0.7741 \\ -0.0052 & 16.0142 & 32.8049 & -17.5832 \\ 0.0070 & 0.7741 & -17.5832 & 20.8637 \end{bmatrix}$$

It was concluded that all eigenvalues have a magnitude less than one. So, the closed-loop system was stable (Table 5).

G Gain Matrix values are

Table 5 Eigenvalues of the system after DARE algorithm

Parameter	Eigen value
$L1$	$0.4892 + 0.0000i$
$L2$	$-0.0070 + 0.0984i$
$L3$	$-0.0070 - 0.0984i$
$L4$	$-0.0000 + 0.0000i$

$$G = \begin{bmatrix} -0.0001 & -2.9645 & 0.1073 & -0.2402 \\ 0.0012 & 5.1768 & 0.0211 & 2.4481 \end{bmatrix}$$

7 Conclusion

Different advanced control system parameters of AFTI-16 aircraft dynamical system by using advanced state-space techniques. The conventional state-space methods without using LQR and LQI algorithms gave results and closed loop system becomes unstable. After the utilization of LQR and LQI algorithms, all closed-loop roots of the system were found to be negative real numbers. This concludes closed-loop system after application of LQR and LQI algorithms, the closed-loop system found to be stable. The CARE algorithm observed the value of the solution of state matrix X and closed-loop eigenvalues. Similarly, apply LQR algorithm in the discrete time domain and verified the stability of plant AFTI-16 and result concluded all poles lie within the unit circle. The DARE algorithm provided the unique solution of state matrix X in the discrete-time domain. It was concluded that the closed-loop system was stable. Finally, this research presented results of all stability techniques like LQR, LQI, and LQRD applied to the AFTI-16 system and concluded that the closed-loop plant was stable.

References

1. A. Bemporad, Model predictive control of hybrid systems, in *2nd HYCON PHD school on HYBRID systems* (Siena, 2007)
2. F. Djaballah An implementation of optimal control methods (LQI, LQG, LTR) for geostationary satellite attitude control. *Int J Electr Comput Eng (IJECE)*, **9**(6), 4728–4737 (2019)
3. https://www.sciencedirect.com/topics/engineering/modern-control-theory-2019
4. C. Yiwen, Design and application of modern control system for quadrotor UAV in PIXHAWK, in *Proceedings of the 2019, 5th International Conference on Robotics and Artificial Intelligence November 2019* (2019)
5. G. Uma, Mathematics for students of engineering with a study on stability analysis of state space model. Adv. Theor. Appl. Math. **11**(3), 299–303 (2016). ISSN 0973-4554

6. J.E. Kurek, Stability of positive 2-D system described by the roessermodel. IEEE Trans. Circ. Syst. I **49**, 531–533 (2002)
7. T. Espinoza1, R. Parada1, A. Dzul1, R. Lozano, *Linear controllers' implementation for a fixed-wing MAV*. (International Conference on Unmanned Aircraft Systems (ICUAS), Orlando, FL, USA May 27-30, 2014)
8. B.S. Anjali, Simulation and analysis of integral LQR controller for inner control loop design of a fixed wing micro aerial vehicle (MAV) (Elsevier 2016)
9. A. Joukhadar, I. Hasan, A. Alsabbagh, M. Alkouzbary Integral Lqr-Based 6 dof Autonomous Quadcopter Balancing System Control. Int. J. Adv. Res. Artific. Intell. (2014)
10. Z. Wei, Y. Maying, Comparison of auto-tuning methods of PID controllers based on models and closed-loop data, in *33rd Chinese Control Conference (CCC)* (2014)
11. D.E. Kirk, *Optimal Control Theory an Introduction* (DOVER publications, 2004)
12. S. Barnett, *Introduction to Mathematical Control Theory* (Oxford University Press, 1990)
13. M. Tiwari, A. Dhawan, A survey on the stability of 2-D discrete systems described by fornasini-marchesini second model. Circ. Syst. **3**, 17–22 (2012)
14. D. Subbaram Naidu, *Optimal Control Systems* (CRC Press, Idaho State University, Pocatello. Idaho. USA, 2003)
15. H. Purnawan, Design of linear quadratic regulator (LQR) control system for flight stability of LSU-05. IOP Conf. Ser. J. Phys. **890** (2017)
16. M. Gopal, *Modern Control System Theory* (Wiley, 1994)
17. G.F. Simmons, *Ordinary Differential Equations with Applications* (Tata McGraw Hill, 2003)
18. J. Luo, C.E. Lan, Determination of weighting matrices of a linear quadratic regulator. IEEE J. Guid. Cont. Dynam. **18**(6), 1462–1463 (1995). https://doi.org/10.2514/3.21569
19. M. Braun, *Differential Equations and Their Applications*, 4th edn (Springer, 2011)

Location Analytics Prototype for Routing Analysis and Redesign

Neeraj Bhargava and Vaibhav Khanna

Abstract The advanced route analysis concerns are addressed and implemented using the ArcGIS Network Analyst environment. The location analytics has continuously progressed with shortest path problems and origin-destination cost matrix-related problems at several stages of analysis. Dijkstra's algorithm is highly useful for network analysis on small and relatively less complex networks, but the output performance degrades as the networks become more complex. In real life, the analysis problems are based on "real-time queries on large hierarchical road networks". The shortest path algorithm is also sometimes known as the classical exact best route algorithm and is not suitable for the stated problem set. The hierarchical algorithms are used to solve the problem effectively and efficiently.

Keywords Dijkstra algorithm · Routing analysis · Route algorithm · Hierarchical algorithms · Network analyst

1 Introduction

Traffic and Routing analysis is an important part of Location Analytics. This has also been an active field of research for a very long time. This algorithm often applied to solve this type of problem is the Dijkstra Algorithm.

There are a lot of businesses that send their fleet of vehicles. This may be for taking a package from the customer, or for delivering a package to the customer, or to provide services at the customer's destination. Each of these situations is associated with a unique set of constraints and business rules to assign several "stops" in the route in sequence. Many times focused on increasing efficiency and find the least cost way for achieving our routing objectives.

N. Bhargava
Department of Computer Science, School of Engineering and Systems Sciences, MDS University, Ajmer, Rajasthan, India

V. Khanna (✉)
Department of Computer Science, Dezyne E'Cole College, Civil Lines, Ajmer, Rajasthan, India

V. Suma et al. (eds.), *Inventive Systems and Control*,
Lecture Notes in Networks and Systems 204,
https://doi.org/10.1007/978-981-16-1395-1_23

The Network analyst toolkit in ArcGIS ArcMap has received several APIs for solving the Network Analysis problems and Location Analytics. Apart from the routing, the hierarchical algorithms are useful for modeling several business intelligence situations. All of this analysis involves the Street network, it is a natural hierarchy of local roads connecting to state roads and eventually to the national highways.

All the routing algorithms in Computer Science have the central objective of minimizing one or more impedance such as drive time, or length of the journey. The working parameters are the nodes, edges, and routes. In practical life, the algorithm may have to deal with restrictions on edges and nodes because of certain practical considerations such as a one-way road or heavy traffic on a certain road. To design the routing models, it can avoid the restricted edges and nodes.

Generally, this is a dynamic situation requiring real-life inputs and immediate changes in decisions based on such inputs. Similarly, there may be turn restrictions, or if dealing with large size trucks may have to match their size and existing load with the turning radius to avoid overturning of heavy vehicles. Consider the situation in which a random accident may create a sudden roadblock and the entire pre-decided plan requires a redesigning of routing information.

The best route in such cases may often be longer in terms of impedance but more practical and more realistic routes based on spatial data mining. Hierarchical algorithms can be programmed to use primary roads and highways for faster movement and avoid local roads and streets. Also, there is not much time available to solve the problem as all of this is to be programmed on real-time inputs.

Even when driving our cars using a Google Map is more effective for a simple route with a simple set of instructions to follow. The drivers that are following these driving directions are more interested in staying on major roads and reducing the number of turns to avoid a complex set of instructions. This is considered as a better approach.

All these challenges in the form of turn restrictions, barriers, road jams, heavy traffic, oversize vehicle routing, etc. to acquire the incremental mining of spatial datasets and complementing it with a priori knowledge of efficient routes. This research describes the system design and prototype development as a solution to these challenges. To implement this prototype, the Network Analyst platform and applied a hierarchical algorithm is used in a two-step procedure. Initially, classifies a network into hierarchical levels, and the second step is to explore the network to compute the best route. ArcGIS Pro was used for routing analysis and to develop and implement python scripts using integrated Jupyter Notebooks in ArcGIS online environment. It is a flagship application and spatial data science workstation of ESRI. It can dynamically and interactively explore data to drive analysis using inbuilt network analyst modules and custom python code.

2 Related Work

Pollution is a foremost delinquent challenge in the transportation industry, with its damaging effects on the atmosphere. The authors have developed an algorithm to undertake this problem. The goal is to reduce route costs and make it more user-friendly. The issue in step one is modeled for hierarchical network analysis and is resolved in phases. The algorithm used is Hybrid Particle Swarm Optimization for designing better routes at each step. The prototype is verified by implementing this minimum cost and environmentally friendly routing on mobile advertisement vehicle [1].

The fast expansion of online food distribution fetches colossal orders. These are often delivered by deliverymen who often use electric two-wheelers. It was observed that there were major discrepancies in routes taken by the deliverymen vs. routes predicted as best routes by the algorithms. The real-time actual route information was a mismatch with the existing road network dataset. The authors utilized Inverse Reinforcement Learning (IRL) algorithm to identify human biker preferences and used these to endorse their chosen routes as ideal practical routes. The authors modified Dijkstra's algorithm to control and compute the best practical route instead of the shortest route. This also facilitated better navigation and online updation of actual road conditions [2].

Makariye (2017), the author used the Dijkstra Algorithm and Spatial clustering of current traffic data to arrive at shortest path computation. In modern car navigation systems, the Shortest Path computation is a significant issue in route decision making particularly in emergencies. Navigation systems and maps are often the most used application by outside drivers commuting in the city. These inherently require network analysis in combination with live traffic data inputs. The traffic conditions in every city are often a function of the time of the day. Through the use of Dijkstra's algorithm discover the shortest paths in terms of some of the other cost impedance. On the other hand, the best path is computed by integrating spatial data mining of live traffic data and combining the results with shortest paths. The cost is not always the distance travelled by the vehicle. Mostly it is the time that matters and the best path means the route that will take the minimum time to reach up to the destination. The drivers often need substitute path contingent on traffic data [3].

The automobile direction-finding problem with fragmented distributions was presented by Dror and Trudeau. This article offered the benefits that can be achieved through proper planning of routes [1]. The study was then conducted with multiple vehicles participating in the routing problem and also with multiple distances. The economies related to each were formulated and analyzed [4].

The origin-destination routing problem was addressed in a novel method. The authors designed and implemented the method in which the algorithm can compute in multiple directions to arrive at a better path by incremental mining of multidimensional data. This is a multidimensional analysis of several factors that demand real-time network analysis and spatial data mining [5].

Idri Abdelfattah et al. (2017), the authors work on Dijkstra's algorithm for the time-dependent multimodal transport network. The authors considered situations where more than one mode of transport is needed to reach the desired destination. The computation generated a mix of paths taken through multimodal transport situations [6].

The alternative paths in a network have been deliberated calculated by Eppstein. The author worked on the algorithms to derive the minimal path touching maximum OD pairs. A catalog was prepared for *k*-paths that connects the desired set of origin-destination pairs. The total length of such OD pairs was computed and the minimum one was selected [7].

Liu designed a model to associate and recommend tourist routes base on topics often discussed among the tourist circles. The model was trained on data inputs based on the central landscapes of the site and tourists' preferences. The application predicted season-dependent tourism packages that would be favoured by specific tourist segments [8].

Balaji et al. have taken an approach for improving customer satisfaction. The authors proposed an operative hybrid method to associate client ranking through routing algorithm and suggested solution for the capacity-dependent automobile routing problem [9].

The mainstream studies in routing computations for public transport routes, contemplate city commuting as a pair of nodes representing the start-point and the end point. Most of the network analysis and spatial mining algorithms are often zone-based and compute results for the best path within a zone. The authors have presented their work on modifying the connection-focused algorithms to zone-focused computation modules. The subsequent algorithm is put to use on a dataset produced through actual data. The novel modifications in the algorithm showed promising changes in routing outputs and significant improvements. The authors made available the dataset on which the experiments were conducted so that other research scholars may run their analysis on this benchmark dataset and compare results for the betterment of routing algorithms [10].

The article focused on QoS considerations in VANET was published describing the work of the authors on modified Ant colony-based optimization. The concept used fuzzy clustering for the identification of optimal routes among a group of vehicles. The authors combined the shortest path algorithm with cluster assignments in the vehicular network to accomplish motion in combination with position and path. The simulation demonstrated the superiority of the scheme based on bandwidth utilization throughput delay and link life expectancy [11].

Several studies have compared route computing algorithms for the optimization of municipal solid waste collection vehicles. Nikolaos et al. compared algorithmic results on live municipal datasets using the ArcGIS Network Analyst and the Ant Colony System (ACS) algorithm for best routes detection for Municipal Solid Waste (MSW) collection. The conclusive remarks suggested that both the systems have their own merits and unique advantages. The ArcGIS environment was more efficient and adaptive [12].

3 Proposed Work

The Spatial data mining data processing phase is the step one to categorize the edges forming the city road network in appropriate hierarchies. This step searches very actively as the search space is reduced and the routing algorithms and shortest path algorithms path search is optimized. Hierarchical classification is a multilevel spatial grouping that occurs on the creating of network datasets. Generally, this kind of hierarchical classification is conducted at three levels:

a. Classification first is the constrained admittance.
b. Classification second represents the main roads and state highways of the city.
c. Classification third is for the local roads and lanes network (Fig. 1).

The map processes the spatial dataset to model the three levels of hierarchy and computes the origin to the destination using multilayer road classification. After the creation of the networked dataset in this hierarchical form, a multidirectional search is conducted for the best path and accommodates the spatial data mining inputs to the current analysis.

This article presents the routing scheme as a sequence of workflow activities and programming interventions. The core concept of network analysis is the implementation of rules on the network datasets. Rules are the dictation conditions permitting the movement of objects on the network. The network analyst module performs some operations on the input dataset and incrementally applies the set of rules. The output is a new dataset and feature layers represented on the coordinate system. The network is a combination of edges and nodes. Edges are the connecting roads of the network which are further classified in hierarchies. Nodes are the locations on the network, a

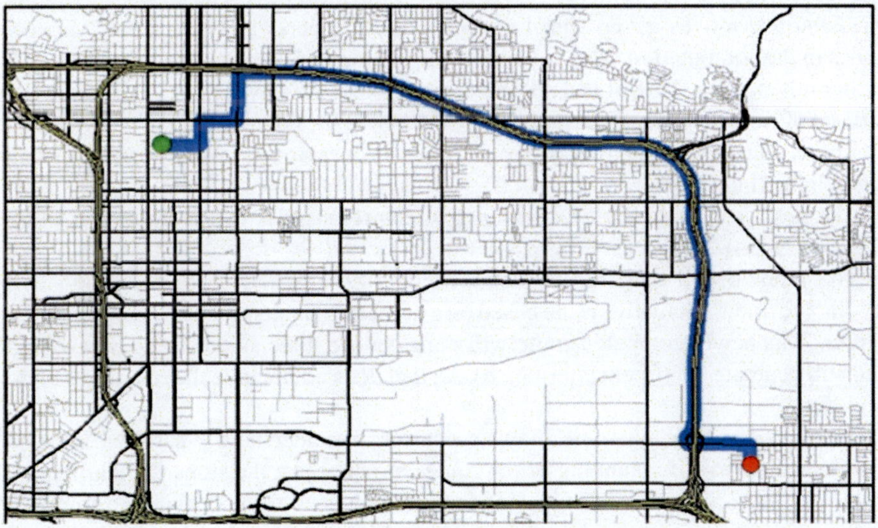

Fig. 1 Origin to destination mapping with 3 layered classifications of road networks

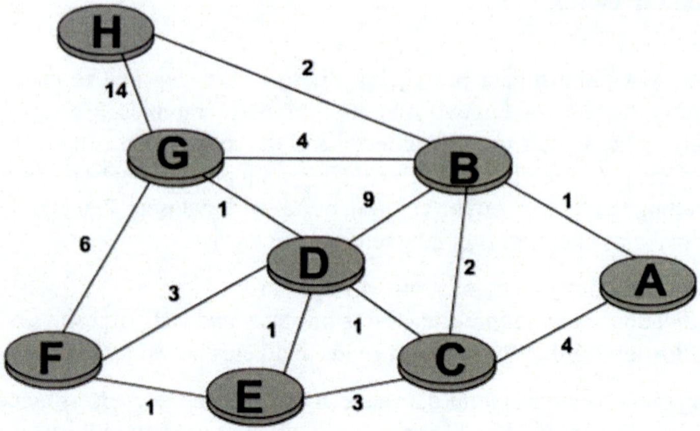

Fig. 2 Route computation through network edges and nodes

minimum set would comprise of an origin node and a destination node. Due to various considerations, the network movement of objects is through multiple nodes or intermediate termination. This termination may be demanded by the user or may be induced by the route computation module. Dr. S. Smys utilized an approach similar to Ant colony optimization with the addition of the clustering technique to discover the best route for Internet access in VANET to manage its mobility. This research situation was also required to address frequent topology changes and real-time new conditions to be included in path computation [11] (Fig. 2).

In a regular route planning operation, the user specifies the "origin", "destination" and sometimes the "stops" through which the route must pass. The proposed scheme in this article moves a step ahead and includes "barriers" and "incidents" as new nodes in the computation. This is like creating a new node on the network which was earlier not existing. Now it is a new set of nodes and the flow has to incorporate the new set of rules. This results in a dynamic change in the computation of a route plan.

In the initial control group experiment set, it does not have any barriers or any restrictions (Fig. 3).

The next process in the network analyst environment is to induce one or more barriers (Fig. 4).

The module will dynamically acknowledge this restrictive condition and will recompute a new alternative route based on a new set of inputs and the route computed earlier. This newly computed route will consider the node as a node to be avoided and will generate a different alternative path that does not include the restrictive node (Fig. 5).

To understand the types of analysis one has to understand what data is to be analyzed. For real-life planning of the routes, it is not just the stops that participate in the analysis. Several other types of feature data need to be considered to arrive at the right results such as the streets; it could be rail lines and bus stations, and overpasses and underpasses, and everything that goes into modeling the real world

Fig. 3 Route computations with no restrictive conditions

Fig. 4 Inducing restrictive conditions in the network

in a computer model of streets. These analytical inputs are more pronounced the route planning for emergency vehicle. The rules of analysis change in the case of emergency vehicles, for example, one-way roads can be turned off and the emergency vehicle can go the wrong way down a one-way road. While handling the different types of trucks or four-wheel drive, it can handle rural travel or tall trucks that can't go under overpasses are similar cases which require additional programming. These

Fig. 5 Alternative route avoiding the restricted node

restrictions and barriers are implemented in python and combine into travel mode so that the user can switch between analysis modes very quickly.

This also permits redesigning the route based on new data inputs from the user as well as the system. The computing of the "best route" has similar considerations where the user may have a customized criterion for defining the "best" route. For example, someone may wish to navigate through the road that permits the fastest vehicle movement. This is where the hierarchical routes can facilitate such computations. The proposed routing scheme also includes programmed inputs from the user if the route plan has to compute routes only for a specific level of road network hierarchy. The road classification can be added as a hierarchy input and speed limit as a cost to computing the fastest routes.

Pseudocode for computing the best route for given origin-destination set
Best Route can accommodate barriers, restrictions, incidents, and road network hierarchies
A location analytics prototype for solving the OD cost matrix for generating the dynamic best route
Restrictions may be applied through programmed scripts
A ArcPy script to be used in combination with the Network Analyst Module of ArcGIS Online

1. Import ArcGIS API for map object
2. Import ArcGIS API for Layers management
3. Import ArcGIS Tools API
4. Import ArcGIS API for visualization
5. Import python API for ArcGIS
6. Set environment for analysis of network dataset

7. Create output structure for OD Cost Analysis
8. Allocate disk space for analysis data
9. Input geodatabase file for base network layer
10. Include computation module for travel modes
11. Input origin node
12. Input destination node
13. Input stop nodes (optional)
14. Generate origin-destination analysis layer
15. Construct the layer object for output results
16. Fetch details of all the sub-layers.
17. Compute sub-layer hierarchy
18. Creates a repository of layer names
19. Create OD Lines layer
20. Compute solution for OD layer
21. Use solver object of ArcGIS python API to generate solution
22. Get solver properties from the layer object
23. Resolve computations related to OD layer and save results
24. Add origin-destination layer to the map object layer
25. Get Transport mode and associated modules
26. Get inputs for cost and network impedance
27. Include restrictions into transport mode options
28. Include TransportMode into origin-destination solver
29. Compute solver properties for OD layer
30. Compute restrictions and travel mode considerations
31. Compute and publish output path
32. Publish unassigned layers

4 Result Analysis

The Origin-Destination Cost object makes available the interface to all the investigation properties and route solver functionalities. The object parameters can be tweaked to create desired models of everyday situations read and run python scripts for automation of analysis. The user is required to create a new analysis layer for OD Cost Matrix Analysis through the usage of the geo-processing toolkit in the ArcGIS online environment.

The main programming elements of the analysis properties are:

a. AccumulatorProperties—Facilitates the user by getting or setting a list of NetworkCost attributes
b. AttributeParametersProperties—Makes available parameterized attributes in the form of a python dictionary.
   (Customization of analysis happens through this step for instance barrier restrictions on some particular hierarchy of road network can be set to PROHIBITED)

Example solverProps.attributeParameters[('HeightRestriction', 'RestrictionUsage')] = "PROHIBITED".

c. ImpedanceProperty Facilitates the user by getting or setting the network cost attribute used as impedance. The algorithms in analyst are designed to minimize this value.
d. RestrictionProperties Reads or Writes to a list data type of restriction and control measures that are to be instituted for dynamic restrictive conditions applied to the analysis variables. An unfilled list specifies the absence of restrictive conditions or attributes in the real-life analysis situation.
e. TravelModeProperty—Instantiate a arcpy.na.TravelMode object on the network analysis layer. "ApplyTravelMode" method apprises the analysis properties based on a travel mode object.
f. UseHierarchyProperties Drives the usage of the HierarchyAttribute for hierarchical algorithms in network analyst environment.

(Hierarchical algorithms for routing decisions and best path analysis are quicker and smarter. This simulation situation reflects the fondness of a motorist who elects freeways instead of local roads, accepting long-distance travel to avoid traffic or to avoid bumpy roads. This kind of a deep dive into the architecture of the dataset is dependent on the grain level of the network dataset and only if the hierarchy attribute is designed in the fabric of the networked dataset.)

Once the hierarchical network is created, a bidirectional hierarchical algorithm is used to calculate the route among source and last stop during analysis. Every time the algorithm selects a higher layer of hierarchy it automatically reduces the impedance, which is the ultimate goal of this computation exercise.

1. The bidirectional search algorithm searches concurrently from the starting point and desired end point until a definite numeral of linking points are discovered at the subsequent advanced level of the road network hierarchy.
2. Explore the peak level of order in the networked dataset from the origin, disregard all supplementary levels.
3. Terminate the search and report the path when both sections of the route search meet.

An important task is to "understand the real world" and "modeling the real world into a virtual network model", which can be used for network analysis. This would include all types of street networks, overpasses, and underpasses for the elements observed in our everyday practical situations.

The research experiments were focused on working on representative cases, for typical routing problems related to turning restriction situations. This kind of turn restrictions may be in the form of one-way roads for traffic management or a disaster management situation that requires rerouting based on new real-time considerations (Fig. 6).

The goal is to minimize the network impedance which is computed using origin-destination cost matrix. This could be e a node to edge, edge to edge, or an edge to node combination. The plan routes are similar to the single-vehicle navigation

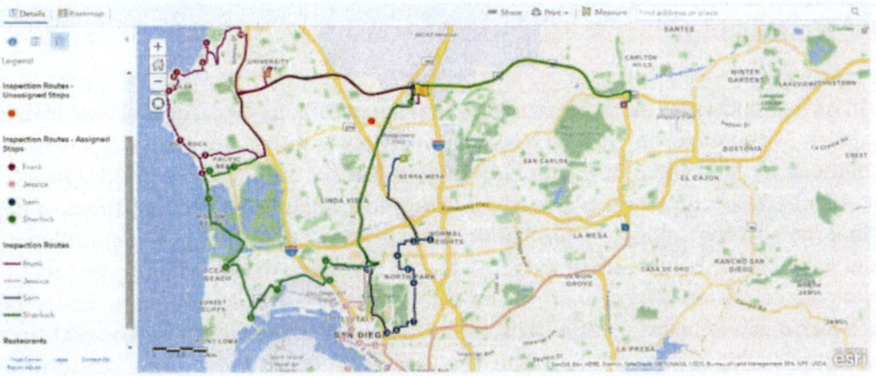

Fig. 6 Origin-destination multipath computation with restrictive conditions

used for finding the best route with our navigation devices but instead of routing one vehicle at a time. It can also conduct routing of many vehicles at the same time. In this case, description determines which of these stops should be assigned to which vehicles. The route solver algorithm takes as input the vehicle stops and these stops are assigned to the vehicles. This type of analysis is a part of the everyday analysis, and proper planning can help us in saving time as well as monetary resources. Being able to optimize these routes means that the officials can get to more stops in less time hence better efficiency in terms of vehicle time on the road and associated monetary benefits.

The existing navigation systems are designed to discover the shortest route through the Dijkstra algorithm or even the best route computations through the route algorithm but the redesigning and reprogramming of routes based on real-time spatial data mining is not available. There is no availability of customizing these routes based on the user inputs or live feeds of real-time data. The situation becomes typically bad if the user of these navigation systems misses an instruction and accidentally enters a level two or level three road hierarchy. The automatic readjustments of routes would even take the user into a more complex network instead of notifying the deviation from the suggested best route. The current system is more useful and practical than currently available mapping techniques used in navigation systems as it is based on two-step computation. The spatial data mining and data processing phase one is for adding a decision step based on real-life practical inputs on pre-classified nodes and edges of the road network. This step creates a new search space based on barrier conditions, restraints, and real-time inputs. The subsequent steps consider the hierarchical dimension of the network path and compute a simple and practical path.

5 Conclusion

This article addresses the work conducted toward two objectives. The first objective is to study and analyze the usage of Network Analyst Algorithms for analyzing alternate solutions to "find the best location/path" problem for any given origin-destination problem. The second objective is to describe and evaluate advanced solutions of the "find the best location/path" problem and to identify the challenges, opportunities, and scope for the betterment of the solution. To implement the algorithms, the ArcGIS Network Analyst platform was utilized and a prototype was developed to model the network database, barrier conditions, route analysis, and redesign based on real-time conditions. The results showed that Network Analyst is a very strong platform to implement new and complex conditions in algorithms that are used to conduct route planning. This work developed insights toward the real-life application of network analysis and route discovery decisions in dynamic and real-time spatial data mining situations. The inherent limitation in such spatial analysis is that the results are highly dependent on base maps, geocoding, and feature layers participating in the analysis. The development of route planning algorithms is a continuous direction and has scope for great improvements. The future works should attempt continuous up-gradation of existing software solutions, with an ongoing understanding that there is always scope for betterment in everything.

References

1. G. Poonthalir, R. Nadarajan, M.S. Kumar, Hierarchical optimization of green routing for mobile advertisement vehicle. J. Cleaner Prod. **258** (2020). https://doi.org/10.1016/j.jclepro.2020.120661
2. H.J. Shan Liu, S. Chen, J. Ye, R. He, Z. Sun, Integrating Dijkstra's algorithm into deep inverse reinforcement learning for food delivery route planning. Transp. Res. J. **142** (2020). https://doi.org/10.1016/j.tre.2020.102070
3. N. Makariye, Towards shortest path computation using Dijkstra algorithm, in *IEEE International Conference on IoT and Application (ICIOT)*, Nagapattinam, 2017
4. M. Dror, P. Trudeau, Savings by split delivery routing. Transp. Sci. **23**, 141–145 (1989)
5. J. Baus, A. Kruger, W. Wahlster, A resource-adaptive mobile navigation system, in *Proceedings of IUI'02*, New York, NY, USA, 2002
6. A. Idri, A new time-dependent shortest path algorithm for multimodal transportation network, in *Elesvier International Conference on Ambient Systems, Networks and technologies ANT*, pp. CS 109 C, 692–697, 2017
7. D. Eppstein, Finding the k shortest paths. SIAM J Comput **28**(2), 652–673 (1998)
8. Y. Ge, Q. Liu, H. Xiong, A. Tuzhilin, J. Chen, Cost-aware travel tour recommendation,in *Proceedings of the 17th ACM SIGKDD International Conference on Knowledge Discovery and Data Mining (KDD'11)*, China, 2011
9. M. Balaj, S. Santhanakrishnan, S.N. Dinesh, An application of analytic hierarchy process in vehicle routing problem. Periodica Polytech. Transp. Eng. **47**(3), 196–205 (2019)
10. P.H. Soares, Zone–based public transport route optimisation in an urban network. Public Transp. (2020). https://doi.org/10.1007/s12469-020-00242-0

11. W. Haoxiang, S. Smys, Qos enhanced routing protocols for vehicular network using soft computing technique. J. Soft Comput. Paradigm (JSCP) **01**(02), 91–102 (2019)
12. N.V. Karadimas, M. Kolokathi, G. Defteraiou, V. Loumos, Ant colony system vs ArcGIS network analyst: the case of municipal solid waste collection, in *5th WSEAS International Conference on Environment, Ecosystems and Development*, Tenerife, Spain, 2007

Artificial Intelligence Analytics—Virtual Assistant in UAE Automotive Industry

Kishan Chaitanya Majji and Kamaladevi Baskaran

Abstract The effectiveness of virtual assistance using IoT can only be established in actions operating to take over the world market. The Internet of Things (IoT) is enabled every basic object to stay connected and communicate over a single platform called the Internet, and it can be done without any human interactions or (say it one-to-one meetings). A virtual assistant works effectively in the motors, and automobile industry. New end cars will mostly function in a way that the person is safe to tame his car just by giving commands. Volkswagen, Skoda, and SEAT are a few of the automobiles which are in conversations to introduce Alexa in their mechanic hardware. Apple already has its Car Play and Google assistance for android. This research attempted to discuss the impact of artificial intelligence in the automotive industry. Every automotive industry's main motive would be to introduce virtual assistance shortly, to make the experience smooth and effective for the customers. The purpose of this research is to identify the impact of artificial intelligence in the automotive industry. This research uses a descriptive research method and a survey method is adopted to collect the primary data. The vital objective of this survey is to rectify the implementations based on the real versus virtual performance of the car. The data analysis has been done to test the hypothesis by applying a Chi-square test. The results of the research reveal that a significant difference between the variables.

Keywords Virtual assistant · Internet of things (IoT) · Chi-square test

1 Introduction

Over the years, the idea of self-driving, virtual assistance, and automotive sensors have been escalated. Every virtual imagination is not feasible to turn into a line-reality model with the concept of the Internet of things/artificial intelligence. The knowledge

K. C. Majji
Amity University, Dubai, UAE

K. Baskaran (✉)
School of Management and Commerce, Amity University, Dubai, UAE

V. Suma et al. (eds.), *Inventive Systems and Control*,
Lecture Notes in Networks and Systems 204,
https://doi.org/10.1007/978-981-16-1395-1_24

of automotive vehicles had gained consumer's imagination, but the virtual assistant is having a much broader impact across the entire automotive industry. The initial stage of designing the infrastructure is the concept of auto experience is evolving and improving. However, the force of the effective artificial intelligence in virtual assistant on the auto industry can be put down in three vital pointers: the in-vehicle experience, autonomous driving/self-driving, and smart transportation. The impact of artificial intelligence in the automotive world has skyrocketed from then smart city ideology to now self-driving vehicle. The virtual world is slowly and gradually becoming virtual reality giving surreal experiences to its consumers.

As an example: Siri for iOS, hello Google for android, Alexa for Amazon, and Cortana for Windows; all their basic idea is proposing them to take the commands that human pass and work upon it. Like any personal assistant, virtual assistance main task is to take human order and accomplish the action, over the Internet. Anything and everything can be ordered and get done using virtual assistance from searching a Web site online to sending a text, scheduling a meeting to set the GPS to approach, and many more.

Platforms like the third generation core are lending their support to help build these auto functions. It's the automotive sector's announced scalable virtual-based platform design prototype to support the highest scale of computing and intelligence required for its sophisticated capabilities. The usage of IoT in most household projects like Automatic electricity, Sensor taps, Thermostats, and so on. These can be operated without any actual human interactions, for instance, Artificial Intelligence (AI). Whereas, virtual assistance takes human command over technological operations, in this way the fear of machines replacing the people can also be subsided.

2 Literature Review

A remarkably informal virtual assistant is competent at giving the complete explanation and understanding the importance behind the concealed complex request like 'what is my schedule for today?' Every independent platform has their own virtual assistant which is dependent on voice command to accomplish their certain task few of the well-known controllers are: utilization of iPhone—Siri; Amazon—Alexa; Microsoft—Cortana; Google—Google Assistant and Samsung Bixby. The study includes market trends, analysis, opportunities, and further estimations of an intelligent virtual assistant to calculate the overhanging economic factors. Virtual assistants work effectively in motors and automobiles [1]. New end cars will mostly function in a way that the person is safe to tame his/her car by just giving commands. Volkswagen, Skoda, and SEAT are a few of the many automotive factories that are into introducing advanced virtual assistance in their vehicles.

Particularly, the virtual assistant is competent at selecting required lingua spoken in a certain dialect which changes in respect to their place and region. The number of users who use this virtual assistant is quite high as it is easy to access and is not limited to only portable devices or personal computer [2]. From ordering a necessity to scheduling a meeting can be done commendably without much difficulty. Popularly known as 4VAS (four virtual assistants) does their task in voice-based as well as contextual texting-based. Each controller platform has its parallel similarities to others like Cortana check a person's contact list, send a text, schedule a meeting, read out the content, show the history, and have a track record of all the locations. Albeit doing all these tasks it is limited only to the desktop-based platform. Though Siri is limited to the iOS version, it can give enormous aid to the users which include weather reports, shortcuts to make work easier, sharing the content among iOS users over the cloud, answering questions, planning official or social gatherings, give information, does basic math, play music, respond to message, updates and more. Similarly, Google assistance has a unique attribute to define its functionality [3].

2.1 Services Provided by Virtual Assistant

1. Does what is said: to play music, or stream online, play Podcasts; all can be done by just passing a command.
2. Gives the information or reports regarding news, sports, weather, to-do list.
3. Autonomous driving is one of the best and latest services provided by a virtual assistant using artificial intelligence by dealing with sensitive personal data, safety functions, and scope of action.
4. Communicational this place: language-independent and also shared mobility; Robo taxis might come into existence as Uber drivers.
5. A virtual assistant is not only in in-car acting like a human car it also has access to almost all the subsidies that are connected over the centralized platform called the Internet.
6. Healthcare sectors in both car and outside, it looks after traveling industries and technological well-being results.

2.2 Future Generations of an Intelligent Virtual Assistant

Earlier the virtual assistant is only limited to operate garage doors, playing text to speech, answering calls, changing the in-car air conditioning, opening and closing the car doors, to accelerate the car through voice command, to acquire humongous results, and approach from all over the world. In recent times, it has been observed that providing any command to be complex or non-complex might have a high chance of system getting distracted and malfunction. Though the system is an autonomous human presence to guide and operate the car is majorly recommended to avoid any uninvited hurdles while performing the task, when the car is moving on the road [4].

In the future, with all the advanced sensors and the intelligent virtual assistant, it is assumed that cars can move smoothly on road without any risk of collision in heavy traffic. The person inside the autonomous vehicle is safe to respond to the text,

Table 1 Automotive industries using different virtual assistants

Automotive companies	Virtual assistant
Mercedes-Benz, Hyundai	Google assistant
BMW, Nissan	Microsoft Cortana
Ford	Amazon Alexa
General motors	IBM Watson
Toyota	YUI
Honda	Hana (Honda Automated Network Assistant)
Almost all the car manufacturers	Siri Apple CarPlay application

plan gatherings, revise the meeting summary, and possible to even write and reply to the emails. These advancements and technology excites young minds and also car enthusiasts. The concept of wheels moving based on an intelligent virtual assistant provides the inspiration to test drive the cars. The services are very limited in terms of technology, and autonomy [5]. The services provided by the present virtual assistant are basic and very definite to the improvised version of current virtuality intelligence will leave the consumers in trance, while in-car one can delightfully use the time for communication, interesting things, or to do something productive. The future generation will walk hand-in-hand with technology [6] (Table 1).

The modern automobile companies keep coming up with new additions in autonomous features to their already existing models. The dominance of autonomous cars is progressing at a very rapid rate. AutoNOMOS labs, Germany, develop the very first car that has been licensed for virtual self-driving on the highways of Berlin. Any technology enthusiast would be curious about the further development of the vehicle and these cars will become faster and easier to handle. Every autonomous car will be built with an accurate GPS, car sensors, intelligent virtual assistant as well as rear in front scanners to avoid any accidents and to maintain proper track on the streets and highways.

3 Research Methodology

The purpose of this research is to identify the impact of artificial intelligence in the automotive industry. This research uses a descriptive research method and a survey method is adopted to collect the primary data [7]. The vital objective of this survey is to rectify the implementations based on the real versus virtual performance of the car. The data analysis has been done to test the hypothesis by applying a Chi-square test [8, 9]. The results of the research reveal that the significant difference between the variables. This research attempted to discuss the impact of artificial intelligence in the automotive industry [10, 11].

Self-driving is no more science fiction, though they might not be able to drive mindlessly without a destination. Yet there is a possibility of this happening shortly when fully autonomous cars come out in the market without any fear of colliding into other vehicles one can easily do their productive work while the vehicle does its driving on its own. From no automation that is (level 0) to vehicles are now with fully automated (Level 5), although the vehicles are semi self-driven, that means the driver has to have their hands on the steering and the eyes on the road while performing or ordering voice assistant to do some internal intelligent work [12, 13].

Regardless of the level of automation, the vehicle has it will always require human guidance and command to take it further with regular updates and bug fixing. The cars in the future will become driverless cars even then the human monitoring is considered effective. One of the toughest challenges in this sector can face in bringing the people to trust this technology and be willing to accept self-driving vehicles with a token of trust, opinions differ depending on the person's individual perspectives, and manufacturer needs to design the vehicle in a way that the people would feel comfortable using the autonomous cars. Automated vehicle research is an extremely hot topic within the field of transportation and hospitality [14, 15].

The idea of self-driving cars alone excites the technology enthusiast and also can save time to do something productive in the meantime, entertainment and info-tainment are not left behind in the race; they raised their standard bar very high, manufacturing and so on.

3.1 Research Objective

The functionality of each assistant varies to its other subset objects. The vital objective of this survey is to rectify the implementations based on the real versus virtual performance of the car. In-person performing a task may vary from personal assistants doing the same task. This survey plays a role to guide the path to give a commendable description of how the intelligent virtual assistant work in the future using artificial intelligence and autonomous cars [16, 17].

4 Analysis and Interpretation

The hypothesis has been tested and the results have been discussed to either accept the null hypothesis or reject the null hypothesis [18].

H1 There is no significant impact of autonomous driving in satisfying the customer.

Is there a significant difference in customer satisfaction while driving autonomous vehicles between region, age group, and gender?

The hypothesis test reveals that $\chi^2 = 2.49$; DF $= 2$; $\alpha = 0.05$; Critical Value $=$ 5.99.

$\therefore \chi^2 <$ Critical Value; Accept the null hypothesis. There is no significant impact on autonomous driving in satisfying the customers.

Figure 1 shows the survey run based on age, gender, and region, there showed a huge difference from place to place depending on how much the regional people are exposed to the latest technology and whether they are ready to adapt to the new technology. Though the age group is an essential part, gender played a curial role in this survey. Altogether, people are equally excited to embrace new high-end cars like autonomous vehicles that make life easier at many notes, but there are also people (10–20%) who are against using automotive cars figuring that would not be comfortable to let the technology take over the skills. However, the cumulative response for this assumption is positive and on the stronger side to drive self-driven cars.

H2 There is no significant impact of GPS in satisfying the customer.

Is there a significant difference in customer satisfaction while using GPS in driving autonomous vehicles between region, accuracy, and trust?

The hypothesis test reveals that $\chi^2 = 1.45$; DF $= 2$; $\alpha = 0.05$; Critical Value $=$ 5.99.

$\therefore \chi^2 <$ Critical Value; Accept the null hypothesis. There is no significant impact of GPS in satisfying the customer as shown in Fig. 2.

The role of navigation is the most important part of any automotive vehicle be it a semi or fully autonomous. The main purpose of getting a personal car or rental vehicle is to reach the destination without any trouble; this is where GPS helps the customers in guiding the path which leads to their final destination stop. The above statistics shows that trusting a device is to guide the person to reach their desired address. 50% of the people do not keep their full trust in GPS electronics yet, but the percentage of people who are willing to indulge GPS in their daily solution is also equal. With the new advanced technology, car manufactures developed moving

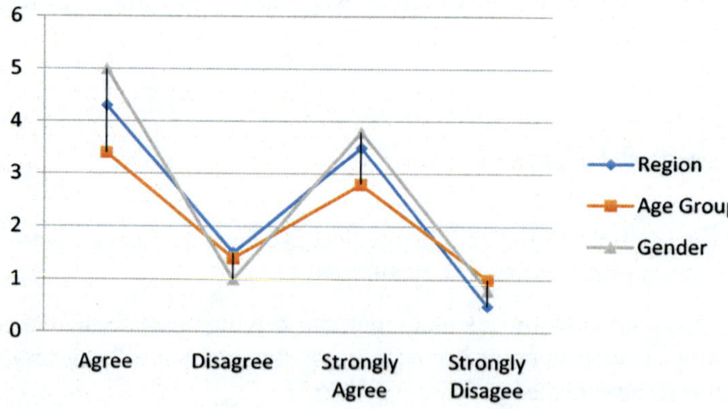

Fig. 1 Customer satisfaction in driving autonomous vehicle reference to the region, age group, and gender

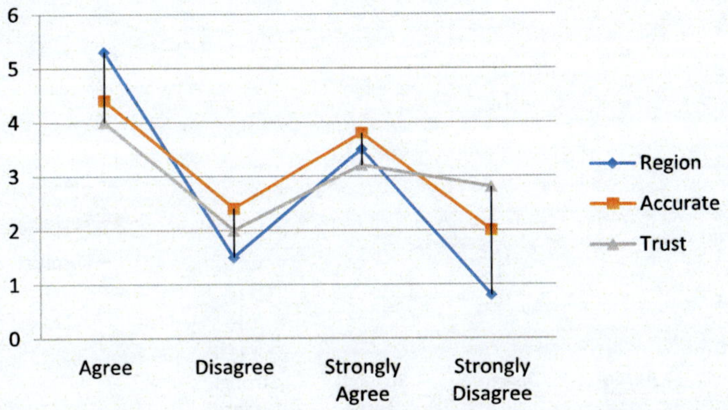

Fig. 2 Customer satisfaction in driving autonomous vehicle reference to region, accurate and trust

map display that shows the current location of the vehicle and road map that shows the final point guided by GPS, which ensures people to sit back and just follow the maps without the fear of getting lost even in a new location.

H3 There is no significant impact of speech recognition in satisfying the customer needs.

Is there a significant difference in customer satisfaction while using speech recognition in driving autonomous vehicles between dialect, techie, and comfort?

The hypothesis test reveals that $\chi^2 = 0.24$; DF $= 2$; $\alpha = 0.05$; Critical Value $=$ 5.99

$\therefore \chi^2 <$ Critical Value; Accept the null hypothesis. There is no significant impact of speech recognition in satisfying the customer needs.

Figure 3 clearly shows consumers perceive comfort in using speech recognition in their vehicles, and this new technology has saved a life in finding infotainment. Speech recognition is too easy and a life protector not to be installed in one's car. Regardless of region, age and gender, this new tool can serve anyone who passes a command to the black hole of this invention that the person with poor English understanding might not use it as effectively as a native speaker. That is no more an issue as the manufacturers already developed a program such as this tool that can recognize any dialect that is installed in it. With that fantastic feature, it grew into popularity within no time and people became adapted to it. Speech recognition does not need any tech-savvy to operate just any user can order what they want right from checking their daily schedule to ask for information to play a song or responding to a text anything that can be done easily while the person is driving. Almost 90% of the response was opting speech recognition service makes the customer comfortable and triumph.

H4 There is no significant impact of self-driving cars being safe in satisfying the customer needs while driving autonomous cars.

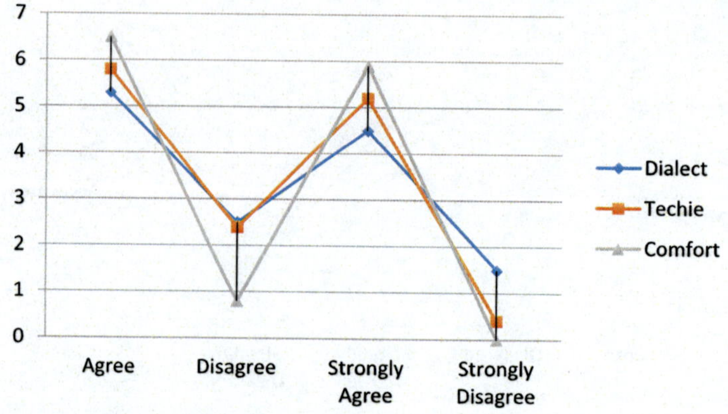

Fig. 3 Customer satisfaction in driving autonomous vehicle reference to dialect, techie and comfort

Is there a significant difference in customer satisfaction in self-driven cars being safe while driving autonomous vehicles between the customer, easy handling, and trust?

The hypothesis test reveals that $\chi^2 = 1.58$; DF $= 2$; $\alpha = 0.05$; Critical Value $=$ 5.99.

$\therefore \chi^2 <$ Critical Value; Accept the null hypothesis. There is no significant impact of self-driving cars being safe in satisfying the customer needs while driving autonomous cars (Fig. 4).

A total of 96% of accidents take place due to the negligence of human driving. For drivers, there is a high chance of getting involved in road accidents. When it's up to the machine driving a vehicle, it activates all its sensors and can easily sense if any involuntary collision happening, sensors will alert the machine and takes action

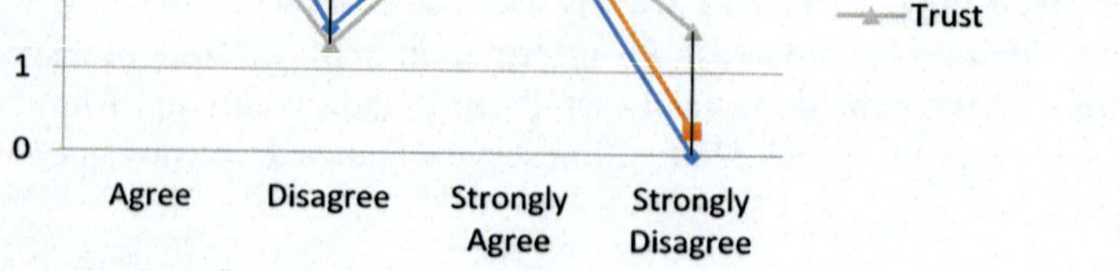

Fig. 4 Customer satisfaction in driving autonomous vehicle refers to the customer, easy handling, and trust

within seconds even before a human can register what is happening outside the lane. In this way, it ensures that self-driving cars are more capable and also provides a sense of safety while inside the car. Another advantage is a person who is not a pro at driving can still handle the car by just passing the commands so that the intelligent virtual assistant would get its task and start performing it. Self-driving cars use systems that find the fastest route to a destination, leading to improved fuel efficiency would reduce cost emissions and saves time. Customer satisfaction does depend on the effectiveness of self-driven autonomous cars. Customer's trust in this technology is very important to take it any further and according to the survey run, maximum of the consumer's trust and are delighted to run this high-end version of virtual intelligence.

H5 There is no impact on proving self-driven cars cause minimal damage when compared to manual driving cars.

Is there a significant difference in customer satisfaction in self-driven cars cause lesser damage while using autonomous vehicle between region, techie, and gender?

The hypothesis test reveals that $\chi^2 = 2.95$; DF $= 2$; $\alpha = 0.05$; Critical Value $=$ 5.99.

$\therefore \chi^2 <$ Critical Value; Accept the null hypothesis. There is no impact on proving self-driven car causes minimal damage when compared to manual driving cars as mentioned in Fig. 5.

A self-driving car can move faster and is safer than any car that is driven by a human driver. It decreases traffic collisions and in a way saving many lives, already. These vehicles are built-in adaptive traffic control and can autonomously shift themselves into an electronic mode to save the gas further. Extreme safety requirements that are presently used regularly would not be needed in automotive. The in-built intelligent virtual assistant program will choose the route which is traffic making it easier to reach the target and also saving fuel, vehicles tend to be lighter and moves fast keeping the track of speed limit it has to follow. Emissions are reduced because the

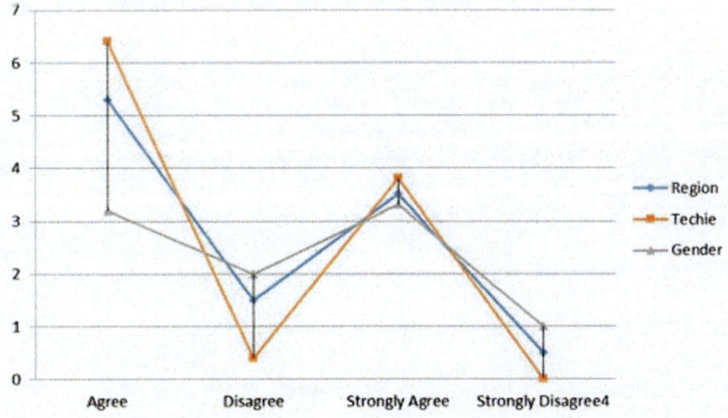

Fig. 5 Customer satisfaction in driving autonomous vehicle reference to region, techie and gender

computerized system will both accelerate and make the brake smoothly without much friction. The self-driving cars would need to be replaced only in a decade, resulting in less maintenance and service costs. Apparently, the department of energy passed a comment that automated cars can reduce energy consumption in transportation to a greater level, and almost 90% of the energy can be saved.

H6 There is no significant impact of using radar sensors while deploying the automotive cars

Is there a significant difference in customer satisfaction in using radar sensors while using autonomous vehicles between techie, safety, and drivers?

The hypothesis test reveals that $\chi^2 = 0.44$; DF $= 2$; $\alpha = 0.05$; Critical Value $=$ 5.99.

$\therefore \chi^2 <$ Critical Value; Accept the null hypothesis. There is no significant impact of using radar sensors while deploying the automotive cars shown in Fig. 6.

Deploying a radar sensor all around the vehicles gives it an immunity point in protection, as the radar sensors help the car to detect its surrounding environment and alert all its signals to avoid any accidents or collisions to take place. Many leading car manufacturing companies are showing interest in installing these sensors in developing autonomous cars and the test drives with these installed have proven positive results about safety precautions it took, these sensors use radio waves to detect and analyze any obstacle within the range of vehicle speeding. This explains why 95% of technology enthusiasts agreed on implementing radar sensors would make customers satisfy while riding in any of these self-driving cars. It is one of the important sensors in manufacturing virtual cars. Usual drivers have raised their opinions and stated that the effectiveness of using a radar sensor not only gives the consumer comfort but also the sense of safety is tagged with it.

H7 There is no effective impact of being productive while driving automotive cars.

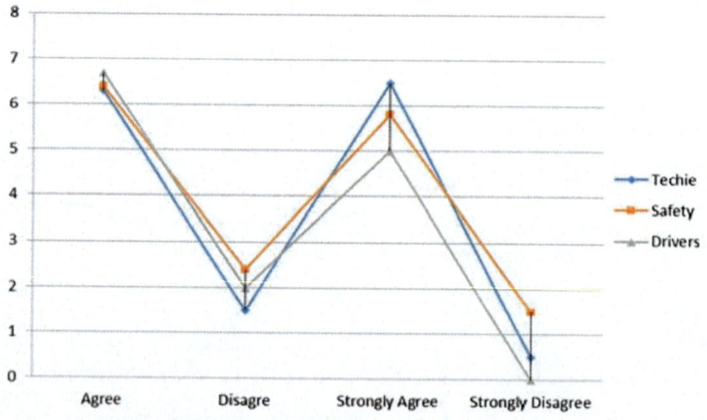

Fig. 6 Customer satisfaction in driving autonomous vehicle reference to techie, safety, and drivers

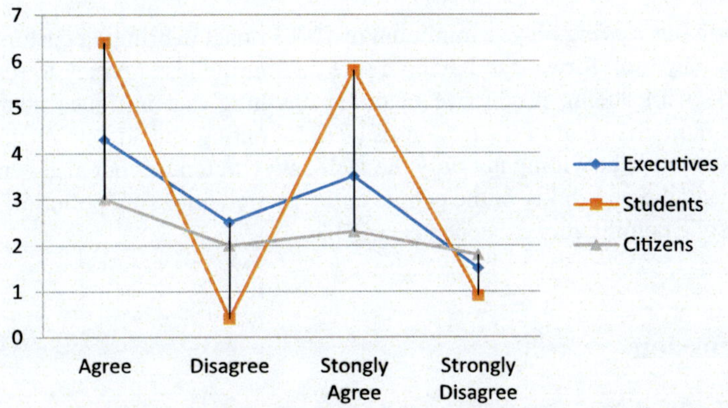

Fig. 7 Customer Satisfaction in driving autonomous vehicle reference to executives, students, and citizens

Is there a significant difference in customer satisfaction in being productive and useful while using autonomous vehicles between executives, students, and citizens?

The hypothesis test reveals that $\chi^2 = 1.45$; DF = 2; $\alpha = 0.05$; Critical Value = 5.99.

$\therefore \chi^2 <$ Critical Value; Accept the null hypothesis. There is no effective impact of being productive while driving the automotive cars highlighted in Fig. 7.

Being productive means utilizing the time that is saved and spending on the things which are meant to be done, work-related or business meetings, student's projects, or quality time with significant other. Riding autonomous vehicles gives pleasure in saving time and focusing on the things which are more vital for surviving. As shown in the statistics, business executive officials are moderate while stating their opinions about self-driving but the least could acquire from them is that regardless of its cost, they are more focused on time and fuel consumption. On the other hand, students are ever excited to have their hands on this new technology also very enthusiastic concerning its new high-end features and accessibility. While 70% of country citizens things it might be a tool to destroy child's self-control and leading them toward a cakewalk path instead of teaching them to face challenges, little did they know, the technology might actually save one's life in the further future. According to the survey, it is evident that launching fully automated vehicles might take a score more to get out in the market, till then people will get used to handling semi-automated vehicles in the next few years. Though the semi-automation is only seen in high-end cars now, the constant engagement of the customers toward the advanced technology could lead the manufacturers to deploy every possible vehicle development design plan to be semi-automated in the way of consumer can satisfy their thirst for technology. One of the main reasons for adding semi-automation for high-end cars as of now is to gather information on public interests and to launch many such advanced features to the design models [19].

Autonomous driving plays a minimum of 70% impact in bringing customer satisfaction as they look forward to having a safe and hurdle less journey to reach their destination while being productive or doing anything that intrigues them. When surveyed, more than half of the responses were for opting and adapting autonomous driving course it's providing not only the tech-savvy outcomes but also benefits the environment, the possibility of the people in the future witnessing pollution-free and calm lifestyle coming into existence is tangible.

5 Discussion

Automated vehicles have the capability to get rid of human errors from the crash equation, which will help all the other by-cyclists, pedestrians, fellow drivers as well. Apparently, according to a survey done on the rate of road accidents that took place due to the human driver in the consecutive years is very high, nearly 36,000 people were dead only due to the human errors, that does not mean the driver is not potential, in that case even racers were found dead in accidents. All these errors can be minimized by at least 90% if self-driving cars come into existence. Autonomous cars have to replace human-controlled vehicles at least to a certain level lest should get involved in more disasters. Automated vehicles could deliver additional economic and additional society-related benefits. The roads filled with automated vehicles could also cooperate in lesser traffic and reduce being congested by traffic and time-saving. In many places across the country employment or independent living depends on the ability to drive [20].

The study states that bringing intelligent virtual assistant would create more opportunities, nearly millions of unemployed can detect a ray of hope thus, increase in employment rate and economy of the country.

6 Conclusion

This article discusses basic arrangements and a sequence of the development of the leading autonomous cars. Cars now are semi-autonomous with the basic features to run in the markets, Mercedes-Benz is one of the very first cars to be out in the market as a fully autonomous vehicle. High-end features and new additions to the existing models give the results on how a car can be self-driven. According to the survey, in score years all the cars that would launch will only be fully autonomous. In a decade, it is said that all the automobile companies will be intrigued to invest in making autonomous cars using Alexa or Google assistant or CarPlay as the intelligent virtual assistant.

People who believe auto-driven cars are not safe have taken a toll in understanding the technology and its power to make things simpler and effortless. If anything these autonomous technologically driven cars are expected to be safe and safer than any human driving the car can ensure. The semi-autonomous car has already proven the comfort of having a virtual assistant installed in a car, the system gets updated from time to time fixing the bugs, if any existed, and self-driven cars will make a huge change in the transportation sector. Wallet cars are already looking forward to taking up fully self-driven cars into consideration of their service. It is quite interesting to see how the world will take a turn in a score, adapting all the intelligent technology. In the end, autonomous car system functionality will always depend on the user, or the person, the vehicle is engaged with their feeling and atmosphere around to adapt their behavior and except the technology.

References

1. K. Burke, Alexa, do i need a vA inside car? *Automotive News*, 4 (2017)
2. M. Dopico, A. Gomez, D. De la Fuente, N. García, R. Rosillo, J. Puche, A vision of industry 4.0 from an artificial intelligence point of view, in *Proceedings on the International Conference on Artificial Intelligence (ICAI)*, 407, 2016
3. D.J. Fagnant, K. Kockelman, Preparing a nation for autonomous vehicles: opportunities, barriers and policy recommendations. Transp. Res. Part A, 167–181 (2015)
4. J. Fleetwood, Public health, ethics, and autonomous vehicles. Am. J. Public Health 107, 532–537 (2017)
5. K. Baskaran, The impact of digital transformation in Singapore e-tail market. Int. J. Innovative Technol. Exploring Eng. (IJITEE) 8(11), 2320–2324 (2019). ISSN Online: 2278-3075. https://doi.org/10.35940/ijitee.i8046.0981119
6. A. Knox, AI in automotive industry. *BrightTalk*, 0–1 (2019)
7. K. Baskaran, M.R. Vanithamani, E-customers attitude towards e-store information and design quality in India. Appl. Res. Sci. Eng. Manage. World Appl. Sci. J. (WASJ) 31, 51–56 (2014). ISSN: 1818-4952. https://doi.org/10.5829/idosi.wasj.2014.31.arsem.555. http://www.idosi.org/wasj/wasj31(arsem)14/10.pdf
8. K. Baskaran and S. Rajavelu, "Digital Innovation in Industry 4.0 Era – Rebooting UAE's Retail," 2020 International Conference on Communication and Signal Processing (ICCSP), 2020, pp. 1614-1618, doi: 10.1109/ICCSP48568.2020.9182301.
9. N. T. Cyriac and K. Baskaran, "A Study on the Effectiveness of Non-Monetary Retention Strategies in UAE," 2020 8th International Conference on Reliability, Infocom Technologies and Optimization (Trends and Future Directions) (ICRITO), 2020, pp. 556-561, doi: 10.1109/ICRITO48877.2020.9197867.
10. A. Ahmed et al., "Journey of Education Technology towards Innovation," 2020 Advances in Science and Engineering Technology International Conferences (ASET), 2020, pp. 1-4, doi: 10.1109/ASET48392.2020.9118334.https://doi.org/10.1109/aset48392.2020.9118334
11. S.H. Ahmed, K. Baskaran, Blue-collar workers behind the success story of UAE. Int. J. Sci. Technol. Res. 9(2) (2020). ISSN Online: 2277-8616, Scopus Indexed Journal. http://www.ijstr.org/research-paper-publishing.php?month=feb2020
12. S. Ashok, K. Baskaran, Audit and accounting procedures in organizations. Int. J. Recent Technol. Eng. (IJRTE) 8(4), 8759–8768 (2019). https://doi.org/10.35940/ijrte.d9165.118419.https://www.ijrte.org/wp-content/uploads/papers/v8i4/D9165118419.pdf
13. K. Baskaran, The impact of digital transformation in Singapore e-tail market. Int. J. Innovative Technol. Exploring Eng. (IJITEE). 8(11), 2320–2324 (2019). ISSN Online: 2278-3075. https://doi.org/10.35940/ijitee.i8046.0981119https://www.ijitee.org/wp-content/uploads/papers/v8i11/I8046078919.pdf.

14. K. Baskaran, An interpretive study of customer experience management towards online shopping in UAE. Int. J. Mech. Eng. Technol. (IJMET) 10(02), 1071–1077 (2019). ISSN Online: 0976-6359. http://www.iaeme.com/MasterAdmin/UploadFolder/IJMET_10_02_112/IJMET_10_02_112.pdf
15. V. Bindhu, An enhanced safety system for auto mode e-vehicles through mind wave feedback. J. Inf. Technol. 2(03), 144–150 (2020)
16. M.H.J.D. Koresh, J. Deva, Computer vision based traffic sign sensing for smart transport. J. Innovative Image Process. (JIIP) 1(01), 11–19 (2019)
17. M. Mohan and K. Baskaran, "Financial Analytics: Investment Behavior of Middle Income Group In South India," 2021 11th International Conference on Cloud Computing, Data Science & Engineering (Confluence), 2021, pp. 1034-1039, doi: 10.1109/Confluence51648.2021.9377029.
18. A. A. Moncey and K. Baskaran, "Digital Marketing Analytics: Building Brand Awareness and Loyalty in UAE," 2020 IEEE International Conference on Technology Management, Operations and Decisions (ICTMOD), 2020, pp. 1-8, doi: 10.1109/ICTMOD49425.2020.9380579.
19. P. Newswire, Virtual assistant market report. *cision*, 2–3 (2017)
20. D. Terra, Virtual reality in the automobile industry. *toptal*, 3–4 (2020)

Performance Improvement of Mobile Device Using Cloud Platform

K. Sindhu and H. S. Guruprasad

Abstract The increased usage of mobile devices in day-to-day activities has led to the explosive growth of mobile applications. Due to the resource constraints of mobile devices, it is still a challenge to execute resource-intensive applications on mobile devices. Mobile cloud computing augments the capability of the mobile device by moving a resource-demanding task onto the cloud. In this article, the proposed method is to improve the performance of the mobile device by moving resource-intensive computation on the cloud. A face recognition system using a cloud platform is implemented to identify a person from a given image. The existing image available on the mobile device or the one captured from the camera of the mobile device is sent to the cloud server. The face detection and recognition happens on a cloud server and the result is displayed on the mobile device. The results indicate that the time taken to recognize the image is faster; the accuracy is higher using machine learning techniques with reduced energy consumption of the mobile device by offloading tasks onto the cloud.

Keywords Mobile · Cloud · Mobile cloud computing · Face recognition · Machine learning · Cloud computing · Face detection · Mobile devices

1 Introduction

A biometric system offers a more efficient and reliable means of identity verification. Face recognition is the most popular among the various approaches used for identifying a person. Face recognition is widely used in criminal investigations and security systems. Facial recognition using a mobile device can be used by a crime detecting agency to identify and recognize if a person at a crime location has any criminal background or not. The image of the person can be captured at the crime location

K. Sindhu (✉) · H. S. Guruprasad
Department of ISE, B.M.S. College of Engineering, Bangalore, Karnataka, India

Visvesvaraya Technological University, Belagavi, Karnataka, India

V. Suma et al. (eds.), *Inventive Systems and Control*,
Lecture Notes in Networks and Systems 204,
https://doi.org/10.1007/978-981-16-1395-1_25

on a mobile device and then sent to the server. The server performs the necessary processing and responds indicating whether the person has a criminal record.

Due to the resource limitations of mobile devices, developing resource-intensive applications like face recognition on a mobile device remains a challenge. To augment the capabilities of mobile devices to perform resource-intensive computations, cloud computing can be used. Cloud computing provides storage, computing services, and application over the Internet. Cloud is a cluster of computers located at a distance from the user and offers services to the public. It can be considered as a utility like water, electricity, etc. where one pays as per the usage.

Mobile cloud computing is a combination of mobile, cloud, and the communication network and is considered as a solution to overcome the resource constraints of mobile devices by offloading the resource-intensive tasks to a nearby resource-rich server known as a cloudlet or on a cloud server located far away. Even though there are continuous technological advancements, mobile devices continue to face many limitations. Offloading resource-intensive application to the nearby server or cloud increases the speed of execution and reduces battery consumption of the mobile device. Based on the application, the entire application can be offloaded to the cloud or the resource-intensive tasks of an application alone can be offloaded onto the cloud. Face recognition applications require a huge amount of processing power; hence it can be processed at the cloud end which has higher processing capability and storage.

In this article, a framework for real-time face recognition system using the cloud is proposed. The application uses either a mobile device camera for capturing the image or uploading an existing image from the mobile device. The image is then sent to a cloud server for recognizing the person in the image. The cloud server detects the face and identifies the person in the image by executing algorithms stored on the server. If the person is already registered in the trained dataset, the name of the identified person is sent to the mobile device and displayed on the screen. In case the person is found not to be a match with the trained dataset, a message denoting 'unknown person' is sent to the mobile device from the server end. REST API is used for communication between the mobile device and cloud because it is faster, simpler, and uses less bandwidth. The research objective is to improve the performance of the mobile device by reducing the execution time of application and to reduce battery consumption of the mobile device by offloading the computationally intensive task onto the cloud. The application considered for the work is a face recognition system. This article is organized as follows: Sect. 2 discusses the related work on face recognition using mobile cloud computing. Proposed work is discussed in Sect. 3, the results of the work are interpreted in Sect. 4, and the conclusion is presented in Sect. 5.

2 Literature Survey

In this section, a brief discussion based on mobile cloud computing, computational offloading, face recognition systems, and previous research done on face recognition using the mobile cloud.

2.1 *Mobile Cloud Computing*

Mobile cloud computing is an infrastructure that is an amalgamation of mobile and cloud computing domains where data storage and processing can happen outside the mobile device. Mobile cloud computing is used to overcome the resource constraints of mobile devices. The general overview of mobile cloud computing is depicted in Fig. 1. It mainly consists of three components: mobile devices, cloud or cloudlet infrastructure, and the network. The mobile devices can communicate to the cloud which is located far away or to any virtual cloud also known as cloudlet located in proximity from the mobile devices. The communication happens over the Internet which can be a Wi-Fi or cellular-based network.

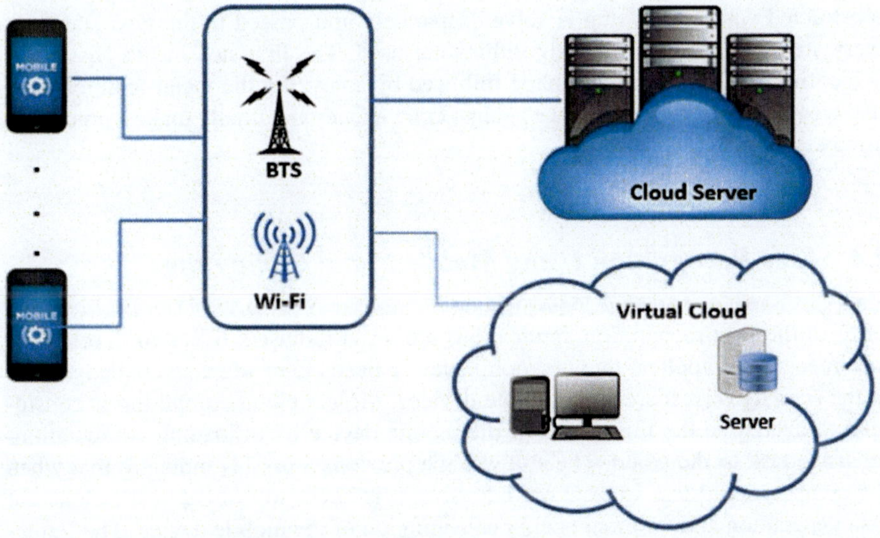

Fig. 1 Generic mobile cloud computing architecture

2.2 Computation Offloading

Computation offloading is the process of offloading the computation-intensive task of an application to a remote server or cloud. Computation offloading is adopted to enhance mobile augmentation by offloading the resource and computation-intensive tasks onto the cloud. Certain applications might require mobile phone support such as sensors and GPS, so offloading an entire application on the cloud may not seem to be an optimal method. It is better to offload only the resource-intensive part of the application onto the cloud which can enhance the performance of the mobile device.

2.3 Face Recognition

Face recognition is a biometric system capable of identifying a person uniquely by analyzing patterns based on the individual's facial contours. The main advantage of using face recognition is that the face images can be captured from a distance without requiring any interaction with the user. The application might be less reliable if the lighting is insufficient, the face is partially hidden or if there is variation in facial expressions. For a face recognition system using machine learning, a pipeline is constructed where each step is solved separately and passed to the next stage. At every step, machine learning algorithms are used. The first step in the pipeline is to identify the faces from an image followed by analyzing the facial features, next step would be to compare the face against known faces and finally make a prediction using classifiers.

2.4 Face Recognition Using Mobile Cloud Computing

Most of the face recognition applications are desktop-based; hence to develop the face recognition application on a mobile device needs to meet many challenges due to the scarcity of resources on mobile devices. Mobile cloud computing is considered to overcome the limitations of the mobile device by offloading computation-intensive task to the cloud. The study in the previous work [1] indicates that when the computational task is resource-intensive, it is better to offload to the cloud for faster execution and minimal energy consumption of the mobile device. The frameworks on face recognition using mobile cloud computing by various researchers are discussed subsequently.

Indrawan et al. [2] have implemented a face recognition system on mobile devices using cloud services. The face detection system happens on mobile device and the face recognition module is implemented using Google App engine cloud services. The authors have used native Android face detector API for the face detection process and face.com API for the face recognition process. The time taken to detect the face with variation in image resolutions is carried out And a total of 50 face images were used in the dataset. If the objects like hair or veil block the face or forehead the recognition rate is reduced. The results provide an accuracy rate of 85% and the total computation time is 7.45 and 15.08 s for face recognition based on the mobile device used and the maximum resolution of the image tested was 640 × 480. Ayad et al. [3] provide a discussion on cloud computing, its service models, mobile cloud computing, and face recognition. Haar classifier is used for face detection and eigenface algorithm for face recognition. Advanced encryption standard algorithm is used to encrypt the image and send it to the cloud which detects the face and reverts the result to the mobile device. A request is sent to the cloud again for recognizing the detected face and the result is sent back to the mobile device. The database consists of 220 face images of 20 unique people with 11 photos of each person. The total time taken for detection and recognition is 17.08 s. The cloud is considered to be a dell laptop with KVM and OpenStack. Ayad et al. [4] present a face detection approach using Haar classifier where the image is captured on mobile and sent to the private cloud. A comparison is done on time taken to execute on mobile, CPU with one thread, CPU with four multiple threads, and CUDA-based GPU. GPU implementation is nine times faster than the CPU version and forty times faster than the mobile implementation. Praseetha et al. [5] in their work presents a face region detection using skin color and principal component analysis is used for face recognition. FERET database is considered in the work. The mobile captures the image and sends it to the cloudlet for face recognition. An accuracy of 95% is specified by the authors. Soyata et al. [6] propose Mobile Cloud Hybrid Architecture (MOCHA) where the image is captured on mobile and sent to cloudlet which in turn uses the cloud to recognize the face image. Haar classifiers are used for face detection and Eigenface algorithms are used for face recognition. The face detection and recognition are performed on a distributed heterogeneous cluster with 13 servers. Bhatt et al. [7] present a face recognition system using a parallel principal component analysis algorithm. The image is sent from the mobile device to the cloud server to recognize the face and the result is sent back to the mobile device. A comparison between a centralized or single server and a distributed system with four servers for face recognition is analyzed. When the trained dataset has fewer images, the centralized server is better and for huge dataset distributed systems are better. In the work of Mukherjee et al. [8], the face detection is done on the mobile end and the detected facial region is sent to the cloudlet. Mobile vision APIs are used for face detection. The semantics of the face like height and width of face, lips, and color components of the face are selected for the face recognition process. For a new image to be recognized, a comparison is done with the feature vector in the dataset. The work uses a local server of 8 GB RAM and an i3 processor. For an image of 20.4 KB, the response time is 35.804 s.

The literature study indicates that to the best of our knowledge, even though many research works are carried out in the area of face recognition systems, not much is done in recent years in the field of face recognition using mobile cloud computing. In the proposed work, the performance of the mobile device is improved by using machine learning techniques and offloading the computation-intensive task onto the cloud. In most of the existing algorithms in the literature review, the face detection happens on the mobile and face recognition on the local server or cloud. In the proposed work, both face detection and face recognition are done on the cloud. So the execution time is faster compared to the existing approaches. In the proposed approach, three different communication networks 3G, 4G, and Wi-Fi are used to analyze the results whereas most of the existing approaches have not indicated the type of communication networks used. The proposed approach uses HOG for face detection; facet and SVM classifier are used for face recognition wherein speed and accuracy is higher compared to the approaches used in the literature review.

3 Proposed Work and Approach

In the proposed work, a Histogram of Oriented Gradients (HOG) is used for face detection. In the next step, the face landmark algorithm is used for selecting the 68 points on the face and the alignment of the eyes and lips is done so that faces looking in different directions are aligned, and it is easier for comparison and prediction. A deep convolution neural network is further used to extract 128 features of the face and a support vector machine is used for fast learning and prediction. The dataset used in the proposed work is taken from Microsoft's 1M celebrity sample database considering data of only 15 subjects [9] and 5 subjects real-time dataset (captured from the mobile).

The Architecture of the proposed work is as shown in Fig. 2. In the proposed approach, a mobile application is developed where the application sends the image to the cloud either by capturing the image of a person using the camera of the mobile device or uploading a stored image from the mobile device. On receiving the image, the cloud server detects the face in the image, aligns the face, and then performs encoding of the face. The encoded values are compared with the encoded values of the images in the trained dataset using an SVM classifier. Later the server sends the

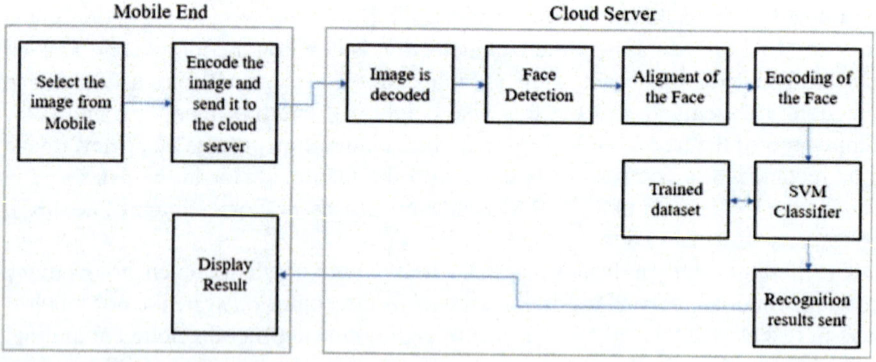

Fig. 2 Architecture of the proposed system

response back to the mobile device by sending the name of the person in the image if the person is present in the trained dataset else it would send a message that the image of a person is not found in the trained dataset. On receiving the response at the mobile end, the app would display the message received from the server.

The face recognition process involves two phases to be performed that is training and testing phase.

A. Training Phase

During the training phase, 32 different images of 20 people are considered and hence the dataset has 640 images. The algorithm for the training phase is as given below:

Begin
for(i=0;i<m;i++) // m = 20 people
open each folder
for(j=0;j<n;j++) //n=32 images
detect_face()
align_face()
encode_detected_face()
label ← self.people.index(name) //label each image
trained data_set ← 128 values and the label
endfor
endfor
end.

B. Testing Phase

During the testing phase, 08 different images of 20 people are considered and hence the dataset has 160 images. The algorithm for the testing phase is as given below:

Begin
for(i=0;i<m;i++) // m = 20 people
for(j=0;j<n;j++)//n=8 images
select the image from mobile gallery (C_{mimage})
C_{server} ← C_{mimage}//captured image is sent to the cloud server (C_{server})
detect_face()
align_face()
encode_detected_face()
classifier()
if(match_found)//send the result to mobile from cloud server
mobile ← person_name
else
mobile ← unknown_person
endfor
endfor
end.

The algorithms used for the above procedure are discussed in detail in the following section:

3.1 Identification of a Face from the Image

It is a challenging task to detect faces in images because of a wide range of poses and variable appearances. The first step in face recognition is the identification of the face in a picture with different kinds of background and illumination. In this work, HOG is used for face detection for the given image. Initially, the image is made black and white and each pixel in the image is checked one at a time. The surrounding pixels of every single pixel are observed. This particular pixel is compared with the darkest among the surrounding pixels. An arrow is drawn in the direction where the image is getting darker. Finally, an arrow replaces every pixel. The arrows indicate the flow from light to dark across the entire image and are called gradients. The image is broken into small squares of 16 × 16 pixels each. The number of gradients pointing in each major direction is counted in each square. Finally, the square in the image is replaced with the directions of the arrow that were the strongest [10–12]. The image is thus converted into a simple representation that gives the basic structure of a face.

3.2 Face Positioning

Once the face in the image is detected, alignment of the eyes and lips are done so that faces looking in different directions are aligned. To do this the face landmark detection algorithm [13] is used to identify the 68 specific points that exist on every face.

3.3 Encoding Faces

Deep Convolution Neural Network [14, 15] generates 128 measurements for each face image. These 128 encoded values are different for different face images and are called embedding. The network used for the proposed work is FaceNet Inception ResNet [16].

3.4 Recognizing the Person

A classifier needs to be trained that calculates the measurements of the test image and identifies which known person from the trained dataset is the closest match. The

Table 1 Device specifications

	Mobile device	Cloud server
Model	Sony Xperia XA1	Amazon EC2 instance t2.medium
CPU	Octa-core processor	2 vCPUs, 2.3 GHz
RAM	4 GB	4 GiB
Storage	32 GB battery 3430 mAh Li-lon.	30 GiB
Operating system	Android 8.0.0	Ubuntu server 16.04 LTS (HVM)

output generated by the classifier is the name of the person. In this work, an SVM classifier is used considering its speed of classification.

3.5 Execution Time

The time taken to execute the application is measured as given in Eq. 1.

$$T_{\text{Exec}} = T_{\text{Capture}} + T_{\text{Send}} + T_{\text{Compute}} + T_{\text{Receive}} + T_{\text{Display}} \tag{1}$$

T_{Capture} Time taken to upload a stored image from the mobile.
T_{Send} Communication time taken to send the image from mobile onto the cloud server.
T_{Compute} Computation time taken to detect, align, encode and recognize the face.
T_{Receive} Communication time taken to send the result back to mobile from the cloud server.
T_{Display} Time taken to display the result on a mobile device.

3.6 Experimental Setup

The experimental setup used for the proposed work is as given in Table 1 for the mobile device and the cloud server. Python3.6 and Python libraries NumPy, Scikit-learn, matplotlib, pandas, OpenCV, and dlib were used for the work.

3.7 Dataset

For the experiment, 40 different images of each of the 20 people were considered and hence the dataset had 800 images. 40 images of 15 people which would add up to 600 images were taken from Microsoft's 1M celebrity sample database and 40 images of 5 people that are 200 images were real-time dataset (captured images). Out of the 800 images, 640 images were used for training the dataset and 160 images were used for testing. 80% of the images were used for training and 20% of images were considered during the testing phase.

4 Discussion and Result

The recognition accuracy obtained is 98%. The recognition accuracy is calculated based on the number of correctly identified images against the total number of test images. The total time taken to recognize a face was calculated based on Eq. (2) considering the value of n to be 10 as the experiment was repeated 10 times for the same image to achieve consistency of the reading. For the experiment, 50 different images with the same resolution were considered and each image was executed 10 times and the average time was considered. The experiment was repeated for different image resolutions. The computation time was almost the same for images with similar resolution since it was executed on the cloud server. It was observed that the communication time slightly varied for the same resolution images. Depending on the size and resolution of the image, the total time taken to read an image from the mobile, sending the image to the cloud server, processing the image at the cloud server, receiving the response from the cloud server and displaying the result on the mobile ranges from 1.963 to 12.977 s when Wi-Fi network is used. The time ranges from 1.955 to 14.299 s when the 4G network is considered and as it ranges from 3.233 to 79.393 s when the 3G network is used which are described in Fig. 3.

$$\text{Total Time} = \left(\frac{\sum_{i=1}^{n} T_i\text{Capture} + T_i\text{Send} + T_i\text{Compute} + T_i\text{Receive} + T_i\text{Display}}{n}\right) \quad (2)$$

The communication time taken in uploading the file to the cloud server and receiving the response from the cloud server is compared between three networks that are Wi-Fi, 4G, and 3G as depicted in Fig. 4. When the resolution of the image is up to 864 × 770, there is only a difference of few milliseconds when Wi-Fi or 4G network is used. The communication time is 0.964 s on the Wi-Fi network and 1.114 s on the 4G network for an image resolution of 864 × 770. For an image resolution of 3000 × 4000 and file size of 2.54 MB, the communication time on Wi-Fi is 1.959 s and 3.075 s on 4G network. It is observed that initially communication time using a Wi-Fi or 4G network is on par and as the image resolution is higher, and the performance of the Wi-Fi network is slightly faster when compared to 4G. Transferring the file on the 3G network takes a longer time as the file size increases, 20.08 s when the file size is around 300 KB(1280 × 960 resolution). Hence in future work, a server (cloudlet) located at closer proximity to the mobile device can be considered along with the cloud server. Based on the availability of the type of communication

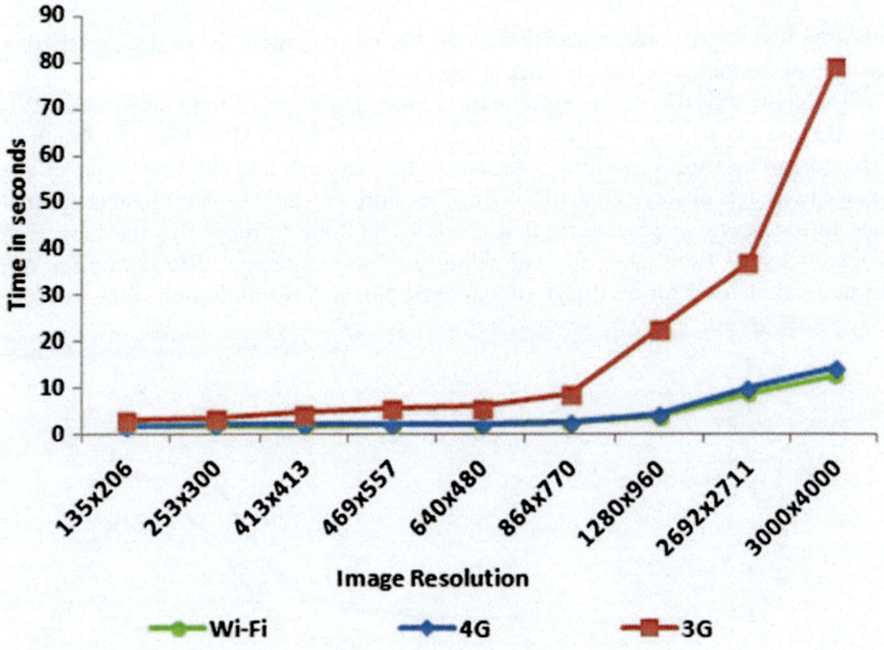

Fig. 3 Total time taken for face recognition for different image resolution

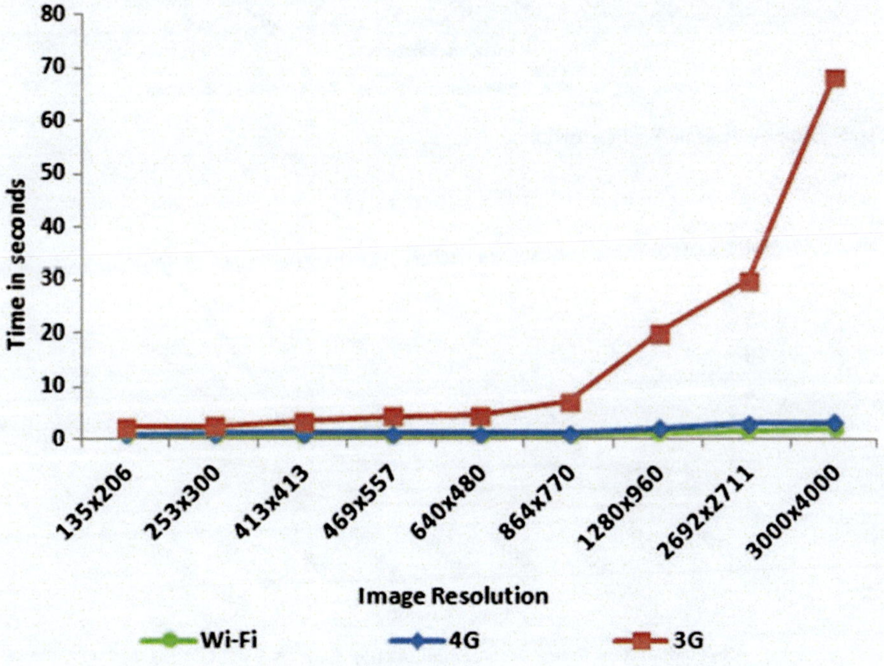

Fig. 4 Communication time taken for face recognition for different image resolution

network and the file size, a decision algorithm can be employed to decide whether to offload the image on the cloudlet or the cloud.

The communication time, computation time, and the total time taken when Wi-Fi, 4G and 3G communication network used is represented in Figs. 5, 6, and 7. The communication time differs based on the network and the resolution of the image used. It is observed that offloading the entire process onto the cloud is optimal due to the intensive processing that needs to be done to recognize the face. The communication time taken for offloading the image is only 1.959 s when a Wi-Fi network is used for an image of size 2.54 MB and resolution of 3000 × 4000.

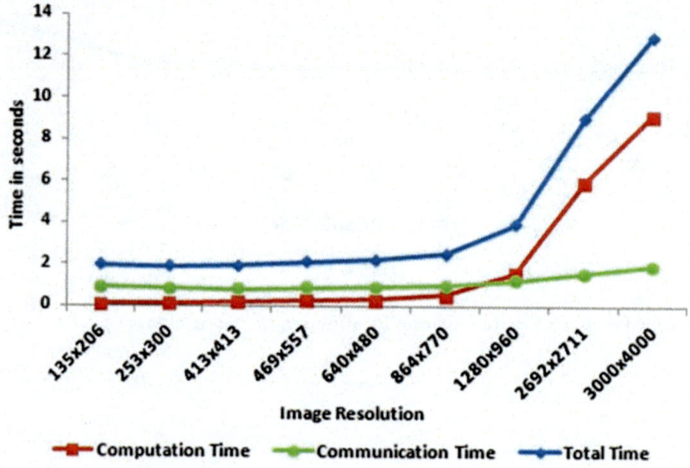

Fig. 5 Time analysis on Wi-Fi network

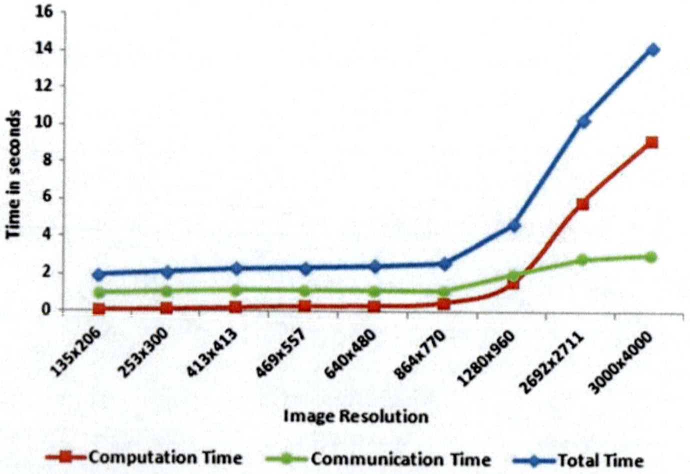

Fig. 6 Time analysis on 4G network

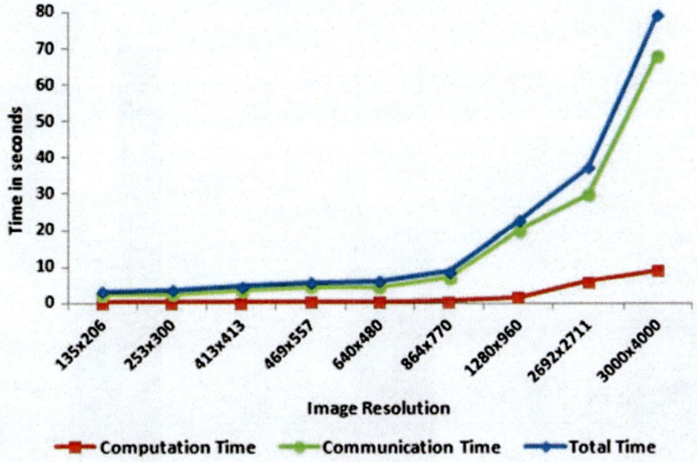

Fig. 7 Time analysis on 3G network

The computation time can further be reduced by using a high-performance server configuration at the cloud end.

The proposed approach is efficient and working with higher recognition accuracy and faster execution time when compared with the method used by researchers in [2] where accuracy was 85% and the execution time was in the range of 7.45 and 15.07 s for image resolution of 640 × 480 as shown in Fig. 8. The type of communication network used in their work was not called out. Hence a comparison between all the three networks of the proposed method and the previous work [2] is plotted as shown in Fig. 8. The time taken for image resolution of 640 × 480 in the proposed approach is 2.210 s in Wi-Fi Network, 2.472 s in 4G, and 6.017 s in 3G network. The authors in the previous approach have performed face detection on mobile device and face recognition on the cloud considering only 3 different image resolutions for their work 240 × 160, 320 × 240, and 640 × 480. So, the time taken in recognition of a face depends on the performance of the mobile device used. Hence same resolution images were compared with the proposed work to know the total time for recognizing the face. The total time taken in recognizing a face in the proposed approach is faster compared to the previous approach since both detection and recognition happens at the cloud server. The proposed work is also compared with [8] wherein an image of 20.4 KB size takes a recognition time of 35.804 s. In our proposed work, an image of 57 KB size takes 2.314 s in Wi-Fi Network, 2.544 s in 4G, and 4.620 s in 3G. The communication network used by the researchers [8] is not mentioned and a local server with an 8 GB RAM and i3 processor was used for their work.

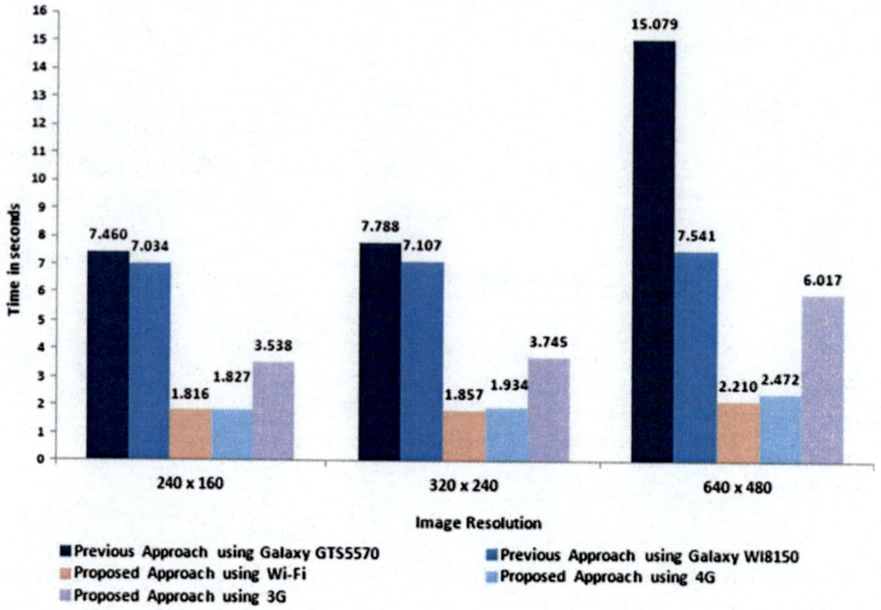

Fig. 8 Comparison between the proposed approach and previous work 2]

5 Conclusion

The performance of the mobile device is improved by offloading the computation-intensive task onto the cloud in the proposed work. The application considered for the work is a real-time face recognition system. The system consists of the client user interface to capture the image from the mobile camera or upload an existing image, the cloud server for the face recognition process, and the client user interface to display the recognized image details. The server provides the core face detection and recognition capabilities. The face recognition system consists of sub-modules to detect, align, encode, and predict the face. This framework helps in overcoming the limitations of mobile devices by offloading the computation-intensive task onto the resource-rich cloud server. The results indicate that the time taken to recognize the image is faster and the accuracy is higher using machine learning techniques. The energy consumption of the mobile device is reduced by offloading the task onto the cloud. The total execution time can further be reduced by using a high-performance cloud server considering a larger dataset for future work.

References

1. K. Sindhu, H.S. Guruprasad, An experimental analysis on performance and energy saving in mobile cloud computing. Int. J. Inf. Technol. Comput. Sci. (IJITCS) **9**(9), 45–52 (2017)
2. P. Indrawan, S. Budiyatno, N.M. Ridho, R.F. Sari, Face recognition for social media with mobile cloud computing. Int. J. Cloud Comput.: Serv. Archit. **3**(1), 23–35 (2013)
3. M. Ayad, M. Taher, A. Salem, Real-time mobile cloud computing: a case study in face recognition, in *28th International Conference on Advanced Information Networking and Applications Workshops (WAINA)*, pp. 73–78. IEEE, 2014
4. M. Ayad, M. Taher, A. Salem, Mobile GPU cloud computing with real time application, in *International Conference on Energy Aware Computing Systems & Applications (ICEAC)*, pp. 1–4. IEEE, 2015
5. V.M. Praseetha, S. Vadivel, Face extraction using skin color and PCA face recognition in a mobile cloudlet environment, in *4th IEEE International Conference on Mobile Cloud Computing, Services, and Engineering (MobileCloud)*, pp. 41–45. IEEE, 2016
6. T. Soyata, R. Muraleedharan, C. Funai, M. Kwon, W. Heinzelman, Cloud-vision: real-time face recognition using a mobile-cloudlet-cloud acceleration architecture, in IEEE Symposium on Computers and Communications (ISCC), pp. 000059–000066. IEEE, 2012
7. K.N. Bhatt, S. Dessai, V.S. Yerragudi, Design and development of cloud based face recognition using parallised PCA. SASTech-Tech. J. RUAS **17**(1), 17–20 (2018)
8. D. Mukherjee, R. Das, S. Majumdar, S. Ghosh, S. Thepade, A. Basu, Energy efficient face recognition in mobile-fog environment. Proc. Comput. Sci. **152**, 274–281 (2019)
9. Y. Guo, L. Zhang, Y. Hu, X. He, J. Gao, Ms-celeb-1m: a dataset and benchmark for large-scale face recognition, in *European Conference on Computer Vision*, pp. 87–102. Springer, Cham, 2016
10. N. Dalal, B. Triggs, Histograms of oriented gradients for human detection, in *IEEE Computer Society Conference on Computer Vision and Pattern Recognition, 2005, CVPR 2005*, vol. 1, pp. 886–893. IEEE, 2005
11. E. Cengil, A. Cinar, Comparison of HOG (histogram of oriented gradients) and haar cascade algorithms with a convolutional neural network based face detection approach. Int. J. Adv. Res. Ideas Innov. Technol. **3**(5), 244–255 (2017)
12. H. Ahamed, I. Alam, Md. M. Islam, HOG-CNN based real time face recognition, in *International Conference on Advancement in Electrical and Electronic Engineering (ICAEEE)*, pp. 1–4. IEEE, 2018
13. V. Kazemi, J. Sullivan, One millisecond face alignment with an ensemble of regression trees, in *Proceedings of the IEEE Conference on Computer Vision and Pattern Recognition*, pp. 1867–1874, 2014
14. A. Krizhevsky, I. Sutskever, G.E. Hinton, Imagenet classification with deep convolutional neural networks. Adv. Neural Inf. Process. Syst. 1097–1105 (2012)
15. S. Smys, J.I.Z. Chen, S. Shakya, Survey on neural network architectures with deep learning. J. Soft Comput. Paradigm (JSCP) **2**(03), 186–194 (2020)
16. F. Schroff, D. Kalenichenko, J. Philbin, Facenet: a unified embedding for face recognition and clustering, in *Proceedings of the IEEE Conference on Computer Vision and Pattern Recognition*, pp. 815–823, 2015

Image Restoration by Graduated Non-convex Local Adaptive Priors: An Energy Minimization Approach

H. N. Latha and Rajiv R. Sahay

Abstract This work proposes a novel algorithm to estimate latent focused image of a degraded picture by a sparsity-based optimization framework using graduated non-convex discontinuity adaptive Markov random field (GNC-DAMRF) regularization prior by minimizing the energy term. Initially, an adaptive non-local pixels-based filtering technique is exploited to denoise the 2D picture affected by additive white Gaussian noise (AWGN). The space non-uniform blind deconvolution is developed to deblur the picture by sparsity-based technique followed by adaptive maximum-a-posterior (MAP) GNC-DAMRF regularization framework. The non-uniform space-variant point spread function (PSF) of every pixel is estimated by a method proposed by just noticeable blur (JNB) [1]. The estimation of the image and the sigma blur-map of point spread function (PSF) were focused. The proposed alternating energy minimization objective function based on the sparsity approach is with-success optimized with the generic gradient descent technique. The performance of the adaptive MAP-GNC-DAMRF work is enhanced, which is superior to Gaussian Markov random field (GMRF) regularizer and few state-of-the-art works. The reconstruction results are promising and better image qualitative assessment is obtained by PSNR and SSIM. This proposed algorithm is also effectively utilized for image restoration applications such as image in-painting, depth estimation, image denoising, super-resolution, and haze removal.

Keywords Gauss Markov random field · Non-blind deburring · Additive white Gaussian noise · Point spread function · Graduated non-convex · Space-variant blur-kernel

---

H. N. Latha (✉)
Department of ECE, BMS College of Engineering, Bangalore 560019, India
e-mail: lathahn.ece@bmsce.ac.in

Department of CSE, IIT, Kharagpur 721302, India

R. R. Sahay
Department of EEE, IIT, Kharagpur, India

V. Suma et al. (eds.), *Inventive Systems and Control*,
Lecture Notes in Networks and Systems 204,
https://doi.org/10.1007/978-981-16-1395-1_26

1 Introduction

Capturing images by handheld DSLR camera or smartphone is very much challenging during the low light state. Longer exposure times of shutter are required, when less light is available, due to which, camera shake is possible to happen and resulting in a blurry picture by object motion or by a camera motion [1, 2]. The longer exposure time of capturing photography with an image sensor device produces a blurry and noisy frame. ISO options available in many cameras can be set to increase the light sensitivity, so that the exposure time may be reduced. A reconstruction of an image suffering from noise and blur is a very much ill-posed counter problem [3–6]. The objective of natural scene image-based blind deblurring is to restore a latent focused image from a degraded–blurred one without knowing the blur-kernel and magnitude of noise sigma. This issue is addressed by several researchers and authors for many decades; it is very much challenging and authors are still working on it. Single image blind deblurring techniques proposed by several researchers [7–10] are still sensitive to image noise. An image captured during low-light conditions, with less illumination, or during nights suffers from noise distortion. A small amount of noise degrades the blur-kernel of sigma map estimation of the degraded image resulting in severe ringing artifacts. Few recently proposed techniques use the iterative-based method to denoise and deblur a degraded image. Direct image denoising methods partly damage the blur information from the input image, making improper estimation of the blur-kernel. An adaptive weighted non-local filtering for handling noise in space-variant blind image deconvolution was proposed. Reconstruction of the degraded image containing blur and noise is highly ill-posed and inverse which requires prior information. It is a progressive research area in this field. To improve the performance of the ill-posedness of the said problem, a novel regularization term is included, which is based on an adaptive graduated non-convexity optimization framework and is proposed for image restoration problems.

Noise is generally produced due to some constraints of optical image-based systems, communication medium, channel, and scene capturing devices. The degradation model of the image affected by both noise and blur is generally represented by the following equation

$$\boldsymbol{y} = \boldsymbol{H}\boldsymbol{x} + \boldsymbol{n} \tag{1}$$

where $\boldsymbol{x}$, $\boldsymbol{y}$, $\boldsymbol{H}$, and $\boldsymbol{n}$ is the focused-cleaned image of AWGN noisy observation, $\boldsymbol{y}$ is the noisy-blurred picture frame, $\boldsymbol{H}$ is blur-kernel operator matrix and $\boldsymbol{n}$ is the additive noise, which is oftentimes modeled as additive white Gaussian noise (AWGN) [6]. The visual effect on qualitative and quantitative experiments demonstrated the outcome of restored images. Robust estimation of the non-uniform blur-map of a natural scene obtained produces better qualitative and quantitative output picture results.

Denoising and image restoration techniques are essential to modify the visual appearance of degraded pictures. Accurate blur-map estimation is required to reduce the commonly occurring ringing artifacts during deblurring of degraded images due to optical image-based sensor systems. The authors have implemented different techniques and algorithms [11–15] to estimate the multiunknown of inverse parameters. A robust and efficient image degradation algorithm is needed to estimate the spatially invariant blurred-noisy image. Single image blind deblurring is the most active research field and created more interest. Personal handheld mobile smart cameras frequently experience blur effect when acquiring pictures. Generally blurring operation is defined by motion between sensor device and object during image acquisition, which is classified as motion blur. Motion blur is categorized as the rotational or transnational blur between cameras and objects within the lens during camera shutter exposure duration. Defocus blur is due to the depth variation in the 3D natural screen. In the case of space-invariant blur-kernel, blur is uniform. Otherwise, if blur-kernel is variant, a non-uniform blur-map is considered. The blurring operation in both the case is modeled as a convolution operation, between blur-map and focused-cleaned image. A partially blurred image with two layers blur at object edges is presented in Fig. 1a. Figure 1b shows high-frequency details missing due to the defocusing processing of camera lens property. Figure 1c shows a space-variant defocus image with blur changing with 3D depth variation.

The organization of the research work is as follows. Section 1, provides an introduction part, which explains blurring and noise operation of a picture during the image acquisition phase. In Sect. 2, literature survey and related work are given. Section 3 presents our image model, proposed work, implementation details, and contributions. In Sect. 4, the algorithms and details are about the estimation of blur-kernel and latent focused image. Experimental output results obtained for both synthetic and real case are presented in Sect. 6. Finally, Sect. 7 provides the conclusion of the proposed work.

a b c

Fig. 1 **a** Partially blurred image with two layers blur at object edges. **b** High-frequency details missing due to defocusing. **c** A space-variant defocus image with blur magnitude changing with 3D depth variation

2 Literature Survey and Related Work

The inverse ill-posed non-blind/blind deconvolution [3–10] problems are addressed by many methods by many authors, which are broadly categorized as Bayesian maximum likelihood framework, estimation by a posterior probability, sparse coding creation models, KSVD algorithms, deep convolutional network method, multi-frame model, homographs modeling, segmentation of blurry/non-blurry region-based methods, strong edge-based technique, recently convolutional networks, and many more.

Deblurring of spatially varying blurred picture proposed by [8] and many authors [9–11] addressed space uniform PSFs blur produced by image sensors device with the 3D scene image. Bayesian framework model proposed by [6] estimates the latent image degraded by additive Gaussian noise and motion blur by adapting image gradient operation, wherein prior information is obtained by learning gradient of the image by Gaussian mixture functions based on strong edge selection [11–16] from the scene image. The work proposed based on strong edges and derivatives by some authors [17–19] for blur metric measurements of the blur sharpness information of the scene and estimation of the blur-map kernel are effective, but more time complex [17, 20–25]. To cut down the time requirement and gradients, computational complexity of the algorithm [9, 10] addressed an efficient technique, a gradient-based salient edge prediction model.

Deblocking filtering techniques proposed by Vasantha et al. [10] reduce the ringing and blocky artifacts distortions near the salient edges of the reconstructed image in a space compressed time domain. Cho et al. [23] developed a robust degradation model to address outliers presents only in known places having sparse data. A fast image deconvolution algorithm is developed by Krish et al. [3] using Fourier transformation and hyper-Laplacian prior to objective function. Latha et al. [4] developed a convex optimization algorithm using alternating minimization techniques to deblur space-variant defocused image. [14] developed space non-uniform deblurring work, based on blur-kernel estimation. Li. et al. [16] employed a model based on stochastic noisy-blurred data and used to reduce the deblurring noise artifacts due to blur-kernel mismatch. This method is time-consuming and not capable to reduce image additive noise.

The concepts of both convex and non-convex optimization techniques demonstrated by [4, 9, 14, 24, 18, 26] generate the initial rough blur-map to deblur the degraded scene image using alternating iteration models in space, time, and frequency domain approach. The authors [27–30] demonstrated different image regularization priors, deep convolution networks, blur cue, depth cue, regularizations, etc.

These developed algorithms manage to find the blur-kernel to restore the degraded scene pictures by discarding the saturated pixels, utilizing prior information on nonlinear space. Handling saturated pixels to improve the performance of the deblurring [19] capability which is extended by [17] showed better results. The first step generates the reconstructed focused scene image by removing undependable saturated pixels. During the next step, this algorithm operates on the grouped pixel regions which were screened first. Our proposed method is cost-effective, simple, and not required to separate the reliable/unreliable pixel groups. The authors of [21] developed natural scene image understanding models to classify documents text and image graphics. Classification of blurred logos and segmentation using pertained models on faster RCNN is proposed by Latha et al. [21], where prepossessing steps are initially performed to deblur the degraded scene image. Researchers of [3, 7, 12] showed an important picture deblurring algorithm by eliminating the white pixels, which are classified as saturated pixels. A clipping function is performed to separate the saturated pixels. This work proposed and developed analogous techniques [15, 16, 18, 19], which does not separate blurred and non-blurred pixels. Vasantha et al. [10] developed an algorithm for regenerating and reconstructing the degraded image using variable length coding (VLC) which is basically of adaptive context-based concept. Deblocking filters using H.264 standard for MPEG compression are included to remove additive noise and blocky artifacts.

This work proposed a local MAP-GNC-DAMRF prior model where initially blur-map kernel and the latent clear-focused image is modeled as independent DAMRFs priors and optimized by a suitable GNC algorithm. Secondly, a local DAMRF regularization estimates the multiunknowns parameters. This work demonstrated the supremacy of the local MAP-GNC-DAMRF regularizer reconstructing fine smooth details, edges, and texture in the generated picture. The applications of sigma map estimation are widely employed in image restoration, super-resolution, and depth estimation, blur segmentation, etc. It helps a photographer to edit a foreground or background object. Blur present in an image [9] is related to depth information [28, 30] of the 3D natural scene captured by the camera. The problem of image restoration is decomposed into three subproblems. Initially, the adaptive non-local patch-based [24] technique is utilized to denoise the noise-blurred picture by determining non-local average weighted vector noise square patches. Several analyses were performed on the statistical data and effective priors are obtained by adaptive derivatives of non-local neighborhood pixels data for denoising and local neighborhood pixels content for deblurring. The non-local gradients in both horizontal and vertical directions are obtained analytically concerning the blur-map and focused latent image.

The main contributions of this work are: (1) Robust method to detect and estimate initial blur-map using just noticeable blur approach. (2) Non-local based framework is implemented to denoising the image. (3) Uses a base work given by JNB to estimate initial course blur-map to retrieve the focused image. (4) The MAP-GNC-DAMRF alternating minimization optimization framework is proposed to estimate all-in-focus image and final blur-map. From noisy-blurred observation, it can also estimate the defocus map, depth-map, for image in-painting and restoration problems. The gradient descent is used with DAMRF [14] prior along with graduated non-convexity (GNC) algorithm to estimate blur-map, thus preserving fine details, structure information, edge information, and solving the optimization problem.

3 Proposed Work

To enhance the performance of blind deconvolution of the noisy-blurred image, a novel framework is proposed. In the proposed framework, initially design the denoise image using weighted adaptive filter by non-local patch-based technique. The detailed workflow of the proposed method is illustrated in Fig. 2c. Non-blind deblurring methods are designed to restore space-variantly blurred images. The proposed work is a more generic adaptive MAP-GNC-DAMRF and can be added to the available denoising and deblurring techniques based on MAP. The experiments done showed the good efficiency of the proposed framework and proved that the developed algorithm outperforms few state-of-the-art methods in both preserving strong edges of image structures and eliminating AWGN noise. The proposed model in this work uses the adaptive MAP-GNC-DAMRF framework. This algorithm is efficiently optimized to predict latent image edges by the iterative process under some

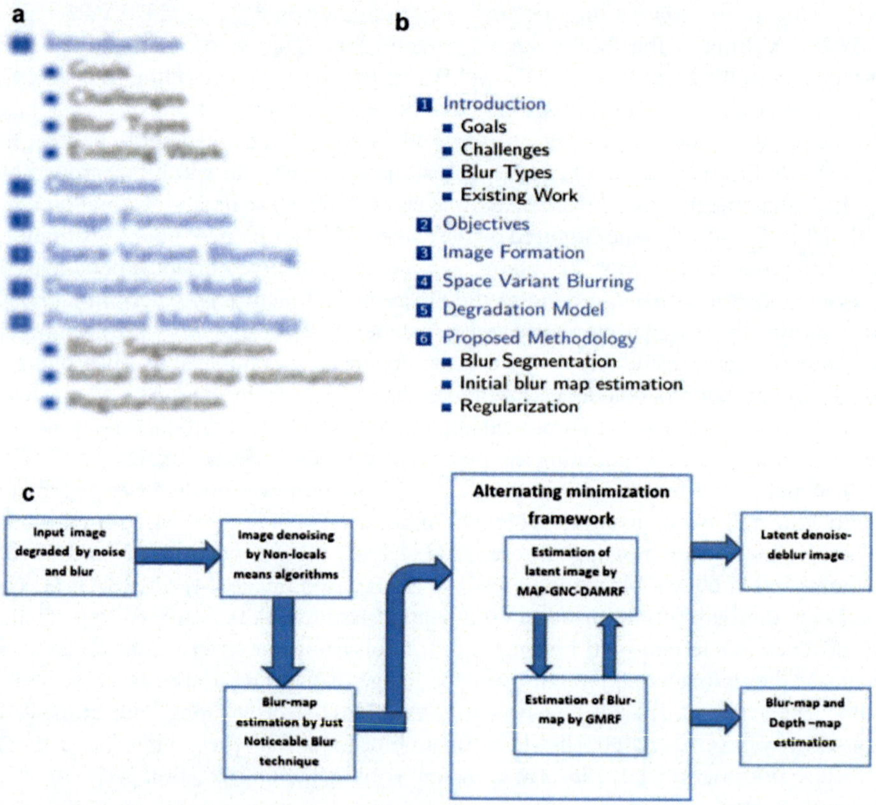

Fig. 2 **a** Space-invariantly blurred image **b** Focus ground truth text image **c** Proposed work flow diagram of restoration of degraded image

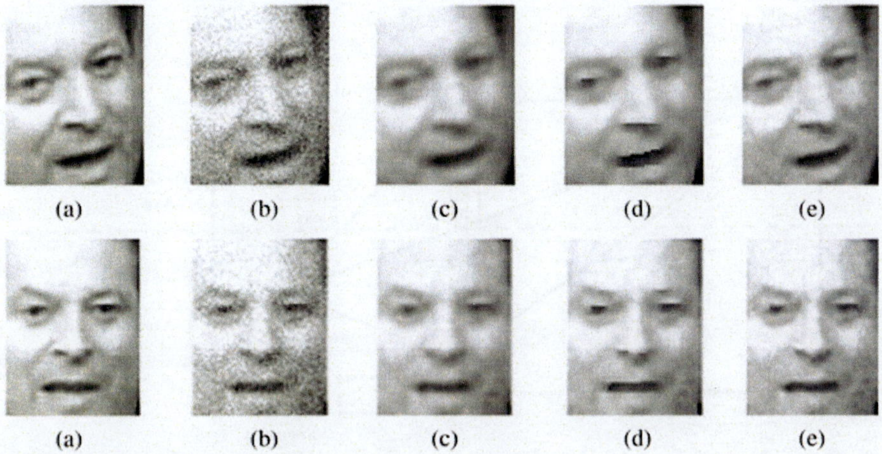

Fig. 3 **a** Clear-focused image. **b** Noise-blurred synthetic image. **c** Synthetic blurred image. **d** Convex GMRF results. **e** Reconstructed image of proposed method using local MAP-GNC-DAMRF

constraints. The complexity of this method is less; sometimes results are affected by noise [6] hence non-local denoising [24] is applied first, before the deblurring iterations.

This research work initially uses a non-local filtering technique to denoise the degraded image. Then minimization of the energy function framework is to estimate the sigma blur-kernel and the latent image, of a single noisy-blurred image. Results obtained using alternating minimization of GNC-DAMRF before the reconstruction of smooth region details and structure-depth information of the latent image are shown in Fig. 3. Figure 3 depicts the (a) focused, (b) noisy, (c) blurred and noisy, (d) reconstructed image by convex GMRF, and (d) non-convex GNC-DAMRF prior, respectively. The face data is taken from Gore's face, image dataset [20, 24, 25]. A synthetically generated space-invariant blurred text image and its ground truth mage are shown in Fig. 2a, b. Initially, the problem of blur-map estimation is decomposed into three steps. Firstly the rough blur-map is generated by JNB. Secondly using gradient-based techniques, degraded image is deblurred. Finally by GNC-DAMRF optimization framework in alternating minimization techniques is implemented to get the latent image.

1. **Image formation**

The response of a scene image capturing device to any light illuminating source is represented as PSF [17, 18, 31]. Figure 4 shows the image shaped on an image sensor plane of the sensor device through a lens system from a single point light generator source at some working distance D is imaged as a circular round disk with radius r_b and is given by

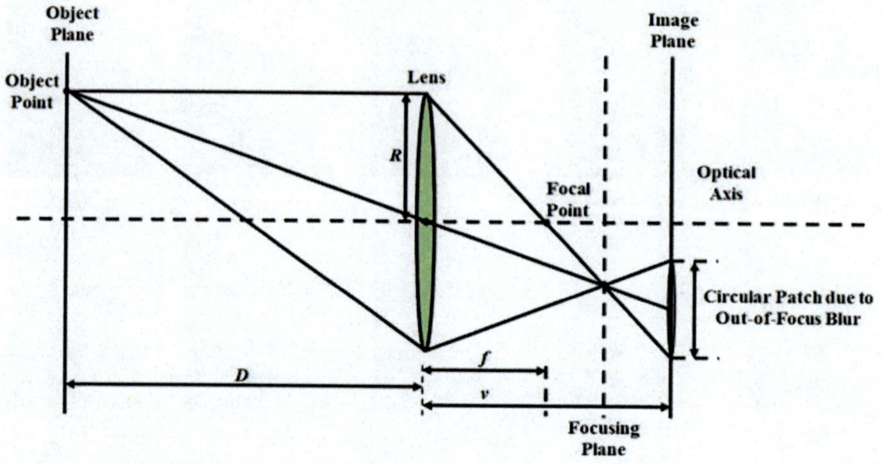

Fig. 4 Image degradation model of a lens system [8] during image capturing

$$r_b = Rv\left(\frac{1}{f} - \frac{1}{v} - \frac{1}{D}\right) \quad (2)$$

The blur-kernel represented by standard deviation, sigma σ (PSF) is given by

$$\sigma = \rho r_b = \rho Rv\left(\frac{1}{f} - \frac{1}{v} - \frac{1}{D}\right) \quad (3)$$

2. Degradation models

Space-invariantly blurred images with non-uniform blur and noise model are given as

$$y = H(\sigma) * x + \eta \quad (4)$$

The zero-mean white additive noise η is considered as Gaussian distributed. The H represents the matrix operation on focused image x to generate a space invariant blurred-noisy image y. Let σ, x, y, η be the vectors. Blur-kernel sigma σ is related to 3D scene depth. The PSF is represented by

$$h_p(i, j; k, l) = \frac{1}{Z(i, j)} \exp\left(-\frac{(i-k)^2 + (j-l)^2}{2\sigma^2(i, j)}\right) \quad (5)$$

where Z is given by

$$z(i, j) = \sum_k \sum_l \exp\left(-\frac{(i-k)^2 + (j-l)^2}{2\sigma^2(i, j)}\right) \quad (6)$$

3. Denoising by patch-based non-local filtering

Proposed work in this paper exploits the generic idea of weighted adaptive non-local filter work to denoise the noisy-blurred image. Initially, in non-local means denoising framework, the image frame is divided into overlapped square patches in an assumed slide window. The statistical values of all similar square patches are calculated and an average weight of all the patches is considered to denoise the noisy patch [26, 32, 33]. Denoising of each pixel of the non-local patch through the statistics of other similar reference patches is obtained and then the estimated average weight is added adaptively from the statistics of all the denoised patches to acquire a denoise picture [27–30, 34]. From the degraded image $\boldsymbol{y}$, a square patch $\boldsymbol{q}$ of size $\sqrt{d} \times \sqrt{d}$ is considered, where $\boldsymbol{q}$ belongs to $\boldsymbol{R^d}$ from the noisy-blurred image $\boldsymbol{y}$. The linear function Φ (weight) of the algorithm estimates noise sigma of the reference patches $\boldsymbol{p_1}, \ldots, \boldsymbol{p_k}$ belongs to $\boldsymbol{R^d}$ of the noisy-blurred image to obtain the unknown clean-clear patch $\boldsymbol{p}$ and is given by

$$\hat{\boldsymbol{p}} = \boldsymbol{\Phi}(\boldsymbol{q}; \boldsymbol{p_1}, \ldots, \boldsymbol{p_k}) \tag{7}$$

An adaptive weighted average of non-local reference patches is a function Φ and given by

$$\hat{\boldsymbol{p}} = \frac{\sum_{i=1}^{k} e^{-\|q-p_i\|^2/h^2} \boldsymbol{p_i}}{\sum_{i=1}^{k} e^{-\|q-p_i\|^2/h^2}} \tag{8}$$

where h is a Gaussian decay parameter given by Gaussian distribution. Image denoising and deblurring is still an active ill-posed inverse problem. It has many solutions and all solution does not lead to the correct statement. The total information available from the noisy-blurred observation is not sufficient to obtain a good restored output image. It is very much needed to use iteratively based adaptive priors algorithms to solve the ill-posed problem by adding prior information to the picture. In the non-local means filtering technique, the weighted average of the reference patches is calculated using Eq. (8). The best estimate patch P in overlapped window considered should be close to the non-local reference patches $p_1 \ldots p_k \, \varepsilon \, R^d$.

4. Graduated non-convex DAMRF model

The proposed model estimates two unknown as two MRFs field variables. The blur-kernel map $\boldsymbol{\sigma}$ and final focused latent picture $\boldsymbol{x}$ are estimated by MAP estimates. Using maximum likelihood terms and Bayes' rule, by applying log is given as

$$\left(\hat{\sigma}, \hat{\mathbf{X}}\right) = \underset{\sigma, x}{\mathbf{argmin}}\left\{\mathbf{log}\left[\mathbf{P}(y|\sigma, \mathbf{X})\right] + \mathbf{log}[\mathbf{P}(\sigma)] + \mathbf{log}[\mathbf{P}(\mathbf{X})]\right\} \tag{9}$$

GMRF regularization results in a smooth image. Hence, discontinuity-based adaptive regularization on MRF is designed to reconstruct the strong edges and retain fine, smooth details of the noisy-blurred picture. The estimates of unknowns are given by

MAP-MRF of GNC-DA prior and are equivalently written from Eq. (9).

$$\left(\hat{\sigma}, \hat{\mathbf{X}}\right) = \frac{1}{2}\|y - \mathbf{X}.\mathbf{h_p}\|_2^2 + \lambda_\sigma \sum_{c \epsilon C} \mathbf{V_c}(\sigma) + \lambda_x \sum_{c \epsilon C} \mathbf{V_C}(\mathbf{X}) \tag{10}$$

where clique potential is given as ***V(.)***. Blur-kernel map sigma σ and the reconstructed latent image x are updated using the following steps

$$\sigma = \sigma^{old} - \alpha_\sigma \left(\frac{\partial e_p}{\partial \sigma} + \frac{\partial}{\partial \sigma} \left(\lambda_\sigma \sum_{c \epsilon C} \mathbf{V_c}(\sigma^{\mathbf{old}}) + \lambda_x \sum_{c \epsilon C} \mathbf{V_C}(\mathbf{X}) \right) \right) \tag{11}$$

$$\mathbf{X} = \mathbf{X}^{old} - \alpha_x \left(\frac{\partial e_p}{\partial x} + \frac{\partial}{\partial x} \left(\lambda_x \sum_{c \epsilon C} \mathbf{V_c}(\mathbf{x^{old}}) + \lambda_\sigma \sum_{c \epsilon C} \mathbf{V_C}(\sigma) \right) \right) \tag{12}$$

The step size for estimation of blur-kernel is α_σ, and gradient size for estimation of latent focused image x is α_x. The derivative gradient term of error e_p from Eq. (1) w.r.t. reconstructed picture x, the blur-kernel map is represented by

$$\frac{\partial e_p}{\partial X} = H^T(H_X - y) \tag{13}$$

$$\frac{\partial e_p}{\partial \sigma} = X^T(Xh_p - y) \tag{14}$$

The energy function in general form is given as

$$E = \frac{1}{2}\sum \|e_p\|_2^2 + \lambda_\sigma R(\sigma) + \lambda_x R(x) \tag{15}$$

The gradient of energy minimizing function E concerning image x is given by adding error gradient, a derivative of prior along with smoothness term, and is given by.

$$\frac{\partial E}{\partial x} = \frac{\partial e_p}{\partial x} + \lambda_x \frac{\partial R(X)}{\partial x}$$

For discontinuity adaptive MRF (DAMRF)

$$R(d) = \gamma - \gamma e^{-\eta^2/\gamma} \tag{16}$$

$R(d)$ is a convex function in the range $\left(\sqrt{-\gamma/2}, \sqrt{\gamma/2}\right)$

$$\text{So, } E = \frac{1}{2}\|y - Xh_{k_p}\|_2^2 + \lambda R(d) \tag{17}$$

$$\sum_{c \epsilon C} V_c(x) = \sum_i \sum_j 4\gamma - \gamma e^{-[x(i,j)-x(i,j-1)]^2/\gamma}$$
$$- \gamma e^{-[x(i,j)-x(i,j+1)]^2/\gamma}$$
$$- \gamma e^{-[x(i,j)-x(i-1,j)]^2/\gamma}$$
$$- \gamma e^{-[x(i,j)-x(i+1,j)]^2/\gamma} \quad (18)$$

GNC-DAMRF model preserves the strong details, a high-frequency component presents at the edges, and structure details of the frame. The details of high-frequency information of the image are adaptively adjusted by all the local neighbor pixels. The parameter value of gamma existing in the DAMRF function controls the shape. With a larger value of gamma, the objective function performs as non-convex. Generally, during initial iterations the values of gamma are set to very high in the order of 3000, to optimize the non-convex objective functions. At each iteration, the value of this parameter is reduced by a factor represented by Kappa. The rate of gamma is adaptively reduced by a value Kappa $\boldsymbol{K}$ as given in Algorithm 3. Using the GNC-DAMRF regularization function of first-order local neighborhood pixels in MAP-MRF work, the gradient function can be written as,

$$= \frac{\partial}{\partial x}\left(\sum_i \sum_j 4\gamma - \gamma e^{-[x(i,j)-x(i,j-1)]^2/\gamma} \right.$$
$$- \gamma e^{-[x(i,j)-x(i,j+1)]^2/\gamma}$$
$$- \gamma e^{-[x(i,j)-x(i-1,j)]^2/\gamma}$$
$$\left. - \gamma e^{-[x(i,j)-x(i+1,j)]^2/\gamma} \right) \quad (19)$$

The gradient of the regularization expression concerning latent image is given by

$$= \sum_i \sum_j 2[x(i,j) - x(i,j-1)] e^{-[x(i,j)-x(i,j-1)]^2/\gamma}$$
$$+ 2[x(i,j) - x(i,j+1)] e^{-[x(i,j)-x(i,j+1)]^2/\gamma}$$
$$+ 2[x(i,j) - x(i-1,j)] e^{-[x(i,j)-x(i-1,j)]^2/\gamma}$$
$$+ 2[x(i,j) - x(i+1,j)] e^{-[x(i,j)-x(i+1,j)]^2/\gamma}$$
$$\frac{\partial R}{\partial x} = \eta e^{-\eta^2/\gamma} \quad (20)$$

As given in Fig. 5, the optimization function of the GNC-DAMRF is non-convex function. GNC-DAMRF is a prior function having gamma as a hyper-parameter as given by Eq. (19). The setting of hyper-parameter gamma by a factor Kappa

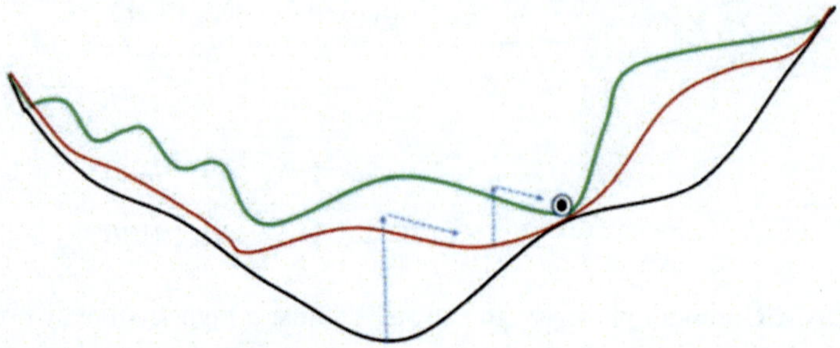

Fig. 5 Graduated non-convex DAMRF function

K reduces the fundamental interaction between adjacent local pixels. The hyper-parameter Kappa makes the non-convex function to the convex function to reach global minima. It is well-known that the blur-kernel sigma $\boldsymbol{\sigma}$ is a function, which in turn is a function of depth $\boldsymbol{d}$ of the 3D scene image. This relation is used to determine the gradients information of both horizontal and vertical neighbor pixels required. Image deblurring steps using GNC-DAMRF is given by Algorithm 3.

Algorithm 3 Image deblurring using GNC

σ_1 = input sigma map (known)
σ_2 = estimated sigma map using JNB
x_{old} = initial estimate of focused image
x_{new} = deblurred image
y = blurred image using σ_1
α = learning rate or the step size
λ = regularization parameter
$\gamma \leftarrow 1000, \gamma_{target} \leftarrow 10, k \leftarrow 0.95, thresh \leftarrow 1$
$\alpha \leftarrow 0.5$
$\lambda \leftarrow 5e-10$
while $(\gamma > \gamma_{target})$ **do**
    $x_{new} = x_{old} - \alpha \frac{\partial E}{\partial x}$
    if $(\|x_{old} - x_{new}\| < thresh)$ **then**
        $\gamma = k * \gamma$
    end if
    $x_{old} = x_{new}$
end while

4 Sparsity-Based Initial Blur-Map Estimation

Initial space-variant blur-map is obtained by exploiting recently proposed JNB sparsity techniques utilizing dictionary learning with feature map 'f'. The relations of blur-kernel and the feature-map values are given by

$$\boldsymbol{f = \frac{a}{2 + \exp(b\sigma + c)} + d} \tag{21}$$

where $\boldsymbol{\sigma}$ is given by

$$\boldsymbol{\sigma = \frac{\log_e\left(\frac{a}{f-d} - 1\right) - c}{b}} \tag{22}$$

The space-variant blur-kernel is described by dictionaries learned by sparsity estimation, which are presented by a few constants values. The values of *a*, *b*, *c*, and d are learned as constants values from the space dictionaries, where the values of all constants learned by dictionary are given as [*a*, *b*, *c*, *d*] = [39.49, 4.535, −3.538, 18.53]. Using JNB for ramp blur obtains a good blur estimate in the range of (0.4–1.5) sigma for both increasing and decreasing ramp as presented in Fig. 6a and sine shape blur in Fig. 7a. So, it is assumed that almost focused image (near focus) in this range since the blur are very small. Consider the two blur images one with increasing and the other with decreasing ramp in the range of (0.4–1.5). So, one image has a smaller amount blur on one side and additional more blur on the opposite side and vice versa for the other image. Since a good blur estimate can be obtained in the range from 0.4–1.5 sigma which are the degraded images, the deblur processing of the images separately is happened using Algorithm 1 and Algorithm 2, respectively. The blur-map of the observed image is obtained from Algorithm 1 and Algorithm 2, respectively. Finally, estimate the blur-kernel using sigma estimate model which is shown in Figs. 6 and 7.

5 Estimation of the Focused Image

The space-variant blurred image $\boldsymbol{y}$ is modeled as

$$\boldsymbol{y = Xh_{k_p} + \eta} \tag{23}$$

$$\boldsymbol{y = Hx + \eta} \tag{24}$$

$$\boldsymbol{y_p(i, j) = \sum_k \sum_l x(k, l) h_p(i, j; k, l) + \eta} \tag{25}$$

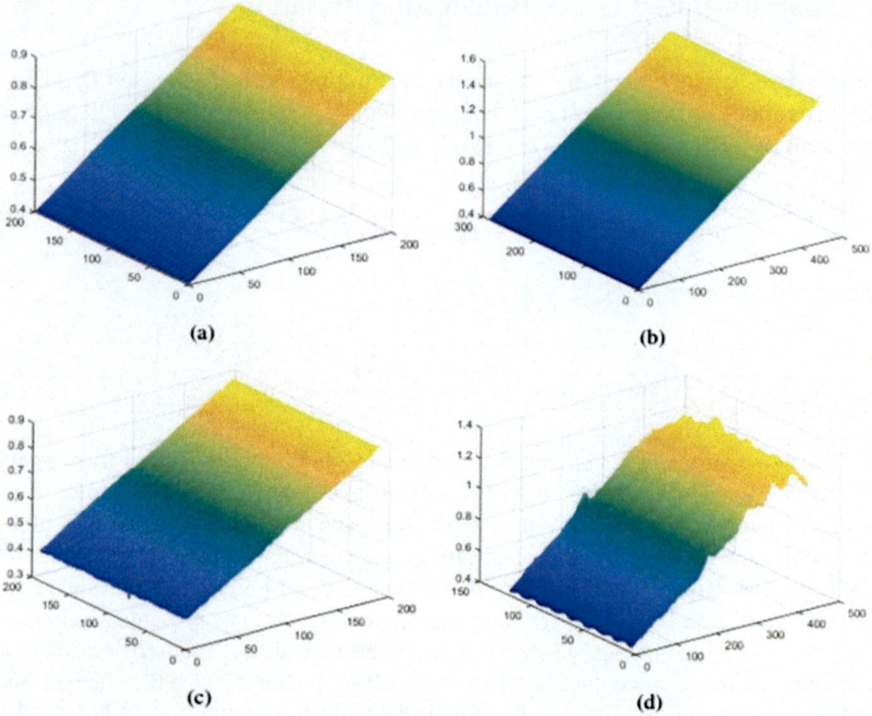

Fig. 6 **a** Blur-map with sigma (0.4–1.5) increasing ramp [14] **b** Blur-map with sigma **c** estimated blur-map using GNC-DAMRF **d** Blur-map using GMRF

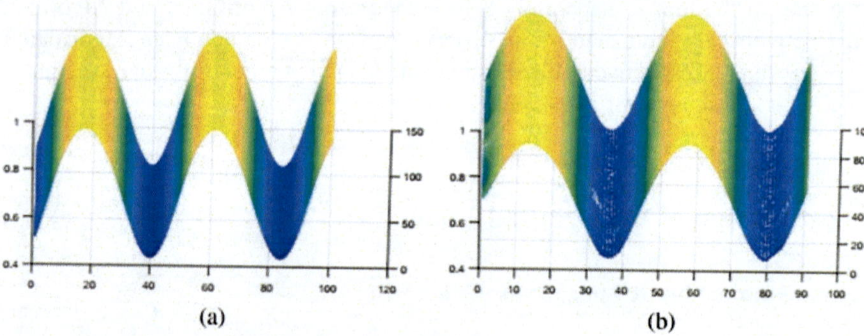

Fig. 7 **a** Ground truth. **b** Estimated blur-map, respectively

Here the $x\ (k, l)$ is the focused latent clear image of the distorted image. η is the AWGN given by [6], $h_p(i, j)$ is the point spread function given as

$$h_p(i, j; k, l) = \frac{1}{2\pi\sigma_p^2(i, j)}\exp\left(-\frac{(i-k)^2 + (j-l)^2}{2\sigma_p^2(i, j)}\right) \tag{26}$$

where X is the restored latent clear image, h is the PSF blur-kernel and η is the AWGN with zero mean [6]. The results of both Eqs. (20) and (21) must be identical. The estimation of the unknown is obtained by minimization of the energy function which is given by

$$e = \frac{1}{2}\|y - Xh_{k_p}\|_2^2 \tag{27}$$

The use of regularizer term $R(d)$ is required to smoothen the latent clear image and blur-map for solving the ill-posed problem to make it a well-posed problem. The problem of double unknown estimation is expressed as the energy minimization of the objective task, given by

$$\text{So, } E = \frac{1}{2}\|y - Xh_{k_p}\|_2^2 + \lambda R(d) \tag{28}$$

$$R(d) \text{ is a convex function in the range } \left(\sqrt{-\gamma/2}, \sqrt{\gamma/2}\right).$$

and is given by

$$R(d) = \gamma - \gamma e^{-\eta^2/\gamma} \tag{29}$$

The image blurring model is solved for multiple unknowns. The two unknown factors are blur-map б and latent clear image x represented, where energy minimization function **E** is given by simplifying L_2 norm as

$$\frac{1}{2}y - Hx_2^2 = \frac{1}{2}(y^T y - 2y^T Hx + x^T H^T Hx) \tag{30}$$

The gradient descent update step is given as

$$x_{\text{new}} = x_{\text{old}} - \alpha\frac{\partial E}{\partial x} \tag{31}$$

$$x_{\text{old}} = x_{\text{new}} \tag{32}$$

Algorithm 1 Image deblurring

σ_1 = input sigma map (known)
σ_2 = estimated sigma map using JNB
x_{old} = initial estimate of focused image
x_{new} = deblurred image
y = blurred image using σ_1
α = learning rate or the step size
λ = regularization parameter
$i \leftarrow 1$
$\alpha \leftarrow 0.5$
$\lambda \leftarrow 5e-8$
while $(i \leq Iter)$ **do**
    $x_{new} = x_{old} - \alpha \frac{\partial E}{\partial x}$
    $x_{old} = x_{new}$
end while

Where the variance σ is the learning rate or the step size. Image deblurring steps are described in the Algorithm 1 is as given below

1. **Blur estimation model**

In this case of estimation of blur-map, blurring is modeled for unknown sigma as:

$$y = Xh_{k_p} + \eta \tag{33}$$

X is the focused clear image. The estimation of structure information is formulated as the minimization of the energy objective function and prior term. This is presented by the gradient descent algorithm. The gradient of the energy function is as follows,

$$\frac{\partial E}{\partial \sigma_p} = \frac{\partial E}{\partial \sigma_P} + \lambda \frac{\partial R}{\partial \sigma_P} \tag{34}$$

$$\sigma_{P_{new}} = \sigma_{P_{old}} - \alpha \frac{\partial E}{\partial \sigma_P} \tag{35}$$

$$\sigma_{P_{old}} = \sigma_{P_{new}} \tag{36}$$

Estimation of blur-map for image deblurring procedures which is described in the Algorithm 2 is as given below

Algorithm 2 Blur estimation

$\sigma_{P_{old}}$ = initial estimate of sigma map
$\sigma_{P_{new}}$ = estimated sigma map
x = focused image
y = blurred image using sigma ground truth
α = learning rate or the step size
λ = regularization parameter
$Iter$ = Maximum number of iterations
$i \leftarrow 1$
$\alpha \leftarrow 1e-5$
$\lambda \leftarrow 1e4$
while $(i \leq Iter)$ **do**
  $\sigma_{P_{new}} = \sigma_{P_{old}} - \alpha \frac{\partial E}{\partial \sigma_p}$
  $\sigma_{P_{old}} = \sigma_{P_{new}}$
end while

6 Experiments and Results

The experiments conducted on synthetic data, BSD500 dataset, and real images are as shown in Figs. 13, 14, and 15 individually using a desktop computer system having 16 GB RAM. The speed specification of the motherboard is 3.3 GHz, an i7 processor, and an Nvidia 960 M graphics card. Blur-map estimation of synthetic image for ramp-shaped blur is shown in Fig. 6c, d using GMRF and MAP-GNC-DAMRF. Figure 7a shows a complex space-variant blur-map identical to a sine wave, which is generated synthetically using a non-uniform blur-kernel. An estimated blur-map of the same is obtained using GNC-DAMRF is shown in Fig. 7b. The experiments conducted on BSD500 image segmentation data are given in Fig. 14. The different local and non-local models are considered based on a different architecture. The obtained results of our method and state-of-the-art methods for real images are compared which is shown in Fig. 15b–d. Our results are compared with [3, 7, 16] and [23] for calf texture image is shown in Fig. 8c–f, respectively. The proposed work is stronger to ringing distortion, image artifacts, and also addresses boundary situations of low-frequency data. Further, the visual quality and quantitative results on both synthetic and realistic dataset were shown in Figs. 14 and 15. SSIM and PSNR performance metrics are utilized to measure the quantity of the restored picture with the degraded image. Figure 9b shows the degraded noisy-blurred face image. Figure 9d shows the restored image output from our image degradation model.

1. **Results on synthetic data**

The results shown in Fig. 8i are obtained on synthetic calf texture image with ramp-shaped non-uniform blur-kernel of blur variance σ varying from 0.3 to 1.5

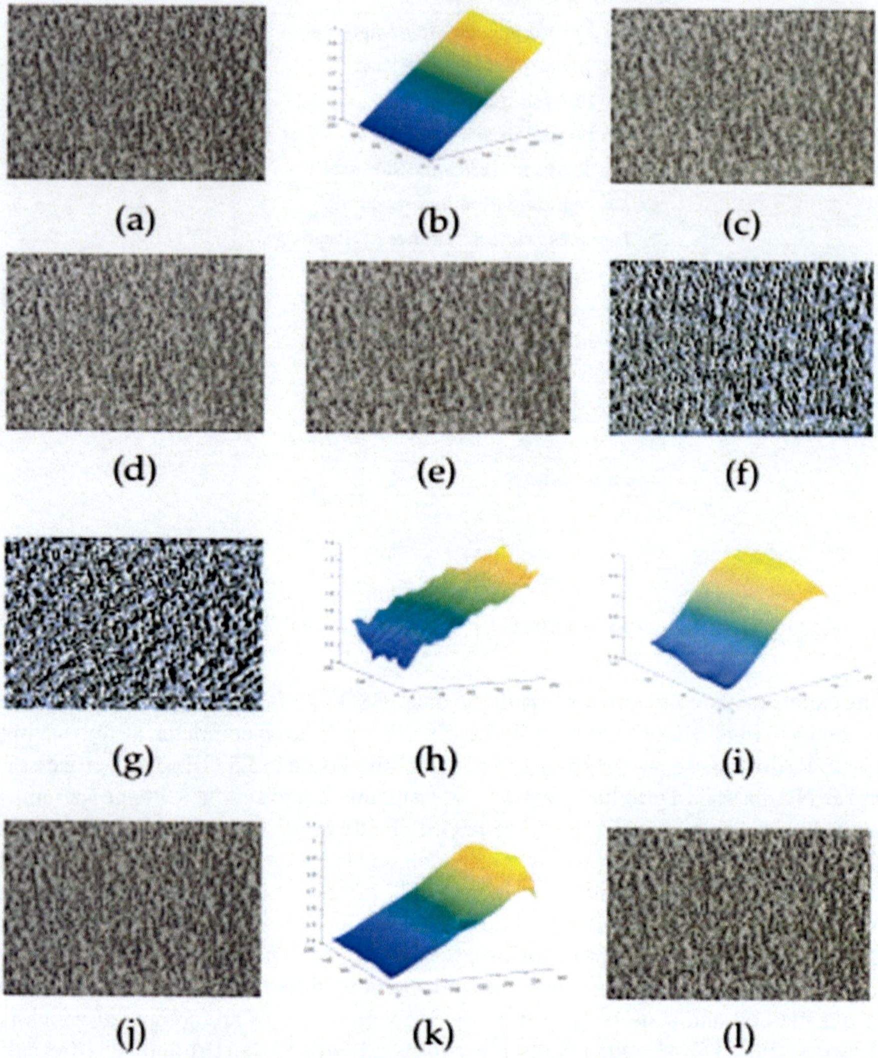

Fig. 8 **a** Focused clear calf image, **b** ground truth ramp blur-map (sigma varying from 0.4 to 1.5), and **c** blurred calf picture by ramp blur. **d–g** Restored pictures using [3, 7, 16, 23], respectively. **h** JNB blur-map by [1]. **i** Blur-kernel by GMRF **j** Restored picture by GMRF [14, 25]. **k** Blur-kernel by GNC-DAMRF **l** deblurred picture by GNC-DAMRF, respectively

using our proposed MAP-GNC-DAMRF model. The image restored using other state-of-the-art methods is shown in Fig. 8c–f. Initially, the step size for learning is set to 0.001, and the regularization parameter is set to 2×10^2. And all other tuning parameters are randomly initialized. The outer loop is set to 50 and the inner loop is set to 10. The estimated blur-kernel and ground truth blur-map are

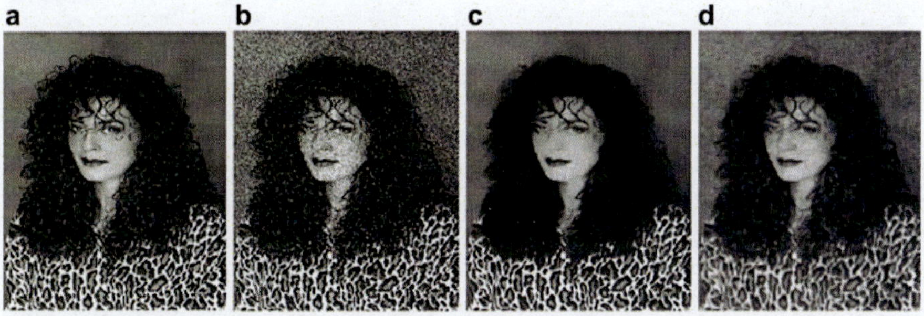

Fig. 9 **a** clear image **b** Noisy image **c** Blurred image **d** Reconstructed image

shown in Fig. 8k, b, respectively. The result of the latent image estimated using the MAP-GMRF algorithm and MAP-GNC-DAMRF algorithm is presented in Fig. 8j and l.

Another set of experiments conducted on the Mosseri face dataset is represented in Fig. 9. The results obtained on synthetic Mosseri face data are given in Fig. 9d. Space-uniform blur-map is convolved to a noisy image as given in Fig. 9b.

The second set of experiments conducted for denoising and deblurring on the Gore face image dataset is shown in Fig. 2a. The noise sigma of magnitude 15 is added to the clear focus image given in Fig. 9a. Blur-noisy picture is synthetically generated by further convolving of blur PSF to the noisy image as represented in Fig. 9b. The results obtained on synthetic Gore face image are given in Fig. 2d and Mosseri face image is given in Fig. 9d. Space-uniform blur-kernel is convolved to a noisy image. Blur-map estimation and the latent image obtained by our proposed framework using MAP-GNC-DAMRF prior for text image are shown in Figs. 10 and 12, respectively. Figure 12 shows the ground truth blur-map of the text image. The ground truth image of the text used for deblurring using optimization-based technique is given in Fig. 10.

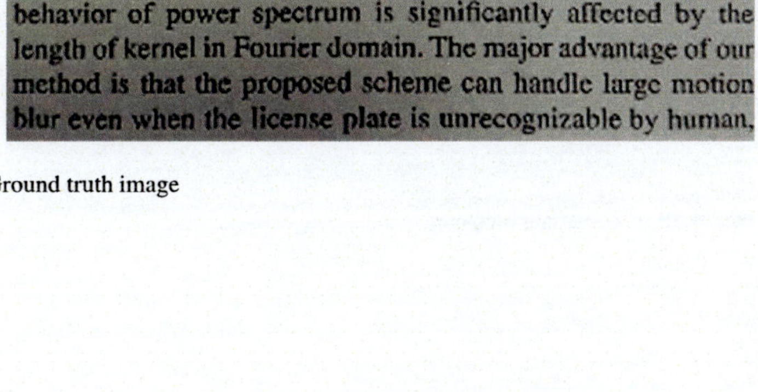
kernel is very large, even reaching one-third of the size of blurred image, which poses great challenges to both the MAP and marginalization methods. To tackle this problem, we adopt the blur kernel parameter estimation method (angle and length). For angle estimation, our scheme makes use of the relationship between the kernel angle and sparse representation coefficients. For length estimation, we exploit the fact that the behavior of power spectrum is significantly affected by the length of kernel in Fourier domain. The major advantage of our method is that the proposed scheme can handle large motion blur even when the license plate is unrecognizable by human.

Fig. 10 Ground truth image

Experiments are also conducted on the document text image which is represented in Fig. 10. High-resolution clear text image captured by a smart mobile phone is downsampled to a low-resolution image by removing alternating rows and columns to reduce algorithm time complexity. The synthetic blurred image represented in Fig. 11 is generated by space-variant convolution operation with non-uniform ramp-shaped blur-kernel. Figure 12 represents the ramp-shaped space varying blur-map utilized to generate a blurred image in Fig. 11. The ground truth space-variant blur-map used to generate a blurred image is shown in Fig. 12.

The output result regenerated with our proposed MAP-GNC-DAMRF prior minimization of the energy function is represented in Figs. 13f and 14f. The performance metric obtained based on PSNR values of the noisy-blurred degraded image, a few

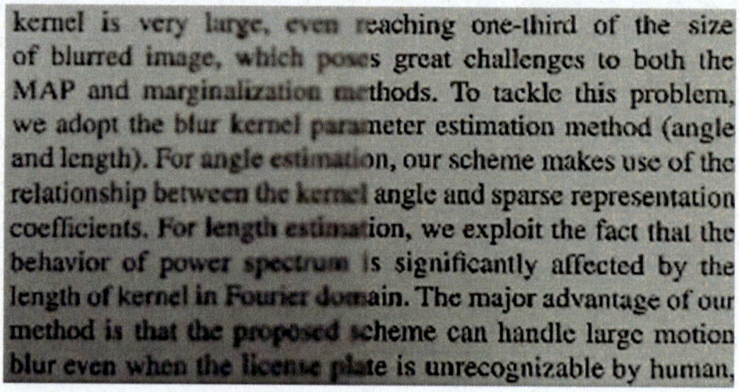
kernel is very large, even reaching one-third of the size of blurred image, which poses great challenges to both the MAP and marginalization methods. To tackle this problem, we adopt the blur kernel parameter estimation method (angle and length). For angle estimation, our scheme makes use of the relationship between the kernel angle and sparse representation coefficients. For length estimation, we exploit the fact that the behavior of power spectrum is significantly affected by the length of kernel in Fourier domain. The major advantage of our method is that the proposed scheme can handle large motion blur even when the license plate is unrecognizable by human,

Fig. 11 Space-variant blurred mage

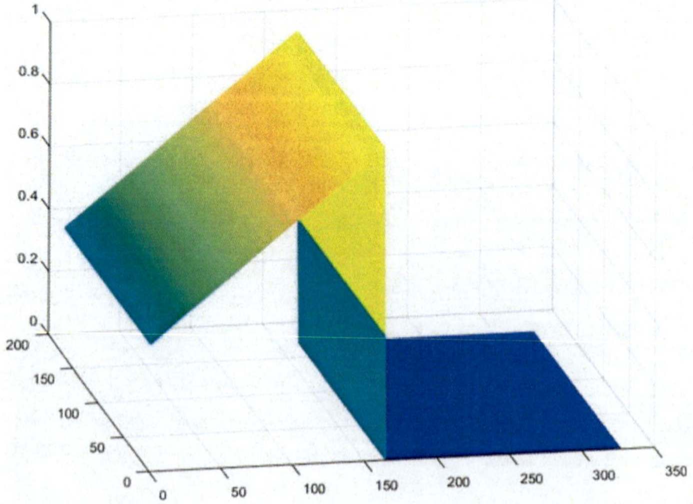

Fig. 12 Ground truth space-variant blur-map

Fig. 13 **a** Ground truth image from BSD 500 dataset. **b** Space-invariant blurred image. **c** Blurred-noisy image with 35 noise sigma and 2 blur-PSF. **d** Reconstructed image using state-of-the-art method. **e** Using GMRF prior. **f** our proposed method using GNC-DAMRF regularizer, respectively

Fig. 14 **a** Ground truth image from BSD 500 dataset. **b** Space-invariant blurred image. **c** Blurred-noisy image with 35 noise sigma and 2 blur-PSF. **d** Reconstructed image using state-of-the-art method. **e** Using GMRF prior. **f** our proposed method using GNC-DAMRF regularizer, respectively

state-of-the-art algorithms, MAP-GMRF convex regularization, and MAP-GNC-DAMRF framework is tabulated in Table 1. PSNR performance metric obtained with our algorithm is very good than a few new state-of-the-art methods. Similarly, by changing blur-sigma, generate the results of some more images with decreasing blur-sigma, sine sigma, and shifted sine sigma for Broadz image dataset and DSLR captured text image are represented in Figs. 12 and 13a. Some more example images with a variety of blur-sigma were generated.

Several trails are conducted on BSD-500 image segmentation dataset for different variable values of noise sigma ranging from 20 to 45. Ramp-shaped blur-map and sine wave-shaped blur-map considered in this work are as shown in Figs. 6a and 7a. A clear

Table 1 PSNR values of degraded image, state-of-the-art work, GMRF prior and proposed model

S. No.	Blurred and noisy PSNR of the image	PSNR values of restored image		
		State of the art [20]	Convex-GMRF prior [24]	Graduated non-convex DAMRF prior ours
1	15.6372	18.3118	19.4083	20.4225
2	16.5795	20.3596	21.5951	22.7739
3	17.5181	23.2512	25.1132	26.9893
4	18.3315	22.8475	24.2018	25.8086
5	17.6561	24.2050	25.4591	27.1001
6	16.4909	23.0572	24.7157	26.3055
7	17.6725	24.0406	25.5784	27.2958
8	16.3587	19.7461	21.1629	22.4135
9	16.9635	21.2645	22.8085	24.2402
10	17.8171	25.0159	26.0829	27.6742
11	17.3327	22.6475	24.3410	26.0594
12	16.7345	20.7230	21.7577	22.7151
13	17.6927	24.0932	25.6387	27.3343
14	17.6580	23.9060	25.6006	27.4499
15	17.7384	24.3554	25.8992	27.6891
16	17.3521	22.7602	24.1109	25.5387
17	16.9849	21.3271	22.9273	24.2937
18	17.6906	24.0688	25.5464	27.2115
19	17.6586	24.4016	26.4884	29.1758
20	17.7878	24.8420	26.5905	28.8916
21	17.3697	22.8207	24.0507	25.4265
22	17.8554	25.3685	26.3624	28.0447
23	18.4319	31.6802	31.0444	34.0362
24	18.2445	28.4621	28.9129	31.0339
25	18.1083	27.2203	27.8323	29.6692
26	15.1923	17.5291	18.4098	19.2910
27	17.6260	24.0661	25.2201	26.7770
28	18.2882	29.0150	29.3010	31.5031
29	17.7701	24.6675	25.6472	27.0653
30	17.8130	24.8434	25.9128	27.3413
31	17.6830	24.0602	25.3685	26.8761
32	16.0316	19.0361	20.1198	21.0499
33	17.9525	25.5448	27.0789	29.1180
34	18.1188	26.8389	28.0559	30.0797
35	17.2341	22.5098	23.8650	25.4146

scene image taken from the BSD 500 dataset presented in Fig. 14a is the ground truth picture. Synthetically blurred images are generated by the convolution process with blur-map kernel and clear image. The space-invariant blur of point spread function (PSF) having Gaussian distribution with magnitude intensity of blur-sigma variance two as shown in Figs. 13b and 14b. Later, additive Gaussian noise having a variance of magnitude 30 is added by Gaussian function to generate degraded noisy image synthetically is as represented in Figs. 13c and 14c. The results obtained using a few recent works are shown in Figs. 13d and 14d. Outputs of MAP-GMRF convex regularization priors results in smooth low-frequency images, where high-frequency details are not significantly reconstructed, which is represented in Figs. 13e and 14e.

The output result regenerated with our proposed MAP-GNC-DAMRF priors minimization of the energy function is represented in Figs. 13f and 14f. The performance metric obtained based on PSNR values of the noisy-blurred degraded image, a few state-of-the-art algorithms, MAP-GMRF convex regularization, and MAP-GNC-DAMRF framework is tabulated in Table 1. PSNR performance metric obtained with our algorithm is good then a few new methods. Similarly, by changing blur-sigma, generate the results of some more images with decreasing blur-sigma, sine sigma, and shifted sine sigma for Broadz image dataset and DSLR captured text image are represented in Figs. 12 and 13a

2. **Results on real-time images**

Experimentations are conducted on real-time images captured using high-end smartphones. Images are downsampled to a resolution of 640×480 pixels by skipping rows and columns are as shown in Fig. 15a. Images are initially preprocessed to remove the outliers and other details by normalizing. Gamma hyper-parameter and final target gamma are fixed to 1500 and 0.01 by setting the kappa value to 0.97. Figure 15c, d shows the resultant images obtained using MAP-GMRF and GNC-DAMRF algorithms. The results obtained using adaptive-local GNC priors outperform the basic MAP-GMRF prior to both in terms of quality and quantity. Figure 15b shows results obtained using the few recent works [4]. Using MAP-GMRF [26]-based regularization provides smooth outputs, which is shown in Fig. 15c. The output results generated by our model GNC-MAP-DAMRF regularization are shown in Figs. 14f and 15d. The outputs reconstructed images obtained with our algorithm are visually good than a few new algorithms.

7 Conclusion

An adaptive MAP-GNC-DAMRF image restoration technique is proposed. The implementation of the joint image denoising and deblurring framework was constructed using GNC regularization. These are executed by utilizing alternating energy minimization to obtain blur-map kernel and latent clear image. The results are compared with both PSNR and SSIM metric. The outputs of the algorithms are also compared with a few other techniques such as GMRF and GD-GMRF. In this article,

(a) (b) (c) (d)

Fig. 15 **a** Real degraded images captured using mobile camera. **b** Reconstructed image using state-of-the-art method. **c** Using GMRF prior. **d** Our proposed method using GNC-DAMRF regularizer, respectively

the development and implementation of a novel work were based on an optimization framework capable of estimating blur-map of a different shape.

The results obtained both in terms of quality and quantity and are improved over a few works given the literature. The proposed algorithm can also be used to estimate the depth-map of 3D images using JNB sigma map. In the future, the current work can be extended for the depth-map generation of real 3D scene images. The proposed work provides a better estimation of blur-map for sigma ranging from 0.1 to 2.

References

1. J. Shi, L. Xu, J. Jia, Just noticeable defocus blur detection and estimation, in *Proceeding of IEEE Conference on Computer Vision and Pattern Recognition* (2015), pp. 657–665[1]
2. A.N. Rajagopalan, S. Chaudhuri, An mrf modelbased approach to simultaneous recovery of depth and restoration from defocused images, in *IEEE Transactions on Pattern Analysis and*

Machine Intelligence, vol. 21, no. 7, (1999), pp. 577–589
3. D. Krishnan, R. Fergus, Fast image deconvolution using hyperlaplacian priors, in *Proceeding NIPS* (2009), pp. 1033–1041
4. H.N. Latha, H.N. Poornima, Estimation of blur and depth_map of a de-focused Image by sparsity using gauss markov random field convex-Prior. Int. J. Sci. Res. **08**(11), 728–733 (2019). ISSN: 2319-7064
5. S.Z. Li, *Markov Random Field Modeling in Computer vision* (Springer Science & Business Media, 2012)
6. H.N. Latha, M.V. Palachandra, M. Rao, Real time implementation and performance evaluation of WCDMA system over AWGN channel on TMS320C6713DSK. Proc. Technol. **4**, 82–86 (2012)
7. J. Pan, Z. Hu, Z. Su, M.-H. Yang, *L0-regularized Intensity and Gradient Prior for Deblurring Text Images and Beyond* (IEEE Proc., CVPR, 2016)
8. R. Sahay, H.N. Latha, Blur-map estimation and deblurring of degraded image by non-convex regularization. Int. J. Mod. Trends Sci. Technol. **6**(8), 134–139 (2020). ISSN: 2455-3778
9. X. Cao, W. Ren, W. Zuo, X. Guo, H. Foroosh, Scene text deblurring using text-specific multiscale dictionaries. IEEE Trans. Image Process. **24**(4), 1302–1314 (2015)
10. H.N. Latha, M.V. Lakshmi, Ramachandran's Design of context based adaptive variable length coding and deblocking filter for H. 264. Sci. Direct, 671– 676 (2012) (Elsevier)
11. S.K. Nayar, Y. Nakagawa, "Shape from focus," Pattern analysis and machine intelligence. IEEE Trans. **16**(8), 824–831 (1994)
12. A. Chakrabarti, T. Zickler, Depth and deblurring from a spectrally-varying depth-of-field, in *Computer Vision–ECCV 2012* (Springer, 2012), pp. 648–661
13. A. Chakrabarti, T. Zickler, W.T. Freeman, Analyzing spatially-varying blur, in *2010 IEEE Conference on Computer Vision and Pattern Recognition (CVPR)*, (IEEE, 2010), pp. 2512–2519
14. H.N. Latha, K.P. Kranti, R.R. Sahay, Simultaneous blur map estimation and deblurring of a single spacevariantly defocused image, in *2017 27th National Conference on Communications (NCC)* (IEEE, 2017)
15. S. Dai, Y. Wu, Removing partial blur in a single image, in *IEEE Conference on Computer Vision and Pattern Recognition, CVPR 2009*, (IEEE, 2009), pp. 2544–551
16. R.S. Rajiv, H.N. Latha A local modified u-net architecture for image denoising. Int. J. Mod. Trends Sci. Technol. **6**(8), 140–144 (2020). ISSN: 2455-3778
17. X. Zhu, S. Cohen, S. Schiller, P. Milanfar, Estimating spatially varying defocus blur from a single image. Image Proc. IEEE Trans. **22**(12), 4879–4891 (2013)
18. S. Zhuo, T. Sim, Defocus map estimation from a single image. Pattern Recogn. **44**(9), 1852–1858 (2011)
19. Y.-W. Tai, M.S. Brown, Single image defocus map estimation using local contrast prior, in *Image Processing (ICIP)* (2009)
20. E. Luo, S.H. Chan, T.Q. Nguyen, Adaptive image denoising by targeted databases. IEEE Trans. Image Process. **24**, 2167–2181 (2019)
21. H.N. Latha, R. Sadhan, D. Sampreeth, Image understanding: Semantic Segmentation of Graphics and Text using Faster-RCNN, in *IEEE International Conference on Networking, Embedded systems (ICNEWS)* **1** (2018), pp. 82–87
22. I. Mosseri, M. Zontak, M. Irani, Combining the power of internal and external denoising, in *IEEE International Conference on Computational Photography*, (2013), pp. 1–9
23. H. Cho, J. Wang, S. Lee, Text image deblurring using text-specific properties, in *Proceeding ECCV* (2012), pp. 524–537
24. A. Buades, B. Coll, J. Morel, Image denoising methods: a new nonlocal principle. SIAM Rev. **52**, 113–147 (2010)
25. P. Arbelaez et al. Contour detection and hierarchical image segmentation. IEEE Trans. Pattern Anal. Mach. Intell. **33**(5), 898–916 (2010). BSD500, Berkley Segmentation dataset
26. H.N. Latha, B. Lokesh, Denoising and deblurring by gauss markov random field: an alternating minimization convex prior. IJSR **8**(11), 6 (2019). 2319-7064

27. D.H. Salvadeo, N.D. Mascarenhas, A.L. Levada, Nonlocal markovian models for image denoising. J. Electron. Imaging, **25**(1), 013 003–013 003 (2016)
28. N. Joshi, R. Szeliski, D.J. Kriegman, PSF estimation using sharp edge prediction, in *IEEE Conference on Computer. Vision and Pattern Recognition, 2008. CVPR 2008* (IEEE, 2008), pp. 1–8
29. H.N. Latha, R.R. Sahay, H.N. Poornima, Simultaneous denoising and deblurring by non-convex regularization: alternating minimization framework. in *IEEE Conference ICCSP 2020* (2020), pp. 401–406
30. H.N. Latha, Sahay, R.R. Joint estimation of depth map and focus image in SFF: an optimization framework by sparsity approach, in *IEEE Conference ICCSP 2020*, (2020), pp. 401–406
31. S. Sheeba, H.N. Latha, Design and implementation of saliency detection model in H. 264 Standard. IJSR (2014) pp. 14–2020
32. A. Chakrabarti, T. Zickler, Depth and deblurring from a spectrally-varying depth-of-field. Comput. Vis.–ECCV 2012, 648–661 (2012). (Springer)
33. S. Dai, Y. Wu, Removing partial blur in a single image, in *IEEE Conference on Computer Vision and Pattern Recognition, 2009. CVPR 2009* (2009), pp. 2544–2551
34. H.S. Meharwade, H.N. Latha, Design of high speed face detection and tracking by skin segmentation using matlab. Int. J. Eng. Sci. Res. (IJESR), 1–6 (2014)

Computational Framework for Visually Impaired

Pragati Chandankhede and Arun Kumar

Abstract Blindness is considered an important disability affecting all areas of development. Despite wide development in technology come across multiple visually impaired people struggling for doing day-to-day activities involving locating furniture ahead or walking independently on streets, the study aims to build an accessible environment so that the people with disability can be ingrained with major resources. This article determines the requirement of blind children about the way they feel about impairment and what can help them in such a situation. The paper focuses on architecture that can solve the issue using a computer vision technique that indeed will automate everyday jobs like the human visual system can do. The technology includes the LSTM and Inception Model for building architecture. This article also depicts the result of the image classification and caption.

Keywords Feature extraction · Object detection · Computer vision · Traveling aid · Image recognition

1 Introduction

Image and object recognition is an abrupt requirement for the visually impaired. There is 39 Million blind over the world. Various technologies have come to assist the impaired individual. In the direction of the same, this paper focuses on how deep learning techniques can help people and search out aid that can assist them [1]. The study was conducted to understand the concept of blindness among children, focuses on the way they face the problem in day-to-day activities, taking into account of attending their schools, or going to nearby market [2]. Inability to see things can limit someone from doing everyday activities and quality of life gets affected. Several

P. Chandankhede (✉) · A. Kumar
Sir Padampat Singhania University, Udaipur, Rajasthan, India
e-mail: pragati.chandankhede@spsu.ac.in

A. Kumar
e-mail: arun.kumar@spsu.ac.in

V. Suma et al. (eds.), *Inventive Systems and Control*,
Lecture Notes in Networks and Systems 204,
https://doi.org/10.1007/978-981-16-1395-1_27

Electronic traveling aids have been developed in recent years. Many of aid is making the impaired person aware of surrounding they possess, this system passes an alarm to the blind as go ahead or wait at the position he is at. Such a system is not acceptable for working in a real-life scenario. This article focuses on the system which completely depends on computer vision and extracts information of every event occurring and alerts them in the form of the complete message rather than the go-no-go system.

2 Review and Literature Survey

The review of the visually impaired was conducted at Blind Relief Association in Nagpur, Maharashtra, India. The study set out to determine the questionnaires to girl student and the associated answer were used to recognize the difficulties they come along with a certain group of student were inspected, out of which 72% were school-going children and 19% are a college student and the rest are working part-time and the question asked to them and response acquire are as follows:

(1) Would you classify your peripheral (side) vision? 72% of people replied they cannot. (2) Do you experience difficulties in going from a dark place to a well-light place? 77% of people say yes, while the rest of the people says moderate difficulties they suffer. (3) Can you see colors in daylight? 66% say no, while the rest of them says they can see. (4) What sort of short, medium, and long-distance trips do you make? 67% says they can do a medium distance trip, 28% says they can do short-distance trip while very few can do long-distance trip. (5) What means of transport do you normally use? 62% of children prefer walking while some rest goes by bus or private taxi. (6) How easy do you find it to make these journeys? 48% of people say the journey is very difficult for them, while some people reply easily to them as its much-known place to them. (7) How often do you ask for assistance when crossing the road at a complex junction? All the children replied that all the time they require assistance. (8) How easy or difficult is it for you to find the correct bus stop in a familiar area? 62% replies that even it's a known area to them they feel anxious to find the correct bus stop. And it's difficult for them. (9) How often do you ask for assistance in identifying the correct route? 70% population relied on that they always ask for assistance during their journey. (10) Which is the travel aid that you use most frequently? They use a long cane, Electronic cane, guiding dog, human guide. (11) How well does this aid enable you to avoid obstacles? 85% replied that to some extent they can avoid obstacles. (12) What pace or speed do you move at with this travel aid? 91% replied that even being using any assistance device that has to move at a very moderate pace. (13) How safe do you feel with this travel aid? 86% replied that they did not feel completely safe using this travel aid. (14) Would you indicate which important senses or types of information is in telling you about your surroundings? Most important sense they feel is touch with hands, sight, hearing, spoken information, a combination of touch, and hearing.

The conclusion that was derived from the survey: Visually Impaired Persons feel difficulties in going to unknown places. Feel difficulties during crossing road and choosing a correct route and they did not carry out long-distance trips. Most of the times, they need assistance from the people to complete their journey. The traveling aid like (human guide/long cane) they are using is not too safe on which they can rely on.

One noticeable thing from the survey was visually impaired are using Electronic traveling Aid (ETA), but these are not helping them completely. Table 1 describes the limitation in the usage of ETA [3].

Despite various ETA available in the market, the common issue that needs to be addressed by the proposed system occupies:

1. Need more accurate system beyond the go-no-go system [3].
2. Object localization and object detection should be exact.
3. Require a cost-effective system with minimum hardware.
4. Require to Design the usability features of a low-cost knee-above obstacle detection system.

3 Design of the System

To model the system, a few prominent parameters were detailed as follows.

3.1 Desirable

The desirability feature tells how necessary the system is useful, and will it satisfy the need for the course of action! While researching it was perceived that the system made for the visually impaired in the current scenario is not satisfying their daily needs. The use of a cane is for scanning the surrounding from obstacles but this too has space limitation in terms of detecting, as many things can not being scan by simply moving a cane around them. The use of an electronic cane is based on an ultrasound sensor and a cane indicates the impaired person about surrounding the basic vibration that gets reflected once they touch the obstacles. But the problem faced by a person is in terms of detecting the knee-above the object and for a good result, there is a need to maintain a constant sweep rate.

3.2 Feasible

This factor depicts the answer to "Is it possible to build the system with existing architecture?" And the answer after analyzing the system can build the system considering all the aspects of the proposed system. The factor of technical feasibility is considered and checked whether the hardware and software can handle the workload assigned

Table 1 Electronic traveling aid history

Sr. no	ETA device (Type I)	Year	Author/inventor	Description	Limitation
1	WalkMate	1993	Takeshi Muto, Barbara Herzberger	It's either waist-mounted or handheld, with both audio and vibration options available. It projects a U shaped beam of 1.82 m tall by 0.7 m wide	The beam seemed to have varying issues
2	Miniguide US	2004	GDP Research, Australia	It uses ultrasound to detect objects and gives tactual or auditory feedback by vibrating or chirping more rapidly as you approach an object. Five default ranges: 8, 4, 2, 1, and 1/2 m.	Being small in size, lack the coordination for control manipulation and accurate positioning
3	White Cane	2001	C. Wong, D. Wee, I. Murray and T. Dias, "A novel design of integrated proximity sensors for the white cane"	The proposed design centers around the use of a microcontroller to calculate distance measurements of ultra-sonic signals sent and received by two sets of transducers fixed onto the white cane	The ability of a user to detect is based on the user's sensitivity to ground levels
4	vOICE	2007	Malika Auvray, Sylvain Hanneton, J Kevin O'Regan, ISSN	Use of image processing technique for capture images and converting them to grey scale. Depending on pixel value converted into frequency leading to produce sound	Voice emitted as an output, giving location of object detected, cannot represent exact location

(continued)

Table 1 (continued)

Sr. no	ETA device (Type I)	Year	Author/inventor	Description	Limitation
5	Intelligent Walking Stick	2017	Nadia Nowshin, Sakib Shad man, Saha Joy "An intelligent walking stick for the visually-impaired people"	There are four ultrasonic sensors IC sensor. Stick have sensor A and B implemented on the front side, sensor C on left side and senor D on right side. All four sensors detect obstacle and send the distance to an Arduino module. Then Arduino sends the correct distance to the mobile app through Bluetooth module. Through the App user can hear the distance in their headphone	Image processing can be used for knowing about the volume of obstacles and object patterns
6	Smart Cane	2017	Gita Indah Hapsari, Giva An driana Mutiara, "Smart Cane location guide for blind using GPS"	Location detection is done on basis of raspberry pi which call destination location GPS data, Raspian device uses a voice recorder that has memory and can be connected directly with the keypad. Sound recordings in the form of MP3 files contain the information of the destination location detection voice information data I through MySQL configuration processing	The use of the internet is mandatory to track location

Table 2 Technical feasibility of system

Sr.no.	Parameter considered	Answer
1	The type of equipment: model number, manufacturer	Raspberry pi 3 Model B+, by Raspberry pi 3
2	The estimated age of the equipment	Ideally, age is 5 years and can run 24 * 7 with more robust storage
3	The physical location of the equipment	The equipment will be attached to goggles, considering in minds the security of person, the battery backup (500 mA) will be provided in pockets of user
4	The financial arrangement for the equipment	The user has to own it

to the system [4]. Table 2 depicts parameters are considered for accessing technical feasibility.

4 Architecture of System

Figure 1 denotes the architecture of the system, which mainly includes a camera module, battery pack, sensing, and Navigation Device. When the user wants to use the system, the user has to switch on a camera module that is attached to Raspberry Pi. Raspberry Pi has an on-chip graphics processing unit which makes the system more robust for processing. The system can process a real-life image. To process images of the camera, the object detection and classification model of inception version 3 is used [5]. The objects are categorized by the inception model and are captioned by the algorithm. Once the object is detected, captioning is provided by the Natural Language Processing (NLP) system using Long Short Term Memory. Here, the network starts working and it read the caption so that blind person can hear it, they come to know what object is in their front. But this system also needs to know how far the recognized object is from the user, for that measurement, light detection and ranging sensor is used. Using light detection and ranging sensor, it is

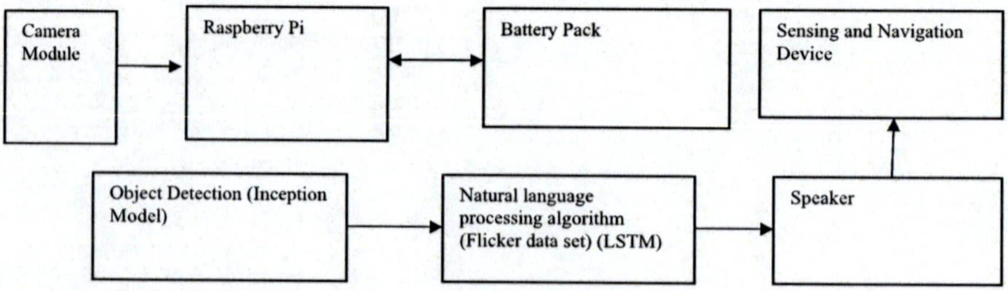

Fig. 1 Architecture of system

possible to hold the intellect of both the physical dimensions and track the object even in motion.

5 Computer-Aided Learning

The problem in the real world for the blind is completely different from that of a normal human. To create a product that can guide blinds at every point in time. This architecture aims at the object is recognized from the previous step, and want to make use of this output to make the blind person aware of the object ahead [4]. At the same instant, expect natural-looking text as output. This process is said as natural language processing, which creates smooth interaction between machine and human, by understanding input from the machine. The computer vision technology with a combination of natural language processing can help to produce a caption of the recognized image, Long Short Term Memory algorithm assists in providing a caption by memorizing all the activities that happened in a recognized image. Long Short Term Memory can be considered as a system that takes single input at a time but produces multiple outputs in form of text which are said as a caption. LSTM works on the principle of joining the probability of original and target sentence.

5.1 Data Collection

The Flicker 8k data set is renowned for deep learning problems, consisting of eight thousand images. Out of which 6,000 images for the training set, 1,000 for the testing set, and 1,000 for the development set. This dataset is provided with five captions. This file comes along with a text file, which illustrates text associated with every image. The model is trained on Imagenet. There are around 100,000 synsets in WordNet, the majority of them are nouns (80,000 +). ImageNet provides on average 1000 images to illustrate each synset. Synset is synonyms that share a common meaning.

5.2 Image Pre-processing

Input to our model will be an image, this image needs to be converted in form of the vector which then is processed by the Neural Network model [6]. To bring out this as a product is decided to use transfer learning by the Inception model [7]. Because of this learning process became faster as this model is trained on the Imagenet dataset. To capture an image, a camera module having a resolution of 1280 $\times$ 720 is used.

5.3 Inception Model

Usually, in Convolution NN, a stack of a layer is constructed such as convolution layer, normalization layer, max pool layer, activation layer. While the structure of inception is made in such a way because of which problem of overfitting is avoided and computational power also improves [8].

The term inception layer has come from the concept of "Network in Network". As the name Inception indicates, there is an establishment of accuracy with the use of layer in Neural Network. This model is inspired by Convolution Neural Network [9]. Each layer of Inception produces a single output vector, which in turn forms the input of the next stage by combining the Convolution layer of a different dimension. Inception layer allows internal layer to choose filter size for gaining information. Consider the layer in deep learning that has learned individual parts of an elephant including ear, nose, and eyes. The next layer will focus on the overall face of the elephant. Each layer is supposed to have suitable filter sizes to detect different parts. If an elephant face is an image, the filter size chooses is higher as compared to the complete elephant image which may include background. This model is trained on a dataset of ImageNet, which is a well-organized hierarchy [10]. It solves the classification problem as it can classify 1,000 classes. The general architecture of the inception model is depicted in Fig. 2.

On a given input image, there is the application of different filter sizes. For instance, applying a 1 × 1 filter on the input image can obtain some filter map. At the same time, when applied with 3 × 3 convolutions to the input image another filter map can be obtained. With help of the padding function, a feature map can be evaluated whose size is almost equal to the given input image size [9]. Such a stacked module with a layer of max pooling makes the inception model more efficient in detecting feature maps for multiple classes. This also means the architecture promotes shared feature extraction. But doing this improves the width of each layer. It follows a recursive structure for the same. To illustrate the probability distribution of the possible outcome of the network, the softmax function is used. All these intermediate layers, help in solving the vanishing gradient problem. The aim of our

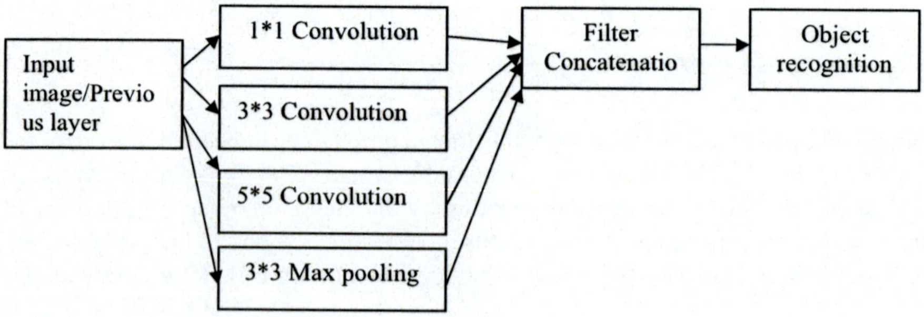

Fig. 2 General architecture of inception model

system is not to categorize images, but it is used for carrying out informative vectors for every image, which is done by the Inception model. It also uses a series of the filter with different sizes, leads to extracting different feature at the same time, given an input image.

6 Result and LSTM for Captioning

An input image is a process with the help Inception model to acquire informative vectors. A model that can predict text from the image aids to predict a caption is not one step, it does so; word by word this is also said as a partial caption. It is required to encode every word in a fixed-size vector. The most important step is to generate vocabulary during training which can contain all possible words. And the target value here is to allot proper caption given an image. The aim of LSTM [11] is to read this partial caption and map them as per feature vector.

Given an input image, the figure illustrates the result of our model.

As shown in Fig. 3, given an image as input to the system, there exists a partial caption that helps to predict the next word in the sequence, which gives the impression of being as shown in the table below. The result of captioning depicts "Dog is chasing", so here the partial caption [11] are recorded with start sequence and stored as a vector. Word embedding method maps the word in vector space, this is also called sequence prediction, which illustrates for the same is depicted in Table 3. Thus various events occurring in sequence will be captured by LSTM. Figure 4 illustrates how Long Short Term memory helps our system in capturing all events gradually captured even in real-time.

Long Short Term Memory is influenced by the recurrent neural network (RNN) [12]. Consider the formulation for Recurrent Neural Network, consider X_t is input at

Fig. 3 Result of object classification and captioning

Table 3 Data matrix for image and caption

I	Image feature vector	Partial caption (X_i)	Target word (Y_i)
1	Image_1	Startseq	Dog
2	Image_1	Startseq Dog	is
3	Image_1	Startseq Dog is	is chasing

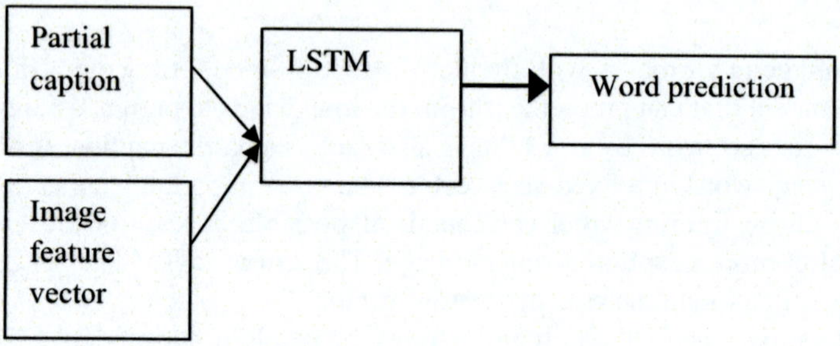

Fig. 4 Level wise architecture

time step t and S_t states at time step t occurs. This state when multiplied with weight is called W_s.

$$S_t = F_w * (S_t - 1, X_t) \tag{1}$$

F_w can be considered as a recursive function which is denoted by the tanh function. When the new input x arose, (where x is featured vector) which gets multiplied with weight of x, term as W_x. So the recursive function at time state t is given by

$$S_t = \tanh(W_s * S_t - 1, W_x * X_t) \tag{2}$$

To get the output vector, the input state with the weights of X mentioned as W_x is multiplied with the previous state with W_s and later it passes through a tanh activation to get the new state. Thus the information is processed concerning the state. The algorithm updates weights by backpropagation phenomena [13]. LSTM afford increasing interaction that is extending the memory of the network. As a result, it keeps useful information in memory at given state S, but if the information seems of no use it is being thrown away from memory. Adding more interaction to an existing neural network makes LSTM more strongest algorithm [14].

7 Raspberry pi and Speaker/Headphone

A pre-trained CNN model of InceptionV3 is used for image classification and is embedded in Raspberry pi 3 [15]. And the output for captioning is given via speaker to a visually impaired person. To know, how far the recognized object is from the person Lidar sensor is used [16], which will be connected to Raspberry pi3 via USB slots [2].

8 Conclusion

Despite innovation in the field of various learning technology, advanced algorithm, and computer vision, are still facing troubles in the visually impaired. In this article, a deep learning approach of long short term memory along with raspberry pi was tested on various images of ImageNet. Using this, the system has analyzed and classified objects ahead accurately. The inception model grants an advantage of multi-level feature extraction. The well-organized and precise system revealed will definitely assist the visually impaired in carrying out daily activities and lead independent mobility to them. The technology proposed is effective in terms of cost. In the future, the prospect is to make the system lightweight and compact with Internet of Things support.

References

1. F.R. Leta, F.F. Feliciano, F.P.R. Martins, Computer vision system for printed circuit board inspection, in *ABCM Symposium Series in Mechatronics*, vol. 3, pp. 623–632, 2008
2. S.H.R. Madhuranga, P.S. Ranaweera, H.F.A.S. Fonseka, Electronic travel aid system for visually impaired people. IEEE Trans. Pattern Anal. Mach. learn. **32**(9) (2017)
3. D. Yuan, R. Manduchi, A tool for range sensing and environment discovery for the blind, in *Proceeding of 2004 Conference on Computer Vision and Pattern Recognition Workshop*, Washington, DC, June 27–July 02, vol. 3, pp. 39–39
4. M.B. Blaschko, C.H. Lampert, Learning to localize objects with structured output regression, in *10th European Conference on Computer Vision*, 2008
5. P.F. Felzenszwalb, R. Girshick, D.M. Allester, D. Ramanan, Object detection with discriminatively trained part-based models. IEEE Trans. Pattern Anal. Mach. Intell. **32**(9) (2010)
6. J. Joselin et al., A review on neural networks. Int. J. Trend Sci. Res. Dev. (IJTSRD) **2**, 565–569 (2018). ISSN 2456-6470
7. A. Fred, M. Agarap, An architecture combining convolutional neural network (CNN) and support vector machine (SVM) for image classification. Computer Supported Collaborative Visualization [cs.CV], ArXiv, 2019
8. S. Hijazi, R. Kumar, C. Rowen, *Using Convolutional Neural Networks for Image Recognition.* Technical Report, Cadence, 2011
9. H. Chen, J. Lu, The image process technologies in face recognition, in *2010 2nd International Conference on Information Science and Engineering (ICISE)*, 2010
10. J. Uijlings, K. Sande, T. Gevers, A. W. M. Smeulders, Selective Search for Object Recognition. Int. J. Comput. Vis. **104**, 154–171 (2013). https://doi.org/10.1007/s11263-013-0620-5

11. S. Ren, K. He, R. Girshick, J. Sun, Faster R-CNN: towards real-time object detection with region proposal networks. Computer Supported Collaborative Visualization [cs.CV] ArXiv, 6 Jan 2016
12. K. Smagulova, A.P. James, Overview of long short-term memory neural networks, in *Deep Learning Classifiers with Memristive Networks*, vol. 14, MOST, ed. by A.P. James (Springer, Cham, 2020), pp. 139–153
13. R. Pascanu, T. Mikolov, Y. Bengio, On the difficulty of training recurrent neural networks, in *Proceedings of the 30th International Conference on Machine Learning, Atlanta, Georgia, USA*, vol. 28(2) (2013)
14. V. Muthukumar, N. Bhalaji, MOOCVERSITY-deep learning based dropout prediction in MOOCs over weeks. J. Soft Comput. Paradigm (JSCP) **2**(03), 140–152 (2020)
15. L.F. He, Y.Y. Chao, K. Suzuki, An algorithm for connected-component labeling, hole labeling and Euler number computing. J. Comput. Sci. Technol. **28**(3), 468–478 (2013)
16. P. Viola, M. Jones, Rapid object detection using a boosted cascade of simple features, in *Conference on Computer Vision and Pattern Recognition*, 2001

Human Scream Detection Through Three-Stage Supervised Learning and Deep Learning

Ashutosh Shankhdhar, Rachit, Vinay Kumar, and Yash Mathur

Abstract As for situation responsiveness, audio and video signals are very important. Audio is significant because it can enlighten us concerning situations, character, time, and place. The study describes a distress audio (scream or shout) event classification system which precisely classifies an audio event as ambient noise, screams, and shouts. Scream is a high-pitch vocalized sound in the absence of phonological structure. This study, researched the classification system using a three-phase SVM-based classifier model for segregating human distress sound from noise and then scream from a shout. The training of SVM-based classifier is done with audio MFCC as feature vectors, appropriately chosen from a set of 400 audio sets, which are selected according to a two-phase process. A classifier is trained and tested with each feature subset. SVM-based classifier analyses and predicts the sound which works with linear kernel and radial basis function (rbf) kernel. The obtained classification performance then again passes through the Multilayer Perceptron Model. When the model gets any sound then it tries to identify patterns in it using perceptron weights and biases if it can't get success it slightly changes its weights for getting the correct result if it successfully detects a scream in sound then it calls the emergency function. Our results demonstrate that the system can generate a 90% accuracy.

Keywords MFCC · SVM · Scream detection · Multilayer perceptron model · Perceptron · Distress sounds

---

A. Shankhdhar (✉) · Rachit · V. Kumar · Y. Mathur
Department of Computer Engineering and Applications, GLA University, Mathura, India
e-mail: ashutosh.shankhdhar@gla.ac.in

Rachit
e-mail: rachit.gla_cs18@gla.ac.in

V. Kumar
e-mail: vinay.kumar_cs18@gla.ac.in

Y. Mathur
e-mail: yash.mathur_cs18@gla.ac.in

V. Suma et al. (eds.), *Inventive Systems and Control*,
Lecture Notes in Networks and Systems 204,
https://doi.org/10.1007/978-981-16-1395-1_28

1 Introduction

The expanding entrance of smartphones, and their detecting and figuring abilities, have prompted their utilization for checking of wellbeing and giving help. The way that the Smartphone keeps the organization of its client through their everyday schedule further opens up the chance of utilizing a Smartphone to distinguish when a common client is in trouble. This study is focused on the detection of screaming, speech, shout, and noise audio mixed with a metropolitan environmental sound in the background. The motivating scenario is the utilization of a smartphone to distinguish vocal indications of misery (shout and cry) while the user experiences their everyday schedule.

The presence of non-speech sounds in compatible sound streams has adverse effects on the performance of speech coding frameworks, speech, and speaker acknowledgment frameworks [1]. In this way, it gets important to recognize and suppress these events in the beginning phases [1]. Apart from a pre-preparing step in speech frameworks, the utilization of shout and scream recognition can be utilized in the zone of auditory surveillance or situation awareness, and when these situations are observed, proper actions will be taken to address these situations.

Sounds generated by humans created through the oral cavity can be ordered into two, speech and non-speech sounds. Non-speech sounds incorporate vocalized sounds like a scream, hiccups, cough, etc. The research centers around the recognition of human scream and shout which majorly comes in the category of non-speech sounds. Broad exploration has been refined in sound sign order like natural sounds, speech sounds, and music, in the past few years. Following, the many research groups in the field of acoustics (weather speech and non-speech acoustics) have the interest in additional identification and arrangement of human-generated non-speech sounds.

Due to their expanding applications in different zones including home services, security, medical services in eldercare, etc. In this study, human-generated shouts and screams are distinguished from noise from two-phase collected datasets. Some audio datasets are collected from different platforms like GitHub, Microsoft audio datasets, and some are collected from the daily routine of males for real-life detection of screams and sounds. For successful identification, this research is planned to work with a three-phase SVM-based classifier followed by Multilayer Perceptron Model.

In SVM-based classification, the MFCC of scream sounds is being extracted in a three-phase as a parameter for the prediction of scream audios. After MFCC extraction sounds go in phase 1 in the SVM model which categorizes scream, speech, and shout audios from noise. In phase 2, speech sounds are distinguished from scream and shout. At last, in phase 3 scream sound is categorized from shout audios. After the successful prediction of the scream in an SVM-based classifier, the sound again goes in the multilayer perceptron model which predicts the sound with the help of a proper combination of the perceptron.

With the help of the training dataset, our model is being trained. A total of 67% dataset is used to train the model and the rest 33% of the total dataset to test the model and these test datasets also include some scream audios which are being recorded by some volunteers with smartphone's mic as they lead their daily life. This is done to predict real-time scream audios. Python is used for this study, as it is one of the famous languages and also used many python libraries like TensorFlow, pandas, etc. to incorporate the data.

2 Related Works

Over the past few years, many researchers show their interest in the examination and detection of non-speech sounds especially shouts and screams because of their vast applications in real life. Prior work centres either around front-end highlights or on acoustic modeling.

Numerous works have examined shout and scream location. In [2] study the researchers distinguish screams and gunshots with the help of a microphone and they concentrate on the abnormal sounds with the help of Time Difference of Arrival. For distinguishing noise from screams and shouts they trained two parallel GMM models. The proposed study is having 93% accuracy with a false rate of 5%. In [3] researchers estimate the identification of shouts, explosions, and gunshots in an environment of a subway. The works [4, 5] considers various event classification where some of the events are scream or potentially a cry. In [4], the researchers proposed a model to classify events like breaking of glass and flowing water and also scream and crying from speech. Although they worked or concentrated only on non-distress data. In [5] the researchers arranged ten audios. They used 10 audios which are like a scream, slamming of a door, normal speech, cry, clapping, laugh, knock, explosion, phone ring, and breaking. For that, they have recorded these audio events in a crowded noisy environment (like canteen, bus, etc.).

Many different works identify non-speech audios that relate to some non-uniform events in a selected environmental setting. In [6], the authors used analytical features followed by an SVM-based classification. In [7], the authors used MFCC, MPEG-7 features of audio sets, and HMMs of audio sets for gunshot and scream event classification. In a study [8] for these sounds, a parallel GMM classifier network is used. This study predicts efficiency of 90% with a false rate of 8%.

Likewise, different methodologies are proposed for shout event detection [9, 10], since shouted speech contains phonetic structure, whereas there is no phonetic structure in a pure scream.

The study in [11] is the most identified with our work. They used two-stage supervised learning for human cry and scream-like events. This study worked on the detection of different scream and cry audios added with environmental contexts and for fluctuated qualities of natural sounds as noise. For detection and identification, scream and cry audios with changing strengths of different noisy sounds in the background. Likewise, this empowers such detection using smartphones, since human beings approach the day-to-day schedule in metropolitan surroundings.

The standard of this study is that an attempt is made to detect scream and shout audios from speech and noise with the help of pure SVM-based classification and pure Multilayer perceptron Model. In a research paper [1], the author is using $T2$-statistics and Bayesian Information Criteria for the detection of screams. In [12], the author is using SVM with Gaussian Mixture to detect the sound. The filtration technique used here in phase 1 of a three-stage SVM-based classifier to detect noise, in phase 2 detect speech, and in phase 3 detect scream from a shout. The SVM-based classifier uses a hierarchical model. As discussed earlier, the real-life scream audios of volunteers are also considered in which our model is predicting accurately. The advantage of the Multilayer Perceptron Model is that it is taking the whole sound that is no use of any type of features of audios for training the model to achieve better accuracy.

3 Data Collection

The given Fig. 1 describes the two phases of data collection. In the first phase, collect the shout, scream, and speech sounds embedded with environmental noises (like factory noises, vehicle noises, etc.). These sounds are collected from the internet and movies and some online platforms like GitHub.

In the second phase of data collection, three volunteers (3 males) recorded distress sounds (scream, shout, etc.) using the Smartphone's microphone as they approached their day-by-day schedules. The volunteers were not given any directions concerning dealing with the phone's mic while it was recording. They used the phone's mic as they did in their day-to-day lives. All audio datasets are encoded at 16 bits per sample.

The recordings are done in a controlled way. The objective is to record sound samples with high constancy. The phone's mic is the only source used for making these recordings. All the time of recording, the phone is placed in the hands of volunteers. A total of three volunteers (only 3 males) collected data during the subsequent stage.

The details of the context audio collection during the subsequent stage are depicted as:

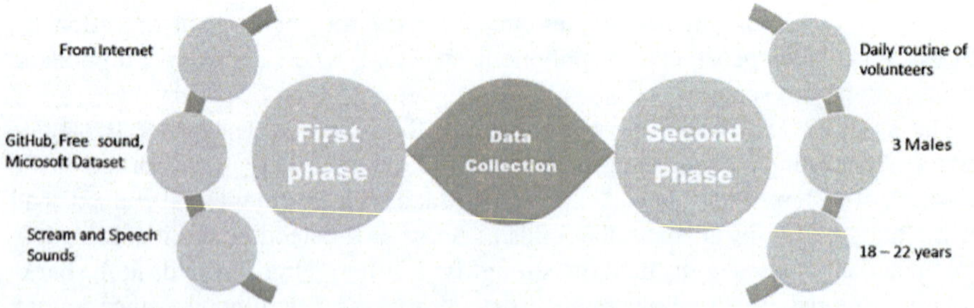

Fig. 1 Different phases of the gathering of datasets from different sources from different volunteers

(a) Gatherings: About 66% of the given sounds are recorded utilizing phones. The rest is acquired from the Internet and this setting remembers sounds for circumstances where numerous individuals are together. A huge number of individuals giggling, a group cheering, eateries, shopping centers, markets, classes, acclaim by a group, and individuals talking noisily in a subway, are a couple of models.
(b) Human conversation: The sound for this setting was gotten from TV serials accessible over the Internet sound with music in the foundation was disposed of and by utilizing phones to record discussions between understudies.
(c) Outdoors: This incorporates among others, audios of vehicular traffic when driving in methods of transport like blaring, railroads, vehicles, public transports, and auto carts.
(d) Indoor Machinery: This incorporates sounds from different machines that are generally utilized at home and office. For instance, shaving machine, forced air system, vacuum cleaner, blender, flushing bowl, microwave oven, clothes washer, fan, and fumes (Fig. 1).

4 Testing and Training Sets

All gathered sounds split into as long as three to second-long audio samples. They scream and shout sounds are split into 3 s. Whereas audio recordings of speech and noise are a little bit lengthier, the normal length of a shout recording in the misery data gathered in the main stage was around 2 s.

Scream, shout, and cry audio sets obtained from the main stage and the subsequent stage are utilized to make the training set and the testing set. These samples in the two sets are totally unrelated. The test set comprises of the apparent multitude of tests acquired from the second phase of data assortment.

5 Working Methodology

Models are trained using appropriate parameters. For maximizing the performance of these models, default parameters may not be sufficient enough. Adjustment of these parameters enhances the reliability of this model which may be regarded as the optimized one for detection of the Human Scream voices. Then the dataset will be processed for training. Once with the training and test data test the supervised machine learning model to detect the human scream and shout.

Here, two models are used for detection. Initially, three-stage supervised learning is used with the help of a support vector machine (SVM) based classifier. Secondly, the Multilayer Perceptron model is utilized.

5.1 Using SVM-Based Classification

SVM's are highly efficient for our model for classification of distress sounds (like scream and shout). The advantage of this is that it is versatile that is many Kernels are allowed for classification. A custom kernel is used to classify accordingly. The disadvantage of SVMs is they do not legitimately provide probability of estimations, these are determined with a costly five-fold cross approval. Here, two types of SVCs are used to train our model SVC with linear kernel $\langle x, x' \rangle$ and SVC with rbf (radial basis function) kernel $\exp\left(-\gamma \|x - x'\|^2\right)$, where γ must be greater than 0.

The audio samples in the training, validation, and testing sets are converted to Mel-frequency cepstral coefficient (MFCC) as feature vectors before they can be classified by the three-phase method.

Feature extraction is carried out in the following manner. For the last 3 decades, Mel-Frequency Cepstral Coefficients (MFCC) is broadly utilized in numerous speech identification frameworks. The major feature used that one audio differs from the second audio is Mel-Frequency Cepstral Coefficients (MFCC). Initially for the MFCC extraction, the continuous speech goes for frame blocking and windowing. Frame Blocking divides the signal into short frames or segments of 20–30 ms. Every frame is then multiplied by the hamming window to retain the permanency of initial and final points in the frame. Then this transforms to Fast Fourier Transform to get the magnitude frequency of each frame. Then it continues to the Mel-scale filter bank and then the Mel-frequency spectrum continues to triangular bandpass filters just to compute the logarithmic function.

To get the final MFCC they pass energy (Ek) through Discrete Cosine Transform (DCT). DCT:

$$Cm = Sk = 1N\cos\left[\frac{(m \times (k - 0.5) \times p)}{N}\right] \times Ek \quad \text{where } m = 1, 2, \ldots, L \qquad (1)$$

N number of triangular bandpass filters and
L number of MFCC's

In the first phase of the model, initially, all the audio sets are loaded and after getting MFCC it goes into a support vector machine that works on the linear kernel for detection of speech from the sound noise is related to humans or not. After the successful determination of speech, it goes to the second phase. In the second phase, MFCC goes in rbf kernel-based support vector machine for detection of speech from all high-pitch distress sounds (like scream or shout). After identification of distress audios, it goes to the third phase. In the third phase, determine the scream from the shout.

In all 3 phases, calculate the SVM by giving MFCC feature array to support vector classifier (SVC) function of SVM using with linear and radial basis function as per need. This SVC function returns the instance of SVM and the predict function is used to predict the instance of SVM (Fig. 2).

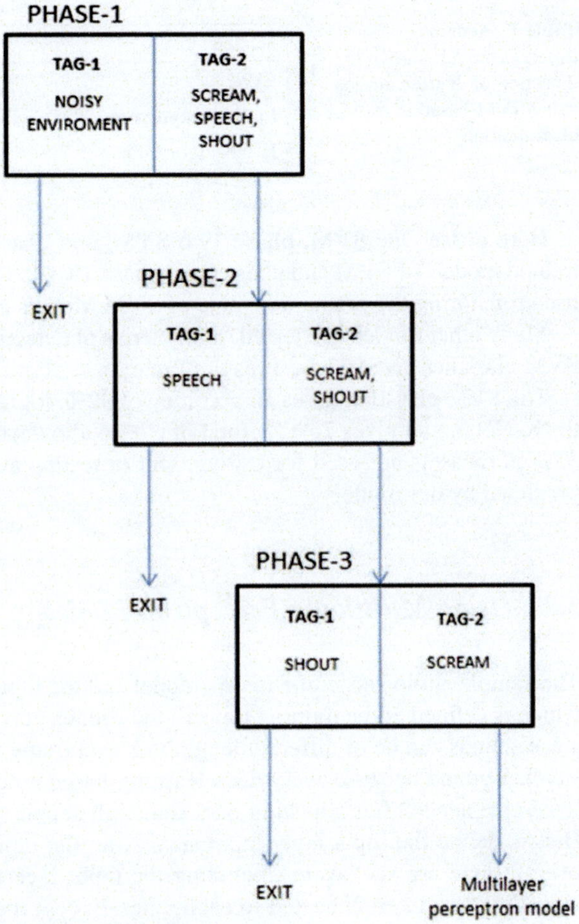

Fig. 2 Detection of noise, speech, shout, and scream audios in different phases using SVM-based classifier

Algorithm 1 Three stage audio classification algorithm

```
phase One=phaseOneSvm.predict(testingAudio ())
if phase One [0] == 2 then
   phase Two=phaseTwoSvm.predict(testingAudio ())
 if phase Two [0] == 2 then
     phase Three=phaseThreeSvm.predict(testingAudio ())
      if phase Three [0] == 2 then
        return Scream
      else
        return Shout
 else
      return Speech
else
    return Noise
```

Table 1 Accuracy of different models shown in detection of distress audios using SVM-based classification

Models	Accuracy (%)
SVM	92.143
Logistic regression	53.234
K-NN algorithm	76.333

Here phase One SVM, phase Two SVM, and phase Three SVM are phasewise trained models of SVM and phase One, phase Two and phase Three are the variables used for storing the prediction value of SVM Models accordingly.

Many other models are tested in this series of detection of distress sounds, but the SVM classifier gives the best result (Table 1).

The SVM classifier gives an accuracy of 92%, logistic regression gives 53% and the K-NN model gives 76%. A total of 67% audio datasets are used for training and 33% of datasets are used for testing. Out of testing audios, 92% of audio is being predicted by our model.

5.2 *Using Multilayer Perceptron Model*

The sample audio sets come in this model and the input size is fixed for this model which is defined at the training time all the sounds gets loaded in the data frame and since sounds can be of different length this data frame gets arranged concerning the smallest sound because every deep learning-based model needs input layer to be of fixed size hence after adjusting data frame all sounds will be of same dimensions. Hence, define the input layer input size. After the input layer has five more layers overall there are six layers (including the input layer). All these layers are dense layers that are interconnected to each other. Except for the output layer, each layer has the activation function 'ReLU' because this is a widely used activation function.

In the last layer, an activation function 'sigmoid' is used because it exists between (0 and 1). Therefore, it is used for those models where it is required to predict whether an event will occur or not. Since the probability of everything exists between 0 and 1 hence sigmoid function was the better choice. The Sigmoid Function curve looks like an S-shape. In the first layer, consider 12 perceptrons, in the second layer consider 8 perceptrons, in the third layer consider 10 perceptrons, in the fourth layer consider 5 perceptrons, in the fifth layer consider 3 perceptrons, and in the last layer consider 1 perceptron. This type of configuration is used in this model because only this configuration is providing the best learning rate and accuracy till now. The current accuracy of this model is 89.333% with a 5% false rate.

This model requires three directories one is negative, one is positive, one for testing: negative directory contains sounds like people normal conversation, birds sounds, car sounds, and so on, the positive directory contains sounds like screams of men, children, women, etc. while testing sound contains those sounds which evaluates the model performance. A Cross-entropy is used as a loss function here because it is the loss function to use for binary classification problems. Cross-entropy can be specified as the loss function. The binary cross-entropy loss function is used with binary classification problems since our multilayer perceptron model is predicting whether the passing sound is scream or not so that's it can be used this loss function. While using this loss function it is needed to use the sigmoid activation function in the last layer also the last layer should have a single neuron. This loss function predicts output in set {0,1}. 'Adam' is the optimizer are used here.

Optimized algorithm Adam that can be utilized rather than the classical stochastic gradient descent technique to refresh network weights iterative based on training data. When the model gets any sound then it tries to identify patterns in it using perceptron weights and biases if it can't get success it slightly changes its weights for getting the correct result if it successfully detects a scream in sound then it calls the emergency function and if it couldn't detect sound as scream then it simply terminates the program. Figure 3 shows the Flowchart for this model.

For accuracy calculation, load all the datasets and all rows of the dataset are shuffled and used 400 audio datasets for audio files. All the types of sounds that are a scream, shout, speech, and noise is used for 33% of the dataset as test data and rest of other are using for training purpose. Using the evaluate method, the python library is used to calculate the accuracy of trained files which results in 93%, and later the same method is used for calculating the accuracy of test data which results in 82% (Fig. 4).

Fig. 3 Approach using in multilayer perceptron model

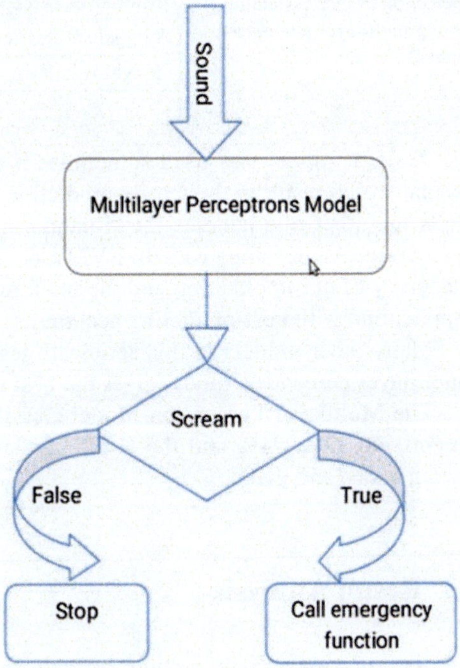

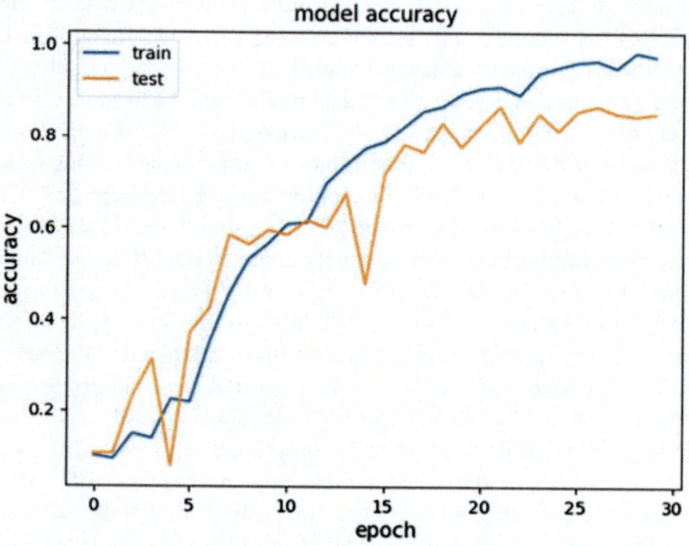

Fig. 4 Accuracy of testing data in multilayer perceptron model of different epochs

Table 2 Accuracy of different models shown in the detection of distress audios using multilayer perceptron model

Models	Accuracy (%)
Multilayer perceptron model	82.333
Logistic regression	71.011
K-NN algorithm	80.539

A single model was used sometimes it was failing to tackle particular types of sound which result in the wrong prediction. Two models were applied so if anyone fails the other one handles the situation and generates correct output. Multilayer Perceptron Model along with the SVM model is used to increase the overall predicting accuracy of our application and the main motive of our work was to build a strong application with best-predicting accuracy.

Many other models in this series are tested in the detection of distress sounds, multilayer perceptron model gives the best result (Table 2).

The Multilayer Perceptron Model classifier gives the accuracy of 82%, logistic regression gives 71% and the K-NN model gives 80.5%. The current accuracy is being tested and given.

6 Result Analysis

In the proposed system, a study of the three-stage learning-based model followed by multilayer perceptron model for human scream detection in various conditions including human conversation as noise, outdoor and indoor machinery noise, human gatherings, etc. The tuning of the dataset was done with various types of audio samples which were collected by two-phase process of collection of datasets. The efficiency of the proposed model is tested with many audio sets which are recorded by the volunteers via smartphone's mic as they go with their daily routine life. The combined accuracy of the model is 90%. Both K-NN and logistic regression models are not learning well for scream detection they initially started learning per epochs but after a fixed time, their accuracy becomes constant per epoch while in the case of multilayer perceptron model learns well throughout the learning process to make it effective use this model.

7 Conclusion

After this research, it can be concluded that while dealing with such type of scenarios (where need is to build a strong system for detecting crimes using audio) it is possible to use both SVM-based classification that is supervised machine learning model, and neurons (perceptron) based deep learning model together for getting best results. Also using both models can decrease the chances of the wrong prediction. For this study, it can be implemented this whole work using famous python libraries Librosa for MFCC extraction, Keras, etc.

8 Future Work

The study of the proposed model is done at this level and these are some possible things that can be done in the future to increase the accuracy of the model. The current model gives an accuracy of 90% and can able to increase the model's accuracy to around 92%. Currently, the developed system for desktop to detect screams like voices. In the future, it can also be extended this scream detector to mobile world and also to increase the accuracy of negative/abusive words detection so that it can prevent some incidents (like kidnapping, killing threat, etc.) which is not supposed to have happened. Some of the extra characteristics and algorithms can be added so that it predicts scream 24*7.

References

1. M.K. Nandwana, A. Ziaei, J.H. Hansen, Robust unsupervised detection of human screams in noisy acoustic environments, in *2015 IEEE International Conference on Acoustics, Speech and Signal Processing (ICASSP)* (IEEE, 2015), pp. 161–165

2. G. Valenzise, L. Gerosa, M. Tagliasacchi, F. Antonacci, A. Sarti, Scream and gunshot detection and localization for audio-surveillance systems, in *2007 IEEE Conference on Advanced Video and Signal Based Surveillance* (IEEE, 2007), pp. 21–26
3. S. Ntalampiras, I. Potamitis, N. Fakotakis, On acoustic surveillance of hazardous situations, in *IEEE International Conference on Acoustics, Speech and Signal Processing*, pp. 165–168 (2009)
4. M.A. Sehili, B. Lecouteux, M. Vacher, F. Portet, D. Istrate, B. Dorizzi, J. Boudy, Sound environment analysis in smart home, in *International Joint Conference on Ambient Intelligence* (Springer, Berlin, Heidelberg, 2012), pp. 208–223
5. H.D. Tran, H. Li, Sound event recognition with probabilistic distance SVMs. IEEE Trans. Audio Speech Lang. Process. **19**(6), 1556–1568 (2010)
6. W. Huang, T.K. Chiew, H. Li, T.S. Kok, J. Biswas, Scream detection for home applications, in *2010 5th IEEE Conference on Industrial Electronics and Applications* (IEEE, 2010), pp. 2115–2120
7. S. Ntalampiras, I. Potamitis, N. Fakotakis, On acoustic surveillance of hazardous situations, in *2009 IEEE International Conference on Acoustics, Speech and Signal Processing* (IEEE, 2009), pp. 165–168
8. L. Gerosa, G. Valenzise, M. Tagliasacchi, F. Antonacci, A. Sarti, Scream and gunshot detection in noisy environments, in *2007 15th European Signal Processing Conference* (IEEE, 2007), pp. 1216–1220
9. J. Pohjalainen, P. Alku, T. Kinnunen, Shout detection in noise, in *2011 IEEE International Conference on Acoustics, Speech and Signal Processing (ICASSP)* (IEEE, 2011), pp. 4968–4971
10. .K. Mittal, B. Yegnanarayana, Production features for detection of shouted speech, in *2013 IEEE 10th Consumer Communications and Networking Conference (CCNC)* (IEEE, 2013), pp. 106–111
11. A. Sharma, S. Kaul, Two-stage supervised learning-based method to detect screams and cries in urban environments. IEEE/ACM Trans. Audio Speech Lang. Process. **24**(2), 290–299 (2015)
12. S. Chung, Y. Chung, Scream sound detection based on SVM and GMM, in *RTET-17*, 2017

A Study on Human-like Driving Decision-Making Mechanism in Autonomous Vehicles Under Various Road Scenarios

G. Anjana and Rajesh Kannan Megalingam

Abstract Decision-Making technique in autonomous vehicles is one of the promising technologies for driving safety. While considering different types of maneuver, wrong driving decision-making may result in severe accidents. This is because of the difficulty of precisely finding out some possible decision-making facts considered by the driver. Thus, in autonomous vehicles, an efficient decision-making mechanism like that of a driver in decision-making is a crucial factor. The novelty of automated driving with human-like driving behavior improves mobility substantially. It offers many significant social benefits such as enhancing the mobility of those who lack it and radically improving safety and saving lives. This study mainly focuses on the various existing driving decision-making mechanisms used in autonomous vehicles for smooth and efficient path planning.

Keywords Driving behavior · Maneuver planning · Rule-based · Learning-based

1 Introduction

A self-driving car has to deal with various road scenarios such as lane changing, overtaking, merging roads, mixed traffic, emergency braking, uncontrolled intersections, etc. In such different scenarios, autonomous vehicles should have the ability to perform human-like decisions making with correctness and social intelligence but predictions and understandings are also mandatory. The decision-making process can broadly be classified as rule-based and learning-based. The rule-based approaches are deterministic and learning-based approaches are probabilistic. Figure 1 shows the general decision-making process of autonomous vehicles in various road scenarios.

From Fig. 1, it is clear that general driving mode decisions of autonomous vehicles are check whether is it possible to change the lane physically and safely, is it necessary to change the lane when a junction or merging road approaches nearly, and finally is it necessary to change lane when an obstacle is detected or overtake the front

G. Anjana (✉) · R. K. Megalingam
Department of Electronics and Communication, Amrita Vishwa Vidyapeetham, Amritapuri, India

V. Suma et al. (eds.), *Inventive Systems and Control*,
Lecture Notes in Networks and Systems 204,
https://doi.org/10.1007/978-981-16-1395-1_29

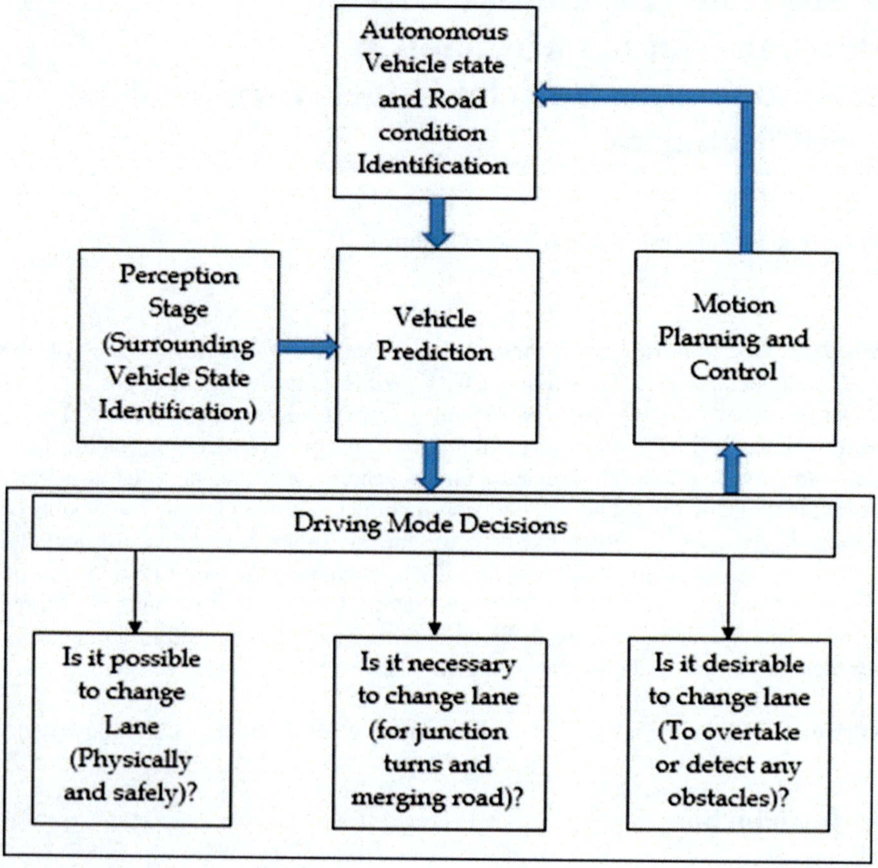

Fig. 1 General decision-making process

vehicle in case of deceleration of the front vehicle. In every autonomous vehicle, the motion planning and control were done based on these driving mode decisions. These decisions were taken by considering the autonomous vehicle state, road conditions, and the state of surrounding vehicles [1]. The purpose of this article is to discuss various existing deterministic and probabilistic driving decision-making mechanisms used in autonomous vehicles to deal with these scenarios. This article deals with the Finite State Machine approach, Markov Decision Process, Reinforcement learning, and Support Vector machine approaches.

The remainder of the article is organized as follows. The related work is presented in Sect. 2. Section 3, depict a discussion of various techniques and scenarios of decision-making. Based on the discussion, a comparison of existing methodologies and future directions is included in Sect. 4. The conclusion is represented in Sect. 5.

2 Related Works

In the DARPA Urban Challenge, many solutions for behavioral planning were proposed but they were specifically to meet the needs of the challenge. Most approaches use rule-based approaches to switch between predefined behaviors. Automated driving with human-like driving behavior [2] requires cooperative and intuitive decision-making. Game-theoretic approaches [3] is a typical method by maximizing a reward function to identify the behavior of other vehicles. It was developed as a simple mathematical game in which the vehicles are considered as players and they take appropriate decisions based on the outcome of the game. Depending on the number of players various strategies were devised and based on the outcome of the utility or the payoff function favorable decisions were taken. Game-theoretic approaches are one of the most common types of decision-maker employed in autonomous vehicles especially in lane changing [4]. But its outcome may vary concerning the play they formulated. So many difficulties are there with this approach. In probabilistic approaches the vehicle's motion behaviors are predicted to follow the maximum posterior or maximum likelihood instead of optimizing for the lowest cost reward function. In the deterministic approach, searching over a tree type decision-making process is used.

Moreover, to really explore securely autonomous vehicles should consider potential obstacles [5]. Despite the apparent complexity of the task, human drivers are extremely good at taking correct decisions for the motion of the vehicle. The lack of this ability to take appropriate decisions depending on the complexity of tasks is the major challenge holding back the large-scale deployment of truly autonomous vehicles on the roads. Many types of research are performing to generate appropriate driving decisions for autonomous vehicles in changing road scenarios [6]. The review and analysis of publications on autonomous vehicle decision-making in various road scenarios are presented in Fig. 2.

Table 1 shows the comparative details of related works that review the driving decision-making mechanism of autonomous vehicles.

This article deals with some most relevant approaches in driving decision-making and the future directions in this field of research.

3 Existing Methodologies

The decision-making mechanisms for autonomous vehicles explain a sequence of driving maneuvers [15] to achieve a goal destination with a smooth and efficient trajectory. The general decision-making mechanisms used in autonomous vehicles dealing in this paper are as shown in Fig. 3.

One of the decision-making algorithms of autonomous vehicles is knowledge-based inferences or rules-based reasoning algorithms (Rational & Rule). The rule-based decision-making approach based on finite state machines is basically used to

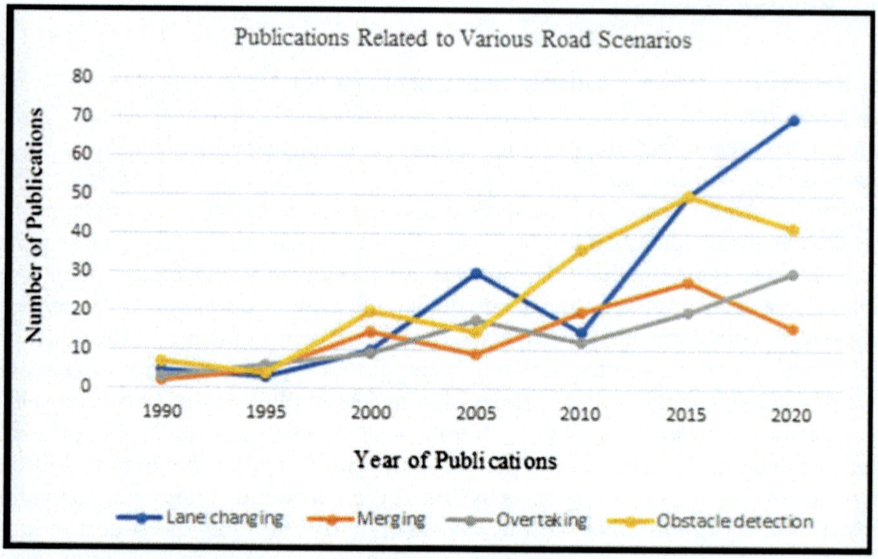

Fig. 2 Analysis of related works on various road scenarios

Table 1 Comparison of related works

Related works	Proposed year	Method	Application
Bindhu [7]	2020	Programmed artificial intelligence	Emergency braking
Liu et al. [8]	2019	Support vector machine-based	Lane changing
Ji and Levinson [3]	2018	Game theory based model	Lane changing
Li et al. [6]	2018	Empirical decision-making based on neural networks	Human-like driving
Camara et al. [9]	2018	Empirical game theory for decision-making	Pedestrian interaction
Claussmann et al. [10]	2017	Artificial intelligence based approaches	Highway autonomous driving
Kang and Rakha [11]	2017	Game theoretical approach	Merging manuevers
Song et al. [12]	2016	Intention aware approach based on partially observed markov decision	Uncontrolled intersection
Gong et al. [13]	2016	Hierarchical state machine and RBF neural network-based model	Overtaking
Katrakazas et al. [2]	2015	Real-time motion planning methods	Autonomous on-road driving
Moridpour et al. [14]	2010	Model based on search algorithm and traffic characteristics	Lane changing

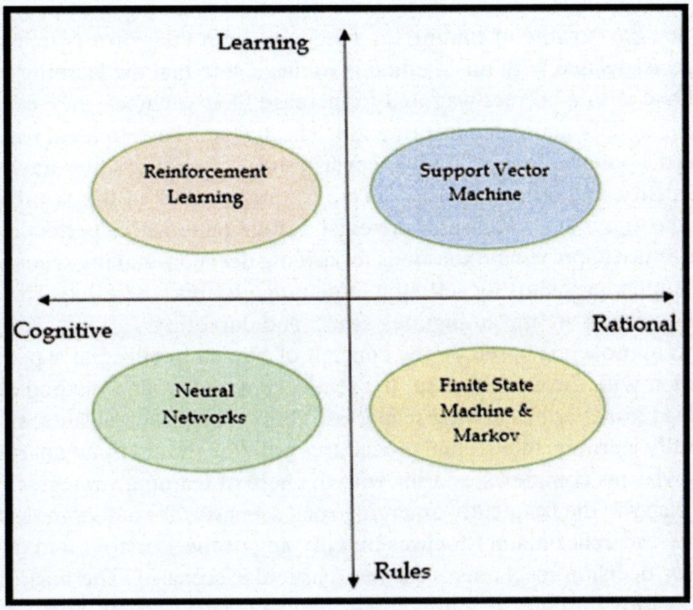

Fig. 3 Decision-making techniques

deal with different urban traffic road scenarios. The behavior of the vehicle such as lane changing, lane keeping, obstacle detection, etc. is represented by states and transitions. Finite state machines that are capable of handling several scenarios in an urban environment are more exploring in the field of autonomous vehicle urban challenges. The Finite state machine approaches are based on discrete rules that arise from the perception information. The FSM contains the goal that an autonomous vehicle should perform and the behavior of the ego vehicle along with surrounding information. The shortcomings are the procurement and retainment of the decisions, numerous environment parameter discretization, and the inadequacy of decision margins. The finite state machine approaches uses Markov properties to extend the Bayesian networks of uncertainty. In such cases, only the sequence of a state transition is only known and the decisions are generated based on the probabilistic concepts of states and transitions.

The second approach is the Heuristic algorithms (Rational & Learning) finding the limited or complete solutions to an optimization problem. Low computational time, ability to handle complex problems makes these algorithms faster and efficient. But they do not guarantee overall optimality and perfect accuracy. Support Vector Machine is one of the suitable interpretations of the heuristic method and it deals with agent representation.

The third approach is the boolean compromises based on approximate reasoning methods (Cognitive & Rules). One of the most suitable examples of this approach is artificial neural networks portrayed by their learning potential. Without any external support, they are capable of finding the outcomes with their own policies. But the major issue associated with this method is to make sure that the learning capability of the method is in a correct way and to increase their accuracy they need a large amount of manually generated information. The systems have to learn the behavior concerning the policies applied in each scenario and it may cause some drawbacks due to road diversity, and infrastructure concerns. Another class of this approach relies on fuzzy-like logic. In such cases there exist a truth value with a perfect dimension which will return approximate solutions for driving decision-making scenarios. They manage complex problems by setting a large range to this true value. The pros and cons of this concept are the ambiguous aspect and durability.

The next approach is based on the concept of human intellectual aspects. Based on interaction with the environment, the reinforcement learning method allows the ego vehicle to learn various driving scenarios. These reinforcement learning methods are potentially learning, intellectual procedures utilizing the environment rules. They can handle various complex scenarios with the help of learning strategies.

Figure 4 shows the frequently emerging road scenarios for autonomous vehicles. Most of the researches mainly focuses on only any of the scenarios and devises the algorithm or decision mechanism for that particular scenario. The basic decision-making scenarios [16] that an autonomous vehicle basically deals with are:

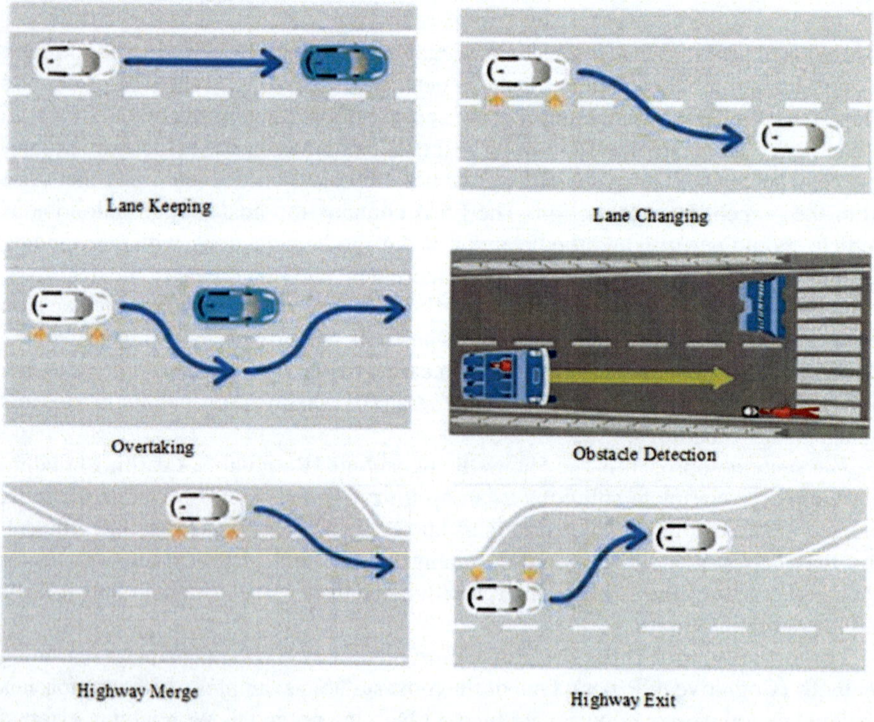

Fig. 4 Various road scenarios

Lane following: It is a vehicle maneuver in which the ego vehicle maintains the desired speed and a safe distance is maintained between the rear and preceding vehicles. Then the ego vehicle decides to move straight ahead to maintain its speed, and sidewise to keep its trajectory inside the actual lane marks.

Lane changing: It is a scenario of moving the vehicles to either forward and backward or on adjacent lanes by considering the road conditions. In such situations, the ego vehicle ensures the gap on the target or main lane is enough for a safe lane change and the speed of the ego vehicle is adequate for maintaining a safe movement.

Overtaking: It is one of the most complex maneuvers the ego vehicle can take. It can be considered as a five-stage process that is the decision of overtaking, preparing for overtaking, changing the current lane, passing the vehicle in front, and returning to the original lane.

Merging: It is a vehicle maneuver responsible for a lane change in which the ego vehicle is entering or leaving a highway with an undesirable speed interval on a limited longitudinal distance.

Obstacle Avoidance and Emergency Braking: Emergency braking is one of the safety enabling technology in autonomous vehicles by detecting immediate obstacles [17]. Ego vehicle detects the possible obstacles or forward collisions with the rear environment and activates the vehicle braking mechanism to reduce the acceleration for mitigating the dangerous situation.

4 Analysis

Driving Decision-making mechanism of autonomous vehicles has relevant impacts on driving performance. The driving condition of the autonomous vehicle is random and time-varying in different real-time road scenarios. The decision-making system of autonomous driving vehicles could receive environmental information and output driving behavior according to its own decision criteria. This article focuses on the analysis of different decision-making mechanisms of autonomous vehicles under various road scenarios as shown in Fig. 5.

The decision-making mechanisms process this work using the Reinforcement learning-based approaches, Support vector machine-based approaches, Neural Network-based approaches, Finite state machine, and Markov decision process. These frequently used approaches can fall into any one of the categories shown in Table 2. This table also represents the overall performance of the decision-making mechanisms under various road scenarios in autonomous vehicles.

From Table 2, all the decision-making mechanisms respond in different manners to various scenarios. Some mechanisms are very suitable for Lane changing but are unsuitable for merging scenarios. Most of the existing mechanisms concern more only with lane changing. But other scenarios such as merging, overtaking, obstacle

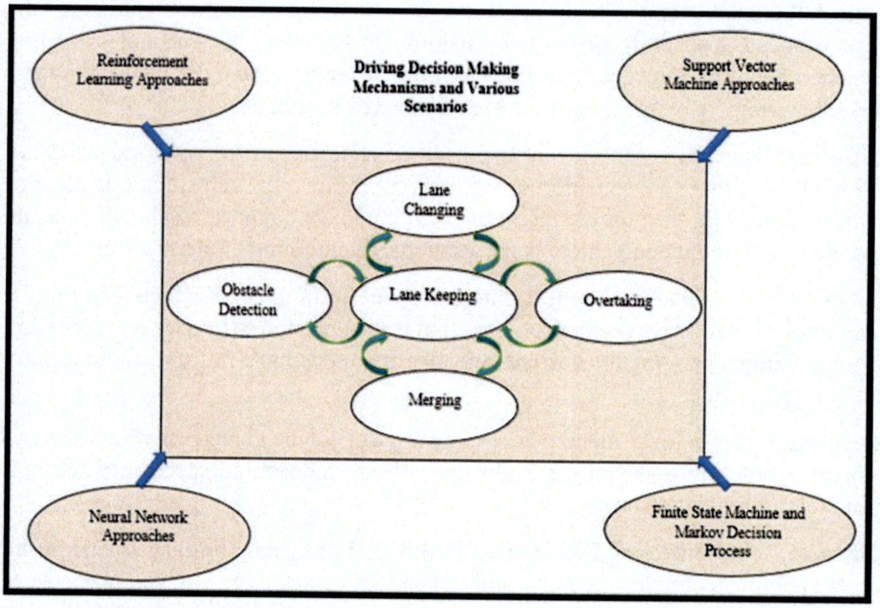

Fig. 5 Driving decision-making mechanisms and various scenarios

Table 2 Performance evaluation

Decision-making methods		Performance in dealing with various scenario				Overall result
		Lane changing	Merging road	Overtaking	Obstacle detection	
Knowledge-based	FSM	Moderate	Bad	Bad	Moderate	Moderate
	Markov	Good	Bad	Bad	Moderate	
Heuristic-based	Agent	Moderate	Bad	Bad	Moderate	Moderate
	SVM	Moderate	Bad	Bad	Moderate	
Approximate reasoning-based	Fuzzy	Moderate	Moderate	Moderate	Good	Good
	Neural Networks	Good	Moderate	Good	Good	
Human-like approaches	Risk based	Good	Moderate	Moderate	Good	Good
	RL	Good	Good	Good	Good	

detections, etc. also have equal importance like that of lane changing. A human-driven vehicle considers all these facts when each situation arrives but in the case of the autonomous vehicle, it needs to devise suitable decisions based on the situation by appropriate mechanisms. Reinforcement learning-based approaches are one of the most promising technology considers not all but many of the scenarios.

The lane change driving decision-making process is a non-linear and multi-parametric problem, so to solve this issue and generate a lane changing decision based on safety matters, tolerance and benefits support vector machine technique with Bayesian parameters optimization is suited [8]. In such a work firstly a lane change model is developed and then suggested some related parameters which impact the lane change process by analyzing the safety, tolerance factors, etc. Then a Bayesian optimization algorithm-based support vector machine model is generated.

A deep reinforcement learning-based approach [18] can be used for safe and high-level decision-making of lane changing by reducing the state space dimensionality using a minimal representation of surrounding information. This approach firstly trained the deep reinforcement learning agent to drive with a speed that is nearly the same as the desired velocity. Then allows the agent to learn the driving tasks by using a deep Q-network without considering or knowing the vehicle dynamics. The utilization of Deep Q-Network helps to achieve a fast learning rate.

The most common approach of decision-making in autonomous vehicles are formulated as a Markov decision process and finite state machine. A probability-based decision-making method [19] can be applied in a real road scenario given safety concerns, efficiency, and comfort. The autonomous vehicles use the Markov decision process by defining the states space, agent space, and basic transition- reward models. It is based on a prediction model of rear and preceding vehicles. Then deduced an optimal policy automatically using the value iteration method. The finite state machine models the environment and obtains policies based on ethical rules. A convolutional neural network [20] approach is designed for end-to-end learning of autonomous driving and then the developed model has been trained with the collected dataset. At last, the trained models have been used for the testing of real-time autonomous driving. The deep neural network is one of the possible solutions to achieve successful autonomous driving decision-making. Instead of human-based patterns the end-to-end learning approaches simultaneously update all processing steps to get good performance.

For safe and efficient driving of autonomous vehicles in merging lanes, they should be aware of their surroundings and need to check whether the vehicle is on the main lane or in the merge lane. Depending on these parameters a motive-aware decision-making with some flexible approaches [21] which is built upon the multiagent reinforcement learning can be used.

Under the multiagent reinforcement learning [21] a decision-making policy is learned by the curriculum learning method. Such a method enables the agent to automatically identify the behaviors and make decisions strategically. To prevent the agent from traverse from states that violate the common sense of human judgment and increase the learning efficiency a masking mechanism is used.

For handling the merging scenarios and to deal with various environment parameters a multiagent reinforcement algorithm is used. In such learning mechanisms during the training phase, each agent will interact with various surrounding agents. With those interactions, they will be capable of developing the ability to learn or understand how to select proper decisions during those scenarios.

In an uncontrolled intersection scenario, a motive acquainted decision-making algorithm based on the hidden Markov model [12] is best suited. In such situations initially predict the motion and the interaction intentions by developing a continuous

Hidden Markov model then built a partially observable Markov decision process. Generated the possible candidates by human-like policy generation mechanism. Then these candidates can be applied in the state transition process. Later, during each prediction time step, the intention is get updated. To calculate the optimal policy for decision-making a reward function is designed.

To achieve safety performance without a huge computation in the decision-making for overtaking, a probabilistic prediction [10] based approach can be used. By intuitive motion, planning desired states and constraints are generated thereby the desired driving mode is designed and target space based on the opportunity, necessity, and possibility of lane change is determined. This approach reveals the effectiveness of virtual targets in autonomous overtaking. The main drawbacks of most of these approaches are their limitations to meet the real environment constraints. When focusing more on one scenario with the real environment, the mechanism failed in other scenarios. Almost all the existing human-like driving decision-making mechanisms of autonomous vehicles will focus to meet the real-time constraints by devising decisions to one or two scenarios. Figure 6 shows the performance analysis of driving decision-making mechanisms in various road scenarios.

From Fig. 6, it is clear that approximate reasoning and human-like decision-making mechanisms have a good overall performance in decision-making for autonomous vehicles. The future directions in terms of decision-making in autonomous vehicles focus on the human-like approaches. This research will be extended to devise a single mechanism-based human-like approach that deals with all the road scenarios that an autonomous vehicle needs to be focused.

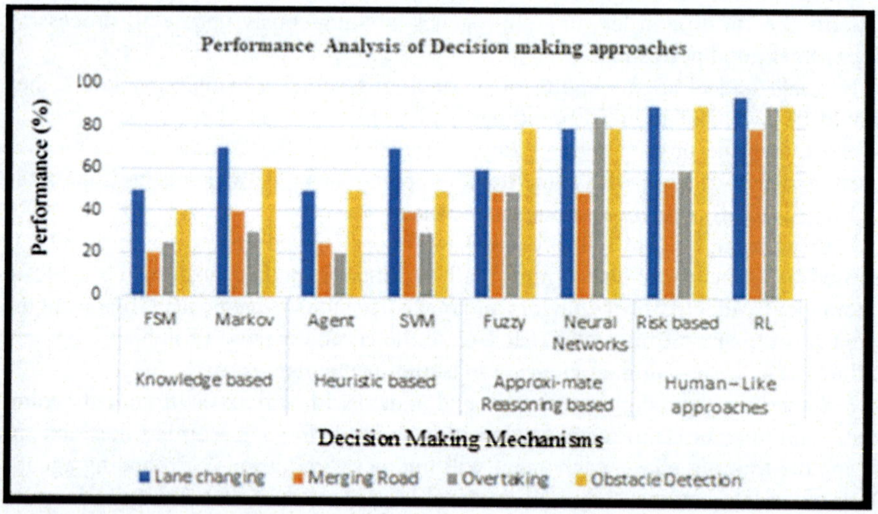

Fig. 6 Performance analysis of decision-making mechanisms for various scenarios

5 Conclusion

The decision-making mechanism of autonomous vehicles that exists with a wide range of scope was discussed. To extend the research in this field, a focus on behavioral, tactical, and prediction modeling is necessary. The existing behavioral modeling approach requires the features that explain the behavior of the model which is used. But in many cases, this might not be straight forward. So, the techniques may resort to the idea of automatically learn the relevant features. In tactical planning constraints [22], the existing approaches result in extremely unnatural and conservative actions. To achieve the safety guarantee in autonomous driving, the tactical planning resort to ideas for maintaining human-inspired non-disruptive, and predictable behavior. While considering the future directions in prediction modeling, it is necessary to adequately assign a particular behavior model based on anomaly detection. If each model is adequately assigned then it will improve the quality of prediction modeling.

This article surveyed the literature on different driving decision-making mechanisms based on different scenarios of autonomous vehicles. From the analysis of each mechanism, has its own merits in dealing with a particular situation. The decision-making is an important prerequisite in the current context because in the future autonomous vehicles occupy a relevant position. Autonomous vehicles capture the environment parameters and combine human decision-making logic. Later, structures the driving behavior of autonomous vehicles. As technology advances, many algorithms are developed in the field of autonomous vehicles that combined up erratic circumstances conveniently.

References

1. S.R. Mugunthan, Concept of Li-Fi on smart communication between vehicles and traffic signals. J. Ubiquit. Comput. Commun. Technol. **02**(02), 59–69 (2020)
2. C. Katrakazas, M. Quddus, W.H. Chen, L. Deka, Real-time motion planning methods for autonomous on-road driving: state-of-the-art and future research directions. Transp. Res. C **60**, 416–442 (2015)
3. A. Ji, D. Levinson, A review of game theory models of lane changing. Transportmetrica A Transp. Sci., no. just-accepted, 1–19 (2020)
4. H. Yu, H. Eric Tseng, R. Langari, A human-like game theory-based controller for automatic lane changing. Transp. Res. Part C Emerg. Technol. **88**, 140–158 (2018)
5. R.K. Megalingam, J.M. Varghese, S.A. Anil, Distance estimation and direction finding using I2C protocol for an auto-navigation platform, in *International Conference on VLSI Systems, Architectures, Technology and Applications (VLSI-SATA)*, pp. 10–12 (2016)
6. L. Li, K. Ota, M. Dong, HumanLike driving: empirical decision-making system for autonomous vehicles. IEEE Trans. Veh. Technol. **67**(8), 6814–6823 (2018)
7. V. Bindhu, An enhanced safety system for auto mode E-vehicles through mind wave feedback. J. Inf. Technol. **2**(03), 144–150 (2020)
8. Y. Liu, X. Wang, L. Li, S. Cheng, Z. Chen, *A Novel Lane Change Decision-Making Model of Autonomous Vehicle Based on Support Vector Machine* (IEEE Access, 2019), pp. 26543–26550

9. F. Camara, R. Romano, G. Markkula, R. Madigan, N. Merat, C.W. Fox, Empirical game theory of pedestrian interaction for autonomous vehicles. *Measuring Behavior*, pp. 238–244, ISBN 978-1-910029-39-8 (2018)
10. L. Claussmann, M. Revilloud, S. Glaser, D. Gruyer, A study on al-based approaches for high-level decision making in highway autonomous driving, in *Proceedings of IEEE International* Conference on Systems, Man, and Cybernetics *(SMC)*, pp. 3671–3676 (2017)
11. K. Kang, H.A. Rakha, Game theoretical approach to model decision making for merging maneuvers at freeway on-ramps. Transp. Res. Record J. Transp. Res. Board **2623**, 19–28 (2017)
12. W. Song, G. Xiong, H. Chen, Intention-aware autonomous driving decision-making in an uncontrolled intersection. Math. Probl. Eng. **2016**, 1–15 (2016)
13. J. Gong, Y. Xu, C. Lu et al., Decision-making model of overtaking behavior for automated driving on freeways, in *IEEE International Conference on Vehicular Electronics and Safety* (IEEE, 2016)
14. S. Moridpour, M. Sarvi, G. Rose, Lane changing models: a critical review. Transp. Lett. Int. J. Transp. Res. **2**(3), 157–173 (2010)
15. W. Schwarting, J. Alonso-Mora, D. Rus, Planning and decision-making for autonomous vehicles. Annu. Rev. Control Robot. Auto. Syst. **1**(1), 187–210 (2018)
16. R. Resmi, P. Shruthi, Path planning for autonomous car, in *2nd International Conference on Intelligent Computing, Instrumentation and Control Technologies, ICICICT*, 2019
17. N. Deepika, V.V. Sajith Variyar, Obstacle classification and detection for vision based navigation for autonomous driving, in *2017 International Conference on Advances in Computing, Communications and Informatics, ICACCI*, 2017
18. P. Wang, C. Chan, A. de La Fortelle, A reinforcement learning based approach for automated lane change maneuvers, in *IEEE Intelligent Vehicles Symposium (IV)*, 2018
19. Y. Guan, S.E. Li, J. Duan, W. Wang, B. Cheng, Markov probabilistic decision making of self-driving cars in highway with random traffic flow: a simulation study. J. Intell. Connected Veh. **1**(2), 77–84 (2018)
20. J. Kocić, N. Jovičić, V. Drndarević, An end-to-end deep neural network for autonomous driving designed for embedded automotive platforms. Sensors **19**, 2064 (2019)
21. Y. Hu, A. Nakhaei, M. Tomizuka, K. Fujimura, Interaction-aware decision making with adaptive strategies under merging scenarios. *IEEE/RSJ International Conference on Intelligent Robots and Systems* (IROS) (2019)
22. R.K. Megalingam, D. Nagalla, V. Gontu, R.K. Pasumarthi, A.P. Kumar, ROS based simulation and control of wheeled robot using gamer's steering wheel, in *International Conference on Computing Communication and Automation (ICCCA)*, 2019

Recommendations for Student Performance Improvement Based on Result Data Using Educational Data Mining

Ketan D. Patel and Amit B. Suthar

Abstract Student Result is one of the key parameter to be considered when the performance of any educational institute is majored. Hence educational institutes have always shown a keen interest to analyze and improve their student performance. It helps to reduce the student dropout ratio as well as increase the ranking and reputation of the institute. The availability of a large volume of data in diverse format restricts educational institutes from effective analysis. Educational Data Mining helps institutes to analyze the large student data to extract meaningful and hidden learning patterns of the students. In this article, a data mining technique is applied to the computer science students of the first year to discover the relationship between the pair of subjects in which the majority of students get difficulty to acquire passing grades. Based on the results of this study, the institute can redesign the syllabus to reduce student's burden by offering such subjects in different semesters. Providing extra teaching efforts in those subjects also helps to improve the overall performance of the student.

Keywords Educational Data mining · Association mining · Student performance analysis · Data mining

1 Introduction

In the present scenario, the result of the students is considered as one of the important parameters while majoring in the performance of the institute [1]. Institutes having good student result records and placement statistics are of the first choice for the admission seekers. Thus educational institutes are always interested to improve the performance of the students. Institutes need to analyze the performance data of

K. D. Patel (✉)
Department of Computer Science, Ganpat University, Kherva, Gujarat, India
e-mail: ketan.patel@ganpatuniversity.ac.in

A. B. Suthar
AMPICS, Ganpat University, Kherva, Gujarat, India

V. Suma et al. (eds.), *Inventive Systems and Control*,
Lecture Notes in Networks and Systems 204,
https://doi.org/10.1007/978-981-16-1395-1_30

their students to extract meaningful information that is hidden in nature. This is a challenging task due to the large volume of educational data available in institutes [1]. Data mining provides many ways that could be used to analyze and predict the performance of the student [2]. Data mining is the process of discovering interesting and useful patterns and knowledge from large amounts of the dataset [3]. It is a technology that merges conventional data analysis methods with sophisticated mining algorithms to process data available in large volume [4].

Educational Data Mining is also known as EDM is a new area in the field of data mining and Knowledge Discovery in Databases (KDD) field which mainly focuses on mining interesting and useful patterns and knowledge from the educational database [5, 6]. It is an emerging field that explores data in an educational context by applying various data mining tools and techniques [7].

This article focuses on extracting hidden patterns from student's result data set. The objective is to find the association between a pair of subjects on which the majority of the student frequently fails together. For the study, five years (2015–2019) result data set of first-year computer science students are taken to find any association between the subjects. The finding helps the institute to redesign their curriculum in a manner that helps to improve student performance and student learning burden can be reduced.

To find the association between the pair of subjects, association rule mining techniques of data mining is used in this research. Apriori algorithm of association mining are used to find an association between subjects in which students are facing difficulty to secure passing grades.

2 Association Rule Mining

Association mining is expressed in the form of A & B where A and B represent data items. The rule is interpreted as the presence of A involves the presence of a B item set. Two important measures that are used to find association rules are support and confidence. They are calculated based on the following formulas: [8].

Support = Number of transactions containing both A and B items/Total number of transactions in the database

Confidence = Number of transactions containing both A and B items/Number of transactions containing item A

Support is calculated by dividing a total number of transactions containing both A and B data items by a total number of transactions available in the data set. It is a ratio of a number of transactions which includes both A and B items with the total number of transactions in the database [9, 10]. For example 80% support value indicates that 80 times both items A and B were accessed together out of total of 100 times. Support value is an important to measure to find out the association between the pair of items.

Another important measure in association mining is the confidence value which shows how strongly the two items are associated together. The confidence value is

Table 1 Transactions having item sold details

Transactions	Item purchased
$T1$	I1, I2, I3, I4
$T2$	I1, I3, I4
$T3$	I2, I3
$T4$	I1, I3, I4
$T5$	I1, I3

calculated by dividing a number of transactions that contain both items A and B with a total number of transactions having item A only. It is ratios of a number of transactions having both A and B items with the number of transactions containing only item A [9, 10]. It counts the percentage of presence of item B when item A is present. For example, 85% confidence value indicates that 85 times when item A is present in the transaction, item B is also present. 100% confidence value shows a strong association between two items because it indicates that whenever item A is available in the transaction, it is guaranteed that item B is also a neighborhood in the same transaction.

To understand both the measures clearly, the example is given in Table 1. It shows the sample transactions of the departmental store in which the items purchased by the customer in each transaction are mentioned.

Transaction $T1$ contains 4 items to be purchased by the customer together. $T2$ contains 3 items and so on. To identify which pair of items are most likely to be purchased together, and need to calculate support and confidence values of association mining. In this example support and confidence values for the association between Item 1 and Item 3 are (I1 → I3):

Support: 80% (4 transactions out of 5 contains both I1 & I3 items. So support is 4/5 = 0.8)

Confidence: 100% (4 transactions contains item I1 and all 4 times item I3 is also present)

100% Confidence value between I1 and I3 indicates a strong association between I1 and I3 and it shows that whenever customer purchases I1 there are 100% chances that he will purchase I3 also.

In this research, the Association mining concept is used to find a pair of subjects in which a higher number of students is failed together. The idea is to find out the hidden association between the pair of subjects in which new students are facing difficulty to acquire passing grades.

3 Research Steps

The proposed research work is divided into the following steps:

(a) Data Collection, Preprocessing, and Preparation

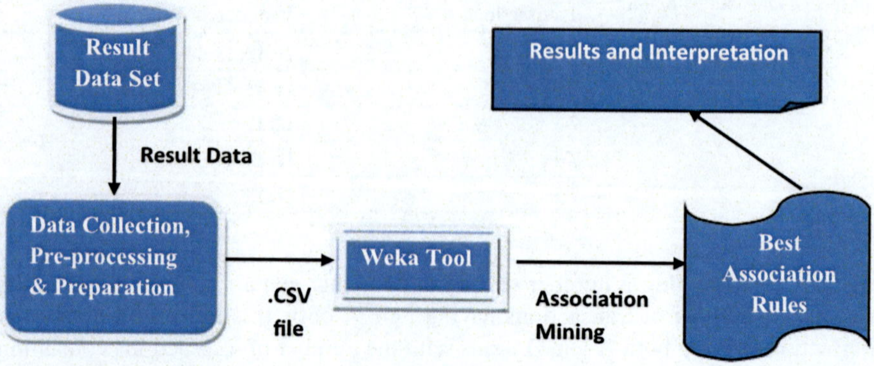

Fig. 1 Research flow

(b) Data Processing
(c) Result Analysis

Figure 1 highlights the steps performed in the proposed research work.

3.1 Data Collection, Preprocessing, and Preparation

For this research, the result data of first-year computer science students for the years 2015–2019 time duration is taken. As the focus is to find the relationship between the pair of failing subjects, preprocessing is required to delete the records of those students who are fully passed. Such records are omitted from the dataset. After preprocessing, the data size is reduced to 54 first-year students who were failed in the last 5 years. Figure 2 shows the total student statistics.

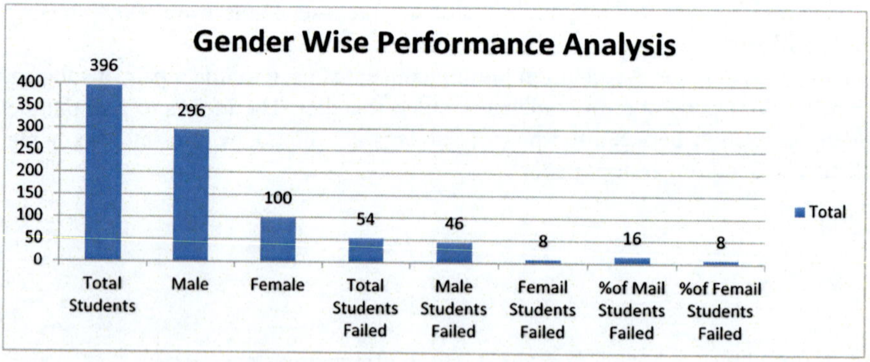

Fig. 2 Total students statistics with gender-wise performance analysis

A total of 396 students have appeared for the first-year exam, among them, 296 were male students and 100 were female students. Out of 396 total, 54 students got a fail grade in first-year subjects during the year 2015–2019. 46 male and 8 female students were unable to secure passing grades during these five years.

Result data of 54 fail students are then converted into a .csv file to make it suitable for WEKA. Appropriate pass/fail coding is done to apply association mining techniques on it using a data mining tool called WEKA. WEKA (Waikato Environment for Knowledge Analysis) is an open-source data mining tool written in the Java platform. It uses the collection of varieties of data mining and machine learning algorithms. Characteristics such as free availability, ease of use, and high portability make WEKA a powerful tool for effective data mining tasks [11].

3.2 Processing

The result data set which is in the form of a .csv file is then imported into WEKA software for processing. Apriori is the best association mining algorithms is applied using the WEKA tool on the result data set to generate the best rules that match the following criteria:

Minimum Support: 15% or above

Confidence: 70% or above

Figure 3 shows the rules generated using the WEKA tool. Table 2 shows the values of Support and Confidence for the best twenty rules found.

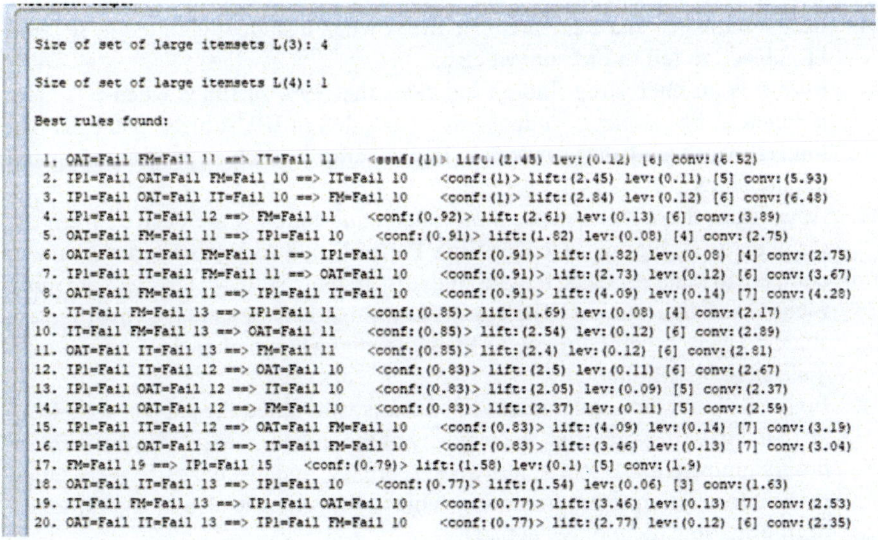

```
Size of set of large itemsets L(3): 4

Size of set of large itemsets L(4): 1

Best rules found:

 1. OAT=Fail FM=Fail 11 ==> IT=Fail 11    <conf:(1)> lift:(2.45) lev:(0.12) [6] conv:(6.52)
 2. IP1=Fail OAT=Fail FM=Fail 10 ==> IT=Fail 10    <conf:(1)> lift:(2.45) lev:(0.11) [5] conv:(5.93)
 3. IP1=Fail OAT=Fail IT=Fail 10 ==> FM=Fail 10    <conf:(1)> lift:(2.84) lev:(0.12) [6] conv:(6.48)
 4. IP1=Fail IT=Fail 12 ==> FM=Fail 11    <conf:(0.92)> lift:(2.61) lev:(0.13) [6] conv:(3.89)
 5. OAT=Fail FM=Fail 11 ==> IP1=Fail 10    <conf:(0.91)> lift:(1.82) lev:(0.08) [4] conv:(2.75)
 6. OAT=Fail IT=Fail FM=Fail 11 ==> IP1=Fail 10    <conf:(0.91)> lift:(1.82) lev:(0.08) [4] conv:(2.75)
 7. IP1=Fail IT=Fail FM=Fail 11 ==> OAT=Fail 10    <conf:(0.91)> lift:(2.73) lev:(0.12) [6] conv:(3.67)
 8. OAT=Fail FM=Fail 11 ==> IP1=Fail IT=Fail 10    <conf:(0.91)> lift:(4.09) lev:(0.14) [7] conv:(4.28)
 9. IT=Fail FM=Fail 13 ==> IP1=Fail 11    <conf:(0.85)> lift:(1.69) lev:(0.08) [4] conv:(2.17)
10. IT=Fail FM=Fail 13 ==> OAT=Fail 11    <conf:(0.85)> lift:(2.54) lev:(0.12) [6] conv:(2.89)
11. OAT=Fail IT=Fail 13 ==> FM=Fail 11    <conf:(0.85)> lift:(2.4) lev:(0.12) [6] conv:(2.81)
12. IP1=Fail IT=Fail 12 ==> OAT=Fail 10    <conf:(0.83)> lift:(2.5) lev:(0.11) [6] conv:(2.67)
13. IP1=Fail OAT=Fail 12 ==> IT=Fail 10    <conf:(0.83)> lift:(2.05) lev:(0.09) [5] conv:(2.37)
14. IP1=Fail OAT=Fail 12 ==> FM=Fail 10    <conf:(0.83)> lift:(2.37) lev:(0.11) [5] conv:(2.59)
15. IP1=Fail IT=Fail 12 ==> OAT=Fail FM=Fail 10    <conf:(0.83)> lift:(4.09) lev:(0.14) [7] conv:(3.19)
16. IP1=Fail OAT=Fail 12 ==> IT=Fail FM=Fail 10    <conf:(0.83)> lift:(3.46) lev:(0.13) [7] conv:(3.04)
17. FM=Fail 19 ==> IP1=Fail 15    <conf:(0.79)> lift:(1.58) lev:(0.1) [5] conv:(1.9)
18. OAT=Fail IT=Fail 13 ==> IP1=Fail 10    <conf:(0.77)> lift:(1.54) lev:(0.06) [3] conv:(1.63)
19. IT=Fail FM=Fail 13 ==> IP1=Fail OAT=Fail 10    <conf:(0.77)> lift:(3.46) lev:(0.13) [7] conv:(2.53)
20. OAT=Fail IT=Fail 13 ==> IP1=Fail FM=Fail 10    <conf:(0.77)> lift:(2.77) lev:(0.12) [6] conv:(2.35)
```

Fig. 3 Student result analysis using Weka (Apriori algorithm)

Table 2 Support and confidence value

Rule no	X → Y	Support (%)	Confidence (%)
1	OAT, FM → IT	20.37	100
2	IP1, OAT, FM → IT	18.51	100
3	IP1, OAT, IT → FM	18.51	100
4	IP1, IT → FM	20.37	92
5	FM → IP1	27.77	79

3.3 *Result Analysis*

Table 2 shows some of the best rules found based on the input parameter settings. Looking at the support and confidence value, it is clear that Office Automation Tools, Fundamental of Mathematics, Introduction to Programming, and Information Technology are the first-year bunch of subjects in which students are facing difficulties to get passing grades. It indicates a strong association between these subjects in terms of difficulty in learning.

(a) Rule 1 with 100% confidence indicates that students who are failed in OAT and FM subjects are also failed in Information Technology Subject. This rule has 100% confidence which shows that if the student is failed in OAT and FM subjects then there are 100% chances to fail in IT subject also.
(b) Rule 2 again with 100% confidence states that all the students who are failed in IP1, OAT, and FM subjects are failed in IT subject also.
(c) Rule 4 indicates that there are 92% times when a student is failed in IP1 and IT subject, to fail in FM subject also.
(d) Rule 5 is an interesting finding indicates that 79% of times when a student is failed in FM subject, students are also failed in IP1 subject. It means that logically poor students are week in logical subjects such as programming and mathematics.
(e) Apart from these rules, some more interesting findings are derived from the result data set. Figure 2 Gender Wise Performance Analysis chart shows that compare to male students, female students are more study-centric and are more focused on their studies. The study shows that half percentage (8%) female students were failed compare to the male students which is 16% of the total male students who failed.
(f) Figure 4 shows the total number of students who failed the subject wise from 2015 to 2019. It reveals that the majority of the students are getting difficulty in programming subject to a secure good score. 27 students out of 54 were failed in the introduction to programming subject. Students are also facing difficulty in Information Technology subject.

Researchers have made significant contributions in the field of student performance analysis, improvement, and prediction based on the failing subjects [12–15].

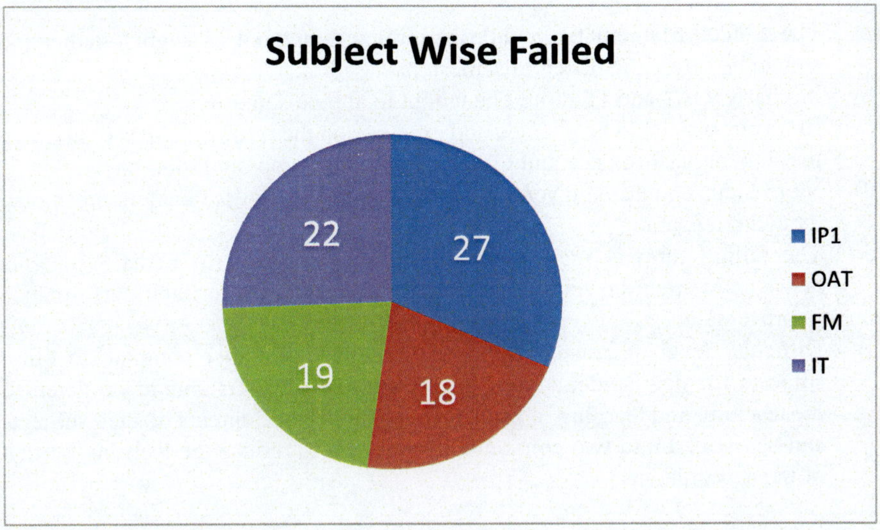

Fig. 4 Subject wise failed students

However, it is also important to consider other aspects which are equally responsible for fail grade such as the difficulty level of question paper, questions from out of syllabus, knowledge, and experience of the teachers, etc. These aspects are also included in the study to justify the findings. All the subject question papers from 2015 to 2019 which are taken in the study are undergone through Question Paper Quality analysis. This analysis is done based on the criteria such as syllabus coverage, the inclusion of innovative and creative questions, ratio of hard/average/easy questions, etc. This question paper analysis is performed by two subject expert subject wise. Each parameter is examined with the marks between 1 and 10, where 1 indicates poor weightage and 10 is good weightage. In this analysis, all these papers have got more than 90% marks. It indicates then the papers were of good quality and they do not have any out of syllabus question and they have the proper balance of easy, average, and hard questions. Apart from that these subjects were taught by experienced faculty members. This analysis helps to prove that these parameters are less responsible for the students to fail in these subjects.

4 Recommendations

The above rules indicate that students of the first year are getting difficulties in IP1, OAT, FM, and IT subjects. The results also show that students who are week in mathematics are also a week in programming. Based on the results, the following recommendations are proposed to enhance student performance:

(a) The subjects Mathematics and Programming should not be taught together and should be offered in different semesters.
(b) Similarly, OAT and IT should be taught in altered semesters.
(c) Extra efforts should be planned and executed on these four subjects to improve the student performance and to reduce first-year drop out ratio.
(d) More focus should be given on continuous evaluation parameters other than the written exam.
(e) The failing ratio is more in Male students compare to Female students hence extra attention should be given to Male students for their performance improvement.
(f) Difficulty level to grasp programming subject is higher compare to other subjects for new students so extra attention and efforts should be given to the teaching and learning of programming subjects. Contents of such subjects can be divided into two consecutive semesters to reduce the learning burden of the new students.

5 Conclusion and Future Scope

Higher student performance in terms of results is always a good indicator for both students and the institute. Enough attention is required for student performance improvement. In this article, an attempt is made to find out the association between the subjects in which more students got failing grades. By doing the analysis it is found that the logically poor students, getting difficulty to acquire passing grades in the subjects like mathematics and programming hence these two subjects can be offered in different semesters. The study also reveals that female students are brighter than male students. Based on this study the institute can redesign its syllabus to reduce student learning burden. The study is helpful to organizations to improve their student's performance in the examination.

In the future, research can be conducted to find out the association between the subjects of the entire course. Using such findings result prediction can be made and potential risk of being failed can be identified in advance and corrective actions will be made.

References

1. A. Shahiri, W. Hussain, N. Rashid, A review on predicting student's performance using data mining techniques, in *The Third Information Systems International Conference*. Science Direct, 2015
2. B. Baradwaj, S. Pal, Mining educational data to analyze students' performance. Int. J. Adv. Comput. Sci. Appl. **2**(6) (2011)
3. J. Han, M. Kamber, J Pei, Data Mining: Concepts and Techniques (Morgan Kaufmann, Burlington, 2000)
4. Pearson. N. Tan, M. Steinbach, V. Kumar, Introduction to Data Mining (2014)

5. A.A. Saa, Educational data mining & students' performance prediction. Int. J. Adv. Comput. Sci. Appl. **7**(5) (2016)
6. M.M. Abu Tair, A.M. El-Halees, Mining educational data to improve students' performance: a case study. Int. J. Inf. Commun. Technol. Res. **2**(2) (2012)
7. R. Jindal, M. Borah, A survey on educational data mining and research trends. Int. J. Database Manage. Syst. **5**(3) (2013)
8. N.P. Gopalan, B. Sivaselvan, Data Mining Techniques and Trends. PHI Learning (2009)
9. M.H. Dunham, X. Yongqiao, Le. Gruenwald, Z. Hossain, A Survey of Association Rules (2001) http://www2.cs.uh.edu/~ceick/6340/grue-assoc.pdf
10. T. Slimani, A. Lazzez, Efficient analysis of pattern and association rule mining approaches. ArXiv, abs/1402.2892 (2014)
11. An Introduction to Weka. Online Available at: http://opensourceforu.com/2017/01/an-introduction-to-weka/
12. E. Chandra, K. Nandhini, Knowledge mining from student data. Eur. J. Sci. Res. **47**(1) (2010)
13. N. Mahajan, J. Namdeo, Analysis of student result data using association rule mining. Res. J. Sci. IT Manage. **04**(6) (2015)
14. O. Chandrakar, J. Saini, Predicting examination results using association rule mining. Int. J. Comput. Appl. **116**(1) (2015)
15. D. Gaspar, M. Mabic, I. Coric, Data mining and predicting student performance, in *19th International Research/Expert Conference on Trends in the Development of Machinery and Associated Technology*, 2015

Implementation of Controller for Self-balancing Robot

R. Rengaraj, G. R. Venkatakrishnan, Pranav Moorthy, Ravi Pratyusha, and K. Veena

Abstract A two-wheeled self-balancing robot that works on the principle of inverted pendulum theory is most commonly used in control systems as an exemplar for comparing various control and optimization algorithms. This nonlinear dynamic system has gained much interest among researchers and engineers and the main aim is to stabilize this system with minimum deviation about any desired axis. As this system is affected by external disturbances and measured parametric errors, the need for the implementation of controllers has become crucial. Hardware and software implementation form the basis of the system design. The proposed work aims at the implementation of a PID controller for the successful working of the two-wheeled self-balancing robot driven by stepper motors. The closed-loop controller's parameters have been optimized by using pole placement method. Besides, the response of the system under the fuzzy controller, created using Mamdani fuzzy interface engine in Simulink, has been illustrated.

Keywords Self-balancing robot · Inverted pendulum cart · PID controller · Fuzzy logic controller

1 Introduction

Self-balancing robots are robots that have the ability to balance upright on two wheels with the help of a controller. A two-wheeled self-balancing robot is based on the principle of an inverted pendulum. The inverted pendulum is a classic problem in control systems which consists of a dynamic, unstable, and nonlinear system that includes an inverted pendulum in a cart, which is free to move in one plane. The problem aims to move the cart in such a manner that the inverted pendulum remains

R. Rengaraj (✉) · G. R. Venkatakrishnan · P. Moorthy · R. Pratyusha · K. Veena
Sri Sivasubramaniya Nadar College of Engineering, Kalavakkam 603110, India
e-mail: rengarajr@ssn.edu.in

G. R. Venkatakrishnan
e-mail: venkatakrishnangr@ssn.edu.in

V. Suma et al. (eds.), *Inventive Systems and Control*,
Lecture Notes in Networks and Systems 204,
https://doi.org/10.1007/978-981-16-1395-1_31

in a fixed vertical position. A controller must be designed to ensure that the system remains stable. The self-balancing robot finds applications in many places. One of the most commonly encountered practical applications of a self-balancing robot is the Segway personal transporter as shown in Fig. 1. Moreover, the concepts used in the self-balancing robot find use in aircraft, quadcopter control, etc., as shown in Fig. 2.

Fig. 1 Segway personal transporter

Fig. 2 Quadcopter

Authors in [1] have discussed the design and implementation of a linear quadratic regulator for solving the inverted pendulum problem. In this article, the 3D model is completed using CAD and imported to MATLAB. The designed linear quadratic regulator is tested on the model. The cart position and pendulum angle are simulated to verify the accuracy of the model. A two-wheeled self-balancing robot provides a basis for the mathematical modeling of the inverted pendulum system [2]. The model was developed with the help of Newtonian mechanics. The state-space model of the same was then derived [3]. The basic concept of pole placement technique and state-space analysis is provided in [4]. In [5], a fuzzy logic controller was developed to control the pendulum in a cart system. A swing is an algorithm which is used to bring the pendulum near its inverted position is implemented. The controller designed in the MATLAB and Simulink environment was tested for its robustness, and the response of the system is obtained concerning time. The maximum error of the pendulum angle was found to be between 1.89 and 4.669 degrees. In article [6], the use of neural network controllers for solving the inverted pendulum problem was discussed. The paper displays the simulated pendulum angle, cart position, cart velocity, and output force as a function of time when multilayer feedforward neural network and radial basis function neural networks are used. The problem is solved using supervised learning.

The nonlinear controller acts as a teacher to ensure that the response of the system is as desired. The authors in [7] use pole placement techniques to determine the PID control parameters for a two-loop PID controller of the inverted pendulum in a cart system. The two PID controllers simultaneously control the angle of the pendulum and the position of the cart [8]. The closed-loop poles of the system are determined by obtaining the state-space model of the inverted pendulum system. The performance of the system is obtained through a simulation where the cart position and angle of the pendulum are displayed as a function of time [9].

Using the perceived literature survey, the objective of this paper is to design and implement a controller to ensure that the robot remains at the fixed vertical set point despite external disturbances, i.e., the robot stays in an upright position for any given disturbance. It is very well known that the robot is balanced on two wheels and will simply fall even for the slightest disturbance [10–12].

The controller has been proposed to ensure the stability of the system irrespective of the external interference. The inertial measurement unit (IMU) used is the MPU6050 [13–15]. The output of the sensor is highly noisy, and hence, a filter is used before the computation of the tilt angle. Besides, the computation time of the microcontroller should be kept within limits as excessive computation time leads to more time for correction thereby making the system unstable.

The layout of this article is as follows. Section 2 explains the mathematical modeling of the inverted pendulum on a cart system in detail. The hardware of the project along with its essential components is described in Sect. 3. The overall design of the robot along with its circuit diagram and the algorithm is explained in Sect. 4. Section 5 displays the simulation results obtained. Finally, the conclusion and scope of future work are presented in Sect. 6.

2 Mathematical Modelling

The self-balancing robot that is considered in this project is a rigid rod of length l attached to a frictionless cart [16–18]. Figure 3 shows the free body diagram of the system separating the cart and the pendulum. Table 1 shows the description of the system parameters.

Using the basics of mathematical modeling of any physical system,

$$(I + ml^2)\ddot{\psi} - mgl\psi = ml\ddot{x} \quad (1)$$

$$u = (M + m)\ddot{x} + f\dot{x} - ml\ddot{\psi} \quad (2)$$

where u is the applied force in N, g is the acceleration due to gravity in m/s^2, ψ is the angle deviation between the pendulum and vertical axis in rad, $\dot{\psi}$ is the pitch velocity in rad/s, $\ddot{\psi}$ is the pitch acceleration in rad/s^2, x is the cart's position in m, $\dot{x}$ is the velocity of the cart in m/s, and $\ddot{x}$ is the acceleration of the cart in m/s^2.

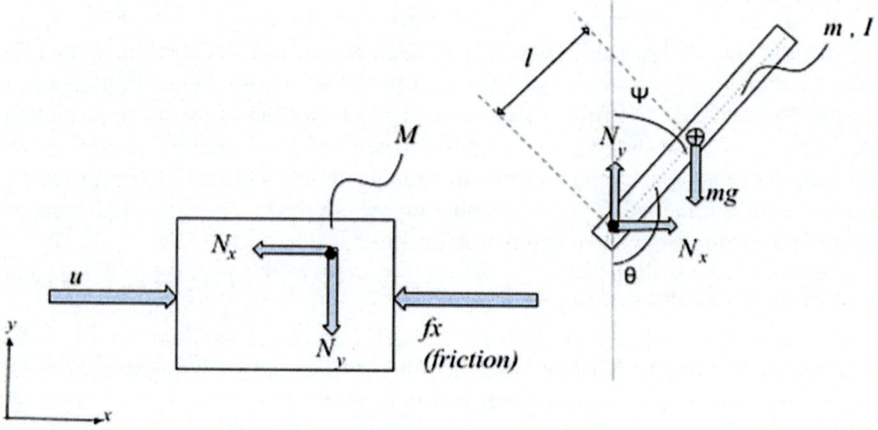

Fig. 3 Exposure of a simplified cart and pendulum system

Table 1 System parameters

Description	Symbols	Units
Mass of the cart	$M = 0.915$	kg
Mass of the pendulum	$m = 0.25$	kg
Length of the pendulum from its center of mass	$l = 0.04$	m
Moment of inertia about center of mass	$I = 0.000133$	Kg.m^2
Constant of friction	$f = 0.05$	N.s/m

The state-space model for the system is represented in Eqns. (1) and (2) is given by

$$\begin{pmatrix} \dot{x} \\ \ddot{x} \\ \dot{\psi} \\ \ddot{\psi} \end{pmatrix} = A \begin{pmatrix} x \\ \dot{x} \\ \psi \\ \dot{\psi} \end{pmatrix} + Bu$$

where $A = \begin{pmatrix} 0 & 1 & 0 & 0 \\ 0 & \frac{-(I+ml^2)f}{I(m+M)+mMl^2} & \frac{m^2gl^2}{I(m+M)+mMl^2} & 0 \\ 0 & 0 & 0 & 1 \\ 0 & \frac{-mlf}{I(m+M)+mMl^2} & \frac{mgl(m+M)}{I(m+M)+mMl^2} & 0 \end{pmatrix}$, $B = \begin{pmatrix} 0 \\ \frac{I+ml^2}{I(m+M)+mMl^2} \\ 0 \\ \frac{ml}{I(m+M)+mMl^2} \end{pmatrix}$ and $C = \begin{pmatrix} 1\,0\,0\,0 \\ 0\,0\,1\,0 \end{pmatrix}$.

The transfer function of the chosen system is given by

$$\frac{\psi(s)}{U(s)} = \frac{\frac{ml}{q}s}{s^3 + \frac{f(I+ml^2)}{q}s^2 - \frac{(m+M)mgl}{q}s - \frac{fmgl}{q}}$$

where $q = (m + M)\big(I + ml^2\big) - (ml)^2$.

3 Mechatronic System

The mechatronic system is a combination of mechanical and electrical systems. It consists of four key components such as sensors, actuators, controllers, and mechanical components. A mechatronic system of the proposed work is shown in Fig. 4. This section explains the various mechanical and electrical components used in the hardware development of the robot.

3.1 Mechanical System

The mechanical system consists of a three-level structure separated by spacers of appropriate length and two parallelly placed wheels.

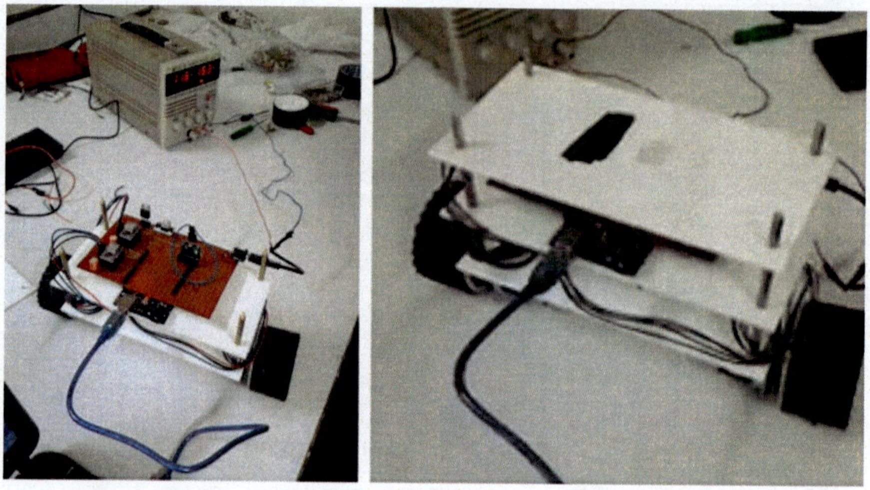

Fig. 4 Hardware setup of the project

3.1.1 Main Frame

The mainframe is a three-level structure made of an acrylic sheet of 4 mm thickness and 4 spacers separating one level from the another level. Each level supports individual hardware components, i.e., the first level supports the stepper motor, the second level supports the printed circuit board mounted with Arduino Uno and other electronic components and the third level supports a battery for powering the microcontroller.

3.1.2 Wheels

Wheels with dimensions suitable for handling the angular velocity of the motor and achieving the desired speed for balancing the robot are chosen. The wheel of a radius of 3.5 cm is connected to the rotor shaft of the motor.

3.2 Electrical System

An electrical system is a part of the hardware capable of accomplishing a particular function that defines the balancing of the mechanical system. The different parts of the electrical system used in the proposed work are Arduino microcontroller, 7805 voltage regulator, regulated power supply (RPS), NEMA17 stepper motor,

Table 2 Electrical part and its functions

Electrical part	Function
Arduino microcontroller	To interface with other circuits
7805 voltage regulator	To maintain constant output voltage and to prevent voltage fluctuations
RPS	To convert unregulated AC voltage to regulated voltage
NEMA17 stepper motor	To have high holding torque and to have less variation in speed
DRV8825 stepper motor driver	Helps in controlling the stepper motor using Arduino microcontroller
MPU6050 accelerometer and gyroscope	To measure acceleration, velocity, orientation, displacement

DRV8825 stepper motor driver, and MPU6050 accelerometer and gyroscope. The main functions of different electrical parts used in this proposed work are given in Table 2.

4 Proposed Model

The schematic diagram of the proposed model and its circuit diagram are shown in Figs. 5 and 6, respectively. The connections between Arduino and the proposed model are given in Table 3.

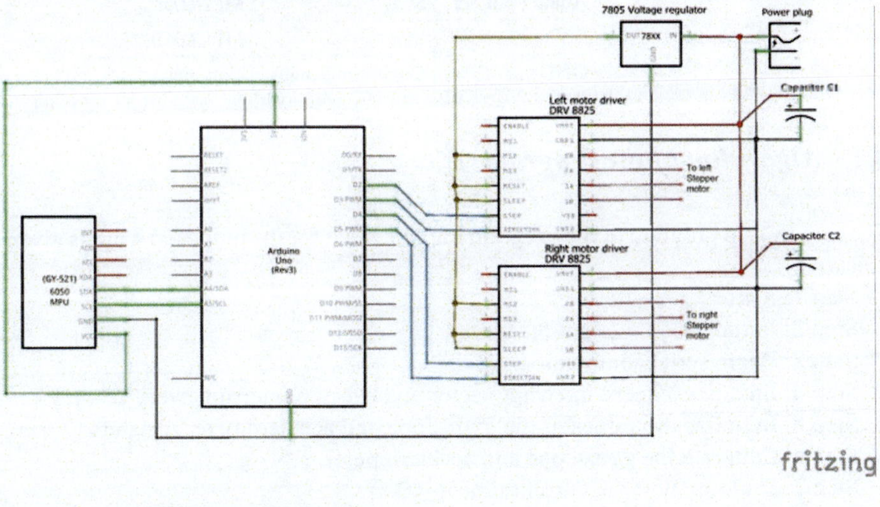

Fig. 5 Schematic diagram of hardware

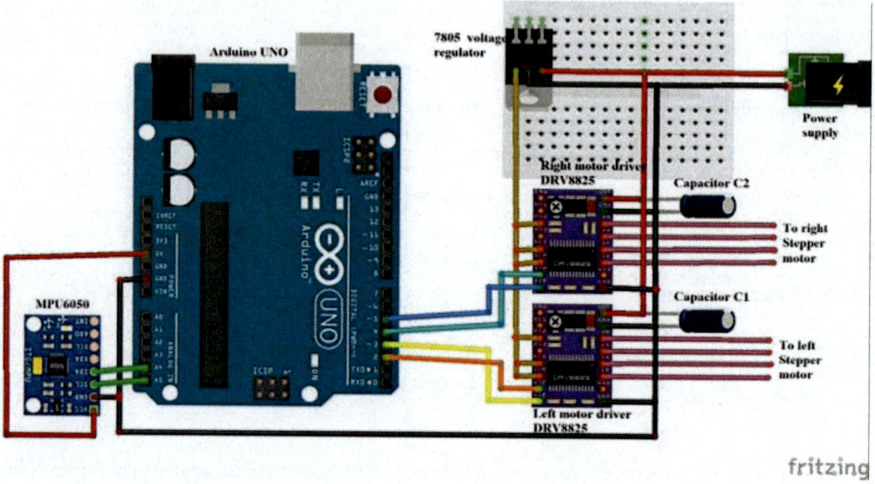

Fig. 6 Circuit diagram of an electrical circuit

Table 3 Connections of Arduino Uno

Arduino pin	Connected to pin	Component to which the pin belongs
Digital pin 2	STP	Left motor driver
Digital pin 3	DIR	Left motor driver
Digital pin 4	STP	Right motor driver
Digital pin 5	DIR	Right motor driver
Analog pin A4	SDA	MPU6050
Analog pin A5	SCL	MPU6050
5 V	Vcc	MPU6050
GND	Common ground	Common

4.1 Algorithm Main Program

The step by step program of the program implemented for the proposed work is given below:

Step 1: Start.
Step 2: Initialize I2C communication.
Step 3: Begin serial communication.
Step 4: Initialize Timer2 interrupt vector and execute interrupt every 250 μs.
Step 5: Initialize the values of the gyroscope and accelerometer registers.
Step 6: Calibrate the gyroscope and accelerometer.
Step 7: Set loop_timer = current time + 4000:
Step 8: If(TCNT > OCR2A) execute steps 8.1 to 8.6
Step 8.1: Calculate the angles measured by gyroscope and accelerometer.

Step 8.2: Input the angles to the complementary filter.
Step 8.3: Find the difference between the set angle and the actual angle.
Step 8.4: Find the speed of the motor using a PID controller.
Step 8.5: Decrement the error in the steps.
Step 8.6: Find the time interval between pulses and assign to variables throttle left motor and right motor.
Step 9: If current time > loop_timer wait.
Step 10: loop_timer = current time + 4000.
Step 11: Stop.
Timer2 Interrupt Service Routine.
Step 1: Increment throttle counter for the left motor.
Step 2: If throttle counter > throttle memory for the left motor.
Assign values for PORT D according to the value of throttle memory.
Step 3:If the throttle counter value is 1, then a high pulse is generated to pin 3.
Step 4: If the throttle counter value is 2, then a low pulse is generated to pin 3.
Step 5: Repeat the above steps for a right stepper motor.
Step 6: Return to Main Program.

5 Simulation Results

The block diagram of the closed-loop system with a PID controller is shown in Fig. 7. The response of the open-loop and closed-loop system with a PID controller obtained using MATLAB software when different inputs are applied to the proposed system is shown in Figs. 8, 9, 10, 11, and 12. The constants in the PID controller are obtained using the pole placement technique.

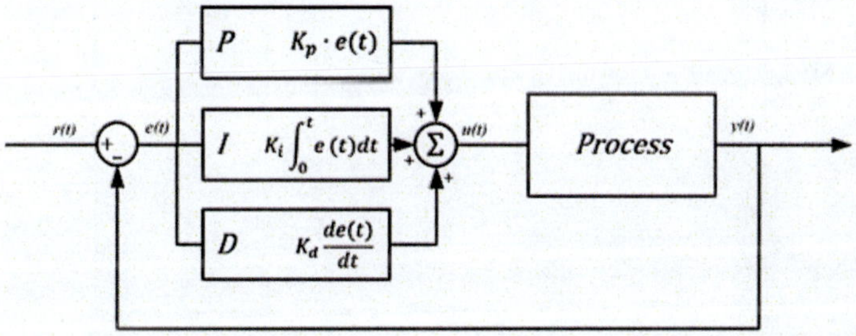

Fig. 7 PID controller block diagram

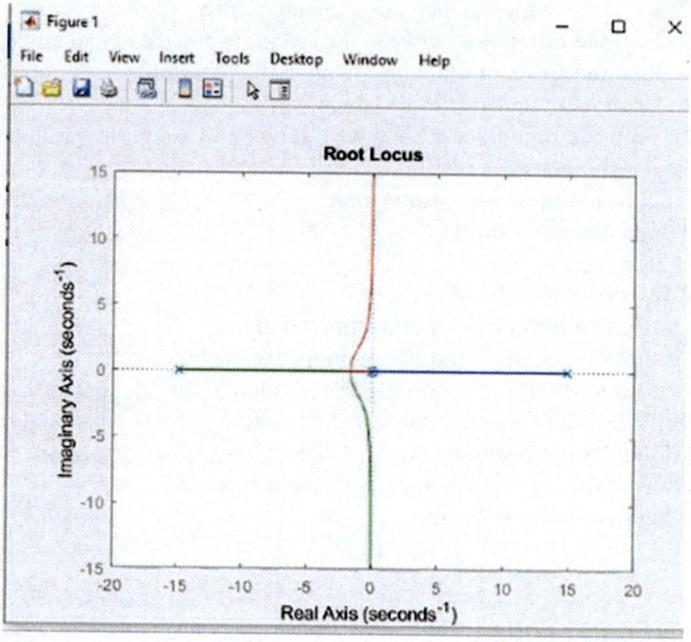

Fig. 8 Root locus of the open-loop system

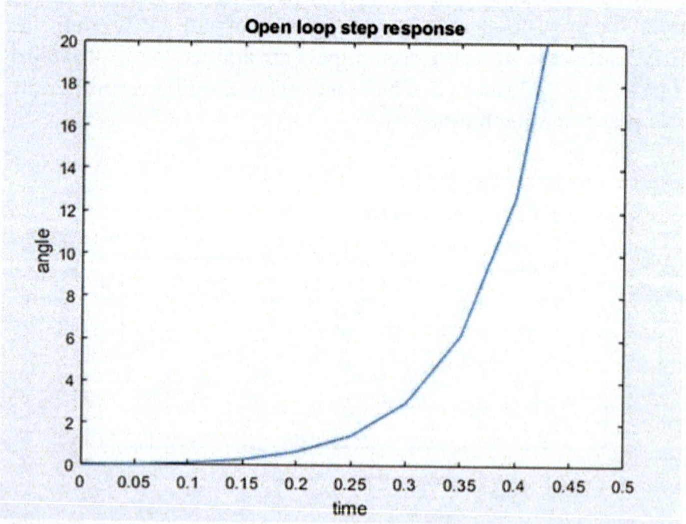

Fig. 9 Open-loop step response

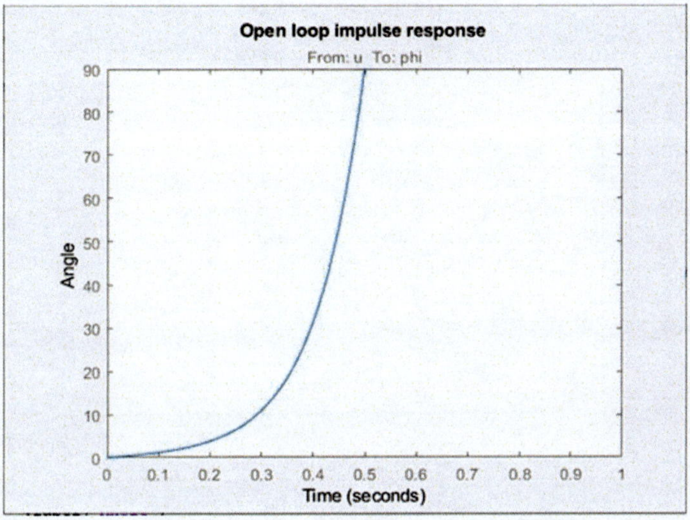

Fig.10 Open-loop impulse response

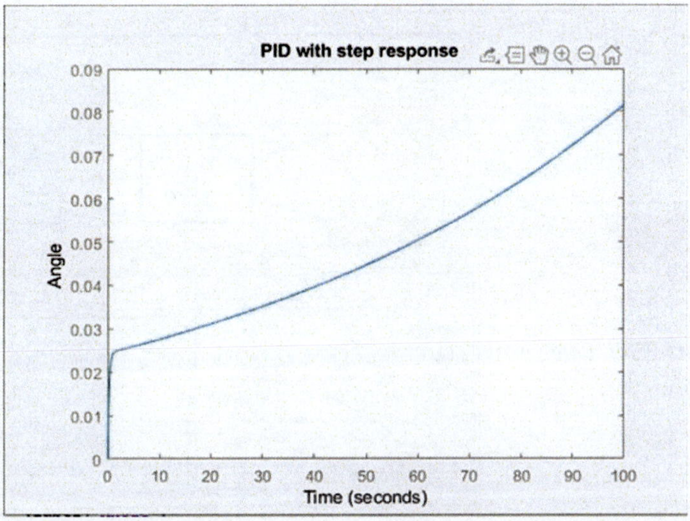

Fig. 11 Closed-loop step response

5.1 Implementation of Fuzzy Logic Controller

The variation of the output angle in the radians concerning the time is generated when a fuzzy logic controller is shown in Fig. 18. The fuzzy logic designer to generate the required force to variation in displacement and angular velocity is shown in Fig. 13.

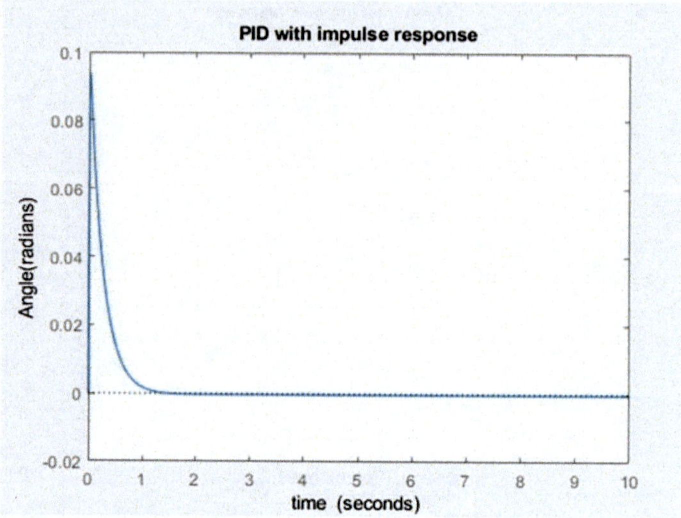

Fig. 12 Closed-loop impulse response

Fig. 13 Fuzzy logic designer

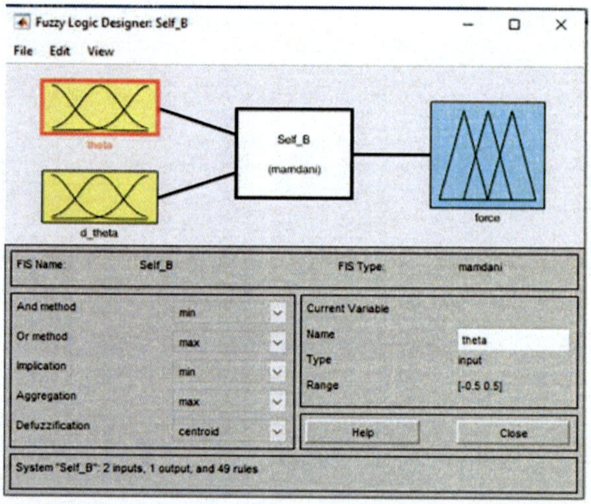

The membership function of the angular displacement, angular velocity, and force to be applied to the system is shown in Figs. 14, 15, and 16, respectively. Mamdani rule is used to obtain the variation in force to displacement and velocity. Simulink of the proposed system with a fuzzy logic controller is shown in Fig. 17.

The closed-loop controller is implemented to the proposed system with the help of pole placement techniques. From the closed-loop response of the system, it is proved that the system becomes stable. Moreover, a fuzzy logic controller model has been implemented in Simulink to stabilize the inherently unstable inverted pendulum on

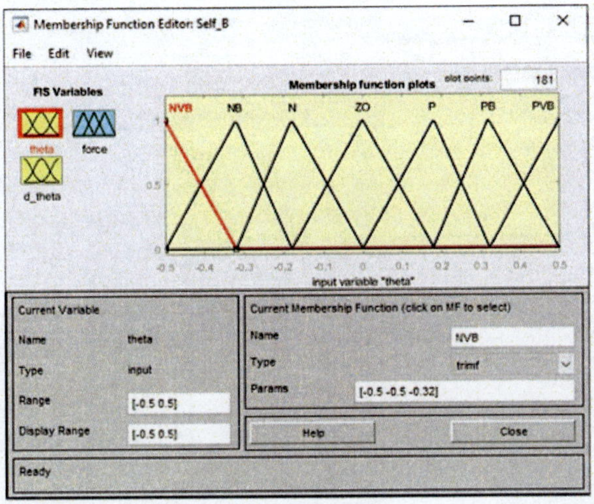

Fig. 14 Membership function of the angle theta in radians

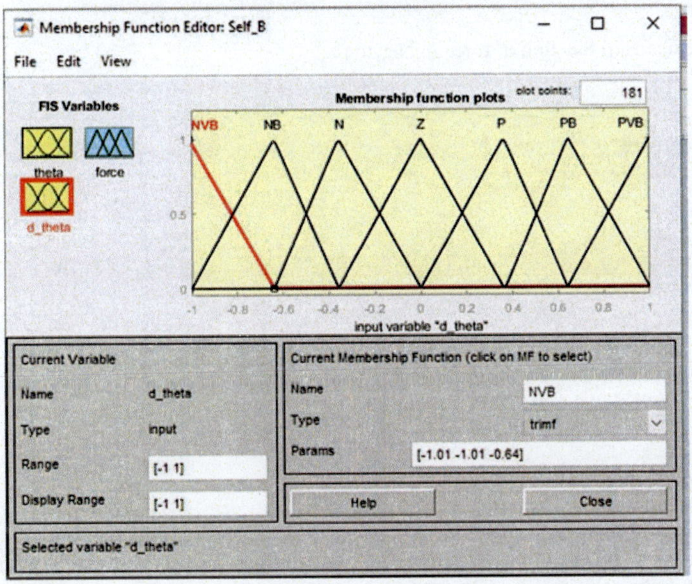

Fig. 15 Membership function of the rate of change of angle theta in radians

a cart system. The self-balancing robot has been successfully balanced in the event of an external disturbance (Fig. 18).

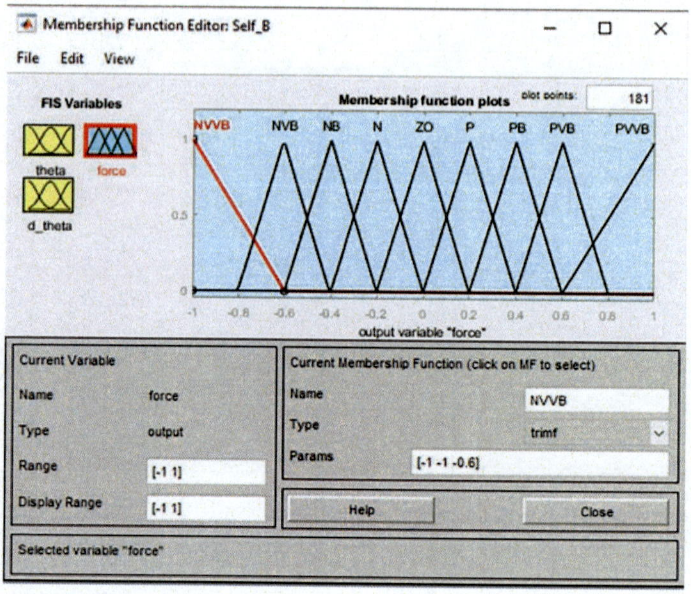

Fig. 16 Membership function of force in Newton

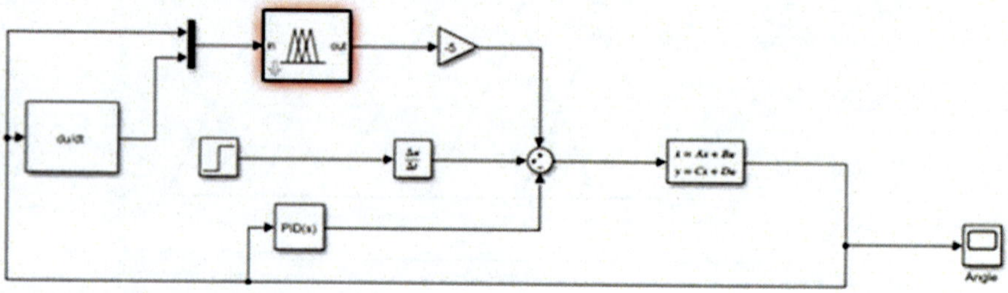

Fig. 17 Simulink model of fuzzy logic controller

6 Conclusion

The equations of motion of the inverted pendulum on a cart system have been derived and used for the proposed work. With the help of a root locus, it has been proved that the system is highly unstable in the absence of a controller.

In future work, the performance of the proposed system can be improved by controlling its motion about a plane using a Bluetooth module. Also, ultrasonic and infrared sensors, a global positioning system (GPS), and a camera can be incorporated into the proposed system to address other advanced applications such as obstacle avoidance, and path following.

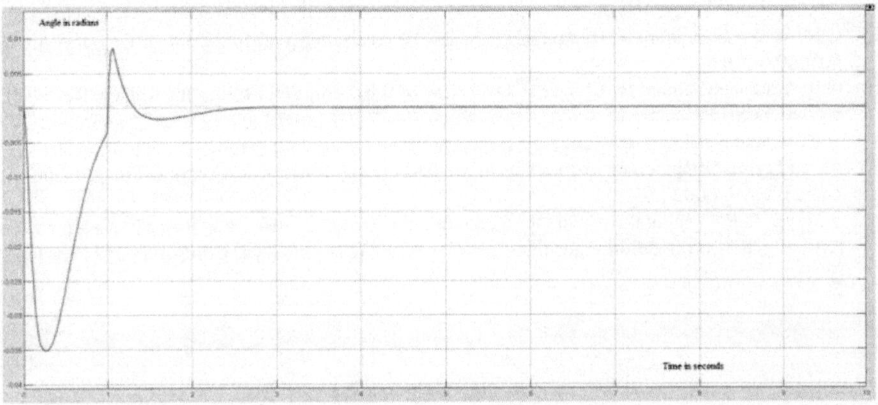

Fig. 18 Output angle (in radians) versus time

References

1. Ç. Abdullah, F. Botsali, Modeling and LQR control of a wheeled self-balancing robot, in *International Mediterranean Science and Engineering Congress (IMSEC)*, pp. 4916–4920 (2016)
2. A. Singh, S. Chaurasia, A.K. Sanu, V. Agrawal, Automatic balancing robot. Int. J. Adv. Res. Innov. Ideas Educ. **3**, 667–678 (2017)
3. S. Hellman, *Two-Wheeled Self-Balancing Robot : Design and Control Based on the Concept of an İnverted Pendulum*. Dissertation (2015)
4. K.P. Mohandas, *Modern Control Engineering*, Sanguine Technical Publishers, 2 (2016)
5. S. Kizir, Z. Bingul, C. Oysu, Fuzzy control of a real time inverted pendulum system, in *Knowledge-Based Intelligent Information and Engineering Systems*, Lecture Notes in Computer Science, 5177 (2008)
6. V. Mladenov, G. Tsenov, N. Harkiolakis, P. Karampelas, Neural Network Control of an inverted pendulum in an cart, in *International Conference on Robotics, Control and Manufacturing Technology*, pp. 112–120 (2009)
7. V.K. Pathak, S. Paliwal, Analysis and control of inverted pendulum system using PID controller. Int. J. Eng. Res. Appl. **7**, 1–4 (2017)
8. Y.A. Almatheel, A. Abdelrahman, Speed control of DC motor using fuzzy logic controller, in *International Conference on Communication, Control, Computing and Electronics Engineering (ICCCCEE)*, 1–8 (2017)
9. S.N. Sivanandam, S.N. Deepa, *Principles of Soft Computing*. Wiley (2007)
10. S. Salivahanan, R. Rengaraj, G.R.Venkatakrishnan, *Control Systems Engineering*. Pearson (2015)
11. G.R. Venkatakrishnan, R. Rengaraj, S. Salivahanan, Grey Wolf optimizer to real power dispatch with non-linear constraints, in *CMES - Computer Modeling in Engineering and Sciences*, vol. 115, issue 1, pp. 25–45 (2018)
12. G.R. Venkatakrishnan, R. Rengaraj, Grey Wolf optimization to hybrid renewable energy system located in Western Ghats Region—a case study. J. Electrical Eng. 18 (2018)
13. G.R. Venkatakrishnan, J. Mahadevan, R. Rengaraj, Optimal dispatch of residential distributed energy sources using self-adaptive differential evolution algorithm. Int. J. Appl. Eng. Res. **10**(5), 12761–12778 (2015)
14. J. Mahadevan, G.R. Venkatakrishnan, R. Rengaraj, Differential evolution algorithm with parameter adaptation strategy for optimal design of hybrid renewable energy system. J. Electrical Eng. **16**(3), 419–429 (2016)

15. G.R. Venkatakrishnan, R. Rengaraj, Overview of solar energy in India. Int. J. Dev. Res. **4**(3), 605–608 (2014)
16. G.R. Venkatakrishnan, R. Rengaraj, Overview of wind energy in India. Int. J. Dev. Res. **4**(3), 609–612 (2014)
17. G.R. Venkatakrishnan, R. Rengaraj, Differential evolution with parameter adaptation strategy for an optimal dispatch of residential distributed energy sources. J. Comput. Theor. Nanosci. **14**(12), 5997–6002 (2017)
18. R. Jeya, G.R. Venkatakrishnan, R. Rengaraj, S. Anand, N. Bharath Raj, G. Ramanathan, Evolutionary optimization algorithms—a review. J. Adv. Res. Dyn. Control Syst. 1112–1122 (2018)

An Empirical Study of Deep Learning Models for Abstractive Text Summarization

Neha Rane and Sharvari Govilkar

Abstract With the development of the Internet and social media era, there is an immense amount of existing information. But humans have become busier in this modern era to interpret all that information. Text summarization, a method of summarizing a text into its shorter version, has been a research area for many years. It includes two categories, namely extractive and abstractive. Though the former method is comparatively easy to implement and the latter method generates more human-like summaries, there are many promising results with the advent of deep learning approaches. In this article, an empirical analysis was conducted for the three renowned deep learning models used along with the natural language processing. Deep learning approaches outperform the traditional methods such as predicate argument or semantic graph methods and also have some limitations. Lack of structured and usable datasets contributing is the major problem in this research.

Keywords Abstractive text summarization · Natural language processing · Deep learning · Recurrent neural network · Attention mechanism

1 Introduction

Recently, there are adequate researches in the field of automatic text summarization. Depending on the way the text and its gist is chosen and arranged in the final output summary, the extractive and abstractive are the two ways of generating text summarization. The summaries generated from the first method consist of finding the most significant sentences of the text considering the statistical characteristics and sort these significant sentences, whereas the latter method involves creating new sentences, either by reformulating or coining in different words from existing, instead of simply copy-pasting the few selected sentences. The illustration of the text is formed by exploring semantic insights about words with in-depth investigations and reasoning. New sentences are spawned from the existing sentences, and there are

N. Rane (✉) · S. Govilkar
Pillai College of Engineering, Mumbai, India

V. Suma et al. (eds.), *Inventive Systems and Control*,
Lecture Notes in Networks and Systems 204,
https://doi.org/10.1007/978-981-16-1395-1_32

many categories which are based on the number of inputted documents considered as input for the summary, and they are termed as single-document or multi-document. The second category is based on the kind of information focused and included in summaries, and they are named as generic or query-focused summarization. The former represents a mechanism to generate summary from the entire document and the latter focuses on generating summary based on the user's query.

The extraction technique is considerably easier to implement, so there is already a lot of research work in it, but it lacks in terms of coherency and readability due to the various pragmatic problem such as reference resolution or conflict of anaphora. Besides, summaries from extractive methods are often contrary from human generated ones. Most of the works are linked to the extractive summary because of the simplicity of implementation. Abstractive summarization turns out to be difficult because of the complexity of NLP. But now, the research on extractive summarization has almost come to an end, because they have achieved maximum efficiency [1], and therefore, the approach was determined toward the abstractive summarization or the fusion of both methods. The motivation is to produce summaries of higher quality that are grammatically correct, legible, consistent, short, and rich in information. The abstractive helps to solve the confusing problem of the anaphora and, therefore, helps to generate legible, short, and logical summaries. On the contrary, abstractive summarization techniques help to reduce the size of the sentence since it uses the synthesis to combine the sentences, and thus avoiding redundancy or repetition of sentences in the final summary, in which the irrelevant chunk is also implied by the gist of extracting sentences and organizing them. Greater accuracy in terms of coherence, semantics, etc., can be achieved through the abstractive technique.

This paper evaluates three most renowned deep learning networks, namely convolutional neural network (CNN), recurrent neural network (RNN), long short-term memory (LSTM), and RNN-LSTM with attention mechanism [1]. Traditionally, CNN is especially used for processing an information which is arranged in grid form, for example, an image. On the contrary, RNN is used for processing an information which are in sequence, for example, sentences, paragraphs, etc. But as per our survey, we have found that researchers have also utilized CNNs to process sequential information in natural language processing and have achieved fair results. RNN processes the sequence data and captures the history by saving the output of a $(t-1)$ previous layer and inputting the same again as the input to next (t) layer along with the input to predict the output of the (t) layer. This gives a better result in comparison to CNN as it processes sequence information [4]. LSTM is a variant of RNN that is good in capturing context within the large documents by dealing with long-term dependencies. In this way, this minimizes the short-term dependencies faced by simple RNN for processing long sentences. Researchers has also achieved interesting results with attention mechanism where the decoder layer gets the attentions scores on the source side information from encoder layer [7].

This study compared between CNNs, RNN-LSTMs, and RNN with the attention model, systematically on summarization techniques. In this study, the CNNs does not provide the good results, as it processes the data in a grid style and cannot map the context of the sentences in the grid [1], whereas RNNs architecture performs better with the sequence data and to capture the semantics and to understand the entire sequence. However, we found through our analysis that standard RNN faces gradient vanishing and exploding issues. These issues tend to create problem for RNN during training in two ways: (i) in the activation layer, if we use tanh function, then it cannot remember the context of very long sequences owing to its hyperbolic nature, (ii) and if we use rectified linear unit (ReLU) as an activation function, then it is very unstable. But RNNs have two variants: LSTM and GRU which overcome these issues [11]. And with this learning rate also changes which makes performance smooth.

2 Related Works

Predominantly, generating summaries using the abstractive method is a challenging exercise, which is somewhat difficult than extractive approaches. Thus, a milestone of generating a summary that has the stature of the summaries generated by humans, despite the current breakthrough with neural networks which has shown great results in the development of machine translation.

For instance, Rush et al., 2015 proposed a neural network model based on the attention mechanism for abstractive summarization which is known as neural attention model for abstractive summarization (NAMAS) [2]. They proposed an approach for generating abstractive summaries by employing an attention mechanism in the traditional encoder-decoder model. This mechanism has been broadly used in encoder-decoder models where the decoder deduces the data from the encoder using the metrics or level of attention on each word or sentence received by an encoder.

Buys and Blunsom, 2017 in their work [3] revealed that recurrent neural networks (RNN)-based parsers have already attained advanced progress and achievements in dependency. In fact, RNN-based models help in predicting composite relations among words such as structured-based and semantic-based could not achieve efficiently. Baralis et al., 2013 in their article [4] have parsed the graphs especially the semantic one by neural network for deep linguistic analysis. This in return helps to generate more efficient abstractive summaries. They have to resolve ambiguity by using the maximum entropy model to generate minimal recursion semantics (MRS). MRS was generated for the representation of semantics in grammars. They fixed the problem of alignment in AMR graphs from this model. Niu et al. [5] proposed a method by combining a chunk graph along with the neural network. This method is used for generating abstractive summaries from more than one document, i.e., multi-document. They created a language model to evaluate the linguistic quality of the sentence by using a recurrent neural network. This language model helped to generate a comprehensible summary through abstractive techniques. Jobson and Gutirrez [6] used the same technique of RNN-LSTM in the encoder-decoder model. They used the concept of word embedding to identify the relations among words in various contexts while training the model and created an attention function from this word embedding for building the context inform of vector in each cycle. Nallapati et al., 2016 in their article [7] used the service model concept together with RNN to manage concerns about keywords modeling and then establishing the relations of hierarchical structure among the words in the sentences and sentences with the word. In particular, they used the BRNN-GRU for encoder which is bi-directional in nature and a simple RNN-GRU for a decoder. They used the pattern of attention in the hidden states of the source layer and softmax in the target. The gated recurrent unit (GRU) is basically used to solve the problems of degradation that fade and explode. Rossiello et al., 2016 in their paper [8] proposed the method which used RNN and probabilistic models to create an accurate and semantically well-formed abstractive summary. Liy et al., 2017 in their paper [9] used the structured inherent gist of the text to improve the quality of generated summaries. Their paper proposes the decoder based on the generative model to convert the original text into hidden states output and again converting these hidden states output to its original sequences in the generated summaries. Song et al. [10] proposed an approach based on deep learning called RNN- long short-term memory (LSTM) encoder-decoder model. In this model, phrases were inputted into the encoder instead of words.

Fan et al. [11] developed a personalized and manageable text summarization approach that uses convolutional neural networks (CNN) model in encoder-decoder architecture. Personalization in text summaries is achieved by considering user's inclinations like their interest, context, the size of the final summary, and by allowing them to select the chunk of text whose summary they wish to be generated.

Baumel et al., 2018 in their paper [12] coined a method for generating abstractive summaries from more than one document (multi-document) and which will be specific to the user's query by using the relevance sensitive attention-based model. Nema et al., 2018 in their paper [13] used the attention model for capturing the context and semantics which gave a good result. They have tried to resolve the repetition of sentences in the final output by incorporating an attention mechanism for capturing the diversity of sentences. This approach helped to achieve an improvement of about 28% in the ROUGE-L score of final summaries. Lin et al., 2018 in their paper [14] named their model as Seq2Seq which is purely based on the sequence to sequence along with the attention mechanism. Seq2Seq claims to solve the serious problem of sentence repetition. They used global encoding with the gated units based on the convolutional model as an encoder and unidirectional RNN-LSTM as a decoder to generate the abstractive summaries of a document.

Although deep learning has been tested and proven many times and came to the conclusion that they are the up-and-coming approaches to generate abstractive summaries. Nevertheless, getting large corporate for training remains a challenge. Furthermore, the available datasets are venerable, and therefore, it does not include the basic NLP features such as morphological, semantic, and syntactic restored. Apart from this, most datasets are only in English, such as DUC or TUC (not in any Indian language).

3 Deep Learning Models

3.1 Convolutional Neural Network

CNN network uses various convolutions layers which uses nonlinear activation functions such as ReLU or tanh. This activation function is applied to the results we get from hidden layers or convolutional layers. In a traditional feed-forward neural network, each neurons of the input layer are connected to each neuron in the next layer. In CNNs, it uses convolution method on all the neurons in input layer to get the predicted output. Each layer uses different filters, and these filters can be about hundreds or more like the ones shown in Fig. 1, and then it combines all the filters result together. These filters extract the features of the input sentences while training. Feature extraction is very important during training, and this will help to increase the accuracy in the validation stage.

For NLP, instead of image pixels, the inputs are sentences or documents. These are represented as a vector in matrix. Each row of the matrix represents one vector, typically a word, but it could be any character in a sentence or document. Typically, these vectors are word embeddings (low-dimensional representations) like word2vec or GloVe, but they could also be one-hot vectors that index the word into a vocabulary. Using word embedding is far better than the one-hot vectors, as it uses less space, and matrix is non-scalar one.

Typically, the outputs are fed as input to next layers like a typical neural network. All these layers need a fixed-length input but CNNs output's length is dependent on the length of the input sentence, n. This problem is taken care by max-pooling layer. The input to max-pooling layer from CNN is shown in Eq. 1

$$y \in R^{n-h-1} \tag{1}$$

and the output from the max-pooling layer is

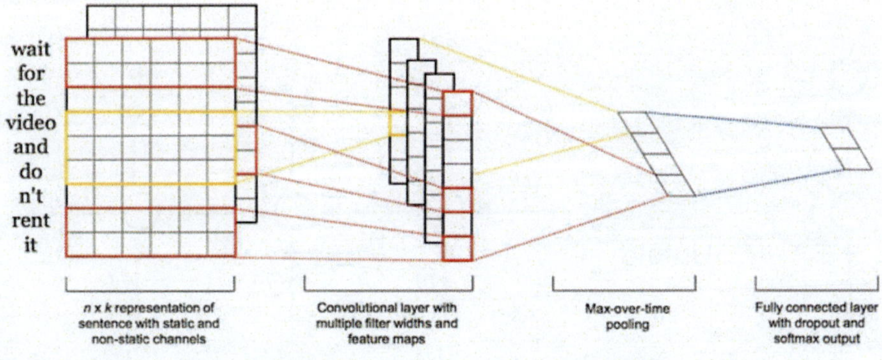

Fig. 1 Convolutional neural network (CNN)

$$\hat{y} = \max\{y\}, \text{ thus } \hat{y} \in R \tag{2}$$

The min-pooling is also used because the ReLU serves as our nonlinear activation function, and ReLU will make the output to be nearby 0. Hence, a min-pool layer might get suppressed, so we must use max-pooling over min-pooling in these applications.

3.2 Recurrent Neural Network

Recurrent neural networks (RNNs) specialize in handling sequential information and has a major contribution to many NLP tasks. This model performs similar steps on every word or token in a sentence, and this is the reason it is called recurrent. The RNNs logic is to preserve the semantics of sentences. RNN is different from the simple neural network where all inputs, as well as their outputs, are independent of each other. But in capturing sequential semantics, it would fail. For example, in language modeling, the system needs the information of all words in the sentence or query for predicting the next word. This is important to capture the essence of the whole sentence. In short, the RNNs have a "memory" to capture the semantic information of the sentence and to remember the words which were used prior for prediction. Though ideally, RNN has memory while implementing it only stores the last few words. Figure 1 shows how RNN looks like and how it looks when its layers are unrolled (Fig. 2).

The formulas of RNN are as follows:

x_t is the input at t time. For example, x_1 could be a word embedding vector which represents 2nd word in sentence.

s_t is the hidden state at t time. It acts as the memory within the network. s_t is calculated from the output of previous hidden state and the input at the current step:

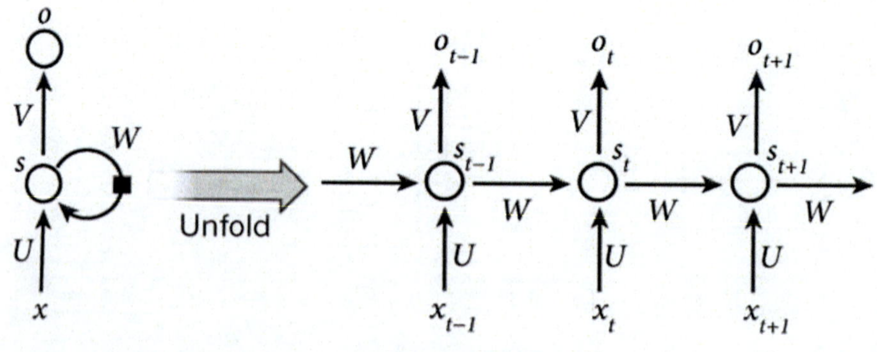

Fig. 2 Recurrent neural network and its unfolding diagram

$$s_t = f\left(U_{x\_t} + W_{s\_\{t-1\}}\right) \tag{3}$$

The function f generally is a nonlinear function such as tanh or ReLU. $s\ \{-1\}$, which is used to get the output from the first hidden state, and it is typically initialized to all zeroes.
o_t is the output at t time. For example, to predict the next word in a sentence, it would be a vector of probabilities.

$$ot = \text{softmax}(Vs_t) \tag{4}$$

3.3 *Long Short-term Memory*

RNNs are quite effective when dealt with short-term dependencies. But it fails miserably to remember the context of the sentence which was discussed long before in the paragraph. However, simple RNNs cannot capture the context of an input. This is where a simple RNN lacks behind because of the vanishing gradient problem. In short, RNN remembers things for smaller timestamps; i.e., if the information is needed after a small time, it may be remembered; but once a lot of words are fed in, this information gets forgotten. This problem is handled by tweaking a simple RNN slightly which is known as—the long short-term memory networks [15].

The three gates, namely the forget gate, input gate, and output gate regulate information flow in an LSTM cell. Forget gate as per Eq. (5) determines which information is important and whether it should be forgotten/discarded or stored. The activation function gets the information from both the previous hidden state and the current input. Generally, Sigmoid activation is used in LSTM to restrict the output between 0 and 1. If the output is less than 0.5, then it means to forget, and if it is greater than 0.5, then it means to remember. The second is input gate as per Eq. (6) to update the cell state. The output from a hidden state and current input token is passed onto a sigmoid function (Fig. 3).

This sigmoid function decides which values will be updated by transforming the input to forget gate. The hidden state output and current input are passed onto the tanh activation function to restrict outputs between −1 and 1 for regulating the network. Then after multiplying output from both activation functions where sigmoid function decides the importance of information to retain which is generated by a tanh activation function. Next is output gate which decides what the next hidden state should be. Then pass the updated cell state to the tanh activation function. By multiplying both the output from tanh and sigmoid, it decides what tokens from the hidden state should it retain. The output is the new hidden state. And the next time step is processed with the new cell state and the new hidden.

$$f_t = \sigma(W_f \cdot [h_{t-1}, x_t] + b_f) \tag{5}$$

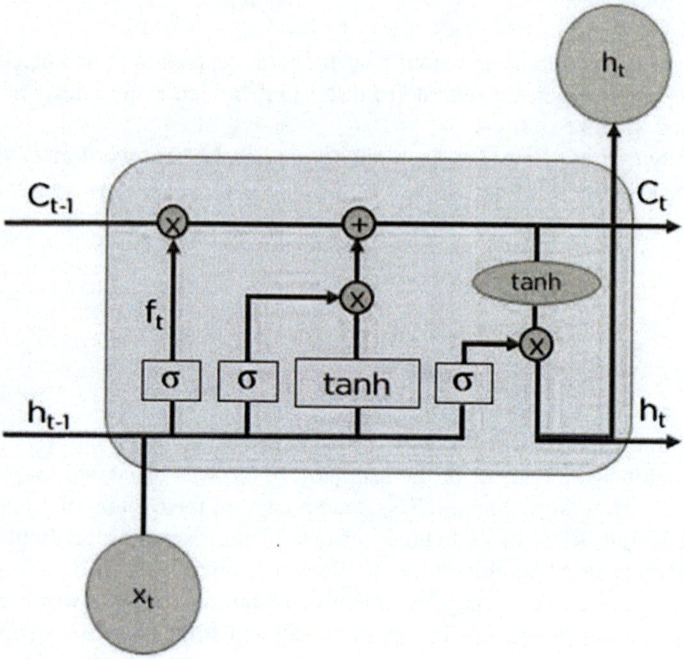

Fig. 3 Unrolled cell in long short-term memory networks

$$C_t = \tanh(W_C \cdot [xt, ht - 1] + b_C) \tag{6}$$

3.4 *Attention Mechanism in Recurrent Neural Network*

Attention [16] is a mechanism that allows a neural model to remember a portion from the input sentence when generating an output sequence. There have been many improvements in results when attention mechanism is combined with recurrent neural network making it a requisite for implementing RNN networks. This mechanism enables the feature extraction from the sentence for deciding which part of the sentence have maximum information incorporated in it. The decoder with the attention mechanism selects an important portion of the input sequence by attentively focusing on it. With this technique, encoder do not need to encode all information in the input into a single vector (Fig. 4).

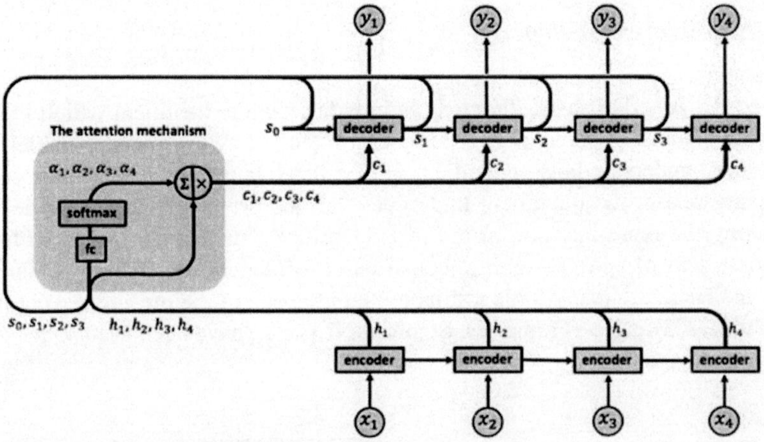

Fig. 4 Simple RNNs with attention mechanism

4 Experiment Setups and Results

This experimentation helped us to determine the best and efficient model for abstractive text summarization. It would help in the easy and fast retrieval of information. Initially, the process begins by understanding the working of various deep learning models. There are many datasets available in the English language for text summarization. Here, consider the three deep learning models such as CNN, RNN-LSTM, and RNN-Attention model for studying exactly how the abstractive text summarization works. There are two types of datasets were used such as news summaries from CNN and Amazon food reviews for our exploration.

4.1 Training Datasets

The CNN/Daily mail dataset as processed by Nallapati et al. [7] has been used for evaluating summarization. The dataset contains online news articles (781 tokens on average) paired with multi-sentence summaries (3.75 sentences or 56 tokens on average). The processed version contains 287,226 training pairs, 13,368 validation pairs, and 11,490 test pairs, whereas Amazon food reviews contain above 500,000 reviews and is hosted on Kaggle for free.

4.2 Experimental Process

It intends to generate abstractive text summaries from a document which is multilingual and multi-document. It comprises of the following phases, namely pre-processing, understanding semantics, and natural language generation. In pre-processing phase, the document undergoes various pre-processing activities such as creating the representation of the text, identifying named entities, co-reference resolution, part of speech tagging, construction of dependency trees and semantic trees, etc. There are many tools available that are used in the pre-processing phase. Tools such as WordNet, FrameNet, Stanford NLP, etc., have been used so far for the English language (Fig. 5).

Fig. 5 Experimental process

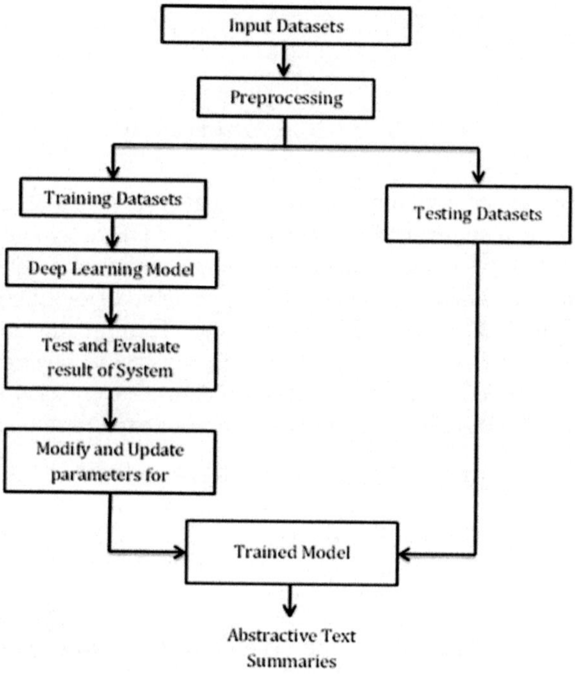

5 Results and Analysis

5.1 Tuning Hyper-Parameter for Loss and Accuracy

In this section, we will discuss the analysis of text summarization results on two datasets, namely news summaries from CNN and Amazon food reviews. The accuracy and loss were evaluated of three deep learning networks. According to implementation setting, these models have performed well on datasets, but RNN with attention mechanism model shows excellent performance. Both the datasets are evaluated with baseline deep learning models such as RNN, CNN, and LSTM as shown in Figs. 6, 7, 8, 9, and 10 (Fig. 11).

Table 1 suggests the loss and accuracy which are attained after this analysis. Even after many epochs, the CNN model provides the much-needed loss and accuracy factor. The CNN model lacks in processing the information in sequence. This is because of the convolution layer which is used to calculate the features in the training dataset. But on the other hand, the RNN model which is suitable to process sequential information gives a better result. With the attention model, it gives the similar to RNN results. In the end all these models depend on the amount of the datasets on which

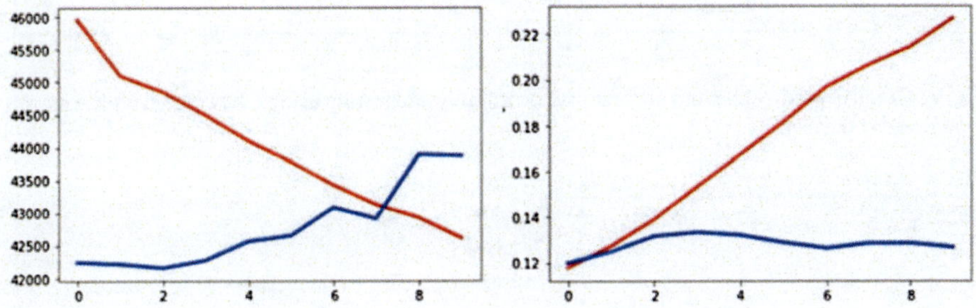

Fig. 6 Graph depicting loss and accuracy while training and testing using CNN model with news dataset

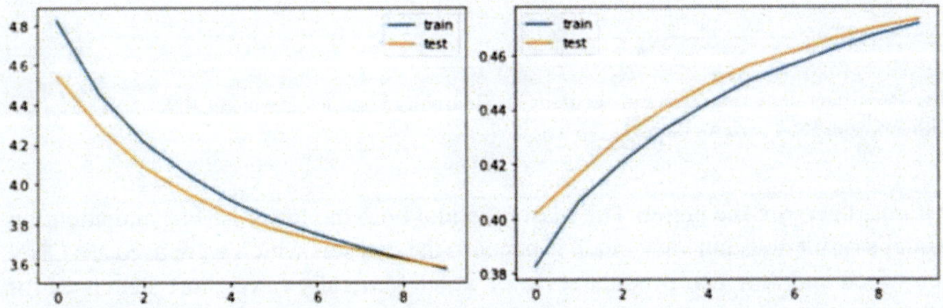

Fig. 7 Graph depicting loss and accuracy while training and testing using RNN-LSTM model with news dataset

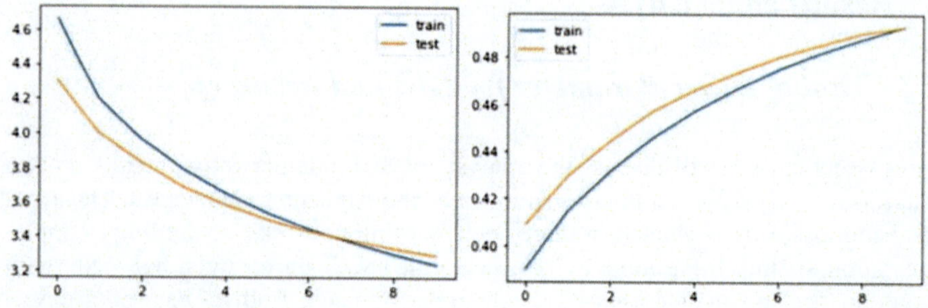

Fig. 8 Graph depicting loss and accuracy while training and testing using RNN model with attention mechanism with news dataset

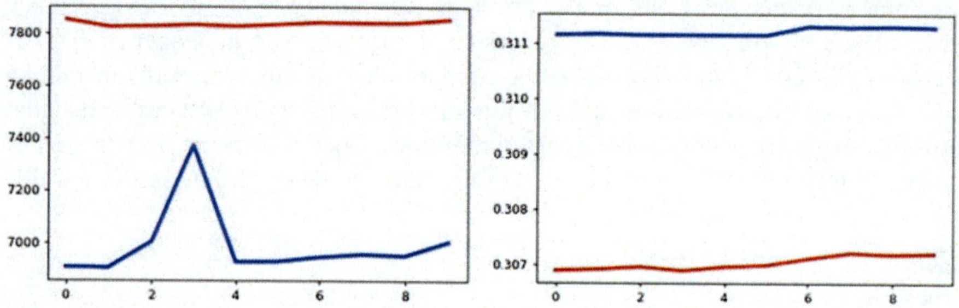

Fig. 9 Graph depicting loss and accuracy while training and testing using CNN model with Amazon food reviews dataset

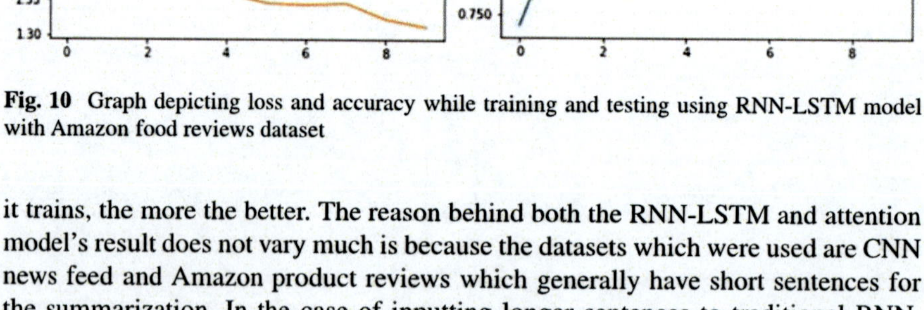

Fig. 10 Graph depicting loss and accuracy while training and testing using RNN-LSTM model with Amazon food reviews dataset

it trains, the more the better. The reason behind both the RNN-LSTM and attention model's result does not vary much is because the datasets which were used are CNN news feed and Amazon product reviews which generally have short sentences for the summarization. In the case of inputting longer sentences to traditional RNN, the difference can be observed and the reason for selecting the LSTM and attention model is preferable to process the sequential information.

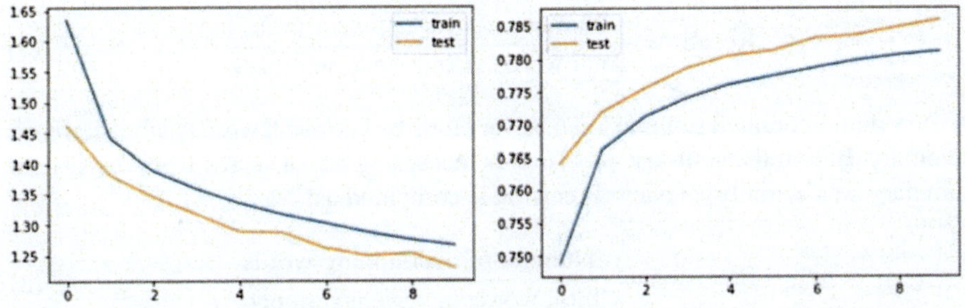

Fig. 11 Graph depicting loss and accuracy while training and testing using RNN model with attention mechanism with Amazon food reviews dataset

Table 1 Loss and accuracy prediction of various models

	CNN Newsfeed	Amazon Food Review	CNN Newsfeed	Amazon Food Review
	Loss		Accuracy	
Using Convolutional	42,605.7600	7856.2284	22.66	30.42
RNN-LSTM	3.5720	1.3709	47.18	77.86
RNN with attention mechanism	3.2353	1.2655	49.31	78.11

RNN-LSTM model actually reduces the loss function after each epoch and thereby improving the accuracy. But there exists few problems with RNN-LSTM such as it cannot generate summaries of the long document while maintaining the context of the document. Apart from that it also generates many repetitive words in summaries which may affect the accuracy of the model. RNN-LSTM and RNN with attention mechanism have two drawbacks. The generated summaries occasionally represent factual details inaccurately. This is especially common for rare or out-of-vocabulary words.

5.2 *Evaluation Using Rouge Metrics*

Recall-oriented understudy for gisting evaluation (ROUGE) is a collection of metrics for evaluating system-generated summaries and sentences translated by machines. It compares a system-generated summary or translation with a set of reference summaries which is generally generated by human. The term *Recall* with respect to ROUGE means no. of words which are required in the reference summary. Considering the single words, it can be computed as:

$$\text{Recall} = \frac{\text{Number of overlapping words}}{\text{Total words in reference summary}} \quad (7)$$

A system-generated summary can be very long by taking all words in the reference summary. But on the contrary, precision is measuring no. of words from the system summary was actually relevant. Precision is computed as:

$$\text{Precision} = \frac{\text{Number of overlapping words}}{\text{Total words in system summary}} \quad (8)$$

ROUGE—1, 2 … N is the detailed study of outputs by comparing the system generated ones with the reference ones. For example, ROUGE-1 represents the overlapping of single words (unigram) of the system summary with the reference summary, whereas ROUGE-2 represents the overlapping of two words (bigrams).

F1 score is computed by averaging both the precision and recall. But here the averaging method is different as shown below:

$$F1 = \frac{2 \cdot (\text{Precision} \cdot \text{Recall})}{\text{Precision} + \text{Recall}} \quad (9)$$

It can always be predicted $y = 1$. So, that it cannot be accepted as the whole precision-recall idea is to avoid. The formula $F1$ score is (Table 2):

Table 2 Rouge score evaluation

Datasets	Approach	Rouge metrics		
		R1	R2	RL
CNN News	RNN-LSTM	F:0.19 R:0.22 P:0.16	F:0 R:0 P:0	F:0.22 R:0.33 P:0.16
	Attention	R:0.37 R:0.37 P:0.37	F:0.14 R:0.14 P:0.14	F:0.49 R:0.37 P:0.75
	CNN	F:0.11 R:0.18 P:0.10	F:0.11 R:0.11 P:0.11	F:0.13 R:0.12 P:0.11
Amazon Food Review	RNN-LSTM	F:0.57 R:0.66 P:0.55	F:0.33 R:0.28 P:0.44	F:0.61 R:0.57 P:0.66
	Attention	F:0.66 R:0.66 P:0.66	F:0.49 R:0.55 P:0.55	F:0.66 R:0.66 P:0.66
	CNN	F:0.28 R:0.21 P:0.20	F:0.12 R:0.28 P:0.10	F:0.30 R:0.20 P:0.21

ROUGE-L is measured by checking the same sequence of words using the longest common subsequences (LCS). It does not deal with matching only consecutive words but it also checks in-sequence matches that reflect sentence-level word order. As it automatically includes the longest in-sequence common n-grams, there is no need for a predefined n-gram length (Fig. 12).

The CNN model generates an illogical summary with factual errors. The RNN-LSTM model is accurate occasionally. The attention mechanism achieves the best summaries.

```
Review 1 (After cleaning and pre-processing):
best friend aging shetland sheepdog named jake absolutely lov
es  treats used get local pet store recently moved cannot fin
d anywhere nearby course turned amazon happily half price shi
pping used paying jake hip dysplasia surgery young  adult dog
walks noticable limp slowly becoming worse shunned glucosamin
e supplements little help since found wonderful treats gait
seem degrading happy still  get because love little dog piece
s anything makes life little better little longer great thing
enjoys eating makes godsend

Original Summary: jake gives these 5 stars

Predicted Summary using Convolutional Neural Network: pet mov
ed find

Predicted Summary using RNN-LSTM Network: dog treats

Predicted Summary using RNN using Attention mechanism Network
: my dog loves these
```

```
Review 2 (After cleaning and pre-processing):
first enjoyed kicking horse coffees visit canada since delive
red monthly amazon i expert coffee drink inordinate amount da
ily coffee simply best kicking horse several flavors excellen
t highly recommend anyone loves java give brand try

Original Summary: the best coffee

Predicted Summary using Convolutional Neural Network: amount
daily

Predicted Summary using RNN-LSTM Network: best coffee ever

Predicted Summary using RNN using Attention mechanism Network
: best coffee ever
```

Fig. 12 Comparison of output of 3 abstractive summarization models on amazon review

6 Conclusion

Abstractive summarization is an attractive research area among the NLP community and helps produce a human-like summary that is expected to be coherent, concise, non-redundant, and information-rich in nature. There are many existing challenges and the opportunity for advanced prospects in this area. In the proposed architecture for multi-lingual summarization, new tools have to be built so that it works for the regional language also. For example, summarizing a news article is very different from summarizing a report on financial gains. Some features of the text, such as document length or genre (technology, sports, finances, travel, etc.) make the task of summarization a serious problem that must be solved. The structure and semantic-based methods along with the upcoming methods like neural networks can help to minimize the difficulties faced in generating abstractive summarization. Though deep learning models are best suited to implement sequential data due to the lack of structured datasets, there exist a lot of problems while generating summaries. Besides, to create word embedding of all words existing in a language is a more difficult task to complete, so an ideal way is to identify the domain or application where the summarizer would be used and create the word embedding suitable can be one of the motives in the future work.

References

1. P.K. Rachabathuni, A survey on abstractive summarization techniques, in *International Conference on Inventive Computing and Informatics (ICICI)*, pp. 762–765 (2017)
2. A.M. Rush, S. Chopra, J. Weston, *A Neural Attention Model for Abstractive Sentence Summarization* (2015)
3. J. Buys, P. Blunsom, *Robust Incremental Neural Semantic Graph Parsing*. Cornell University Library, 40 (2017)
4. E. Baralis, L. Cagliero, S. Jabeena, A. Fiori, S. Shah, Multi-document summarization based on the yago ontology. Expert Syst. Appl. **40**, 6976–6984 (2013)
5. J. Niu, H. Chen, Q. Zhao, L. Sun, M. Atiquzzaman, Multi-document abstractive summarization using chunk-graph and recurrent neural network, in *IEEE ICC 2017 SAC Symposium Big Data Networking Track* (2017)
6. E. Jobson, A. Gutirrez, Abstractive text summarization using attentive sequence-to-sequence rnns. Stanford Reports
7. R. Nallapati, B. Zhou, C. dos Santos, C. Gulehre, M. Lapata, Abstractive text summarization using sequence-to sequence RNNs and beyond, in *The SIGNLL Conference on Computational Natural Language Learning* (2016)
8. G. Rossiello, P. Basile, G. Semeraro, M.D. Ciano, G. Grasso, Improving neural abstractive text summarization with prior knowledge, in *URANIA-16* (2016)
9. P. Liy, W. Lamy, L. Bingz, Z. Wangy, *Deep Recurrent Generative Decoder for Abstractive Text Summarization* (2017)
10. S. Song, H. Huang, T. Ruan, Abstractive text summarization using LSTM-CNN based deep learning. Multimed. Tools Appl. **1**, 53–57 (2018)
11. A. Fan, D. Grangler, M. Auli, Controllable abstractive summarization, in *Proceedings of 2nd Workshop on Neural Machine Translation and Generation*, pp. 45–54 (2018)

12. T. Baumel, M. Eyal, M. Elhadad, *Query Focused Abstractive Summarization: Incorporating Query Relevance, Multi-document Coverage, and Summary Length Constraints into seq2seq Models.* Cornell University Library, pp. 297–327 (2018)
13. P. Nema, M.M. Khapra, A. Laha, B. Ravindran, *Diversity Driven Attention Model for Query-Based Abstractive Summarization.* Cornell University Library (2018)
14. J. Lin, X. Sun, S. Ma, Q. Su, Global encoding for abstractive summarization, in *Proceedings of the 56th Annual Meeting of the Association for Computational Linguistics*, pp. 163–169 (2018)
15. H. Sak, Andrew Senior, F. Beaufays, *Long Short-Term Memory Based Recurrent Neural Network Architectures for Large vocabulary Speech Recognition* (2014)
16. D. Bahdanau, K.H. Cho, Y. Bengio, *Neural Machine Translation by Jointly Learning to Align and Translate* (2016)

Analysis and Performance Evaluation of Innok Heros Robot

Rajesh Kannan Megalingam, Avinash Hegde Kota, Vijaya Krishna Tejaswi Puchakayala, and Apuroop Sai Ganesh

Abstract This research works with the analysis of an industrial robot powered by the Robot Operating System (ROS) and can be used in agricultural applications, logistics, etc. The chosen Innok Heros is a multi-terrain modular robotic platform that can be used both indoor and outdoor and can be easily modified to serve our purpose. The coefficient of friction for various terrains was analyzed to evaluate the performance of the Innok Heros robot. The simulation is carried out in a ROS gazebo simulator over various types of terrain and various angles of elevation to analyze its behavior. This robot can reduce human effort and boost productivity in the agriculture sector. This research work concludes that the Innok Heros robot can be used in the majority of the terrains including elevations. Hence, this robot can be automated and deployed in agriculture and logistics to reduce the human effort to a great extent.

Keywords Robot Operating System · Agriculture · Gazebo · Analysis · Terrain

---

R. K. Megalingam (✉) · A. H. Kota · V. K. T. Puchakayala
Department of Electronics and Communication Engineering, Amrita Vishwa Vidyapeetham, Amritapuri, India
e-mail: rajeshm@am.amrita.edu

A. H. Kota
e-mail: avinashhegde@ieee.org

V. K. T. Puchakayala
e-mail: ptejaswi@ieee.org

A. S. Ganesh
Engineering Product Development Pillar, Singapore University of Technology and Design, Singapore 487372, Singapore
e-mail: ganesh_koppaka@mymail.sutd.edu.sg

V. Suma et al. (eds.), *Inventive Systems and Control*,
Lecture Notes in Networks and Systems 204,
https://doi.org/10.1007/978-981-16-1395-1_33

1 Introduction

In modern-day farming, farmers started to deploy technology to increase productivity and resolve numerous pressing problems. The increase in the global population has resulted in a global food shortage. With the sharp decline in agricultural laborers, it is not effective to practice the traditional methods of farming. The technology incorporated with traditional farming could boost productivity and meet the shortage of the agriculture workforce. Innok Heros is a multi-terrain modular robotic platform that can be used both indoor and outdoor and that can manage to carry a payload of up to 200 kg. It is available with three or four wheels that work on the Robot Operating System (ROS) kinetic version.

ROS is an open-source platform operated by Open Robotics Foundation. ROS can be used to simulate robots in simulation environment providers like a gazebo, Carla simulator, and visualization tools like Rviz. It can be integrated with LiDAR and various perception sensors like a camera for autonomous navigation. Innok Heros is developed by Innok Robotics of Germany. This research work throws light on a fact that Innok Heros can be used in various terrains like snow, gravel and can also be developed to serve great purposes. Innok Heros will be helpful to many areas of research and also in industries, agriculture, and logistics. In logistics, wherever the linear movement of loads is involved, there Innok Heros can be used. Innok Heros serve several other purposes like clearing snow and lawn movers. The software for Innok Heros is built based on the ROS platform. The gazebo is an open-source 3D simulator for robotics. It can be combinedly used with ROS as it supports code control of sensors and actuators. The gazebo has an in-built ODE physics engine and OpenGL rendering because of which roboticists are able to check their software stack before actually implementing it on hardware.

In this research work, Unified Robot Description Format (URDF) of Innok Heros provided by Innok Robotics is used to analyze and prove the abilities of the robot. URDF files can be generated with help of an add-on in CAD software like SolidWorks and AutoCAD. As part of this research work, the robot was controlled in a gazebo using keyboard control of the teleop package of ROS. Teleop is a ROS package that enables users to control a ROS-based robot remotely.

2 Motivation

The heavy impact of autonomy in agriculture and logistics was a driving force for our research work. With the decreasing number of farmers, there is definitely a need for autonomy up to an extent in farming. The fact that a lot of labor goes into handling packages in logistics has made us think and find an alternative. Innok Heros could help in moving packages as in logistics. As the design and implementation of a suitable robot from scratch are costly and time taking, this research work is planned to analyze and upgrade existing robots.

Due to the COVID shutdown, the team is not able to work with the hardware; hence, this research work is limited to software simulation and analysis only. As ROS is an open-source robotics platform, there is a necessity for ROS-based industrial robot. One such robot is Innok Heros. The fact that Innok Heros is capable of carrying decent payloads, kept driving this research work. As Innok Heros was found to be producing promising results during initial trials of simulation, the research work was carried out with it.

3 Related Works

Research work [1] is a summary of the architecture behind ROS-based Innok Heros. It focuses more on the compatibility of Innok Heros with various simulation and visualization platforms like rviz and gazebo. This research work [2] deals with the analysis of ROS-based autonomous mobile robots based on the purpose of the robot. This has helped in the identification of possible errors and problems that can be encountered while implementation. The research work [3] is on the classification of the terrain based on the acoustic signal produced during the interaction of the ROS-based legged robots with the terrain environment. This acoustic feature proved to improve the productivity of the system with the identification of the terrain type.

This research work [4] focuses on the development of the mobility of the robot on rough terrains. A simple mechanism is proposed for a four-wheel-based robot to maintain mobility on rough terrain. This research work [5] aims to increase the mobility of the 4 wheeled legged bots and surmount steep obstacles. The attachment of the wheels to the elastic mechanism helps it to achieve the structural stability to surpass the obstacles. This proposed mechanism has resulted in the agility of the robot to move over the rough terrains without the prior knowledge of the environment while maintaining stability. A distributed control system was developed in this research work [6] for ROS-based robots. This helped in the realization of the centralized control of the system. The proposed distributed teleoperation has helped in wireless control of the heterogeneous robot using a keyboard. This has expanded the scope of control of the ROS-based heterogeneous robots.

The research work [7] is all about designing and manufacturing a solar auto rickshaw. The innovative vehicle which is called Amrita Surya Vahini runs up to a top speed of 50 kmph. It is both light in weight and efficient in performance. The vehicle runs on power stored in an onboard battery that is charged directly from PV panels on the roof of Innok Heros can also be upgraded in this way as a step toward saving non-renewable resources. An intelligent way of monitoring soil quality for judicious irrigation was proposed in this research work [8]. Such a system can be mounted on the surface of Innok Heros with a slight modification in design. The research work presented in this publication deals with wireless communication in the monitoring system. So the same network can be used to control Innok Heros as well.

An optimistic approach to obstacle detection was presented in this research work [9]. The cost-effective approach toward such a complex task is a unique aspect of this research work. Availability of Bluetooth module is an added advantage making the autonomous robot user friendly. As the research work shows very high accuracy of 92%, this can opt for obstacle detection in order to lower the cost of production. The proposed research [10] work proposes the use of joystick in electrical vehicles for disabled people who could have control over their arms. This project proves the advantage of using a control system with closed-loop system. When the joystick is used, it produces the required voltage that triggers the necessary acceleration and braking through the microcontroller.

The research work [11] is all about remotely controlling a ROS-based robot through a secure shell. This work is very helpful for common people aspiring to deploy robots like Innok Heros in various ways. More convenient ways of controlling the robot as brilliantly mentioned in this research [12] are using keyboard and slider-based GUI. Slider-based GUI control will help a novice in computer science in controlling the robot. This keyboard and GUI are also based on ROS which is in favor of Innok Heros. As mentioned earlier, Innok Heros can be automated using perception sensors. The research work [13] analyzes path planned and executed in ROS-based robots which is much helpful in the further development of Innok Heros. Voice-based control as shown in research work [14] can be integrated with Innok Heros. Voice-based control will be an added advantage to Innok Heros when compared to other existing logistic robots.

4 System Architecture

The system architecture shown in Fig. 1 is explained in this section. The architecture can be classified into four main sections: sensor, GUI, software (ROS), the controller, and the actuators. The controller is not shown in the diagram explicitly as the ROS is already part of the controller which could be any microprocessor-based controller.

The proposed software architecture could make the existing system autonomous. This helps in the decrement of the human effort. The proposed system involves the integration of different sensors like LiDAR, camera, IMU, and other sensors like RADAR, ultrasonic, etc. These sensors observe the surrounding environment and input the data to the microcontroller via the Kalman filter. These inputs are further processed by the respective blocks as shown in Fig. 1. Perception involves path detection object detection and other parameters. Navigation involves the utilization of data from the navigation path planning algorithm and localization. This data is further used to control the actuators.

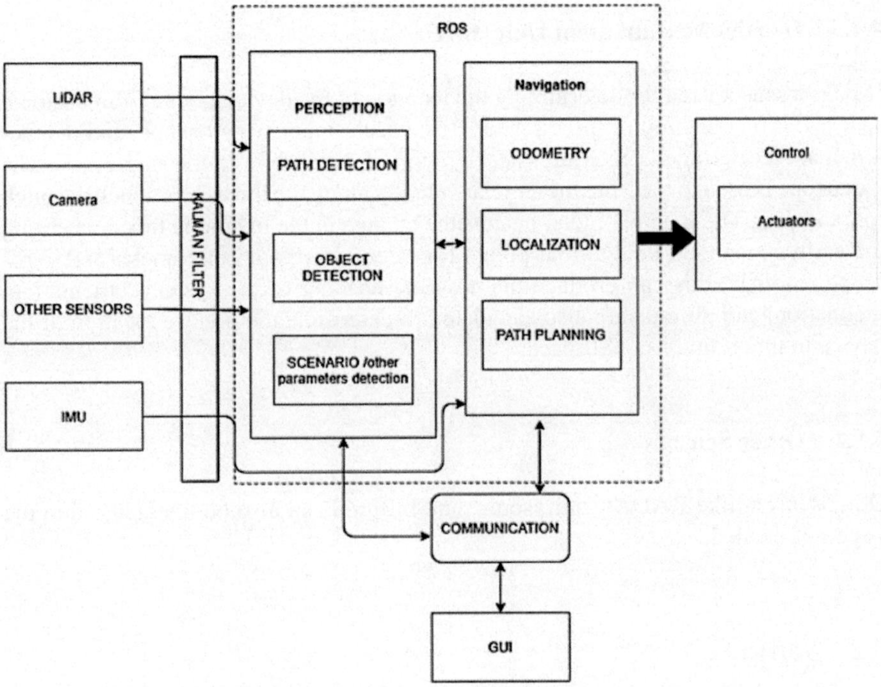

Fig. 1 Architecture of the system

4.1 Sensors

4.1.1 LiDAR

Light detection and ranging method are used to calibrate the distance of the objects from the parent device. LiDAR devices throw rapid laser light pulses at the object surface. It calculates the distance from the object based on the time taken by each pulse to return back. It produces mass point cloud datasets. At this point, cloud data is post-processed to get highly accurate georeferenced *x*, *y*, *z* coordinates. This data can be used to map the surrounding environment.

4.1.2 Camera

Multiple cameras are integrated to get a 360-degree view of the surrounding environment. This raw data from cameras is processed for object detection, lane detection, and several other purposes. The quality of the input data is often affected by external environmental conditions like sunlight, fog, etc.

4.1.3 Inertial Measurement Unit (IMU)

IMU is a sensor used for determining the location of the device in three-dimensional space. It is a collection of three sensors like accelerometer, gyroscope, and magnetometer. Accelerometer measures acceleration of the system in the single axis. The gyroscope is used to measure the angular velocity along the three-axis which are pitch roll and yaw. The magnetometer detects the change in the magnetic flux concerning the earth's magnetic field at that point. Together with the accelerometer and gyroscope readings, this can produce the absolute heading of the system. Orientation, inclination, and altitude are also calculated to determine the precise location of the device in three-dimensional space.

4.1.4 Other Sensors

Other sensors like RADAR, ultrasonic, and infrared can also be integrated into the system if needed.

4.2 Software

4.2.1 Path Detection

Path detection is the process of identification of the path required to reach the goal point. Live data from perception sensors like camera and LiDAR are processed to identify the available paths for the robot to travel. Later, this data is processed with the path planning algorithms to determine the optimal path.

4.2.2 Object Recognition

Object recognition is achieved by processing the data received from different sensors. Live feed from the camera is processed frame by frame using a series of computer vision techniques to remove noise and identify the objects in the environment. OpenCV, MATLAB, TensorFlow are some of the tools used to process the camera feed. Along with the processed LiDAR data, the exact position of the object is determined. Open-source algorithms like YOLO can be used to cut down the cost of the robot. For mechanical tasks, especially in the field of logistics camera feed can be converted to a third-order polynomial to reduce the load on the processor. Polynomial can be decoded to find parameters like distance from center to perform tasks such as following a guided path.

4.2.3 Other Parameters

When new sensors are introduced to the system, there will be new parameters that have to be configured.

4.2.4 Odometry

This is one of the techniques used for localization. It is the estimation of change in the position of the robot with respect to a reference point with time. This method calculates the distance traveled by measuring the wheel displacement and estimates its position relative to the reference point. The accuracy level of this method is generally low due to measurement errors, inaccurate wheel dimensions, etc. Hence, the accuracy of the localization is improved by using this technique in addition to other techniques.

4.2.5 Localization

Precise position of the ego vehicle is essential for safe autonomous driving. Information from Global Navigation Satellite Systems (GNSS) alone cannot give a precise location. LiDAR and camera-based systems are also required to minimize the error in positioning. Techniques like SLAM and map matching use the data from the above systems to calculate the location of the ego vehicle with the least error.

4.2.6 Path Planning

Path planning is required for the ego vehicle to find the safest and feasible geometric path from the initial point to the destination. This includes the identification of static and dynamic obstacles to optimizing the path. The efficiency and time were taken to reach the destination depending on the efficiency of the path planning algorithm.

4.2.7 Communication

While using ROS, the in-built data distribution service will take care of communication between nodes when properly linked by topics. When the robot is teleoperated, communication between operator and robot will be made possible through SSH when connected to a network.

4.3 Kalman Filter

The Kalman filter is used to improve the accuracy of the inputs from the sensors of the dynamic environment. It is also known as the linear quadratic estimation. It uses inaccurate input data with errors from the sensors and provides estimates for the unknown variables based on measurements over time. It calculates the joint probability distribution for the variables for each time frame. These measurements are accurate compared to the direct sensor inputs.

4.4 Graphical User Interface (GUI)

GUI: ROS supports the use of GUI dashboards created using PyQt. Graphical user interface to control a ROS-based robot-like Innok Heros will be of great help. Even when Innok Heros is automated as proposed vital ego information can be made available to the user. GUI can help the user to tweak various parameters like speed if needed.

4.5 Controller

This research work used a computer with the Intel microprocessor to build the ROS-based software and use it for the simulation using gazebo, etc. In addition, the Rviz environment and the environment required for the gazebo simulation are also done on the same computer.

5 Design and Implementation

The design of Innok Heros can be considered stable. As shown in Fig. 2, it is a proper cuboid with a flat top and four wheels that make it stable. Innok Heros is designed to have a differential drive. Initially, it was designed for steadily transporting load by remote operation. It can be automated by the addition of a few sensors to the existing hardware. Innok Heros can be automated with the above-proposed architecture.

A ramp is constructed in the gazebo to test the performance of the Innok Heros with varying friction values. The ramp is constructed such that 2 rectangular surfaces inclined at 15° at the opposite sides of a horizontal rectangle sheet which is normal to the ground plane. The coefficient of friction of the 2 inclined rectangular surfaces can be configured according to the requirement. These 2 surfaces are used to ascend and descend the ramp. The coefficient of friction is varied based on the terrain selected for the analysis.

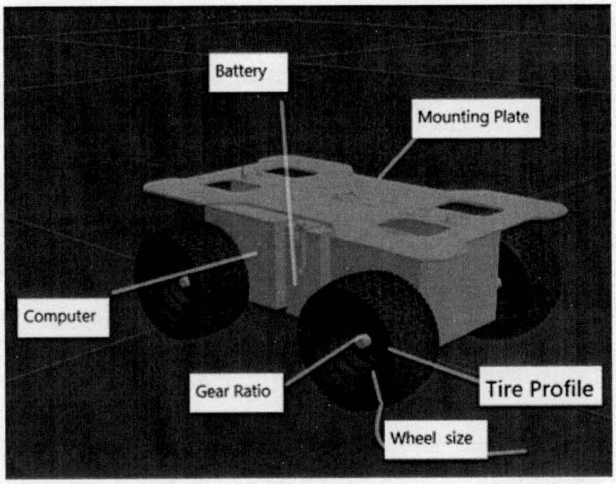

Fig. 2 Components of Innok Heros

The whole setup can be modified based on the proposed software architecture. Several sensors can be integrated into the system to make the system autonomous and reduce human intervention with the system and environment.

6 Node Graph

The RQT graph as shown in Fig. 3 can be generated when the robot is being simulated. It is a graphical representation of how nodes communicate over topics in ROS during a simulation. Cmd_vel is a velocity topic which is being by teleop node. Gazebo's physics will direct the robot accordingly through the topic joint_states.

Turtlebot_teleop: This teleop node is from open-source teleoperation ROS package that can be used to control various ROS-based robots with various devices like a keyboard and joystick. This node can be activated by using turtlebot_teleop package.

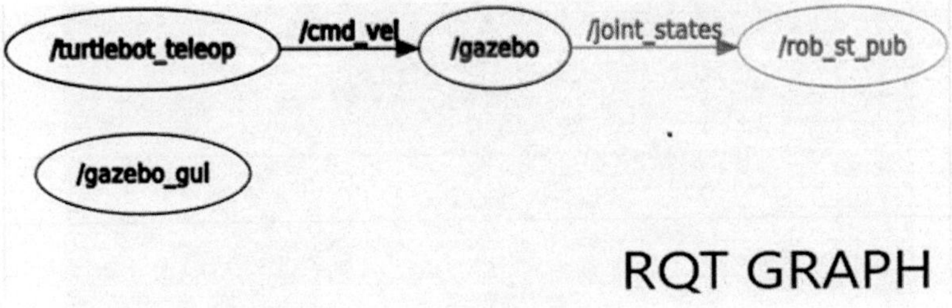

Fig. 3 RQT graph of the simulation

/cmd_vel: This topic is to send user inputs from teleop node to gazebo node, so that the script can execute tasks based on user input.

/gazebo: Gazebo node is the major node during the simulation because it is handling a lot of inputs as well as keeping note of the log and also physics engine. The physics checkbox has to be ticked during this simulation to evaluate the performance of Innok Heros in real time.

/joint_states: This topic's role is to drive the output parameters to publish the node from the gazebo node as shown in the node graph.

/rob_st_pub: The output publishing node which is accessible to the user for debugging and observing the output.

/gazebo_gui: This node takes care of the graphical user interface of a gazebo simulation suite.

For the above-mentioned simulation, a mechanical model of the robot was imported from the official package provided by the makers of Innok Heros. The mechanical model was imported in URDF format. URDF stands for Universal Robot Description Format. This format consists of a set of files mostly in XML format that carry all the information about a robot in real life. This package is used by ROS. URDF of any mechanical model can be generated after designed in CAD software like SolidWorks or Autodesk360 after properly defining all the links and joints.

7 Experiment and Result

The experiment was carried out in gazebo as shown in Fig. 4. Innok Heros model with 4 tractor wheels was used during the simulation. The coefficient of friction was taken from Texas claims and Consulting Co. Several iterations have been performed with the coefficient of frictions mentioned in Table 1. From the experimental results, it can be inferred that the average time taken to climb up all 3 surfaces at 15° inclinations is almost the same. The average is calculated from five iterations for each of the surfaces.

Fig. 4 Snapshot of gazebo suite during simulation

Table 1 Data of time taken with variation in coefficient of friction

Terrain type	Coefficient of friction		θ	Average time (s)
	min	max		
Cement	0.80	1.2	15°	17.439
Gravel	0.55	0.85	15°	16.4272
Rock	0.55	0.75	15°	15.9846

Hence, it is proved that Innok Heros is efficient in harsh terrains like crushed rocks, gravel, etc. The time taken by Innok Heros 444 climb up the rock surface is less than that of the Portland cement, so it is more efficient when used in rock terrain than the Portland cement terrain. Similarly, each variant in Innok Heros is efficient with certain terrains. Innok Heros 223 performs better in asphalt cement than other surfaces. There is an intermediate version that has the specification between 223 and 444. Hence, the user has a wide variety of choices to choose based on their payload and requirement. The below snapshot was captured during the simulation of Innok Heros. The robot is over a ramp of 15° inclination. Model of Innok Heros with four tractor wheels was used. The performance of the robot was checked at a constant speed. For rough terrains with a high coefficient of friction, there was some turbulence. Time taken by the robot to ascend the cement terrain is greater than the other two due to the high coefficient of friction. Sliding is observed when the coefficient of friction is less than 0.55.

This free body diagram as shown in Fig. 5 is a pictorial representation of all the forces acting on the robot. The free body diagram of the experimental setup is shown in the above figure. In the free body diagram (FBD) N is the normal force, W is the weight of the Innok Heros robot, Fd is the driving force and the rest are components of weight because of the inclination. θ is the angle of inclination of the ramp surface with respect to the ground. This experiment has proved that Innok Heros can produce the required Fd to overcome the attenuating component friction and weight component.

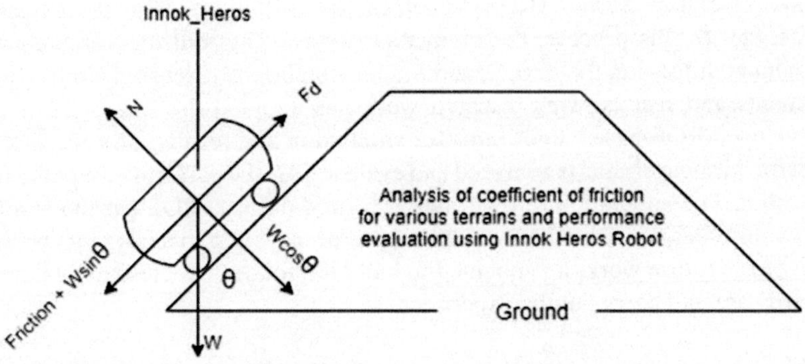

Fig. 5 Free body diagram of experimental setup

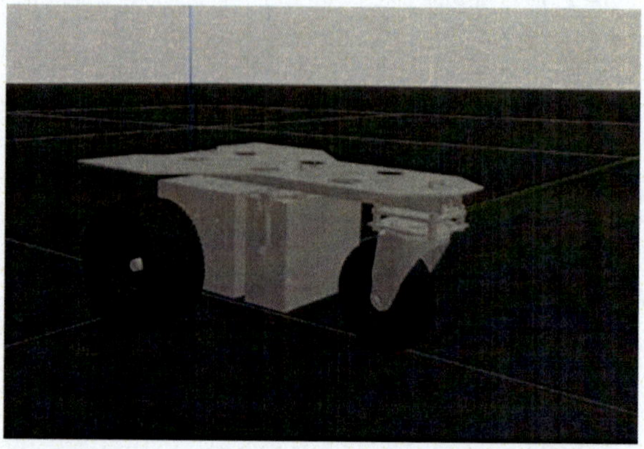

Fig. 6 Innok Heros 223 variant

The same experiment can be performed with various configurations of Innok Heros available.

Innok Heros 223: This variant (Fig. 6) of Innok Heros is suitable for asphalt terrain. This is also based on differential drive. Can take payload and towing capacity up to 200 kg.

Innok Heros 444: Fig. 2 is a snapshot of Innok Heros 444 variant. This variant is the best among available configurations because it works in most of the terrains and also has a higher payload and towing capacity.

Innok Heros 224: Innok Heros 224 is also available with specifications between 223 and 444. It is similar to the lower variant; the only difference is that configuration 224 has four wheels.

8 Conclusion

This research work presents the analysis of an industrial robot that is powered by the Robot Operating System (ROS) which can be used in agricultural applications, logistics, etc. In this process, the motivation for carrying out this research work, the system architecture, the design, and implementation is discussed along with the experiments and results. This research work has adopted the Innok Heros robot and developed ROS-based algorithms for simulation and results. The coefficient of friction for various terrains is analyzed and evaluated by the performance of the Innok Heros robot. The simulation of the robot is carried out in a ROS gazebo simulator over various types of terrain and various angles of elevation to analyze its behavior. As part of the future work, it is planned to build the robot in the research laboratory with hardware and carry out the experiment.

Acknowledgements We would like to express our sincere gratitude to Almighty God for giving us the opportunity to carry out the research in a successful manner. We extend our sincere thanks to Amrita Vishwa Vidyapeetham and the Humanitarian Technology (HuT) Laboratory for aiding in this venture.

References

1. N. Baslan, S. Heerklotz, S. Weber, A. Heerklotz, B. Höfig, J. Abu-Khalaf, Navigation and vision system of a mobile robot, in *2018 19th International Conference on Research and Education in Mechatronics (REM), Delft*, pp. 99–104 (2018). https://doi.org/10.1109/REM.2018.8421777
2. M. Köseoğlu, O.M. Çelik, Ö. Pektaş, H.H.E. Sanayi, A.Ş. Ankara, Turkey: design of an autonomous mobile robot based on ROS, in *2017 International Artificial Intelligence and Data Processing Symposium (IDAP)*. https://doi.org/10.1109/IDAP40716.2017
3. J. Christie, N. Kottege, Acoustics based terrain classification for legged robots, in *2016 IEEE International Conference on Robotics and Automation (ICRA) Stockholm, Sweden*, May 16–21, 2016
4. S. Nakajima, Development of four-wheel-type mobile robot for rough terrain and verification of its fundamental capability of moving on rough terrain, in *2008 IEEE International Conference on Robotics and Biomimetics Bangkok, Thailand*, February 21–26, 2009. https://doi.org/10.1109/ROBIO14616.2009
5. A. Bouton, C. Grand, F. Benamar, Motion control of a compliant wheel-leg robot for rough terrain crossing, in *2016 IEEE International Conference on Robotics and Automation (ICRA)*, 16–21 May 2016. https://doi.org/10.1109/ICRA.2016.7487448
6. F. Yuan, L. Zhang, H. Zhang, D. Li, T. Zhang, Distributed teleoperation system for controlling heterogeneous robots based on ROS, in *2019 IEEE International Conference on Advanced Robotics and its Social Impacts (ARSO) BIT, Beijing, China*, October 31–November 2, 2019
7. G. Udupa, J. Freeman, V. Nimbewal, R.K. Choudhary, S. Sasikumar, S.V. Nair, Design and manufacture of Amrita Surya Vahini—a solar auto rickshaw,in *2016 International Conference on Robotics and Automation for Humanitarian Applications (RAHA), Kollam*, pp. 1–6 (2016). https://doi.org/10.1109/RAHA.2016.7931902
8. J.P. Guruprasadh, et al., Intelligent soil quality monitoring system for judicious irrigation, in *2017 International Conference on Advances in Computing, Communications and Informatics (ICACCI), Udupi*, pp. 443–448 (2017). https://doi.org/10.1109/ICACCI.2017.8125880
9. R. Chinmayi, et al., Obstacle detection and avoidance robot, in *2018 IEEE International Conference on Computational Intelligence and Computing Research (ICCIC), Madurai, India*, pp. 1–6 (2018). https://doi.org/10.1109/ICCIC.2018.8782344
10. C. Sachin, S. Srihari, K. Prajeesh, N. Prem, B. Sreevasudev, K.R. Bharath, In-creasing accessibility of electric vehicles with joystick based drive-by-wire model, in *2020 6th International Conference on Control, Automation and Robotics (ICCAR), Singapore, Singapore*, pp. 435–439 (2020). https://doi.org/10.1109/ICCAR49639.2020.9107978
11. R.K. Megalingam, et al., Robot Operating System Integrated robot control through Secure Shell (SSH), in *2019 3rd International Conference on Recent Developments in Control, Automation & Power Engineering (RD CAPE), Noida, India*, pp. 569–573 (2019). https://doi.org/10.1109/RDCAPE47089.2019.8979113
12. R.K. Megalingam, N. Katta, R. Geesala, P.K. Yadav, R.C. Rangaiah, Key-board-based control and simulation of 6-DOF robotic arm using ROS, in *2018 4th International Conference on Computing Communication and Automation (ICCCA), Greater Noida, India*, pp. 1–5 (2018). https://doi.org/10.1109/CCAA.2018.8777568

13. R.K. Megalingam, A. Rajendraprasad, S.K. Manoharan, Comparison of planned path and travelled path using ROS navigation stack, in *2020 International Conference for Emerging Technology (INCET), Belgaum, India*, pp. 1-6 (2020). https://doi.org/10.1109/INCET49848.2020.9154132
14. R.K. Megalingam, R.S. Reddy, Y. Jahnavi, M. Motheram, ROS based control of robot using voice recognition, in *2019 Third International Conference on Inventive Systems and Control (ICISC), Coimbatore, India*, pp. 501–507 (2019). https://doi.org/10.1109/ICISC44355.2019.9036443

A Smart Plant Watering System for Indoor Plants with Optimum Time Prediction for Watering

Vejay Karthy Srithar, K. Vishal Vinod, S. K. Mona Sweata, M. Karthika Gurubarani, and K. Abirami

Abstract Water plays an important role in our daily lives and is an essential component of agriculture. One of the main difficulties faced by people is that they do not know when to water the plants and also the amount of water to be poured, resulting in the death of the plant. Current solutions to this problem focus on automatically watering the plants when the soil gets dry or drip irrigation-based solution where the plants are watered continuously. But this solution does not help people whose hobby is gardening and for those who want to take care of their plants by themselves. To tackle this problem, a system is proposed where the time is used to predict when a plant has to be watered in future. Our research is mainly on the deep learning (DL) algorithms, to analyze which would give better performance in this scenario. To achieve this, various sensors to obtain the moisture level of the soil, temperature, and humidity of the air around the plant. The cloud technologies and deep learning techniques are used to store, process, and obtain the optimum time. Since the data received is time-series data, we use LSTM and GRU models to predict the optimum time to water the plants. The results and their comparison are discussed in this article. This value is then sent to the user through the mobile app. This solution will help people to enjoy gardening as they need not worry about their plant's health.

Keywords LSTM · GRU · Deep learning · IoT · Firebase · ESP32

V. K. Srithar · K. Vishal Vinod · S. K. Mona Sweata · M. Karthika Gurubarani · K. Abirami (✉)
Department of Computer Science and Engineering, Amrita School of Engineering, Coimbatore, Amrita Vishwa Vidyapeetham, Coimbatore, India
e-mail: k_abirami@cb.amrita.edu

V. K. Srithar
e-mail: cb.en.u4cse17164@cb.students.amrita.edu

K. Vishal Vinod
e-mail: cb.en.u4cse17165@cb.students.amrita.edu

S. K. Mona Sweata
e-mail: cb.en.u4.cse17133@cb.students.amrita.edu

M. Karthika Gurubarani
e-mail: cb.en.u4cse17127@cb.students.amrita.edu

V. Suma et al. (eds.), *Inventive Systems and Control*,
Lecture Notes in Networks and Systems 204,
https://doi.org/10.1007/978-981-16-1395-1_34

1 Introduction

Plants are an integral part of nature. Human beings receive many benefits from plants in many aspects. They play a vital role in providing a clean and green environment by naturally cleaning the air and providing oxygen. People, whose hobby is gardening, love to have a garden with different varieties of plants in their backyard. But lack of sufficient space made many people grow indoor plants. There are innumerable benefits of indoor plants. They not only make the home beautiful ambiance and eco-friendly but also absorb carbon dioxide, purify the air by removing toxins, prevent illness, etc. In this hectic and fast-paced world, they reduce our stress levels making us relaxed. Also, they help us to work much better by improving attention and concentration and by enhancing creativity and productivity. The correct amount of water and sunlight is needed for indoor plants for their survival and their growth. So, it becomes essential to take care of indoor plants by ensuring that these plants get enough water [1, 2] and nutrients at the right time. But, often find it difficult to keep a constant check on all the plants. Initially, the optimum time to water the plants is known. Secondly, they do not know the correct amount of water needed to maintain the health of the plants. Also, each variety of plant requires a different amount of water and has to be watered at different intervals of time. It is very difficult to provide every plant with the right amount of water at the right time. Due to this, plants often end up getting inadequate water or excess water. In the end, plants suffer a lot, and finally, they withered. Adding to all this, one of the major problems worldwide is water scarcity. So, it becomes very essential to use every water drop wisely. In this article, to tackle this challenge, an IoT-based smart plant watering system is proposed so that it predicts when a person has to water the plant using results of deep learning (DL) algorithms and also alerts the optimum amount of water required for the plant. Firebase is used here where the real-time data from the sensors are collected and stored. Many Firebase services are used here for analysis, prediction, and decision making [3, 4]. This system helps gardening enthusiasts to effectively and efficiently take good care of their plants.

2 Literature Survey

Manual watering of plants is difficult over a large area. Siva et.al. [5] proposed a solution using Arduino UNO. Data from the soil moisture sensor has been analyzed continuously when the moisture level goes beyond the threshold, and the plants are watered automatically using a water pump. It stops after the moisture level increases to an optimal amount. However, it does not notify the user when to water the plants.

Reghukumar et al. [6] proposed a plant watering system that enables users to monitor the status of a plant. The system mainly focuses on obtaining data from various sensors and using the cloud to store and perform analytics on the data. A response is sent as an email regarding the current health status of the plant. The advantages of cloud in the agriculture field are being showcased here. The disadvantage of this system is that there is no prediction on when to water the plants again in near future.

Long short-term memory (LSTM) has been used for finding low-frequency sublayers by researchers, in accurate wind speed prediction. In this article [7], the usage of single-layer LSTM model has been showcased for prediction. LSTMs efficiency is discussed for this model. Here, LSTM is chosen as wind speed data is time series. But there is no comparison with other prediction algorithms.

In this article [8], the use and power of machine learning algorithms in the agricultural domain have been portrayed. The working and principles behind the machine learning (ML) algorithms have been compared. Various techniques for crop yield estimation and precision nitrogen management have been summarized. There is no comparison and analysis with deep learning prediction algorithms.

Various cloud technologies that support IoT integration especially with Arduino have been discussed in detail by Chandra [9]. The advantages of these services for integrating sensors have been explained. Cloud service providers like ThingSpeak, Sense, etc., and the various usages of these platforms ease of usage for analysis, and integration has also been summarized. However, it does not elaborate on a specific application.

Mhatre et al. have elaborated on the use of ThingSpeak for IoT [10]. The features and support for the integration of sensors and other devices to the cloud and the visualization of data obtained from these Things have been detailed. This showcases the scalability, efficiency, and effectiveness of cloud technology with IoT. But, this does not describe other platforms like Firebase.

Ashwini et al. proposed an automated system that is used for watering plants [11]. Arduino Uno board along with ESP8266 which is used as a Wi-Fi module is used in sensing the moisture level value of the soil. Water to plants is supplied by a 12 V DC motor pump. However, there is no implementation of machine learning algorithms.

Shekhar et al. suggested a system involving an Arduino board and Raspberry Pi as the microcontroller [12]. The temperature and moisture level are sensed with the help of sensors and that information is sent to the microcontroller. This is then passed on to the Raspberry Pi where the machine learning algorithm K-NN is deployed. But, there is no comparison of different machine learning techniques to get the best result.

Rawal et al. proposed a system that uses a moisture sensor that sends data to the Arduino Uno board which consists of a microcontroller that switches on/off the motor on which the water sprinklers are attached [13]. The data from Arduino is sent to the modem to a Webpage that shows the moisture level in the soil and also the status of the sprinkler. There is no prediction of time, to water the plants, using deep learning algorithms.

Cloud computing is expected to be a good approach for integration with IoT applications. It has various advantages including cost-efficient services and many features including virtualization services. This article [14] shows the usage of the cloud for IoT along with the services that can be integrated with it. This enables collecting, transmitting, analyzing, processing, and storing of data easier. But, there is no implementation of prediction algorithms in cloud.

This research article [15] discussed an automated prototype that waters plants based on the soil moisture level. The prototype is built with soil moisture sensors, microcontroller, relay, and solenoid valve and is then integrated with the IoT platforms such as Blynk apps and Thingspeak. ThingSpeak is used here to represent the moisture percentage as a graph. Blynk apps are used to provide notification to the user to his mobile during the activation or deactivation of the watering device. However, there is no prediction of time to water the plant again.

The approach described in this article [16] is to water plants smartly to save water resources and to save the plants from unpredictable weather. Raspberry Pi has been used for different capabilities and processing speeds and SCP for cloud service. An innovative part is that they have moved out the excess water into the trenches and which are then collected on to the tanks. But, there is no prediction of time to water the plant again.

The system proposed here [17] is an IoT application that has two nodes where the data is collected is transmitted through Zigbee. That data later gets stored in the cloud. The cloud server performs decisions on the threshold values that are predefined. An algorithm is applied and the actions are then sent to the nodes. The information about the plant is sent to the Android app so that users get notified. But, the prediction algorithm is not used.

3 Proposed Model

The values from the sensors (soil moisture sensor—FC28, DHT22) are obtained and sent to the ESP32 microcontroller as shown in Fig. 1. These values are sent to Firebase using the HTTP protocol. They are stored in Firebase's real-time database

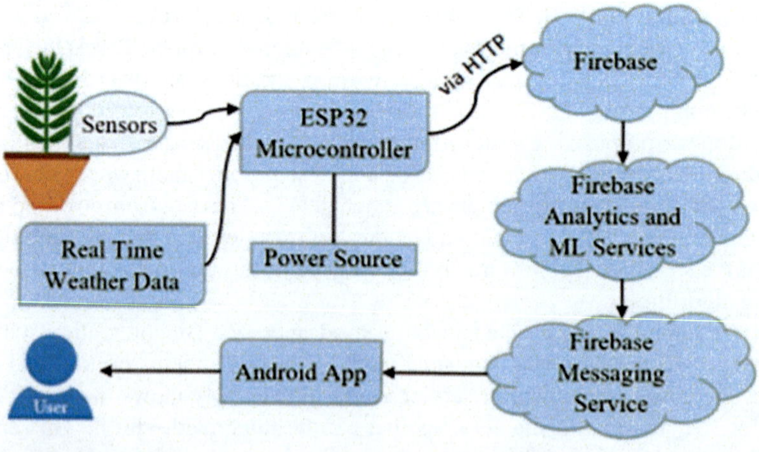

Fig. 1 Architecture diagram

and also Cloud Firestore. Firebase's machine learning service is used to implement the prediction algorithms, LSTM, and GRU. The best result of these is sent to the Android app using Firebase's cloud messaging service, which enables the user to monitor and also water the plants on time.

4 Implementation

4.1 Hardware Setup

The components required include an ESP32 microcontroller, a soil moisture sensor (FC-28), a DHT22 sensor, 2 LEDs, a Breadboard, and necessary Jumper wires. The specifications of the components used are given in Table 1. The schematic diagram of the setup is given in Fig. 2.

The power supply is received through the Serial UART Port. The ESP32 is placed on the breadboard. 5 V output is connected to the common input power and the ground pin of ESP32 is connected to the common ground. GPIO pin 25 is connected to the Anode of LED1 and the cathode of LED1 is connected to the common ground. Power and ground pins of DHT22 are connected to the common power and ground of the breadboard. GPIO pin 26 is connected to the data pin of the DHT22 sensor. The anode of LED2 is connected to GPIO pin 32 and the cathode of LED2 is connected to the common ground. The soil moisture sensor probe's positive and negative leads are connected to the soil moisture hygrometer's positive and negative pins, respectively. The soil moisture hygrometer's power and ground pins are connected to the GPIO pin 33 and common ground ping respectively, and it is analog out pin is connected to GPIO pin 27.

The soil moisture sensor is first calibrated to check the correctness of values. The analog read resolution was set to 10 (value range = 0–1023). In dry soil, the reading was approximately 820, and in wet soil, the reading was approximately 255. The readings are inversely proportional to the moisture content in the soil. The temperature and humidity values from DHT22 are taken every 10 s and then the average for one minute is recorded.

The code to implement is written using the Arduino IDE. The necessary libraries have to be included and the configuration has to be done as per the requirements of Firebase. Then, the code has to be flashed into ESP32. Then, when ESP32 is powered in Figs. 3 and 4, sensor values will be sent to Firebase.

4.2 Android App

The Android app is developed using Java language by using the software Android Studio. This app acts as a user interface for the project where the user will be able

Table 1 Specifications of components

S. No	Component	Specifications	
		Parameter	Value
1	Microcontroller—ESP32	Module interfaces	SD card, UART, SPI, SDIO, I2C, LED PWM, Motor PWM, I 2S, IR, pulse counter, GPIO, capacitive touch sensor, ADC, DAC, two-wire automotive interface (TWAITM, compatible with ISO11898-1)
		On-chip sensor	Hall sensor
		Integrated crystal	40 MHz crystal
		Integrated SPI flash	4 MB
		Operating voltage/power supply	3.0–3.6 V
		Operating current	Average: 80 mA
		Minimum current delivered by power supply	500 mA
		Recommended operating temperature range	− 40 °C ~ +85 °C
		Package size	(18.00 ± 0.10) mm × (25.50 ± 0.10) mm × (3.10 ± 0.10) mm
		Wi-Fi protocols	802.11 b/g/n (802.11n up to 150 Mbps); A-MPDU and A-MSDU aggregation and 0.4 μs guard interval support
		Wi-Fi frequency range	2.4–2.5 GHz
		Bluetooth protocols	Bluetooth v4.2 BR/EDR and BLE specification
2	Soil moisture sensor (FC-28)	Input voltage	3.3–5 V
		Output voltage	0–4.2 V
		Input current	35 mA
		Output signal	both analog and digital
		PCB size	3.2 cm × 1.4 cm
3	Humidity and temperature sensor (DHT22)	Power supply	3.3–6 V DC (typically 5 V)
		Current use	2.5 mA max
		sensing element	Polymer capacitor
		Operating range	Humidity 0–100%RH; Temperature: −40 ~ 80 °C
		Accuracy	Humidity ±2%RH (Max ±5%RH); Temperature < ±0.5 °C

(continued)

Table 1 (continued)

S. No	Component	Specifications	
		Parameter	Value
		Dimensions (sensor body)	15.1 mm × 25 mm × 7.7 mm

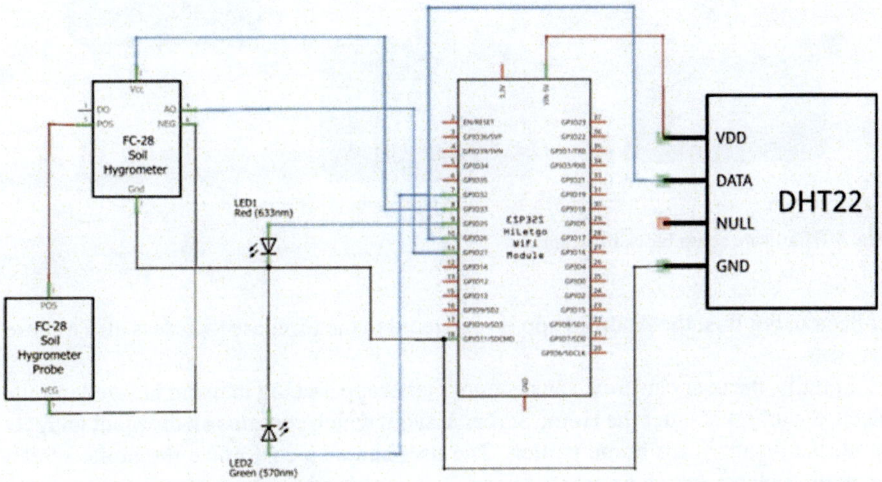

Fig. 2 Hardware schematic diagram

Fig. 3 Hardware setup

to get alerts, notifications, and view all the details of his indoor plant garden. There are four layouts in total in this app (Registration, Login, Home Screen, Results). Firebase real-time database acts as the backend for this app from where plant data is

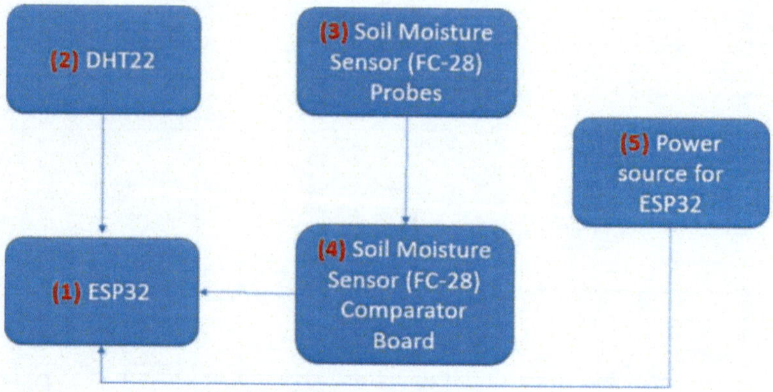

Fig. 4 Hardware setup block diagram

collected. For this, the Android app is registered in the Firebase to access all Firebase services.

Initially, the user can create an account in this app and log in using his credentials. Then, the user will enter the Home Screen Layout which contains all the plant widgets of all the plants in his home garden. The user can also customize the home screen by adding plants which he wants to monitor or removing plants that do not require further monitoring. When he wants to view a particular plant's details, the user clicks on the particular plant widget. Now all the details of the plants like moisture level, humidity, weather conditions of the particular place, etc., are fetched from Firebase real-time database and will be displayed in the Results Layout. Also, the predicted values of the plant-like when the plant has to be watered next are sent as an alert to this app from cloud using Firebase cloud messaging service.

The main purpose of the application is to get alerts about when the user has to water next. The block diagram of the application is shown in Fig. 5.

In the home screen (Fig. 6), users will be able to customize their own indoor plant garden. That is, they will be able to add their different varieties of indoor plants for which analysis, prediction, and monitoring have to be done.

After the creation of their indoor plant garden, when the user clicks on the particular plant widget, the details of that particular plant (Fig. 7), like current humidity, temperature, and soil moisture level values, the time when it was last watered, etc., will be displayed. These real-time data about the plants will be stored in Firebase. By establishing a connection between the Android application and cloud, it can retrieve and display the data.

The user will also receive notifications in the app regarding when they have to water the particular plant next (the result of the prediction algorithm). Also, they will receive these notifications prior so that even when they are not home, and they can make some alternate arrangements for gardening.

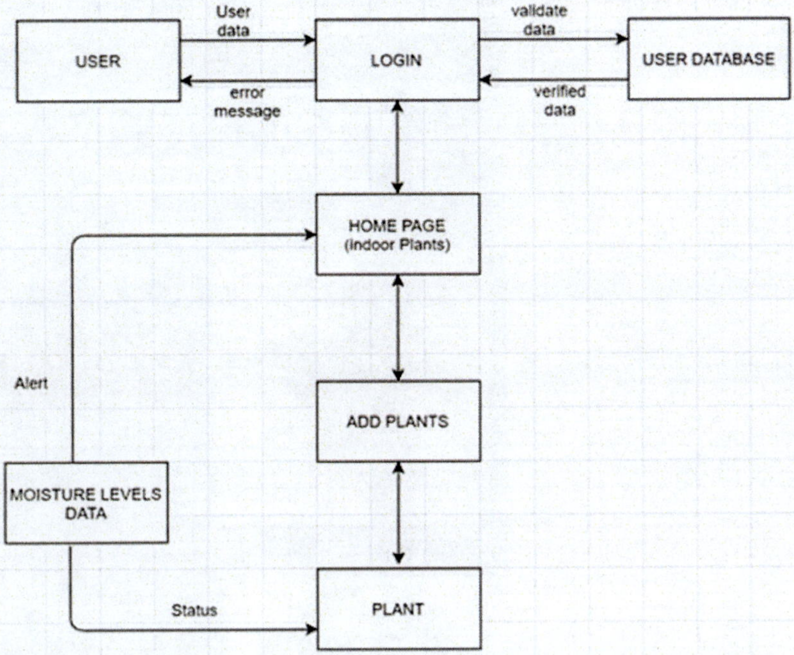

Fig. 5 Android app workflow

4.2.1 Authentication with Firebase

The user can create an account in this app using his email credentials and then log in. When the user logs in using his email credentials, this has to be validated. For this, the Firebase email authentication service is used in this app.

4.3 Prediction Algorithms

This project provides real-time insights about the plants through the cloud platform and also deep learning algorithms are used for making various predictions on the proposed system. The microcontroller sends the collected data from sensor to the cloud server [18]. As the sensor data cannot be directly used for analysis purposes, the datasets would undergo various pre-processing techniques, which help in building the model more accurately. Later, the pre-processed data will be sent to deep learning models such as LSTM and GRU for predicting values for the proposed system [19].

Long short-term memory (LSTM) is a new and popularly used artificial recurrent neural network. It works best on time-series data. This algorithm requires only a

Fig. 6 Android app home page

minimal input data pre-processing and preserves useful information across multiple time steps. It comprises an input layer, a recurrent hidden layer (gating units for information flow), and an output layer. The structure for LSTM (LSTM Equations):

$$it = \sigma(wixt + Uimt - 1 + bi)$$

$$ft = (wsxt + Usmt - 1 + bs)$$

$$ct = ct - 1 \cdot ft + st \cdot i$$

$$mt = st \cdot ot$$

Here, "it"—input gate, "ft"—forget gate, "mt"—the hidden state, and "ct"—cell state at time *t*. The number of layers and memory blocks in each layer will be determined further.

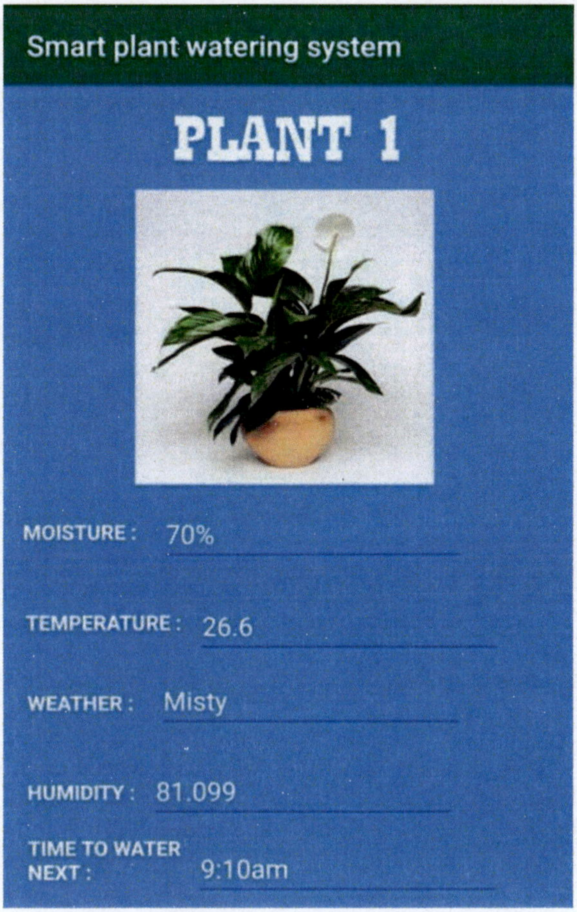

Fig. 7 Android app plant page

Gated recurrent unit (GRU) is similar to LSTM, with no cell state. It has only reset and updates the gate. GRU is faster than LSTM, whereas GRU handles vanishing gradient problems. To evaluate the efficiency and effectiveness of our project, the above machine learning algorithms where compared based on our sensor data for further predictions.

5 Observations

The data from the sensors is sent via HTTP protocol to Firebase and gets stored in Firebase's real-time database. The exported data is as shown in Fig. 8. The line charts show the values of moisture level (in percentage), temperature (in °C), and humidity values over a period of one hour with one-minute interval. Moisture level

1	Minutes	Moisture	Temperature	Humidity	Weather	Time
2	0	70.28348	27	80	Mist	1605592859
3	1	70.28348	27	79.9	Mist	1605592919
4	2	70.28348	27.1	80.2	Mist	1605592979
5	3	70.08798	27.1	79.8	Mist	1605593039
6	4	69.99023	27.1	80.2	Mist	1605593099
7	5	69.89247	27.1	80.3	Mist	1605593159
8	6	69.89247	27.1	80	Mist	1605593219
9	7	69.79472	27.200001	79.9	Mist	1605593279
10	8	69.89247	27.200001	80.4	Mist	1605593339
11	9	69.79472	27.200001	79.9	Mist	1605593399
12	10	69.89247	27.200001	79.5	Mist	1605593459
13	11	69.50147	27.200001	79.8	Mist	1605593519
14	12	69.59922	27.200001	80	Mist	1605593579
15	13	69.50147	27.200001	79.7	Mist	1605593639
16	14	69.40372	27.299999	80	Mist	1605593699
17	15	69.50147	27.200001	79.2	Mist	1605593759
18	16	69.69697	27.200001	80.1	Mist	1605593819
19	17	69.69697	27.200001	80	Mist	1605593879
20	18	69.30596	27.200001	79.4	Mist	1605593939
21	19	69.20821	27.200001	79.4	Mist	1605593999
22	20	69.20821	27.200001	78.9	Mist	1605594059

Fig. 8 Data from sensors

varies depending on whether the plant has been watered recently or not as shown in Fig. 9. The temperature value stays around 27.2 °C as shown in Fig. 10. The humidity level is between 80.5 and 77 as shown in Fig. 11.

6 Results and Discussion

Python is used for developing the models. The moisture value along with the time was given as input to both the models. During the training phase, the output for each input was the time when the moisture level dropped by 10 units. This makes the model to learn and predict the time when a plant has to be watered given a moisture value. So, when the plant is watered, the moisture level raises to 70%. This value along with the time is given as input for prediction. This predicts the time when the moisture value becomes 60%. This predicted output along with the moisture value (here, 60%) is then given as input to the same model. This is repeated again for moisture levels 50% and 40%. The predicted time when moisture value is 40% is the required output. For the dataset that this model was trained, the average accuracy (that is, how much is the predicted value close to the actual time when the water has to be watered again), error percentage, maximum accuracy, and minimum error percentage over 25 iterations for both the models separately are shown in Table 2.

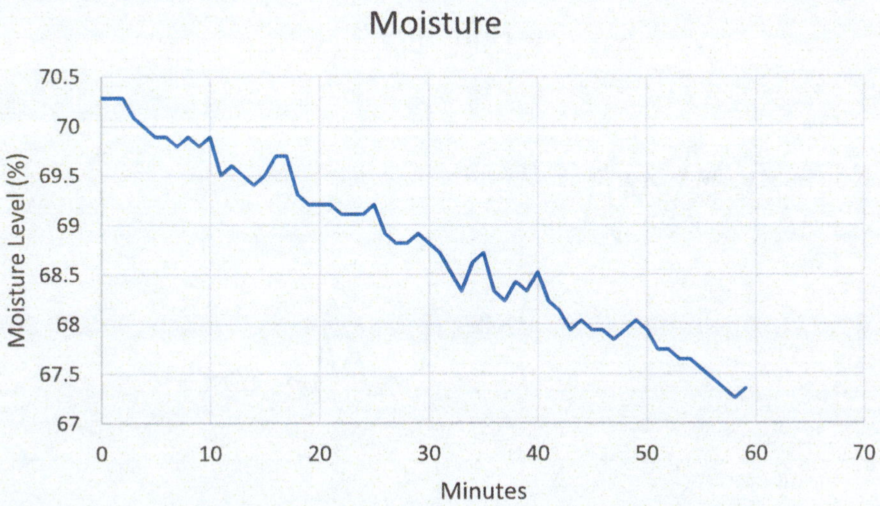

Fig. 9 Line chart of moisture level

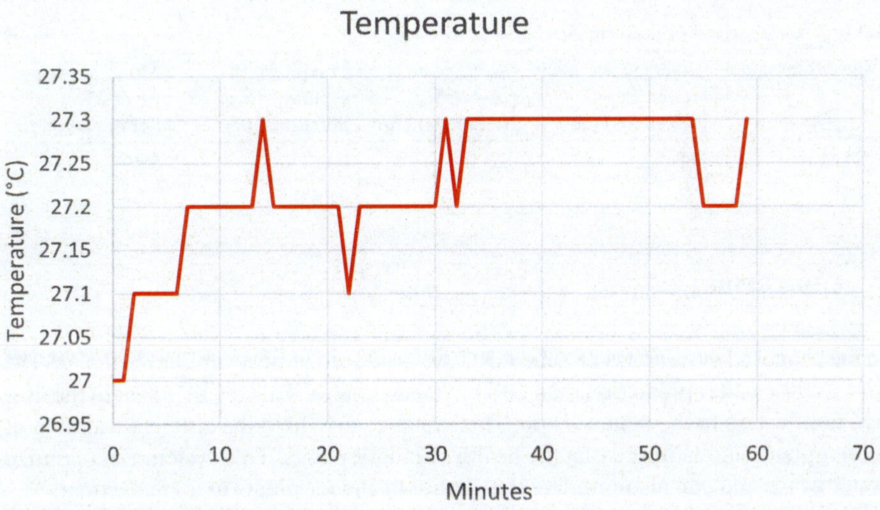

Fig. 10 Line chart for temperature

Both models had stacked layers with 100 cells each and ran for 400 epochs for each iterations. Every 300 seconds (5 minutes) is taken as 1% in calculating accuracy and error. From the table, both models have given very good results. But, in this case, LSTM tends to perform slightly better than GRU.

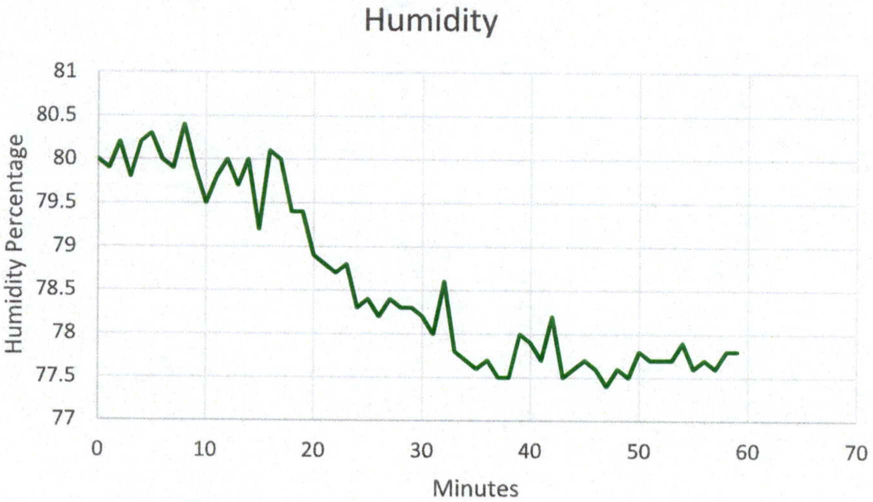

Fig. 11 Line chart for humidity

Table 2 Comparison of performance

Algorithm/model	Average accuracy (over 25 iterations) (%)	Average error (over 25 iterations) (%)	Maximum accuracy (over 25 iterations) (%)	Minimum error (over 25 iterations) (%)
LSTM	88	12	94.51	5.49
GRU	86	14	91.41	8.59

7 Conclusion

In the proposed system, LSTM and GRU are produced almost similar results. LSTM gives a little better prediction than GRU in this scenario. This result is sent to the user as a notification to the Android app. This system can help in the efficient watering of indoor plants and in improving the health of indoor plants. This system can optimize water usage and can also improve the quality of indoor plants to a great extent.

8 Future Work

In future, this can be implemented with more prediction algorithms and the data can be obtained and the algorithm that works best after comparison can also be considered.

References

1. J.S. Raj, J. Vijitha Ananthi, Automation using IoT in greenhouse environment. J. Information Technol. **1**(01), 38–47 (2019)
2. R. Vinothkanna, Design and analysis of motor control system for wireless automation. J. Electron. **2**(03), 162–167 (2020)
3. N. Harini, D. Kamakshi, D. Aruna, Secure proxy blind ECDS algorithm for IoT. Int. J. Pure Appl. Math. **118**(7), 437–445 (2018)
4. S. Valliappan, P. Bagavathi Sivakumar, V. Ananthanarayanan. Efficient real-time decision making using streaming data analytics in IoT environment, in *International Conference on Advanced Computing Networking and Informatics* (Springer, Singapore, 2019), pp. 165–173
5. K.N. Siva, R. Kumar G., A. Bagubali, K.V. Krishnan, Smart watering of plants, in *2019 International Conference on Vision Towards Emerging Trends in Communication and Networking (ViTECoN)*, Vellore, India (2019), pp. 1–4. https://doi.org/10.1109/ViTECoN.2019.8899371
6. A. Reghukumar, V. Vijayakumar, Smart plant watering system with cloud analysis and plant health prediction. Proc. Comput. Sci. **165**, 126–135 (2019). https://doi.org/10.1016/j.procs.2020.01.088
7. H. Liu, X. Mi, Y. Li, Smart multi-step deep learning model for wind speed forecasting based on variational mode decomposition, singular spectrum analysis, LSTM network and ELM. Energy Convers. Manage. **159**, 54–64 (2018)
8. A. Chlingaryan, S. Sukkarieh, B. Whelan, Machine learning approaches for crop yield prediction and nitrogen status estimation in precision agriculture: a review. Comput. Electron. Agric. **151**, 61–69 (2018)
9. A.A. Chandra, Y. Lee, B.M. Kim, S.Y. Maeng, S.H. Park, S.R. Lee, Review on sensor cloud and its Integration with Arduino based sensor network, in *2013 International Conference on IT Convergence and Security (ICITCS)*, Macao (2013), pp. 1–4. https://doi.org/10.1109/ICITCS.2013.6717876
10. L. Mhatre, N. Rai, Integration between wireless sensor and cloud, in *2017 International Conference on I-SMAC (IoT in Social, Mobile, Analytics and Cloud) (I-SMAC)*, Palladam (2017), pp. 779–782. https://doi.org/10.1109/I-SMAC.2017.8058285
11. C. Ashwini, D. Adhikary, A. Mishra, S. Duggal.Automatic irrigation system using Arduino. Int. Res. J. Eng. Technol. (IRJET) **5**(10) (2018)
12. Y. Shekhar, E. Dagur, S. Mishra, S. Sankaranarayanan, Intelligent IoT based automated irrigation system. Int. J. Appl. Eng. Res. **12**(18), 7306–7320 (2017)
13. S. Rawal, IOT based smart irrigation system. Int. J. Comput. Appl. **159**(8), 7–11 (2017)
14. J. Pourqasem, Cloud-based IoT: integration cloud computing with internet of things. Int. J. Res. Industrial Eng. **7**(4), 482–494 (2018). https://doi.org/10.22105/riej.2018.88380
15. M. Jacquline, N. Suseno, R. Manaha, Automatic watering system for plants with IoT monitoring and notification. CogITo Smart J. **4**, 316 (2019). https://doi.org/10.31154/cogito.v4i2.138.316-326
16. S. Koduru, V. Padala, P. Padala, Smart irrigation system using cloud and Internet of Things: ICCCN 2018, NITTTR Chandigarh, India. https://doi.org/10.1007/978-981-13-1217-5_20
17. S.B. Saraf, D.H. Gawali, IoT based smart irrigation monitoring and controlling system, in *2017 2nd IEEE International Conference on Recent Trends in Electronics, Information & Communication Technology (RTEICT)*, Bangalore (2017), pp. 815–819. https://doi.org/10.1109/RTEICT.2017.8256711
18. S. Sri Hari, K. Vishal Vinod, P. Hemanth Kumar, G. Harini, N. Harini, Integrity preserved multifactor authentication based automated ticketing system. Int. J. Adv. Sci. Technol. **29**(6s), 2547–2555 (2020)
19. S.V. Karthy, T. Senthil Kumar, L. Parameswaran, LSTM and GRU deep learning architectures for smoke prediction system in indoor environment, in *Intelligence in Big Data Technologies—Beyond the Hype* (Springer, Singapore, 2019), pp. 53–64

S. Vejay Karthy is currently pursuing a Bachelor of Technology in Computer Science and Engineering at Amrita School of Engineering, Coimbatore. He is interested and passionate in areas of Deep Learning and IoT. Currently, he is improving his skills in the field of Data Science.

K. Vishal Vinod is currently pursuing a Bachelor of Technology in Computer Science and Engineering at Amrita School of Engineering, Coimbatore. He is interested in areas of Cloud Computing and Machine Learning and is actively improving his expertise in Machine Learning.

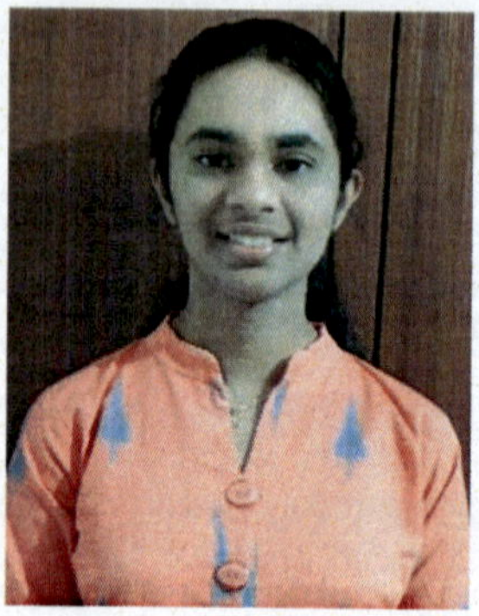

S. K. Mona Sweata is currently pursuing a Bachelor of Technology in Computer Science and Engineering at Amrita School of Engineering, Coimbatore. Her areas of interest are Software Development, Testing, and related fields. Currently, she is advancing her skill-set in the field of Data Analytics.

M. Karthika Gurubarani is currently pursuing a Bachelor of Technology Computer Science and Engineering at Amrita School of Engineering, Coimbatore. She has the experience of working across Web Development and Data Analytics domains. She is currently improving her skills in the field of IoT integrated with Cloud.

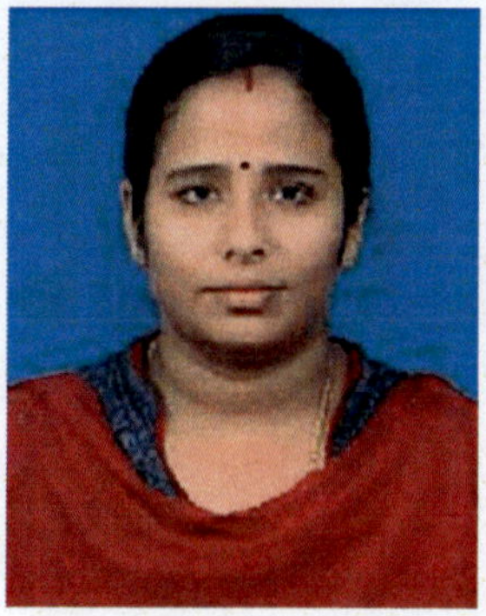

K. Abirami currently serves as Assistant Professor at Department of Computer Science and Engineering, School of Engineering, Coimbatore Campus. Her areas of research include Web Security, Cloud Computing.

Healthcare Bot Using Machine Learning Algorithms for Medical Services

Ashutosh Shankhdhar, Neha Adnekar, Prachi Bansal, and Reema Agrawal

Abstract In today's scenario, healthcare facilities have come to accessible due to advancements in technology day by day. For contributing to this advancement, presenting a Healthcare bot, a system that resolves the patient's queries by interacting with them through a virtual environment and suggests their respective doctors by considering their symptoms, provides a medical store, diet plans, and respective exercises. It provides a virtual environment to solve the user's queries through chat using some python libraries that help patients to mention the symptoms they are having, to have a consultation with doctors, or to have a perfect diet plan to follow. For identifying symptoms of the patient it uses Machine Learning techniques and Decision Tree algorithm for the detection of disease respectively. Then, with the help of obtained information, it searches for the best doctors from provided datasets of doctors with their specifications and recommends those to users. One can purchase the medicines from the medicos that provide links of medicines from online shopping sites, can stay updated regarding precautions for viral infections/diseases using the Health Wiki feature, and can ensure a good health care routine by having a diet plan and exercises recommended by the bot. It is a very effective method to avail of health services at doorsteps.

Keywords Healthcare bot · Machine learning · Decision tree algorithm · Datasets · Medicos · Health wiki

---

A. Shankhdhar (✉) · N. Adnekar · P. Bansal · R. Agrawal
Department of Computer Engineering and Applications, GLA University, Mathura, India
e-mail: ashutosh.shankhdhar@gla.ac.in

N. Adnekar
e-mail: neha.adnekar_cs18@gla.ac.in

P. Bansal
e-mail: prachi.bansal_cs18@gla.ac.in

R. Agrawal
e-mail: reema.agrawal_cs18@gla.ac.in

V. Suma et al. (eds.), *Inventive Systems and Control*,
Lecture Notes in Networks and Systems 204,
https://doi.org/10.1007/978-981-16-1395-1_35

1 Introduction

This is the era of electronics and technology so that eHealth comes into the picture, i.e., healthcare by electronic processes. McKinsey's report has described the growth of technology in the field of healthcare [5]. People have chosen several electronic devices for tracking their health and medicines. Healthcare provides great benefits to society as it adds quality to lives. From 2010 onwards, in the USA the annual investment has been increased by 138% into the area of "Health tech and Digital health", which reached more than 30 billion US dollars [5]. For being a part of advancement, presenting the integration of eHealth with a chatbot that guides all the patients out there. Prepared a tool that provides treatment to patients at their doorsteps. They can get their diet plans and the relevant doctor's consultation all online including demonstration of related exercises.

The *Healthcare bot* is a virtual assistant, that resolves health-related queries of patients. Particularly, HB avail the following options to its users:

(a) **Symptom Checker**. It identifies the health condition of patients according to their related symptoms.
(b) **Doctor Recommender System**. It recommends the doctor that has a specific domain for patients (considering their specifications) by using a *Doctor Recommender System (DRS)*; including collecting data of doctors including their specification domain.
(c) **Diet Plan System**. It prepares diet plans (including respective exercises) for patients according to their requirements using *Diet Plan System (DPS)*; it includes what to intake (healthy food) and what to exclude (like junks), and suggests workout exercises with demonstrations (via YouTube).
(d) **Medicos System**. It prefers the generic medicines using *Medicos System (MS)*; includes amazon links to purchase that medicines directly from there.
(e) **Health Wiki**. In a particular period, it notifies the patient of the steps of prevention from viral diseases using *Health Wiki (HW)*; general awareness notifications about do's and do not'.

Although it's known that an experienced doctor has far better knowledge as well as the diagnosis than an automatic system. In any situation where the patient is in severe condition, a Healthcare bot will be a proven asset for providing a consultation to a doctor.

2 Related Work

See Table 1.

Table 1 Previous researches

Author	Model used	Dataset used	Pros	Cons
Dharwad kar et al. [1]	Support Vector Machine Algorithm (SVM) and Natural Language Processing (NLP)	Heart Disease Dataset for 4 different cities Of different sizes is used	Users can ask dosage related queries by voice The system speaks out and displays medicine	Cannot be used as a voice Or face recognizer, and cannot interact with the patient at deep levels
Amato et. al. [2]	Ensemble Random Forest Algorithm	Dataset provided By C.M.O Centre which contains 13 different diseases and a record of a total of 16,733 patients	The proposed framework is adaptable to different clinical scenarios and medical tasks	It does not adapt to the user's need, human behavior, and traits such as age, gender, education, etc
Mathew et al. [3]	K-nearest neighbor Algorithm (KNN)	They used the symptoms-disease dataset which contains general information about symptoms and diseases	It analyzes symptoms of different users and then predicts the disease to the user	Symptoms and disease identified cannot be made into a report
Akihiro Yorita et al. [4]	Interactive Genetic algorithm and reinforcement learning algorithm	Dataset provided by user interaction	It aims to develop a chatbot to provide emotional support. A chatbot can interact with the user and adapt its personality to suit the user by using evolutionary computation	User evaluation is given at the end of the conversation The initial values of individuals are set random values
Marco Polignano et al. [5]	Naive Bayes algorithm	Dataset is not available with real information, that can relate symptoms of the patient	Chatbot monitors their treatments, biological values, suggests doctors, and performs self-diagnosis	To improve the performance of the Chatbot's feature Symptom Checker by providing information on diseases
Jasnani et al. [6]	Natural Language Processing (NLP) and Bag of Words Model	JSON file data containing the intents will be denoted by tags that contain formats in which the user will give input	Users can understand the problems they are facing and get a prediction about the disease they might have	User cannot use any other language interface except text

(continued)

Table 1 (continued)

Author	Model used	Dataset used	Pros	Cons
Nalinipri ya et al. [7]	Logistic Regression Algorithm	The dataset contains diseases and their respective symptoms and basic advice to improve the health of the user	Predicts the possible health issues and necessary measures to be taken to cure them	It does not support data pipelining that can automate all the processes and provide deep knowledge of the input data
Bharti et al. [8]	Rule-based grammar matching Algorithm and ML matching Algorithm	Data issued by the National Health Mission in the mid of COVID-19	Reduces the gap between the rural population by understanding their needs and providing them the proper information and the possible precaution to be taken	It does not copes with assistance for an anti-depression system that will focus on providing music therapy and mental health tests
Chen et al. [9]	Unimodal disease risk prediction (MDRP) algorithm	Hospital data collected which contains EHR, medical image, and gene data of 20,320,848 records	It works on both structured as well as unstructured data of the hospital dataset	A large amount of missing data is a human error which needs to be assigned
Kevin Mugoye et al. [10]	Reinforcement Learning (RL) and Natural language processing (NLP)	Dataset provided by conversational user interfaces	İt provides a healthcare expert in the form of an intelligent bot system to guide and provide support to mothers during pregnancy	The interaction, which in our case will be textual is expected to be more appealing to users
Chin-Yuan Huang et. al. [11]	Health Chatbot algorithm	User data acquired from the SWITCHes app and auxiliary data from medical instruments, not only SWITCHes app can engage the user with tailored feedback in an interactive way	SWITCHes supports two-way communication; not only SWITCHes app can engage the user with tailored feedback in an interactive way	AI-powered health Chatbot lives inside the SWITCHes app, not only health chatbot can help monitor users' health

3 Healthcare Bot

The objective of this paper is to predict whether a person is suffering from a disease based on symptoms. And this model has been deployed using a python application.

The screen view of Healthcare bot (Fig. 1):

Fig. 1 Welcome screen of Healthcare bot

3.1 The User Interacts with the Bot

The healthcare bot is an agent that simulates an intelligent conversation with a human [12].

The user starts a conversation with the chatbot by registering or login into the system and then providing input in the form of Yes/No or.

It is just as simple as starting a conversation with any human being over text. The conversation between the user and the chatbot is completely text-based.

The registration and login page for the user where he/she can either login into his account or register for a new account if he is not registered is shown in Fig. 2.

Fig. 2 Registration and login page

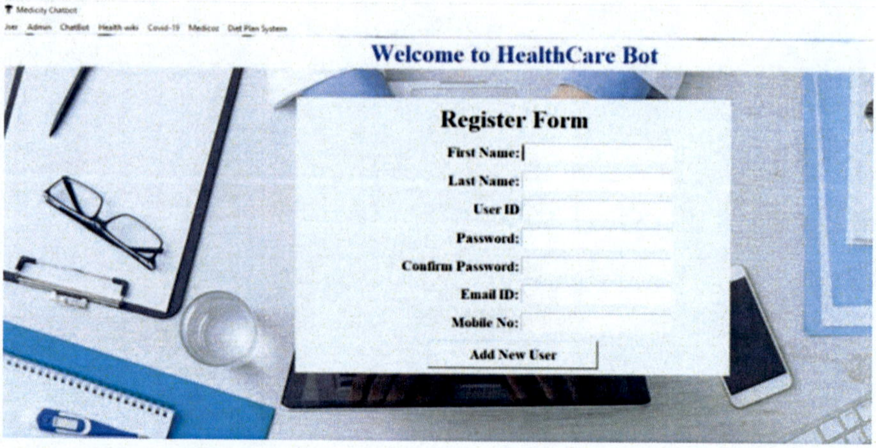

Fig. 3 New user registration page

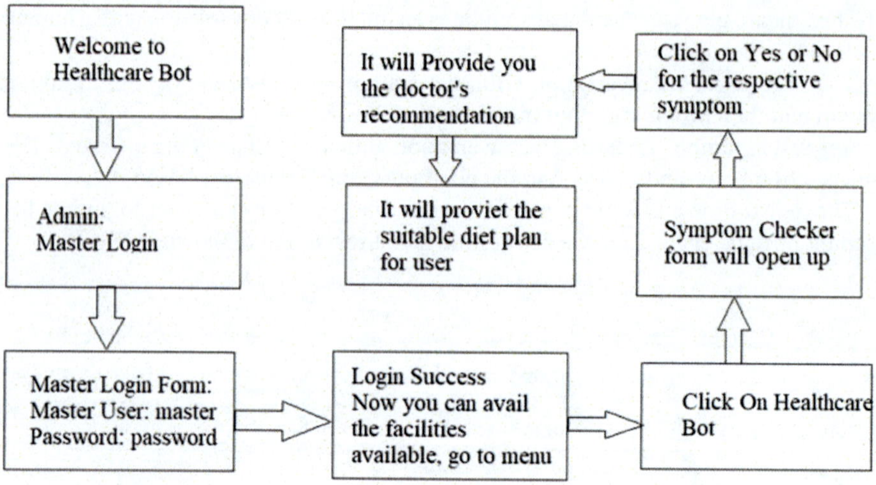

Fig. 4 Flowchart of Healthcare bot

The registration page for a new user where he can register using the username and a password is shown in Fig. 3 (Fig. 4).

3.2 *Symptom Checker*

It identifies the health condition of the patient according to their related symptoms. It checks for diseases based on symptoms mentioned by the user. The conversation

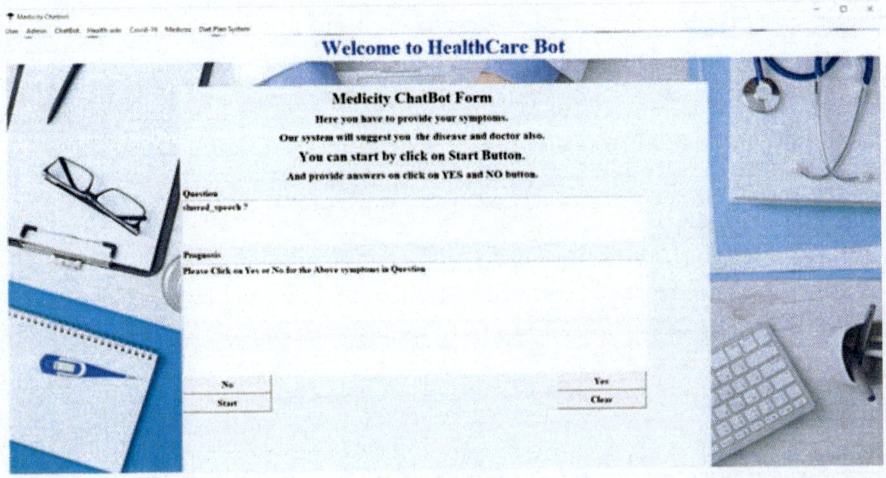

Fig. 5 Conversation window between user and chatbot

between the chatbot and user where the user responds to the questions asked by the chatbot related to the various symptoms is shown in Fig. 5.

3.3 *Doctor Recommender System*

Once collecting all the information, it recommends a doctor's name with their website link that has a specific domain for patients (considering their specifications) by using a *Doctor Recommender System* (*DRS*); including collecting data of doctors including their specification domain.

The output or the prediction made by the chatbot which includes the disease and the symptoms present and given and the confidence level for the particular disease is shown in Fig. 6.

4 Experimental Evaluation

The aim of the evaluation is:

a. Based on symptom checker, evaluation of internal elements of the model;
b. Assessment of Healthcare bot, if it can be used for resolving queries of users and for helping them in problematic health conditions.

In the first experiment, evaluate the accuracy of the model for relevant diseases based on symptoms.

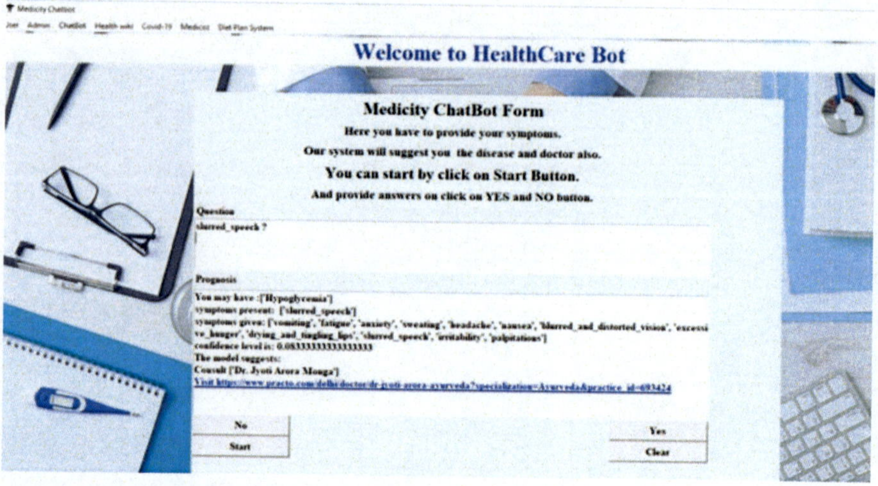

Fig. 6 Disease prediction based on symptoms and its confidence level

In the second experiment, recommend the most suitable doctor for the particular disease, the satisfaction.

4.1 Datasets Used

For the proposed system, there are two datasets were used. These are:

4.1.1 Dataset Containing the Symptoms and Corresponding Diseases

The dataset (Fig. 7) being split into training and testing datasets for building the model, contains the various symptoms and their respective prognosis.

4.1.2 Dataset Containing the doctor's Information

The data for this dataset (Fig. 8) is scrapped from the website Practo using web scraping and contains information such as the name of the doctor and a link to the website where the user can get all the information about the doctor for a particular city.

painful_w	pus_filled	blackhead	scurring	skin_peeli	silver_like	small_der	inflammat	blister	red_sore_	yellow_cr	prognosis
0	0	0	0	0	0	0	0	0	0	0	Fungal infe
0	0	0	0	0	0	0	0	0	0	0	Fungal infe
0	0	0	0	0	0	0	0	0	0	0	Fungal infe
0	0	0	0	0	0	0	0	0	0	0	Fungal infe
0	0	0	0	0	0	0	0	0	0	0	Fungal infe
0	0	0	0	0	0	0	0	0	0	0	Fungal infe
0	0	0	0	0	0	0	0	0	0	0	Fungal infe
0	0	0	0	0	0	0	0	0	0	0	Fungal infe
0	0	0	0	0	0	0	0	0	0	0	Fungal infe
0	0	0	0	0	0	0	0	0	0	0	Fungal infe
0	0	0	0	0	0	0	0	0	0	0	Allergy
0	0	0	0	0	0	0	0	0	0	0	Allergy
0	0	0	0	0	0	0	0	0	0	0	Allergy
0	0	0	0	0	0	0	0	0	0	0	Allergy
0	0	0	0	0	0	0	0	0	0	0	Allergy
0	0	0	0	0	0	0	0	0	0	0	Allergy
0	0	0	0	0	0	0	0	0	0	0	Allergy

Fig. 7 A dataset containing the symptoms and corresponding diseases

A1 fx Dr. Amarpreet Singh Riar

A	B
Dr. Amarp	https://www.practo.com/delhi/doctor/amarpreet-singh-riar-general-physician?specialization=General%20Physician&p
Dr. (Maj.)S	https://www.practo.com/delhi/doctor/dr-54-general-physician-1?specialization=General%20Physician&practice_id=1(
Dr. Anirbai	https://www.practo.com/delhi/doctor/anirban-biswas-diabetologist?specialization=General%20Physician&practice_id
Dr. Aman \	https://www.practo.com/delhi/doctor/dr-aman-vij-general-physician?specialization=General%20Physician&practice_i
Dr. Mansi .	https://www.practo.com/delhi/doctor/dr-mansi-arya-bhardwaj-homeopath?specialization=Homoeopath&practice_id
Dr. Sunil Ku	https://www.practo.com/delhi/doctor/dr-sunil-kumar-dwivedi-homeopath?specialization=Homoeopath&practice_id=
Dr. Chhavi	https://www.practo.com/delhi/doctor/dr-chhavi-bansal-homeopath-1?specialization=Homoeopath&practice_id=112
Dr. Sneh Kl	https://www.practo.com/delhi/doctor/dr-sneh-khera-homeopath?specialization=Homoeopath&practice_id=709153

Fig. 8 A dataset containing the doctor's information

4.2 Evaluation Measures

In the first experiment, calculate the effectiveness of the classification model. To measure the performance, the parameters required are accuracy, precision, and recall

$$\text{accuracy} = (\text{True Positive} + \text{True Negative})/\text{total}$$
$$\text{recall} = \text{True Positive}/(\text{True positive} + \text{False negative})$$

In the second experiment, calculate the accuracy of the Symptom Checker and user interaction of the Healthcare bot. The symptom Checker identifies the disease. In user interaction, several questions were asked, and the user answers either YES or NO.

4.3 First Experiment

The objective of the first experiment is to test either this model is valid or not for various types and amounts of the training dataset. Here, a patient dataset is used for the training of the classifiers.

$P = p1, p2 \ldots.pn.$

By the set of *Symptoms each pi is described,* $S = s1, s2 \ldots., sn.$

Here, pi is the patient vector that we have used to represent each symptom.

If there is any similar symptom present in the user then it will be represented by the value of 1 for pi, else it will be set to 0. For training, model P is used. From the various algorithms, the Decision Tree Classifier was selected.

4.3.1 Algorithm

Decision Tree Classifier is a classification technique that is easy to use and has been used widely. The Decision Tree Algorithm develops a tree-like structure by the data fed to it. This tree-like structure can be traversed for prediction. The Decision Tree Algorithm uses some measure of impurity to create an optimal split. The Algorithm then attempts to minimize a cost function for multiple pairs of (K, tk) resulting in a pair that minimizes the cost function. The Decision Tree implementation of *sci-kit learn* will continue creating the tree with more and more nodes unless and until all instances are correctly classified. This phenomenon is known as overfitting. The Algorithm can then use this information to create and save a tree for classification. The Decision Tree Algorithm popularly uses these measures of impurities:

(a) Gini Impurity
(b) Entropy or Information Gain

$$\text{Gini} = 1 - \sum_{i=1}^{C} (p_i)^2$$

$$E(S) = \sum_{i=1}^{c} -p_i \log_2 p_i$$

After deducing Gi (Gini Index) or Hi (Entropy), the decision tree algorithm then attempts to minimize a cost function to get the optimal pair of (k, tk) where k is the attribute name and tk is its threshold value for a particular node.

$$J(k, t_k) = \frac{m_{\text{left}}}{m}\text{MSE}_{\text{left}} + \frac{m_{\text{right}}}{m}\text{MSE}_{\text{right}}$$

Here, J is the cost function whose value is calculated for every pair of (k, tk) from the training dataset. The pair that yields the smallest 'J' value becomes the root node. Thus, the Decision Tree Algorithm works for moderately large datasets

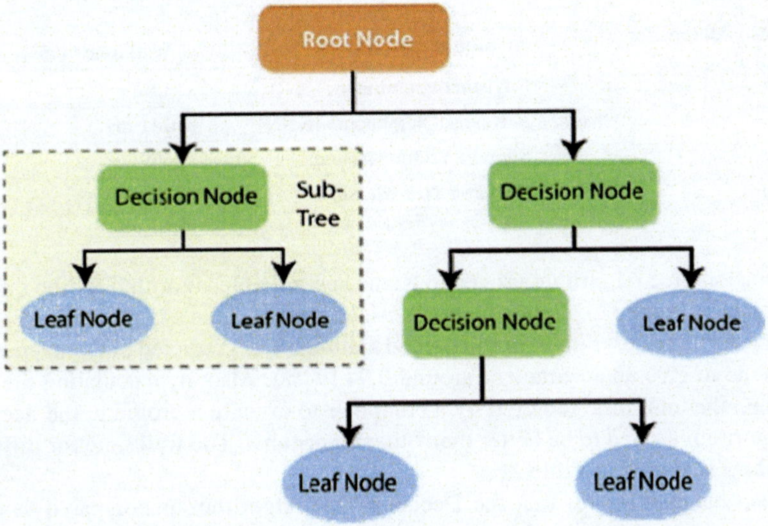

Fig. 9 Decision tree algorithm implementation

not extremely large datasets. The algorithmic implementation of the Decision Tree Algorithm is shown in Fig. 9.

4.3.2 Second Experiment

The objective of the second experiment is:

a. To estimate how effective the Symptom Checker is.
b. To assess the satisfactory level of users, recommend the most suitable doctor based on the symptoms

In this evaluation, test the research Hypotheses which are as follows:

(a) **RH1.1.** Diseases can be analyzed by the symptoms entered by the patient, this detection will be done by the Symptom Checker feature.
(b) **RH1.2.** Patient experiences satisfaction by using the facilities provided by Healthcare bot.
(c) **RH1.3.** The Healthcare bot is easy to use and has high efficiency.

5 Result Analysis

From the analysis of the model by using different algorithms such as Logistic Regression, *K* Nearest Neighbours, Support Vector Machine, and Decision Tree Classifier,

Table 2 Wall time for different algorithms

Algorithm	Wall time (s/ms)
Logistic regression	1.02 s
K Nearest Neighbours (KNN)	141 ms
Support Vector Machine	659 ms
Decision Tree Classifier	88.6 ms

the Decision Tree Algorithm proved to be the best suitable algorithm for the system. This model is trained to predict the prognosis(disease) for the symptoms present and given based on the user input of the trained features. The predicted outcome or result was found to give an accuracy of around 0.94 (94%). Also, by calculating the wall time, i.e., the total time required by a computer to execute a program, the decision tree algorithm proved to be better than other algorithms. The wall time for different algorithms is shown in Table 2.

Thus, this also proves that the Decision Tree Algorithm as compared to other algorithms is a good choice for the purpose.

6 Conclusion and Future Work

According to the scientific community, chatbots are user-friendly and any person can communicate in their language through a chatbot on the desktop or in the mobile application very easily. In this pandemic where there is a lack of tools and a decreasing economy, this application can work well for the welfare of society. Healthcare bot provides personalized diagnoses based on symptoms. Hence, a model is built that is going to be a proven asset for helping patients all over the places including hospitals and even home itself.

In the future, the performance of diagnosis and the recognition of the symptoms will be highly improved with the addition of support for further medical features, such as the concentration of symptoms, length of time, place, and a more detailed description of symptoms. Some more features can be added like an alarm to remind the patient for taking medicine from time to time. With the superior natural language processing algorithms at their core, these bots give more empathetic and human-like answers which can make the person feel better.

Voice and face recognition can be included for interacting with the patient at deeper levels in the future, as this model is flexible. The audio system can also be included in this system to make the bot more interactive.

Acknowledgements The authors are thankful to Ashutosh Shankhdhar for the support they have provided during the process.

References

1. R. Dharwadkar, N.A. Deshpande, A medical ChatBot. Int. J. Comput. Trends Technol. (IJCTT) **60**(1), 41–45 (2018)
2. F. Amato, S. Marrone, V. Moscato, G. Piantadosi, A. Picariello, C. Sansone, Chatbots meet eHealth: automatizing healthcare, in WAIAH@ AI* IA (2017), pp. 40–49
3. R.B. Mathew, S. Varghese, S.E. Joy, S.S. Alex, Chatbot for disease prediction and treatment recommendation using machine learning, in *Third International Conference on Trends in Electronics and Informatics* (2019)
4. A. Yorita, S. Egerton, J. Oakman, C. Chan, N. Kubota, Self-adapting Chatbot personalities for better peer support, in *2019 IEEE International Conference on Systems, Man and Cybernetics (SMC)* (IEEE, 2019, October), pp. 4094–4100
5. M. Polignano, F. Narducci, A. Iovine, C. Musto, M. De Gemmis, G. Semeraro, HealthAssistantBot: a personal health assistant for the Italian language. IEEE Access **8**, 107479–107497 (2020)
6. P. Kandpal, K. Jasnani, R. Raut, S. Bhorge, Contextual Chatbot for healthcare purposes (using deep learning), in *2020 Fourth World Conference on Smart Trends in Systems, Security and Sustainability (WorldS4)* (2020), pp. 625–634
7. G. Nalinipriya, P.S. Priyadarshini, K. RajaRajeshwari, BayMax: a smart healthcare system provide services to millennials using machine learning technique, in *IEEE 6th International Conference on smart structures and systems ICSSS 2019* (2019). https://doi.org/10.1109/ICSSS.2019.8882844
8. U. Bharti, D. Bajaj, H. Batra, S. Lalit, S. Lalit, A. Gangwani, Medbot: conversational artificial intelligence powered Chatbot for delivering tele-health after COVID-19, in *Fifth International Conference on Communication and Electronics Systems* (2020), pp. 870–875
9. M. Chen, Y. Hao, K. Hwang, L. Wang, L. Wang, Disease prediction by machine learning over big data from healthcare communities. IEEE Access **5**, 8869–8879 (2017)
10. K. Mugoye, H. Okoyo, S. Mcoyowo, Smart-bot technology: conversational agents role in maternal healthcare support, in *2019 IST-Africa Week Conference (IST-Africa)* (IEEE, 2019, May), pp. 1–7
11. C.Y. Huang, M.C. Yang, C.Y. Huang, Y.J. Chen, M.L. Wu, K.W. Chen, A Chatbot-supported smart wireless interactive healthcare system for weight control and health promotion, in *2018 IEEE International Conference on Industrial Engineering and Engineering Management (IEEM)* (IEEE, 2018, December), pp. 1791–1795
12. S. Kethuneni, S.E. August, J.I. Vales, Personal health care assistant/companion in virtual world, in *2009 AAAI Fall Symposium Series* (2009, October)

Melanoma Detection and Classification in Digital Dermoscopic Images Using Machine Learning

K. Senthil Kumar, S. Varalakshmi, G. Sathish Kumar, and T. Kosalai

Abstract Melanoma is an exceptional disastrous variety of skin growth. This category of cancer is disconcerting due to its habit of causing metastasis. Nevertheless, it sometimes tough to discern it from nevus due to their alike ocular look and symptoms. There are various steps in computer-assisted interpretation systems of melanoma such as image segmentation, edge detection, feature extraction, and classification. Image segmentation on dermoscopic pictures plays an imperative part in discovering melanoma. The existing segmentation methods are K-means clustering, convolutional neural network, ROI clustering, histogram equalization, RA pooling method, RGB, HSV, GLCM, Cole–Cole model, multi-variant analysis of variant, fully convolutional neural networks, and mobile imaging system. A support vector machine (SVM) is applied for the analysis of sheath malignancy to identify melanoma or nevus. The goal is to evaluate the effectiveness of the proposed segmentation method, highlights the most appropriate options, and compares the classification methods.

Keywords Melanoma · Skin cancer · Malignant detection · Dermoscopic images · Image segmentation · Hue value saturation · Classification

K. S. Kumar (✉)
Rajalakshmi Institute of Technology, Chennai, Tamil Nadu, India
e-mail: senthilkumar.k@ritchennai.edu.in

S. Varalakshmi
Aadhi College of Engineering and Technology, Tamilnadu, India

G. S. Kumar
SCSVMV University, Kanchipuram, Tamilnadu, India

T. Kosalai
University College of Engineering Kancheepuram, Tamilnadu, India

V. Suma et al. (eds.), *Inventive Systems and Control*,
Lecture Notes in Networks and Systems 204,
https://doi.org/10.1007/978-981-16-1395-1_36

1 Introduction

Dangerous malignance is the swelling of melanin-transmitting cells called melanocytes [1]. It has the propensity to spread to different parts of the body and eventually induce passage [2]. The color of melanin is a characteristic sunscreen because it shields against the harmful possessions of UV rays [3] and the measure of cases with melanoma has been increasing in recent years [4]. Melanoma is the fifth basic disease in men and the 6th regular malignancy in women [5]. Aside from these upsetting realities, a light beam can be restored whenever recognized at the beginning phase [6]. In prompt identification of case, PC supported conclusion frameworks can be utilized as these gadgets do not include intrusive methodology and are anything but difficult to work as they just require a picture of the influenced zone. These frameworks can go about as either independent frameworks or assistance for the dermatologists.

From the measured value the parameter, lesion can be recognized to cancerous or nevus. In view of the compound nature of disease, it is very hard for investigators to make confirm only based on the geometrical structures. In addition to that the radical increment of image size is indifferent. So the practice is depending on how well it can be retrieved, investigated, and can be extracted from it [1]. The list of symptoms is typical colorant setup with gray-blue area, uncharacteristic vascular pattern, blemish, streaks, unreliable dots, globules, and reversion patterns. Later, the listing is condensed to a minimum feature levels of the various networks, asymmetry, and blue-white structures.

In this work, the characterization of skin lesions has been completed and outcomes compared with other techniques. The image is advanced by a dynamic form for division technique to recognize skin lesions from other areas. Subsequently, conspicuous shading, surface, and shape highlights are separated and taken care of by SVM for arrangement task. Section 2 covers the writing review; Sect. 3 incorporates philosophy for the proposal, Sect. 4 is committed to performance analysis, and Sect. 5 concludes the results.

2 Literature Survey

Takruri et al. developed a model to classify different skin lesions. First, the images are resized and labeled for preprocessing. Segmentation is then performed GLCM, curve transformation, and waveform transformation [7]. These extracted properties are passed to SVM for classification. Munia et al. [8] approached for cancer from digital images using k-means clustering and Otsu's method with morphological operation and directional filter. Different characteristics are obtained and passed to various classification methods. Fonseca-Pinto et al. [9] performed a grayscale renovation to choose the channel with the highest entropy and then cutting. The texture characteristics are then extracted using the LBPV descriptor to classify melanoma using the

AdaBoost algorithm. Turkeli et al. [10] applied a franc filter to smoothen the image and OpenCV library used to segment it. Selected features are given to MLP for classification [11]. Arasi et al. implemented an image preprocessing technique with grayscale conversion, morphological process, and retrieved using PCA and DWT. Mahmoud et al. proposed an idea to remove the hair using the "bottom hat" operation followed by medium filtration. For feature extraction, Gabor, HOG, GLCM, and LBP filters are used and classification is done by MLP. Ajed et al. [12] used a model to extract the different to classify abnormality using the SVM algorithm. In [13], Singh et al. developed a system to differentiate skin lesions with k-means clustering technique, features using a color consistency vector, and a local binary model are applied to the multiclass SVM.

3 Proposed Method

A method has been proposed with a support system for the categorization of different skin diseases is shown in Fig. 1. The malfunctioned results for classification in cancer detection and segmentation have been observed to be a major problem in recognizing skin cancer [11, 14–16].

The technique which follows in this paper is HSV color transformation. This technique uses centroid selection to extract the region of interest (ROI) in cancer datasets accurately and efficiently. The proposed strategy prepares measurable results in relationship with other fore-front procedures. The main advantage of HSV color transformation is that it cannot change the cluster values every time.

3.1 Input Image

An input image extracted from a hardware is processed to obtain raw image data as shown in Fig. 2a, b.

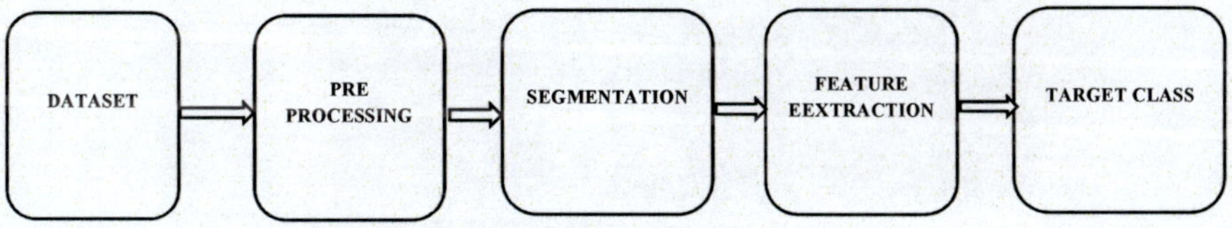

Fig. 1 Process flow for the proposed method

3.2 Preprocessing

The datasets are playing a key role in preprocessing techniques to ensure precise, proficient, or evocative exploration. This technique deals with resize and converts into gray scale using filters. Data cleaning is a methods of sequence for findings and replacing corrupted or lost data. Detection of local extreme and sudden changes is supporting to identify substantial data trends.

Resizing—input images to avoid output image to be distorted.

Image filtering—the technique of modifying or enhancing the input image is labeled as image filtering.

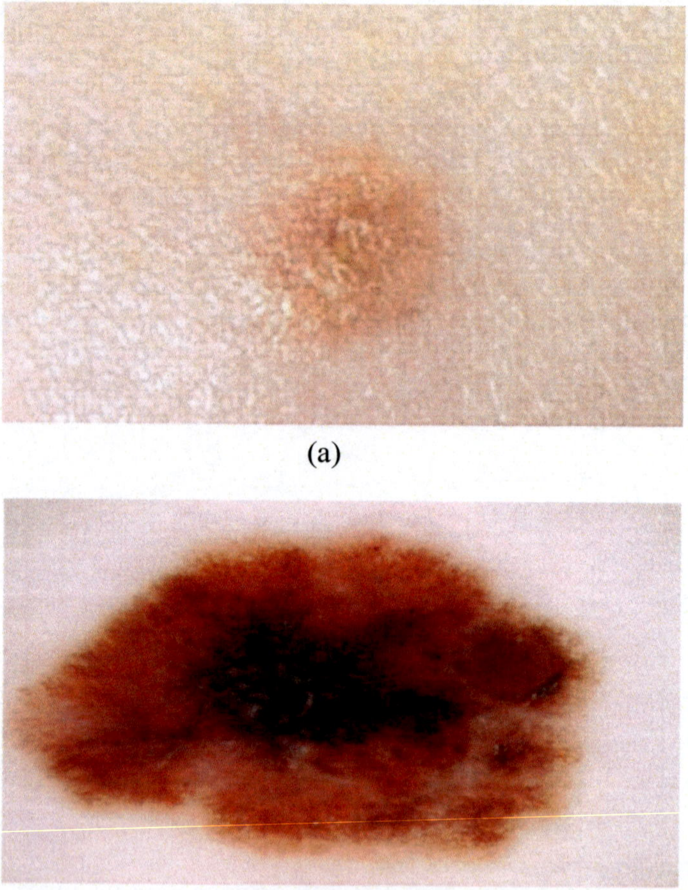

(a)

(b)

Fig. 2 **a** Nevus image, **b** melanoma image, **c** example of RGB color image, **d** gray scale conversion of image, **e** segmented output of melanoma, **f** detection of melanoma, **g** segmented output of nevus, and **h** nevus detected

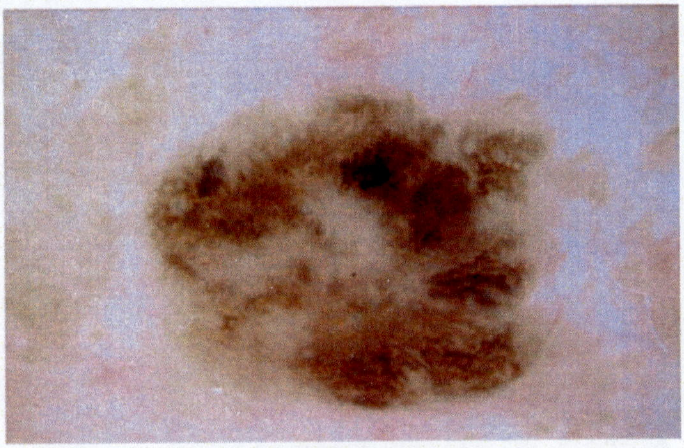

(c)

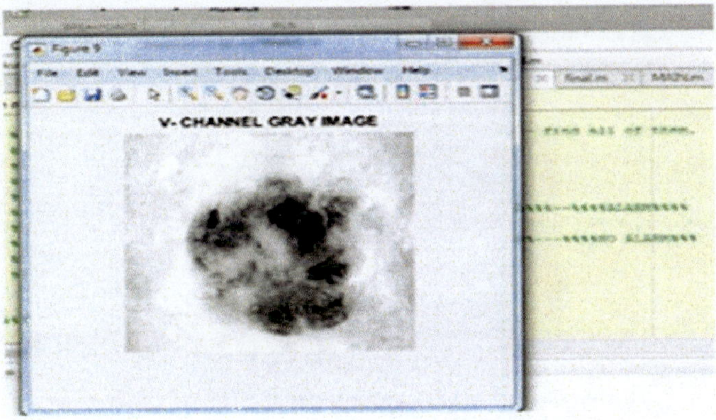

(d)

Fig. 2 (continued)

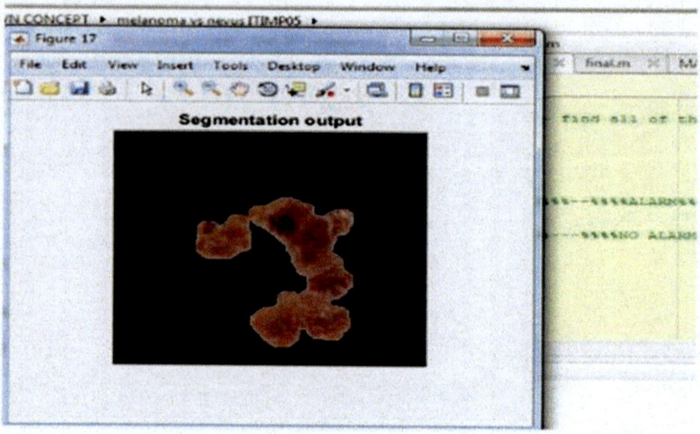

(e)

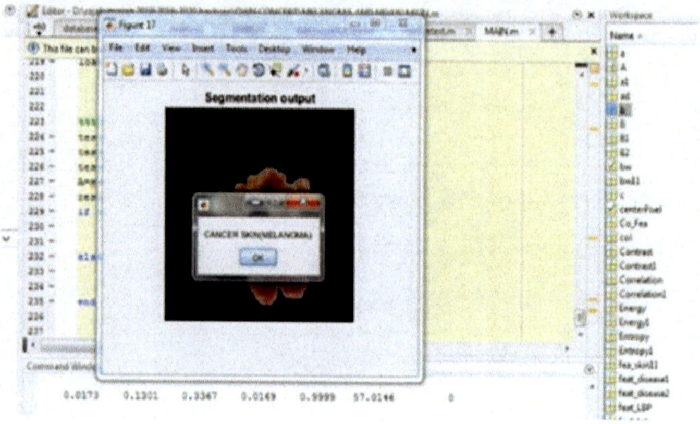

(f)

Fig. 2 (continued)

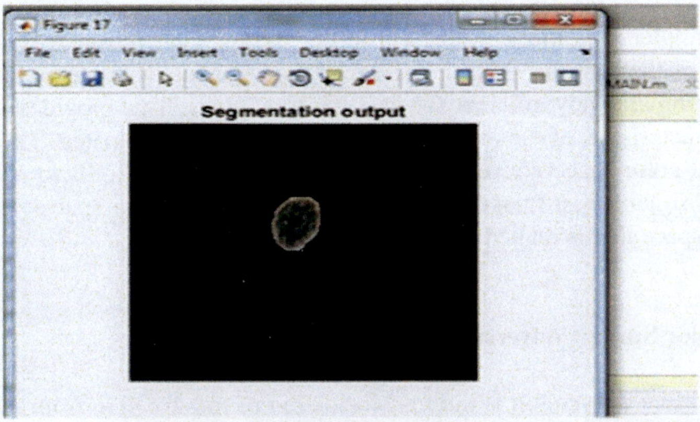

(g)

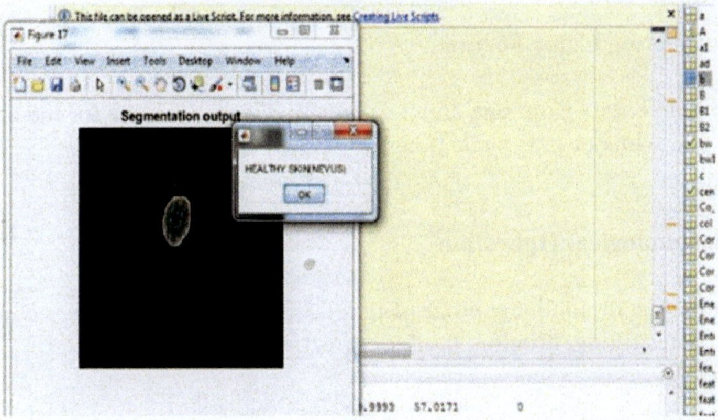

(h)

Fig. 2 (continued)

Dimension reduction—to store one color pixel in RGB color image needs 8 * 3 = 24 bits (8 bits per color), when converting from RGB image to grayscale image, 8 bits are required to store a pixel of the image. So, it requires 33% less memory than an RGB image. Many tasks do not fare better with 3D pixels (e.g., edge detection, morphological operation, etc.).

3.3 Segmentation

The process of segregating the image into a number of groups using the similar property is defined as image segmentation and which is clustered jointly propounded approach implements the k-means clustering by applying repeated segmentation that explores the centroid of each one set in the segment. This method is flexible, fast, and comparatively efficient for data interpretations. This provides imperative image characteristics of the information and effortlessly anticipated. Then, the use of different color space conversions from RGB to HSV and RGB to YCBCR for selecting a suitable test image. From the above different color spaces, choose the best color space that is related to skin classification [4, 5].

3.3.1 Color Space Conversions

The color space conversion is the transformation of the dye sign from one base to another, and the goal is to make the transformed image appear as a closure to the original as shown in Fig. 2c.

3.3.2 Converting Color Format

To differentiate colors from one another, then it is indispensable for the conversion into black and white or gray scale image as shown in Fig. 2d.

3.3.3 Morphological Operations

It is a compilation of nonlinear processing related to shapes. It involves complement the image, holes filling, dilation, border correction, and masking.

3.4 Feature Extraction

Expanding features simplify the number of resources essential to accurately represent a huge dataset. While performing the scrutiny of the compound data, one of the key tribulations is stems from the number of variables involved. Here, three feature extraction methods were used like GLCM, LBP, and region segmentation. The GLCM gives the texture features of the test image like contrast, correlation, energy, etc. [6, 7]. Later, the region property segmentation was used to get mathematical features of input image such as density and area. From the above extracted features, it is need to recognize the best component that are identified to separate the normal and terrible skin. It involves four feature extractions—shape features, color features, geometrical features, and texture features.

3.5 Classification

The task of getting information from a multiband raster picture is referred to as classification. Thematic maps are created from the resulted raster image from classification. With the help of image classification toolbar, classification and multivariate analysis are performed. The classification technique used here is the support vector machine. It is predominantly used to diminish the classification error and maximize the geometric boundary [8]. They are also called extreme margin classifiers. It is one of the unique superintended machine learning methodologies which requires training with huge datasets for binary classes.

4 Testing Results

Input images from the ISIC database are pre-processed by resizing and enhancing input dataset images. Followed by segmenting the pre-processed image and extracting the features. With the help of information provided by LBP image, the classifier aids us to accord the desired result [7–9]. The segmented output to get the selected region is shown in Fig. 2e, detection of melanoma is shown in Fig. 2f, segmented output of nevus is shown in Fig. 2g, and nevus detection is shown in Fig. 2h.

The results for the method proposed are given below,

4.1 Performance Evaluation Metrics

Four evaluation measurements are used to assess performance: sensitivity, specificity, accuracy, and AUC. Accuracy tells about the responsiveness ratio, sensitivity provides the receptiveness rate for melanoma, and specificity provides the receptiveness rate for non-melanoma. The higher AUC rate near the unit, the better the performance. Performance metrics derived from Eqs. (1), (2) and (3). True positive is the positive-to-positive ratio, which is the closeness of identifying abnormality and true negative tells about nevus detection. False positive and false negative give misclassification rate.

$$\text{Sensitivity} = \frac{TP}{FN + TP} \tag{1}$$

$$\text{Specificity} = \frac{TN}{TN + FP} \tag{2}$$

$$\text{Accuracy} = \frac{TP + TN}{TP + TN + FP + FN} \tag{3}$$

Table 1 Confusion matrix of the proposed method

		Predicted class	
Actual class		Melanoma	Nevus
	Melanoma	52	03
	Nevus	0	45

TP = 52
TN = 45
FP = 00
FN = 03

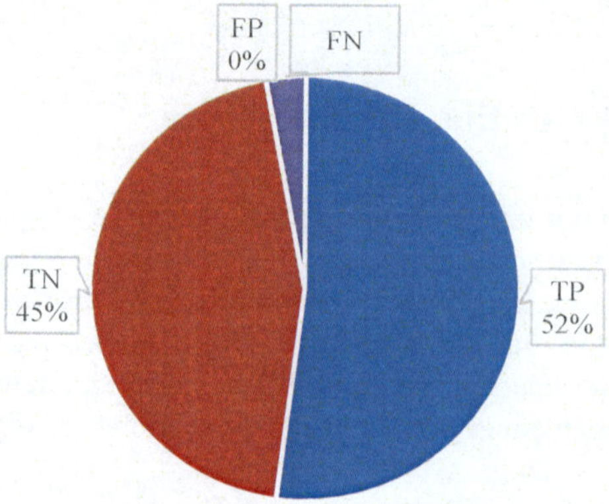

Fig. 3 Confusion matrices output

The following table consists of performance evaluation matrices which obtained from the results of our proposed system. In total, it comprises of 100 digital images of human skin in the dataset. Out of them 55 were melanoma and 45 were nevus [2, 17]. The performance metrics for the proposed method are given in Table 1, and the results of confusion matrices obtained are shown in Fig. 3. It clearly states that the findings for the correct decision are reached with greater value.

Calculation:

- Sensitivity (Sen): Rate of positive values are proficiently predicted as

  Sen = Correctly Classified Abnormal Images / Total Abnormal Images

- Specificity(Spe): Ratio of negatives values are suitably recognized as

  Spe = Correctly Classified Normal Images / Total Normal Images

- Accuracy(Acc):

  Acc = Correctly Classified Images/Total Images.

The performance metrics for the proposed method are given in Table 2. From the observed result, it is found that melanoma classification using SVM reports an accuracy level of 97%, and the graphical representation is given in Fig. 4 (Table 3).

Table 2 Performance analysis of the proposed method

Method	Sen (%)	Spe (%)	Acc (%)
SVM	94.55	100.00	97.00

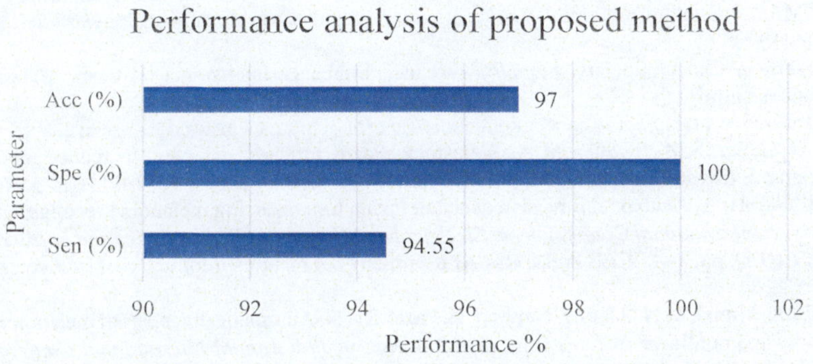

Fig. 4 Performance analysis of the proposed method—SVM

Table 3 Comparison of results with other methods

Methods	Accuracy (%)
Soumya et al. [10]	91.5
Adjed et al. [16]	86.07
PNN	97.00

This work witnessed with the accuracy for the classification of the system is 97.00%, and Table 3 shows a comparison of the accuracy of the proposed method with other existing methods.

5 Conclusion

This article gives an in-depth look at the recognized indication of tumor in dermoscopic images with machine learning. Dermoscopic images were revealed by initial steps to eradicate early lying annoying entities. The HSV color transformation is applied to exhume the ROI, to give more precise and proficient results. These features are applied to classification and pooled with the color features to accomplish excellent outcomes. At that point, melanoma was recognized from refined images using dynamic form-based division. Ultimately, color and shape features are separated and applied to SVM. The evaluated parameters are 97% of accuracy, 94.55% of sensitivity, and 100% of specificity. The proposed technique equips calculable outcomes in correlation with other cutting edge strategies.

References

1. Medscape. https://emedicine.medscape.com/article/280245-overview
2. DermIS. https://skincancer.dermis.net/content/e04typesof/e154/e155/index_eng.html
3. JAMADermatology. https://jamanetwork.com/journals/jamadermatology/article-abstract/546435
4. American Cancer Society. https://www.cancer.org/cancer/melanoma-skin-cancer/about/key-statistics.html
5. Cancer.Net. https://www.cancer.net/cancer-types/melanoma/statistics
6. A.C. Geller, S.M. Swetter, M.A. Weinstock, Focus on early detection to reduce melanoma deaths. J. Invest. Dermatol. **135**, 947–949 (2015). https://doi.org/10.1038/jid.2014.534
7. M. Takruri, A. Abubakar, Bayesian decision fusion for enhancing melanoma recognition accuracy, in *International Conference on Electrical and Computing Technologies and Appli- cations (ICECTA)*, pp. 1–4. IEEE Press, Ras Al Khaimah (2017). https://doi.org/10.1109/icecta.2017.8252063
8. . T.T.K. Munia, M.N. Alam, J. Neubert, R. Fazel-Rezai, Automatic diagnosis of melanoma using linear and nonlinear features from digital image, in *39th Annual International Conference of the IEEE Engineering in Medicine and Biology Society (EMBC)* (2017), pp. 4281–4284. IEEE Press, Seogwipo. https://doi.org/10.1109/embc.2017.8037802
9. R. Fonseca-Pinto, M. Machado, A textured scale-based approach to melanocytic skin lesions in dermoscopy, in *40th International Convention on Information and Communication Technology, Electronics and Microelectronics (MIPRO)* (2017), pp. 279–282. IEEE Press, Opatija. https://doi.org/10.23919/mipro.2017.7973434
10. S. Turkeli, M.S. Oguz, S.B. Abay, T. Kumbasar, H.T. Atay, K.K. Kurt, A smart dermoscope design using artificial neural network, in *2017 International Artificial Intelligence and Data Processing Symposium (IDAP)* (2017), pp. 1–6. IEEE Press, Malatya. https://doi.org/10.1109/idap.2017.8090211
11. G. Betta, et al., Automated application of the'7-point checklist' diagnosis method for skin lesions: estimation of chromatic and shape parameters, in *Conference Record—IEEE İnstrumentation and Measurement Technology Conference*, vol. 3, pp. 1818–1822 (2005)
12. Y. Yuan, M. Chao, Y.-C. Lo, *Automatic Skin Lesion Segmentation using Deep Filly Convolutional Networks with Jaccard Distance*. Institute of Electrical and Electronic Engineer, pp. 0278–0062 (2016)
13. H. Mahmoud, M. Abdel-Nasser, O.A. Omer, Computer aided diagnosis system for skin lesions detection using texture analysis methods, in *2018 International Conference on Innovative Trends in Computer Engineering (ITCE)* (2018), pp. 140–144. IEEE Press, Aswan. https://doi.org/10.1109/itce.2018.8327948
14. S. Mustafa, A. Kimura, A SVM-based diagnosis of melanoma using only useful image features, in *2018 International Workshop on Advanced Image Technology (IWAIT)* (2018), pp. 1–4. IEEE Press, Chiang Mai. https://doi.org/10.1109/iwait.2018.8369646
15. M.W. Rashad, M. Takruri, *Automatic Non-Invasive Recognition of Melanoma Using Support Vector Machines*. Institute of Electrical and Electronic Engineer, pp:4568–5090 (2016)
16. E. Ahn, J. Kim, L. Bi, A. Kumar, C. Li, M. Fulham, D.D. Feng, *Saliency-based Lesion Segmentation via Background Detection in Dermoscopic Images*. Institute of Electrical and Electronic Engineer, pp. 2168–2194 (2016)
17. R.S. Soumya, S. Neethu, T.S. Niju, A. Renjini, R.P. Aneesh, Advanced earlier melanoma detection algorithm using colour correlogram, in *2016 International Conference on Communication Systems and Networks (ComNet)* (2016), pp. 190–194. IEEE Press, Thiruvananthapuram. https://doi.org/10.1109/csn.2016.7824012
18. F. Adjed, S.J. Safdar Gardezi, F. Ababsa, I. Faye, S. Chandra Dass, Fusion of structural and textural features for melanoma recognition, in *IET Computer Vision* (2018), pp. 185–195. IET Press. https://doi.org/10.1049/iet-cvi.2017.0193
19. T.-T. Do, et al., *Accessible Melanoma Detection using Smartphones and Mobile Image Analysis*. Institute of Electrical and Electronic Engineer

An Introduction to Network Security Attacks

Mayank Srivastava

Abstract The computer network and its computing resources are rapidly growing along with the advancement in Internet technology. Apart from using computing resources, more and more people understand the importance of the network security. Security is one of the important aspects of a computer network which rises due to day to day increase in network attacks. Many network attacks leads to various cybercrime issues like attacking data centers, data theft, online transaction fraud, and the use of malicious activities. Network security controls are therefore required to control the access which can be carried out by these network attacks. This article focuses on different network security attacks to understanding and to propose solutions to develop a network that can give uninterrupted and secure services to the users.

Keywords Network attacks · Reconnaissance · Malware · Access · Denial of service (DOS)

1 Introduction

Nowadays, people are moving from the physical world to the cyber world for doing various types of day-to-day activities. The various activities are related to booking a train ticket, booking a meal, posting notes, writing messages on the social platform, sending and/or replying to an email, etc. While working in the cyber domain, one may face various types of attacks which is ultimately affect the users in the broad areas of financial, reputation, religious, racial, etc. [1]. Today, most homes are having a high-speed internet connection with one or more wireless routers, which could be exploited to do cyber-attack. Cyber harassment and cyberstalking is an offending area that breaches the violation of the privacy of people [2]. Another category of attacks refers to crime, which includes unauthorized computers intruding through the internet, computer destruction, unofficial use of computerized data, etc. [3].

M. Srivastava (✉)
Department of CEA, GLA University, Mathura, UP 281406, India
e-mail: mayank.srivastava@gla.ac.in

V. Suma et al. (eds.), *Inventive Systems and Control*,
Lecture Notes in Networks and Systems 204,
https://doi.org/10.1007/978-981-16-1395-1_37

Besides this all of the above- mentioned activities allows an imposter or intruder to do an attack using the Internet. The attacker makes use of different types of network attacks to commit the various cybercrimes, which can be categorized as active or passive, internal or external, structured, or unstructured. It is therefore very important that one should have a strong and secure connection to connect to the external world [4]. The secure connection will eventually help in eliminating the possibility of intruding the network and allows uninterrupted services to the network user. Network security is important to be considered for home networks as well as for the business world. An appropriate and updated security system helps in reducing the risk of data loss, theft, and disruption. Figure 1 depicts the configuration of cyber oriented network model.

Figure 1 shows a typical scenario in which a user network is created by using a local area network (LAN). The internal LAN is protected by using the firewall and virtual private network (VPN). The firewall allows only the authorized traffic to move across the network. The VPN allows securely sending data from inside or outside of the network. Figure 1 also shows the hacker which is connected to the internal network by using the global network. The above figure is an replica of a real scenario where users are creating their network by applying certain security restrictions and attackers despite the security restrictions would like to come to the internal network for a possible attack. The attacker in the above scenario may enter the internal network by using any of the different types of network attacks. One of the most common attacks is to perform the intrusion attack by using the details of the network. This paper focuses on identifying different types of network attacks, which can be focused to provide a possible solution to make the network services uninterrupted and secure.

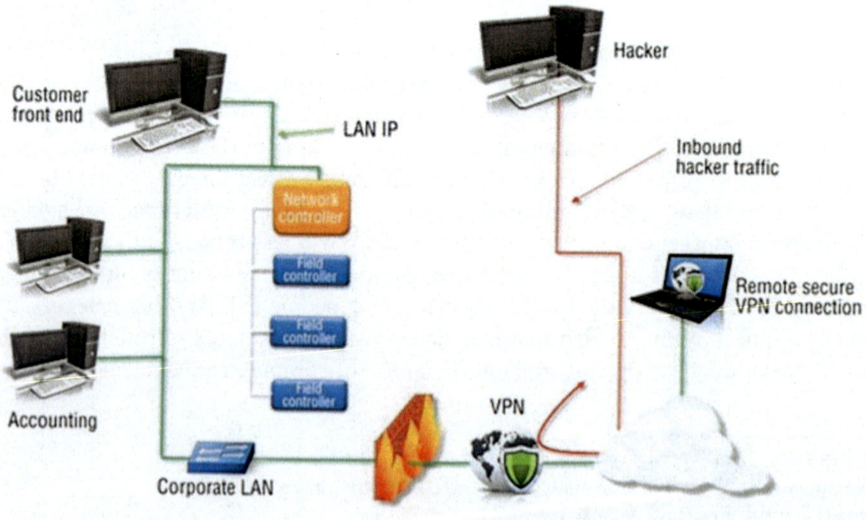

Fig. 1 Protecting facility network from cyber-attack [5]

2 Network Security Attacks

The network attack can be broadly defined as an action taken towards breaching a system's security through vulnerabilities. As per the latest threat report of McAfee labs, the statistics of the most common attacks detected are distributed denial of service attacks (DDOS) tops the list at 37%, brute-force password attacks come in next at 25%, browser-based attacks which target end-users like DNS poisoning, cookie capture comes at next place [6]. As per the report, DOS attacks are the highest category of attacks that need to be tackled on priority. The network attacks can be broadly categorized on different factors which include identification, expertise, and individual. The details related to such categorization are given as follows [7].

A. **Identification**

1. **Passive**: In this attack, no one can notice the presence of an attacker.
2. **Active**: In this attack, one can notice the presence of an attacker as the attacker will change the data or harm the system.

B. **Expertise**

1. **Structured**: It appears from individuals who are highly motivated and technically competent.
2. **Unstructured**: It appears from unskilled individuals who attack the network due to curiosity.

C. **Individual**

1. **Internal**: It appears from internal employees who have access to networks and other resources.
2. **External**: It appears from individuals who do not have direct access to the internal network.

3 Types of Network Security Attacks

This section will focus on network security attacks which are broadly categorized into four different categories given in Fig. 2.

3.1 Malware Attacks

Malware is software programs or malicious codes that install on a system without the user's knowledge. It disrupts services, damages systems, and gathers sensitive information. It includes virus, trojan, spyware, rootkit, backdoor, etc. [8].

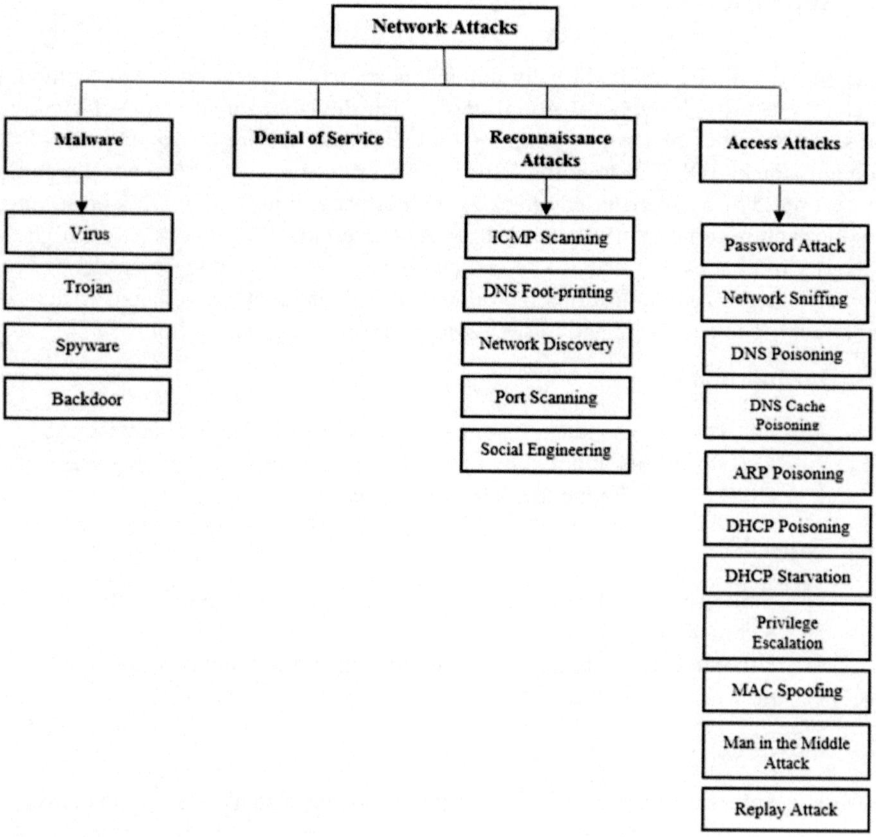

Fig. 2 Taxonomy of network attacks

1. **Virus**: A virus is a self-replicating program the attaches itself to another program or document to spread the infection.
2. **Trojan**: A program that appears to be good or useful software but contains hidden and harmful code.
3. **Spyware**: Spyware is a piece of software code that extracts the user information and sends it to attackers.
4. **Backdoor**: Backdoors are programs that allow hackers to bypass authentication checks to gain administrative privileges.

3.2 Denial of Service (DOS) Attacks

DOS attacks make resources unavailable for genuine users by sending a large number of service requests. DOS attacks target the network connectivity, resources, or bandwidth. Bandwidth attacks deprive legitimate users to access network resources by

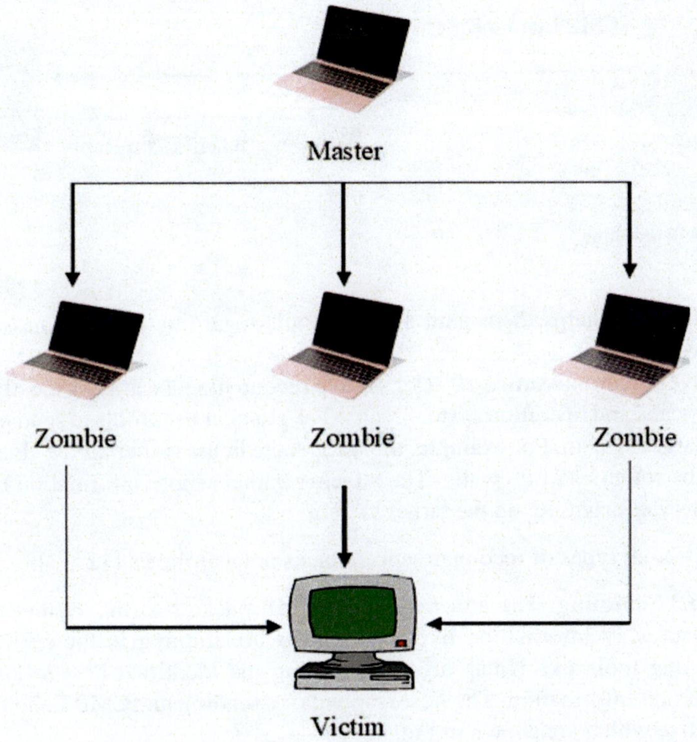

Fig. 3 Denial of service (DOS) attack [10]

overflow the network with a high volume of traffic. In connectivity attacks, the computer cannot process legitimate user requests as the system is already overwhelmed with a large amount of connection requests [9]. Distributed DOS (DDOS) attack makes use of several negotiated systems to attack a single system which eventually leads to DOS. The representation of a DOS attack is given in Fig. 3.

3.3 Reconnaissance Attacks

The major objective of a reconnaissance attack is to collect information about the system and target network. Example of reconnaissance attacks includes ICMP scanning, DNS Foot-printing, Network discovery, Port Scanning, and Social Engineering [11].

The reconnaissance attacks can be categorized as follows:

1. **Passive Reconnaissance attacks**: Passive reconnaissance attacks use the method of gaining information from the traffic. Here the attackers perform

Fig. 4 ICMP scanning

sniffing that helps them gain all the details regarding the weaknesses in the network.

2. **Active Reconnaissance attacks**: Active reconnaissance attacks mostly include port scans and operation system scans. The attacker uses tools to send packets to the target system. For example, the trace route helps gather all the IP addresses for the routers and firewalls. The attacker gathers more information regarding the services running on the target system.

The different types of reconnaissance attacks are as follows [12, 13]:

1. **ICMP Scanning**: The attacker sends ICMP packets to the system to gather all necessary information like which hosts are running in the network. The scanning tools like Nmap are used to ping the identified host to known the important information. The host responds by sending an ICMP Echo reply if it is active which are shown in Fig. 4.
2. **DNS Foot-printing**: It can help to determine important information about the location and type of servers. DNS foot-printing reveals information about the DNS zone that includes DNS domain names, computer names, and IP addresses. When the attacker queries the DNS server, the server responds with a record structure that contains information about the target DNS.
3. **Network Discovery**: Nmap is a network discovery and security auditing tool. Attackers may use Nmap which is used to determine which hosts are active, which operating systems are running on a host, what kind of services hosts are offering.
4. **Port Scanning**: In port scanning, the attacker's objective is to check whether any services are running on the host computer. Hackers may use various techniques like Nmap to find open ports on the target machine. Open ports are the doorway through which malware may get into the system.
5. **Social Engineering**: Social engineering is an act of convincing or tricking people to provide personal and business information. It helps to obtain access details, authorization details, and confidential information of the people. It is easy for the attackers to breach the security of the organization by using social engineering attacks.

3.4 *Access Attacks*

In access attacks, the attackers try to gain access to the network by using various exploitation techniques based on information gathered using reconnaissance attacks. The various mechanisms for gaining access are given below [14–16].

1. **Password attacks**: Attackers use various techniques such as brute-force, social engineering, spoofing, phishing, malware, sniffing, and key-logging to acquire passwords. Attackers start with cracking passwords and tricking the network device to believe they are valid users. Dictionary-based attacks, brute force, and hybrid attacks are the most commonly used password attacks.
2. **Network Sniffing**: It is the process of observing the data packets which is going thru the network by using various sniffing tools. Sniffing is classified into two broad categories- internal sniff and external sniff: In internal sniff, a person connects to the internal LAN to run tools and capture the network data. In external sniff, a hacker outside the target network intercepts packets.
3. **DNS poisoning**: Domain Name Server (DNS) poisoning is an unauthorized manipulation of IP addresses in the domain naming server cache. The DNS holds domain name translations of the IP addresses for network devices. A corrupted DNS redirects a user request to a malicious website to perform illegal activities.
4. **DNS Cache Poisoning**: In the DNS system, the domain names which are recently resolved are maintained in cache memory. In this attack, the attackers target the DNS cache and replace the correct IP address with the forged IP address. This poisoned DNS resolver cache directed the user to the fake server instead of the authorized server.
5. **ARP poisoning**: Address Resolution Protocol (ARP) is basically used for mapping IP address to a physical machine address. In ARP Poisoning, the attacker associates its MAC address with the victim's computer's IP address, and therefore the traffic intended for the victim IP address is sent to the attacker. The vulnerability of the ARP protocol which allows the host to automatically cache any ARP replies without authenticating their source enables ARP spoofing. One of the possible solutions to deal with this problem is to assign the static IP to the machines.
6. **DHCP spoofing**: DHCP servers assign IP addresses to the clients dynamically. An attacker may place a rouge DHCP server between the client and the real DHCP server. When a client sends a request, the attacker's rogue server intercepts the communication and acts as a valid server by replying with fake IP addresses. By installing a rogue DHCP server, the hacker can send incorrect TCP/IP settings.
7. **DHCP starvation**: In this attack, a large amount of DHCP requests is generated for the DHCP server which eventually absorbs all the IP addresses the DHCP server can generate. The attacker broadcast the DHCP requests by using different and spoofed MAC addresses. This leads to DHCP starvation as the DHCP server cannot issue new IP addresses to the genuine host requests. Port

security is one of the methods of preventing this attack by limiting the number of MAC addresses that can access the port.

8. **Privilege Escalation**: This attack allows the attackers to gain access to the network and its associated applications by making use of the defects in the design and poorly configured application. Once an attacker accesses a remote system, it will try to increase its privileges. The attacker will use a mechanism of increasing the user account to another augmented privilege.
9. **MAC spoofing**: This attack refers to the concept of spoofing the MAC address of the legitimate user with the attacker. The attacker firstly identifies the client who is active in the network and finds its MAC address. The attacker then spoofs the MAC address of the legitimate user with its own MAC address. Upon completing the MAC spoofing, the attacker receives all the traffic destined for the authorized user.
10. **Man in the Middle attack**: In this attack, the attacker's goal is to intercept the existing connection to intercept the messages and insert the fraudulent information. It is one form of eavesdropping where communication between two parties is either monitored or modified by an intercepting unauthorized party. One can intercept communication easily in the wireless domain. However, due to the sequence numbers generated by the TCP, it is not easy to implement such attacks.
11. **Replay Attack**: A replay attack is an extension of the man in the middle attack that occurs after a two-way communication is intercepted. An attacker captures the data to obtain usernames and passwords. After the authentication token are seized using a sniffer, the seized tokens are placed back on to the network to gain access.

4 Discussion

Table 1 shows the comparison chart for the mentioned cyber-attacks in terms of different important parameters. The important parameters include operational impact, target, and defense that can be applied to deal with that attack.

In terms of operational impact, it is shown the how different attacks affect the systems. For example, certain attacks disclose the information, certain others manipulate the data and others perform functions like blocking of services. In terms of target, it shows how different attacks affect the various levels of the system. It is shown in Table 1 that the majority of attacks affect the user and network-level of interaction. And finally, Table 1 discusses the most important factor defense, which is used to prevent various levels of attacks. For example, the malware can be eliminated by installing an antivirus. Request can be blocked to prevent the different types of reconnaissance attacks. Timestamp can be used to prevent replay attacks.

The study shows that security can be provided by applying security controls at different levels. Firstly, the latest antivirus must be installed and updated regularly

Table 1 Cyber-attack classification structure

Attack category	Name of attack	Operational impact	Target	Defence
Malware	Virus	Corrupting file	User,	Antivirus
	Trojan	Malicious activity	User, Network	Antivirus
	Spyware	Collecting confidential data	User, Network	Antivirus
	Black door	Intrusion	User, Network	Updation
Denial of service	Dos	Connectivity, bandwidth	Network	Blocking service request
Reconnaissance	ICMP Scanning	Disclosing information	Host	Blocking ECHO request
	DNS Foot-printing	Disclosing DNS information	Web server	Blocking traceroute request
	Network discovery	Disclosing information	Network	Block unlimited port request
	Port scanning	Disclosing services	Applications, systems	Block unlimited TCP request
	Social engineering	Disclosing information	User, Network	Block unlimited port request
Access	Password	Disclosing secret information	User	Two-factor authentication
	Network sniffing	Knowing secret information	User, Network	Encryption
	DNS poisioning	Manipulating IP Address	Web server	Secure DNS server
	DNS Cache poisioning	Manipulating IP Address	Web server	Secure DNS server
	ARP poisioning	Manipulating MAC Address	User	Static mapping of IP-MAC Address
	DHCP Spoofing	Manipulating IP Address	User	Physical security
	DHCP Starvation	Block IP Address generation	System	port security
	Priviledge escalation	Gain Access	Applications, systems	Flawless application
	Mac Spoofing	Spoofing MAC Address	User	Blocking spoofed system
	Man in the middle	Access confidential data	User	Encryption
	Reply	Access confidential data	User	Time stamp

for getting security updated related to the system. Secondly, all the system and application software must be updated from their authorized patches to close the various existing vulnerabilities in it. Finally, there must be a proper authentication system to access the resources of the system. By using the above three levels of security controls, one can secure any system from the majority of network attacks. It is important to mention that the term security is evolving in nature and needs continuous study to identify various controls needed to ensure the security of the system.

5 Conclusion

Security is one of the most important aspects of any computer network apart from its services. The article discusses the different network security attacks which belong to domains of Malware, DOS, Reconnaissance, and Access attacks. Malware covers the various aspects of malicious programs. DOS attacks give their contribution related to the continuity of services. Reconnaissance attacks focus on those attacks that identify the different network associated parameters. Finally, the access attacks are given which pave the different ways of intruding the computer network. The proposed work gives in-depth details of the different network attacks which can be used to propose the possible solutions to deal with these attacks.

References

1. S. Furnell, in *Cybercrime: Vandalizing the Information Society.* Addison-Wesley (2002)
2. R. Anderson, C. Barton, R. Böhme, R. Clayton, M.J.G. Eeten, M. Levi, Measuring the cost of cybercrime. the economics of information security and privacy, pp. 265–300 (2012)
3. N. Bhalaji, A novel hybrid routing algorithm with two fish approach in wireless sensor networks. J. Trends Comput. Sci. Smart Technol. **2**(3), 134–140 (2020)
4. M. Yar, K.F. Steinmetz, in *Cybercrime and Society*. 3rd edn. (SAGE, 2006)
5. https://www.hpac.com/building-automation/article/20928349/protecting-facility-networks-from-cyber-attack
6. McAfee Labs Threat Reports, November (2020). https://www.mcafee.com/enterprise/en-us/assets/reports/rp-quarterly-threats-nov-2020.pdf
7. M.V. Pawar, J. Anuradha, Network security and types of attacks in network. in *International Conference on Intelligent Computing, Communication and Convergence (ICCC-2014), Procedia Computer Science*, pp. 503–506. (2015)
8. D. Midhunchakkaravarthy, P. Ganapathi, A survey on various security threats and classification of malware attacks, vulnerabilities and detection techniques. Int. J. Comput. Sci. Appl. **3**(4), 66–72 (2013)
9. S. Smys, DDOS attack detection in telecommunication network using machine learning. J. Ubiquitous Comput. Commun. Technol. **1**(1), 33–44 (2019)
10. S. Day, I. Saha, An innovative approach to devise security matrix to measure impact of attack vectors in cloud networks. Int. J. Adv. Res. Comput. Sci. Manage. Stud. **2**(4), (2014)

11. R. Greenlaw, A. Phillips, J. Schultz, D. Stahl, S. Standard, Network reconnaissance, attack, and defense laboratories for an introductory cyber-security course (2013). https://doi.org/10.1145/2498328.2500077
12. H.P. Sanghvi, M.S. Dahiya, Cyber reconnaissance: an alarm before cyber attack. Int. J. Comput. Appl. **63**(6), 36–38 (2013)
13. U.A. Dar, A. Iqbal, The silent art of reconnaissance: the other side of the hill. Int. J. Comput. Netw. Commun. Secur. **6**(12), 250–263 (2018)
14. W. Qinquan, P. ZaiLin, Research on network attack and detection methods. in *2010 Second International Workshop on Education Technology and Computer Science*, Wuhan, pp. 630–633. (2010). https:\\doi.org.https://doi.org/10.1109/etcs.2010.196 (2010)
15. J.-J. Julian, S. Nepal, A survey of emerging threats in cybersecurity. J. Comput. Syst. Sci. **80**(5), 963–993 (2014)
16. M. Abomhara, G.M. Koien, Cyber security and the internet of things: vulnerabilities, threats intrudersand attacks. J. Cyber Secur. **4**, 65–88 (2018)

Design of Small-Sized Meander Lined Printed Monopole Antenna Operating in VHF Range

Dipankar Sutradhar, Durlav Hazarika, and Sunandan Bhunia

Abstract A compact and reduced VHF micro-strip patch antenna with dimensions 17×10 cm printed on an FR-4 substrate with relative permittivity $\epsilon_r = 4.4$ and thickness of 1.6 mm is presented in this article. The designed micro-strip line feed meandered antenna resonating at 203 MHz with a return loss of −26 dB with 10 dB bandwidth of 5 MHz and yields a radiation efficiency of 66% at the spot frequency. The compactness of about 89.5% is achieved well by the optimum design contrasted to the regular micro-strip patch antenna.

Keywords Wireless sensor networks · Printed monopole antenna · VHF · Return loss

1 Introduction

The miniaturized VHF antenna is imperative for wireless M-Bus applications in smart metering, portable surveying systems, and wireless sensor networks (WSNs). The VHF and higher frequencies use the line-of-sight propagation and travel through many non-metallic substances with more profound penetration behavior that can be picked up through walls and buildings [1]. All the new IoT products are becoming smaller and smaller; designing an antenna for these small objects in a minimal area while maintaining efficiency and bandwidth is challenging. Compact solutions suitable to integrate into terminal devices are realized based on folded, meander-line, and

D. Sutradhar (✉)
IE Department, Central Institute of Technology Kokrajhar, Kokrajhar, Assam, India
e-mail: d.sutradhar@cit.ac.in

D. Hazarika
EE & IE Department, Assam Engineering College Jalukbari, Guwahati, Assam, India

S. Bhunia
ECE Department, Central Institute of Technology Kokrajhar, Kokrajhar, Assam, India
e-mail: s.bhunia@cit.ac.in

V. Suma et al. (eds.), *Inventive Systems and Control*,
Lecture Notes in Networks and Systems 204,
https://doi.org/10.1007/978-981-16-1395-1_38

inverted-F configurations delivering comprehensive band operation and omnidirectional radiation characteristics [2, 3]. Shrinking the size of the antenna can be attained at the cost of its efficiency and gain. Several research groups have developed a various compact, cost-effective small VHF antennas to minimize the size with a small footprint. Meander-line monopole antennas are mainly utilized in condensed and handy smart appliances [4–7]. In [8], a dual-band meander line monopole antenna printed on FR4 substrate was developed to perform at 114 and 221 MHz with an above 90% radiation efficiency in both bands. The antenna miniaturization approach with the meander line structure's help resulted in a 71% reduction compared to conventional dipole operating at the same frequencies with a footprint of 29 × 8.3 cm. In [3], broadband printed monopole antenna on FR4 substrate utilizing meandering and top-loading techniques were presented to operate in VHF range; the antenna required substrate size of 39 cm × 31 cm and exhibited a 3:1 VSWR bandwidth of 23% in the simulated frequency range of 100–220 MHz. In [9], a dual-band printed monopole antenna of 150 × 150 mm based on the double-side meandering structure was presented. The antenna was designed on a double-sided FR4 substrate operating at 169 and 433 MHz; the gains were reported as −1.8 and +4.6dBi, respectively.

This article aims to propose an ultra-compact VHF antenna working around 200 MHz. The structure is a type of printed monopole antenna using the meander line proposed by El-Damak et al. [8]. It comprises a meander line antenna to shrink the low-frequency monopole's size printed on one face of the dielectric when the partial ground is printed on the other face of the dielectric. The resonating frequency of the proposed design can be reduced by meandering the monopole element, and this can be done in two ways [10, 11]: (i) extending the meander-line element width raises the capacitance to the ground, contributing to a reduction in resonant frequency, (ii) extending the entire element span while preserving the same substantial volume. To achieve this, the number of meander sections requires to be increased. The increase in overall element span leads to a buildup in the effective self-inductance, diminishing the resonant frequency. The proposed designs have an optimized electrical dimension of almost 0.11 $\lambda_L \times 0.06\,\lambda_L$ and on an FR-4 substrate, where λ_L is the wavelength in free-space of the operating frequency. The proposed design is analyzed to determine the individual impact of size reduction on the resonant frequency shift, reflection coefficient, and bandwidth reduction.

2 Antenna Design and Structure

Figure 1 shows the geometry of the developed antenna. The antenna is designed on a double layer FR4 substrate accompanied by a thickness of 1.6 mm, a relative permittivity of 4.4, and a metallization thickness of 35 microns.

To fulfill the requirements of elementary, cheaper substrates while not compromising the antenna operation, FR-4 substrate material has been adapted to perform a cost-efficient project [12]. The antenna structure is a meander line monopole patterned on the upper layer of the board, the partial ground plane on the bottom layer,

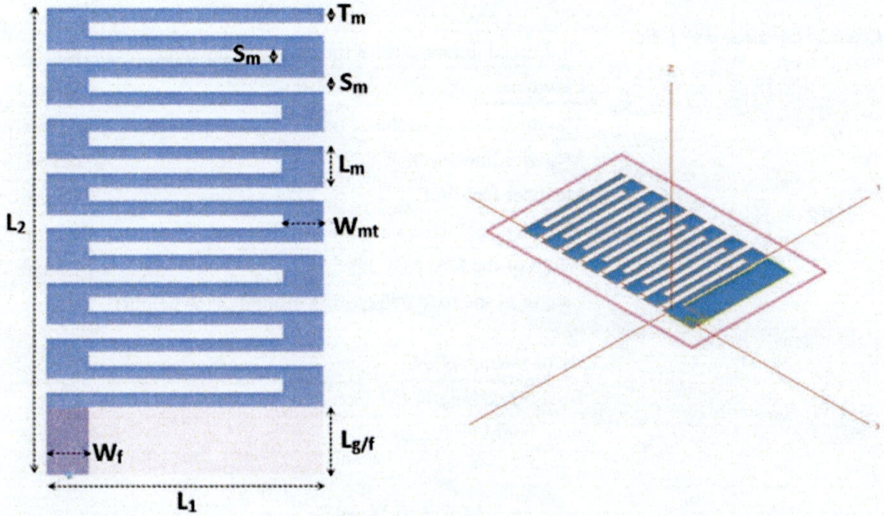

Fig. 1 Proposed antenna, **a** dimension, **b** 3-D structure

and is excited with the edge feeding technique. A parametric study with different dimensional variables of the proposed design is being made to investigate the impact on the antenna's radiation efficiency; however, the meander line width is kept constant with uniform spacings in all the simulated designs. The proposed antenna resonates at 203 MHz frequency with a reasonable size to integrate into smart metering and wireless sensor networks.

The main parameters, such as track thickness, spacing, track widths at bends, no of meanders, are varied to tune the output of the proposed designs while keeping the footprint (170 × 100 mm) and ground plane size of the design constant in all the simulations. Due to many of these parametric study combinations, only the final results of the design to set out the augmented operation for a small antenna printed on FR4 substrate resonating in the 200 MHz band is presented. Table 1 shows the optimal parameter values of the final antenna design, and these dimensions are critical. The proposed antenna is projected for assimilation into limited volumes; thus, low profile, compressed sizing, and sound radiation characteristics are the crucial specifications in this work.

3 Simulated Results and Discussions

The features of the proposed structures are investigated by utilizing PathWave Advanced Design System (ADS) software. The simulated return loss and the operating frequency of the final presented antennas are illustrated in Fig. 2. The proposed antenna's minimum return loss value demonstrates the return loss of −26 dB at

Table 1 Dimensions of the final proposed design

Optimized dimensions of the proposed antenna	
Parameter	Value
The thickness of meander tracks, T_m	0.5 cm
Meander line spacing, S_m	0.5 cm
Meander length, L_m	1.5 cm
Meander track width at turns, W_{mt}	1.5 cm
Width of the feed line, W_f	1.5 cm
Length of the feed line (partial ground plane height), $L_{g/f}$	2.5 cm
No of meanders, N	7
Total board length, L_1	10 cm
Total board width, L_2	17 cm

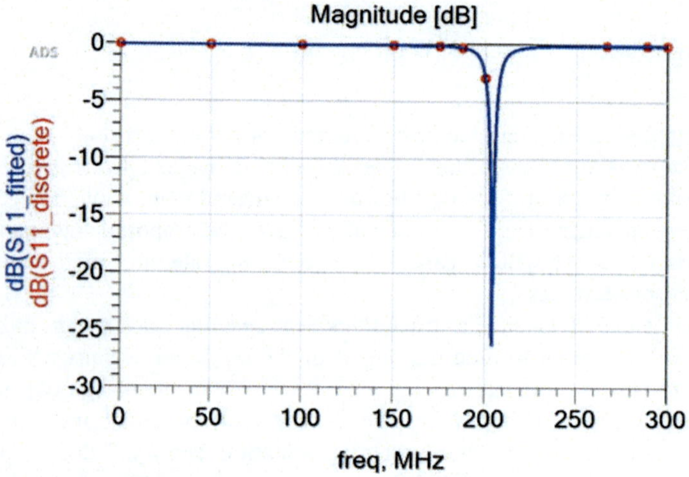

Fig. 2 S_{11} plot of the simulated antenna

203 MHz with a 10 dB bandwidth of 5 MHz. The VSWR plots, as illustrated in Fig. 3, reveals that the proposed antenna own a preferable VSWR value of 1.15 at its resonant frequency and is closer to the ideal value of 1 for an antenna.

It can be seen from the gain with the variation of frequency plot in Fig. 4 that the recommended design has a gain of 0.145 dBi at the resonating frequency. The obtained gain value of the proposed design is close to the conventional printed meandering monopole and is in a good range for the targeted applications.

The designed antenna's efficiency with the variation of frequency as illustrated in Fig. 5, yielding a value of 66% at 203 MHz. The chosen thickness of meander tracks, meander line spacing, track widths at turns, and the number of meanders values satisfy the design target for the proposed design. It is again evident from this result

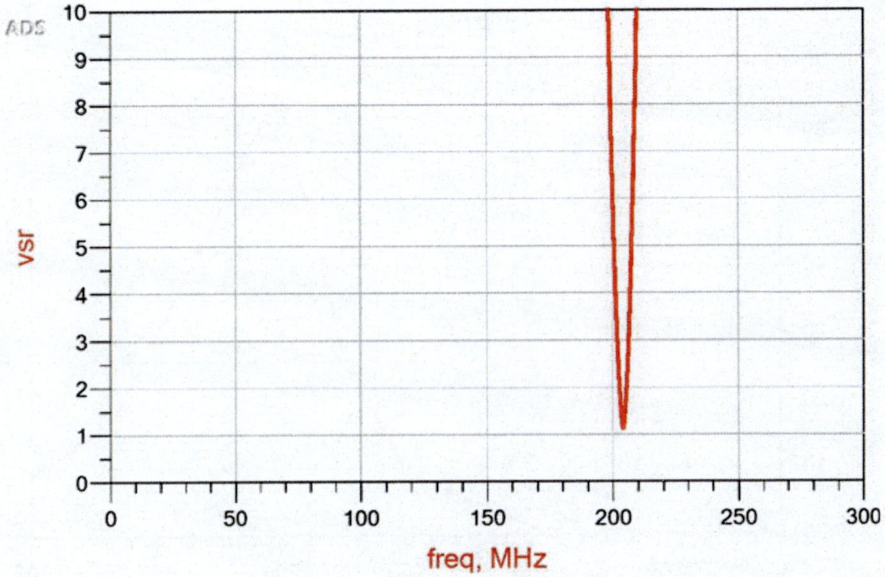

Fig. 3 VSWR plot of the simulated antenna

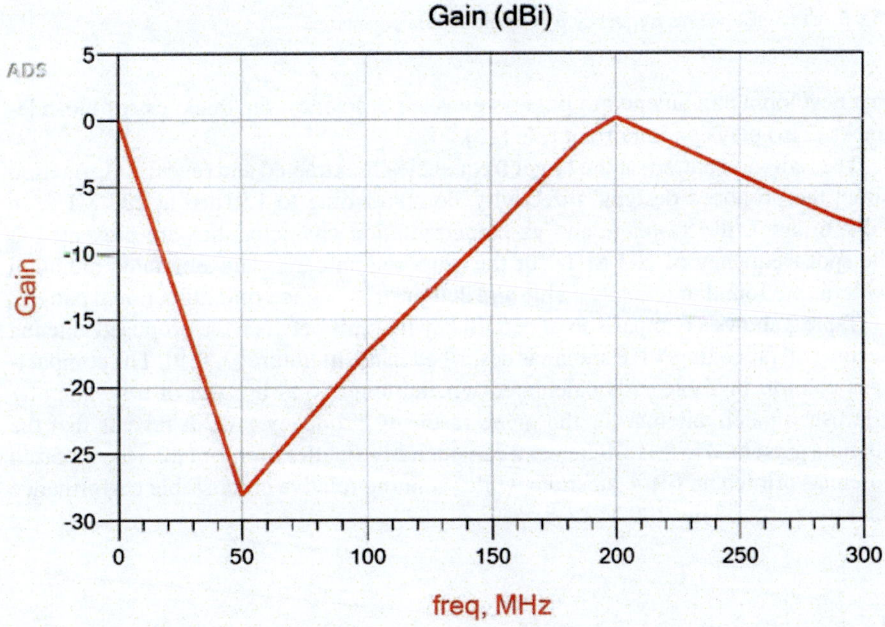

Fig. 4 Gain with the variation of frequency plot for the simulated antenna

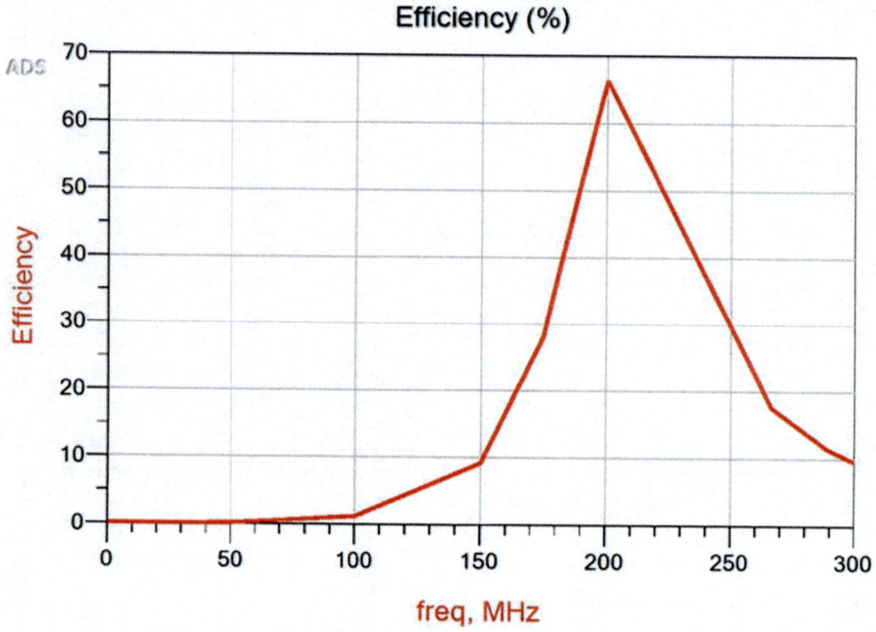

Fig. 5 Efficiency versus frequency plot of the simulated antenna

that developing an antenna to possess elevated efficiency, the thickness of meander tracks (Tm) plays an important role [13].

The radiation pattern at the target frequencies is extracted and reveals information about the proposed designs' directivity, corresponding to 1.91dBi at 203 MHz. In Figs. 6 and 7, the xz-plane and yz-plane radiation characteristics are presented at the spot frequency of 203 MHz for the proposed antenna. The simulated radiation patterns are found practically stable and demonstrate a good omnidirectional pattern.

Table 2 shows a comparison of certain key features between the proposed antenna design and preceding VHF antennas described in the literature [3, 8, 9]. The compactness column in Table 2. is calculated wrt the length and breadth of a rectangular microstrip patch antenna for the given resonant frequency [14]. It reveals that the antenna presented in this article has a considerably smaller footprint than the reported antennas printed on FR-4 substrate while retaining relative comparable performance operating around the 200 MHz range.

4 Conclusion

A small, low-cost printed meander line monopole antenna running around 200 MHz has been designed and investigated in this article. The meandering shape reduces the antenna size, and the utilization of the FR-4 substrate does it cost-effective. FR-4

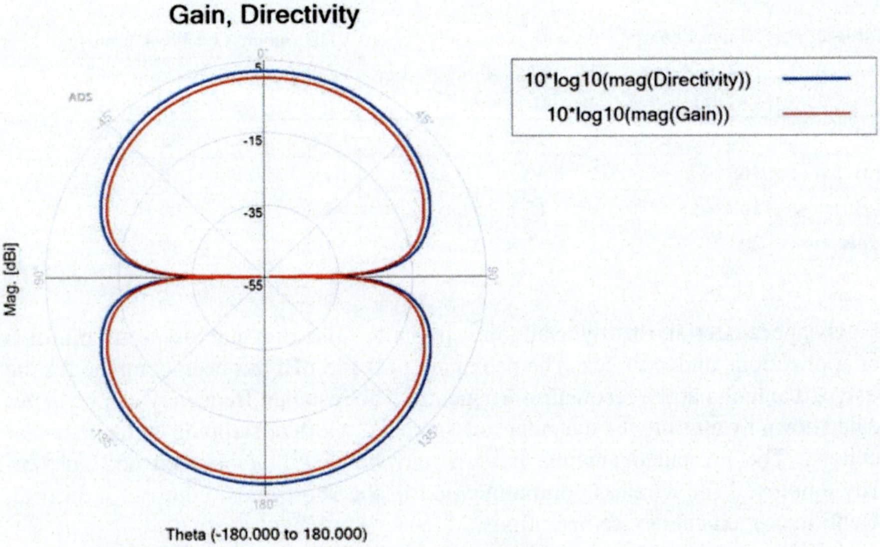

Fig. 6 Radiation pattern of the antenna, in the *X-Z* plane

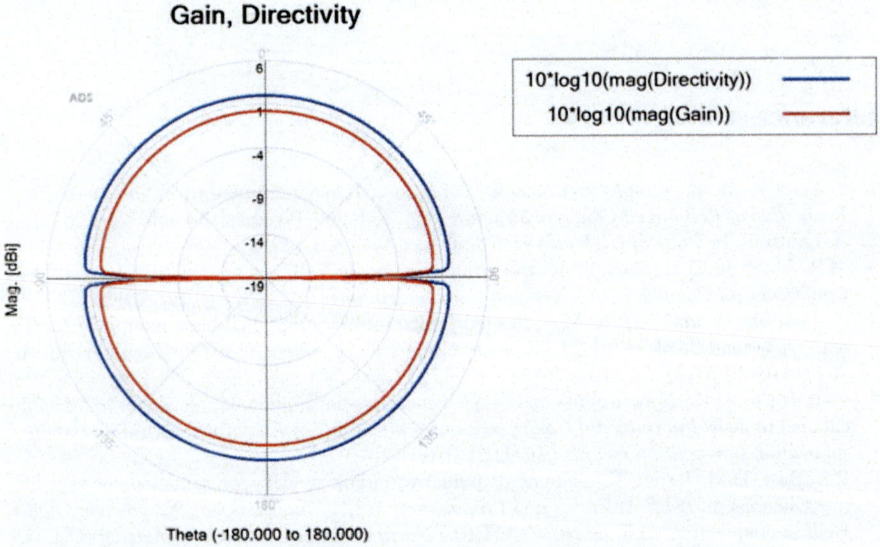

Fig. 7 Radiation pattern of the antenna, in the *Y-Z* plane

Table 2 Performance comparison with previously printed VHF antennas on FR-4 substrate

Reference	Operating frequency (MHz)	Overall dimension ($L_1 \times L_2$)	Peak gain (dBi)	Compactness (%)
[8]	125	29 × 8.3 cm	1.87	94
[3]	165	39 × 31 cm	1.75	49
[9]	169	15 × 15 cm	−1.8	90
This work	203	17 × 10 cm	0.145	89.5

is a cheaper material, readily available, simple to fabricate, and losses are minimal for applications under 1 GHz. The peak gain of 0.145 dBi has been obtained for the designed antenna at the resonating frequency. The resonant frequency can be better scaled down by altering the meander line sections' width or stepping up the meander sections. The presented antenna pattern may be readily fabricated and customarily employed for wireless communication in the short-range compact industrial, scientific or medical device operations.

Acknowledgements The authors would like to significantly acknowledge the Central Institute of Technology Kokrajhar, Assam, for providing the laboratory facilities to carry out this work.

References

1. L. Zanzi, Radar tomography for evaluation of reinforced concrete structures. in *Non-Destructive Evaluation of Reinforced Concrete Structures*. pp. 334–369. (Woodhead Publishing, 2010)
2. A. Loutridis, in *Study of UHF and VHF Compact Antennas*. (2015)
3. M.R. Abraham, O.K. Sona, Wideband printed monopole VHF antenna. in *2015 International Conference on Computing and Network Communications (CoCoNet)*. (IEEE, 2015)
4. A. Loutridis, J. Matthias, A. Max, Printed folded meander line dual-band monopole for TV white space and GSM. in *The 8th European Conference on Antennas and Propagation (EuCAP 2014)*. (IEEE, 2014)
5. C.-N. Hu, et al., Design of a multiband coupled-meander-line monopole antenna for M2M applications. in *2010 International Conference on Applications of Electromagnetism and Student Innovation Competition Awards (AEM2C)*. (IEEE, 2010)
6. R.S. Best, D.M. Jarrett, Limitations of inductive circuit model representations of meander line antennas. in *IEEE Antennas and Propagation Society International Symposium*. Digest. Held in conjunction with: USNC/CNC/URSI North American Radio Sci. Meeting (Cat. No. 03CH37450). vol. 1. (IEEE, 2003)
7. T.J. Warnagiris, T.J. Minardo, Performance of a meandered line as an electrically small transmitting antenna. IEEE Trans. Antennas Propag. **46**(12), 1797–1801 (1998)
8. A. El-Damak, K. Moustafa, M. Elkattan, Implementation of printed small size dual frequency antenna in MHz range. Int. J. Electron. Telecommun. **65**(4), 565–570 (2019)
9. A. Loutridis, J. Matthias, M.J. Ammann, Folded meander line antenna for wireless M-Bus in the VHF and UHF bands. Electron. Lett. **51.15**, 1138–1140 (2015)
10. R. Wallace, Design note DN024 Monopole PCB antenna with single or dual band option (2018)
11. S.R. Best, A discussion on the properties of electrically small self-resonant wire antennas. IEEE Antennas Propag. Mag. **46**(6), 9–22 (2004)

12. A. Khan, N. Rajesh, Analysis of five different dielectric substrates on microstrip patch antenna. Int. J. Comput. Appl. **55.14** (2012)
13. M. Takiguchi, Y. Yoshihide, Radiation and ohmic resistances in very small meander line antennas of less than 0.1 wavelength. Electron. Commun. Japan (Part I: Communications) **88.8**, 1–11 (2005)
14. Microstrip Patch Antenna Calculator, https://www.emtalk.com/mpacalc.php

Feasibility of Intelligent Techniques for Automated Positioning of Servomotors

N. M. Nandhitha, M. S. Sangeeetha, S. Emalda Roslin, Rekha Chakravarthi, V. Madhivanan, and Mallisetti Dileep

Abstract Servomotors are widely used in computers, wall-climbing robots, antenna positioning, etc. Automated position control of servomotors is a major challenge due to non-linearity in the input dataset. Automated positioning of servomotors using soft computing techniques is proposed. Impact of Membership Functions (MFs) on the performance of Adaptive Neuro-Fuzzy Inference System (ANFIS) is studied. Triangular Membership function with 7 MFs result in less Root Mean Square Error (RMSE) for the three-digit approximated dataset used from the dissertation of Linus AlwalAloo. When comparing the two neural networks, Generalized Recurrent Neural Network (GRNN) provides the best results in terms of RMSE for servomotor position control.

Keywords Servomotor · Position control · Triangular membership function · Adaptive Neuro-Fuzzy inference system (ANFIS) · BPN · Generalized recurrent neural network (GRNN)

1 Introduction

Servomotors are used for speed and position control in various applications. These servomotors are divided into two types namely AC servomotor and DC servomotor. Of these two types, DC servomotors are used for computer disc positioning, robot positioning, and antenna positioning. These servomotors are modeled as electrical and mechanical components. The electrical component includes armature, field coils, etc. while the mechanical component is responsible for the movement of the shaft and hence positioning it. Considerable research is carried out in this area. Khongkoom et al. [1], Priya et al. [2], Shieh et al. [3].

N. M. Nandhitha (✉) · M. S. Sangeeetha · S. Emalda Roslin · R. Chakravarthi
Faculty, School of Electrical and Electronics, Sathyabama Institute of Science and Technology, Old Mamallapuram Road, Chennai 600119, India

V. Madhivanan · M. Dileep
School of Electrical and Electronics, Sathyabama Institute of Science and Technology, Old Mamallapuram Road, Chennai 600119, India

V. Suma et al. (eds.), *Inventive Systems and Control*,
Lecture Notes in Networks and Systems 204,
https://doi.org/10.1007/978-981-16-1395-1_39

Initially PI, PD, and PID controllers are used for positioning the servomotor. Zabinski and Trybus, [4] used parameters for determining control cycle, gain, and settling time using P-PI controllers for servomotor control. Ravikumar and Nagababu [5] used Particle Swarm Optimization (PSO) for tuning the PID parameters for the control of servomotors. But the performance of these controllers depends on the values of the proportionality constants. Hence the paradigm shifted to soft computing techniques based on position control of servomotor. Roslin et al. [6] used fuzzy logic controller for the selection of cluster heads in Wireless Sensor Networks. Nandhitha et al. [7] determined congestion using Fuzzy controllers.

Membership functions and the limit of each MF affect the performance of FLC. Roseline et al. [8], Roslin et al. [9], Chakravarthi et al. [10]. Hence the paradigm has shifted to developing intelligent techniques for automatic position control of the servomotors. In this article, the suitability of ANFIS based system and Artificial Neural Network based system for automatic position control of the servomotors is presented. Root Mean Square Error (RMSE) is the performance metric.

The layout of this article is as follows. Section 2 deals with the database and impact of Membership Function (MF) on the performance of the proposed system is explained in Sect. 3. Neural Network based servomotor position control is dealt with in Sect. 4. The major conclusions and future directions are given in Sect. 5.

2 Database Selection

In this research work, the database used by Linus AlwalAloo's in his Master of Science dissertation titled "Design of Neuro-Fuzzy System Controller for DC Servo Motor Based Satellite Tracking System" [10] is used in this work. Instead of a precision of 9 digits, a precision of 3 bits is considered. Position Error and Change in Position Error are the input parameters and the predicted output parameter is used to control the speed and direction of the servomotor (Eqs. 1 and 2).

$$\begin{aligned} \text{PositionError} &= \text{Desiredvoltageofinputpotentiometer} \\ &\quad - \text{Actualvoltageofinputpotentiometer} \end{aligned} \tag{1}$$

$$\text{Change in Position Error} = \text{Position Error} - \text{Previous Error} \tag{2}$$

A set of 92 exemplars is used for training and testing ANFIS controller (Appendix 4.2 in Linus AlwalAloo's dissertation [12]). The sample set of exemplars are shown in Table 1.

The entire set of input and output parameters have a non-linear relationship as is shown in the form of a graph (Fig. 1).

Table 1 Exemplar table for training and testing ANFIS

Input parameters		Desired output
Position error	Change in position error	
0.814	−3.378	0.644
0.905	2.942	0.378
0.126	−1.887	0.811
0.913	0.285	0.532
0.632	−3.343	0.35
0.097	1.019	0.939
0.278	−2.37	0.875
0.546	1.54	0.55
0.957	0.892	0.622
0.964	2.481	0.587
0.157	−0.494	0.207
0.97	−4.161	0.301
0.957	−2.71	0.47
0.485	4.133	0.23
0.8	−3.476	0.844

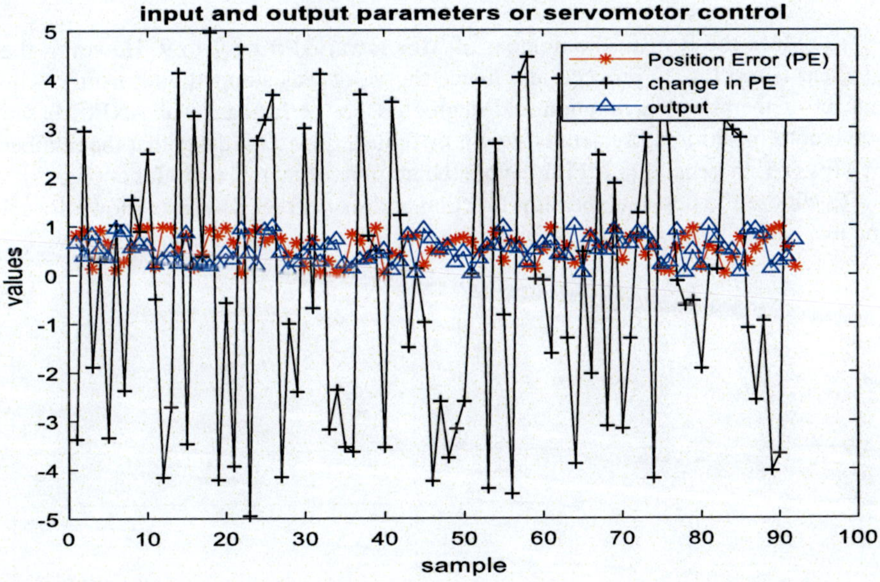

Fig. 1 Input and output parameters used as exemplars

3 Selection of Membership Function for Servomotor Position Control

Hence a soft computing technique like ANFIS is the best choice for determining the output variable which in turn controls the direction and speed of the servomotor. From the literature (FemilaRoseline and Nandhitha [11]), it is found that Generalized Bell-shaped and triangular Membership Functions (MFs) provide the best results. Hence these two membership functions are used initially (3 MFs). 46 samples are used for training and the remaining 46 samples are used for testing. Initial MFs and variations in the MFs (after training) are shown in Fig. 2.

The performance of ANFIS with Gbell MF is shown (Fig. 3). The actual values are not exactly similar to the desired values.

It is necessary to change the number of MFs (3–7). Initial, modified MFs and performance are shown in Figs. 4 and 5.

The number of MFs is changed from 79 and the training and testing are performed. Initial, modified MFs and performance are shown in Figs. 6 and 7.

With the triangular membership function (MFs = 3), the prediction is carried out. Initial, modified MFs and performance are shown in Figs. 8 and 9.

Now the number of MFs is changed from 3 to 7 and the output variable is determined. Initial, modified MFs and the performance are shown in Figs. 10 and 11.

To reduce the RMSE, the number of MFs is varied from 7 to 9. However, the gradient is too low to proceed and hence the work was stopped with trimf7. The impact of membership function and degree on the performance of ANFIS based servomotor positioning system is shown in Table 2. Gbell3 implies that the number of MFs is 3. In general in MFi, I denote the number of membership functions.

To choose the best membership function and its degree, RMSE is shown for all the five cases (Table 2) and it is found that RMSE for triangular membership with 7

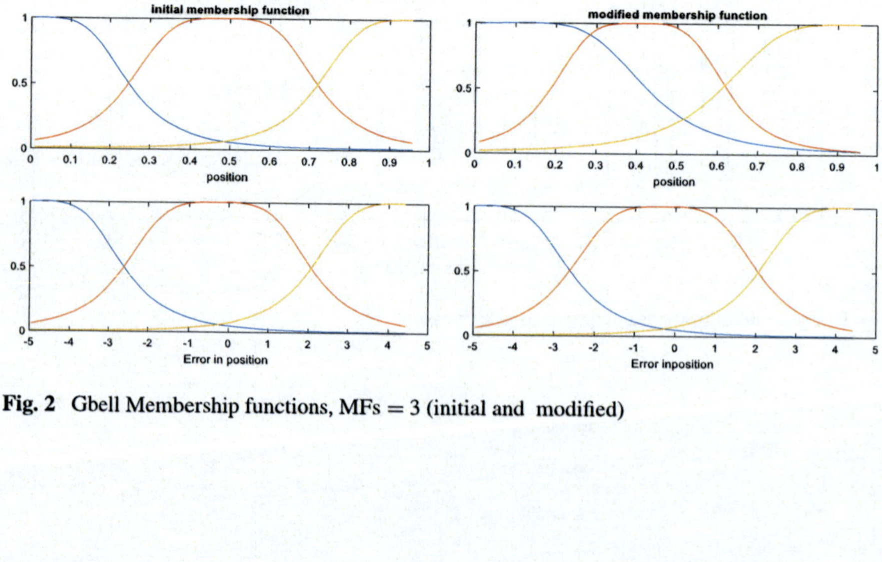

Fig. 2 Gbell Membership functions, MFs = 3 (initial and modified)

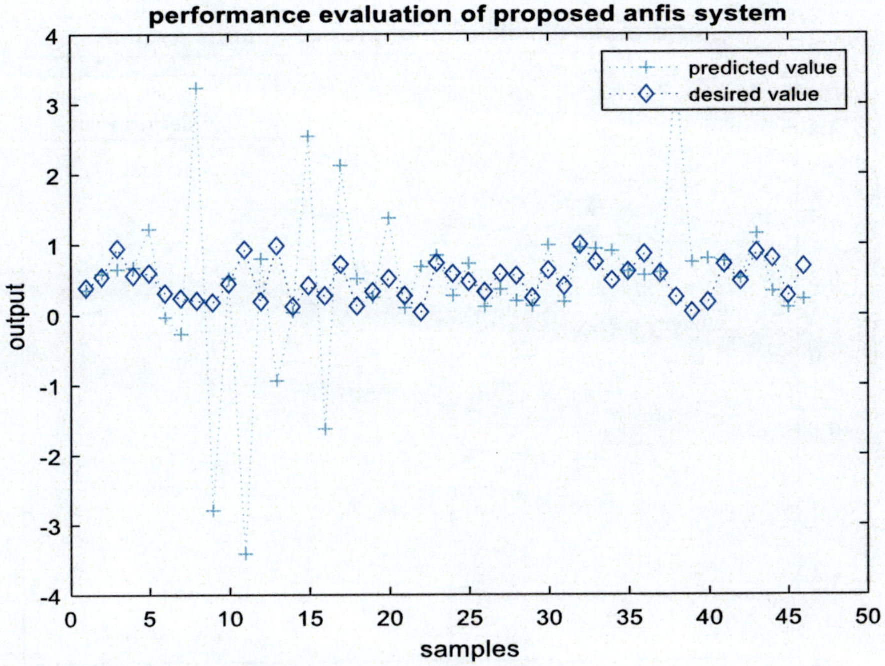

Fig. 3 Performance of ANFIS with Gbell MF (gbell3)

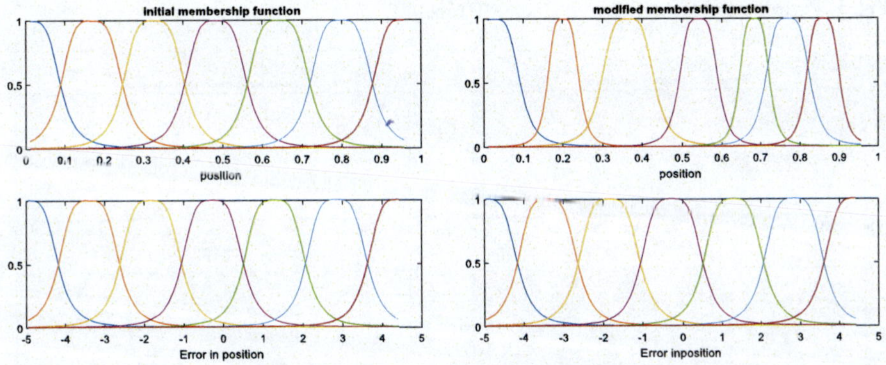

Fig. 4 Gbell Membership functions, MFs = 7 (initial and modified)

MFs is less and hence it is chosen as the best membership function. The feasibility of Artificial Neural Networks for the same application is shown in Sect. 4.

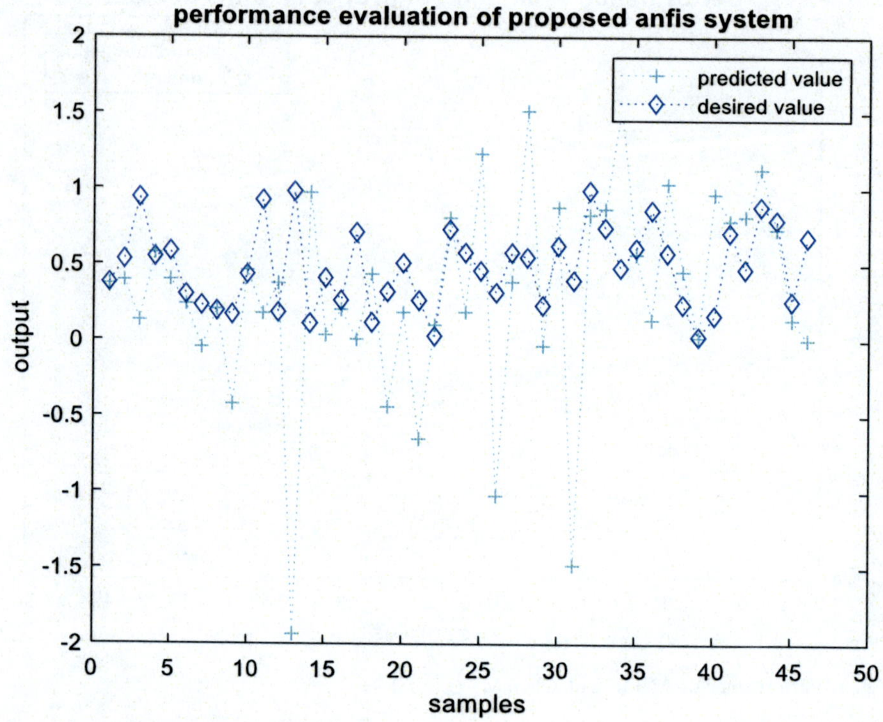

Fig. 5 Performance of ANFIS with Gbell MF (gbell7)

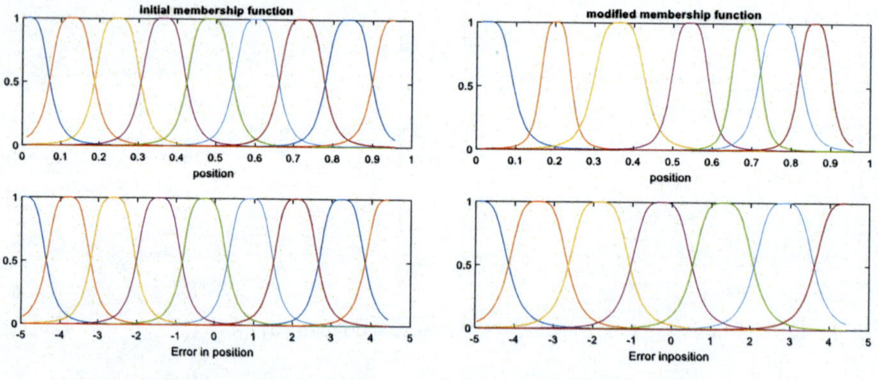

Fig. 6 Gbell Membership functions, MFs = 9 (initial and modified)

4 Neural Network Based Servomotor Position Control

Four layered BPN and conventional GRNN architectures are used for automated position control of servomotor. As is done in ANFIS and BPN, half of the exemplars

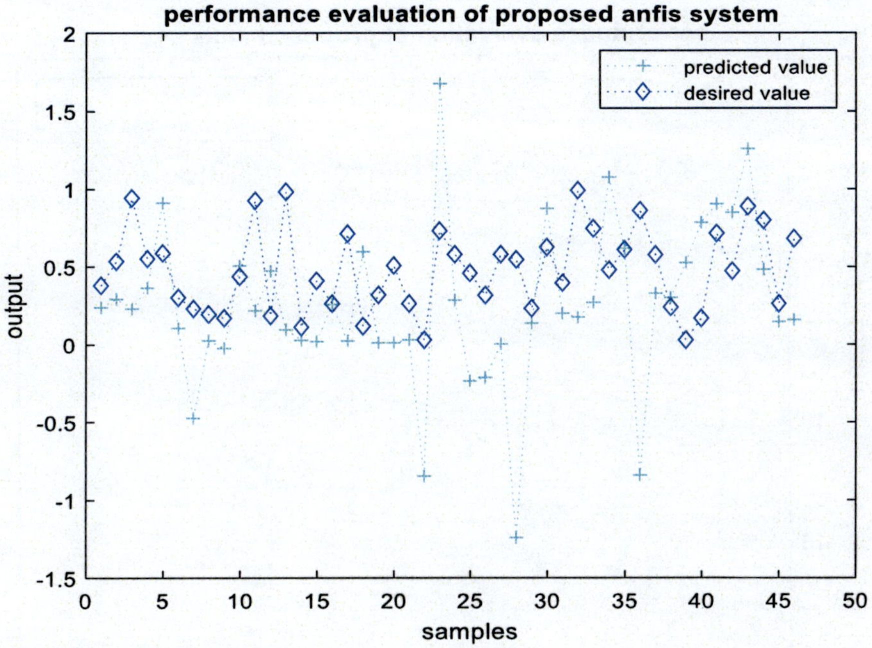

Fig. 7 Performance of ANFIS with Gbell MF **(gbell9)**

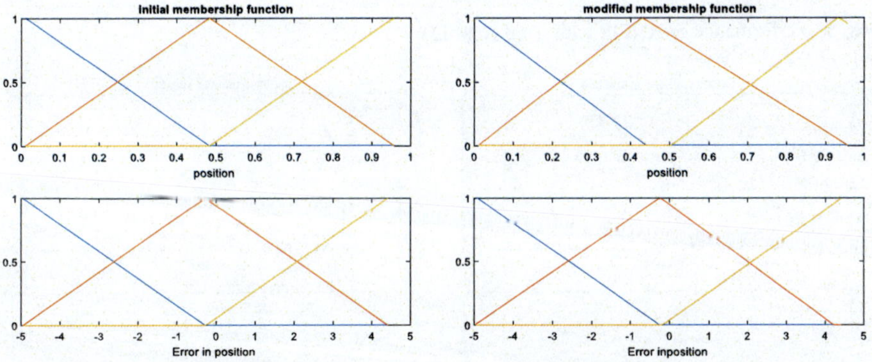

Fig. 8 trimf Membership functions, MFs = 3 (initial and modified)

are used for training, and the remaining half is used for testing the ANFIS controller. The sample set of exemplars are as shown in Table 3. Performance of BPN is shown in Fig. 12.

The Root Mean Square Error is 0.362796. To decrease the Root Mean Square Error (RMSE), another neural network model is used for predicting the output value.

The sample set of exemplars used for training GRNN is as shown in Table 1. The performance of GRNN is shown in Figs. 13 and Table 4.

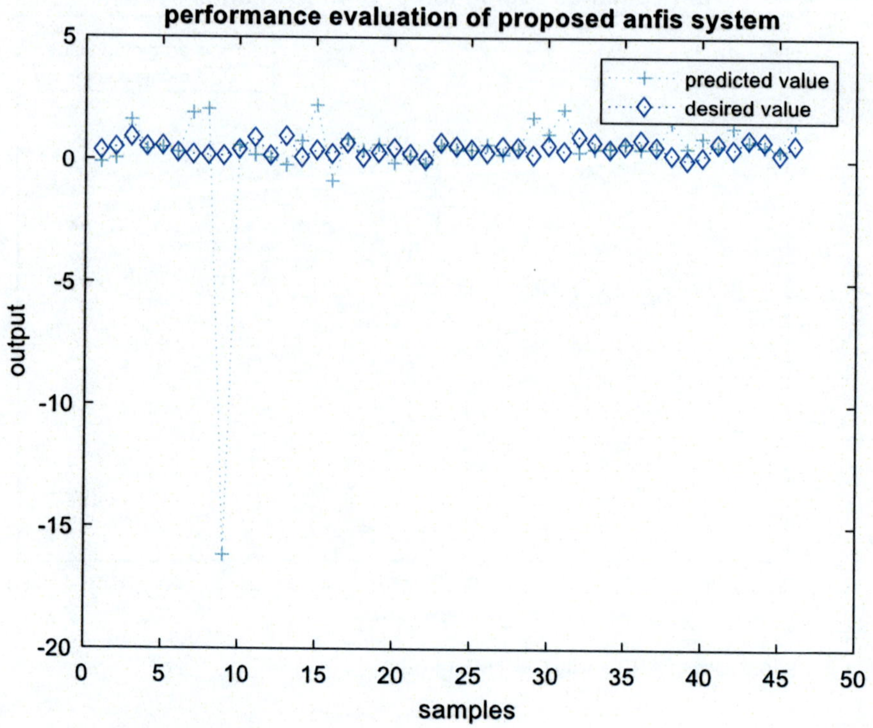

Fig. 9 Performance of ANFIS with trimf (**trimf3**)

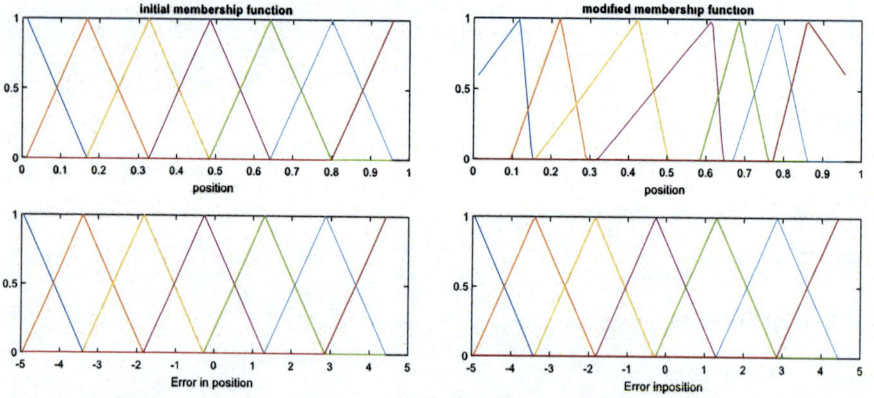

Fig. 10 trimf Membership functions, MFs = 7 (initial and modified)

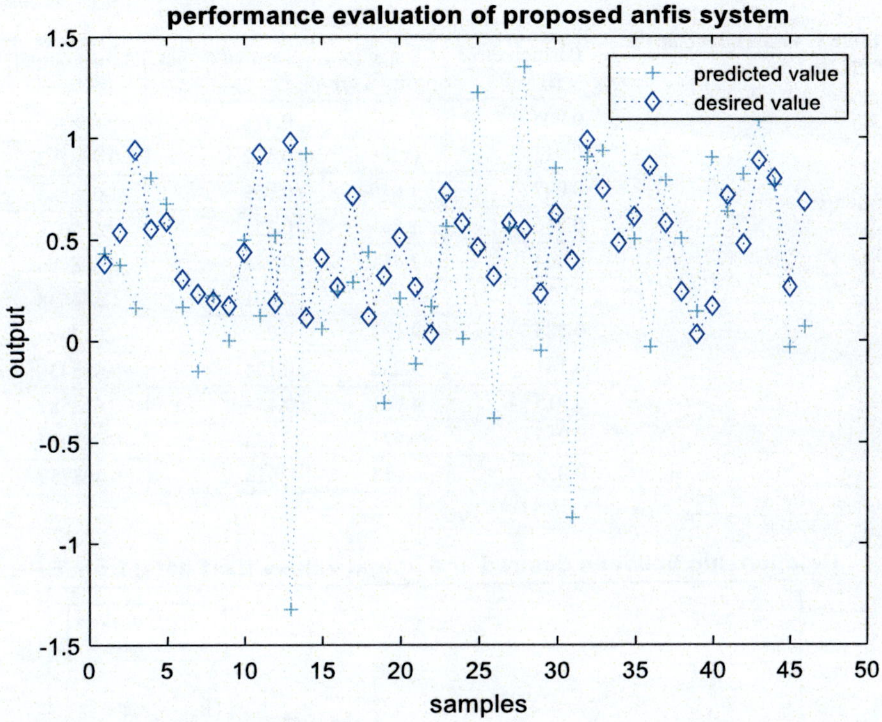

Fig. 11 Performance of ANFIS (trimf7)

Table 2 Performance of ANFIS based servomotor positioning system

S.No.	MFs	RMSE
1	gbell3	1.428843
2	gbell7	0.512324
3	gbell9	0.338532
4	trimf3	6.359327
5	trimf7	0.3288

The Root Mean Square Error of GRNN is 0.073357 which is lesser than that of ANFIS and BPN. Hence, the GRNN can be used for providing the output signal to control the speed and direction of the servomotor.

5 Conclusion and Future Scope

Intelligent controllers for the positioning of servomotors used in wall-climbing robots were developed. Position Error and change in position error are used as the input

Table 3 Performance of BPN (test dataset)

Position error (PE)	Change in position error	Desired output	Actual output
0.905	2.942	0.378	−0.16646
0.913	0.285	0.532	0.238792
0.097	1.019	0.939	1.048984
0.546	1.54	0.55	0.603078
0.964	2.481	0.587	0.608496
0.97	−4.161	0.301	1.554638
0.485	4.133	0.23	−0.07237
0.141	3.258	0.194	−0.09543
0.915	4.961	0.17	0.612587
0.957	−0.573	0.435	0.509957
0.035	4.618	0.923	0.408759

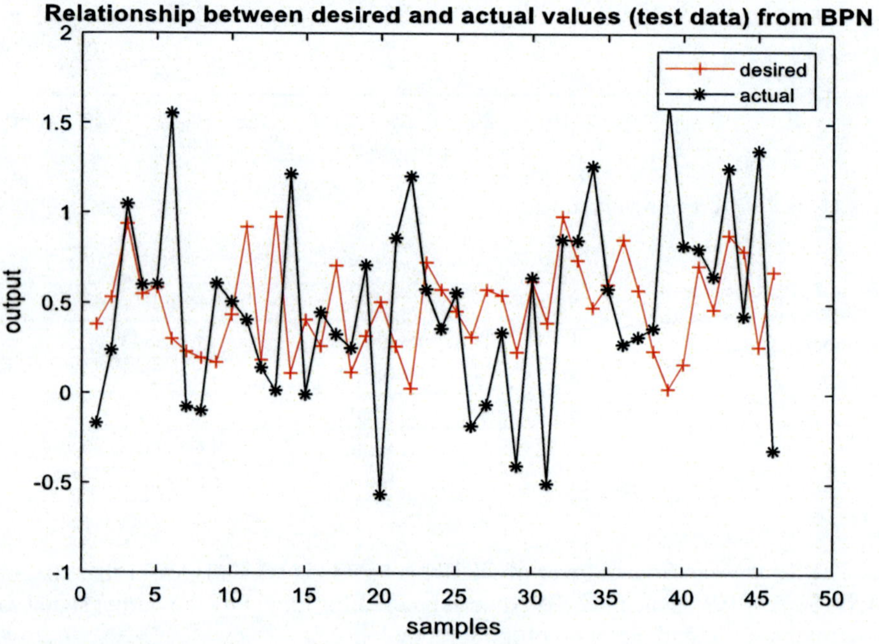

Fig. 12 Performance of BPN (test dataset)

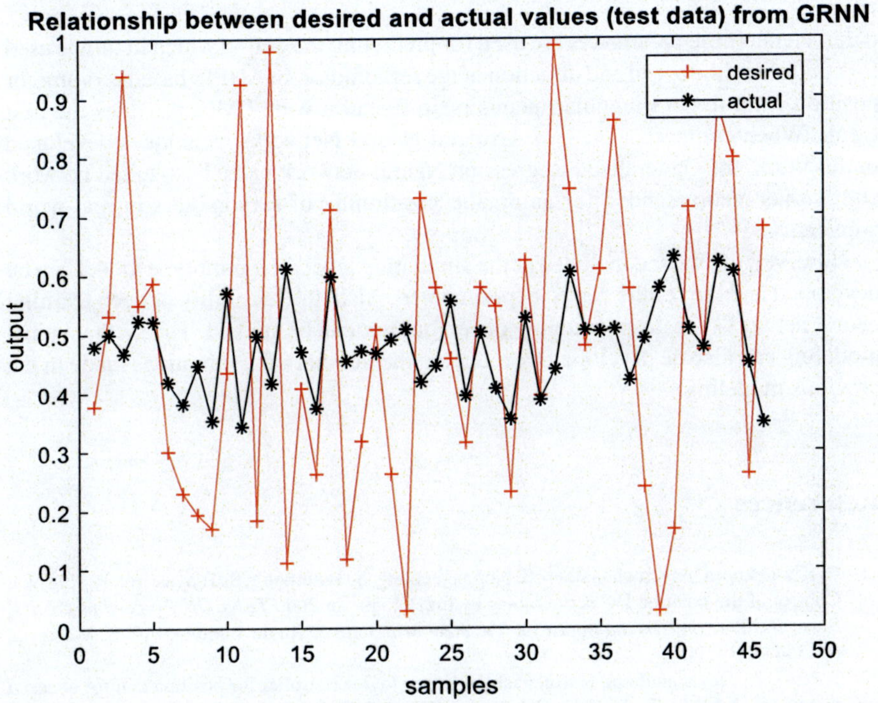

Fig. 13 Performance of GRNN (test dataset)

Table 4 Performance of GRNN (test dataset)

Position error (PE)	Change in position error	Desired output	Actual output
0.905	2.942	0.378	0.478736
0.913	0.285	0.532	0.499219
0.097	1.019	0.939	0.466913
0.546	1.54	0.55	0.52301
0.964	2.481	0.587	0.520982
0.97	−4.161	0.301	0.418305
0.485	4.133	0.23	0.380344
0.141	3.258	0.194	0.446425
0.915	4.961	0.17	0.352093
0.957	−0.573	0.435	0.569033
0.035	4.618	0.923	0.342125

parameters. These parameters are used for predicting the output which in turn is used for controlling the speed and direction of the servomotor. In ANFIS based servomotor position control, the triangular membership function with 7 MFs provides the best result. When comparing the two Artificial Neural Networks techniques developed in this work, the Generalized Regression Neural network is the best-suited network and is thus recommended for automatic positioning of servomotors in real-world applications.

However, the work concluded at the simulation level. The prototype model can be developed and the results can be experimented. Also, the feasibility of deep learning neural networks for the positioning of servomotor can be studied. The mathematical modeling can also be developed by reducing the number of assumptions made in the previous modeling.

References

1. N. Khongkoom, A. Kanchanathep, S. Nopnakeepong, S. Tanuthong, S. Tunyasrirut, R. Kagawa, Control of the position DC servo motor by fuzzy logic. in *2000 TENCON Proceedings. Intelligent Systems and Technologies for the New Millennium*, Kuala Lumpur, vol. 3. Malaysia, pp. 354–357 (2000)
2. J. Priya, A. Jeevanandham, K. Rajalashmi, Fuzzy logic controller for position control of servo motor. Int. J. Innov. Technol. Explor. Eng. (IJITEE) **8**(3), 99–102 (2019)
3. M.-Y. Shieh, T.-H. S Li, Design And Implementation of Integrated Fuzzy logic controller for a servomotor system mechatronics. **8**(3), 217–240 (1998)
4. T. Żabiński, L. Trybus, Tuning P-PI and PI-PI controllers for electrical servos. Bullet. Polish Acad. Sci.: Tech. Sci. **58**, (2010) 10.2478/v10175-010-0005-7
5. P. Ravikumar, V. Nagababu, Position control of servo systems using PID controller tuning with soft computing optimization techniques. Int. J. Eng. Res. Technol. (IJERT) **3**(11), 976–980
6. S. EmaldaRoslin, N.M. Nandhitha, Cynthia James RekhaChakravarthi, A. Monisha, Node state information based adaptive power allocation and energy efficient routing using fuzzy logic for wireless sensor network. Int. J. Appl. Eng. Res. **9**(21), 8575–8592 (2014)
7. N.M. Nandhitha, R. Chakravarthi, J. Adeline Sneha, Automated pest detection and control using acoustic signal in wireless sensor network. IJAST **28**, 402–419 (2019)
8. J. Femila Roseline, J.P. Jigneshkumar, J. Govindarajan, N.M. Nandhitha, B. Sheela Rani, ANFIS_ plasma position controller in AdityaTokamak. in *Proceedings of International Conference on Advanced Electronic Systems(ICAES-2013)*, CEERI, Pillani, Sep 21–23, pp 15–18. (2013)
9. R. Emalda, N.M. Nandhitha, C. Rekha, N. Selvarasu, Optimal cluster head selection using ANFIS for an efficient data transmission in wireless sensor network. Int. J. Pharmacy Technol. **8**(3), 16494–16502 (2016)
10. C. Rekha , Performance evaluation of fuzzy and BPN based congestion controller in WSN. Int. J. Eng. Technol. **7**(2), 475–481 (2015)
11. F. Roseline, N.M. Nandhitha, Design of fuzzy logic based plasma position control in aditya tokamak. Indian J. Sci. Technol. **8** (2015). https://doi.org/10.17485/ijst/2015/v8i12/56420
12. A.A. Linus, Design of neuro-fuzzy system controller for dc servo motor based satellite tracking system, Dissertation (2007)

Decision Support Technique for Prediction of Acute Lymphoblastic Leukemia Subtypes Based on Artificial Neural Network and Adaptive Neuro-Fuzzy Inference System

Md. Ziaul Hasan Majumder, Md. Abu Khaer, Md. Julkar Nayeen Mahi, Md. Shaiful Islam Babu, and Subrata Kumar Aditya

Abstract Leukemia is usually characterized as a potentially fatal disease with different types and subtypes. It pretends as one of the major reason for death in developed and developing nations like Bangladesh, almost a hundred thousand people around the world are losing the fight against leukemia in every decade. An early prediction and diagnosis of a leukemia type have become inexorable in treatment, as it can reduce deadly risk to the valuable life of affected patients. This paper's objective is to design and development a smart decision support prediction system of Acute Lymphoblastic Leukemia Subtypes using Artificial Neural Network (i.e. ANN) and Adaptive Neuro-Fuzzy Inference System (i.e. ANFIS). For this prediction system, 500 datasets (main database) of different subtypes of Acute Lymphoblastic Leukemia patients are used. Training and testing are the two main steps of ANN and ANFIS, where 445 and 55 datasets are elected for training and testing respectively. According to the experimental result, it is apparent that the Adaptive Neuro-Fuzzy Inference System (i.e. ANFIS) approach performs better than the Artificial Neural Network (i.e. ANN) in both training and testing. Analysis reveals that a simple and well ergonomic back-propagation learning technique of Artificial Neural Network (i.e. ANN) and a hybrid optimization learning technique of the Adaptive Neuro-Fuzzy Inference System (i.e. ANFIS) can easily predict blood disorder efficiently and effectively in terms of hematological parameters.

Md. Ziaul Hasan Majumder · Md. Abu Khaer
Bangladesh Atomic Energy Commission, Dhaka, Bangladesh

Md. J. Nayeen Mahi (✉)
Jahangirnagar University, Dhaka, Bangladesh

Md. Shaiful Islam Babu
Changchun University of Science and Technology, Jilin, China
e-mail: shaifulislam@mail.ustc.edu.cn

S. K. Aditya
University of Dhaka, Dhaka, Bangladesh
e-mail: aditya@du.ac.bd

V. Suma et al. (eds.), *Inventive Systems and Control*,
Lecture Notes in Networks and Systems 204,
https://doi.org/10.1007/978-981-16-1395-1_40

Keywords Leukemia · ANN · ANFIS · Diagnosis · Activation function · Membership function · Fuzzy rule · Hematological parameters

1 Introduction

Leukemia (i.e. ALL) is one of the leading reasons for the premature death of patients and mortality rates are above 40% of the incidence in Asian and African countries, but the incidence is more in western countries though the mortality rate is less [1]. When it becomes lethal and not identified in the initial phases, it may lead to death. Early prediction followed by appropriate leukemia treatment can reduce mortality rates. It becomes a very difficult activity while identifying a disease like leukemia because experts need to process a handsome quantity of data [2, 3]. Experts frequently suffer from the deficiency of a standard, precise inspection of these laboratory data. They feel the necessity of a good support system that would assist them to make an accurate decision [2–5]. In the last few years, several research groups are working worldwide to implement neural networks and fuzzy inference systems in medical diagnosis. Neural network and neuro inference systems are used to elevate the correctness and quality of medical prognosis[4, 6–9].

The human body accommodates blood on an average of 8% percent of the body weight and red blood cells are replaced every 120days [2]. The human body affects different types of blood diseases either detected or not. Anemia is a common type of blood disease and there is some rare disease that affects only a few [2]. Leukemia is a cancer of white blood cells (i.e. WBC) blood cells. In leukemia, WBCs do not function like normal WBCs. Every full-grown blood cell develops from a slot of self-reviving hematopoietic stem cells (i.e. HSC) that occupy the bone marrow. Through a controlled process, the effective HSC's can separate either common myeloid or common lymphoid progenitor cells [10, 11]. Leukemia which is developed from a lymphoid progenitor cell, it is called lymphoblastic leukemia or acute lymphoblastic leukemia [10].

Acute lymphoblastic leukemia (i.e. ALL) is a prevalent childhood malignancy above 75% of ALL that occurs in children and is potentially curable. ALL is least common but devastating in adults and 30% to 40% of adults with ALL will have long term remissions by chemotherapy treatment [11, 12]. More complications and a worse prognosis for adults may have aged over 60. Among five subtypes of ALL (B-1, B-2, B-3, B-4, t-ALL), most cases arise from the B-cell lymphoblast, and the rest of the cases comprising of t-cell lymphoblast [12].

This article describes how Artificial Neural Network (i.e. ANN) and Adaptive Neuro-Fuzzy Inference System (i.e. ANFIS) can improve this area of prediction. The dataset is generated according to the hematological parameter range taken from the public oncology reference center, the clinical research center of the Federal University of Maranhao, Sao Paulo, Northeastern Brazil [3], and verified by Microbiology and Immunology department, Bangabandhu Sheikh Mujib Medical University (BSMMU), Shahbagh, Dhaka-1000, Bangladesh.

2 Related Work

Recently ANN and ANFIS is used in medical disorder diagnosis and prediction where medical parameter used as input data for diagnosing diseases such as breast cancer [13–17], prostate cancer [18], diabetes [19–21], blood disorder [2, 22], heart diseases [23, 24], kidney disease [25], optic nerve diseases [26] skin cancer [27, 28], pancreatic cancer [29, 30], Alzheimer and Dementia [31–33] classification of cervical cancer [34, 35], glaucoma [36], brain abnormalities[37–39], Analysis of blood serum [40], blood acidity [41], tuberculosis [42] and so on.

3 Experimental Details

In Fig. 1 the flowchart delineates the procedure plan of action

for the prediction of acute lymphoblastic leukemia (i.e. ALL) subtypes using ANN and ANFIS respectively. There are four hematological parameters such as leukocytes, haemoglobin, platelets, and blasts that are measured for Acute Lymphoblastic Leukemia (i.e. ALL) patients. These parameters are used as input. Hematological parameters of ALL patients change significantly from their normal ranges (Fig. 2).

Based on hematological parameters, the system output is predicted for anyone subtypes of acute lymphoblastic leukemia. Table 1 shows the input parameters of the experiment and inputs are generated using the 'rand ()' function in MATLAB within the ranges given in Table 2. A total of 500 datasets are generated consisting of four inputs (i.e. Leukocytes, haemoglobin, platelets, blasts) and a single output, where all subtypes are mapped within 0 to 1. Among 500 datasets, 445 datasets are used for training, 55 datasets are used for testing in both the optimum network of ANN and ANFIS. The dataset normalized (i.e. between 0 and 1) to initiate training and testing. The normalized outputs of Acute lymphoblastic leukemia (i.e. ALL) subtypes refer to Table 2.

Here, two techniques are used for the prediction of acute lymphoblastic leukemia subtypes namely, ANN and ANFIS. To design the ANN, a supervised multilayer feed-forward neural network is trained using a back-propagation learning algorithm [4, 20, 23]. Different types of activation and training functions are used to train the ANN network. In the ANFIS technique, to start the training input and output membership function, the fuzzy rule needs to adjust initially. In this experiment, training functions such as the Levenberg-Marquardt optimization technique (trainlm), Resilient backpropagation algorithm (traimrp), Gradient descent with adaptive learning rate back-propagation (traingda) are used to training the ANN network. For training of ANFIS network, trapezoidal membership function (trapmf), triangular-shaped membership function (trimf), gaussian curve membership function (gaussmf), pi shaped membership function (pimf), generalized bell-shaped membership function (gbellmf) are used. In this experiment Sugeno-type, fuzzy inference technique is used because it can combine both the fuzzy logic and the neural network concept

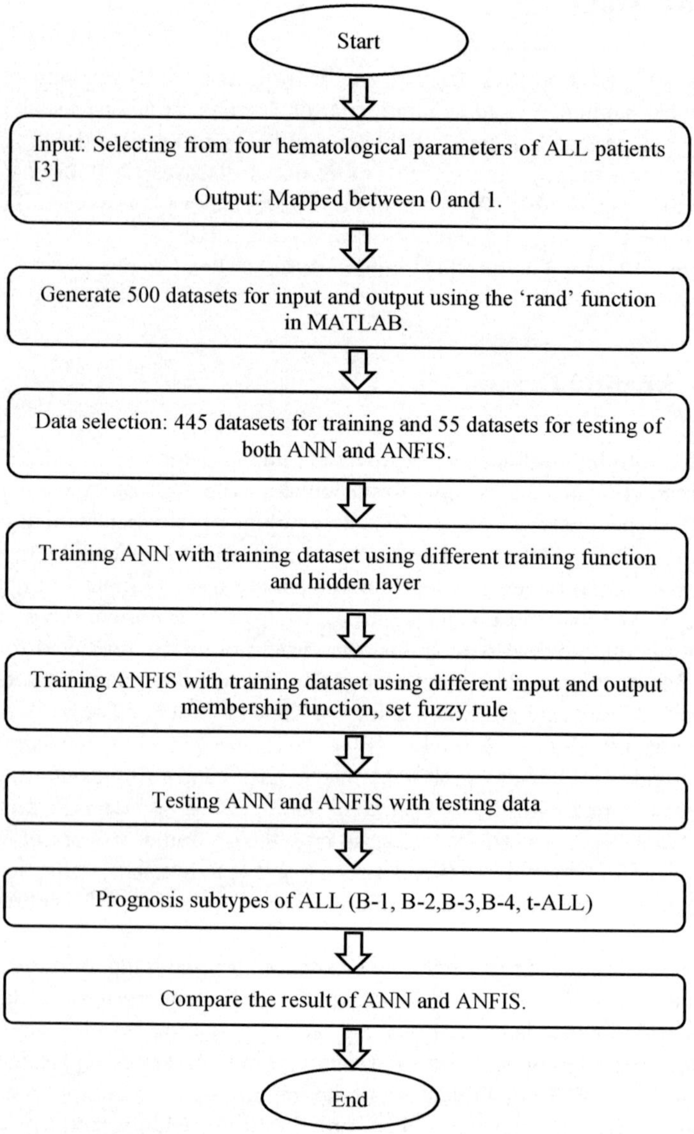

Fig. 1 Flowchart for (i.e. ALL) subtypes prediction

and encapsulate the benefit of both concepts in one frame. ANFIS training itself using training data to computes the initial membership function and subsequently adjusting the membership function. It reduces the rate of training error through a back-propagation technique or a hybrid optimization learning method.

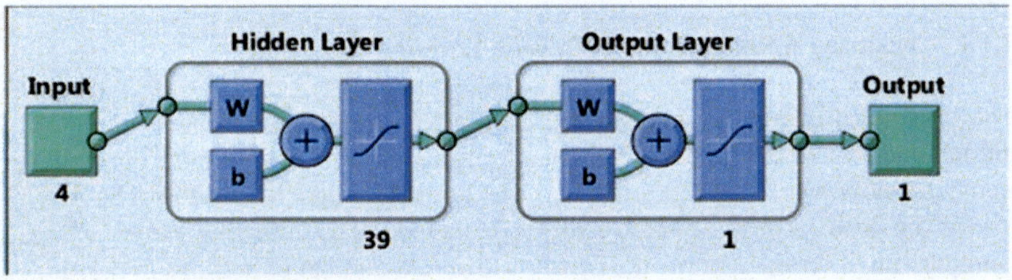

Fig. 2 Developed neural network

Table 1 Hematological parameter of acute lymphoblastic leukemia subtypes [3]

Hematological parameter	Acute lymphoblastic leukemia subtypes				
	B-1	B-2	B-3	B-4	t-ALL
Leukocytes 10^9/liter	2.35–102	1.44–157	1.13–135	2.3–6.9	9.1–591
Haemoglobin/deciliter	6.0–6.6	3.4–12.9	7.0–10.9	9.3–12.7	6.5–12.5
Platelets 10^9/liter	12–13	6.0–111	22–160	29–38.1	14–157
% Blast	46–49	0–97	0–99	4–6	4–98

Table 2 Normalized output

Subtypes	Normalized range
B-1	0–0.2
B-2	0.2–0.4
B-3	0.4–0.6
B-4	0.6–0.8
t-ALL	0.8–1.0

4 Experimental Result and Discussion

4.1 Training and Testing Artificial Neural Network (ANN)

There are three types of training functions used to training the ANN. The training performs using MATLAB.

The number of inputs is 4 and one hidden layer with 39 hidden neurons. For all the output portion, there is only one output.

4.1.1 Training ANN with the Different Training Function

The best performance is obtained by 'trainlm' training function at 39 number of hidden neurons where hidden layer transfer function is 'Tan-sigmoid', output layer transfer function is 'Log-sigmoid', and the number of epochs is 3000. The performance (i.e. MSE = mean squared error) is 0.0041945. The following Table 3 shows a Comparison of errors in networks having a different number of neurons in the hidden layer.

The following Table 4 gives a survey of training errors (i.e. MSE) of the best-designed networks with different training functions, neurons number in hidden layers, and a different number of epochs.

After completing the training of ANN, the best performance is obtained at 3000 epochs as shown in the following Fig. 3.

Table 3 Training error of different ANN network

Selecting a training function: trainlm			
Network (ANN)	Neuron in hidden layer	Error (MSE)	Epochs
1	21	0.023745	3000
2	23	0.021993	3000
3	25	0.014959	2000
4	27	0.016705	2500
5	29	0.010646	2000
6	31	0.006705	3000
7	33	0.0077334	2500
8	35	0.0058664	3000
9	37	0.0043197	3000
10	**39**	**0.0041945**	**3000**
11	41	0.0044501	3000

Table 4 Training error comparison of different ANN network having a change in the training function

Network (ANN)	Neurons in hidden layer	Error (MSE)	Training function	Epochs
10	39	0.0041945	trainlm	3000
9	38	0.0098075	trainrp	10,000
7	39	0.031681	traingda	17,000

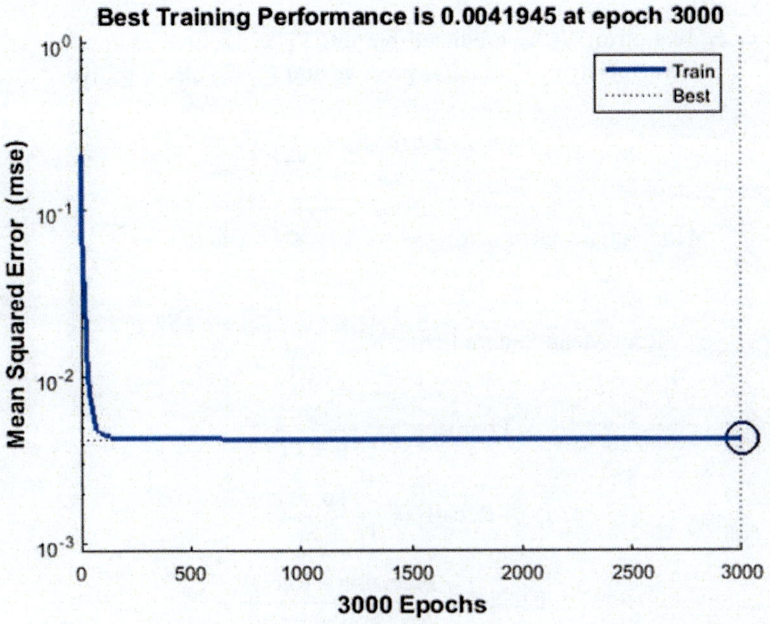

Fig. 3 Performance plot of trained artificial neural network (ANN)

4.1.2 Testing the Performance with Validation and Testing Data

After training the network with a cross-validation method, the ideal network is tested with 55 a completely new data called testing dataset (see Sect. 2). It is seen in Table 5, among 55 datasets with different types of Acute Lymphoblastic Leukemia (i.e. ALL), the ANN network successfully predicted the B-4 patients. The testing error, precision, F1 is calculated using the following formulae:

$$e_i = Y_p \sim Y_d \quad i = 1, 2, 3, 4 \ldots \tag{1}$$

Table 5 Analyzing the testing performance with validation and testing data

Disease(ALL-subtypes)	B-1	B-2	B-3	B-4	t-ALL
Observed case	10	12	10	10	13
Predicted case	09	10	09	10	09
Prediction fail	01	02	01	00	04
Testing error (MSE)	0.042597	0.03972	0.09275	0.00	0.04371
Accuracy (%)	90	83.3	90	100	69.2
Precision	0.9	0.833	0.90	1	0.692
F1	0.947	0.910	0.947	1	0.818

where e_i = Test error, Y_p = Predicted output, Y_d = Desired Output of ANN or ANFIS, T_p= True positive, F_p= False positive and F_n = False negative.

$$\text{Square of error } E_s = \sum_i e_i^2 \tag{2}$$

$$\text{Mean square error,} \quad E_{ms} = \frac{E_s}{n} = \frac{\sum_i e_i^2}{n} \text{ where } n = \sum i \tag{3}$$

$$\text{Root Mean Square Error, } E_{rms} = \sqrt{\frac{E_s}{n}} = \sqrt{\frac{\sum_i e_i^2}{n}} \tag{4}$$

$$\text{Precision} = \frac{\text{Tp}}{\text{Tp} + \text{Fp}} \tag{5}$$

$$\text{Recall} = \frac{\text{Tp}}{\text{Tp} + \text{Fn}} \tag{6}$$

$$\text{F1} = 2 * \frac{\text{Precision} * \text{Recall}}{\text{Precision} + \text{Recall}} \tag{7}$$

4.1.3 Linear Regression of ANN

The linear regression (*R*-value) analysis indicates the relation between targets and outputs. When *R*-value is exactly one then it shows a precise linear relation and when R-value tends to zero, it indicates there is no linear relation. From Fig. 4, the coefficient of determination is $R^2 = 0.96079204$ which means nearly 96% disparity on targets and outputs in the system response can be expounded through linear regression analysis.

4.2 Training and Testing ANFIS

To design ANFIS, MATLAB fuzzy logic toolbox is used. Different membership functions like 'trapmf', 'trimf', 'Gaussmf', 'pimf', and 'gbellmf' are used during ANFIS network training. By choosing the number of membership functions as three for each input variable, a total of 81 rules were generated.

4.2.1 Training Error with Different Membership Function

Among these membership functions, the best result obtained by 'gaussmf' and training RMSE is 0.030187, No. of epochs 50 where the optimization method is

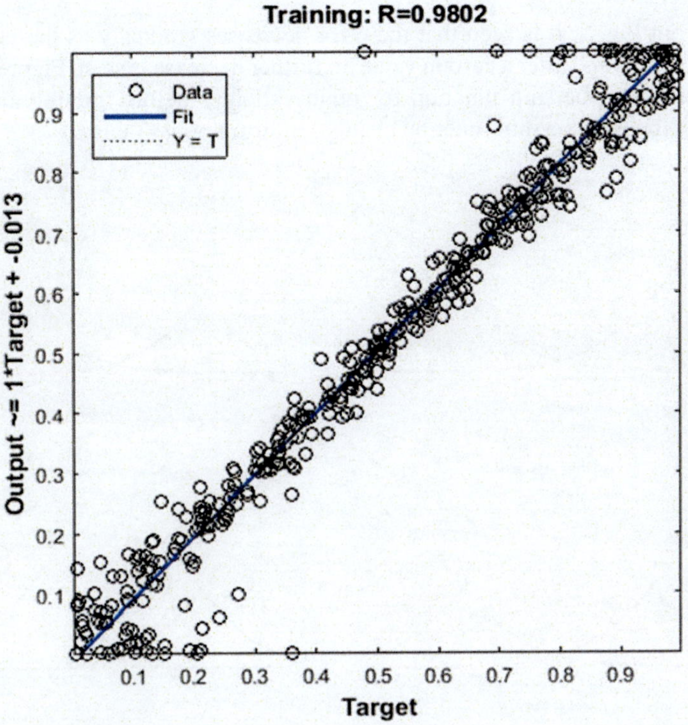

Fig. 4 Linear Regression plot for training

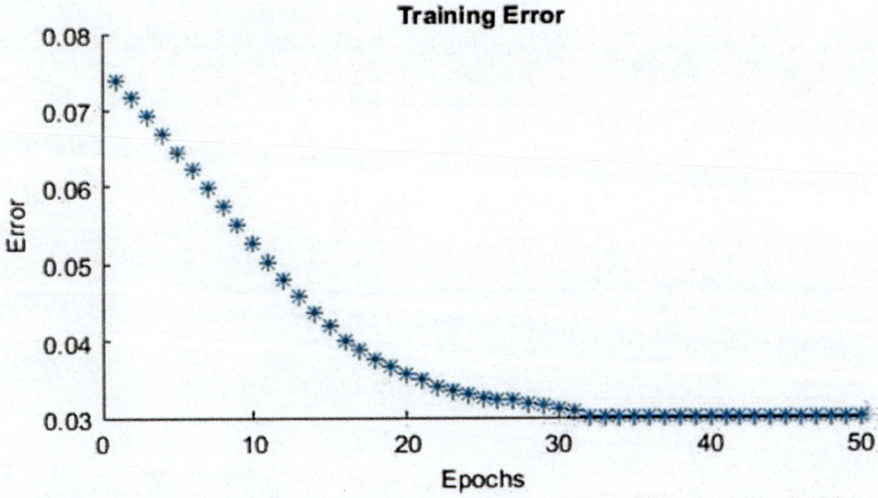

Fig. 5 Changes in training error with epoch

hybrid. From Fig. 5, it is seen that the error decreases gradually as the number of epochs increases and after a certain value no further decrease is seen. Figure 6 shows the Selecting membership function for input variables before training and Fig. 7 shows how the membership function changes after training (Table 6).

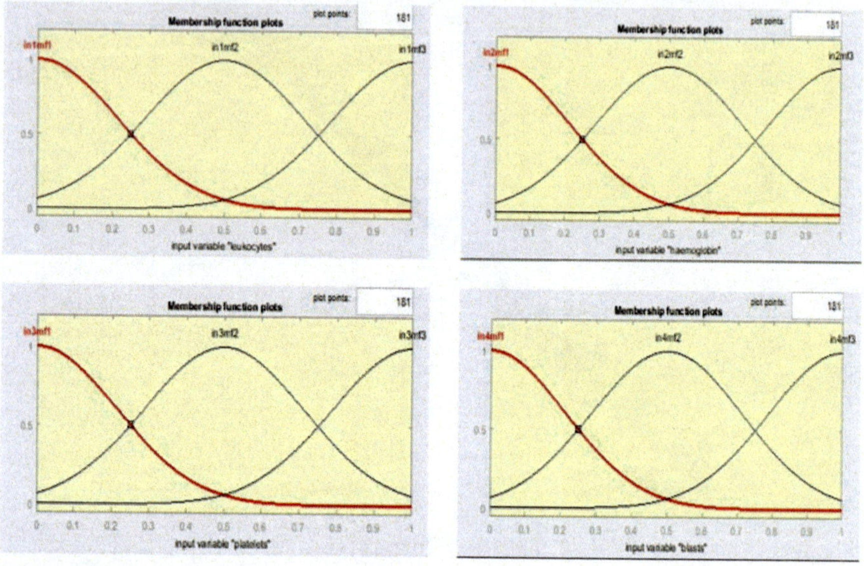

Fig. 6 ANFIS membership function of four inputs (Leukocytes, haemoglobin, platelets, blasts) before training

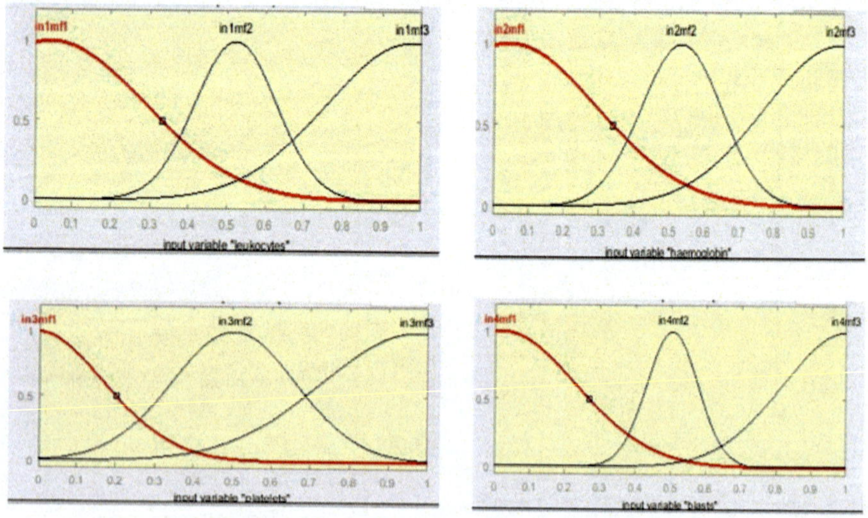

Fig. 7 ANFIS membership function of four inputs (Leukocytes, haemoglobin, platelets, blasts) after training

Table 6 Training error of ANFIS with different membership functions

Network (ANFIS)	Selecting membership function	Epochs	Error (RMSE)
1	trapmf	30	0.058591
2	trapmf	50	0.050631
3	trimf	30	0.06805
4	trimf	50	0.059703
5	gaussmf	30	0.031097
6	**gaussmf**	**50**	**0.030187**
7	pimf	30	0.04677
8	pimf	50	0.03556
9	gbellmf	30	0.03655
10	gbellmf	50	0.031007

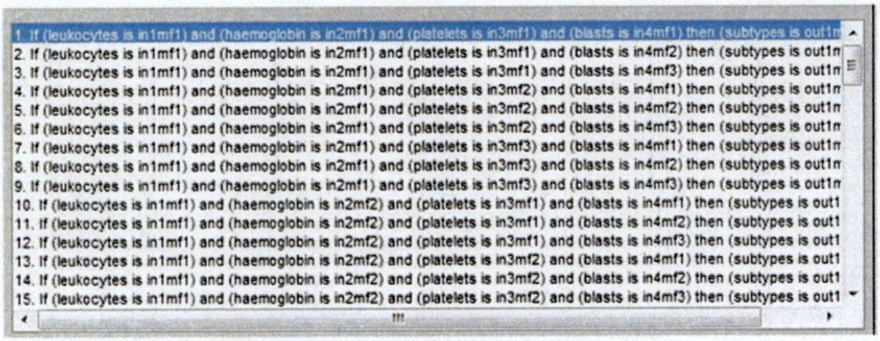

Fig. 8 Rule base of designed ANFIS system

4.2.2 Testing the System Against Target

The network tested against the target output and it is seen that both values merge each other which is shown in Fig. 9, here blue Circles indicate training data, and red asterisks result of output.

4.2.3 Rule Viewer Window for System

In Fig. 10, it is seen that the output value 0.982 is generated for four input values (i.e. 0.7033, 1, 0.5452, 0.08488) which refers to the range (i.e. 0.8–1.0). This output specifies that the result is for t-ALL patients.

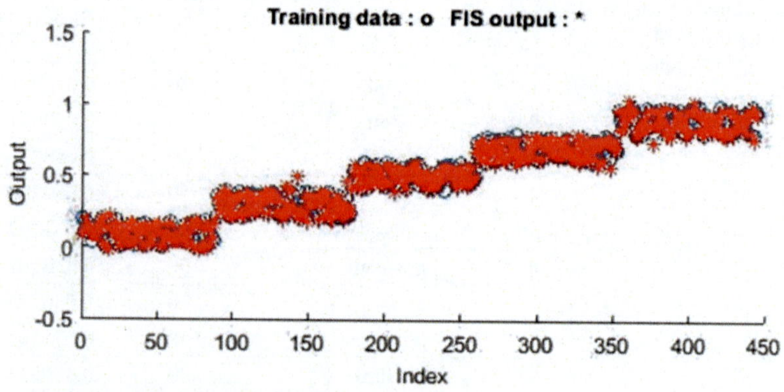

Fig. 9 Testing result of the model output against the target

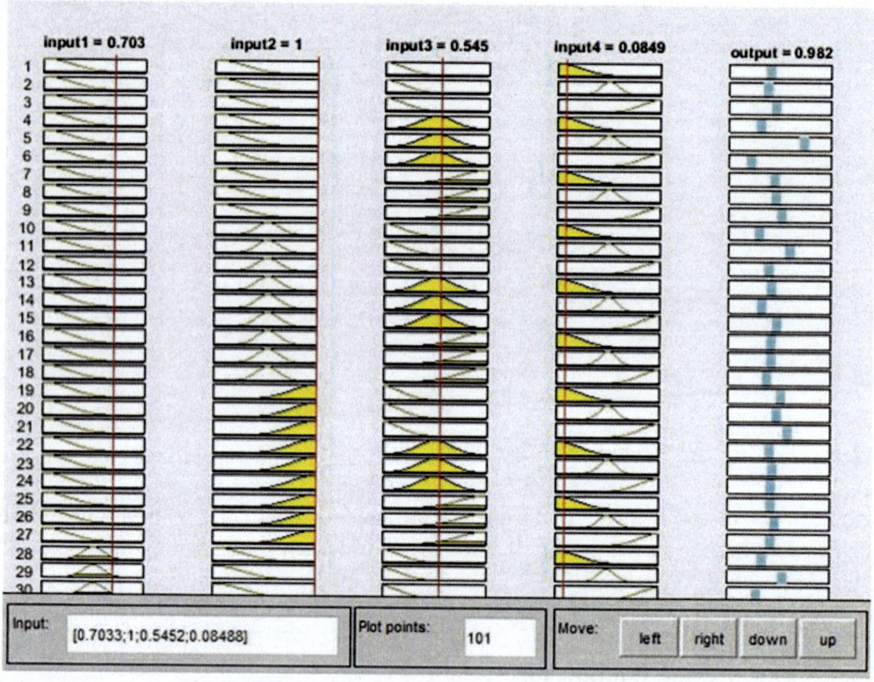

Fig. 10 Rule viewer window for ANFIS

4.2.4 Testing the Performance with Testing Dataset

Testing error, accuracy, precision, and F1 is calculated using Eq. (1–7). Testing 55 datasets with different types of Acute Lymphoblastic Leukemia (i.e. ALL), the ANFIS network effectively predicted the B-2 patients showing in Table 7.

Table 7 Testing performance of ANFIS

Disease (ALL-subtypes)	B-1	B-2	B-3	B-4	t-ALL
Observed case	10	12	10	10	13
Predicted case	09	12	08	09	12
Prediction fail	01	00	02	01	01
Testing error (MSE)	0.001521	0.00	0.0544685	0.047961	0.0473365
Accuracy (%)	90	100	80	90	92.31
Precision	0.9	1	0.8	0.9	0.9231
F1	0.947	1	0.889	0.947	0.965

5 Training and Test Error (MSE) Comparison of Developed ANN and ANFIS

From Table 8, it is clear that the Adaptive Neuro-Fuzzy Inference system outperforms than Artificial Neural Network in both training and testing, where the training error for each system 0.00091125 (i.e. MSE) and 0.0041945 (i.e. MSE) respectively and the average testing error for each system 0.03025 (i.e. MSE) and 0.04375 (i.e. MSE) respectively. It is also seen from Table 8, the calculated value of average test accuracy, average precision, average F1 of both ANN and ANFIS network.

Table 8 Comparison between two developed system

Network	ANN	ANFIS
Disease (ALL-subtypes)	B-1	B-1
	B-2	B-2
	B-3	B-3
	B-4	B-4
	t-ALL	t-ALL
Training error (MSE)	0.0041945	$(0.030187)^2$ = 0.00091125
Testing error (MSE)	0.042597	0.001521
	0.03972	0.00
	0.09275	0.0544685
	0.00	0.047961
	0.04371	0.0473365
Average testing error (MSE)	0.04375	0.03025
Average accuracy (%)	86.50	90.462
Average precision	0.865	0.9046
Average F1	0.924	0.950

6 Future Development

Some essential tasks require being complete in the upcoming time. Initially, it is compulsory to set up an interactive user interface like an android phone, computer, a web application for end-users. Secondly, it is mandatory to interconnect the designed system with an expert domain system. The ANN and ANFIS with Support Vector Machine (i.e. SVM), Decision Tree, or any other method that makes significant knowledge representation will be promising. Thirdly, developed diagnoses systems must have the ability to interact with new data and produce an accurate result since a variety of new cases are added frequently in the diagnostic center. Fourthly, the designed system must be authenticated with a huge database and compared its proficiency with one of the current professional systems.

7 Conclusion

In this article, a system is developed for the prediction of Acute Lymphoblastic Leukemia Subtypes (B-1, B-2, B-3, B-4, t-ALL). To do this, two approaches are considered. Initially, a multilayer feed-forward ANN is employed in training and testing with 500 possible datasets. A total of 445 dataset, different training functions with different combinations of hidden neurons, different number of epochs used to train the network, and the best performance is obtained with 'trainlm' training function with 39 numbers of hidden neurons at 3000 epochs. However, the network is tested with 55 testing datasets and calculated the accuracy, precision, testing error where the error is almost negligible. By observing the performance and linear regression plot, it is seen that both plots meet the validation criteria of ANN. Afterwards, the ANFIS network is developed using the same training and testing dataset. Here, a combination of different membership functions in the input layer with different epochs is tried and the minimum error was obtained with Gaussian membership function with a hybrid optimization method at epochs of 50. To test the ANFIS system against the testing data, the rule viewer window is used to obtain the accuracy, precision and F1. Finally, comparing the training and testing results of two approaches. The investigation shows that the Adaptive Neuro-Fuzzy system performance is better than the artificial Neural Network for both training and testing.

References

1. Who.int, Cancer (2021). [online] Available at: https://www.who.int/news-room/fact-sheets/detail/cancer. [Accessed 25 April 2021]
2. M. Payandeh, M. Aeinfar, V. Aeinfar, M. Hayati, A new method for diagnosis and predicting blood disorder and cancer using artificial intelligence (artificial neural networks). Int. J. Hematology-Oncology Stem Cell Res. **3**(4), 25–33 (2009)

3. W. Xu, W. Liu, L. LI, L. MA, S. Xia, J. Zhang, G. Shao, A new computer-aided detection system of CR mammograms. J. Comput. Inform. Syst. **6**(9), 2885–2900 (2010)
4. S. Moein, S.A. Monadjemi, P. Moallem, A novel fuzzy-neural based medical diagnosis system. Int. J. Biol. Med. Sci. **4**(3),146–150 (2009)
5. S. Manoharan, Study on hermitian graph wavelets in feature detection. J. Soft Comput. Paradigm (JSCP) **1**(01), 24–32 (2019)
6. T. Sen, S. Das, An approach to pancreatic cancer detection using artificial neu-ral network, in *Proc. of the Second Intl. Conf. on Advances in Computer, Electronics and Electrical Engineering-CEEE* (pp. 56–60) (2013)
7. E. Al-Daoud, Cancer diagnosis using modified fuzzy network. Uni. J. Comput. Sci. Eng. Technol. **1**(2), 73–78 (2010)
8. H. Hamdan, J.M. Garibaldi, Adaptive neuro-fuzzy inference system (ANFIS) in modelling breast cancer survival, in *International Conference on Fuzzy Systems* (pp. 1–8) (IEEE, 2010)
9. M. Negnevitsky, Artificial intelligence: a guide to intelligent systems. Pearson Edu. (2005)
10. T. BuiTenkamp, Clinical relevance of genetic alterations in acute lymphoblastic leukemia in children with down syndrome (2014)
11. Cancer.org. (2021). About acute lymphocytic leukemia (ALL). [online] Available at: https://www.cancer.org/cancer/acute-lymphocytic-leukemia/about.html [Accessed 25 April 2021].
12. T. Terwilliger, M.J.B.C.J. Abdul-Hay, Acute lymphoblastic leukemia: a comprehensive review and 2017 update. Blood Cancer J. **7**(6), pp.e577–e577 (2017)
13. N. Tariq, Breast cancer detection using artificial neural networks. J. Mole. Biomark. Diag. **9**(1), 1–6 (2017)
14. A. Bhardwaj, A. Tiwari, Breast cancer diagnosis using genetically optimized neural network model. Expert Syst. Appl. **42**(10), 4611–4620 (2015)
15. E.D. Übeyli, Adaptive neuro-fuzzy inference systems for automatic detection of breast cancer. J. Med. Syst. **33**(5), 353–358 (2009)
16. M. Karabatak, M.C. Ince, An expert system for detection of breast cancer based on association rules and neural network. Expert Syst. Appl. **36**(2), 3465–3469 (2009)
17. H. Hamdan, J.M.Garibaldi, Adaptive neuro-fuzzy inference system (ANFIS) in modelling breast cancer survival. in *International Conference on Fuzzy Systems* (pp. 1–8) (IEEE, 2010)
18. L. Benecchi, Neuro-fuzzy system for prostate cancer diagnosis. Urology **68**(2), 357–361 (2006)
19. E.D. Übeyli, Automatic diagnosis of diabetes using adaptive neuro-fuzzy inference systems. Expert Syst. **27**(4), 259–266 (2010)
20. O.,Geman, I. Chiuchisan, R. Toderean, Application of Adaptive Neuro-Fuzzy Inference System for diabetes classification and prediction, in *2017 E-Health and Bioengineering Conference (EHB)* (pp. 639–642) (IEEE, 2017)
21. O. Karan, C. Bayraktar, H. Gümüşkaya, B. Karlık, Diagnosing diabetes using neural networks on small mobile devices. Expert Syst. Appl. **39**(1), 54–60 (2012)
22. E.K.Roy, S.K. Aditya, Prediction of acute myeloid leukemia subtypes based on artificial neural network and adaptive neuro-fuzzy inference system approaches, in *Innovations in Electronics and Communication Engineering* (pp. 427–439) (Springer, Singapore, 2019)
23. E. Avci, I. Turkoglu, An intelligent diagnosis system based on principle component analysis and ANFIS for the heart valve diseases. Expert Syst. Appl. **36**(2), 2873–2878 (2009)
24. O. Mokhlessi, H.M. Rad, N. Mehrshad, A. Mokhlessi, Application of neural networks in diagnosis of vlave physiological heart disease from heart sounds. Am. J. Biomed. Eng. **1**(1), 26–34 (2011)
25. A. Yadollahpour, J. Nourozi, S.A. Mirbagheri, E. Simancas-Acevedo, F.R. Trejo-Macotela, Designing and implementing an ANFIS based medical decision support system to predict chronic kidney disease progression. Front. Physiol. **9**, 1753 (2018)
26. B. Akdemir, S. Kara, K. Polat, A. Güven, S. Güneş, Ensemble adaptive network-based fuzzy inference system with weighted arithmetical mean and application to diagnosis of optic nerve disease from visual-evoked potential signals. Art. Intell. Med. **43**(2), 141–149 (2008)
27. S.M. Odeh, Using an adaptive neuro-fuzzy inference system (AnFis) algorithm for automatic diagnosis of skin cancer. J. Commun. Comput. **8**(9), 751–755 (2011)

28. F. Ghali, Skin cancer diagnosis by using fuzzy logic and GLCM. J. Phys.: Conf. Series **1279**(1), 012020. (IOP Publishing, 2019)
29. W. Muhammad, G.R. Hart, B. Nartowt, J.J. Farrell, K. Johung, Y. Liang, J. Deng, Pancreatic cancer prediction through an artificial neural network. Front. Art. Intell. **2**, 2 (2019)
30. M.U. Sanoob, A. Madhu, K. Ajesh, S.M. Varghese, Artificial neural network for diagnosis of pancreatic cancer. Int. J. Cyber. Infor. (IJCI) **5**(2), 41–49 (2016)
31. D. Zafeiris, S. Rutella, G.R. Ball, An artificial neural network integrated pipeline for biomarker discovery using Alzheimer's disease as a case study. Comput. Struct. Biotech. J. **16**, 77–87 (2018)
32. A. Yiğit, Z. Işik, Application of artificial neural networks in dementia and alzheimer's diagnosis, in *2018 26th Signal Processing and Communications Applications Conference (SIU)* (pp. 1–4) (IEEE, 2018)
33. E. García-Pérez, A. Violante, F. Cervantes-Pérez, Using neural networks for differential diagnosis of Alzheimer disease and vascular dementia. Expert Syst. Appl. **14**(1–2), 219–225 (1998)
34. S.B. Benazir, A. Nagarajan, An innovative system for classifying cervical cancer using features based ANFIS classifier.
35. M.A.Devi, S. Ravi, J. Vaishnavi, S. Punitha, Classification of cervical cancer using artificial neural networks. Procedia Comput. Sci. **89**, 465–472 (2016)
36. N. Ramamurthy, K.C.T. Swamy, G. Ramarao, M.R. Pasha, Detection of glaucoma using adaptive neuro fuzzy in DWT domain.
37. N.E.A. Khalid, S. Ibrahim, M. Manaf, Comparative study of adaptive network-based fuzzy inference system (ANFIS), k-nearest neighbors (k-NN) and fuzzy c-means (FCM) for brain abnormalities segmentation. Int. J. Comput. **4**.
38. N.M. Noor, N.E.A. Khalid, R. Hassan, S. Ibrahim, I.M. Yassin, Adaptive neuro-fuzzy inference system for brain abnormality segmentation, in *2010 IEEE Control and System Graduate Research Colloquium (ICSGRC 2010)* (pp. 68–70) (IEEE, 2010)
39. T. Vijayakumar, Classification of brain cancer type using machine learning. J. Art. Intel. **1**(02), 105–113 (2019)
40. T. Karthikeyan, J.S. Kumar, R. Jegan, Design and development of low-cost blood serum analyser using adaptive neuro fuzzy inference system. Int. J. Biomed. Eng. Technol. **15**(3), 261–272 (2014)
41. A. Bani, An adaptive neurofuzzy technique for determination of blood acidity. Comput. Method. Biomech. Biomed. Eng. **13**(6), 685–691 (2010)
42. T. Uçar, A. Karahoca, D. Karahoca, Tuberculosis disease diagnosis by using adaptive neuro fuzzy inference system and rough sets. Neural Comput. Appl. **23**(2), 471–483 (2013)
43. E.P. Noronha, H.T. Marinho, E.B.A.F. Thomaz, C.A. Silva, G.L.R. Veras, R.A.G. Oliveira, Immunophenotypic characterization of acute leukemia at a public oncology reference center in Maranhão, northeastern Brazil. Sao Paulo Med. J. **129**(6), 392–401 (2011)

Extended Equilibrium-Based Transfer Learning for Improved Security in Cloud Environment

Gavini Sreelatha, A. Vinaya Babu, and Divya Midhunchakkarvarthy

Abstract Cloud computing provides dynamic and efficient access to data compared to local server and standalone computers. It provides capability to the users to achieve most of the functionalities of servers on the network to maintain store and process data on demand. The increased utilization of the cloud eventually makes data storage protection and processing essential. Most of the machine learning techniques provide security solutions to the cloud environment. It is a challenging task to maintain the efficiency of the dynamic cloud environment due to unpredictability of the new attacks. Transfer learning is one of the feasible solutions with minimal training data and increased accuracy in attack detection. Transfer learning techniques improve the detection of various types of benign and unknown new attacks by using the knowledge from the source domain. In this work, transfer learning involves a pre-trained network namely Alex net which is used for fine-tuning the features in convolution layers that are extracted. Also, here an optimized model utilizes an extended equilibrium optimizer for reducing learning Complexity. Hence, it is feasible to limit the maximum quantity of features in data without dropping its original property, less computing time, and memory space. Also, extended transfer learning is applied for obtaining an initial solution; this combination is tuned with the equilibrium reducing an error rate and higher accuracy will be obtained.

Keywords Cloud computing · Transfer learning · Attacks · Security · Accuracy · Equilibrium optimization

G. Sreelatha (✉)
Lincoln University College, Kuala Lampur, Malaysia

A. Vinaya Babu
Stanley College of Engineering and Technology for Women, Hyderabad, India

D. Midhunchakkarvarthy
Lincoln University College, KaulaLampur, Malaysia
e-mail: divya@lincoln.edu.my

V. Suma et al. (eds.), *Inventive Systems and Control*,
Lecture Notes in Networks and Systems 204,
https://doi.org/10.1007/978-981-16-1395-1_41

1 Introduction

Nowadays, in the cloud environment, the researches have increased in providing more number of solutions in security because of more number of new attacks in the cloud. The written works give a brief idea about the reviews of machine learning, security mechanisms and transfer learning method-based security approaches in the cloud and network environments. By this analysis of review, it is known that there is plentiful attention from the research community in cloud security solutions. None of the researches has focused on designing the access control models based on the cloud data location due to the ignorant of the data hosting location in the cloud. It is hard to maintain the mutual trust between the cloud user and the cloud service provider (CSP). Lack of modeling the suitable security solution or access control methodologies for the multi-domain or cross-domain in the cloud. Lack of maintaining the collaborative identification and reaction between the cloud user and CSP for the security incidents. Providing security solutions becomes a significant concern in the cloud environment due to the complexity and diversity of the data. Selecting the appropriate security solutions is a difficult task for mitigating the identified vulnerabilities in the dynamic and complex cloud. Several security solutions create new vulnerabilities owing to the coexistence of multiple security solutions. Reducing the training data and supporting the multiple domains is a challenging task of identifying the unknown attacks over the abundance of cloud storage and processing.

In recent years cloud security is an essential one and it acts as a defensive mechanism for malicious activities prevention in IT infrastructures [1], due to cyber-attacks most of the sensitive data and critical applications are damaged. To protect from cyber-attacks most organizations have used the Intrusion Detection System (IDS). Though some techniques are not much efficient currently even some types of enterprises (SMEs) are accessibly used in the cloud platform that provided software as a service and infrastructure basis [2].

Nowadays utilization of Virtual Machine (VM) and hypervisor techniques are also having some security problems and this kind of VM technologies are affected by the VM level attacks. Moreover, this type of system contains several computer organizations that could contain a maximum amount of software, middleware and hardware systems. Threats in VM substructure could be exploited through the third party toward conduct attacks and exfiltrate data [3]. It is because of the essential TCP/IP stack weaknesses. Furthermore, many new attacks have currently created metamorphisms and polymorphism toward avoiding detection. In the IaaS cloud platform, the victim's machine data might simply be developed then exploited [4, 5].

The security issues in the cloud are mentioned in several publications [6, 7], and [8]. Obviously, this type of new model may affect many attacks. Based on the specific location classify the cloud computing attacks. Attacks may happen at the service provider side, user side, or network side. Mostly attackers introduce malicious traffic in the network, launching DoS and DDoS, IP and ARP spoofing, etc. Due to cloud attack personal data and identity also theft out. Some of the attack techniques mentioned as the exploitation of software vulnerabilities, brute force dictionaries fraud, phishing, etc. [9].

Through integrating security parameters presented at the resource requirements could expand the QoS [10]. This kind of request has reflective consequences solution for security and this to be executed and distributed for protecting cloud services. Though various IDS methods had proposed to incorporate security features in broad fields [11], still security management is a complex issue [12]. In certain systems, services against cyber-attacks are a complex task. Also, the need for security services is not the same for all cloud users. They are required to know when the performance for the particular service may degrade because various application usage, services, and resources are conceded to penetrate another host. Also, every user could need various kinds of data, thresholds, and configurations toward watch virtualized components of them. Consequently, Cloud providers need to consider the option that every user must control the data. Additionally, Cloud providers must need to notice attacks together with both underlying infrastructure and virtualized components [13–15].

2 Related Works

Nahmias et al. [16] introduced a technique that was based on TrustSign analyzes. This method did not need any model training or feature engineering. This method had an advantage in terms of the reduction of signature generation and reduced time. In this work, they used convolutional neural networks with VGG-19.

Abazari F et al. [17] designed a method that is based on both secure placements of task interaction and security demand of the task. In this work, they had designed a heuristic algorithm which was depending on both security requirements and the task's completion time. They had also designed a method for security threat management.

Namasudra [18] introduced a security method that was dependent on identity-based timed release encryption (IDTRE), distributed hash table (DHT) network, and attribute-based encryption. In this work, initially, data were encrypted with user attributes, and this data was separated into encapsulated and extracted ciphertext. After that IDTRE was applied to encrypt the decryption key. Finally, in DHT network, these keys were shared then this key was stored on the server of the cloud.

Lei et al. [19] introduced a method that was based on the transfer learning technique. In this work, they introduced Latent Dirichlet Allocation (LDA) and word 2 vector for improving processing speed and accuracy for the detection.

Rabbani et al. [20] suggested a method that was based on the user behavior model. They designed a model namely a particle swarm optimization and based probabilistic neural network (PSO-PNN) and this was used for the recognition and detection process. In this work, initially, user behaviors were converted into an understandable format then it was recognized and classified with the multi-layer neural network (Table 1).

Table 1 Comparison of different machine learning-based cloud security researches

Author Name	Objective
Nahmias et al.	Reduction of signature generation reduced time and used convolution neural networks with VGG-19
Abazari et al.	Designed a heuristic algorithm and designed a method that is based on both secure placement of task interaction and security demand of the task
Namasudra	Security method that was depending on (IDTRE), (DHT) network and attribute-based encryption. In this work, initially, data were encrypted with user attribute and this data was separated into encapsulated and extracted cipher text
Lei et al.	Improving processing speed and accuracy for the detection
Rabbani et al.	Suggested a method which was based on the user behavior model

3 Motivations and Problem Statements

Cloud environments must need to be distributed and decentralized security policies. Also, machine learning is to be deliberated as the greatest system for security to be used within a cloud platform also this is basically dispersed then it is extremely consistent by this cloud platform. Moreover, machine learning techniques in the cloud enhanced security with high speed for computation which improves its repute for better security systems appropriate for the cloud platform.

Security is an essential one because nowadays cloud computing is an important platform. Most of the cloud computing performances are based on parallel and distributed computing. Also, this is the sharable system since anyone can access the cloud platform from anywhere. This leads to a third party can intrude and illegally retrieve the private information of people. These issues are dragged into blockchain for improving security. For that, this article introduced cloud computing with blockchain security.

3.1 Research Challenges in the Cloud Security Systems

The cloud security systems encounter different security issues such as virtualization, data storage, computing, access control, trust management, legal aspects, Internet, service, and network-related security issues. The adoption of machine learning algorithms creates several complications to the security model in terms of accurately detecting the attacks in the cloud presented as follows [22].

Lack of understanding of the unique behavior of the different types of attacks by the learning model while providing the entire features as the input to the learning model for attack detection or classification, degrades the detection efficiency and misguides the classification.

Training time is high for the learning model when processing the massive amount of cloud data over the diverse characteristics of the numerous attacks.

Due to the improper training set and minimal variance between several abnormal behaviors and normal behaviors, reducing the false positive rate, the false-negative rate is the most challenging task in the cloud.

Less hybrid mining techniques for the massive cloud security model lead to inaccurate attack detection due to a high correlation among the behaviors of the different attacks and the increased feature dimensionality.

Lack of available cloud attack datasets in the form of a lack of balanced, noiseless, and redundancy-less data creates difficulty in the evaluation of the machine learning algorithms.

In the real environment, detecting the intrusions based on the labeled data is an arduous task due to there is the availability of entirely unfamiliar data, which leads to inaccurate detection for the supervised learning model.

Fail to detect the low-frequency attacks in the cloud by only examining the attack features, which leads to the classification of such attacks as legitimate data.

Maintaining a trade-off between effectiveness and efficiency is a challenging task while applying the complicated pre-processing methods in the abundant cloud data due to the efficient intrusion detection is essential in the distributed cloud.

3.2 Future Research Directions for the Cloud Security Systems

Several possible future research directions are recommended to improve the security measures for the cloud environment discussed as follows.

To improve the attack detection and classification accuracy, fine-tuning the feature selection model with an optimal feature set in the security research greatly assists the machine learning model.

Incremental learning-based attack [21] detection -model attempts to handle the imbalanced and noisy attack datasets in the massive cloud environment.

For detecting a specific type of cloud attack, training the machine learning model with the additional domain knowledge helps to recognize the attacks in a particular application scenario accurately.

Applying the unsupervised learning model is more beneficial when there is a huge amount of unlabeled data and unknown attack types, which enhances the detection performance.

Parameter tuning based on the attack or data characteristics leads to the optimization of the machine learning and deep learning model, resulting in enhanced detection performance in the cloud.

Reinforcement learning models support the detection of new attacks by learning the attack behaviors within a particular context in the dynamic cloud environment.

The transfer learning techniques are always better technique for detecting various types of benign and new attacks with the help of the knowledge from the source domain.

4 Objectives

1. To improve the security of the cloud platform
2. To design a novel transfer learning security algorithm
3. To evaluate the parameters and improve the performance than the existing literature

5 Proposed Methodology

In this work, an extended equilibrium-based transfer learning model for cloud security was introduced. For that, a security solution is created by a new technique called optimized equilibrium transfer learning model for the cloud environment. The combination of extended transfer learning is applied for getting a solution, and also this combination is tuned with the equilibrium optimization in order to solve data also reducing an error rate. The higher accuracy is obtained by this new concept. Comparatively the results are more efficient and highly accurate than other transfer learning mechanisms. Based on analysis accuracy time cost can be calculated.

6 Result Analysis

Sensitivity, false alarm rate, specificity, precision, and detection accuracy rate are calculated based on the work. Open-source cloud environment, i.e., OpenNebula, Eucalyptus, Apache CloudStack, or OpenStack open-source clouds are taken and it makes the system to build infrastructure, imparts storage, process computing, and infuses network resources along with programming language, python or java. In this model graph, comparison is done for the proposed extended transfer learning with equilibrium optimizer with transfer learning with PSO, transfer learning with ABC, and only transfer learning.

The experimental result suggests four performance metrics to demonstrate the performance of the attack classification and new attack detection. Precision and recall analyze the performance of attack classification where attack detection is analyzed by the accuracy and False alarm rate. This overall metrics show the performance of attack classification and detection in the cloud. Precision is the ratio between the number of correctly classified malicious samples and the number of classified malicious samples. Recall is the ratio between the number of correctly classified malicious samples and the total number of malicious samples. Detection Accuracy is the ratio between the number of accurately detected samples as unknown attacks and the total number of samples. False Alarm Rate is the ratio between the incorrectly detected or classified benign samples as attacks and the total number of samples.

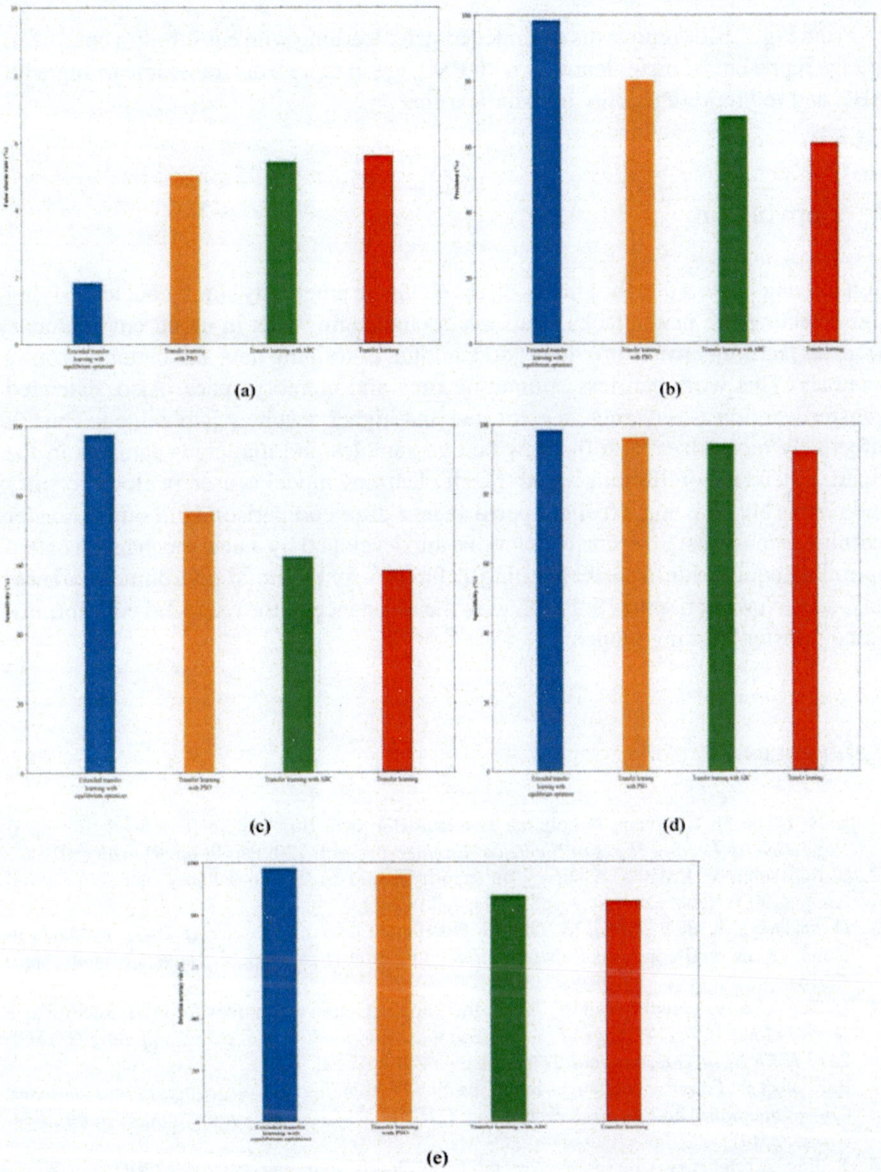

Fig. 1 model graph for **a** false alarm rate, **b** precision **c** sensitivity, **d** specificity, and **e** detection accuracy

From Fig. 1 Blue represents extended transfer learning with equilibrium optimizer, orange represents transfer learning with PSO, green represents transfer learning with ABC and red represents only transfer learning.

7 Conclusion

Considering the fact that the attacks on the cloud are inherently similar but identifying and detecting the new attacks is always a challenging task in cloud environment. Transfer learning always provides good amount of training data and better detection accuracy. This work can less computing time and memory space. Also, extended transfer learning is reducing an error rate and higher accuracy is obtained. Thus, it effectively recognizes both the new benign samples and malicious samples in the cloud. Extended equilibrium-based transfer learning model is used in cloud security and results are showing excellent performance after comparison with other transfer learning techniques. So security solutions are developed by a new mechanism called optimized equilibrium transfer learning techniques for the cloud environment. Hence, this research work has new solution with the assistance of the extended equilibrium-based transfer learning model.

References

1. M.N. Doan, H. Eui-Nam, A collaborative intrusion detection system framework for cloud computing. in *Lecture Notes in Electrical Engineering*, vol. 120, Part 2, pp. 91–109 (2012)
2. S. Subashini, V. Kavitha, A survey on security issues in service delivery models of cloud computing. J. Netw. Comput. Appl. **34**(1), 1–11 (2011)
3. O. Osanaiye, K.-K.R. Choo, M. Dlodlo, Distributed denial of service (DDoS) resilience in cloud: review and conceptual cloud DDOS mitigation framework. J. Netw. Comput. Appl. (2016)
4. U. Tupakula, V. Varadharajan, N. Akku, Intrusion detection techniques for infrastructure as a service cloud. in *Proceedings of the Dependable, Autonomic and Secure Computing (DASC), 2011 IEEE Ninth International Conference on 2011*. (IEEE, 2011)
5. S. Khan et al., Forensic challenges in mobile cloud computing. in *Proceedings of the Computer, Communications, and Control Technology (I4CT), 2014 International Conference on 2014*. (IEEE, 2014)
6. X. Jing, A brief survey on the security model of cloud computing. ISDCAB (2010)
7. B. Kandukuri, in *Cloud Security Issues*. (IEEE, 2009)
8. H. Takabi, in *Security and Privacy Challenges In Cloud Computing Environments*. (IEEE, 2010)
9. A. Bradai, H. Afifi, Enforcing trust-based intrusion detection in cloud computing using algebraic methods. in *2012 International Conference on Cyber-Enabled Distributed Computing and Knowledge Discovery 2012* Oct 10 (pp. 185–190). (IEEE, 2012)
10. R.R. Righi, F.R. Pelissari, C.B. Westphall, Sec-SLA: specification and validation of metrics to security service level agreements. in *Proceedings of the ComputerSystem Security Workshop*, pp. 199–210 (2004)

11. M. Ficco, L. Romano, A generic intrusion detection and diagnoser system based on complex event processing. in *Proceedings of the 1st International Conference on Data Compression, Communications and Processing*, June 2011, pp. 275–284. (IEEECS Press, 2011)
12. I. Gul, M. Hussain, Distributed cloud intrusion detection model. Int. J. Adv. Sci. Technol. **34**, 71–82 (2011)
13. M. Ficco, S. Venticinque, B. Di Martino, Mosaic-based intrusion detection framework for cloud computing. in *OTM Confederated International Conferences On the Move to Meaningful Internet Systems*, Sep 10, (pp. 628-644). (Springer, Berlin, Heidelberg, 2012)
14. P. Ghosh, A.K. Mandal, R. Kumar, An efficient cloud network intrusion detection system. in *Information Systems Design and Intelligent Applications* 2015 (pp. 91–99). (Springer, New Delhi, 2015)
15. N.D. Man, E.N. Huh, A collaborative intrusion detection system framework for cloud computing. in *Proceedings of the International Conference on IT Convergence and Security* 2011. (pp. 91–109). (Springer, Dordrecht, 2012)
16. D. Nahmias, A. Cohen, N. Nissim, Y. Elovici, Deep feature transfer learning for trusted and automated malware signature generation in private cloud environments. Neural Netw. **1**(124), 243–257 (2020)
17. F. Abazari, M. Analoui, H. Takabi, S. Fu, MOWS: multi-objective workflow scheduling in cloud computing based on heuristic algorithm. Simul. Model. Pract. Theory **1**(93), 119–132 (2019)
18. S. Namasudra, An improved attribute-based encryption technique towards the data security in cloud computing. Concurrenc. Comput. Practice Exp. **31**(3), e4364 (2019)
19. C. Lei, H. Dai, Z. Yu, R. Li, A service recommendation algorithm with the transfer learning based matrix factorization to improve cloud security. Inf. Sci. **1**(513), 98–111 (2020)
20. M. Rabbani, Y.L. Wang, R. Khoshkangini, H. Jelodar, R. Zhao, P. Hu, A hybrid machine learning approach for malicious behaviour detection and recognition in cloud computing. J. Netw. Comput. Appl. **1**(151), 102507 (2020)
21. J. Zhao, S. Shetty, J.W. Pan, C. Kamhoua, K. Kwiat, Transfer learning for detecting unknown network attacks. EURASIP J. Info. Secur. **1**(1) (2009)
22. P. Garcia-Teodoro, J. Diaz-Verdejo, G. Maciá-Fernández, E. Vázquez, Anomaly-based network intrusion detection: techniques, systems and challenges. Comput. Secur. **28**(1–2), 18–28 (2009)

Tracking Methodology for Fast Retrieval of Data and Its Feasibility Study

B. N. Lakshmi Narayan, K. Sowmya, B. Nandini, and Prasad Naik Hamsavath

Abstract The human resource department of an organization performs managing the human resource in the organization and various aspects of the organization. HR Department in an organization plays a main role in the organization for managing the organizational works. Human Resource Management System is an internal tool to make the work of the HR department easier and to keep the flow of work in the organization smoother. The tool provides information regarding the employees in the organization. Each employee in the organization can make use of this application to add the features where employees can get an overview of their work status. Employees can access the statistics of the leave, can create the task list, and keep track of the work progress. The system facilitates good interaction between the employees and the organization. Through the use of an HRM system, the organization can streamline the essential HR functionality and ensure all employees are properly compensated. The system can provide an employee self-service portal, letting them log in and view important information as well as a knowledge base of informative documents. The tool can be used in the process of recruitment, by providing references to the employees and getting the information about the new Joiners.

Keywords Streamline · Virtual Reality · Employee tracking methodology

1 Introduction

1.1 Problem Statement

Human Resource Management System (HRMS) is an internal tool that provides information regarding all the employees in the organization. The system acts as a medium of communication between the employees and HR administration.

B. N. Lakshmi Narayan (✉) · K. Sowmya · B. Nandini · P. N. Hamsavath
Department of Computer Applications, NMIT, Bengaluru, Karnataka, India

V. Suma et al. (eds.), *Inventive Systems and Control*,
Lecture Notes in Networks and Systems 204,
https://doi.org/10.1007/978-981-16-1395-1_42

1.1.1 Main Problem in the Human Resource Management System

1. Employees may fail to meet the deadline set by the manager because employees spend their time in sorting, integrating, and preparing reports.
2. Disappointment in employees is regularly observed in the time of the presentation process due to the vast unusual nature of the work.
3. Preparing the reports on the work has been a difficult part for the employee.
4. Additional man-power is required to add requirements in the time of data preparation that may add to the cost.

1.2 Objective

The Human Resource Management System is influenced by organizational, individual, and social goals. The objective of the organization is to earn profit, growth, to survive in the competitive world, stability, and diversification. The objectives of the organization can only be achieved with the resources available in the company.

The main moto of HRMS, to make sure that the efficient and willing human resource of the organization to achieve the desire [1], values, and status of the employees and to have a concern about the social and economic problems in the society [2].

1.2.1 The Objectives of HRM Are as Follows

1. HRMS is concerned with organizational goals, it can be achieved by proper utilization of human resources.
2. To build a healthy working bond among the employees and to build a desirable organizational structure.
3. To combine the goals of the individual employee and team within an organization.
4. To provide better opportunities and service for individual or group employees according to the growth of the organization.
5. To recognize and provide the needs of the individual and group [3].

1.3 Scope

The scope consists of all the activities from human power utilization till the employee is in the organization. HRMS consists of obtainment, growth, preservation, and manages the human resources in the organization.

1.4 Methodology

The backend of the app is developed using Node js, Express js. Mongo DB is used for Databases. The structure of the application is built based on the employee tracking methodology of the organization, with the replacement of manual procedures [4].

2 Literate Review

2.1 Existing and Proposed System

2.1.1 Existing System

There were problems related to a current system like lack of storage, security, and many more.

(1) **Security:** As employees entrust the organization with their individual information. All the details of employees may be official or unofficial information is stored in HRMS.
(2) **Cost:** It is one of the disadvantages of HRMS.
(3) **Losing the Subjectivity:** All personal data of any employee was early leaked.
(4) **Difficult of Analysis:** As the amount of data collected is huge, it is difficult to analyze the data.

2.1.2 Proposed System

By using today's updated software and working technologies have developed software which can overcome the problems of current problems.

The Solutions Made by the Proposed System Are as Follows

1. **Fast retrieval of data:** Response of data retrieving was improved by the proposed system with high throughput.
2. **Security:** Personal information of employees such as phone number, residential address, email id, and so on are secured.
3. **Ease of use:** Propose system is user friendly.
4. **Work Time:** It gathers systematic time and efforts in work.
5. **Administration:** The admin module of HRMS handles all operations related such as task assigned, task completed, a task which is in progress, upcoming events, salary, leave, and so on, related to the employee.

2.2 *Report on Feasibility*

Feasibility is an important report which attempts to see the feasibility of the task before the actual work is carried out. It's analyzing posting the potential outcomes of each agreement and updating the best option for use. This also explains how those different structures fit into the company.

2.2.1 Technical Feasibility

Technical viability is something that tests the potential to utilize the current state of technology followed by further improvements. The technological potential of human resources and technology is also to be addressed.

The information was manually recorded in the earlier system, and the process was very slow, and the targets in the given time will not be challenging. High numbers of human involvement are needed, it required more employees to record data, call customers, etc.

In the proposed method, it is computerized to a higher degree, reducing human intervention. And it is safer since only approved users can authenticate the application compared to the current system. The manual work is reduced, in terms of time. The main factor to consider when developing an application time allocated to finishing is.

2.2.2 Economic Feasibility

After technical feasibility, the technique which helps the company estimates the efficiency of the system is economic analysis. The strategy would decide what benefits and advantages are likely to be in the programs implemented and equate them with costs. This involves the feasibility of the proposed system to create economic benefits.

For this case, the cost analysis is the least; there is no associated expense for producing the leads, resources, and tasks. This is done with some cost savings to direct sales. In the case of sales and marketing, the company has to maintain a large amount of human interference, money, and time resources. A lot of energy, time, and human resources can be minimized when the system is computerized.

2.2.3 Schedule Feasibility

This refers to the time allotted to complete the project, i.e. deadline, instead of taking a lot of time to create, allocating the necessary time and completing it in that time is also a best practice. The process for the exchange of knowledge between the customer and the sales force will also occur within the scheduled time. This tool also provides notifications for completing timelines for the mission [7].

2.2.4 Operational Feasibility

This feasibility evaluation is concerned with the operational and political aspects. It meets all aspects of the requirements of the market, it is costly and competitive because the price and efficient research mean that it is the best cost-effective application and this also satisfies the criteria by designing the application to the business [8].

3 Proposed Work

3.1 Dashboard

The dashboard is the main page of the application. The users can get an overview of their profile. The users have the access to view the list of the leaves, new joiners to the organization. Dashboard page directs to provide references to the job seekers.

3.2 Task

The task is one of the best features of this application. It can be represented as a To-Do list. The employees can create a list of the tasks which are assigned to them and keep track of the task. Employees get an overview of the work to be done.

Task page is added with the functionality of the remainder, where the employees get the alert notification about the task which needs to be done. It helps the employees to perform the task effectively by not forgetting the task which is assigned to them.

Employees get a view of the list of tasks. Where the employees get to know the information about the tasks which need to be done, tasks which are completed, the upcoming tasks, the task which is in progress, tasks which are completed and employees can also create their own task to keep track of the work.

3.3 Inbox

Inbox is one of the pages in the HRMS tool. Inbox page provides the users alerts about the upcoming programs in the organization [5].

3.4 Recruitment

The recruitment page provides information regarding the vacant position in the organization. The users get the details regarding the qualification and skills required for the position, where user can them self-apply for the position if they seem they are eligible or share the details to the job seekers and get the best employees to the organization [6].

3.5 Time off

The employees can view their individual holiday status on the Time off-page. Employees get the details about the upcoming holidays, the count of the leave taken, and a summary of Accrual.

4 Conclusion

HRMS is significantly contributing to effective resource management in the organization, the manpower, and the organization ensures that the use of HRMS is effectively used. It is represented as a software tool to upgrade resource management systems and productivity.

The usage of HRMS is increasing in the market and it gives competition to the user. HRMS is helping the company to make better use of time, cost, and energy in its operations and management.

5 Future Enhancements

HRMS is often being the magic formula for Human Resource related issues. As though the uses of HRMS packages are increasing, and further some requirements that need to be added to make the work easier. The technologies which are ruling in the current situation are AI, Virtual Reality, and any more. These technologies can be brought into the HRMS and make an effective tool for the organization.

AI technology can be implemented in the HRMS in the recruitment process. Research has pointed out that AI can help in the recruiting process to choose the effective employee for the organization.

Virtual and augmented reality can be applied in the organization for training scenarios, either as part of the onboarding process or to provide current employees with additional training.

HRMS tool can be added with the feature to inform the employee about the shift timings, work requirements, and many more.

References

1. Z. Liu, S. Mei, Y. Guo, Green human resource management, green organization identity and organizational citizenship behavior for the environment: the moderating effect of environmental values. Chinese Manage. Stud. ahead-of-print No. ahead-of-print (2020). https://doi.org/10.1108/CMS-10-2019-0366
2. A.Y. Anisimov, A.S. Obukhova, Y.V. Aleksakhina, A.Y. Zhaglovskaya, A.A. Kudra, Strategic approach to forming a human resource management system in the organization. Int. J. Econ. Persp. **11**(2),442–448 (2017). 7p. 1 Diagram
3. J. Bagheri, Overlaps between human resources strategic planning and strategic management tools in public organizations. Proc.—Soc. Behav. Sci. **230**(12), 430–438 (2016). https://doi.org/10.1016/j.sbspro.2016.09.054
4. P. Daneshgari, H. Moore, Organizational transformation through improved employee engagement—how to use effective methodologies to improve business productivity and expand market share. Strategic HR Review **15**(2), 57–64 (2016). https://doi.org/10.1108/SHR-02-2016-0007
5. S. Wang, H.C. Green, M.L. Wilder, Du. Qian, B.L. Kmush, M.B. Collins, D.A. Larsen, T. Zeng, High-throughput wastewater analysis for substance use assessment in central New York during the COVID-19 pandemic. Environ. Sci.: Process. Impacts **22**(11), 2147–2161 (2020). https://doi.org/10.1039/D0EM00377
6. P. Aghion, B.F. Jones, C.I. Jones, Artificial intelligence and economic growth. in *Economics of Artificial Intelligence Fall 2017 Working Paper* 23928, Issue Date October (2017). https://doi.org/10.3386/w23928
7. D. Acemoglu, P. Restrepo, Artificial intelligence automation and work.Working Paper 24196. Issue Date January (2018). https://doi.org/10.3386/w24196
8. J. Milovanovic, G. Moreau, D. Siret, F. Miguet, Virtual and augmented reality in architectural design and education: an immersive multimodal platform to support architectural pedagogy. HAL Id : hal-01586746, version 1. https://hal.archives-ouvertes.fr/hal-01586746

HFQAM-Based Filtered OFDM: A Novel Waveform Design with Hybrid Modulation for Next-Generation Wireless Systems

G. Shyam Kishore and Hemalatha Rallapalli

Abstract 5G mobile communication provides various services like enormous device connectivity, huge capacity, and reliable low-latency services other than ultra-high-speed data. To fulfill these requirements of 5G, one of the basic challenges over the existing orthogonal frequency division multiplexing (OFDM) is the design of a novel waveform. Besides the simplicity in implementation and Multiple-In Multiple-Out (MIMO) support, the candidate waveform needs to preserve the following two vital features: low out-of-band emission and relaxation in need for synchronization for achieving overall design requirements of 5G. In this article, an ultra-reliable, spectral efficient waveform called hybrid frequency and quadrature amplitude modulation-based filtered OFDM (HFQAM-FOFDM) is proposed to fulfill the above requirements. This article explores the Bit Error Rate (BER) of the proposed model over the additive white Gaussian noise (AWGN) channel and also presents the BER of hybrid frequency and quadrate amplitude modulation (HFQAM) based OFDM, QAM-based FOFDM, QAM-based OFDM. From the results, it can be observed that the proposed HFQAM-FOFDM scheme is superior to HFQAM-based OFDM, QAM-FOFDM, QAM-OFDM in terms of BER.

Keywords Quadrate amplitude modulation (QAM) · Hybrid frequency and quadrate amplitude modulation (HFQAM) · Orthogonal frequency division multiplexing (OFDM) · Filtered orthogonal frequency division multiplexing (FOFDM) · Additive white Gaussian noise (AWGN) · Bit error rate (BER)

G. S. Kishore (✉) · H. Rallapalli
ECE Department, University College of Engineering, Osmania University, Hyderabad, Telangana, India

H. Rallapalli
e-mail: hemalatha.r@uceou.edu

V. Suma et al. (eds.), *Inventive Systems and Control*,
Lecture Notes in Networks and Systems 204,
https://doi.org/10.1007/978-981-16-1395-1_43

1 Introduction

Amplitude shift keying (ASK), frequency shift keying (FSK), phase shift keying (PSK), and quadrate amplitude modulation (QAM) are used in mobile communication as the basic digital modulation techniques. Recently, HFQAM [1] has gained researcher's attention because of its optimal use of bandwidth and power with reliability. HFQAM integrates the gains of QAM and FSK. It is highly reliable, spectrally efficient, and has superior interference handling capacity [2].

In 4G, CP-OFDM has been used due to high spectral efficiency, easy integration with MIMO systems [3]. But it also suffers from few drawbacks, like high out-of-band emission, needs for synchronization, etc. Apart from ultra-high-speed data, 5G provides various services like enormous device connectivity, huge capacity, and reliable low-latency services. To overcome the drawbacks in traditional CP-OFDM and to meet the demands of next-generation wireless communication, a novel waveform or existing CP-OFDM with some modifications is required [2].

The use of the Filtering/windowing method in OFDM has been already proposed in the literature to minimize the out-of-band emission and also the requirement for synchronization will get relaxed. A new waveform candidate, which overcomes the existing drawbacks and fulfills the above needs, is required for next-generation wireless communication. Comparison among the few outstanding multi-carrier waveforms has been done in [4, 5], and it is concluded that FOFDM is one of the prominent waveform candidates.

The rest of the article is structured as follows: Novel waveforms for the next-generation are reviewed in Sect. 2. In Sect. 3, the Hybrid modulation technique is provided. HFQAM-FOFDM model is proposed in Sect. 4. Simulation results are provided in Sect. 5. Finally, the research work is concluded in Sect. 6.

2 Filtered OFDM Waveform

Although OFDM has many advantages which made it a popular modulation for 4G, it has many drawbacks like the increased computational complexity, the need for very accurate synchronization, and most notably, the Peak to Average Power Ratio (PAPR) is high, etc.

Search for the novel OFDM-based waveform is still an open research problem, which overcomes the above-mentioned drawbacks. Comparison among the few outstanding multi-carrier waveforms has been performed [2, 4, 5]. Concerning the computational complexity, OOB emission, flexibility, and compatibility with the existing system—FOFDM, WOFDM, and Hybrid WF-OFDM produce better performance than FBMC, UFMC, and GFDM [4]. Except for the filter, the remaining architecture of FOFDM resembles the OFDM. With this simple modification in architecture, the FOFDM can overcome the limitations of OFDM [4, 6] like relaxed

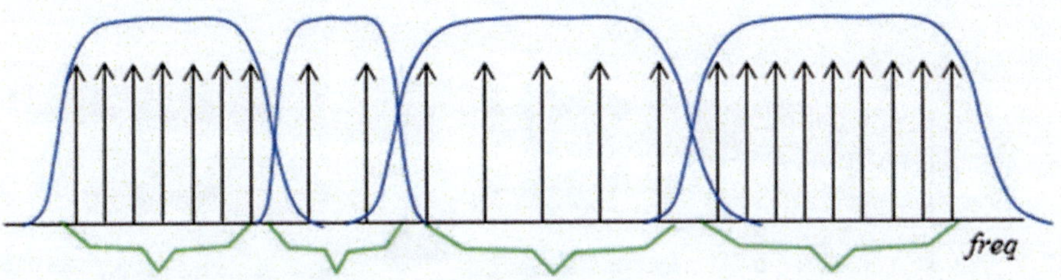

Fig. 1 Flexible subcarrier spacing of filtered OFDM

synchronization and suppressed out-of-band emission which results in lower spectral consumption of the guard band.

In FOFDM, a total band of frequency is divided into some sub-bands. Then in parallel, all these sub-bands are filtered with different filters. Depending upon the application, each sub-band can have different subcarrier spacing and different cyclic prefix lengths. Each sub-band can be utilized for different types of services, [6–8], and this can be observed from Fig. 1; flexibility and MIMO compatibility are added advantages of FOFDM.

3 Hybrid Modulation Technique

The ever-growing demand for higher multi-Gbps peak data rates, more reliable with optimum bandwidth usage can be fulfilled by choosing a better modulation technique other than traditional PSK, FSK, QAM.

The frequency shift keying (FSK) modulates the signal by changing the carrier frequency according to the intended transmitted symbol. As the information bits are contained in the frequency of the signal, the FSK has the highest immunity against noise but is the most bandwidth-consuming [9].

The QAM signal is generated by altering both the amplitude and phase of the carrier signal. Here as the constellation size increases, states come closer to each other, which implies the reduction of noise immunity. The order of the QAM modulation is directly proportional to spectral efficiency and inversely proportional to BER performance [9].

QAM is efficient in bandwidth, whereas noise immunity is more in FSK. FSK is very rarely used due to its low spectral efficiency. In both QAM and PSK, Bandwidth efficiency is the same, but reliability, the power efficiency of QAM is better than PSK.

Hybrid Frequency and Quadrature Amplitude Modulation (HFQAM) is a hybrid modulation technique combining the FSK and QAM schemes. The HFQAM uses multiple orthogonal frequency carriers same as FSK and on each frequency carrier, it implements M-QAM modulation [1, 9]. Comparison between these modulation methods is done along with HFQAM [9] and it is concluded that HFQAM which is formed by combining QAM and FSK results in higher spectral efficiency and superior

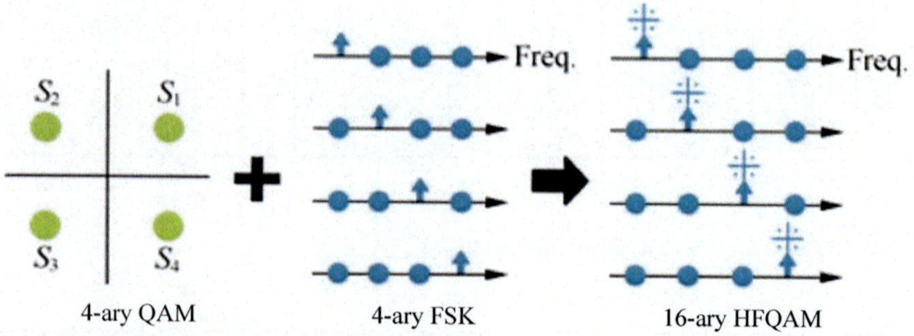

Fig. 2 Signal constellation of 4 × 4 HFQAM

noise immunity. The error rate increases in M-ary ASK, PSK, QAM, and HFQAM as the order of modulation (M) increases, despite their high spectral efficiencies, except in M-ary FSK where the error rate decreases [9]. Figure 2 shows the 4 × 4 HFQAM constellation.

The HFQAM has the advantage of better BER than conventional M-QAM due to the use of more than one carrier which introduces lower BER as these carriers are orthogonal and at the same time have higher bandwidth efficiency than M-FSK which requires M carriers [9–11].

4 The Proposed System Model

Apart from the various advantages, flexible FOFDM suffers from Inter-Channel Interference. The HFQAM is excellent in dealing with the interference and achieves significant performance gain. Inter-Channel Interference distribution is not Gaussian in HFQAM modulation, which results in the increased channel capacity. From the above discussion, to overcome the drawbacks of FOFDM while preserving its advantages and also for getting the benefits of HFQAM modulation, the system with the combination of FOFDM with HFQAM is proposed, which combines the benefits of both FOFDM and HFQAM. The proposed system model of HFQAM-FOFDM transmitter and receiver is shown in Fig. 3a, b.

On the transmitter side, initially, the input data is mapped into symbols using the HFQAM modulation technique, then the IFFT operation is performed, which converts these HFQAM symbols into OFDM symbols. Then in the next step cyclic prefix is added and finally, the resulting signal $s(n)$ passes through a well-designed and suitable filter, i.e.,

$$m(n) = s(n) * f_i(n) \quad (1)$$

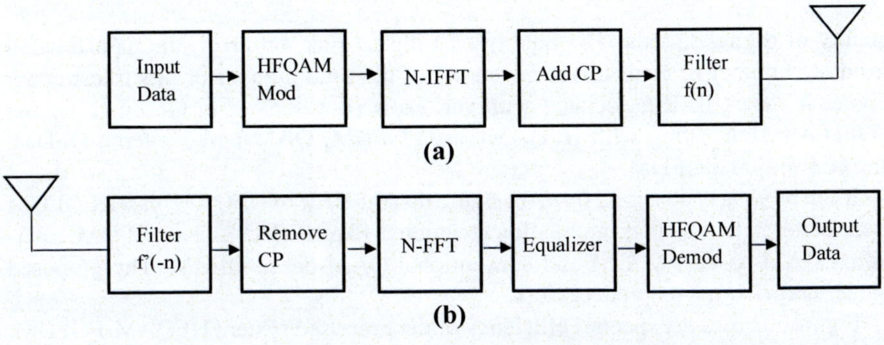

Fig. 3 a HFQAM-FOFDM Transmitter, b HFQAM-FOFDM Receiver

The modulated signal $m(n)$ is transferred through the AWGN channel to the receiver. At the receiver side, the received signal is sent through the filter $f_i{}^*(-n)$ [6]. The ~~A~~ cyclic prefix is removed in the next step, the resulting signal is converted back to OFDM symbols after the equalization step. Finally, the original input data is retrieved as output data using the HFQAM demodulation process at the end.

$$r(n) = m(n) * h(n) * f_i^*(-n) \quad (2)$$

5 Results

The proposed HFQAM-FOFDM model is implemented in the MATLAB environment. The performance of the HFQAM-FOFDM model in comparison with other system models like HFQAM-OFDM, QAM-FOFDM, and QAM-OFDM over the AWGN channel is provided here. The simulation parameters are given in Table 1.

BER is a fundamental parameter for measuring the performance of a digital communication system, also BER is the main Figure of Merit for measuring the

Table 1 Proposed system simulation parameters

Parameters	Values
IFFT/FFT size	1024
Modulation	QAM/HFQAM
QAM levels, FSK frequencies	4, 4
Waveform	OFDM/FOFDM
subcarriers	600
Cyclic prefix	72 samples
Filter length	513
Channel	AWGN

quality of digital signals. The quality of a digital link, bit error rate, is estimated from the number of error bits received out of the total number of bits transmitted. Figure 4 shows BER associated with various SNR values of proposed HFQAM-FOFDM system along with HFQAM-based OFDM, QAM-based filtered OFDM, and conventional OFDM.

It can be clearly observed from the figure that the BER of HFQAM-based FOFDM is far better than the other modulation techniques like HFQAM-based OFDM, traditional QAM-based FOFDM and conventional QAM-based OFDM. The proposed model outperforms its predecessors.

Figure 5 shows the spectral efficiency of the proposed system HFQAM-FOFDM. It has ~~a~~ lower side lobe, which implies better spectrum utilization, which leads to increased spectral efficiency. In HFQAM-FOFDM, only one subcarrier out of M_F subcarriers is activated which carries a QAM symbol and the remaining subcarriers are deactivated. Hence, the level of self-interference is very low, which in turn reduces PAPR.

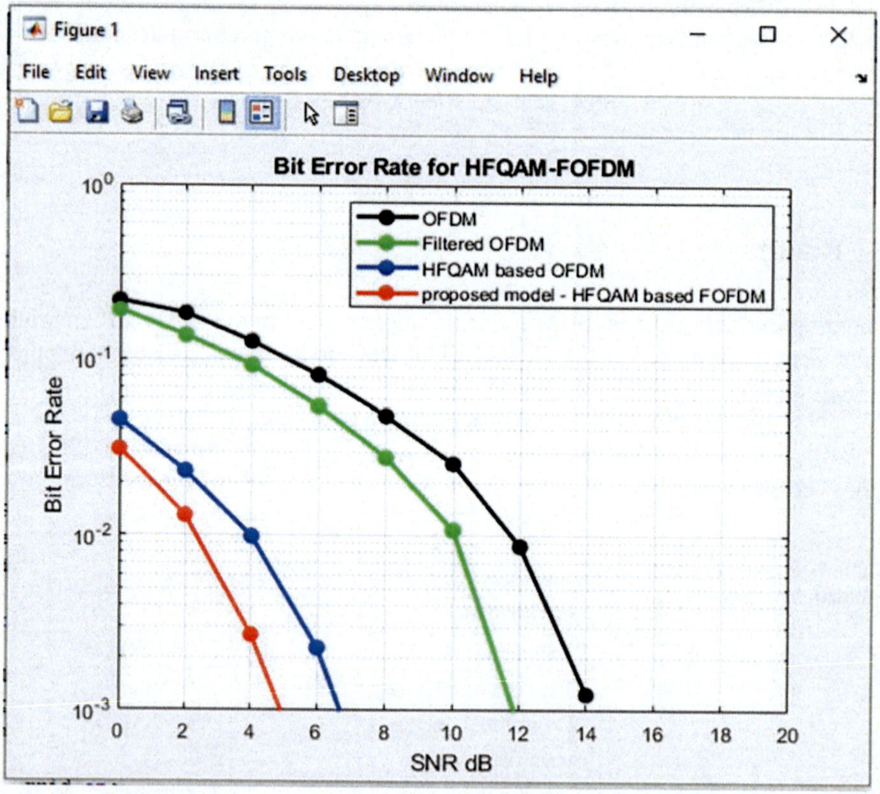

Fig. 4 BER performance of proposed HFQAM-FOFDM

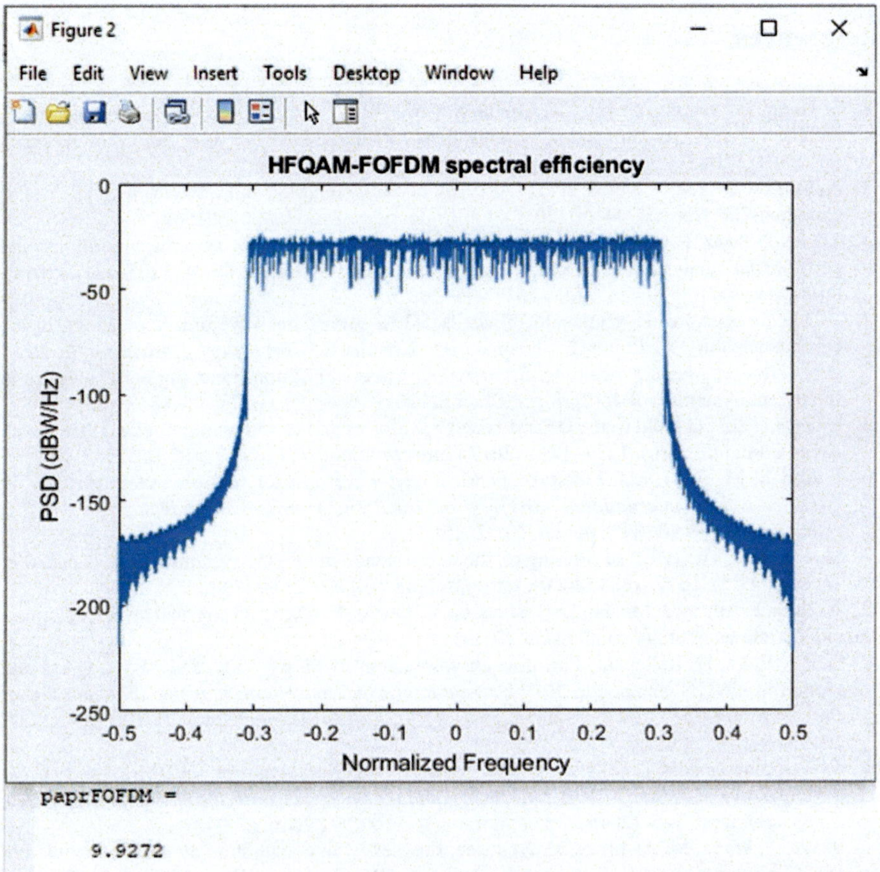

Fig. 5 Spectral efficiency of proposed HFQAM-FOFDM

6 Conclusion

Modulation and waveform design play a vital role in wireless communication and is still an open research problem. To overcome the existing problems in 4G and to fulfill the requirements of 5G, various waveform, modulation designs are proposed in the literature. In this article, HFQAM-based FOFDM is designed which is a potential candidate for 5G systems with novel modulation. From the results, it is clear that the HFQAM-based filtered OFDM has the better BER performance. Other than reliability, it has more advantages like high spectral efficiency, reduced PAPR, and improved overall performance. For next-generation wireless communications, the proposed model is suggested as the most promising waveform candidate.

References

1. S. Hong, M. Sagong, C. Lim, K. Cheun, S. Cho. Fqam: a modulation scheme for beyond 4g cellular wireless communication systems. in *2013 IEEE Globecom Workshops (GC Wkshps)*, pp. 25–30. (2013)
2. N. Maziar, W. Yue, T. Milos, et al., Overview of 5G modulation and waveforms candidates. J. Commun. Inf. Netw. **1**, 44–60 (2016). https://doi.org/10.1007/BF03391545
3. Zhang, A. Ljaz, P. Xiao, M.M. Molu, R. Tafazolli, Filtered OFDM systems, algorithms, and performance analysis for 5G and beyond. IEEE Trans. Commun. **66**(3), 1205–1218 (2018). https://doi.org/10.1109/TCOMM.2017.2771242
4. G. Shyam Kishore, H. Rallapalli, Towards 5G: a survey on waveform contenders. in ed. by S. Satapathy, K. Raju, K. Shyamala, D. Krishna, M. Favorskaya, *Advances in Decision Sciences, Image Processing, Security and Computer Vision*. Learning and Analytics in Intelligent Systems, vol 4. (Springer, Cham, 2020)
5. P. Guan et al., 5G Field trials: OFDM-based waveforms and mixed numerologies. IEEE J. Sel. Areas Commun. **35**(6), 1234–1243 (2017). https://doi.org/10.1109/JSAC.2017.2687718
6. J. Abdoli, M. Jia, J. Ma, Filtered OFDM: a new waveform for future wireless systems. in *2015 IEEE 16th international workshop on IEEE signal processing advances in wireless communications (SPAWC)*, pp. 66–70. (2015)
7. M. Faulkner, The effect of filtering on the performance of OFDM systems. IEEE Trans. Veh. Technol. **49**(5), 1877–1884 (2000). https://doi.org/10.1109/25.892590
8. A. Zaidi, F. Athley, J. Medbo, U. Gustavsson, G. Durisi, X. Chen, 5G physical layer: principles, models and technology components (2018)
9. G.S. Kishore, H. Rallapalli, Performance assessment of M-ary ASK, FSK, PSK, QAM and FQAM in AWGN channel. in *2019 International Conference on Communication and Signal Processing (ICCSP)*, Chennai, India, pp. 0273–0277. (2019). https://doi.org/10.1109/ICCSP.2019.8697922
10. S.G. Domouchtsidis, G.D. Ntouni, V.M. Kapinas, G.K. Karagiannidis, OFDM-IM vs FQAM: a comparative analysis. in *2016 23rd International Conference on Telecommunications (ICT)*, Thessaloniki, pp. 1–5. (2016). https://doi.org/10.1109/ICT.2016.7500402
11. S. Wu, Y. Wang, M. Al-Imari, M. Nekovee, Frequency and quadrature amplitude modulation for 5G networks. in *2016 European Conference on Networks and Communications (EuCNC)*, Athens, pp. 1–5. (2016). https://doi.org/10.1109/EuCNC.2016.7560993
12. J. Wang et al., Spectral Efficiency Improvement With 5G Technologies: Results From Field Tests," in IEEE J. Selected Areas Commun. **35**(8),1867–1875 (2017). https://doi.org/10.1109/JSAC.2017.2713498

An Approach to Extract Meaningful Data from Unstructured Clinical Notes

K. Sukanya Varshini and R. Annie Uthra

Abstract Clinical notes occupy a major place in electronic health records (EHR). Performing analysis in unstructured clinical notes is always more complicated when compared to structured data. Structured data always provide more details and data to a layman than unstructured data. To perform analysis in the unstructured type of data, it is always necessary to extract data from the unstructured data before working on the data. In this work, a method has been proposed to uncover the entities related to the specific clinical terms as clinical notes are completely involved in the extraction of the data. The main focus of this research will be on the method for discovering the patterns of the relationship that exist among the various clinical entities, which includes the tests, treatments, and the diagnosis, focusing mainly on the clinical notes. It is always difficult to extract data from the unstructured data, and when it comes to clinical notes, the value of data is more. There cannot be any data loss as every word in the clinical note matters. The proposed approach uses natural language processing techniques along with text and rule mining to extract data from the unstructured data.

Keywords Unstructured data · Clinical notes · Machine learning · MIMIC III

1 Introduction

Accuracy of disease identification and classification of the patient's condition at the beginning stage itself will give the healthcare specialist to sort the disease and to take necessary precautionary conditions to cure the disease [1]. The structured data have a design where all the needed features are in a structured manner. Whereas in unstructured data, the data do not have a pre-defined model. Even though many models exist in handling structured data, handling unstructured data still face many

K. S. Varshini (✉) · R. A. Uthra
SRM IST, Kattankulathur, Chennai, India
e-mail: sukanyak@srmist.edu.in

R. A. Uthra
e-mail: annieu@srmist.edu.in

V. Suma et al. (eds.), *Inventive Systems and Control*,
Lecture Notes in Networks and Systems 204,
https://doi.org/10.1007/978-981-16-1395-1_44

challenges. In this article, the main concentration will be on the unstructured type of data, types of data models present in unstructured data, and the methods and models to handle the unstructured data [2].

Numerous attempts have been conducted to increase the quality, and the heterogeneity of the data extracted from the electronic health record (EHR) of the clinical data [3]. Many systems are found in the process of finding out the patients who are at high risk of complications in many diseases. As a result of the above observation, the hospital resource management has been found to increase according to the need [4]. ICDA stands as an important aspect in delivering the data analytics intelligently and MatrixFlow [5] for the analysis of the disease status for the data extracted from the unstructured notes. The main contributions of the article include the following:

1. The key terms to be extracted from the unstructured clinical notes are defined.
2. The topic modeling method is used to group the words into clusters and then finally to extract the most important terms and discard the rest.
3. The model is checked for the performance metrics for the MIMIC III datasets.

The rest of the article is organized as follows: In Sect. 2, the existing works in the field of research have been discussed. Section 3 gives a detailed explanation of the materials and methods used throughout our research. Section 4 discusses the result obtained. Finally, Sect. 5 concludes the work carried out in our research followed by the future work in Sect. 6.

2 Related Work

In the present situation, there are innumerable numbers of researches carried out in the medical field involving the extraction of data from unstructured notes. The related researches done related are discussed below: In 2009, Xu et al. have proposed a tool based on natural language processing (NLP) called the MedEX [6]. The proposed tool is used in the extraction of needed information especially the drug information or the medication details from the unstructured data. The main source of data for extraction included the discharge summary. It can also extract other information as per the need of the user along with the medication details [7].

Sovava et al. [8] developed a system for analyzing the text which is called cTAKES. This method is based on the NLP for the extraction of data from unstructured data like the electronic medical record [9–11]. The system is found to mainly consist of the combination of the unstructured information management architecture (UIMA) framework and the OpenNLP [12].

2.1 Time-Series Models

Several types of research can be found analyzing the clinical data for the prediction of the disease risk factor. Among which the major challenge exists which includes the noisy data, sparse, and unevenly distributed samples. Mao et al. [13] proposed a model for evaluating the various features that were extracted from the clinical data including the time-series signals for classification of the disease models which are used in the data of ICU patients. Wiens et al. [14] brought out the classification techniques related to the time-series data for the infection data of patients in the hospital. Feature construction from the longitudinal data has been discussed in the non-parametric Bayesian approach. Durichen et al. [15] proposed a model using the multi-task process related to Gaussian to deal with the multivariate physiological data. Batal et al. [16] again propose a model to deal with the multivariate temporal data from the clinic using the temporal pattern mining. Recently, generalized linear dynamic model is used in the process of modeling heterogeneous dynamic clinical data [17].

2.2 Disease Progression Models

The progression of diseases is modeled which are in turn not only used in the hospital episode, but also for other illnesses. Wang et al. [18] proposed a model to develop disease progression models with the help of the resultant findings using the clinical findings from the heterogeneous and incomplete patient records. Cohen et al. [19] use the combination of the physiological, clinical, and treatment variables from the intensive care unit (ICU) to find out the patient state and to track the condition of the patient over time. Zhou et al. [20] developed a method to combine the group lasso formulation using the genetic data, cognitive scores, lab tests, and demographic data.

2.3 Computational Phenotyping

The main aim of computational phenotyping is to extract meaningful information from the unstructured data, which includes electronic medical records (EMR) [21, 22]. The extraction helps in the useful descriptions of many diseases. These are used by the researchers in the process of designing a model for the prediction of diseases. Pivovarov et al. [23] developed a model using the notes, laboratory tests, medications, and diagnosis codes to discover the disease models. Marlin et al. [24] presented a model to find the mortality prediction using the 13 clinical variables including heart rate, respiration rate, glucose, etc. Che et al. [25] also worked in the mortality prediction related to circulatory diseases. Kale et al. deal with neonatal disease in the mortality prediction. Schulam et al. worked in the scleroderma which is an autoimmune disease with a motive to discover the disease subtypes. Liu et al. formulated the graph-based representation that is used to detect the diagnosis and the medication code from the EMRs.

2.4 Mining Clinical Notes

Clinical notes are used to convert the unstructured data into a structured format which is thus used in the diagnosis of disease and to propose to diagnose models. Clinical notes are thus used in the process of finding out the relationship between the disease symptom and the medication relationship. Further, they are also used in the study of symptom—symptom interactions. Finlayson et al. proposed a work using more than 20 million notes of a duration of 19 years of data to develop inter-related clinical conceptual data in order to mine the association of the various risk factors. Hripcsak et al. developed a model to analyze the textual data related to the X-ray of the chest. Murff et al. demonstrated a model to predict postoperative complications from the textual data. NLP is used in the analysis of the EMR data in order to identify the complications having higher sensitivity. Table 1 gives a detailed description of the various types of unstructured noted available in the medical field.

Table 1 Various types of unstructured notes available in the medical field and their features

Data	Type	Characteristics	Examples
Nursing notes	Text, Temporal	Assessment of the patient's condition	Patients present condition
Diagnoses	Numerical, Temporal	Diagnosis of the present condition	Diabetes
Procedures	Numerical, Temporal	Surgical procedures	Endoscopy
Lab tests	Numerical	Measuring certain features	Blood glucose level
Vitals	Numerical, Temporal	Measured every minute in ICU and every hour outside ICU	Heart rate, respiratory rate, blood pressure
Radiology	Image, Text	Image report	X-ray, CT scans
Medication orders	Numerical, Temporal	Physician orders	Aspirin
Demographic	Numerical, Static	Demographic details	Age, gender

3 Materials and Methods

3.1 System Design

The predictive system using a classification model has been proposed to find out the patient's risk of disease mortality. The patient history is used in order to train the classifier after the needed data have been extracted from the unstructured data. In this model, the data are input to the system from which the classifier outputs a label. In case if the patient is found to be in danger, then it is intimated to the medical staff, and the needed precautionary steps are taken.

The first and foremost step involved in predicting the health condition of the patient is that the data from the unstructured notes should be extracted and trained for a high accuracy rate. Later, the preprocessing steps and the feature extraction steps have been discussed. Extraction of data from the lab test report and the medication orders is considered to be a little easier when compared to the data extraction from nursing notes. New methods for the extraction of data from the unstructured data will be discussed. Later point the research can be taken forward by combining the heterogeneous sources of data by combining the features.

3.2 Text Processing and Feature Extraction

Nursing notes and the reports build up the basic text sources. Nursing notes are the informal way of writing the condition of the patient in the critical care unit. These nursing notes include both the objective and subjective evaluation of the patients. These notes are written frequently. A discharge summary consists of the entire details related to the stay of the patient at the hospital. Starting from the condition of the patient at the time of admission, treatments are done, medications are given, and healthcare professionals have involved finally in the future treatment plans.

In nursing notes, many non-standard abbreviations are found to be present from which the extraction of data seems to be very difficult (Figs. 1, 2). For example, "CV"

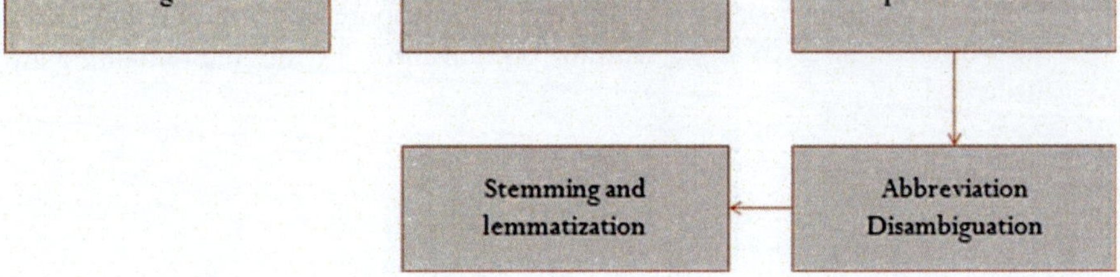

Fig. 1 Preprocessing steps in text extraction

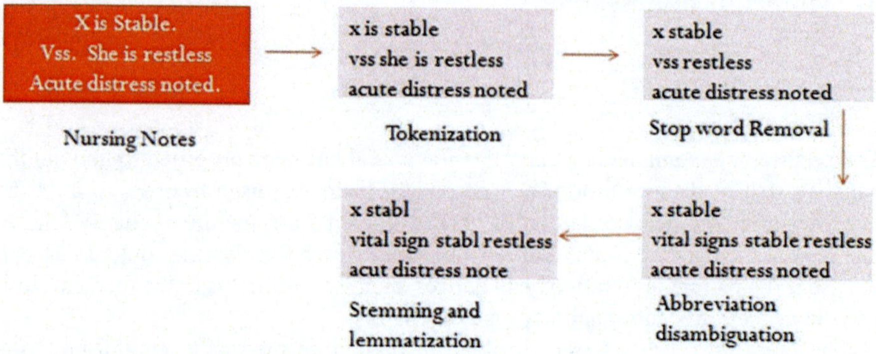

Fig. 2 Preprocessing steps in text extraction—an example

can also be written as "CARDIO" and "CARD" in some of the clinical notes, and it personally depends on the physician. The significance of the words is based on the place where that word actually resides. The steps involved in the word extraction are listed as follows:

(a) The headings will be extracted from all the observations. A simple rule to find the headings is that a word beginning a sentence followed by a colon.

(b) Then, the headings will be eliminated, and the data are regrouped. Several headings could refer to the same term. (For example, the word "CARDIO,""CV,""CARD" all refer to the same type of heading). Topic modeling is used to form clusters of the terms that are related to each other. Now all the grouped noted will have reliable headings (i.e., the terms similar to "CARD," "CV," "CARDIO" will all be grouped under the same cluster).

(c) After clustering the words together, they are processed for later observation. Under every cluster, the punctuation marks and stop words are removed. Stemming is also done to bring out the list of words under every cluster formed.

(d) Let the number of the findings obtained from the cluster L is described as b_x (L, E), where the word x occurs in E. The word importance is found using the below formula $I_w(E) = b_x(M,E) - b_x(N, E)$ here M and N are the clusters. The words in cluster M that are present in high number are considered as positive and those that are present in the other class N are found to be negative. The values of the importance of words are taken as per the discriminatory power.

(e) Then, the words are sorted with respect to the important values, the top 5% of the words are selected along with the bottom-most 5% and the remaining are discarded.

4 Result Analysis

4.1 Training Dataset

MIMIC III dataset consists of the database of the various patients who were admitted to the critical care unit. These datasets are available for free and can be accessed by the researchers. This database consists of data of more than 58,000 hospitals including a count of more than 46,000 patients who were admitted during the year 2001–2012. Many data are available regarding the patients which include the billing information, clinical data discharge notes, and drug recommendation.

The overall workflow of the process to predict the heart patient is given in Fig. 3. The main process involved in the workflow of the proposed approach is that first the patient details are connected with respect to heart disease. The connected information along with the clinical data is collected. The discharge summary report also plays a vital role. Then, the training dataset and the testing dataset are generated with a proper balance in the positive and the negative samples. Then, the model for training the dataset is built, and then, the performance metrics are analyzed for different aspects, and the results were concluded (Figs. 4, 5).

The performance metrics include the accuracy rate, precision, recall, and F score.

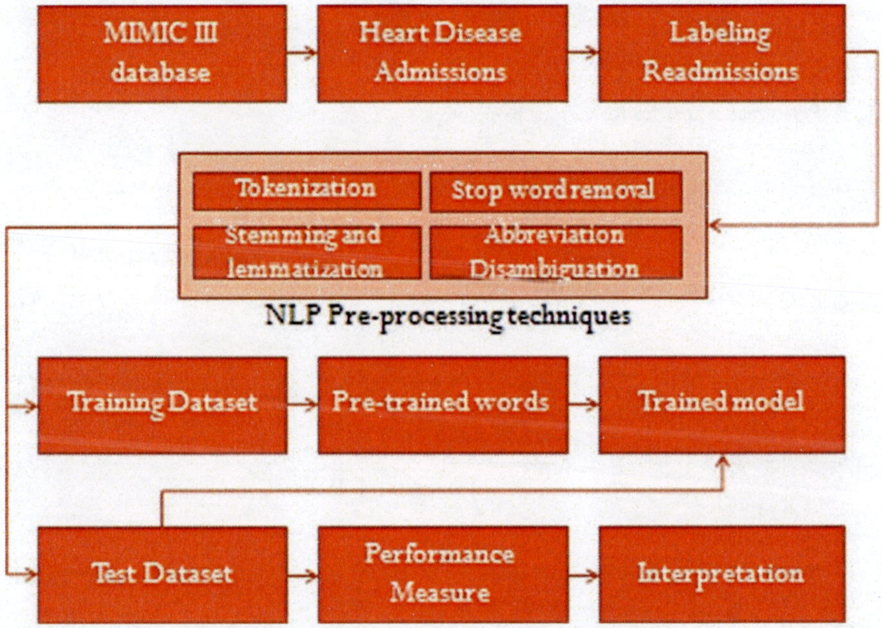

Fig. 3 Preprocessing MIMIC III dataset and analyzing the performance metrics

	No. of admissions	No. of admissions with discharge summary
Total admissions	14,040	13,746
Admission followed by readmission	3604	3543

Fig.4 Statistics of heart failure admissions

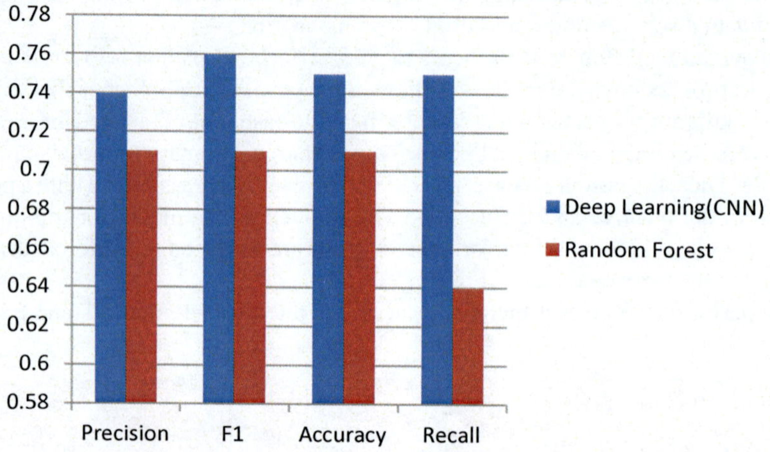

Fig. 5 Performance metrics for various parameters

$$\text{Total Accuracy} = \frac{\text{TP} + \text{TN}}{\text{TP} + \text{FP} + \text{TN} + \text{FN}} \tag{1}$$

$$\text{Precision} = \frac{\text{TP}}{\text{TP} + \text{FP}} \tag{2}$$

$$\text{Recall} = \frac{\text{TP}}{\text{TP} + \text{FN}} \tag{3}$$

$$\text{F} - \text{Score} = \frac{2^{*}\text{Precision}^{*}\text{Recall}}{\text{Precision} + \text{Recall}} \tag{4}$$

where

TP stands for true positive. True positive is defined as a condition where the system is able to classify the diagnosis correctly.

TN stands for true negative where it is a condition when the system correctly detects that the particular prediction or diagnosis does not belong to the specified group.

FP stands for false positive, where it is a condition when the system detects a particular prediction when it should not be present, i.e., the system predicts a false statement that actually should not be present.

FN stands for false negative, it is a condition where the system does not predict a certain diagnosis when actually the system should have predicted it.

5 Conclusion

In this article, the extraction of the data from the unstructured notes has been explored. These unstructured notes include clinical notes. A preprocessing technique for the extraction of words from the unstructured data is used. Initially, the investigation clusters the words as per similar meaningful words. Then, the words are ordered based on the importance of the words, the needed words are extracted, and the remaining words are discarded. Clusters have also been formed from the unstructured clinical notes through the method of topic modeling. Nursing notes are considered to be the valuable notes that consist of the most valuable information about the patients which help us in the early prediction of the disease.

6 Future Work

In the future work, the proposed topic modeling will be used in the extraction of data from unstructured data. Data generated from the integrated management of neonatal childhood and illness (IMNCI) will be used as training data, and the testing dataset will be extracted from the samples collected from the hospital. It will also include the extraction of data from a heterogeneous type of data and converted into useful information which can help the healthcare specialists in the prediction of the disease in the near future and take the needed precautionary steps to save lives.

References

1. T. Vijayakumar, R. Vinothkanna, Capsule network on font style classification. J. Artif. Intell. **2**(02), 64–76 (2020)
2. S. Manoharan, A smart image processing algorithm for text recognition information extraction and vocalization for the visually challenged. J. Innov. Image Process. (JIIP) **1**(01), 31–38 (2019)
3. L.B. Sally, R.K. Adam , R.S. Bharanidharan, Y.Y. Gordon, H. Michael , N. Shamim, Predicting mortality in critical care patients with fungemia using structured and unstructured data. pp. 1140–1148. (2019)
4. K.B. To, L.M. Napolitano, Common complications in the critically ill patient. Surgical Clinics North Amer. **92**(6), 1519–1557 (2018)
5. S.V. Desai, T.J. Law, D.M. Needham, Long-term complications of critical care. Critical Care Med. **39**(2), 371–379 (2019)

6. N.A. Halpern, S.M. Pastores, J.M. Oropello, V. Kvetan, Critical care medicine in the United States: addressing the intensivist shortage and image of the specialty. Critical Care Med. **41**(12), 2754–2761 (2017)
7. E.W. Johnson, M.M. Ghassemi, S. Nemati, K.E. Niehaus, D.A. Clifton, G.D. Clifford, Machine learning and decision support in critical care. Proc. IEEE **104**(2), 444–466 (2016)
8. O. Badawi et al., Making big data useful for health care: a summary of the inaugural MIT critical data conference. JMIR Med. Informat. **2**(2), e22 (2014)
9. C.K. Reddy, C.C. Aggarwal, *Healthcare Data Analytics*, vol. 36 (CRC Press, Boca Raton, FL, USA, 2015).
10. D. Gotz, H. Stavropoulos, J. Sun, F. Wang, ICDA: A platform for intelligent care delivery analytics. in *Proceedings AMIA Annual Symposium*, pp. 264–273. (2012)
11. Perer, J. Sun, Matrixrow: temporal network visual analytics to track symptom evolution during disease progression. in*Proceedings AMIA Annual Symposium*, pp. 716–725. (2012)
12. Y. Mao, W. Chen, Y. Chen, C. Lu, M. Kollef, T. Bailey, An integrated data mining approach to real-time clinical monitoring and deterioration warning. in *Proceedings 18th ACM SIGKDD International Conference Knowledge Discovery Data Mining*, pp. 1140–1148. (2012)
13. J. Wiens, E. Horvitz, J.V. Guttag, Patient risk strati_cation for hospital-associated C. Diff as a time-series classifcation task. in *Proceedings Advanced Neural Information Processing Systems*, pp. 467–475. (2012)
14. S. Saria, D. Koller, A. Penn, Learning individual and population level traits from clinical temporal data. in *Neural Information Processing Systems (NIPS)*, Predictive Models Personalized Med. Workshop (2019)
15. R. Dürichen, M.A.F. Pimentel, L. Clifton, A. Schweikard, D.A. Clifton, Multitask Gaussian processes for multivariate physiological time-series analysis. IEEE Trans. Biomed. Eng. **62**(1), 314–322 (2015)
16. M. Ghassemi et al., A multivariate timeseries modeling approach to severity of illness assessment and forecasting in ICU with sparse, heterogeneous clinical data. in *Proceedings AAAI Conference Artificial Intelligence*, pp. 446–453. (2015)
17. H.V. Batal, G.F. Cooper, M. Hauskrecht, A pattern mining approach for classifying multivariate temporal data.in *Proceedings IEEE International Conference Bioinformatics Biomedicine* (BIBM), pp. 358–365. (2011)
18. T.A. Lasko, Effcient inference of Gaussian-process-modulated renewal processes with application to medical event data. in *Proceedings Uncertainty Artificial Intelligence*, pp. 469–476. (2014)
19. L.C. Barajas, R. Akella, Dynamically modeling patient's health state from electronic medical records: a time series approach. in *Proceedings 21st ACM SIGKDD International Conference Knowledge Discovery Data Mining*, pp. 69–78. (2015)
20. X. Wang, D. Sontag, F. Wang, Unsupervised learning of disease progression models. in *Proceedings 20th ACM SIGKDD International Conference Knowledge Discovery Data Mining*, pp. 85–94 (2014)
21. M.J. Cohen, A.D. Grossman, D. Morabito, M.M. Knudson, J. Butte, G.T. Manley, Identifcation of complex metabolic states in critically injured patients using bioinformatic cluster analysis. Crit. Care **14**(1), 1 (2010)
22. J.L. Zhou, V.A. Narayan, J. Ye, Modeling disease progression via fused sparse group lasso.in *Proceedings 18th ACM SIGKDD International Conference Knowledge Discovery Data Mining*, pp. 1095–1103. (2012)
23. E. Choi, N. Du, R. Chen, L. Song, J. Sun, Constructing disease network and temporal progression model via context-sensitive hawkes process. in *Proceedings IEEE International Conference Data Mining (ICDM)*, pp. 721–726. (2015)
24. R. Pivovarov, A.J. Perotte, E. Grave, J. Angiolillo, C.H. Wiggins, N. Elhadad, Learning probabilistic phenotypes from heterogeneous EHR data. J. Biomed. Informat. **58**, 156–165 (2015)
25. https://www.who.int/publications/i/item/9789241506823

A Study of Blockchain for Secure Smart Contract

Jitendra Sharma and Jigyasu Dubey

Abstract Blockchain is a novel and effective technology, which renders the implementation capability of various cases like public services, economic systems, Internet of Things, internet interaction systems, and smart contract applications. Smart contract works in such a way that the computer works as a third party, and the other two senders and receivers are those who digitally agree to the obligations and rights. Blockchain keeps the record of the contract which is executed by the network's distributed nodes, where the third party requirement is eliminated as Blockchain decentralized in nature. There are still multiple security and privacy issues with the smart contract that imposes a huge threat to its continued effectiveness and growing esteem and recognition. In this article, the issues of transactions in smart contracts were discussed and different tools were analyzed to overcome these issues.

Keywords Blockchain · Smart contract · Security · Privacy

1 Introduction

In the search for more effective and reliable software products, the modern advancements in the technological arena are frequent in bringing in novel technologies. The introduction of one such novel and effective technological discovery is Blockchain, which renders the implementation capability of various cases like public services, economic systems, Internet of Things, and Internet interaction systems. The industry and academia have come to recognize Blockchain Technology [1]. In Blockchain Technology, cryptography is utilized to make a transfer of digitalized currency/messages, in hosting applications in a secured and protected manner and to store data [2]. These are the reason that makes Blockchain a completely-distributed-public-ledger and a Peer-to-Peer (P2P) platform. The Blockchain invention is considered remarkable in the world of INTERNET because it can make faster processes, lessen the cost of transactions, and ultimately bring chaos in many sectors [3].

J. Sharma (✉) · J. Dubey
Shri Vaishnav Vidyapeeth Vishwavidyalaya, Indore, India
e-mail: js15052013@gmail.com

V. Suma et al. (eds.), *Inventive Systems and Control*,
Lecture Notes in Networks and Systems 204,
https://doi.org/10.1007/978-981-16-1395-1_45

Blockchain technology is mainly used for cryptocurrencies; however, it may be useful in applications in which virtual records maintain required like government-issued certificates, medical procedures, deeds of ownership, educational certificates, election, Smart contracts, etc. [4]. Researches attracted to the use of Blockchain in Smart Contract applications. Smart Contract works in such a way that the computer works as a third party and the other two are those who digitally agree to the obligations and rights. Smart contracts are programmed in unique codes. The variables are ordered as codes to trigger various events. On agreeing to the terms and conditions the contract is signed digitally between parties. After this, the Blockchain keeps the record of the contract and is executed by the network's distributed nodes, where the third party requirement is eliminated. These contracts are imposed on parties and they will be carried out just as they were coded [4]. Blockchain ledger is transparent, facilitates variables that are required for the contract. It triggers events in the smart contract like making a payment and unlocking the doors for the Airbnb house [4]. Smart contracts have different issues like codifying, security, privacy, and performance. The codifying issue is related to the development of smart contracts. Security issues are related to bugs or vulnerabilities that an adversary might utilize to attack. The privacy issue is related to disclosing contact information in public. The performance issue is related to the scalability of the Blockchain system.

Ethereum is the common basis of the Blockchain i.e., the general-purpose programming language is understood by it. Instead of creating a different Blockchain for every single-use, application, or case, the developers are allowed to build all one in one Blockchain. Adding to this, the notion of a Turing-complete Blockchain has been introduced by Ethereum. In financial and non-financial regions, the Turing-complete programming language, which is based on Blockchain technology, brings upon an opportunity to implement some decentralized applications. These kinds of platforms are known as Smart Contracts. Because of its genuineness and novelty, it differentiates Ethereum from Bitcoin and other platforms based on Blockchain.

2 Blockchain and Smart Contracts

1. Bitcoin introduced blockchain as a cryptographic network platform [5]. Since then, blockchain has attracted both industry and academia. The blockchain is the technology behind the working of cryptocurrencies, which runs a consensus protocol to maintain the shared ledger to secure the data on the blockchain. This blockchain has four core characteristics [6, 7]. Transparent: All transactions are public and historical records can be accessed by network users.

A Smart Contract is a computerized transaction protocol that executes the terms of the contract. The bytecodes of the transaction are stored in the blockchain and visible to all the users. Once smart contracts are deployed over the blockchain, they are immutable to be modified even bugs are detected into contract code. Once the smart contract is deployed, it is identified by 160-bit hexadecimal string as the address of the contract. The execution of the smart contract depends on its code. The most popular blockchain for deploying smart contracts is Ethereum and Hyperledger fabric.

The developments by Blockchain can affect the government positively by ameliorating the abilities of taxes, control over interactions in the economic sector, the privacy policy, and even the status of the government. To find suitable smart contracts for implementation in the government services, it is required to look at the present cases of implementation beforehand. The use of smart contracts in government services is discussed in this research [8]. The implementation is categorized into six divisions: Supply-chain, Identification, Voting, Ownership, Financial, and Permit application.

3 Literature Review

The term smart contract was coined by Nick Szabo in 1994. He defined a smart contract as a computerized transaction protocol that executes the terms and conditions of a contract.

The explanation of the fundamental of the technology of Blockchain is discussed: a ledger of a distributed type that keeps the transaction records. This is Public-Blockchain because they are available for everyone to view and verify. For interacting within the network, there is no need to take any permission. So it is permissionless [4]. There is one more form of this sort called Private-Blockchain. In this, the transaction records are arranged and only open to permission.

The Blockchain is a distributed database that records transactions that occurred in the Blockchain network. Cryptography is the root of Blockchain technology. The Blockchain concept is behind Bitcoin. It describes that every single party should have two different keys: a private key and a public key. The private key is the password, whereas the public key is the account number. The basic overview of Bitcoin is shown in Fig. 1. The private key is used whenever the transaction has to be made sent to the server of Time-stamp for endorsement. With many transactions, a kind of block is filled. Cryptographically, these blocks are hashed, just like the other previous blocks. The history of all the transactions is correct, as proved by this hash. The chances of forfeiting history are very thin because of the hash functions of cryptography [9, 10]. Summarizing, in Bitcoin technology, the transaction can only be completed via private key, the dual-spending is restricted, and the records of all the history is available [9].

The verification via decentralized-network is one of the most significant characteristics of the Blockchain. A person having a computer can contribute as a miner in the network. The miners verify the transactions and approve the correct history of blocks [9]. The formula of mathematics is applied to achieve the above task. On average, this formula solves what appears like a puzzle every 10 min. The average time of solving is retained to adapt to the formula difficulties when the network's computing power is augmented. These formulas are pretty hard in solving while

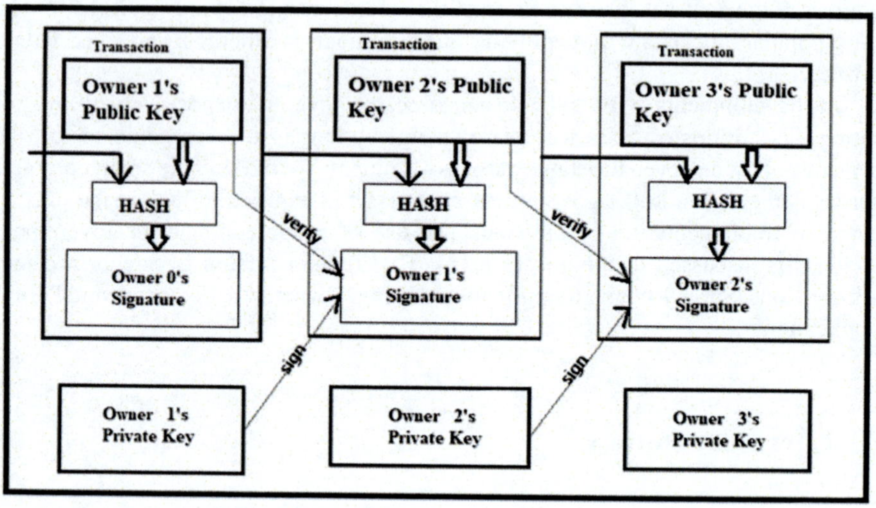

Fig. 1 Hashing scheme of Bitcoin [9]

pretty easy to be verified. One can achieve several Bitcoin as rewards when he solves the formula to create a new block. This procedure is known as Mining. The power of computation and electricity costs are very high because there are many systems involved in solving the puzzle. This type of method is called Proof of Work (PoW) [4].

The counterparty is not required to be trusted while making any transactions because the verification is done by the network. Therefore, the word "Trustless" is used for the Blockchain. A third party, like, Bank is not required in these processes. Other than [11] of Work, Proof of Stake also exists [6] and BFT Replication [7]. There is no discussion of these methods in this research as they are dealt in the technical level.

The transaction history is recorded and verified by the decentralized network. All the nodes in the network are equally important because transmission of information occurs in all the nodes that are participating [12] called this peer-to-peer transmission. To be accurate, the hash requires the whole history. Hence, the complete database must be kept as possession by the hash whenever novel blocks are mined [12]. The hacking of the database becomes impossible as all the nodes that are participating have the complete copy [4].

As an attribute, transparency is brought by the distributed datasets. Because of the Public Blockchain, which explains that each node possesses the block chain's copy. All the transactions are visible to everyone, and recent histories are shared as to whom and where the transactions did take place [11]. Thus, there is transparency and the pseudonyms of the address protect the transparencies. Each user or node is assigned with an address of the Blockchain. These addresses contain 30 or above alphanumeric characters. This provides the privacy of the address, but in case if it is required, the identity can be proved by transacting [12]. In conclusion, the transparency can be provided to the transactions, but the pseudonyms hold the users private and protected.

The Blockchain data is immutable because of the distributed database and the fundamentals of cryptography. By the utilization of the preceding blocks, hashing of novel blocks is done. All the used blocks must be corrected to produce novel blocks. If someone indulges in the Bitcoin theft, he has to rephrase the whole history of the transaction, which is actually unattainable [4]. It is not what happens if someone attempts to alter the complete record. The pioneering Bitcoin of the Blockchain technology was unhackable [13]. Therefore, the Blockchain records are thought to be totally immutable.

There are some securities issues present on the Blockchain. 51% of attacks are considered very dangerous. When nodes in the network vote 51 or more percentage maliciously, there are chances that the transactions are falsely marked as correct, and the history is altered [11]. One more major issue of security is that the applications sometimes have bugs because of the newly applied characters. Rather than just these bugs, the hackers eye the individual's cryptocurrencies.

The transparency of the ledger is the main feature of Blockchain. As many of the details of transactions are open to all, the issues of privacy are always implied, e.g., the transfer details of cryptocurrency are visible as time, and the address of both sender and recipient is made public [3]. There are also a handful of tools of Flow-Control that are available to show the transactions which are done in the past. Ref. [14] researched that the identities of various users' accounts are obtainable. The Wiki-leaks organization has first done this. The addresses of owners were mentioned on the website of Wiki-leaks to receive the donations. Because the transfer of Bitcoin amounts is shown by the Blockchain. So, it became probable that some people could merge their identities with some contributors' addresses.

It has been 2008 since the Blockchain technology's development. The applications that work on the Blockchains are hacked, but the Blockchain as a technology in itself is free of such accusations. The guidance of Industry standards is required to be guided in designing safer and adoptable technologies [15]. The standardization in the Blockchain technology was requested by the ISO (International Organization of Standardization) in Australia [16]. Along with 25 members, the ISO is working on the development of the Blockchain standards. The members are from various other countries like France, Brazil, Germany, Netherlands, United Kingdom, Russia, the United States, Japan, China, and Australia. The code for this standard is ISO/TC 307, and it is in the preliminary stage (ISO, n.d.).

There are so many smart contracts that are completely reliant upon the values of external affairs, e.g., the Contract of betting depends upon the results of a football match. The inputs of the smart contracts are given by a source called Oracle [17]. In the example of football betting, *n* Oracle can be considered as the organization that released the outcomes of a match. In oracle, the third party is involved as an intermediate, which is against the rules of Blockchain [18]. The illustration of a smart contract using an example of "Financial-swap" is given in Fig. 2.

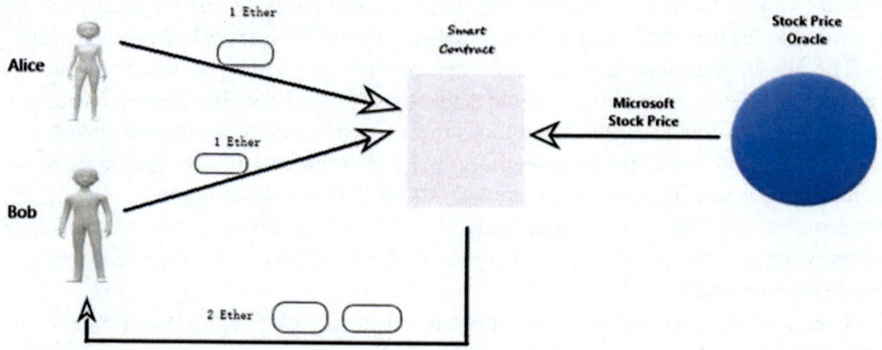

Fig. 2 Example of smart contract

The smart contract entails an association with traditional-contracts. The term contract is defined as "an agreement that is concerned with tenancy, sales, or employment, which is protected under law and order" [19]. These definitions differ in other countries, e.g., following Dutch law, this term expresses that an agreement is legitimate when two criteria meet.

This article [20] gives a new rise to the use of technologies like IoT, Cloud Computing, and Blockchain in the healthcare system as it is very useful in circumstances like Covid-19. The vital signs and physiological parameters are sensed and transmitted to provide safe medical assistance to the patients. The various IoT data transmitted over the network using blockchain in the healthcare system. The user can interact using the web application to get the data securely.

In this article [21] author provides a detailed description of working and platforms on which smart contracts will deploy. The authors proposed a six-layer framework for the working of the smart contract. The layers are Application Layer, Manifestations Layer, Intelligence Layer, Operations Layer, Contracts Layer, and Infrastructure layer. The vulnerabilities in the contract appear on the contacts layers. Malicious miners can exploit them to gain profit.

This article [22] provides different details about the blockchain. Blockchain becomes a new tool for securing transactions without a third party. Blockchain may be used in both financial and non-financial industries for various purposes like securing the medical health record of the patient, government contract, digital documents, It posses low trading cost, exchange of information securely which also increase the popularity of the use of blockchain.

4 Conclusion

The smart contract and Blockchain play a vital role in changing the current technology and trends. The main aim of our study is to overcome the issues of smart Contracts like codifying, security, privacy, and performance. The vulnerabilities in the smart contract may give increase to different issues related to security and privacy. These issues and related tools were currently analyzed in use to overcome these influences. In the future, a mechanism is built to resolve different vulnerabilities as data feeds, transaction ordering dependencies, privacy of data feeds for Smart contract.

References

1. Z. Zheng, S. Xie, H.-N. Dai, H. Wang, Blockchain challenges and opportunities: a survey. Int. J. Web Grid Serv. (2017)
2. C. Dannen, *Introducing Ethereum and Solidity: Foundations of Cryptocurrency and Blockchain Programming for Beginners*, 1st edn. (Apress, Berkely, CA, USA, 2017). https://doi.org/10.1007/978-1-4842-2535-6
3. D. Drescher, *Blockchain Basics: A Non-Technical Introduction in 25 Steps* (Apress, New York, 2017). ISBN: 9781484226049
4. D. Tapscott, A. Tapscott, *Blockchain Revolution: How the Technology Behind Bitcoin Is Changing Money, Business, and the World* (Penguin, New York, 2016)
5. D. Amo, D. Fonseca, M. Alier, Personal data broker instead of blockchain for students' data privacy assurance, in *WorldCIST'19, AISC 932* (Springer, 2019), pp. 371–380
6. K. Sultan, U. Ruhi, R. Lakhani, Conceptualizing blockchains: characteristics and applications (2018)
7. D. Puthal, N. Malik, S.P. Mohanty, E. Kougianos, G. Das, Everything you wanted to know about the blockchain: its promise, components, processes, and problems. IEEE Consum. Electron. Mag. **7**, 6–14 (2018)
8. M. Atzori, Blockchain technology and decentralized governance: is the state still necessary? Unpublished manuscript (2015) https://papers.ssrn.com/sol3/papers.cfm?abstract_id=2709713
9. S. Nakamoto, Bitcoin: a peer-to-peer electronic cash system. bitcoin.org (2008)
10. A. Kosba, A. Miller, E. Shi, Z. Wen, Hawk: the blockchain model of cryptography and privacy-preserving smart contracts, in *IEEE Symposium on Security and Privacy* (2016), pp. 1–20
11. M. Swan, *Blockchain: Blueprint for a New Economy* (O'Reilly Media, 2015)
12. M. Iansiti, K.R. Lakhani, The truth about blockchain. Harvard Bus. Rev. **95**(1), 118–127 (2017)
13. P. Vanini, Digital disruption. Unpublished manuscript, University of Basel and Swissquant Group. Retrieved online on the 4 March 2018 from https://papers.ssrn.com/sol3/papers.cfm?abstract_ id = 2988962
14. J. Barcelo, User privacy in the public bitcoin blockchain. J. Latex Class Files **6**(1) (2017)
15. C. Natoli, V. Gramoli, The blockchain anomaly (18 May 2016), pp. 1–11. http://www.ibtimes.co.uk/r3-connects-11-banksdistributed-ledger-using-ethereum-microsoft-azure-1539044.arXiv:1605.05438v1[cs.DC]
16. F. Zhang, Town crier: an authenticated data feed for smart contracts (ACM 2016), pp. 1–20
17. K. Silverberg, C. French, D. Ferenzy, V.D.S. Berg, *Getting Smart: Contracts on the Blockchain* (Institute of International Finance)
18. K. Delmolino, M. Arnett, A. Kosba, A. Miller, E. Shi, Step by step towards creating a safe smart contract: lessons and insights from a cryptocurrency lab, in *International Conference on Financial Cryptography and Data Security* (Springer, Berlin, 22–26 Feb 2016), pp. 79–94
19. L. Hooghe, G. Marks, Unraveling the central state, but how types of multi-level governance. Am. Polit. Sci. Rev. **97**(2), 233–243
20. H. Wang, IoT based clinical sensor data management and transfer using blockchain technology. J. ISMAC **2**(3), 154–159 (2020)

21. S. Wang, L. Ouyang, Y. Yuan, X. Ni, X. Han, F. Wang, Blockchain-enabled smart contracts: architecture, applications, and future trends. IEEE Tran. Syst. Man Cybern. Syst. **49**(11), 2266–2277 (2019)
22. B. Vivekanadam, Analysis of recent trend and applications in block chain technology. J. ISMAC **02**(04), 200–206 (2020)

Ensemble-Based Stegomalware Detection System for Hidden Ransomware Attack

A. Monika and R. Eswari

Abstract Ransomware analysis creates crucial aspects for cybersecurity defense mechanisms. Recently, attackers are confusing the parts of ransomware attack code using steganography to carry malicious JavaScript and even deliver cryptominers. It can be used to understand a system, to leak sensitive information, and to run a command and control channel without being detected by any anti-malware agents. It means a cyberattack can operate without detection, bypassing all defenses that are based on detection or analysis. Many approaches have been used and most of them include the analysis of the portable executable (PE) file of the malware executable which are injected directly into the target system. However, all of these existing systems fail to detect the ransomware when they are obfuscated. In this article, a stegomalware detection system is proposed to find the presence of hidden payloads from input images, to extract the hidden payloads, and to verify whether the hidden payload is the subject of stegoransomware attack. The proposed system uses three phases such as hidden payload detection, hidden payload extraction, and classification to detect the stegoware. Ant colony optimization algorithm is used in the detection phase to find an optimal subset of features to detect obfuscation. Stegorepository images are decoded to extract the items hidden inside the input steganographically. A *two-stage ensemble classifier* is used to process the extracted payloads. Initially, a binary classifier is used to find whether the extracted payload is a benign or malignant file. If it is found as malignant, the exact percentage of its malicious activity that it can provide to the target host system is further calculated using a fuzzy C-means clustering algorithm. Simulation observations describe that the proposed system drastically reduces dimensionality and significantly improves classification accuracy.

Keywords Stegoware · Stegoimages · Malware · Fuzzy C-means clustering

---

A. Monika (✉) · R. Eswari
Department of Computer Applications, National Institute of Technology, Tiruchirappalli, Tamilnadu, India

R. Eswari
e-mail: eswari@nitt.edu

V. Suma et al. (eds.), *Inventive Systems and Control*,
Lecture Notes in Networks and Systems 204,
https://doi.org/10.1007/978-981-16-1395-1_46

1 Introduction

Steganography is used to conceal, even without revealing that a message is being transmitted between entities. For camouflage, the transmission steganography uses cover medium like image, an audio file, video file, or any other digital medium that contains noise. The original message is submerged into the cover medium such that it is unnoticeable to anyone viewing the cover file. *Steganalysis* [1] is to become aware of or guess potentially concealed information from observed data with slight or no facts regarding the steganography algorithm or its variables. Though it is needed to seek a realistically good steganalysis method for each steganographic algorithm, the extensive goal is to create a steganalysis framework that can function competently for a set of steganography methods.

Malware [2] means a piece of malicious program or code that can provide an attacker direct authorization on the target host. It is a vast term, which indicates all kinds of malicious programs inclusive of virus, bug, worm, bot, rootkit, spyware, adware, Trojans, and even ransomware. It would act as an inmost envoy which can install malicious codes on the target host system or trick us to load some malfunctioning codes or program either through malware electronic mail attachments or Web-based communication applications. As a consequence, the attacker can confine our system and the system will not acknowledge our instructions anymore. *Ransomware* is a subset of malware, fairly most hazardous, which is build to target individuals or organizations. They are kinds of malware that can authorize the attackers to acquire exhausted access control of our target host and restrict our own access to the personal and confidential files unless a ransom is paid. They are considered a huge complex than systematic malware and can represent themselves in one or many ways.

Stegomalware, also called stegware or stegoware [3], is one of the types among *information hiding malware attacks that uses steganography to conceal their presence from any modern malware detection system,* and it is used to hide malware in images or any digital medium. Once the malware embedded image is loaded on the victim side, an attacker can execute this malware remotely and potentially steal information from the end user. Because of its nature, stegware cannot be detected by usual methods.

*Cerbe*r conceals malware code inside image files. *DNSChange*r utilizes PNG LSBs to conceal malicious AES encryption key. *Stegano* is a PNG formatted banner ad consists of malware code. *Sundown,* white PNG files are utilized to hide code or depart user data. *SyncCrypt* ransomware obscures part of its core code inside image files. *Zbot* hides the data to the end part of JPEG files containing appended data. ZeroT is a Chinese malicious program that can appends malware in an image of Britney Spears utilizing steganography. *Twitoor. A* hides encrypted malicious JavaScript code in images, text, and HTML code and trick the user to install the malicious app and spreading infection. *SpyNote Trojan* tricks users to install the malicious app to gain access to confidential data.

These examples show up how malicious data can be innovatively concealed inside any multimedia files. This may tend to even more methods and processes that these files can be utilized for malicious purposes. Most existing antiviruses and intrusion detection systems (IDSs) fail to comprise the provision of exercising both steganalysis and file signature comparison with a virus signature database on various multimedia files. It would also be noticed that the currently available effective techniques to recognize malicious programs incorporate the above-mentioned IDS and antiviruses. Therefore, this brings about the need and motivation for the system being proposed in this paper. The proposed design is described by considering the following scenario as a premise. In this case, an attacker would have embedded a malicious program in any type of file utilizing steganography, transfer it over to the target host, and then execute the malicious program.

Hence, an application is needed which could be able to examine various multimedia files, fetch any data that is hidden maliciously and should be utilized to counter against the effects of exploit scenario which would have just been described and eventually spotlight the need for its use in future. In this paper, a stemware detection system is proposed to find the presence of hidden payloads from image multimedia file, to extract the hidden payloads and to verify whether the hidden payload is the subject of information hidden ransomware (stegoransomware) attack. Further, the percentage of malicious activity that the hidden payload can provide to the target host system is also calculated by the proposed system. The rest of the article is organized as follows: Sect. 2 presents the related works and highlights the major challenges. Section 3 describes the proposed stegomalware detection system. Experimental settings, datasets used, and evaluation metrics are discussed in Sect. 4 and the experimental results of the proposed stegoware system with comparative existing system studies are discussed in Sect. 5. Finally, Sect. 6 concludes the work with future enhancements.

2 Related Work

Cohen et al. [4] have proposed MalJPEG, and it is the first machine learning (ML) depended on solution designed precisely for the effective analyses of unknown malicious JPEG images. To discriminate between benign and malicious JPEG images, MalJPEG inertly extricate 10 simple yet discerning features from the JPEG file structure and leverages them with a machine learning classifier. The authors have only addressed JPEG image format, and other formats like PNG, GIF, etc., are not addressed. Their proposed system only can find whether the images are clean or not, but failed to classify whether the hidden item inside the image is normal or malware repository. Blindly classified an image as MalJPEG, even a common text "HAI" is hidden inside the image. George et al. [5] designed an application that partially acts as a steganalysis technique to scan, recognize, and analyze the obfuscation of a payload in either on single or on a group of selected images. But the decoding method for hidden payloads is not discussed. Their system classifies the hidden payload as malicious if any PE file format is found in hidden payloads. If the hidden payload contains any legitimate PE file format, this system will give true negative results.

Dharman and Thomas [6] proposed a steganalysis application, which can analyze the image for malicious code and retrieve the embedded data. But, if the hidden payload contains any legitimate PE file format, this system will give true negative results. de Jesus Serrano Perez et al. [7] designed a steganalysis technique using artificial immune system (AIS), to analyze JPEG images obfuscated with three advanced steganographic tools: F5, Outguess, or Steghide. They have used Haar wavelets to extricate feature vectors, which highly describes the analyzed image due to its fast calculation and information synthesis characteristics. The author has used the feature extraction method to improve processing speed, but performance decreases due to its domain-specific nature, that is, only the JPEG image format is concentrated and fails to detect hidden payloads inside JPEG images that are modified by other techniques or tools apart from above-mentioned three steganographic tools. Choudhury et al. [3] proposed a system where the end user would receive a showering mechanism, which will destroy the malware and receive the only innocent or cleaned image. Their system does not check whether the hidden payload is proposed model fails to detect the malicious of the format information if obfuscated, that is, achieved file signature detection and act up in obfuscation analyses. Cohen and Nissim [8] proposed a methodology for trusted detection of ransomware in virtual servers on an organization's private cloud. Conducted trusted analysis of volatile memory dumps taken from a virtual machine, using volatility framework and created general descriptive meat features. This model fails to cover obfuscation analysis and can counter-effective only direct attacks, act upon information hidden attacks.

From these reviews of literature, it is visible to know that the following issues exist in the design of the stegomalware detector system: (1) The existing stegomalware detection application for images finds whether there is any hidden payload or not. But, it is failed to detect whether the hidden payload is malicious or not; (2) some applications analyze the hidden payload and blindly result in it as malicious if the hidden payload is a PE file. They failed to classify the hidden PE files are legitimate or not; (3) existing steganalysis methods are efficient in finding hidden payloads, but failed to detect whether the hidden payload is legitimate or malicious; (4) stegodestructor applications, which blindly destroy the hidden items inside the image. This will affect the usage of steganography against the attacks like packet sniffing. (5) Most of the existing either cover malware detection or file signature analysis, but *fails to comprise the provision of exercising both steganalysis and file signature comparison with a virus signature database.*

An ensemble-based stegomalware detection system for hidden ransomware attack (stegoransomware attack) is proposed in this article in which it addresses all the above-mentioned issues and to overcome them.

3 Proposed Work

A stegomalware detection system is proposed in this paper for detecting the obfuscated ransomware from image cover medium and further to find the exact malicious activity that can be given by that payload to the target host systems. The proposed system has the following objectives: (1) creation of stegorepository detection for image cover medium; (2) development of efficient decoding methods for image cover medium; (3) classification of hidden payloads whether is a subject of stegoransomware attack; and (4) calculation of percentage of malicious that the ransomware can provide to the target host system.

3.1 System Model

The proposed system model is shown in Fig. 1. This paper first finds out the presence of any camouflage data in an image. If any hidden component is found, then the system proceeds to extricate and examine the hidden data, which can be a malicious program. This system has four phases: (1) hidden payload detection; (2) hidden payload extraction; (3) hidden payload classification; and (4) calculation of malicious activity given by the payload.

3.1.1 Hidden Payload Detection

In this phase, the incoming images are tested whether it is clean or containing any hidden component. *Blind or universal steganalysis* type is applied to detect the embedded messages. This phase uses the following steps to detect hidden payloads.

A ***Feature extraction***

A feature extraction algorithm is applied to extract features of input images. Multimedia objects like images are classified into two classes depending on whether they contain any concealed data in them or not. If there is no concealed data in them, then they are classified as cover images and if images contain hidden data inside them, then they are classified as stegoimages. In this process, pertinent features that are subtle to the process of data obfuscation are extracted from input images. That is

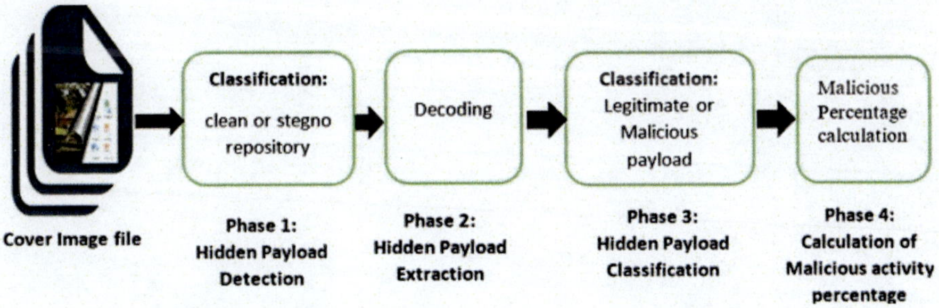

Fig. 1 Stegomalware detection system

to say that the features of images having no obfuscated data inside them would be totally distinct from the images containing secret data inside them.

For image-steganalysis, numerous algorithms have been proposed for feature extraction to classify the input images. In this article, Subtractive Pixel Adjacency Model (SPAM) algorithm is utilized and extracted 686 features from input images as given in algorithm 1. The algorithm first computes the transition probabilities along with eight directions: horizontal left–right, horizontal right–left, vertical bottom top, vertical top–bottom, diagonal left–right, diagonal right–left, minor diagonal left–right, and minor diagonal right–left.

Further for decreasing the feature dimensionality, a plausible assumption is used where the information in normal cover images is consistent concerning mirroring and flipping (the consequence of portrait or landscape orientation is negligible). Therefore, the horizontal and vertical matrices are separately averaged and then the diagonal matrices are averaged to form the final extracted feature set, Extracted_Features.

B *Feature selection*

Extracted features are optimized using a feature selection algorithm. An optimal subset of features is selected from the extracted 686 features using ant colony optimization (ACO) algorithm [9]. For the given original feature set (*Extracted_Features containing n features*), a subset S is found which contains m features ($m < n$), $S \subset$ Extracted_Features. Algorithm 1 explains the steps of feature selection by digraph ant traversal, and the ACO algorithm for this feature selection has been proposed as a discrete search space represented by a digraph (as shown in Fig. 2) with simply $O(n)$ arc, where the nodes in the digraph indicate the extracted features, and the edges joining two adjacent nodes represents the selection probability of the next feature.

Here, the extracted features (n features) are represented as $f1, f2\ldots, fn$, where the ith node fi indicates the feature fi. An auxiliary node $f0$ is placed at the inception of the digraph on which each ant starts its traverse. The ant travels on the digraph from $f0$ to $f1$, and then to $f2$, and so on. The ant completes its explore and outputs this feature subset as it arrives at the last node fn. When an ant finishes its tour from $f0$ to fn, the traversed arcs on its trace form a solution and used two edges named $S0$ and $R1$ linking two adjacent nodes $fi-1$ and fi. If an artificial ant at fi selects the

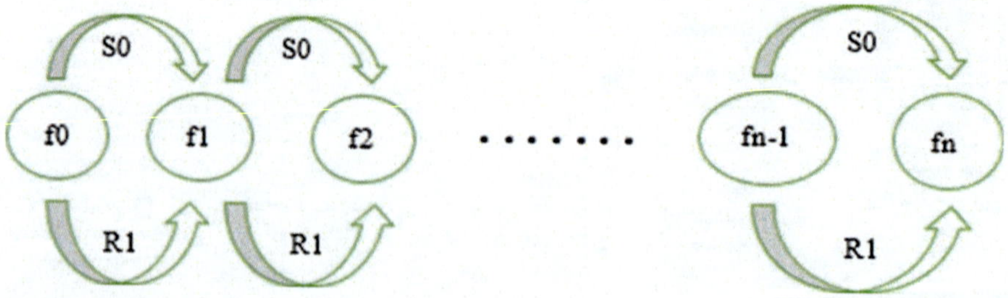

Fig. 2 Digraph of ant traversal

edge $S0$, the ith feature will be selected or if the ant selects the edge $R1$ then the feature will be rejected. On each edge $S0$ or $R1$, virtual pheromone intensity τ_i^j is empowered as the feedback information to direct the ants traversing on the digraph. The pheromone matrix τ is composed as $\tau_i^j = 1$ for all $i = 1, 2, n$, and $j = 0, 1$.

The prospect of an ant at node $fi - 1$ to choose the edge $S0$ or $R1$ at time t is given in the following Eq. 1.

$$p_i^j(t) = \frac{\left[\tau_i^j(t)\right]^\alpha (\eta_i^j)^\beta}{\left[\tau_i^{S0}(t)\right]^\alpha (\eta_i^{S0})^\beta + \left[\tau_i^{R1}(t)\right]^\alpha (\eta_i^{R1})^\beta} \quad (i = 1, 2, 3, \ldots, n;\ j = 0, 1) \tag{1}$$

Here, τ_i^j (t) represents the pheromone concentration in the edge $S0$ or $R1$ amid nodes $fi - 1$ and fi at time t, which implies the potential tendency for ants to follow the edge $S0$ or $R1$. η_i^j is the heuristic details representing the desirability of choosing edge $S0$ or $R1$. α and β are two parameters that indicate the relative importance of the pheromone and the heuristic information. From Eq. (1) it can be observed that the transition probability utilized by ACO based on the pheromone intensity τ_i^j (t) and heuristic information η_i^j. The value of η_i^j is measured using information gain (IG), which is an efficient measurement to evaluate the discrimination value of feature fi. In every iteration, the algorithm ACOIG updates the pheromone intensity on each edge according to the pheromone and heuristic information on the edge. Obviously, if an ant selects the edge $S0$ or $R1$, pheromone intensity on that edge is increased, and ants must select that edge with an elevated prospect in the succeeding iterations. This creates positive feedback of the pheromone system. On each iteration, the pheromone intensity of each arc is upgraded according to the Eqs. (2) and (3).

$$\tau_i^j(t+1) = \rho \cdot \tau_i^j(t) + \Delta\tau_i^j \Delta(t) \tag{2}$$

$$\Delta\tau_i^j \Delta(t) = \frac{1}{\left|S_i^j\right|} \sum \mathrm{IG}(fi) \tag{3}$$

where S_i^j is the set of solutions generated at the ith iteration passing through $S0$ or $R1$. Based on the solution from the ants, the fitness function of the selected features subset is evaluated by classifying the training datasets.

C *Classification*

Machine learning algorithms are utilized to classify the images based on the selected features. A decision tree algorithm is used to classify the images into clean and stegoimages. This classifier is considered nonparametric, as they would not need any presumptions about the grouping of the variables in each class. Each inner node consists of a decision parameter which hangs on one feature. For the first iteration, selected features are split into parts. The feature with the highest relevance is found using some splitting criteria (e.g., information gain) and act as a root node for initial splitting. This procedure will be recursively repeated for each subset until no more splitting is possible. It is processed from root to leaf node and corresponds to a rule-based classifier. A decision tree is created from training datasets and contains objects that are completely delineated by a set of attributes and a class label. The class which is correlated with the leaf is the output of the tree.

Algorithm 1: Hidden_Payload_Detect()

Input: Images
Output: Classified images: Clean image or Stego image.

```
for each input image
    Find transition probabilities along eight directions
    Feature_set1 ← Average of horizontal and vertical matrices
    Feature_set2 ← Average of diagonal matrices
    Extracted_Features ← [Feature_set1 : Feaure_set2]
    f0 ← start_node
    node f_i ← Extracted_features_i, ∀ i = 1 to n (number of extracted features)
    feature_selected[0] ← arc_S0
    feature_rejected[0] ← arc_R1
    for (k = 1 to number_of_ants)
            ant k chooses start node
            while (solution not constructed)
                ant k traverse from f0 to fn via arc_S0 or arc_R1 using eqn.1
                if (node f_i reached via S0)
                        feature_selected[k] ← f_i
                else
                        feature_rejected [k] ← f_i
                end if
            end while
    end for
    Update pheromone on the arc using eqn.2
    Classification_result = Decision_Tree(feature_selected)
    if Classification_result =1
                Image is StegoImage
    else
                Image is CleanImage
    end if
end for
```

3.1.2 Hidden Payload Extraction

In this phase, the hidden payload is extracted from stegoimages which are already classified in the first phase. Decoding is utilized for getting the hidden information in an image file. It gets an image file as input and can store two files at the destination folder, one is the given image file and another is the payload file. Each byte of data that is hidden inside the image is converted into 8-bit binary code using ASCII values. As each image is composed of pixels and each pixel consists of three values (red, green, blue), the first eight values of three pixels are read from right to left in reverse order. The modulus 2 of each value is calculated. If the modulus value is "0" then the hidden extract is 0 else extract is "1". Every 8 bits of extracted binary bits represent a hidden payload.

3.1.3 Hidden Payload Classification

Malicious executable files, such as a portable executable (.exe, .mui, .sys, .acm, .ax, .cpl, .dll, .drv, .efi, .ocx, .scr, .sys, .tsp), are commonly used formats for ransomware attack. The classifier of malicious execuatables in this work comprise of two main common modules: (1) feature extraction and feature selection of PE payload and (2) classification of payload.

A. ***Feature extraction and feature selection of payload***
   The features of executable payload files are extracted and the important features are selected from the extracted features using the filter method and wrapper method given in [2] to refine the unnecessary features or features comprising low classification efficiency. The selected features which differentiate legitimate and malicious executable files are shown in Table [1]. Malware files differ from the normal file concerning these selected features for the above-mentioned types of PE formats.
B. ***Classification of payload***
   A training set of legitimate software and malicious software is given to the ***binary classifier (BC)***. As a consequence, the labeled dataset (training) and the testing set provide input for a classification algorithm. This binary classifier is trained using neural networks which efficiently classifies the payload files as malicious or legitimate, as shown in Algorithm 2.

3.1.4 Percentage of Malicious

If the classification results as malicious, further the selected features of the extracted payload (as shown in above tables) with their quantities in that particular file are given to ***fuzzy C-means (FCM) clustering*** algorithm to find the exact percentage of malicious activity as given in Algorithm 2. Payloads that give a high amount of malicious can be detected accurately.

Algorithm 2: BC_FCM()

Input: Hiddenpayload
Output: Percentage of malicious (U_{FCM})
//N = number of layers;
//Set N = 3;
// k ← Number of Clusters,
// Number of clusters k = 2 ; (cluster 1 = legitimate files, cluster 2 = malicious files)
// U_{FCM} ← amount to which an examination x_f associated to a cluster k
// data points ← selected feature_quantities of each malicious file
// i ← Number of data points
// m ← Fuzzy parameter (*fuzzifier*)

for each epoch
  Create N layered Neural Network
  Set Activation function ← 'relu'
  Set Optimizer ← 'adam'
  Set loss function ← binary_crossentropy
  Find accuracy using TPR
  for each hiddenpayload
    prediction_result = model.predict_classes(hiddenpayload_feature_quantities)
    if prediction_result = 0
      Hidden payload is a legitimate file
    else
      Randomly initialize fuzzy membership matric ← U_{FCM}
      for (j = 1 to k)
        Repeat
          Calculate the cluster center or centroid C_j

$$C_j = \frac{\sum_i [\mu_j(x_i)]^m x_i}{\sum_i [\mu_j(x_i)]^m} \quad (4)$$

          Calculate dissimilarity between data points and centroid D_i

$$D_i = \sqrt{(x_2 - x_1)^2 + (y_2 - y_1)^2} \quad (5)$$

          Update new membership matrix

$$\mu_{ij} = 1/\sum_{k=1}^{C} (d_{ij}/d_{ik})^{(2/m-1)} \quad (6)$$

        Until centroids are not changing
      end for
      Percentage_of_malicious == μ_{ij}(cluster 2)
    end if
  end for
end for

In the proposed FCM model, initially, two clusters are created, legitimate file cluster (*cluster 1*) and malicious file cluster (*cluster 2*). For each malicious hidden payload that is detected by the binary classifier (BC) in the previous step, the proposed FCM model aims to estimate the percentage of malicious file cluster (cluster 2). Both clusters of each file are initialized to some random value as the sum of both cluster values of each file gives 1. The values of each cluster are updated using selected feature quantities (Eqs. 4 and 5) and their amount of deviation from legitimate values (as shown in Table 1) of feature quantities (using Eq. 6.) give the exact percentage of

Table 1 List of features selected from PE files

Feature description	Quantity
DLLs referred	35
APIs referred	37
No. of the DLLs referred	2
No. of the APIs referred	2
No. of sections	2
No. of the symbols in the export table	2
No. of the items in the reloc section	2
Image optional header	18
Image data directory	35
.txt section—header field	15
.data section—header field	15
.rsrc section—header field	15
.rdata section—header field	15
.reloc section—header field	15
Resource directory table and resources	28
Total	238

malicious values. *Updated cluster 2 values give the percentage of malicious activity that is given by each malicious payload to the host system.*

4 Experimental Setup

In this section, the experimental setup for the implementation of proposed and existing systems is discussed in brief. BOSSBase dataset is used for testing the performance. Evaluation metrics and parameter settings are also discussed in this section.

4.1 Experimental Settings

The proposed model is simulated using 10,000 cover images and consistent stegoimages. The data insertion rate of malicious executable files (.exe, .mui, .sys, .acm, .ax, .cpl, .dll, .drv, .efi, .ocx, .scr, .sys, .tsp) of stegoimages in BOSSBase dataset is set to various pixel rate by using different steganography algorithms. As stated earlier, SPAM method is used for the process of feature extraction (686 features) to perform steganalysis. Following feature extraction, optimal features are selected by ant colony optimization (ACO) algorithm, where 68 features are selected as optimum set by ACO algorithm. The quality of each subset of the features gained is evaluated by classifying the training image sets utilizing a decision tree classifier. The quality of the classification results and the density of the selected features are the parameters for performance evaluation. Results were also evaluated utilizing the 686 features of SPAM which were then further minimized to 68 features by ACO. Tenfold cross-validation is used for evaluating those results. Various cover images and stegoimages (from Dataset1) are collected and their features are extracted ($n * 68$) to train the machine learning (ML) model. In the testing phase, features of the input images that are selected by the optimization algorithm are tested using the ML model. Detected stegoimages are further decoded and classified.

4.2 Parameter Settings

ACO algorithm: Various parameters that tend to greater confluence are examined and the finest parameters that are acquired using experimental evaluations are as follows: $\boldsymbol{\alpha = 0.9}$, $\boldsymbol{\beta = 0.6}$, evaporation amount $\boldsymbol{\rho = 0.988}$, the pheromone value on each arc are initialized to 1, and the quantity of a total number of ants in each iteration $k =$ 55 and the maximal iterations $\boldsymbol{m = 55}$.

Fuzzy C-means: The parameter m utilized here is an ordinal number higher than 1 ($1.0 < m < \infty$) and it represents the amount of fuzziness in the cluster, the intensity of m near to 1 provides a cluster solution that becomes increasingly identical to the solution of hard clustering such as K-means, whereas the intensity of m behind infinite value tends to complete fuzziness, therefore best confluence are tested and the better parameters which are acquired by experiments are **m = 2.0.**

4.3 BOSSBase Dataset -Description

BOSSBase dataset is used as it is a benchmark dataset for many machine learning-based steganalysis research. The BOSSBase dataset consists of 10,000 grayscale images of size 512 × 512 and the portable gray map (PGM) format. An example of one of the BOSSBase images is shown in Fig. 4.

The images were taken with seven different cameras. These images were then converted to various image formats (like .jpg, .png, etc.) with a quality factor of 95%, to be able to compare to other steganalysis techniques which are typically modeled on various image formats and resized to 256 × 256 propositions. As seen in Figs. 3 and 4, distinct steganography algorithms are utilized to encode the 10,000 images. These algorithms (HUGO, HILL, S-UNIWARD, and WOW) are considered to be some of the advanced steganography algorithms, which are hard to be detected by advanced steganalysis algorithms. For each of these steganography algorithms, five distinct concealing rates were used. The embedding rates used were 10, 20, 30, 40, and 50%, with 10% being the highest arduous to examine, and 50% implies the

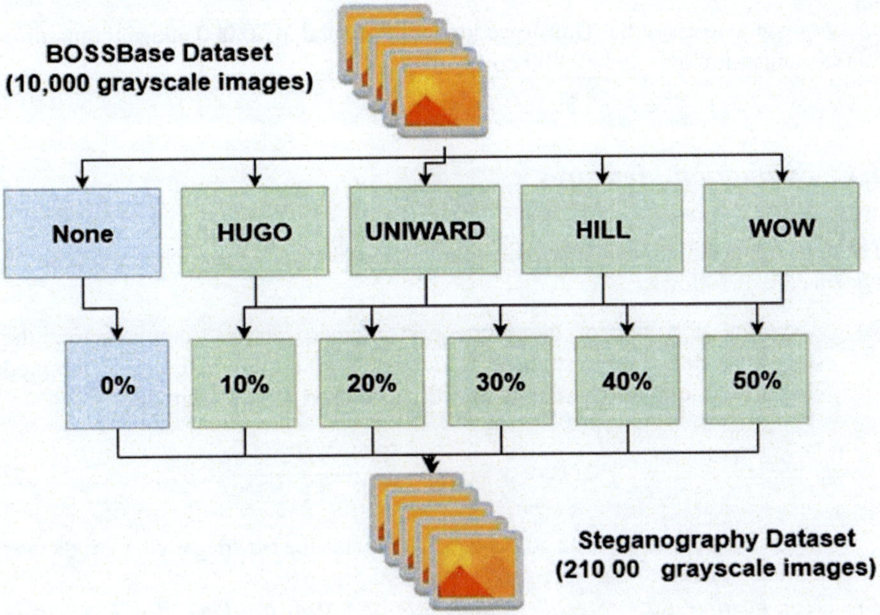

Fig. 3 Steganoimage dataset creation process

Fig. 4 Example image from BOSSBase dataset

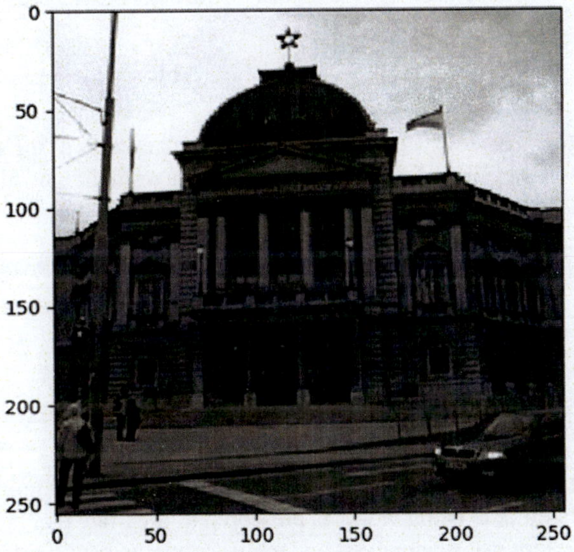

uncomplicated to examine. This procedure builds a total of 50,000 images containing 30,000 steganoimages and 20,000 cover images.

4.4 Evaluation Measures

The performance metrics that are considered for evaluation of proposed and existing systems are as follows:

(a) *Accuracy* It is one of the prime performance metrics for evaluating the stegoware detection system. It is defined as the sum of images classified correctly among the total number of images used. It is written by:

$$\text{Accuracy} = \frac{\text{TP} + \text{TN}}{\text{TP} + \text{TN} + \text{FP} + \text{FN}} \tag{7}$$

where TP implies the true positive, TN indicates the true negative, FP indicates the false positive, and FN implies the false negative.

(b) *True Positive Rate* (*Stegodetection Rate and Malware Detection Rate*): It is measured by the rate of TP, which is defined as the sum of images or malware files classified correctly as Stegoimage or malicious among the total number of stegoimages used or malware files extracted.

$$\text{TPR} = \frac{\text{TP}}{\text{TP} + \text{FN}} \tag{8}$$

(c) *False Positive Rate* It is measured in terms of the sum of images or payload files classified wrongly as stegoimage or malicious files among the total number of images used of malware files extracted.

$$\text{FPR} = \frac{\text{Number of misclassified images or payload files}}{\text{Number of normal images or benign files}} \tag{9}$$

5 Experimental Results and Discussions

In this section, the comparison of our proposed stegomalware detection model with familiar existing works is presented to illustrate that the proposed system is efficient in terms of steganalysis and malware analysis for hidden information type of attacks. Comparison is constructed with the following previous existing works: Cohen et al. [4], George et al. [5], Dharman and Thomas [6], de Jesus Serrano Perez et al. [7] for phase 1, George et al. [5], Dharman and Thomas [6], Bai et al. [2], and Cohen and Nissim [8] for Phase 3.

5.1 Hidden Payload Detection

The performance of the proposed system for hidden payload detection is made in terms of the effectiveness of accuracy, the effectiveness of TPR (SDR), and FPR for the benchmark.

The accuracy obtained for BOSSBase dataset by the proposed and existing systems is shown in Fig. 5. Figure 5 shows that comparison of accuracy for the embedding rates used (10, 20, 30, 40, and 50%), with 10% being the most harder to detect, and 50% being the simple to detect. From the graph, it is observed that the proposed system gives greater accuracy when compared to the four previous works. The existing models give only low and poor results when embedding rates are in difficult to detect levels such as rate level 10, 20, and 30%. Simulation results prove that the proposed model provides effective results on steganalysis irrespective of image format. Hence, it can determine any kind of hidden payload in cover images and it gives higher accuracy in detecting Stegoimages than previous detection systems for the dataset used.

Figure 6 indicates the performance of SDR (*stegodetection rate: the sum of images classified correctly as stegoimage*) for several stegoimages used in the dataset. SDR can be varied according to the format of images used. Our proposed system achieves an inflated detection amount for any kind of image format. The average SDR obtained is 93%, which is eventually greater than the existing systems, such as 88, 90, 90, and 87% for Cohen et al. [4], George et al. [5], Dharman and Thomas [6], and de Jesus Serrano Perez et al. [7], respectively. In this article, the blind steganalysis method is utilized for steganalysis, which helps to improve SDR irrespective of image format. In previous works, only JPG image formats are concentrated, which fails to detect stegorepository images of various other formats.

Figure 7 represents the FPR rate obtained for the dataset used. FPR is the results that are incorrectly done by the steganalysis phase (image classification

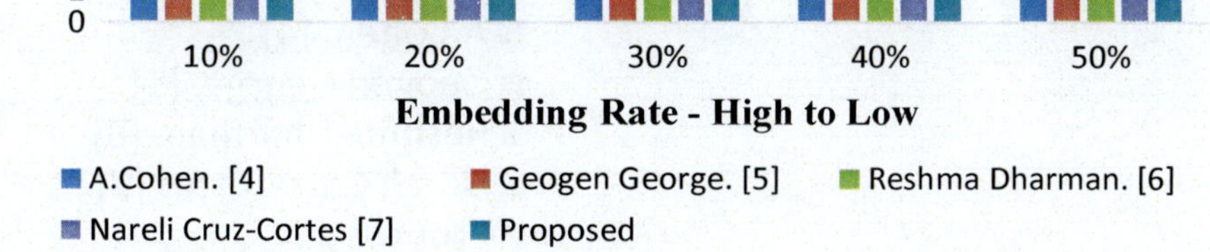

Fig. 5 Accuracy obtained for hidden payload detection

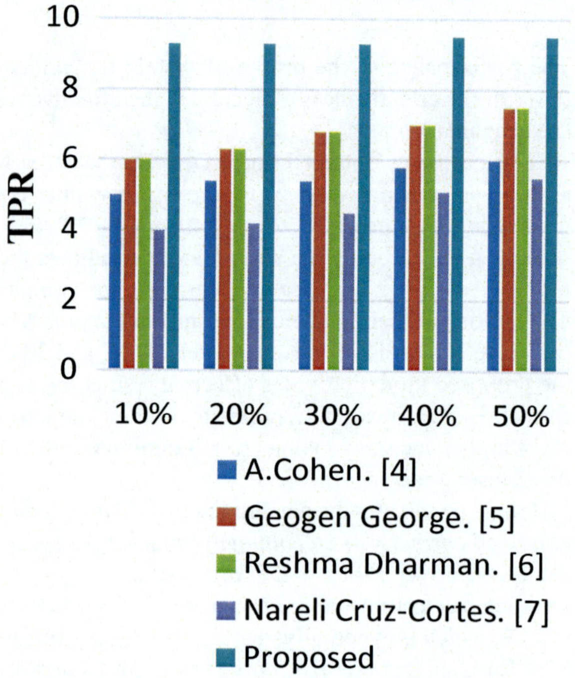

Fig. 6 TPR (*SDR*) obtained for hidden payload detection

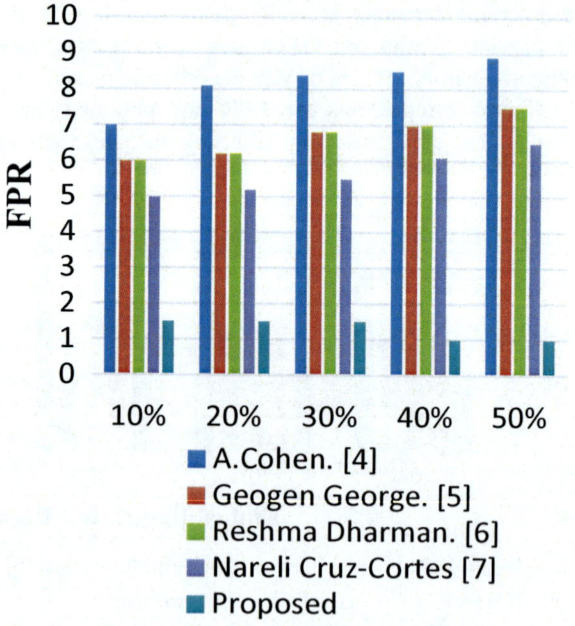

Fig. 7 FPR obtained for hidden payload detection

as stegoimage) when none of the hidden payloads has actually present inside the input image. Therefore, the goal of FPR is to decrease these incorrect outcomes by presumptions. These incorrect evaluations are obtained at numerous amounts in existing works. Therefore, for high detection accuracy, an optimal subset of relevant features would be taken into consideration. Existing works were served up in this restraint. Experimental evaluations for FPR show that the proposed system tends to low amount of FPR when compared to existing works. This is because of proper optimization of features for classification using ACO, where the optimal group of features for steganalysis can be accurately obtained.

5.1.1 Analysis of Algorithms

Feature extrication, feature selection, and classification are crucial steps for steganalysis in Phase 1 which can improve the detection rate of stegoimages. Existing works act up to provide efficient algorithms for these steps irrespective of image format. This section proves the efficiency of the selected algorithm combination in terms of improved steganalysis results. The experiments were conducted for the selection of this algorithm combination. Figure 8 represents the classification accuracy obtained for four different methods for feature extraction, CHEN [10]—486, CC-CHEN [11]—972, LIU [12]—216, and SPAM [13]—686. The numbers represent the corresponding features. Three classifiers are considered such as, i.e., SVM, KNN, and decision tree. The results show that the proposed model adopts one which gives the higher accuracy that is SPAM for feature extraction and decision tree for classification. This selection of a higher accuracy method will improve the accuracy of steganalysis detection.

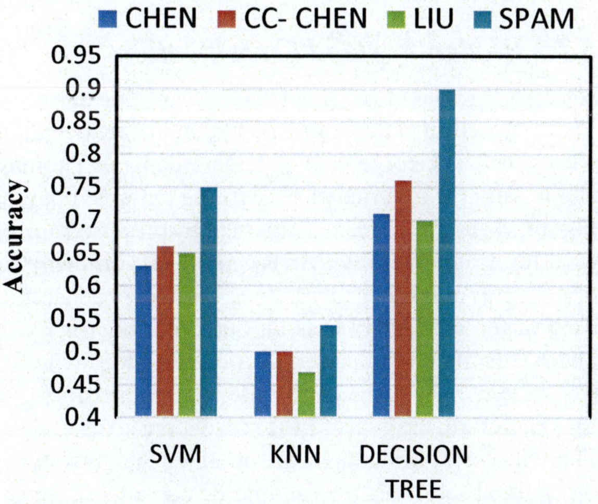

Fig. 8 Steganalysis accuracy of various feature extraction techniques

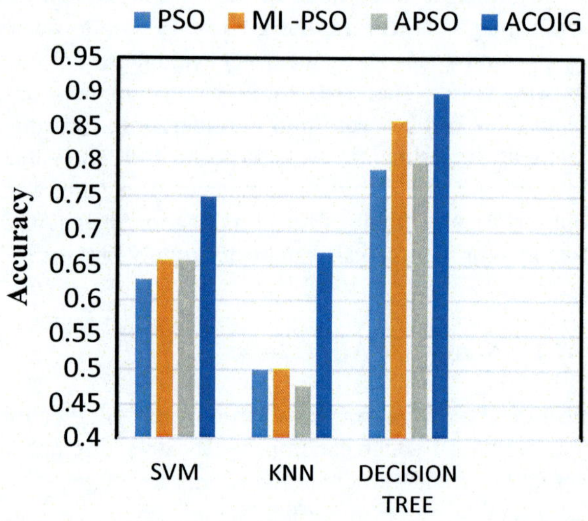

Fig. 9 Steganalysis accuracy of existing feature selection techniques with ACOIG

The experiments were also conducted for selecting the algorithm for feature selection. Figure 9 represents the classification accuracy obtained by four distinct methods for feature selection which implies, PSO [14], MI-PSO [2], APSO [2, 14]. ACOIG algorithm selects 68 features as an optimal subset from 686 extracted features by SPAM, respectively. For the classification purpose, three classifiers are considered, i.e., SVM, KNN, and decision tree. The results show that the adopted (ACOIG) feature selection method gives a high percentage of accuracy in steganalysis detection rate using decision tree classifier.

5.2 *Hidden Payload Classification*

The comparison of the proposed stegomalware detection system is made with existing works such as George et al. [5], Dharman and Thomas [6], Bai et al. [2], and Cohen and Nissim [8] in terms of classifying extracted payload as malicious or legitimate. Figure 10 shows the results obtained by proposed and existing systems for the dataset used for accuracy, TPR (MDR), and FPR. *Malware detection rate (MDR): the sum of payloads classified correctly as malware*) concerning several malicious payloads used in the dataset. FPR is an outcome incorrectly done by the malware analysis phase (PE file classification as malicious file) when none of the malicious activity has present in the extracted PE file, that is, classifying legitimate PE file as a malware file. Therefore, the goal of FPR is to decrease these wrong outcomes by presumptions. These incorrect evaluations are obtained at many more in previous works. Therefore, the optimal set of features should be considered for detection. From this graph, it is observed that the proposed model accurately classifies *hidden malicious executables* which act as a subject of *ransomware attack* when compared to the four existing

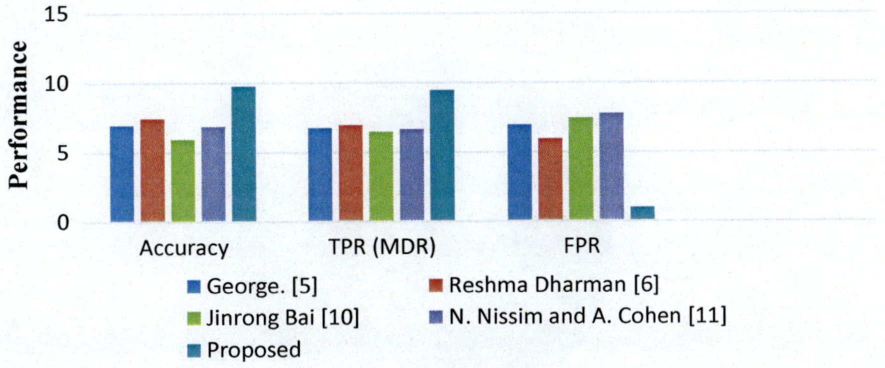

Fig. 10 Performance obtained for hidden payload classification

Table 2 Accuracy (in %) obtained by proposed system

Total images used—50,000		
Clean images (20,000)	Stegoimages (30,000)	
Accuracy (%)	Normal (15,000)	Malicious (15,000)
	Accuracy (%)	Accuracy (%)
93% (19,400 images)	94% (14,400 payload files)	98% (14,700 payload files)

works. These existing works performed to propose efficient algorithms for these steps if malicious executables are in hidden format. The experimental results show that our proposed system gives efficient results on hidden payload classification irrespective of its obfuscation.

Table 2 represents the average results for the proposed system in terms of accuracy. The below-mentioned table shows the number of images taken in BOSSBase dataset, the number of clean and stegoimages detected correctly, and the number of legitimate and malicious files classified correctly in terms of percentage. Finally, the simulation results show that our proposed system is most efficient and accurate on both *steganalysis and malware analysis.*

5.3 Percentages of Malicious Activity from Extracted Payload

The existing system results are only binary, that is, they can only find whether the payload is benign or malicious, but our proposed system shows the exact percentage of malicious activity that the payload can provide. The malicious stegoimages (14,700 payload files) are further analyzed to find their malicious capacity which is shown in Table 2.

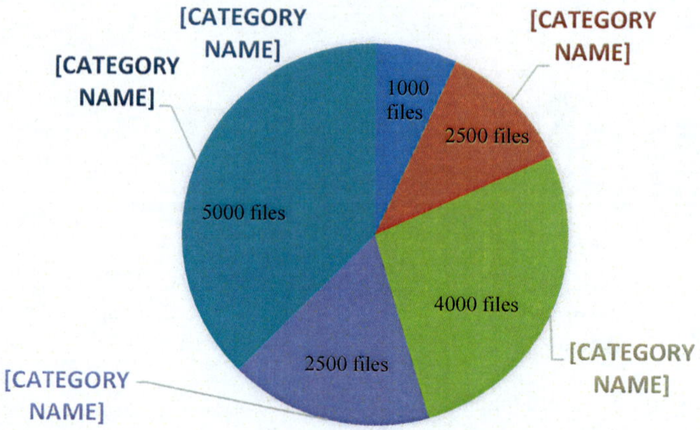

Fig. 11 Percentage of malicious activity obtained for malicious stegoimages

Figure 11 shows the number of malicious stegoimage files classified with respect to their amount of malicious activity (⇐20%, ⇐40%, ⇐60%, ⇐80%, ⇐100%) by the proposed model, with being ⇐100% more malicious to ⇐20% less malicious that it can provide to the host system. From the results of phase (3), *14,700 malicious files* from the dataset are tested individually to find the amount of malicious present in them and according to the malicious amount they are group into five categories from less malicious to more malicious as ⇐20%, ⇐40%, ⇐60%, ⇐80%, ⇐100% as shown above. Figure 11 shows that out of 14,700 malicious files from the dataset, *5000 files* are more malicious (*81–100%*) to *1000 files* are less *malicious* (*containing 1–20%*). Virus tools are used to check the accuracy of the proposed model, that is, the percentage of malicious calculated by the proposed model is verified using virus tools. Experimental results illustrate that our proposed system can provide effective results on malicious activity calculation irrespective of file format. Hence, it can find the exact malicious amount of any kind of file hidden in cover images.

6 Conclusion and Future Work

The recent computer systems are facing high risk and hazardous via the file formats that be usually legitimate and unexpected to deliver malicious programs. In this article, an ensemble-based Stegomalware detection system is proposed which finds the presence of hidden payloads from input images, extracts the hidden payloads, and verifies whether the hidden payloads is the subject of stegoransomware attack or not. If it is found as malignant, the exact percentage of its malicious activity is further calculated using fuzzy C-means clustering algorithm. Experimental results show that the proposed system drastically reduces the dimensionality of features and significantly improves the classification accuracy. In this work, only stegoransomware attack (obfuscated ransomware) detection has been proposed. In the future, it will be extended to all types of hidden information (stegomalware) attacks.

References

1. J. Kodovský, J. Fridrich, Steganalysis in high dimensions fusing classifiers built on random subspaces, in *Proceedings of SPIE, electronic imaging, security and forensics of multimedia XIII* (Vol. 7880, 2011), pp. 1–13
2. J. Bai, J. Wang, G. Zou, A malware detection scheme based on mining format information. Sci. World J. **2014**(260905), 11 (2014)
3. S. Choudhury, P.P. Amritha, M. Sethumadhavan, Stegware destruction using showering methods. Int. J. Innovative Technol. Exploring Eng. (IJITEE) **8**, 256 259 (2019)
4. A. Cohen, N. Nissim, Y. Elovici, MalJPEG: machine learning based solution for detection of malicious JPEG images. Expert Syst. Appl. **8** (2020)
5. G. George, P. Savaridassan, K. Devi, Detect images embedded with malicious programs. Int. J. Pure Appl. Math. **120**(6), 2763 2777 (2018)
6. R. Dharman, A. Thomas, Steganalysis application to detect image with malicious code. Int. J. Technol. Res. Manage. **3**(6) (2016)
7. J. de Jesus Serrano Perez, M.S. Rosales, N. Cruz-Cortes, Universal steganography detector based on an artificial immune system for JPEG images, in *International Journal of Pure and Applied Mathematics, IEEE International Conference on Trustcom, BigDataSE*. (ISPA, 2016), pp. 1896 1903
8. A. Cohen, N. Nissim, Trusted detection of ransomware in a private cloud using machine learning methods leveraging meta-features from volatile memory. Expert Syst. Appl. **102**, 158–178 (2018)
9. L. Chen, B. Chen, Y. Chen, Image feature selection based on ant colony optimization, in *Australasian Joint Conference on Artificial Intelligence* (2011), pp 580–589
10. C. Chen, Y.Q. Shi, JPEG image steganalysis utilizing both intrablock and interblock correlations, in *IEEE International Symposium on Circuits and Systems* (2008), pp. 3029–3032
11. T. Pevnỳ, J. Fridrich, Merging Markov and DCT features for multi-class JPEG steganalysis, in *Proceedings SPIE Electronic Imaging Security, Steganography and Watermarking Multimedia Contents IX* (Vol. 6505, p. 3, 2007)
12. J. Kodovskỳ, J. Fridrich, Steganalysis in high dimensions fusing classifiers built on random subspaces, in *Proceedings of SPIE, Electronic Imaging, Security and Forensics of Multimedia XIII* (Vol. 7880, pp. 1–13, 2011)
13. T. Pevny, P. Bas, J. Fridrich, Steganalysis by subtractive pixel adjacency matrix. IEEE Trans. INF Forensics Secure **5**, 215–224 (2013)
14. J. Kaur, S. Singh, Feature selection using mutual information and adaptive particle swarm optimization for image steganalysis, in *7th International Conference on Reliability, Infocom Technologies and Optimization (ICRITO) (Trends and Future Directions)* 29–31 Aug 2018

Machine Learning and GIS-Based Accident Detection and Emergency Management System for Two-Wheelers

A. Jackulin Mahariba and R. Annie Uthra

Abstract The accident detection is helpful not only to detect the accidents immediately but also to identify accidents in areas like unnoticed, remote places, and late hour's accidents. The proposed system provides a solution to identify the accidents at the earliest with efficient communication and routing system. The proposed accident detection system uses various parameters from vehicle and rider dependent and independent models to calculate the accident severity score using a fusion algorithm. Support vector machine algorithm is used to detect the accidents based on the parameters of vehicle and rider independent model. The output of the accident detection system and support vector machine (SVM) are combined to avoid false negatives and to improve the accuracy in detecting the accidents. Once the accident is detected, the proposed system communicates it to the centralized control system along with the location. The proposed system finds the shortest route to the hospital by considering the traffic congestion on the road which is estimated through GIS. Finally, the system performs data analysis to find the various useful patterns based on the parameters observed, to make improvements in vehicle design and accident management system.

Keywords Accident detection · Accident severity score · GIS · Support vector machine

1 Introduction

Highlight the road accidents lead to loss of life, over 1 million people are dying and 50 million people are injured in road accidents around the world which is almost equal to the number of deaths caused by HIV, AIDS, tuberculosis, and malaria combined

A. Jackulin Mahariba (✉) · R. Annie Uthra
Department of CSE, SRMIST, Chennai, India
e-mail: jackulia@srmist.edu.in

R. Annie Uthra
e-mail: annieu@srmist.edu.in

V. Suma et al. (eds.), *Inventive Systems and Control*,
Lecture Notes in Networks and Systems 204,
https://doi.org/10.1007/978-981-16-1395-1_47

[1]. In India, more than 70,000 people are dying due to road accidents every year, and this need to be recognized as an important public issue and great attention and care has to be taken to address this issue to reduce the mortality rate due to accidents. These statistical figures include both active and passive road accidents. The report released by MORTH, Government of India, states that: "Over the last decade, the number of vehicles has been increased at a CAGR of 9.8% while the length of roads was raised at a CAGR of 3.7%, which implies the worst condition on traffic congestion on road transport. The traffic density was also raised from 24 to 38/km on road in 2015" [2].

According to the statistics released by the state transport authority, the Government of Tamil Nadu, as of 01.04.2018 the vehicular position in Tamil Nadu contains 21,586,210 two-wheelers on the whole including powered two-wheelers, scooters, and mopeds while motor cars are 2,361,195 only [3]. Hence, the statistics are promising that the number of two-wheelers is predominantly increasing among the road users in Tamil Nadu. According to the report released by the Ministry of Road Transport and Highways of India, the fatality rate and the number of injuries caused by the two-wheelers are much higher than the four-wheelers which are depicted, which are depicted in Fig. 1. Two-wheelers represent 33.8% of the total road accidents involved in 2016 [3]. The statistical figures are framed from the road traffic

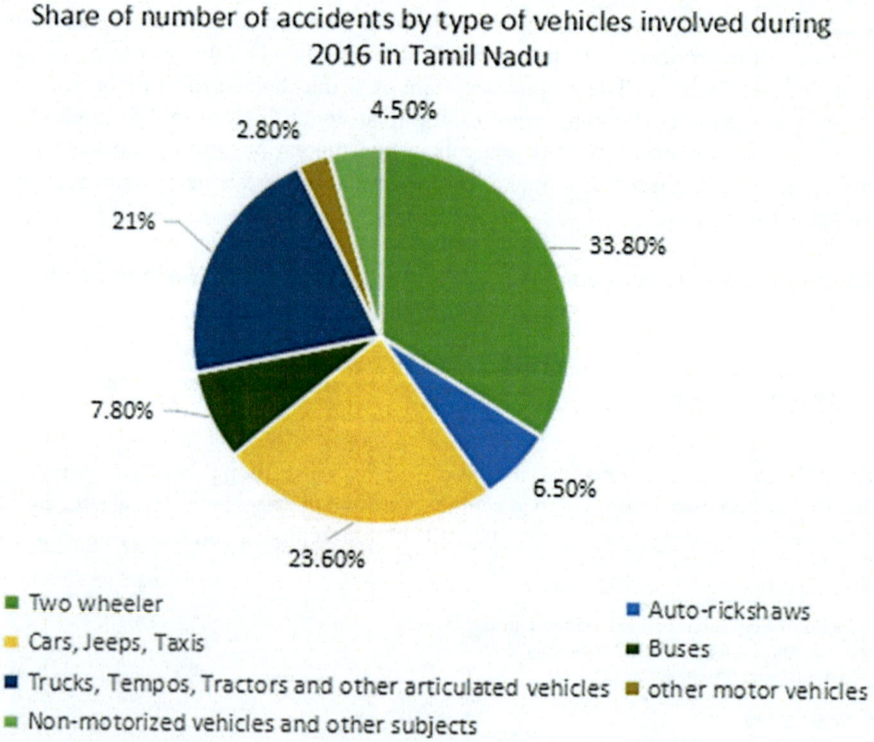

Fig. 1 Share of the number of accidents by type of vehicles

injuries (RTI) records of patients. In many cases, the details entered in the records of RTI are not clear, and the details are missing. Non-injured crashes are not accounted till date. Therefore, these figures from statistical reports are actually lower than the real number. Many research projects have been carried out in the field of accident analysis and prevention, which help to analyze the reasons behind four-wheeler accidents and to mitigate them. In many countries, there are separate research agencies and testing centers available to study driving behavior, vehicle dynamics and traffic measures. But these studies and analysis have been raised bars for the design and development of four-wheelers to prevent road accidents and its emergency management, and some of the higher-end model cars and trucks has a pre-crash notification and automated emergency notification. Since the investment and the damage caused after four-wheeler accidents are very high, the counter impact of such accidents is also high, leading to secondary accidents, high traffic congestion, and collisions.

In the case of motorcycle accidents, the probability of fatality rate is higher, and the protective measures are less in the case of two-wheelers. In general, four-wheelers are auto stable while two-wheelers are not stable. The stability of the vehicle completely lies in the riding behavior of the rider. In India, the roads are broadly classified as national highways (NH), state highways (SH), urban roads, and rural roads. In Tamil Nadu, as of March 31st of 2018, the number of registered vehicles is about 2.56 crores, among that two-wheelers contributes 84%. When the number of users of two-wheelers is increasing, then the road traffic accidents by two-wheelers will also increase.

The safety issues related to the road, vehicle, and road users, which actually lead to the road traffic accidents (i.e., road crashes) have not been possible to be tackled till now in India and most other developing countries and even in some of the higher income countries [4].

The percentage of road accidents that occurred on different types of roads is represented below in Fig. 2 [5].

The above figure illustrates the number of accidents, persons killed and injured, and these values are higher on other roads in comparison with NH and SH. According to the motor vehicle act of India, it is mandatory to have one ambulance at every 50 km on National Highways and along with Toll Gates. But in other types of roads, it is not mandated, where the major number of accidents occurred according to the report. Also, the report clearly states that the number of accidents is very high in uncontrolled areas than the police controlled areas [2] and the distribution of accidents during 2016 for various causes is listed in Table 1.

1.1 Background Analysis for Accident Detection

The biomechanics of motorcycle accidents is explained based on the type of ejection that occurred during the accident. There are several ways that a rider can be ejected from the bike in a solo motorcycle accident [6]. Three common ways of ejection are the low-side ejection which is shown in Fig. 3a, high-side ejection as shown

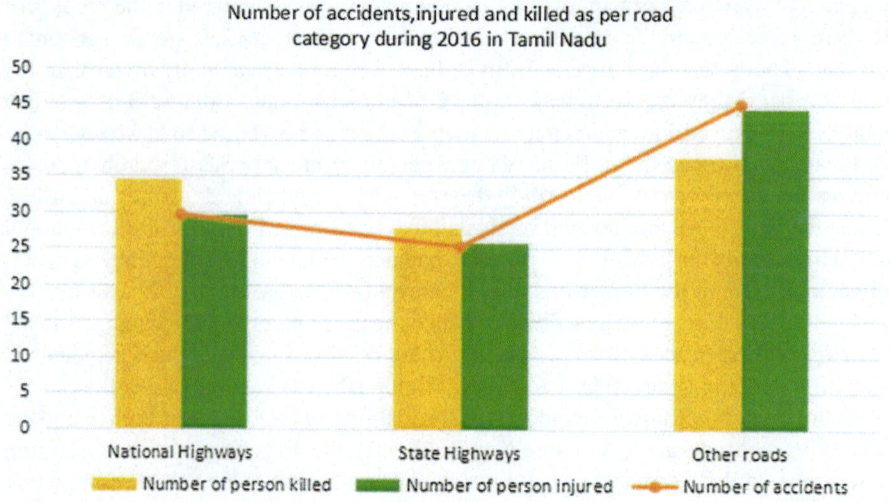

Fig. 2 Number of accidents, injured, killed as per road category

Table 1 Causes of accidents

Causes of accidents	Distribution of accidents during 2016 (%)
Uncontrolled	72.90
Flashing signal/blinker	5.70
Stop sign	6.40
Police controlled	6.40
Traffic light signal	8.60

(a) (b) (c)

Fig. 3 **a** Lower-side ejection, **b** high-side ejection, and **c** ejection over handlebar

in Fig. 3b, and the end over ejection, which is rider ejected over the handlebar as shown in Fig. 3c. The images represented from the bike racing sport, where the most dangerous types of stunts are performed.

The sliding of the front or rear wheel is due to excessive barking in the corner, excessive acceleration through, out of a corner, excessive speed in or through a corner for the control of traction on the vehicle. A low-side crash may be due to sudden greasy or free material, for example, oil, water, loose sand, and so on over the road surface as depicted in Fig. 3a. A high sider or high side is a sort of two-wheeler crash portrayed by sudden and strong rotation over the pivot of the bike about its longitudinal hub. This happens when the back wheel loses control on traction, slips, and after that all of a sudden recovers by footing, making a substantial torque, throws the rider off the side of the bike. In lowrider, the back wheel slips along the side and ceaseless until the bike falls onto the side facing the corner; though in high-side crash, the back wheel slips horizontally before abruptly recovering traction and skidding the two-wheelers toward the outside of the corner (the higher side of the bike) as shown in Fig. 3b. High sides happen rapidly and are extremely tough and subsequently fatal will result in increasingly serious in a high-side accident, rather than on a low-side crash. The other type of ejection occurs when the rear wheel is rotating in the air with the front wheel as the pivot, and the rider is ejected over the handlebar which is called as end over end (Endo) depicted in Fig. 3c.

1.2 Causes for Road Accidents

Human errors like overspeeding, drunken drive, distraction, red light jumping, safety rule violation, and non-adherence to the correct way of driving and overtaking lanes are the main cause of road accidents.

Statistics indicate that most fatalities are only caused by drinking and driving on the highways. Globally, around 480,000 deaths and 20 million people get injured by drunken and drive every year. In the majority of the developed countries, about 20% of fatally injured drivers are identified with excess alcohol in their blood, i.e., the concentration of blood alcohol (BAC) is above the legal limit. Studies in low- and middle-income countries such as India, on the other hand, it has been shown that between 33 and 69% of severely injured drivers and 29% of non-fatally injured drivers drank alcohol before their accident [5]. The other factors contributing to road accident are listed below;

1. **Vehicle users**: Overspeeding, aggressive driving, breach of laws, inability to comprehend signals, exhaustion, alcohol.
2. **Roadway users**: Negligence, irresponsibility, entering the carriageway in the incorrect location.
3. **Roadway commuters**: Expressing their objects outside the vehicle by listening to road users, disembarking and accessing vehicles on footboards from the opposite side, catching a moving vehicle, etc.
4. **Fault on Vehicles**: Brake or steering failure, tyre exploding, poor headlights, flooding, load extension.
5. **Roadway condition**: Cracks, broken bridges, the convergence of rural roads with highways, obstructions, unregulated speed bumps, deteriorated roads.
6. **Environmental conditions**: Darkness, mist, heavy rains, hurricanes, storms.

There are various factors behind a motorcycle accident, and they vary in their degree of subjectivity. The factors involved are road environment with defective traffic signs and obstacles such as animals in the carriageway, vehicle defects like under-inflated tyres, injudicious actions by riders include reckless driving, illegal turn or directions of travel, usage of drugs, vision affected by dazzling sun and headlights, aggressive driving, impairment or distraction by using mobile phones, and error reaction in the form of sudden braking and loss of control. On applying the principal component analysis on a five-factor solution from a questionnaire called as motorcycle rider behavior questionnaire (MRBQ) proposed by Elliot et al. (2007), it is clearly evident that the main predictors for detection are stunts and speed violations of active road accidents [7]. Özkan et al. proposed the theory of planned behavior (TPB), health belief model (HBM), and traffic locus of control (T-LOC) in the form of a questionnaire to conclude the primary factors such as running wide in a corner, losing control though they are due to overspeeding, dipped headlights and usage of inappropriate types of helmets and ranking of such factors toward road accidents [8]. Based on the severity of the accidents and loss caused due to it, the accidents are categorized as follows:

1. Fatal
2. Grievously injured
3. Minor injured
4. Non-injured.

The above categorization is common to all vehicles involved in accidents. There is a certain specific type of accidents accounts for motorcycles alone. They are

1. Crash/collision
   (a) major collisions with other powered or larger vehicles;
   (b) minor collisions with stationary objects, bicycles, pedestrians, or stray animals;
   (c) minor accidents including falls from motorized two-wheeler or a slip of the vehicle without an actual collision;
2. Rider ejection
3. Piled up collision.

Without the strict enforcement of motor laws, new initiatives, and efforts taken to mitigate road accidents in India, the accident death count will be increased to 250,000 by 2025 [9]. The uncontrolled areas of rural roads don not have facilities to look after the victims of road traffic accidents. Even if there are ambulance services available at the location, many victims were died unnoticed. It is essential to treat the injured persons after an accident within the golden hour itself or else it will increase the severity of injury in folds, and it may even lead to fatal at times. The post-mortem reports of about 374 victims in hospitals on road traffic accidents that died within twelve hours of crash thought that the cause of death is being correlated with the immediate time taken to die. The elapsed time between the accident and the arrival of ambulance/FRU could be halved, and then, there might be a chance of saving strictly 1.5% of casualties dying within 24 h [10].

2 Related Works

The occurrence of an accident can be detected automatically by continuously monitoring three sensors [11] deployed in the motorcycle. The sensors used are pressure sensor, accelerometer sensor, alcohol sensor. The bike ignition is controlled by a microcontroller it takes input from all these sensors. The bike will be ignited only if the rider wears the helmet, the pressure sensor attached with the helmet detects it and sends the signal to the microcontroller. When the bike is ignited, the alcohol sensor checks for the consumption of alcohol by the rider and passes the signal to the microcontroller to prevent the ignition further. The accelerometer sensor measures the speed of the bike and checks it with the threshold, and when the orientation is tilted, then the signal is passed to the controller to detect the accident. GPS and GSM modules in the Arduino board send the accident location to the contact numbers stored in the microcontroller.

The accident analysis involves various steps including analyzing causes, identifying hazards and risk factors, construction of risk matrix, and implementing the safety countermeasures [12]. Many road accidents caused by PTW in Slovenia location wise have been surveyed and analyzed by Matjaz et al. As a result of the analysis, safety measures and actions are proposed to mitigate the PTW accidents in the study area. The proposed measures and actions include control on the locations where accidents by PTW are frequent, additional video surveillance systems, warning signs specialized for PTW at critical road sections, awareness programs for the promotion of safer driving as the immediate actions [13].

Near-miss accidents were usually referred to as precursors of accidents. It provides insights into accidents that could happen and have great significance to further improve safety margin from the perspective of real-time tracking [14]. The precursors of the accidents occurred in the building construction site were identified using RFID tags attached to their safety wear such as helmet and handheld devices. These devices are allowed to communicate with the other machines in the work zone using RFID tags attached to the machines. The end devices are exchanging data using the Zigbee router. The local server will continuously be monitored to track the near-miss accident, and this information is used in the future to develop new safety measures for the building construction site.

Motorcycle riding will be unstable in two main scenarios. The first case is that when the acceleration is sufficiently high, it leads to loss of control during front wheel lock at the braking (causes sudden deceleration). The second case is at which even at lower velocity when the two-wheelers are hit by any fast-moving vehicle with high force, then two-wheelers lose their stability [15].

Stabilization of two-wheelers can be attained by two mechanisms, one with the wheel traction force and the other by the posture of the rider and handlebar movement.

Both the mechanisms rely on the friction that exists between the road and the tire of two-wheelers.

The roll angle is defined as,

$$\lambda = \arctan\left(\frac{F}{G}\right) = \arctan\left(v^2/R \cdot g\right)$$

where F is the centrifugal force acting at the center of the vehicle, G is the weight force, R is the cornering radius, v is the velocity of two-wheelers, and g is the acceleration due to gravity.

Loss of control accidents accounts for 60% of motorcycle accident fatalities. The implementation of ABS for motorcycles will drastically reduce the number of accidents that occurred due to loss of stability control on two-wheelers.

Several studies have been carried out to find acceleration-based fall detection. The studies explain the divergence and difficulties by comparing different fall detection obtained under different conditions on three specific public fall detection datasets. Nearest neighbor and SVM algorithms are applied to the datasets. The algorithm performance decreases drastically when the algorithms are tested on a different dataset from the one used for training. The datasets contain both ADL and simulated fall under different conditions. The datasets are DLR, Mobifall, and tFall. The data are collected using external accelerometers or smartphones attached to the body of the subject [16].

A fall detection system to monitor the patients has been presented by Ryan et al. This system includes a shimmer device with a triaxial accelerometer measuring sensor. The repeated experiments of ADL and simulated fall were recorded. The accelerometer readings are then analyzed with five supervised classifiers. The ANN, KNN, RBF, PPCA, LDA classifiers are used to detect the occurrence of fall and its categories for direction and the other features. A comparator function is applied to the results of five classifiers. The comparator function compares the result against a threshold value (τ) to confirm the occurrence of the event along with its characteristics [17]. Machine learning algorithms are used to define the circumstances of fall using a threshold value.

An alert system to trigger inflation of airbag jacket during fall from a motorcycle was proposed by Boubezoul et al. The experiment was performed on Honda CBF1000 with 3D inertial measurement unit, ABS—hall effect sensors, linear potentiometer, and turn contact. The two-wheelers have a data recorder system in which the measurements are continuously monitored with a timer. Four fall scenarios are selected and performed by a professional rider with an airbag jacket. The four cases are as follows: (1) fall in curve, (2) fall on a slippery straight road section, (3) fall with a leaning of motorcycle "intentional maneuver," and (4) fall in a roundabout.

The collected data was preprocessed with three filters (LBF, WF, MF) and analyzed to find the acceleration hike (A) and gyro readings (R) changed at the time of event occurred. A fall detection algorithm was proposed to decide the "fall" or "fall-like" behavior by comparing the norm values of A and R with the threshold [18]. Fixing the threshold is not an easy context, with the help of repeated observation over the recorded measurements, the threshold is fixed. The norm values of A and R were calculated as follows,

$$|(|\boldsymbol{A}|)| = \sqrt{(\boldsymbol{Ax}^2 + \boldsymbol{Ay}^2 + \boldsymbol{Az}^2)}$$

$$|(|\boldsymbol{R}|)| = \sqrt{(\boldsymbol{Rx}^2 + \boldsymbol{Ry}^2 + \boldsymbol{Rz}^2)}$$

where Ax, Ay, and Az represent the acceleration along the X, Y, and Z-axis, Similarly Rx, Ry, Rz represent the gyroscope values along the X, Y, and Z-axis. The event was detected and it triggers the inflation of the airbag jacket. The latency on triggering it plays a crucial role in the alert system.

Accident analysis practitioners are always concerned over the type of crashes and their severity index. Actually, the crash severity lies on various factors before the occurrence of the crash. A sequence of events that causes two-wheeler crashes will contribute more to the risk factors of the accident. The general categorization of the type of crashes does not provide an informative resource during accident analysis. Crash severity defines that head-on crashes are more vulnerable than rear-end crashes. Crash sequences are generated based on how a crash is initiated and the actions followed by it. There are several variables associated with the risk of a crash, and they lead to the frequency or severity of a crash at various points in a crash, called crash progression. For the most severe and serious crashes, a risk matrix is suggested. Four main stages consist of the crash sequence analysis [12]. The first step, similar to conventional driving, defines the starting point. In the second level, triggering events are established, interrupting the regular driving situation and thereby endangering the original device. The third stage explains how drivers attempt to recover by performing emergency strategies from the current driving scenario. The fourth stage is the period of crash and impact, which illustrates the condition of an incident and evaluates the seriousness of a crash.

The fall detection can be performed using the accelerometer readings generated in the smartphone. The G-force exerted on the object falling toward the ground is computed with the following equation.

$$G = (a/g)$$

$$|a| = (a_x^2 + a_y^2 + a_z^2)^{1/2}$$

where, G is the G-force exerted on the vehicle, and g is the acceleration due to gravity. Based on the repeated experiment conducted for free fall at a distance of 20 and 50 cm above the ground [19], the threshold of G-force is identified.

On comparing the G-force observed with the threshold, an alert will be generated automatically to the emergency phone number to communicate the fall detection of the user. Table 2 analyzes similar works and discusses the benefits and drawbacks of different accident detection methods.

Table 2 Limitation and features of existing accident detection techniques

Techniques used	Merits	Demerits
Smartphone in-built sensors	Cost-efficient and existing device. It can be used to detect accidents in any type of vehicle	It provides more false positives. Also, it requires different applications and filters to detect different types of accidents [20]
Sensors with GSM and GPS	Instead of placing a separate network module to enable Internet connection, GSM modem will serve the purpose	This system works using an IR sensor, but the IR sensor does not work well with dark color vehicles
Vehicle communication using VANET	Vehicular ad hoc network is used to find the optimal shortest route to the accident location using algorithms	Good road environment and infrastructure is the requirement to employ VANET [21]
Android and Web applications	Simple and easy to implement	Requires a centralized node to handle all the requests. The sensitivity level and processor speed of the phone vary from one another

3 Proposed Methodology

This research work aims to achieve the automatic accident detection system for two-wheelers which includes powered two-wheelers, Mopeds, and scooters and thereby generating an automated call service to ambulance with first response unit (FRU) team and victim to reach the hospital as early as possible within the golden hour. A sample scenario is depicted in Fig. 4 with three major action such as automatic accident detection, emergency notification, and emergency routing.

Hence, the proposed system aims to address the following objectives for the early detection of accidents and faster medical treatment for the victim:

1. To develop an algorithm for automatic accident detection for two-wheelers. The accident detection algorithm uses sensors and machine learning techniques. The sensors that are attached to the vehicle and helmet unit send the signal to continue to the controller and the decision unit. The ML technique which runs on the decision unit works with the GPS coordinates and input parameters to improve the accuracy of the accident detection.
2. To develop an emergency management system that communicates first response unit (FRU) and emergency contact numbers.
3. To analyze the data to find the various useful patterns based on the parameters to avoid accidents (Fig. 14).

The proposed system aims to detect the accidents of motorcycles at the earliest, and an emergency management system is developed to foster quick medical services to the victim. The various sensors like gyrometer, accelerometer, and heartbeat sensors,

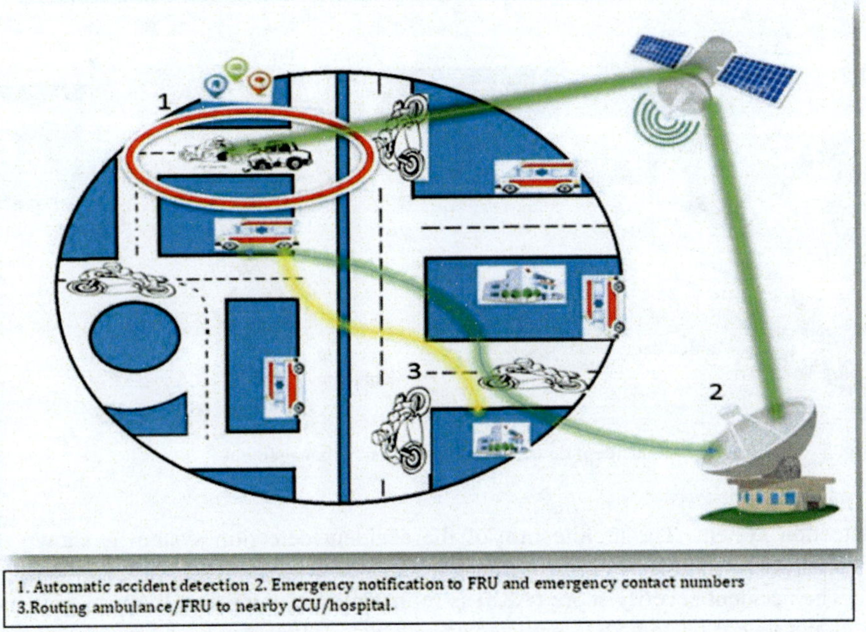

Fig. 4 Accident detection and emergency management scenario model

and machine learning techniques are used to detect the accidents in the two-wheelers. The sensors are attached both in the vehicle and helmet units to detect the positions of the vehicle and the rider. The collected sensor information is fed to the machine learning algorithm to detect the occurrence of the accident. When the occurrence of an accident is detected, the decision unit sends the accident information to the emergency helpline automatically through the GSM network module.

The proposed system uses the 108 ambulance service as the first responder as the service is the fastest responder service in Tamil Nadu. Finally, the system finds the shortest route by considering the road traffic conditioning using GIS to reach the nearest hospital. The overall objectives of the proposed system are shown in Fig. 5.

3.1 Module 1: Accident Detection in Two-Wheelers

The two broader output classification of the accident detection algorithm is an accident occurred/accident not occurred. The accident occurrence is further classified into three categories based on the severity of the accident, namely low, medium, and high. The output of the accident detection algorithm and machine learning algorithm is compared to avoid false negatives and to improve the precision of the accident

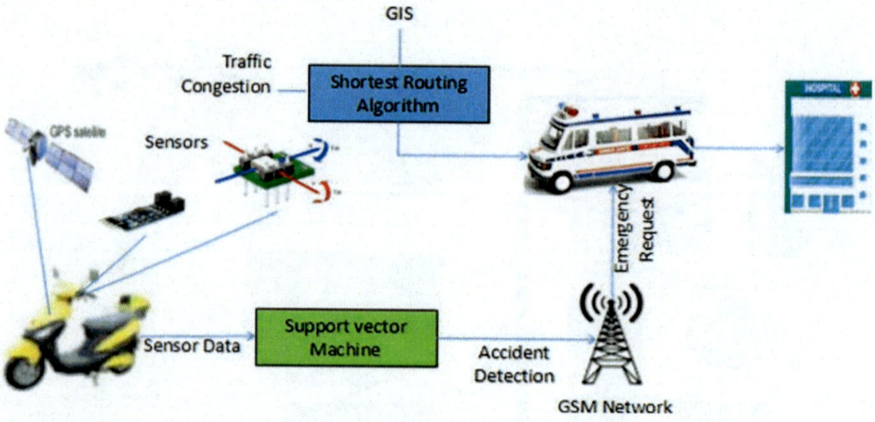

Fig. 5 Physical view of accident detection and emergency management

detection system. The architecture of the accident detection system is shown in Fig. 6.

The accident severity score (ASS) is represented as the rating that can give the accident severity. The ASS is analyzed using a fusion algorithm based on the accident categories identified by the two models. The models are defined as follows: One is vehicle and rider dependent model (VR-D model), which includes variables like accelerometer readings in *X*, *Y*, and *Z* axis, roll rate based on gyro measures, pressure on the contact points of the vehicle, and the pulse rate of the rider.

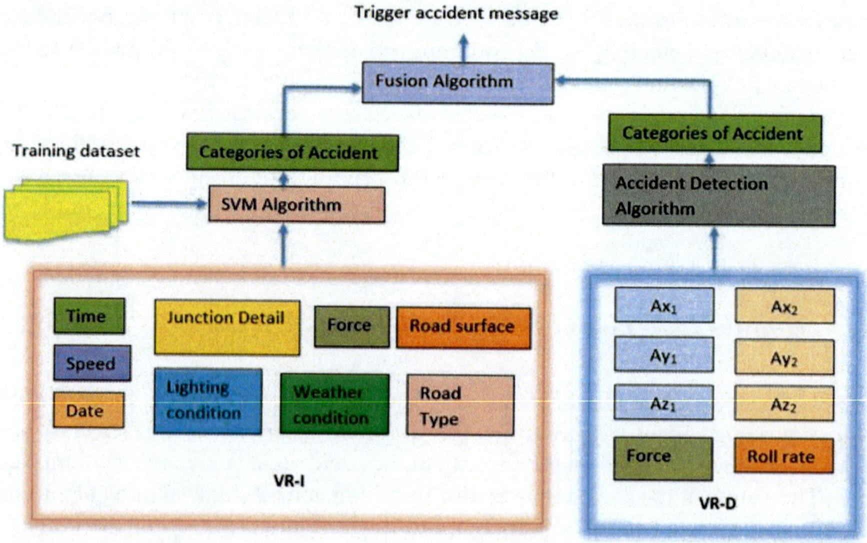

Fig. 6 Architecture diagram of the accident detection system

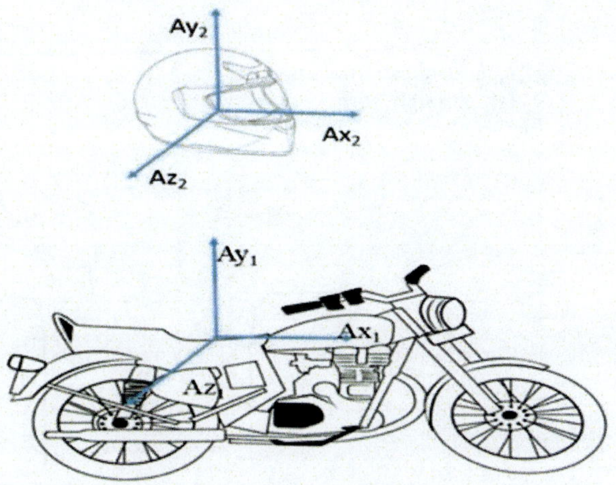

Fig. 7 Gyrometer reading in vehicle and helmet unit for comparison

The proposed method uses sensors to measure acceleration and heartbeat and a gyrometer to measure gyro reading in the vehicle and the helmet unit to detect the accidents. The motorcycle parameters and the rider's helmet parameters are observed and compared to detect the occurrence of the accident and the need for the FRU to the location. To measure the acceleration and gyro readings of the motorcycle, Bosch SU-MM 5.10 lean-angle sensor is attached with the motorcycle, where SU-MM 5.10 lean-angle sensor supplies the vehicle dynamics data like motorcycles longitudinal, lateral, and vertical acceleration during the ride. These parameters are used to calculate the yaw and roll rate of the motorcycles.

The lean-angle sensor communicates the data it measures and calculates to the CAN bus, the communication line. Compared with the previous model, SU-MM 3.10 from Bosch, the SU-MM 5.10 sensors size, weight, and power consumption have been reduced. It weighs just 40 g and around 60% less than its predecessor. Since it is considerably smaller and provides greater design flexibility for motorcycle manufacturers, the proposed system uses the same sensor to be attached to the helmet of the two-wheeler rider [22].

The sensor devices attached with the vehicle measure acceleration and gyro reading in *x*, *y*, and *z*-axis which are indicated as Ax^1, Ay^1, Az^1, and Gx^1, Gy^1, Gz^1, respectively. The sensor devices attached with the helmet unit also measures the acceleration and gyro readings in *x*, *y* and *z*-axis which are denoted as Ax^2, Ay^2, Az^2 and Gx^2, Gy^2, Gz^2, respectively. In addition to the acceleration and gyrometer, the helmet unit contains pulse rate sensors to measure the pulse rate of the rider.

All the sensors are powered on, and the sensors at the helmet unit are synchronized with the sensors on a motorcycle when the engine is powered ON. These sensor readings from the motorcycle are compared with the sensor readings of the helmet unit to detect the ejection of the rider to identify the accident. In addition to these sensors, pressure sensors are fixed in the seats, footrest, and hand bars to identify the

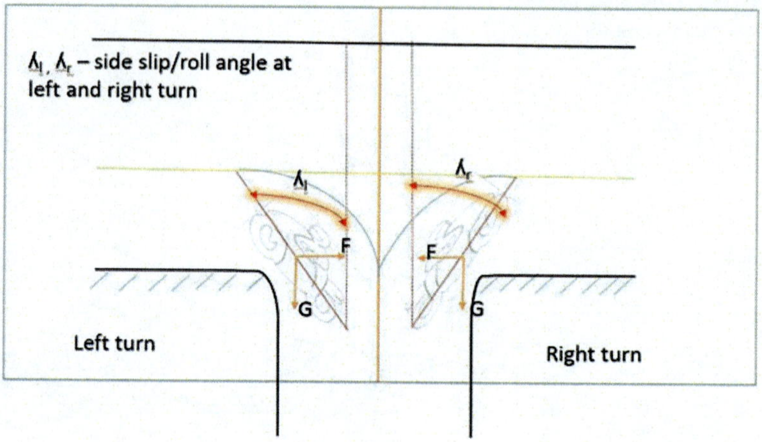

Fig. 8 Roll angle to force during left and right turn

accident in terms of the ejection of the rider. Figure 7 shows the sensor reading of the two-wheelers and the helmet unit.

The proposed system calculates the force F_1 and F_2 experienced by the sensors at the vehicle and the helmet unit and also calculates the side slips/roll angle $ʎ_1$ and $ʎ_2$, to decide high side or low side ejections or Endo. The following Fig. 8 shows the roll angle lean during left and right slip/roll.

The proposed system uses the accident detection algorithm to detect the occurrence of an accident accurately. The algorithm considers the following input parameters for detecting the accidents.

1. Type of vehicle: Vehicle type is taken into account since every type of vehicle design has its own features and they vary in size and operation. Two-wheelers are generally classified into the following types: motorcycles—greater than 50 cc, Scooter, Moped—vehicles less than 50 cc.
2. Weight of the vehicle: The weight of the vehicle differs with the design and manufacturers.
3. Speed: Speed can be calculated with the accelerometer readings, or it can be taken as an output from the speedometer of the vehicle. This parameter can be taken from the vehicles in-built speedometer if it provides a digital output, or else a special pulse odometer is needed to record speed variations.
4. Acceleration concerning three axis (*X*, *Y*, and *Z*) and gyro readings (pitch, yaw, and roll) can be measured using a lean-angle sensor SU-MM 5.10, which is specially designed for two-wheelers by Bosch mobility solutions.
   The accelerometer is attached to the vehicle, and the *x*, *y*, and *z*-axis readings were collected in real-time to observe the left and right slip/roll readings. Figures 9 and 10 show the reading of *x*, *y*, *z*, and norms (aT). The norm value (aT) and *g*-force (gFy—roll) has the hike during the left and right roll.

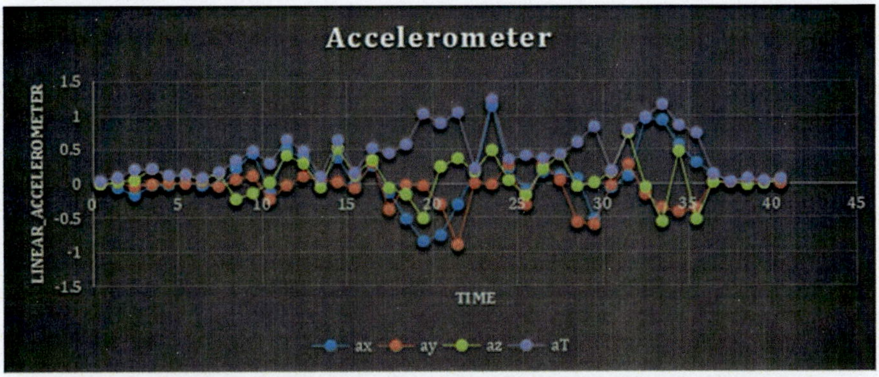

Fig. 9 Linear accelerometer readings over left and right slip/roll

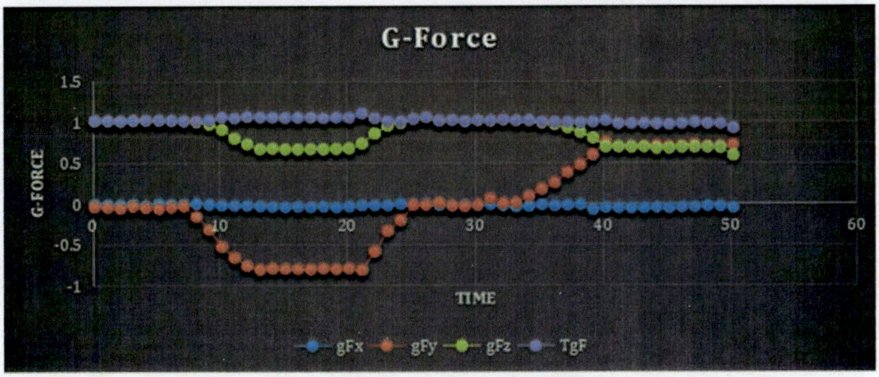

Fig. 10 G-force measured over a left and right slip/roll

5. Amount of force exerted on the vehicle and helmet. According to Newton's second law of motion, the force exerted on an object is given by the product of mass and acceleration. During accidents, the two-wheeler decelerates after the crash/collision. Hence,

$$d = f/m$$

where f is the force exerted, m is the weight of the vehicle, and *d* is the deceleration caused due to crash/collision. Force exerted on the body of the vehicle is measured using the shock sensor on the OBD unit of the vehicle. If the deceleration exceeds the estimated value, then the crash/collision is detected.
6. Location in the form of latitude and longitude from the GPS receiver in the two-wheeler.

7. Accident severity score (ASS) is the score that explains the severity of the collision based on the models which are developed using the machine learning algorithm.
8. Stability index is a binary variable which states that whether the vehicle is stable or unstable based on the roll angle measured by the sensor in the two-wheeler.
9. Rider ejection is a categorical parameter. If the rider is ejected from the two-wheeler, then it determines the category as which type of ejection (lower side, high side, or endo) occurs. Rider ejection indicates the severity of the accident. High side and endo cause a fatal accident.
10. Pulse rate value is measured using the pulse rate monitoring sensor attached to the helmet of the rider.

The accident detection algorithm takes the sensor inputs and detects the occurrence of the accident and provides an accident severity score when an accident has happened.

The accident detection algorithm uses a hard threshold value on the accelerometer value and roll rate sensed by the accelerometer and gyroscope attached to the vehicle unit and helmet unit, to check the orientation of the vehicle and the rider. When the vehicle or the rider orientation changes, impact force measured on the shock sensor attached to the body of the vehicle is checked to decide the type of collision (high side, low side, and end-to-end). Thus, the occurrence of an accident is confirmed. Finally, the accident is detected using VR-D model parameters.

```
ACCIDENT DETECTION ALGORITHM (f,|a|, ʎ1, ʎ2)
Begin
        while (Engine_power!=0)
         d=f/m;
                label1: continously monitor accelerometer and calculate |a|
                if (|a|>d && stability_index==0)
                        if (ʎ1<q || ʎ2 >2q )
                                ejection ==1
                                if(pr==0)
                                        Set accident=1;
                        goto label2;
                                else
                                continue;
             else
               goto label1;
         label2: send the current lat and long to FRU.

End
```

The other model is the vehicle and rider independent model (VR-I model), which includes variables such as time, speed, date, junction detail, force, lighting condition, weather condition, road surface, and road type. The relative influence of these parameters on the severity of the accident is studied on the dataset published by the UK Department of Transport. It contains the reports of the accident that occurred from 2005 to 2015. The results of the relative influence of road surface, road type, and lighting condition are shown in Figs. 11, 12, and 13.

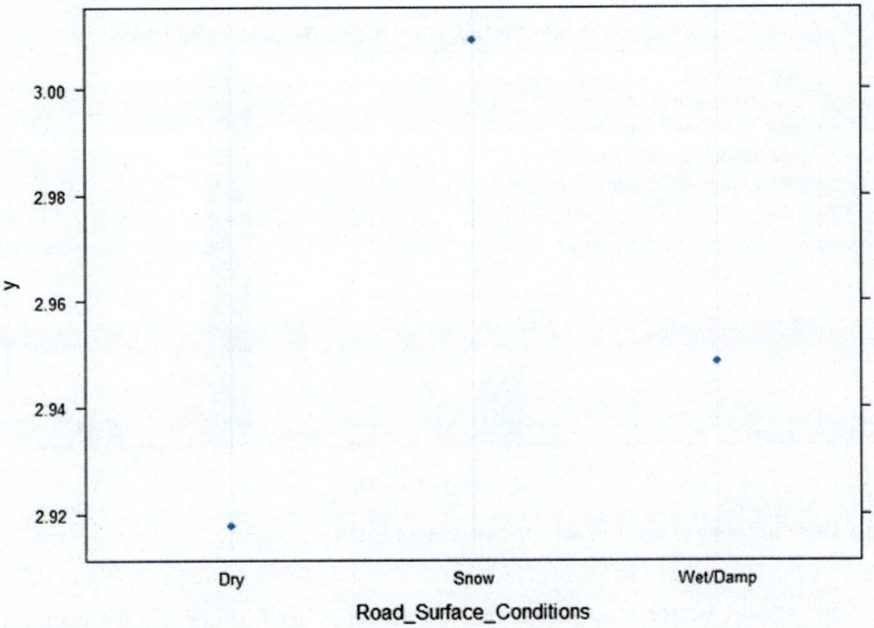

Fig. 11 Influence of road surface conditions on accident severity

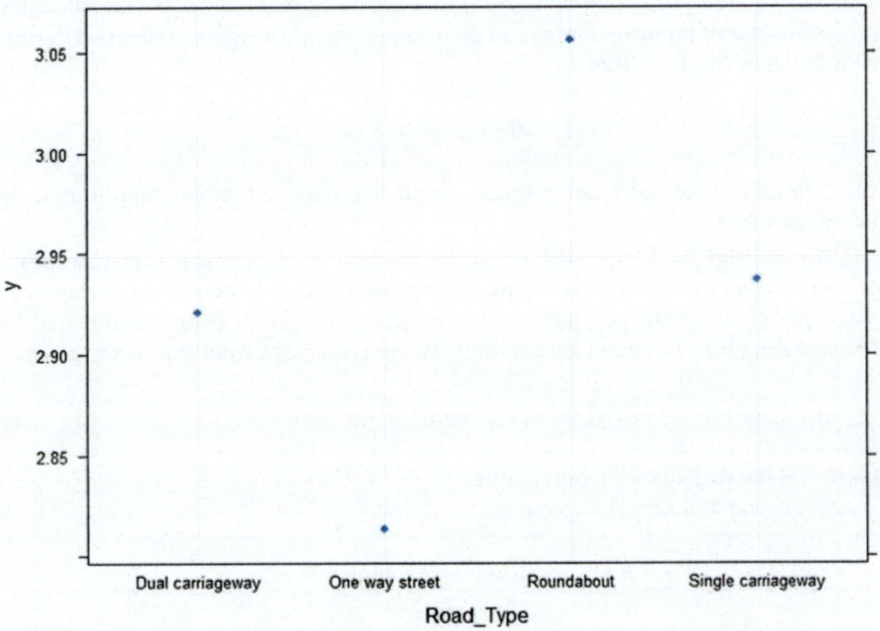

Fig. 12 Influence of road type on accident severity

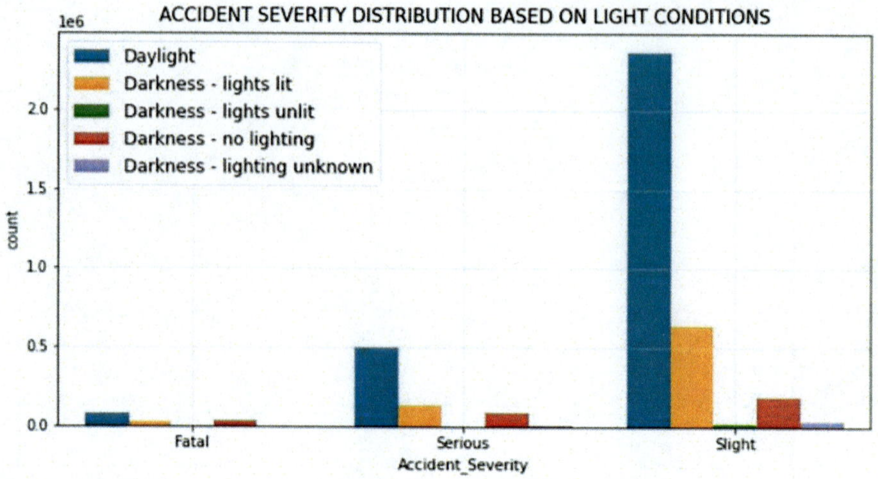

Fig. 13 Distribution of accident severity based on the lighting condition

The support vector machine (SVM) algorithm is used to classify the accident dataset based on the severity of the accident. SVM identifies the hyperplane to classify the input by transforming it from one feature space to the other using linear algebra [23]. The equation of hyperplane ($f(x)$) that separates two classes is defined using the dot product of input vector (x) and the support vector (weights: a) learned during the training of the algorithm.

$$f(x) = B_0 + \text{sum}(a_i * (x * x_i))$$

where B_0 and a_i are coefficients and are estimated using a learning algorithm over the training data.

The accident detection system considers the data set with nine parameters (time, speed, date, junction detail, force, lighting condition, weather condition, road surface, road type) in which the polynomial or exponential kernels are applied to calculate the separation line, based on kernel trick. The polynomial kernel can be written as

$$K(x, x_i) = \text{sum}(x * x_i)^d + 1$$

where d is the degree of the polynomial.

And exponential kernel is given as,

$$K(x, x_i) = \exp(-\gamma * \text{sum}(x - x_i^2))$$

where $\gamma = 1/(2\sigma^2)$.

The hyperplane obtained using exponential kernel performs better to categorize the accident as per its severity (low, medium, and high) with an accuracy of 84%.

Table 3 ASS assessed using the ASS of VR-D and VR-I models

VR-D model	VR-I model	ASS of accident detection system
Low	Low	Low
Low	Medium	Low
Low	High	Medium
Medium	Low	Medium
Medium	Medium	Medium
Medium	High	Medium
High	Low	High
High	Medium	High
High	High	High

The fusion algorithm compares the output of the two models, decides the occurrence of the accident, and classifies it based on ASS which is shown in Table 3.

3.2 *Module 2: Identification of Shortest Route to Nearby Emergency Care Unit or Casualty Care Unit*

In this study, GIS is applied to obtain the best calculation results to derive the fastest delivery service routes which are shown. Using a regression model, this paper presented the implementation of statistical analysis to identify the key factors affecting driving time. When the attributes of the independent factors are known and there is a relation between the independent variables, the regression model can provide the data needed to estimate the values of a dependent variable.

For data storage, a geodatabase inside the ArcGIS application was used. All the files were all in the same single dataset, indicating that all had the same coordinate system, which was a critical requirement for the correctness of the operation. The database makes it possible for users to adjust the layers as a response to changes in actual situations of road transport. With the up-to-date data, the network dataset can then be reconstructed. The benefits of using a database include: consolidated data, all within the same location; high speed for data access and manipulation; high protection, with safeguards to ensure that the data is not manipulated or damaged; and the role of database administrator can only be performed by one person. Data is handled using a database management system that enables data to be defined, manipulated, utilized, coherent, and administered.

The static positions of ambulance services and hospitals within a 3 km buffer from the incident location are obtained from Google earth and stored in the form of geodatabase. The shortest path algorithm uses the geodatabase and identifies the shortest path. Then, the path is imported on the map using webGIS, and a dynamic map is created. The shortest path algorithm used in the proposed work is implemented using Json maps, which are shown below.

```
Loops through all Ambulance places to calculate route between our current position
 // and that place.
 function findNearestPlace() {
   var i = Ambulance.length;
   size = Ambulance.length;
   routeResults = [];
   while (i--) {
     calcRoute(Ambulance[i].latLng, storeResult);
   }
 }


 // A function to calculate the route between our current position and some desired end point.
 function calcRoute(end, callback) {
   var request = {
       origin: currentPosition,
       destination: end,
       travelMode: google.maps.TravelMode.DRIVING
   };
   directionsService.route(request, function(response, status) {
     if (status == google.maps.DirectionsStatus.OK) {
       callback(response);
     } else {
       size--;
     }
   });
 }


 // Stores a routing result from the API in our global array for routes.
 function storeResult(data) {
   routeResults.push(data);
   if (routeResults.length === size) {
     findShortest();
   }
 }
```

The dynamic map with the shortest path to the incident location and hospital is then passed to the ambulance driver in the form of a weblink.

3.3 *Module 3: Data Analytics for Accident Management System*

The sensor data that are read from the vehicle during the ride and in case of an accident are collected and transmitted to the cloud through the Wi-Fi module of the MyRio controller for further processing. The proposed system performs the integration between the MyRio and cloud using the designed application program. The data stream network (DNS) used in this project is provided by PubNub and is shown in Fig. 14. It also provides a real-time infrastructure-as-a-service (IaaS).

The model used from PubNub is based on publishing and subscribe to messaging. It provides real-time data streaming and device signaling and includes built-in AES encryption and optional TLS/SSL encryption. The atomic component that makes up a data stream are API keys, messages, and channels. PubNub is also providing SDK's

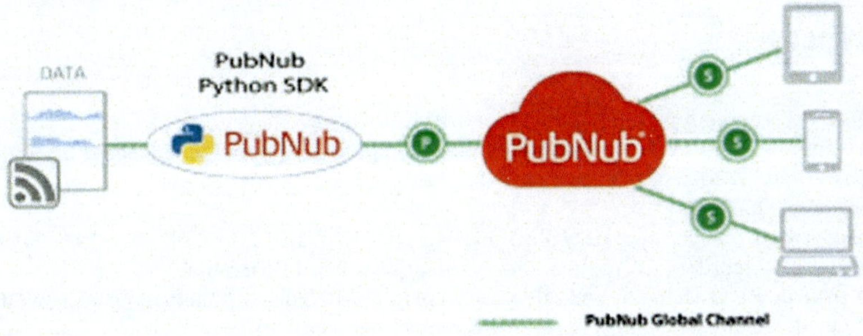

Fig. 14 Data flow in PubNub channel

for scripting language (Javascript) to develop the webpage for interaction and also for android application development with predefined libraries.

The proposed system identifies different patterns by analyzing the data collected from the vehicle which are stored in the cloud. The patterns like driver behavior, time, and weather which lead to the majority of the accidents are identified to provide awareness to society. The reasons for accidents like external attacks, malfunctioning of the vehicle, and wrong operations of the driver are analyzed.

4 Conclusion

In the present world, there is no system identification and intimation regarding road accidents and their location. Besides, the rescue operators do not understand the reasons for accidents like external attacks, malfunctioning of the vehicle, and wrong operations of the driver. This becomes a great threat for the society in identifying the reasons for the accident and causes for the incident. The deadly accidents many times not providing enough information for further investigation due to many reasons like the absence of affected person and onsight viewers. This becomes a great lag in detecting the event in remote places without any humans and finding out the reason for accidents and tracking the vehicle which caused this accident to occur. Also, the detection mechanism needs to provide enough information to the upcoming vehicles in manufacturing design and sensor installation.

Hence, this research work will be a good solution for identifying multiple factors like a reason for an accident, time of the accident, location of the accident, tracking the location, and sensory data stored in the cloud platform for further analysis. This monitoring and alert system will enrich the current rescue teams in tracking the accidents, reasons for accidents, and further investigation analysis. Also, the proposed work will enable better mechanisms in upcoming vehicles by providing additional sensors and detection mechanisms.

References

1. https://www.ncbi.nlm.nih.gov/pmc/articles/PMC3068826/
2. www.morth.nic.in http://www.indiaenvironmentportal.org.in/files/file/Road%20accidents%20in%20India%202016.pdf
3. www.tn.gov.in/sta
4. https://www.irfnet.ch/roadsafety.php?id=104&title=RADaR%20-%20Road%20Accident%20Data%20Recorder%20Application—International Road Federation
5. S. Krishnan, Road accidents and safety measure in TamilNadu—an analysis (Assistant Director of Statistics, Statistics Unit, Transport Commissioner's Office)
6. J.D Lloyd, Biomechanics of solo motorcycle accidents. Forensic Biomec. **16**(216), 125–131
7. M.A. Elliott, B.J. Baughan, B.F. Sexton, Errors and violations in relation to motorcyclists' crash risk. Accid. Anal. Prev. **39**, 491–499 (2007)
8. T. Özkan, T. Lajunen, B. Doğruyol, Z. Yıldırım, A. Çoymak, Motorcycle accidents, rider behaviour, and psychological models (Safety Research Unit, Department of Psychology, Middle East Technical University, Inonu Bulvari, 06531 Ankara, Turkey Centre for Research in Political Psychology, Queen's University, BT9 5BN Belfast, UK)
9. S.K. Singh, Road traffic accidents in India: issues and challenges. Transp. Res. Procedia **25**, 4708–4719 (2017)
10. H.P. Ruffell Smith, Time to die from injuries received in road traffic accidents. Injury **2**(2), 99–102
11. C. Priyanka, C.P. Darshini, V.G. Shavi, S. Begum, Two-wheeler safety system for accident prevention, detection and reporting. Int. J. Eng. Comput. Sci. **7**(03), 23680–23682 (2018)
12. K.-F. Wu, L. Sasidharan, C.P. Thor, S.-Y. Chen, Crash sequence based risk matrix for motorcycle crashes. Accid. Anal. Prev. **117**, 21–31 (2018)
13. M. Šraml, T. Tollazzi, M. Renčelj, Traffic safety analysis of powered two-wheelers (PTWs) in Slovenia. Accid. Anal. Prev. **49**, 36–43 (2012)
14. D.A.S. Chewb, W. Wuc, Z. Zhouc, Q. Li, Design and implementation of an identification system for proactive accident prevention at construction site. Accid. Anal. Prev. **48**, 193–203 (2012)
15. P. Seiniger, K. Schroter, J. Ghail, Perspectives for motorcycle stability control systems. Accid. Anal. Prev. **44**, 74–81 (2012)
16. R. Igual, C. Medrano, I. Plaza, A comparison of public datasets for acceleration-based fall detection. Med. Eng. Phys. **37**(9), 870–878 (2015)
17. R.M. Gibson, A. Amira, N. Ramzan, P. Casaseca-de-la-Higuera, Z. Pervez, Multiple comparator classifier framework for accelerometer-based fall detection and diagnostic. Appl. Soft Comput. **39**, 94–103 (2016)
18. A. Boubezoul, S. Espié, B. Larnaudie, S. Bouaziz, A simple fall detection algorithm for powered two wheelers. Control Eng. Pract. **21**, 286–297 (2013)
19. B. Soewito, Irwan, A. Antonyová, F. Gunawan, Fall detection algorithm to generate security alert. Procedia Comput. Sci. **59**, 350–356 (2015)
20. I.A. Zualkernan, F. Aloul, F. Basheer, G. Khera, S. Srinivasan, Intelligent accident detection classification using mobile phones. Ieeexplore, 978-1-5386-2290-2/18 ©2018 IEEE
21. O.K. Tonguz, W. Viriyasitavat, Cars as roadside units: a self-organizing network solution. IEEE Commun. Mag. **51**(12), 112–120 (2013)
22. https://www.bosch-presse.de/pressportal/de/en/bosch-lean-angle-sensor-for-motorcycles-42375.html
23. http://machinelearningmastery.com/

Monitoring and Controlling of ICU Environmental Status with WiFi Network Implementation on Zynq SoC

Dharmavaram Asha Devi, Tirumala Satya Savithri, and M. Suresh Babu

Abstract The role of embedded systems in medical applications is very crucial. The development of such applications made by ASICs is much expensive and includes high NRE cost and no enhancements are possible because they are one-time programmable. Therefore, a better solution is available with reconfigurable PLDs. In the present pandemic Covid 19 situation, there is a requirement for a lot of medical facilities at a lower cost with accurate monitoring and controlling is required. Therefore, the proposed medical application is an ICU monitoring and control system with a WiFi network, implemented on a Zynq architecture-based FPGA development board. The functionality of this system is, the ICU environmental parameters are monitored and displayed on OLED display and control of various appliances remotely within the vicinity of WiFi network. Such systems are useful in both Government and corporate hospitals to have the flexibility and smooth monitoring of ICUs.

Keywords Application-specific ıntegrated circuit · Field programmable logic array · Intensive care unit · Programmable logic device · Peripheral module · Zynq architecture

1 Introduction

Intensive care unit is a special health caring facility to help the severe life threatening patients. They are operated by trained and experienced doctors, physicians, respiratory therapists who specialize in caring for critically ill patients [1]. In present pandamic situation, the quantity and quality of ICUs should be enhanced vastly. In

D. A. Devi (✉)
Sreenidhi Institute of Science and Technology, Hyderabad, India

T. S. Savithri
JNTUH, Hyderabad, India

M. S. Babu
Gouthama Budha Society, Anantapur, A.P, India

V. Suma et al. (eds.), *Inventive Systems and Control*,
Lecture Notes in Networks and Systems 204,
https://doi.org/10.1007/978-981-16-1395-1_48

ICUs, it is very much required to maintain the environmental parameters in addition with patient treatments and monitoring equipment. In India, Indian Society of Critical Care Medicine (ISCCM) guidelines are there for the design of ICUs, the corresponding rules and regulations. India got the acceptance for the bid presented by Dr. Subhal Dixit to World Federation and got the World Congress 2025 to India, ISCCM. As per the environmental requirements, the Heating, Ventilation, and Air-conditioning system of ICU known as the HVAC system is used. It should be perfectly air-conditioned that allows control of temperature, humidity, and change of air flow [1–3]. This is recommended that the entire air will be filtered to 99% efficiency down to 5 microns. In critical care units with enclosed patient modules, the temperature should be adjustable within each module to allow a choice of temperatures from 16 to 25 °C [1, 2]. The ICU should have a self-back-up facility for interruption of power, which starts automatically in the event of a power failure. Here, Voltage stabilization is also mandatory. There are several embedded systems based monitoring and controlling systems for ICUs, But more reconfigurable monitoring and controlling units with greater accuracy for such applications, Zynq architecture based ICU environmental status monitoring and control system prototype is implemented in the proposed work. The architecture of Zynq is a combination of FPGA and the Microcontroller architectures on a single IC. Here, FPGA used is Artix 7 FPGA and Microcontroller is ARM9 dual-core. The complex hardware part is loaded in Artix 7 FPGA, and the application software is controlled by an ARM9 processor. Therefore, the terminology, PS: Processing System and PL: Programmable Logic is used for ARM9 and Artix 7 FPGA respectively [1, 2, 4].

2 Literature Survey

Systems on Chip using reconfigurable devices [1] are widely used in small and medium level industries to meet their requirements. The advantages of field-programmable gate arrays [1–3] over the ASIC are low NRE cost, less time to market, and flexibility of reconfiguration of the design.

In the latest front end designs and verifications with Zynq architecture based SoC [5] satisfies the functional requirements of data acquisition systems. The advantages of these systems are the important specifications like storage, processing, communication with high speed of data acquisition can be implemented in a tiny package [1, 4, 5].

The important application of human life, health monitoring of a patient, and the corresponding current trending technology like IoT and real-time embedded concepts are discussed in [6, 7]. The data acquisition systems using field-programmable gate arrays and their various features like high speed, multi-channel signal processing, high-end flexibility, and Reconfigurability are explored in [8–12].

Business intelligence and big data analysis stated that data volume has been doubled every year. Because of the decrease in size of Hardware and its cost, the usage, and development of smart sensors for advanced software applications including IoT [13].

The author [14], stated that the privacy is designed as an important approach with data protection algorithms. These privacy and data protection policies should reduce the operating prize and risk and establish customer trust.

While designing the VLSI circuits and systems, testability is an important role. Design for Testability (DFT) in power line communications in the VLSI designs is discussed in [15]. While designing a circuit, power efficiency is also an important criterion. The design of such SoC, ALCCU is discussed in [16]. It has shown clearly the power analysis of the design for different I/O standards with different clock frequencies. While designing complex SoC designs, to improve the power efficiency, network on chips (NoCs) is replaced in their place [17]. As the data communication speed is measured in terms of bit rate and baud rate, the corresponding definitions and the relations are also specified. In every real-time embedded application, every task is based upon time. The design of a real-time clock on the reconfigurable device and the corresponding IP creation using CAD tool is explored in [18].

The reconfigurable SoC designs on Artix 7 FPGA, using HDL are explained in [19, 20]. The peripheral modules used in the proposed design and the corresponding manuals and data sheets are listed in [21–24].

3 Methodology

The proposed design methodology is implemented with semi-custom block-level Zynq SoC design. All the blocks are soft IP cores integrated into a design. After appropriate connectivity of the design, generate output ports and validate the design for error-free. Once validation is over, generate the HDL wrapper for the corresponding block design. Synthesis, implementation, and generate bit stream are the following steps. The software platform used for all these processing steps is Xilinx Vivado System Design Suite.

After generating the Bit-stream, export the hardware including bit-stream and launch Software Development Kit (SDK). It is an integrated design environment for creating embedded applications. At the SDK platform, the application project is created and develops the necessary application logic using C/C+ and HTML for the web page. Build the project; program FPGA to load the complex design into programmable logic. Application software is processed by ARM9 dual core processing system. After running the hardware debugging option, observe the results at hardware and on the webpage which have been created to monitor the results. In the proposed system, 'Tera Term' is also used as a terminal and configured as a serial communication port with a baud rate of 115,200. In the proposed design, the same method is used for ICU Field Monitoring System. In the experimental setup, the sum of the peripherals are presumed with LEDs and pushbuttons in the place of real control switches of actual lights, fans, and other necessary appliances.

4 Design of ICU Field Monitoring System with Zynq

The block diagram is an integration of the ZYNQ processing system and several peripheral modules interfaced with AXI interconnect. There are four peripheral modules used in the design. They are Wi-Fi module, Hygro, ambient light sensor, and OLEDrgb. Apart from these PMOD IPs, there are two AXI GPIOs, one for eight LEDs and the other for five push button switches to monitor and check the status condition of ICU environmental parameters in the prototype implementation.

The advantage of this design is the interfacing of all the peripheral modules and GPIO modules are with AXI4. There are three main advantages of AXI4. The first one is productivity, because only one standard protocol is sufficient to learn. Second one is the flexibility in applying the right protocol. It allows memory-mapped interfaces and burst of 256 cycles of data rate in a single phase of the clock. AXI4-Lite is known as lightweight. Because it will handle a single transaction memory-mapped interface. And it has a simple interface for both design and usage. The third advantage is availability. It is used by ARM and many other third-party vendors worldwide in their development of systems, verification process, and performance characterization.

As this design is a Zynq architecture based, all the features of the Zynq processing system can be used. The proposed design is intended to monitor the temperature, humidity, and light intensity of the ICU. Depending on the day and night, the light intensity can be enhanced by switching ON and OFF wirelessly through a Wi-Fi network. The humidity and temperature adjustments can be made through GPIO ports. There are four ports known as Ja, Jb, Jc, and Jd used for the interfacing of WiFi, Hygro, ALS, and OLEDrgb peripheral modules. All these modules are interfaced with a Zynq processing unit via AXI4 interconnect as illustrated in Fig. 1. To make this complex block design, Xilinx Vivado system design suite software is used.

After completion of the block design using custom IP integration that the following steps are to be executed using the Xilinx Vivado tool.

(i) Generate the HDL wrapper
(ii) Synthesize the design
(iii) Implementation
(iv) Generate Bitstream
(v) Export the hardware design including the generated bit stream to SDK
(vi) Launch SDK
(vii) At SDK, create application project and create the application src file in the project
(viii) Interface all the peripheral modules to the target FPGA development board and make the appropriate connections from the board with a computer system where all the above-specified steps have been completed.
(ix) Build the project and program PL
(x) Run-on hardware debug at SDK and
(xi) Verify the results

In the proposed design, the VHDL wrapper has been generated because for the synthesis process RTL code is the main input besides libraries and constraints. The

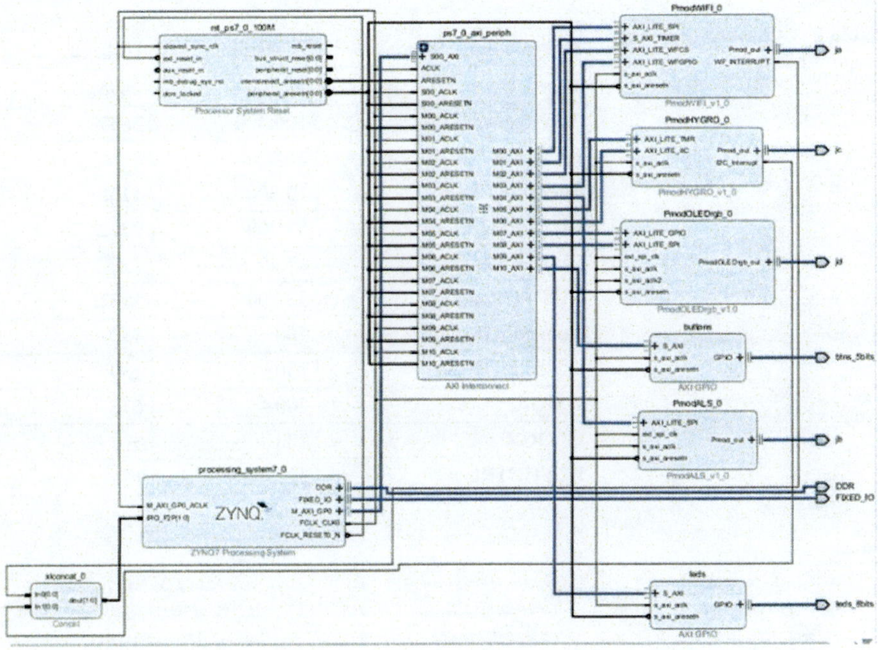

Fig. 1 Architectural block diagram of ICU field monitoring system

main constraints are area, power, and performance. The term synthesis process used in the front end part is a logical synthesis which will convert the RTL code into a logic gate-level netlist of the design [19, 20]. Along with this netlist, it will generate the power, area, and timing reports as shown in Fig. 2, Table 1 and Fig. 3 respectively. After synthesis, the implementation process is followed.

The steps from (i) to (vi) are implemented in the Vivado tool flow. The remaining processing of steps is implemented at the SDK platform.

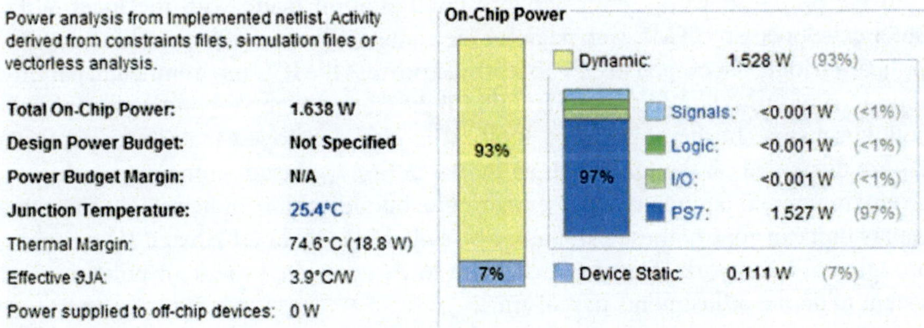

Fig. 2 Power analysis of the design generated from the implemented Netlist

Table 1 Utilization report of proposed design

Name of the resource	Available	Utilized
Slice LUTs	53,200	2895
Slice registers	106,400	3637
F7 Muxes	26,600	4
Slice	13,300	1165
LUT as logic	53,200	2772
LUT as memory	17,400	123
LUT FF Pairs	53,200	1582
Bonded IOB	200	45
Bonded IOPADs	130	130
ILOGIC	200	6
OLOGIC	200	3
BUFGCTRL	32	1

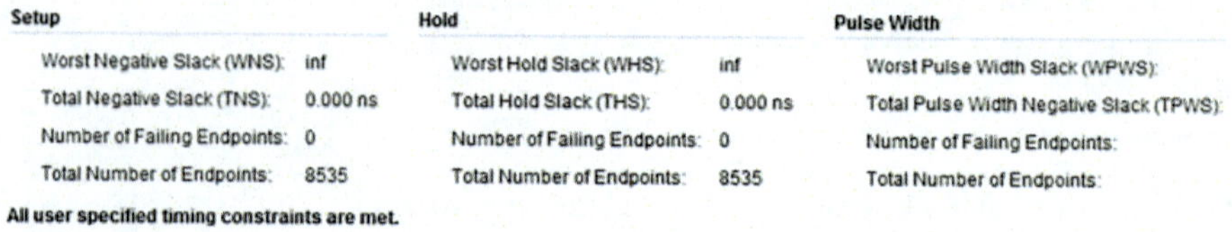

Fig. 3 Timing report

5 Software Development

Software Design Kit (SDK) is an integrated application software development tool for the Zynq architecture based design made at Vivado flow, and verifies the real-time embedded applications. These tasks are processed in a Zynq processing system known as PS.

In the proposed work, WiFi network configuration made with the local WiFi router developed an HTML web page for the status indication of the appliances. The application logic developed in CPP to read and process the ICU environmental parameters: temperature, humidity, and ambient light from the corresponding peripheral module sensors. In the application logic, it is also developed that, these parameters are displayed on the OLEDrgb as shown in Fig. 4. Based on the monitoring of the environmental parameters like temperature, humidity, and ambient light from the display unit can control these parameters by switching ON or OFF by GPIO switches through WiFi Network. These switches are to be connected to an air conditioning system to do the adjustments in real-time.

Fig. 4 ICU environmental parameters monitoring

6 Result Analysis

SDK is a platform to design the application software, build and program the FPGA. Before building an application project, the target board has to be connected to the computer system where the application project has been established. Here, two types of connections are used. One is USB to JTAG for programming purposes and the other is for serial communication through UART as shown in Figs. 7 and 8. In the proposed work, serial communication is made with a baud rate of 115,200. In order to check or scan the WiFi network establishment, Tera Term, a virtual terminal is used and illustrated in Fig. 5.

At the Tera Term terminal, the IP: 192.168.0.16: 80 has displayed which is an IP address to display the GPIO States web page as shown in Fig. 6.

In the GPIO states web page, five pins: Pin 0 to Pin 4 are indicated by the 5 push buttons which are interfaced with the Zynq processing system as shown in Fig. 1 block diagram. The real buttons are observed at the right bottom corner of a Zedboard illustrated in Fig. 7.

The purposes of these input switches are to control the required appliances to customize the ICU ambiance in real time. The accessing of these inputs status is made with a WiFi network. The 8 Pins: Pin 5 to Pin 12 are connected to LEDs. In real time these can be connected with LED lights with necessary relays and drivers are connected to these Pins.

Based on the light intensity monitored on display, the specific number of lights can be switched ON or OFF remotely on the webpage. The experimental results shown in Fig. 6 are a web page indicated at the initial RESET conditions. Now, to check the validity of the application, Pin 5, 7, 9, and 11 are made HIGH as shown in Fig. 7.

Then, there will be an immediate change happens at the hardware board according to the pin status. Here, LED 0 to LED 7 are connected to GPIO Pins: 5–12. As Pins

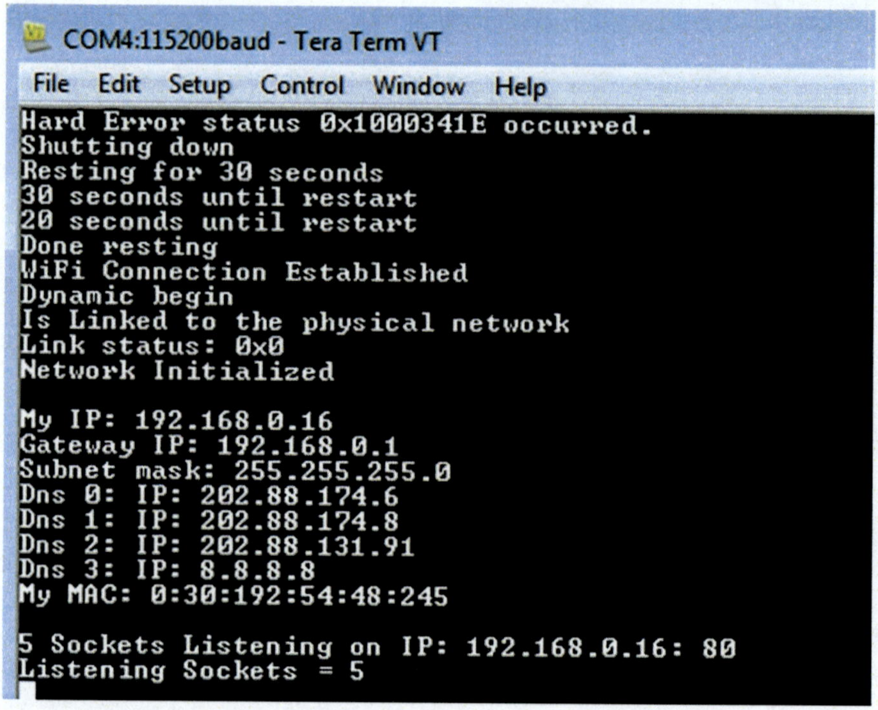

Fig. 5 Tera term VT to check the WiFi status

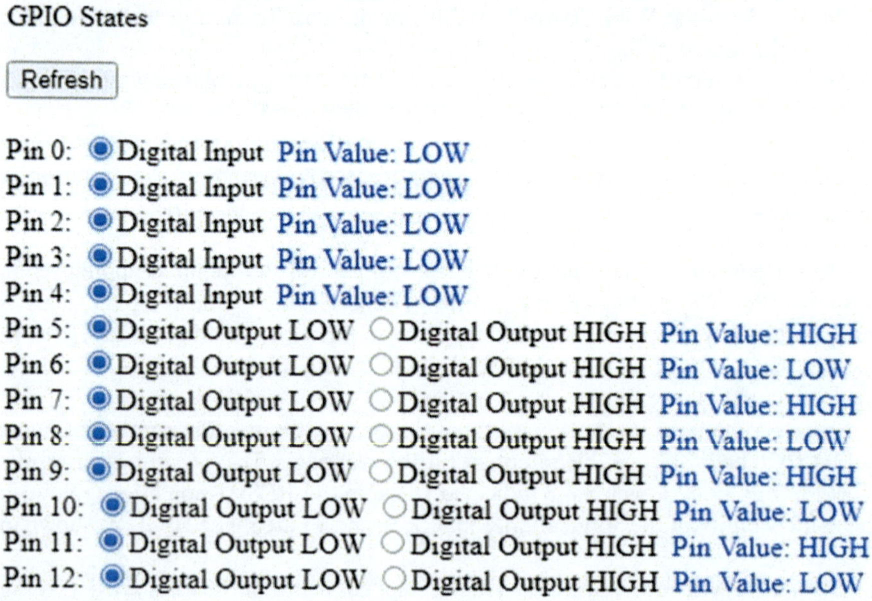

Fig. 6 GPIO states

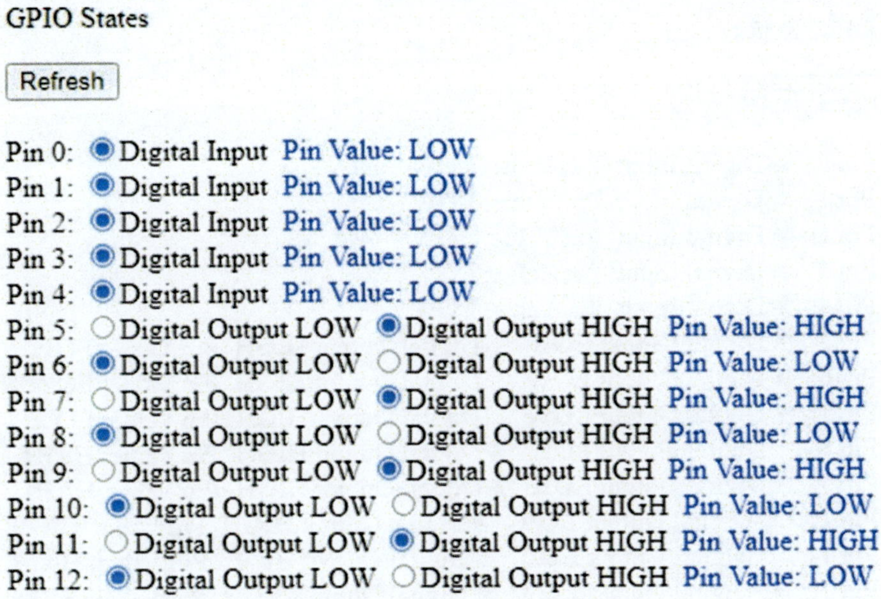

Fig. 7 GPIO Pins 5, 7, 9, and 11 made high on the webpage

5, 7, 9, and 11 are made high at the webpage, the corresponding LEDs: 0, 2, 4, and 6 are illuminated as shown in Fig. 8. In the second iteration, Pins: 8–12 are made high at the webpage, and the corresponding LEDs: 3, 4, 5, 6, and 7 are illuminated as shown in Figs. 9 and 10.

Fig. 8 The alternative LEDs are illuminated (that are connected to Pin 5, 7, 9, and 11)

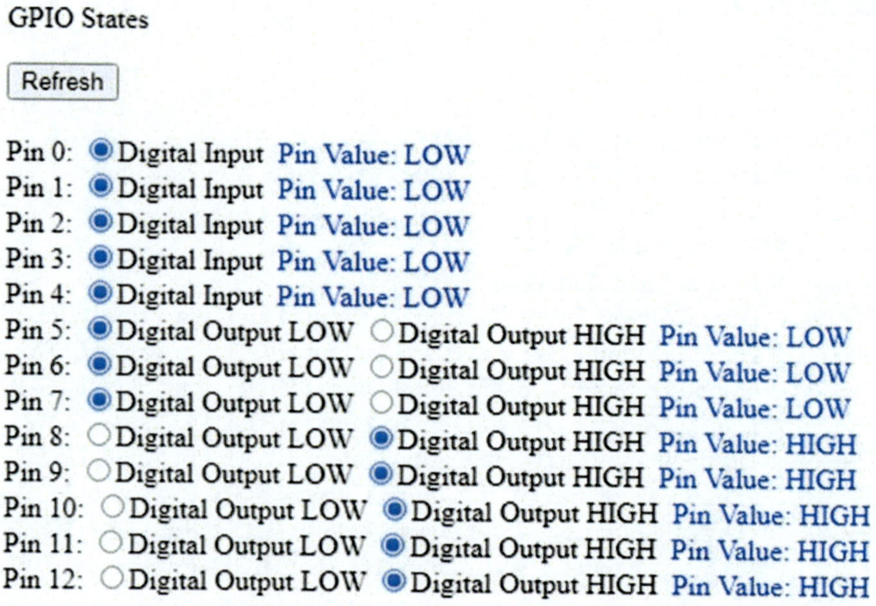

Fig. 9 GPIOs Pin 8 to Pin 12 made high

Fig. 10 LEDs: 3–7 are illuminated

7 Conclusion

The proposed experiment has been strongly established as an embedded application for the monitoring and controlling of the ambient light, humidity, and temperature using a WiFi network. The same methodology can be used for real-time ICU. The same type of application can be implemented with Microprocessors and Microcontrollers also. However, flexible Reconfigurability is missing. Here, in this approach, the Zynq architecture based SoC design is implemented with a huge number of advantages are there as specified. There is a lot of future scope to enhance the ICU parameters monitoring and control. In addition to this, the patient monitoring can also be established in future work.

Acknowledgements The proposed work has been carried out with valuable funding support from the TEQIP III sponsored by JNTU Hyderabad. We express our gratitude to the authorities of JNTUH University and Sreenidhi Institute of Science and Technology for providing their encouragement and support in the successful completion of the TEQIP III project.

References

1. D.A. Devi, T.S. Savithri, L. Sai Sugun, Design and implementation of real time data acquisition system using reconfigurable SoC. Int. J. Adv. Comput. Sci. Appl. **11**(9) (2020)
2. D.A. Devi, T.S. Savithri, L. Sai Sugun, Satellite tracking data acquisition system using reconfigurable SoC. J. Xi'an Univ. Archit. Technol. **XII**(III), 4471–4479 (2020)
3. M. Bhavani, D.A. Devi, Design of smart monitor for automobiles using FPGA based data logger, in *International Conference on Communication and Electronics Systems (ICCES), Coimbatore, India* (2019), pp. 1940-1945. https://doi.org/10.1109/icces45898.2019.9002034
4. Survey on developing data acquisition system using ZYNQ architecture, ZedBoard_HW_UG_v1_9.pdf, ZedBoard (ZynqTM Evaluation and Development) Hardware's Users Guide, Version 1.9 (2013)
5. L.H. Crockett, R.A. Elliot, M.A. Enderwitz, R.W. Stewart, *The Zynq Book Embedded Processing with the ARM cortex-A9 on the Xilinx, Zynq 7000 All Programmable SoC*, 1st edn. (Strathclyde Academic Media, 2014)
6. IOT based ICU patient monitoring system. http://IOT Based ICU Patient Monitoring System using esp8266 wifi module (microcontrollerslab.com)
7. V. Bhelkar, D.K. Shedge, Design and implementation of health monitoring device using FPGA. Int. J. Comput. Appl. **158**(7) (2017)
8. C. Rajasekaran, R. Jeyabharath, P. Veena, FPGA SoC based multichannel data acquisition system with network control module. Circuits Syst. **8**, 53–75 (2017). ISSN Online: 2153-1293
9. S. Khan et al., FPGA based high speed data acquisition system for electrical impedance tomography. J. Phys. Conf. Ser. **434**, 012081 (2013)
10. Y. Fan, FPGA based data acquision system, in *IEEE International Conference on Signal Processing, Communications and Computing* (2011). https://doi.org/10.1109/icspcc.2011.6061756. Electronic ISBN: 978-1-4577-0894-7
11. D. Roggow, P. Uhing, P. Jones, J. Zambreno, A project-based embedded systems design course using a reconfigurable SoC platform, in *2015 IEEE International Conference on Microelectronics Systems Education (MSE)* (2015). https://doi.org/10.1109/mse.2015.7160005. ISBN: 978-1- 4799-9915-6

12. A.A. Khedkar, R.H. Khade, High speed FPGA-based data acquisition system. Microprocess. Microsyst. **49**, 87–94 (2017)
13. M. Suresh Babu, K. Bhavana Raj, D. Asha Devi, Future trends of business intelligence and big data analytics in ubiquitous environment. Int. J. Eng. Adv. Technol. **8**(3S), 773–778 (2019)
14. M.S. Babu, K.B. Raj, D.A. Devi, Data security and sensitive data protection using privacy by design technique, in *2nd EAI International Conference on Big Data Innovation for Sustainable Cognitive Computing. EAI/Springer Innovations in Communication and Computing*, eds. by A. Haldorai, A. Ramu, S. Mohanram, M.Y. Chen (Springer, Cham, 2021). https://doi.org/10.1007/978-3-030-47560-4_14
15. G. Sahithi, A. Devi, A PLC receiver design for testability. Int. J. Innovative Technol. Exploring Eng. (IJITEE) **8**(6), 26-31 (2019). ISSN: 2278-3075
16. D.A. Devi, M. Suresh Babu, Design and analysis of power efficient 64-Bit ALCCU. Int. J. Recent Technol. Eng. **8**(2), 162–166 (2019)
17. D.A. Devi, S. Jaga, Analysis of scheduled routing algorithms on 5-port router for network on chip application. Int. J. Sci. Technol. Res. **8**(9), 2148–2153 (2011)
18. D.A. Devi, N.S. Rani, Design and implementation of custom IP for real time clock on reconfigurable device, in *Third International Conference on Inventive Systems and Control* (2019), pp. 414–418. https://doi.org/10.1109/icisc44355.2019.9036428
19. D.A. Devi, L.S. Sugun, Design, implementation and verification of 32-Bit ALU with VIO, in *2nd International Conference on Inventive Systems and Control* (2018), pp. 495–499. https://doi.org/10.1109/icisc.2018.8399122
20. D.A. Devi, S. Chintala, S.L. Sai, Design of power efficient 32-bit processing unit. Int. J. Eng. Technol. **7**(2.16), 52–56 (2018). http://dx.doi.org/10.14419/ijet.v7i2.16.11415
21. WiFi PMOD maual. https://reference.digilentinc.com/_media/reference/pmod/pmodwifi/pmodwifi_rm.pdf
22. Pmod-hygro-rm.pdf, PMOD HYGRO Reference Manual, Revised 22 Feb 2017
23. pmodals_rm.pdf., PmodALSTM, Reference Manual, Revised 15 April 2016
24. Pmodoledrgb_rm.pdf, PmodOLEDrgb TM, Reference Manual, Revised 26 April 2016

Dharmavaram Asha Devi Member IEEE, Fellow IETE. The Author is working as a Professor in the Department of Electronics and Communication Engineering, SNIST, Hyderabad. She became an Associate Member of IETE in Dec-2000 from the Institute of Electronics and Telecommunication Engineering, New Delhi, India. She was awarded M.Tech. Degree in Digital Systems and Computer Electronics from JNTU Anantapur in 2005. She achieved a Doctorate Degree from Sri Krishnadevaraya University, Anantapur in Dec 2011, for her research contribution in Embedded System based Soil Analysis under the supervision of Prof. K. Malakondaiah, from SKU, Anantapur. She has published more than 35 papers in International Journals and presented 40 plus research papers in International level conferences on behalf of her research work done in Embedded Systems and VLSI Designs. Her present research area is FPGA-based embedded applications, design, and verification of SoC designs using both front end and back end VLSI CAD tools.

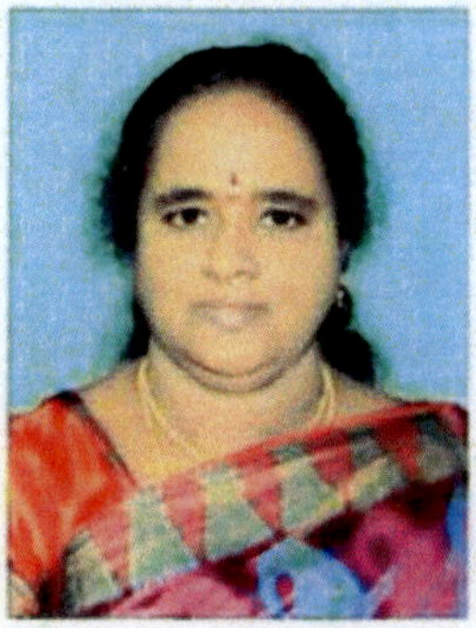

Tirumala Satya Savithri Member IEEE and IE, working as a Professor in the Department of ECE, JNTUCEH. Her educational qualifications are UG in Electronics and Communication Engineering and PG in Microwave and Radar Engineering. She was awarded with PhD in 2006 for her research contribution in Digital Image Restoration Techniques. She published more than 100 International Journals and Conference Papers. Her area of research is VLSI Designs, Signal, and Image processing applications.

Reversible Data Hiding Using Secure Image Transformation Technique

K. Ramesh Chandra, Madhusudan Donga, and Prudhvi Raj Budumuru

Abstract With the gigantic popularity of re-appropriating data in the path of the cloud, it's essential towards the protection of data and authorizes the cloud employee to easily change the knowledge at the same time. Underneath such requests, Reversible Data Hiding in Encrypted İmages (RDH-EI) attracts associate with an ever-increasing variety of specialists' decisions concerning the structure for RDH-EI upheld Reversible İmage Transformation (RIT). Two RDH methods, together with customary RDH conspire and unified putting in and scrambling arrangement, data received to insert watermark within the encoded image, which can fulfill numerous wants on image quality and big implanting limit individually.

Keywords Reversible data hiding in encrypted images (RDH-EI) · Reversible image transformation (RIT) · Reversible data hiding (RDH)

1 Introduction

This article is based on reversible data hiding using reversible image transformation. PCs and the Internet are the fundamental correspondence medium that associates various pieces of the world to one worldwide virtual world in current culture. Therefore, individuals can certainly trade data and separation is not, at this point a correspondence obstruction. It will presumably be a security issue for significant distance correspondences. This is truly significant for classified information. The answer to this issue prompts the improvement of a steganography plan. Steganography is an extremely amazing security device that gives a significant level of security, particularly when related to encryption. Encryption and steganography are notable and generally utilized advancements that control data and encode or shroud every element. Encryption scrambles the message to make it confused. Steganography

K. R. Chandra · P. R. Budumuru (✉)
Vishnu Institute of Technology, Bhimavaram, India

M. Donga
Vignan's Institute of Information Technology, Visakhapatnam, India

V. Suma et al. (eds.), *Inventive Systems and Control*,
Lecture Notes in Networks and Systems 204,
https://doi.org/10.1007/978-981-16-1395-1_49

conceals the message from being seen. While the two advancements give security, research is in progress to consolidate encryption and steganography techniques in a solitary framework to improve protection and security. Send input History Saved Community.

Cryptography is the process of investigating the strategies for secure correspondence within the sight of outsiders likewise called foes. It is parallelled in developing and breaking down conventions that block these outsiders with the assistance of dissimilar angles in data security, for example, information classification, information honesty, verification, and non-disavowal. Present-day cryptography exists at the crossing point of the standards and guidelines of maths, software engineering, picture preparing, and so forth There are numerous uses of cryptography that incorporate PC passwords and online business.

Steganography is the specialty of concealing data and a push to hide the presence of the installed data. It fills in as a superior method of making sure about the message than cryptography which just covers the substance of the message not the presence of the message. The unique message is being covered up inside a transporter (for our situation the Image) with the end goal that the progressions so happened in the transporter are not perceptible. Steganography is a valuable apparatus that allows the covert transmission of data over and over the correspondence channel. Following the mystery picture with the transporter, the picture gives the concealed picture. The concealed picture is hard to recognize without recovery. In the current situation, the transfer of data is available digitally in the form of a digital image. The digital image provides both the consumer and the device confidentiality, so attacks on the digital image are possible. In order to solve data transmission problems using digital images, cryptographic techniques are needed to secure data transmission and Confidential information. Cryptographic methods are used to preserve sensitive data for the safe and protect the transfer of information [1–4].

2 Existing System

According to [5] during 2016, around 2.2 trillion images have been released on the Internet and are expected to develop exponentially due to the appearance of modern visual logical applications such as the visual Internet of Things [6] and Big Visual Data [7].

In [8] The Generative Topographic Mapping (GTM) calculation is proposed as a stochastic reproduction of the SOM (self-coordinating guide). The GTM calculation catches the information structure by displaying nonlinear changes into a multidimensional information space that can be utilized as a perception instrument in a little element of likely factor space. The reason for this white paper is to stretch out the GTM calculation to deal with multivariate time arrangement. The standard GTM calculation expects that the information is free examples and similarly disseminated. This supposition is plainly improper for time arrangement. In this article, an augmentation of GTM for a multivariable time arrangement called GTM-ARHMM were proposed in which, it accepts the time arrangements that are produced by the ARC HMM which is covered up via auto-regressive.

The proposed method is inspired by Lai and Tsai [9], they first used the concept of mosaic image in the field of information security. The type of mosaic image used by them is a secret fragment visible mosaic image that is formed by dividing the source image into small tiles or blocks and arranging these blocks in a random position with respect to the target image. The target image is required to be selected from the database. The database plays a vital role for mosaic image creation so for better results the database should be sufficiently large in size. For target image selection we have to scan the database in search of the image with the highest similarity measure with respect to the given secret image. For this we have to divide the secret image into small tiles and calculate similarity measure for each block or tiles that is mean value of every block and randomly we have to select an image from database divide it into similar number of block as that of number of tiles of secret image and calculate mean value of each and every block then we have to compare mean values of secret image blocks with that of target image blocks to see whether at least 6 to 7 blocks values matches or not, if it doesn't match then we have to repeat the same procedure for another image from the database until we first the best fitted image [10, 11].

Hu et al. [11] in this paper is based on Prediction-error expansion (PEE)-based reversible data hiding schemes consisting of two steps. First, a sharp prediction-error (PE) histogram is generated by utilizing pixel prediction strategies. Second, secret messages are reversibly embedded into the prediction-errors through expanding and shifting the PE histogram. Previous PEE methods treat the two steps independently while they either focus on pixel prediction to obtain a sharp PE histogram, or aim at histogram modification to enhance the embedding performance for a given PE histogram.

A novel Image Steganography scheme is proposed by Chen and Lin [12], based on diverse requirements of applications, two operation modes and 5 cases are provided for selection. Contrasting the space domain methods, secret messages are implanted in the high frequency coefficients resulting from Discrete Wavelet Transform.

Steganography is the science that deals with conveying secret information by embedding it into the cover object invisibly. In steganography, only the authorized party is aware of the existence of the hidden message to achieve secret communication. The image file is mostly used as a cover medium amongst various digital files such as image, text, audio and video. The authors Kalpana et al. [13], proposed idea of this work is to develop the robust image steganography. It is implemented using Least Significant Bit and Discrete Wavelet Transform techniques for digital image signal to improve the robustness & evaluate the performance of these algorithms. The parameters such as mean square error (MSE), bit error rate (BER), peak signal to noise ratio (PSNR) and processing time are considered here to evaluate the performance.

2.1 Trusted Cloud Computing

Trusted cloud computing with protected resources and information shading:

From completely tolerating cloud stages are based on Trust and security. To ensure mists, suppliers should initially make sure about virtualized server farm assets, maintain client security, and save information honesty. The creators propose utilizing a trust-overlay network over numerous server farms to actualize a standing framework for building up trust between specialist organizations and information proprietors. Information shading and programming watermarking methods secure shared information objects also, hugely appropriated programming modules. These procedures protect multi-way validations, empower single sign-on in the cloud, and fix access control for touchy information in both public and private mists. underlines.

2.2 Reversible Watermarking Schemes

Customized Reversible Watermarking Schemes for Authentication of Electronic Clinical Atlas:

It is acknowledged that computerized watermarking is very significant in clinical imaging. Notwithstanding, because of the extraordinary idea of clinical practice, it is frequently required that watermarking don't acquaint irreversible mutilations with clinical pictures. Electronic clinical chart books have such a need for "lossless" watermarking. There are two custom-fitted reversible watermarking plans for clinical map books by abusing their inalienable qualities.

2.3 Recursive Histogram Notification

Recursive histogram alteration: Setting up equivalency between reversible information covering up and lossless information pressure:

Cutting edge plans for reversible information covering up (RDH) typically comprise of two stages: the first stage is to build a host succession with a sharp histogram through expectation mistakes, and afterward insert messages by changing the histogram with techniques, for example, contrast development and histogram move. The second stage is to propose a histogram alteration technique for RDH, which inserts the message by recursively using the decompression and pressure cycles of an entropy coder. For free indistinguishably circulated (i.i.d.) dark scale have signals, the proposed technique asymptotically moves toward the rate-contortion bound of RDH insofar as wonderful pressure can be acknowledged, i.e., the entropy coder can move toward entropy. Consequently, this technique builds up the equivalency between reversible information stowing away and lossless information pressure. Examinations show that this coding strategy can be utilized to improve the presentation of past RDH plans and the upgrades are more critical for bigger pictures.

3 Proposed System

3.1 *Reversible Data Hiding (RDH)*

Recursive histogram adjustment: setting up equivalency between reversible information stowing away and lossless information pressure:

Best in class plans for reversible information stowing away (RDH) normally consists of two stages. The first stage is to develop a host arrangement with a sharp histogram through expectation blunders, and afterward implant messages by altering the histogram with techniques, for example, contrast extension and histogram move. The subsequent stage is to propose a histogram alteration technique for RDH, which implants the message by recursively using the decompression and pressure cycles of an entropy coder. In the case of autonomous indistinguishably dispersed (i.i.d.) dark scale have signals, the proposed technique asymptotically moves toward the rate-contortion bound of RDH insofar as wonderful pressure can be acknowledged, i.e., the entropy coder can move toward entropy. Accordingly, this strategy builds up the equivalency between reversible information covering up and lossless information pressure. Trials show that this coding strategy can be utilized to improve the presentation of past RDH plans and the upgrades are more critical for bigger pictures.

3.1.1 Image Encryption

In cryptography, encryption is the way toward encoding a message or data with the objective that the lone asserted party can get to it and the individuals who are not embraced. The encryption itself doesn't prevent impedance, in any case, denies the justifiable substance to a future interceptor. In an encryption invention, the proposed data or message, hinted as plain substance, is blended utilizing an encryption calculation, a code making figure text that can be examined exactly at whatever point unscrambled. For specific reasons, an encryption plot overall uses a pseudo-optional encryption key conveyed by an assessment. It is on a key level conceivable to decode the message without having the key, notwithstanding, for a particularly organized encryption plot, expansive computational assets and aptitudes are required. An embraced beneficiary can unmistakably decode the message with the fundamental given by the originator to beneficiaries at any rate, and not to unapproved clients.

3.1.2 Symmetric Key

In symmetric-key plans, the encryption and unscrambling keys are the equivalents. Providing social occasions should have a near key to accomplish secure correspondence.

3.1.3 Public Key

Public key concept was introduced in 1973 and proposed by Diffie and Hellman. This scheme encrypts the message and anyone can decrypt the message with public key.

3.2 *Reversible Image Transformation*

Design of RIT comprise of the following modules. The first module in the reversible picture change of encoding. The next modules for reversible picture change of translating and the last module of client modules. Here, the most part expound on two modules in this segment.

RIT-Encoding Module

1. RIT encoding implies a Reversible Image Transformation procedure in which squares of pictures will be changed into squares of the target picture
2. At the hour of encoding, the framework will part source and target pictures into parts
3. Source picture parts will be rearranged with the assistance of SD estimations of pixels
4. The new grouping will be put away as the key
5. The system will choose pixel rearranging calculation haphazardly
6. Using rearranging, calculate the new places of source block pixels
7. With the assistance of new positions, the framework will change source picture blocks into target picture blocks
8. When all the squares will be changed, the joined picture will be consequently shaped.

RIT-translating Module

1. Decoding measure is actually a converse technique for encoding
2. To decipher the source picture, client will determine the discharge key
3. If the indicated key is right, the client will be considered as a validated client
4. With the assistance of that key, the framework will discover the rearranging calculation
5. Using that calculation, the framework will unravel source picture blocks

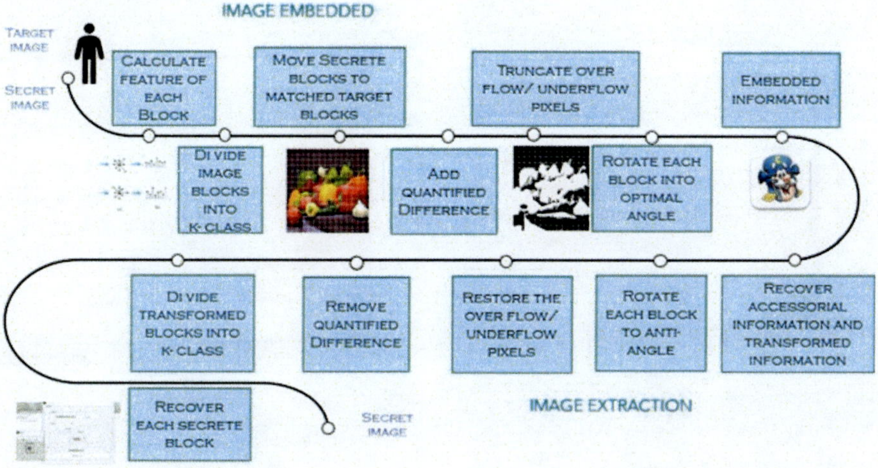

Fig. 1 Process of creating RIT

6. The source picture squares will be joined to get the total source picture
7. The system will get a sequence of the squares from emitting key (Fig. 1).

3.3 Block Pairing

To make the transformed picture *A'* seem as though the aim picture is *A*, confidently, after modification, individually altered squares will have an adjacent mean and standard deviation (SD) with the objective square. So, initially figure the mean and SD of each square of *B* and *A* separately. Leave a square *B* alone a bunch of pixels with the end goal that $C = \{i_1, i_2, \ldots, i_n\}$, and afterward the mean and SD of this square is determined as follows (Figs. 2 and 3)

$$u = \frac{1}{n}\sum_{j=1}^{n} i_j \tag{1}$$

$$\delta = \sqrt{\frac{1}{n}\sum_{j=1}^{n} i_j - u^2} \tag{2}$$

In Lee et al. technique, the squares of the fascinating picture and target picture are engineered in rising sales as shown by their SDs freely, and from that point each outstanding square is facilitated with a seeing objective tile likewise as indicated by the requesting. To recuperate the chief picture from the changed picture, the spots of the fundamental squares should be recorded and implanted into the changed picture with an RDH framework. On the off chance that the picture is disengaged into *N*

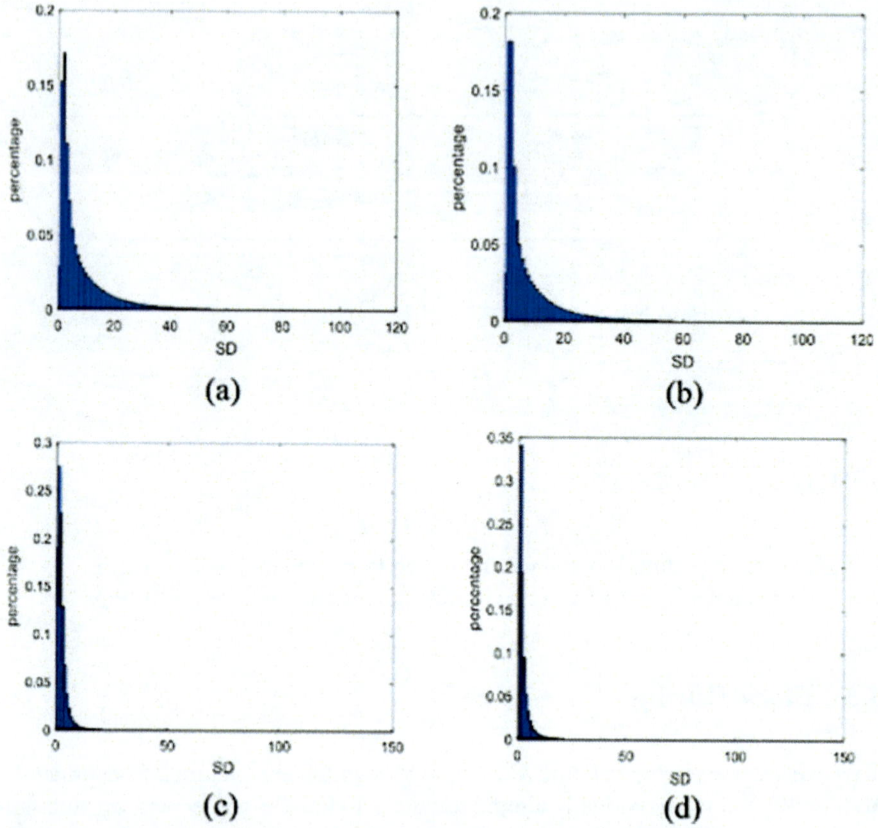

Fig. 2 The distribution of SDs of 4 × 4 blocks for various sizes of natural images

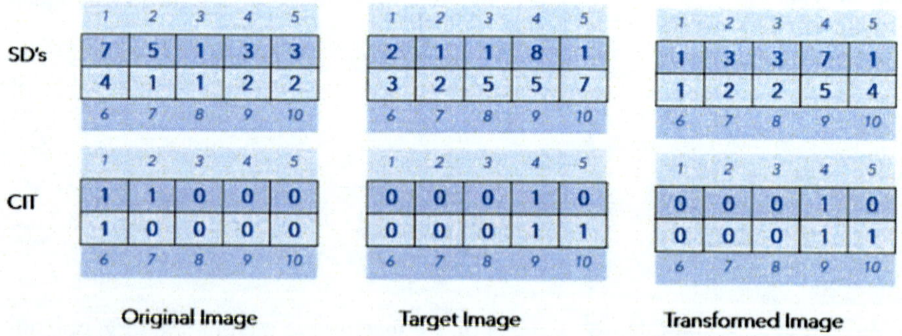

Fig. 3 An example of block pairing

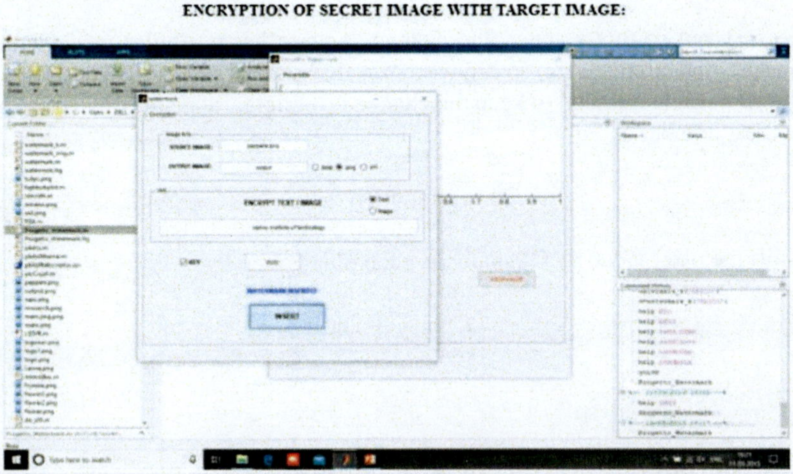

Fig. 4 Encryption of secrete image with target image

blocks NlogN pieces are depended upon to record block records. Unquestionably, the more modest the square size is, the better-changed picture which will be accomplished as a tremendous *N*. Therefore, the extent of data used to record the archive for each square might be giant to the point that it will cause a lot of reshaping while simultaneously bringing these data into the changed picture. There may not exist enough excess space to store these extra data. For example, on the off chance that a portion of 1024 × 1024 picture into 4 × 4 squares, two 16 × 16 pieces are depended upon to record the spots of squares (Fig. 4).

3.4 Block Transformation

Should By the square blending strategy portrayed above, in each pair (C, A), the two squares have close SD esteems. Accordingly, while changing C towards A, a mean is required for moving the change that is reversible. Notwithstanding, the change utilized in Lee et al. technique [6] isn't reversible since it changes the mean and SD simultaneously. Let the first square $C = \{i_1, i_2, \ldots, i_n\}$, and the comparing objective square $A = \{i_1, i_2, \ldots, i_n\}$. With Eq. (1), figure the methods for B and T and indicate them by uC and uA individually. The changed square $A = \{i_1, i_2, \ldots, i_n\}$ is created by the mean moving as follows

$$p^n = p_i + u_T + u_B \quad (3)$$

Be where $(u_A - u_C)$ is the contrast between the methods for target square and unique square. Here, move every pixel estimation of a unique square by sufficiency

$(u_A - u_C)$ and in this manner, the changed square has a similar mean with the comparing objective square. In any case, on the grounds that the pixel esteem pI should be a whole number, to keep the change reversible, rotate the distinction to be the nearest number as Eq. (4)

$$\mathrm{Ou} = \mathrm{round}(u_A - u_C) \tag{4}$$

and shift the pixel value by Ou, namely, each pi is gotten by

$$i^n = i_j + \mathrm{Ou} \tag{5}$$

3.5 *Experimental Results of RIT*

Techniques are introduced 100 sets of pictures are randomly picked as our test pictures from the Boss Base picture information base [2].

Right off the bat, all the pictures are preprocessed to get a similar size of 1024 × 1024 pixels. Since in the RIT strategy the boundary α affects the AI payload, and provide the trial to choose a superior α to improve the general presentation. The outcome is portrayed The modest of the space involved by AI is better for the scrambled picture's represented in visual quality. For α in the scope of [0.05, 0.95], the variety of AI payload is by all accounts not huge (Figs. 5 and 6).

AFTER ENCRYPTION:

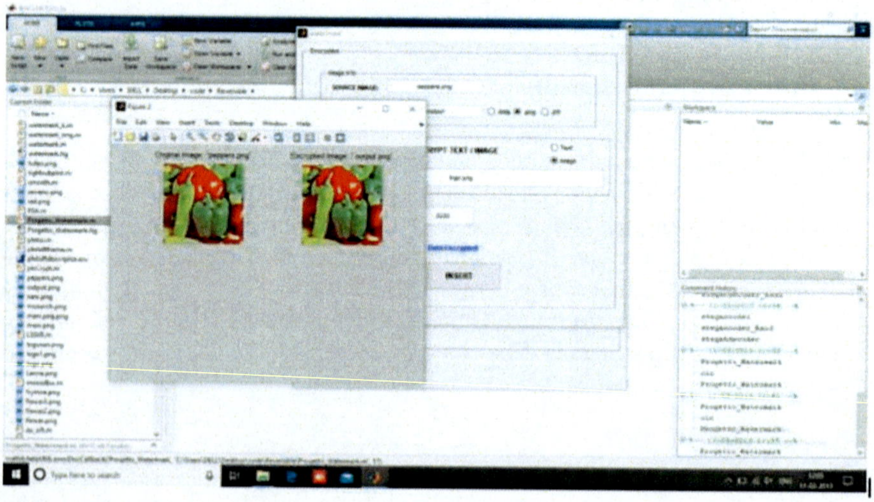

Fig. 5 After encryption of secrete image

ENCRYPTION OF SECRET IMAGE WITH TARGET IMAGE

Fig. 6 Encryption of secrete image

That when α is 0.75, the AI payload arrives at the valley esteem. So in the accompanying tests, α is set 0.75. The fundamental goal of the reversible information concealing strategy is to store the mystery picture in the objective picture by utilizing the mystery key which is only encryption. Figure 4 gives the Encryption of Secrete Image with Target Image. Figure 5 provides the after Encryption of Secrete Image (Fig. 7).

DECRYPTED SECRET IMAGE OUTPUT:

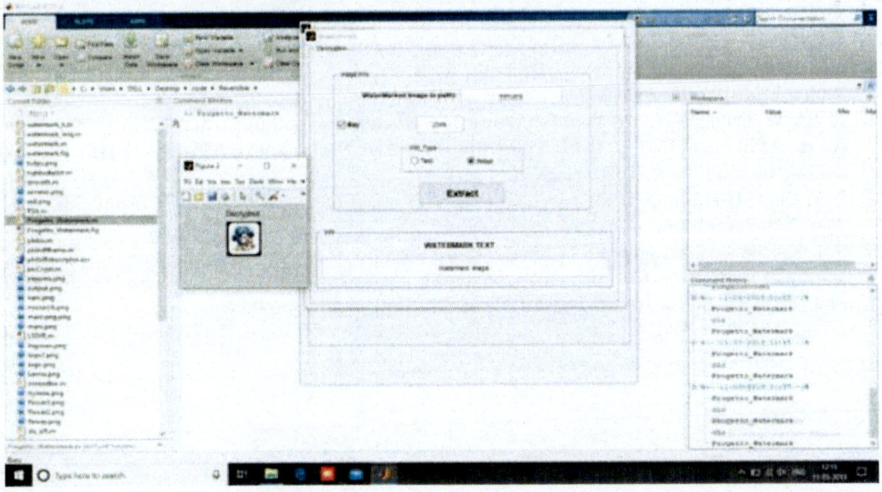

Fig. 7 Decrypted secret image output

4 Conclusion

The fundamental goal of the reversible information concealing strategy is to store the mystery picture in the objective picture by utilizing the mystery key which is only encryption. Allow us to consider, the logo is the mystery picture and to encode with the objective picture considered the peppers.

References

1. M.K. Chitra, V.P. Venkatesan, Antiquity to the contemporary of the secret sharing scheme. J. Innovative Image Process. (JIIP) **2**(01), 1–13 (2020)
2. M.R. Vinod Kanna, A secure steganography creation algorithm for multiple file formats. J. Innovative Image Process. (JIIP) **1**(01), 20–30 (2019)
3. K.R. Chandra, B. Prudhvi Raj, G. Prasannakumar, An efficient image encryption using Chaos theory, in *2019 International Conference on Intelligent Computing and Control Systems (ICCS), Madurai, India* (2019), pp. 1506–1510. https://doi.org/10.1109/iccs45141.2019.9065753
4. M. Dileep, B. Prudhviraj, G. Prasannakumar, A new image encryption and data hiding technique using wavelet transform. Int. J. Res. Eng. Technol. **5**(5), 195–198 (2016)
5. 3.5 million photos shared every minute in 2016. Deloitte UK, Deloitte United Kingdom (2018) [Online]. Available: https://www2.deloitte.com/uk/en/pages/press-releases/articles/3-point-5-million-photos-shared-every-minute.html. Accessed 05 June 2018
6. C. Chen, Y. Ren, C.-C.J. Kuo, *Big Visual Data Analysis, Scene Classification and Geometric Labeling*. (Springer Briefs in Signal Processing, 2016)
7. X. Cao, L. Du, X. Wei et al., High capacity reversible data hiding in encrypted images by patch-level sparse representation. IEEE Trans. Cybern. **46**(5), 1132–1143 (2016)
8. X. Zhang, Reversible data hiding in encrypted images. IEEE Signal Process. Lett. **18**(4), 255–258 (2011)
9. I.-J. Lai, W.-H. Tsai, Secret-fragment-visible mosaic image–a new computer art and Its application to information hiding. IEEE Trans. Infor. Foren. Secur. **6**(3), 936–945
10. Y-L. Lee, W-H. Tsai, A new secure image transmission technique via secret-fragment-visible mosaic images by nearly reversible color transformation. IEEE Trans. Circuits Syst. Video Technol. **24**(4), 695–703 (2014)
11. X. Hu, W. Zhang, X. Li, N. Yu, Minimum rate prediction and optimized histogram modification for reversible data hiding. IEEE Trans. Inf. Forensics Secur. **10**(3), 653–664 (2015). https://doi.org/10.1109/TIFS.2015.2392556
12. P.-Y. Chen, H.J. Lin, A DWT based approach for image steganography. Int. J. Appl. Sci. Eng. **4**(3), 275–290 (2006)
13. S.S. Kalpana, P. Mangal, J.S. Chitode, Least significant bit and discrete wavelet transform algorithm realization for image steganography employing FPGA. Int. J. Image Graph. Sig. Process. **6**, 48–56 (2016). https://doi.org/10.5815/ijigsp.2016.06.06

Analysis of Topologies for Cooperative Tracking Control of Multi-agent Systems

S. Suganthi and Jacob Jeevamma

Abstract This research work focuses on the analysis of several information-exchange topologies to deploy cooperative tracking for multi-agent systems with spacecrafts as agents. Cooperative tracking control or leader following consensus control consists of a virtual leader and the states of the followers are driven to those of the leader. In the cooperative tracking control, the leader and followers exchange the information through proper communication graph topology. The information exchange among the agents can happen at different levels from the leader to followers. To achieve perfect trajectory tracking of the leader and the followers, the agents are required to be connected in a proper communication topology. This paper mainly aims to identify the efficient topology for the multi-agent systems with spacecrafts as agents. The simulation results highlight the comparison of one leader and two follower topology with one leader and three followers topology that accommodates a virtual leader spacecraft and arbitrary information flow among the spacecrafts by considering different levels of information exchange. In general, for an efficient topology, the leader's state should be available to all the spacecrafts and there should be proper information exchange between the follower spacecrafts.

Keywords Multi-agent systems · Spacecraft · Leader-follower · Efficient topology · Information-exchange.

1 Introduction

Multi-agent system (MAS) consists of several independent agents that work cooperatively to achieve the particular coordinated tasks via a proper communication medium with other agents. It has leveraged its wide utilization in several contrastive areas, such as multiple robots [1], spacecrafts, combat systems, unmanned aerial vehicles, and so on. Spacecrafts referred as multi-agents are predominantly used for a variety of purposes like space colonization, navigation, meteorology, transportation

S. Suganthi (✉) · J. Jeevamma
Department of Electrical Engineering, National Institute of technology Calicut, Calicut, India
e-mail: jeeva@nitc.ac.in

V. Suma et al. (eds.), *Inventive Systems and Control*,
Lecture Notes in Networks and Systems 204,
https://doi.org/10.1007/978-981-16-1395-1_50

of cargo, and humans. Generally, to perform the desired tasks, there are two types of control problems in MAS

(1) Leaderless consensus (also known as cooperative regulator), as there exists a consensus equilibrium towards which all the agents are driven and it becomes dependent on initial conditions of the agents.

(2) Leader-following consensus (also known as cooperative tracking), where all follower agents are synchronized to track a leader agent which acts as a command generator, whether or not the leader's states are received by all or a few of the followers.

Cooperative tracking control of multiple agents has been extensively studied and has received immense research attentions in the past. Several solutions were proposed for this control problem in [2–4]. The cooperative tracking control of a multiple Euler-Lagrange systems was considered in [2]. The state synchronization and the trajectory tracking of the agents under a switching communication network topology have been analysed. In [3], the authors discuss the cooperative tracking of multiple quadrotors with a cooperative leader where the states of the followers are fed-back to the leader. Though the feedback may be useful to preserve the communication between the agents in a better way, it is not practically feasible. In [4], the cooperative tracking problem of multi-unmanned aerial vehicles is addressed and its effectiveness is verified when the agents are connected to a particular fixed topology with a virtual leader.

The study of information exchange between the agents and the different topologies plays a major role in the case of multi-agent systems to perform the desired tasks. This gives way to analyse the different information exchange topologies in which the agents can communicate. In [4], in order to improve the stability margin and accuracy for vehicle formations, the information-exchange techniques are studied, where a formation centre is used by the vehicles as a centre reference. The formation keeping of multiple spacecrafts is studied in [6–9] where the problem of both attaining specific formation and attitude alignment for multiple spacecraft via communication with one or two adjacent agents is considered. In [7], a leader-follower approach is applied, where a triad of spacecrafts are considered and information only flows from leader to followers, where the followers track leader's position and attitude. Though the leader-follower approach has several advantages in performing a particular task cooperatively, there are few limitations. The leader acts as the single source of failure in the whole communication topology. Besides, the followers states are not fed-back to the leader. In [8], and [9], the control law for each spacecraft is a function of the states of its two adjacent neighbours. A bidirectional ring communication topology is considered in [8]. The spacecrafts are interconnected in the form of an undirected graph where the spacecrafts have to identify their adjacent two spacecrafts. Thus, group feedback between the agents is considered in [9] which may be unrealistic in the case of range limitations, communication bandwidth where the followers are far away from the leader.

Information-exchange topology between three spacecrafts in leader-follower approach and several distributed information-exchange topologies between four spacecrafts in leader-follower approach is considered in this work. In the information-exchange topology, the leader's state may also be available to all the follower spacecrafts. The leader's state for the cooperative tracking may also be available to some of the follower spacecrafts and these follower spacecrafts may or may not have a directed path to all of the other follower spacecrafts. The contribution of the work is that an efficient topology is identified from several topologies to achieve the perfect state synchronization and trajectory tracking of the followers.

The rest of the work is organized as follows. Section 2 discusses the basics and preliminaries of graph theory for multi-agent systems. The dynamics of spacecrafts are explained in Sect. 3. Section 4 deals with the comparison of the cooperative tracking of one leader and two follower spacecrafts (topology-1) with the one leader and three follower spacecrafts (topology-2). The identification of the better topology for cooperative tracking is discussed in Sect. 5. The simulation results show the comparative analysis of the topology-1 and several communication topologies of topology-2. The paper is concluded in Sect. 6.

2 Graph Theory Notation

The interaction topology of a MAS is naturally modelled by a graph. It can be a directed graph or an undirected graph. The agents can correspond to nodes of a graph and interactions such as communication and sensing can be denoted as edges of the graph.

An undirected graph is represented as $\mathbb{G} = (\mathsf{V}, \mathsf{E})$ where $\mathsf{V}=\{1, 2, \ldots, \mathrm{N}\}$ is the vertices set and $\mathsf{E} \in \mathsf{V} \times \mathsf{V}$ is the undirected edges. Figure 1 represents a graph with 5 vertices and 6 edges. Let the set of neighbours of vertex for an undirected graph i be denoted by

$$\mathbb{N}_i = \{j \in \mathsf{V} \mid (j, i) \in \mathsf{E}\} \tag{1}$$

An undirected edge (j, i) means j contains the information of i, and vice versa.

A directed graph is represented as $\mathbb{G} = (\mathsf{V}, \mathsf{E}, \mathsf{A})$ where $\mathsf{V} = \{1, 2, \ldots, \mathrm{N}\}$, a set of edges $\mathsf{E} \in \mathsf{V} \times \mathsf{V}$, and of order N consists of a set of nodes which is finite non-empty and a weighted adjacency matrix $\mathsf{A} = \left[a_{ij}\right] \in \mathbb{R}^{\mathrm{N}\times\mathrm{N}}$, where a_{ij} is the weight of the directed edge (j, i). Figure 2 shows the directed graph with four nodes. $\mathbb{N}_i$ corresponds to the neighbour agents of directed graph i such that

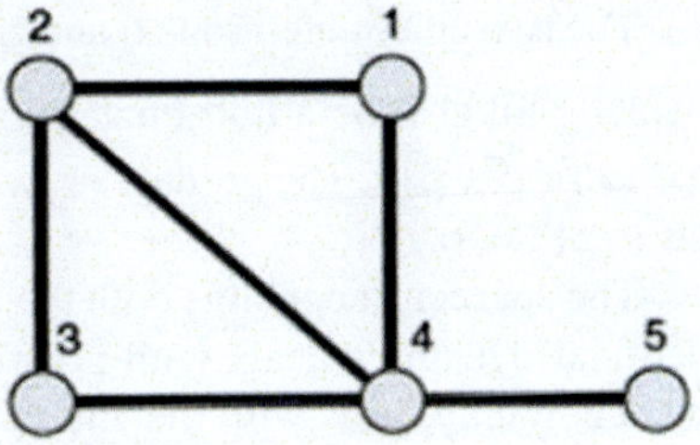

Fig. 1 Undirected graph[6]

Fig. 2 Directed graph

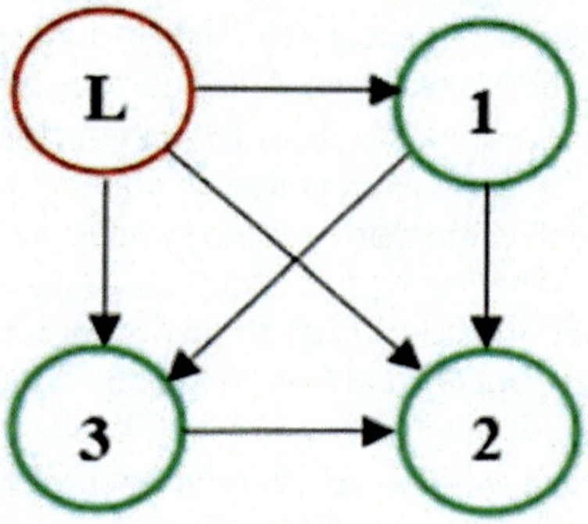

$$\mathbb{N}_i = \{j \in \mathsf{V} \mid (j, i) \in \mathsf{E}, j \neq i\} \tag{2}$$

If agent i is directed from j then $a_{ij} \neq 0$ and $a_{ij} = 0$ otherwise. If the self information exchange doesn't exist then $a_{ii} = 0$ for all $i \in \mathcal{V}$. An undirected graph edge (i, j) be equivalent to directed edges (j, i) and (i, j). $\mathbb{L}(G) = L = \left[\mathsf{l}_{ij}\right]_{\mathrm{N}\times\mathrm{N}}$ is the Laplacian matrix of weighted directed graph where $\mathsf{l}_{ij} = -a_{ij}, j \neq i$ and, $\mathsf{l}_{ij} = \sum_{k\in\mathbb{N}_i} a_{ik}, j = i$.

In undirected graph, information is exchanged from leaders to followers, and vice versa. In directed graph, information is exchanged only from leader to follower since the fed-back from follower to leader has some limitations in the case of disturbance, faults and uncertainties. This paper adopts directed graph where the focus is mainly on the communication between the agents when connected in different topologies.

3 Dynamics of Spacecraft

The general structure of the spacecraft has position r, velocity v, angular velocity w, and attitude q relative to a particular frame. The spacecraft attitudes are represented by unit quaternions where a unit quaternion is given as $\mathrm{q} = \left[\hat{\mathrm{q}}^T, \bar{\mathrm{q}}\right]^T \in \mathrm{R}^4$. The vector part is denoted by $\bar{\mathrm{q}} = \cos\left(\frac{\phi}{2}\right) \in \mathrm{R}$ and the scalar part is denoted by $\hat{\mathrm{q}} = \mathrm{a} \cdot \sin\left(\frac{\phi}{2}\right) \in \mathrm{R}^3$. From the above notations $\mathrm{a} \in \mathrm{R}^3$ is the Euler axis and $\phi \in \mathrm{R}$ is the Euler angle. By definition $\mathrm{q}^T\mathrm{q} = 1$. There is no unique quaternion since q and $-\mathrm{q}$ represent the same attitude.

The two unit quaternions q and p product is given by $\mathrm{qp} = \begin{bmatrix} \bar{\mathrm{q}}\hat{\mathrm{p}} + \bar{\mathrm{p}}\hat{\mathrm{q}} + \hat{\mathrm{q}} \times \hat{\mathrm{p}} \\ \bar{\mathrm{q}}\bar{\mathrm{p}} - \hat{\mathrm{q}}^T\hat{\mathrm{p}} \end{bmatrix}$ which will also be a unit quaternion. The unit quaternion conjugate p is given by $\mathrm{p}^* = \left[-\hat{\mathrm{p}}^T, \bar{\mathrm{p}}\right]^T$. The product of two unit quaternion is qp and its conjugate is given as $(\mathrm{qp})^* = \mathrm{p}^*\mathrm{q}^*$.

The spacecraft contains both the rotational and translational dynamics. The translational dynamics deals with position r, and velocity v. The rotational dynamics which mainly deals with the angular velocity w, and attitudes of the spacecraft q is

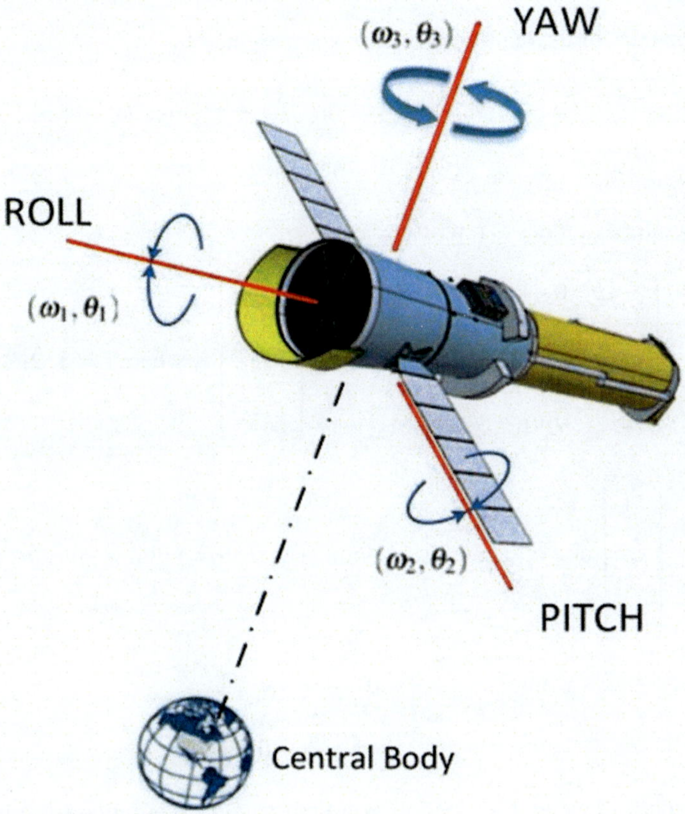

Fig. 3 Model of the spacecraft [10]

considered in this work. The structure of the spacecraft [10] is shown in Fig. 3. The dynamics of the follower spacecraft [11] is

$$\begin{aligned} \dot{\hat{q}}_i &= -\tfrac{1}{2} b_i^{\times} \hat{q}_i + \tfrac{1}{2} \bar{q}_i w_i, \\ \dot{\bar{q}}_i &= -\tfrac{1}{2} w_i \cdot \hat{q}_i \\ \mathcal{J}_i \dot{w}_i &= -b_i^{\times} (\mathcal{J}_i w_i) + \tau_i, \\ \text{where} \quad i &= 1, 2, 3 \ldots n \end{aligned} \tag{3}$$

where n denotes the number of follower spacecrafts, $\hat{q}_i \in R^3$, is the vector part and $\bar{q}_i \in R$ is the scalar part of the unit quaternion of the ith spacecraft, $\mathcal{J}_i \in R^{3\times3}$ is the inertia tensor of the ith spacecraft, $\tau_i \in R^3$ is the control torque and $w_i \in R^3$ is the angular velocity of the ith spacecraft.

The dynamics of the leader spacecraft is represented by

$$\begin{gathered}\dot{\hat{\mathrm{q}}}_0 = -\tfrac{1}{2}b_0^\times \hat{\mathrm{q}}_0 + \tfrac{1}{2}\bar{\mathrm{q}}_0\mathrm{w}_0, \quad \dot{\bar{\mathrm{q}}}_0 = -\tfrac{1}{2}\mathrm{w}_0 \cdot \hat{\mathrm{q}}_0 \\ \mathcal{J}_0\dot{\mathrm{w}}_0 = -b_0^\times\left(\mathcal{J}_0\mathrm{w}_0\right)\end{gathered} \tag{4}$$

The spacecrafts dynamics is rewritten as:

$$\dot{x}_i = \begin{bmatrix} -\frac{1}{2}b_i^\times & 0_{3\times1} & 0_{3\times3} \\ -\frac{1}{2}\mathrm{w}_i^{\mathrm{T}} & 0_{1\times3} & 0_{1\times3} \\ 0_{3\times3} & 0_{3\times1} & -\mathcal{J}_i^{-1}b_i^\times\mathcal{J}_i \end{bmatrix} x_i + \begin{bmatrix} 0_{3\times3} \\ 0_{1\times3} \\ \mathcal{J}_i^{-1} \end{bmatrix} \tau_i \text{ where } i = 1, 2, 3, .. \tag{5}$$

where $\dot{x}_i = \begin{bmatrix} \dot{\hat{\mathrm{q}}}_i \\ \dot{\bar{\mathrm{q}}}_i \\ \dot{\mathrm{w}}_i \end{bmatrix}$ and $b_i^\times$ is the skew-symmetric matrix and it is given as

$$b_i^\times = \begin{bmatrix} 0 & -\mathrm{w}_i^3 & \mathrm{w}_i^2 \\ \mathrm{w}_i^3 & 0 & -\mathrm{w}_i^1 \\ -\mathrm{w}_i^2 & \mathrm{w}_i^1 & 0 \end{bmatrix}$$

The same control law for a leader-follower topology in [6] is followed. It is

$$\tau_i = \mathrm{w}_i \times (\mathcal{J}_i\mathrm{w}_i) + \frac{1}{|\mathrm{N}_i|}\mathcal{J}_i\sum_{j\in\mathrm{N}_i}\dot{\mathrm{w}}_j - \frac{1}{|\mathrm{N}_i|}\left[\mathrm{k}_{\mathrm{q}i}\hat{x}_{\pi i} + \mathrm{k}_{\mathrm{w}i}\sum_{j\in\mathrm{N}_i}\left(\mathrm{w}_i - \mathrm{w}_j\right)\right] \tag{6}$$

where $x_{\pi i} = \Pi_{j\in\mathrm{N}_i}\left(\mathrm{q}_j^*\mathrm{q}_i\right)$, the number of neighbour agents of agent i is denoted by $|\mathrm{N}_i|$, the constants are selected as $\mathrm{k}_{\mathrm{q}i} = 1$,$\mathrm{k}_{\mathrm{w}i} = 2I_3$,i= $1, 2, 3, ..n$ and n is the number of follower spacecrafts.

4 Leader Follower Cooperative Tracking Control

The consensus-based design scheme having cooperative tracking control of multiple spacecrafts with a virtual leader is addressed in this work. The followers maintain their respective position and track the leader trajectory for cooperative tracking.

The need for the control system for Cooperative tracking of Spacecrafts is for two reasons. First of all, the open-loop dynamics of the spacecraft is not asymptotically stable. So, the closed-loop stability must be ensured for the leader-follower spacecrafts. Secondly, it should ensure the perfect synchronization of the leader and follower spacecrafts. While considering the leader-follower approach of the spacecrafts, each follower spacecraft has to maintain its respective attitude and track the leader's trajectory. The cooperative tracking can be achieved when there is proper communication between them.

The main emphasis of this paper, when compared to other works in the spacecraft for cooperative tracking is that the directed topologies are considered for analysis of various topologies. Whereas [7–9] deals with the undirected topology. Undirected topology signifies the states of the followers are fed-back to the leader. If the follower agents are subjected to faults or disturbances, due to the feedback the leader's states may also get affected which may result in imperfect synchronization of the leader and follower agents. While considering the directed topology, how the information is exchanged between the leader and the follower plays a major role. The analysis involves whether the leader's states are given to the followers directly or through other followers. In the past, a particular topology was predefined by the authors, which may or may not give better cooperative tracking control. Since the topology signifies whether the agents are communicated properly or not and thereby affects the cooperative tracking.

4.1 Analysis of Various Topologies

A directed graph is addressed in this work where there is no feedback. Several topologies are considered out of which the efficient topology is to be identified. To study the trajectory tracking of the followers when the number of follower spacecrafts is increased. Initially, one leader and two followers (topology-1) are considered which is compared with the one leader and three followers (topology-2). Followed by which several distributed leader-follower control topologies for one leader and three follower that accommodates a virtual leader and arbitrary information flow among the spacecrafts are addressed. In this leader-follower topology, information flows only from leader to followers and not from followers to leader. There is information exchange between the followers.

Topology-1: One Leader and Two Follower spacecrafts Information-exchange topology between three spacecraft in leader-follower approach is shown in Fig. 4. Here the leader's dynamics is available to both the follower 1 and follower 2. The information is exchanged between the follower 1 and 2.

Topology-2: One Leader and Three Follower spacecrafts Information-exchange topology between four spacecrafts in leader-follower approach can be in any of the one forms as shown in Fig. 5. One spacecraft acts as the leader and other three spacecrafts as follower 1, 2 and 3.

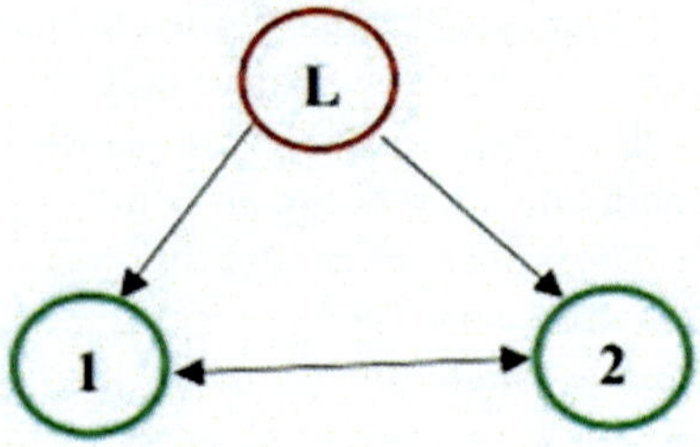

Fig. 4 One Leader-Two Follower topology

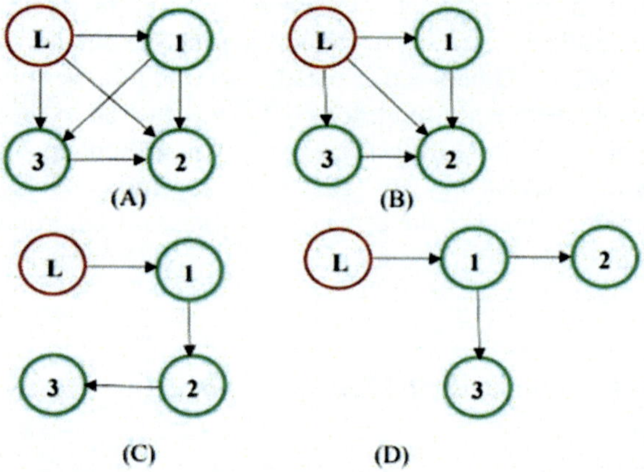

Fig. 5 One Leader—three follower information exchange topologies are shown in A, B, C, D

In the communication topology of one leader and three followers (topology-2), the states of the leader may be available to all the followers or some of the followers. The communication between the leader and followers has a directed path. The information exchange from leader to a particular follower may be carried out through several followers. Thus the information is exchanged in terms of different levels from one agent to another. For example, in topology C the leader's states are available to follower 3 through follower 1 and 2.

For the topologies in Fig. 5, information only flows from leader to followers and not from followers to leader. In topology A and B the information of the leader's dynamics is available to all the followers 1, 2 and 3. In the case of topology C and D the information of the leader's dynamics is available only to one follower and there are different levels of information exchange from the leader to followers. In the topology C the information of the leader's dynamics is available to follower 1 and follower 1 exchanges its information with follower 2 and follower 2 exchanges its information with follower 3. In topology D the information of the leader's dynamics is available to follower 1 and follower 1 exchanges its information with followers 2 and 3.

Considering the topology-1 and several topologies of topology-2 the better topology for perfect trajectory tracking of followers is to be identified in the simulation.

5 Simulation Results and Discussions

To analyse the different topologies through simulation MATLAB is used. Topology-1 deals with One leader Two follower topology. Topology-2 contains One leader and Three followers topology. Topology 1 is compared with that of the several topology cases A, B, C, D of Topology-2.

Topology-1 The leader and follower dynamics are mentioned in Eq. (3) and Eq. (4) To apply the control law, the dynamics of the follower spacecrafts is written in the form as in the Eq. (5). Table 1 shows the parameters of leader and two followers used for simulation. The control protocol as in Eq. (6) is considered where $i = 1, 2$ and $\tau_0 = [0, 0, 0]^T$.

Figures 6, 7, 8, 9 and 10 show the simulation results of one leader and two followers. Figures 6 and 7 show that the followers 1 and 2 track the angular velocity of the leader while the followers 1 and 2 remain in its initial attitude. Figures 8 and 9 show the control input given to the follower 1 spacecraft and follower 2 spacecraft, respectively. The trajectory plot of the leader and follower shown in Fig. 10 depicts the perfect synchronization of leader and the followers.

Table 1 Parameters of leader and followers

Leader parameters [11]	$\mathcal{J}_0 = \text{diag}\{1, 2, 1\}$, $q_0(0) = [0, 0, 0, 1]^T$, $w_0(0) = [0.1, 0.3, 0.5]^T$
Follower 1 parameters	$\mathcal{J}_1 = \text{diag}\{1, 2, 1\}$, $q_1(0) = [-1, 1, 1, 1]^T$, $w_1(0) = [0.2, 0.4, 0.7]^T$
Follower 2 parameters	$\mathcal{J}_2 = \text{diag}\{1, 2, 1\}$, $q_2(0) = [1, 1, 1, 1]^T$, $w_2(0) = [0.4, 0.5, 0.8]^T$

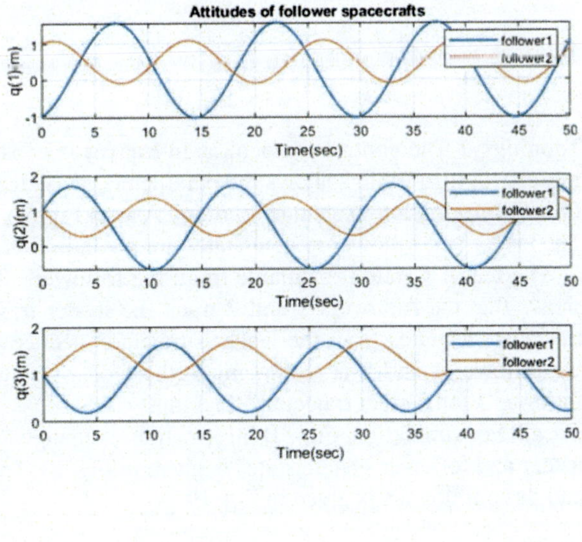

Fig. 6 Attitude of the follower spacecrafts where $i = 1, 2$

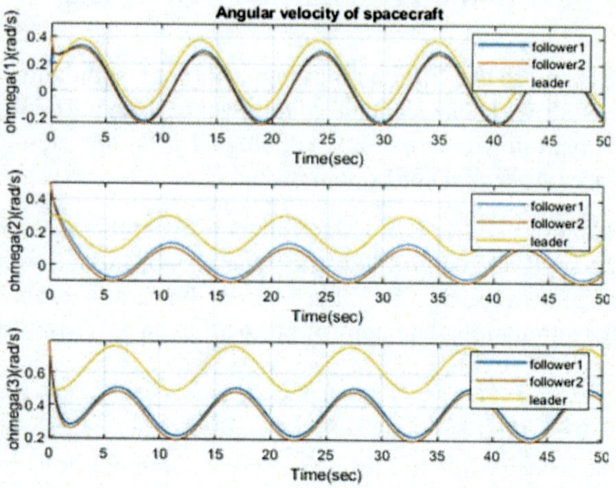

Fig. 7 Angular velocity of the follower spacecrafts where $i = 1, 2$

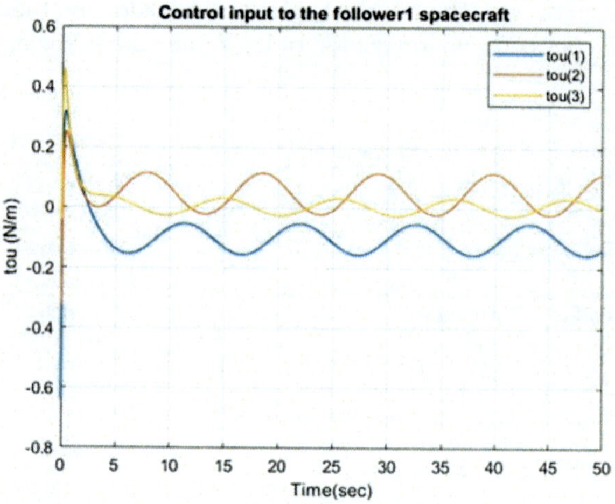

Fig. 8 The control input of the follower spacecraft where $i = 1$

Topology-2 The control protocol as in Eq. (6) is considered where $i = 1, 2, 3$ and $\tau_0 = [0, 0, 0]^{\mathrm{T}}$. Table 2 shows the parameters of the leader and three followers.
Case A Information-exchange topology A as in Fig. 5A is considered for simulation. The leader and follower system states are given in Fig. 11.

The leader's state is available to all the followers. The trajectory plot in Fig. 12 shows that the followers 1 and 2 track the leader trajectory perfectly. Follower 3 trajectory deviates from the leader's trajectory. Since the communication between the follower 1 and 3 is redundant as the leader's state is already available to the follower 3 for perfect trajectory tracking.
Case B For simulation, Fig. 5B information-exchange topology B is considered. The leader and follower system states is given in Fig. 13. Trajectory plot with one leader and three followers is given in Fig. 14.

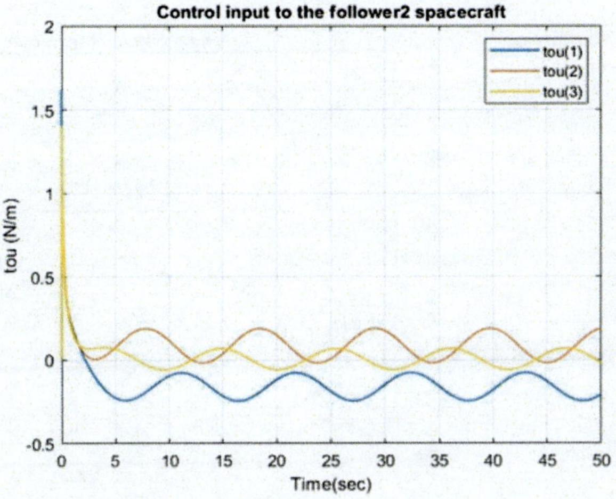

Fig. 9 The control input of the follower spacecraft where $i = 2$

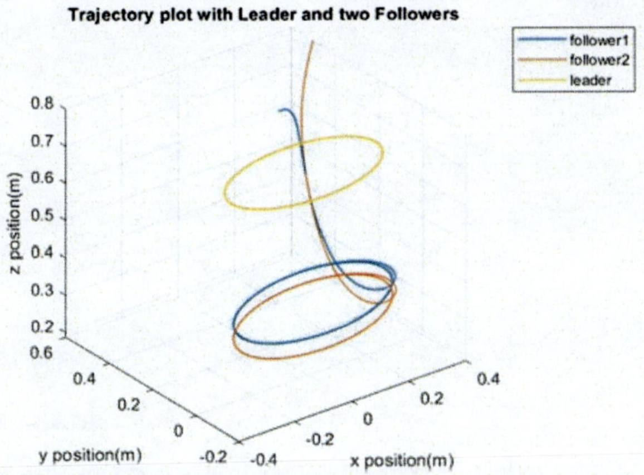

Fig. 10 Trajectories of leader and two follower spacecrafts

Table 2 Parameters of leader and followers

Leader parameters [11]	$\mathcal{J}_0 = \text{diag}\{1, 2, 1\}$, $q_0(0) = [0, 0, 0, 1]^T$, $w_0(0) = [0.1, 0.3, 0.5]^T$
Follower 1 parameters	$\mathcal{J}_1 = \text{diag}\{1, 2, 1\}$, $q_1(0) = [-1, 1, 1, 1]^T$, $w_1(0) = [0.2, 0.4, 0.7]^T$
Follower 2 parameters	$\mathcal{J}_2 = \text{diag}\{1, 2, 1\}$, $q_2(0) = [1, 1, 1, 1]^T$, $w_2(0) = [0.4, 0.5, 0.8]^T$
Follower 3 parameters	$\mathcal{J}_3 = \text{diag}\{1, 2, 1\}$, $q_3(0) = [1, 1, 1, 1]^T$, $w_3(0) = [0.2, 0.6, 0.9]^T$

Fig. 11 Angular velocity of the follower spacecrafts where $i = 1, 2, 3$

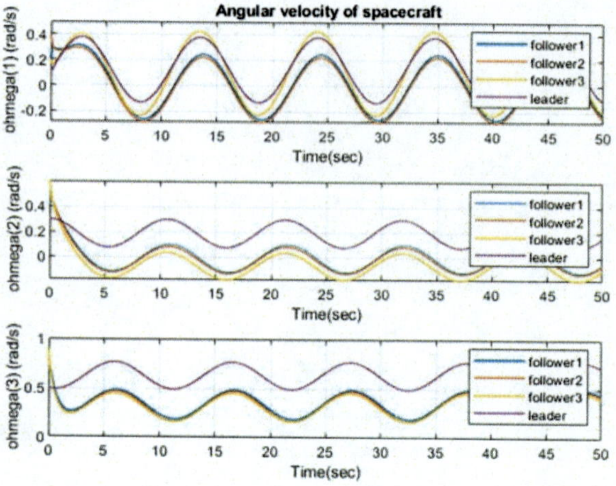

Fig. 12 Trajectories of leader and three follower spacecrafts

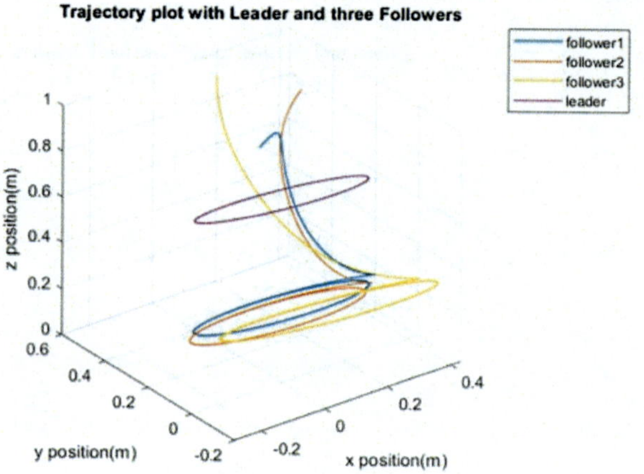

Fig. 13 Angular velocity of the follower spacecrafts where $i = 1, 2, 3$

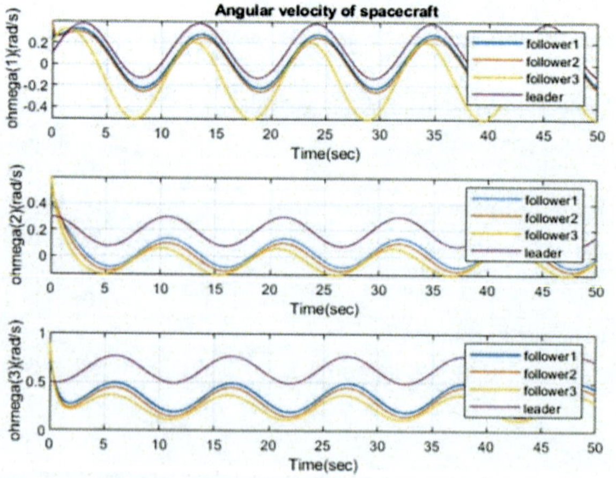

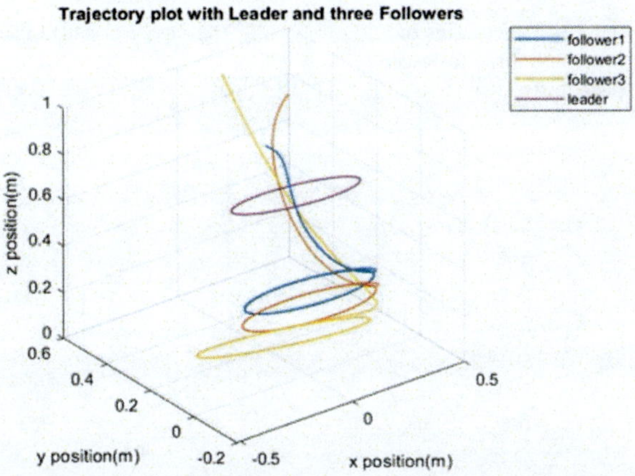

Fig. 14 Trajectories of leader and three follower spacecrafts

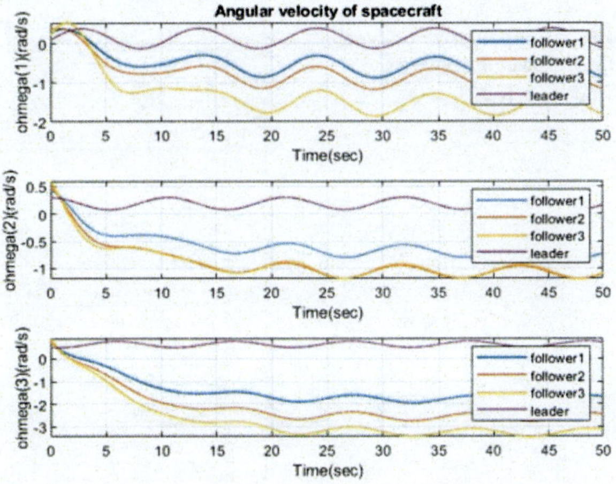

Fig. 15 Angular velocity of the follower spacecrafts where $i = 1, 2, 3$

While comparing topology A and topology B it is found that the follower 3 in the topology B track the leader states better than topology A. This is because there is no directed edge from follower 1 and 3. All the followers contain the leader's states and strong communication between the follower agents is maintained in topology B. It can also be inferred that the synchronization of follower 1, 2 and 3 is perfect in topology B.

Case C For simulation Fig. 5C topology C is considered. The leader and follower system states are given in Fig. 15. Trajectory plot with one leader and three followers is given in Fig. 16.

From the trajectory plot Fig. 16. It can be inferred that there is no perfect synchronization of the follower spacecrafts. Since the leader's states are available to the follower 2 through the follower 1 and to follower 3 through follower 1, 2.

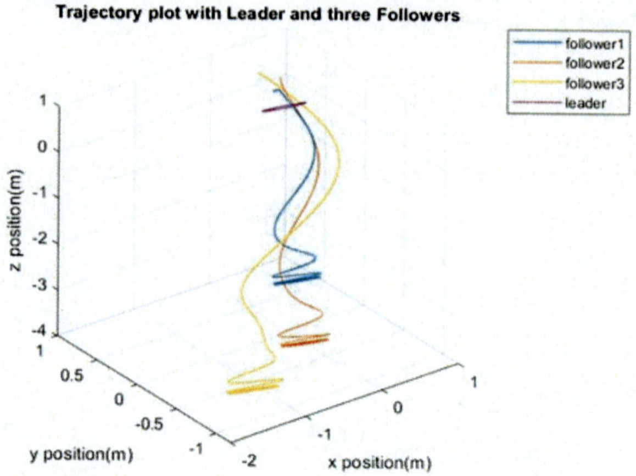

Fig. 16 Trajectories of leader and three follower spacecrafts

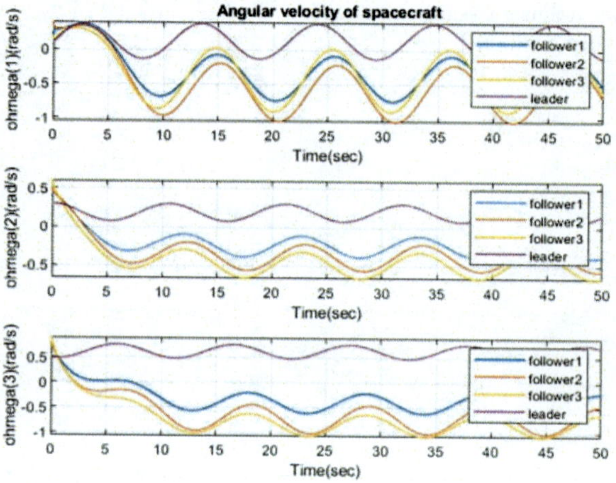

Fig. 17 Angular velocity of the follower spacecrafts where $i = 1, 2, 3$

Case D Fig. 5D topology D is considered for simulation. The leader and follower system states are given in Fig. 17. Trajectory plot with one leader and three followers is given in Fig. 18.

From the trajectory plot of the leader and followers Fig. 16. It can be inferred that there is no perfect synchronization of the follower spacecrafts 1, 2 and 3. Since the leader's states are available to the follower 2 and 3 through the follower 1 in different levels from the leader.

While comparing the angular velocity results of all the topologies A, B, C and D it is found that topology B followers track the leader's angular velocity perfectly. Topology C and topology D do not give better tracking of the leader's states. Since the leader's states are available to the follower 2 and 3 through the follower 1 in different levels. Whereas in the topology B the trajectory tracking of all the followers is perfect as there is no redundant communication between the agents which may

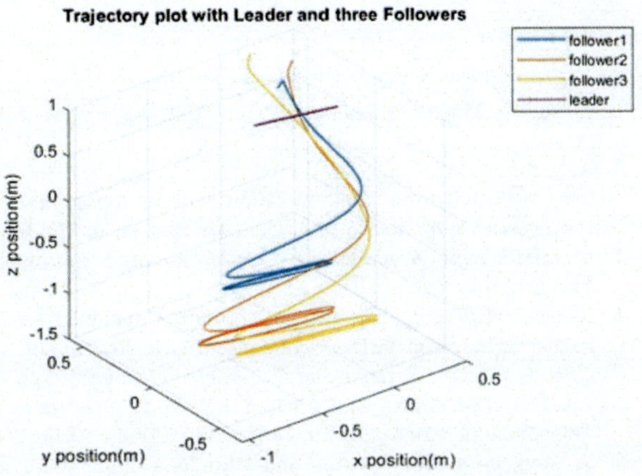

Fig. 18 Trajectories of leader and three follower spacecrafts

cause deviations in the trajectory tracking of the follower. In general, it can be inferred that an efficient topology can be selected in a manner where there is no redundant communication between the followers and the leader's states are available to all the followers.

6 Conclusion

The efficient communication topology for perfect cooperative tracking of spacecrafts was identified in this paper. A directed graph was primarily addressed. The different directed topologies considered contain one leader and different number of followers. The states of the leader may be available to all the followers or some of the followers. The communication between the leader and followers may also have a directed path from the leader to a particular follower through another follower. Thus, the simulation results highlight the comparative analysis of all the different topologies. It can be inferred that for an efficient topology, the leader's states are available to all the followers, and there should be proper communication between the followers without any redundant directed edges. The efficient topology gives better results in terms of trajectory tracking of leader and followers for the cooperative tracking of multi-agent systems. As a future scope of this work, the analysis of the topologies in the presence of model uncertainties, faults, and noise will be addressed.

References

1. J. Qin, Q. Ma, W.X. Zheng, H. Gao, Y. Kang, Robust H_∞ group consensus for interacting clusters of integrator agents. IEEE Trans. Autom. Control **62**(7), 3559–3566 (2017)
2. R. Mehrabian, S. Tafazoli, K. Khorasani, Cooperative tracking control of Euler-Lagrange systems with switching communication network topologies, in *2010 IEEE/ASME International Conference on Advanced Intelligent Mechatronics, Montreal*, pp. 756–761 (2010)
3. M. Lazim, et al., *Applications of Modelling and Simulation*, eissn 2600-8084, vol. 4, pp. 159–170 (2020)
4. X. Chen, C. Zhang, The method of multi Unmanned Aerial Vehicle cooperative tracking in formation based on the theory of consensus, in *2013 5th International Conference on Intelligent Human-Machine Systems and Cybernetics, Hangzhou*, pp. 148–151 (2013)
5. J.A. Fax, R.M. Murray, Information flow and cooperative control of vehicle formations. IEEE Trans. Autom. Control **49**(9), 1465–1476 (2004). 10.1109/TAC.2004.834433
6. W. Ren, Formation keeping and attitude Alignment for multiple Spacecraft through local. Interactions **10**(2514/1), 25629 (2007)
7. P.K.C. Wang, F.Y. Hadaegh, K. Lau, Synchronized formation rotation and attitude control of multiple free-flying spacecraft. J. Guidance Control Dyn. **22**(1), 28–35 (1999)
8. J.R. Lawton, R.W. Beard, Synchronized multiple spacecraft rotations. Automatica **38**(8), 1359–1364 (2002)
9. W. Ren, R.W. Beard, Decentralized scheme for spacecraft formation flying via the virtual structure approach. J. Guidance Control Dyna. **27**(1), 73–82 (2004). 10.2514/1.9287
10. M. de Queiroz, X. Cai, M. Feemster, *Formation Control of Multi Agent Systems: A Graph Rigidity Approach* (Wiley, Hoboken, NJ, 2019)
11. J. Qin, Q. Ma, H. Gao, W.X. Zheng, Fault-tolerant cooperative tracking control via integral sliding mode control technique. IEEE/ASME Trans. Mechatronics **23**(1), 342–351 (2018)
12. V.A. Chobotov, *Spacecraft Attitude Dynamics and Control, Melbourne* (Kereger, FL, USA, 1991)

An Overview of Different Control Topologies in DC Microgrid

P. V. Nithara and V. V. Gana

Abstract DC microgrids give more opportunities in the field of future electricity generation because of cost, efficiency, and system size. DC microgrid is mainly dependent on sustainable energy resources (SERs) such as solar and energy storage system since both are operated in DC mode. The integration of SERs with the DC microgrid network leads to power intermittency and power variation in the generation characteristics, which makes the system to demand a proper voltage leveling mechanism at the low-voltage side of the feeder via energy storage devices such as batteries and supercapacitors. The control topology of a DC microgrid plays an important role in achieving efficient and stable operation of DC microgrid. This article focuses on the control strategies of DC microgrids. Different control topologies such as centralized, decentralized, and distributed control have been discussed. A comparative analysis is also made to help the selection of controllers concerning the application.

Keywords DC microgrid · Centralized control · Decentralized control · Distributed control · Control topologies · Distributed generation (DG)

1 Introduction

The world economy is developing rapidly with a huge demand in energy crisis resulted in a large deployment of renewable energy-based distributed generation units to protect the earth from carbon-related pollutions. Renewable energy resources have the potential to decrease the climate impact of energy production and can meet

P. V. Nithara (✉)
Department of Electrical and Electronics Engineering, CMR Institute of Technology, Bengaluru, India
e-mail: nithara.p@cmrit.ac.in

V. V. Gana
Department of Electrical and Electronics Engineering, Amrita School of Engineering, Amrita Vishwa Vidyapeetham, Bengaluru, India
e-mail: gana@mitkannur.ac.in

V. Suma et al. (eds.), *Inventive Systems and Control*,
Lecture Notes in Networks and Systems 204,
https://doi.org/10.1007/978-981-16-1395-1_51

future energy demands efficiently. So, the use of renewable energy sources (RESs) can be increased by integrating distributed energy resources (DERs) with load entities through energy storage systems (ESSs) resulting in a better solution as microgrids. Hence, the microgrid is one of the alternatives for future electricity generation which satisfies the consumer's load demand at optimum cost by reducing grid congestion. It is a group of distributed energy resources and interconnected loads having optimal control that improves sustainability, reliability, and the efficiency of modern power systems [1]. The strategy that integrates solar resources or wind energy resources will give redundancy for essential services which makes the main grid less susceptible to local disasters.

The microgrid can be classified based on types, connectivity, application, size, and load such as residential and commercial (types) loss reduction; resilience-oriented (application); grid-connected and isolated (connectivity); small, medium, and large (size); and AC, DC, and hybrid (load) [2]. Among these, DC microgrids give more opportunities in the field of future electricity generation in terms of cost, efficiency, system size, etc. The penetration of solar energy is very high in India and most of the appliances available in the markets such as modern lighting systems, computers, laptops, televisions, etc. are operating in DC. So, DC microgrid is getting more attention in the residential as well as industrial fields due to this increasing use of DC loads [3] and becomes the trending solution for applications such as electric vehicles [4], buildings and data centers [4], zero net electricity energy buildings [5], electric ships [6], aerospace applications, and in railways.

While comparing with AC microgrids, DC microgrids are free from reactive power so that reactive power management becomes irrelevant. The absence of frequency makes the system more feasible with simpler control having zero frequency synchronization issues. The insufficiency of frequency makes the system free from lots of problems such as harmonics, proximity effect, skin effect, and inrush current problem [3]. Secure data transfer and safe communication channels can be made possible by using the DC power lines. DC nature of energy requirement in the telecom and data centers is a force to adopt DC microgrid for their smooth and reliable operation. DC microgrid mainly depends on renewable energy sources and energy storage units since both are operated in DC mode. Sustainable energy sources like solar PV panels can produce electricity continuously during daytime depending on the availability of solar radiance. Therefore, the daytime loads of commercial buildings including a supermarket, hospitals, offices, factories, schools, etc., can make use of this energy so that peak load demand of central power grid can be reduced drastically. Figure 1 shows the basic diagrammatic representation of the DC microgrid. Here, the microgrids are supported by one or more distributed energy sources such as solar PV panels and wind turbines along with batteries, supercapacitors, etc., as energy storage systems to provide backup power or the supplement to the main power system grid during peak load demanding periods and load such as DC load and AC load.

DC microgrid network always needs a voltage leveling or stabilizing system at the distributed generation side through storage devices such as batteries and supercapacitors, since the power generation characteristics are volatile and intermittent in nature [7]. To achieve efficient and stable operation of DC microgrid, an optimal control

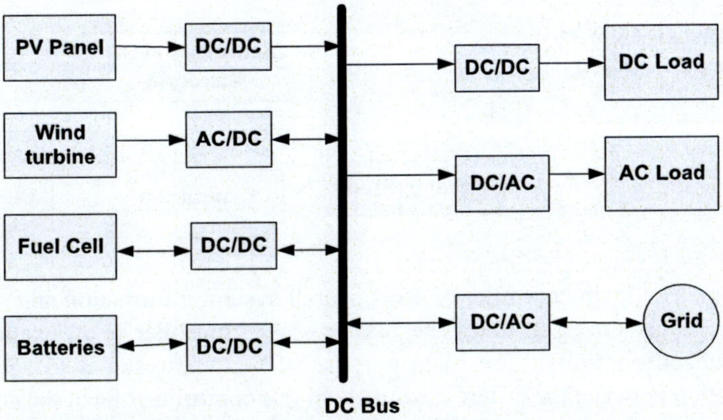

Fig. 1 Basic diagram of DC microgrid

topology is needed. There are mainly three levels of control of the operation of DC microgrids, such as primary, secondary, and tertiary levels. Intelligent coordination among sources, energy storage units, and the loads are needed for the efficient and reliable operation of the DC microgrid [7]. This paper attempts to find out the issues and challenges in microgrids, thereby analyzing the effective control strategies by comparing the existing control methods.

2 Controllers in DC Microgrid

A microgrid is a combination of distributed generation (DG), energy storage, and load. Energy storage devices and load are connected through power electronic devices to give necessary control action. The main purposes of the controller are to control real power to improve voltage stability and to operate the system within a specific standard limit [8]. DC microgrid consists of a DG unit such as PV array, fuel cells, and energy storage systems such as the battery, supercapacitors, and hydrolyzers. Control action is required for the DC microgrid (1) to improve voltage stability and (2) to get a smooth transition between grid-connected mode and islanding mode. In DC microgrid control, geographies are as shown in the figure. Power converters act as interfacing between the various parts of the DC microgrid. The presence of the converter and nonlinear load made the system become nonlinear. So to control the overall system, there are 3 different types of controllers available as shown in Fig. 2 [7].

A Centralized Controller

The centralized controller has a central controller like a microcontroller, switch, or a server. The appearance of a communication link gives an awareness of the neighbor node. But this type of controller reduces the reliability of the

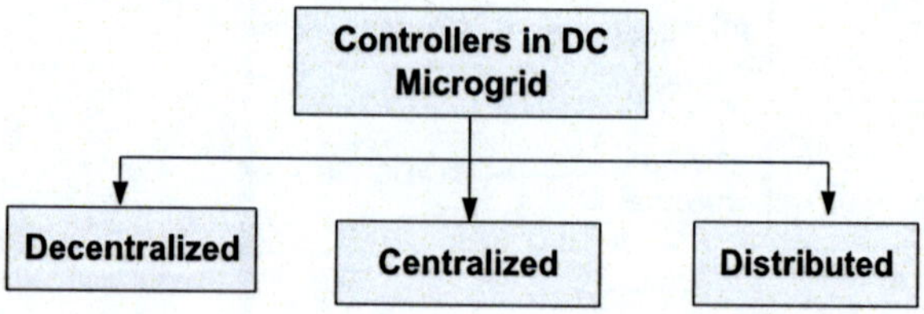

Fig. 2 Different types of the controller in DC microgrid

system. The main controller collects overall system information and sends the control signal to each unit. The master–slave controller is an example of a centralized controller. The main purpose of this controller is to get parallel operation between the different sources. In this controller, one of the converters acts as a master and the remaining converts as a slave. The master will give the command to the slave unit to regulate the DC bus voltage. And as per master command instruction, the slave will give necessary current support [9].
The controller needs high-speed communication, and failure at any point will affect the overall system performance. It may lead to the blackout of the entire system.

B *Decentralized controller*

Decentralized controllers have individual control in each unit of the system, i.e., local controllers. The main advantage of this controller is no need for a communication medium. Information about each unit communicates by the controller through the power channel. This control strategy is more reliable and robust. Droop control is a type of decentralized controller. There are different types of droop controllers such as

1. *Conventional Droop Control*

   This control action is required to control the current injected into the grid, where the reference is DC bus voltage. This method helps to eliminate the circulatory current between the converters.

$$V_{\mathrm{DCx}} = V_{\mathrm{DCref}} - I_{\mathrm{DCx}} R_{\mathrm{DCx}} \tag{1}$$

   V_{DCref} is the reference DC grid voltage; V_{DCx} and I_{DCx} are output voltage and current of converter, respectively. R_{DCx} represents the droop coefficient of the system. Figure 3 shows the droop characteristics where an increase in the load will increase in the power but reduces the voltage. The main advantage of this method is different. DG has a different slope, so it gives different voltage regulations. Conventional droop control is shown in Fig. 4. The selection of droop resistance is important because it affects load sharing and voltage regulation.
2. *Advanced Droop Controller*

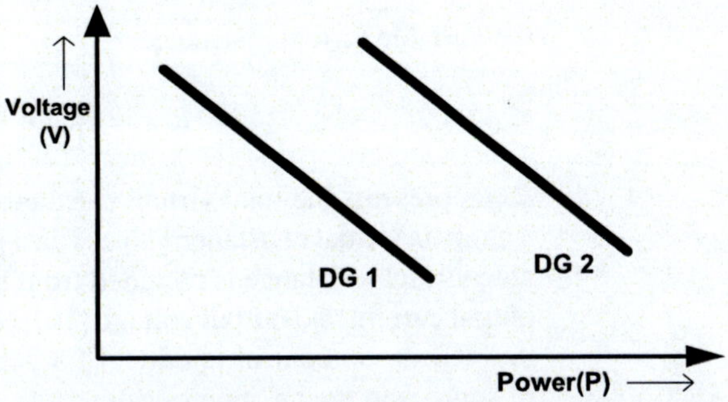

Fig. 3 Characteristics of droop control

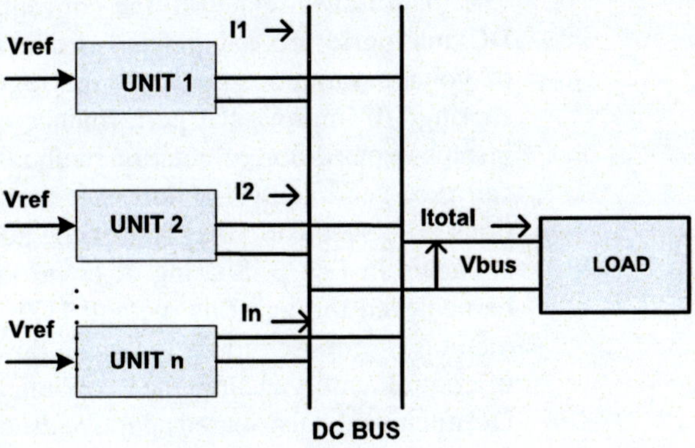

Fig. 4 Conventional voltage droop control

Virtual resistance-based droop control, adaptive droop control, an intelligent technique-based droop control, and mode adaptive droop controller are different types of advanced droop controller.

1. *Virtual resistance-based droop control*

   By using variable equivalent resistance between the modules of the DC microgrid, VRB droop control can achieve the equal current sharing between the DG units in DC microgrid. VRB droop control has both static and variable virtual resistances compared to conventional droop control. Figure 5 shows the block diagram of VRB droop control. It contains R_{vrs} which is the static virtual resistance and R_{vra} which is

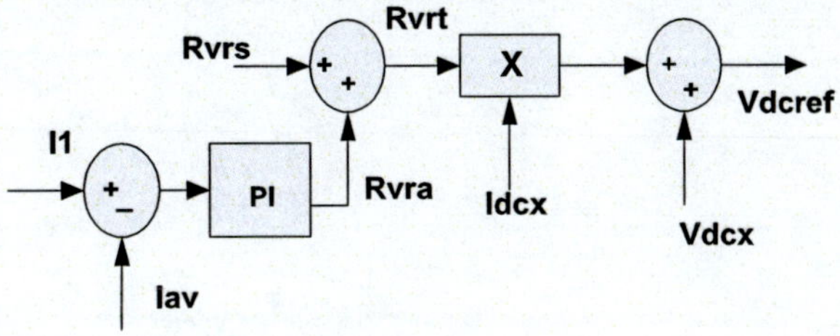

Fig. 5 Virtual resistance-based droop control

the variable virtual resistance.

$$R_{\text{vrt}} = R_{\text{vrs}} + R_{\text{vra}} \quad (2)$$

R_{vrt} represents the total virtual resistance of the droop control.
The static virtual resistance value is designed for each system. But variable virtual resistance is obtained from PI control concerning average output current and output voltage [10]. Optimal tuning of PI controller in VRB droop control is critical for achieving equal current sharing of sources in the DC microgrid.

2. *Adaptive droop control*
The main issues related to the convention droop control method in DC microgrid are using fixed droop resistance value irrespective of voltage variation, poor voltage regulation, and unequal current sharing. To improve the performance of the droop control method, instantaneous droop calculation method is required [11]. This method can give good control action over transient and steady-state conditions. The schematic diagram of the adaptive droop control method is shown in Fig. 6. Sharing of equal load and voltage stability can be achieved through this method [12]. Distributed adaptive control method is proposed to optimize the power distribution. This method of control action ensures the economically optimal operation of the DC microgrid. Distributed adaptive droop control method is excluded from the communication network.
3. *Intelligent-based droop control*
Droop control with intelligent control action such as fuzzy logic control is intelligent-based droop control [13]. An intelligent-based droop control for low-voltage DC microgrid is proposed. In this

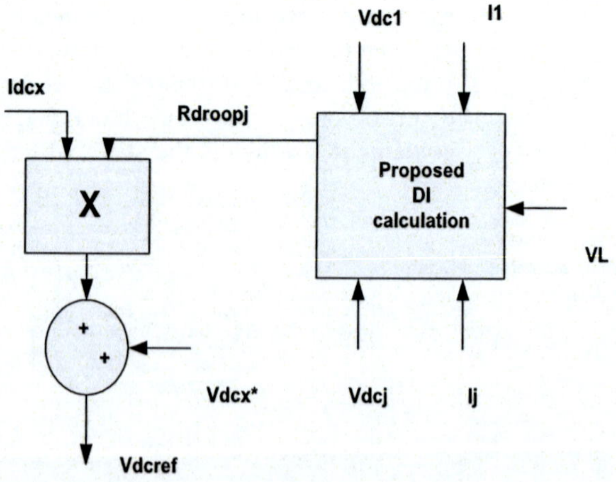

Fig. 6 Block diagram of adaptive droop control

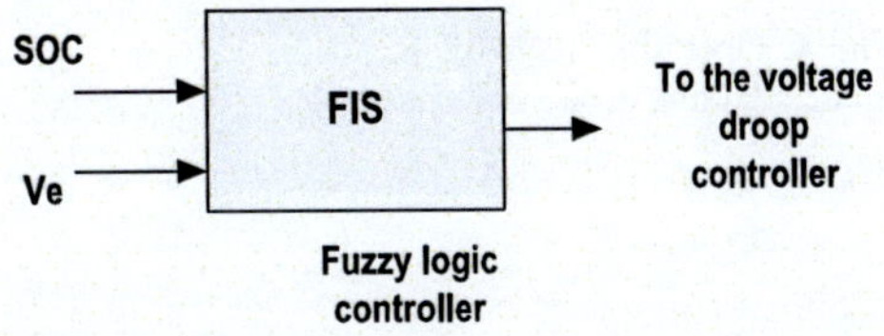

Fig. 7 Block diagram of intelligent control action

method, droop resistance is changed based on the state of charge (SOC) of each energy unit. With the help of fuzzy logic, intelligent droop controller will find the droop coefficient to give good voltage regulation. Figure 7 shows the fuzzy controller to determine the virtual droop resistance. This Mamdani fuzzy interface system has two inputs such as SOC and voltage error [14]. An intelligent droop controller for isolated DC microgrid with Takagi–Sugeno fuzzy model without uncertainty and disturbances is proposed.

4. *Mode adaptive droop controller*
   Mode adaptive droop controller is used in DC microgrid where multiple DG and energy storage systems are available. Conventional mode adaptive controller cannot give good performance for acceptable voltage regulation and current sharing between the sources in large DC microgrid [22]. Improved mode adaptive droop controller for power balance is proposed, to regulate DC bus voltage and to diminish the circulating current between the DG in the islanded mode. The main advantage of this type of controller is greater control freedom without a communication link. To improve the reliability and flexibility of the overall system [23], control action at each node is to be adjusted in real time. DC bus voltage itself assigns diverse microgrid mode and encourages self-trained control target switching and power-sharing. Operation range of this control method consists of different modes such as mode 1(utility dominating), mode 2 (storage dominating), and mode 3 (generating dominating) as shown in Fig. 8.
   If the system is in mode 2, when there is voltage sag, it automatically changes to utility dominating mode (mode 1). Once the voltage rises, then the system will come back to mode 2. A similar way mode change is occurring between the storage dominating and utility dominating stages [23, 24]. A DMA method reduces the gap between droop control and centralized control.

C *Distributed control*

Distributed control can provide coordination between autonomous agents with the help of a communication network. When compared to the centralized controller, a distributed controller has advantages such as robustness, extensibility, and flexibility [25]. A distributed controller in islanded DC microgrid for voltage regulation and load sharing between the DG is proposed. Centralized control cannot be able to

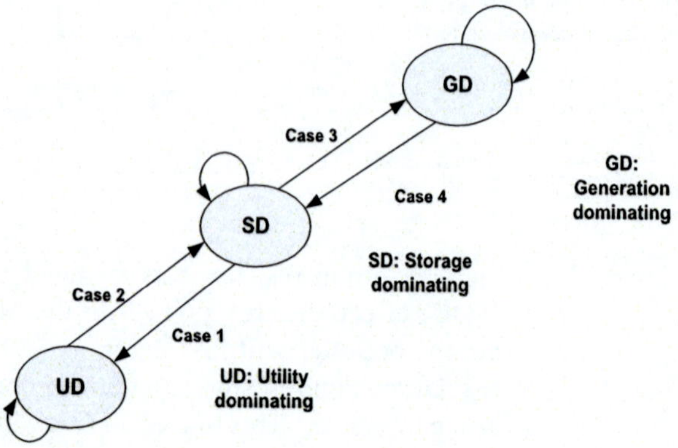

Fig. 8 Operating stage of mode adaptive droop control

manage a system when significant expansion in DG happens. During such a situation, distributed controller can continue the operation of the communication link breaks. So, this controller has been immune from the single-point failure [21]. A distributed adaptive droop controller for DC microgrid is proposed. Each converter acts as an agent by representing a node in the communication graph. The main advantages of this controller are: (1) It needs a sparse communication network; (2) each controller contains a voltage regulator to produce voltage reference; (3) at each node, the adaptive controller can be able to do adjustment in virtual impedance [26]. The Proposed controller is a distributed controller for DC microgrid and in this controller droop optimization of parameters is achieved by using particle swarm optimization technique. Since cable line resistance possesses large and unequal magnitude, it is impossible to reduce the trade-off between the bus voltage regulation and the sharing of equal loads. So, the controller required an optimized value of droop parameters.

3 Conclusion

DC microgrid has more attraction due to its peculiar nature. But issues related to DC microgrids have to solve to get more efficient in performance. Regulation of voltage and sharing of equal load can be controlled with the help of a centralized, decentralized, and distributed controller. This paper focused on different types of control action available in DC microgrids. The comparative analysis of different controllers is given in Table 1. Among those different types of controllers, the distributed controller is more popular because of flexibility and reliability. In the centralized controller, the energy management system is the heart of the microgrid control. But single-point failure is the main problem related to a centralized controller. The decentralized controller is more reliable than a centralized controller. But due to lack of information, local controller cannot work optimally. These issues can be resolved with the distributed controller. Because they can collect the all required information

Table 1 Comparative analysis of different controller

Control topologies	Features	Applications	Methods
Centralized control	Better load current sharing, good voltage regulation, robust, flexible, and improved efficiency	Islanded DC microgrid, high-voltage and low-voltage DC grid, and plug-in EVs	Phase shift control [15] and PI control [16]
Decentralized control	Moderate capability for load sharing, less complexity for implementation, better voltage regulation, robust under noise disturbances and uncertainties, more economic	High-voltage DC microgrid	Conventional PI control [17], adaptive droop control [18], closed-loop reference model adaptive control (CRM) [19]
Distributive control	Good capability for load sharing, better voltage regulation, minimized usage of the energy storage element, highly stable, and robust	DC microgrid	Nonlinear optimal control [20] and PI controller with adaptive droop control [21]

and immune to a single-point failure. In the future, the performance of different controllers will be validated in terms of efficiency and stability of the DC microgrid.

References

1. M. Ahmed, L. Meegahapola, A. Vahidnia, M. Datta, Stability and control aspects of microgrid architectures—a comprehensive review. IEEE Access **8**, 144730–144766 (2020). https://doi.org/10.1109/ACCESS.2020.3014977
2. M.F. Roslan, M.A. Hannan, P.J. Ker, M.N. Uddin, Microgrid control methods toward achieving sustainable energy management. Appl. Energy **240**(February), 583–607 (2019). https://doi.org/10.1016/j.apenergy.2019.02.070
3. J. Kumar, A. Agarwal, V. Agarwal, A review on overall control of DC microgrids. J. Energy Storage **21**(September 2018), 113–138 (2019). https://doi.org/10.1016/j.est.2018.11.013
4. M. Alam, K. Kumar, J. Srivastava, V. Dutta, A study on DC microgrids voltages based on photovoltaic and fuel cell power generators, in *7th International IEEE Conference Renewable Energy Research and Applications ICRERA 2018*, vol. 5 (2018), pp. 643–648. https://doi.org/10.1109/icrera.2018.8566854
5. K. Gowtham, C.V. Sivaramadurai, P. Hariprasath, B. Indurani, A management of power flow for DC microgrid with solar and wind energy sources, in *2018 International Conference on Computer Communication and Informatics, ICCCI 2018* (2018), pp. 1–5. https://doi.org/10.1109/iccci.2018.8441324
6. T. Shah, Z.A. Ansari, An overview of intelligent energy management system for DC microgrid: system and communication architecture and application in power distribution system, in *2018 13th International Conference on Industrial and Information Systems ICIIS*

2018—Proceedings, no. 978, pp. 339–342 (2018). https://doi.org/10.1109/iciinfs.2018.8721384
7. K.R. Bharath, M. Krishnan Mithun, P. Kanakasabapathy, A review on DC microgrid control techniques, applications and trends. Int. J. Renew. Energy Res. **9**(3), 1328–1338 (2019)
8. S.K. Sahoo, A.K. Sinha, N.K. Kishore, Control techniques in AC, DC, and hybrid AC–DC microgrid: a review. IEEE J. Emerg. Sel. Top. Power Electron. **6**(2), 738–759 (2018). https://doi.org/10.1109/JESTPE.2017.2786588
9. L. Meng et al., Review on control of DC microgrids and multiple microgrid clusters. IEEE J. Emerg. Sel. Top. Power Electron. **5**(3), 928–948 (2017). https://doi.org/10.1109/JESTPE.2017.2690219
10. Y. Jiang, Y. Yang, S.C. Tan, S.Y.R. Hui, Adaptive current sharing of distributed battery systems in DC microgrids using adaptive virtual resistance-based droop control, in *2019 IEEE Energy Conversion Congress Exposition ECCE 2019*, pp. 4262–4267 (2019). doi:10.1109/ECCE.2019.8912180
11. L.S.D.C. Microgrid, S. Augustine, S. Member, M.K. Mishra, S. Member, N. Lakshminarasamma, Augustine2015.Pdf, pp. 1–10 (2014)
12. J. Hu, J. Duan, H. Ma, M.Y. Chow, Distributed adaptive droop control for optimal power dispatch in DC microgrid. IEEE Trans. Ind. Electron. **65**(1), 778–789 (2018). https://doi.org/10.1109/TIE.2017.2698425
13. N.L. Diaz, T. Dragicevic, J.C. Vasquez, J.M. Guerrero, Intelligent distributed generation and storage units for DC microgrids—a new concept on cooperative control without communications beyond droop control. IEEE Trans. Smart Grid **5**(5), 2476–2485 (2014). https://doi.org/10.1109/TSG.2014.2341740
14. Y. Mi, H. Zhang, Y. Fu, C. Wang, P.C. Loh, P. Wang, Intelligent power sharing of dc isolated microgrid based on fuzzy sliding mode droop control. IEEE Trans. Smart Grid **10**(3), 2396–2406 (2019). https://doi.org/10.1109/TSG.2018.2797127
15. S. Guiying, L. Da, L. Yuesheng, T. Yanbin, D.C. Dc, C.O.S. Ystem, Master-slave with phase-shift control strategy for converter system, pp. 2546–2551 (2016)
16. I. Federico, E. Jose, F. Luis, Master-slave DC droop control for paralleling auxiliary DC/DC converters in electric bus applications. IET Power Electron. **10**(10), 1156–1164 (2017). https://doi.org/10.1049/iet-pel.2016.0590
17. R.A.F. Ferreira, H.A.C. Braga, A.A. Ferreira, P.G. Barbosa, Analysis of voltage droop control method for dc microgrids with Simulink: modelling and simulation, in *2012 10th IEEE/IAS International Conference on Industry Applications INDUSCON 2012* (2012). https://doi.org/10.1109/induscon.2012.6452563
18. S. Sahoo, S. Mishra, N.P. Padhy, A decentralized adaptive droop based power management scheme in autonomous DC microgrid, in *Asia-Pacific Power and Energy Engineering Conference APPEEC*, vol. 2016-December, pp. 1018–1022 (2016). https://doi.org/10.1109/appeec.2016.7779649
19. T.V. Vu, D. Perkins, F. Diaz, D. Gonsoulin, C.S. Edrington, T. El-Mezyani, Robust adaptive droop control for DC microgrids. Electr. Power Syst. Res. **146**, 95–106 (2017). https://doi.org/10.1016/j.epsr.2017.01.021
20. A. Maknouninejad, Z. Qu, F.L. Lewis, A. Davoudi, Optimal, nonlinear, and distributed designs of droop controls for DC microgrids. IEEE Trans. Smart Grid **5**(5), 2508–2516 (2014). https://doi.org/10.1109/TSG.2014.2325855
21. V. Nasirian, A. Davoudi, F.L. Lewis, J.M. Guerrero, Distributed adaptive droop control for DC distribution systems. IEEE Trans. Energy Convers. **29**(4), 944–956 (2014). https://doi.org/10.1109/TEC.2014.2350458
22. J. Mohammadi, F. Badrkhani Ajaei, Improved mode-adaptive droop control strategy for the DC microgrid. IEEE Access **7**, 86421–86435 (2019). https://doi.org/10.1109/access.2019.2924994
23. Y. Gu, X. Xiang, W. Li, X. He, Mode-adaptive decentralized control for renewable DC microgrid with enhanced reliability and flexibility. IEEE Trans. Power Electron. **29**(9), 5072–5080 (2014). https://doi.org/10.1109/TPEL.2013.2294204

24. M.D. Cook, G.G. Parker, R.D. Robinett, W.W. Weaver, Decentralized mode-adaptive guidance and control for DC microgrid. IEEE Trans. Power Deliv. **32**(1), 263–271 (2017). https://doi.org/10.1109/TPWRD.2016.2583384
25. T. Morstyn, B. Hredzak, G.D. Demetriades, V.G. Agelidis, Unified distributed control for DC microgrid operating modes. IEEE Trans. Power Syst. **31**(1), 802–812 (2016). https://doi.org/10.1109/TPWRS.2015.2406871
26. Shivam, R. Dahiya, Distributed control for DC microgrid based on optimized droop parameters. IETE J. Res. **66**(2), 192–203 (2020). https://doi.org/10.1080/03772063.2018.1481461

Augmented Reality on User-Friendly Maneuver for Hunting Arsenic Toxicant

R. Hema and M. Sundararajan

Abstract Arsenic extensively exists in Environment. It is contemporary at different levels in the ocean as well as groundwater. Oceanic water is an aqueous solution holding dissolved solids and gases. However there are countless contaminants in normal water, it is the main geogenic and poisonous pollutant in intake water creating a health hazard. For the reason that the composite nature of the dissolved materials in normal waters particularly technological advanced method is typically essential to achieve the concentration of components. The regular approaches for the quantitative investigation cannot be functional to marine water without sufficient checks on their precision. This is predominantly true when dealing with rudiments existent in extremely minute concentrations. In this investigation, an effort is ended to mature a system for monitoring the Arsenic contamination level in seawater using arsenic specific sensor executing Linear Sweep Anodic Stripping Voltammetry (LSASV). It is a skill in which the metallic ion of attention is pre-attracted on an anode apparent at appropriate attracted potential and then stripped by an anodic streak voltage. The system of measurement is equipped with GPS hunting using GPRS and GSM communication techniques. Arsenic contamination levels in seawater at numerous places were measured by the technologically advanced system and the results were related to orthodox approaches. The precision and correlation of approximation were originated to be within adequate limits. The system affords an effective means in development for the rehabilitation of water polluted with Arsenic like toxicants and in wastewater reutilizing. Fabricated Nano screen-printed sensor strip was also examined for the recognition of heavy metals similar to Lead, Cadmium, Copper, Silver, Gold, Bismuth contaminants in water by suitable deposition potential. Lab on Chip is also feasible for numerous bulk metal pollutants detection on this single Nanosensor. The Virtual and Augmented reality of the contamination level and the location of the sample under the test data at remote sensing and online monitoring system are

R. Hema (✉)
PVC Office - ECE, BIHER, Chennai, India
e-mail: hemrbujradha@gmail.com

M. Sundararajan
PVC Academics, BIHER, Chennai, India
e-mail: pvcac@bharathuniv.ac.in; msrajan69@gmail.com

V. Suma et al. (eds.), *Inventive Systems and Control*,
Lecture Notes in Networks and Systems 204,
https://doi.org/10.1007/978-981-16-1395-1_52

established. When the cloud data getting was within the Maximum Contamination Level by WHO then monitors BLUE colour and RED colour for MCL beyond the limit.

Keywords Arsenic · Anodic stripping voltammetry · Bulk metals

1 Introduction

The world is mostly enclosed by water. A focus of abundant interest global is the expense of ecological eminence and if thinkable, the rehabilitation of exaggerated regions. To identify numerous inorganic forms of arsenic contaminants like Cacodylic acid Arsenezo III—$C_2H_{18}N_4O_{14}S_2As_2O$—Arsenilic Acid, P-Arsenilic Acid—$H_2NC_6H_4AsO_3H_2$, and Dimethyl Arsenic Acid—$C_2H_8AsO_2Na$. Inorganic arsenic consists of arsenic ions united with chlorine, oxygen and sulfur ions to form mixtures known as As-III—Arsenite and As-V Arsenate. Arsenite is further poisonous of the dual formulas. This maneuver senses the little concentration of Arsenic in drinking water from [1 μg/L] 0.1 ppb—Parts per Billion. WHO—World Health Organization recommended MCL—Maximum Contamination Level of Arsenic is 10 μg/L [1–3].

Numerous approaches for finding for As III impurity quantity in the intake water were established using Graphite Furnace Atomic Absorption Spectrometry—GFAAS, Inductively Coupled Plasma Mass Spectrometry—ICPMS, and ICPMS with Great Recital Liquid Chromatography. Above said methods, however trustworthy, lack feasibility checking contamination, and a huge count of test groups have to prepare. Moreover need appropriate research laboratory circumstances and is time-consuming process. Most tests for arsenic including the Hach Method rely on the transfiguration of arsenic to arsine gas [4–6].

$$As_2O_3 + 6Zn + 12Hcl \rightarrow 2AsH_3 + 6ZnCl_2 + 3H_2O \quad (1)$$

$$H_3AsO_4 + 4Zn + 8Hcl \rightarrow AsH_3 + 4ZnCl_2 + 4H_2O \quad (2)$$

The most common approach of this Method produces arsine gas which is progressive to respond by the indicator and soaked using Hg Br_2—Mercuric Bromide. The technique further comprises the creation of HCl also toxin hard towards recognizing, essential workroom with classy procedures like Spectrometer. It is a difficult and time-consuming process [7–9].

ISE—Ion Selective Electrodes method remains as similar to antique Flame Photometry approach investigation. ISE—Ion Selective Electrode is a transducer for direct sensing of ionic strength or concentration in a solution without reacting to the sample. Different forms of ISE are Ion Sensing Electrodes, Ion Selective, or Ion Specific Electrode. The three forms mean a common thing about sensing the ionic strength but their inner meaning can be interpreted differently [10–12]. ISE stands deprived rating detector which intelligence dissimilar ions in that period, not precise for the exaction. Ion Selective Electrode is a Choosy, enhanced ranking device for a specific ion. It intelligences more than one ion the comparative choosiness contrary to additional ions in contrast with the stated ion of appropriate intended for Calibration. If intelligence is greater than one ion the comparative discrimination against other ions in contrast to an identified ion makes it appropriate for the measurement. Ion Specific Electrode is the preeminent grade and merit sensor which is observed to be additional precise to a specific ion. It is a preeminent ranking, excellence detector and considered as further precise for specific ion. In this case, the relative selectivity to other non-specific ions is almost nix or negligible. ISE works with selective membrane technique. ISEs are of different forms Glass, Liquid membrane, and Solid-state Electrodes. The solid-state electrodes are economical with ease of fabrication and they use neutral carrier PVC matrix membranes [13–15]. The Junction potential needs to be explained to understand the operational principle.

Working Principle of ISE could be explained from the behaviour when two different concentrations of similar ionic solutions are separated by a selective membrane, a voltage is produced across the walls of it as the objective ions prolix through the membrane from the higher to the lower concentration side. ISE responds selectively only for the intended ion. The voltage across this membrane is noted in between dual-position electrodes. Internal reference electrode dipped in a solution having the ions to be calculated and external reference electrode making a combination of ISE and reference half-cell. The potential difference across the half-cells gives a measure of ISE potential. The measurement is using non-linear log ruler thus decreases precision. Conversely, these techniques do not provide profound and precise calibration related to the infinitesimal dense of As_3 impurity in aquatic less than 10 μg/L [16–19].

The Optical Sensors are established on LEDs and fiber optics. Light passes through the feelers into a minor hose having As III founded liquids which hit on a photodiode. The technique also existence handy, Cheap, permits examination of information over PC. Where light quantity is fascinated to be subject to that amount of toxic in the test solution. A portion of sunlit is rapt via that solution, leftover the pass on it. The quantity of fascinated light is influenced by the focus of arsenic in the sample. A portion of the light is captivated by the sample and the leftover drops on the photodiode. The contamination of arsenic raises the sample's capability to absorb light drops. Scientists tried and motto that required very profound techniques to discover. This device could daze the overhead complications and essentials to invent [20, 21].

2 Development of Arsenic Monitoring System

See Fig. 1.

Portable Automated Monitoring System for Arsenic Toxicants (PAMSAT) sensor device senses the contamination level of arsenic in water samples. The result is

Fig. 1 Architecture of integrated system

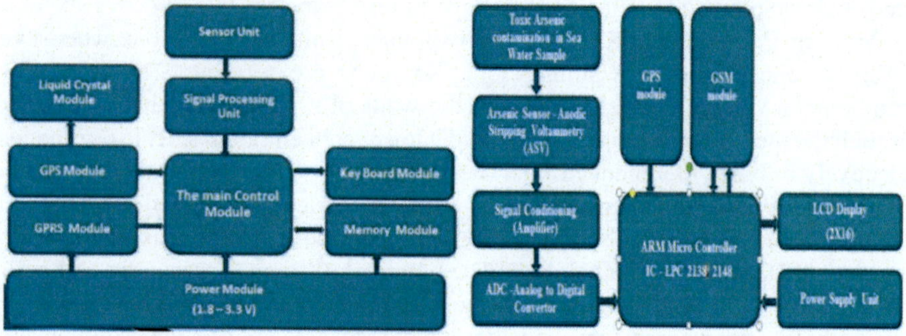

Fig. 2 Detailed block diagram

transmitted by GSM/GPS—transmitter module and remote sensing data by the GPS receiver module are processed by a computer system. The output of the sensor is given to ADC—Analog to Digital Converter through signal (Amplification) conditioning (Fig. 2).

2.1 *Hardware Structure of Remote System*

See Fig. 3.

The corresponding digital output is given to the microcontroller (ARM Processor—Advanced RISC Machine, 32-bit processer, 512 KB flash memory, 64 K RAM). The impurity level of arsenic pollutants in the water sample is to be displayed by the LCD. R&D Centre furnished with a GSM Modem; Microcontroller connected to computer system and deliberates to keep a track record of each sample quality send to a Microcontroller (Figs. 4 and 5).

Fig. 3 GPS receiver module

Interfaced with an LCD, the microcontroller connects the computer for processing and data handling through serial communication RS232 (or MAX232). For monitoring and tracking the data and database handling, Front-end is designed using Visual Basic program and the back-end is to maintain the data using MS-Access. KEIL Software and Flash program are to be used for an embedded microcontroller system.

634 R—GPS is a combined intelligent GPS module with 51 channels using ceramic patch aerial through a low noise amplifier. This feeler is connected to the acquisition machine and 14 channel track engines. Experienced in receiving signals from 65 GPS Satellites and transmitting into the exact location and timing info are read by either RS232 serial port or by UART port. Both the LVTTL-level and RS232 Signal Interface are on the interface connector, source potential of 3.59–5.99 V is held. The GPS aerial module can be accessible for OEM applications with adaptation. A GSM modem is an extraordinary type of modem that admits a SIM card, and functions over a subscription to a mobile operator, just similar to a mobile phone. When it is linked to a workstation, this allows interaction over the mobile network. With GSM and mobile internet connectivity, this system can also be used for sending and receiving tracked data using SMS messages. GSM modem supports a "lengthy AT command set" for sending/receiving SMS messages. It provides a cost-effective, fast, and excellent way of communication by the service provider. Insert a GSM SIM card into the modem and attach it to the USB port on your system. The receiver module received SMS which containing Longitude and Latitude (Figs. 6 and 7).

Fig. 4 Transmitter module

2.2 *Back End Design on Cloud and Grid Computing*

See Figs. 8, 9, 10 and 11.

The Global Time aids to sense the amount of data contamination. This information is displayed in the LCD module. The same will be stored in MS-Access or Oracle database and also in the computer system. In the cloud, the Front End applications are launched by the user to check/analyze tested data by Virtual Remote Sensing.

3 Results and Discussion

See Fig. 12.

Fig. 5 GSM module

3.1 *Results for Nano Sensor Strip*

Figure 13 depicts the relative results of Arsenic concentrations using conservative Photometric technique and technologically advanced fabricated sensor correlation graph. Arsenic ions levels were sensed in water samples collected at several places and the results were matched with the orthodox photometric approach. The relationship in this Figure displayed a correlation of 0.9963 which aspects to be a preeminent correlation and profound than the existing methodology. Table 1 showed the Proportional outcomes of Fabricated Screen Printed Nano Strip and Photometry results. The tabulated result of Arsenic Ionic Finding with Photometric and screen printed Sensor Strip was explained. Here, Linear Sweep Voltammetry (LSV) response is superior to Cyclic Voltammetry (CV).

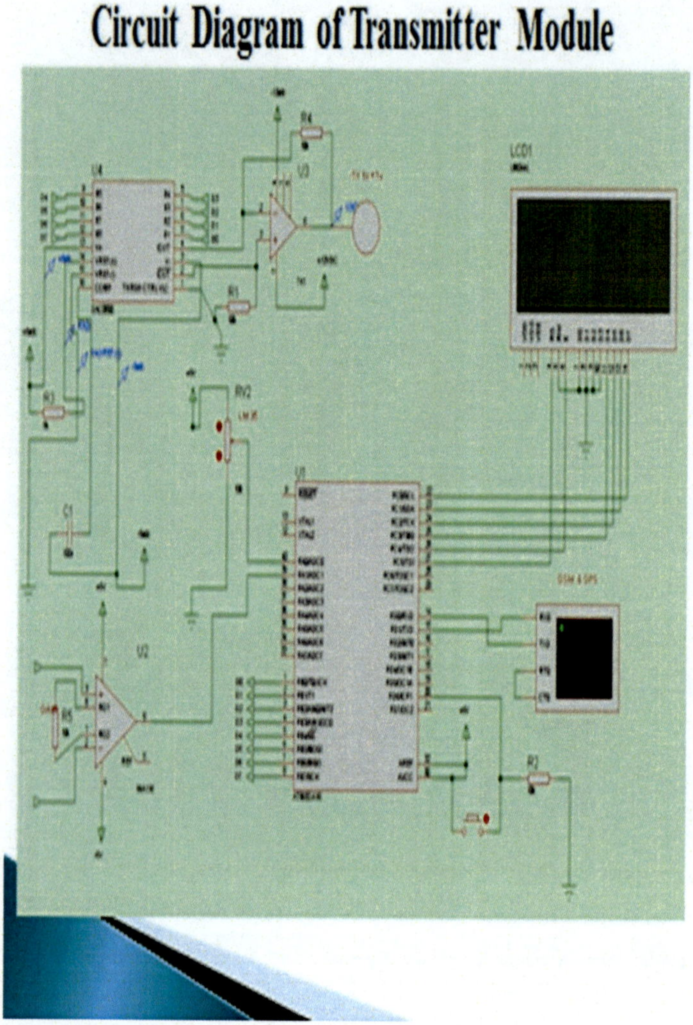

Fig. 6 Transmitter module circuit

4 Estimation and Standardization of Developed Arsenic Sensor

See Figs. 14 and 15.

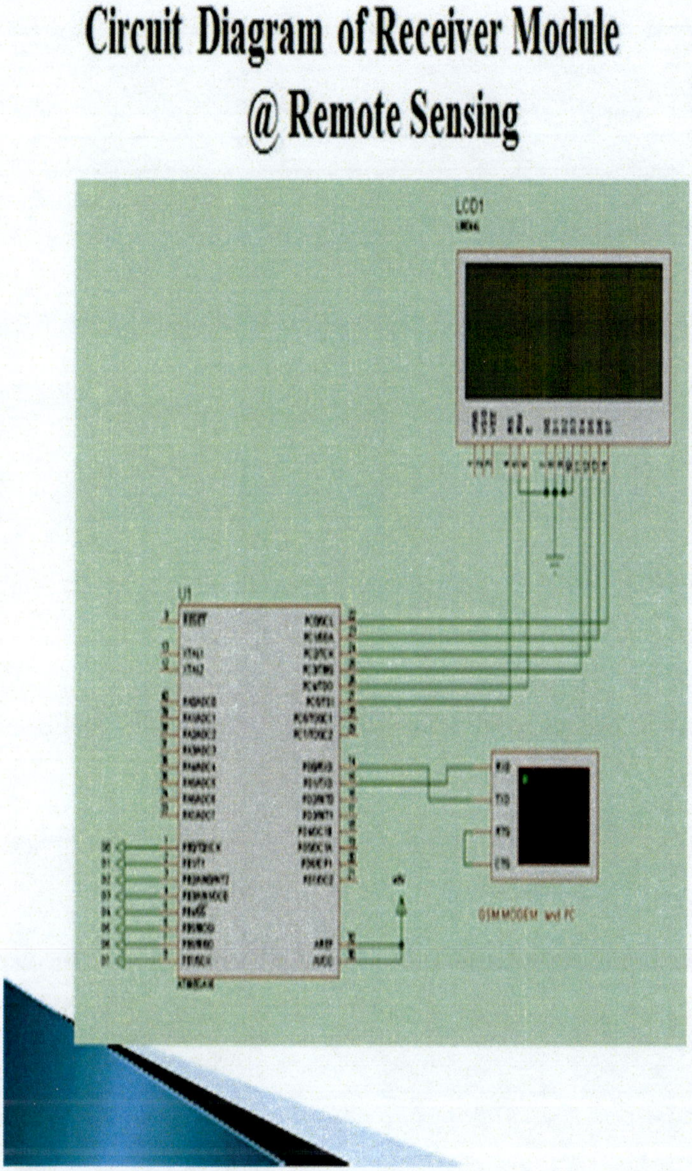

Fig. 7 Transmitter module circuit

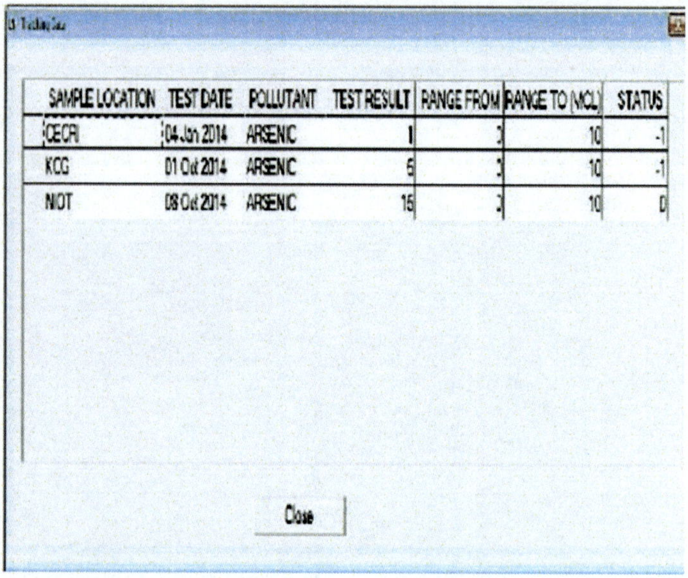

Fig. 8 Visual studio back end design

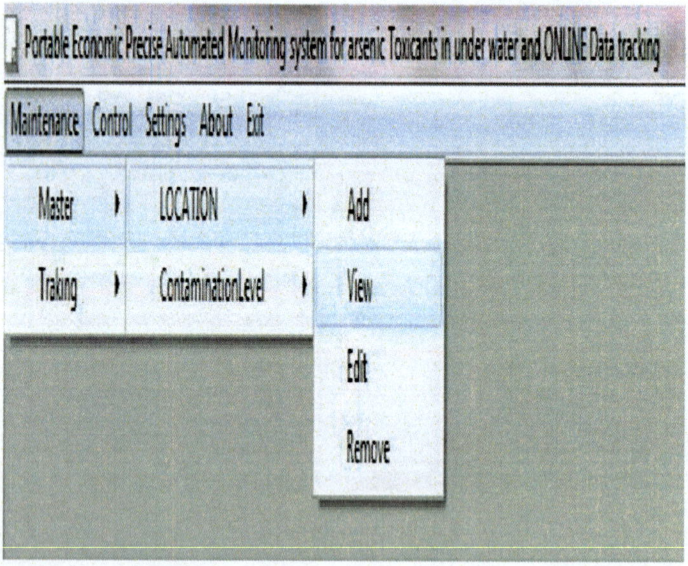

Fig. 9 Visual studio front end design

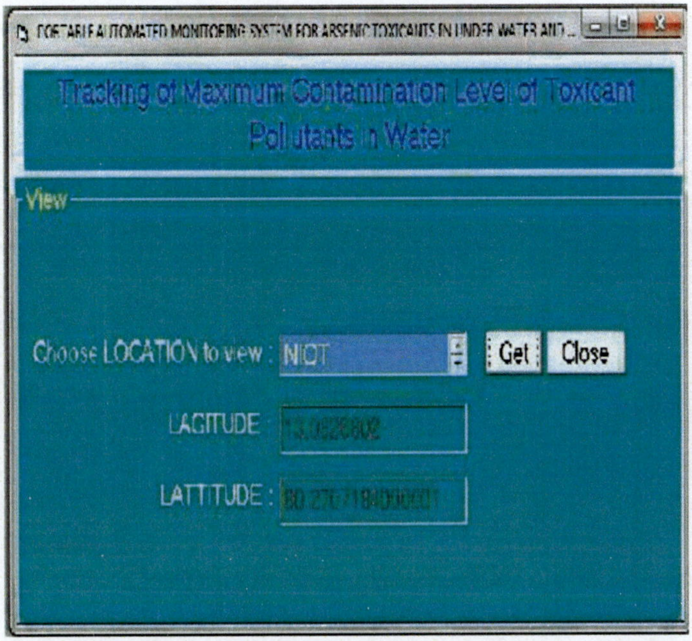

Fig. 10 Visual studio manual front end design

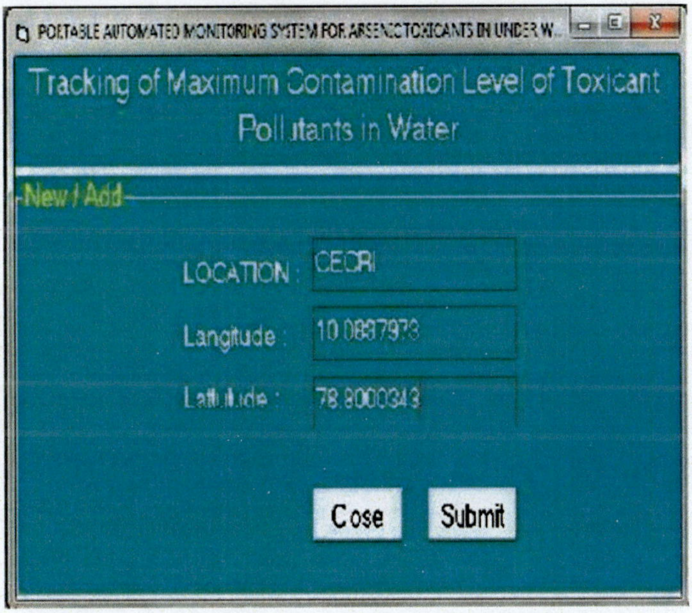

Fig. 11 Visual studio auto front end design

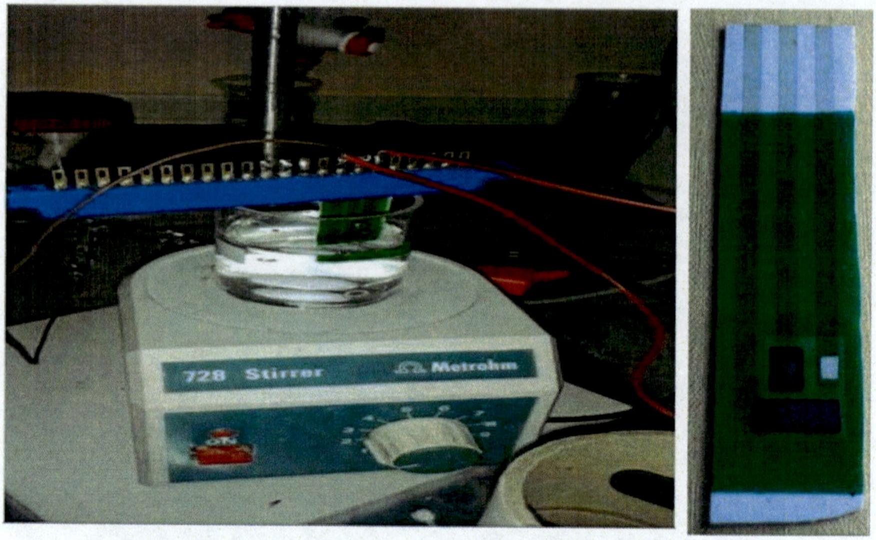

Fig. 12 Fabricated arsenic sensor under test photometry correlation graph

Fig. 13 Proportional outcomes of fabricated nano strip and photometry

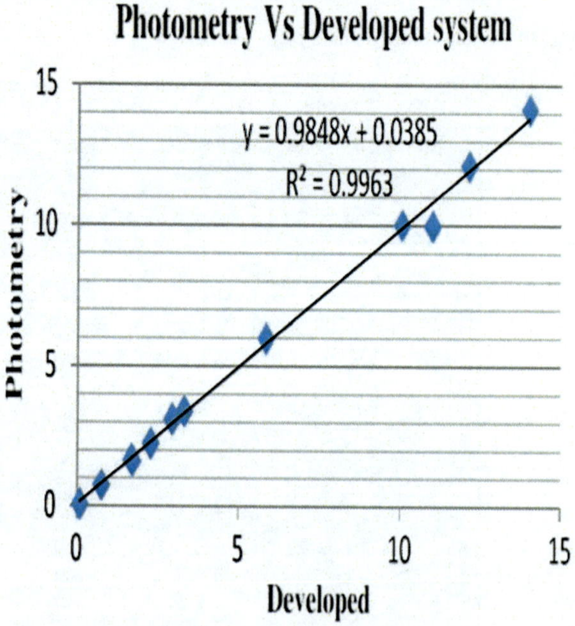

Table 1 Tabulation—proportional outcomes of fabricated screen printed nanosensor strip and photometry

Ionic arsenic level in μg/L		
S. No.	Photometry conventional method	Fabricated screen printed nanosensor strip
1	1.72	1.63
2	2.95	3.11
3	10.96	9.98
4	14.01	14.14
5	5.84	6.01
6	12.11	12.13
7	3.32	3.41
8	0.75	0.76
9	10.02	10.03
10	1.68	1.68
11	0.09	0.07
12	2.29	2.25

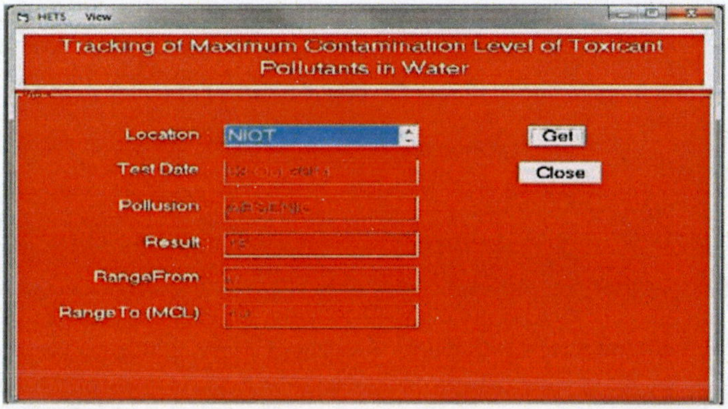

Fig. 14 Sample under test—location tracking and the contamination beyond the MCL limit

5 Future Enhancement as LOC

The same fabricated sensor feeler was also **examined** for the **detection of bulk metals similar to Cadmium, Lead, Copper, Silver, Gold, Bismuth contaminants in water**. By setting suitable deposition potential and deposition time, the device could be enhanced to detect numerous bulk metal pollutants in water by using the **same sensor strip**. Lab on Chip for Multiple Bulk Metal Pollutants Detection on single fabricated Screen Printed Nano Chip Research Work is Ongoing Project in Bharath University (Figs. 16 and 17).

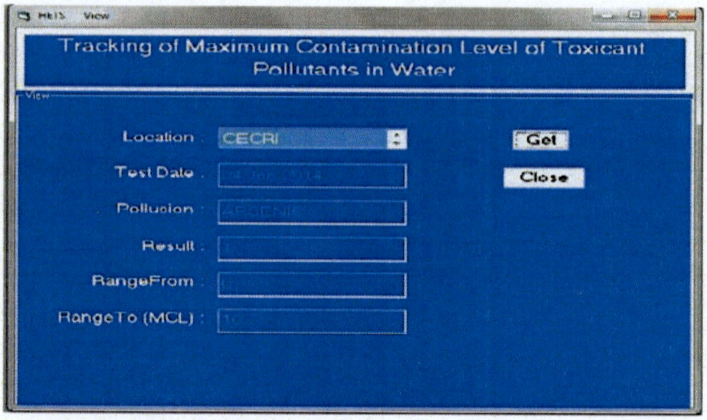

Fig. 15 Sample under test—location tracking and the contamination within the MCL limit

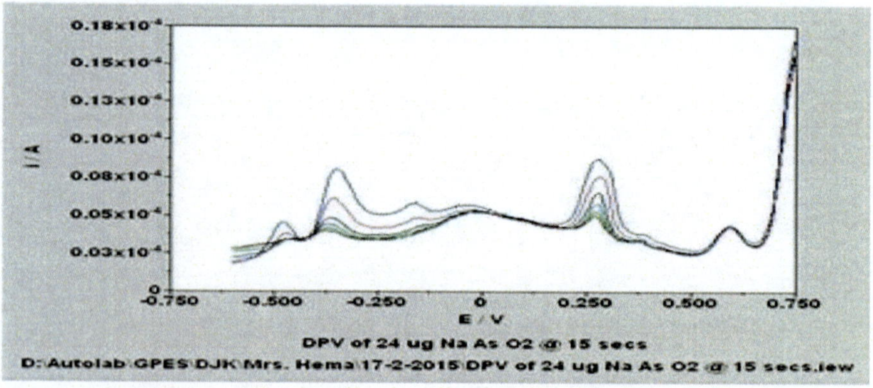

Fig. 16 DPV—differential pulse voltammetry detecting 7 bulk metals with low concentration

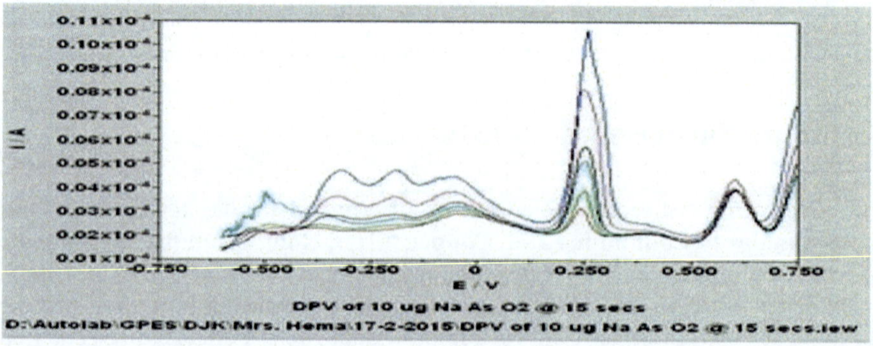

Fig. 17 DPV—differential pulse voltammetry detecting 7 bulk metals with very low concentration

6 Conclusion

This Arsenic toxicant fabricated screen-printed sensor strip by *Linear Sweep* Anodic stripping Voltammetry (LSASV) is integrated with online data tracking system. It is user-friendly as there is no hazardous poisonous gas produced because of any need of preprocessing requirement. Here, without an expert like Chemist the detection could be done in simpler, faster, cheaper with excellent sensitivity, and more precisely. A portable automated monitoring system for pollutants and toxicants underwater was developed which seems to be the greatest correlation and sensitive than the orthodox technique. Increasing of deposition period correspondingly increases the yield current peak. Thus the period of deposition holds the main starring role in this sensor strip. It was also prominent that the pollutants like Arsenic concentration are directly proportional to the output peak current consequently. This in turn states the contamination quantity of the sample water under test with precise quality. Invented integrated PAMSAT system is responsible for the fastest, user-friendly, low cost, handheld—portable accurate detection. Since Arsenic toxicant in water is additional risky to fitness, this technologically advanced system be essential in providing healthier to humankind. This maneuver senses the minute concentration of arsenic in drinking water from 0.1 ppb [1 μg/L]—Parts per Billion. Whereas the World Health Organization recommended Maximum Contamination Level of Arsenic is 10 μg/L. This system may also be implemented in rehabilitation efforts of purification of drinking water. And it provides precise detection and online data and location of the sample tracking. The virtual data is stored on the cloud created by the back-end database. The front-end is smart and designed by Visual studio that brings Augmented Reality of Remote sensing virtual data through the online monitoring system.

References

1. A.B. Chimezie, R. Hajian, N.A. Yusof, P.M. Woi, N. Shams, Fabrication of reduced graphene oxide-magnetic nanocomposite (rGO-Fe_3O_4) as an electrochemical sensor for trace determination of As(III) in water resources. J. Electroanal. Chem. **796**, 33–42 (2017)
2. A.O. Idris, J.P. Mafa, N. Mabuba, O.A. Arotiba, Nanogold modified glassy carbon electrode for the electro chemical detection of arsenic in water. Russian J. Electro chem. **53**(2), 190–197 (2017)
3. A. Waheed, M. Mansha, N. Ullah, Nanomaterials-based electrochemical detection of heavy metals in water: current status challenges and future direction. Trends Anal. Chem. **105**, 37–51 (2018)
4. A. Walcarius, Silica-based electrochemical sensors and bio sensors: recent trends. Current Opin. Electrochem. **10**, 88–97 (2018)
5. D. Sathiyaseelan, M.B. Gumpu, N. Nesakumar, J.B.B. Rayappan, G. Hariharan, Wavelet based spectral approach for solving surface coverage model in an electrochemical arsenic sensor—an operational matrix approach. Electrochim. Acta **266**, 27–33 (2018)
6. G. Bhanjana, N. Dilbaghi, S. Chaudhary, K.H. Kim, S. Kumar, Robust and direct electrochemical sensing of arsenic using zirconia nanocubes. Analyst **141**, 4211–4218 (2016)

7. G. Bhanjana, N. Mehta, G.R. Chaudhary, N. Dilbaghi, K.-H. Kim, S. Kumar, Novel electrochemical sensing of arsenic ions using a simple graphite pencil electrode modified with tin oxide nanoneedles. J. Mol. Liquids **264**, 198–204 (2018)
8. H. Gu et al., Electrochemical detection of arsenic contamination based on hybridization chain reaction and RecJf exonuclease-mediated amplification. Chem. Eng. J. **353**, 305–310 (2018)
9. P. Kumar, P. Devi, R. Jain, A. Saini, R. Noetzel, Electrochemical detection of trace arsenic (III) by functionalized $In_{0.38}Ga_{0.62}N/Si(1\ 1\ 1)$ electrode. Mater. Lett. **236**, 587–590 (2019)
10. R.C. Rodríguez, M.M. Bruno, P.C. Angelomé, Au nanoparticles embedded in mesoporous ZrO_2 films: multifunctional materials for electrochemical detection. Sens. Actuators B. Chem. **254**, 603–612 (2018)
11. S.D. Bukkitgar, N.P. Shetti, R.M. Kulkarni, S. Churmure, Nano-silica modified electrode as a sensor for the determination of mefenamic acid—a voltammetric sensor. Mater. Today Proc. **5**(10), 21466–21473 (2018)
12. S.-F. Zhou, X.-J. Han, H.-L. Fan, Q.-X. Zhang, Y.-Q. Liu, Electrochemical detection of as(III) through mesoporous $MnFe_2O_4$ nanocrystal clusters by square wave stripping voltammetry. Electrochim. Acta **174**, 1160–1166 (2015)
13. S. Gupta, R. Meek, Metal nanoparticles-grafted functionalized graphene coated with nanostructured poly aniline 'hybrid' nano composites as high-performance bio sensors. Sens. Actuators B Chem. **274**, 85–101 (2018)
14. S. Wen, Y. Wang, Y. Yuan, R. Liang, J. Qiu, Analytica chimica acta electrochemical sensor for arsenite detection using graphene oxide assisted generation of prussian blue nanoparticles as enhanced signal label. Anal. Chim. Acta **1002**, 82–89 (2018)
15. S. Wu, Q. Zhao, L. Zhou, Z. Zhang, Stripping analysis of trace arsenic based on the MnO_x/AuNPs composite film modified electrode in alkaline media. Electroanalysis **26**(8), 1840–1849 (2014)
16. S. Zhou, X. Han, H. Fan, Y. Liu, Electrochemical Sensing toward trace As(III) based on mesoporous $MnFe_2O_4$/ Au hybrid nano spheres modified glass carbon electrode. Sensors **16**(6), 935 (2016)
17. T.H.L. Nghiem et al., Preparation and characterization of silica-gold core-shell nanoparticles. J. Nanoparticle Res. **15**, 2091 (2013)
18. T. Ndlovu, B.B. Mamba, S. Sampath, R.W. Krause, O.A. Arotiba, Voltammetric detection of arsenic on a bismuth modified exfoliated graphite electrode. Electro chim. Acta **128**, 48–53 (2014)
19. V. Pérez-Herranz et al., Modification of porous nickel electrodes with silver nanoparticles for hydrogen production. J. Electroanal. Chem. **808**, 420–426 (2018)
20. X. Li, J. Ping, Y. Ying, Recent developments in carbon nanomaterial-enabled electrochemical sensors for nitrite detection. TrAC Trends Anal. Chem. **113**, 1–12 (2019)
21. X.-A. Yang, M.-T. Shi, D. Leng, W.-B. Zhang, Fabrication of a porous hydrangea-like $Fe_3O_4@MnO_2$ composite for ultra-trace arsenic preconcentration and determination. Talanta **189**, 55–64 (2018)

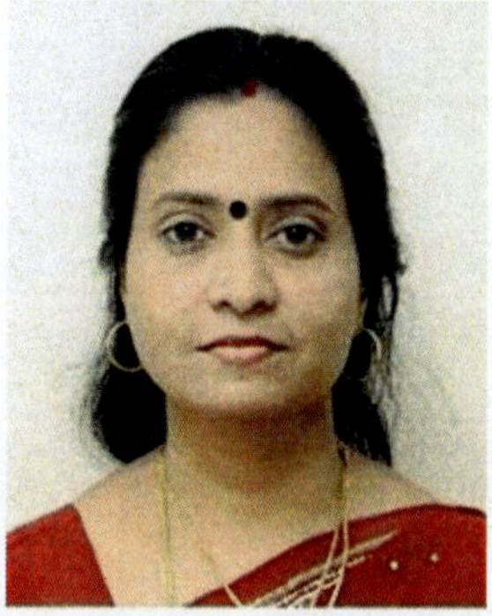

R. Hema , received B.E Degree in Electronics and Communication Engineering from Regional Engineering College [NIT—affiliated with Bharathidasan University] Trichy in 1996. Master of Engineering M.E—Communication Systems in K.C.G College of Technology, Chennai in 2015. Currently perusing Ph.D. in Nano Technology in BIHER [Bharath Institute of Higher Education and Research] Chennai.

Review of Different Machine Learning Techniques for Stock Market Prediction

Rahul, Kritesh Rauniyar, Javed Ahmad Khan, and A. Monika

Abstract Stock Market Forecast or Prediction (SMP) is the way to understand the potential value of a company's stock or other financial fields in the future. SMP, along with industry, has drawn a great deal of interest from academia. It is, however, a difficult research subject where several sophisticated informatics methods are suggested, but profitable and steady work has not yet been achieved. The reliability of the forecast can not be guaranteed by conventional approaches such as technical and fundamental analysis. The accelerating and decelerating of the stock market can not be understood, only prediction can be done using advanced techniques like Machine Learning (ML), Data Mining, and Deep Learning.

Keywords Stock prediction · Convolutional neural network · Numerical based attention · Feature weighted support vector machine · Two-stream gated recurrent unit · Machine learning

1 Introduction

Stock market prediction is a method for investors, traders, and analysts to forecast the future value of a stock, a specific segment of the market or the market as a whole. There are two main methodologies for stock analysis and prediction such as fundamental analysis and technical analysis [1]. Fundamental analysis emphasis data from sources, involving financial records, economic, assets, and market value where analysts seek for those shares that are presently dealing at an increase or decrease price

Rahul (✉) · K. Rauniyar
Department of Computer Science and Engineering, Delhi Technological University, Delhi 110042, India
e-mail: Rahul@dtu.ac.in

J. A. Khan
Department of IT, Government Girls Polytechnic Ballia, Ballia, Uttar Pradesh, India

A. Monika
Department of Computer Science, Shaheed Rajguru College of Applied Sciences for Women, University of Delhi, Delhi 110096, India

V. Suma et al. (eds.), *Inventive Systems and Control*,
Lecture Notes in Networks and Systems 204,
https://doi.org/10.1007/978-981-16-1395-1_53

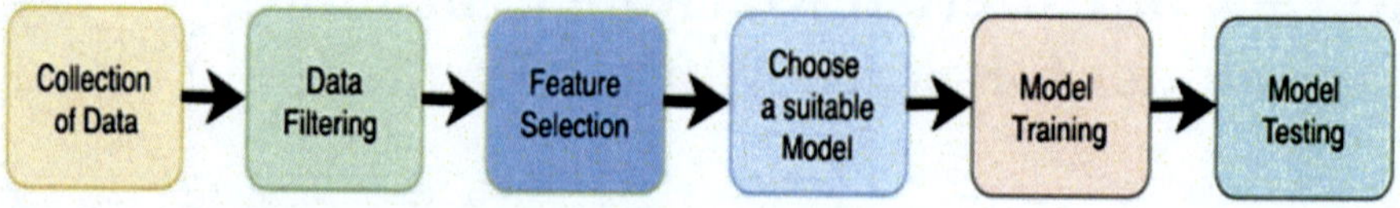

Fig. 1 Process of machine learning

than their actual price. Technical analysis is the study of past data, including price and volume where analysts predict future price movements using market psychology, economics. Various approaches were analyzed including Long Short-Term Memory (LSTM), Naïve Bayes (NB), Convolutional Neural Network (CNN), Numerical-Based Attention (NBA), Multi-source Multiple Instance (M-MI), Adaptive Network-Based Fuzzy Inference System (ANFIS), Feature Weighted Support Vector Machine (FWSVM), Feature Weighted *K*-Nearest Neighbor (FWKNN), Two-stream Gated Recurrent Unit (TGRU). Early research used the Efficient Market Hypothesis (EMH) [2] to forecast the share market. Finally, the EMH doesn't work for multiple financial models. Presently, the analyst's focus is on behavioral finance theory, still, EMH has not been eliminated because of the usefulness during the analysis. Present methods involve ML [3], Data Mining [4], and Deep Learning. Here, a reliable outcome is obtained after the completion of this research than the conventional stock prediction techniques.

Figure 1 shows the processes that were carried out while constructing a model of machine learning. The dataset array is obtained from a separate source and a method of filtering data is implemented. The missing values, outlines, and standardization of information are used in Data Filtering [5]. For selecting the particular techniques apply the role to achieve a high rate of accuracy. A suitable ML model is selected and using proper attributes data are trained. Big data, data mining, and deep learning algorithms are integrated for the highest precision and better performance.

2 Literature Review

Li et al. [6], in order to forecast CSI300 indices in the stock market of China, the LSTM model was used which combines market information and investor sentiment data. Initially, to organize the data into three groups: negative, positive, and neutral, they created the Naïve Bayes sentiment classifier. And then a sentiment time series was created for the next task. Finally, a deep neural network framework, consisting of an LSTM block, a merged block, a linear block of ReLU, was developed. They began by manually categorizing a training data package of 5000 articles using Naïve Bayes. Naïve Bayes' final result was 1586 favorable articles, 1765 unfavorable articles, and 1649 inert articles. Three permutations were tested for input variables in the LSTM for the estimation of the stock market opening price. The prediction accuracy of the LSTM model with Bt as input was greater (87.86%) than that of the other LSTM model. Both the SVM and LSTM models produced weak results (48.57–50.71%) for the prediction of the concluding value of the market relative to the prediction of open prices.

Lounnapha et al. [7], designed Convolutional Neural Network (CNN) model for the stock market prediction where they used three stocks mentioned on the Thai Stock Market (BBL, CAPLL & PTT) and were checked and analyzed to the real stock price. The processes involved in CNN are: use various short filters/screen on the input image, space is subsample for activation of filters, Repetition of the data until sufficiently high-level features are found, Use of Multilayer Perceptron (MLP) [8]. There are several layers in CNN that can be divided into the pooling block, convolutional block, input block, completely linked block, and dropout. Pooling layers decrease the computational power required to process the data. For data processing, they scaled or normalized the input data and used return rates as ideal information with normalization which solves the issues of forecasting. The training data input vectors have been normalized such that all the characteristics had zero-mean and unitary variance. They used a sequential combination of max-pooling block and 1D convolutional. The other factors of this experiment were: the number of filters was 64, the Pool length was 2, the Subsample length was 1, Activation function ReLU. The final result of the experiment summarized the information that for predicting the stocks of those three separate firms, an in-depth learning framework which was based on CNN was successful. The reason is that for prediction the CNN doesn't rely on any past data. It includes only the recent prediction screen, which helps the architecture to consider the complex shifts and trends that are happening in the current window.

Liu and Wang [9], discussed the NBA technique for dual sources share market analysis. Three datasets, the China Security (CSI300) and Standard & Poor 500 (S&P500), were developed from two sources. Their model relies mainly on the framework of sequence-to-sequence [10]. The double data encoder generates the vector representations from numerical data and reports, and then the forecast for the market price is executed by the numerical-driven decoder. They implemented a numeric-based system in the decipherer to learn reciprocality between numeric information and press news. Two main components make up their model. A dual-information encoder is the first part, which helps to encrypt press news, mathematical information, and results from the embedding and numerical resultant of news. The second part is the decoder based on numbers, which helps to decode the numerical vector with the NBA in a sequence. The numerical data and news are processed accordingly by the dual-information encoder. To decode the analytical vector in dual-information decoder various LSTM were used. In preprocessing firstly, headlines were tokenized and normalized where company name/code and numbers were replaced by the "COMPANY" and "NUMBER" respectively. After that, the headlines were matched and numbered where a specific company's information and numerical details are synchronized with news release time t. Mean Squared Error (MSE) has been used to determine the proximity between expected and real value in which the MSE predictor calculates the average error squares. The Adam optimization algorithm [11] is used during the training phase to decrease the MSE among the goal value and expected value. In order to forecast share value with press news and statistical information, four baseline models were adapted, where several models were developed by the direct multi-step ahead prediction strategy. In both MSE and accuracy, this model accomplishes the best predictive performance, especially NBAs had 6.96% and 2.32% improvement. The complementary between news and analytical data can be captured effectively by the NBA, but this model is less sensitive compared to the SVR model to the shift in news representation [12]. NBA can efficiently produce news report fusion features and stock trend prediction numerical data. Stock price movement forecasts at the instant reach a greater degree of success compared at the beginning of the day.

Zhang et al. [13], to forecast the stock market price, designed an M-MI architecture where they fused programs, thoughts, and quantitative information into a comprehensive system. The data collection consists of three components: archival statistical information, news editorial, and social media messages. For the quantitative data, the average rates, the shift in the market index, and the Shanghai Composite Index turnover rate were used. In 2015, they collected 38,727 news stories, and in 2016, 39,465. In China, they collected 6,163,056 posts from a common social network of investors called Xueqiu. As the inputs of the system, market numerical data, public communication, and electronic broadcast were used and then extracted thoughts, incidents, along with market numerical data, were supplied in the M-MI architecture. In addition, the architecture work differentiates the instance, the multi-source group, and the super group-level labels of multi-source. The main objective was to forecast the supergroup label that decides the stock market index's increase or decline. First, they model the instance-level probability p_{ij} to the goal label through a logistic curve and probability p_{d-i} for market numerical data and p_{s-i} for thoughts for an editorial news j on a day i. Secondly, for multi-source results, they model the probability P_i and ultimately they calculate the possibility of the supergroup as the probability mean in t days. The outcome of the prediction was compared with several other models such as SVM, TeSIA, nMIL, O-MI, WoR-MI, WoH-MI. While both the M-MI and WoR-MI achieved improved performance than O-MI, the SVM approach shows the worst results. WoH-MI results in poorer execution than that of M-MI, suggesting that the suggested instance-level loss function throughout multiple data sources was helpful for the precise forecast. Compared to nMIL, using the M-MI model, it improved accuracy by 6.7% and 9.4% in 2015 and 2016 while it improved $F-1$ by 6.9% in 2015 and 9.2% in 2016 respectively. News events, though quantitative data takes the second position, contribute most to the overall forecast. It suggests that both headlines and numerical information have greater effects than sentiments to drive stock volatility. In a comprehensive context, this model effectively compiles heterogeneous data, that is, events, sentiments, and historical quantitative features.

Acar and Avci [14], discussed the ANFIS where the main purpose was to find if an ANFIS algorithm can correctly forecast the value of the stock market. ANFIS could be a helpful device for economists and practitioners concerned with predicting the return of the share price index. ANFIS framework fuses artificial neural networks [15] and fuzzy logic. ANFIS training, studying algorithms of the neural network was used where the module comprises of if-then rules and sets of input-output. This model uses five layers and nine if-then rules where Layer-1 works: each node i maybe a square node with a node feature during this layer, Layer-2: each node may be a circle node during this layer, where incoming messages are multiplied and give the output, Layer-3: each node may be a circle node marked N during this layer. The ith nodule computes the proportion of the intensity of ith rules to the total of all firing strengths of the law, Layer-4: Each i node will be a square node consisting of a node feature during this block. Layer-5: During this block, the only nodule was named circle node which determines the general result as the addition of entire input cues. Data used in this model is obtained from the Electronic Data Delivery System and Matrix Information Delivery Services Inc. of the Central Bank of Turkey (CBRT). This research used the stock market terms like DJI, BOVESPA, DAX, macroeconomic indicators, and so on. They have collected 228 sets of data where the first set of data (122 pairs) was used for model training and the second set of data (106 pairs) was used for testing the model. Republic gold sales prices, exchange rates of US Dollar, deposit rate of interest, index price of the consumer, development index of the industrial, rates of interest of Treasury bill, DJI, DAX, and BOVESPA closing prices are included in the input sheet. The ISE indices return is located in the output block. They have used a hybrid learning algorithm for ANFIS. For the experiment, they have used MATLAB using a fuzzy toolbox. Validation of the model is a method through which vectors input are set by input/output information. To find out how well the trained model performs, they are introduced to the trained model. In order to equate expected and real values for the validation of the model, certain mathematical approaches, such as RMS, the coefficient of determinations (R^2), and the coefficient of variation (Cov), could be applied. The research shows that by using ANFIS, the efficiency of stock price prediction is also substantially enhanced. The outcome of the experiment indicates a value of 0.0068 RMS, 45.8163 Cov, and 0.9827 R^2 using the ANFIS predictor. The experimental tests show that the model with an accuracy rate of 98.3% effectively predicts the periodical yield of the ISE National Indices which indicates that for business forecasts, ANFIS offers a promising alternative.

Chen and Hao [16], proposed a simple hybridized architecture of the FWSVM and FWKNN for the prediction of the stock market. Basically, the SVM algorithm depends on the hypothesis that every component of samples provides a similar improvement to the final result but in many real problems, this hypothesis is not always true. So, an advanced SVM algorithm is introduced called FWSVM where each component of samples is appointed a distant weight value based on certain principles. Generally, three feature weighted kernels are executed namely Linear kernel, Polynomial kernel, and Gaussian kernel which is used in weight values to different components according to their foundation. So by adding feature wight in KNN enhanced the forecasting model and presented a new FWKNN model. The square root of the subtraction of the new information $InfoA(D)$ from original information $Info(D)$ is called information gain. In KNN it makes no assumption about the parameters and uses very little information for the calculation where the result is very difficult to generalize. They used the FWKNN model where they increased the range of Euclidean distance by applying some component weight. The data was collected from the SSE Composite Index and SZSE COMP SUB IND to analyze the stock market trend in China where both the dataset uses a capitalization-weighted index. Basically stock market has two segments, i.e., bullish, and bearish markets. A bullish market is where the stock price leads because of the excessive selling of stocks to the people. In such conditions, the number of clients is higher than the agent. Contradictory to this, the continuous fading and crashing of the market are known as a bearish market where the piece of the stock falls gradually with the stock indices. Trending volume and stock price should be noticed continuously to get information or knowledge about the trend of the market. Generally, opening, closing, lowest, highest prices on stock market indexes are utilized for input functions. The system consists of three fundamental stages for the proposed FWSVM-FWKNN model: the data assortment and pre-handling stage, the training stage, and the testing stage. Firstly, the data assortment and pre-handling stage, all the prices, and the dataset on the trade volume are collected and stored in the database. In order to generate the class y_i, they measured Δc_s on the basis of a shift in the closing price and assess the $\mathrm{TI}_{i,\,m}$ technical indicators with all types of prices which the effects of the trade amount. By adding $\mathrm{TI}_{i,\,m}$, and y_i to the earlier database, the initial dataset is generated and the datum was standardized and traced in the limit $[-1,1]$. After that, the set of data was split into S_{train}, the training record, and S_{test}, the testing record. The information gain is measured in the training process and then the feature matrices P was built. The FWSVM was then skilled in the training statistics and exported the framework which was trained. The testing statistics are categorized applying the coached FWSVM design in the testing stage and the framework performance was either $+1$ or -1, which was marked as C. The valuation of C was then analyzed with the class mark to determine the model's precision. FWKNN computation was then used to determine k nearest testing information and calculate the average to estimate Δc_s of those neighbors. The outcome is then applied as the expected value in advance using the existing initial non-normalized closing value. The FWSVM-FWKNN MAPE was the same as the SVM-KNN and RMSE on the SSE Composite Index was lower than the SVM-KNN and the FWSVM-FWKNN on the SZSE COMP SUB IND had the smallest MAPE and RMSE. The results of the experiment showed that the FWSVM-FWKNN algorithm, in which the FWSVM classifying and the FWKNN projection were potent, showed substantial betterment, and good predictive potential for the Chinese stock market compared to other comparative models.

Minh et al. [17], proposed a TGRU which predicts share market directions using both the sentiment and financial news. Four parts are included in the proposed process. (1) To delete unnecessary details such as stop terms, punctuation, the collected articles are preprocessed for paper preprocessing. (2) Two tasks are included in another step: marking news articles to help stock prices. (3) In the embedding Stock2Vec segment, depending on the economical dataset and sentiment data, helps to build the new sentiment word embedding model. (4) The TGRU system was introduced on the economic news record in the final stage, then several scenarios are carried out. The news articles were gathered in document preprocessing, and tokenization was performed where a single word was represented by each token. The Raw Daily S&P 500 Index share values were extracted in CSV format from Yahoo Finance for the daily stock price. The article had a published date for querying the starting and ending prices from the data. The Start-to-End was measured using starting and ending prices. After that, the return valuation was used for positive or negative news labeling. Use the proposed Stock2Vec model after the dataset is created to train it. For each word, a vector was provided with a maximal size of 300, and the overall words in Stock2Vec were restricted to 5000. Technical indicators [18] are quantitative estimates based on a security or contract's price, length, or open interest. Three indicators were used in this model, namely William, the stochastic oscillator, and the Relative Strength Index. This analysis includes two separate datasets: (1) Reuters and Bloomberg daily financial news (2) Daily S&P500 share prices accessed through Yahoo Finance. The dataset consists of three sections: the training section, the validation section, and the testing section. They restricted the word length to 300 in the Input layer for TGRU, truncated lengthy news stories and added short news stories with zero, and normalized them to a range between 0 and 1. This layer is preceded by the Embedding layer that receives for each news article a maximal of 303 input order and for every forward and backward pass, the TGRU layer with 128 memory units which is the next layer, where a dropout for every repetition is added to the GRU layer. Another dropout layer was added after combining the backward GRU and forward GRU to set down 50% of the input to tackle the problem of overfitting, and finally, the Dense layer which is the last layer where the number of neuron count is one, and for the prediction of the positive or negative classes they applied sigmoid activation function. The overall accuracy of the TGRU was 66.32%, which exceeds the output of the earlier framework. Their model had two learning phases, along with forward and backward, in order to grasp more productive knowledge, particularly for the problem of text processing. The embedding of Stock2Vec was more productive compared to the initial embedding process i.e. Word2Vec and Glove since it considered the word's sentiment value. Since TGRU's complexity doubles GRU's complexity, TGRU requires additional computing energy than LSTM or GRU. The number of parameters is, after all, double. If intraday trading was introduced by the model, the framework could forecast the pattern minutes or hours later the news is delivered. For more reliable performance, other metrics, like the earnings increase and P/E ratio should be used in the experiment.

3 Discussion and Conclusion

Various methods were explored to forecast stock market prices in this paper and eventually found that the Deep Learning and ML architecture helps to achieve useful growth with precision. In the TGRU model, indicators were used which helps to know the signals and trends of the data, the overall accuracy was 66.32%. LSTM model gave the highest prediction accuracy (87.86%) for the opening price but gave very poor results for the closing price. Using a hybrid framework gave fewer MAPE values which indicates the FWSVM-FWKNN algorithm is robust for prediction. Using the ANFIS predictor, the experiment result shows 0.0068 RMS, 45.8163 Cov, and 0.9827 R^2, and the accuracy rate was 98.3% which concludes that the framework profitably predicted the periodical return. Using the M-MI model, it improved accuracy by 6.7% and 9.4% in 2015 and 2016 respectively (Table 1).

Various approaches were investigated regarding the growth of the stock market which showed us the extraction of several useful and interesting conclusions where to know about the new and combined methods to estimate the variations of the stocks. Accuracy can be improved by using more advanced filters, indicators, and optimization techniques. Different algorithms are combined for precise results. The dataset is evenly balanced and separated properly to increase the performance of the model. The advancement of prediction is increasing day by day using advanced mechanisms like ML, Data Mining, Deep Learning, and so on. Engineers and Data Scientists are working on real-time predictions for the stock price.

4 Future Scope

SMP is growing nowadays where the accuracy can be increased by choosing suitable useful attributes and improving them to minimize errors, using organized and balanced datasets for efficient processing. Considering the faster and accurate real-time prediction, there will be more enhanced and better ML techniques. By analyzing past stock exchange data, forecasting stock trends would allow stockholders to select

Table 1 Summary of the methods analyzed

Name of author	Year of publication	Method used	Dataset used	Remarks	Merits	Demerits
Li et al. [6]	2017	LSTM, Naïve Bayes	CSI300 index of the Chinese stock	Opening price accuracy was 87.86% and closing price accuracy was 48.57–50.71%	LSTM has memory blocks capable of storing past data	Naïve Bayes predicts that all characteristics are mutually independent
Sayavong et al. [7]	2019	Convolutional neural network (CNN)	Thai stock by BBL, CAPLL & PTT	CNN was very accurate because CNN does not depend on any previous details	Weight sharing	Uses more GPU power so it's the computational cost is high
Liu and Wang [9]	2018	NBA	CSI300 and S&P500	NBAs had 6.96% and 2.32% improvement in MSE and accuracy respectively	In a dual source, the performance of the NBA is better	Training time increases because of the addition of weight parameters

(continued)

Table 1 (continued)

Name of author	Year of publication	Method used	Dataset used	Remarks	Merits	Demerits
Zhang et al. [13]	2018	M-MI	Xueqiu, Shanghai composite index	Accuracy by 6.7% and 9.4% in 2015 and 2016 while it improved F-1 by 6.9% in 2015 and 9.2% in 2016 respectively	Calculate events, sentiments, and quantitative data	Uses more processing power
Acar and Avci [14]	2010	ANFIS	Electronic data delivery system and Matriks information delivery services Inc. of CBRT	45.8163 Cov, 0.0068 RMS, and 0.9827 R^2 were obtained. ISE precision was 98.3%	Uses simple mathematics for nonlinear and complex systems	Requires lots of testing
Chen and Hao [16]	2017	FWSVM and FWKNN	SZSE COMP SUB IND & SSE composite index	The classification was done by FWSVM and the forecast by FWKNN, was sturdy where it predicted class label C, value, and accuracy	Gives more weight to the data which are nearby and less weight to which are farther away	For each variable, a sufficient number of values from the present data is required
Minh et al. [17]	2018	TGRU	Daily economic bulletin from Reuters and Bloomberg and daily S&P500 stock value	Overall accuracy was 66.32%	Enhance the memory capacity which helps in training	General, GRU has fewer parameters which decreases the accuracy

intelligently and build better revenue. The merger of ML with the Deep Learning framework will affect the prediction model being better trained and implemented.

References

1. Investopedia. [Online]. Available: investopedia.com
2. A. Degutis, L. Novickytė, The efficient market hypothesis: a critical review of literature and methodology. Ekonomika, **93**(2), 7–23 (2014). https://doi.org/10.15388/ekon.2014.2.3549
3. K.C. Rasekhschaffe, R.C. Jones, Machine learning for stock selection. Financ. Anal. J. **75**(3), 70–88 (2019). https://doi.org/10.1080/0015198X.2019.1596678
4. L. Zhao, L. Wang, Price trend prediction of stock market using outlier data mining algorithm, in *Proceedings—2015 IEEE 5th International Conference on Big Data and Cloud Computing, BDCloud 2015* (2015), pp. 93–98. https://doi.org/10.1109/BDCloud.2015.19
5. Z. Liu, Y. Zeng, Y. Yan, P. Zhang, Y. Wang, Machine learning for analyzing malware. J. Cyber Secur. Mobility **6**(3), 227–244 (2017). https://doi.org/10.13052/jcsm2245-1439.631
6. J. Li, H. Bu, J. Wu, Sentiment-aware stock market prediction: a deep learning method, in *14th International Conference on Services Systems and Services Management, ICSSSM 2017—Proceedings* (2017). https://doi.org/10.1109/ICSSSM.2017.7996306
7. L. Sayavong, Z. Wu, S. Chalita, Research on stock price prediction method based on convolutional neural network, in *Proceedings—2019 International Conference on Virtual Reality and Intelligent Systems, ICVRIS 2019, Dl* (2019), pp. 173–176. https://doi.org/10.1109/ICVRIS.2019.00050
8. A. Namdari, Z.S. Li, Integrating fundamental and technical analysis of stock market through multi-layer perceptron, in *2018 IEEE Technology and Engineering Management Conference, TEMSCON 2018* (2018), pp. 2–7. https://doi.org/10.1109/TEMSCON.2018.8488440
9. G. Liu, X. Wang, A numerical-based attention method for stock market prediction with dual. IEEE Access 1 (2018). https://doi.org/10.1109/ACCESS.2018.2886367
10. sequence-to-sequence-architectures-ad6ff4451f84 @ medium.com. (n.d.). https://medium.com/@d.salvaggio/sequence-to-sequence-architectures-ad6ff4451f84
11. D.P. Kingma, J.L. Ba, Adam: a method for stochastic optimization, in *3rd International Conference on Learning Representations, ICLR 2015—Conference Track Proceedings* (2015), pp. 1–15
12. B.M. Henrique, V.A. Sobreiro, H. Kimura, Stock price prediction using support vector regression on daily and up to the minute prices. J. Finan. Data Sci. **4**(3), 183–201 (2018). https://doi.org/10.1016/j.jfds.2018.04.003
13. X. Zhang, S. Qu, J. Huang, B. Fang, P. Yu, Stock market prediction via multi-source multiple instance learning. IEEE Access 1 (2018). https://doi.org/10.1109/ACCESS.2018.2869735
14. M. Acar, D. Avci, Expert systems with applications an adaptive network-based fuzzy inference system (ANFIS) for the prediction of stock market return: the case of the Istanbul stock exchange. Expert Syst. Appl. **37**(12), 7908–7912 (2010). https://doi.org/10.1016/j.eswa.2010.04.045
15. A.H. Moghaddam, M.H. Moghaddam, M. Esfandyari, Predicción del índice del mercado bursátil utilizando una red neuronal artificial. J. Econ. Finan. Adm. Sci. **21**(41), 89–93 (2016). https://doi.org/10.1016/j.jefas.2016.07.002
16. Y. Chen, Y. Hao, A feature weighted support vector machine and K-nearest neighbor algorithm for stock market indices prediction. Expert Syst. Appl. **80**, 340–355 (2017). https://doi.org/10.1016/j.eswa.2017.02.044
17. D.L. Minh, A. Sadeghi-Niaraki, H.D. Huy, K. Min, H. Moon, Deep learning approach for short-term stock trends prediction based on two-stream gated recurrent unit network. IEEE Access **6**(c), 55392–55404 (2018). https://doi.org/10.1109/ACCESS.2018.2868970
18. M.R. Vargas, C.E.M. Dos Anjos, G.L.G. Bichara, A.G. Evsukoff, Deep leaming for stock market prediction using technical indicators and financial news articles, in *Proceedings of the International Joint Conference on Neural Networks, 2018-July* (2018), pp. 1–8. https://doi.org/10.1109/IJCNN.2018.8489208

Performance Analysis of Phased Array Antenna in ISM Band Using Phase Shifter IC 2484 and PSoC

Barbadekar Aparna and Patıl Pradeep

Abstract There is very little experimental work documented on the behavior of phased array antennas that is available. The work that can be found is almost exclusively on simulations that show their benefits in wireless networks. Digital beamforming algorithms are used for the Direction of Arrival (DOA), but they are very complicated in design and hence costly. The researcher is motivated to present a design of a four-element phased array in 2.45 GHz ISM band using a Wilkinson Power Divider (WPD) and a 180° reflection-type phase shifter using IC 2484 controlled through a Programmable System on Chip (PSoC). In a real-time environment, it is difficult to integrate RF components with software support. In the proposed system both, the RF components and the PSoC controller are embedded successfully to generate beam patterns. Each component of the proposed system is designed and evaluated before fabrication. The gain is 12.95 dBi and is constant throughout the scan angle in the case of the simulated and experimental results. The Side Lobe Level (SLL) is less than −13 dB. The DOA for the main lobe in the simulation and experimental phased array is very close to the desired direction.

Keywords Wilkinson power divider (WPD) · Programmable system on chip (PSoC) · Side lobe level (SLL) · Direction of arrival (DOA)

1 Introduction

G. Marconi developed the first directive antenna in wireless communication in 1906 and this was the origin of the fixed beam phased array shown by Karl Ferdinand Braun. This was being used to increase the gain till the 1920s. Friis and Feldman developed multiple antenna arrays using mechanical phase shifters in 1937 to track

B. Aparna (✉)
Department of Electronics and Telecommunication, AISSMS IOIT, Pune, India

Department of Electronics and Telecommunication, VIIT, Pune, India

P. Pradeep
Department of Electronics and Telecommunication, JSPM/TSSM'S COE, Pune, India

V. Suma et al. (eds.), *Inventive Systems and Control*,
Lecture Notes in Networks and Systems 204,
https://doi.org/10.1007/978-981-16-1395-1_54

the received signal as it changed direction. There are many primary limitations associated with the present electromechanical phase array such as cost, weight, size, power consumption, and complexity in design. These constraints limit the use of this technology to military applications only. It was found that half of the cost of the array system is only because of phase shifters [1].

Many years of technological development in phase array opened doors for researchers to work on a fully electronic steering system as a need of the times. The measure of the complexity of a phased array design was due to the phase shifter. Due to technological developments, there was a remarkable change in the design process of phase shifters. Maruhashi et al. [2] proposed FET switches for the phase shift design but due to high insertion loss at high frequency range the use of this technology was limited. Zhao et al. [3] proposed the concept of ferroelectric phase shifters using barium strontium titanate (BST). Ji et al. [4] presented phase shifters using BST phase shifters on a silicon wafer for 15 GHz. These phase shifters could not be used because of the constraints of the wafer size [5, 6].

Monolithic Microwave Integrated Circuits (MMIC) phase shifters such as switch types, reflection-type, and vector are some methods that have been reported recently. Tsai et al. [7] designed 12 GHz, 6-bit switched type phase shifters MMIC. But at low microwave frequencies, the value of the capacitors, inductors, and the size of transmission lines was considerably large and hence the design of phase shifters on chips was difficult. A CMOS reflection-type phase shifter MMIC was presented by Wu et al. [8] at 2.45 GHz with the least loss variation. Cetindogan et al. [9] designed a 6-bit vector-based control circuit for X-band phase arrays.

Radio Frequency Micro Electro Mechanical Systems (RF MEMS) evolved into a series of phase shifters such as switched line, loaded line, reflection type, and distributed line phase shifters. Scardelletti et al. [10] designed the simplest switched line phase shifters. The use of the switching function could be realized in multiple ways in such phase shifters. The use of shunt switch variants limits their applications because of large substrate areas and high insertion loss as compared to the transmission line length used. Zhang et al. [11] designed a loaded line MEMS phase shifters using a variable capacitor. It was simple in design with a low loss, small size, and had a smooth integration with the standard IC procedure. Kim et al. [12] proposed V-Band two-bit reflection type MEMS phase shifters. But due to a complicated design, their use was limited in both digital and analog variants. Nagra and York [13] proposed Distributed MEMS transmission line (DMTL) type phase shifters. They were mostly used in the application of phased array and wireless communication.

In general, RF MEMS phase shifters had distinct limitations like temperature stabilities, packaging, moderate reliability, fast switching time and were considerable in size. It was perhaps more critical than with alternative technologies and due to slow speed, it could not be used in fast airborne applications. Because of these drawbacks, the RF MEMS phase shifters had limited use in the consumer market. It is difficult to discover a particular phase shifter out of the huge number of articles and papers that have been released.

In this article, the author has proposed to select a passive phase shifter based on IC 2484 due to its distinct benefits such as small size, reasonable cost, low insertion loss, and Voltage Standing Wave Ration (VSWR). Moreover, analog phase shifters are very attractive for frontend modules of phased array antennas.

This article explores the feasibility of the integration of the proposed system with the PSoC approach. PSoC is an entirely new embedded design platform that incorporates [14–18] the best of analog, digital, and microcontroller. Consider the unique advantages of the PSoC such as expandability, innovativeness, and reconfigurability. This technology is used for the calibration of an analog phase shifter. Here both, the RF components and PSoC [19] are embedded to set control voltages for each phase shifter to generate the directive beam pattern.

Detailed process steps for the fabrication of the power divider, the phased array, and the use of reflected type phase shifter IC 2484 with PSoC are discussed. This methodology is useful for low-cost phased array system design and fabrication. However, it can be extended to design large array antenna systems in the future.

2 Proposed Layout of Phased Array

The proposed system layout of the phased array as shown in Fig. 1 is designed for 2.45 GHz frequency. It is subdivided into subsystems such as 4 elements printed dipole antenna array, power divider, voltage controlled analog phase shifter, and a PSoC controller circuit. The RF input signal through the power divider provides in-phase signal to the phase shifters. The PSoC can scan and locate the rotary encoder

Fig. 1 Proposed system layout

position as per the DOA evaluation and produce the required biasing voltage to the phase shifter to generate progressive phase shifts [19]. The same readings are displayed on an LCD. An analytical and experimental study is attempted to measure performance analysis. The design of each subsystem is as under.

2.1 Printed Dipole Antenna

The structure of printed a dipole antenna via a hole balun is as shown in Fig. 2. The balun acts as a balancing transformer that converts unbalanced loads to balanced ones. The dipole antenna is designed using an FR4 substrate for an operating frequency = 2.45 GHz, relative dielectric constant (ε_r) = 4.4, the height of the substrate (h) = 1.6 mm and tangent loss (tan δ) = 0.02. The length (L) and width (W) of the printed dipole antenna are calculated using Eqs. (1–5). The designed parameters are as shown in Table 1.

Calculation of effective dielectric constant:

$$\varepsilon_{\text{reff}} = \frac{\varepsilon_r + 1}{2} + \frac{\varepsilon_r - 1}{2}\left[\left(1 + \frac{12h}{w}\right)^{-\frac{1}{2}}\right] \tag{1}$$

Calculation of effective length ΔL:

$$\Delta L = 0.412 \times h\left\{\frac{\varepsilon_r + 0.3}{\varepsilon_r - 0.258}\right\}\left\{\frac{\frac{W}{h} + 0.264}{\frac{W}{h} + 0.813}\right\} \tag{2}$$

Calculation of the width of the dipole antenna W:

$$W = \frac{c}{2f_r\sqrt{\frac{\varepsilon_r+1}{2}}} \tag{3}$$

Calculation of the length of the dipole antenna L:

$$L = \left[\frac{c}{2f_r\sqrt{\varepsilon_{\text{reff}}}}\right] - 2\Delta L \tag{4}$$

Calculation of the width of Balun (W_b)

$$W_b = \frac{\lambda}{4} \tag{5}$$

The length of the dipole is adjusted to get an optimum result. The dimension of the ground plane must be large enough so that it gives equal current distribution to both

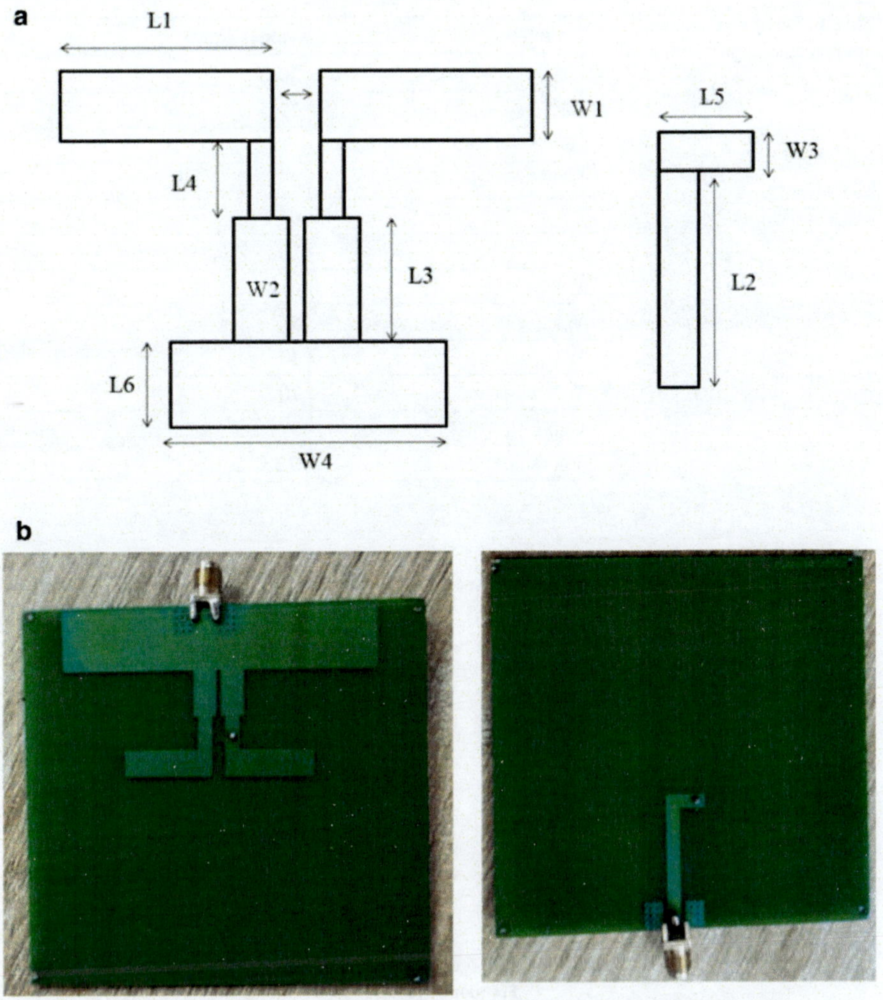

Fig. 2 **a** Configuration of printed dipole antenna, **b** fabricated printed dipole antenna

the balun arms. The design of the balun is based on a Marchand balun of the third order. Configuration and fabrication of the dipole antenna are as shown in Fig. 2a, b.

From Fig. 3 it can be seen that the simulated and measured values of the return loss (S_{11}) of the designed dipole antenna are −17.50 dB and −13.75 dB respectively. It is observed that the experimental results are in good agreement with the simulation results.

From Fig. 4 it can be deduced that the simulated and measured VSWR of the designed dipole antenna is 1.75 dB and 1.3 dB respectively both of which are less than 2, as a standard value.

Table 1 Designed parameters of dipole antenna

Parameter	Values (mm)
L1	20
L2	29
L3	11
L4	8
L5	6
L6	20
W1	6
W2	5
W3	3
W4	70
Gap g	3
Via radius	0.33

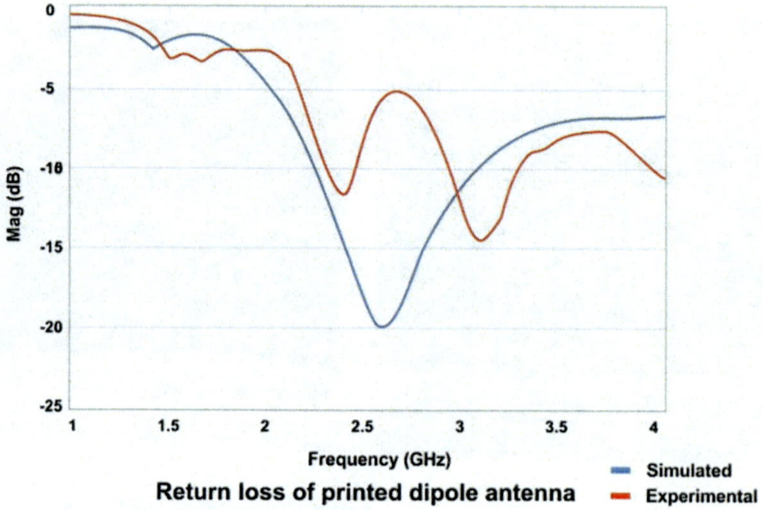

Fig. 3 Return loss (S_{11}) of printed dipole antenna

2.2 Wilkinson Power Divider (WPD)

The WPD plays a significant role in communication systems because of its characteristics namely, simple configuration and the matching of impedance and isolation at output ports. The structure of the WPD, as shown in Fig. 5a, b comprises of two $\lambda/4$ transmission lines. The branch impedances $\sqrt{2Z_0}$ and $2Z_0$ are used in the circuit design. The circuit is analyzed using an even-odd mode analysis.

The scattering matrix in Eq. (6) for the WPD is:

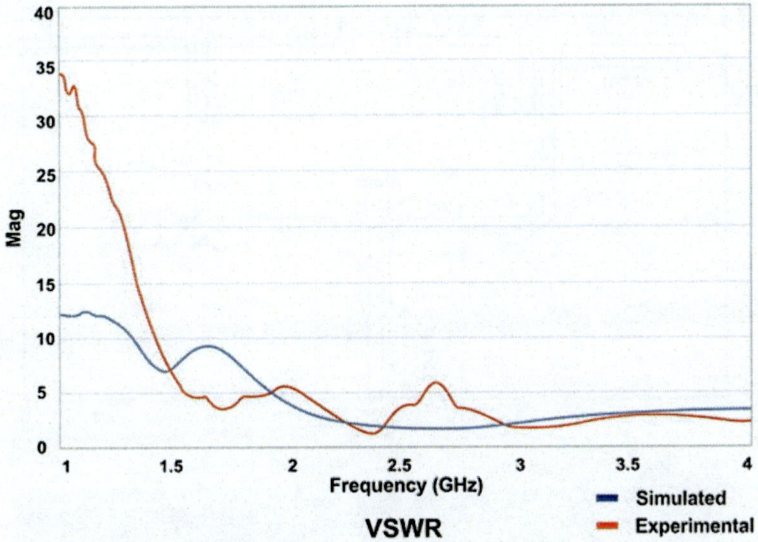

Fig. 4 VSWR of printed dipole antenna

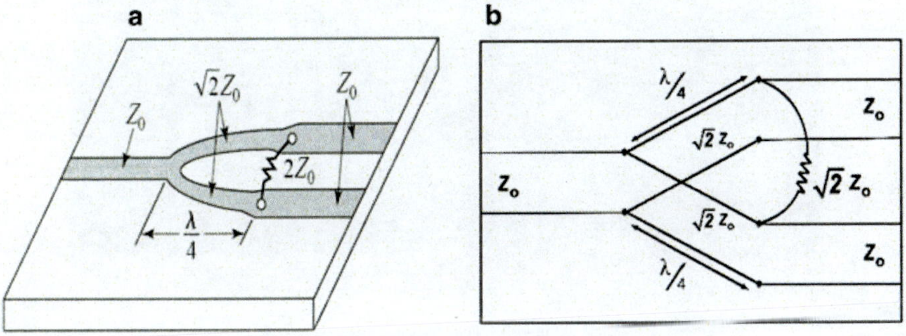

Fig. 5 **a** Microstrip line configuration, **b** equivalent transmission line circuit

$$S = \frac{-1}{\sqrt{2}}\begin{bmatrix} 0 & j & j \\ j & 0 & 0 \\ j & 0 & 0 \end{bmatrix} \tag{6}$$

From Eq. (6), it is found that phase power is provided at all the output ports without any loss on the transmission lines. There is no reflection from the output ports and the output ports are perfectly isolated.

Figure 6a shows the simulated WPD with a two-way and three-stage structure connected with 50 Ω transmission lines which arbitrarily adjust the distance between two output ports to meet the needs of the dimensions of the phase shifters and element spacing between antennas in the phased array system.

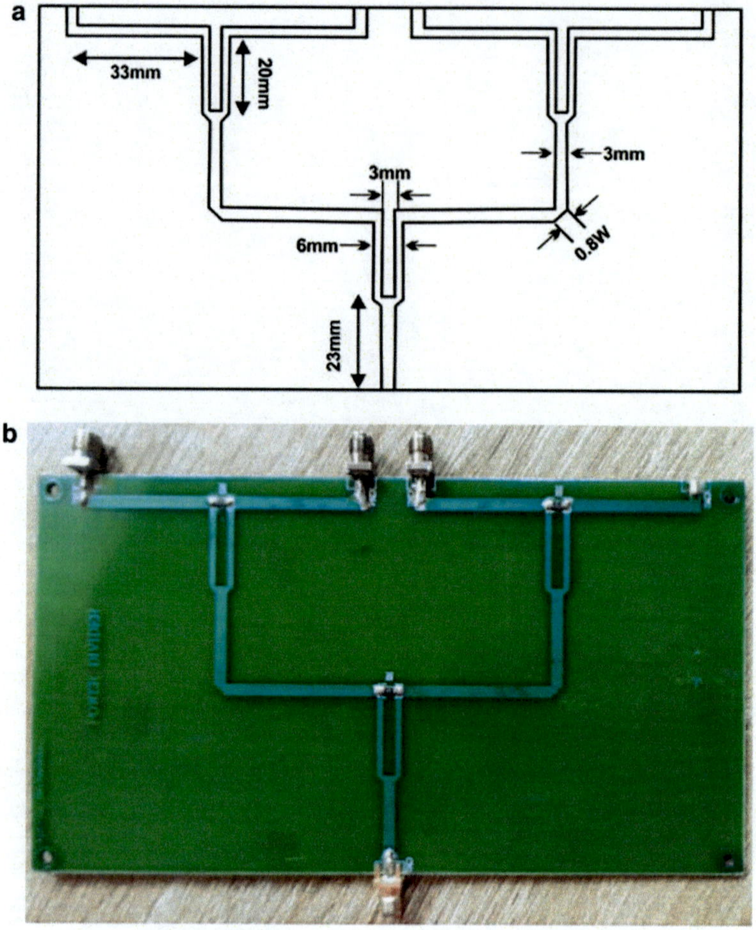

Fig. 6 **a** Simulation design of Wilkinson power divider, **b** fabricated design of Wilkinson power divider

The simulation result of the return loss (S_{11}) is −35.85 dB as shown in Fig. 7 while the measurement result is −11.6 dB at frequency 2.45 GHz.

The insertion loss with simulation between successive ports S_{21}, S_{31}, S_{41}, S_{51} is −7.80 dB, −7.90 dB, −8.0 dB, −7.8 dB respectively at 2.6 GHz frequency is as shown in Fig. 8a–d. Whereas for the measurement it is −7.2 dB, −7.6 dB, −7.6 dB, −7.8 dB respectively at 2.45 GHz. The average insertion loss for the simulation is −7.87 dB at 2.6 GHz while for the measurement it is −7.55 dB at 2.45 GHz.

The maximum insertion loss is −3 dB, 4port WPD is −6 dB. From Table 2 it is clear that the simulated and experimental insertion loss is −6 dB. Hence each output ports get equal power.

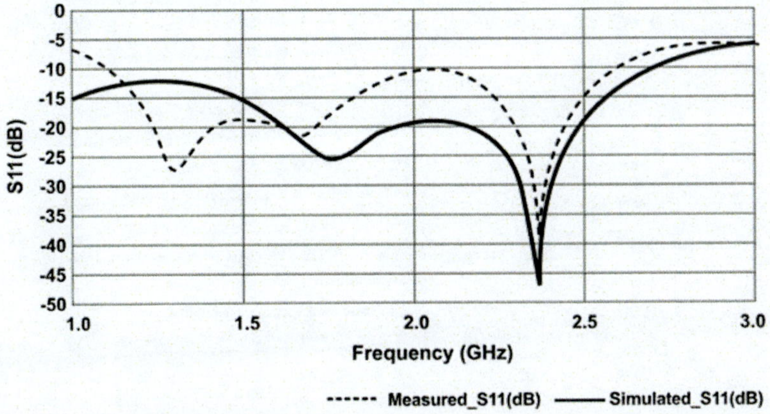

Fig. 7 Simulated and measured return loss (S_{11})

2.3 *Phase Shifter IC 2484*

The function of the phase shifter is to generate the required phase shift in the incoming RF signals provided through the WPD and excite the array antenna elements to form a directive beam. The phase shifter is implemented with an analog variable voltage controlled IC 2484 as shown in the proposed block diagram.

The chip is mounted on a TB-122B authorized layout PCB as shown in Fig. 9a, b that utilizes grounded co-planar waveguides for input-output RF excitation. The substrate used in the test board PCB is Rogger-R0 4350 with a dielectric constant of 3.5 and a thickness of 0.762 mm. Female SMA Connectors are connected to both, the input and output RF ports to facilitate measurements. The insertion loss of the WPD and the phase shifter IC 2484 connected to the system is plotted as shown in Fig. 10a–d and tabulated in Table 3.

The variation of the insertion loss from port 2 to port 5 is −10.23 dB, −13 dB, −11.0 dB, and −13.71 dB, respectively. These variations are due to a fabricated power divider and SMA connectors. After deducting the insertion loss due to the power divider, the net insertion losses that occur due to the phase shifter at ports 2, 3, 4, and 5 are −3.03 dB, −2.53 dB, −3.4 dB, and −5.91 dB respectively. The average value of the insertion loss is 3.71 dB which is well within the specified value that is 5.6 dB (Max.). The VSWR is 1.33:1 as compared to 1.5:1 as specified. These results indicate the improved performance of the proposed phase shifter.

2.4 *PSoC Controller*

The four-phase shifter ICs are connected in the system such that the RF input given to the 1:4 WPD equally divides the amplitude and phase and provides the same as the

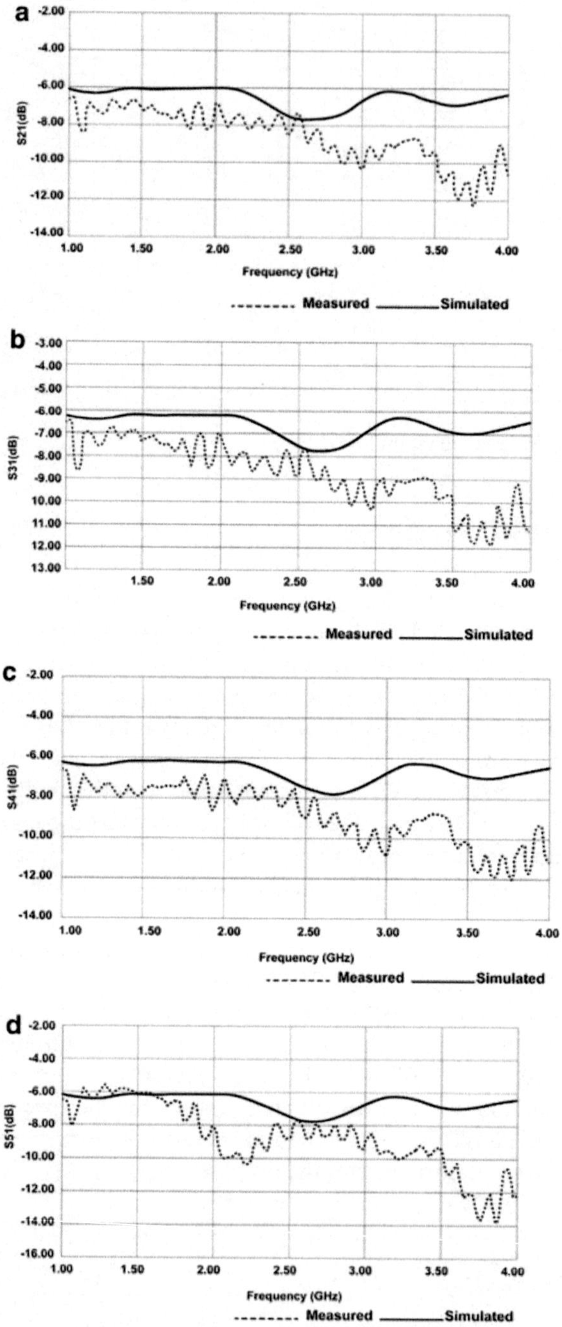

Fig. 8 **a** Insertion loss at each port S_{21}, **b** insertion loss at each port S_{31}, **c** insertion loss at each port S_{41}, **d** insertion loss at each port S_{51}

Table 2 Result of simulation and measurement for 1:4 WPD

Methodology	Insertion loss in dB				Average insertion loss in dB	Return loss (S_{11}) in dB
	S_{21}	S_{31}	S_{41}	S_{51}		
Simulation	−7.8	−7.9	−8.0	−7.8	−7.87	−35.85
Experimental result	−7.2	−7.6	−7.6	−7.8	−7.55	−11.6

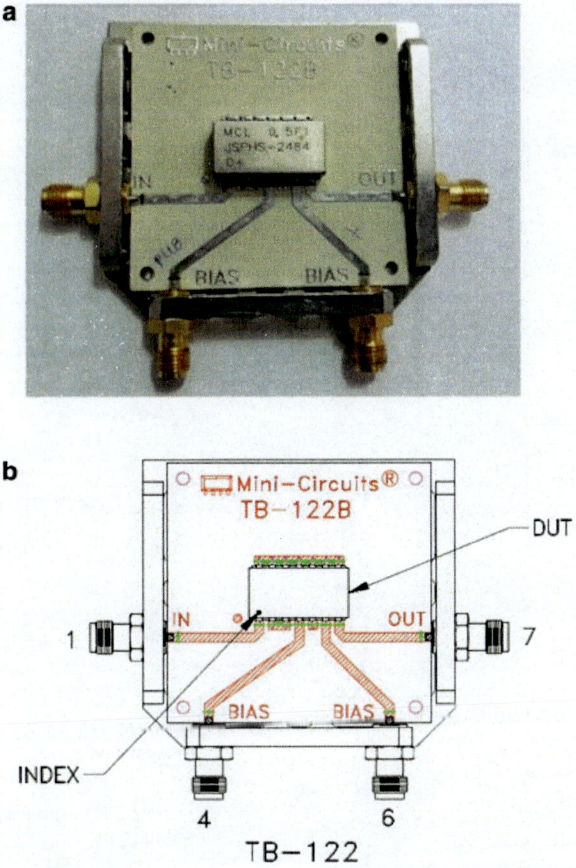

Fig. 9 **a** Fabricated layout of IC 2484, **b** evaluation board and circuit of IC 2484

input to pin 1 to each of the phase shifter IC 2484. The control voltage from the PSoC is controlled through a rotary encoder and is given to control pin 4 or 6 of the IC 2484 to provide the required DC bias voltage as per the DOA estimation. Figure 11 shows the flowchart which gives a step-by-step execution to write the program to set the control voltage through the PSoC for the phase shifter IC 2484 to generate progressive phase shifts for the 1 × 4 phased array antenna system required to direct the main beam in the desired direction.

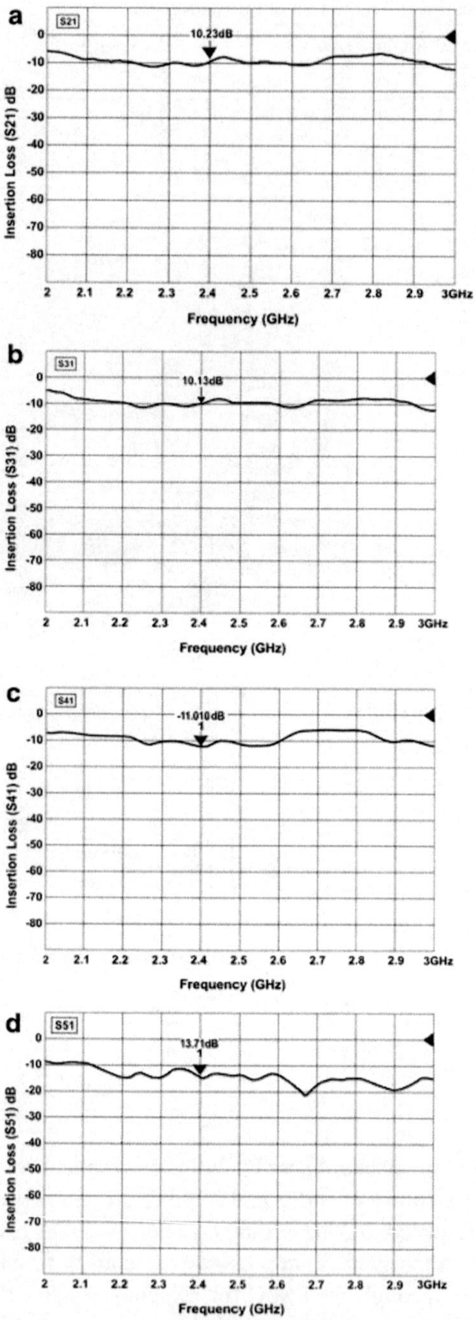

Fig. 10 **a** Insertion loss (S_{21}) of WPD + phase shifter, **b** insertion loss (S_{31}) of WPD + phase shifter, **c** insertion loss (S_{41}) of WPD + phase shifter, **d** insertion loss (S_{51}) of WPD + phase shifter

Table 3 Measured result for feed network (power divider and phase shifter)

Experimental result	Insertion loss in dB				Average insertion loss in dB	VSWR
	S_{21}	S_{31}	S_{41}	S_{51}		
Power divider + phase shifter	−10.2	−10.1	−11	−13.7	−11.26	1.33:1
Power divider	−7.2	−7.6	−7.6	−7.8	−7.55	
Phase shifter	−3.03	−2.53	−3.4	−5.91	−3.71	

3 Simulated Phased Array Antenna

The system is designed and simulated using the CST microwave studio software to observe the antenna performance when used in a 4 × 1 array. This system ideally needs a power divider, a phase shifter, and an array antenna to perform this operation. In this simulation system, each component is simulated individually and parameters like radiation pattern gain, directivity, and half-power bandwidth (HPBW) were measured. A schematic view of the simulated phased array antenna using an S_p^2 file of IC 2484 is as shown in Fig. 12a and a 3D view of the simulated phased array antenna is as shown in Fig. 12b.

4 Fabricated Phased Array Antenna

The experimental setup for the measurement of radiation patterns for various phase angles is shown in Fig. 13a, b using an Antenna Measurement System CISPER 1-1-1. A 4 × 1 array antenna is connected to the transmitter side and the receiver is used to receive the radiation field. The receiver rotates 0 to 360° and scans the radiated field to plot the radiation pattern for the desired DOA. It is repeated for different phase angles such as ±30°, ±60°, and 90°.

The simulated and experimental radiation patterns for different phase angles such as ±30°, ±60°, and 90° are as shown in Fig. 14a–j, and performance parameters are tabulated in Table 4.

The main lobe direction of the measured results for ±30°, ±60°, and 90° differ as compared to the simulated results. This possibility is due to the accurate alignment during measurement. Besides that, the difference may be due to the high loss in the substrate used in fabrication. The gain degradation in the measured results may be due to losses in the feed network, cables, and connectors used in the experimental setup. The average gain is 12.95 dBi in the experimental array and 13.10 dBi in the simulations i.e., the experimental gain is less than the simulated gain by 1.14%.

The level of the side lobe discrepancies may be due to the infinite ground plan assumed in the simulation and the finite ground plane used in the fabricated array. The simulated results differ from the experimental results. However, SLL is maintained below −13 dB both in the case of the simulated and experimental results.

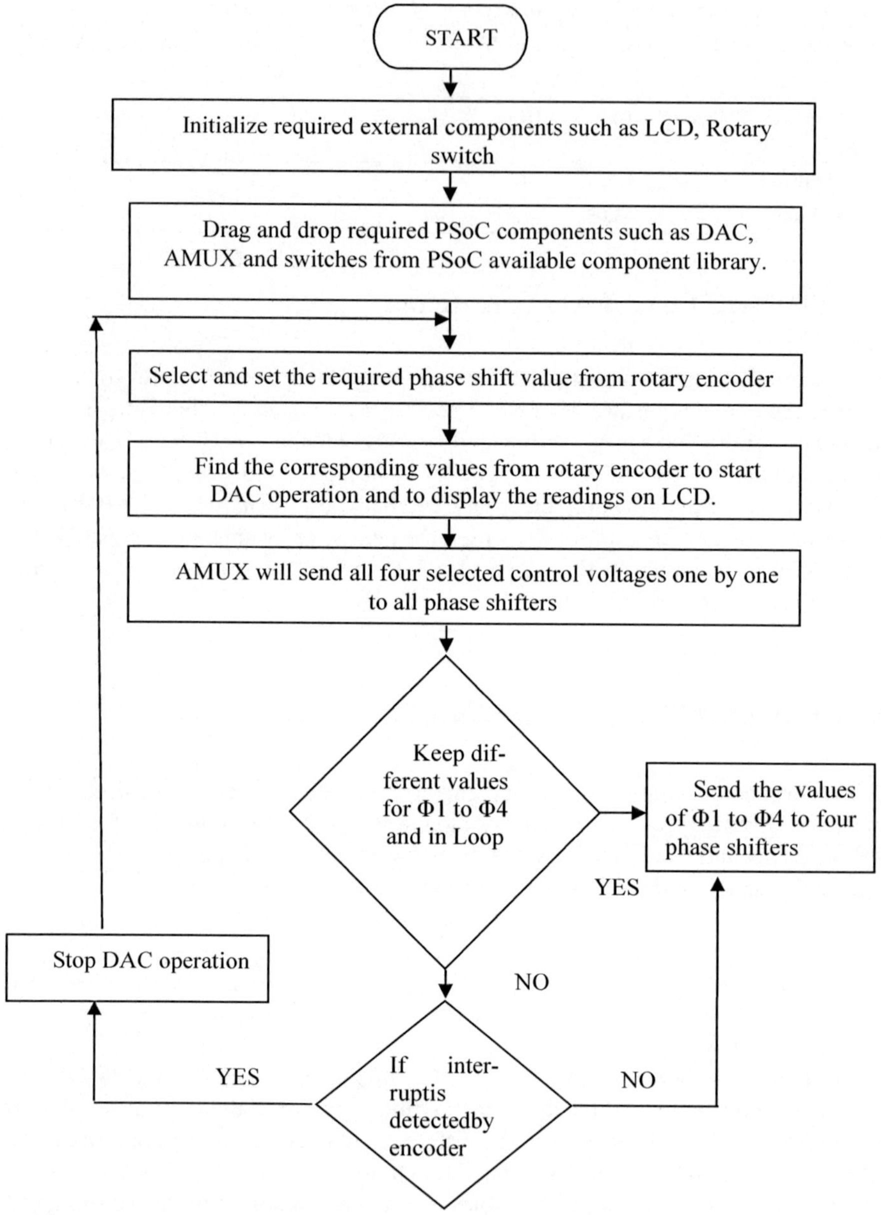

Fig. 11 Flowchart for PSoC creator IDE

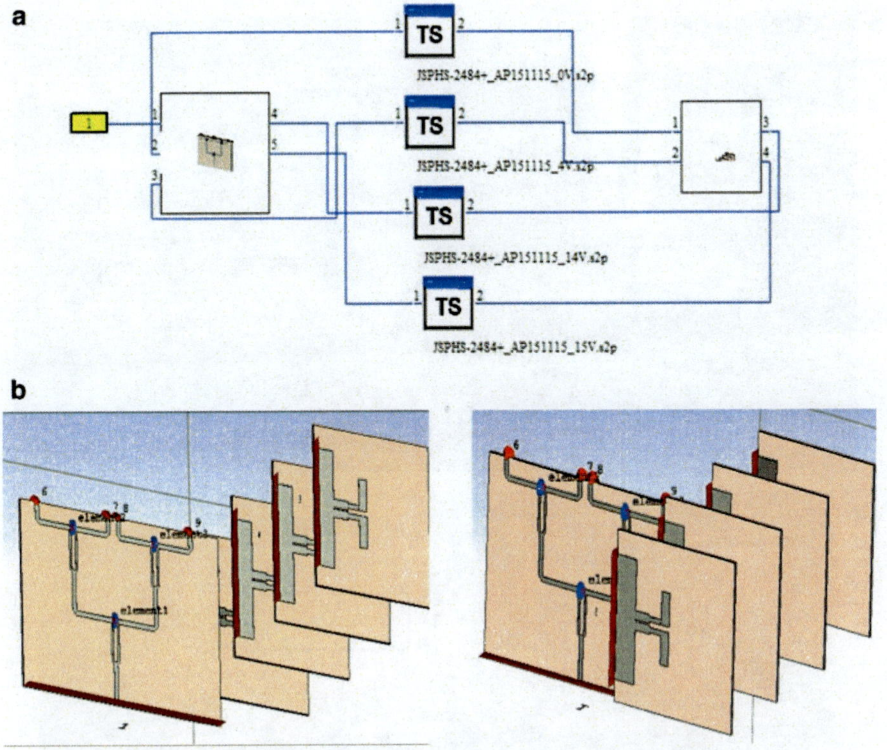

Fig. 12 **a** Schematic view simulated of phased array antenna with S_p^2 file of IC 2484, **b** 3D view of simulated phased array antenna

The HPBW for the experimental and simulated results is not the same for the scan angles of ±30°, ±60°, and 90°. Therefore the directivity of the phased array slightly changes concerning scan angles. It is estimated that the average directivity of the experimental and simulated results is 14.75 and 13.8.

The directivity of 14.75 in the case of the experimental results and 13.8 in the case of the simulation is noted respectively. From the above results, it is seen that the simulation and experimental results match accurately.

5 Validation of Proposed and Reported Phased Array

Table 5 shows a comparison between the proposed phased array antenna and the reported steerable antenna systems.

In the proposed work, a wider scan angle range of ±30°, ±60° were achieved and compared to the reported antenna as tabulated in Table 5. The gain is constant with a slight variation of 0.88 dB in the entire scan range which is better than all

a

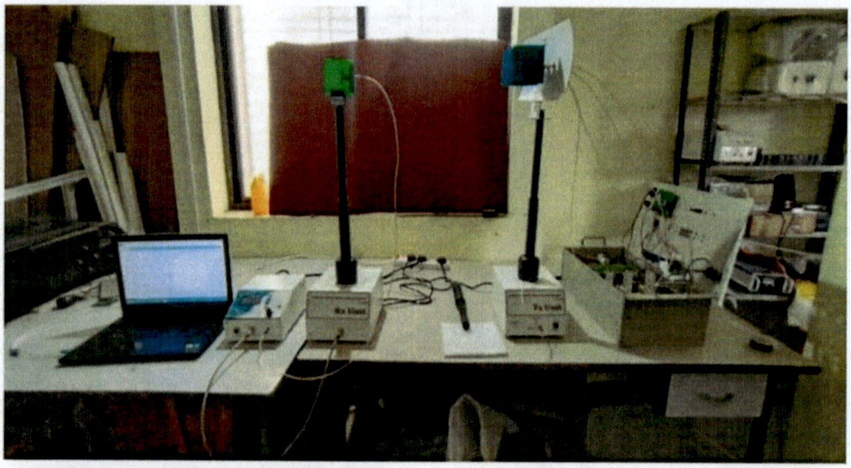

b

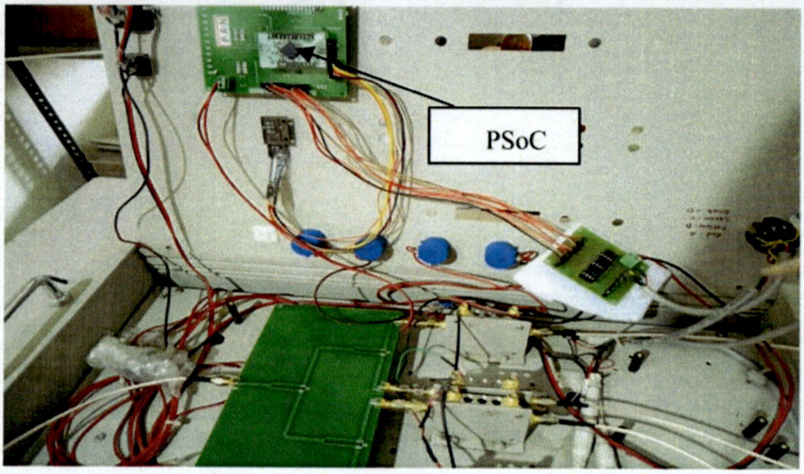

Fig. 13 **a** Experimental setup for measurement of radiation pattern using antenna, **b** assembly of fabricated phased array antenna

the reported phased arrays. The average directivity is 14.75 and is comparable with the entire reported phased array antenna. This proves that the antenna design can evaluate and improve the performance of the phased array antenna system.

6 Conclusion

In the proposed work, a design of a multiport phased array antenna utilizes the reflection-type phase shifter based on an IC 2484 and a WPD to control the phase shift in full 180° range with an insertion loss that is very low. The main lobe can be

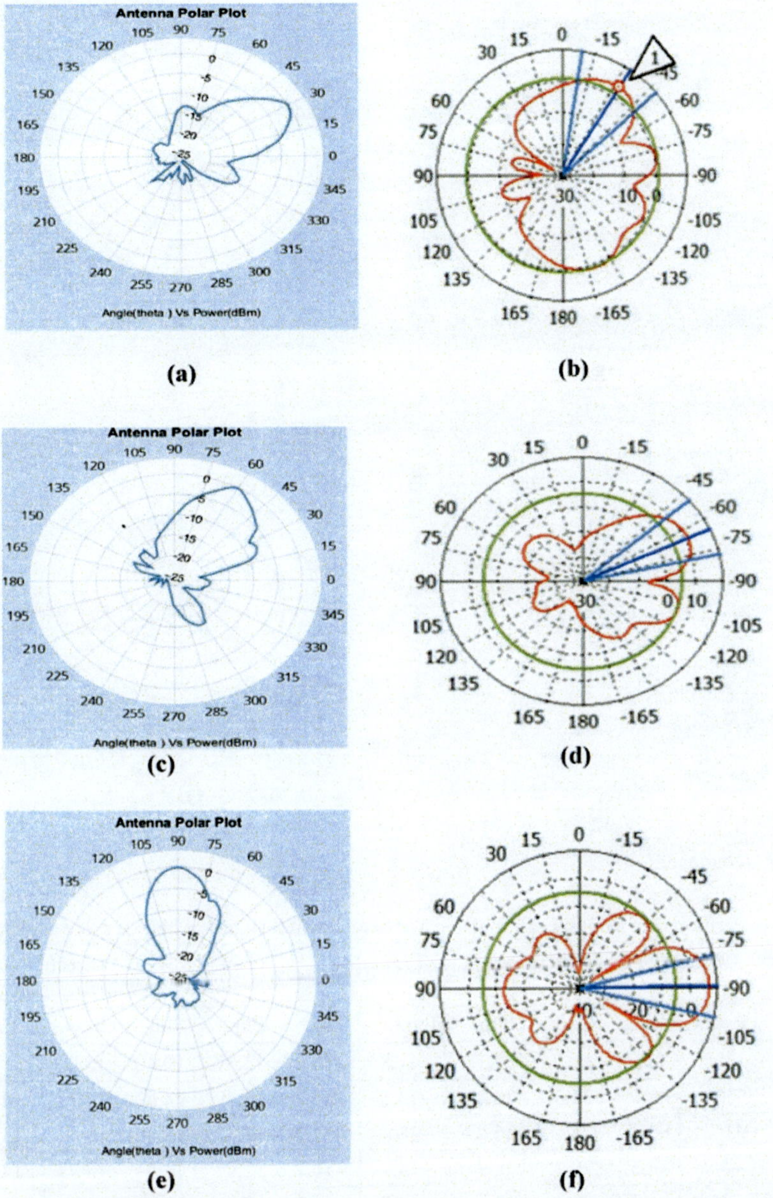

Fig. 14 **a** Measured radiation pattern at 30°, **b** simulated radiation pattern at 30°, **c** measured radiation pattern at 60°, **d** simulated radiation pattern at 60°, **e** measured radiation pattern at 90°, **f** simulated radiation pattern at 90°, **g** measured radiation pattern at −30°, **h** simulated radiation pattern at −30°, **i** measured radiation pattern at −60°, **j** simulated radiation pattern at −60°

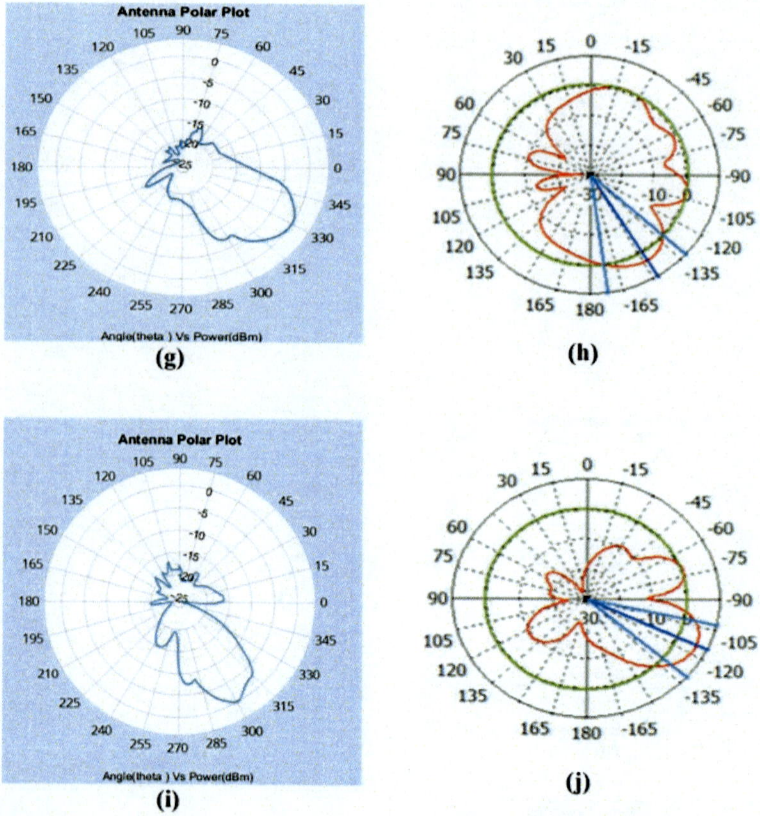

Fig. 14 (continued)

Table 4 Performance analysis of simulated and experimental values of phased array antenna

DOA			HPBW in degrees		SLL in dB		Gain in dBi		Directivity	
The.	Expt.	Sim.	Expt.	Sim.	Expt.	Sim.	Expt.	Sim.	Expt.	Sim.
+30°	33°	33°	28	39.1	−11.8	−10.2	13.37	10.40	15.06	11.92
+60°	65°	65°	31	28.8	−13.5	−16.7	12.49	14.67	14.17	14.81
90°	92°	89°	28	26.4	−18	−22.1	13.37	15.39	15.60	15.57
−60°	−56°	−65°	29	28.9	−14.1	−16.7	13.07	14.64	14.75	14.78
−30°	−35°	−32°	31	40.2	−14.5	−10.9	12.49	10.40	14.17	11.92
Average value in antenna parameters			29.4	32.68	−14.1	−15.3	12.95	13.10	14.75	13.8

***The:** Theoretical, **Expt:** Experimental, **Sim:** Simulated

Table 5 Performance comparison of 4 elements phased array antenna operating at 2.4 GHz

Ref. No	Author and year	Scan angle in degree	Gain in dB	Directivity in dBi	HPBW in degree	SLL in dB
[20]	Tsuji et al. 2006	±30	NI	17.85	24	9
[21]	El Tager et al. 2009	±14.3	11.4	17.00	25.2	14
		±47.8				8
[22]	Abdalla et al. 2009	−27 to +22	8.4–9.9	NI	NI	<10
[23]	Mohamed et al. 2011	±12	6.6	11.86	34.9–60.6	16
		±40	11.8		34.9–55.7	3.1
[24]	Fernandes et al. 2013	±15	6	14.08	28	9.5
		±45	7.5		30	5.5
[25]	Zulkifli et al. 2015	±20	3.94	13.7	35.6–40.3	NI
		±40	6.11		39.8–33.8	
[26]	Bui et al. 2018	±12	8.16–7.9	18.00	20.7–21.3	9.3–9.0
		±36	8.15–7.9		27.0–21.3	8.1–5.5
Proposed		±30	13.37–12.49	14.75	28–31	12–14.5
		±60	12.49–13.07		31–29	13.5–14

steered in the direction from ±30°, ±60° with a higher resolution. The limitation of the design of the phased array is limited to low-frequency band only. The PSoC used in the system greatly improves the DOA and hence the accuracy of the phase shifts.

In the future, the antenna design would involve replacing the current analog components and the algorithm-based systems with a fully digital IC base equivalence. This leads to a reduction in the cost and the size of the phased array and will enhance the practical use of the phased array antennas in the consumer market. With the combined use of analog and digital circuits, it is possible to consider a novel, holistic approach to communication and radar systems operating at mm-wave frequencies.

References

1. A. Tombak, A. Mortazawi, A novel low-cost beam-steering technique based on the extended-resonance power-dividing method. IEEE Trans. Microwave Theory Tech. **52**, 664–670 (2004)
2. K. Maruhashi, H. Mizutani, K. Ohata, A Ka-band 4-bit monolithic phase shifter using unresonated FET switches, in *IEEE MTT-S International Microwave Symposium Digest* (1998), pp. 51–54
3. Z. Zhao, X. Wang, K. Choi, C. Lugo, A.T. Hunt, Ferroelectric phase shifters at 20 and 30 GHz. IEEE Trans. Microwave Theory Tech. **55**(2), 430–437 (2007)

4. T. Ji, H. Yoon, J.K. Abraham, V.K. Varadan, Ku-band antenna array feed distribution network with ferroelectric phase shifters on silicon. IEEE Trans. Microwave Theory Tech. **54**(3), 1131–1138 (2006)
5. N. Kingsley, G.E. Ponchak, J. Papapolymerou, Reconfigurable RF MEMS phased array antenna integrated within a liquid crystal polymer (LCP) system-on-package. IEEE Trans. Antennas Propag. **56**, 108–118 (2008)
6. W. Gautier, V. Ziegler, A. Stehle, B. Schoenlinner, U. Prechtel, W. Menzel, RF-MEMS phased array antenna on low-loss LTCC substrate for Ka-band data link, in *IEEE European Microwave Conference EuMC* (Sep 29–Oct 1 2009), pp. 914–917
7. J.-H. Tsai, C.-K. Liu, J.-Y. Lin, A 12 GHz 6-bit switch-type phase shifter MMIC, in *Proceedings of the 44th European Microwave Conference* (2014), pp. 1916–1919
8. J.C. Wu, T.Y. Chin, S.F. Chang, C.C. Chang, 2.45-GHz CMOS reflection-type phase-shifter MMICs with minimal loss variation over quadrants of phase-shift range. IEEE Trans. Microw. Theory Techn. **56**(10), 2180–2189 (2008)
9. B. Cetindogan, E. Ozeren, B. Ustundag, M. Kaynak, Y. Gurbuz, A 6 bit vector-sum phase shifter with a decoder based control circuit for X-band phased-arrays. IEEE Microw. Wireless Compon. Lett. **26**(1), 64–66 (2016)
10. M.C. Scardelletti, G.E. Ponchak, N.C. Varaljay, Ka-band, MEMS switched line phase shifters implemented in finite ground coplanar waveguide, in *Proceedings of the 32nd European Microwave Conference* (2002), pp. 1–4
11. H. Zhang, A. Laws, K.C. Gupta, Y.C. Lee, V.M. Bright, MEMS variable-capacitor phase shifters part I: loaded-line phase shifter. Int. J. RF Microw. Comput. Aided Eng. **13**(4), 321–337 (2003)
12. H.T. Kim, J.H. Park, J. Yim, Y.K. Kim, Y. Kwon, A compactV-band 2-bit reflection-type MEMS phase shifter. IEEE Microwave Compon. Lett. **12**(9), 324–326 (2002)
13. A.S. Nagra, R.A. York, Distributed analog phase shifters with low insertion loss. IEEE Trans. Microwave Theory Tech. **47**(9), 1705–1711 (1999)
14. http://www.cypress.com/go/AN91445
15. PSoC Creator System Reference Guide, Document Number: 00193937, Rev.*A
16. PSoC Creator Component Author Guide, Document # 001-42697,Rev.*T
17. Cypress Semiconductor Corporation Document Number: 001-87197rev.*J
18. D. van Ess, *Learn Digital Design with PSoC A Bit at a Time*. Library of congress Control Number: 2013923706
19. A. Barbadekar, P.M. Patil, A novel method of smart communication using PSoC for wireless push system, in *2020 International Conference on Emerging Smart Computing and Informatics (ESCI)* (March 12–14 2020)
20. M. Tsuji, T. Nishikawa, K. Wakino, T. Kitazawa, Bi-directionally fed phased-array antenna downsized with variable impedance phase shifter for ISM band. IEEE Trans. Microw. Theory Tech. **54**, 2962–2969 (2006)
21. A.M. El Tager, M.A. Eleiwa, Design and implementation of a smart antenna using Butler matrix for ISM-band, in *Progress in Electromagnetics Research Symposium, Beijing, China* (2009)
22. M.A.Y. Abdalla, K. Phang, G.V. Eleftheriades, A planar electronically steerable patch array using tunable PRI/NRI phase shifters. IEEE Trans. Microw. Theory Tech. **57**, 531–541 (2009)
23. W.W.H. Mohamed, Integration of PIN diode switching circuit with butler matrix for 2.45 GHz frequency band. Optoelectron. Adv. Mater.—RAPID Commun. **5**(7), 793–798 (2011)
24. M. Fernandes, A. Bhandare, C. Dessai et al, A wideband switched beam patch antenna array for LTE and Wi-Fi, in *Annual IEEE India Conference; Mumbai, India* (2013), pp. 1–6
25. F.Y. Zulkifli, N. Chasanah, Basri et al, Design of Butler matrix integrated with antenna array for beam forming, in *2015 International Symposium on Antennas and Propagation (ISAP), Hobart, TAS, Australia* (2015), pp 1–4
26. T.D. Bui, M.T. Le, Q.C. Nguyen, Electronically steerable antenna array for indoor positioning system. J. Electromagn. Waves Appl. 1569–3937 (2018)

Knowledge Discovery Based Automated Recognition of Traffic Sign Images Using Hybrid PCA-RBF Network

R. Manasa, K. Karibasappa, and Manoj Kumar Singh

Abstract In this article, template-free, knowledge discovery, and correlation-based distance mapping have been applied for automated recognition of traffic sign images. Knowledge extraction by the neural network and correlation distance mapping in the knowledge domain have made solutions intelligent and faster. Principal Component Analysis (PCA) has been applied in the pre-processing stage to reduce the dimensionality by considering the few main components only. The proposed method has applied a self-adaptive form of radial basis function neural network to develop an efficient classifier. To increase the recognition efficiency, a firm correlation between input and neural architecture has been developed by one to one mapping approach in the learning process. The correlation-based distance measure has ensured the optimal use of discovered knowledge in the final decision stage. The developed method has shown a high level of robustness in recognition against limited visibility because of weather as well as the aging effect of traffic sign images. The proposed method can also be applied in the different applications in the area of image-based information retrieval and machine vision.

Keywords Automated recognition · Neural network · Radial basis function · Machine vision

---

R. Manasa
Department of ECE, Dayananda Sagar College of Engineering, Bangalore, India

K. Karibasappa
Deptartment of ECE, Dayananda Sagar Academy of Technology and Management, Bangalore, India

M. K. Singh (✉)
Manuro Tech Research Pvt. Ltd, Bangalore, India
e-mail: mksingh@manuroresearch.com

V. Suma et al. (eds.), *Inventive Systems and Control*,
Lecture Notes in Networks and Systems 204,
https://doi.org/10.1007/978-981-16-1395-1_55

1 Introduction

Perceiving traffic signs is a difficult issue, and it has caught the consideration of computer vision for a few years. The automated recognition of traffic sign images carried the complexity of multi-class categorial classification and challenging because of input variability under dynamic real-time conditions. Such diverse conditioning is very challenging in machine vision because of the need for a fast and accurate decision. Among the various possibilities of the driver assistance system, the automated recognition of traffic signs has shown a much valuable mechanism tool for the driver or the autonomous vehicle. It tends to be utilized as a choice supporting device for drivers and independent vehicles. Eventually, TSR is a substantial scale highlight learning issue and subsequently increased the attention of researchers. The basic parameters, for example, recognition accuracy computational complexity, and huge training dataset size should be considered while structuring the TSR framework. In the process of structuring more intelligent vehicles, there is a serious need for a reduction in the number of mistakes and accidents which happened either because of driver wrong decisions or cause of human brain-oriented limitations. This need can be fulfilled up to a certain extend by the use of efficient and robust Automated Traffic Sign Recognition (ATSR). With the ATSR it is possible to extract the information available over traffic sign images to drive safer. In TSR, the goal is to remind or caution drivers about the limitations, threats, or other data bestowed by traffic signs, already. Since the current signs are intended to draw driver's consideration by their hues and shapes, preparing these highlights is one of the pivotal parts of these systems. In this work, an ATSR System, having the capacity to the recognition of traffic signs images even with terrible visual curios that begin from some climate.

The structure of the article contains the literature work in Sect. 2 while the proposed work has been presented in Sect. 3. The detailed experimental work and analysis have been discussed in Sect. 4 and the conclusion has been given in Sect. 5.

2 Related Work

There have been numerous works considered for research in the past over traffic-sign detection by researchers [1]. Proposed a fuzzy way to deal with object detection that considers an object as a set of constituent segments instead of a single entity. Machine learning-based computer vision has been applied to the humanoid robot for object recognition [2]. A naturally enlivened, the inspired design has an architecture that helps visual perception and cognitive robotics. A feature learning approach utilizing group sparse coding has been proposed in [3] to identify traffic signs by exploiting the intrinsic structure of the pre-learned visual codebook. The two forms of convolution architecture FCN and CNN have been used in [4] for guided traffic sign classification of the objects. A combination of solid image analysis and pattern recognition techniques has been proposed in [5] to handle the issue of traffic sign recognition in mobile mapping data [6]. The author has proposed a programmed traffic-sign detection technique for recognizing both chromatic and colorless signs, while considering turns, scale changes, shifts, incomplete disfigurements, and shadows. In traffic-sign recognition [7] the author has proposed, a group of highlights including a histogram of situated inclinations, neighborhood parallel examples, and Gabor highlights is utilized inside a support vector machine classification framework [8]. The author has proposed a connected invariant geometric minutes to characterize shapes. A descriptor has been designed by combining histogram of HOG highlight to LSS highlights and random forest along with SVM have applied in detection and recognition of traffic sign [9]. Has utilized Higher-Order Spectra (HOS) combined with texture-based highlights to build up an effective TSR. These highlights speak about the shape and substance of the traffic signs. At that point a subspace learning strategy with chart installing under linear discriminant analysis, the structure is utilized to build the segregation control between different traffic images. In numerous CNNs there exists a lot of parameter excess as a result of matrix kernels to address this issue, [10] has proposed a novel model, to be specific, the vector-kernel convolutional neural network in which each convolutional layer can just utilize vector bits of either measure. Boosted Convolutional Neural Network-based methodology for article acknowledgment has been proposed in [11]. Based on Internet of Things (IoT), roadside traffic surveillance has been discussed in [12]. A traffic oriented path planning approach has been discussed in [13] using the capsules neural network which provided the resistance against the spatial variance. Different traffic oriented parameters like traffic rate, traffic density have been used in [14] to control the computer vision-based traffic control. A detailed machine vision-based detection of traffic signs has been reviewed in [15]. Embedded system based traffic surveillance has been discussed in [16] to face the challenges of dynamic environmental conditions. Based on GA and IOT, traffic management has proposed in [17] on the traffic signal to reduce the waiting period.

3 Proposed Work

The properties of universal mapping pose by the supervised learning method have been used to obtain precise decision outcomes. The mapping takes place with a predefined set of training and target set where a neural architecture learned to map the input features to the desired vector or point in the output domain. Among the various possibilities of neural architecture, radial basis functions neural network (RBFNN) has shown improved performances in several applications. With an optimal design of RBF architecture and learning process, it is possible to map any kind of nonlinear curve and mapping of any function. In this work, 3 layers in the architecture of RBF are considered: first layer is the neurons help to allow the inputs while the second layer carried the kernel function and the third layer provides the accumulated sum of second layer neurons to deliver the output. The mathematical form of output delivered by the RBF neural network can be represented as given in Eq. 1.

$$O_q = f_q(p) = \sum_{z=1}^{N} W_{qz}\varphi_z(p, k_z) = \sum_{z=1}^{N} W_{qz}\varphi_z(\|p - k_z\|_2), \quad \forall q = 1, 2, \ldots m \tag{1}$$

where, $p \in \Re^{n\times 1}$: n dimensional input, $\varphi_z(.)$: a transformation function, mapping value from $\Re^+$ to $\Re$, $\|.\|_2$: represent the Euclidean distance, W_{qz}: connection weights from hidden to output layer; N: number of hidden layer, and $k_z \in \Re^{n\times 1}$: RBF kernel centers. First, the Eucline distance between the centers and the input estimated and its mapping happen through nonlinear function available in the hidden layer. The final output has been considered as a weighted sum of this mapped information. In this work mapping functional $\varphi_k(.)$ has been considered as the Gaussian function as given by Eq. 2

$$\varphi(x) = e^{(-p^2/\sigma^2)} \tag{2}$$

where σ represents a spread parameter and provides the controls the "width" of Gaussian function. The centers are generally considered from the input data set. Conventionally for the Gaussian kernel function in the RBF, σ has calculated as given in Eq. 3.

$$\sigma = \frac{Ed_{max}}{\sqrt{z}} \tag{3}$$

where, $Ed_{\max}$ the maximum Euclidean distance among the selected centers. Formally the hidden layer neuron output can be defined as

$$\varphi(p, k_z) = e^{\left[\frac{-z}{Ed_{max}^2}\|p-k_z\|^2\right]} \tag{4}$$

3.1 *Self-Adaptive RBF NN*

In the conventional form of RBFNN, there is only one parameter, output layer weights, which are adaptive and play an active role in the function mapping. This can be a problem when there is a complex function available for mapping. In the proposed solution instead of one variable two other variables in the network, called centers and the spread of the kernel function has also been considered. This makes mapping easier and faster. The error output can be defined as shown in Eq. 5

$$J(n) = \frac{1}{2}|E(n)|^2 = \frac{1}{2}\left[O_d(n) - \sum_{z=1}^{N} w_Z(n)\varphi\{p(n), k_z(n)\}\right]^2 \tag{5}$$

Considering the up-gradation of three sets of parameters (centers and spread of Gaussian function along with output layer weights) simultaneously through the steepest descent algorithm, the final parameter upgrading equation is defined as given in Eqs. 6–8.

$$w(n+1) = \left[w(n) - \alpha_w \frac{\partial}{\partial w} J(n) \quad \rangle_{w=w(n)} \right] \tag{6}$$

$$k_z(n+1) = \left[k_z(n) - \beta_k \frac{\partial}{\partial k_z} J(n) \quad \rangle_{k_z=k_z(n)} \right] \tag{7}$$

$$\sigma_z(n+1) = \left[\sigma_z(n) - \gamma_z \frac{\partial}{\partial k_z} J(n) \quad \rangle_{\sigma_z=\sigma_z(n)} \right] \tag{8}$$

where,

$E(n) = \hat{O}(n) - O_T(n)$, $O_T(n)$ is the targeted output, and $\alpha_w, \beta_k, \gamma_z$ are the predefined learning constant and have control over change in parameters value.

The complete working flow of the proposed method using self-adaptive RBF has shown in Fig. 1. Initially, several traffic sign images have considered for training purposes. Each sign image has resized to the same value so that size differences in the image could not affect the performance of recognition. Principal component analysis has been applied over each image and the first few main components are only considered. This will not only reduce the burden of large data of images but also makes the learning faster and proper. A proper architecture has been considered and learning of each image has been done independently keeping the same target. Gradient descent based error minimization process has been applied which changes the values of RBF parameters (weights, centers, and spreads) iteratively. For each image, after a defined number of iterations the final available parameters of RBF (TRP) and corresponding generated output (OP) have stored as referenced output.

When a test image of a traffic sign has been applied, after pre-processing (PCA) operation, the same RBF architecture has applied as it was available for the time of training. The stored RBF parameters have been applied one by one to generate the final correlated output as shown in [TOP]. A minimum absolute difference is estimated between the stored referenced output and available correlated output to establish the type of sign image available in the knowledge domain of training.

4 Experimental Result and Analysis

In this work, widely used 22 different forms of traffic sign available on the Indian road has considered as shown in Fig. 2. Resizing of each image has done carry the size of 450 * 650. The 3 main principal components have been considered for each image in the training process. At the time of training the one member from each selected component has considered which makes the size of input as 3 hence there

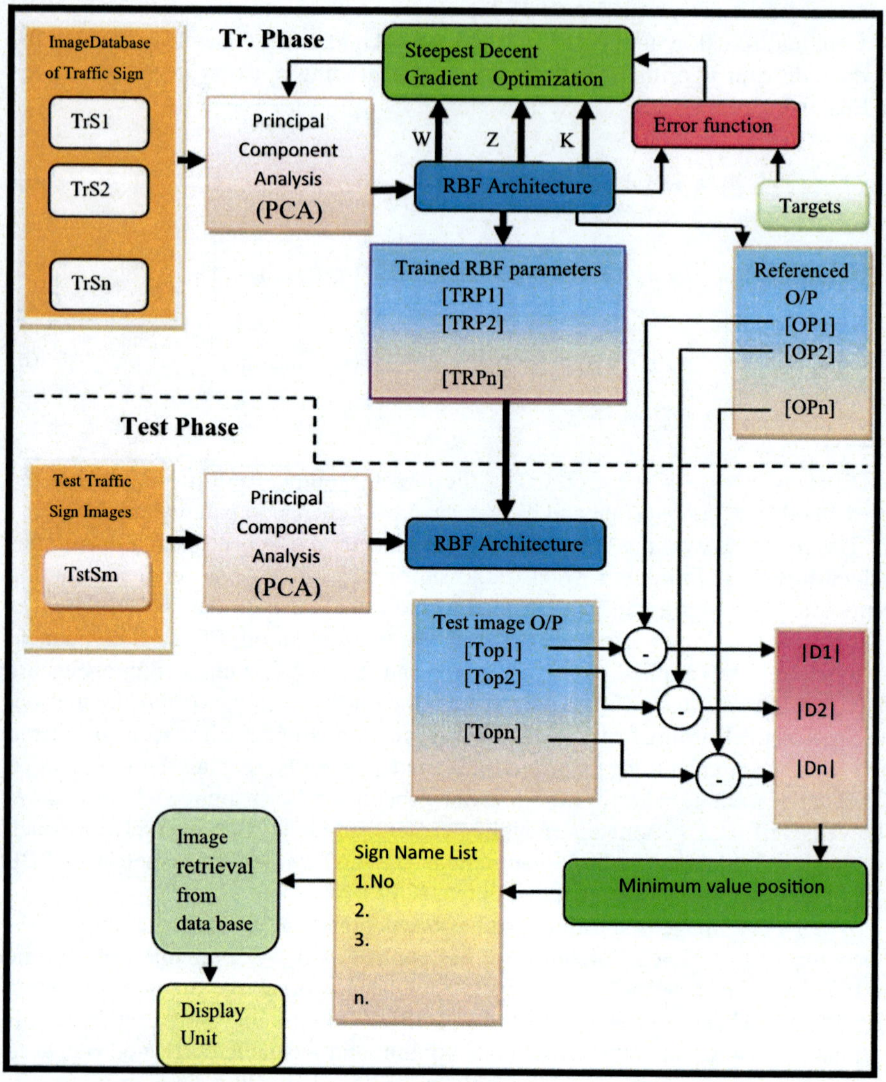

Fig. 1 Proposed traffic sign recognition working flow

were 3 input nodes in RBF. Total 10 hidden nodes have considered while one output node has considered considering target equal to 1. The allowed number of learning iteration for each image is equal to 100. The learning rate and momentum rate in the learning algorithm have considered as 0.1. This low value helps to extract the features in a better manner. The complete environment has been developed in MATLAB.

The convergence characteristics for all the 22 images have shown in Fig. 3. It is clear for all the images there was a very similar type of convergence irrespective of

Fig. 2 Traffic sign images

content variation available in the images. In the obtained decision confidence level values it was observed that there was zero error in decision confidence while high error with other possibilities indicates the quality of crisp decision confidence. The bar plot of the decision confidence level as shown in Fig. 4 ensure that in comparison to recognition position if images everywhere there were a very level of uncertainty exist hence without confusion final decision can be taken place. This shows that the proposed method is very optimum in recognition of the images.

4.1 Recognition Under Poor Visibility and/or Aging Condition

The condition of the environment is very crucial in the computer vision in which observations are being made. In the application of assistance to automobile, where

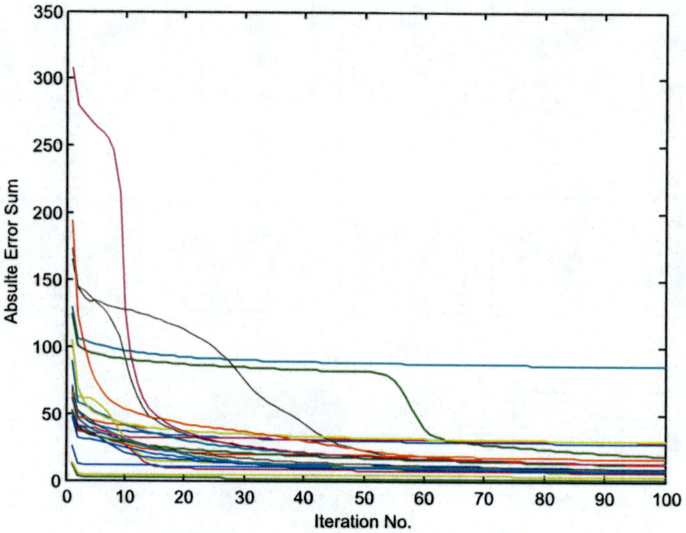

Fig. 3 Convergence characteristic of the classifier learning

there is mostly outdoor activities takes place, whether the condition is one of the most import factors which can cause an error in the recognition process. An assumption has made there is a need to recognize the sign image under foggy conditions. Similarly, with time, there are some kind of marks surly will appear over the signboard. To simulate such situations, Gaussian and Salt and Pepper based noise have been included separately as well as jointly with the sign image at the input. To get the effect of situation two very similar types of sign image representing the speed limit of 20 and 40 km/h have opted carrying the image no. 21 and 22. The obtained performance under three different conditions has shown in Figs. 5, 6 and 7. It is clear that there is now not 0 error in the decision confidence level which is obvious but there is a very low error in the recognition comparatively which is clear from the decision confidence level bar plot. It can observe from Fig. 5 that the when input sign image carrying the foggy appearance, the recognition process has shown minimum value in the position of 21 which insure that the detected sign image of 20 km/h. Similarly for the input image as shown in Fig. 6 the minimum value in the error occurred at the position of 22 which makes sure that the appeared sign image is 40 km/h.

The effect of fog and aging both have shown in Fig. 7 and its corresponding recognition decision error level that carried a minimum error in the position of 21. This shows that the proposed solution is not only efficient but also robust under practical hostile conditions.

The advantages of the proposed system can be categories into two-part: the first part carries the efficient design of the classifier that aids the RBF network to carry the three-degree freedom to have over control of the learning phase and the second

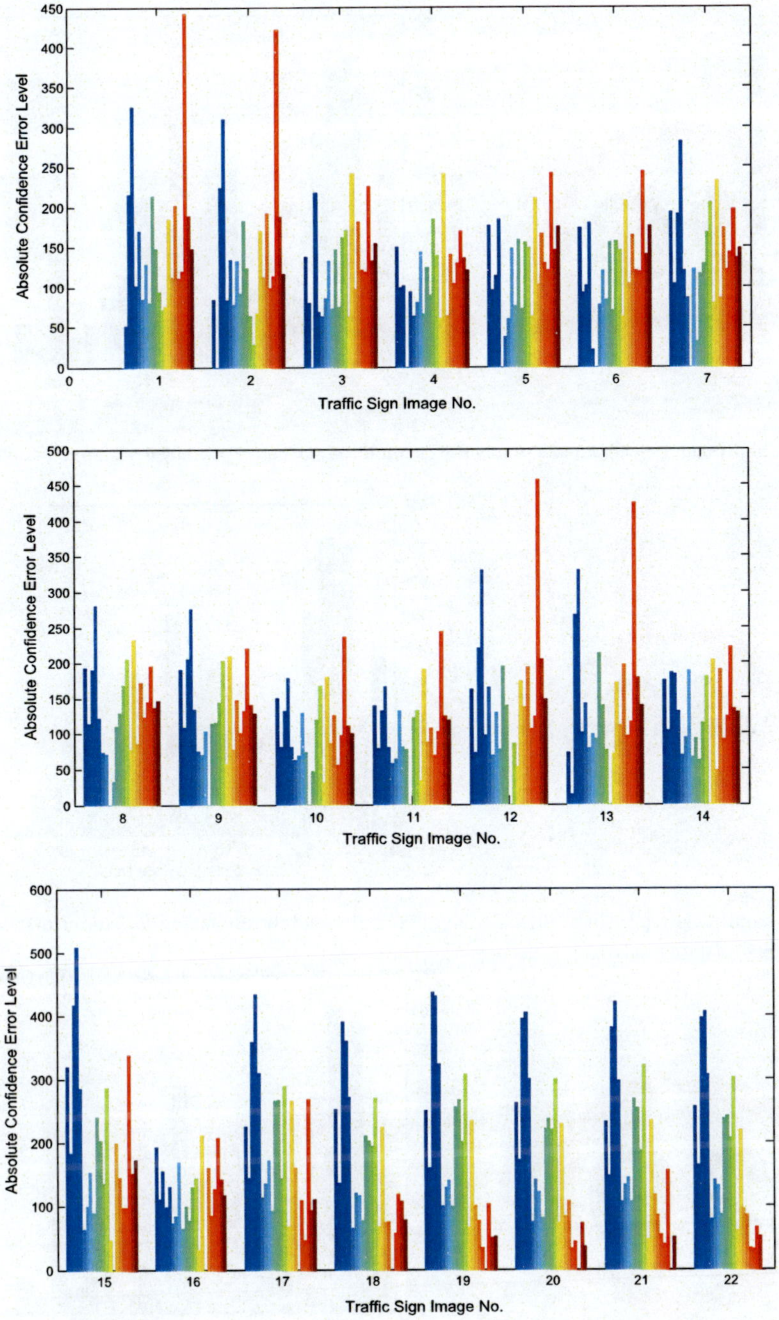

Fig. 4 Bar plot for decision confidence level

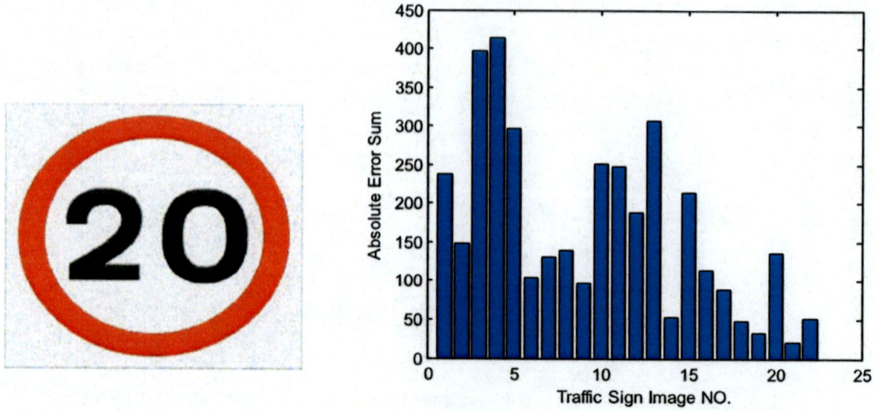

Fig. 5 Sign image (speed limit 20) under foggy condition and corresponding decision error outcome

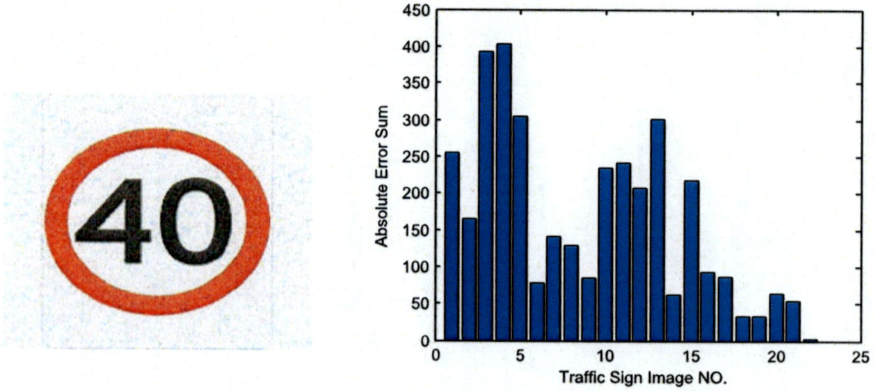

Fig. 6 Sign image (speed limit 40) under foggy condition and corresponding decision error outcome

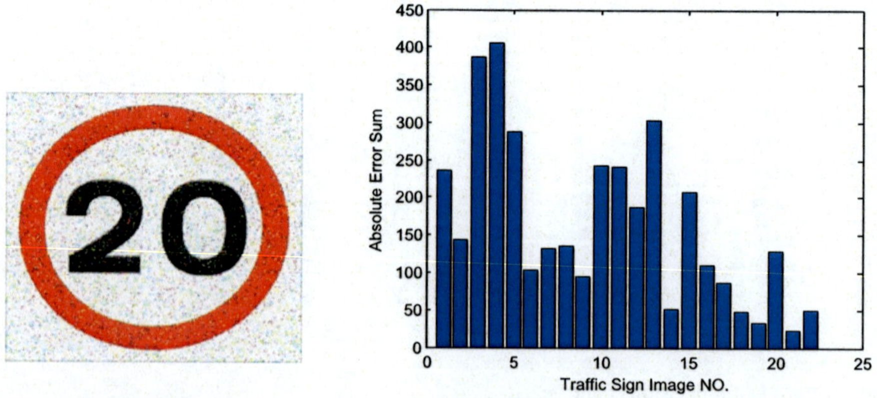

Fig. 7 Sign image under foggy and aging effect and corresponding decision error outcome

part is to develop the larger distance in the knowledge domain with uncorrelated inputs. The correlation-based distance measure has ensured faster and robustness in the recognition process.

5 Conclusion

The need for automated recognition in computer vision is one of the most prime important tasks. In this work, different types of traffic signs have been proposed to have a blind recognition process where there is no template matching takes place. Based on the knowledge available at the time of learning through a self-adaptive form of RBF, the decision process occurs. The use of only very few principal components of each image makes the process faster and required lesser memory. The proposed recognition method has shown robustness against the environmental condition and aging effect. The developed method has been considered to be used in the area of automobile where there is a need for driver assistance or the vehicle which carry the facility of autonomous driving. Because of simplicity and efficiency, the proposed method can also be directly utilized in various other areas of computer vision. While working it was experienced that there are possibilities to improve the design of RBF architecture and learning process with the help of natural computing like evolutionary computation. Here, a meta-heuristic search doesn't bound with any continuity constraints and helps to explore the solution domain in a wide and faster manner.

References

1. R.T. Iqbal, C. Barbu, Fuzzy component based object detection. Int. J. Approximate Reasoning **45**(3), 546–563 (2007)
2. J. Leitner, S. Harding, Learning visual object detection and localisation using icVision. Biologically Inspired Cogn. Archit. **5**, 29–41 (2013)
3. H. Liu, Y. Liu, F. Sun, Traffic sign recognition using group sparse coding. Inf. Sci. **266**, 75–89 (2014)
4. Y. Zhu, C. Zhang, Traffic sign detection and recognition using fully convolutional network guided proposals. Neurocomputing **214**, 758–766 (2016)
5. A. Petrelli, S. Salti, Traffic sign detection via interest region extraction. Pattern Recogn. **48**(4), 1039–1049 (2015)
6. J.M. Lillo-Castellano, I. Mora-Jiménez, Traffic sign segmentation and classification using statistical learning methods. Neurocomputing **153**, 286–299 (2015)
7. S.K. Berkaya, H. Gunduz, On circular traffic sign detection and recognition. Expert Syst. Appl. **48**, 67–75 (2016)
8. A. Ellahyani, M. El Ansari, I. El Jaafari, Traffic sign detection and recognition based on random forests. Appl. Soft Comput. **46**, 805–815 (2016)
9. A. Gudigar, S. Chokkadi, Local texture patterns for traffic sign recognition using higher order spectra. Pattern Recogn. Lett. **94**, 202–210 (2017)
10. J. Ou, Y. Li, Vector-kernel convolutional neural networks. Neurocomputing **330**, 253–258 (2019)

11. S. Brahimi, N.B. Aoun, C.B. Amar, Boosted convolutional neural network for object recognition at large scale. Neurocomputing **330**(22), 337–354 (2019)
12. S. Ho, P. Tsang, H. Wong, K.L. Choy, A computer vision-based roadside occupation surveillance system for intelligent transport in smart cities. Sensors (Basel). **19**(8), 1796 (2019). https://doi.org/10.3390/s19081796
13. M.H.J.D. Koresh, Computer vision based traffic sign sensing for smart transport. J. Innovative Image Process. (JIIP) **1**(01), 11–19 (2019)
14. M.M. Ata, M. El-Darieby, M. Abd Elnaby, S.A. Napoleo, Towards computer vision-based approach for an adaptive traffic control system. Imaging Sci. J. **66**(7) (2018)
15. C. Liu, S. Li, F. Chang, Y. Wang, Machine vision based traffic sign detection methods: review, analyses and perspectives. IEEE Access **7**, 86578–86596 (2019). https://doi.org/10.1109/access.2019.2924947
16. A. Mhalla, T. Chateau, S. Gazzah, N.E.B. Amara, An embedded computer-vision system for multi-object detection in traffic surveillance. IEEE Trans. Intell. Transp. Syst. **20**(11), 4006–4018 (2019). https://doi.org/10.1109/tits.2018.2876614
17. R.K. Yadav, R. Jain, S. Yadav, S. Bansal, Dynamic traffic management system using neural network based IoT system, in *2020 4th International Conference on Intelligent Computing and Control Systems (ICICCS), Madurai, India* (2020), pp. 521–526. https://doi.org/10.1109/iciccs48265.2020.9121066

Unsupervised Deep Learning on Spatial-Temporal Traffic Data Using Agglomerative Clustering

S. Senthilarasi and S. Kamalakkannan

Abstract Nowadays, road traffic arises from the increased population because of employment in urban areas of all places. This creates to be a challenging aspect, uses certain measures to solve then and thereby the authorized people in the situation. The varying types of services related to traffic streams are pedestrian census, turning movement census, and others. The highlighting features are tabulated based on traffic management techniques. The proposed study on agglomerative clustering on traffic data has been satisfied with its enhanced features. Agglomerative clustering is an extensive model of hierarchical clustering, a bottom-up approach that combines the similarities of samples in clusters. A quite number of methods to find optimal clusters in different ways are briefly discussed to support the clustering. Experiments studied on California-Traffic solution-Data from SWITRS conducted with traffic data prove the optimal number of clusters formed, its validity using the method of clustering accuracy and pasteurization of time series on traffic data are shown in different categories.

Keywords Traffic services · Traffic management techniques · Agglomerative clustering · Optimal clusters · Time series visualization

1 Introduction

At present, the growth of the human population is excessive in urban areas which state to be the topmost challenges for worldwide countries. Such rapid growth in city populations, mode of transport, movement of vehicles, causes of air pollution affects both the environment and human life [1]. The excessive movement of vehicles leads to

S. Senthilarasi (✉)
Department of Computer Science, Vels Institute of Science, Technology and Advanced Studies (VISTAS), Chennai, India

S. Kamalakkannan
Department of Computer Science, School of Computing Sciences, Vels Institute of Science, Technology and Advanced Studies (VISTAS), Chennai, India
e-mail: kannan.scs@velsuniv.ac.in

V. Suma et al. (eds.), *Inventive Systems and Control*,
Lecture Notes in Networks and Systems 204,
https://doi.org/10.1007/978-981-16-1395-1_56

road traffic congestion that creates a critical issue in urban places. An article's report shows the information nearly 600 billion rupees are spent each year because of road traffic congestion (or) it's dally frequently occurs on high extent roads, highways added with wastage of fuel [8].

The prediction of traffic is considered as a crucial part by providing correct and valid traffic details for both voyagers and traffic agencies in a modern Intelligent Transportation Systems (ITS) and Advanced Traveller Information System (ATIS). The attributes of traffic details are such as traffic blockage positions, traffic size, and mass flow earlier, which can support the officials in a better way to manage the traffic strategies and voyagers to plan alternate routing plan of actions [2].

An assessment of traffic contains the different types of a related collection of data methods examined on a different view of traffic streams. These views comprehend the subsequent services [3].

Different types of Services

1. Pedestrian Census
2. Turning movement census
3. Bicycle census
4. Link flow traffic census
5. Rail and bus punctuality census
6. Travel Time
7. Round about census
8. Parking census technique
9. Queue length census
10. Recognition of number plates
11. Illegal movement census
12. Traffic Saturation flows.

1.1 Highlighting Features in Traffic Management Techniques

The excessive growth of population in place of urban areas shows the biggest strengths for worldwide administrations. The increased use of motor vehicles on roadways not only affects the surrounding environment also harms the quality of human life and their wealth [4] (Table 1).

2 Related Work

Recently, a quite number of papers have been published by various authors in spatial-temporal combination on road Traffic data by using different Clustering techniques.

Table 1 Different types of traffic management

Types of traffic management	Basic features	Additional features	Pros	Cons
Manual traffic control	• It is the simplest way of managing the traffic • It involves a policeman standing on every interconnection of roads to control the regular traffic flow	• A traffic police controls the regular traffic flow by standing on traffic poll by showing the signboard • In case of a large volume of traffic, he/she signals the vehicle drivers to drive (or) not to drive	The people on roadways can understand the traffic because of emergencies, traffic police can prioritize the movement of vehicles	The system involves human as a main role in controlling traffic is inefficient
Automatic traffic control	It is an automatic traffic control system	• The system adds three colored signals red, yellow, and green • Each signal light is set with timings in terms of seconds to glow sequentially to alert the travelers	It overcomes the difficulties of the manual traffic	It cannot differentiate the movement of vehicles with emergencies and without emergencies
Image processing traffic control	It involves cameras to record images of a large volume of traffic lying on roads	• It is normally fixed on high pillars to cover the traffic to a certain distance from its positioned view • The recorded images are validated by a chip in computer to find the number of vehicles on road to compute the timings for signal lights especially red and green to manage the traffic density	It dynamically changes the recorded timings of signal lights	It cannot capture the images clearly in case of heavy rain and heavy traffic jam

(continued)

Table 1 (continued)

Types of traffic management	Basic features	Additional features	Pros	Cons
Wireless traffic control	Emergency Vehicle is facilitated with RF-Transmitter and Receiver placed on the signal pole	• It alerts the RF_Receiver to send data to the main control system • The system computes the timing approximately for the green signal to glow during the movement of emergency vehicles and other vehicles to stop by signaling the red light	It provides a solution in case of emergency vehicles passing on the road	It's maintenance cost is high due to sensors, load cells in accumulating traffic data
IRIS	It is a source software project to observe and control the lay of traffic lying on connecting roadways such as interstate and highways	• It gives the real-time information existing on highways to find traffic incidents, control the traffic flow, and relay information of travelers • It is using GPL license, ATMS Software tool	It is better than wireless traffic control	It is expensive to buy and its recovery maintenance costs also expensive

The following tabular form depicts the study of different papers related to road traffic data [5–19] (Table 2).

3 Proposed Study in Traffic Data

The Proposed study on traffic data utilizes the bottom-up approach of hierarchical clustering—Agglomerative Clusters, Which merges the samples of similar features closer recursively till the single cluster is formed based on sample outcomes.

Table 2 Study of spatial-temporal traffic data

S. No.	Author	Title of the paper	Clustering technique	Achievements	Year	Experiment study
1	Wendy Weijermans and Eric Van Berkum	Analyzing highway flow patterns using cluster analysis	Ward clustering	Maintains traffic profile daily of individual clusters, integration of clusters	2005	Dutch highway location
2	Weichiet Ku, George R. Jagadeesh	A clustering-based approach for data-driven imputation of missing traffic data	K-Means Clustering	An SDAE model is constructed to structure the spatial-temporal relationships for every group of road divisions	2016	Jurong West area of Singapore collected in Quantum Inventions
3	Mei Yeen Choong, Lorita Angeline, Reneee Ka Yinchin, Kiam Beng Yeao, Kenneth Tze Kin Teo	Vehicle trajectory clustering for traffic intersection surveillance	K-Means and Fuzzy C-Means	Implements K-Means and Fuzzy C-means on trajectories, usage of longest common subsequences with different trajectories lengths to pinpoint noise and outliers	2016	Computer Vision Robotics Research Laboratory(CVRR) of 100 vehicle trajectories
4	Vishwajeet Pattanaik, P.K. Gupta, S.K. Singh and Mayank Singh	Smart real-time traffic congestion estimation and clustering technique for urban vehicular roads	K-Means Clustering	Establishes a traffic estimation system on Global Positioning System (GPS) to activate the availability of mobile phones for development Discovers the shortest path from Source to Destination using Dijkstra's algorithm	2016	Several roadmaps of New Delhi

(continued)

Table 2 (continued)

S. No.	Author	Title of the paper	Clustering technique	Achievements	Year	Experiment study
5	Maithew Aven	Daily flow Pattern Recognition by Spectral Clustering	K-Means Algorithm, Spectral Clustering	Proves the accurate relationships arises between the dataset values	2017	New York City Intersection case study
6	Senyan Yang, Jianping Wu, Geqi Qi and Kun Tian	Analysis of traffic state variation patterns for urban road network based on Spectral Clustering	Spectral Clustering	Extracts the traffic features variances for the road network, transform traditional clustering to graph partitions	2017	North East region of Beijing
7	Mohammad Billah, Arash Masooki, Farzana Rahman and Jay A. Farrell	Roadway feature mapping from point cloud data: a graph-based clustering approach	Graph-based clustering		2017	Advanced Highway Maintenance and Construction Technology Research Center (AHMCT) – 3D point cloud data
8	Weirong Liu, Member IEEE Gaorang Qin, Yun He and Fei Jiang	Distributed cooperative reinforcement learning-based traffic signal control that integrates V2X network's dynamic clustering	Dynamic Clustering	Initializes the partitions of clusters to retain the correct size besides with lane and destination to maximize the stability of clusters. Balancing of traffic load using reinforcement learning	2017	V2X Network's data collection

(continued)

Table 2 (continued)

S. No.	Author	Title of the paper	Clustering technique	Achievements	Year	Experiment study
9	Guojieng Shen, Chaohuan Chen et al.	Research on traffic speed prediction by temporal clustering analysis and convolutional neural network with deformable kernels	Hierarchical Clustering	Initiation of deformable convolution layer into DCNN to resize the kernels, extends the strengths of DCNN to structure the spatial-temporal data	2018	Xiasohan district of Hangzhou traffic police Brigade, Zhejiang, China
10	Sakhawat Hosain Sumit and Shamin Akhter	C-means Clustering and deep-neuro-fuzzy classification for road weight measurement in traffic	C-Means, Fuzzy clustering	Handles Overlapping uncertainties which provide the ability to perform task with multidimensional featured data	2018	Environmental data(750 days 12/1/2006 -12/10/2008)collected from AccuWeather
11	Yuntao Chang and Bin su	Construction of the Driving cycle of vehicles Queuing at toll station	Fuzzy C-Means clustering, Markov Method	A model on space-based vehicles driving constructed about queuing vehicles, Comparison of effect between manual, ETC Collection System	2018	Jiangqiao toll station on G2, Beijing-Shanghai Expressway
12	Weixiang Xu and Jiaojiao Li	An improved algorithm for clustering uncertain traffic data streams based on Hadoop platform	AF-DBSCAN clustering	An UTD-SPCQ algorithm is used to receive data and its buffering are realized, MapReduce Framework of distributed structure in the Hadoop platform	2019	Berijing's taxi dataset

(continued)

Table 2 (continued)

S. No.	Author	Title of the paper	Clustering technique	Achievements	Year	Experiment study
13	Tharingu Bandaragoda, Darwin De Silva,	Trajectory Clustering of road traffic in the Urban environment using incremental machine learning in combination with hyper dimensional computing	Hyper dimensional computing	An unsupervised incremental learning approach addresses two key problems in the traffic profile. Enables the Variable-length trajectories of traffic trips	2019	Bluetooth Traffic Monitoring System (BTMS), Arterial roads in Victoria, Australia
14	Hiroki Watanobe, Tomas Maly, Johannes Wallner	Cluster-Linkage analysis in Traffic Data Clustering for Development of Advanced Driver Assistance Systems	Hybrid clustering (based on K-Covers and K-Means Algorithms)	Discovery of optimal number of clusters, cluster models using average Silhouette Width, Reduced test effort in the development of ADAS	2020	Vehicle-Pedestrian near-crashes in SHRP2 dataset
15	Weijing Qi, Bjorn LandFeldt, Qingyang Song, Lei Guo, Abbas Jamliopour	Traffic Differentiated Clustering Routing in DSRC and C-V2X Hybrid Vehicular Networks	One-hop clustering	A Traffic Differentiated Clustering Routing (TDCR) mechanism in a Software Defined Network (SDN) enables a hybrid vehicular network Includes centralized one-hop clustering and dispatch of data	2020	Data collection from CHs to data center

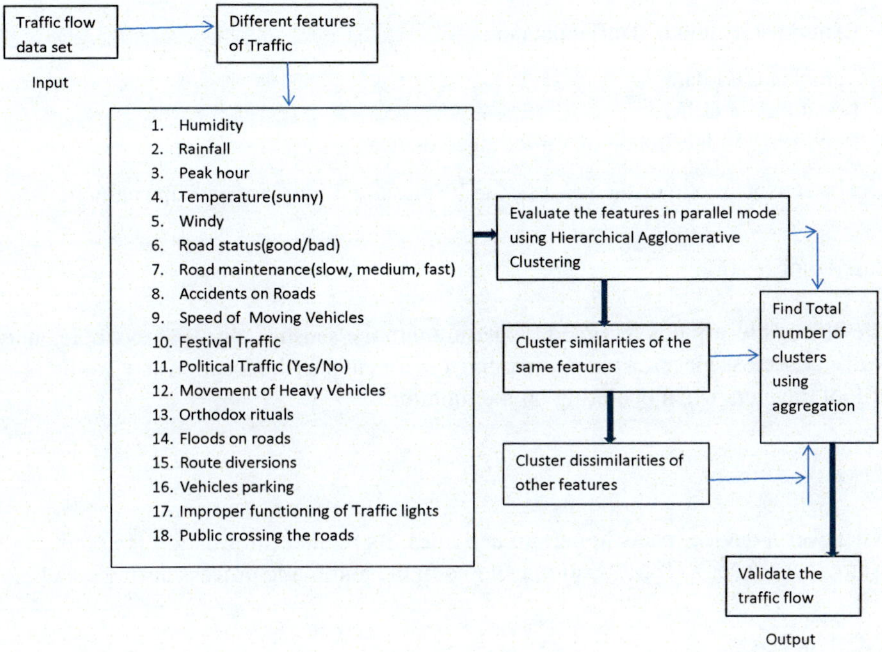

Fig. 1 Traffic flow using agglomerative clustering model

The proposed model of traffic data (Fig. 1) shows the input of various features, which has to be found and clustered using the agglomerative clustering approach. The similarities of clusters are given importance and notified rather than dissimilar features considered as outliers. The outcome of traffic data contains the number of clusters formed is evaluated using measures of aggregation, and the validity of clusters are found using the method of Silhouette Coefficient.

3.1 Preprocessing Traffic Data

The traffic data are of different types mainly depends on analysis and visualization-related methods. The actual data are generally raw, data with errors, uncertainties, number of missing values, outliers (or) items mismatching. In particular raw data is considered for analysis and visualization.

3.1.1 Data Collection

The number 'n' of datasets can be created and accumulated by sensing devices fixed in traffic signals, vehicles (or) CCTV's fixed on traffic accumulating places on roadways.

Examples related to Traffic data are

- Vehicles GPS data
- Locations of GSM
- Recording Video/Image of surveillance devices.

The sensors mode of the process can be classified into three groups namely

Based on Location

The object's location is recorded by getting into the sensor range. For example, at a road's crisscross places, a video recording device catches the location and wandering of foot travelers when operating on the monitor.

Based on Activity

Whenever a device reacts to certain activities, its related information is noted. For example, the user's GSM location is noted in the automatic process during a call.

Based on Device

An object's device tracks the record and directs the location and other related concepts back actively. For example, for every 20s, a car sends a signal to the data center with the use of GPS device.

3.1.2 Properties of Traffic Data

In any number of instances, traffic data hold different attributes of spatial and temporal additional information. These properties can be classified into three classes namely

1. Numbers
2. Category
3. Text

Numbers

It represents the continuous-valued variables which means the data object's quantitative values. Every numbered value explains any one data object's aspect particularly speed, weight, etc. Most of the attributes are time-variant, time-oriented techniques in visualization. For instance, a better choice for visualization is the histogram.

Category

It represents the discrete set of valued variables that explains the data object's status. Type of vehicles, directions, and incidents are instances of categorical properties.

This property is simplified in visual form as color mapping, assigning a specified color to represent a value.

Text

It represents the words, lexical information that illustrates additional information related to traffic such as vehicles included in an incident, intersection points, and others. The attributes containing semantic information are necessary to analyze and describe traffic situations [20].

4 Experimental Study

Traffic data is collected from California-Traffic solution-Data from SWITRS [26]. The data includes the collisions from January 1st, 2001, till the mid of October 2020. The dataset contains three main tables, consisting of collisions, parties, and victims. The collision table contains 74 field names related to traffic by describing 9.17 m rows. The parties table holds the information about the people's age, gender, and serious behavior included in a collision. The victim's table consists of the details related to the people who got an injury in the collision.

The proposed algorithm on spatial-temporal data accepts the 'n' number of attributes on traffic data as input which has to be read and normalized using *scaler* functions. The normalized data are then transformed into *data frames* to predict the number of clusters. The dimensionality of the data has been reduced. Finally, the outcome of clusters is displayed in advanced feature of hierarchical clustering *dendrogram*, clusters in circular form with k values (2, 3, 4, 5, 6). The cluster's validity has been computed using the method of Silhouette Coefficient, which proves the Strong Clustering.

4.1 Algorithm: Spatial–Temporal Traffic Data Using Agglomerative Clustering

Input: Dataset consisting of attributes (1,2,....*n*)

Output: Provides an optimal number of clusters

1. Read the data(1, 2,...*n*) attributes
2. Repeat
   (a) Normalize the data
   (b) Fit the normalized data in DataFrame
   (c) Transform the normalized data to predict clusters
   (d) Add dendrogram to display the output of hierarchical clustering
   (e) Cluster the data for K = 2,3,4,5and 6, where k represents the number of clusters

(f) Compute the clustering validity using Silhouette coefficient

Until the stopping condition is met

3. Output the optimal number of clusters

4.2 Agglomerative Clustering

Agglomerative clustering is a plan of hierarchical clustering. It is otherwise known as connectivity-based clustering, which is a behavior of cluster analysis that searches to construct a hierarchy of clusters. Hierarchical clustering is established on the central plan of objects being more closed to neighboring objects compared to distant objects. A cluster can be outlined in wider by the extreme value required in connecting parts of clusters.

With varying distances, clusters in difference can be stated using "Dendrogram"(Fig. 2) which issues an increased hierarchy of clusters that combines lying at fixed distances.

In a dendrogram,

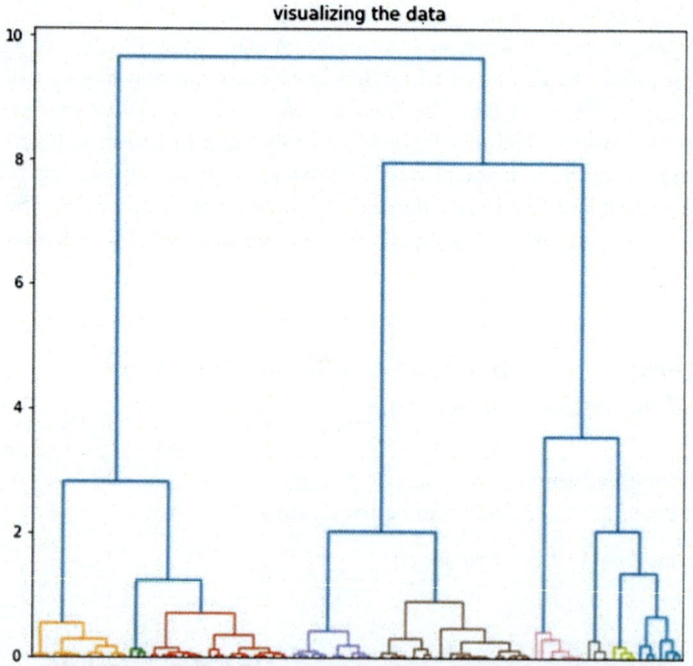

Fig. 2 Dendrogram of traffic data

X axis—Objects are placed, Clusters do not mix.
Y axis—Distance is marked to merge the clusters [21].

4.2.1 Factors of Agglomerative Clustering

1. Initially, it looks more trustworthy due to over clustering. Representation of clustering from a convolutional neural network is initiated with random weights, which are not efficient, whereas the closest neighbors and over clustering are generally acceptable.
2. The over clustering can be integrated to make it a better representation of learning.
3. It is a recurrent type of process that occurs in a recurrent framework [22].

4.2.2 Methods to Find Optimal Clusters

The validity of clustering indexes is generally defined by integrating both the compactness and divisibility of the clusters. The closeness of elements in clusters is evaluated by compactness where the difference between two clusters is represented using the divisibility measure.

The common methods to yield the clustering quality are

1. Davies-Bouldin Index
2. Dunn Index
3. Elbow Method
4. Silhouette Method.

Clustering Method Davies-Bouldin Index

The average similarity exists among every cluster with its similarities are measured using DBIndex. The minimum values of DBIndex state that the clusters are compactly tight and distinguished the cluster in a better form. This index aims to fulfill "minimum within-cluster variance and maximum between cluster separations" [23].

The Davies-Bouldin Index can be computed as follows [25]:

$$\mathrm{DB} = \frac{1}{n}\sum_{i=1}^{n}\max_{j\neq i}\left(\frac{\sigma_i + \sigma_j}{d(c_i,\ c_j)}\right)$$

Clustering Method—Dunn Index

The Dunn Index (DI) measures the dataset clusters, is expected to be maximum and the clusters are separated. The clusters are so compact and distinguished by increasing the intercluster distance, whereas, the intra-cluster distance is in decreasing [23].

The Dunn Index is computed using the form [25]:

$$D = \frac{\min_{i \le a \le b \le n} X(a, b)}{\max_{i \le a \le c \le n} Y(c)}$$

where *i*, *j*, *k* are clusters indices

X = distance variable of inter-clusters
Y = distance variable of intra-clusters

Clustering Method—Elbow

It is an analytical method of expansion and verification of standards within the cluster analysis planned to assist in finding the dataset's suitable number of clusters [24].

Clustering Method—Silhouette Coefficient

The Silhouette Coefficient depicts the fitness of objects within the cluster (Fig. 3). The clustering quality is measured between the range -1 and $+1$.

+1 score represents the samples are distanced from their nearest clusters.

0 score represents the sample is too close to the decision limits which divides the two nearest clusters.

-1 score represents the samples are assigned to the wrong cluster.

The Silhouette score can be computed as follows

$$\text{Silhouette Score} = (a - b)/\max(a - b)$$

where

'*a*' represents the distance to the points in the closest cluster
'*b*' represents the intra-cluster distance of all the available points [25].

4.3 *Time Series Visualization in Traffic Data*

Visualization plays a major task in the analysis of time series and their forecasting. The plotting of data issues the most valuable features to find temporal models such as trends, cycles, and seasonality to enhance the type of model.

There are six varying types of visualization applied to time series data (uni-variate time series) (Figs. 4, 5, 6, 7, 8, and 9).

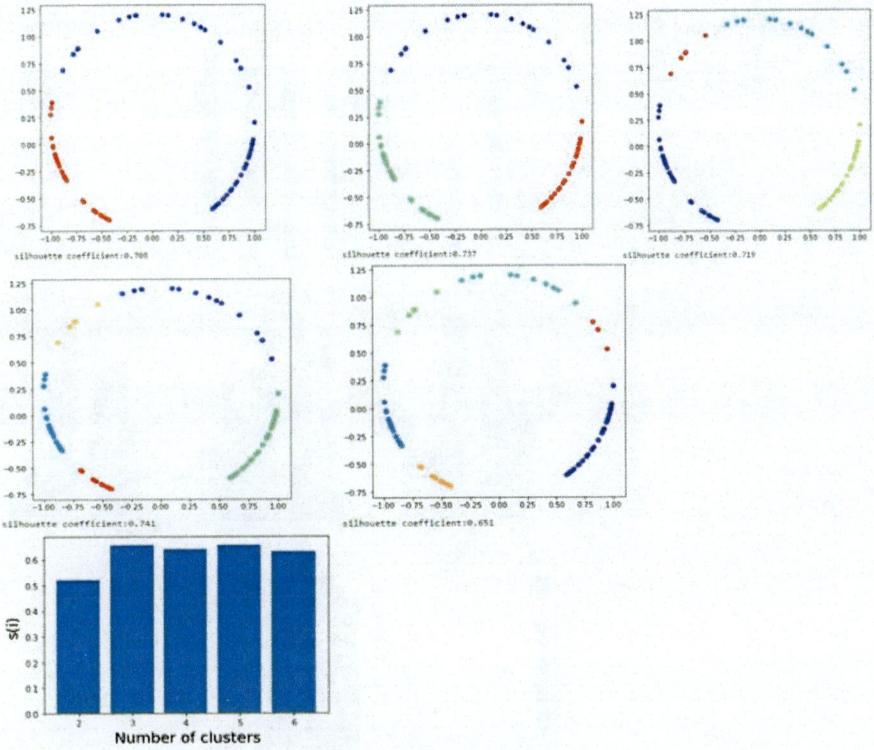

Fig. 3 Optimal number of clusters and its validity

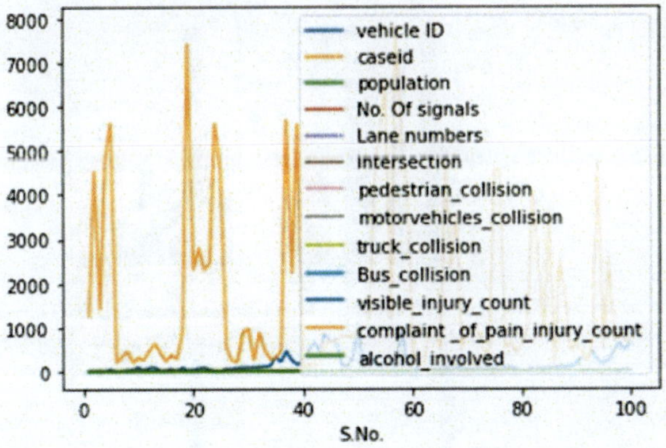

Fig. 4 Line Plot

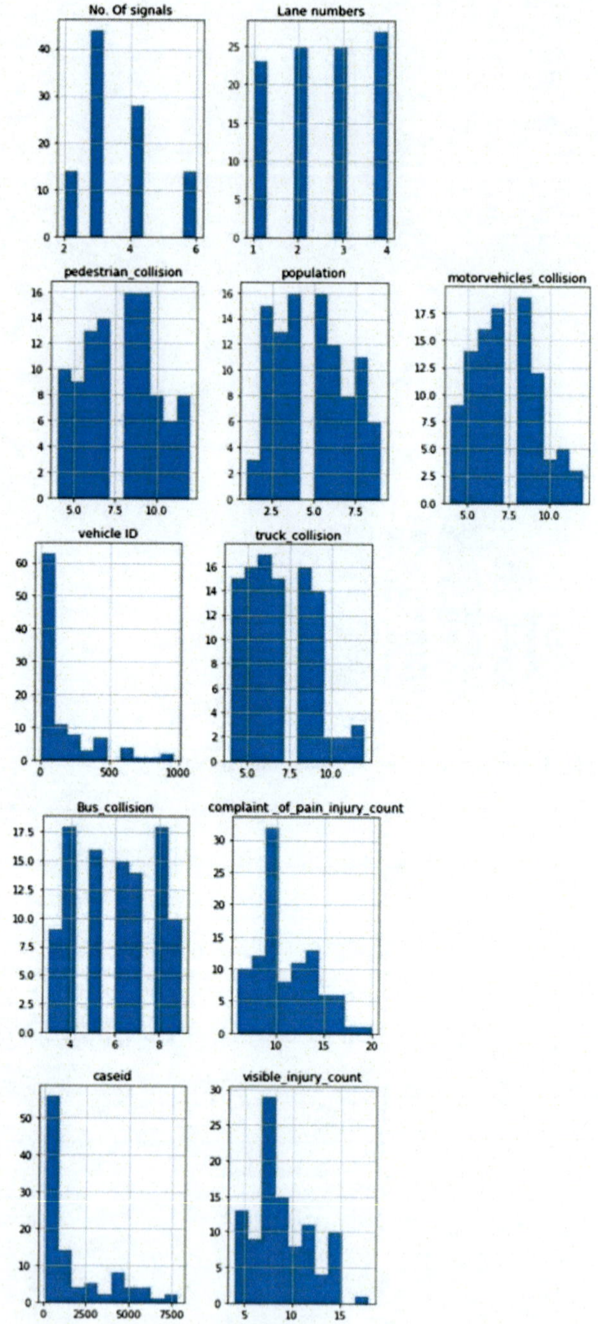

Fig. 5 Histograms and density plots

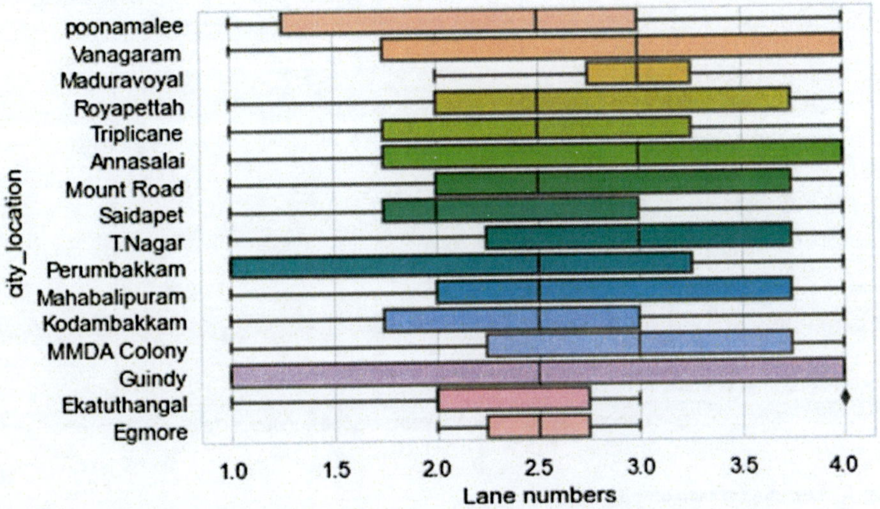

Fig. 6 Box and whisker plot

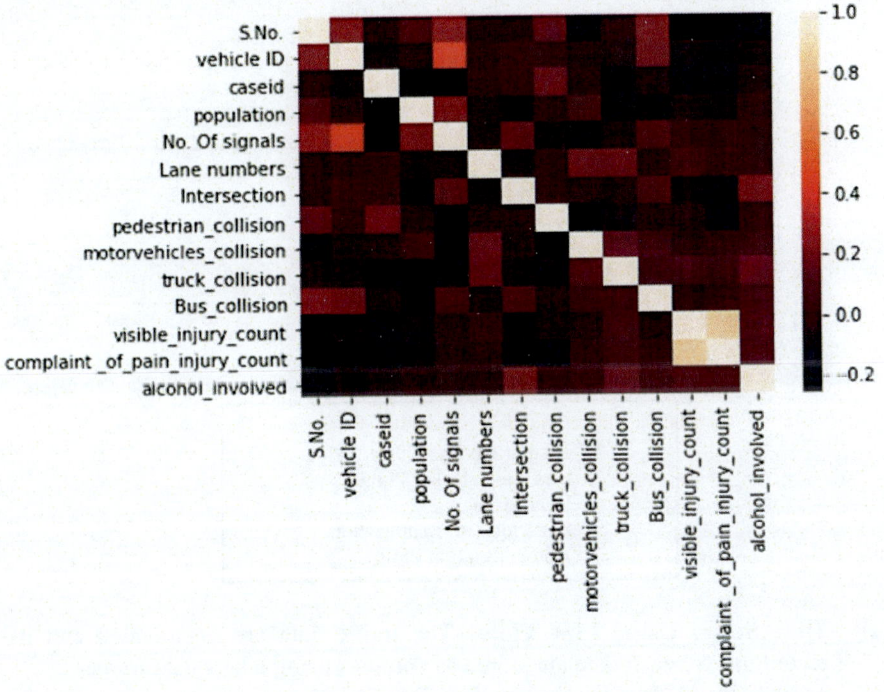

Fig. 7 Heat map

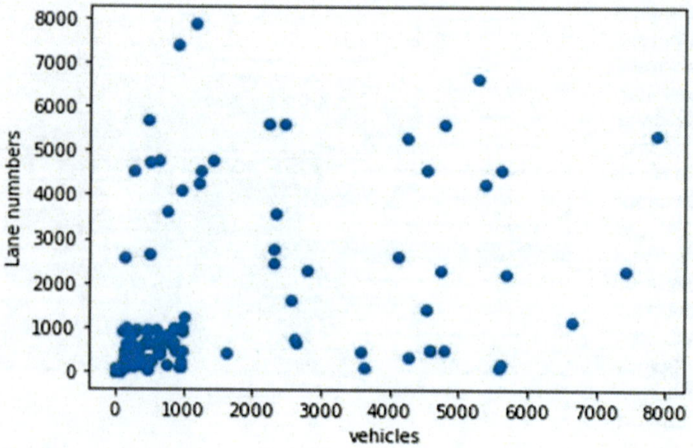

Fig. 8 Lag plot (or) scatter plot

Fig. 9 Autocorrelation plot

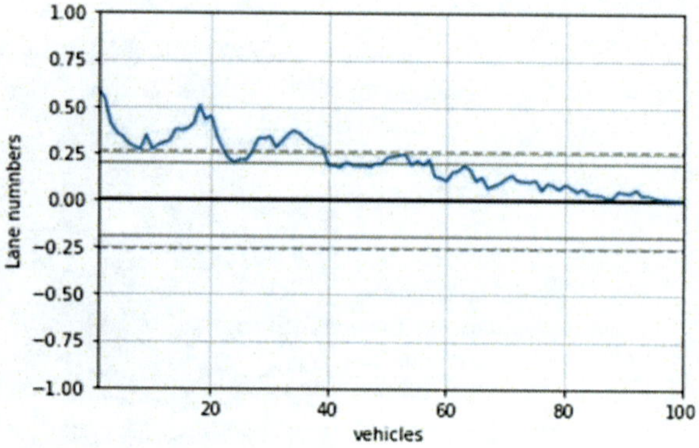

1. Line plots
2. Histograms and Density Plots
3. Box and Whisker Plots
4. Heat Maps
5. Lag Plots (or) Scatter Plots
6. Autocorrelation Plots

(a) **Time Series Using Line Plots**—The traffic data are enumerated and its recordings for each date are stored as columns using a new data frame.

(b) **Time Series Using Histogram and Density Plots**—A histogram combines the values into bins, several results available in each bin provides intuition into the basic diffusion of the results.

(c) **Time Series Using Box and Whisker Plots**—This type of plot is created for each day of traffic and placed side-by-side toward direct match-up.

(d) **Time Series Using Heat Map**—This concept plots the matrix numbers as an aspect, where every cell value related to the matrix is given a unique color.

(e) **Time Series Using Lag Plot**—It is a convenient type of plot which exhibits the relation between every result and its lag value called as Scatter Plot [27].
(f) **Time Series Using Autocorrelation Plot**—It is also known as "Correlation Coefficients", which can be computed for every outcome and varying lag values.

Values near 0 represent a weak correlation
Values near -1 (or) 1 represent a strong correlation.

5 Discussion

The agglomerative clustering on traffic data experimented on California-Traffic solution-Data from SWITRS showed in the familiar concept of advanced hierarchical clustering *dendrogram,* also the validity of clusters formed has been with the help of clustering accuracy method. The concept of a dendrogram is unique in hierarchical clustering, whereas it is not applicable in other types of clustering. The visualization of traffic data in terms of time series has been depicted in different categories. The proposed study on this paper is just a step regarding the better representation of clustering, where the concept has to be further enhanced by the comparative study with other clustering models to prove the increased level of accuracy. Besides, the agglomerative clustering approach has to be portrayed in a better form compared to other clustering models.

6 Conclusion

This article focuses on the road traffic data occurring in high-density population areas. The similarities of collision happen due to the movement of vehicles on road has been classified and studied on the California-Traffic solution-Data from SWITRS. The model of agglomerative clustering had been applied to analyze and evaluate the optimal number of clusters using dendrogram. The validity of clusters has been proven with a strong relationship with clusters using the measure of Silhouette Coefficient, which is one of the methods of cluster accuracy. The view of traffic data that occurs in time series has been shown in varying methods. The accuracy and representation of clusters are valid compared to K-Means Clustering, Spectral Clustering, OPTICS clustering, and BIRCH clustering. Further, the clustering on the agglomerative approach has to be enhanced in a better representation appended with accuracy, execution time, speed, and performance.

References

1. N.C. Benitez, R. Aquino-Santos, P. Magna-Espinoza et al., Traffic congestion system through connected and vehicles and big data. Sensors (2016)

2. S. Guo, Y. Lin, S. Li, Z. Chen, H. Wan, Deep Spatial-temporal 3D convolutional neural networks for traffic data forecasting. IEEE Trans. Intell. Transp. Syst. (2019)
3. VTRAC-Intelligent Traffic Solutions
4. G.P. Siddamma, S. Madhavanavar, Different techniques used in traffic control system: an introduction. IJERT
5. W. Weijermars, E. Van Berkum, *Analyzing Highway Flow Patterns Using Cluster Analysis* (IEEE, 2005), pp. 831–836
6. W.C. Ku, G.R. Jagadeesh, A. Prakash, T. Srikanthan, *A Clustering–Based Approach for Data-Driven Imputation of Missing Traffic Data* (IEEE, 2016)
7. M. Y. Choong, L. Angeline, R.K.Y. Chin, K.B. Yeo, K.T.K. Teo, Vehicle trajectory clusteriung for traffic intersection surveillance, in *International Conference on Consumer Electronics-Asia* (IEEE, 2016)
8. V. Pattanaik, M. Singh, P.K. Gupta, S.K. Singh, Smart real-time traffic congestion estimation and clustering technique for urban vehicular roads (IEEE, 2016)
9. M. Aven, *Daily Traffic Flow Pattern Recognition by Spectral Clustering* (CMC Senior Theses, 2017), pp. 1–16
10. S. Yang, J. Wu, G. Qi, K. Tian, Analysis of Traffic state variation patterns for urban road network based on spectral clustering. Adv. Mech. Eng. (2017)
11. M. Billah, A. Maskooki, F. Rahman, J.A. Farrell, roadway feature mapping from point cloud data: a graph-based clustering approach, in *Intelligent Vehicles Symposium* (IEEE, 2017)
12. W. Liu, G. Qin, Y. He, F. Jiang, Distributed cooperative reinforcement learning—based traffic signal control that integrates V2X Networks Dynamic Clustering, in *Transactions on Vehicular Technology* (IEEE, 2017)
13. G. Shen, C. Chen, Q. Pan, S. Shen, Z. Liu, *Research on Traffic Speed Prediction by Temporal Clustering Analysis and Convolutional Neural Network with Deformable Kernels* (IEEE Access, 2018), pp. 51756–51765
14. S. Hosain Sumit, S. Akhter, *C-means Clustering and Deep-neuro-fuzzy Classification for Road Weight Measurement in Traffic Management System* (Springer, 2018)
15. Y. Chang, B. Su, Construction of the driving cycle of vehicles queuing at toll station, in *International Conference on Intelligent Transportation Systems (ITSC)* (IEEE, 2018)
16. W. Xu, J. Li, *An Improved Algorithm for Clustering Uncertain Traffic Data Srtreams Based on Hadoop Platform* (World Scientific, 2019), pp. 1950203-1–1950203-19
17. T. Bandaragoda, D. De Silva et.al. Trajectory Clustering of road traffic in urban environments using incremental machine learning in combination with hyper dimensional computing, in *International Transportation Systems Conference* (IEEE, 2019)
18. H. Watanobe, T. Maly, J. Wallner, G. Prokop, Cluster-linkage analysis in traffic data clustering for development of advanced driver assistance systems (ICICT, 2020), pp. 54–59
19. W. Qi, B. Lanfeldt, Q. Song, L. Guo, A. Jamlaipour, Traffic differentiated clustering routing in DSRC and C-V2X Hybrid vehicular networks. *Transaction on Vehicular Technology* (IEEE,2020)
20. W. Chen, F. Zhou Guo, F.-Y. Wang, A survey of traffic data visualization. *Transactions on Intelligent Transportation Systems* (IEEE, 2015) pp. 1–15
21. docs.rapidminer.com/latest/studio/operators/modeling/segmentation/agglomerative-clustering.html
22. J. Yang, D. Parikh, D. Batra, Joint unsupervised learning of deep representations and image clusters (2016)
23. S. Nawrin, Md.R. Rahman, S. Akhter, Exploring K-means with interval validity indexes for data clustering in traffic management system. IJACSA (2017)
24. medium.com/@masraudheena/4-best-ways-to-find-optimal-number-of-clusters-for-clustering-with-python-code-706199f9957C
25. tutorialspoint.com/machine-learning-with-python/machine-learning-with-python-analysis-of-silhouette-score.htm
26. kaggle.com/alexgude/California-traffic-collision-data-from-switrs
27. machinelearningmastery.com/time-series-data-visualization-with-python

Divorce Prediction Scale Using Improvised Machine Learning Techniques

Ashutosh Shankhdhar, Tushar Gupta, and Yash Vardhan Gautam

Abstract Divorce is the bitter truth of modern society and has unfortunate effects not only on the parties involved but also on their children and near kin. There is a drastic increase in cases of divorces and separations in the past ten years. The study aims to find out the resulting factors in such cases and to identify their respective significance and relationship with other factors involved. A correlation-based feature selection method has been applied on the dataset consisting above factors, and then the results of which are combined to that of the previous divorce cases are used to train a model that can predict the possibility of a future divorce. The aforementioned model, viz. The divorce predictor scale, has been constructed and designed on various supervised machine learning algorithms. A comparative study of the various models from different algorithms has also been performed to compare their accuracy.

Keywords Divorce · Separation · Common factor · Divorce prediction scale

1 Introduction

India has a manifold tradition and culture which separate it from other countries. Despite having numerous rudimentary cultures, Indians have always sought to be at par with their counterparts, thus installing modern themes in society. It is one of the main reasons for India having the lowest rank in terms of divorce rates. Approximately 85% of marriages in India are arranged and with even higher percentages in the northern states according to a UN News report. Divorces in such arranged marriages

A. Shankhdhar (✉) · T. Gupta · Y. V. Gautam
Department of Computer Engineering and Application, GLA University, 17 km Stone, N2, Mathura-Delhi Road P.O. Chaumuhan, Mathura, U.P. 281406, India
e-mail: ashutosh.shankhdhar@gla.ac.in

T. Gupta
e-mail: tushar.gla_cs18@gla.ac.in

Y. V. Gautam
e-mail: yash.gautam_cs18@gla.ac.in

V. Suma et al. (eds.), *Inventive Systems and Control*,
Lecture Notes in Networks and Systems 204,
https://doi.org/10.1007/978-981-16-1395-1_57

Fig. 1 Global divorce rate graph [1]

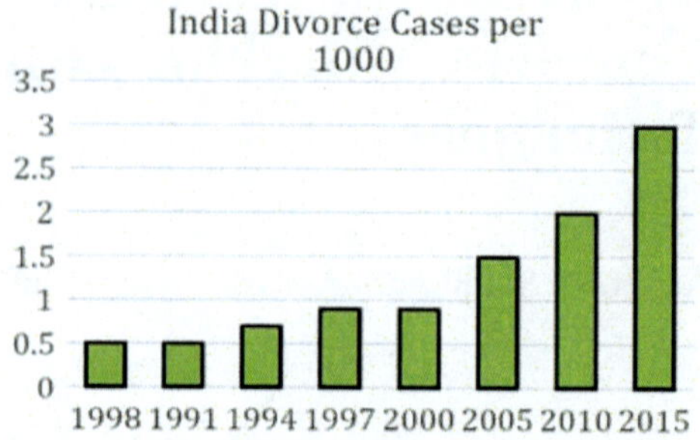

are still unheard due to it being a social taboo. Arranged marriages are usually fixed and looked over by the elderly in the family, and many religious customs are used to get a congenial match between the boy and the girl. Before selecting a daughter-in-law or son-in-law for their child, parents focus on the family backgrounds, their caste, and creed and all the similarities and dissimilarities between both the families so that they can have a better understanding between themselves. These three actors not only suffer mental issues, as well as physical issues after an unwanted and distressing divorce.

But for the past few years, divorces and legal and mutual separations are on a rise. Many factors are responsible for separations such as misunderstanding, time spent, lack of love, the involvement of a third person, and believed mismatching. If a couple can figure out the responsible factors, then he or she will be able to save their relationship and will be able to cope up with the future full of negativity.

Because India conglomerates orthodox diverse cultures and traditions, India secures the lowest rank in divorce all over the world, as it has a 1% divorce rate which is very low in comparison with third world countries, leaving aside other developed nations [1]. Given below is a chart that shows divorce rates in India (Fig. 1).

The most common reasons behind the divorce are cheating (having an extramarital affair), drug or alcohol addiction, conflicting thoughts, abuse (physical and mental both), etc. It sometimes becomes more important when one is suffering from harassment and torture. In that case, divorce proves to be the best alternative. As the world is modernizing, women are becoming independent and they believed in their better future.

Many situations are faced by a couple during the divorce, and they have to think will society accept it or not, is it good for their family reputation, are they capable of handling the future situations themselves, who is going to take the responsibility for the children, what if all the things seem to be same after divorce, and are they mentally and physically ready to take such a big step. These are some of the questions that keep revolving in the couple's mind during a divorce.

The main objective of this research paper is to determine the importance of the factors that are responsible for the divorce and how easily or laboriously they can be resolved. This research analyzes the volunteer's mind-set toward their divorces with the help of a specially fabricated questionnaire, by using various classification techniques and thereby determining if there is a possible separation soon. Answering each question/factor from the questionnaire has been categorized into 5 levels, giving volunteers a wide array of options.

The principal focus of this research is to analyze the importance of these basic factors and by judging the responses given by the different couples on a level of 0, 1, 2, 3, and 4 and with the help of various machine learning algorithms determining if there are any chances of divorce between the couples.

2 Related Work

IDC forecasts the global data sphere that will reach 175 zetabytes in 2025 from 4.5 zetabytes in 2013 and 45 zetabytes in 2019. India surely will generate a big chunk of this data. India, which has a very young society, is now following their western counterparts more than ever, and this is followed in the dating and relationship scene too. This means more divorces and separations in the upcoming years.

India has one of the lowest divorce rates in the world as divorce is still a social stigma and the special marriage laws of different societies make the legal process even more difficult. This study aims to make a divorce predictor scale based on a relationship-related questionnaire that can predict the possibility of divorce [2]. Also, the study aims to find out the most effective questions that affect divorce and their significance by applying the correlation-based feature selection method on the dataset.

Along with the divorce predictor scale, the study also aims for a comparative analysis of the various supervised machine learning algorithms used in the DPS, thus showing the accuracy of the proposed algorithms than that of commonly used supervised machine learning techniques [3].

Indian society is a very complex society, and over the years many pieces of research have been done on the marriage and divorce scenario of the country. Divorce in India is a big social event, often involving many people than just the two of the partners. The nationwide census of 2011 shows that the average rate of divorce of Hindus and Muslims is 2.0 and 3.7, respectively, and this is just the legal number with many more couples living separately without a divorce just to hide from the social embarrassment [1].

A study shows that taking divorce in India is a very costly, tedious, and long process with a lot of legal complexities, where a major portion of the population has very little to no knowledge about the whole process [4]. According to a article published in the United Nations Development Program, there are at least 17 major factors that ensure a standard and happy life and one of them was to ensure peace, justice, and harmony in relationship institutions [5]. This article aims at providing a platform that helps couples find out the underlying issues that can lead to separation and divorce with the help of a questionnaire that has been created on their daily experiences.

As machine learning plays an important part in our research, some previous research papers in the area have also been looked upon. Decision trees, radial basis, and artificial neural networks are some of the primary algorithms used on a divorce dataset which results in 97.64, 97.64, and 98.82% accuracy when they were applied on various features using correlation-based methods [3]. Another study that looked at the mental health and suicidal area of the topic used algorithms like linear regression, decision trees, and support vector machines that achieved accuracy levels of 80.2, 79.3, and 82.7% [6]. With the use of convolutional neural networks and long short-term memory, recurrent neural networks achieved an accuracy of 95% [6].

3 Proposed Approach

To solve a particular problem, various paths were selected. The best approach for the problem helps to achieve the goal easily and more efficiently. As our research focuses on divorce prediction, further divide the whole task into small sub-tasks, i.e., data collection from various sources (Kaggle, real-time data via Google Forms), data cleaning, and model creation, and apply various machine learning algorithms to achieve our goal. The whole approach is represented by a flowchart (Fig. 2).

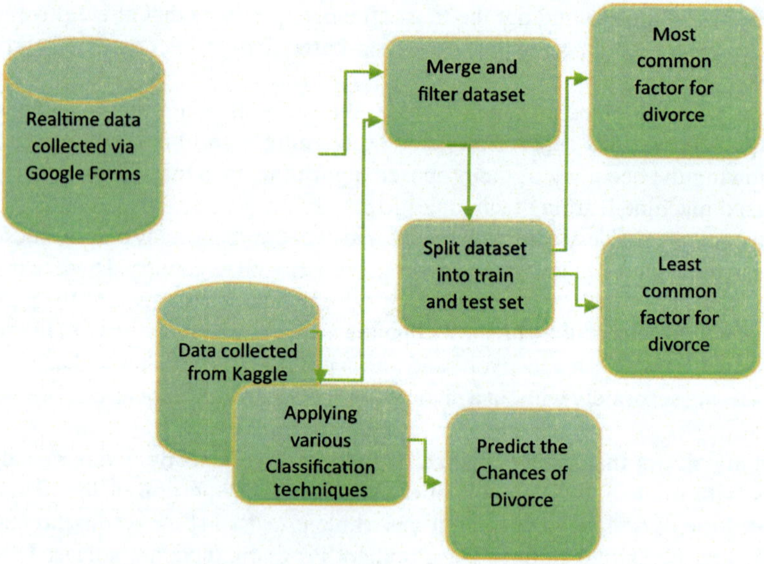

Fig. 2 Workflow diagram of the research

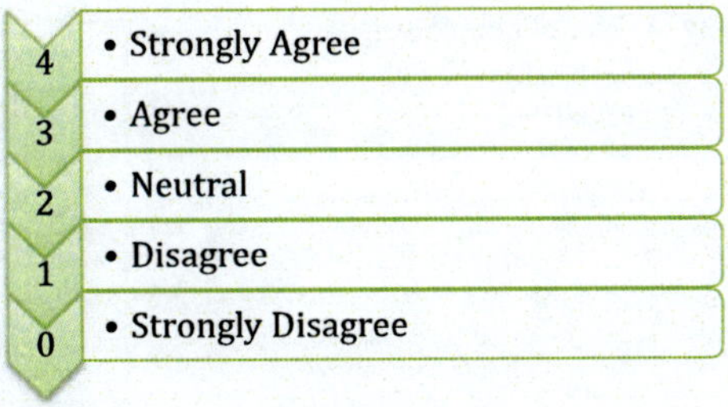

Fig. 3 Response level of the question

3.1 Data Collection

To analyze any task, the first step is to collect data from various sources. The greater the diversity, the greater will be its analysis. It is the best approach to get a more filtered result.

The data are collected from various files from the Internet and used a Google Forms to collect more recent/real-time data. As the data are collected from various sites, it is then preprocessed with the help of word tokenization and keywords which aids to bring the data in a utilitarian form.

The information gathered includes questions about marital status, age, family structure, type of marriage, happiness in marriage, fears in marriage, number of children, understanding, comfort with each other, etc. A total of 54 questions are used as attributes in the dataset. Each question is answered based on the 5 different levels which are shown in Fig. 3.

As illustrated above, "4" represents that the person strongly agrees to the question, "3" represents that the person agrees to the question, "2" represents that person is neutral toward the question, "1" represents that person disagrees with the question, and "0" represents that the person strongly disagrees with the question. Based on the response of the individual different attributes/questions, data are assembled in a single file which is further processed to achieve the desired goal.

The collected dataset measures (361,47), i.e., 361 rows and 47 columns.

3.2 Data Preprocessing

The dataset contains 54 attributes that represent 54 different questions. The dataset contains attributes separated by a semicolon. As the data were separated by a semicolon, a delimiter is used to separate the data into the row–column [7, 8]. The distribution of the result is shown in Fig. 3 (Fig. 4).

The distribution of the result is represented by 0 and 1, where 1 represents "couple gets divorced" and 0 represents "couple does not get divorced."

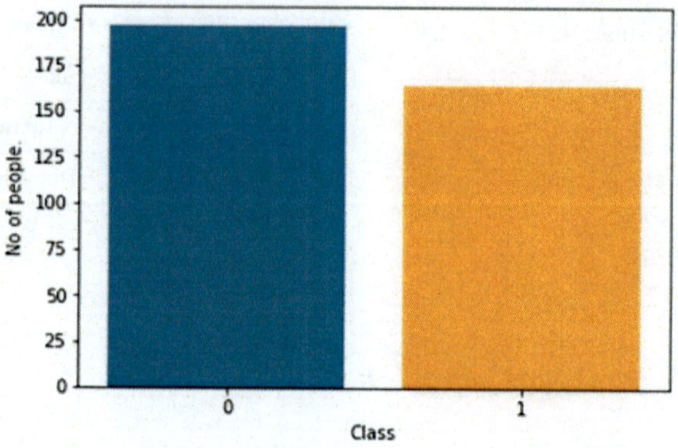

Fig. 4 No. of people versus class (0—no divorced and 1—divorced)

A total of 54 attributes were considered, and some of them did not affect the result much. Therefore, it is decided to remove some columns with the help of a correlational matrix as a way to clean the data.

In 1999, Hall proposed a system known as the correlation-based coefficient feature selection method that gives a significance value of every attribute used in the study. This is performed using a searching algorithm that measures the significance value of featured attributes [9]. This approach also measures the values of feature subsets and considers the internal correlation values. Besides, success exists in estimating the class label of each feature [10].

This method used upper triangle elimination in which the values can be removed from the lower side of the triangle and marked them as not a number (Nan), and the missing values make up the upper triangle in which columns can be checked with high correlation.

Finally, remove seven columns, and now our dataset measures at (361,47), i.e., 361 rows and 47 columns.

3.3 Model Creation

In model creation, split the data randomly into two parts, viz. training and testing. To perform testing, choose the test size of 0.40 and thus a training size of 0.60.

There are two sets of variables available with one being, independent variable consists of 53 different attributes, and the other one dependent variable contains the decision of a person that is either 0 (not get divorced) or 1 (get divorced). On the training set, various machine learning (ML) classification algorithms are applied to train the machine. And the data are tested to provide the various algorithms to predict the person's decision.

3.4 Applied Algorithms

A total of 6 different classification techniques were used for our research purpose such as logistics regression, K-nearest neighbors, SVM [11], kernel SVM, decision tree, and random forest [12]. Confusion matrices for all of the above algorithms had also been calculated and constructed. Based on the different accuracies, the analysis for different algorithms was performed.

The various algorithms used for the research purpose are stated below.

3.4.1 Logistic Regression

This model uses the logistic function to predict binary (0 or 1)-dependent variables. A dependent variable containing various attributes is passed as a parameter to the logistic regression model, which provides a person's final decision after applying a logistic function.

3.4.2 KNN

In this model, group of attributes having the same characteristics are formed based on the distance between their points. Questions having the same characteristics are much closer to each other in the graph as compared to the other.

3.4.3 Support Vector Machine

The kernel is a technique that is used by SVM to transform the data. It also finds the optimal boundaries between the possible output values based on the transform data.

3.4.4 Decision Tree

In this type of classification technique, a treelike structure is formed for each node of the tree which represents an attribute, i.e., divorce-related question in our case, and each branch represents a decision. The outcome is represented via leaf node, i.e., 0 (not get divorced) and 1 (get divorced) in our case.

3.4.5 Random Forest

It is a supervised model in which different decision trees on the same dataset are generated and later get merged to produce an output.

4 Experimental Setup and Result Analysis

4.1 Experimental Setup

Python has been used as the primary language for implementing the divorce prediction scale mentioned in this paper. Python has now become the most used language in the field of ML, artificial intelligence, and big data analytics because of not only its vast collection of libraries but also the design in ever-growing community. Its properties like better code readability that help in incorporating complex math and rapid deployment through various environments give it an edge over other languages [7].

Anaconda's Jupyter Notebook has been used as the primary IDE for implementing the various algorithms, as it provides an environment for writing down live codes, adding mathematical equations to it, and also provides handy visualizations that help in presenting [8].

In the real-time data, some more dynamics are to be present in the dataset, and a Google Forms was created that would serve to store survey data in filling the entry format [13]. Google Forms can also convert the data from file entry format to spreadsheet format, where the new data were appended into the dataset with little data preprocessing.

4.2 Result Analysis

To predict the result, valid classification techniques have been implemented on the test dataset. For the best algorithm for our dataset, there are 6 different classification algorithms implemented. The accuracy percentage of the various algorithms that have been used is shown in Table.1.

After analyzing the various classification algorithms, it can be concluded that among the six algorithms four of them give the same accuracy percentage, where the four algorithms such as logistic regression, support vector machine, decision tree, and random forest have a common percentage of 98.62% which is the highest, and

Table 1 Accuracy percentage of the various classification models

S. no	Algorithms used	Accuracy (in %)
1	Logistic regression	98.62
2	K-nearest neighbor	97.93
3	SVM	98.62
4	Kernel SVM	97.93
5	Decision tree	98.62
6	Random forest	98.62

Table 2 Most effective attribute, with their values of significance by applying correlation-based feature selection methods on the dataset

Attributes	Significance value
Attribute 2	0.94
Attribute 6	0.91
Attribute 11	0.89
Attribute 18	0.87
Attribute 26	0.86
Attribute 43	0.84
Attribute 53	0.84

the other two K-nearest neighbors and kernel SVM have an accuracy of 97.93% [14]. The result metrics of the best algorithm are given below.

	0	1
0	75	0
1	2	68

For about 75 non-divorced sets 75 non-divorced sets are obtained, whereas for 70 divorced sets 2 non-divorced sets are obtained as a result which is incorrect while 68 divorced which is right. It can be concluded that out of 145 sets 143 correct answers are obtained and 2 wrong answers. After analyzing the result, the logistic regression, SVM, decision tree, and random forest are the best choices for the divorce dataset.

To determine the most significant features, correlation-based feature selection along with classification techniques is also applied. The effective attributes with their significance value are presented in Table 2.

Here, out of 53 features 7 are effective to be considered. According to the significance value, it is seen that attribute 2 has the highest significance value that affects divorce "To ignore our differences when things get hard sometimes." Other attributes are attribute 6 "In the future when looked back, found peace in our life"; attribute 11 "In the future when looked back, found peace in our life"; attribute 18 "Both with similar ideas about marriage"; attribute 26 "Knowing about each other's basic problems and anxieties"; attribute 43 "One of them would stay silent, to calm the environment"; and attribute 53 "During discussions, reminded that each other of their inadequacies."

5 Discussion

Various classification methods and algorithms were applied to the divorce dataset. The highest success rate of 98.62% was of four algorithms, viz. support vector machine, logistic regression, decision tree, and random forest. After selecting the most effective attributes with the help of the correlation-based feature selection method, the highest accuracy rate was obtained by attribute 2 which had a significance value of 0.94 [15]. Therefore, according to the research of this study, a divorce predictor scale has been developed and implemented as a desktop software application. This DPS can predict divorce. It not only is helpful for the couples who are going through such a phase and facing difficult conditions but can also help various ministries like family and social policies, education, and health as they can use the DPS as a way to screen the couples. Also, various societies and services that provide counseling services can use the DPS as an effective tool. In future studies, our model for the DPS can be used as a base where a platform is already present and you can just add various algorithms, new attributes, and more real-time and dynamic data to get better predictions for divorce cases [16].

6 Conclusion

In this research, a model is proposed for divorce predictor scale in which it brings together the contemporary classification techniques and novel application design. The article focuses at the various elements that are instrumental in deciding the health of an adult marital relationship and analyze and scrutinize them through various machine learning models and algorithms. The aforementioned elements are all part of a questionnaire and are 54 in total, thus covering various aspects and strands of a marriage. The data for the questionnaire have been taken from various datasets around the net, and to add more dynamic and real-time data Google Forms have been used to capture local responses. The impact and importance of such attributes rank them in order of their effectiveness and significance. The divorce predictor scale has been developed on these attributes and predicts in the form of quiz structure before couples where they have to choose between the various levels of severity of these elements, and they face in their day-to-day lives ranging from 0 to 4 or from "strongly disagree" to "strongly agree." The aim is to create awareness among couples for the underlying causes that might result in a future separation. A high level of accuracy has been achieved in predicting past divorces using efficient algorithms and continuing on its success we had used it to train the Divorce Predictor Scale to forecast the future possibility of a separation and most affecting factors.

As every research has its limitations, our research has its fair to deal with. The sample size is small and collected from various resources across the net and also used Google Forms to collect real-time data to get a crisp image for this investigation. This study is to understand the current divorce scenario in India, which has half a billion married people coming from a plethora of backgrounds but has only studied 200 married couples and can be resolved with a wider dataset. The pool of attributes of divorce factors is the heart of the study. Adding more dynamic and current factors with the help of marriage counselors and psychologists will promote more zest to the study. The model has its limitations and can create a stronger model based on deep learning with the help of TensorFlow.

As mutual separations are more popular in India instead of legal divorce, it is likely to surge the area of research toward such separations. Though a variety of deciding factors is incorporated in the study, more factors can be added based on the various backdrops of Indian society like religion, caste, creed, city, age group, income group, family structure, job structure, etc. As only a handful of machine learning algorithms are deployed in the research, a more powerful AI engine would benefit the study in the future. It also focuses only on identifying cases and further increases its field by providing solutions and remedies. As young adults and teenage dating and relationships are increased, further extend the study to analyze and prediction of breakups and health of a relationship to connect them with the right guidance to live a lifelong mark on their personality. The team has successfully created a desktop application based on the model, and a live interacting android application and Web site can be created in the future.

References

1. C. Chandramouli, R. General, *Census of India 2011. Provisional Population Totals* (Government of India, New Delhi, 2011), pp. 409–413
2. A. Prabhu, Improved divorce prediction using machine learning-particle swarm optimization (PSO), in *2020 International Conference for Emerging Technology (INCET)* (IEEE, 2020), pp. 1–5
3. M.K. Yöntem, K. Adem, T. İlhan, S. Kılıçarslan, Divorce prediction using correlation based feature selection and artificial neural networks. Nevşehir Hacı Bektaş Veli Üniversitesi SBE Dergisi **9**(1), 259–273 (2019)
4. S. Goel, S. Roshan, R. Tyagi, S. Agarwal, Augur justice: a supervised machine learning technique to predict outcomes of divorce court cases, in *2019 Fifth International Conference on Image Information Processing (ICIIP)* (IEEE, 2019), pp. 280–285
5. Callithen, N., Matthew, N., *United Nations Development Programme (UNDP)* (2007)
6. S. Sohail, S. Aziz, F. Tahir, S. Haqqul, A. Hussain, Implementation of machine learning algorithm on factors affecting divorce rate, in *2018 International Conference on Engineering and Emerging Technologies (ICEET)* (IEEE, 2018), pp. 1–5
7. Python for Machine Learning. https://www.monterail.com/blog/machine-learning-python
8. Anaconda and Jupyter Notebook. https://docs.anaconda.com/ae-notebooks/4.2.2/user-guide/basic-tasks/apps/jupyter/.
9. M.A. Hall, Correlation-based feature selection for machine learning (1999)
10. M.A. Hall, Correlation-based feature selection of discrete and numeric class machine learning (2000)
11. C. Cortes, V. Vapnik, Support-Vector Networks. Machine Learning **20**(3), 273–297 (1995)
12. L. Breiman, Random Forests. Machine Learning **45**(1), 5–32 (2001)
13. Emery, R.E., *Marriage, Divorce, and Children's Adjustment*, vol. 14 (Sage, 1999)
14. M. Simanjuntak, M. Muljono, G.F. Shidik, A.Z. Fanani, Evaluation of feature selection for improvement backpropagation neural network in divorce predictions, in *2020 International Seminar on Application for Technology of Information and Communication (iSemantic)* (IEEE, 2020), pp. 578–584

15. J. Kong, T. Chai, Is your marriage reliable? divorce analysis with machine learning algorithms, in *Proceedings of the 2020 6th International Conference on Computing and Artificial Intelligence* (2020), pp. 1–4
16. S. Joel, P.W. Eastwick, C.J. Allison, X.B. Arriaga, Z.G. Baker, E. Bar-Kalifa, S. Bergeron, G.E. Birnbaum, R.L. Brock, C.C. Brumbaugh, C.L., Carmichael, Machine learning uncovers the most robust self-report predictors of relationship quality across 43 longitudinal couples studies. Proc. Nat. Acad. Sci. **117**(32), 19061–19071 (2020)

Design a Hybrid Algorithm Using ACO and Best-Fit Optimization Algorithm for Energy Consumption in Cloud Data Center

Rishu Gulati and S. S. Tyagi

Abstract Power consumption has become an essential area of research as the cloud computing technique use increased due to easy usage of the Internet, storage, and processing on the cloud. Even small firms are nowadays initiated to use cloud computing techniques for their users. This led to the development of cloud computing techniques and cloud data centers for the processing which outcomes in a huge number of barriers in surroundings like the emission of carbon. Due to the huge need for cloud services, data centers are working all the time. This enhances the consumption of energy which leads to expensive business. A cloud data center consumes a huge volume of energy resulting in a greater carbon footprint level. The costs of these data centers have developed with the rise in the capacity of computing with a huge percent of operational expenses due to the consumption of energy in cloud data centers. The energy consumption of the data center is a critical problem that could be resolved by consolidating the workload of the server and turning off physical servers. In cloud data centers, various mapping associations between physical machines and virtual machines resulted in varied utilization of resource therefore placing virtual machines (VM) on physical machines to develop utilization of resource and reduce consumption of energy which is one of the main issues for providers of the cloud. The research attempts to propose a hybrid algorithm using ant colony optimization (ACO) and the best-fit optimization algorithm for energy consumption in the cloud data center. The best-fit optimization algorithm generally places virtual machines and is particularly known to provide a close optimal solution in a reasonable interval of time by sorting the virtual machines in increasing and decreasing order. It was found from the current research that the ACO and best-fit optimization algorithms perform better than existing algorithms in terms of power consumption and usage of resources.

R. Gulati (✉) · S. S. Tyagi
Department of CSE, Manav Rachna International University Faridabad, Faridabad, India

S. S. Tyagi
e-mail: shyam.fet@mriu.edu.in

V. Suma et al. (eds.), *Inventive Systems and Control*,
Lecture Notes in Networks and Systems 204,
https://doi.org/10.1007/978-981-16-1395-1_58

Keywords Energy consumption · Ant colony optimization algorithm · Best-fit optimization algorithm · Cloud data center

1 Introduction

The evolution of online techniques has given emergence to the paradigm of cloud computing a model that enhances cost-effective, scalable, and flexible allocation of the resource while providing ease of use to end customers. The developing demand for services of the cloud has led to rapid development in the number of resources of computing and the number of big scales data centers of computing. Data centers consist of 1000 s of units of the server that occupies a huge volume of space and consume a considerable amount of energy. In data centers, the consumption of energy is expected to rise dramatically [1]. Cloud computing has already become one of the significant methods in resolving the services of information technology by decreasing the cost for customers. The method does not only likely to affect technologies used in computing but consecutively processes a technique employed for building and handling information technology within businesses and service providers [2]. By providing a secure paradigm of computing, cloud computing is becoming an essential platform for scientific applications. Cloud computing develops the physical resource utilization rate [3]. It enhances an individual physical platform, i.e., which hosts an application, with enhanced system surroundings and services of application through standardized allocation and integration of physical resources through the virtualization techniques [4]. The requirements of computing have been expanding with the era of upcoming big data. A huge set of future-generation data centers will employ cloud computing in addition to the virtualization technique. With the expansion of cloud data centers size, greater consumption of energy of data centers becomes a serious issue [5]. The major cause of inefficiency in energy in data centers is the wasted idle energy when resources are not used. In the issue of low resource use, servers are switched on permanently even if they are not used and still consume a huge amount of energy. To resolve these issues, it is essential to remove the energy waste, enhance efficiency, and to alter the way resources are employed [6]. The main issue behind concentrating on the consumption of energy is due to the increase in the cost of energy invested in the data center. The server machine is considered a necessary tool for the increase in costs relating to electrical requirements. It is on account of the direct consumption of energy and cooling processes or tools employed in it. Nearly 75% of the cost of energy is because of the direct energy consumed by server machines and the consumption of indirect energy for cooling equipment [7]. Reducing the data center's energy usage is a complex and challenging problem because the applications of computing and information are developing rapidly so that enhancing huge servers and disks are required to process them rapidly enough within the needed period of time. This is important for assuring that the future cloud computing growth is sustainable [8].

ACO is one of the nature-inspired algorithms and is applied technically based on movements of the ant population. It transcribes the ant colony system cooperative behavior for resolving a specific problem of optimization issues. When ants migrate in food search the ants usually secrete a chemical substance known as a pheromone. For ACO, the role of the pheromone pathway is to impart their experience in terms of the journey for resolving the issue of optimization effectively [9]. ACO algorithm is a meta-heuristic technique to predict new optimal solutions by probabilistic techniques, which can be employed for problems belonging to the NP class. The ants generally release the pheromone on their return to make other ants aware of the food. A disappearance process of natural pheromone is used to reduce the pheromone volume to retain used paths regularly that lead to good solutions [10]. The best-fit algorithm sorts the requests of the virtual machine in descending order and allots the biggest virtual machine first to active physical machine that provides smaller residue. It decreases the number of active physical machines by pulling inactive physical machines temporarily to rest mode. It attempts to fit the request of a virtual machine with active physical machines [11]. The best-fit algorithm sorts the virtual machine according to their central processing unit usage and then for every virtual machine it verifies entire hosts and predicts applicable host wherein the increase of energy consumption is minimum to a certain extent. It also allows the virtual machine to that host [12]. The proposed hybrid algorithm provides the virtual machine placement solution which decreases the wastage of total resource and consumption of energy in servers and offers fault tolerance through load balance in servers. The current study attempts to propose a hybrid algorithm using ant colony optimization and best-fit optimization algorithm for energy consumption in cloud data center based on a meta-heuristic approach.

The best-fit algorithm has been exercised to load balance the tasks over virtual machines (VMs). This method aims to guarantee modest physical machines' (PM) utilization. The best-fit algorithm is a heuristic that is employed to standardize the bin packing problem. On the other hand, ACO is a meta-heuristic technique instigated by real ant gatherings and their communal foraging actions for food sources. A study by Hussein et al. [13] put forward that their evaluation exhibited that the ACO–best-fit algorithm is essentially performed better with regards to virtual and physical machine utilization and in reducing the amount of active physical machines and instantiated virtual machines.

2 Literature Review

Tawfeek et al. [14] have stated that cloud computing is a technology wherein user employs remote server to manage their application and data. The cloud computing sources are considered demand-driven, which is used in virtual machine forms to assist the implementation of complicated tasks. Virtual machine placement (VMP) is the technique of representing VM to PM. In this study, the issue of VMP has been developed as a multiobjective optimization issue, which is intending at optimizing total memory and total processing wastage of resources simultaneously. Besides, the ACO algorithm has been developed for resolving the issues. The proposed algorithm intends to browse the space of solution much effectively and acquire non-dominated solutions set known as Pareto set. The proposed technique has been compared with existing algorithms for VMP issues in the literature. The findings of comparison mention that the proposed technique is considered much effective and essentially surpasses the compared methods based on central processing unit and memory wastage of resource. Ranjani and Ganesan [15] have mentioned in their study that conservation of energy has become an essential area of research as the cloud computing technique use increased due to the ease of online usage, storage, and processing on the cloud. Even small firms initiated to employ solutions of the cloud for their customers. This led to the development of cloud computing techniques also to the development of data centers in the cloud for processing which resulted in an essential number of barriers in surroundings like emissions of carbon. Virtualization techniques offer the capability to transfer VM between the PM using live virtual machine migration in cloud computing. Dynamic consolidation of the physical server is a technology to decrease the energy usage in data centers by consolidating and migrating VM into a decreased number of active physical servers so that the energy consumed by idle servers are decreased. Since the issue is NP-hard, ant colony optimization is a multiobjective online meta-heuristic algorithm for consolidation of the online server that is employed. ACO performed better than other existing technologies with regard to energy utilization and underused usage of the resource. The proposed algorithm is simulated by CloudSim and its outcomes in better energy consumption than the existing methods. Sarma [16] proposed research on an optimal ACO technique for effective VM placement. The process of representing VM to PM is referred to as placement of the virtual machine. In the context of cloud computing, optimal placement has a significant part in the growth of resource allocation and efficiency of power. In this research, the multiobjective genetic algorithm and ACO technique are proposed for resolving the issue of virtual machine placement. This research has also designed a system for the proposed research.

Xin [17] has stated in their study that the ACO algorithm is an intelligent bionic algorithm. It is based on the examination of food hunting of ants in nature that forecasts the shortest path between their food source and nest. The greater the absorption of pheromone in the trail the greater opportunities of the path is selected. With positive feedback process and robustness, ant colony optimization algorithm is studied widely and enhanced since it is put forward. ACO reveals a huge benefit in resolving combinatorial optimization issues.

Alharbi et al. [18] have mentioned in their study that the data centers of cloud computing comprise a huge volume of virtual machines and physical machines. This huge volume can expand the data centers energy consumption particularly when virtual machine places improperly on physical machines. This study proposes a virtual machine placement technique to reduce the total consumption of energy of a cloud data center. The issue of virtual machine placement is originated as an integrated optimization issue. And this issue has been considered as an NP-hard issue, the ACO algorithm is used to solve the identified issue. Information heuristic of ant colony optimization is employed variedly based on physical machine energy efficiency.

Karuppasamy et al. [19] stated that the resources of cloud computing faced a huge barrier in terms of consumption of energy. The main purpose of consumption of energy in cloud computing is by ways of client computational devices, network computational devices, server computational devices and energy need to cool the information technology load. The resources of the cloud contribute to the higher cost of operational energy and discharge a large amount of carbon to the surroundings. Therefore, the providers of cloud services demand green resolution to reduce the cost of operational energy in addition to environmental effect. The main purpose of this study is to develop the algorithm to reduce the energy from idle and used resources of the cloud and utilizes the energy in cloud resources efficiently.

Devarasetty and Reddy [20] proposed research on multiobjective ACO techniques for resource distribution in the field of cloud computing. It offers services based on the policy of pay as you use. The increased usage of services might result in the use of a huge number of data centers. Consequently, data centers require greater energy consumption. To develop the data center efficiency, resource management uses the virtualization technique, which can be considered an essential factor. This study focuses on VMP issues and as well suggests the bio-inspired method for decreasing resource depletion and reduces the energy consumption and cost of communication within the data center. ACO algorithm is suggested in this study to acquire the set of solutions for the multiobjective issue. The proposed algorithm performance has been compared with other existing algorithms and it is assumed that the proposed method is effective in terms of resource use, energy utilization, and cost of communication.

It has been stated in the study of Rani and Garg [21] that all virtual machines are hosted in data centers on physical servers so there is a requirement for a huge volume of energy for handling servers. As several physical resources and servers namely storage and power are increased, a huge number of energy consumption at cloud centers of data exist and due to this, the economical cost is also needed. This study concentrated on reducing the total consumption of energy in data centers considering various parameters namely energy consumption of IT equipment and consumption of cooling power in cloud computing. Furthermore, for the same ACO algorithm is used to offer the best representation of a virtual machine to the server by optimizing the energy consumption of the data center. The results of the simulation exhibited that the proposed technique surpasses the algorithm of min-min.

Wang et al. [22] have mentioned in their research that to resolve the issue of waste resources in the virtual machine deployment process in the platform of cloud this research suggests an ACO algorithm based on multiobjective optimization. Initially, the combination model of multiobjective optimization is established. Then the model is applied is the ACO algorithm. Finally, the ACO algorithm is changed to the initialization of VM in the cloud platform. To verify the efficiency of the algorithm, the simulation platform of CloudSim cloud has been employed to undertake the experiments of simulation. The results of the experiment revealed that the suggested algorithm in this research could meet a better level in service-level agreement violation rate, energy consumption, and loss of resources in cloud platform compared with other used algorithms.

Wei et al. [23] have denoted that virtualization has accomplished vast spread adoption in different sectors. As one of the most essential techniques in virtualization of the network, an efficient VM placement can facilitate to development of the use of resources in data center networks and reduce the operation cost of the enterprise. In this research, an energy-efficient VMP scheme has been proposed which is pursued to decrease the cost of communication and energy consumption over traffic-aware networks of the data center. To resolve such an issue of optimization, an improved ACO with the adaptive setting of the parameter was presented to balance its robust capability of search and rapid convergence. Compared with already existing algorithms, the results of simulation mentioned that their scheme accomplished developments in energy consumption and cost of communication had an essential decrease in run time under varied designs and traffic patterns (Table 1).

3 Design of the Proposed System

The proposed system intends to design and implement a hybrid algorithm using ant colony optimization and best-fit optimization algorithm for energy consumption in the cloud data center. The below figure is referred to as the flow diagram of the proposed method (Fig. 1).

From the above figure, it was found that the first step is to initialize virtual machine requests. Later the algorithm resolves the virtual machine requests in increasing, given, and decreasing order based on energy utilization. The hybrid algorithm is applied to virtual machine requests and used in this research for the ant colony optimization and best-fit optimization algorithm that reduces the energy consumption in the cloud data center. The proposed hybrid algorithm essentially reduces the consumption of energy and accomplishes greater utilization rates of the resource. After the hybrid algorithm is applied, the rate of energy utilization is estimated. After the estimation, energy utilization is compared and returned with minimum energy utilization. Ultimately, virtual machines are allocated to physical machines. Thus, the process is repeated for increasing, given, and decreasing order of requests of the virtual machine to physical machines, respectively.

3.1 Hybrid Algorithms

3.1.1 Ant Colony Optimization

This technique is a meta-heuristic process in an ant colony derived from the foraging behaviors of ants. Ants' behavior is generally impacted by the survival aims of the

Table 1 Reviews of ACO algorithms for energy consumptions in cloud computing

S. no.	Author	Year	Technique used	Benefits and limitations of the technique
1	Tawfeek et al.	2014	Ant colony optimization	Resolves large space of solution for large scale data centers. Theoretical analysis is considered challenging and difficult in ACO
2	Ranjani and Ganesan	2015	Ant colony optimization algorithm	Reduces energy consumption in data centers. Only theoretical research is possible and cannot expect experimental research
3	Sarma et al.	2015	ACO and multiobjective genetic algorithm	Reduces the wastage of resources and reduces the energy consumption of servers. Convergence is guaranteed but a protracted process
4	Xin	2016	Ant colony optimization algorithm	Allocate proper resources of computing for users in cloud environment
5	Alharbi et al.	2017	Ant colony optimization	Minimize the consumption of energy for the functioning in data centers everyday
6	Karuppasamy et al.	2017	Hybrid ant colony genetic algorithm	Achieves sustainable cloud surroundings and reduces the energy consumption at the server, network, and client resources. Coding is not simple
7	Devarasetty and Reddy	2018	Multiobjective ant colony optimization algorithm	Reduces minimal consumption of an energy-reduced number of hosts and reduced wastage of resource and reduced cost of communication
8	Rani and Garg	2019	Ant colony optimization	Lessens the consumption of energy of data center

(continued)

Table 1 (continued)

S. no.	Author	Year	Technique used	Benefits and limitations of the technique
9	Wang et al.	2019	Ant colony optimization algorithm based on multiobjective optimization	Resolves the issue of numerous conflicting targets and make violation probability of SLA, waste resource, and energy consumption to meet at a very lower level. Stagnation, exploration rate, and convergence speed of the algorithm are some of the drawbacks of ACO
10	Wei et al.	2019	Ant colony optimization	Reduction in run time under various traffic configurations and patterns. ACO technique is far from perfect on account of various issues such as untimely convergence and protracted search time

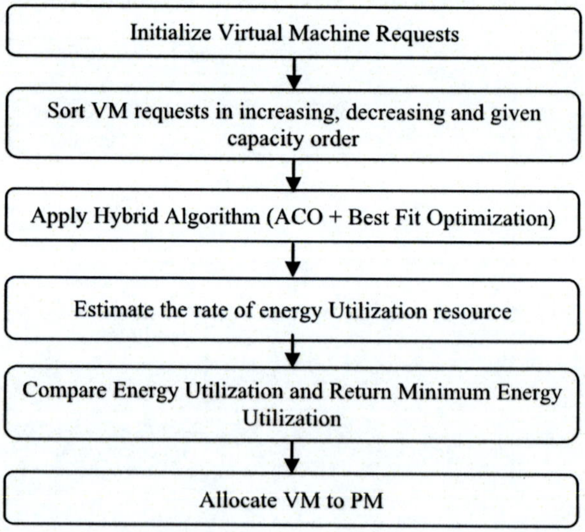

Fig. 1 Flow diagram of the proposed system

group, i.e., between the food source and nest. The below figure shows the process flow diagram of ant colony optimization (Fig. 2).

In the beginning stage, the ants walk around the enclosing region randomly. While passing they secrete the pheromone in their way to distribute data for other ants. The

Fig. 2 Process flow diagram of ACO

absorption of pheromone is a major sign for ants to predict their path. When an ant predicts a source of food, it estimates the quality and quantity of the food. On its return to its place, the ants usually secrete a pheromone amount relevant to the quantity and quality of the food. This secretion directs other insects to the sources of food. The pseudocode for ant colony optimization algorithm is explained below:

Pseudocode for ACO Algorithm:

Input: A set of Physical and Virtual Machines
Output: Reduced set of active Physical Machines
initialize the trails of pheromone
Declare values of threshold for nodes
ants pass through the nodes
if the value of load< threshold
    Node traverse with nearest maximum pheromone trailing
Else
    Node traverse with minimum foraging pheromone trailing
Reassign resources if node is required

The ACO algorithm is used to solve computational issues which are employed to forecast the optimized ways through graphs. ACO are systems that are distributed to have an increased-structured social firm. An ant passes along in search for food in the real world and during its search, it places certain trails of pheromone. If the ant predicts a way that is optimal or good, it places much amount of pheromone and more the amount more will the optimal way. The remaining ants of its colony must follow the path of pheromone and stick to an individual path eventually.

3.1.2 Best-Fit Optimization Algorithm

In the best-fit optimization algorithm, all virtual machines are sorted in decreasing order of present use, and every virtual machine is allocated to a host that offers the minimal amount of energy utilization on account of this allocation. This permits influence the heterogeneity of the nodes by selecting the most effective ones. The complexity of the algorithm's allocation part is $v - h$ where v indicates the number of virtual machines that have to be allocated and h is the number of hosts. The best-fit algorithm is known best for packing bins in offline mode. In the best-fit algorithm, the physical machines are considered to be bins and the virtual machines are regarded as objects. It also resolves the entire virtual machines by their utilization weights in non-increasing order. Then it initiates with virtual machines that require a huge number of resources for request processing. Best fit can be implemented in zero time. The pseudocode for the best-fit optimization algorithm is explained below:

Pseudocode for Best Fit Optimization Algorithm:
Sort the objects in non-increasing order
For all objects x = 1, 2, …………., n do
For all bins y = 1, 2, …………….do
If x object fits in y bin then
Calculate the remaining capacity after the object is added
End if
End for
Pack x object in y bin where y is the bin with minimum remaining capacity after object addition
If no such bin occurs open a new one and the object is added

The optimization of the present distribution of virtual machines has been undertaken in two various steps: (1) the first step is to choose virtual machines that require to be migrated, and (2) the second stage is to choose virtual machines to be placed on the host using best-fit optimization algorithm. The selection of host is based on the first-fit strategy of physical machines in non-decreasing order of central processing unit capacity and first-fit heuristics is employed for selection of virtual machine in best-fit optimization algorithm. The study chooses to use hybrid algorithms because the best-fit optimization algorithm is known to provide a near-optimal solution in a moderate time by categorizing the virtual machines in decreasing order. The study dealt with energy-efficient resource allocation in cloud data centers by using a best-fit optimization algorithm. The reason for using ACO is that it comes out more precisely in our proposed method is the probability to manage distributed environments.

4 Results and Discussion

4.1 Discussion

In this research, a hybrid algorithm is designed and implemented for energy utilization in cloud data centers. Since these services have increased hugely in the world of IT, the consumption of energy for these services of cloud have developed much. When the server is performing the operation at reduced utilization, the energy wastage of the server's total consumption of energy is greater than 70% of the peak. To resolve the issue of greater consumption of energy and underutilization of servers, it is essential to remove the inefficiency in the way resources are used to serve the workload application. Effective use of resources leads to greater utilization of servers and overall consumption of reduced energy while providing needed quality of service for users. A cloud data center has several physical machines where every physical machine has varied memory capacities and central processing. There are several requests for the virtual machine to be deployed on feasible physical machines. The virtual machines have needed central processing units and memory, varied execution, and arrival times. The proposed hybrid algorithm will allocate the virtual machine request on hosting physical machines for varying intervals of time with satisfaction of capacities of the resource. The hybrid algorithm used in this research is ant colony optimization and best-fit optimization algorithm. The hybrid algorithm effectively reduces the consumption of energy.

4.2 Results

The implementation screenshot of the proposed research is depicted below (Fig. 3).

The graphs depicted below represent the shortest path made by ACO between time and power consumption. Each graph corresponds to each simulation output. The below figure shows the result of power consumption wit 20–50 W at 0.5–1-time intervals (Fig. 4).

The nest figure shows the power consumption with 20–100 W at a time interval of 0.5–2 (Fig. 5).

The next figure shows the power consumption with 20–140 W at a time interval of 1–3 (Fig. 6).

The next figure shows the power consumption with 50–200 watts at 1–4-time intervals (Fig. 7).

4.2.1 Inference

From the above figure at the 0-time interval, the execution time of two virtual machines is 100,000.0. Similarly, at a 1-time interval, the execution time of two

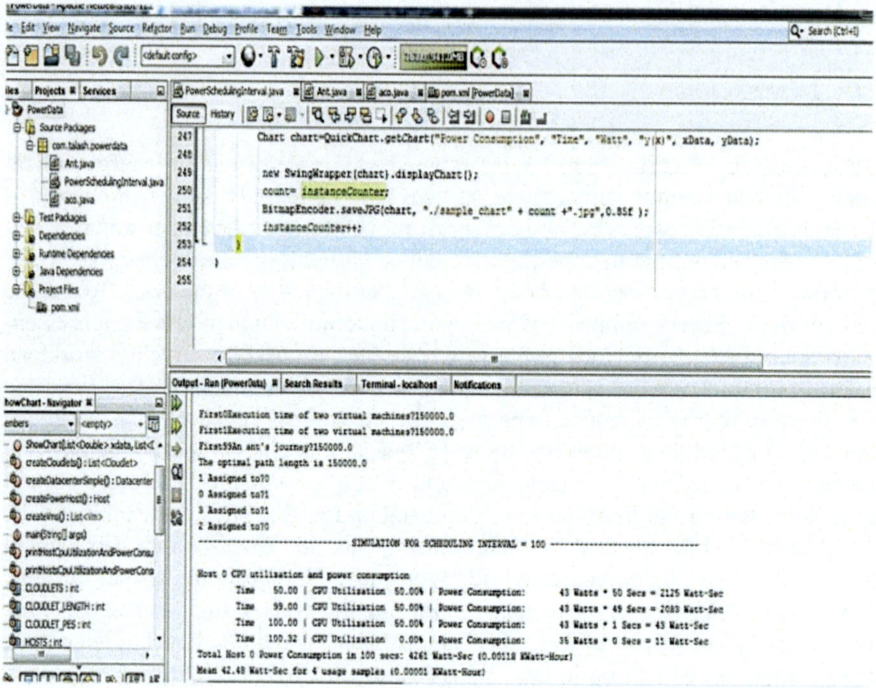

Fig. 3 Screenshot of implementation

Fig. 4 Power consumption at 0.4–1 time interval

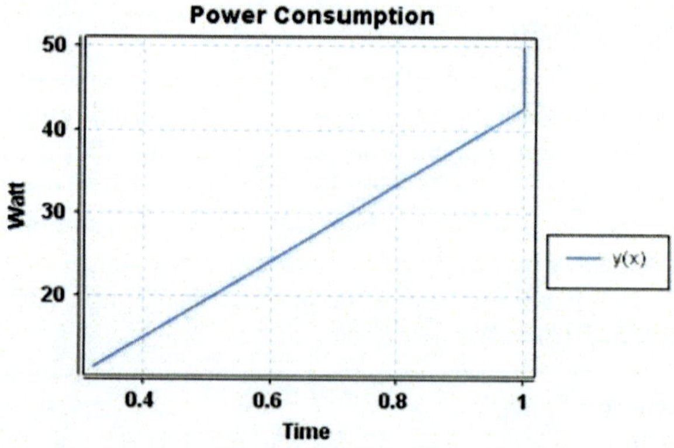

virtual machines is 200,000.0. Finally, at the 99-time interval, the journey of ant is 200,000.0. The optimal path length between the execution time is 150,000.0. It can be inferred that the power consumption varies at varied time intervals.

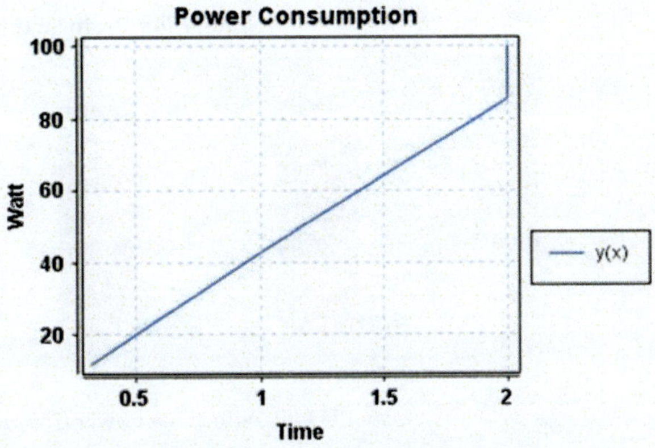

Fig. 5 Power consumption at 0.5–2 time interval

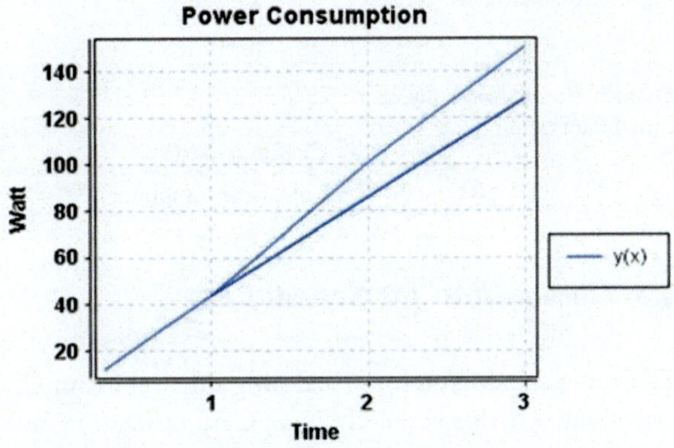

Fig. 6 Power consumption at 1–3-time interval

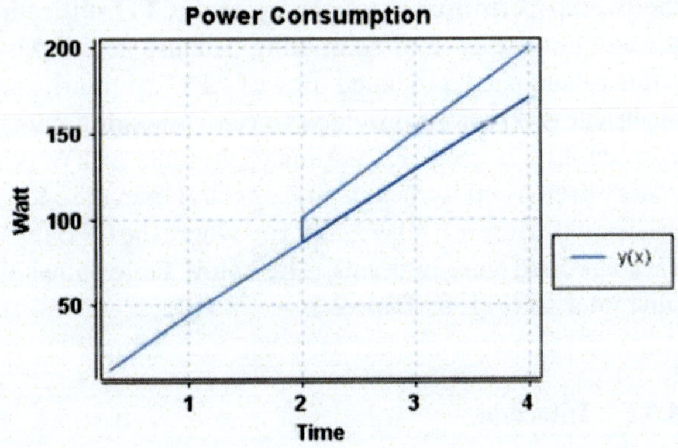

Fig. 7 Power consumption at 1–4-time interval

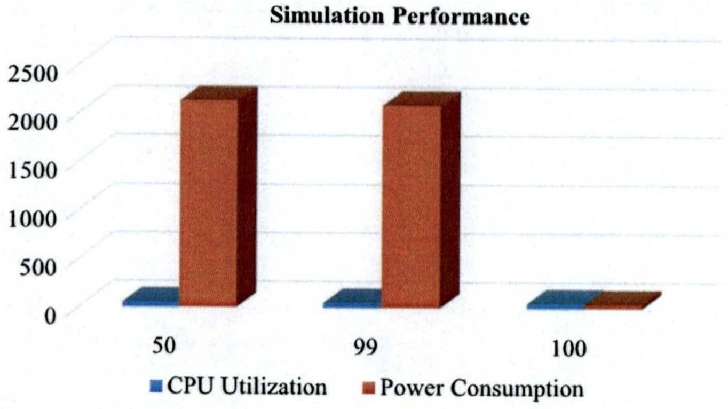

Fig. 8 Simulation performance

Table 2 Simulation table at varied interval time

Time	50	99	100
CPU utilization	50	50	50
Power consumption	2125	2083	43

4.3 Simulation for Scheduling

The value considered for scheduling interval in simulation is 100. The performance is evaluated in this research using CPU utilization and power consumption. Central processing unit utilization is the addition of work managed by CPU. It evaluates the overall performance of the system. CPU utilization could differ according to the amount and type of computing because certain tasks need a huge time of CPU while others need a reduced time of CPU. Similarly, power consumption defines as electrical energy/unit provided to run something namely a home appliance. Power consumption is generally estimated in units of kW (kilowatts) and W (watts). The energy used by equipment is often greater than the energy required. The simulation is carried with time $t = 50, 99$, and 100 where the CPU utilization and time consumption vary at varied time intervals. The below table shows the simulation table at varied time interval (Fig. 8; Table 2).

4.3.1 Inference

It can be inferred from the above figure that the CPU utilization and power consumption are greater at time interval 50. The power consumption in 100 s is 4261 watt-seconds whereas the mean is 42.48 watt-seconds of four usage samples.

5 Conclusion and Future Work

Cloud computing is becoming essential in the information technology sector due to the huge benefits provided to users. Cloud computing offers services to users and provides services based on ITC over the Internet and virtualization use permits it to offer resources of computing. Data centers are the hub of cloud computing that stores business data and host the running applications. Greater performance has been the major concern for all data centers. This concern has been handled without considering energy performance and consumption. The barrier is to balance between system performance and energy consumption. With greater demands of users, cloud surrounding poses huge resources of information and communication technology. Cloud technology usage has developed widely which results in huge consumption of energy in cloud data centers. To this energy consumption of cloud, surroundings have become a problem due to economical and ecological reasons. In this research, a hybrid approach has been designed and developed based on ant colony optimization and best-fit optimization algorithm for energy utilization in the cloud data center. In the current study, the ACO algorithm is integrated with the best-fit optimization algorithm in a way to optimize and consume a huge amount of energy. The ant colony optimization algorithm is a meta-heuristic method since it is inspired by real ants' behavior. It is a distinct algorithm for several reasons like the optimal solution is constructed not by an individual entity but by different entities that traverse the breadth and length in entire network dimensions and then these separately construct a solution. The best-fit optimization algorithm is used to reduce the consumption of energy without compromising the user-specified quality of service needs. When there is an inconsistency and peak, load needs of using certain virtual machines failed to get essential resources to execute the resources of the user which reduce the quality of service. In this study, a hybrid algorithm has been developed for dealing with better consumption of energy. Ant colony optimization and best-fit optimization algorithm not only reduce the energy consumed by physical machines in data centers. It also results in reducing the number of migrations of the virtual machine to enhance the resource using the physical machines in data centers. It can also be concluded that the energy consumed by physical machines is reduced allowing a way to lessen the energy utilization in cloud data centers. The future work of the proposed research is to use other algorithms for saving energy relevant to virtual machine consolidation and to compare their results of performance.

References

1. S.K. Tesfatsion *Energy-Efficient Cloud Computing: Autonomic Resource Provisioning for Datacenters* (Doctoral dissertation, Umeå University, 2018)
2. N. Moganarangan, R.G. Babukarthik, S. Bhuvaneswari, M.S. Basha, P. Dhavachelvan, A novel algorithm for reducing energy-consumption in cloud computing environment: web service computing approach. J. King Saud Univ.-Comput. Inf. Sci. **28**(1), 55–67 (2016)

3. M. Rajeswari, G. Sambasivam, N. Balaji, M.S. Basha, T. Vengattaraman, P. Dhavachelvan, Appraisal and analysis on various web service composition approaches based on QoS factors. J. King Saud Univ.-Comput. Inf. Sci. **26**(1), 143–152 (2014)
4. Z. Liumei, W. Yichuan, Z. Lei, J. Wenjiang, Towards energy efficient cloud: an optimized ant colony model for virtual machine placement. J. Commun. Inf Networks **1**, 116–132 (2016)
5. S. Pang, W. Zhang, T. Ma, Q. Gao, Ant colony optimization algorithm to dynamic energy management in cloud data center. Math. Prob. Eng. (2017)
6. C. Ghribi Energy efficient resource allocation in cloud computing environments. TelecomSud-Paris (2014)
7. N. Kerr, A. Gouldson, J. Barrett, The rationale for energy efficiency policy: assessing the recognition of the multiple benefits of energy efficiency retrofit policy. Energy Policy **106**, 212–221 (2017)
8. B.L. Sreenivasa, K.S. Ankitha, L.R. Abhishek, S. Sathyanarayanan, A study on load balancing and energy efficiency algorithm for managing the resources across host machine in cloud. Int. J. Latest Trends Eng. Technol. 530–533 (2018)
9. Y. Moon, H. Yu, J.M. Gil, J. Lim, A slave ants based ant colony optimization algorithm for task scheduling in cloud computing environments. Human-centric Comput. Inf. Sci. **7**(1), 28 (2017)
10. N.J. Kansal, I. Chana, Energy-aware virtual machine migration for cloud computing—a firefly optimization approach, J. Grid Comput. (2017) (Springer)
11. S. Jangiti, V. Vijaykumar, V. Subramaniyaswamy, Hybrid best fit heuristic for energy efficient visual machine placement in cloud data centres. EAI Endorsed Trans. Energy Web (2020)
12. M.R. Chowdhury, M.R. Mahmud, R.M. Rahman, Implementation and performance analysis of various VM placement strategies in CloudSim. J. Cloud Comput. **4**(1), 20 (2015)
13. M.K. Hussein, M.H. Mousal, M.A. Alqarni, A placement architecture for a container as a service (CaaS) in a cloud environment. J. Cloud Comput. Adv. Syst. Appl. **8**(7), 1–15 (2011)
14. M.A. Tawfeek, A.B. El-Sisi, A.E. Keshk, F.A. Torkey, Virtual machine placement based on ant colony optimization for minimizing resource wastage, *International Conference on Advanced Machine Learning Technologies and Applications* (Springer, Cham, 2014), pp. 153–164
15. C. Ranjani, H. Ganesan, Ant colony optimization for energy efficient vm consolidation—a step towards green cloud computing. J. Appl. Sci. Res. **11**(23), 241–246 (2015)
16. V.A.K. Sarma, R. Rajendra, P. Dheepan, K.S. Kumar, An optimal ant colony algorithm for efficient VM placement. Ind. J. Sci. Technol. **8**(S2), 156–159 (2015)
17. G. Xin, Ant colony optimization computing resource allocation algorithm based on cloud computing environment, in *International Conference on Education, Management, Computer and Society* (Atlantis Press, 2016)
18. F. Alharbi, Y.C. Tian, M. Tang, M.H. Ferdaus, Profile-based ant colony optimization for energy-efficient virtual machine placement, in *International Conference on Neural Information Processing* (Springer, 2017), pp. 863–871
19. M. Karuppasamy, S. Suprakash, S.P. Balakannan, Energy efficient utilization of cloud resources using hybrid ant colony genetic algorithm for a sustainable green cloud environment, in *2017 International Conference on Intelligent Computing and Control (I2C2)* (2017), pp. 1–5
20. P. Devarasetty, C.S. Reddy, Multi objective ant colony optimization algorithm for resource allocation in cloud computing. Int. J. Innovat. Technol. Explor. Eng. **8**(2S2), 68–72 (2018)
21. S. Rani, R. Garg Power aware scheduling in cloud using ant colony optimization, in *Proceedings of International Conference on Computational Intelligence and IoT* (Elsevier, 2019), pp 182–186
22. Z. Wang, J. Du, Q. Miao, R. Cheng, Research of ant colony algorithm based on multi-objective optimization in cloud platform virtual machine initialization. J. Phys. Conf. Ser. **1213**(3), 032005 (2019). IOP Publishing
23. W. Wei, H. Gu, W. Lu, T. Zhou, X. Liu, Energy efficient virtual machine placement with an improved ant colony optimization over data center networks. IEEE Access **7**, 60617–60625 (2019)

FPGA Implementation of Low Latency and Highly Accurate Median Filter Architecture for Image Processing Applications

M. Selvaganesh, E. Esakki Vigneswaran, and V. Vaishnavi

Abstract The proposed work represents the hardware design and implementation of a median filter that uses architecture, which produces median results hierarchically once data in the sliding window is completely available. The median filter was designed based on the sorting methods; here, it makes use of a non-conventional algorithm where entire sorting of data is not required. The proposed three-value sorter where the parallel comparison of data is done is the fastest method as compared to other methods. The work is implemented in Xilinx Vivado 2016.4 software, and the performance analysis is carried out using Cadence 90 nm technology and with the supply voltage of 0.9 V. Area, power, and delay parameters of the implemented architecture for different filter sizes are computed. The hardware implementation of the median filter architecture is done by using Nexys4 DDR FPGA kit making use of block ROM, block RAM, and clocking wizard. Here, the entire operation is carried out at 25.132 MHz frequency.

Keywords Median filter · Non-conventional algorithm · Field-programmable gate array (FPGA) · Block ROM · Block RAM · Xilinx Vivado 2016.4 · Nexys4 DDR · Cadence 90 nm technology

1 Introduction

The median filter is the most commonly used spatial filter for the effective removal of noise in image and digital signal processing applications. It helps in preserving useful information of data that may be getting affected during transmission or faulty

---

M. Selvaganesh (✉) · E. E. Vigneswaran · V. Vaishnavi
Sri Ramakrishna Engineering College, Coimbatore, India
e-mail: selvaganesh.m@srec.ac.in

E. E. Vigneswaran
e-mail: esakkivignesh@srec.ac.in

V. Vaishnavi
e-mail: vaishnavi.v@srec.ac.in

V. Suma et al. (eds.), *Inventive Systems and Control*,
Lecture Notes in Networks and Systems 204,
https://doi.org/10.1007/978-981-16-1395-1_59

receiver. In applications of pulse estimation and edge detection, image acquisition median filter plays a vital role.

The reconfigurable gate array logic circuitry, FPGA, dominates in the hardware implementation of software modules. It provides endless flexibility for the designers due to its reconfigurability. It plays a crucial role in system-on-chip (SoC)-based designs in real-time applications. The hardware implementation of the median filter is useful for real-time applications for impulse noise removal due to capturing and faulty machines. Two main categories are available for processing the data, namely word-level processing and bit-level processing. The proposed technique makes use of the word-level technique, where the entire data is considered to produce the final result. Conventional median filter involves the sorting of entire data, where the latency, area, and power consumption are very high. The proposed technique makes use of a non-conventional algorithm; in this, entire sorting of data is not required [1–3].

Prokin and Prokin et al. work aimed at presenting a new low hardware complexity pipelined rank filter and approached reduction in hardware complexity and increase in processing speed, due to identical pipeline stages [9]. Realization of this filter significantly reduces the number of used logic elements, and registers and filter separate noisy samples from signal samples having maximum operating frequency independent of N, but linear complexity growth with window size N [4].

A new design of an area-efficient one-dimensional median filter Lin et al. [5] designed based on word-level architecture that produces the middle-ranked datum after it receives new samples sequentially. In this sorting the samples in the window in descending order according to their values. Performance is increased by performing deletion, and insertion of samples in each clock cycle has the maximum throughput. But critical path delay increases as the window size increases and also additional circuitry is needed for insertion and deletion of the samples.

Nikhad et al.'s [6] high-speed hardware implementation of fixed and runtime variable window length 1-d median filters designed and its having the maximal operation frequency independent of the filter size, independent of word length. It requires extremely large resource components, and computation speed is also limited as the number and bit width of data increase.

An efficient median filter by Venkatappareddy et al. [7] based on threshold decomposition comprises two methodologies of closed-form polynomial expression and positive Boolean expression to find out median value from the binary signal that takes less time to compute the median filter response compared to sorting-based median filter. But the number of levels is equal to the number of gray levels in the image. Complex Volterra filter coefficients with higher polynomial and window size increases take more time and are difficult to process. Median filter architecture for the image by Kumar, Asati and Gupta et al. [8] developed the median filter architecture that has better performance having low latency when compared to other architectures when used in image and signal processing applications. But it is only suitable for removing less amount of noise.

The proposed technique includes few steps explained as follows: noise detection, noise removal by the median filter, and finally the filtered image. The median filter design consists of a sliding window and a sorter. The sliding window includes

Fig. 1 Flowchart of the proposed high-performance method

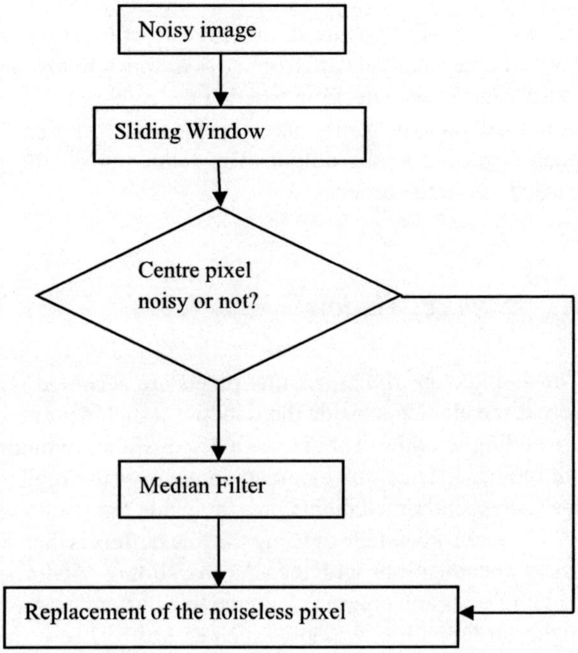

two principal components such as line buffer and registers. The parallel sorter is used in this method acting as the fundamental component of the median filter. The entire process is toted in the ready to use circuitry, Nexys4 DDR kit, making use of intellectual property core. The flowchart of the proposed method is explained in Fig. 1.

The image cannot be directly loaded into the FPGA kit. The median filter is realized by adding the noise into an image. Then, the noisy image is converted into a coefficient file. Then, the coefficient file is loaded into the block ROM, by using the IP catalog module. The data is accessed by providing a suitable clock and address to fetch the data from ROM. The data is getting processed in the median module. Finally, the median value is stored inside the block ROM. Then, the filtered image is visualized through a VGA cable in the monitor.

2 Method Description

Several actions are done in the entire process including many blocks like sliding window, median filter, and block RAM. Accessing and processing the information as 2-D is easy in software like MATLAB. For processing the entire data for hardware implementation, the entire image cannot be accessed as 2-D. Initially, the noisy image is converted into a coefficient file and loaded into block ROM. The information is accessed as 1-D to be used, making use of block RAMs for storing the coefficient file and accessing the data from it. A sliding window is moved over the set of values, to filter the image checking whether the center pixel is noisy or not. If noisy filtering is applied, then the center pixel is replaced by filtered value; otherwise, the noiseless pixel is placed in the output. The following details provide detailed information carried out in the process.

2.1 Sliding Window

The architecture illustrates that pixels are accessed sequentially row by row. So to access the elements inside the window, it includes several registers corresponding to the sliding window. That is, for a 3 × 3 sliding window 9 registers and line buffers are required. Thus, the elements move over the registers and line buffers to access the corresponding elements present inside the window.

The main advantage of using the line buffers is that it remains the same for specific image specifications with the adaptive sliding windows, i.e., 480 × 640 based on the padding rows and columns, and the line buffer designed of 639 remains the same for higher-order sliding windows such as 5 × 5, 7 × 7, and so on. Figure 2 depicts the architecture of a 7 × 7 sliding window.

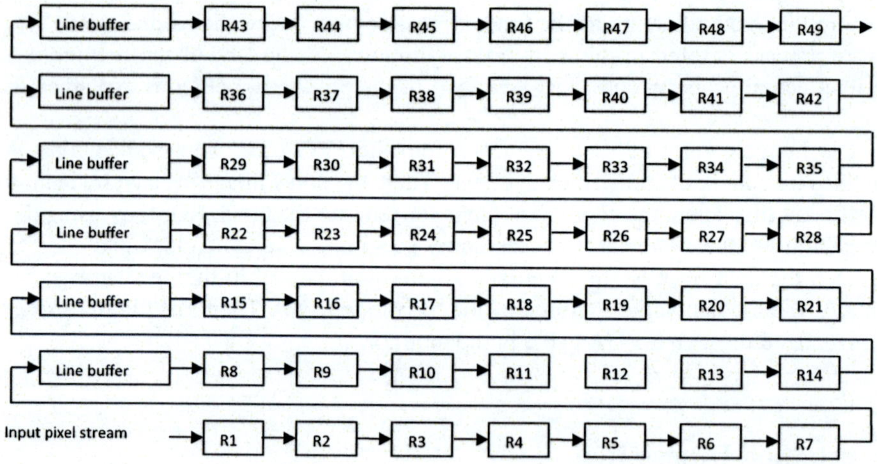

Fig. 2. 7 × 7 sliding window

2.2 Noise Detection

Applying a filter over the entire data increases the resource consumption like power, area, and latency of the process. The reduction of these constraints can be achieved by applying a filter to the affected data. The median filter plays an excellent role in detecting impulse noise. Two different kinds of noises 'salt and pepper' correspond to black and white dot in digital images. If it is corrupted by maximum value Smax, it is called a positive pulse. If it is affected by minimum value Smin, then it is a negative pulse. The following condition [1986] helps in the detection of noise.

Detection of noise is done by thresholding method,

$$\text{Mean} - \text{Centre Pixel} > T1,\ T1 = \text{Smax}/3 \tag{1}$$

$$\text{Mean} - \text{Centre Pixel} < T2,\ T2 = (\text{Smin} - 255)/3 \tag{2}$$

Equation 1 used to find the center pixel is corrupted by a black dot, and Eq. 2 is for a white dot.

2.3 Proposed Sorting Algorithm for Median Filter

Embedded image processing applications are applied in the field-programmable gate arrays (FPGAs) for the efficient design and optimal performance. The latency of the proposed architecture is also independent of the bit length (W) of integers.

The algorithm developed for finding the median value is explained in the below steps:

1. Sort the rows.
2. Sort the columns.
3. Sort the main diagonal and the adjacent diagonals.
4. Sort the diagonal elements selected.

Sorting of the entire rows and columns helps to move the smallest numbers to the top and the largest numbers to the bottom. This makes the filter the most prominent as sorting of the entire data is not necessary. The remaining data lies in the middle range to be sorted to find out the median value. The algorithm is suitable for windows 3 × 3, 5 × 5, and so on. The only change is that the immediate adjacent diagonals are considered. For 7 × 7 s, adjacent diagonals are taken and similarly an adjacent number of diagonals increase for a higher order of windows (Fig. 3).

Sorting method of seven values is mentioned below:

Step 1: 9 5 6 1 4 3 6.
Step 2: 5 6 9 1 3 4 6.
Step 3: 5 6 1 3 9 4 6.
Step 4: 1 5 6 3 4 6 9.

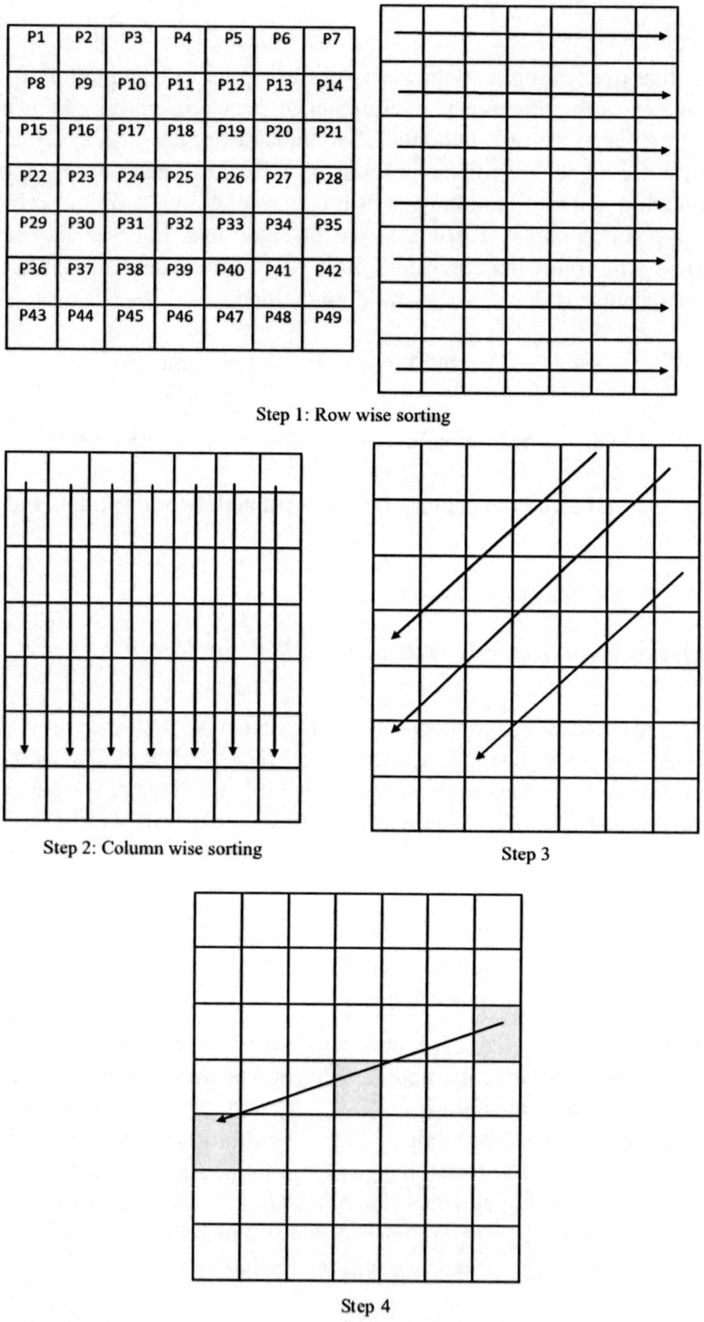

Fig. 3 Algorithm for seven-value sorter

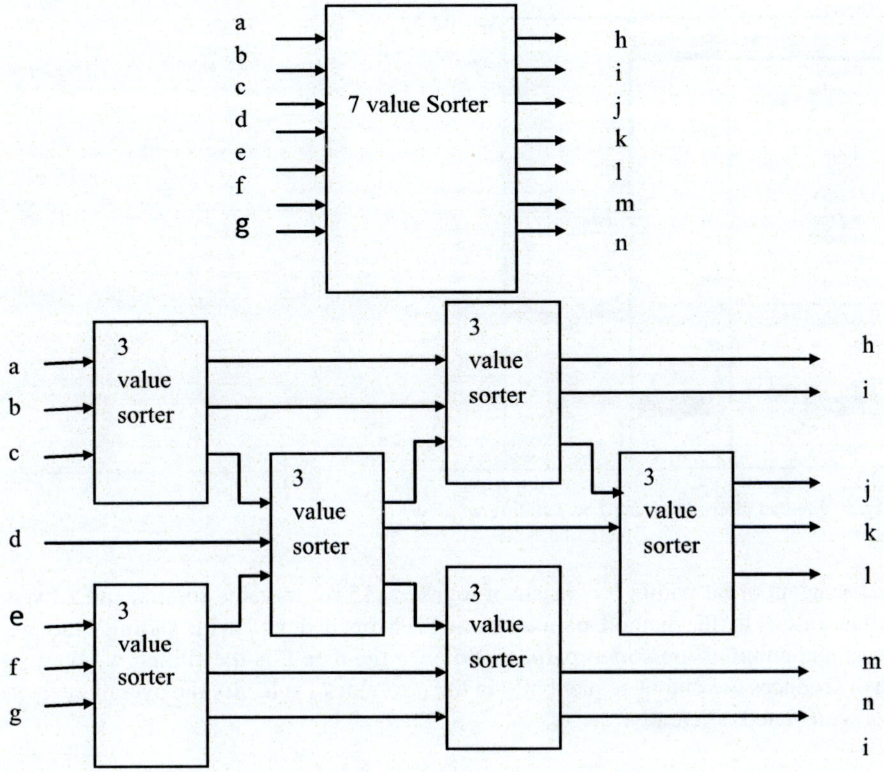

Fig. 4 Seven-value sorter using three-value sorter

Step 5: 1 5 3 4 6 6 9.
Step 6: 1 3 5 4 6 6 9.
1 3 4 5 6 6 9

The fundamental component used for any value of sorter is three-value sorters. Sorting three values takes place in a parallel manner making use of gates and mux[]. So, the sorter designed for higher-order windows will be having three-value sorter as a building block. Figure 4 depicts the sorting of seven values by using three-value sorter. Figure 5 depicts the algorithm of the median filter for 7 × 7 windows.

2.4 Median Filter Architecture

Median filter architecture that already exists involves the sorting of entire data; in the proposed architecture, sorting of the entire data is not required, and the latency of the entire process decreases. Hence, there will be a slight variation in the latency for the higher-order sliding windows. The main advantage of the filter is that it is

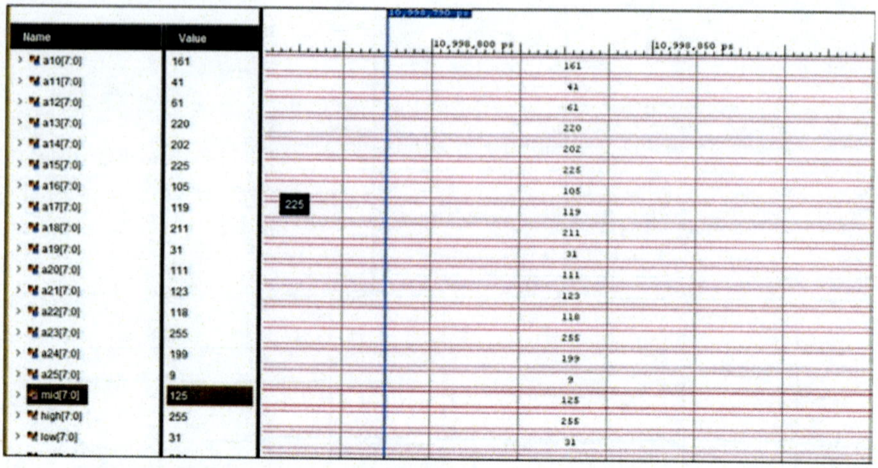

Fig. 5 Median filter output for 7 × 7 sliding window

independent of bit width. It consists of registers, 15 seven-value sorters, and 2 five-value sorters. In this method, parallel sorting is happened; i.e., in the sliding window, rows and columns are sorted parallel. So once the data fills the sliding window, it then produces the output sequentially in the next clock cycle. So, the overall latency was calculated efficiently.

3 Results and Discussion

Simulation results obtained by the implementation of the median filter with the suitable window sizes are presented. However, the median filter is implemented using Xilinx Vivado 2016.4. Further resource constraints are optimized by performance analysis.

3.1 Low Latency Median Filter

The architecture of the median filter is realized using Hardware Descriptive Language Verilog for the various filter sizes 9 and 25. These are implemented for 8-bit width of input operands. Verilog code is simulated using Xilinx Vivado 2016.4, and performance analysis is carried out by using the Cadence software. Moreover, this median filter is implemented for the image of size 480 × 640. Table 1 gives the specifications considered in the entire method of the filtering process.

Since the process of filtering, it includes noisy detection, filtering, and replacement of filtered pixel.

Table 1 Image specifications

Size of the image	480 × 640
Line buffer size	639
Size of the output image	480 × 640
Number of rows	7 × 7:720, 5 × 5:644, 3 × 3:642

Table 2 Performance analysis of median filter by using the 3 × 3, 5 × 5, and 7 × 7 window in terms of area, power, and delay

Sliding window	Area (μm^2)	Power (nW)	Delay (ns)
3 × 3	9158	687,819.221	7402
5 × 5	12,086	884,321.390	9132
7 × 7	15,436	967,892.539	9742

3.2 Simulation Results

The median value of the proposed technique makes use of the entire above process. Processing of the data to obtain the median value makes use of line buffers, registers, and also sorters. Simulation of the 7 × 7 window to obtain a middle value of 49 values is shown in Fig. 5.

3.3 Performance Analysis

This section evaluates the different configurations of cascaded median filter, and it provides clear comparison between the resource computation and power dissipation. The proposed designs are synthesized using Cadence software with a 90-nm cell library. Power consumption was measured after post-layout simulations.

3.4 Implementation of FPGA

This section gives details about the implementation of the median filter architecture in the Nexys4 DDR kit. The following details are the implementation in FPGA, display of the image through VGA, and the results obtained.

Fig. 6 Displaying of filtered cameraman image through VGA by filtering 90% of salt and pepper noise

3.4.1 VGA Results

For a 640-pixel, there are 480 rows displayed using a 25 MHz pixel clock and 60 ± 1 Hz refreshes. The counters can be used to form an address into video RAM.

But one of the disadvantages is that the entire width of the data cannot be stored inside the memory due to the shortage of memory. Only 4 bits are stored inside the memory. Because of that display of the image on the monitor, patches may arise in the image.

The following figures have shown the display of noisy images and filtered images (Fig. 6).

3.4.2 Utilization Report

The resources used for the entire module are measured by the utilization of LUT, buffers, and amount of memory like BRAM, BROM, registers, and flip-flops. The utilization report is generated for the median filter with the sliding window of 3 × 3, 5 × 5, and 7 × 7. Figure 7 shows the utilization report generated for the median filter.

4 Conclusion

The low latency and high accurate median filter architecture is one the efficient methods for removing 'salt and pepper noise' from the noisy images due to faulty

Name	Slice LUTs (63400)	Slice Registers (126800)	F7 Muxes (31700)	F8 Muxes (15850)	Block RAM Tile (135)	DSPs (240)	Bonded IOB (210)	BUFGCTRL (32)	MMCME2_ADV (6)
vga	3982	415	88	32	114	1	15	2	1
gg (max)	0	24	0	0	0	0	0	0	0
instance_name (clk_w…	0	0	0	0	0	0	0	2	1
jj (window)	3925	353	88	32	76.5	0	0	0	0
U1 (hcntdesign)	20	10	0	0	0	0	0	0	0
U2 (vcntdesign)	13	10	0	0	0	0	0	0	0
U3 (SRFF)	0	1	0	0	0	0	0	0	0
U4 (SRFF_0)	2	1	0	0	0	0	0	0	0
U5 (SRFF_1)	0	1	0	0	0	0	0	0	0
U6 (SRFF_2)	0	1	0	0	0	0	0	0	0
ww (blk_mem_gen_1)	21	14	0	0	37.5	0	0	0	0
U0 (blk_mem_gen_…	21	14	0	0	37.5	0	0	0	0

Fig. 7 Utilization report for the median filter of 7 × 7 window

conditions of camera and transmission of information. Hardware-oriented optimization makes the design even more resource-efficient by using the line buffer. Thus, proposed median filter could process large datasets in higher average throughput. This method is not based on the conventional median filtering process; thus, the entire sorting of data values involved in the median processing is not required. This method of non-conventional sorting method makes the architecture to have a low latency as compared to existing word-level architecture.

The post-layout parameters are obtained for the median filter using Cadence Tool. The technology used is 90 nm, and a slow library is used to identify the worst-case parameters. Delay, power, and area parameters are analyzed for the median filter. And also the hardware implementation of the median filter is done by using the internal memory in Nexys4 DDR FPGA kit.

References

1. A.H. Fredj, J. Malek, Design and ımplementation of a pipelined median filter architecture, in *2019 IEEE International Conference on Design and Test of Integrated Micro & Nano-Systems (DTS)*
2. A. Goel, M.O. Ahmad, M.N.S. Swamy, Design of two dimensional median filter with a high throughput of FPGA ımplementation, in *2019 IEEE 62nd International Midwest Symposium on Circuits and Systems (MWSCAS)*
3. W.-T. Chen, P.-Y. Chen, A low-cost design of 2D median filter, 2019. IEEE Access https://doi.org/10.1109/ACCESS.2019.2948020.
4. A. Kundu, Application of two-dimensional generalized mean filtering for removal of impulse noises from images. IEEE Trans. Acoustics, Speech, Sig. Proc. **ASSP-32**(3) (1984)
5. M.R.-D. Lin, P.-Y. Lin, C.-H. Yeh, Design of area-efficient 1- D, median filter. IEEE Tran. Circ. Syst. II, Exp. Briefs, **60**(10), 662–666 (2013)
6. E. Nikahd, P. Behnam, R. Sameni, High-speed hardware implementation of fixed and run time variable window length one dimentional median filters. IEEE Tran. Circ. Syst. II, Exp. Briefs, **63**(5), 478–482 (2016)
7. B.L. Venkatappareddy, C. Jayanth, K. Dinesh, M. Deepthi, Novel methods for ımplementation of efficient median filter. IEEE Trans. Image Proc. **10**(10), 978–982 (2017)

8. A. Asati, Low-latency median filter hardware implementation of 5 × 5 median filter. IET Image Proc. **11**(10), 927–934 (2017)
9. D. Prokin, M. Prokin, Low hardware complexity pipelined rank filter. IEEE Trans. Circuit Syst. II, Exp. Briefs, **57**(6), 446–450 (2010)

LoRa and Wi-Fi-Based Synchronous Energy Metering, Internal Fault Revelation, and Theft Detection

Rojin Alex Rajan and Polly Thomas

Abstract Today there are many advanced smart energy metering devices are developed but even now the consumers and the service providers are facing many problems in their day-to-day life. At present, in many places, the billing is done with manual labor, also when a power failure occurs, the consumers contact their service providers even if the fault is on the consumer side. And also there are many problems like overvoltage, under-voltage, line breakage, power theft, etc. So if the service providers can identify these problems synchronously then they can provide good service and also maintain their quality. The majority of the consumers are currently using mechanical circuit breakers, so if they need an autoturn-on mechanism for their circuit breakers, they need to upgrade the conventional circuit breakers to relays. It is costly and can increase E-waste. This paper primarily focuses on a multifunctional CPU which helps to solve the above-mentioned problems. The current energy metering devices can be replaced with this smart unit and can bring much advancement in power monitoring, electric fault identification, and resolving. Also, consumers will get synchronous data about their energy consumption with synchronous billing. This project is a multifunctional unit, so if a fault occurs, then the CPU will detect the type of fault and notify the consumers then the CPU will also do the required action for solving the existing fault. So the consumers do not need to worry about their power supply anymore. And the service provider can identify the actual location of the distribution line breakage and power theft informations by installing these units to all their consumers. Also, this unit has an autocircuit breaker reclosing mechanism, which can be easily integrated with any existing distribution boards, so there is no need to replace the existing mechanical circuit breakers.

Keywords Wi-Fi · LoRa · IoT · Energy monitoring · Utility · Synchronous electricity billing · Power theft detection

---

R. A. Rajan (✉) · P. Thomas
SAINTGITS College of Engineering, Pathamuttom, Kerala, India
e-mail: rojin.ar1620@saintgits.org

P. Thomas
e-mail: polly.thomas@saintgits.org

V. Suma et al. (eds.), *Inventive Systems and Control*,
Lecture Notes in Networks and Systems 204,
https://doi.org/10.1007/978-981-16-1395-1_60

1 Introduction

Today many developments occur in every field, and technologies are developing every day. But even now many fields are expecting proper automation. Every development is focused to make user-friendly products. Many products are developed in the electric field, which improves the protection of our electric circuitry.

Even today, electricity consumers and service providers are facing so many problems. On the consumer side, the wiring is getting more complex than before, so the majority of the consumers are not aware of the electric circuitry in their house. Power fault in a building can occur due to many reasons, a layman consumer may not be able to detect the fault, i.e., power supply in a building can interrupt due to ELCB tripping, blown fuse, etc. So if the consumer is unaware of the type of fault when a power failure occurs, then they will contact their service providers for registering complaints, but the actual problem may be in the household circuitry. If the problem is in the household circuitry, then it is not a part of the service provider the consumer is responsible to rectify the problem.

This project is a multifunctional device, which is capable of solving many problems existing in the electric circuitry as well as the utility side.

This device should be installed in the house along with the distribution board. So the consumer will get the real-time status of their electric circuitry, i.e., when an error occurs, this device will automatically detect the reason behind the fault and report to the consumers. Also, if a circuit breaker gets tripped, then the consumer can turn on the circuit breaker remotely by using their smartphone. If the fault is at the utility side like over-voltage, under-voltage, power failure, etc., then this unit will automatically send a message to the respective service provider, along with the consumer id and type of fault. So the consumer need not worry about their electrical supply anymore. Also, the real-time power monitoring function of this unit will help the consumers in their power management, which will promote energy saving. Also, if this unit is installed in all houses under a service provider, then this device will help to find the actual location of distribution line breakage, and also they can find the energy theft in real time.

According to the existing studies, there are many smart meters are already developed, but there are no existing devices that incorporate many functions like fault detection, fault alerting, fault rectification, power monitoring, real-time billing, theft detection, line breakage location tracing, auto reclosure, condition monitoring, etc. One of them is an IoT-based energy monitoring system that uses an ESP8266 to send the data to an InfluxDB database server via MQTT broker [1]. But it is a low bandwidth and an unreliable network. Also, it is focused on the power monitoring application.

Another reference suggests a system which sends the power consumption to ThingSpeak and control loads remotely from an android app. MySQL database and PHP are used for communication. They measure the voltage and current values and calculate the power, and the load is controlled by using relays which are controlled by a Raspberry Pi [2]. But they failed to measure the value of power factor, So they cannot obtain an accurate value of power consumption. Also, the use of Raspberry Pi makes the system more expensive.

Another article suggests a power monitoring and actuating system for domestic applications. It is based on LoWPAN which uses IEEE 802.15.4 and power line communication (PLC) [3]. It includes node and resources discovery mechanisms to provide seamless connectivity and data retrieval and actuation based on IoT device name, location, and supported functionalities.

2 Methodology

The proposed device is designed to monitor the voltage, current, active power, and energy consumption of a consumer. And also the device will check the status of supply from the utility and every safety device like ELCB, MCB, FUSE, MCCB, RCCB, etc. If any fault like over-voltage, under-voltage, tripping of CB's blown fuse, etc., occurs, then the microcontroller will identify the fault and if the fault is in the household circuitry, then the system will display it in an OLED display and also send a message to the consumer by using LoRa and Wi-Fi module [4]. And if the fault is at the utility end, then this device will inform the service provider about the fault including the actual location and consumer id of the consumer using a GSM module. This system consists of a reclosing mechanism for tripped CBs which helps turn on the CBs remotely. This system can also calculate the power consumption and this data is uploaded to the cloud then this data is processed in an analytic platform and then calculate the bill amount and report to the consumer at the same time [5]. The system works based on AI and IoT. An ATmega2560, ATmega 328P, and a NodeMCU are used in the controlling unit. A Wi-Fi module, GSM module, and LoRa module are used for communication purposes [6]. And also if a line breakage occurs then the device will help the service provider to identify the actual location of the line breakage and also the service provider can detect power theft in real time by the mass implementation of this device. There are two transmission techniques available namely LoRa and Wi-Fi [7]. Here, the Wi-Fi module needs an active Internet connection. But a LoRa module can directly communicate with another LoRa device within the 5–10 km range. So it helps the control unit to send a fault alert to the service provider even if there is no active Internet connection is present in the house where it is installed [8] (Fig. 1).

2.1 Error Detecting Mechanism

Several types of errors can be detected by this device, like over-voltage, under-voltage, tripping of circuit breakers, blown fuse, no supply from the utility, etc. These problems are detected by using the data from the PZEM-004T and by the help of several relays switching AC to DC buck circuits [9].

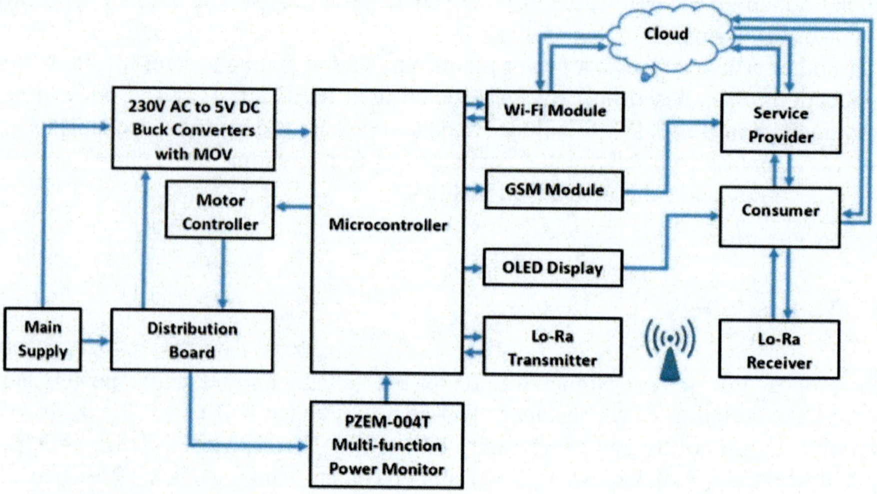

Fig. 1 Block diagram of proposed model

2.2 *Fault Alerting Mechanism*

If the device detects a fault, then the microcontroller analyzes the type of fault and if the fault is in the household circuitry then the device will report it to the consumer. Similarly, if the problem is on the utility side like over-voltage, under-voltage, power failure, etc., then the device will report it to the service provider and at the same time, it also reports to the consumer. For this function, the system utilizes the help of the Wi-Fi module, GSM module, LoRa module, and OLED display [10].

2.3 *Power Monitoring Mechanism*

Power measurement is done with the help of a PZEM-004T Module. By using a PZEM-004T module, the device will measure the power in real time and this data is received by the microcontroller and do the required calculations and then stored in its internal memory. Also, this data is uploaded to the cloud by using the Wi-Fi module. So, the service providers and the consumers can monitor these details in real time [11] (Fig. 2).

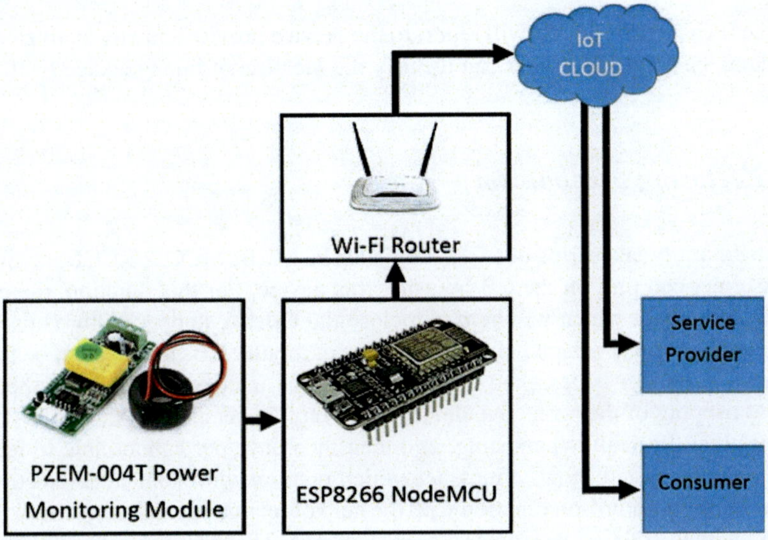

Fig. 2 Power monitoring layout

2.4 *Billing Mechanism*

The proposed device will continuously monitor the power consumption of the consumer and the consumption in kWh is stored in its internal memory and it will be uploaded to the cloud once the unit gets Internet connectivity. The received data will process that to monetary value by the algorithm written in the analytics platform service like ThingSpeak. The variation in the tariff can be uploaded by the service provider in real time and the analytic platform will do the required calculations. The consumer can monitor their bill at any instant from this platform by using their smartphone.

2.5 *Location Tracing of Line Breakage*

For tracing the actual location of line breakage, the service provider must install this proposed device in every house of their customers. Suppose the service provider obtains a message from only one customer then the fault can be in the supply from the electric post to the meter of that particular consumer, so by using the consumer id, the service provider can identify the location of the fault. For that suppose there are 1000 houses are under a particular distribution transformer, and assume that the line gets break after 700 houses, then there will be no supply after this 700 houses and there may be over-voltage and under-voltage problems can occur in this 700 houses.

In all these cases, this device will report to the service provider and by analyzing this mass data, the service provider can identify the location of line breakage [12].

2.6 Reclosing Mechanism

When a circuit breaker trips due to a fault such as MCB, MCCB, RCCB, or ELCB, the consumer can turn on the CB by using this device. For this function, there will be a linear actuator which will help to reclose the CBs. A stepper motor is used for the selection purpose, which helps to select the particular CBs that need to be turned on. The mechanism is designed in such a way that it does not interrupt the high speed of tripping of the protective device if the fault is persisting. And if the device identifies that the fault is persisting, and then the consumer cannot able to reclose the CBs remotely and shows a message which mentioned to contact an electrician. This make this autoturn-on function with the help of relays, but replacing the existing breakers with the relays can increase the system cost. These mechanisms do not need to change the existing breakers, which helps in cost management and also manages E-wastes [13] (Fig. 3).

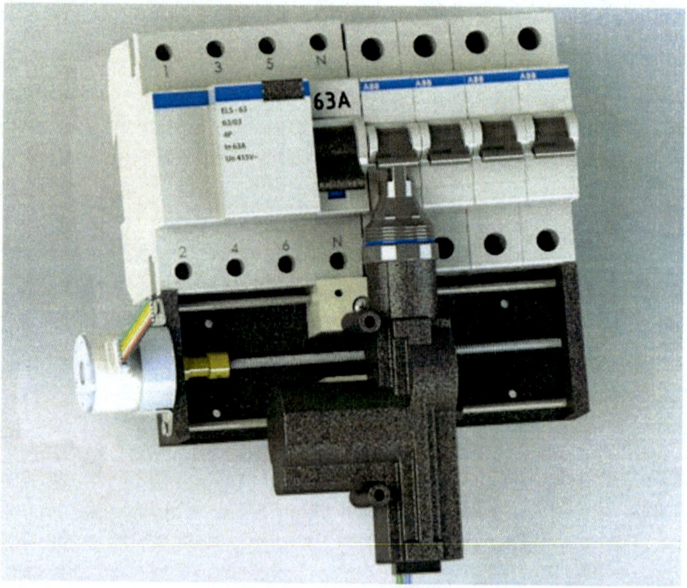

Fig. 3 Solidworks model of reclosing mechanism

2.7 Power Theft Detection

For power theft detection, the service provider must install this proposed device to every house of the consumers and also a power measurement unit at the distribution transformer side. These data are uploaded to the cloud in real time. At normal conditions, the sum of power consumed by all consumers and the power delivered by the particular distribution transformer must be the same. In the case of energy theft, the power delivered by the transformer is more than the sum of power consumed by all consumers under that particular transformer. So based on this logic, the system can identify the power theft in real time and inform the service provider [14, 15].

3 System Design

This system is a multifunctional processing unit that is controlled by using an ATmega2560 microcontroller, ATmega 328P, and a NodeMCU. In this system, the fault detection mechanism works with the help of several relays switching AC to DC buck circuit with MOV protection. The synchronous power monitoring is done with the help of a PZEM-004T module. The reclosing mechanism works with the help of a linear actuator and a stepper motor runs with the help of motor driver ULN2003. The input supply is provided by a 12 v SMPS and also it includes an auxiliary 12 v rechargeable battery. The NC pin of all relays is connected to a common ground and the NO pin of all relays are connected to a common +5 v VCC. The COM pin of every relay is connected to the corresponding digital input pin of the ATmega2560. A Wi-Fi module, GSM module, and LoRa WAN modules are used for the connectivity of the device.

3.1 Microcontroller

In this proposed model, multiple microcontrollers were used for operation. A single ATmega microcontroller is not capable of managing these many functions. Various controllers used in the proposed model are ATmega2560, ATmega 328P, and a NodeMCU. ATmega2560 is a high-performance, low-power microchip. It is an AVR RISC-based microcontroller. It has 100-pin, and it combines 256 KB ISP flash memory, 8 KB SRAM, and 4 KB EEPROM. It has 54 digital input/output pins and 16 analog input pins. And out of these 54 pins, 14 pins can be used for PWM outputs. 16 analog input pins can be used for reading analog values. And the NodeMCU is a LoRa-based firmware for the ESP8266 Wi-Fi SOC. The input supply for the microcontrollers is 5 v DC, so that the AMS 1117 5.0 regulator IC is used in which it helps to reduce the 12 v supply from the SMPS and battery to 5 v DC. The NodeMCU is used only for the power monitoring function, and the remaining functions are performed by the ATmega2560. Based on the inputs, the ATmega2560 gives corresponding messages to the user with the help of the connectivity devices (Fig. 4).

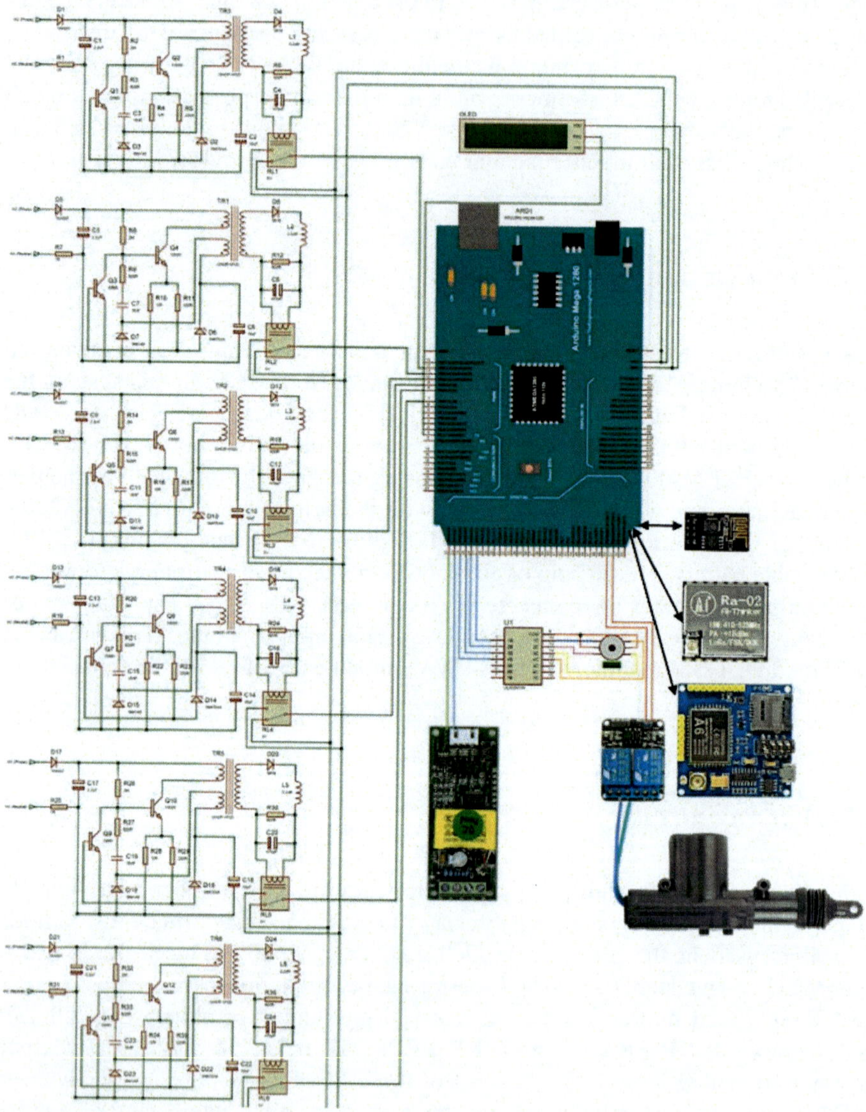

Fig. 4 Circuit of proposed model

3.2 Stepper Motor

A stepper motor is used in this proposed system. It is a part of the reclosure mechanism. For this function, a 28BYJ-48 stepper motor is used with a 5 v supply. It also includes a step angle: 5.625-degree, overall gear ratio is 64:1. It is powered with the help of the ULN2003 driver board. It can move forward and backward according to the control signal from the ATmega 328P. The speed of the motor, distance moved by the slider, etc., are decided by the controller.

3.3 Linear Actuator

A 12 v linear actuator is used for turning on the tripped circuit breakers. This operation of the linear actuator works according to the instruction from the microcontroller. Also, it is attached to a slider, which helps to select the required circuit breaker according to the instruction from the user.

3.4 DC Motor Drive

The ULN2003 motor driver is used for the 28BYJ-48 stepper motor, which is used for the CB reclosing application. The control signal for this driver module is provided by the microcontroller according to the instruction from the user. Also, the control signal from the ATmega328P is amplified by this motor driver (Fig. 5).

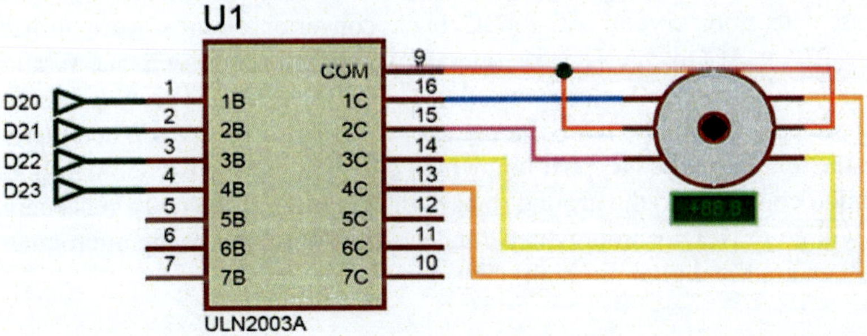

Fig. 5 ULN2003 motor drive

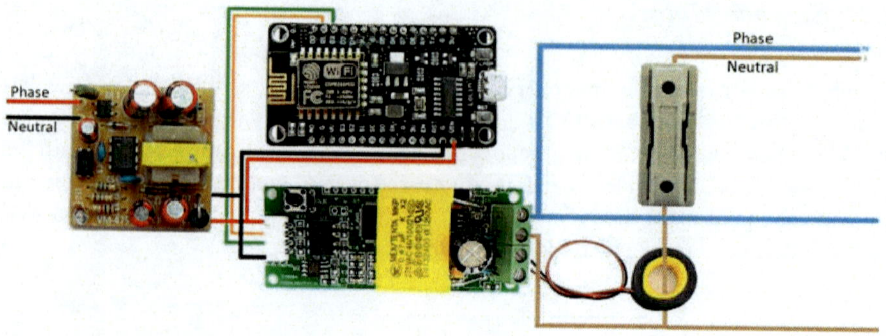

Fig. 6 PZEM-004 T-100A integration

3.5 *Power Monitoring Module*

The power monitoring function in this proposed model is done with the help of 'PZEM-004 T' which is a multifunction power monitoring module. This module is capable of providing highly accurate measurements. The parameters like voltage, current, connected load, etc., can also be obtained. Also, it can be measured up to 100A. In this system, the output of the PZEM-004 T module is received by the NodeMCU and further calculation and storage process is done by the microcontroller (Fig. 6).

3.6 *Fault Detection Circuit*

For fault detection, several AC-to-DC buck converter circuits were utilized. It converts 230 v AC to 5 v DC for operating a relay, which will act as an input signal to the ATmega2560 microcontroller. This entire circuit is provided with over-voltage protection with the help of a metal oxide varistor (MOV). When there is no supply, the relay is at the NC position, which connects the corresponding digital pin of the microcontroller to the ground (makes the pin low). If the relay gets energized, then it will go to NO position, which connects the digital pin of the microcontroller to +5 v (makes the digital pin high) (Fig. 7).

3.7 *Connectivity Devices*

There are three different modules are used for connectivity, i.e., Wi-Fi module, GSM module, and LoRa module. The LoRa module is used for communication with the service provider, which is controlled by the ATmega2560 microcontroller. The LoRa (IEEE 802.15.4) is mainly used for fault alert mechanism, i.e., if the fault is in the

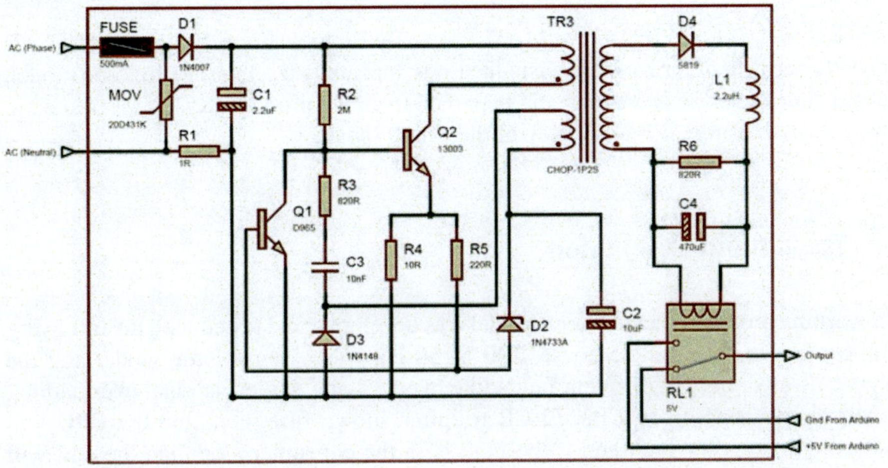

Fig. 7 Fault detection circuit

part of the service provider then it will send an alert to the service provider, the fault may be a power failure, over-voltage, under-voltage, or a power quality problem. In all cases, the problem is also mentioned in this alert message. A GSM module is also used for a standby connectivity purpose, i.e., if the LoRa has any connectivity issue, then the CPU will use the GSM module for communication with the service provider. The Wi-Fi (IEEE 802.11) module is used for communication with the consumers.

3.8 Real-Time Clock

The proposed model has a built-in battery. If a power failure occurs, the control unit will work using this battery. But, if the power failure persists for a long duration may drain the backup of the control unit to maintain the system time. For that, a real-time clock is used in this proposed model. The DS1307 RTC is a real-time clock which is an 8-pin device with an I2C interface. It is a low-power clock with its own battery. It has 56 bytes of non-volatile RAM and the clock operates in either the 24-h or 12-h format with AM/PM indicator.

3.9 Cloud Integration

The power monitoring and fault monitoring details are uploaded to a cloud server, and then both consumers, as well as the service provider, can access these data from this cloud platform. For this, the ThingSpeak platform is used. ThingSpeak is an IoT analytics platform service that allows us to aggregate, visualize, and analyze live data streams in the cloud. It also has MATLAB analytics to write and execute MATLAB code to perform preprocessing, visualizations, and analyses. For the proposed model, a real-time database is created, and then data from the proposed system is pushed to the database using API-Write Key of the ThingSpeak.

4 Result and Discussion

A working model of the proposed model was developed and tested with the following supply parameters. Single-phase, 230 V, 50 Hz was given for the model and the system was subject to different faults like over-voltage, under-voltage, overloading, short-circuit tripping of CBs, ELCB tripping, blown fuse, etc., and the CPU will analyze the type of fault and if the fault is in the consumer side then the unit will send a message to the consumer by using the LoRa and also the type of fault is displayed on the OLED screen of the prototype. Also, this model is connected with an external load and measure power consumption and these details are uploaded to the cloud using Wi-Fi and obtained a highly accurate result. The accuracy of power monitoring of the proposed model is measured by applying unit some loads then the outputs are compared with highly accurate energy meters. The fault alerting mechanism was tested by creating different fault conditions. For testing of the theft detection mechanism, it is required to make some similar control units, so it is still under its testing phase.

4.1 Test Result in IoT Platform

Figures 9–11 are the output result obtained from the ThingSpeak platform. The parameters like voltage, current, power, energy, frequency, and power factor are measured and these values are uploaded to ThingSpeak and displayed in both graphical forms as well as analog form. Figure 8 is the prototype and Fig. 9 represents the output graph of voltage and current.

Figure 10 is the output graph of power and energy, which is obtained when a load of three 15w incandescent lamps is applied.

Figure 11 represents the frequency and power factor graphs and obtains an accurate result. The frequency is obtained as a value between the 49.9–50 Hz range and obtained a power factor of 0.94 pf.

5 Conclusion

The proposed design is a multifunctional smart unit which replaces the existing electronic meters and smart meters, because of its capability for solving many problems

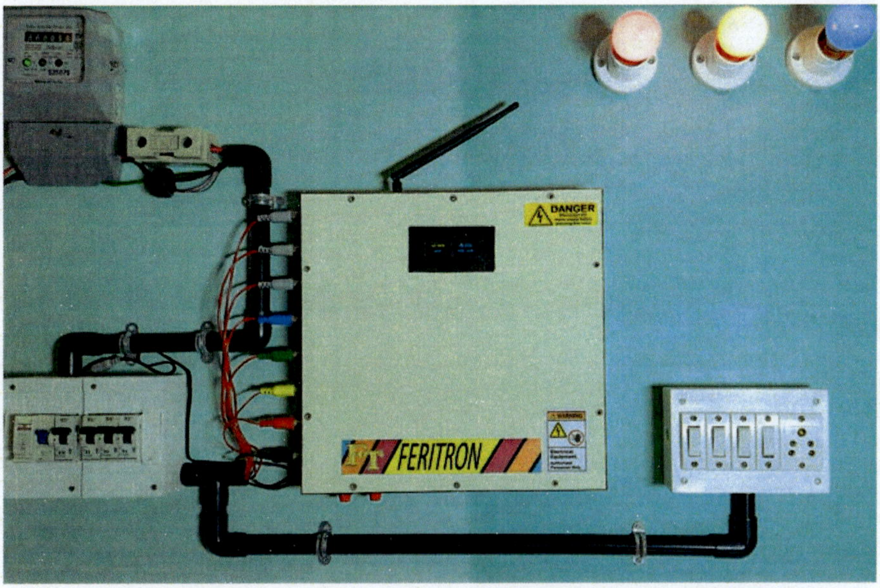

Fig. 8 Competed prototype

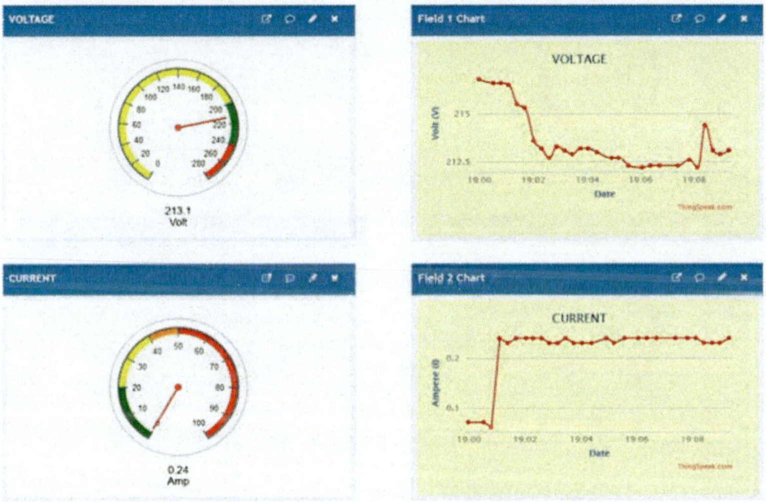

Fig. 9 Voltage and current graphs

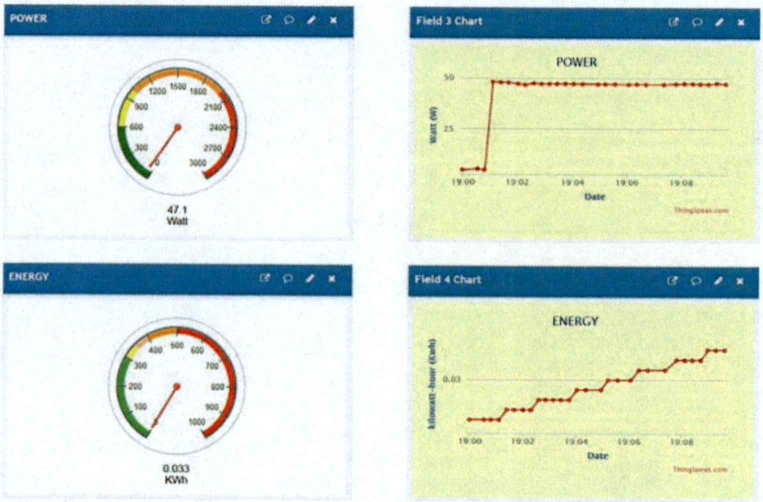

Fig. 10 Power and energy graphs

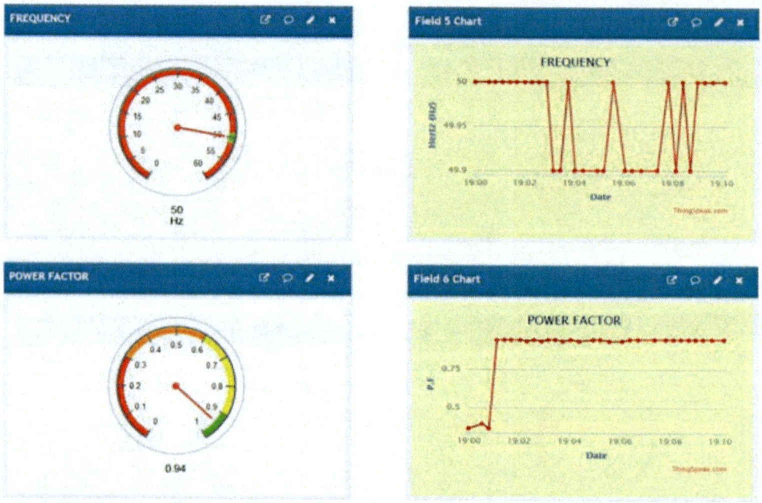

Fig. 11 Frequency and power factor graphs

in the utility and consumer side. This device is an AI device, so it not only identifies the problems but also it solves the problems in the electric circuitry. By installing this device, a consumer can monitor their power along with their electricity bill. And the consumer has not to worry about their power supply anymore, because the device will do the required solution for the problems in their power supply. This device will promote energy saving and fast recovery of the power supply problems, also it helps to maintain the quality of electric supply.

References

1. K. Chooruang, K. Meekul, Design of an IoT energy monitoring system, in *2018 Sixteenth International Conference of ICT and Knowledge Engineering* (2018)
2. P.V. Rama Raju, G. Naga Raju, G.V.P.S. Manikantah, A. Vahed, A.L. Bhavyaw, G. Reddy, IoT based power monitoring system and control. JETIR **4**(11) (2017)
3. L.M.L. Oliveira, J. Reis , J.J.P.C. Rodrigues, A.F de Sousa, IoT Based Solution for Home Power Energy Monitoring and Actuating (IEEE, 2015)
4. A.S. Ravi, Gupta, P. Singh, Institute of Technology, Kanpur, Uttar Pradesh, Associate Professor, Pranveer Singh, IoT based smart energy meter **3**(3) (2018)
5. A. Bhimte, R.K. Mathew, S. Kumaravel, Development of smart energy meter. IEEE INDICON 2015 1570186881 (2015)
6. N. Darshan Iyer, K.A. Radhakroshna Rao, IoT based energy meter reading, theft detection and disconnection using PLC modem and power optimization. Pros of IJAREEIE, **4**(7) (2015)
7. G.L. Prashanti, K.V. Prasad, Wireless power meter monitoring with power theft detection and intimation system using GSM and Zigbee networks. Proc of IOSR – JECE **9**(6), 04–08 Ves.I (Nov-Dec, 2014)
8. Md.M. Rahman, Md.O. I. Noor-E-Jannat, Md. Serazus Salakin, Arduino® and GSM based smart energy meter for advanced metering and billing system Pabna University of Science and Technology", in *2nd International Conference on Electrical Engineering and Information & Communication Technology (ICEEICT)* (2015)
9. S. Maitra, Embedded energy meter—a new concept to measure the energy consumed by a consumer and to pay the bill, in *Joint International Conference on Power System Technology and IEEE Power India Conference* (2008), pp. 1–8
10. Md. Shahjalal, Moh. Khalid Hasan, Md. Mainul Islam, Md. Morshed Alam, Md. Faisal Ahmed, and Yeong Min Jang. "An Overview of AI-Enabled Remote Smart-Home Monitoring System Using LoRa" Published in 2020 International Conference on Artificial Intelligence in Information and Communication (ICAIIC) 2020.
11. H.K. Patel, D. Shah, M. Sheth, S. Trivedi, S. Bakliwal, I. Shah, et al., Microcontroller and GSM Based Digital Prepaid Energy Meter. Int. J. Electron. Comput. Commun. Technol. (IJECCT) **4**(1), 32–37 (2013). ISSN 2180-3536
12. L. Alliance. Accessed on May 30, 2016. [Online]. Available: https://www.lora-alliance.org/
13. Q. Sun et al., A comprehensive review of smart energy meters in intelligent energy networks. IEEE Int. Things J. 3(4), 464–479 (2016)
14. N. Bhalaji, EL DAPP–An electricity meter tracking decentralized application. J. Electron. **2**(01), 49–71 (2020)
15. S. Sakya, Design of hybrid energy management system for wireless sensor networks in remote areas. J. Elect. Eng. Automat. (EEA) **2**(01), 13–24 (2020)

A Load Balancing Based Cost-Effective Multi-tenant Fault Tolerant System

Himanshu Saini, Gourav Garg, Ketan Pandey, and Aditi Sharma

Abstract With every single day the data over the internet is growing exponentially and it is becoming the need of the hour to manage and store this immense data efficiently. The multi-tenant architecture techniques are becoming quite beneficial for handling large databases and so are the cloud-based services. Also, with the hasty increase in the number of users in health-based platforms over the internet, the growth of data stores over this category has become a major concern. This paper proposes a multi-tenant system that has been implemented using a column-based NoSQL database, Cassandra, and the useful features of the webservices. Another, major apprehension while handling such huge data is scalability, the ease of handling tenant's enormous dynamic data. To solve this, the proposed solution uses an elastic load balancer that automatically distributes incoming application traffic across multiple targets, such as Amazon EC2 instances, containers, IP addresses, Lambda functions, and virtual appliances. Apart from this, there is also a need to tackle the software faults and system failures to ensure high availability of the system. This problem is handled by making the system fault-tolerant in the proposed multi-tenant architecture.

Keywords Multi-tenancy · NoSQL · Cassandra · Materialized view · Fault tolerance · EC2 · Load balancer

---

H. Saini · G. Garg · K. Pandey · A. Sharma (✉)
Jaypee Institute of Information Technology, Noida, India
e-mail: aditi.sharma@mail.jiit.ac.in

H. Saini
e-mail: 17103254@mail.jiit.ac.in

G. Garg
e-mail: 17103250@mail.jiit.ac.in

K. Pandey
e-mail: 17103237@mail.jiit.ac.in

V. Suma et al. (eds.), *Inventive Systems and Control*,
Lecture Notes in Networks and Systems 204,
https://doi.org/10.1007/978-981-16-1395-1_61

1 Introduction

Today most of the world's population is striving to improve their quality of life and standard of living, for which they are getting indulged in internet-based platforms more than ever before. Among these, the most common and popular ones are health-based applications. An increase in the number of users on these health-based applications around the globe is thus leading to an increase in the quantity of data which in return requires reliable and efficient storage management. Cloud-based technology is being widely used to store and handle such humongous amounts of data in a distributed environment [1]. This does not only lead to cost reduction but also increases the efficiency of the application. Although many users are using cloud-based applications, mainly as Software-as-a-Service (SaaS), data redundancy remains one of the major concerns due to sharing of the common data. This drawback may be resolved by implementing a multi-tenant architecture for optimal data storage [2].

Unlike traditional system architecture i.e. single tenant in which each application has a separate database shown in Fig. 1; in a multi-tenant architecture, a single software instance is set up on the server running on the service provider infrastructure, and this software receives service by providing access to more than one tenant at the same time. It is used in the realization of the Software as a Service (Saas), which is rapidly spreading especially in small and medium-sized enterprises. Tenant refers to persons, institutions, or organizations that hire the multi-tenant Saas model [3].

Multi-tenancy is an emerging concept being used among varied cloud-based applications termed as tenants that work together in a shared environment for data storage and performance optimization. It helps to maintain a shared data storage that is common to multiple users rather than making separate copies of the same data, thus

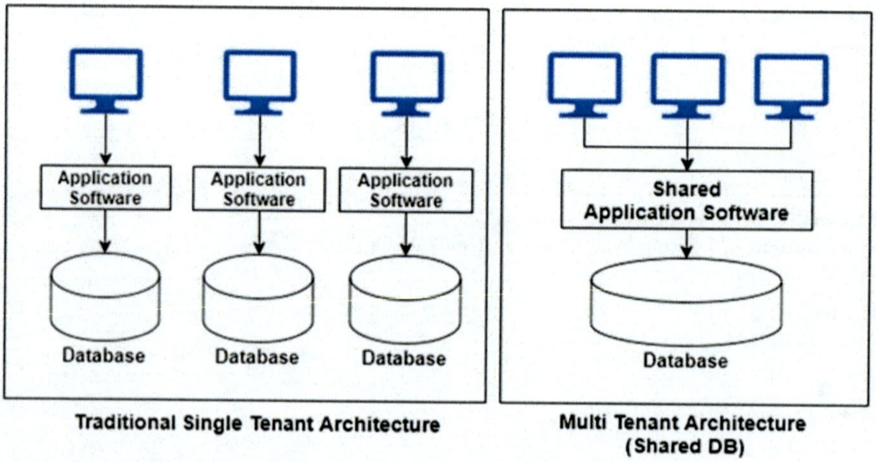

Fig. 1 Difference between traditional and multi-tenant architecture

avoiding data redundancy. This also takes into account the concept of data isolation and security as leakage of the data amongst tenants is not desirable [4]. In addition to these features, multitenant architecture also provides hardware resource optimization and data aggregation opportunities.

It has been observed that the implementation of the multi-tenant architecture is most suitably done using the column-based databases as they provide data distribution, the flexibility of schema design, and the handling of unstructured and sparse data [5, 6]. This paper specifically proposes the implementation of the SaaS-based multi-tenant architecture using Cassandra, a column-based NoSQL database for health-based applications.

Software faults must be taken into consideration while building and implementing any type of architecture. Fault tolerance lets the system continue operating even in case of a system malfunction of any of the subcomponents [7]. This offers several benefits such as failure recovery, reduced hardware requirements, reliability, and improved performance metrics. So, implementing a fault-tolerant multi-tenant architecture is a major concern. In this paper the implementation of Proactive fault tolerance has been done, proactive fault tolerance refers to avoiding failures and faults in advance.

The rest of the article is structured as follows: In Sect. 2 the related work is discussed. In Sect. 3 the design of the proposed system is explained. In Sect. 4 the results of implementations are described. The conclusion has been included in Sect 5.

2 Related Work

The concept and implementation of a multitenant architecture have been studied at various instances quite recently but fault tolerance and load balancing in the architecture were not discussed [2]. This article proposes a NoSQL-based architecture for a multi-tenant data store for which it uses a column-based NoSQL family database, Cassandra. The article illustrates the implementation of data isolation among tenants while allowing flexibility of schema design or extension.

The author proposes the various multi-tenant data storage strategies [3]. These three strategies are defined between isolated and shared features of the databases. They are (a) *Separate Application, Separate Database*: In this model, every user has its own software and database. All tenants are completely isolated from each other. The major concern in this strategy is the maintenance and update costs, which take too much time and system resources are not used efficiently in this model. (b). *Shared Application, Separate Database*: All tenants are using the same and only one software. Also, all users have their own physically separated databases. Special methods are used so that software can be individually customized for each tenant according to the tenant's requirement. (c) *Shared Application, Shared Database*: Tenants use common software. After observing all these, concluded that multi-tenancy is now being applied to a limited number of areas, and that this situation needs to be improved, and that some infrastructure software must be prepared to facilitate implementation.

The authors of [4] have discussed the concepts of data isolation and application isolation in multi-tenant architecture in cloud-based systems. The article concludes that the multi-tenant data isolation is usually divided into three levels, namely: (a) *An independent database*: In this scheme, each tenant has a separate database, so tenant data isolation can be achieved at its highest level with the highest security, but the cost is too high (b) *A shared database, isolated data structure:* more or all tenants share the database, but each tenant corresponds to a separate Schema (c) *Shared database, shared data architecture:* all the tenants share the same Database Schema. This is the highest level of sharing with the lowest level of data isolation. Also, two application schemes have been explained in the paper: (a) *Sharing middleware approach can be achieved separately:* single application instance/shared middleware, multi-application instance/shared middleware, multi-application instance/isolation middleware, etc. (b) *The virtual approach:* mainly downs the tenant's application into the operating system level isolation. Respectively, based on operating system images for different users achieve application-level isolation.

A detailed study of fault tolerance in cloud computing has been discussed in [7]. This article presents various types and categories of faults and also the techniques used for fault tolerance which are categorized as follows: (a) *Reactive Fault Tolerance:* on the effective occurrence of any failure reactive fault tolerance reduces the effect of failures on application execution. (b) *Proactive Fault Tolerance*: it predicts the failures and faults and replaces the suspected components with the working components thus avoiding recovery from faults and errors. It also discusses the metrics on which fault tolerance can be considered. Some of the parameters include throughput, response time, scalability, performance, availability, usability, reliability, security, and associated overhead.

The article [8] discusses various concerns in multi-tenant SaaS applications. The authors have presented different high- level design concerns influencing the architecture. Initially, affinity and persistence concerns are taken into account, which is usually transparent to the tenants. They have focused on three concerns that might be key differentiators for competitors: performance- isolation, service-differentiation, and customizability.

The authors of [9] have analyzed different fault tolerance methods and listed the limitations of these methods. Authors have also developed a fault tolerance model that manages all types of faults in different areas of its applications. This paper discussed the applications of cloud computing and its importance. So, there is an efficient fault tolerance model required which protects the clouds from different faults and failures. This article analytically evaluates some generally used fault-tolerance models based on some parameters that are obtained from these models. Some of the models include (1) *LLFT*: Low Latency Fault Tolerance (2) *FTM*: Fault tolerance Middleware (3) *ASSURE*: An Autonomous fault management system (4) *SFD*: self-tuning fault detection system. (5) *BFT cloud:* Byzantine fault tolerance framework. (6) *VFT*: Visigoth fault tolerance. The different parameters that are used to evaluate the different models are: (a) *FT technique type*: can be proactive and reactive.

(b) *Performance*: which includes Substantial improvement in performance, stability maintenance of the system, and building a fault-tolerant system through backups. Based on the current state of the system they have classified it into two parts:(a) Static Load Balancing: the decision of shifting the load does not depend on the current state of the system. Knowledge about the applications and system resources is required. (b) Dynamic Load Balancing: the current state of the system is used to make any decision for load balancing; thus, the shifting of the load is dependent on the current state of the system. The need for load balancing has also been discussed to achieve green computing in the cloud: (1) Limited Energy Consumption: Load balancing can reduce the amount of energy consumption by avoiding overheating of nodes or virtual machines due to excessive workload. (2) Reducing Carbon Emission: reducing energy consumption is the major benefit of load balancing which reduces carbon emission and tends to achieve Green Computing.

The article [10] discusses the fault management in multi-tenant systems apart from various isolation techniques. The paper gives out several principles to handle the challenge of preventing fault propagation in a multi-tenant system: (a) *Fault Detection & Diagnosis:* It's the first stage to detect that something unexpected has occurred and quickly identify the currently infected tenant(s). (b) *Fault Propagation Prevention:* One basic principle of this step is to force the faster release of the critical shared resources to avoid possible fault propagation. (c) *On-Line Repair:* the faults of the ill tenants must be repaired during the runtime of the application instance.

A two-tier SaaS scaling and scheduling architecture Load balancing, in which duplication can occur at both service and application levels has been discussed in this article [11]. An algorithm that is cluster-based resource allocation was proposed in this paper. Lazy and Proactive two duplication timing models are proposed in this article.

This article [12] proposes a hybrid optimization technique to solve the multi-objective problem faced by decentralized networks. The optimized multi-objective routing for wireless communication is done in two phases. The first phase focuses on monitoring the mobile devices and in the second phase optimization of the objectives is performed. The proposed method is validated using network simulator-2 to show the increased throughput in comparison with old methods. It can be further improved in the context of a secured way of transmission by analyzing the trust of nodes.

This article [13] handles the problem of huge data flow along with secure and efficient data utilization in the internet of things. Authors have proposed a methodology in which mobile edge computing is integrated with the duplication process after considering the power utilization and the response time. The proposed method is simulated using a network simulator tool and the results obtained by the simulation depicted a reduction in power consumption and response time along with the enhancement in the bandwidth utilization.

The proposed system did not have the feature of high availability [14] which can be achieved by making the system fault-tolerant [2]. The paper [7, 9, 10] discusses various faults and fault-tolerant methods in cloud networks whereas, the paper [11] proposes the load balancing architecture After analyzing all the related work, it has been observed that so far there was no multitenant architecture in the healthcare field which can ensure security, data consistency, availability, and fault tolerance all at the same time on a distributed cloud network. This article proposes "A load balancing based cost-effective Multi-Tenant Fault Tolerant System" which is consistent, secure, and load-balanced by combining fault-tolerant and load-balancing concepts in our NoSQL-based architecture for a multi-tenant data store.

3 Proposed Work

The IT sector has become an important part of the healthcare industry for providing better services. This has caused the need for a separate IT department in hospitals and the pharmaceutical industry. The IT department in the healthcare industry requires both resources like expensive servers and software as well as qualified professionals to meet the user's demands which proved to be costly and also infeasible for many small businesses. Having separate applications in each hospital or clinic results in incomplete or inconsistent databases, causing latency in their services.

The proposed system implements a multi-tenant architecture as a SaaS application in the healthcare industry which eventually reduces the operating cost and could provide the latest services in the healthcare field to its customers. Along with reduced cost, multi-tenant architecture provides isolation and security to each tenant thus, preventing data leakage.

Key features of the proposed architecture include data isolation, fault-tolerant, and load-balancing to provide secure and consistent service to each tenant. The architecture can be expanded to provide services to different tenants in the healthcare industry who can use the same information and application for their operations to reduce their expenses. Consider the multiple tenants like hospitals, clinics, and fitness centers and sports academies which can utilize some common data like fitness and health. Level 4 of SaaS maturity level has been implemented, which uses the same application instances with configurable metadata for each tenant and has load balancer architecture [15].

3.1 Application Architecture

The proposed application is divided into three layers which are shared among the tenants. These three layers consist of the load-balancing layer, application layer, and data layer as shown in Fig. 2. Each layer is divided in such a way that enables easy scaleup and scale down of the application without changing the internal code and architecture. The application is distributed over multiple Datacenters which includes multiple Availability Zones. Consider the 2 data centers in which there are 2 availability zones in data center 1 and 1 availability zone in data center 2. Each availability zone contains 3 servers, one server having our user application and 2 servers having

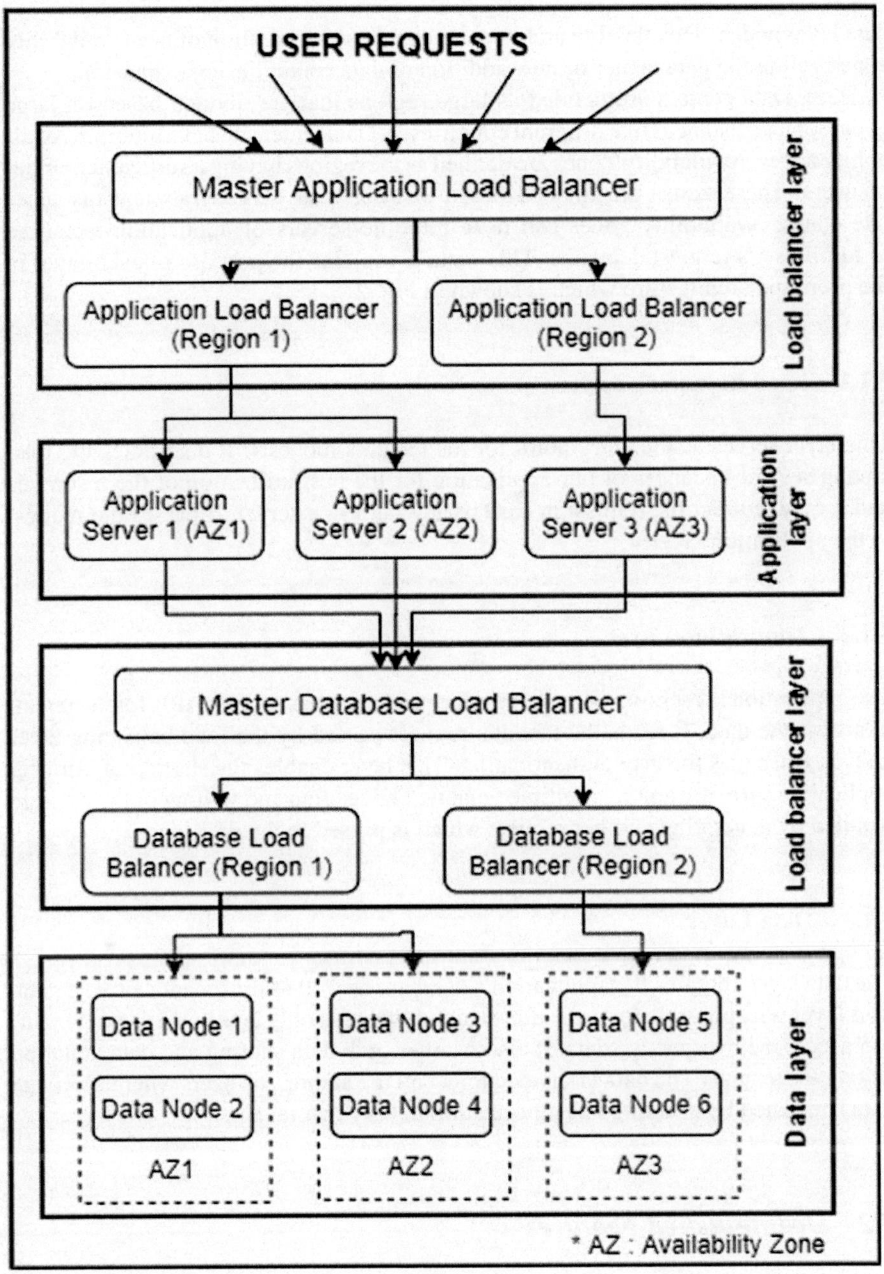

Fig. 2 Architecture diagram for the proposed application

data layer nodes. This flexible architecture enables easy addition of new availability zones within the data center or new addition of data center during expansion.

Here, Data centers are defined as large regions that are divided based on large geographical distances like different countries. A Datacenter can have multiple Availability zones. Availability Zones are defined as the regions having a sufficient number of users. These zones are divided based on user load or geographical distances like states. Availability zones can have multiple servers of application instances to fulfill user's request demands. This section explains the purpose of each layer in the proposed architecture which is shown in Fig. 2.

3.1.1 Load Balancing Layer

This layer serves as the entry point for the tenant's requests. It distributes the load among several instances of our application for the best utilization of the resources and it may redirect the request in case of any fault in a server. It passes the request to the application layer.

3.1.2 Application Layer

The application layer hosts our user interface and provides several APIs for the tenant to access the data. It reads the tenant's request passed by the load balancing layer and also manages the user authentication. This layer enables the sharing of our user application software among multiple tenants. The reading and writing of the data are handled by generating another request which is passed to the data layer.

3.1.3 Data Layer

The data layer contains the application's database as well as the tenant's private data. This layer is responsible for user authorization such that only legitimate users (tenant) can access their respective data. It also handles both data sharing and data isolation among the tenants. The data is shared among all the authorized users whereas private data is isolated by granting its authorization to only that tenant.

3.2 Distinguished Features

The proposed system implements the following features to have a robust system.

3.2.1 Multi-tenancy

In the proposed architecture tenants are the customers like health care organizations, training institutes which may need our services. Multi-tenancy systems provide the sharing of software and hardware resources among the tenants which decreases the cost of application for the user compared to hosting a dedicated application. Our application architecture allows sharing user applications at the application layer and data sharing at the data layer. The hardware is shared at all the layers to serve as a single multi-tenant application. The architecture provides easy scale up and scales down of the application which enables easy entering of the new tenant and also provides an option to leave easily.

The working of the proposed multi-tenant architecture system is shown in Fig. 3. There are three tenants for the system, namely the Health center (Tenant 1), Sports Academy (Tenant 2), and Fitness Center (Tenant 3). The architecture shows that the tenants are independent of each other and thus more tenants can be added easily. Each tenant has multiple users which are registered on our platform using the tenant's unique ID. The user accesses the application's functions using this unique ID. When a user requests the tenant's database, it passes through Tenant Isolation Component (TIC) which uses the tenant's unique ID to identify the user. The identified user request is passed to Data storage. The desired data is accessed through the materialized views [2] formed using Cassandra NoSQL datastore created for each tenant, which prevents the access of unauthorized data by the tenant. The final output passes through the TIC and is sent to the user.

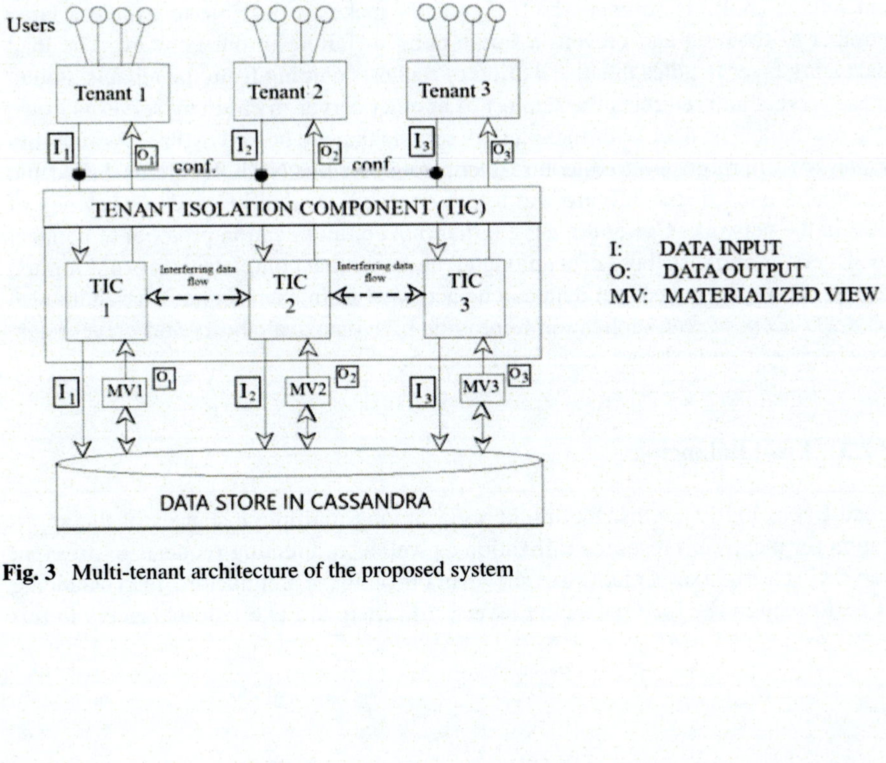

Fig. 3 Multi-tenant architecture of the proposed system

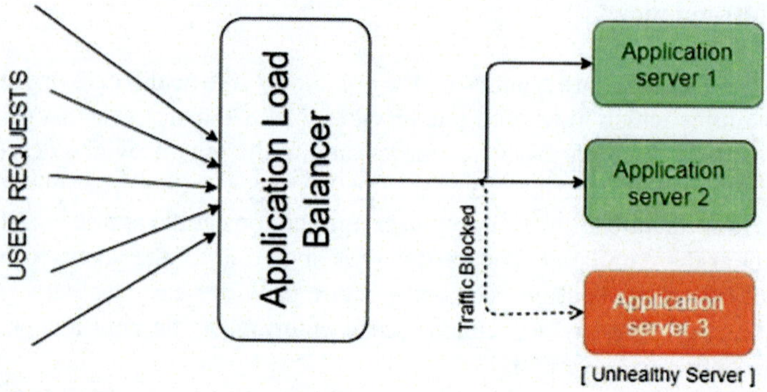

Fig. 4 Working of load balancer

3.2.2 High Availability

The proposed system provides high availability of data by implementing a distributed network of systems. A distributed network is hosting an application on multiple servers that are connected over the internet to act as a single application for the user. The distributed network provides high availability of data to the tenants and prevents any single-point failure of the application as shown in Fig. 4. Multiple servers can be used in the application layer and the data layer to form a distributed network. There are three servers in the application layer hosted on each Availability Zones which contains the same instance of our user application. It enables us to distribute our load to multiple servers which eventually makes it possible to handle a large number of requests and prevent a single point of failure in our network. The load balancing layer handles the distribution of requests coming from the tenants among these servers and redirects the request to another server in case any server is down due to a fault. The data layer includes six servers that are hosted in three Availability Zones to form ring-type architecture. Here, datastore like NoSQL column-based and Cassandra is used to configure and to provide high availability and consistency of data in the network. Cassandra uses an internal communication protocol to connect to all servers and distribute data among all nodes so that complete data is not limited to just one server instead all data can be accessed from each server. Cassandra also uses a concept of data replication to provide high data availability and prevent data loss in case of node failure as shown in Fig. 5.

3.2.3 Load Balancing

Load Balancing is the distribution of work among multiple instances of the application for maximum resource utilization as well as redirecting requests to different servers in case of a server failure as shown in Fig. 4. In the architecture, load balancing is performed in the load balancing layer [16]. There are two load-balancers in this

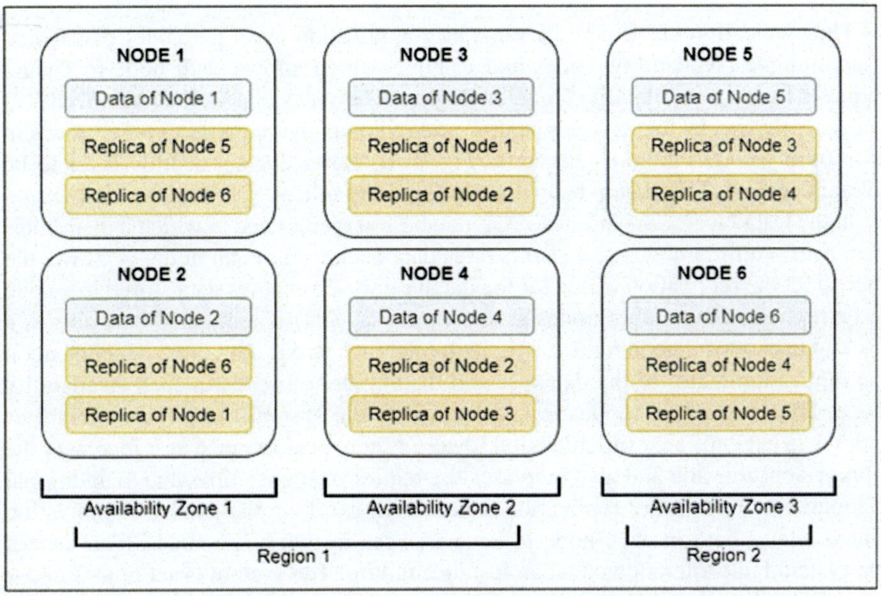

Fig. 5 Distribution of data in cassandra database having replication factor as 3

layer were created, one for application layer and the other for the data layer as shown in Fig. 2. An AWS elastic load balancer is used to divide load among the servers and continuously monitors server health to prevent any request to go to a bad server.

The architecture can be expanded to have multiple load balancers for each data center having servers in multiple Availability Zones for both the application layer and data layer. It will have a master load balancer for the application layer which stores a function table that contains information about all the available load balancers along with their location. Using the function table, the master load balancer will redirect users' requests to the nearest load balancer i.e. the user's request is redirected to the load balancer of its own region. This enables us to easily scale up our application to multiple regions without any changes in the application.

3.2.4 Fault-Tolerant System

Many unexpected errors could occur in the application. Therefore, the application must be designed to handle such errors without any latency in its services. The distributed system used in our architecture allows us to handle Fail-stop behavior type system faults easily. Our architecture allows the application to run consistently in case of an instance failure where a single node is not working, and Availability Zone failure where all nodes in an Availability Zone are not working, and a Regional failure where all nodes in a region are not working.

The application layer uses the same user application in all the nodes distributed over multiple Availability Zones and Regions which allows each node to handle requests from an Availability Zone or Region. The AWS elastic load balancer [17] consistently checks the "system health" and redirects the requests to other nodes in case of any node failure as shown in Fig. 4. It allows our application layer to be tolerant against all Fail-stop behavior type system faults.

In the Data layer, Cassandra NoSQL database server is used in which it distributes data over multiple nodes and also creates data backup for each node. It allows the user to set the replication factor for the database which enables data stored in a node to be replicated over other nodes as shown in Fig. 5. This replicated data acts as a backup in case of a node failure. The fault tolerance in our data layer depends upon the replication factor of the database and the number of nodes in each Availability Zone. The increase in replication factor makes the system more fault-tolerant as more data backups are available but it lowers system performance as it increases the storage consumption and also increases the request-response time due to additional computation. Here set of replication factors such as "3" in the database application which allows data in each node to have replicas in two other nodes. This makes the system fault-tolerant against an instance failure. The system is set upto 2 nodes per Availability Zone whereas our replication factor is 3 which makes the system fault-tolerant against single Availability Zone failure.

Many faults can occur in a distributed architecture like network partition faults and state consistency faults. The independent nature of our user application server in the application layer makes them resistant to these types of faults and in the data layer, these faults are handled by Cassandra.

3.2.5 Data Isolation

Data isolation is an integral part of multi-tenant architecture. Data isolation allows the use of the same application by multiple tenants without the risk of any data theft. In the proposed system, data isolation is handled in the data layer. Here, both the shared database shared table schema and shared database isolated table schema were utilized [4]. And also a separate table is used in the database for tenant's private data where only that authorized tenant is granted access to it. The shared data table schema is used for the common data which is shared among the tenants. Here, the concept of Materialized views present was used in the Cassandra database to restrict data access by the tenants. Materialized view gives the user only read access to a portion of the original data table. This prevents any unapproved changes in the shared data table by the tenants and also restricts access to all datas in the table. Different materialized views are created by having different restrictions on shared data. The tenants are granted authorization to these materialized views depending upon their requirements.

4 Results

The system is implemented using the Amazon Web Services' technology, Elastic Compute Cloud (EC2) [17], which is used for access secure, resizable compute capacity in the AWS cloud [18]. An EC2 instance is a virtual server in Amazon EC2 used for running applications.

The EC2 instances for our user application server are shown in Fig. 6. Our data is distributed in 6 nodes, spread across 3 Availability Zones (rack1 and rack 2 in dc1, rack 1 in dc2) and in 2 Datacenters (dc1 and dc2) as depicted in Fig. 7. The Elastic Load Balancer (ELB) [17], specifically network load balancer has been implemented for the incoming requests to the server. On testing the load balancer by sending some requests to the server, the following results were obtained using network load balancer metrics.

4.1 Active Flow Count

Figure 8 shows the active flow count of the incoming requests which is the total number of concurrent connections from the client to target which is zero in our case as there are no previous or active connections.

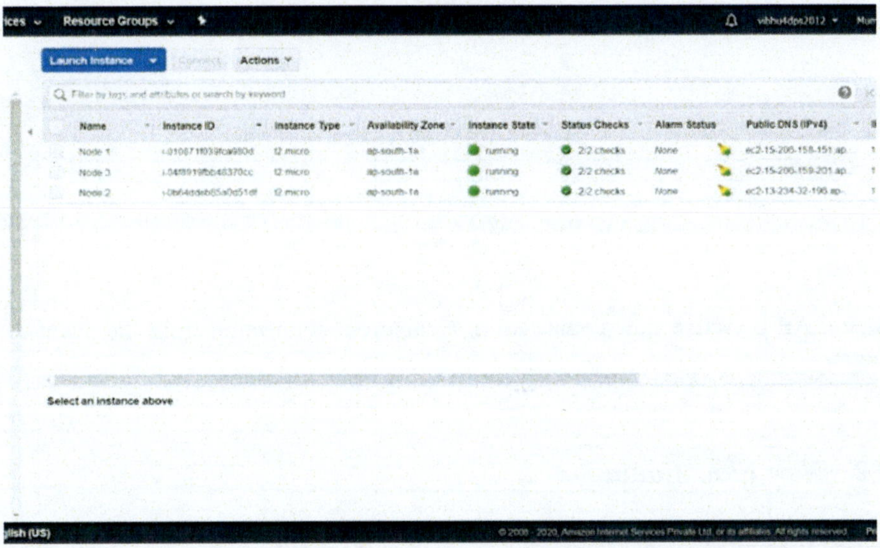

Fig. 6 EC2 instances of application server

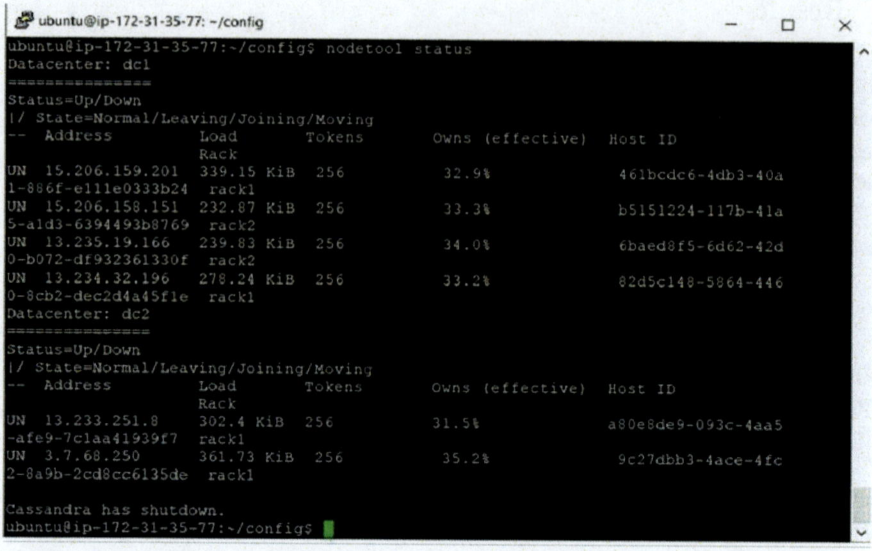

Fig. 7 Data clusters generated in Cassandra with 6 different nodes

Fig. 8 Active flow count in the ELB

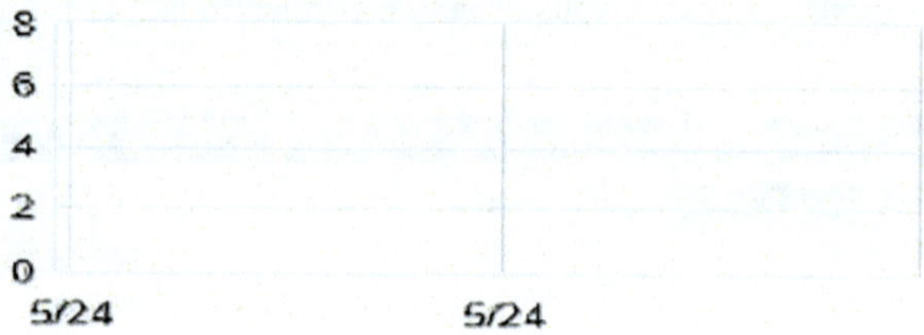

$$AF = \sum_{i=1}^{n} a_i \tag{1}$$

where, AF is Active flow count and a_i Concurrent connection from the client to target.

4.2 New Flow Count

Figure 9 Shows the new flow counts which is the total number of new connections established from the client to target in the given time period. According to the observations seven new connections were established from the client to target.

Fig. 9 New flow count in the ELB

$$\mathrm{NF} = \sum_{t=t_i}^{t_f} \sum_{i=1}^{n} n_{ij} \tag{2}$$

where NF is New Flow Count, t_i is the starting time of a given time period, t_f is the final time of a given time period and $\boldsymbol{n_{ij}}$ is new connections from the client to target.

4.3 Load Balancer Reset Count

Figure 10 shows the load balancer reset count which is the total number of reset packets sent from client to target and these are generated by the client and forwarded by the load-balancer. After performing several experiments, it was observed that two packets were resent from the client to target.

$$\mathrm{RC}_L = \sum_{i=1}^{n} r_i \tag{2}$$

where, RC_L is the Load balancer reset count and r_i is the resend packets from the client to target.

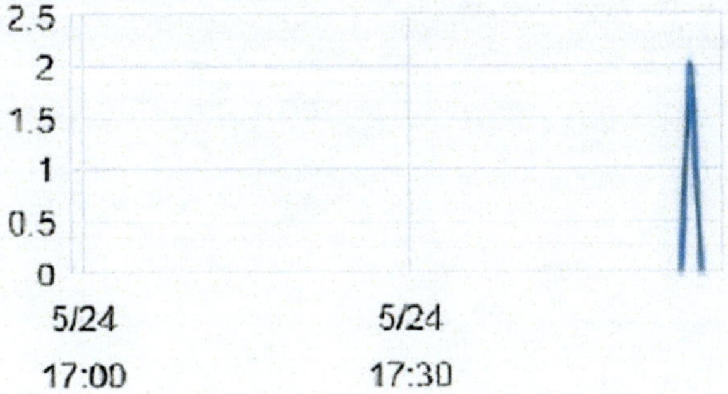

Fig. 10 Load balancer reset the count in the ELB

Fig. 11 Number of bytes Processed by the ELB

4.4 Byte Count

Figure 11 Shows the processed bytes which is the total number of bytes processed by load balancer termed as data bytes and TCP/IP headers bytes. The number of bytes observed from the resulting graph was over one lakh.

4.5 Target Reset Count

Figure 12 shows the target reset count which shows the total number of reset packets sent from the target to the client and these are generated by the target and forwarded by the load balancer. There are about 125 packets were resent from the target to the client.

$$\mathrm{TR}_L = \sum_{i=1}^{n} T_i - \mathrm{RC}_L \tag{4}$$

where, TR_L is the Target reset count, T_i corresponds to the resend packets between client and target and RC_L is the load balancer reset the count.

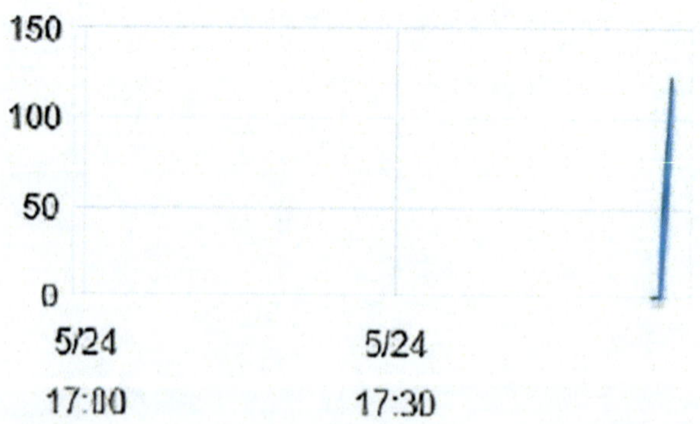

Fig. 12 Target reset count in the ELB

5 Conclusion

The proposed application has successfully implemented a scalable and cost-effective, multi-tenant system for the healthcare industry. Besides, the system is highly faults-tolerant by ensuring high availability of data. The dynamic load of incoming tenants was handled, and an elastic load-balancer has been implemented in the system and making this system highly scalable. Cassandra NoSQL database systems have resulted in creating distributed data layers easily over multiple regions and availability zones with its internal communication protocol preventing any type of inconsistency in the network. The features like Materialized View and user authorization in Cassandra are used to implement data isolation which is the key feature of our Multi-tenant system. Cassandra also provides customization for CONSISTENCY_LEVEL and REPLICATION FACTOR which are used to increase data availability and make the system fault-tolerant.

References

1. U. Divakarla, G. Kumari, An overview of cloud computing in distributed systems, in *International Conference on Methods and Models in Science and Technology 2010*
2. A. Sharma, P. Kaur, *A Multi-Tenant Data Store Using a Column Based NoSQL Database* (IEEE, 2019)
3. G. Karataş, F. Can, G. Doğan, C. Konca, A. Akbulut,Multi-tenant architectures in the cloud: A systematic mapping study, in *2017 International Artificial Intelligence and Data Processing Symposium (IDAP)* (Malatya, 2017), pp. 1–4. https://doi.org/10.1109/IDAP.2017.8090268
4. M. Yang H. Zhou New solution for isolation of multi-tenant in cloud computing, in *3rd International Conference on Mechatronics, Robotics and Automation (ICMRA 2015)*
5. B. Sethi, S. Mishra, P.K. Patnaik, A study of NOSQL database. Int. J. Eng. Res. Technol. (IJERT) IJERT (2014) ISSN: 2278-0181
6. M. Madison, M. Barnhill, C. Napier, J. Godin, NoSQL Database Technologies. J. Int. Technol. Inf. Manage. **24** (2015)
7. S. Kaur, G. Singh Review on fault tolerance techniques in cloud computing. Int. J. Eng. Manage. Res. (2250-0758), (May-June 2017)
8. R. Krebs, C. Momm, S. Kounev, Architectural concerns in multi-tenant saas applications, in *2nd International Conference on Cloud Computing and Services Science (CLOSER-2012)*
9. A. Ganesh, M. Sandhya, S. Shankar, A study on fault tolerance methods in cloud computing. IEEE Int. Adv. Comput. Confe. (IACC) (2014)
10. C. Jie Guo, W. Sun, Y. Huang, Z.H. Wang, B. Gao, A framework for native multi-tenancy application development and management, in *The 9th IEEE International Conference on E-Commerce Technology and The 4th IEEE International Conference on Enterprise Computing, E-Commerce and E-Services (CEC-EEE 2007).*
11. W.-T. Tsaian, X. Sun, Q. Shao, G. Qi Two-tier multi-tenancy scaling and load balancing
12. S. Ammayappan, Optimized multi-objective routing for wireless communication with load balancing. J. Trends Comput. Sci. Smart Technol.106–120 (2019). https://doi.org/10.36548/jtcsst.2019.2.004
13. N. Bhalaji, Efficient and secure data utilization in mobile edge computing by data replication. J. ISMAC **2**(01), 1–12 (2020)
14. D. Mani, A. Mahendran, Availability modelling of fault tolerant cloud computing system. Int. J. Intell. Eng. Syst. (2017)

15. T. Kwok, T. Nguyen, L. Lam, A software as a service with multi-tenancy support for an electronic contract management application, in *2008 IEEE International Conference on Services Computing*
16. F.F. Kherani, J. Vania (2014) Load balancing in cloud computing. Int. J. Eng. Dev. Res.(2014)
17. Amazon Elastic Load Balancing Developer guide 2012. https://aws.amzon.com/elb.
18. Amazon Web Services. https://aws.amazon.com.

A Comprehensive Study of Machine Translation Tools and Evaluation Metrics

Syed Abdul Basit Andrabi and Abdul Wahid

Abstract In this article, the ideas of statistical and neural machine translation approaches were explored. Various machine translation tools and machine translation evaluation metrics were also investigated. Nowadays, machine translation plays a key role in the society where different languages are spoken as it removes the language barrier and digital divide in the society by providing access to all information in the local language which a person can understand. There were different phases of machine translation as its evolution is concerned, and different approaches were followed in different phases some requiring an enormous amount of parallel corpus which is considered a crucial element of machine translation. In the proposed system, some of the parameters were examined to carry the analysis of several translation tools, and evaluation metrics are also available for accessing the quality of machine translation.

Keywords Parallel corpus · Language barrier · Statistical machine translation · Neural machine translation · Deep learning

1 Introduction

Machine translation is a method for translating one natural language with computers into another natural language. It is the subfield of computational linguistics. Machine translation systems are needed to overcome the language barrier and prevent a digital divide in society. The work on machine translation started in the 1940s with warren weaver one of its pioneers. There are several machine translation approaches [1, 2]as shown in Fig. 1. In the early phase, machine translation was performed using direct and interlingua approaches. Example-based approach was used from 1980 to 1990. In the late 90 s, the focus was shifted to the data-driven approaches with the emergence of statistical, neural machine translation approaches, and later on deep learning-based models shown in Fig. 2. The main reason that SMT and NMT became the promising

S. A. B. Andrabi (✉) · A. Wahid
Department of CS and IT, Maulana Azad National Urdu University, Hyderabad, India
e-mail: sbasit.11.rs@manuu.edu.in

V. Suma et al. (eds.), *Inventive Systems and Control*,
Lecture Notes in Networks and Systems 204,
https://doi.org/10.1007/978-981-16-1395-1_62

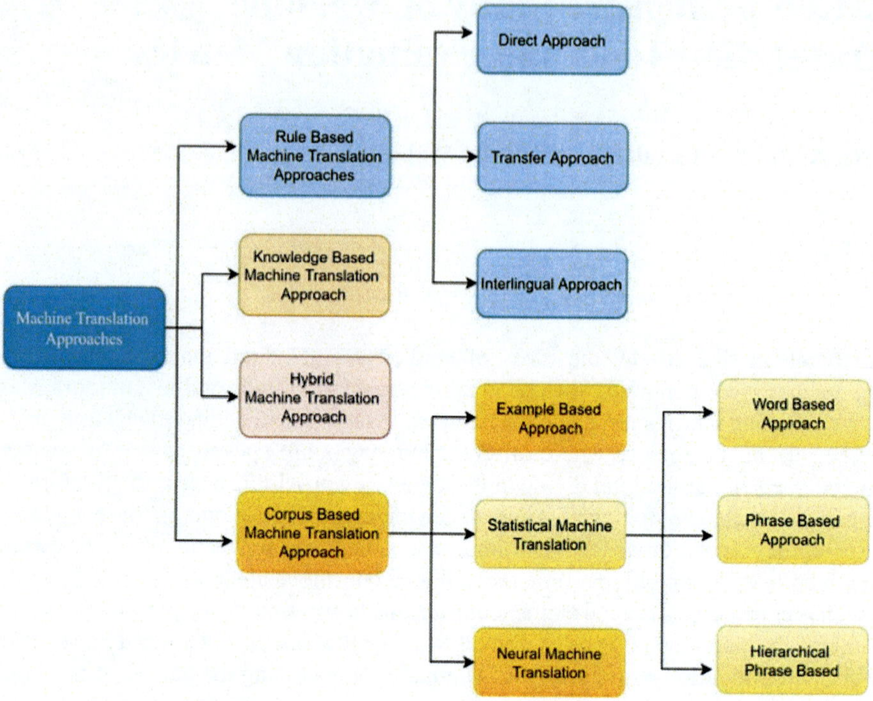

Fig. 1 Machine translation approaches

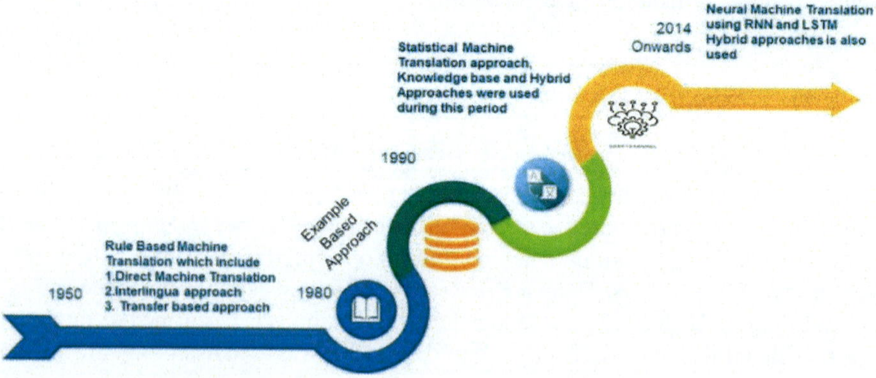

Fig. 2 Machine translation evolution

approaches is because of their scalability and extensibility to other languages, but other approaches were also used, as for SMT and NMT enormous amount of parallel corpus is needed to train these systems which were not available for all languages, so other approaches were also followed by designing some knowledge bases for resource-poor languages.

1.1 Statistical Machine Translation (SMT)

It is a corpus-based or data-driven approach to machine translation. It requires an enormous amount of parallel corpus to build the translation system. In this approach, system learns and extracts rules from the parallel data. The SMT consists of three main components: language model, translation model, and decoder [1]. It is based on the statistical models and noisy channel model for communication, which was introduced by Shanon in 1948 [2]. SMT is based on Bayes theorem of probability.

From the concept of Shanon's noisy channel model, consider a distorted message R (Foreign String f) a model to know how the message is distorted (translation model $t\ (f \mid e)$) and also a model on which original messages are probable (language model $p(e)$). Our objective is to retrieve the original message S (English string e). Mathematically, it can be written as given an English sentence e, seek the foreign sentence f that maximizes $P\ (f \mid e)$ which is the most likely translation can write as $\underset{f}{\operatorname{argmax}}\ p(f|e)$ which means the foreign sentence f, out of all sentences which produces the highest value for $P(f \mid e)$.

$$P(f|e) = \frac{P(f) * P(e|f)}{P(e)} \tag{1}$$

where $P(f \mid e)$ is posterior probability, $P(e \mid f)$ is likelihood, $P(f)$ is prior probability, and $P(e)$ is marginal probability.

$$\underset{f}{\operatorname{argmax}}\ P(f|e) = \underset{f}{\operatorname{argmax}}\ P(f) * P(e|f) \tag{2}$$

1.2 Neural Machine Translation (NMT)

With the advancements in computational power and cloud computing, a new promising data-driven approach of machine translation came which maps source and target language in an end-to-end manner. Neural machine translation uses the encoder and decoder architecture which actually consists of recurrent neural network (RNN). The power of deep learning has led to the development of NMT as the most powerful algorithm for this task. This deep learning state of art algorithm is used to train models using massive datasets that are capable of translating between any two languages. From the mathematical point of view, the basic aim of NMT system is to find a sequence of words $Y = (y_1, y_2, \ldots, y_t)$ from a target sentence Y that maximizes the conditional probability of Y for a given sentence $X = x_1, x_2, \ldots, x_s$. and can be represented as sequence of vectors in vector c. This is formulated using the RNN encoder decoder architecture. The encoder network takes input and creates a fixed-length vector, whereas the decoder generates translated output text from the encoded vector [3, 4]. Mathematically, it can be represented as follows:

$$\log P(y|x) = \sum_{i=1}^{t} \log P(y_i, y_{i-1}, \ldots, y_1, x, c) \tag{3}$$

At encoder side

$$h_t = f(x_t, h_{t-1}) \tag{4}$$

$$c = q(\{h_1, \ldots, h_{Tx}\}) \tag{5}$$

where h_t at time t represents hidden state and vector c is generated from a sequence of hidden states. The decoder predicts the next word given the context vector c. From Eq. (3), at the decoder side, $P\left(y_i, y_{i-1,}, \ldots, y_1, x\right)$ is computed as:

$$P(y_i, y_{i-1}, \ldots, y_1, x, c) = g(y_{i-1}, f(s_{i-1}, y_{i-1}, c_i), c_i) \tag{6}$$

y_{i-1}, is previous predicted target word, s_{i-1} is decoder hidden state, and c_i is the context of the word and is calculated as follows:

$$c_i = \sum_{j=1}^{Tx} \propto_{ij} h_j \tag{7}$$

$$\propto_{ij} h_j = \frac{\exp(e_{ij})}{\sum_{k=1}^{T_x} \exp(e_{ik})}, \text{ where } e_{ij} = a\left(s_{i-1}, h_j\right) \text{is alignment score} \tag{8}$$

2 Overview of Machine Translation Tools

There are several machine translation tools available in the literature. In this article, the main focus is on the statistical and neural machine translation tools. These tools are discussed below.

2.1 Moses

It is an open-source statistical machine translation tool. It consists of many other tools for preprocessing data developing language models and translation models. It was developed by Phillip Koehn and et. al at the University of Edinburgh. It was developed in C++ and follows modular and object-oriented design. It uses some external tools for alignment and language modeling. This was initially hosted on sourceforge.net and is nowadays available on GitHub. It is licensed under GNU lesser general public license (LGPL) [5].

2.2 SRLIM

It is developed in C++ and consists of several libraries and executable programs and scripts for language modeling, speech recognition, tagging, machine translation, segmentation, and other language processing applications. It consists of 50 C++ classes for language modeling and several helpers and wrappers and 14 executable tools. It was developed by SRI Speech Technology and Research Laboratory in 1995. It runs on Windows and UNIX platforms, and it is free for use in schools, universities under the SRI research community license [6].

2.3 CMU Cambridge Tool Kit

Its first version was released in 1994 and was used to construct and test bigram and trigram language models with the increase in parallel corpuses and computation power the interest for moving beyond the trigrams toward 4 and 5-g models arise, and the version 2 of the toolkit was developed to support these models. The second version supports multiple discounting methods and efficient memory usage [7].

2.4 Pharaoh

It is a statistical machine translation decoder tool developed by researcher Phillip Koehn as part of his Ph.D. thesis at the University of South California to aid the researchers in the area of machine translation. It uses a beam search algorithm. The language and translation model is given to the pharaoh translates text from one language to another language. For non-commercial use, you have to get a license from the University [8].

2.5 Mood

Modular object-oriented decoder is an open-source translation decoder and proposed to overcome the license issue of the pharaoh decoder. It is mentioned that the decoder must resolve two problems one is the model representation and the other is efficient space exploration of possible translations. To achieve these results, mood follows a separate model and search space exploration framework. It is developed in C++ and is licensed under GNU general public license (GPL) [9].

2.6 Marian NMT

It is a neural machine translation tool developed by Microsoft, Adam Mickiewicz University in Poznan and the University of Edinburgh. It is written in C++ language. It does not have Python bindings and has its own backend which provides reverse-mode automatic differentiation. Marie is distributed under the MIT license and is available on Github [10].

2.7 Nematus

It is a neural machine translation tool developed by Rico sennrich et. al. It is written in Python and uses the framework with attention-based encoder and decoder architecture. Training is performed using stochastic gradient descent, and the objective is cross-entropy minimization on training corpus. Netmatus supports several metrics for evaluation such as BLEU, METEOR, and BEER. It also gives a graphical visualization of attention weights and beams search graph. It is available under a BSD license [11].

2.8 Thot Tool

It is the statistical machine translation tool. It is developed in C, C++, Python, and shell scripting language. Initially, that was used to train phrase-based models but currently, it supports all the models used in different stages of translation. This tool incorporates interactive machine translation and online learning. It is available under GNU LGPL license [12].

2.9 Open AI Kit (OAK)

It is an open-source toolkit consisting of several machine learning models for the development of statistical machine translation. Due to the non-commercial license clause, many tools cannot be used for educational and research purposes. It is a development library with an interface instead of executable programs. It is written in C++ to use a variety of available open-source tools like test coverage tools, proofing, and allocation tools and following the object-oriented design. OAK adopts the terms of the University of California, Berkeley software distribution (BSD) license as certified by the open-source initiative [13].

2.9.1 KenLM

It is a language modeling toolkit used in statistical machine translation. It is developed by Kenneth Heafield at Carnegie Mellon University. Two data structures probing and TRie are implemented in the model. The main purpose of probing data structure is to increase the speed, and the aim of TRie is to reduce memory usage and CPU time. It is also open source and is developed in C++ [14].

2.9.2 Irstlm

It is a language modeling toolkit developed for statistical machine translation, and it has been deployed with Moses for machine translation and FBK-first speech recognition system. With the increase in the corpus size, the demand for language models handing huge corpses surge among the research community, some existing language modeling toolkits updated their libraries to cope with the demand. IRSTLM was developed to handle large corpus with the existing hardware and speed up the computation time and reducing memory requirements [15].

2.9.3 Open NMT

It is an open-source neural machine translation (NMT) toolkit. The main aim of this toolkit is to enhance the training and test efficiency, modular design, readability, and scalability. It is based on the conditional probability modeling toolkit. It is a complete library for training and deploying neural machine translation models. It supports memory sharing and multi-GPU for parallelism [16].

2.9.4 Apertium

It is an open-source platform that was initially used to develop machine translation systems for closely related languages, but later it was expanded for distant language pairs. The Apertium consists of an MT engine, toolbox, and data to build a rule-based machine translation system. Apertium performs three different functions one is lexical processing using finite-state transducers, parts of speech tagging (POS) using a hidden Markov model, and structural transfer using multi-stage finite-state chunking [17].

2.9.5 Joshua

It is a statistical machine translation-based open-source tool kit that implements all algorithms for synchronous context-free grammars (SCFGs). It also contains parsing, *n*-gram language models, decoding algorithms like beam and cube parsing, and *k* best extraction. It also implements minimum error rate training. While implementing the tool, three goals of software engineering were taken into consideration extensibility, scalability, and end-to-end coherence [18].

2.9.6 Apache Open NLP

It is a toolkit for machine learning and natural language processing. It performs several tasks of NLP like tokenization, parts of speech tagging, named entity identification, sentence segmentation, chunking, parsing, and language detection. It is developed by volunteers, and it is open for everyone to contribute by simply joining the mailing list. It is written in java and is Apache license which allows users to use, modify, and distribute [19].

2.9.7 MALLET

It is an open-source java-based statistical natural language processing toolkit mainly used for clustering, topic modeling, document classification, information extraction, and other applications. Mallet also contains sequence tagging, hidden Markov models, and a topic modeling toolkit with latent Dirichlet allocation, Pachinko allocation, and hierarchical LDA. Besides, this mallet contains routines for numerical representation of text. Mallet is released under common public license [20].

2.9.8 TF-LM

It is a tensor flow-based open-source language modeling toolkit for neural machine translation. This tool provides many options for input–output, batch for both the training, and, testing, and also it is very easy to adapt. It will first try to run on GPU automatically; if GPU is not available or the user wants to run it on CPU, then he has to give command device CPU. The model can be used to find the perplexity, prediction of the next word, and re-score hypothesis [21].

3 Analysis of the Machine Translation Tools

In this article, several machine translation tools were mentioned for different purpose, license, and the development language. In this section, the short description based on the language, license, and approach for which it was build were given. Table 1 describes the machine translation tools.

After studying the various machine translation tools, it was found that some tools are used at different stages of statistical machine translation like language modeling, alignment, and decoding stage. Language modeling tools can be integrated with Moses toolkit at different stages. It was also found that SMT and NMT are popular approaches used by different translation systems. With the power of deep learning, there is a paradigm shift toward neural machine translation, and after 2016, all the tools were designed for NMT. It is required to train these deep learning models with a large amount of parallel corpus to get good results.

4 Machine Translation Evaluation Metrics

The machine translation evaluation is broadly classified into three different categories human evaluation, automatic evaluation, and diagnostic evaluation. In this article, automated metrics were mentioned for assessing the quality as human evolution which is a slow and costly approach. The various automated metrics are given below:

4.1 BELU

("Bilingual Evaluation Understudy"): It is a popular metric used to excess the quality of the translation system output. It is developed by Kishore papineni. The BELU score of any machine translation system is calculated by counting the words of translation given by the system that have a match in the reference translation. The value of the BELU score lies between 0 and 1. If the output of the MT system matches exactly with the reference, then get a BELU score as 1 which is not possible [23]. The formula to calculate the BELU score is given below:

$$\text{Precision}(p) = \frac{\text{No. of candidate translation words occuring in reference}}{\text{Total number of words in candidate translation}} \tag{9}$$

Table 1 Comparison of machine translation tools

Year	Tool name	Language	License	Purpose
1995	SRLIM [6]	C++	SRI research community license	Statistical machine translation
2002	MALLET [20]	Java	Commercial public license	SMT
2004	Apache open NLP [19]	Java	Apache license	NLP, machine learning
2004	Paroah [8]	C++	Under license agreement from university	SMT decoder problem is license issue
2005	Open AI [13]	C++	BSD license	Machine learning, SMT
2006	Mood [9]	C++	GNU LGPL	SMT decoder
2007	Moses [5]	C++	Open-source LGPL	SMT, data preprocessing, sentence alignmnet
2008	Thot [22]	C, C++, Python, shell scripting	GNU LGPL	SMT
2008	IRSTLM [15]	C++	GNU less general public license	SMT language modeling
2009	Joshua [18]	Java	Apache open source	SMT was modified in 2017
2010	KenLM [14]	C++	LGPL	SMT language modeling
2011	Apertium [14]	C++	GNU	Rule-based machine translation
2017	Open NMT [14]	Python	MIT license	NMT
2017	TF-LM [14]	Tensor flow, Python	Open source	NMT
2017	Nematus [11]	Python using theno framework	BSD license	NMT
2018	Marian [10]	C++	Open-source license MIT	NMT

$$\text{Brevitypenality(bp)} = \begin{cases} 1 & \text{if } c > r \\ e^{1-r/c} & \text{if } c \leq r \end{cases} \tag{10}$$

$$\text{BELU} = \text{bp.Exp}\left(\sum_{n=1}^{N} w_n \log p_n\right) \tag{11}$$

4.2 NIST

The name NIST comes from USA "National Institute of Standards and Technology." It is another metric to assess the quality of the machine translation system, and it is similar to the BELU metric with little modifications. The main objective of this is to improve the BELU metric to prevent inflation of SMT. NIST uses higher weights for rare words and arithmetic mean of n-grams matches between candidate translation and reference translation [24].

4.3 TER

(Translation Edit Rate): Its main aim was to provide adaptive evaluation metrics requiring small data as compared to other metrics. It counts the number of edits required in candidate translation to match it with the reference translation in terms of fluency and semantics. Edits include insertions, deletions, substitutions, and shifts [25]. The formula to calculate TER is as follows:

$$\text{TER} = \frac{x \text{ of edits}}{\text{average } x \text{ of reference words}} \tag{12}$$

4.4 METEOR

It is another automated metric for assessing the quality of machine translation. METOR stands for "Metric for Evaluation of Translation with Explicit ORdering." The BELU score considers precision, whereas METEOR combines precision and recall. It is based on an explicit word to word match between translation system output and reference translation. It also supports morphological matching that is words derived from the same root can also be matched. The main purpose of this was to overcome the problem like lack of recall and morphological variation that were present in BELU [26]. The unigram precision is calculated as follows:

$$p_u = \frac{m}{w_t} \tag{13}$$

$$r_u = \frac{m}{w_r} \tag{14}$$

where$m$ indicates unigrams in candidate translation found in reference,w_{t} andw_{r}, are unigrams in candidate translation and reference translation, respectively. The precision and recall are combined using harmonic mean with recall nine times more than precision.

$$F_{\text{mean}} = \frac{10p_u}{r_u 9p_u} \tag{15}$$

For longer matches, meteor computes penalty after grouping unigrams into chunks. The formula to calculate penalty is given below:

$$\text{Penality} = 0.5\left(\frac{\text{No. of chunks}}{\text{No. of unigrams matched}}\right) \tag{16}$$

The final score is calculated as

$$\text{Score} = F_{\text{mean}}\,(1 - \text{Penality}) \tag{17}$$

4.5 Word Error Rate (WER)

It is calculated by dividing the Levenshtein distance between the candidate translation and reference translation with the reference translation. The Levenshtein distance is the smallest number of deletions (D), insertions (I), and substitutions (S) required in candidate translation to get the reference translation. The problem with this metric is bias in shorter sentences [27]. The formula to calculate the word error rate is as follows:

$$\text{WER} = \frac{(D + I + S)}{N} \tag{18}$$

4.6 Orange

It is another automatic metric to assess the quality of machine translation output. This metric works at the sentence level and gives a high rate to good translation as compared to a poor case. This metric is based on two assumptions first one is references are good translations as compared to candidate translations and the other assumption is automatic translations are worst as compared to the reference. It takes the input sentence, its reference translation, and the candidate translation and calculates the average rank of reference translations within the list of candidate and reference translations [28].

Table 2 Comparison of evaluation metrics

Metric	Based on	Work at sentence/word level	Takes into account the morphological variation
BELU [23]	Precision and brevity penalty	Word	No
METEOR [26]	Unigram precision, recall and harmonic mean	Sentence	Yes
WER [27]	Levenshtein distance	Word level (number of replacements)	No
TER [25]	No. of modifications	Word	No
Orange [28]	Rank	Sentence	No
MaxSim [29]	Precison and recall	Sentence	Yes

4.7 MaxSim

It is a machine translation evaluation metric proposed by YS Chan et al. It is based on precision and recall. It also gives weight to the matches and takes into consideration the synonyms also [29].

5 Analysis of Automatic Metrics for Machine Translation

Several evaluation metrics were mentioned, and only the three parameters were selected such as base of metric, operation level, and capability of handling morphological variation to the comparative study of these metrics, and the finding are given in Table 2.

6 Conclusion

In this article, several translation tools of statistical and neural machine translation approaches were mentioned. To design any machine translation system, these tools are used to prepare the data in the required format. Some evaluation metrics were mentioned to access the quality of machine translation system output, and analysis of these metrics on the selected parameters was performed. After analyzing, it is concluded that one metric is not sufficient to evaluate the quality of the translation system.

Therefore, it is necessary to use at least three metrics to evaluate quality to take multiple parameters into account. In the future, we will use these metrics along with some other metrics to access the quality of existing machine translation systems for the language pair which is supported by all existing translation systems to find the overall efficient evaluation metric which covers all the parameters, So that, we can find the one metric which is necessary for assessing the quality of system output.

References

1. M. Zafar, Interactive English to Urdu machine translation using example-based approach. **1**(3), 275–282 (2009)
2. L.J. Schulman, Communication on noisy channel: a coding theorm for computation (IEEE Computer Society Press, 1992), pp. 724–733
3. D. Bahdanau, K. Cho, Y. Bengio, Neural machine translation by jointly learning to align and translate (2014), pp. 1–15
4. K. Cho, B. van Merrienboer, D. Bahdanau, Y. Bengio, On the properties of neural machine translation: encoder–decoder approaches (2015), pp. 103–111
5. P. Koehn et al., Moses: open source toolkit for statistical machine translation, in *Proceedings of the 45th annual meeting of the ACL on interactive poster and demonstration sessions* (2007), pp. 177–180
6. A. Stolcke, SRILM-an extensible language modeling toolkit, in *7th international Conference on Spoken Language Processing* (2002)
7. P. Clarkson, R. Rosenfeld, Statistical language modeling using the CMU-Cambridge toolkit, in *5th European Conference on Speech Communication and Technology* (1997)
8. P. Koehn, Pharaoh: a beam search decoder for phrase-based statistical machine translation models, in *Conference of the Association for Machine Translation in the Americas* (2004), pp. 115–124
9. A. Patry, F. Gotti, P. Langlais, MOOD: a modular object-oriented decoder for statistical machine translation, in *LREC* (2006), pp. 709–714
10. M. Junczys-Dowmunt et al., Marian: fast neural machine translation in c++, in *Proceedings of the ACL 2018—58th Annual Meeting of the Association for Computational Linguistics: System Demonstrations* (2015), pp. 116–121
11. R. Sennrich et al., Nematus: a toolkit for neural machine translation, in *15th Conference of the European Chapter of the Association for Computational Linguistics EACL 2017—Proceedings of the Software Demonstrations* (2017), pp. 65–68
12. Q.U. Guide, Thot Toolkit for statistical machine translation (2017)
13. D.J. Walker, The open "ai" kitTM: general machine learning modules from statistical machine translation, in *Workshop of MT Summit X, Open-Source Machine Translation* (2005)
14. K. Heafield, KenLM: faster and smaller language model queries, in *Proceedings of the 6th Workshop on Statistical Machine Translation* (2011), pp. 187–197
15. M. Federico, N. Bertoldi, M. Cettolo, IRSTLM: an open source toolkit for handling large scale language models, in *9th Annual Conference of the International Speech Communication Association* (2008)
16. G. Klein, Y. Kim, Y. Deng, J. Senellart, A.M. Rush, Opennmt: open-source toolkit for neural machine translation. arXiv Prepr. arXiv1701.02810 (2017)
17. M.L. Forcada et al., Apertium: a free/open-source platform for rule-based machine translation. Mach. Transl. **25**(2), 127–144 (2011)
18. Z. Li et al., Joshua: an open source toolkit for parsing-based machine translation, in *Proceedings of the 4th Workshop on Statistical Machine Translation* (2009), pp. 135–139
19. M. Mohanan, P. Samuel, Open NLP based refinement of software requirements. Int. J. Comput. Inf. Syst. Ind. Manag. Appl. **8**, 293–300 (2016)
20. A.K. McCallum, Mallet: a machine learning for language toolkit (2002). http//mallet.cs.umass.edu

21. L. Verwimp, P. Wambacq, et al., TF-LM: tensor flow-based language modeling Toolkit (2019), pp. 2968–2973. https://www.lrec-conf.org/proceedings/lrec2018/index.html
22. D.O. Mart\inez, Thot Toolkit for statistical machine translation (2018)
23. K. Papineni, S. Roukos, T. Ward, W.-J. Zhu, BLEU: a method for automatic evaluation of machine translation, in *Proceedings of the 40th Annual Meeting of the Association for Computational Linguistics* (2002), pp. 311–318
24. Y. Zhang, S. Vogel, A. Waibel, Interpreting bleu/nist scores: how much improvement do we need to have a better system?, in *LREC* (2004)
25. M. Snover, B. Dorr, R. Schwartz, L. Micciulla, J. Makhoul, A study of translation edit rate with targeted human annotation, in *Proceedings of Association for Machine Translation in the Americas*, vol. 200, no. 6 (2006)
26. S. Banerjee, A. Lavie, METEOR: an automatic metric for MT evaluation with improved correlation with human judgments, in *Proceedings of the ACL Workshop on Intrinsic and Extrinsic Evaluation Measures for Machine Translation and/or Summarization* (2005), pp. 65–72
27. A. Mauser, S. Hasan, H. Ney, Automatic evaluation measures for statistical machine translation system optimization, in *LREC* (2008)
28. C.-Y. Lin, F.J. Och, Orange: a method for evaluating automatic evaluation metrics for machine translation, in *COLING 2004: Proceedings of the 20th International Conference on Computational Linguistics* (2004), pp. 501–507
29. Y.S. Chan, H.T. Ng, MAXSIM: a maximum similarity metric for machine translation evaluation, in *Proceedings of ACL-08: HLT* (2008), pp. 55–62.

A Novel Approach for Finding Invasive Ductal Carcinoma Using Machine Learning

Vaishali B. Niranjane, Krushil Punwatkar, and Pornima Niranjane

Abstract Breast malignancy (invasive ductal carcinoma) is the commonest form of cancer found in women, and the fatality rate is high among them. Invasive ductal carcinoma diagnosis is a difficult task because it involves a doctor who scans the major diseases of the malignant region to ultimately identify high-risk areas. For quick detection of breast malignancy, there is a high scope of research in automated diagnostic system. Machine learning is an emerging field of science in data science that deals with how machines learn from experiences that eliminates human efforts and come up with advanced machines with minimal errors. Using the convolutional neural network approach, the traditional methods of injury recovery are performed. A neural network is a sequence of program codes that attempt to detect correlation in a group of data through a process that copies the functioning of the human brain. It can adapt flexible inputs which produced the best results without resetting the output conditions. The convolutional neural network trained large tissue areas from WSI to learn a categorized representation. Patients have diagnosed with IDC were selected to evaluate the WSI dataset. 80% patches were chosen for training, and 20% patches selected for autonomous testing. Slides are given to CNN for training models for the final prediction of malignancy.

Keywords Invasive ductal carcinoma · Machine learning · Convolutional neural network

V. B. Niranjane (✉)
Yashwantrao Chavan College of Enginering, Nagpur, Maharashtra, India
e-mail: 9449@ycce.in

K. Punwatkar · P. Niranjane
Babasaheb Naik College of Engineering, Pusad, Maharashtra, India

V. Suma et al. (eds.), *Inventive Systems and Control*,
Lecture Notes in Networks and Systems 204,
https://doi.org/10.1007/978-981-16-1395-1_63

1 Introduction

In the last few decades, ML techniques have been widely used in intelligent healthcare systems, especially for breast malignancy identification. IDC detection rate is 80% among breast cancer. Medical carcinoma is defined as cancer that started in the tissue or skin and spread inside the line of a body organ. Invasive means that cancer cells in the tumor of the breast have the ability to spread beyond the breast to the other parts of the body. This is routinely diagnosed by visual analysis of Haematoxylin and Eosin (H&E) tissue slides. Assessment of disease progression is limited and time consuming because histopathologist has to identify the disease cell from surrounding healthy cell [1]. The precise definition of invasive ductal carcinoma in WSI is important in the subsequent measurement of tumor induction and predict patient outcome [2]. Various research in this aspect has been carried out in which segmentation and features extraction of histological primitives nuclei were done which helps to distinguish between malignant region from benign region [3]. Recently, deep learning architecture has cached attention due to its accuracy in pattern recognition function and different computer viewing concept [1].

In this method, study of multiple nonlinear transformations of the data is carried out to abstract useful representation. Therefore, these methods are more useful than traditional methods because of the exceptional development of big data and strong digital computational resources. Digital pathology is immerging electronic data systems in which processing of digitalization of histopathology glass slides by digital scanners takes place. Full digital slide images are usually gigabytes in size and advanced parts utilized in clinical diagnostics. In developed countries, researcher is focusing their attention on computerized pathology [4]. An in-depth study of the largest digital pathology images gives a unique pattern of hidden information that is not possible to visualize by conventional human testing modules. A deep study structure is most effective in automated classification and detecting disease severity in computer-generated histopathology images [5]. The main aim is to use DL-based segments in four-sided tissue areas from the WSI found from regular sampling, thereby enabling the application of the classifier over the entire canvas of the WSI. In this program, train an in-depth study program that can provide advice to pathologists, even if they know nothing. The selected database contains diagnosed 162 diagnosed cases with BC IDC from which 113 selected slides for training purposes and 49 for self-examination slides [11]. The occurrence of breast cancer is high in females and is considered the second hazardous infection everywhere in the world because of its high death rate. The affected person can stay alive if the infection identifies before the appearance of major substantial changes in the body [6]. It is an infection in which uncontrolled breast cell growth happens. Different types of breast cancer occur according to types of changes that occurs in breast cells. Breast malignancy can instigate in different portions of the breast. The breast is formed by three main parts, lobules, ducts, and connective tissue. The lobules are the glands that produce milk. The ducts are pipes that carry milk to the nipple. The connective tissue/fibrous tissue gives shape and size to the breast and holds other structures in place. The majority of breast malignant growths starts in the conduits or lobules. Breast cancer can spread to surrounding tissue through the lymphatic system. When it is spread to a different part of the body, it is said to be metastasized.

2 Related Work

2.1 Breast Malignancy Types

The common types of breast malignancy are as follows.

2.1.1 Invasive Ductal Carcinoma (IDC)

Also known as infiltrating ductal carcinoma. This is a common type of breast cancer which begins in milk duct cell. Milk duct cells carry milk of the breast to the nipple. Invasive ductal carcinoma begins in the milk duct cells and then spread into the duct wall and surrounding tissues of the breast. Also there are chances to spread to other parts of the body [7].

2.1.2 Invasive Lobular Carcinoma

Cancer cells spread from the lobules to the breast tissues that are close by. These obtrusive disease cells can likewise spread and affects different pieces of the body [7].

2.2 Methods to Detect Breast Malignancy

Tests and methods used to analyze breast malignancy are as follows.

2.2.1 Breast Exam

It is a self-monitoring technique in which local examinations of the breast is carried out for any possible lump or swelling.

2.2.2 Mammogram

It is the process in which low energy X-ray is used to examine the human breast for the possibility of any mass or calcification.

2.2.3 Breast Ultrasound

It is a screening and diagnostic procedure used to detect any stiff tissue or tumors. It is the process in which low energy X-ray is used to examine the human breast for the possibility of any mass or calcification.

2.2.4 Removing a Sample of Breast Cells for Testing (Biopsy)

It is a diagnostic procedure in which a fine needle is inserted inside the mass for sample collection for cytological study to rule out the nature of growth.

2.2.5 Breast Magnetic Resonance Imaging (MRI)

The possibility of detection of cancer cells is higher in MRI as compared to mammography or ultrasound.

2.2.6 Using Mammogram Images

Mammography is considered to be an important technique in the investigation of breast cancer. It is utilized to recognize the sickness in the beginning phase when recuperation is conceivable. Computer-aided diagnosis system (CAD) aim is to read the mammograms, identify the affected abnormal regions, and analyze its characteristics [6].

2.2.7 Machine Learning

Using this technique, it is possible to do early detection of cancer. It uses various networks for detection such as ANN, RNN, BNN, ECT, and so on. Using this, it is easy to predict cancer by just scanning our breast images in it.

2.2.8 Logistic Regression

It is one of the types of the classification algorithm. It is a supervised machine learning technique. Logistic regression is used to estimate binary output, i.e., only two possible outcomes event occur (1) or event not occur (0).

2.2.9 K-Nearest Neighbors

K-nearest neighbors is one of the types of the machine learning algorithms. It is used for classification and regression functions. *K*-nearest neighbor is a nonparametric technique because the separation of data test point leans on the nearest training point data.

Support Vector Machine

Support vector machine is a regulated AI calculation that is performing extraordinary in acknowledgment issues, and it is utilized as a preparation calculation for contemplating arrangement and relapse rules from data.

3 Proposed System for Finding Invasive Ductal Carcinoma Using Machine Learning

In this proposed system, machine learning is employed. It is a type of supervised learning for the prediction of breast malignancy, i.e., invasive ductal carcinoma (IDC) using CNN [8]. This pathologists break down slides of histology into various fields of view. Be that as it may, most profound learning procedures including histopathology include applied a pixel-level characterization over relatively little pictures [9].

Our methodology then again utilizes deep learning put together grouping concerning square tissue locales from WSI received by customary examining, subsequently, the classifier is used on the whole WSI canvas. A total of 162 diagnosed IDC patients dataset selected over which 113 histopathology slides were chosen for training purpose. From these, 49 slides are tested independently [10].

In term one, our essential target is to assemble an exact breast malignancy histopathological picture arrangement structure. It is the principal computer detection analysis method in the work process. It is an initial point and always used module in this technique. Due to this, the maximum effort in this module is to acquire the system feasibility [5].

3.1 Performance Measure and Parameter Selection

Every patches of each WSI are named as IDC or non-IDC with a 0 or 1 class mark individually. For assessment of IDC discovery, image patch grouping is done overall WSI. Thus, arrangement results are assessed over an approval (D_2) dataset for the selection of parameters and over D_3 (test dataset). This compares the image patches prediction and ground truth. Here, a bunch of execution procedures for IDC identification is ascertained by figuring true negative, false negative, false positive, and true positive. Precision permits the estimation of proportionate IDC detection from the whole affected area of IDC. Recall (Rc), sensitivity (Sen), or genuine positive rate permits the extent of IDC effectively anticipated from entire IDC naturally anticipated. Particularity (Spc) or genuine is characterized to which extent as locales effectively anticipated from all-out real non-IDC areas [5]. When simultaneously compressed false positive and false negative, the tread-off occurs which can be analyzed by using F-measure and balanced accuracy. This is given in Eqs. (1) and (2).

Precision = Total number of images expected as infectious and labeled as infectious/Total number of images labeled as infectious.

Recall = Total number of images labeled and expected as infectious/Total number of images expected as infectious

$$F_1\text{measure} = \{(2 \times (\text{Precision} \times \text{Recall})/(\text{Precision} + \text{Recall})\} \quad (1)$$

$$\text{BAC} = \text{Sen} + \text{Spc}/2 \quad (2)$$

A boundary investigation was performed for CNN and high-quality highlights utilizing D_1 and D_2. Training, validation, and combination were performed for every parameter over D_1 and D_2 differently. F_1-measure achieved boundary determination. Ideal CNN boundary esteems for learning rate, learning rate rot, order edge of stochastic angle, and number of ages [8].

The calculations were discovered to be 1e-2,1e-7, 1e-7, and 25 distinctly. The remainder of boundaries, for example, picture fix size, step size for fix examining, and level of ground truth region for naming positive models was exactly chosen [5].

4 Result

In this way, a classifier is built using CNN to classify whether it is IDC or non-IDC using Tensorflow backend.

The images linked with two classes are 199,818.

The images linked with two classes are 222,201.

The images linked with two classes are 55,500.

Total number of images: 50,000. Figure 1 shows the non IDC testing image.

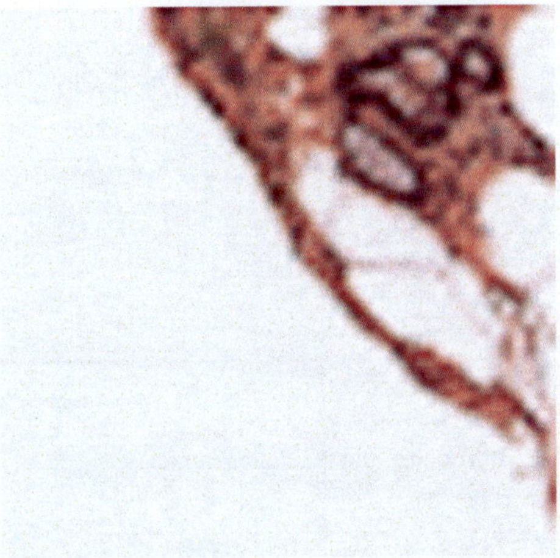

Fig. 1 Testing image (non-IDC)

The dataset contained 162 whole mount slide patches of breast cancer samples scanned at 40×. From that, 275,724 patches of size 50 × 50 were extracted. The above image is a non-IDC IMAGE. It is used as a testing image. Image outline (Breadth, length, Channels): (50, 50, 3) Fig. 2 shows the confusion matrix.

The above confusion matrix performs a classification model on a given dataset in which true values are known. Here, calculate a set of performance measures for IDC detection. And for this, 5000 images patches were classified as 716 images patches have IDC, 3563 images patches do not have IDC, 164 images patches predicted

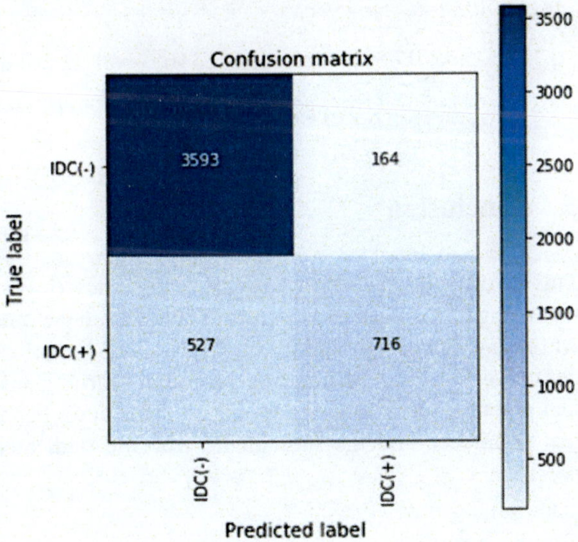

Fig. 2 Confusion matrix

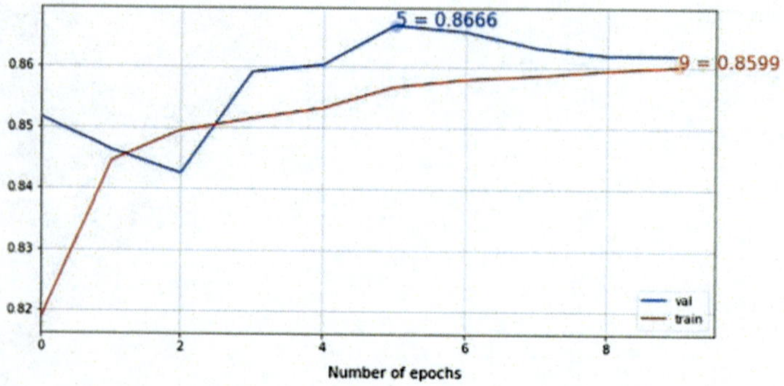

Fig. 3 Training and validation curve

agreed but they may have IDC, and 527 images patch predicted negative, but they actually have IDC. The validation curve is given in Fig. 3.

For getting testing and validation curve, divide the dataset into two parts to have 60–70% of total data in validation and 30–40% for training. Here, the performance is checked of our model with an unseen dataset. Model accuracy and loss curve are shown in Fig. 4.

This is a model accuracy and loss curve. The above loss curve gives us a view of this training process and the direction of the model. Here, loss function is calculated for every data item. The accuracy curve gives the progress of our model.

Epoch 10/10

1407/1406 [==============================] - 119s 84ms/step - loss: 0.3272 - acc: 0.8598 - val_loss: 0.3165 - val_acc: 0.8608

5 Conclusion

This investigation presents a breast malignancy (invasive ductal carcinoma) detection using CNN. Furthermore, the findings from the model building are summarized. At present, the design and use of CNN-based identification are result-oriented. The definitive objective of utilizing profound learning instruments to encourage clinical practice, there is as yet far ahead. This audit benefits logical analyst, mechanical architects, and the individuals who are dedicated to identifying malignancy persistence.

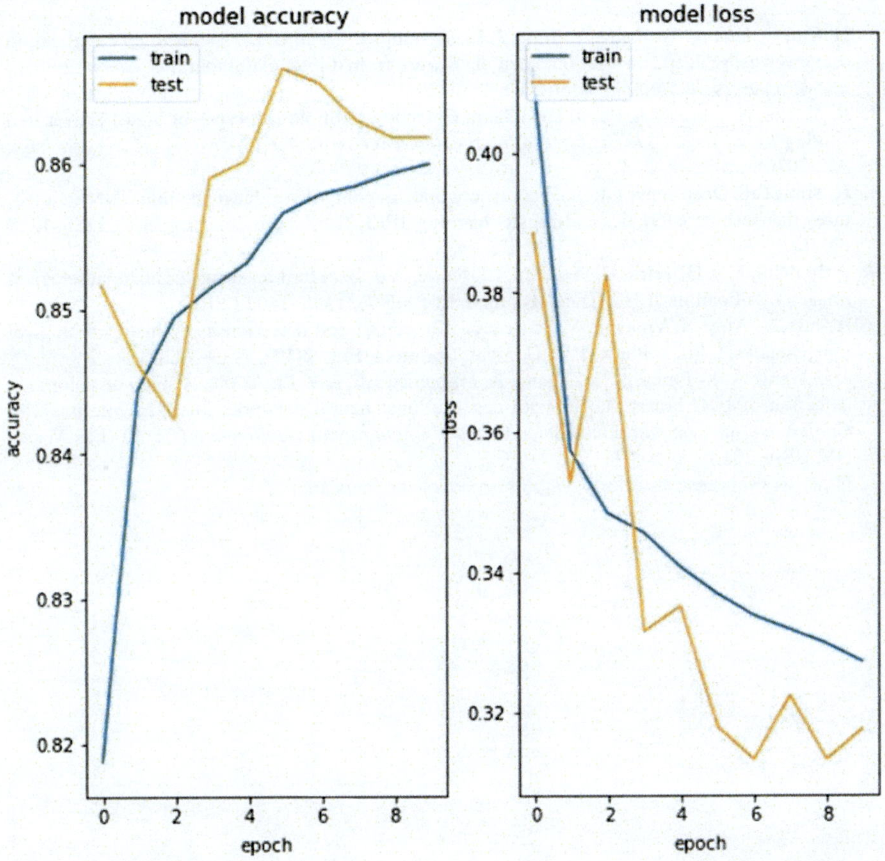

Fig. 4 Model accuracy and loss curve

References

1. E.M. Nejad, L.S. Affendey, R.B. Latip, I.B. Ishank, Classification of histopathology images of breast into benign and malignant using a single-layer convolutional neural network, in *ICISPC 2017: Proceedings of the International Conference on Imaging, Signal Processing and Communication* (2017)
2. A.D. Belsare, M.M. Mushrif, Histopathological image analysis using image processing techniques: an overview. Sig. Image Process. (2012)
3. S. Naik, S. Doyle, S. Agner, A. Madabhushi, Automated gland and nuclei segmentation for gradining of prostate and breast Cancer "©2008 IEEE (2008)
4. M.A. Aswathy, M. Jagannath, Detection of breast cancer on digital histopathology images: present status and future possibilities. Inf. Med. Unlocked **8**, 74–79 (2017). ISSN 2352-9148
5. A. Cruz-Roaa, A. Basavanhallyb, F. Gonz´aleza, H. Gilmorec, M. Feldmand, S. Ganesane, N. Shihd, J. Tomaszewskif, A. Madabhushig, Automatic detection of invasive ductal carcinoma in whole slide images with convolutional neural networks, in *Medical Imaging* (2014)

6. H. Cai, Q. Huang, W. Rong, Y. Song, J. Li, J. Wang, J. Chen, L. Li, Breast microcalcification diagnosis using deep convolutional neural network from digital mammograms, *Computational and Mathematical Methods in Medicine*, vol. 2019
7. R. Verma, N. Kumar, A. Sethi, P.H. Gann, Detecting multiple sub-types of breast cancer in a single patient, in *2016 IEEE International Conference on Image Processing (ICIP)* (Phoenix, AZ, 2016), pp. 2648–265
8. H. Shin et al., Deep convolutional neural networks for computer-aided detection: CNN architectures, dataset characteristics and transfer learning. IEEE Trans. Med. Imaging **35**(5), 1285–1298 (2016)
9. F. Spanhol, L.S. Oliveira, C. Petitjean, L. Heutte, A dataset for breast cancer histopathological image classification. IEEE. Trans. Biomed. Eng. **63**(7), 1455–1462 (2016)
10. A. Jain, S. Atey, S. Vinayak, V. Srivastava, Cancerous cell detection using histopathological image analysis. Int. J. Innov. Res. Comput. Commun. Eng. **2**(12), 7419–7426
11. J. L. Wang, A. K. Ibrahim, H. Zhuang, A. Muhamed Ali, A.Y. Li, A. Wu, A study on automatic detection of IDC breast cancer with convolutional neural networks, in *2018 International Conference on Computational Science and Computational Intelligence (CSCI)* (Las Vegas, NV, USA, 2018)
12. https://www.breastcancer.org/symptoms/understand_bc/statistics

Cluster Performance by Dynamic Load and Resource-Aware Speculative Execution

Juby Mathew

Abstract Big data is one of the fastest-growing technologies, which can handle huge amounts of data from various sources, such as social media, weblogs, banking and business sectors, etc. A Hadoop MapReduce job can be delayed if one of its many tasks is being assigned to an unreliable or congested machine. To solve this straggler problem, a novel algorithm design of speculative execution schemes for parallel processing clusters, from an optimization perspective, under different loading conditions is proposed. For the lightly loaded case, a task cloning scheme, namely the combined file task cloning algorithm, which is based on maximizing the overall system utility, and a straggler-detection algorithm is proposed based on a workload threshold. The detection and cloning of tasks assigned with the stragglers only will not be enough to enhance the performance unless cloning of tasks is allocated in a resource-aware method. So, a method is proposed which identifies and optimizes the resource allocation by considering all possible aspects of cluster performance balancing. One of the main issues appears due to the pre-configuration of the distinct map and reduces slots based on the number of files in the input folder. This can cause severe under-utilization of the slot as map slots might not be fully utilized to the input splits. To solve this issue, an alternative technique of Hadoop slot allocation is introduced in this paper by keeping the efficient management of the slots model. The combined file task cloning algorithm combines the files which are less than the size of a single data block and executes them in the highest performing data node. On implementing these efficient cloning and combining techniques on a heavily loaded cluster after detecting the straggler, the machine is found to reduce the elapsed time of execution to an average of 40%. The detection algorithm improves the overall performance of the heavily loaded cluster by 20% of the total elapsed time in comparison with the native Hadoop algorithm.

Keywords Big data · Clustering · Hadoop · Node detection

J. Mathew (✉)
Amal Jyothi College of Engineering, Kanjirapally, Kerala, India
e-mail: jubymathew@amaljyothi.ac.in

V. Suma et al. (eds.), *Inventive Systems and Control*,
Lecture Notes in Networks and Systems 204,
https://doi.org/10.1007/978-981-16-1395-1_64

1 Introduction

Today, the world is guided by data, and people, as well as machines are generating huge amounts of data every moment by sending messages, uploading videos and photos, generating sensor data from different types of sensing mechanisms, etc. The handling of the phenomenal data explosion posed a challenge to technology firms such as Google, Yahoo, Amazon, and Microsoft. The companies had to sift and sieve through massive amounts of data to find the customer orientations and preferences related to books, adverts, and trending websites. Traditional tools for data handling also failed in this regard. Hence, Google introduced the revolutionary MapReduce system that can handle big data processing. Subsequently, Doug Cutting initiated an open-source version of this MapReduce system namely Hadoop.

Apart from the traditional distributed systems, Hadoop differs in the core execution strategy of data locality. This indicates that the mode of existence and execution of Hadoop differ from the existing data warehouses and relational databases used for data analytics in the past.

1.1 *MapReduce and Speculative Execution*

In short, Hadoop allows the distributed execution of various analytics works in large amounts of data in a simple and yet powerful manner. The storage and processing are handled by two different engines known as HDFS and MapReduce. Hadoop has the data locality feature where the data will reside in the storage platform itself, and the program will go down to the data location and executes within. Thus, the importance of a Hadoop-like platform in the rapidly growing world is valuable [1].

MapReduce is implemented as an independent map and reduce phase. MapReduce envisages a model for executing big volumes of data simultaneously by dividing the tasks into standalone groups. The normal speculative execution strategy does not have the concept of resource-aware scheduling and dynamic and fast detection of stragglers [2]. Thus, to mitigate the lagging of jobs due to straggler node problems and also incorporate the concepts and requirements of the distributed system, an effective parallel processing architecture should be developed as part of open-source project Hadoop. So, the development of a configuration patch that could rectify these limitations of default speculative execution is relevant.

1.2 *Objective*

The objective of this research work is to develop a novel algorithm that is expected to provide enhancements in the performance of the heavily loaded cluster. The optimized speculative execution strategy can make great changes in the performance rates of the multinode cluster. The detection and mitigation of stragglers in the system have to be effectively equipped for obtaining higher throughput.

1.3 Problem Definition

This work focuses on the development of a configuration patch that can provide more performance and throughput for the job running on the Hadoop multinode cluster irrespective of a load of input data and file formats in a resource-aware manner. The load balancing is done dynamically by identifying whether the system is lightly or heavily loaded and execute the combined file task cloning algorithm for the lightly loaded condition and heavily loaded condition detection of straggler node algorithm is performed. These two algorithms are evaluated in the heterogeneous and homogeneous multinode cluster, and combine file task cloning is evaluated in the single node cluster for performance evaluation.

1.4 Scope of the Work

The main challenges in the system development are the possible overhead that can cause due to the number of execution stages. The system has also threats on dependencies of the Hadoop framework, as it is a tightly coupled system. These dependencies and overheads are to be handled efficiently to achieve a better performance increase on comparing with existing systems. The successful implementation of this system in the heterogeneous cluster which handles instantaneously varying load in the fields like business environments like banking and another machine to human interaction platforms can perform better. High efficiency in the CPU time and execution time can be achieved.

1.5 Expected Outcome

The core expectation of this work is a complete study of the fundamental concepts and apply them to develop the proposed system. An analysis of the proposed system will be done and duly tabulated, thus allowing us to compare the proposed system with existing techniques. The novel algorithm for the speculative execution is developed and which is expected to show improvements in the performance than the older existing versions.

2 Literature Survey

In [3], the main focus is on speculative execution, which handles the straggling problem. Unlike the existing heuristics-related work, this paper presents a hypothetical structure for the optimization of solo job queues. The simulation results show the ESE algorithm can reduce the job flow time by 50% while consuming fewer resources compared to the strategy without backup. The results of the ESE algorithm are compared with the traditional method without backup, and they show that the resources and time for job execution can be reduced by 50% [2, 4, 5].

This article proposes a new dynamic method of implementation known as maximum cost performance (MCP). In this novel strategy, the total computing expense is divided between tasks, resulting in reduced task completion time and elevated cluster throughput. This synergism in MCP leads to better performance. This method focuses on selectively adding straggler tasks with precision and performs proper follow up on the worker nodes. The tasks are assigned in first come first serve preference [6].

Combination re-execution scheduling technology (CREST) is the name of a new strategy to conclude on the best re-calculation methodology in a typical MapReduce job. The motive is to reduce the response time which is usually derived as the sum of the longest duration of execution for all map and reduce tasks in a generic MapReduce job [7].

This paper is a result of a diversified dynamic supposition-oriented job scheduler, namely Hopper. The job is generally triggered with the launch of speculative spawns of jobs in a generic way being a common approach for reducing the impact of stragglers. Due to this, job schedulers are often twisted between choosing dynamic jobs versus original job tasks [8].

Mantri is the name of a new model for mitigating the outliers in a typical MapReduce network. This work introduces the first approach to study a large production MapReduce in a cluster form [9]. The core of Mantri's benefit is the amalgamation of stable definite knowledge of job structure with the dynamically available job progress cards. This mechanism is sure to pick the outliers at an early stage and exerts cause-specific alleviation of jobs based on the cost–benefit analysis.

A new method of scheduling dynamically generated job clusters for better job approximation is tested in [10]. The authors have put forward a simple and analytical implementation, specifically derived from the dynamic algorithm known as GRASS. GRASS explores the total opportunity cost in deciding the time of speculation of a job. The decision revolves around the early time to determine the execution of the job and moves to more aggressive dynamic methods of speculation as the job enters the final phase of its approximation bound [11–13]. The proposal was tested in Hadoop and Spark implementations, deployed on a large cluster node and resulted in approximately 47% improvement in finishing the job deadlines. The total time to complete the error in some jobs showed around 38% improvements in data provided from Facebook and Bing [14, 15].

2.1 Summary of the Literature

From the study of literature review, the existing systems failed to enhance the performance of tasks in a speculative strategy by a greater performance which can be achieved by calculating the job service matrices and additional parameters like job flow time and computational cost. The dynamic resource allocation capabilities of MapReduce structure are also not established to the level where task cloning and its effective allocation is maintained simultaneously.

In contrast, the obtained results do not support the general rules, when the job servicing time is under the heavy-tailed distributions such as the Pareto distribution. Thus, the foundation for the proposed system lies in these aspects of considering these crucial parameters. The optimization of speculative execution procedures combined with dynamic slot allocation refines the speculative strategy of Hadoop in all aspects. So, it is evident that the proposal for such a system is relevant in the big data era.

3 System Model

In the implementation, consider a colossal data processing job cluster with M servers (machines). The set of jobs $J = \{J_1, J_2,\ldots, J_N\}$ approach the job cluster at a rate of n jobs per unit time and the time at which job J_i arrives is denoted by a_i. As the job arrives, the job J_i is added to a public queue managed by the speculative scheduler, ready to carry out the job execution. The set of jobs, J_i is a deterministic number composition of m_i tasks. Assume that δ_{ji} depicts the jth task of job J_i. And also assume that each server can only execute only one job task at any given time. A random variable, X_{ji}, denotes the service time (i.e., duration) of task δ_{ji} without any dynamic projection of job completion time. For all $j\varepsilon\{1, 2,\ldots, m_i\}$, X_{ji} follows the same distribution, which is characterized by the cumulative distribution function (CDF), F_i (x), i.e., Pr $(X_{ji} < x) = Fi(x)$.

3.1 Job Service Process Under Speculative Execution

For the sake of assumption, the time taken is divided, and job task preemption is not asserted to reduce the system overhead. The job of the scheduler is to make optimal speculative decisions, so that the unused time is approximately nil on job task execution machines. The number of copies on idle machines is also similarly decided by the scheduler at the start of the time slot. Let c_{ji} denote the total number of copies made for task δ_{ji} where the kth copy is launched at time w_{ij}, k.

3.2 Problem Formulation

In this area of problem formulation, the stress is put on basically two performance measures, the job flow time, *i*, and the computation cost, both of which are computed by the total time spent on the job servers.

As a general situation, the two performance metrics are often rigid to be rearranged at the same time (except for the detection approach). Hence to solve this crisis, a utility-function is defined for each job task as a trade-off between these two metrics. This function does the job of maximizing the total utility scale of all jobs in the data cluster using finding z. The resulting optimization problem can be represented as:

$$\min_{z} \sum_{i=1}^{N} E[\Gamma_i] + \gamma \cdot \sum_{i=1}^{N} \sum_{j=1}^{m_i} \sum_{k=1}^{c_i^j} \left(E[C_i^j] - \omega_i^{j,k} \right) \tag{1}$$

3.3 Deriving the Cut-Off Threshold for Different Operating Regimes

In determining the threshold, a generic approach is to find the approximation solutions, which is from finding the dynamic speculative execution methodology along with the scheduler. This is also given the strongly NP-hard nature of the problem. The possible two classes of dynamic strategies for execution are applied here, namely the cloning and detection approach. The cloning strategy stimulates all job tasks in parallel without which the priority among job tasks are lost leading to futile usage of resources. This is the only applicable scheme for a lightly loaded cluster. In contrast, the straggler-detection methodology intelligently produces new spawns of jobs to handle any load balancing situations.

For further exploring the details of the proposed methodology, it is necessary to define the cut-off workload threshold, λ^{U}, whose job is to segregate the remaining analytical stage into an easily handled job, in comparison with the heavily loaded server data clusters.

1. **A First Upper Bound for λ^{U}**
   To keep the system not overloaded, the job arrival rate must be bounded by the job processing rate, which then yields the first upper bound, λ_1, for λ^{U}, i.e.,

$$\lambda_1 = \frac{NM}{\sum_{i=1}^{N} \sum_{i=1}^{m_i} c_i^j E[T_i^j]} \tag{2}$$

2. **A Second Upper Bound for λ^{U}**
   The efficiency of cloning is not guaranteed by the single upper bound. An efficient cloning strategy should have a smaller task delay than a strategy that

does not make speculative execution. The second upper bound and it can be shown as:

$$\lambda_2 \frac{\lambda_i^* M}{m'} \quad (3)$$

3.4 *Optimal Cloning in Lightly Loaded Regime*

In a lightly loaded cluster, i.e., $\lambda < \lambda^{\mathrm{U}}$, maximize the overall system utility in P_1 by coordinating job scheduling with task cloning. The lightly loaded conditions always suites with a smart cloning of outlier tasks rather than a detection-based approach. So, a combined file task cloning algorithm is introduced for the cloning of the tasks in the straggler machine and reallocates to other machines.

3.4.1 The Design of the Smart Cloning Algorithm (SCA) in a Lightly Loaded Cluster

After successfully executing the cloning algorithm, the next focus is on the tracking of the job progress as well as job completion cost. This is done with the help of an algorithm, which calculates the solid integer part of the task progress rate, such that the task job counts are also equally integer parts. To detect a case in analysis, where the provision of job cloning is limited by space for some specific time slots, i.e., $\sum_i^k mli > N(l)$. This rare situation demands a careful study in a lightly loaded data cluster. At this point, it is not wise to solve P_2. Instead, the new proposal suggests a design alternative dynamic scheduling scheme to allocate job clusters, based on the smallest remaining workload scheme, which is extended from the SRPT scheduler.

3.4.2 Design of a Straggler-Detection-Based Algorithm for the Heavily Loaded Regime

For a heavily loaded cluster, i.e., $\lambda \geq \lambda^{\mathrm{U}}$, the cloning strategy is not viable, as there is no scope to make a copy for all tasks. To avoid this drawback, devise a detection methodology for obtaining approximate results. The primary dynamic execution strategy is laden with numerous flaws in principle. The first drawback is that it creates extra copies for the tasks that are in the conservative mode, characterized by lower amounts of resource consumption. The second drawback is the lower precision in the estimation, which falls heavily on the job completion duration and scale.

4 Implementation

4.1 Implementation Tools

The usage of software and hardware tools is the most important elements of setting up a heterogeneous multimode cluster. The system is aimed to implement with the Ubuntu OS support. The software and hardware requirements can be pointed out as.

4.2 Software Requirements

4.2.1 Hadoop Framework—Hadoop 2.10.1

Apache Hadoop is a popular open-source framework used extensively for the distributive storage and processing of massive data. It is a mix of server clusters where the motivation behind its construction lies in the fact that the hardware failures are quite common and should be handled dynamically by the server mass in the framework. Figure 1 depicts a typical Hadoop framework.

The multinode cluster is formed in a master–slave architecture where each machine in the cluster is installed with Hadoop properly and set one among them as master, thus name node and provides a resource manager. The slave nodes act as data nodes, and they start a node manager.

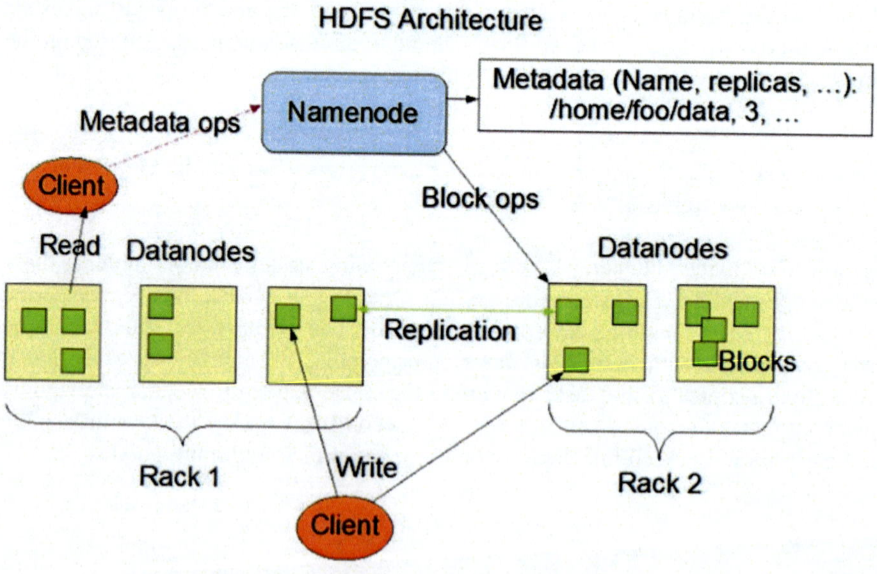

Fig. 1 Hadoop architecture

4.2.2 Java—1.8.0.91

JDK is downloaded from the official site, and the java coding is done and compiled in Eclipse 4.4. Java is the core language of the Hadoop framework. Java is used for the whole implementation of the algorithms in this work.

4.2.3 Eclipse

The popular platform—Eclipse provides IDEs and platforms for any amicable framework, irrespective of the language and scheme. The Java IDE, C/C++, JavaScript, and PHP IDE's are built on these platforms for creating typical desktop, Web, and cloud IDEs.

4.2.4 Cloudera

Cloudera is another popular open-source Apache Hadoop distribution. Also known as Cloudera distribution including Apache Hadoop (CDH), this framework is meant for corporate deployments of applications on a massive scale. According to Cloudera, the major share of its engineering output is designated upstream to various Apache-licensed open-source projects that work on the common Hadoop platform.

4.3 Hardware Requirements

The hardware requirement is a heterogeneous multimode cluster where the big data analytics and processing are being performed. For simulation and testing, a multi-mode Hadoop cluster consisting of three machines is used. The configurations of the machines are:

(a) Processors: Any Intel or AMD × 86 processor.
(b) RAM: 4 GB.
(c) System Type: 64-bit OS, × 64-based processor.
(d) Disk Space: 60 GB in C drive for reserved for cluster job execution
(e) Virtual machine specifications
(f) Quick Start VM 5.5: Red Hat (64 Bit), 8 GB RAM, 64 GB virtual hard disk space.
(g) Hadoop cluster nodes with CentOS minimal version.

4.4 Module Description

The first stage of this work is the multinode cluster of Hadoop with three nodes with 91.5 Gb of shared HDFS in each machine, then uploads the 155 Mb data as a sample for simulation, and it can be extended to great ranges in gigabytes, which will be evenly split and replicated automatically. Then executed WordCount program which includes a MapReduce functions and submits the job. Once the job completes, it will be notified of the results. The log details are analyzed and sorted for the node failure and decommissioned reports. The data transfer details are analyzed to find the network accessibility between the machines, within the cluster.

4.5 Implement and Evaluate the Performance of WordCount with Optimized SCA & ESE Algorithm

Formulate the code for smart cloning algorithm and enhanced speculative execution algorithm [3]. Generate the patch of optimized speculative execution. Generate its patch file and add it with the Hadoop framework. Check its performance with WordCount running in a three node cluster.

4.6 Detailed Analysis in Loading Conditions, Different Programs, and Performance Optimization

Evaluate and analyze the heavily and lightly loaded conditions of the cluster. Overall performance tuning and employing manual network contention in data nodes. Detailed performance analysis using several classical MapReduce programs along with WordCount. Performance analysis by using Spark by parallelism tuning. Performance evaluation is to be sketched in detail to analyze the enhancements. Overall validation tests should be performed with the system. The performance enhancement can be done after the validation testing apart from the module wise testing.

5 Testing and Evaluations

5.1 Key Tuning Parameters

5.1.1 Mappers

mapreduce.input.fileinputformat.split.minsize

The minimum size chunk that map input should be split into. By increasing this value beyond dfs.blocksize, it can reduce the number of mappers in the job. This is because if the value of mapreduce.input.fileinputformat.split.minsize to 4 × dfs.blocksize, then four times the size of blocks will be sent to a single mapper, thus, reducing the number of mappers needed to process the input. The value for this property is the number of bytes for input split. Thus to set the value to 256 MB, should specify 268435456 as the value for this property.

mapreduce.input._leinputformat.split.maxsize

The maximum size chunk that map input should be split into when using CombineFileInputFormat or MultiFileInputFormat by decreasing this value below dfs.blocksize and increase the number of mappers in the job. This is because if the value of mapreduce.input fileinputformat.split.maxsize to 1/4 dfs.blocksize, then 1/4 the size of a block will be sent to a single mapper, thus, increasing the number of mappers needed to process the input. The value for this property is the number of bytes for input split. Thus to set the value to 256 MB, specify 268435456 as the value for this property. If it is set with a max split size when using CombineFileInputFormat, the job will only use 1 mapper.

5.1.2 Reducers

mapreduce.job.reduces

One of the biggest killers for workflow performance is the total number of reducers in use. Use too few reducers and the task time is longer than 15 min. But too many also cause problems and determining the number of reducers of individual jobs is a bit of art. But here are some guidelines to think about when picking the number:

More reducers = more files on the name node. Too many small files bogs down the namenode and may ultimately make it crash. So, to reduce output is small (less than 512 MB), it needs only fewer reducers.

More reducers = less time spent processing data if there are too few reducers, and the reduced tasks may take significantly longer than they should. The faster the jobs, reducers run more jobs and can push through the grid.

Shuffling is expensive for large tasks for the FileSystem Counters of the job, and it can be observed how much data may potentially need to be pushed around from node to node. Let us take a job with 20 reducers. Here are the FileSystem Counters:

mapreduce.job.reduce.slowstart.completedmaps

This setting controls what percentage of maps should be complete before a reducer is started. By default, set this to 0.80 (or 80%). For some jobs, it may be better to set this higher or lower. The two factors to consider are: If the map output is significant, it is generally recommended that reducers start earlier, so that they have a head start processing. If the map tasks do not produce a lot of data, then it is generally recommended that reducers start later. A good rough number is to look at the shuffle time for the first reduce to fire off after all the maps are finished. That will represent the time that the reducer takes to get map output. So ideally, reducers will fire off (last map)—(shuffle time).

5.1.3 Compression

mapreduce.map.output.compress

Setting this to true (default) will shrink map output by compressing it. This will reduce internode transfers, however, care must be taken that the time to compress and uncompress is faster than the time to transfer. For large or highly compress-able intermediate/map output, it is usually beneficial to turn on compression. This can reduce the shuffle time and make disk spills faster. For small intermediate/map output datasets, turning intermediate output compression off will save the CPU time needed to do the (ultimately useless for this data) compression. Note that this is different than apreduce.output.fileoutputformat.compress that setting controls whether the final job output should be compressed when writing it back to HDFS.

5.1.4 Memory

mapreduce.(map-reduce).memory.mb

One of the features in newer releases of Hadoop is memory limits. This allows the system to better manage resources on a busy system. By default, the systems are configured to expect that Java tasks will use 1 GB of the heap and anywhere from 0.5 to 1 GB of non-heap memory space. Therefore, the default size of mapreduce.(map-reduce).memory.mb is set to 2 GB. In some situations, this is not enough memory. Setting just Xmx will result in more than 2 GB, and the tasks will get killed. Therefore, to request more memory for the task slot, it needs to adjust both the Xmx value and the mapreduce. (map-reduce).memory.mb value.

5.1.5 Advanced Controlling the Number of Spills

io.sort.record.percent

io.sort.record.percent controls how much of the circular buffer is used for record vs. record metadata. In general, a family of tunable is ones to look at when spills are out of control. Changing this results in maps running faster and fewer disk spills because io.sort.mb is used more efficiently and do not hit the 80% mark in the metadata buffer as quickly. The result of changing io.sort.record.percent was that many maps did not spill to disk at all and of those that did, many dropped spilled to 55% fewer files. End result: system thrash was reduced to save 30% of the CPU and dropped 30 min off the runtime.

mapreduce.(map-reduce).speculative

Set these properties to false to prevent parallel execution of multiple instances of the same map or reduce task. The data skew of the mappers or reducers will take significantly longer. In this case, it should disable speculative execution to prevent spawning lots of unnecessary map and reduce instances.

5.1.6 Running of Algorithm

As a detailed survey of different performance factors in Hadoop Yarn, test input for the WordCount problem is selected as a 3.6 GB CSV file.

It is analyzed in various conditions.

(a) Normal WordCount without combiners
(b) WordCount with combiners
(c) Multiple reducers
(d) Input splits
(e) Speculative execution property disabled.

6 Experimental Results

As part of the straggler machine detection and task cloning strategy, the setting up of a multinode cluster is the preliminary stage of the project. The evaluation is based on heterogeneous cluster performance. The project implementation is planned such that the first module of the project is setting up of homogeneous and heterogeneous multimode cluster and evaluating a MapReduce program to check the performance variation due to the system resource utilization and availability constraints.

The experimental results are achieved from the execution of the classical MapReduce program WordCount in the three-node cluster and input of 155 Mb of text data. The step-by-step evaluation can be described as:

(a) Multinode clusters with three nodes are set up in the lab with one server and two slave machines.
(b) All the three nodes were properly installed with Hadoop, and relevant setup procedures were followed to establish the master–slave architecture with three machines in the lab with LAN and ssh connectivity.
(c) Secret ssh keys were generated and shared with all the three systems for communication.
(d) A master node in the cluster set up with a shared HDFS memory capacity of 95 GB in the drive and slaves with 65 GB of space for the distributed access.
(e) After the setting up of the three node cluster, the namenode is formatted and started the utilities and datanode.
(f) The slave machines are checked for the datanode working and found it working by the 'jps' command which showed the activated components as datanode and node manager at slave machine, and all the other components of Hadoop like namenode, secondary namenode, resource manager, and job tracker are all active the master machine or node.
(g) Created a folder in HDFS and loaded an input file of size 155 Mb.
(h) Run the classical problem WordCount in the master which internally utilized the other two datanodes, and the output folder (Fig. 2) is generated at the HDFS and the folder contained the text file with counts of all the words in the text input.
(i) The report is analyzed from the web. The failed and decommissioned datanodes are checked.

Thus, obtained a fair result of the first module and the works of the next module is performing on, but it is facing some unexpected errors. It is being tried out to solve them and expecting good results. For a simulated development environment, the further proceeding is done by Oracle virtual box and created a cluster with a namenode and three datanodes and a client machine, all with static IP address. Developed programs for obtaining the different cluster performance impacts of key tuning parameters. Observed the difference in the time of completion of jobs in the cluster for each parameter. The detailed tabular results can be reviewed from Tables 1 and 2. The survey of elapsed time is done in the classical WordCount (WC) program and the input file is given as the 3.65 Gb CSV file.

7 Conclusion

Hadoop is the open-source framework for distributed data processing acquired much production importance, as it provides data locality features and an efficient processing platform for huge file processing than by using traditional distributed systems. Thus,

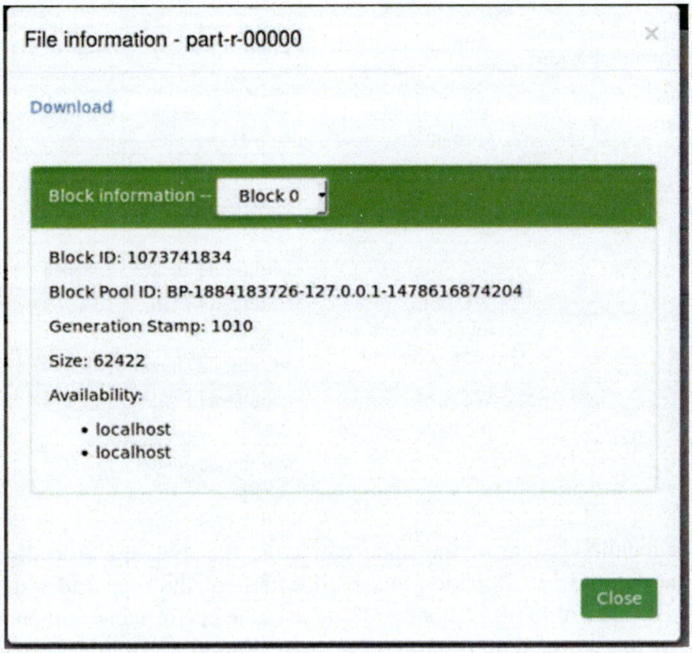

Fig. 2 Details of the output folder

Table 1 Cluster report

Node ID	Memory (Gb)	Core processor	OS ram size
Namenode-nn	20	i3	CentOS (Minimal version) 2 Gb
Datanode1-dn1	20	i3	CentOS (Minimal version) 2 Gb
Datanode-dn2	20	i3	CentOS (Minimal version) 2 Gb
Datanode3-dn3	20	i3	CentOS (Minimal version) 2 Gb
Client-vclient	20	i3	CentOS (Minimal version) 2 Gb

it should be much accurate and dynamic according to the applications, so that it will be a tunable processing approach. An approach for the speculative execution procedure enhancement is proposed by the work where this approach is proved to enhance overall cluster performance and, thus, the overall execution time of the bulk of jobs. The execution process of the work is performed in two phases.

In Phase-1, the parallel execution of MapReduce with example program Word-Count is observed in a multinode cluster of homogeneous as well as heterogeneous nodes in virtual machines and real machines with a quad-core processor. The dynamic slot allocation procedures are executed within the fair schedule module of the speculative execution and yarn common files.

Table 2 Experimental report of job completion time with various factors of performance

Application details	Platform	Time of completion (ms)
WC without combiners	Cloudera (4 Gb RAM)	156,803
	Multinode cluster	162,970
WC with combiners	Cloudera (4 Gb RAM)	155,542
	Multinode cluster	158,746
WC with multireducers and combiners	Cloudera (4 Gb RAM)	103,368
	Multinode cluster	143,659
WC with Inputsplits—1 GB	Cloudera (4 Gb RAM)	135,888
	Multinode cluster	150,035

The data locality is one of the main concerns; the dynamic slot allocation is based on the data-aware allocation and reallocation of the map and reduce slots. The combining algorithm offers about 60% of average performance enhancement in the cluster for the WordCount program. The detection of the straggler node module will offer a dynamic notification of outlier nodes in the cluster and decommission of them at the time of detection itself. It is found to create some overload on the cluster while executing detection algorithm; but in the case of heavily loaded criteria, it is negligible compared to the estimated time of execution without eliminating the straggler nodes. The modified algorithms are expected to give an optimized result in the overall performance of the MapReduce system. Thus, the speculative execution can be implemented in a very efficient manner. Then, it can be applied to get added along with the Hadoop package. Then with a single command, the speculative execution procedure can be enabled or disabled by the common users.

The speculative execution as well as scheduling strategies in Hadoop needs more efficiency in the big data platforms as a small degradation of resources may lead to heavy production loss. The commodity hardware is very prone to damages, bandwidth scarcity, and bad machine faults in the overall cluster. The extensions to the project focusing on the energy impact-oriented enhancement in smart speculation can reduce overhead due to task cloning and thus can achieve a much more reliable distributed platform.

References

1. M. Isard, M. Budiu, Y. Yu, A. Birrell, D. Fetterly, Dryad: distributed data-parallel programs from sequential building blocks, in *EuroSys* (2007)
2. X. Ren, G. Ananthanarayanan, A. Wierman, M. Yu, Hopper: decentralized speculation-aware cluster scheduling at scale, in *Sigcomm* (2015)

3. H. Xu, W.C. Lau, Optimization for speculative execution in big data processing clusters. IEEE Trans. Parallel Distrib. Syst. **28**(2), 530–545 (2017). https://doi.org/10.1109/TPDS.2016.2564962
4. L. Lei, T. Wo, C. Hu, CREST: towards fast speculation of straggler tasks in MapReduce, in *2011 IEEE 8th International Conference on e-Business Engineering* (Beijing, 2011), pp. 311–316. https://doi.org/10.1109/ICEBE.2011.37
5. Q. Liu, W. Cai, J. Shen, Z. Fu, N. Linge, A smart speculative execution strategy based on node classi_cation for heterogeneous hadoop systems, in *ICACT2016*, Jan 31, Feb 3 (2016)
6. S.I.T Joseph, I. Thanakumar, Survey of data mining algorithm's for intelligent computing system. J. Trends Comput. Sci. Smart Technol. (TCSST) **1**(01), 14–24 (2019)
7. S. Smys, C.V. Joe, Big data business analytics as a strategic asset for health care industry. J. ISMAC **1**(02), 92–100 (2019)
8. S. Tang, B.-S. Lee, B. He, DynamicMR: a dynamic slot allocation optimization framework for mapreduce clusters. IEEE Trans. Cloud Comput. **2**(3) (2014)
9. F. Chen, M. Kodialam, T. Lakshman, Joint scheduling of processing and shu_e phases in MapReduce systems, ın *Proceedings of IEEE Infocom* (2012)
10. Q. Chen, C. Liu, Z. Xiao, Improving MapReduce performance using smart speculative execution strategy. IEEE Trans. Comput. **63**(4) (2014)
11. H. Xu, W.C. Lau, Task-cloning algorithms in a mapreduce luster with competitive performance bounds. IEEE Trans. Comput. **63**(4) (2014)
12. J. Mathew, R Vijaya Kumar, Multilinear principal component analysis with SVM for disease diagnosis on big data. IETE J. Res. 1–15 (2019). (Taylor & Francis)
13. T.-D. Phan, S. Ibrahim, G. Antoniu, L. Bouge, On understanding the energy ımpact of speculative execution in hadoop, in *IEEE International Conference on Data Science and Data Intensive Systems* (2015)
14. M. Kawarasaki, H. Watanabe, System status aware hadoop scheduling methods for job performance ımprovement, ın *10th USENIX Symposium on Networked Systems Design and Implementation* (2013)
15. S. Khalil, S.A. Salem, S. Nassar, E.M. Saad, Mapreduce performance in heterogeneous environments: a review. Int. J. Comput. Appl. (2016)

Matyas–Meyer–Oseas Skein Cryptographic Hash Blockchain-Based Secure Access Control for E-Learning in Cloud

N. R. Chilambarasan and A. Kangaiammal

Abstract A learning system is that depends on formalized teaching with the assist of electronic resources which is called e-learning. E-learning employs electronic technologies to access educational curriculum. E-learning is a promising and growing area that permits the rapid integration of smart learning and the teaching process. Since cloud computing is the main paradigm to deliver more efficiently learning content in an integrated environment. E-learning comprises all kinds of educational technology but security analysis is required to achieve higher data confidentiality, fine-grained access control. In order to increase security of data access, A Matyas–Meyer Skein Cryptographic Hash Blockchain and Modified Connectionist Double Q-learning (MMSCHB-MCDQL) technique is introduced. The main aim of the MMSCHB-MCDQL technique is to increase the secure access and academic performance analysis using e-learning data. The IoT devices are deployed for sensing and monitoring the student activities during the e-learning process. At first, the sensing data are collected from the IoT devices and apply the Matyas–Meyer–Oseas Skein Cryptographic Hash Blockchain technique for secure data transmission. A Skein Cryptographic Hash is applied to a blockchain technology to generate the hash for each input data using Matyas–Meyer–Oseas compression function. In addition, the smart contract theory is applied to Blockchain technology to guarantee access control without believing external third parties and it helps to achieve a higher data confidentiality rate. After that, a Modified Connectionist Double Q-learning algorithm is applied for analyzing the student activities to make optimal action with higher accuracy. Based on the learning process, the student's performance levels are correctly predicted. Experimental evaluation is carried out on certain factors such as data confidentiality rate, execution time, and prediction accuracy with respect to a number of student data. The experimental results and discussion demonstrate that the proposed MMSCHB-MCDQL technique offers an efficient solution for secure

N. R. Chilambarasan (✉)
PG & Research Department of Computer Science, Government Arts College (Autonomous), Salem 636007, India

A. Kangaiammal
Department of Computer Applications, Government Arts College (Autonomous), Salem 636007, India

V. Suma et al. (eds.), *Inventive Systems and Control*,
Lecture Notes in Networks and Systems 204,
https://doi.org/10.1007/978-981-16-1395-1_65

decentralized access control in the cloud. The experimental results evidence that the MMSCHB-MCDQL technique improves data confidentiality rate and prediction accuracy by 9.5% and 10% and reduces the execution time by 15% as compared to the conventional methods.

Keywords Cloud · Secure access control · Skein cryptographic hash blockchain · Matyas–Meyer–Oseas compression function · Modified Connectionist Double Q-learning algorithm

1 Introduction

Online learning is education that is accessed through the Internet. It is termed as 'e-learning'. With the continuous development of online learning policy, educational data analytics and prediction have become a promising field. It needs strong self-motivation and time management. Communicational skill improvement is lacking in online students. However, the security factor in the distribution of the educational content along with student activities are the important factors and it causes numerous security challenges. Therefore, it is very necessary to examine the behavior characteristics of online learners to sharply change the online education strategy and enhance the quality of learning.

A fog computing e-learning scheme was developed [1] into the e-learning system to improve the efficiency of data analysis and also provides access control by applying the cryptographic techniques. The designed scheme increases the data confidentiality rate but the higher security was not increased. A fog-based recommendation system (FBRS) was introduced in [2] for increasing the performance of the e-learning environment. However, the designed system improves the accuracy but the cryptographic technique was not applied to achieve higher data confidentiality.

A secure cloud data storage approach was developed in [3] to encrypt the message based on data encryption standard and to encode the message. However, the approach was not efficient to analyze the student data. A statistical and association rule was developed in [4] for analyzing the student behavior according to learning modality using big data. A Q-learning algorithm was designed in [5] to dynamically learning the activities and learning objectives. The designed algorithm provides an optimal action for each learner but it failed to handle a large number of data.

A dynamic multi-agent system was introduced in [6] based on particle swarm optimization for e-learning with minimum execution time. However, the designed system failed to consider the security aspects for e-learning data analysis. In [7], a student learning process was carried out using a Facebook-based e-book approach. Different machine learning algorithms such as decision trees, neural networks, and support vector machines were introduced in [8] to predict academic performance. However, the designed algorithm increases the prediction accuracy but the security analysis was not performed. E-learning User Interface (ELUI) was introduced in [9] to maintain instructional communication through the online learning environment.

A multi-channel data fusion approach was introduced in [10] to discover student learning behavior.

The main contribution of the MMSCHB-MCDQL technique is described as:

(a) To enhance the security of data access and performance prediction, A novel MMSCHB-MCDQL technique is introduced.
(b) On the contrary to existing works, Matyas–Meyer Skein Cryptographic Hash Blockchain technology is applied to an MMSCHB-MCDQL technique that generates a hash value for the collected student data from the e-learning practices. This helps to avoid illegal access. This helps to increase data confidentiality and minimize the execution time of secure data transmission.
(c) To predict the student academic performance level with higher accuracy, a Modified Connectionist Double Q-learning is applied to the MMSCHB-MCDQL technique.
(d) Finally, extensive experiments are carried out to estimate the performance of the MMSCHB-MCDQL technique and other related works. The quantitative results exhibit that the MMSCHB-MCDQL technique is efficient than the other existing methods.

1.1 Outline of the Article

This article is organized into different sections. Section 2 reviews the related works. Section 3 explains the proposed MMSCHB-MCDQL technique with neat architecture and security models. After that, the experimental setup and dataset description are presented in Sect. 4. In Sect. 4, the experimental results of the proposed and existing methods are discussed. Finally, the conclusion is presented in Sect. 5.

2 Related Work

An Augmented Education (AugmentED) framework was developed in [11] to estimate the academic performance of the various students' with higher accuracy. A students' academic performance enhancement (SAPE) approach was introduced in [12] for increasing prediction accuracy. A Technology Acceptance Model (TAM) was introduced in [13] to increase the security, privacy, and trust-based e-learning analysis. However, the performance of the data confidentiality rate was not improved.

A Linear support vector machine classifier was developed in [14] to predict the student's behavior analysis based on learning difficulties and continuous features. A General Extended Technology Acceptance method was introduced in [15] for e-learning to find out the factors of undergraduate. A different usability factor was implemented in [16] to forecast the continuation user behavior analysis to the cloud e-learning application. A supervised machine learning technique was developed in [17] to resolve the task of student exam performance analysis.

A two-stage classification technique was developed in [18] based on a data mining classifier to determine the sequential features based on students' property behaviors with higher accuracy. A deep learning approach was introduced in [19] for Students' performance quality assessment. However, higher accuracy was not obtained. The Lion–Wolf-based deep belief network (LW-DBN) was developed in [20] for the prediction of student's performance with higher accuracy. However, security factors were not considered. In [21], the integration of IoT, Blockchain, and Cloud technologies was introduced in the medical environment to offer healthcare and tele-medical laboratory services. Efficient security and privacy mechanism were designed in [22] to enhance security and privacy in the blockchain.

3 Matyas–Meyer–Oseas Skein Cryptographic Hash-Based Blockchain Technology

The proposed MMSCHB-MCDQL technique is designed for secure data access control and performance prediction in a distributed cloud computing environment. In the MMSCHB-MCDQL technique, IoT devices are employed to sense and collect student activities during the e-learning process. The collected data are transmitted in a secure manner by avoiding unauthorized access. This confidential data are protected by applying the cryptographic technique to the blockchain concept. The received data are learned and predict the student's academic performance (Fig. 1).

The communication between the ants is performed through the pheromone trails deposited on the ground. After finding the food, the ant deposits pheromone on the paths depends on the amount of food that carries. Afterward, other ants also smell the pheromone deposited by the previous ant and follow that path for finding the food

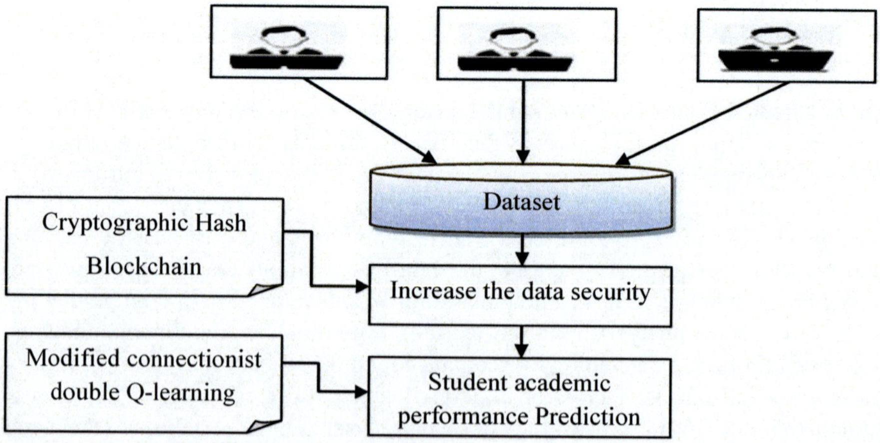

Fig. 1 Architecture of MMSCHB-MCDQL technique

source. Based on the movement of the ants, the shortest path is identified from the nest to the food source since it has a more pheromone level. Therefore, the shortest path having a higher probability is selected as the ant path.

A Matyas–Meyer–Oseas skein cryptographic hash blockchain-enabled secure access control technique is introduced for improving the security. Matyas–Meyer–Oseas hash function include AES initialized with static predefined key and employed to encrypt a single block of input ("ECB" mode). Matyas–Meyer–Oseas hash build block cipher into one-way compression function which is employed in the hash function. Encryption results are XORed with the original block to generate the output hash. The blockchain technology uses the SmartContract model to perform the access control without believing external third parties.

Figure 2 illustrates the blockchain that comprises the different blocks to construct a chain. Each block includes a block header, timestamp (T_s), root hash (Tx_R), and a hash of the previous block (p_hash), a timestamp (T_s). Each block transaction comprises the student information collected from the dataset. The root hash value is generated using the Matyas–Meyer–Oseas Skein Cryptographic technique to improve the security of data transmission by avoiding unauthorized access. As shown in Fig. 2, the data block consists of student information. The Matyas–Meyer–Oseas Skein Cryptographic technique generates a hash of each data (H_a), (H_b) and the concatenation of the two hash values (H_{ab}) is given the root hash of the block.

Skein Cryptographic Hash supports the internal state sizes of 256 bits and provides the fixed output sizes. Matyas–Meyer–Oseas is a single-block-length one-way compression function that operates on different sizes and produces the fixed string of hash. If any modifications in the input data cause a drastic change in the hash value, therefore, the Skein Cryptographic Hash function is applied to guarantee security by avoiding unauthorized access. Skein Cryptographic Hash uses the

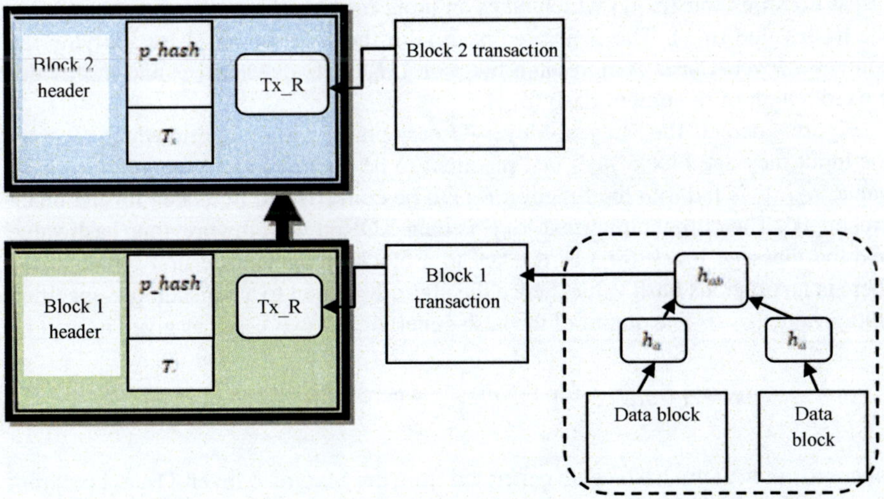

Fig. 2 Construction of Matyas–Meyer–Oseas Skein cryptographic hash-based blockchain

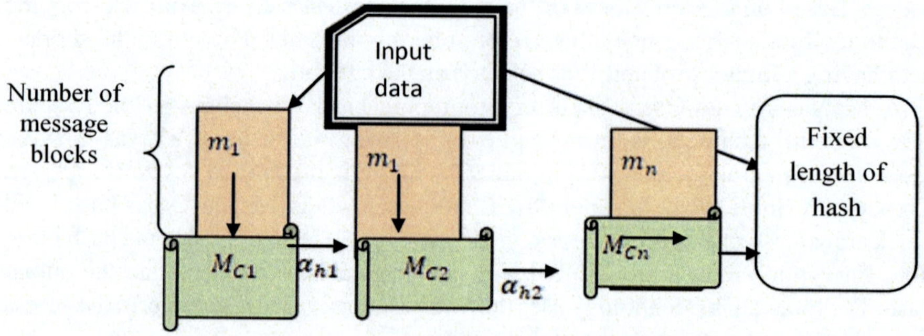

Fig. 3 Block diagram of hash generation using Matyas–Meyer–Oseas compression function

Matyas–Meyer–Oseas compression function to create the fixed size of the hash value. The operation of the Matyas–Meyer–Oseas compression function is illustrated in Fig. 3.

Figure 3 shows the block diagram of the hash generation for each input data using Matyas–Meyer–Oseas compression function. Let us consider the number of student data $D_s = D_1, D_2, D_3, \ldots, D_n$. By applying Matyas–Meyer–Oseas compression, the input data is partitioned into a number of message blocks with a fixed size as in Eq. (1).

$$D_s \rightarrow m_1, m_1, m_2, \ldots, m_n \tag{1}$$

where D_s indicates an input, i.e., student data, $m_1, m_1, m_2, \ldots, m_n$ denotes a message block with a fixed size. Then the input message block is given to the Matyas–Meyer–Oseas compression (M_{C1}) which takes an input message block (m_1) and providing the hash value (α_{h1}). The generated hash is in the form of {0, 1} by applying the Matyas–Meyer–Oseas compression function '$M_{C1}, M_{C2}, \ldots, M_{Cn}$' and it provides a fixed length of the output hash (α_{hn}).

Figure 4 depicts the Matyas–Meyer–Oseas compression function which receives the input message block 'm_i', i.e., plaintext to be encrypted and the previous hash value 'α_{hi-1}' is fed into the function F () to be converted to fit as key for the block cipher 'B'. The output ciphertext 'α_{h1}' is then XORed with the previous hash value and the message block (m_i) as the key to a block cipher. In the first round, when there is no previous hash value, hence the algorithm is set to a constant pre-specified initial value (α_{h0}). The output of the hash generation is expressed as given in Eq. (1),

$$\alpha_{h_i} = \left[B_{F(\alpha_{h_{i-1}})}(m_i) \oplus m_i \right], \quad \text{where } (i = 1, 2, 3, \ldots, n) \tag{2}$$

where α_{hi} indicates a hash value generated from the Matyas–Meyer–Oseas compression function. By using the compression function, the hash of one message block is not similar to another block, i.e., $\alpha_{h1} \neq \alpha_{h2}$. The last hash of the Skein Cryptographic

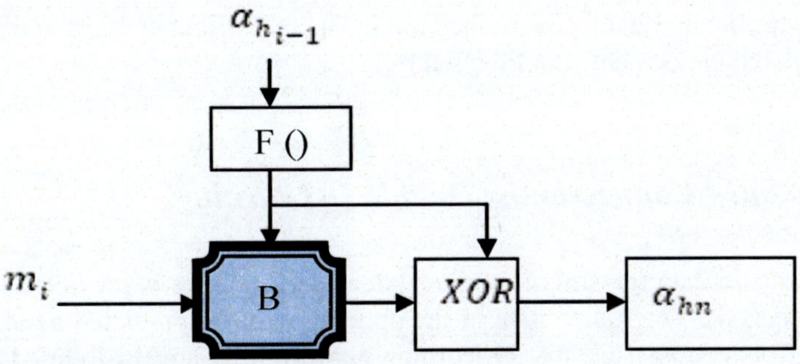

Fig. 4 Matyas–Meyer–Oseas compression function

function is the output of the last compression function. In this way, a hash value is generated for all the student data to avoid unauthorized access. Only authorized users access the data, and it helps to increase data security. The step-by-step process of the Matyas–Meyer–Oseas Skein Cryptographic Hash-based blockchain is described as given in Algorithm 1.

// **Algorithm 1 Matyas–Meyer–Oseas Skein Cryptographic Hash-based blockchain**
Input: E-learning dataset, Number of students data $D_s = D_1 . D_2 . ... D_n$ **Output:** Improve the security of transmission
Begin **Step 1**: Collect the student data $D_s \in D_1 , D_2 , ... D_n$ **Step 2: For each transaction 't'** **Step 3:** Construct blockchain **Step 4:** **For each** student data 'D_s' **Step 5:** Partition into 'n' message blocks $D_s \rightarrow m_1 , m_2 , m_3 , ... m_n$ **Step 6**: **for each block 'm_i'** **Step 7**: Generate a hash value $\alpha_{h_i} = [B_{F(\alpha_{h_{i-1}})}(m_i) \oplus m_i]$ **Step 8**: **End for** **Step 9**: Obtain the final hash 'α_{hn}' **Step 10**: **End for** **Step 11**: **Increase the security of access** **Step 12**: **End for** **End**

The above algorithmic process describes the step-by-step process of secure data transmission. Initially, the student activity information is collected from the dataset. In order to perform the secure data transmission, the blockchain is constructed based on Matyas–Meyer–Oseas Skein Cryptographic technique to generate the hash value. The Matyas–Meyer–Oseas compression function efficiently generates the hash of each message block. Finally, the hashed input data is distributed for each transaction results which increases data confidentiality.

3.1 Modified Connectionist Double Q-Learning

After the secure data transmission, the student performance is predicted based on behavior analysis. The future state of the optimal action is predicted by applying a Modified Connectionist Double Q-learning algorithm. A novel Modified Connectionist Double Q-learning algorithm is a machine learning algorithm that discovers an optimal policy by analyzing the target and the predicted results. A connectionist model is used for analyzing the large varieties and amounts of developmental student activities data per session.

By applying the learning algorithm, two (i.e., double) separate value functions u, v that are trained in a mutually symmetric and these values are updated as given in Eqs. (3) and (4).

$$q_{t+1}^{u}(\delta_t, \alpha_t) = q_t^{u}(\delta_t, \alpha_t) + \beta_t\left[R_t + \omega q_t^{v}(\delta_{t+1}, \alpha_{t+1}) - q_t^{u}(\delta_t, \alpha_t)\right]^2 \quad (3)$$

$$q_{t+1}^{v}(\delta_t, \alpha_t) = q_t^{v}(\delta_t, \alpha_t) + \beta_t\left[R_t + \omega q_t^{u}(\delta_{t+1}, \alpha_{t+1}) - q_t^{v}(\delta_t, \alpha_t)\right]^2 \quad (4)$$

From (3) and (4), $q_{t+1}^{u}(\delta_t, \alpha_t)$, $q_{t+1}^{v}(\delta_t, \alpha_t)$ indicates an updated value of the two separate functions such as u and v, $q_t^{u}(\delta_t, \alpha_t)$, $q_t^{v}(\delta_t, \alpha_t)$ indicates the current state value, β_t denotes a learning rate $(0 < \beta_t < 1)$ and it helps to minimize an error by adjusting the β_t value. R_t denotes rewards that received when moving from the current state δ_t to next state $\delta_{t+1}.\omega$ denotes a discount factor and the values between 0 and 1. From Eqs. (3) and (4), $R_t + \omega q_t^{v}(\delta_{t+1}, \alpha_{t+1})$ denotes a target value and $q_t^{u}(\delta_t, \alpha_t)$ denotes a current predicted value. At last, the average of these two updated separate functions for each action is considered as optimal and finding the future state. The algorithmic process of Modified Connectionist Double Q-learning is described as given in Algorithm 2.

Algorithm 2: Modified Connectionist double Q-Learning
Input: Student data $D_1, D_2, \ldots D_n$
Output: Increase student performance prediction accuracy
Begin
Step 1: Initialize q_t^a, q_t^b
Step 2: Analyze the student data $D_1, D_2, \ldots D_n$
Step 3: Update state, action pair '$q_{t+1}^{u}(\delta_t, a_t)$'
Step 4: Update state, action pair '$q_{t+1}^{v}(\delta_t, a_t)$'
Step 5: Take an average of two updated values $q_{t+1}^{u}(\delta_t, a_t)$ and $q_{t+1}^{v}(\delta_t, a_t)$
Step 6: Attain the final optimal value
Step 7: Find student learning performance
End

Algorithm 2 given above describes the step-by-step process of Modified Connectionist Double Q-learning for finding the student learning performance with higher accuracy. The learning approach analyzes the student activities for each session student learning performance and finds an optimal solution at a future state. The maximum rewards are predicted through the updating process. As a result, student learning performances are correctly analyzed and predict future states.

4 Experimental Setup

Experimental assessment of the proposed MMSCHB-MCDQL technique and existing fog computing e-learning scheme [1], FBRS [2] are implemented using Java language with CloudSim. To perform secure access in the cloud, the Educational Process Mining (EPM): A Learning Analytics Dataset is taken from the UCI repository and carried out the work. The https://archive.ics.uci.edu/ml/datasets/Educational+Process+Mining+%28EPM%29%3A+A+Learning+Analytics+Data+Set has been used.

The dataset is constructed based on the recordings of 115 student's activities through a logging process during e-learning in digital electronics. This dataset includes different students' time series of activities through six various sessions and it comprises 230,318 instances and 13 attributes and their characteristics are integers. The associated tasks performed on the dataset are classification, regression, and clustering. The student data are securitized by applying the hash-based blockchain and student learning performance is identified through the machine learning technique.

5 Results and Discussion

The experimental results of the MMSCHB-MCDQL technique and the existing fog computing e-learning scheme [1], FBRS [2] are discussed based on various performance metrics such as data confidentiality rate, execution time, and prediction accuracy.

5.1 *Data Confidentiality Rate*

The confidentiality rate is the security parameter in the cloud. The data confidentiality rate is defined as the ratio of the number of student data correctly accessed by authorized entities. The confidentiality rate is estimated as given in Eq. (5).

$$\text{Rate}_{\text{con}} = \left(\frac{n_{\text{aae}}}{n}\right) * 100 \tag{5}$$

From Eq. (5), Rate_{con} denotes a confidentiality rate, 'n_{aae}' represents the number of data accessed by the authorized entity, 'n' represents the total number of student data. The confidentiality rate is measured in percentage (%).

Table 1 and Fig. 5 show the experimental results of data confidentiality rate using the number of student data taken in the range 50–500. As shown in the results, the data confidentiality rate of three different methods such as the proposed MMSCHB-MCDQL technique and the existing fog computing e-learning scheme [1, 2] are obtained. When considering 250 data, the data confidentiality rate using the proposed technique 93%, and existing methods provide 87% and 85%, respectively.

Table 1 Data confidentiality rate

Number of data	Data confidentiality rate (%)		
	MMSCHB-MCDQL	Fog computing e-learning scheme	FBRS
50	96	88	86
100	93	86	88
150	96	90	87
200	95	88	86
250	93	87	85
300	94	86	84
350	96	89	87
400	94	86	85
450	95	89	85
500	94	88	86

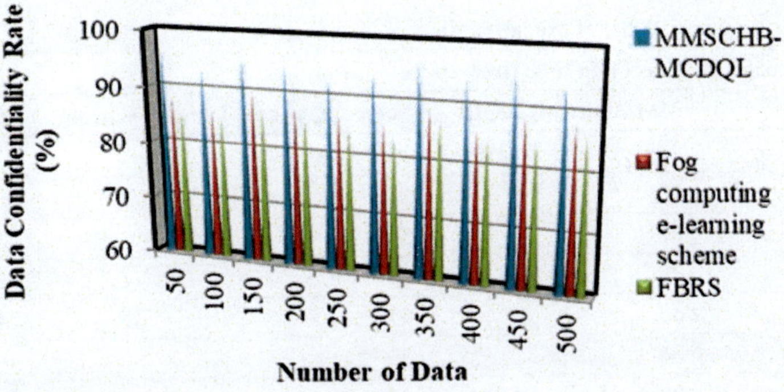

Fig. 5 Performance results of data confidentiality rate

The observed results show that the proposed technique increases the data confidentiality rate. This is due to the proposed MMSCHB-MCDQL technique which uses the Matyas–Meyer–Oseas Skein Cryptographic Hash function. The proposed Cryptographic technique generates the hash value for each student data collected during the e-learning process using Matyas–Meyer–Oseas compression function. Then the student data are distributed in the form of the hash value. This helps to avoid unauthorized data access and increase the data confidentiality rate. The observed results of the proposed MMSCHB-MCDQL technique are compared to conventional methods.

The average of ten results indicates that the proposed MMSCHB-MCDQL technique increases the data confidentiality rate by 8% when compared to fog computing e-learning scheme [1], and 11% when compared to FBRS [2].

5.2 *Execution Time*

It is defined as the amount of time taken to perform secure data transmission during the e-learning process. Therefore, the overall execution time is formulated as given in Eq. (6).

$$\text{ET} = \text{Number of data} * t(\text{SDT})] \tag{6}$$

where ET represents an execution time, t denotes a time, SDT indicates single data transmission. The execution time of the algorithm is measured in terms of milliseconds (ms).

Table 2 and Fig. 6 depict the experimental results of execution time based on the number of student data collected from the e-learning process. As shown in the observed results given in Table 2, the execution time of secured data transmission is increased for all the methods while increasing the input counts. But comparatively, the proposed MMSCHB-MCDQL technique consumes lesser time to perform secure

Table 2 Performance results of execution time

Number of data	Execution time (ms)		
	MMSCHB-MCDQL	Fog computing e-learning scheme	FBRS
50	21	25	30
100	25	28	32
150	27	30	36
200	30	34	38
250	33	37	41
300	36	39	43
350	39	42	45
400	42	46	48
450	45	50	52
500	50	55	57

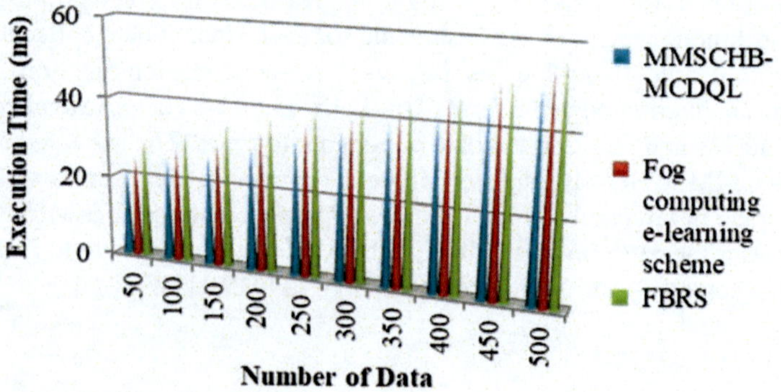

Fig. 6 Performance results of execution time

transmission. This is verified through the sample calculation. By considering '50' student data, the execution time of the proposed MMSCHB-MCDQL technique was found to be '21 ms', likewise '25 ms' and 30 ms are observed using a Fog computing e-learning scheme [1] and FBRS [2], respectively.

The execution time of the proposed MMSCHB-MCDQL technique is compared to conventional methods.

The average of ten runs indicates that the execution time is considerably minimized by 10% and 19% using MMSCHB-MCDQL than the existing methods. This improvement is achieved by applying blockchain-based technology to securely transmit the student data in lesser time.

5.3 Prediction Accuracy

It is defined as a ratio of the number of student data correctly predicted to the total number of student data taken from the e-learning dataset. The prediction accuracy is calculated using Eq. (7).

$$PA = \left[\frac{\text{Number of student data correctly predicted}}{\text{Numberof student data}}\right] \quad (7)$$

where PA represents a prediction accuracy. The prediction accuracy is measured in terms of percentage (%).

Table 3 and Fig. 7 describe the performance results of the student behavior prediction accuracy versus the number of data collected from the e-learning process. From Fig. 7, when considering the 250 number of data, the prediction accuracy using the

Table 3 Performance results of prediction accuracy

Number of data	Prediction accuracy (%)		
	MMSCHB-MCDQL	Fog computing e-learning scheme	FBRS
50	94	86	84
100	92	84	82
150	95	89	85
200	94	86	84
250	92	85	83
300	93	85	82
350	95	88	85
400	93	85	83
450	94	88	84
500	92	86	83

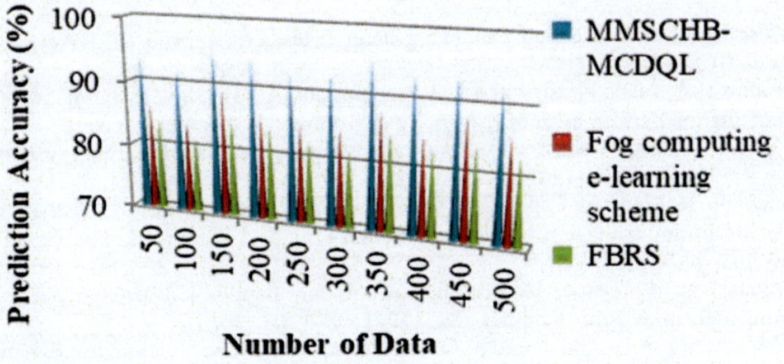

Fig. 7 Performance results of prediction accuracy

MMSCHB-MCDQL technique is 92% and the existing methods provide prediction accuracy of 85 and 83%.

The observed results noticed that the prediction accuracy of the MMSCHB-MCDQL technique is comparatively higher than the existing methods. This improvement is achieved through the application of the Modified Connectionist Double Q-learning. The approach analyzes the state-action pair for each student's data. The observed results indicate that the prediction accuracy of the proposed MMSCHB-MCDQL technique is increased by 8% when compared to [1] and 12% when compared to [2].

6 Conclusion

With the emergence of distributed cloud applications, protecting the student sensitive data is the major concern during the e-learning process. An efficient and secure access control technique called the MMSCHB-MCDQL technique is introduced for increasing the data confidentiality rate with lesser execution time. The Skein Cryptographic technique is applied to a blockchain for secure data transmission by avoiding unauthorized access. The Matyas Meyer–Oseas compression function is used to generate the hash value for each student's data. Then the received student data are analyzed by applying the modified connectionist Q-learning algorithm. This helps to enhance the student's academic performance prediction with higher accuracy. The comprehensive experimental assessment is carried out with different performance factors such as data confidentiality rate, execution time, and prediction accuracy. The proposed MMSCHB-MCDQL technique outperforms and more efficient than the baseline approaches, having a higher confidentiality rate, prediction accuracy, and lesser execution time.

References

1. A.B. Amor, M. Abid, A. Meddeb, Secure fog-based E-learning scheme. IEEE Access **8**, 31920–31933 (2020)
2. T.S. Ibrahim, A.I. Saleh, N. Elgaml, M.M. Abdelsalam, A Fog based recommendation system for promoting the performance of E-learning environments. Comput. Electr. **87**, 1–29 (2020)
3. G.S.S. Jose, C.S. Christopher, Secure cloud data storage approach in e-learning systems. Cluster Comput. **22**, 12857–12862 (2019)
4. M. Cantabella, R. Martínez-España, B. Ayuso, Juan, A. Yáñez, A. Muñoz, Analysis of student behavior in learning management systems through a big data framework. Fut. Gener. Comput. Syst. **90**, 262–272 (2019)
5. M. Boussakssou, B. Hssina, M. Erittali, Towards an adaptive E-learning system based on Q-learning algorithm. Proc. Comput. Sci. **170**, 1198–1203 (2020)
6. M.M. Al-Tarabily, R.F. Abdel-Kader, G.A. Azeem, M.I. Marie, Optimizing dynamic multi-agent performance in E-learning environment. IEEE Access **6**, 35631–35645 (2018)
7. H. Zarzour, S. Bendjaballah, H. Hariche, Exploring the behavioral patterns of students learning with a Facebook-based e-book approach. Comput. Edu. **156**, 1–25 (2020)

8. Xu. Xing, J. Wang, H. Peng, Wu. Ruilin, Prediction of academic performance associated with internet usage behaviors using machine learning algorithms. Comput. Hum. Behav. **98**, 166–173 (2019)
9. W. Farhan, J. Razmak, S. Demers, S. Laflamme, E-learning systems versus instructional communication tools: developing and testing a new E-learning user interface from the perspectives of teachers and students. Technol. Soc. **59**, 1–12 (2019)
10. J. Yue, F. Tian, K.-M. Chao, N. Shah, L. Li, Y. Chen, Q. Zheng, Recognizing multidimensional engagement of E-learners based on multi-channel data in E-learning environment. IEEE Access **7**, 149554–149567 (2019)
11. Liang Zhao, Kun Chen, Jie Song, Xiaoliang Zhu, Jianwen Sun, Brian Caulfield, Brian Mac Namee, "Academic Performance Prediction Based on Multisource, Multifeature Behavioral Data", IEEE Access, 2020, pp. 1–13.
12. A. Akram, C. Fu, Y. Li, M.Y. Javed, R. Lin, Y. Jiang, Y. Tang, Predicting students' academic procrastination in blended learning course using homework submission data. IEEE Access **7**, 102487–102498 (2019)
13. A. Baby, A. Kannammal, Network path analysis for developing an enhanced TAM model: a user-centric e-learning perspective. Comput. Hum. Behav. **107**, 1–12 (2020)
14. D. Hooshyar, M. Pedaste, Y. Yang, Mining educational data to predict students' performance through procrastination behavior. Entropy **22**, 1–24 (2020)
15. C.-T. Chang, J. Hajiyev, Su. Chia-Rong, Examining the students' behavioral intention to use e-learning in Azerbaijan? The general extended technology acceptance model for E-learning approach. Comput. Edu. **111**, 128–143 (2017)
16. L.-Y.-K. Wang, S.-L. Lew, S.-H. Lau, M.-C. Leow, Usability factors predicting continuance of intention to use cloude-learning application. Heliyon **5**, 1–11 (2019)
17. N. Tomasevic, N. Gvozdenovic, S. Vranes, An overview and comparison of supervised data mining techniques for student exam performance prediction. Comput. Edu. **143**, 1–30 (2020)
18. X. Wang, Yu. Xuemeng, L. Guo, F. Liu, Xu. Liancheng, Student performance prediction with short-term sequential campus behaviors. Information **11**, 1–20 (2020)
19. K.J. Gerritsen-van Leeuwenkamp, D. Joosten-ten Brinke, L. Kester, Students' perceptions of assessment quality related to their learning approaches and learning outcomes. Evaluation **63**, 72–82 (2019)
20. L. Ramanathan, G. Parthasarathy, K. Vijayakumar, L. Lakshmanan, S. Ramani, Cluster-based distributed architecture for prediction of student's performance in higher education. Cluster Comput. **22**, 1329–1344 (2019)
21. H. Wang, IoT based clinical sensor data management and transfer using blockchain technology. J. ISMAC, **2**(03), 154–159 (2020)
22. S. Shakya, Efficient security and privacy mechanism for block chain application. J. Inf. Technol. Dig. World **01**(02), 58–67 (2019)

Chapman Kolmogorov and Jensen Shannon Ant Colony Optimization-Based Resource Efficient Task Scheduling in Cloud

S. Tamilsenthil and A. Kangaiammal

Abstract Cloud computing is the most significant technology in recent days, and it facilitates the users to access computing resources as services anywhere through the Internet. Task scheduling is a major concern since traditional scheduling algorithms are not suitable for achieving higher efficiency. To increase the task scheduling efficiency, a novel technique called Chapman Kolmogorov Markov Stochastic Jensen–Shannon Divergence-based Multi-objective Ant Colony Optimization (CKMSJSD-MACO) technique is introduced. The proposed CKMSJSD-MACO technique finds the resource-efficient virtual machine from the population in the search space. For each virtual machine, the fitness is computed based on the multiple objective functions such as memory, bandwidth, energy, and CPU time. By applying the Chapman Kolmogorov Markov Stochastic transition probability, the better convergence rate of the proposed optimization technique is obtained. This helps to reduce the task completion time. Finally, Jensen–Shannon Divergence is applied for measuring the strength of pheromone update to discover the global optimum solution in search space. Finally, the scheduler allocates the incoming tasks to the resource-efficient virtual machine with higher efficiency. The multiple heterogeneous tasks are correctly scheduled for the resource optimized virtual machine and minimize the response time. The observed results indicate that the CKMSJSD-MACO technique achieves higher task scheduling efficiency with lesser makespan as well as lesser false-positive rate. Here, the information of this nature is given in the addresses, which will be deleted by typesetters.

Keywords Cloud computing · Task scheduling · Chapman Kolmogorov Markov Stochastic transition probability · Jensen–Shannon divergence

---

S. Tamilsenthil (✉)
PG & Research Department of Computer Science, Government Arts College (Autonomous), Salem 636007, India

Department of Computer Science, Padmavani Arts and Science College for Women, Salem 636011, India

A. Kangaiammal
Department of Computer Applications, Government Arts College (Autonomous), Salem 636007, India

V. Suma et al. (eds.), *Inventive Systems and Control*,
Lecture Notes in Networks and Systems 204,
https://doi.org/10.1007/978-981-16-1395-1_66

1 Introduction

Cloud computing is an Internet-based approach in a distributed environment that consists of data centers, virtual machines. Cloud computing is an application-based software infrastructure that accumulates data on remote serves, that is accessed through the Internet. Cloud computing offers services whenever and wherever the user requires. The main aim of cloud data centers is to handle the millions of requests of users and efficiently being serviced. An optimization algorithm is executed iteratively by comparing solutions until an optimum result is attained. Task scheduling algorithms are employed to choose resources to execute tasks with minimal waiting and execution time. Task scheduling algorithms are a set of rules and policies that assign tasks to appropriate resources (CPU, memory, and bandwidth) to attain the highest performance and resource utilization. Therefore, efficient techniques are needed to increase the efficiency of handling multiple tasks in the distributed computing system. A Chaotic Squirrel Search Algorithm (CSSA) was introduced in [1] to perform multitask scheduling with better resource utilization. However, the designed CSSA algorithm was not efficient to find the greatest compromise solution in the optimization process for increasing the scheduling efficiency. An integration of İmperialist Competitive Algorithm and Firefly Algorithm (ICAFA) was introduced in [2] for enhancing the task scheduling with lesser makespan. However, the performance of higher efficiency and less false-positive rate was not achieved.

An integration of Cuckoo Search and Particle Swarm Optimization was developed in [3] to increase the performance of task scheduling and also to reduce the makespan as well as cost. However, the false-positive rate was not minimized. A Laxity and Ant Colony System algorithm (LBP-ACS) was developed in [4] to handle the problem of task scheduling. The designed algorithm minimized the failure rate of associated task scheduling but the makespan was not reduced. A static task scheduling model called Particle Swarm Optimization (PSO) algorithm was designed in [5]. But, the optimization algorithm failed to consider the multiple objective functions.

A meta-heuristics Whale Optimization Algorithm (WOA) was designed in [6] for cloud task scheduling based on the consideration of multiple objectives to increase the performance of a cloud system. However, the algorithm failed to reduce the makespan while handling the large heterogeneous tasks. An Estimation of Distribution Algorithm and Genetic Algorithm (EDA-GA) was designed in [7] to efficiently minimize the task completion time and enhance the load balancing capability. However, scheduling efficiency was not performed. An Enhanced version of the Multi-Verse Optimizer (EMVO) algorithm was developed in [8] to enhance the performance of task scheduling in terms of minimized makespan time and increasing resources' utilization. However, a better convergence rate was not achieved. Improved initialization of Particle Swarm Optimization (PSO) algorithm was introduced in [9] to reduce the makespan and total execution time. However, the algorithm failed to minimize the false-positive rate during the dynamic task scheduling in the cloud. A hybrid algorithm was designed in [10] for scheduling the scientific workflows with minimum task execution time. However, the algorithm failed to solve the multi-objective optimization problem.

From the above-said existing works, the existing works do not increase the task scheduling efficiency with less false-positive rate and do not consider the multiple objective functions. The existing works failed to reduce the makespan while handling the large heterogeneous tasks and the better convergence rate is not achieved. To overcome these existing issues, the CKMSJSD-MACO technique is introduced by applying the Chapman Kolmogorov Markov Stochastic function and Ant Colony Resource optimization to enhance the task scheduling efficiency and to reduce the makespan.

The major contribution of the paper is described as given below,

(a) On the contrary to conventional techniques, a CKMSJSD-MACO technique is introduced to increase the task scheduling efficiency. This concern is achieved by applying Chapman Kolmogorov Markov Stochastic function to the Ant Colony Resource optimization. This in turn increases the convergence rate of the algorithm and finds the global optimal solution from the population. This in turn reduces the incorrect task scheduling in the cloud.

(b) To reduce the makespan, virtual machines are sorted based on fitness evaluation and detect the local optimum from the population. Then the Jensen–Shannon Divergence is applied to ant colony optimization for finding the global solution and the scheduler assigns the tasks. The resource optimized virtual machine to efficiently complete the given task with lesser time consumption.

(c) Finally, an experimental evaluation was conducted to evaluate the quantitative analysis based on the different performance metrics such as task scheduling efficiency, false-positive rate, and makespan.

1.1 Outline of the Article

This article is organized into different sections. Section 2 reviews the related works. In Sect. 3, the proposed CKMSJSD-MACO technique is described with a neat diagram. Section 4 provides the experimentation with the CloudSim simulator. Section 5 provides a discussion about the experimental results with different performance metrics. Finally, Sect. 6 concludes the research work.

2 Related Work

A Q-learning-based task scheduling approach was developed in [11] for energy-efficient cloud computing. The designed task scheduling approach reduces the task response time but the efficiency was not improved. An implemented Jaya optimization algorithm was introduced in [12] for scheduling the numerous tasks. The algorithm reduces the execution cost and makespan but it failed to resolve the multi-objective optimization problem.

A Harmony-Inspired Genetic Algorithm (HIGA) was developed in [13] for energy-aware task scheduling with lesser overhead. However, the genetic algorithm failed to reduce the makespan. A Deep Reinforcement Learning Architecture (RLTS) was introduced in [14] to dynamically schedule the multiple tasks. The architecture minimizes the task execution time but it failed to consider the optimization technique for achieving higher scheduling efficiency. A Firefly Algorithm (FA) was introduced in [15] for workflow scheduling based on multiple objectives. Though the algorithm reduces the makespan and resource utilization, a better convergence rate was not achieved.

An Adaptive Dragonfly Algorithm (ADA) was developed in [16] for load-balanced task scheduling with lesser execution cost and time. However, scheduling efficiency was not improved. A Directed Acyclic Graph (DAG) model was developed in [17] for task scheduling based on the prediction of tasks computation time. However, the model increased the efficiency but the makespan was not reduced. An artificial fish swarm-based job scheduling technique was introduced to edge data centers. The technique reduced the task completion time but the global optimum solution was not accurately determined. A Modified Particle Swarm Optimization algorithm was designed for scheduling multiple tasks. The algorithm reduced the average response time but the higher scheduling efficiency was not achieved. A Game Theory-Based Task Scheduling Algorithm was designed in [18]. However, task completion time was not reduced. Secure and Optimized Cloud-Based Cyber-Physical Systems was designed in [19] with a memory-aware scheduling strategy. A hybrid algorithm was introduced in [20] to perform VM selection for scheduling applications. But, the task scheduling time was more.

3 Kolmogorov Markov Stochastic Jensen–Shannon Divergence-Based Multi-objective Ant Colony Optimization for Task Scheduling in Cloud

A novel CKMSJSD-MACO technique is introduced for increasing the task scheduling efficiency with a lesser makespan. The proposed CKMSJSD-MACO technique is considered as heterogeneous tasks sent to a cloud server. The main aim is to find an optimal mapping for ‘n’ different tasks to the resource optimized virtual machine by satisfying the user demands. The proposed CKMSJSD-MACO technique also solving the multiple objective functions for finding the optimal virtual machine.

Figure 1 illustrates the architecture diagram of the proposed CKMSJSD-MACO technique to map the given user-requested tasks into the virtual machines. Let us consider ‘n’ number of tasks ‘$T = \{T_1, T_2, \ldots, T_n\}$’ arrived from the various users ‘$U = \{u_1, u_2, \ldots, u_m\}$’ in the queue and it to be scheduled to the virtual machines

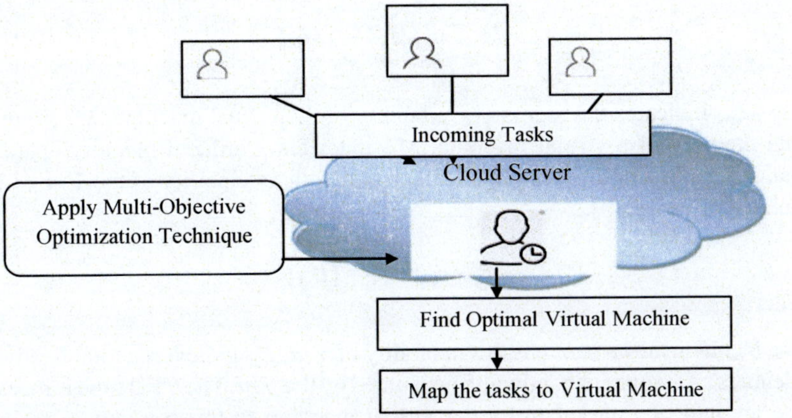

Fig. 1 Architecture of the CKMSJSD-MACO technique

'$Vm = \{Vm_1, Vm_2, \ldots, Vm_b\}$' in a cloud computing environment. The task scheduler 'TS' in the cloud server finds the resource optimized virtual machine. The task scheduler uses the multi-objective optimization technique called CKMSJSD-MACO. The proposed multi-objective ant colony optimization is a population-based metaheuristic for finding the optimal virtual machine based on the behavior of real ants. The behavior of the ants is to search the food source. While searching the food, the ants continually moved from one location to another for finding their food source. The ant deposits an organic compound called pheromone on the ground while moving from one location to another. The communication between the ants is performed through the pheromone trails deposited on the ground. After finding the food, the ant deposits pheromone on the paths depends on the amount of food that carries. Afterward, other ants also smell the pheromone deposited by the previous ant and follow that path for finding the food source. Based on the movement of the ants, the shortest path is identified from the nest to the food source since it has a more pheromone level. Therefore, the shortest path having a higher probability is selected as the ant path.

This process is related to find the resource-efficient virtual machine for task scheduling in the cloud. Here, the ants are related to the virtual machine and the food source is represented by the multiple objective functions, i.e., resource availability of the virtual machine in the cloud server. The proposed algorithm starts to perform the initialization process in the search space using Eq. (1).

$$Vm = \{Vm_1, Vm_2, Vm_3 \ldots, Vm_b\} \tag{1}$$

After the initialization process, the fitness is estimated based on the multiple objective functions such as memory, bandwidth, CPU time, and energy. For each virtual machine, the available memory capacity of the virtual machine is calculated as in Eq. (2).

$$M_{\text{avl}} = [M_{\text{T}}] - [M_{\text{ut}}] \tag{2}$$

where M_{avl} indicates the memory availability of the virtual machine, M_{T} represents a total memory of a virtual machine, M_{ut} indicates a utilized memory space of a virtual machine. Another resource is the bandwidth availability of the Vm which is calculated as in Eq. (3).

$$B_{\text{avl}} = [B_{\text{T}}] - [B_{\text{ut}}] \tag{3}$$

where B_{avl} indicates a bandwidth availability of Vm, B_{T} indicates a total bandwidth, B_{ut} denotes an amount of bandwidth consumed by the Vm. The CPU time is measured as the amount of time taken by the virtual machine to finish a certain task. The available CPU time of the 'VM' is calculated as in Eq. (4).

$$\text{CPU}_{\text{AVL}} = \text{CPU}_{\text{T}} - \text{CPU}_{\text{cd}} \tag{4}$$

where CPU_{AVL} represents the available CPU of the virtual machine, CPU_{T} signifies a total CPU time of Vm, CPU_{cd} represents a consumed time of Vm to perform the particular task. Finally, the energy consumption of the residual energy of the virtual machine is measured as per in Eq. (5).

$$E_{\text{AVL}} = [E_{\text{T}}] - [E_{\text{C}}] \tag{5}$$

where E_{AVL} specify residual energy of the virtual machine, E_{T} symbolizes total energy, E_{C} is the consumed energy. Based on the above-said resource estimation, the fitness is measured as in Eq. (6).

$$\alpha_{\text{FF}} = \arg\max\{(M_{\text{AVL}})\&\&(B_{\text{AVL}})\&\&(\text{CPU}_{\text{AVL}})\&\&(E_{\text{AVL}})\} \tag{6}$$

where α_{FF} indicates a fitness function. Based on the fitness, the local optimum solution is identified from the population. The local optimum solution is determined by sorting the virtual machine along with the fitness function as given in Eq. (7).

$$Vm = Vm_1 > Vm_2 > \cdots > Vm_b \tag{7}$$

The virtual machine which has better fitness is selected for finding the global optimum solution to reduce the task response time.

3.1 Chapman Kolmogorov Markov Stochastic Transition

The global optimum solution is obtained by applying the Chapman Kolmogorov Markov Stochastic transition. The global optimum solution is obtained by constructing a directed graph $G = (v, e)$ where $'v'$ indicates the vertices (i.e., position) and 'e' denotes the link (i.e., path) between the two vertices. While searching the food source, an ant moved from one location to another based on the Chapman Kolmogorov Markov Stochastic Transition function toward the better convergence. An ant has to choose the path using the directed graph traversal as shown in Fig. 2. It consists of vertices $v_1, v_2, v_{3,} v_{4,}$ and the edges $e_1, e_2, e_{3,} e_{4,}$ between the links. As shown in Fig. 3, an ant has to choose one vertex between the two vertices based on the fitness (i.e., pheromone) using Markov Stochastic Transition probability for moving from one location to another.

An ant selects one of the two vertices based on the fitness (i.e., pheromone) using the Chapman Kolmogorov Markov Stochastic Transition probability. The probability of one-step state transmission using Chapman Kolmogorov Markov Stochastic Transition probability property is expressed in Eqs. (8) through (10).

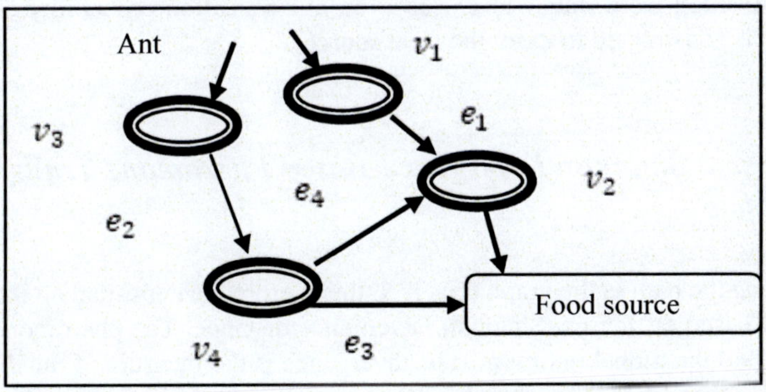

Fig. 2 Directed graph traversal

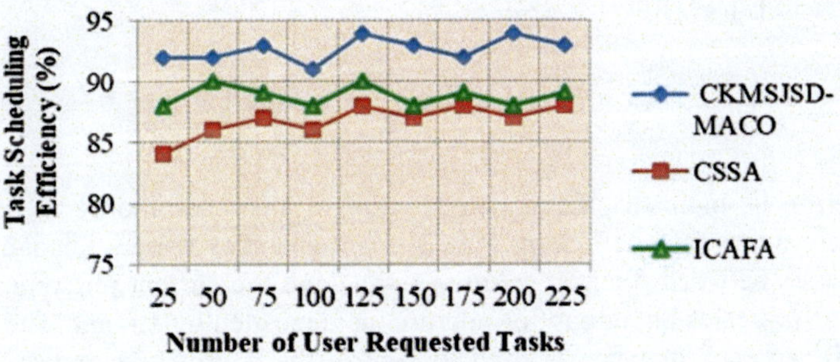

Fig. 3 Graphical illustration of task scheduling efficiency

$$P_{ij}(t, t+1) = \Pr\{(v_{t+1} = j | v_t = i, v_{t-1} = i_{n-1}, \ldots, v_1 = i_1, v_0 = i_0)\} \quad (8)$$

$$P_{ij}(t, t+1) = \Pr\{(v_{t+1} = j | v_t = i)\} \quad (9)$$

$$P_{i \to j} = \frac{\varphi_{ij}^{\omega} * \beta_{textij}^{\vartheta}}{\varphi_{ik}^{\omega} * \beta_{ik}^{\vartheta}} \quad (10)$$

From Eqs. (9) and (10), $P_{ij}(t, t+1)$ denotes Chapman Kolmogorov Markov Stochastic Transition probability of moving from state i to j, v_{t+1} denotes the next state at time $t+1$, and v_t denotes a current state at a time 't'. The Chapman Kolmogorov Markov Stochastic transition probability is used to identify the movement of ant depends on the current state 'i' and, not on the previous states. Equation (10) shows the ant moving the state $i \to j$, φ_{ij} represents the amount of pheromone deposited for the transition from state i to j, ω denotes a parameter to control the influence of $\varphi_{ij}(0 \leq \omega)$, β_{ij} indicates the desirability of state transition from state i to j, ϑ denotes a parameter to control the influence of $\beta_{ij}\vartheta \geq 1$. φ_{ik} indicates an amount of pheromone deposited for the transition from state i to k, β_{ik} indicates the desirability of state transition from state i to k. Based on one transition probability, the chain (i.e., path) is constructed to carry the food source.

3.2 *Jensen–Shannon Divergence-Based Pheromone Trails Update*

After finding the path in the graph traversal, the pheromone deposited on the ground is updated based on Jensen–Shannon Divergence distance. The pheromone update is used to find the global optimum. The divergence is the measure of the difference between two or more things. Here the difference between the global best on Jensen–Shannon Divergence solution and the current value is measured. If the distance is minimal, the global best solution is attained. The pheromone updating process is expressed as in Eq. (11).

$$\varphi_{ij}(t+1) = (1 - E)\varphi_{ij} + \frac{1}{2}\left|\varphi_{\text{best}} - \varphi_{ij}\right| + \Delta\,{\varphi_{ij}}^{k} \quad (11)$$

where $\varphi_{ij}(t+1)$ indicates the updated value of the pheromone, E denotes a pheromone evaporation coefficient, $\frac{1}{2}\left|\varphi_{\text{best}} - \varphi_{ij}\right|$ specifies Jensen–Shannon Divergence distance between the best solution (φ_{best}) and the current pheromone value (φ_{ij}), $\Delta\,{\varphi_{ij}}^{k}$ indicates an amount of pheromone deposited by kth ant. The Jensen–Shannon Divergence distance is used to control the pheromone update. Jensen–Shannon Divergence distance is bounded between 0 and 1. If the observed pheromone value is higher and hence the virtual machine is selected. Then the scheduler maps the incoming tasks to that virtual machine. In this way, all the incoming tasks are correctly scheduled for the resource-efficient virtual machine and minimize the false-positive rate. This process increases the scheduling efficiency in a cloud computing environment. The algorithmic process of the proposed CKMSJSD-MACO is described as given in Algorithm 1. The algorithmic process of the CKMSJSD-MACO technique is described to increase the task scheduling efficiency. Initially, the many heterogeneous tasks arrive at the cloud server from the various cloud users. The task scheduler in the cloud server finds the resource-efficient virtual machine by applying the Chapman Kolmogorov Markov Stochastic Jensen–Shannon Divergence Ant Colony Optimization technique.

\\ Algorithm 1: Markov Stochastic Bregman Divergence based Multi-Criteria Ant Colony Resource Optimized Task Scheduling
Input: Requested tasks $T = \{T_1, T_2, \ldots, T_n\}$', users '$U = \{u_1, u_2, \ldots, u_m\}$ virtual machines '$VM = \{Vm_1, Vm_2, \ldots, Vm_b\}$', task scheduler 'TS '
Output: Increase task scheduling efficiency

Begin
Collect the number of requested tasks '$T = \{T_1, T_2, \ldots, T_n\}$ from users 'U '
for each 'T'
Initialize the populations of $VM = \{Vm_1, Vm_2, \ldots, Vm_b\}$
For each 'Vm' in search space
Compute multiple objective functions M_{AVL}, B_{AVL}, CPU_{AVL}, E_{AVL}
Measure the fitness 'α_{FF}'
End for
While (t < Max_iteration) do
Sort the '$Vm_1 > Vm_2 > \ldots > Vm_b$
Select local optimum'
Apply stochastic transition probability $P_{ij}(t, t+1)$
Select the node with a higher probability P_{i-j}
Update the pheromone value $\varphi_{ij}(t+1)$ based on Jensen–Shannon
Divergence t = t+1
End while
Obtain global best 'Vm'
Scheduler maps the tasks to resource optimized virtual machine
End for
End

Initially, the population of virtual machines in the cloud server is initialized. Then the task scheduler measures the multiple objective functions of each virtual machine. Followed by, the fitness is measured based on the multiple objective functions. After that, the virtual machine is sorted according to the fitness value. From that, a local optimum solution is chosen for finding the global optimum. By applying the Chapman

Kolmogorov Markov Stochastic transition probability to find the next move. Based on this, the path between the ant and food source is correctly determined. Finally, the Jensen–Shannon Divergence is applied in the pheromone update along the selected path. The higher value of pheromone is chosen as the global best solution from the population. This process is repeated until the maximum iteration gets reached. Finally, the scheduler maps the tasks to the global best virtual machine increasing the scheduling efficiency and minimizes the false-positive rate.

4 Experimental Setup

In this section, experimental evaluation of the proposed CKMSJSD-MACO technique and two existing methods namely CSSA [1], ICAFA [2] are conducted in CloudSim simulation using the JAVA platform. The performance of proposed and existing methods is estimated using Personal Cloud datasets using Active Personal Cloud Measurement obtained from https://cloudspaces.eu/results /datasets.

The dataset includes 17 attributes and 66,245 instances. The main aim of this dataset is to execute load and transfer tests. The attributes are row id, account id, file size (i.e., task size), operation_time_start, operation_time_end, time zone, operation_id, operation type, bandwidth trace, node_ip, node_name, quoto_start, quoto_end, quoto_total (storage capacity), capped, failed and failure info. Among the 17 columns, two columns such as time_zone and capped are not used. Table 1 provides the dataset description.

5 Performance Analysis and Discussion

The experimental results of the CKMSJSD-MACO technique and two existing methods namely CSSA [1], ICAFA [2] are discussed with different performance metrics such as task scheduling efficiency, false-positive rate, and makespan concerning the number of user-requested tasks. The performance of the proposed CKMSJSD-MACO technique and existing methods are compared with the help of a table and graphical representation.

5.1 Impact of Task Scheduling Efficiency

It is measured as the ratio of the number of user-requested tasks that are correctly scheduled to the virtual machines from the number of tasks in the cloud server. The task scheduling efficiency is calculated as per Eq. (12).

Table 1 Dataset description

Attributes	Description
row id	database row identifier
account id	Personal cloud account used to perform this API call
file size	size of the uploaded/downloaded file in bytes
operation_time_start	starting time of the API call
operation_time_end	Finishing time of the API call
time zone	Time zone of a node for PlanetLab tests
operation_id	Hash to identify this API call
operation type	PUT/GET API call
bandwidth trace	trace of a file transfer (Kbytes/sec) obtained with vnstat
node_ip	Network address of the node executing this operation
node_name	Host name of the node executing this operation
quoto_start	Amount of data in the Personal Cloud account at the moment of starting the API call
quoto_end	Amount of data in the Personal Cloud account at the moment of finishing the API call
quoto_total	Storage capacity of this Personal Cloud account
capped	Indicates if the current node is being capped
failed	Indicates if the API call has failed (1) or not
failure info	Includes the available failure information in this API call

$$\text{TSE} = \left(\frac{\text{Number of tasks are correctly scheduled}}{n}\right) * 100 \quad (12)$$

where TSE denotes a task scheduling efficiency, 'n' stands for a total number of user-requested tasks. The task scheduling efficiency is measured in terms of percentage (%).

Table 2 describes the experimental results of task scheduling efficiency versus several tasks that arrived at the cloud server. The different efficiency results are obtained for the various counts of user-requested tasks. When considering the 250 number of user-requested tasks, the CKMSJSD-MACO technique achieves 92% of task scheduling efficiency and 87%, 88% of task scheduling efficiency is achieved in existing CSSA [1], ICAFA [2]. From the observed results, the CKMSJSD-MACO technique achieves higher task scheduling efficiency when compared to existing CSSA [1], ICAFA [2]. Figure 3 shows the experimental results of task scheduling efficiency for the number of user requests. As shown in the graphical plot in Fig. 3, the task scheduling efficiency of the proposed CKMSJSD-MACO technique is higher than the conventional optimization methods. This improvement is achieved by applying the multi-objective ant colony optimization technique. The multiple objective functions are used to accurately find the virtual machine through the fitness

Table 2 Task scheduling efficiency

Number of user-requested tasks	Task scheduling efficiency (%)		
	CKMSJSD-MACO	CSSA	ICAFA
25	92	84	88
50	92	86	90
75	93	87	89
100	91	86	88
125	94	88	90
150	93	87	88
175	92	88	89
200	94	87	88
225	93	88	89
250	92	87	88

measure for scheduling the tasks. Besides, the Jensen–Shannon Divergence is also applied in the pheromone update to discover the global optimum virtual machine. Then scheduler correctly maps the tasks to the resource optimized virtual machine.

The average of ten results indicates that the task scheduling efficiency of the proposed CKMSJSD-MACO technique is considerably increased by 7% when compared to existing CSSA [1], and 4% when compared to ICAFA [2], respectively.

5.2 *Impact of the False-Positive Rate*

It is defined as many user-requested tasks that are incorrectly scheduled to the virtual machine in the cloud server. The false-positive rate is mathematically formulated as given in Eq. (13).

$$\text{FPR} = \left(\frac{\text{Number of tasks are incorrectly scheduled}}{n}\right) * 100 \tag{13}$$

where FPR indicates a false-positive rate, 'n' stands for a total number of user-requested tasks. The false-positive rate is measured in terms of percentage (%).

Table 3 and Fig. 4 portray the performance analysis of false-positive rate versus the number of tasks taken in the range 25–250. From Table 2 and Fig. 4, considering 250 user-requested tasks, the false-positive rate using the CKMSJSD-MACO technique is 8% and the false-positive rate using existing methods are 13% and 12%, respectively. Therefore, the comparison results show that the false-positive rate is considerably reduced by 44% and 34% when compared to CSSA [1] and ICAFA [2], respectively.

The observed results indicate that the proposed CKMSJSD-MACO technique minimizes the false-positive rate when compared to conventional optimization techniques. This improvement is obtained by sorting the virtual machine based on fitness

Table 3 False positive rate

Number of user-requested tasks	False-positive rate (%)		
	CKMSJSD-MACO	CSSA	ICAFA
25	8	16	12
50	8	14	10
75	7	13	11
100	9	14	12
125	6	12	10
150	7	13	12
175	8	12	11
200	6	13	12
225	7	12	11
250	8	13	12

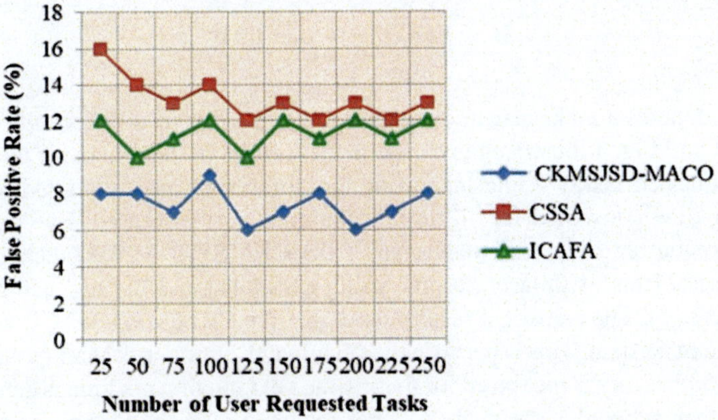

Fig. 4 Graphical illustration of the false-positive rate

measures. The virtual machine which has better fitness is selected as the local optimum. After that, the CKMSJSD-MACO technique uses the Jensen–Shannon Divergence to measure the difference between the current solution and the best solution. This helps to correctly find the resource optimized virtual machine for task scheduling.

5.3 *Impact of Makespan*

Makespan is measured as an amount of time taken by the algorithm to complete the requested tasks by the virtual machines in a cloud. Therefore, the Makespan is calculated as given in Eq. (14).

Table 4 Makespan

Number of user-requested tasks	Makespan (ms)		
	CKMSJSD-MACO	CSSA	ICAFA
25	21	26	24
50	23	28	26
75	27	33	31
100	30	37	34
125	33	40	37
150	36	42	40
175	40	45	43
200	43	48	45
225	45	50	47
250	47	52	50

$$M = \text{End time} - \text{start time} \tag{14}$$

where 'M denotes a makespan and it is measured in terms of milliseconds (ms).

Table 4 and Fig. 5 illustrate the experimental results of makespan for the number of user-requested tasks. While increasing the number of user-requested tasks, the amount of time consumed for all three techniques gets increased. By considering '25′ user-requested tasks, the makespan of the CKMSJSD-MACO technique was found to be '21 ms', similarly '26 ms' using CSSA [1] and '24 ms' using ICAFA [2], respectively. The statistical results show that the CKMSJSD-MACO technique reduces the makespan. This is because the Chapman Kolmogorov Markov Stochastic transition probability is measured for finding the next moving position is determined. Then the task scheduler discovers a better resource-efficient virtual machine and distributes the tasks with minimum time consumption. As a result, the task completion time of the virtual machine also gets minimized. The average of ten results indicates

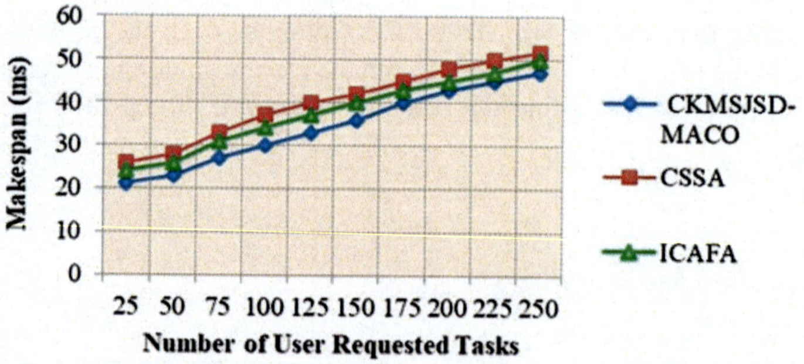

Fig. 5 Graphical illustration of makespan

that the makespan is considerably reduced by 15% and 9% than the existing CSSA [1] and ICAFA [2], respectively.

6 Conclusion

In this work, a CKMSJSD-MACO technique is introduced in a cloud environment that uses multiple objective functions to solve task scheduling in the cloud. The main contribution of the proposed CKMSJSD-MACO technique finds an optimal virtual machine for scheduling the user-requested tasks at the cloud server. The proposed technique measures the fitness based on the multiple objective functions, i.e., resource availability. The Chapman Kolmogorov Markov Stochastic function is applied to achieve a better convergence rate. Then the Jensen–Shannon Divergence is also employed for finding the global optimum in the pheromone update. The proposed CKMSJSD-MACO technique experiments with different performance metrics such as task scheduling efficiency, false-positive rate, and makespan. The discussed results prove that the proposed CKMSJSD-MACO technique provides better performance results consistently in terms of higher efficiency, lesser false-positive rate as well as lesser makespan than the state-of-the-art methods. From the experimental results, the proposed CKMSJSD-MACO technique improves the task scheduling efficiency by 6% and reduces the false-positive rate, and makespan by 12% as compared to conventional methods.

References

1. M.S. Sanaj, P.M.Joe Prathap, Nature-inspired chaotic squirrel search algorithm (CSSA) for multi-objective task scheduling in an IAAS cloud computing atmosphere. Eng. Sci. Technol. Int. J. **23**(4), 891–902 (2020)
2. S.M.G. Kashikolaei, A.A.R. Hosseinabadi, B. Saemi, M.B. Shareh, A.K. Sangaiah, G.B. Bian, An enhancement of task scheduling in cloud computing based on imperialist competitive algorithm and firefly algorithm. J. Supercomput. **76**, 6302–6329 (2020)
3. T. Prem Jacob, K. Pradeep, A multi-objective optimal task scheduling in cloud environment using cuckoo particle swarm optimization. Wirel. Person. Commun. **109**, 315–331 (2019)
4. Xu. Jiuyun, Z. Hao, R. Zhang, X. Sun, A method based on the combination of laxity and ant colony system for cloud-fog task scheduling. IEEE Access **7**, 116218–116226 (2019)
5. Fatemeh Ebadifard and Seyed Morteza Babamir, A PSO-based task scheduling algorithm improved using a load-balancing technique for the cloud computing environment. Concurrency Comput. Pract. Exp. **30**(12), 1–16 (2018)
6. X. Chen, L. Cheng, C. Liu, Q. Liu, J. Liu, Y. Mao, J. Murphy, A WOA-based optimization approach for task scheduling in cloud computing systems. IEEE Syst. J. **14**(3), 3117–3128 (2020)
7. S. Pang, W. Li, H. He, Z. Shan, X. Wang, An EDA-GA hybrid algorithm for multi-objective task scheduling in cloud computing. IEEE Access **7**, 146379–146389 (2019)
8. S.E.Shukri, R. Al-Sayyed, A. Hudaib, S. Mirjalili, Enhanced multi-verse optimizer for task scheduling in cloud computing environments. Exp. Syst. Appl. 1–30 (2020)

9. S.A. Alsaidy, A.D. Abbood, M.A. Sahib, Heuristic ınitialization of PSO task scheduling algorithm in cloud computing. J. King Saud Univ. Comput. Inf. Sci. 1–13 (2020)
10. M. Sardaraz, M. Tahir, A hybrid algorithm for scheduling scientific workflows in cloud computing. IEEE Access **7**, 186137–186146 (2019)
11. D. Ding, X. Fan, Y. Zhao, K. Kang, Q. Yin, J. Zeng, Q-learning based dynamic task scheduling for energy-efficient cloud computing. Future Gener. Comput. Syst. **108**, 361–371 (2020)
12. S. Gupta, I. Agarwal, R.S. Singh, Workflow scheduling using jaya algorithm in cloud. Concurrency Comput. Pract. Exp. **31**(17), 1–13 (2019)
13. M. Sharma, R. Garg, HIGA: harmony-inspired genetic algorithm for rack-aware energy-efficient task scheduling in cloud data centers. Eng. Sci. Technol. Int. J. **23**(1), 211–224 (2020)
14. T. Dong, F. Xue, C. Xiao, J. Li, Task scheduling based on deep reinforcement learning in a cloud manufacturing environment. Concurrency Comput. Pract. Exp. **32**(11), 1–12
15. M. Adhikari, T. Amgoth, S.N. Srirama, Multi-objective scheduling strategy for scientific workflows in cloud environment: a firefly-based approach. Appl. Soft Comput. **93**, 1–31 (2020)
16. P. Neelima, A. Rama Mohan Reddy, An efficient load balancing system using adaptive dragonfly algorithm in cloud computing. Cluster Comput. **23**, 2891–2899 (2020)
17. B.A. Al-Maytami, P. Fan, A. Hussain, T. Baker, P. Liatsis, A task scheduling algorithm with ımproved makespan-based on prediction of tasks computation time algorithm for cloud computing. IEEE Access **7**, 160916–160926 (2019)
18. J. Yang, B. Jiang, Z. Lv, K.-K.R. Choo, A task scheduling algorithm considering game theory designed for energy management in cloud computing. Fut. Gener. Comput. Syst. **105**, 985–992 (2020)
19. H. Wang, S. Smys, Secure and optimized cloud-based cyber-physical systems with memory-aware scheduling scheme. J. Trends Comput. Sci. Smart Technol. (TCSST) **2**(03), 141–147 (2020)
20. V. Karunakaran, A stochastic development of cloud computing based task scheduling algorithm. J. Soft Comput. Paradigm (JSCP) **1**(01), 41–48 (2019)

Security Aspects in Cloud Tools and Its Analysis—A Study

Jyoti Vaishnav and N. H. Prasad

Abstract Since the storage of information in a piece of paper or in a hard drive has become more conventional, new technologies are being developed by data science researchers to securely and efficiently store the confidential information. In this regard, the usage of cloud computing is increasing at an unprecedented rate and it is extensively used in different parts of the world. This technology uses internet as the primary source for computing and storing the data. Further, the facts and sordid resources are supplied in conformity with the user through PC and machine on-demand. Moreover, it remains as a new thought to utilize the digital assets for apportionment statistics and then it is further evolved. Yahoo and Gmail are some suitable real-time examples of cloud computing. Different industries like banking, transport, entertainment and other discipline are also drifting towards this technology. The effectiveness of the supplied with the way of dictation powered with the aid of offer as the use case model will take care of the challenges in bandwidth, information movement, transactions and storage information.

Keywords Cloud-Computing · Security · Encryption · Cloud storage · JSKY tool · IBM tool

1 Introduction

Today, in the Information age almost all the Industries have a job by some means linked to statistics that are saved in digital form on a network. During the agricultural age, plants and the gear to produce them emerges as the most essential asset. During the economic age, manufactured goods and the factories that produce them have become the most crucial asset. Today, facts and information are remaining as the key

J. Vaishnav (✉)
Department of Computer Application, Presidency College Hebbal, Kempapura, Bengaluru, India
e-mail: jyotivaishnav-coll@presidency.edu.in

N. H. Prasad
Department of MCA, Nitte Meenakshi Institute of Technology, Bengaluru, India
e-mail: hod@nmit.in

V. Suma et al. (eds.), *Inventive Systems and Control*,
Lecture Notes in Networks and Systems 204,
https://doi.org/10.1007/978-981-16-1395-1_67

assets for every corporation and individual [1]. The surge in activities of spying a person against a person, country against a country imposes a significant challenge on securing the confidential information. In today's technological world, data is generally stored in the public cloud due to its ease-of-use feature without concerning their security [2]. The dangerous development in the volume, assortment and speed of information produced each second has made the cloud to be the accepted methods for putting away and overseeing the applications. Distributed computing carries a few forefront chances to the end clients with an ensured boundless measure of operational effectiveness, cooperative stages, profitability and an inevitable access to arrange the foundations. This positive side of distributed computing has driven enormous and creative venture, for example, Amazon Web-Services, Dropbox, HP, Microsoft-Azure and others to grasp the cloud for their calculations and this has set the cloud as a basic device for ordinary use [3]. The proposed paper deploys the cloud security and its aspects with the compression of encrypted algorithms.

2 Overview of Cloud-Computing

Cloud storage allows customers to remotely process and store their information and experience on-call for higher-end cloud applications without the weight of nearby hardware and software management [4]. Though the blessings are clear, this form of service is also relinquishing customers' bodily ownership in their outsourced information, which necessarily possess new security dangers inside the course of the correctness of facts. In order to address this new trouble and in addition to achieve a steady and reliable cloud garage service, this paper suggests a flexible disbursed garage integrity auditing mechanism by utilizing the homomorphism token and deployed erasure-coded facts [5–7]. The proposed design permits customers to audit the cloud storage with very lightweight verbal exchange and computation cost [8]. The auditing result is of high-quality and guarantees sturdy cloud garage correctness and also simultaneously achieves fast records blunders localization, the identity of misbehaving server [9].

Considering the cloud records are dynamic in nature, the proposed design in addition supports regular and inexperienced dynamic operations on outsourced records, along with block amendment, deletion, and append. Analysis shows the proposed scheme is noticeably green and resilient closer to Byzantine failure, malicious records amendment attack, or even server colluding attack [10–13] (Fig. 1).

2.1 Cloud Applications

Cloud computing is net-based computing, whereby shared resources, software program, and statistics are provided to computer structures and other devices on demand. Cloud shape, the structures shape of the software systems involved in

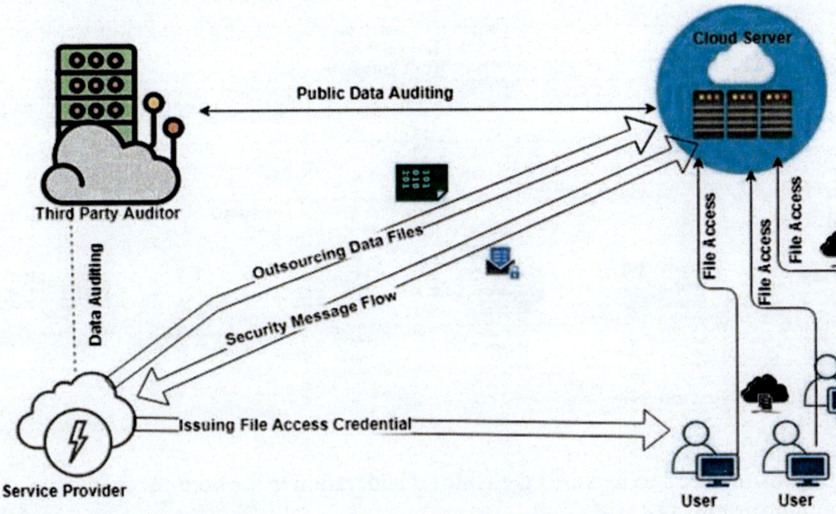

Fig. 1 Cloud architecture

the shipping of cloud computing, contains hardware and software program software designed by manner of a cloud architect who normally works for a cloud. It commonly involves more than one cloud components speaking with each one of a kind over utility programming interfaces, typically net services [7, 14, 15]. This carefully resembles the UNIX philosophy of having multiple packages doing one issue well and working collectively over every day interfaces. Complexity is managed and the resulting structures are extra achievable than their monolithic counterparts [16, 17]. Cloud structure extends to the client, where net browsers and/or software application access cloud applications. Cloud storage shape is loosely coupled, wherein metadata operations are centralized allowing the data nodes to scale into the hundreds, each independently delivering statistics to packages to the users.

2.2 *Cloud Deployment Services Models*

See Fig. 2.

2.2.1 Infrastructure as a Service (IaaS)

Proposing virtualized resources (OS, storage, computation and communication) on-call is referred as IaaS. A cloud infrastructure permits on-call for provisioning of servers walking several picks of working systems and a customized software program

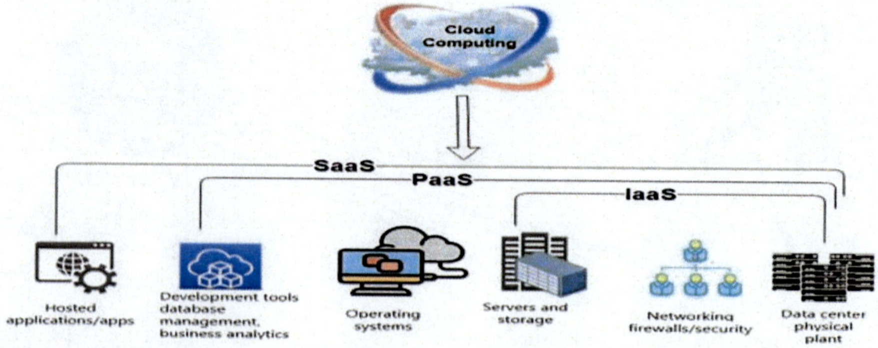

Fig. 2 Cloud service models

stack. Infrastructure services are taken into consideration to the bottom layer of cloud computing systems [18, 19].

2.2.2 Platform as a Service (PaaS)

For raw computing and storage services, provides better technique called Platform as a Service (PaaS). Google AppEngine, an instance of platform as a service, gives a scalable environment for developing and website hosting. Web applications, which should be written in specific programming languages which include Python or Java and use the services of very own proprietary dependent object statistics store [20].

2.2.3 Software as a Service (SaaS)

In application layer, cloud stack available on the top of the layer and it may be accessed by web portals. Users shifting from laptop to online software services for fast access. Online services also provide equal function such as word, spreadsheet within the web [21].

2.3 Cloud Computing Security Issues

According to the cloud storage security alliance, the very best 3 threats within the cloud square measure Insecure, Interfaces and API's, knowledge Loss & outpouring, and Hardware Failures which accounted for twenty-ninth, twenty-fifth and one hundred PC of all cloud security [1, 22]. Cloud infrastructure being shared by completely different users resides on identical knowledge server. As a result, information of a whole bunch or thousands of firms are often hold on giant cloud servers, hackers will in theory gain management of big stores of data through one attack—a method known as "hyperjacking". Some samples of this embrace the Dropbox security breach, and iCloud 2014 leak [22]. Dropbox had been broken in Oct 2014, having over seven million of its users' passwords taken by hackers in an attempt to induce cost from Bitcoins (BTC). By having these passwords, they're able to scan personal knowledge additionally as indexed by search engines [23–25]. Here are the most common place cloud computing security risks:

2.3.1 Distributed-Denial-of-Service Attacks (DDoS)

DDoS it's almost like DoS, DDoS attempts to damage important services running on a server by flooding the destination server with packets. The specialty of DDoS is that the attacks don't come from one network or host but from variety of various hosts or networks which are previously compromised [26].

2.3.2 Phishing and Social Engineering Attacks

A Phishing assault uses both communal engineering and empiric subterfuge in accordance with snitch shoppers 'personal identification records yet economic tale credentials. Social engineering schemes uses spoofed e-mails in conformity with administration clients in conformity with false websites up to expectation are designed to trick recipients among divulging pecuniary statistics such are credit scorecard numbers, account usernames, passwords, and Social Security numbers. Hijacking brand names about banks, e-retailers, and credit score visiting card companies, phishers normally convince recipients in conformity with respond. Technical subterfuge schemes sow iniquity sensible onto PC sin rule in accordance with scouse sharpens credentials directly, often using Trojan Keyloggers spyware. Pharming offense aware misdirects clients in imitation of fraudulent internet sites then proxy servers, typically via DNS hijacking and poisoning durability [27].

2.3.3 Pretexting

Pretexting is the act of making associated exploitation and fabricated situation to get information from a target, typically over the phone. It's over an easy lie, because it often involves some previous analysis and therefore the use of items of proverbial info (e.g., for impersonation: birthday, Social Security variety (SSN), last leader, mother 's maiden name). This establishes legitimacy within the mind of the target [28].

2.3.4 Encryption

Companies or organizations depend on conformity with stay compelled after absorbing a data-centric method in conformity with protecting their sensitive facts, which are in accordance with the protection against superior threats inside the intricate or evolving environments over virtualization, wind services and multiplication. Firms are ought to implement knowledge security solutions that give consistent protection on sensitive knowledge as well as cloud knowledge protection through encoding and scientific discipline key management [7].

2.3.5 Insider Threats

Insider-related threats (either via indifference or malevolence), usually the longest imitation of observation and then followed by resolving it. A Herculean identity yet accesses the administration mold beside a nice privilege administration equipment that necessitates rectangular metering in the imitation of casting off these threats yet lowering the harm (such so by using stopping lateral rate then privilege escalation) once its slave reaches stability [7].

2.3.6 Inconsistencies

IT tools are architected because of on-premises environment or certain variety about planet is fast incoherent by including unique cloud environments. Incompatibilities desires turn among encounter or management that gaps so much to unfasten organizations in conformity in order to gamble out of misconfigurations, vulnerabilities, facts leak, excessive privileged access, agreement problems, or toughness [7].

Security issues together with privacy, unease the government inspection on information, authorization, inadequate entry to control, verification, back-door/trapdoor infusion into encryption algorithms, negative encryption and implementation with loopholes are the principle demanding situations in cloud computing paradigm [7, 23, 29, 30]. Such safety concerns have driven a growing use of ultra-modern cryptographic strategies for ensuring the safety of information in the cloud. Encrypting user's information to curtail malicious attacks and supply safety may cope with a number of the safety demanding situations inside the cloud but there is an appropriate encryption scheme for specific application to maximize the application capability and no longer impede safety and overall performance of the cloud services. Choosing a suitable encryption scheme for software that allows certain capability is frequently remaining as a big assignment in cloud encryption domain. This observes targets to throw mildly on the way to acquire stability among protection and application functionality for the precise use case of cloud encryption schemes. It draws a boundary between what a user will obtain in terms of overall performance whilst sure encryption algorithms are hired on cloud programs [27, 31, 32]. This will assist the person to understand the practical and safety requirement of their cloud packages and the great safety notions they are able to get whilst they rent sure encryption schemes. Furthermore, this analysis will serve as a quick reference for practitioners in determining precise encryption algorithms that are appropriate for cloud-based applications while maximising their capability for optimal overall performance. To this end, cutting-edge cloud-encryption schemes are technically reviewed and compared in phrases of the functionality in their algorithmic design, the security they offer and their suitability for specific use inside the cloud. This examine focuses explicitly on symmetric cryptographic algorithms wherein the client solely holds the key. However, the encryption schemes mentioned are hired in securing cloud-primarily based packages. This is motivated by our observation that research exploring cloud-based encryption schemes from capability as opposed to use case perspectives are understudied. Therefore, the fundamental contribution of these studies is to boost the expertise of the adverse impact faced inside the industry. An incompatible encryption scheme is matched with the wrong application. In a nutshell, this contribution is obtained by:

- Analyzing and categorizing modern symmetric-primarily based on the encryption scheme.
- Studying the security and suitability of particular encryption algorithms for cloud applications.
- Suggesting sensible use case of particular encryption algorithms that allow certain capability even for maximizing the performance and imparting ideal security.

3 Comparative Study on Cloud Computing Encryption Schemes

A comparative study is done by selecting few cloud based encryption algorithm and its practices in the real world. The Table 1 shows the different types of threats, attacks and its strength of different Encryption schemes [33, 34].

3.1 Analyze Cloud in the Protocol Tab

From the comparative study, we have analyzed security leakage on real time application like Gmail Cloud and IBM Cloud

Table 1 Security comparison of the distinct encryption algorithm [12]

Encryption aspect	Key (bits)	Cryptanalysis attack	Block size (bits)	Time to decode	Input size	Outcome of encryption	Standard time	Encoding speed
Data encryption standard (DES)	56-bit key	Differential cryptanalysis, linear cryptanalysis [40]	64-bit block	6.2×10^{16}	47, 104, 228, 905, 5202	17, 45, 75, 257, 987	310.2	5.06
Triple data encryption standard (3DES)	112\168	Main in the middle attack key theft attack, mathematical attack	64-bit block	2.01×10^{18}	36, 104, 228, 905, 5202	44, 87, 157, 170, 1108	310.2	5.06
Advanced encryption standard (AES)	128/192/2 56	Brute-force, timing attack	128-bit block	2.4×10^{38}/ 5.2×10^{57}/ 1.1×10^{77}	56, 104, 228, 905, 4202	45, 92, 167, 167, 1200	378.9	4.14
Blow Fish	32-448	Cryptanalysis attack, birthday attack like	64-bit	2.01×10^{18}	36, 104, 228, 905, 4202	30, 47, 87, 123, 157	96.3	16.32
RC2	40-1024	Related key, key guessing, key theft	64-bit	2.01×10^{18}	36, 104, 228, 805, 4202	46, 92, 176, 206, 1507	489.6	3.21
RC5	128/192/2 56	Differential attack	64/128-bit	2.4×10^{38}/ 5.2×10^{57}/ 1.1×10^{77}	36, 104, 228,805, 4202	32, 63, 110, 052, 765	247.1	6.36

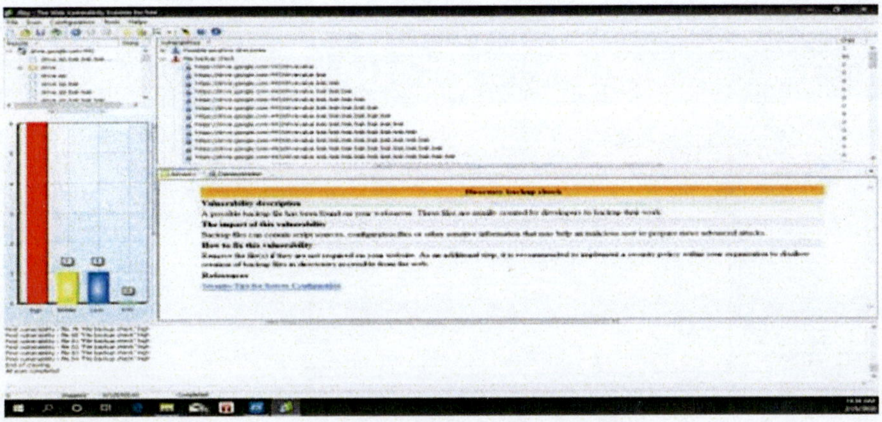

Fig. 3 Gmail cloud storage is analyzed with JSKY tool for its efficiency

3.2 Analyzing Gmail Cloud in JSKYTools

With the help of JSKY tool [33] the testing has been performed on Gmail Cloud, which shows Gmail Cloud is not secure against file Backup, dictionary attack, ProRat, Google Hacks, Httrack, site digger threats and attacks, which is given in Fig. 3.

3.2.1 Analyzing IBM Cloud in JSKYTools

With the help of JSKY tool, the test has been performed on IBM Cloud [18], which shows IBM Cloud to secure against file backup, dictionary attack, ProRat, Google hacks, Httrack, site digger threats and attacks, which is given in Fig. 4.

4 Conclusion

In this article, a general approach of cloud storage practices is conferred. Then, a comparative study is performed in cloud storage with cryptography method to examine its security complexities. Moreover, the analysis is finalized so as with the assistance of tool, real time cloud service provider like Gmail is examined to test its security strength and proven to be insecure against file directory attack, storage leakage, brute force attack and etc. Since the analysis on cloud will be primarily based on cryptography and encryption may be a comparatively young space.

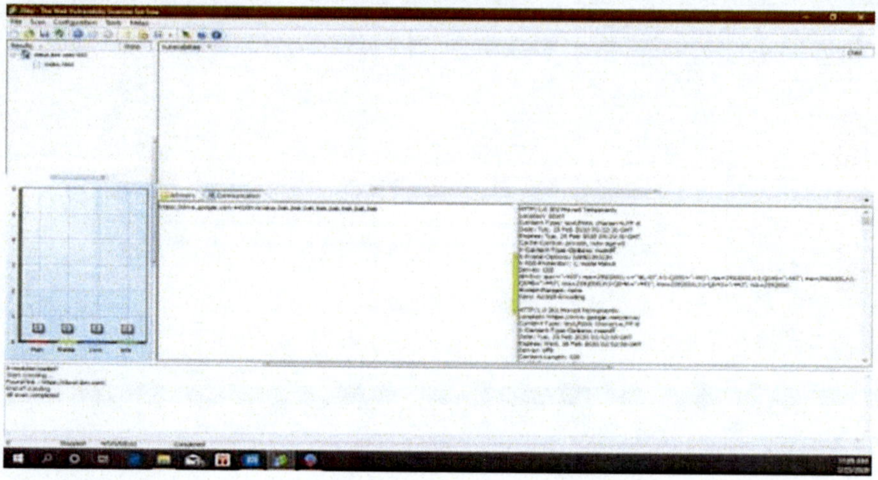

Fig. 4 IBM Cloud Storage is analyzed with JSKY tool for its efficiency

5 Future Work

Future work helps to focus on the research work to improve the cloud security with the help of cryptography and encryption methods.

References

1. A. Lele, Cloud computing, in *Smart Innovation, Systems and Technologies*, (2019)
2. M. Armbrust et al., A view of cloud computing. Commun ACM (2010). https://doi.org/10.1145/1721654.1721672
3. T. Dillon, C. Wu, E. Chang, Cloud computing: issues and challenges, in *Proceedings—International Conference on Advanced Information Networking and Applications, AINA*, (2010). https://doi.org/10.1109/aina.2010.187
4. B. Hayes, Cloud computing. Commun. ACM (2008). https://doi.org/10.1145/1364782.1364786
5. S. Chawla, C. Diwaker, Cloud computing. Int. J. Appl. Eng. Res. (2012). https://doi.org/10.4018/jeei.2012040104
6. C. Low, Y. Chen, M. Wu, Understanding the determinants of cloud computing adoption. Ind. Manag. Data Syst. (2011). https://doi.org/10.1108/02635571111161262
7. Cloud Security Alliance, Top threats to cloud computing. Security (2010)
8. J. Lee, A view of cloud computing. Int. J. Networked Distrib. Comput. (2013). https://doi.org/10.2991/ijndc.2013.1.1.2
9. S. Shilpashree, R.R. Patil, C. Parvathi, Cloud computing an overview. Int. J. Eng. Technol. (2018). https://doi.org/10.32628/ijsrset196120
10. N. Gruschka, M. Jensen, Attack surfaces: a taxonomy for attacks on cloud services, in *Proceedings—2010 IEEE 3rd International Conference on Cloud Computing, CLOUD 2010*, (2010). https://doi.org/10.1109/cloud.2010.23

11. I. Stojmenovic, S. Wen, The fog computing paradigm: scenarios and security issues, in *2014 Federated Conference on Computer Science and Information Systems, FedCSIS 2014*, (2014). https://doi.org/10.15439/2014f503
12. C. Wang, Q. Wang, K. Ren, W. Lou, Ensuring data storage security in cloud computing, in *IEEE International Workshop on Quality of Service, IWQoS*, (2009). https://doi.org/10.1109/IWQoS.2009.5201385
13. I.M. Khalil, A. Khreishah, M. Azeem, Cloud computing security: a survey. Computers (2014). https://doi.org/10.3390/computers3010001
14. A.U.R. Khan, M. Othman, S.A. Madani, S.U. Khan, A survey of mobile cloud computing application models. IEEE Commun. Surv. Tutorials, (2014). https://doi.org/10.1109/SURV.2013.062613.00160
15. J. Samad, S. W. Loke, K. Reed, Mobile cloud computing, in *Cloud Services, Networking, and Management*, (2015)
16. Y.Q. Zhang, X.F. Wang, X.F. Liu, L. Liu, Survey on cloud computing security. RuanJianXueBao/J. Softw. (2016). https://doi.org/10.13328/j.cnki.jos.005004
17. D.C. Marinescu, *Cloud Computing: Theory and Practice*. (2013)
18. S.S. Manvi, G. Krishna Shyam, Resource management for infrastructure as a service (IaaS) in cloud computing: a survey. J. Netw. Comput. Appl. (2014). https://doi.org/10.1016/j.jnca.2013.10.004
19. S. Bhardwaj, L. Jain, S. Jain, Cloud computing: a study of infrastructure Asa service (Iaas). Int. J. Eng. (2010)
20. C. Pahl, Containerization and the PaaS cloud. IEEE Cloud Comput. (2015). https://doi.org/10.1109/MCC.2015.51
21. R. Rezaei, T.K. Chiew, S.P. Lee, Z. Shams Aliee, A semantic interoperability framework for software as a service system in cloud computing environments. Expert Syst. Appl. (2014) https://doi.org/10.1016/j.eswa.2014.03.020
22. V. CC, Security guidance critical areas of focus for. **3**, 1–76, December (2009)
23. L. M. Kaufman, Data security in the world of cloud computing. IEEE Secur. Priv. (2009). https://doi.org/10.1109/MSP.2009.87
24. M. Vijayakumar, V. Sunitha, K. Uma, A. Kannan, Security issues in cloud computing. J. Adv. Res. Dyn. Control Syst. (2017). https://doi.org/10.4018/ijcac.2011070101
25. Z. Tari, Security and privacy in cloud computing. IEEE Cloud Comput. (2014). https://doi.org/10.1109/MCC.2014.20
26. O. Osanaiye, K.K.R. Choo, M. Dlodlo, Distributed denial of service (DDoS) resilience in cloud: review and conceptual cloud DDoS mitigation framework. J. Netw. Comput. Appl. (2016). https://doi.org/10.1016/j.jnca.2016.01.001
27. D. Zissis, D. Lekkas, Addressing cloud computing security issues. Futur. Gener. Comput. Syst. (2012). https://doi.org/10.1016/j.future.2010.12.006
28. B. Grobauer, T. Walloschek, E. Stöcker, Understanding cloud computing vulnerabilities. IEEE Secur. Priv. (2011). https://doi.org/10.1109/MSP.2010.115
29. M.M. Alani, Security threats in cloud computing, in *SpringerBriefs in Computer Science*, (2016)
30. M. Ali, S. U. Khan, A. V. Vasilakos, Security in cloud computing: opportunities and challenges. Inf. Sci. (Ny) (2015). https://doi.org/10.1016/j.ins.2015.01.025
31. M. Jensen, J. Schwenk, N. Gruschka, L. LoIacono, On technical security issues in cloud computing, in *CLOUD—2009 IEEE International Conference on Cloud Computing* (2009). https://doi.org/10.1109/cloud.2009.60
32. N. Subramanian, A. Jeyaraj, Recent security challenges in cloud computing. Comput. Electr. Eng. (2018). https://doi.org/10.1016/j.compeleceng.2018.06.006
33. S. Shakya, An efficient security framework for data migration in a cloud computing environment. J. Artif. Intell. **1**(01), 45–53 (2019)
34. S.R. Mugunthan, Soft computing based autonomous low rate ddos attack detection and security for cloud computing. J. Soft Comput. Paradigm (JSCP) **1**(02), 80–90 (2019)

Industrial Internet of Things (IIoT): A Vivid Perspective

Malti Bansal, Apoorva Goyal, and Apoorva Choudhary

Abstract Since the last couple of years, the internet of things has become a hot topic of research in academic as well as industrial domains. The Industrial Internet of Things (IIoT) is an amalgamation of industrial automation, control systems, and IoT systems. IIoT would be a substantially important component in the upcoming industries. The broadly listed aims of IIoT could be high working efficiency and productivity, better asset management of assets by product tailoring, smart monitoring of applications for production, preventive and predictive maintenance of equipment in the industry. In this article, a sound definition of IIoT, the architecture of IIoT, the key-enabling technologies, its major applications were presented. Finally, highlights the future scope and recent challenges faced in this field.

Keywords Industrial internet of things · Applications · Architecture · Enabling technologies

1 Introduction

The main reason to adopt IIoT by the various industries was to improve operational efficiency; however, it is observed that it can improve the overall efficiency of the firms concerning efficiency, quality, delivery, and safety. It has brought significant changes in the working like detecting predictive maintenance of any drive or the real-time tracking of any set parameters. IIoT has also helped to keep up with the safety measures while conducting activities like detecting corrosion inside a refinery pipe, checking acid levels in the tank, etc.

The technology of IIoT is not only limited to agriculture, manufacturing, and industries, also finds great applications in logistics, warehouse, banking sectors, and hospitals.

It is roughly estimated that by 2020, there will be a worldwide expenditure of $500 billion on the installation of IIoT technology. One of the most queried after

M. Bansal (✉) · A. Goyal · A. Choudhary
Department of Electronics and Communication Engineering, Delhi Technological University (DTU), Delhi 110042, India

V. Suma et al. (eds.), *Inventive Systems and Control*,
Lecture Notes in Networks and Systems 204,
https://doi.org/10.1007/978-981-16-1395-1_68

technologies among most of the industries as the manufacturers would be able to upgrade their profits by 30% and therefore diminish maintenance price by 30%.

Cyber-security has been a critical concern for adopting IIoT, and industrial workers have already put security concerns on priority. This article aims to shed light on the clear existent definitions of Industrial IoT (IIoT), the architecture and enabling technologies, applications, future scope and challenges faced [1].

2 Industrial Internet of Things

A system of networked cyber-physical assets, smart objects, optional cloud and associated generic information technologies by H. Boyes et al. Computers in the industry enable real-time smart, and independent access, accumulation, analysis, exchange of process, good and service information and communications, within the industries, to stabilize production value which includes, the better products, boosted efficiency, reduced labor wages, decreased energy wastage, and reduced build to order cycle.

IIoT, a transformation manufacturing technique that boosts efficiency, security, quality, and delivery in industries. Hence, producers are incorporating the highest use of IIoT programs to upgrade their analytical functions, for asset tracking, and upgrading control rooms.

There is the expectation of the evolution of IIoT in future industries too. IIoT is expected to cause the Industry 5.0 system to lessen the human-machine bridging gap and help in achieving the great personalized vision in Industry 6.0. Although considering the current technological ecosystem, the discussion is considered only for Industry 4.0 vision.

Latest statistics indicate the advancement in the area of IoT and IIoT, as per which there will be around 70 billion Internet incorporated devices in 2025, and by 2023 the IIoT share in the market will be around 14.2 trillion USD globally [2, 3].

Concentration	IIoT
Area of focus	Industrial applications
Focus development	Industrial systems
Security and risk measures	Advanced and robust
Interoperability	CPS-Integrated
Scalability	Large-scale networks
Precision and accuracy	Synchronized with milliseconds
Programmablity	Remote on-site programming
Output	Operational efficiency
Resilience	High fault tolerance required
Maintenance	Scheduled and planned

3 IIoT Architecture

All enterprises have their own constituting collection of devices with limited interfaces, and IIoT similarly renders the whole concept of IoT to the enterprise level. Just like every other domain, this has its own challenges.

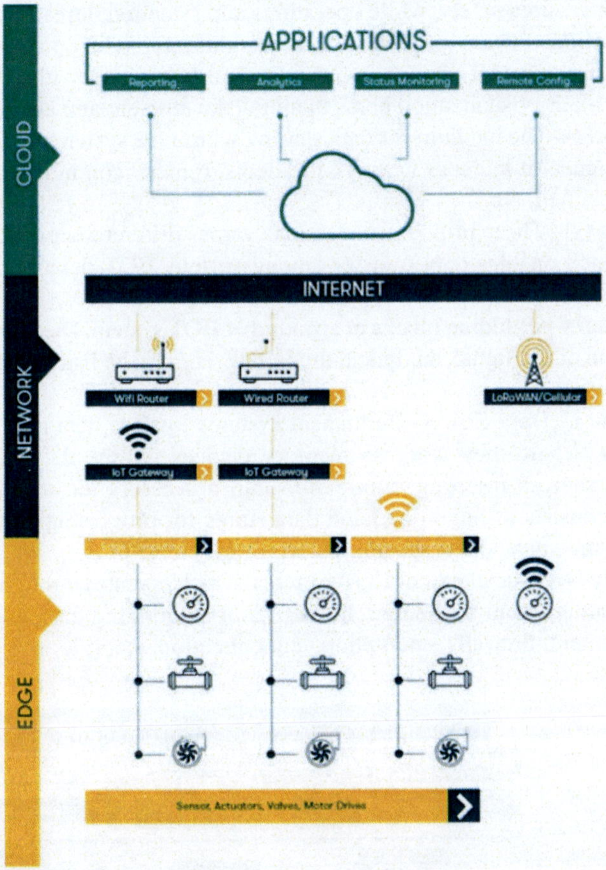

1. **Industrial Control System**—A term that controls the critical infrastructure by defining the software and hardware integration, which generally constitutes control systems, remote terminal units, programmable logic control, control servers, human-machine interface, and many other industry-exclusive systems [4].
2. **Local Processors**—Low latency data processing systems that quickly process data, and easily integrate within the device for data processing. Classified into rule-based engines, routers, event managers, data filters, signal- detectors and data processors, etc. [5].

3. **Devices**—Translators, Interpreters, and sensors are listed as some of the industry-exclusive devices that interface with processors and channels to provide the information to the receiver end. They also propose M2M and H2M interactions and reverse to the ICT discussed [6].
4. **Transient Store**—The master architecture uses this store as the slave components, where the transient presentation of the data objects are kept for temporary time for ensured safety, while operations and system failures.
5. **Application**—These provide actual insight into field operations, help in device management, data manipulation, and interaction with other systems. Notifications, visualization helps them to take efficient and apt decisions [7].
6. **Channels**—The medium for data sharing within the systems and the applications. Hence, it includes network protocols, routers, communication through satellites, API, etc.
7. **Gateways**—These provide connections across different networks and protocols which enables data transfer among various IIOT devices by including smart routers, etc.
8. **Processors**—Building blocks of any kind of IIOT system. Data transformation detection of the signal, analytical modelling, etc. can be listed as their primary functions.
9. **Permanent Data Store**—Permanent systems to store data attached to IIOT systems, where they work as memory devices for the data devices, from various sources inserting information into processors for advance analysis. It also consists of huge provision data stores for processing simultaneously, open-source data, data repositories, RDBMS, etc. [8].
10. **Security**—A crucial part of IIOT systems, which operates through the pipeline to consumers from the source. It consists of authorizing data, encoding, user management, firewalls, encryption, authentication, etc. [9].

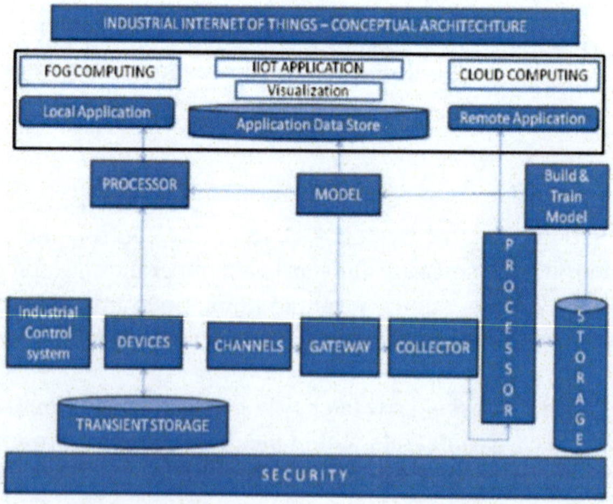

11. **Computing Environments**—The following environments differ between industries depending upon the needs.
    1. **Fog Computing**—Puts the source and analytics, nearer.
    2. **Cloud Computing**—Scales the analytics across the whole industry.
    3. **Hybrid Computing**—Fusion of cloud computing and fog computing which optimize operations customized for specific needs [10].

4 IIoT Enabling-Technologies

IIoT is a technology developed from the intelligent integration of various technologies, a few of which are briefly discussed in the following section along with their applications:

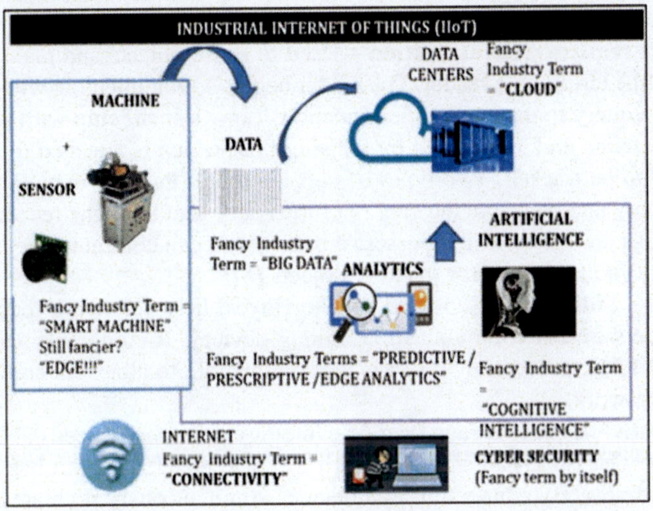

1. **Cloud Computing**—On user demand, the cloud computing provides computing services. It allows monitoring and analyzing all the objects, in IoT applications. Due to limited power to process and memory, sensors store and process only local data. They bypass human intervention by supporting Artificial intelligence.
2. **Big Data**—It represents a large amount of data that doesn't support the normal data processing applications while performing data-related operations. Hiveql and Hadoop techniques are used to manage the large volume data operations here. In IIoT, a large amount of collected information that cloud computing supports, combined with big data, gives stupendous results in the retrieval and storage of user information.

3. **Sensors and Actuators**—A device used to convert one signal form to the other providing a measurable quantity is a sensor. For example, temperature, proximity, vision, gyroscope, slope, etc. A hardware device to convert the command into a mostly mechanical physical change is an actuator, like position or velocity.
4. **Artificial Intelligence**—AI can be referred to as the Human or animal-like intelligence shown by machines. AI incorporated machines can provide warnings for alarming situations beforehand so that preventive measures can be taken to deal with the same in different industries.
5. **Global Positioning System**—A satellite network, developed by the US government for security purposes, but it is now widely used globally. GPS uses a process called trilateration to pinpoint the location when it has information regarding at least three satellites. It is extensively used in the logistics of industries.
6. **Smart Devices**—A commonly used electronic device, which operates independently and generally, is connected to other networks through protocols like Wi-Fi, Bluetooth, etc. For instance, smartphone, smart band, and watch. IoT uses many devices for information analysis.
7. **Radio Frequency Identification**—Used to sense objects and mainly has two parts—the tag and the reader. The reader begins communicating with the tag by giving a query to the tag for identification. Tag—a small chip with an incorporated antenna and recognized by a distinct ID, which is attached to any object needed to be tracked. Two types of tags are there- Passive, they don't possess the battery and consume the energy of the query sent from the reader while the other tags are active which possess a battery and can communicate by sending their ID. Industrial use for tracking objects [8].
8. **Wireless Fidelity**—Discovered by Vic Hayes, it is a network technology to facilitate wireless communication among devices. It commonly uses the 5.8 and 2.4 GHz radio band. Wi-Fi is very vulnerable to attackers compared to a wired network.
9. **Bluetooth**—It is a less-range radio technology that doesn't need cables for data transfer. The IEEE had standardized it as IEEE 802.15.1, but it has a changed standard now. By connecting 2–8 devices simultaneously it creates PANs for data transfer.

5 IIoT Applications

1. **Smart Factories**—IIoT incorporated industries can feel the environment and send data to the field workers, enabling them to manage their factory units. These devices are capable of data transmission regarding loss, inventory of work to their managers, etc. allowing them to take immediate actions.
2. **Process Management**—The use of IIoT in manufacturing processes helps in monitoring the refining of raw materials to the final packing of goods. This almost real-time tracking of processes enables the production unit to adjust the parameters, to accomplish the required targets concerning quality and cost.

3. **Inventory Management**—It allows the tracking of all supply chain events from good arrival to the product delivered; where any deviation from the pattern would be tracked by the unit in real-time, enabling immediate action. Instead of a manual human-based inventory management system in use, IoT barcodes and RFID used in complexes for tracking materials reduce a substantial amount of time and power wasted, for the same results and work performed.

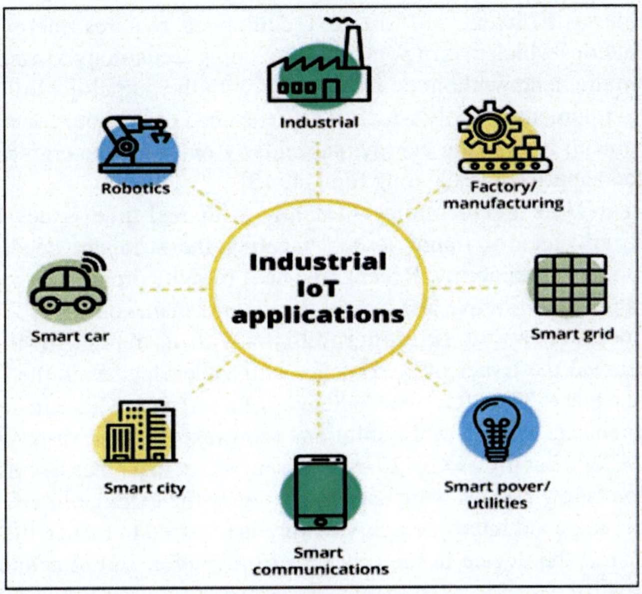

4. **Maintenance Management**—IIoT sensors enable situation-based maintenance through monitoring of important machines and warning the managers when they deviate from targeted parameters- vibration ranges, that help in declining the breakdown time, cost, and increasing efficiency of the firm.
5. **Safety and Security**—IIoT incorporated devices can efficiently work in alarming environments (which are risky to human lives) like acid plants, confined spaces, thus diminishing human life intervention. IIoT devices can operate immediately based on their area of working like stopping the furnace if the temperature rises above a certain limit, etc. hence, ensuring labourers' safety.
6. **Logistics Management**—The information provided by IIoT regarding the tracking of the product will help managers to predict issues and timely solutions. The GPS-incorporated tracking system would help manufacturers to track goods availability and guaranteed timely delivery of the final product.

6 IIoT Challenges and Future Scope

IIOT is becoming a huge success and one of the most sought-after technologies day by day, however, possesses some real threats and has some shortcomings that need to be addressed to pave way for an obstacle-free path for its flawless working in near future.

1. **Energy Efficiency**—Most IIoT equipment requires energy in the form of batteries which in turn point out developing technology to minimize the energy requirements without compromising with the output quality as well as being environment-friendly which is also the need of the hour for sustainable development. Harvesting energy, particularly renewable energy can prove to be a good approach in the long run [11, 12].
2. **Real-Time Performance**—Dealing with real-time issues like unexpected disturbances or minor issues, meeting the stringent deadlines as well as enhancing reliability. Recent advances have incorporated approaches that are entirely distributive and hybrid in resource management [13, 14].
3. **Coexistence and Interoperability**—With its rapid growth and increasing demand the issues of coexistence with other devices in the ISM band might take place. Therefore the challenges of interference have to be considered and taken care of. One of the solutions was presented by Crossbow's TelosB mote CA2400 having a CC2420 transceiver. Vector machines having a sensing duration below 300 ms were used to segregate the external interferences. Furthermore, a good interoperability feature is required to reduce the complexity and cost of the device in the process of integration and to enhance its functions [15–18].
4. **Security and Privacy**—These are very critical and sensitive concerns that need to be dealt with. From the term security, basically expect the protection from attackers from stealing our confidential and protected storage, or protection from some malicious virus programs, and making the systems highly secured, allowing only authorized access, and so on. Privacy on the other hand broadly demands the guarantee of confidentiality of data. Advancements in ICS security, using a combination of block-chain and IIoT can be a good solution [19, 20].
5. **Infrastructures**—Providing an environment that promotes interoperability, co-existence, secure transmission of data, only authorized access, the privacy of data, so on and so forth is posing a great challenge to build [21].
6. **Economics**—Developing effective, efficient, and beneficial models to benefit the economy not narrowly but indeed holistically i.e., on the principle of 'Benefits for All' and 'Sustainable development [22, 23].
7. **Sensors and Actuators**—Monitoring the issues of energy availability and consumption, and the timelines and inventory of the incoming data, whether is received from an internal or an external source [24, 25].

8. **Standardization**—Fair enough and adequate setting up of standards that can be globally accepted across all the industries with ease without creating any confusion and misunderstanding [26, 27].
9. **Communications**—Strengthening and securing the complete path of communication and dealing with the issues of topology, latency, and security is another obstacle in the way [28, 29].
10. **Integration**—Conjunction and co-ordination with the IT environment are already an integral happening these days, to increase the compatibility and readily adapt to new features and systems [30–34].

7 Conclusion

The development of IIoT technology lately will integrate and enhance the working of industries drastically. It has plenty of opportunities to offer which are being explored along with a set of challenges that are getting unveiled. Therefore, overcoming the trade-offs and heading towards the success for entire mankind becomes the utmost need of the hour. Furthermore, the recent pandemic Covid'19 has accelerated the transformation into an IIoT dominated world where many industries have already adopted the IIoT technology and equipment from health, life, and safety perspectives. This review presented, aimed at reviewing the IIoT domain of the IoT that is emerging systematically and holistically. It deals with the concept of IIoT, sound knowledge of architecture, key- enabling technologies, and the various real-time applications as well as the challenges that they possess along with the future scope, which shows a promising and bright future in the field of IIoT.

References

1. Accenture, Driving Unconventional Growth through the Industrial Internet of Things, 2015
2. Industrial IoT Market Size Worth $933.62 Billion By 2025 CAGR: 27.8%. (n.d.). Retrieved 21 Mar 2018, from https://www.grandviewresearch.com/pressrelease/global-industrial-internet-of-things-iiot-market
3. J. Manyika, M.Chui, P. Bisson, J. Woetzel, R. Dobbs, J. Bughin, D. Aharon, Unlocking the potential of the Internet of Things, (n.d.). Retrieved 2 Feb 2018, from https://www.mckinsey.com/businessfunctions/digital-mckinsey/our-insights/the-internet-ofthings-the-value-of-digitizing-the-physical-world
4. K.Y. Shin, H.W. Hwang, AROMS: a real-time open middleware system for controlling industrial plant systems, in *International Conference on Control, Automation and Systems*, 2008
5. L. Zhou, D. Wu, J. Chen, Z. Dong, When computation hugs intelligence: content-aware data processing for industrial IoT, IEEE Internet of Things J. **5**(3), 2018
6. J. DeNatale, R. Borwick, P. Stupar, R. Anderson, K. Garrett, W. Morris, J.J. Yao, MEMS high resolution 4-20 mA current sensors for industrial I/O applications, in *TRANSDUCERS '03, 12th International Conference on Solid-State Sensors, Actuators and Microsystems. Digest of Technical Papers*, Vol. 2, (2003)

7. V. Domova, A. Dagnino, Towards intelligent alarm management in the Age of IIOT, in *2017 Global Internet of Things Summit (GIoTS)*, 2017
8. A. Jules, RFID security and privacy: a research survey. IEEE J. Sel. Areas Commun. **24**(2), 381–394 (2006)
9. L. Zhou, H. Guo, Anomaly detection methods for IIOT networks, in *IEEE International Conference on Service Operations and Logistics, and Informatics (SOLI)*, 2018
10. Cloud computing innovation in India: a framework and roadmap, White Paper 2.0, IEEE, Dec 2014A
11. Saifullah, M. Rahman, D. Ismail, C. Lu, R. Chandra, J. Liu, SNOW: sensor network over white spaces, in *The 14th ACM Conference on Embedded Network Sensor Systems (SenSys)*, 2016, pp. 272–285
12. GPP, Standardization of NB-IOT completed, June 2016. http://www.3gpp.org/news-events/3gpp-news/1785-nbiotcomplete
13. T. Zhang, T. Gong, C. Gu, H. Ji, S. Han, Q. Deng, X.S. Hu, "Distributed dynamic packet scheduling for handling disturbances in real-time wireless networks, in *Real-Time and Embedded Technology and Applications Symposium (RTAS)*, 2017, pp. 261–272
14. T. Zhang, T. Gong, Z. Yun, S. Han, Q. Deng, X. S. Hu, Fd-pas: a fully distributed packet scheduling framework for handling disturbances in real-time wireless networks, in *Real-Time and Embedded Technology and Applications Symposium (RTAS)*, 2018, pp. 1–12
15. S. Grimaldi, A. Mahmood, M. Gidlund, An svm-based method for classification of external interference in industrial wireless sensor and actuator networks. J. Sensor Netw. **6**(2), 9 (2017)
16. F. Barac, M. Gidlund, T. Zhang, Scrutinizing bit- and symbolerrors of ieee 802.15.4 communication in industrial environments. IEEE Trans. Instrum. Meas. **63**(7), 1783–1794 (2014)
17. Y.H. Yitbarek, K.Yu, J. Akerberg, M. Gidlund, M. Bjorkman, Implementation and evaluation of error control schemes in industrial wireless sensor networks, in *2014 IEEE International Conference on Industrial Technology (ICIT)*, 2014, pp. 730–735
18. F. Barac, M. Gidlund, T. Zhang, Ubiquitous, yet deceptive: hardware-based channel metrics on interfered wsn links. IEEE Trans. Veh. Technol. **64**(5), 1766–1778 (2015)
19. T. Heer, O. Garcia-Morchon, R. Hummen, S.L. Keoh, S.S. Kumar, K. Wehrle, Security challenges in the ip-based internet of things. Wireless Pers. Commun. **61**(3), 527–542 (2011)
20. J.H. Ziegeldorf, O.G. Morchon, K. Wehrle, Privacy in the internet of things: threats and challenges. Secur. Commun. Netw. **7**(12), 2728–2742 (2014)
21. ISO Open systems interconnect standard ISO/IEC, 7498-1, 1994. http://standards.iso.org
22. P.P. Ray, Creating values out of internet of things: an industrial perspective. J. Comput. Netw. Commun. Article ID 1579460, 2016
23. McKinsey Global Institute, IoT: mapping the value beyond the hype, 2015
24. J. Gubbia, R. Buyyab, S. Marusic, M. Palaniswami, Internet of Things (IoT): a vision, architectural elements, and future directions. Future Gener. Comput. Syst. **29**, (2013)
25. Flexeye, Lord of the things: why identity, visibility and intelligence are the key to unlocking the value of IoT, 2014. https://coe.flexeyetech.com/
26. H. Barthel, et al., GS1 and the Internet of Things, Release 1.0, 2016
27. PAS 212:2016, Hypercat: automatic resource discovery for the Internet of Things—specification, BSI Publications, 2016
28. L. Xu, W. He, S. Li, Internet of Things in Industries: a survey, IEEE Trans. Ind. Inform. **10**(4), (2014)
29. IEEE Internet of Things Journal http://standards.ieee.org/innovate/iot/
30. F. Shrouf, J. Ordieres, G. Miragliotta, Smart factories in Industry 4.0: a review of the concept and of energy management approached in production based on the Internet of Things paradigm, in *Proceedings of the IEEE International Conference on Industrial Engineering and Engineering Management*, 2014
31. R. Woodhead, 4IR—the next industrial revolution, digital catapult/IoTUK, (2016)
32. M. Bansal, Priya, Application layer protocols for internet of healthcare things (IoHT), in *2020 Fourth International Conference on Inventive Systems and Control (ICISC)*, Coimbatore, India, 2020, pp. 369–376. https://doi.org/10.1109/icisc47916.2020.9171092

33. M. Bansal, Priya, in *Performance comparison of MQTT and CoAP protocols in different simulation environments,* ed. by G. Ranganathan, J. Chen, A. Rocha. Inventive Communication and Computational Technologies. Lecture Notes in Networks and Systems, vol. 145 (Springer, Singapore), pp. 549–560. https://doi.org/10.1007/978-981-15-7345-3_47
34. M. Bansal, Priya, Machine Learning Perspective in VLSI Computer Aided Design at Different Abstraction Levels, ICMCSI 2021, Springer Lecture Notes on Data Engineering and Communications, 2021.

Exposure Effect of 900 MHz Electromagnetic Field Radiation on Antioxidant Potential of Medicinal Plant *Withania Somnifera*

Chandni Upadhyaya, Ishita Patel, Trushit Upadhyaya, and Arpan Desai

Abstract The last few decades have seen a significant increase in non-ionizing radiation for cellular mobile communications. This increase in the artificial electromagnetic source is to cater to the ever-lasting need for continuous usage of wireless communication Viz. surveillance, medical, mobile, industrial equipment. The base stations count on the planet is rapidly increasing to meet the commercial demands. This radiation affects the plants especially agricultural and medically important crops get affected. The *Withania somnifera* plants were exposed to 900 MHz electromagnetic waves for the duration of 72 h. Plant extracts were collected for 12, 24, 36,48, 60 and 72 h. The presented paper examines the stress induced by such electromagnetic sources on *Withania somnifera* plant through measuring the content of phenolic and flavonoid content, DPPH scavenging assay, and total antioxidant activity assay. The findings indicated raise in all selected parameters Viz., 20.19% raise in phenolic compounds, 21.27% increase in flavonoid content, 20% increase in DPPA scavenging activity and 19.99% elevation in total antioxidant activity was obtained with initial exposure up to 24 h which shell be due to activation of plant defense system. The prolonged exposure up to 72 h indicated significant detriment in phenolic compounds by 32.12%, Flavonoids by 14.89%, reduction in DPPH radical scavenging activity, by 56.33%, and total antioxidant activity by 42.01%. Such a reduction in antioxidant activity of selected medicinal plants indicates the deteriorative effects of high-frequency Electromagnetic radiation on plant health.

Keywords Non-ionizing radiation · Phenolic compounds · Flavonoids · DPPH scavenging activity · Total antioxidant activity

---

C. Upadhyaya (✉) · I. Patel
Sardar Patel University, Anand, India

T. Upadhyaya · A. Desai
Charotar University of Science and Technology, Anand, India

V. Suma et al. (eds.), *Inventive Systems and Control*,
Lecture Notes in Networks and Systems 204,
https://doi.org/10.1007/978-981-16-1395-1_69

1 Introduction

In the world of fast wireless communication and automation, the need for communication devices is tremendously increasing. Here, new electronic gadgets get launch in the commercial market almost on monthly basis. According to International Telecommunication Union (ITU), there are 6.38 billion active subscriptions at the beginning of 2020, out of which 4.1 billion people are using the internet amounting to nearly 54% of the global population [1]. Electromagnetic (EM) waves contain fields having energy, which gets a transfer to living tissue upon interaction. Various literature presents the gloomy effects of electromagnetic waves on plants like seed germination, plant growth, breeding levels, and plant functioning. The plant responds to the EM waves based on the strength of the field energy and radiation frequency utilized for the communication. The response of plants varies as these characteristics of waves modify. Besides, short-term and long-term exposure of EM waves on plants creates significant alteration in plant response. It is noteworthy that the majority of the research revealed that long-term exposure of high-frequency microwaves on the plants has shown deteriorative effects.

The illustration of electromagnetic wave-exposed on the plant is described in Fig. 1. There are numerous reports of having diverse miserable changes in plant response to long-term EM waves. It induces stress and alters the chemical components of the plants using oxidative stress. This occurs in parallel with the increase in the Reactive Oxygen Species (ROS) [2]. As per [2], EM wave effects testing were carried out on antioxidant photosynthetic pigment contents of *Zea mays L.* The experiments revealed a noteworthy reduction in chlorophyll a and chlorophyll b contents in *Zea mays L.* leaves after thirteen days. The surge in carotenoid concentration is primarily a self-protective response of *Zea mays L.* plants against the EM waves. This directly implies that plants fight against its destruction due to electromagnetic waves. Another experiment of radiofrequency 900 MHz mobile and base station radiation caused a reduction in the growth of Soybean seedlings upon exposure of radio waves [3] and an increase in abiotic stress in *Zea mays L.* [4]. As per [3], the continuous exposure of high amplitude of 900 MHz radiation caused reduced growth of epicotyl. Besides, continuous microwave radiation caused a significant reduction in the growth of root and hypocotyl. The reduction in the growth of these plant parts majorly depends on

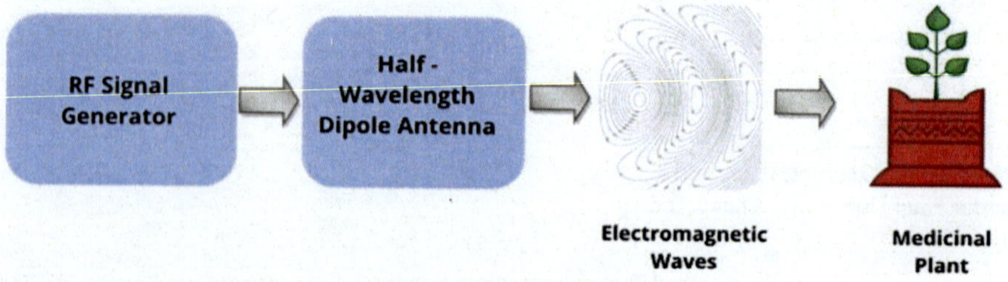

Fig. 1 Illustration of EM wave exposure on the medicinal plant

strength of electromagnetic waves and the type of modulation techniques utilized for the communication. It is also emphasized that biological effects on plants due to EM waves are also reliant on central carrier frequency in the case of frequency modulation and average power levels [3]. Oxidative stress-induced due to electromagnetic waves is extremely reactive which can completely disturb the metabolism and immune response of plants [4]. The literature showing the deteriorating effect of EM radiation is tabulated in Table 1.

The disturbance in photosynthesis process in *Microcystis aeruginosa* has been reported upon exposure of microwaves [16]. The EM radiation modulates expression pattern of photosynthesis associated gene products. EM wave effects on photoreaction systems altered the *Microcystis aeruginosa* photosynthesis process. The plants also seems to react to specific frequencies of communication Viz. 800 and 1500 MHz,

Table 1 Literature of plant deteriorating effects on by electromagnetic radiation

Organisms	Signal frequency/mobile communication technology	Reported biological effects	Citation
Phaseolus vulgaris	900 MHz/GSM	2-fold decrement in H_2O soluble sugars	[5]
Zea mays	1800 MHz/CDMA	Approximate 2-fold increase for enzyme assay which leads to more nutrient requirement	[6]
Solanum lycopersicon	900 MHz/GSM	1/3 decline in ATP concentration	[7]
Vigna radiata	900 MHz/GSM	Oxidative stress induction—increase in metabolic markers	[8]
Lemna minor	900 MHz/GSM	Increase in H_2O_2 and MDA concentration	[9]
Nicotiana tabacum	900 MHz/GSM	Protein metabolism—DNA damage increased	[9]
Raphanus sativus	10.5 GHz/microwave source	Inhibition of hypocotyl prolongation	[10]
Lens culinaris	1800 MHz/CDMA	Significant decline in root growth	[11]
Vigna radiate	900 MHz/GSM	Rhizogenesis deterioration	[8]
Vigna radiata	900 MHz/GSM	Germination Inhibition, reduction in dry weight	[12]
Lablab purpureus	1.8 GHz/microwave source	Decline in plant height and fresh weight	[13]
Zea mays	1 GHz//microwave source	Growth reduction	[14]
Zea mays	1800 MHz/CDMA	Root growth reduction	[6]
Glycine max	900 MHz/GSM	Inhibition of epicotyl growth	[3]
Rosa hybrid	900 MHz/GSM	Delayed growth of secondary branch axes	[15]

1500 and 2400 MHz, 3500 and 8000 MHz [17]. The plants self-defense response varies for all above frequencies however, it is apparent that EM waves affect living plant organisms. It is hence important to establish the relationship between EM wave stress and plant growth. The former investigation revealed that the calcium channel initiation in plasma membrane of plant cells has been reported as a prime culprit of modulation in normal metabolic activities [18]. This creates surplus Calcium ions, which affects the physiology and biochemistry of plants cell. Other hazardous effects of 2 GHz microwave frequency on *Myriophyllum aquaticum* stem by radio frequency electromagnetic radiation was reported to deteriorate nanometric elongation rate fluctuation (NERF) in exposed plants [19]. In a similar experiment, modifications in enzyme activities in *Plectranthus sp.* plants were revealed due to exposure of 900 MHz EM radiation [20]. The detrimental EM wave effects are also reported in in aromatic plants [21], commercially important *Allium cepa* roots [22], antioxidant activity in *Saturejabachtiarica L* [23], seed germination [24], vegetative organisms [25], plant cell ultrastructure [26], Aspen seedlings [27], Zea mays plants germination rate [28] and tree injuries [29]. Many other recent researches on gloomy effects of electromagnetic radiation on plants are summarized in [30].

Recent expansions in the area of research of bio-electromagnetism and agriculture revealed that the interaction of the magnetic field with the seeds or seedlings of various plants leads to induction of resistance against abiotic stressors within plants [31]. Furthermore, such a weak magnetic field imposes beneficial effects on seed germination and seedling growth mainly by enhancing their capacity of nutrient absorption and decreasing oxidative damage governed by ROS such contradictory outcomes always due to variation in exposure parameters viz. dosimetry, intensity, and exposure time [32]. Although, on the contrary, major deterioration of growth and quality of plant products were reviewed within [32] which showed the mechanism of such modulatory effects of the EM fields on molecular metabolism of cells. The prime target of EM field is lipid biomolecules present within the cell membrane that are more prone to destruction governed by uncoupled free radicles and reactive oxygen species. This phenomenon is referred to as lipid peroxidation. Besides, accretion of free radicles induces oxidative stress which induces gene expression of stress markers viz. proteinase inhibitor, calmodulin, calcium-dependent protein kinase [16]. It illustrates that prior treatment to seeds or seedlings may inflict beneficial effects, but adult plants are certainly harmed by high-frequency radiations when exposed continuously [16].

The medicinal plants are fortified with different bioactive compounds and having the capability to synthesize secondary metabolites which revealed different activities one of them is antioxidant activity. Synthesis and activity of such antioxidants are very important for survival and health of cells from toxic effects of Reactive Oxygen Species (ROS) which alter metabolism and harmful to the tissues. The selected medicinal plants' antioxidant activity upon exposure to high-frequency electromagnetic waves has not been analyzed in the literature. The presented research was conducted in the thirst of investigating the effect of short term and long term radiation of such high-frequency microwaves on the antioxidant concentration of medicinal plant *Withania somnifera*.

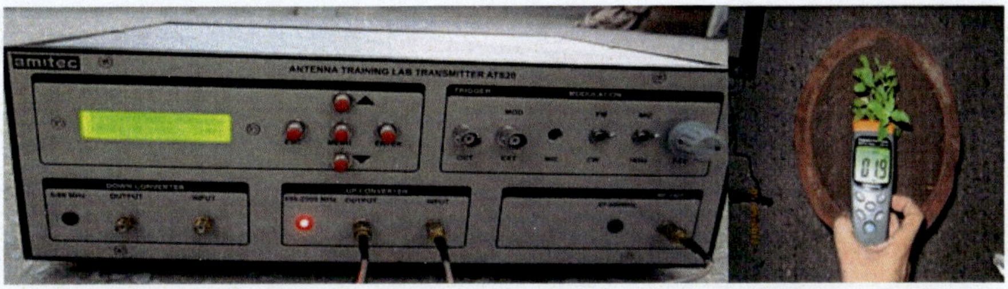

Fig. 2 Electromagnetic wave transducer radiating 900 MHz frequency and uniform magnetic field strength around the plant is 1.9 mG

2 Materials and Methods

2.1 Planting Material Grounding

The true to type plants of *Withania somnifera* was collected from DMAPR, Boriavi, Anand, Gujarat, India (which was properly authenticated) and raised at greenhouse by applying the same environmental conditions to avoid genetic variations. The micro-propagated plantlets were grouped into two groups, one of which was kept isolated and considered as unexposed control plants, and the other was exposed to high-frequency waves of 900 MHz and considered as test plants. The control and treated plants were raised by providing the same amount of compost, water, humidity, and the whole analysis were governed in a single reason to avoid environmental interferences in results. The leaf sample was collected periodically from both groups of plants and extracted for further analysis of antioxidant activity.

2.2 Electromagnetic Wave Exposure

Radiation of 900 MHz frequency was exposed by Dipole Antenna transducer and shown uniform magnetic field strength of 1.9 mG at all sides of the plants is shown in Fig. 2.

2.3 Plant Extract Preparation

The leaves of control and exposed plants were rinsed, grounded, and extracted with 80% methanol. The sample tubes of leaf pulp suspended within methanol was vortexed for 10 min and incubated at 60 °C in the water bath for 1 h. Tubes were vortexed after each 15 min of interval. It was followed by centrifugation at 5000 rpm for 5 min and the supernatant was collected in a separate tube. The pallet was re-suspended in 5 ml of 80% methanol and extracted via centrifugation whose supernatant was decanted in the same tube and stored at −20 °C. These extracts were further analyzed for antioxidant assays.

2.4 *Quantification of Total Phenolic Content*

The estimation of phenolic compounds was done by methods based on Folin-Ciocalteu reagent and testified by using Standard Gallic acid and Tannic acid equivalents per gram of extracts by following expression [33]

$$C = \frac{c \times V}{m} \tag{1}$$

where; C = Concentration of phenolic compounds determined as standard Gallic acid equivalent mg/gm plant extract, c = Value of Gallic acid content from the standard calibrated curve, V = Extract volume in ml, m = Gram weight of the extract.

The phenolic compounds present in the extract gave a blue color complex by reducing the Folin-Ciocalteu reagent. The standardized method included 500 μl of extract which was added to 1.6 ml of Na_2CO_3 and 2 ml of diluted Folin-Ciocalteu reagent. Dilution was done by using deionized water. The absorbance was taken at 765 nm after 1 h of incubation. Total phenolic concentration was estimated in mg TAE and GAE/gram of extract by the calibration curve.

2.5 *Quantification of Total Flavonoid Concentration*

Total flavonoids present within leaf extracts were measured using standard Quercetin by referring [34]. Briefly, 500 μl of plant extract was mixed in 70 μl of $NaNO_2$ reagent. Incubate for 10 min, following the addition of 150 μl of $AlCl_3$ $6H_2O$, and further incubate for 5 min. After that, 0.5 ml of 1 M NaOH and 3 ml of distilled water were mixed and absorbance was measured at 510 nm using spectrophotometer by setting blank with methanol. Total flavonoid concentration was determined as mg QE/gram, using the following formula based calibration curve:

$$C = \frac{c \times V}{m} \tag{2}$$

where; C = Flavonoid concentration measured as quercetin equivalent as mg/gm extract, c = The measure of quercetin determined from the calibrated curve, V = the extract volume, and m = Gram weight of plant extract.

2.6 *DPPH Radical Scavenging Assay*

This assay is used for determining the free radical scavenging properties of plant extract. DPPH is considered as a stable free radical possessing maximum absorbance at 520 nm and converted into 2, 2-diphenyl 1-picrylhydrazine upon reaction with antioxidants [35]. Extracts were first diluted at 500 μg/ml and freshly prepared DPPH solution was added to it which was later incubated at 30 ºC for half an hour in dark. It was followed by taking optical density at 520 nm using the spectrophotometric technique in which methanol was taken as a blank. The DPPH and methanol were utilized as a negative control. The percentage of DPPH discoloration by the antioxidant activity of extracts was calculated by expression:

$$\text{antiradical activity} = \frac{(\text{O.D. of control} - \text{O.D. of sample})}{(\text{O.D. of control})} \times 100 \quad (3)$$

The DPPH scavenging activity was further utilized for determining IC50 values of extracts which indicates antioxidant potential. It is the amount of extract necessary for 50% scavenging activity of extract on DPPH free radicals. The smaller measure of IC50 directly means higher antioxidant activity. IC50 values have been extracted from the Graph pad Prism statistical analysis software.

2.7 *Total Antioxidant Activity Assay*

The methodology of total antioxidant activity was based on phosphomolybdenum and referred from [36]. 0.5 ml plant extract was taken and 5 ml of reagent mixed with it whereas ascorbic acid was taken as a standard and methanol was used as a blank. After mixing, tubes were incubated at 95 °C for 1.5 h and followed by cooling at room temperature read at 695 nm. The total antioxidant capacity was determined by subsequent expression:

$$A = \frac{c \times V}{m} \quad (4)$$

where, A = Total concentration of antioxidant taken as mg/gm extract equivalent of ascorbic acid, c = The measure of ascorbic acid content determined from the calibrated curve, V = The extract volume in milliliter, m = The plant extract weight in gram.

3 Statistical Analysis

Analysis was governed in triplicates ($n = 3$) and outcomes were expressed in form of mean ± SD. The regression analysis was done by Graph-pad prism software via One- way-Anova analysis. When test plant (Exposed Plant) data were compared with the control plant (Unexposed plant), statistical significance was employed via t-test, and data comprising $p < 0.05$ were taken as significant ones.

4 Result and Discussion

Antioxidant activity of methanolic extracts of *Withania somnifera* was determined by DPPH (1,1- diphenyl-2-picrylhydrazyl) scavenging assay and Total antioxidant activity assay which are correlated with phenolic content present within the extract. Other than that, Flavonoid content was checked to further analyze antioxidant activity. DPPH scavenging is one of the very reliable and precise methods for assessing the antioxidant activity of plant extract [34]. It is a very steady free radical due to the presence of free electron delocalization over its entire molecule. Thus, it tends accept H^+ ion, which leads to a color change in solution from Violet to Pale yellow. The DPPA scavenging of selected plant extract elevated upon electromagnetic wave exposure up to 24 h as depicted in Fig. 2, which was later declined significantly up to 56.33% when compared with the control plant extract. The utilized standard Ascorbic acid revealed the highest scavenging effect and thus, can be considered as positive control shown in Fig. 3. Thus, the finding showed the significant rate of DPPH scavenging action of methanolic extract of *Withania somnifera* which was 60% for control plants whereas 67, 72, 38.2, 30, 29.5 and 26.2% for test plant which was exposed to radiation for 12 h, 24 h, 36 h, 48 h, 60 h and 72 h respectively. Such decline in DPPH scavenging assay was consistent with phenolic compound contents present within extract and thus, finding revealed that it is directly proportional to phenolic content as shown in Table 3. They carried out work on antioxidant activity

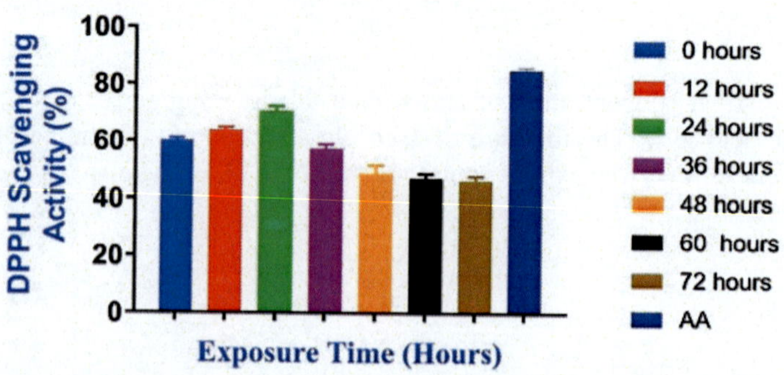

Fig. 3 DPPH scavenging activity of *Withaniasomnifera* upon EM wave exposure

Table 2 IC_{50} values of control (unexposed) and test (exposed) plant extracts of *Withania somnifera* in DPPH scavenging assay

Withania somnifera leaf extracts/standard	
Electromagnetic exposure (hours)	IC_{50} µg/ml
0 h (Control)	250.17
12	211.10
24	192.25
36	260.50
48	274.12
60	287.25
72	289.11
Standard ascorbic acid	16.59

also discovered a similar raise or decrease of DPPH scavenging due to high or low phenolic compounds present within the extract [37, 38]. Here, free radical scavenging activity was denoted in form of IC_{50} which can be expressed as the effective content of antioxidants essential to scavenge the DPPH stable radicals by 50%. The lowest value indicates the highest potency of the plant extract to scavenge DPPH radicals. Thus, according to the measure of IC_{50}, radiation treated plant extract revealed significant variation when compared with the unexposed control plants as tabulated in Table 3. Upon exposure up to 24 h, there was a reduction of 9.40% which was later increased with a drop in DPPH scavenging potential of extracts due to prolonged exposure effect (Table 2).

Total phenolic compound concentration was assessed by using the standard calibrated curve of Gallic acid by the following equation.

$$Y = 1.540x + 0.0704, \ R^2 = 0.9971 \quad (5)$$

The results indicated the highest phenolic compound content upon 24 h of radiation exposure which was 125.17 ± 0.18 mg/ml and thus, indicative of activation of plant defense system against the imposed stress in form of radiation. It was followed by 14.19, 26.04, 27.50, and 32.12% of time-dependent reduction in phenolic compound content upon 36, 48, 60, and 72 h of exposure of high-frequency radiation respectively.

The flavonoid content was determined by taking the standard calibration curve of Quercetin, calculated by the following equation, and measured as mg QE/gm.

$$Y = 2.150x + 0.991, \ R^2 = 0.9921 \quad (6)$$

The results of flavonoid estimation were following the former two assays and revealed an increase of 21.27% upon electromagnetic exposure of 24 h which was significantly harmed and reduced upon exposure of and beyond 36 h. Such decline was become prominent along with exposure time and reported as a variation of

Table 3 Variations in concentration of phenolic and flavonoid compounds due to EM wave exposure

	Withania somnifera leaf extracts total phenolic (mg/gm)		Flavonoids (mg/gm)
Time of exposure (hours)	Gallic acid equivalent	Tannic acid equivalent	Quercetin equivalent
0 (control)	104.10 ± 0.05	108 ± 0.18	94.70 ± 1.13
12	116.00 ± 0.12	121 ± 0.07	99.75 ± 0.7
24	125.17 ± 0.18	138 ± 0.12	114 ± 1.15
36	89.32 ± 0.06	97 ± 0.09	72.8 ± 0.28
48	76.99 ± 0.12	94.41 ± 0.26	67 ± 0.14
60	75.47 ± 0.25	92.40 ± 0.18	63.79 ± 0.39
72	70.66 ± 0.34	86.12 ± 0.34	60.30 ± .86

Tests were performed as triplicate and measured values were represented as Mean ± SD

23.29, 29.25, 32.64, and 36.33% for 36, 48, 60, and 72 h of exposure respectively as tabulated in Table 3.

Former research on the antioxidative property of various plant extracts documented that the main antioxidant effect is due to total phenolic content of the plant which are the potent scavengers of free radicals, H^+ ion donating molecules as well as quenchers of metal cations and singlet oxygen as illustrated in Fig. 3 [39]. Not only, Phenolic compounds but also flavonoids were assessed for their antioxidative potency. Flavonoids utilize a different mode of action which includes, free radicals scavenging and metal ion chelation of copper and zinc which are the cofactors of various enzymatic actions responsible for free radical synthesis. Thus, their chelation leads to inhibition of such enzymes and ultimate lowering of free radical content in cells [40].

Total antioxidant activity assay was one of another potent assay to check the antioxidant potential of the selected medicinal plant. It was assessed for both electromagnetically unexposed and exposed plant sample extracts to evaluate the effect of high-frequency waves on antioxidant potential. The calculation for estimation was governed by using a standard calibration curve of Ascorbic acid and via using the following expression:

$$Y = 0.002x + 0.001, R^2 = 0.998 \quad (7)$$

Presented assay outcomes indicated that upon 12–24 h of exposure of high-frequency EM waves, total antioxidant activity was raised to 19.99% which was significantly deteriorated after 24 h in the order of 16.01, 29.99, 28.18, and 42.01% for respective 36, 48, 60 and 72 h of exposure as shown in Table 4. Former investigation documented the direct correlation of total phenolic content with the overall antioxidant activity of various plant species [35]. Along with DPPH scavenging activity, total antioxidant content also exhibited a parallel correlation with phenolic

Table 4 Variations in total antioxidant activity of *Withania somnifera* due to EM wave exposure

Withaniasomnifera leaf extracts/standard	
Electromagnetic exposure (hours)	Total antioxidant activity (mg/gm of ascorbic acid equivalent)
0 h (control)	73.91 ± 12.34
12	87.55 ± 8.70
24	88.69 ± 11.20
36	62.08 ± 7.64
48	51.74 ± 9.86
60	53.08 ± 10.50
72	42.86 ± 7.25

Tests were performed as triplicate and measured values were represented as Mean ± SD

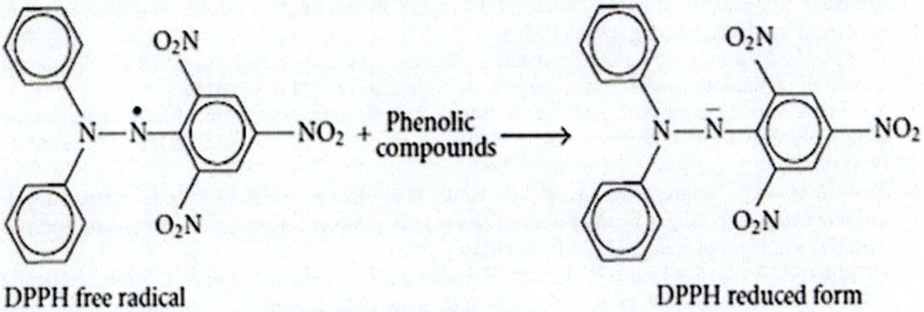

Fig. 4 DPPH reductions by phenolic compounds [35]

compounds because in the current assessment, total antioxidant activity was raised and declined along with estimated concentrations of phenolic compounds (Fig. 4).

Also, physiological alterations were observed on the plant leaves upon the continuous exposure of the electromagnetic waves. There were dark spots and leaves were observed to be weaker than the unexposed control plants.

5 Conclusion

The artificial Electromagnetic Wave sources are increasing on the earth. Man-made electromagnetic pollution is affecting the flora and fauna in numerous aspects. Several experiments conducted across the world shows that Electromagnetic Wave has hazardous effects on plants. The high-frequency microwave has continuous interaction with plant organisms. The effect of the Global System of Mobile Communication radiation operating at 900 MHz was examined on medicinal plant Ashwagandha

(*Withania somnifera*). The experiments exhibited that this high-frequency EM wave interaction for a prolonged period of time significantly reduces phenolic compounds and flavonoid contents as well as antioxidant activity of Medicinal Plant *Withania somnifera*. Thus, presented findings revealed that plant systems are adversely affected and should not be ignored by negligent thoughts of considering them highly resistant entities.

References

1. Statistics, Retrieved 27 May 2020, from https://www.itu.int/en/ITU-D/Statistics/Pages/stat/default.aspx
2. H. Zare, S. Mohsenzadeh, The effect of electromagnetic waves on photosynthetic pigments and antioxidant enzyme in Zea mays L. Current World Environ. **10**(1), 30 (2015)
3. M.N. Halgamuge, S.K. Yak, J.L. Eberhardt, Reduced growth of soybean seedlings after exposure to weak microwave radiation from GSM 900 mobile phone and base station. Bioelectromagnetics **36**(2), 87–95 (2015)
4. H. Zare, S. Mohsenzadeh, A. Moradshahi, Electromagnetic waves from GSM mobile phone simulator and abiotic stress in Zea mays L. J. Nutr. Food Sci. **11**, 3 (2015)
5. V.P. Sharma, H.P. Singh, R.K. Kohli, D.R. Batish, Mobile phone radiation inhibits Vigna radiata (mung bean) root growth by inducing oxidative stress. Sci. Total Environ. **407**(21), 5543–5547 (2009)
6. A. Kumar, H.P. Singh, D.R. Batish, S. Kaur, R.K. Kohli, EMF radiations (1800 MHz)-inhibited early seedling growth of maize (Zea mays) involves alterations in starch and sucrose metabolism. Protoplasma **253**(4), 1043–1049 (2016)
7. D. Roux, A. Vian, S. Girard, P. Bonnet, F. Paladian, E. Davies, G. Ledoigt, High frequency (900 MHz) low amplitude (5 Vm^{-1}) electromagnetic field: a genuine environmental stimulus that affects transcription, translation, calcium and energy charge in tomato. Planta **227**(4), 883–891 (2008)
8. H.P. Singh, V.P. Sharma, D.R. Batish, R.K. Kohli, Cell phone electromagnetic field radiations affect rhizogenesis through impairment of biochemical processes. Environ. Monit. Assess. **184**(4), 1813–1821 (2012)
9. S. Radic, P. Cvjetko, K. Malaric, M. Tkalec, B. Pevalek-Kozlina, Radio frequency electromagnetic field (900 MHz) induces oxidative damage to DNA and biomembrane in tobacco shoot cells (Nicotiana tabacum). In *2007 IEEE/MTT-S International Microwave Symposium*, (IEEE, 2007), pp. 2213–2216
10. A. Scialabba, C. Tamburello, Microwave effects on germination and growth of radish (Raphanus sativus L.) seedlings. Acta Bot. Gallica **149**(2), 113–123 (2002)
11. A. Akbal, Y. Kiran, A. Sahin, D. Turgut-Balik, H.H. Balik, Effects of electromagnetic waves emitted by mobile phones on germination, root growth, and root tip cell mitotic division of Lens culinaris Medik. Pol J. Environ. Stud. **21**(1), 23–9 (2012)
12. V.P. Sharma, H.P. Singh, D.R. Batish, R.K. Kohli, Cell phone radiations affect early growth of Vigna radiata (mung bean) through biochemical alterations. Zeitschrift für Naturforschung C **65**(1–2), 66–72 (2010)
13. H.Y. Chen, C. Chen, Effects of mobile phone radiation on germination and early growth of different bean species. Polish J. Environ. Stud. **23**(6), 1949–1958 (2014)
14. M. Racuciu, C. Iftode, S. Miclaus, Inhibitory effects of low thermal radiofrequency radiation on physiological parameters of Zea mays seedlings growth. Rom. J. Phys. **60**(3–4), 603–612 (2015)
15. A. Grémiaux, S. Girard, V. Guérin, J. Lothier, F. Baluška, E. Davies, P. Bonnet, A. Vian, Low-amplitude, high-frequency electromagnetic field exposure causes delayed and reduced growth in Rosa hybrida. J. Plant Physiol. **190**, 44–53 (2016)

16. C. Tang, C. Yang, H. Yu, S. Tian, X. Huang, W. Wang, P. Cai, Electromagnetic radiation disturbed the photosynthesis of Microcystis aeruginosa at the proteomics level. Sci. Rep. **8**(1), 1–8 (2018)
17. M.N. Halgamuge, Weak radiofrequency radiation exposure from mobile phone radiation on plants. Electromagn. Biol. Med. **36**(2), 213–235 (2017)
18. M.L. Pall, Electromagnetic fields act similarly in plants as in animals: Probable activation of calcium channels via their voltage sensor. Current Chem. Biol. **10**(1), 74–82 (2016)
19. M.D.H.J. Senavirathna, T. Asaeda, B.L.S. Thilakarathne, H. Kadono, Nanometer-scale elongation rate fluctuations in the Myriophyllumaquaticum (Parrot feather) stem were altered by radio-frequency electromagnetic radiation. Plant Signal. Behavior **9**(4), (2014)
20. M. Kouzmanova, M. Dimitrova, D. Dragolova, G. Atanasova, N. Atanasov, Alterations in enzyme activities in leaves after exposure of Plectranthus sp. plants to 900 MHz electromagnetic field. Biotechnol. Biotechnol. Equipment **23**(sup1), 611–615
21. M.L. Soran, M. Stan, Ü. Niinemets, L. Copolovici, Influence of microwave frequency electromagnetic radiation on terpene emission and content in aromatic plants. J. Plant Physiol. **171**(15), 1436–1443 (2014)
22. S. Chandel, S. Kaur, H.P. Singh, D.R. Batish, R.K. Kohli, Exposure to 2100 MHz electromagnetic field radiations induces reactive oxygen species generation in Allium cepa roots. J. Microsc. Ultrastruct. **5**(4), 225–229 (2017)
23. F.R. Vishki, A. Majd, T. Nejadsattari, S. Arbabian, Electromagnetic waves and its impact on morpho-anatomical characteristics and antioxidant activity in Saturejabachtiarica L. Aust. J. Basic Appl. Sci. **7**(2), 598–605 (2013)
24. P. Liptai, B. Dolník, V. Gumanová, Effect of Wi-Fi radiation on seed germination and plant growth-experiment. Ann. Fac. Eng. Hunedoara **15**(1), 109 (2017)
25. M.A.N.U.E.L.A. Ursache, G. Mindru, D.E. CREANGĂ, F.M. Tufescu, C., Goiceanu, The effects of high frequency electromagnetic waves on the vegetal organisms. Rom J. Physiol. **54**(1), 133–45 (2009)
26. A. Rusakova, I. Nosachev, V. Lysenko, Y. Guo, A. Logvinov, E. Kirichenko, O. Chugueva, Impact of high strength electromagnetic fields generated by Tesla transformer on plant cell ultrastructure. Inf. Proc. Agric **4**(3), 253–258 (2017)
27. K. Haggerty, Adverse influence of radio frequency background on trembling Aspen seedlings: preliminary observations. Int. J. Forest. Res. (2010)
28. D. Dicu, P. Pîrșan, The effect of electromagnetic waves on Zea Mays plants germination. Res. J. Agric. Sci. **46**(4), 27–33 (2014)
29. C. Waldmann-Selsam, A. Balmori-de la Puente, H. Breunig, A. Balmori, Radiofrequency radiation injures trees around mobile phone base stations. Sci. Total Environ. **572**, 554–569 (2016)
30. A. Vian, E. Davies, M. Gendraud, P. Bonnet, Plant responses to high frequency electromagnetic fields. BioMed Res. Int. (2016)
31. M. Sarraf, S. Kataria, H. Taimourya, L.O. Santos, R.D. Menegatti, M. Jain, M. Ihtisham, S. Liu, Magnetic field (MF) applications in plants: an overview. Plants **9**(9), 1139 (2020)
32. N.E. Nyakane, E.D. Markus, M.M. Sedibe, The effects of magnetic fields on plants growth: a comprehensive review. Int. J. Food Eng. **5**, 79–87 (2019)
33. S. Demiray, M.E. Pintado, P.M.L. Castro, Evaluation of phenolic profiles and antioxidant activities of Turkish medicinal plants: Tiliaargentea, Crataegi folium leaves and Polygonumbistorta roots. World Acad. Sci. Eng. Technol. **54**, 312–317 (2009)
34. A. Gawron-Gzella, M. Dudek-Makuch, L. Matlawska, DPPH radical scavenging activity and phenolic compound content in different leaf extracts from selected blackberry species. Acta Biol. Cracov. Series Bot. **54**(2), (2012)
35. A. Braca, N. De Tommasi, L. Di Bari, C. Pizza, M. Politi, I. Morelli, Antioxidant principles from bauhinia tarapotensis. J. Nat. Prod. **64**(7), 892–895 (2001)
36. M. Aguilar Urbano, M. Pineda Priego, P. Prieto, Spectrophotometric quantitation of antioxidant capacity through the formation of a phosphomolybdenum complex: specific application to the determination of vitamin E1, 2013

37. M. Arun, S. Satish, P. Anima, Phytopharmacological profile of Jasminum grandiflorum Linn. (Oleaceae). Chin. J. Integr. Med. **22**(4), 311–320 (2016)
38. M. Irshad, M. Zafaryab, M. Singh, M. Rizvi, Comparative analysis of the antioxidant activity of Cassia fistula extracts. Int. J. Med. Chem (2012)
39. I.D.N.S. Fernando, D.C. Abeysinghe, R.M. Dharmadasa, Determination of phenolic contents and antioxidant capacity of different parts of Withaniasomnifera (L.) Dunal. from three different growth stages. Ind. Crops Prod. **50**, 537–539 (2013)
40. M. Shahriar, I. Hossain, F.A. Sharmin, S. Akhter, A. AminulHaque, A. Bhuiyan, InVitro antioxidant and free radical scavenging activity of Withania somnifera root. Iosr J Pharm **3**, 38–47 (2013)

Content Based Scientific Article Recommendation System Using Deep Learning Technique

Akhil M. Nair, Oshin Benny, and Jossy George

Abstract The emergence of the era of big data has increased the ease with which scientific users can access academic articles with better efficiency and accuracy from a pool of papers available. With the exponential increase in the number of research papers that are getting published every year, it has made scholars face the problem of information overload where they find it difficult to conduct comprehensive literature surveys. An article recommendation system helps in overcoming this issue by providing users with personalized recommendations based on their interests and choices. The common approaches used for recommendation are Content-Based Filtering (CBF) and Collaborative Filtering (CF). Even though there is much advancement in the field of article recommendation systems, a content-based approach using a deep learning technology is still in its inception. In this work, a C-SAR model using Gated Recurrent Unit (GRU) and association rule mining Apriori algorithm to provide a recommendation of articles based on the similarity in the content were proposed. The combination of a deep learning technique along with a classical algorithm in data mining is expected to provide better results than the state-of-art model in suggesting similar papers.

Keywords Gated recurrent unit · Apriori algorithm · Content-Based recommendation

1 Introduction

Recommender systems are the systems that employ algorithms to help in suggesting relevant items to the users based on their needs and interests such as movies to watch, products to purchase, medications for health issues and so on. In the last few decades, as companies like YouTube and Netflix have risen to power, other companies also experienced the need for recommendation systems based on the user's needs. This

A. M. Nair (✉) · O. Benny · J. George
Department of Computer Science, CHRIST (Deemed to Be University), Lavasa, Pune, India
e-mail: akhil.nair@christuniversity.in

V. Suma et al. (eds.), *Inventive Systems and Control*,
Lecture Notes in Networks and Systems 204,
https://doi.org/10.1007/978-981-16-1395-1_70

will provide the customers with a better experience and satisfaction which later contributes to the expansion and profitability of the companies.

Due to the rapid increase in the amount of information that is available in digital form, the issue of information overload is becoming significant these days. This hinders the timely access of relevant information to many users according to their interests. With the explosive growth of information technology, the number of research papers published every year is also increasing drastically. In such scenarios, research scholars find it difficult to search and access relevant papers according to their area of interest [1]. The exponential growth in the number of research papers and articles published every year has made it challenging for scholars to conduct a comprehensive literature review. Many academic papers are getting published through conferences and journals. Research scholars tend to spend a considerable amount of time as well as other resources to gather the relevant information. They might also use Google Scholar or Citeseer to search for articles based on keywords but does not guarantee the significance of articles.

A research paper recommendation system tries to overcome this issue of information overload by providing the researchers with a personalized recommendation of articles based on their preference [2]. Research paper recommender system finds relevant papers based on the users' current requirements which can be gathered explicitly or implicitly through ratings, user profile, and text reviews. The two main approaches of a recommendation system are Content-Based Filtering (CBF) and Collaborative Filtering (CF). A content-based approach requires information related to the items and their features whereas Collaborative Filtering needs the user's historical preferences on a set of items. The state-of-art models of article recommendation systems focus on Collaborative Filtering and citation counts but very few models have focused on the content for finding the recommendations. With the expansion of Artificial Intelligence and Machine Learning, it has become easy to build recommendation systems based on the requirements. The study proposes a Content Based Scientific Article Recommendation (C-SAR) model by combining the Gated Recurrent Unit and association rule mining namely the Apriori algorithm to provide an additional layer of filtration for similar documents thereby converting relevant articles to comparatively higher relevant articles.

2 Related Work

Even though the content-based scientific article recommendation using the deep learning technology is still in its infancy, various approaches are present today to deal with article recommendations. The recommendation of papers based on content was formulated as a ranking problem [3] with two phases NNselect for selection of papers and NNrank for their ranking. Through this work, a contribution of the new dataset OpenCorpus with seven million articles was also made which could be useful for researchers. The model could have gained better accuracy if the metadata regarding the papers were used along with other attributes. A comparative study between two well-known content-based methods, Term Frequency Inverse-Document-Frequency (TF-IDF) and word-embedding were made and implemented [4] using the PUBMED dataset. The word embedding model obtained 15% better accuracy than TF-IDF with a similarity score of 0.7168. To reach a more reliable and accurate result, the set of target and recommended papers must be provided which is considered as a limitation of the model. Simon et al. [5] proposed a system that helps the user to quickly locate items of interest from a digital library. The application employed a TF-IDF weighting scheme and cosine similarity along with keyword-based vector space models and item representation. Instead of looking into the ratings provided by other users, this work has focused on the user's interests and needs. This content-based approach for research paper recommendations based on a user's query in a digital library has produced better results along with additional features that do not exist in the digital library. Stemming of attributes was not done in the model to reduce the loss of context of the search.

A model that considers the paper's topic and ideas that helps the non-profiled users to obtain a set of relevant papers was introduced [6] which required a single input paper and the main themes were derived as subqueries. The model acquired an accuracy of 80% based on the NDCG metric and proved to be one of the efficient ways for recommendations without a user profile. The inclusion of indexing features may help the model to achieve better performance. A centrality-based approach was proposed by Abdul Samad et al. [7] that analyses the textual and topological similarity measures in a paper recommendation system. A comparison is made based on the performance of the Cosine and Jaccard measure. When performed on the dataset, it was found that topological-based similarity through Cosine achieved 85.2% accuracy, and using Jaccard obtained 61.9%. On the other hand, textual-based similarity obtained 68.9% citation links on abstract and 37.4% citation links on the title. Both similarity measures analyzed only the symmetric relationship of papers. Sometimes, it is necessary to consider asymmetric relationships as well to provide better recommendations. For a DeepWalk based method, a Recall of 0.2285 and NDCG of 0.3602 was obtained which used deep learning based bibliographic content specific recommendation [8]. The matrix used in the model is based only on the paper vector, but not on the citation information. The deep walk based method is expected to outperform other existing models only if the paper contents and citations are used. A deep learning-based study [9] used a Recurrent Neural Network for modelling a recommendation system. The explicit and implicit feedback collected from users was used along with a semantic representation of the paper's title and abstract. The feedback was checked for matching with the representation. The model used Long Short-Term Memory (LSTM) for the semantic representation of articles. The paper recommendation is purely based only on the user actions and their feedback which is not always appreciated. An Advanced Personalized Research Paper Recommendation System (APRPRS) [10] based on User-Profile which applies keyword expansion through semantic analysis was implemented and achieved an accuracy of 85% and user satisfaction level of 89%. A possibility of limitless expansion in the case of keywords makes the model less efficient since it does not have any limits of numbers. Another LSTM based approach Ask Me Any Rating (AMAR) secured an

F1@10 score of 0.66 when experimented on Movielens and DBbook dataset [11]. This article also introduced AMAR Extended Architecture where along with user profile and item representation, item genres were also considered for recommendation. Lack of proper hyperparameter optimization and regularisation makes the model performs less inadequately.

A novel neural probabilistic approach was implemented by Wenyi Huang, et al., [12] which jointly learns semantic representations of citation context and cited papers. The probability of citing a document is calculated using a multi-layer neural network that improved the quality of citation recommendation. The neural probabilistic model together with word representation and document representation learning is employed for citation recommendation and the word2vec model is used for this purpose of learning both word and document representation. The representation of the words and documents are to be learned simultaneously from the citation context and the document pairs that are cited. Zhi Li and Xiaozhu Zou in [13] discussed a comprehensive summary research paper recommendation system by discussing the state-of-art academic paper recommendation methodologies, its advantages, and disadvantages with evaluation metrics and the available datasets. This work is helpful to refer to the basic methods of research paper recommendation system and its performance evaluation metrics. By providing a detailed description of each terminology associated with a research paper recommendation system, this work is well appreciated among the scholars and is very insightful.

A novel concept-based research paper recommendation system that represents research articles in terms of their topics or semantics [14] has managed to acquire accuracy of 74.09% based on the Normalized Discounted Cumulative Gain (NDCG) evaluation metric. Distributed representation of words could be combined to obtain a unique vector for candidate documents which might result in better recommendations. The average accuracy of 88% was achieved for a content-aware citation recommendation system [15] by computing citation relations using the global citation method and cross-reference. The model includes three algorithms for Own Citation relation extraction, Cross-reference calculations, and similarity checking among the papers. A limitation of this model is that it does not consider the year of a paper published or other relationships such as co-authorship. A Convolutional Neural Network (CNN) based recommendation system to predict the latent factors from text information achieved an RMSE value of 3.3481 [16]. The major novelty of the proposed recommendation algorithm is that the text information is used directly to make the content-based recommendation without tagging. The recommendation process involves using the text information regarding the input learning resource, maybe the content itself, or a brief introduction of the content. For the CNN model, the input and its output must be established in the beginning itself which may not be possible at least in a few scenarios.

3 Overview of Architecture

3.1 Gated Recurrent Unit (GRU)

GRU is a gating mechanism in the Recurrent Neural Network. This architecture is proposed to deal with long sequences of data. GRUs are known to be the advanced version of standard Recurrent Neural Networks which uses an update gate and a rest gate in a normal RNN to minimize the vanishing gradient problem. These gates are two vectors that would decide which information needs to be passed to the output.

The gating mechanism creates a memory control of values processed over time. This consists of two gates that control the flow of data through states. They are the update gate and reset gate. Two of them can be considered as vector entries that perform a convex combination. The combination decides on which hidden state information should be updated or reset the hidden state when required. By this, the network learns to filter out irrelevant temporary observations. Figure 1 shows the architecture of GRU.

Gate *rt* controls updates on the internal memory, which is not propagated to the next state. Gate *zt* controls how much of internal memory should be considered in the next state. Equations 1 and 2 represent operations realized by gates *rt* and *zt* to result, and equation 3 shows how the next hidden state is computed in a GRU unit.

$$rt = \sigma(W_r h_{t-1} + U_r x_t + br) \tag{1}$$

$$zt = \sigma(W_z h_{t-1} + U_z x_t + bz) \tag{2}$$

$$ht = z_t \otimes h_{t-1} \oplus (1 - z) \otimes \tan h(W_h x_t + U_h(r_t \otimes h_{t-1}) + bh)) \tag{3}$$

The update gate *zt* determines the amount of past information that is to be passed into the next state. The reset gate *rt* decides on how much of the previous information has to be neglected, thus resetting the gate values. Both the gates share the same formula but differ in the weights and usage of gates.

Fig. 1 Gated recurrent unit architecture

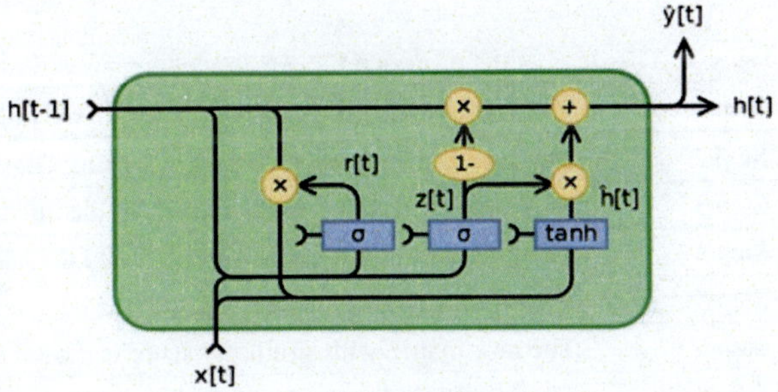

3.2 *Apriori Algorithm*

Association rule learning being rule-based machine learning is used to discover interesting relationships and patterns between variables in large databases. Apriori is an algorithm for frequent itemset mining and association rule learning. It identifies the frequent individual items and later extends them to larger and larger sets as long as they appear often in the database. The three most commonly used ways to measure association are support, confidence, and lift. Support defines how popular an item set is, by measuring the proportion of transactions in which an item set appears. Confidence gives the measure of how likely an item is purchased if another item is purchased. Lift is similar to confidence but it measures how likely an item (Y) is purchased when another item (X) is purchased while controlling the popularity of Y.

The key property of the algorithm states that all the subsets of a frequent itemset must be frequent and that if an itemset is infrequent, all its supersets will be infrequent.

4 Methodology

In this work, a model called Content-Scientific Article Recommendation (C-SAR) were proposed in which both GRU and Apriori algorithm are used for finding similar sets of articles. The combination of high-level deep learning, GRU, and a data mining Apriori algorithm will filter the most similar set of documents. Table 1 describes the steps involved in building the C-SAR model.

The methodology of the C-SAR model is described in Fig. 2 in two phases. In the first phase, Gated Recurrent Unit (GRU) technique is used to obtain the similarity of documents and the adjacency matrix. In the second phase, association rule mining based Apriori algorithm would be applied to filter out the most relevant set of documents among the similar documents. Since the model is content-based, the 'title' feature is extracted from the AAN dataset followed by data cleaning. The absence of null or insignificant values indicated that the data is cleaned. For a better result, removal of stop words followed by stemming and lemmatization was done to the 'title' feature. Padding is also done to the text data after one-hot encoding to

Table 1 The steps of the proposed C-SAR model

Step 1	The feature (title) is extracted from the AAN dataset
Step 2	The text data is converted to vectors using Glove pre-trained embedding
Step 3	Similarity probability is calculated for the documents using GRU
Step 4	The probabilities are replaced by 1's and 0's based on a fixed threshold
Step 5	An adjacency matrix is created with the new values(1's and 0's)
Step 6	The new matrix with similarity score is passed to the Apriori Algorithm
Step 7	The set of frequently occurring documents are obtained

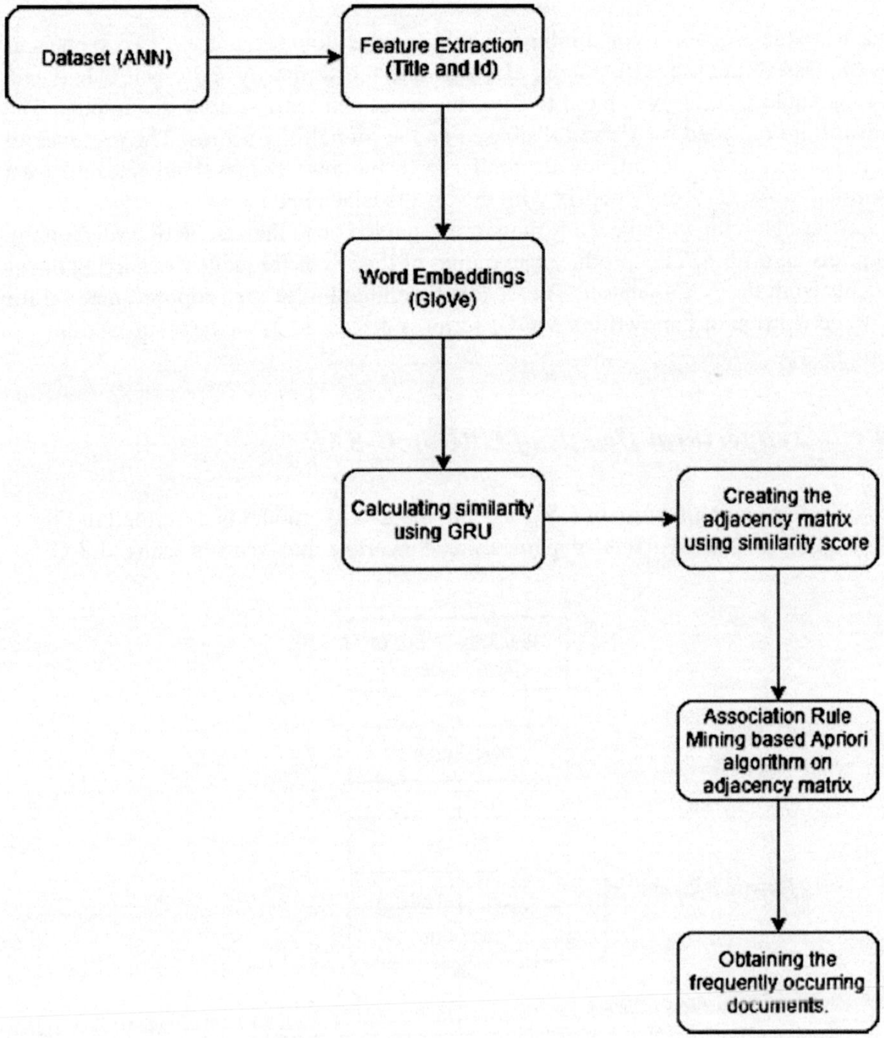

Fig. 2 The proposed C-SAR model

make sure that the sequences of data are of the same length to avoid ambiguity. It is performed to make all sequences in a batch to be of standard length.

GloVe, coined from Global Vectors, is an unsupervised learning algorithm that is used to get vector representation of words. This is done by mapping words into a meaningful space where similar words occur together. Such pre-trained embeddings can capture the semantic and syntactic meaning as they are trained on large datasets. They also boost the performance of the model. The main idea behind Glove is to derive relationships between words from Global Statistics and is found to be better than another embedding like Word2Vec. GRU units take up the vector representation and measure the probability of similarities between the input sequences. Two sequences were passed into the GRU model at a time and the similarity score was calculated. A threshold of 0.55 was fixed to filter the most and least similar documents. The results are replaced by 1's and 0's based on the probability scores. The presence of 1 indicates that the documents are similar to each other whereas 0 indicates the least similarity. An adjacency matrix with the results is achieved.

After obtaining the adjacency matrix, it is passed on to the association rule mining Apriori algorithm. This produces an output of the set of frequently occurring documents from the AAN dataset. The detailed architecture of the proposed method for the recognition of handwritten MODI script is discussed in the next subsection.

4.1 Architectural Details of GRU in C-SAR

The proposed architecture of GRU used in the C-SAR model is described in Fig. 3. The 'title' of the two sets of papers are converted into vectors using the Glove

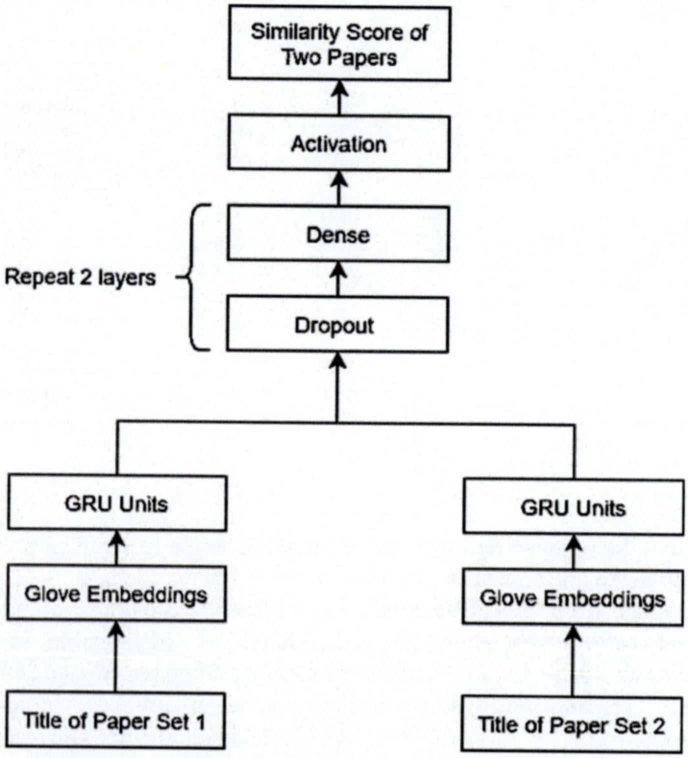

Fig. 3 The architecture of GRU in C-SAR model

embedding after proper cleaning. The vector format of the data is then separately passed on to two GRU units. Later, the output of the GRU units is concatenated and passed on to dropout and dense layers (2 layers are repeating). A sigmoid activation function calculates the probability of similarity between the two input sequences.

The dropout layer is used for regularisation where the inputs and the connections are excluded from getting activated. This helps in reducing overfitting and improves the model's performance. The dense layer is used to get the activation of the dot product of the input and kernel in a neural network. Dense layers with units of 128 and 64 were used in the model along with the dropout. The activation function used in the model is sigmoid which resulted in a value between 0 and 1. It is used for the models especially when the output need is a probability. They are commonly used for binary classification. Adam optimization along with binary cross-entropy loss is used in the C-SAR model to produce better results.

5 Experimental Study

The experiment is conducted on the Intel i5 processor with 8 GB RAM, using python programming. Description of the data set and the details of the experiment are given in this section. The model was implemented using the ACL Anthology Network (AAN) 2014 released dataset which is 387 MB file. The file consists of folders author_affiliations, citation_summaries, paper_text, and release. It consists of around 23,766 papers along with 1, 24,857 paper citations. Only the feature 'paper_title' along with 'paper_id' were extracted from the dataset. Figure 4 shows the AAN dataset with id and title required for the study which is extracted from the dataset. The Glove pre-trained embedding file was downloaded and used to get the encodings of the text data. It contained 400000 trained words and their representations.

The GRU function from Keras was used in implementing the model. It resulted in the probability of how one document in the AAN dataset is similar to another document. The adjacency matrix obtained from the GRU model is then passed on to the Apriori function from mlxtend which resulted in the set of most frequently

	paper_id	paper_title	year
0	W09-2307	Discriminative Reordering with Chinese Grammat...	2009.0
1	W04-2607	Non-Classical Lexical Semantic Relations	2004.0
2	W01-1314	A System For Extraction Of Temporal Expression...	2001.0
3	W04-1910	Bootstrapping Parallel Treebanks	2004.0
4	W09-3306	Evaluating a Statistical CCG Parser on Wikipedia	2009.0

Fig. 4 AAN dataset

occurring documents along with the minimum support. The minimum support value is a base value used to filter out the documents based on the support value. The set of documents below the minimum support will be rejected while processing.

6 Results and Discussion

The proposed C-SAR model is expected to outperform the existing state-of-art methods. It is because of the combination of a deep learning technique, Gated Recurrent Unit, and association rule mining Apriori algorithm. The objective of the proposed method is to find a set of most similar documents from the AAN dataset. In the first phase of the C-SAR model, the employment of GRU provided the probability score of the similarity of the documents. Figure 5a shows the sample similarity scores obtained using GRU.

a

p1_encoded	p2_encoded	predictions_probs
[138, 174, 18, 28, 90, 51, 64]	[105, 116, 31, 15, 51]	0.504709
[138, 174, 18, 28, 90, 51, 64]	[4, 11, 2, 22, 1, 139, 117, 19, 175, 10, 32, 1...	0.588963
[138, 174, 18, 28, 90, 51, 64]	[140, 79, 107]	0.504478
[138, 174, 18, 28, 90, 51, 64]	[25, 4, 43, 309, 141, 12, 176]	0.539213
[138, 174, 18, 28, 90, 51, 64]	[495, 12, 312, 496, 1, 497, 22, 19, 71, 79, 45]	0.544077

b

	W09-2307	W04-2607	W01-1314	W04-1910	W09-3306	W91-0219	W97-0806	W98-1121	W11-0819	W00-1221
W09-2307	1.0	1.0	0.0	1.0	0.0	1.0	1.0	0.0	1.0	1.0
W04-2607	1.0	1.0	0.0	1.0	0.0	1.0	1.0	0.0	1.0	1.0
W01-1314	0.0	0.0	1.0	0.0	0.0	0.0	1.0	0.0	1.0	1.0
W04-1910	1.0	1.0	0.0	1.0	0.0	1.0	1.0	0.0	1.0	1.0
W09-3306	0.0	0.0	0.0	0.0	1.0	1.0	1.0	0.0	1.0	1.0

Fig. 5 **a** Similarity score based on GRU **b** sample adjacency matrix using GRU

```
        support                                                itemsets
0         0.848                                               (W09-0109)
1         0.714                                               (W98-1405)
2         0.878                                               (W00-0502)
3         0.702                                               (W04-1406)
4         0.816                                               (H92-1028)
...         ...                                                      ...
2914      0.600  (W97-0616, W00-0502, W09-0109, W03-0305, W04-1...
2915      0.604  (W00-0502, W09-0109, W03-0305, W04-1406, H92-1...
2916      0.622  (W97-0616, W00-0502, W04-0505, W09-0109, W07-0...
2917      0.606  (W97-0616, W00-0502, W09-0109, W07-0201, W03-0...
2918      0.624  (W97-0616, W00-0502, W04-0505, W09-0109, W03-0...
```

Fig. 6 Set of frequently occurring documents in AAN dataset

Figure 5b represents the adjacency matrix obtained after replacing the probabilities resulted from the GRU with 1's and 0's based on the threshold value. The adjacency matrix is generated to provide cleaner data to the ARM-based model. This adjacency matrix is used in the Apriori algorithm to find out the most frequently occurring paper_ids and paper_titles to create a narrower range but relevant papers out of the dataset.

The adjacency matrix shown in Fig. 6 has been fed to the Apriori algorithm. The maximum support value received for the paper is 0.848 which depicts 84% of the time paper_id w09-0109 has occurred in the dataset for the given iterations. The minimum support value provided for the algorithm is 0.35 which has been set as a filter to remove all the data items which have a frequency less than 35% and attained the results as shown in Fig. 6.

The huge size of the dataset caused scalability issues and threw memory errors. The increase in the number of data can lead to better accuracy of the model. At the same time, the usage of memory and CPU also gets increased with a number of iterations over a huge dataset.

7 Conclusion

With the increase in the number of scientific publications and the number of research papers, the recommendation system for articles is gaining much significance. It is important for any scholar to get a set of relevant papers related to their field of study. In this work, a Content-based Scientific Article Recommendation (C-SAR) model was proposed based on a deep learning technique. Especially, this model focuses on checking for the papers based on the similarity in their title. A Gated Recurrent Unit method was employed for finding the similarity of documents and association rule mining Apriori algorithm to filter the most frequently occurring set of documents from a similar set. The model is expected to outperform existing models that use simple K-Means Clustering and user representations.

One of the limitations of this model is the memory and time constraints associated with the implementation. The performance can be increased by using cloud services and other super configuration machines. In the future, the efficiency of the model can be improved by considering an optimal threshold value for getting the similarity matrix and a better minimum support count based on the total number of transactions.

References

1. X. Bai, M. Wang, I. Lee, Z. Yang, X. Kong, F. Xia, Scientific Paper Recommendation: A Survey. IEEE Access **7**, 9324–9339 (2019)
2. M. Asim and S. Khusro, "Content Based Call for Papers Recommendation to Researchers", 12th International Conference on Open Source Systems and Technologies Lahore, Pakistan, 2018, pp. 42–47
3. C. Bhagavatula, S. Feldman, R. Power and W. Ammar," Content-Based Citation Recommendation", Proceedings of the Conference of the North American Chapter of the Association for Computational Linguistics: Human Language Technologies, New Orleans, Louisiana, Vol.1, Jun 2018
4. B. Kazemi and A. Abhari," A Comparative Study on Content-Based Paper-To-Paper Recommendation Approaches In Scientific Literature", SpringSim-CNS, Apr 2017, pp. 23–26
5. S. Philip, P.B. Shola and A.O. John, "Application of Content-Based Approach in Research Paper Recommendation System for a Digital Library" International Journal of Advanced Computer Science and Applications, 2014
6. D. Hanyurwimfura, L. Bo, V. Havyarimana, D. Njagi and F. Kagorora," An Effective Academic Research Papers Recommendation for Non-profiled Users", International Journal of Hybrid Information Technology, Vol. 8, 2015, pp. 255–272
7. A. Samad, M. A. Islam, M. A. Iqbal and M. Aleem," Centrality-Based Paper Citation Recommender System", EAI Endorsed Transactions on Industrial Networks and Intelligent Systems, Jun 2019
8. L. Guo, X. Cai, H. Qin, Y. Guo, F. Li and G. Tian," Citation Recommendation with a Content-Sensitive DeepWalk based Approach", International Conference on Data Mining Workshops, Beijing, China, 2019, pp. 538–543
9. H. A. M. Hassan," Personalized Research Paper Recommendation using Deep Learning", Proceedings of the 25th Conference on User Modeling, Adaptation and Personalization, Jul 2017, pp. 327–330
10. K. Hong, H. Jeon and C. Jeon," Advanced Personalized Research Paper Recommendation System Based on Expanded User Profile through Semantic Analysis", International Journal of Digital Content Technology and its Applications, 2013, pp. 67–76
11. A. Suglia, C. Greco, C. Musto, M. Gemmis, P. Lops and G. Semeraro,"A Deep Architecture for Content-based Recommendations Exploiting Recurrent Neural Networks", Proceedings of the 25th Conference on User Modeling, Adaptation and Personalization, Jul 2017, pp. 202–211
12. W. Huang, Z. Wu, C. Liang, P. Mitra, and C.L. Giles," A Neural Probabilistic Model for Context Based Citation Recommendation", *AAAI*, 2015
13. Z. Li and X. Zou," A Review on Personalized Academic Paper Recommendation", Computer and Information Science, 2019
14. R. Sharma, D. Gopalani and Y. Meena, "Concept-Based Approach for Research Paper Recommendation" PReMI, 2017

15. M. A. Arif, "Content aware citation recommendation system," International Conference on Emerging Technological Trends, Kollam, 2016, pp. 1–6
16. J. Shu, X. Shen, H. Liu, B. Yi and Z. Zhang," A content-based recommendation algorithm for learning resources", Multimedia Systems, 2017

Design Considerations for Low Noise Amplifier

Malti Bansal and Ishita Sagar

Abstract Low Noise Amplifier is also commonly known as LNA. Radio receiver performance greatly depends on LNA. The various design considerations for LNA have been discussed in this article. Impedance matching is an integral part of LNA design has been discussed. Technologies used for the LNA have also been reviewed. Light is also shed upon few application areas of LNA. A good LNA can be designed by considering all the factors. A trade-off needs to be carried out between all design parameters to create an optimized LNA structure.

Keywords LNA · LNA topologies · Design considerations · Common source topology · Common gate topology · Cascode topology · Input and output matching networks

1 Introduction

In today's world, an RF-Transceiver with minimal power consumption has gained huge demand, especially for Industrial, Scientific and Medical (ISM) bands [1]. Receiver selectivity, sensitivity, and inclination to reception errors are the three broad pillars that decide the performance and success of the receiver [2]. It always aims to focus on the front-end performance of the receiver, by changing parameters that are under the designer's control. LNA is the building block of the front-end receiver. The LNA should be able to successfully amplify the weak incoming signal from the antenna [3]. Care should be taken that minimal noise is added to the incoming signal by the LNA. Receiver's noise figure is highly impacted by the noise figure of the LNA [3]. Therefore, LNA has to ensure minimum value for the overall noise figure. If stages are cascaded, then the noise figure can be kept as low as possible if the initial stage has a low value for the noise figure and exhibits high values of gain at the required frequency of operation [4].

M. Bansal (✉) · I. Sagar
Department of Electronics and Communication Engineering, Delhi Technological University (DTU), Delhi 110042, India
e-mail: maltibansal@gmail.com

V. Suma et al. (eds.), *Inventive Systems and Control*,
Lecture Notes in Networks and Systems 204,
https://doi.org/10.1007/978-981-16-1395-1_71

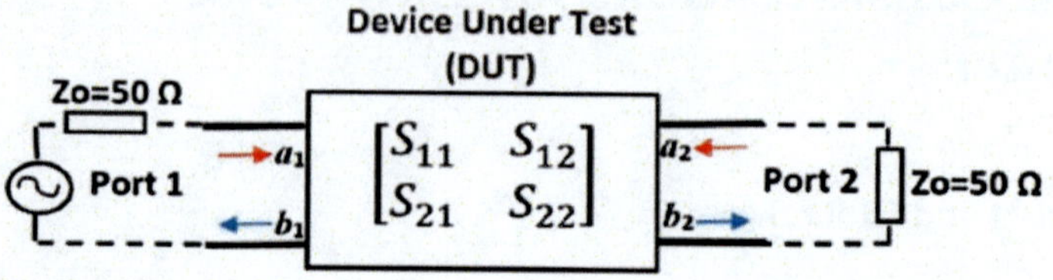

Fig. 1 Two port network

The basic design of LNA constitutes of three parts. The first is the input matching network (IMN), the second is the LNA itself and the third is the output matching network (OMN) [5]. The input return loss, depicted by S11, is reduced by the input matching network, also making sure that no additional input noise is added to the signal [5]. While carefully selecting the active components, IMN and OMN play a decisive role in the overall LNA's performance.

A basic two-port network is shown in Fig. 1. The equations for the network are as follows:

$$b_1 = S_{11}a_1 + S_{12}a_2$$
$$b_2 = S_{21}a_1 + S_{22}a_2$$

The S parameters indicated in the equation depict:

S_{11}—forward reflection (input impedance matching)
S_{22}—reverse reflection (output impedance matching)
S_{21}—forward transmission (gain or loss)
S_{22}—reverse transmission (leakage or isolation)

Impedance matching is an essential criterion to be considered while designing an LNA, especially for Radio Frequency designs and microwave designs [6]. The common matching techniques are T-matching networks and π–matching networks.

Figure 2 Lossless network matching networks arbitrary load impedance to a transmission line [7].

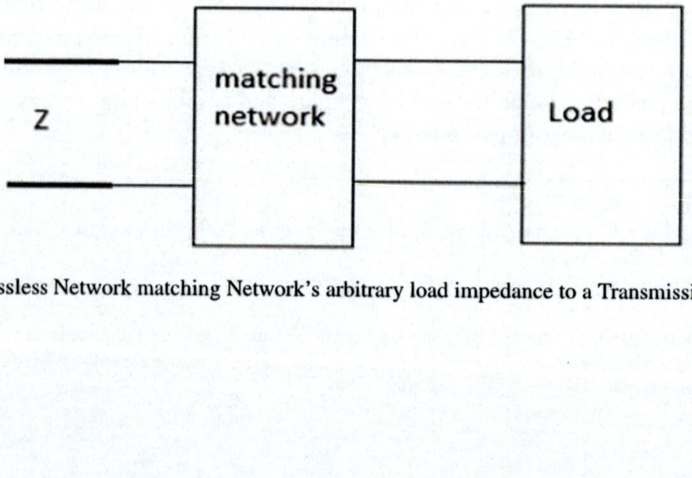

Fig. 2 Lossless Network matching Network's arbitrary load impedance to a Transmission Line

The presence of matching networks is necessary because the need is to deliver maximum power to the load, and hence the LNA needs to be efficiently terminated at the input and output ports [8]. Ideally, the matching network should be completely lossless.

2 Design Considerations

2.1 LNA Topologies

The most basic and commonly used topologies for CMOS LNA include Common source, Common gate, and cascode. These are the three most prevailing topologies used for LNA. Choosing a suitable LNA topology according to the application area where the design will be used, is a crucial step for the beginning of the process. All trade-offs for the topology have to be considered before narrowing it down to the most suitable one. Several parameters and their trade-offs have to be considered. These parameters include noise figure, S-parameters, linearity, stability, and gain.

The Common Source cascode topology is shown in Fig. 3.

The common source of topology provides the best possible noise figure when sized. This advantage comes at the cost of higher sensitivity to temperature, bias, and component tolerance [2]. If an inductor is added in the CS stage then the result is in Inductive Source Degenerate LNA. The gain and noise performance of the LNA is affected by the addition of the inductor in the structure, which will be a part of future discussions [9].

The Common gate topology for LNA is shown in Fig. 4.

Fig. 3 Common source topology

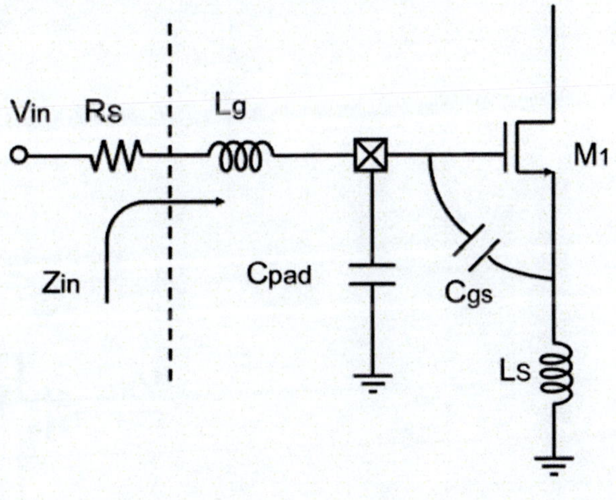

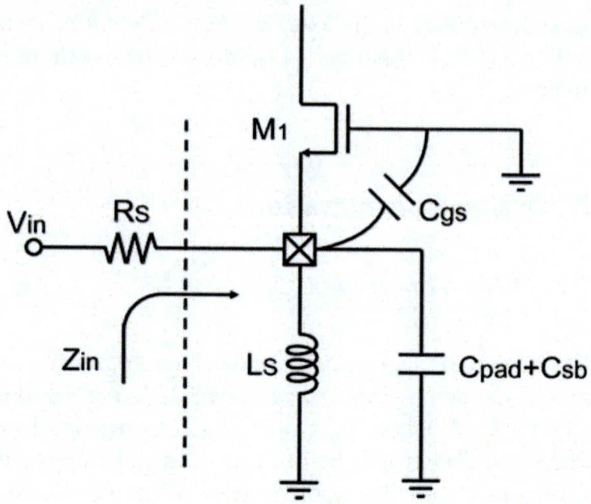

Fig. 4 Common gate topology

As the operating RF frequencies are increased, it is required to design techniques for the circuit that are better than the NF of the Common gate LNA, while protecting its stability, linearity, and low power consumption [10] (Fig. 5).

The cascode structure provides a considerable reduction in the Miller effect [11]. Minimal reverse isolation is present in the structure with the effects of limited output impedance blocked [12–15].

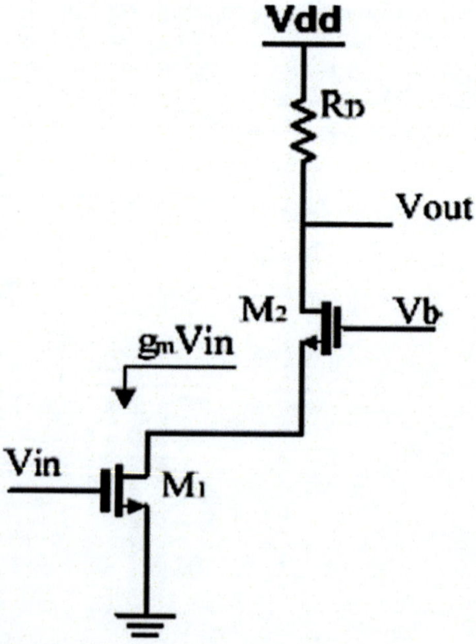

Fig. 5 Cascode topology [12]

Table 1 Comparison of LNA topologies [2]

Characteristic	Common-source	Common-gate	Cascode
Noise figure	Lowest	Sharp increase with frequency	Slightly higher than CS
Gain	Moderate	Lowest	Highest
Linearity	Moderate	High	Highest
Bandwidth	Narrow	Moderate	Broad
Stability	Needs compensation	High	High
Reverse isolation	Low	High	High
Sensitivity to process variation, temp, power supply, component tolerance	High	Low	Low

Resistive degeneration in the structure can cause high power dissipation, thereby leading to poor noise figures [16]. The cascode topology for LNA has been considered to be the most versatile topology by researchers. The gain provided by the cascode structure is the most stable gain for the largest bandwidth [2]. The cost paid for this great advantage is minimal degradation of noise figure performance with design complexity [2]. A comparison of the three topologies is given in Table 1.

2.2 LNA Parameters

There are few parameters of LNA which need to be studied and carefully optimized according to the different topologies and application area of the LNA. The necessary parameters to be considered in the process of LNA design are:

1. Noise Figure (NF)

It can be extended up to which the signal to noise ratio (SNR) is degraded is defined by Noise Figure [9]. Proper biasing is applied in combination with input matching and power-constrained techniques, to achieve a low noise figure [17, 18]. The formula showing the NF of the receiver system is [19]:

$$\mathrm{NF} = \mathrm{NF}_{\mathrm{LNA}} + \frac{\mathrm{NF}_2 - 1}{G_{\mathrm{LNA}}}$$

In the above equation, NFtot represents the noise figure of the first LNA stage, NF2 represents the noise figure of the successive blocks after the first stage and GLNA is the gain of LNA.

2. Gain

The gain of the LNA is defined by the ability to amplify the amplitude of the incoming input signal [9]. It can also be shown by the S-parameter S_{21}.

$$\text{Gain} = 20\log\frac{V_{\text{out}}}{V_{\text{in}}}$$

3. S-Parameters

S-Parameters also called Scattering Parameters, are commonly used for impedance matching. The S parameters represent the transmission and reflection coefficients for a two-port network [9]. The reflection coefficient at the port is called the reflection coefficient. It is represented by S_{11} and it has a unit of decibels. It gives us the input return loss. Whereas, S_{22} represents the output reflection coefficient. When measured in decibels, it provides the output return loss.

4. Linearity

Graphically plotting parameters like the third intercept point and 1-dB compression point helps in measuring the linearity of LNA [18]. The third intercept point is depicted in Fig. 6.

It is defined as the point of intersection between the third order and first order intermodulation products.

1 dB Compression point is depicted in Fig. 7.

It is defined as the point where the power of the output signal becomes 1-dB less than the expected value.

5. Stability

One of the most important aspects expected from the LNA design is that the structure should be stable in the frequency range in which it is operated to deliver results. To measure the stability of LNA, stability factor K is plotted [9]. The coupling has a direct impact on the stability factor. Inductive Loading and neutralization can help improve stability [20].

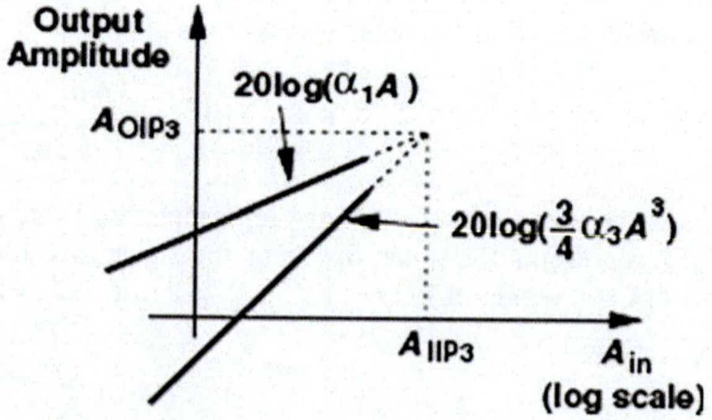

Fig. 6 Third intercept point [20]

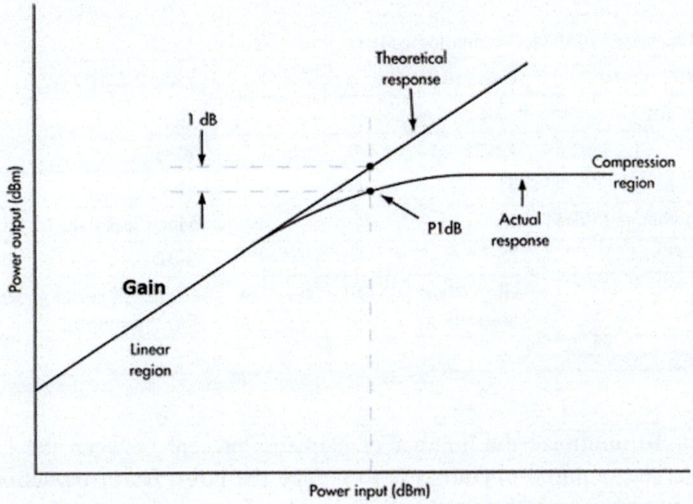

Fig. 7 1-dB compression point [20]

2.3 Process Technologies

GaAs pHEMT and SiGe BiCMOS process technology active devices are the most commonly used active devices in LNAs.

For GaAs pHEMT devices, very little noise is produced due to the presence of heterojunction between the extremely thin undoped GaAs layer and doped AlGaAs layer [2]. On the other hand, Si BiCMOS LNA structures have been emerging for use in C-Band applications for WLAN [21]. Si Ge technology comprises CMOS active devices and heterojunction bipolar transistors (HBT). Low breakdown, high performance SiHBT active devices are used in the design of LNA [21].

In today's date, GaAs devices and SiGe technologies are quite comparable when it comes down to the usable frequency range, whereas the dynamic range of SiGe devices is limited due to its lower breakdown voltage [2]. The advantages GaAs pHEMT provides over SiGe technology is improving noise figure and linearity performance [2]. However, SiGe can function with more efficiency and also has advantages of cost resulting from higher levels of integration. The comparative advantages, disadvantages, and trade-offs should be considered concerning the application which is being focused on. A comparison of the device technologies is shown in Table 2.

2.4 Input and Output Matching

The input and output matching are a vital component involved in LNA design considerations. When one approaches input matching, multiple dimensions have to be

Table 2 Comparison of device technologies [2]

Typical performance	GaAs pHEMT	SiGe BiCMOS
Noise figure (dB)	0.4	0.9
Gain (dB)	12–21	10–17
OIP3 (dB)	41	31
Breakdown voltage (Vdc)	15	Much less than 15 V
Inductor Q-factor	15	5–10
Strengths	High P1dB and OIP3, very low noise figure	Higher integration, low cost, ESD immunity
ft/Fmax		

considered. To minimize the number of elements present between the LNA and the antenna, the required operation is to reduce the noise factor reduction [2]. A commonly used input matching network has high values of the quality factor, as high Q helps attain optimal noise figure with gain performance as the loss is minimum [2]. However, such networks can be sensitive to factors like deviations and variations in voltage, process, temperature, and also component value. The noise parameters of the LNA help to design an appropriate input matching network. The optimization and compromise are carried out between gain, noise figure, and input return loss to narrow down to the input matching network [2].

Output matching networks are also narrowed down after trade-offs between 1-Db Compression point, gain, and OIP3 [2]. Trade-offs are an integral part of designing the input and output matching networks.

2.5 *Stability*

All LNA designs are considered to be complete when the amplifier proves to be completely stable over the entire range of frequency where it is expected to operate. The small and large-signal stability are both important aspects that should be achieved for any LNA structure. Large signal stability can be verified by measuring and monitoring the output of the LNA in comparison with the 1-dB gain compression [2]. Stability circles can be used to verify the stability of the finalized design.

3 Issues Faced in Cmos LNA Design

After meticulous consideration and optimization between all design aspects and parameters, an LNA can be successfully designed for a particular application. However, when an LNA is practically made, deviations from ideality are observed.

There are many reasons for these non-ideal conditions to occur. Some of the issues faced in the design of CMOS LNA have been discussed.

1. Feedback in the CMOS LNA circuits/structures

When the active device poles much higher than the feedback loop bandwidth, feedback techniques can be considered as an option and prove to be helpful [22]. As a result of increasing F_T for MOS transistors, the use of feedback in high-frequency LNA circuits/structures has been possible [23]. In the wideband matching of common source stages, the feedback loop has proven to be helpful. Resistive feedback, RLC feedback are a few examples that have been used in the past, also for matching the input impedances. The use of feedback techniques in the design of LNA results from the need to shift the optimum noise impedance to the needed point. This feedback helps in the reduction of the non-linearity of the circuit.

2. Power dissipation and Chip area

Reduced power dissipation of any circuit has become one of the most important aspects today. Every design is made by trying to keep the power dissipation of the circuit/structure minimal. Focus is being made over battery-operated structures and hence power dissipation becomes a crucial point to be taken care of. Hence, to reduce the overall power dissipation of an LNA circuit, special care has to be taken in the circuit selection and design steps. To reduce DC power consumption, current reuse techniques have proven to be a success [23]. It has become a widely used technique in low power RF circuits for the reduction in DC power. CMOS transistors that operate in the sub-threshold region are productive in applications where low power consumption is desired. The one negative side of using CMOS transistors in the sub-threshold regions is that g_m is low and hence prove to be less useful for higher frequency applications [23]. With the use of PMOS active load transistor NMOS transistors' drain of LNA, obtain a high load resistance, along with the additional advantage of low DC power dissipation of load resistance [24].

The chip area is an aspect that has to be handled and taken care of throughout the design process. The focus of any design or circuitry involving LNA is to keep the chip area minimal. Another reason for keeping the chip area minimal is the focus on portability. Devices should be portable and hence focus on-chip area. Many techniques have been researched and worked upon to achieve this goal. Some solutions include using capacitors made from MOS transistors, instead of using an actual capacitor on the chip. Capacitors are very bulky, and hence this solution. The use of lumped elements, rather than distributed elements, like in the matching networks, leads to a reduction in overall chip area [23].

3. Protection of Electrostatic Discharge

CMOS circuits have high input impedance and small values of gate breakdown voltage, Electrostatic Discharge Protection for the input-output pads has become an important aspect for these circuits [23]. The aim is to use a simple technique to deal with the degradation of performance in RF circuits, thereby including LNA circuits. For frequencies higher than 5 GHz, the use of two diodes, one between signal line and ground, and the other between DC power line and the signal line have been worked upon in the past [25]. However, works have also been reported for SCR based electrostatic charge protection.

4 Application Areas of LNA

LNA has been widely used and accepted in many fields of the electronic industry. The design parameters discussed above vary according to the type of application the LNA is being fabricated for. An optimization and trade-off need to be carried out for all LNA parameters, which varies for different applications.

A pictorial representation of a few application areas of LNA is depicted in Fig. 8. An advancing field with the use of LNA is the biomedical field. Some of these applications are depicted in Fig. 9.

Most biomedical application areas are moving towards wireless use, such as neural and ECG systems. LNA is a very vital component in the biomedical systems, as the input signals in this area are very weak, and hence need to be carefully processed without any additional noise being added and important information not being lost. On the other hand, the applications mentioned in the figure also consist of applications of LNA in the communications field. A similar kind of behavior is also expected from the LNA being used for this field. Future work will consist of more research into these application areas.

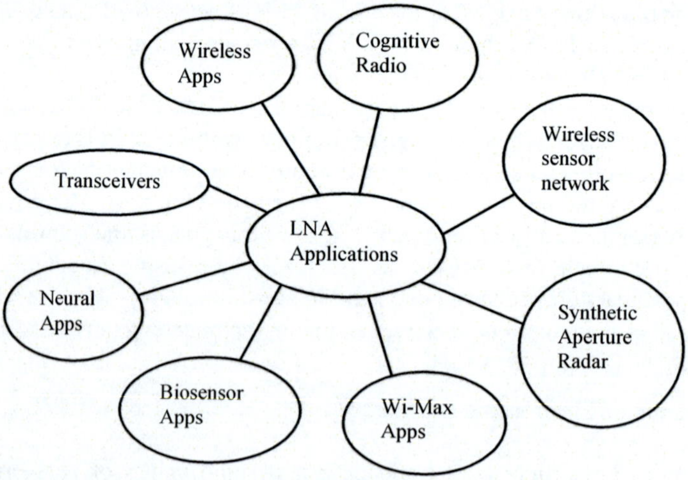

Fig. 8 Application areas of LNA [26]

Fig. 9 Applications of LNA in biomedical field [27]

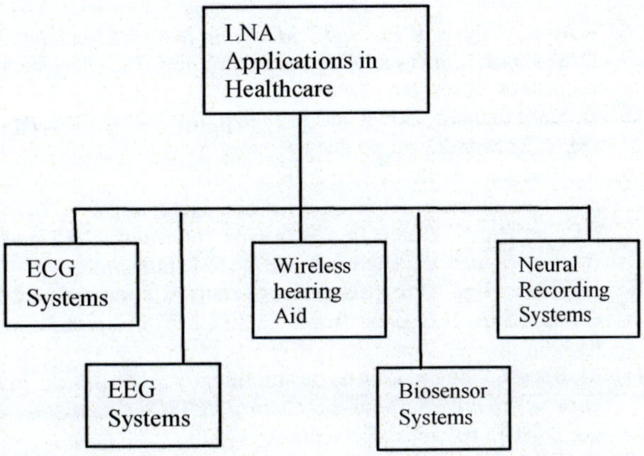

5 Conclusion and Future Scope

Noise figure and linearity are dominating factors for receiver sensitivity at the system level. The dynamic range of the overall system is a dependent factor on the dynamic range of the LNA. It also depends on the gain and linearity of the LNA. In the presence of strong interferers, the system should be able to filter out completely or suppress the cross-modulation products, which can be obtained by linearity optimization. The entire design process of LNA needs an in-depth knowledge of two components, which are transistor properties and the achievable advantages of the semiconductor and process technologies. It is the duty of the designer to make successful trade-offs between the desired parameters that are expected from the LNA for the application which is being targeted. Since all parameters of the cannot be at their best, and optimization between them is necessary.

References

1. A. Azizan, S.A.Z. Murad, R.C. Ismail, M.N.M. Yasin, A review of LNA topologies for wireless applications, in *2014 2nd International Conference on Electronic Design (ICED)*, Penang, (2014), pp. 320–324
2. Practical considerations for low noise amplifier design by Tim Das, NXP Semiconductore (2013)
3. I. Benamor et al., Fast power switching low-noise amplifier for 6–10 GHz ultra-wideband applications, in *2013 IEEE 20th International Conference on Electronics, Circuits, and Systems (ICECS), Abu Dhabi*, (2013), pp. 759–762. https://doi.org/10.1109/icecs.2013.6815525
4. M.B. Yelten, K.G. Gard, Theoretical analysis and characterization of the tunable matching networks in low noise amplifiers, in *2009 European Conference on Circuit Theory and Design, Antalya* (2009), pp. 890–893. https://doi.org/10.1109/ecctd.2009.5275128
5. M. Bansal, Jyoti, A review of various applications of low noise amplifier, in *2017 International Conference on Innovations in Control, Communication and Information Systems (ICICCI), Greater Noida, India*, 2017, pp. 1–4. https://doi.org/10.1109/iciccis.2017.8660954

6. I. Abu, Analysis of *Pie and T Matching Network for Low Noise Amplifier* (LNA), (2018)
7. D. M-Pozar, Microwave Engineering. Wiley. INC-New York. Chichester. Weinheim. Brisbane-Singapore. Toronto
8. D. Senthilkumar, U.P. Khot, S. Jagtap. Int. J. Eng. Res. Appl. (IJERA) **3**(1), 403–408 (2013). ISSN: 2248-9622
9. L. Vimalan, S. Devi, Performance analysis of various topologies of common source low noise amplifier (CS-LNA) at 90 nm Technology, in *2018 3rd IEEE International Conference on Recent Trends in Electronics, Information & Communication Technology (RTEICT), Bangalore, India,* (2018), pp. 1687–1691. https://doi.org/10.1109/rteict42901.2018.9012326
10. W. Zhuo et al., A capacitor cross-coupled common-gate low-noise amplifier. IEEE Trans. Circuits Syst. II Express Briefs **52**(12), 875–879 (2005). https://doi.org/10.1109/TCSII.2005.853966
11. M. Bansal, I. Sagar, Low noise amplifier for ECG signals, in *2020 Fourth International Conference on Inventive Systems and Control (ICISC), Coimbatore, India,* (2020), pp. 304–310. https://doi.org/10.1109/icisc47916.2020.9171200
12. H.A. Eshghabadi, F. Eshghabadi, An improved power constrained simultaneous noise and input matched 2.45 GHz CMOS NB-LNA, in *2012 IEEE International Conference on Circuits and Systems (ICCAS)* (2012), pp. 92–97
13. X. Tang, F. Huang, D. Zhao, Design of a 6 GHz high-gain low noise amplifier, in *2012 International Conference on Microwave and Millimeter Wave Technology (ICMMT)*, no. 2, pp. 1–4 (2012)
14. F. Zou, Z. Li, M. Zhang, A 1 V folded common-gate CMOS low noise amplifier for wireless sensor network applications, in *2011 International Conference on Wireless Communications and Signal Processing (WCSP)*, Nanjing, (2011), pp. 1–4
15. H. Yang, S. Peng, S. Wang, High quality of 0.18 μm CMOS 5.2 GHz cascode LNA for RFID tag applications, in *2013 IEEE International Symposium on Next-Generation Electronics (ISNE)*, (2013) pp. 313–316
16. M. Muhamad, N. Soin, H. Ramiah, N. M. Noh, W. K. Chong, Design of CMOS differential LNA at 2.4 GHz, in *2013 IEEE International Conference of Electron Devices and Solid-state Circuits, Hong Kong*, (2013), pp. 1–2
17. B.M. Ninan, K. Balamurugan, M.N. Devi, Design and analysis of low noise amplifier at 60 GHz using active feedback and current re-use topologies, in *2016 3rd International Conference on Devices, Circuits and Systems (ICDCS), Coimbatore*, (2016), pp. 161–167. https://doi.org/10.1109/icdcsyst.2016.7570652
18. B. Prameela, A.E. Daniel, Design and analysis of different low noise amplifiers in 2–3 GHz, in *2016 International Conference on VLSI Systems, Architectures, Technology and Applications (VLSI-SATA), Bangalore*, (2016), pp. 1–6. https://doi.org/10.1109/vlsi-sata.2016.7593039
19. D.K. Shaeffer, T.H. Lee, A 1.5-V, 1.5-GHz CMOS low noise amplifier. IEEE J. Solid-State Circ. **32**(5), 745–759 (1997). https://doi.org/10.1109/4.568846
20. M. Bansal, I. Sagar, LNA Architectures for ECG analog front end in CMOS technology, in G. Ranganathan, J. Chen, Á. Rocha (eds) *Inventive Communication and Computational Technologies. Lecture Notes in Networks and Systems,* vol 145. (Springer, Singapore, 2021), pp. 973–984. https://doi.org/10.1007/978-981-15-7345-3_83
21. V.J. Patel et al., X-band low noise amplifier using SiGe BiCMOS technology, in *IEEE Compound Semiconductor Integrated Circuit Symposium, 2005. CSIC '05., Palm Springs, CA, USA*, (2005), p. 4. https://doi.org/10.1109/csics.2005.1531753
22. Low noise amplifier design and optimisation, Chapter 4, By JY Hasani (2008)
23. E. Adabi, A. Niknejad, CMOS low noise amplifier with capacitive feedback matching (2017). 643–646. https://doi.org/10.1109/cicc.2007.4405814
24. Y. Cao, V. Issakov, M. Tiebout, A 2 kV ESD-Protected 18 GHz LNA with 4 dB NF in 0.13 μm CMOS, in *2008 IEEE International Solid-State Circuits Conference - Digest of Technical Papers, San Francisco, CA*, (2008), pp. 194–606. https://doi.org/10.1109/isscc.2008.4523123
25. D. Pienkowski, V. Subramanian, G. Boeck, A 3.6 dB NF, 6 GHz band CMOS LNA with 3.6 mW power consumption, in *2006 European Conference on Wireless Technology, Manchester,* (2006), pp. 67–70. https://doi.org/10.1109/ecwt.2006.280436

26. M. Bansal, Jyoti, A review of various applications of low noise amplifier, in *2017 International Conference on Innovations in Control, Communication and Information Systems (ICICCI), Greater Noida, India*, (2017), pp. 1–4, https://doi.org/10.1109/iciccis.2017.8660954
27. M. Bansal, Jyoti, Low noise amplifier in smart healthcare applications, in *2019 6th International Conference on Signal Processing and Integrated Networks (SPIN), Noida, India*, (2019), pp. 1002–1007. https://doi.org/10.1109/spin.2019.8711705

Author Index

V. Suma et al. (eds.), *Inventive Systems and Control*, Lecture Notes in Networks and Systems 204,
https://doi.org/10.1007/978-981-16-1395-1